TOPIC	SYMBOL	MEANING	PAGE
FUNCTIONS	$f(a)$	value of the function f at a	97
	$f:A \to B$	function from A to B	97
	$f_1 + f_2$	sum of the functions f_1 and f_2	98
	$f_1 f_2$	product of the functions f_1 and f_2	98
	$f(S)$	image of the set S under f	99
	$\iota_A(s)$	identity function on A	102
	$f^{-1}(x)$	inverse of f	102
	$f \circ g$	composition of f and g	103
	$\lfloor x \rfloor$	floor function of x	105
	$\lceil x \rceil$	ceiling function of x	105
	$f(x)$ is $O(g(x))$	$f(x)$ is big-O of $g(x)$	132
	$n!$	n factorial	137
	$f(x)$ is $\Omega(g(x))$	$f(x)$ is big-Omega of $g(x)$	140
	$f(x)$ is $\Theta(g(x))$	$f(x)$ is big-Theta of $g(x)$	141
	$\sim$	asymptotic	144
	$\min(x, y)$	minimum of x and y	160
	$\max(x, y)$	maximum of x and y	161
	a_n	term of $\{a_i\}$ with subscript n	225
	$\sum_{i=1}^{n} a_i$	sum of $a_1, a_2, \ldots, a_n$	229
	$\sum_{\alpha \in S} a_\alpha$	sum of a_α over $\alpha \in S$	232
	$\prod_{i=1}^{n} a_n$	product of $a_1, a_2, \ldots, a_n$	237
	$\approx$	approximately equal to	357
INTEGERS	$a \mid b$	a divides b	153
	$a \nmid b$	a does not divide b	153
	$\gcd(a, b)$	greatest common divisor of a and b	159
	$\text{lcm}(a, b)$	least common multiple of a and b	161
	$a \textbf{ div } b$	quotient when a is divided by b	161
	$a \textbf{ mod } b$	remainder when a is divided by b	161
	$a \not\equiv b \pmod{m}$	a is not congruent to b modulo m	161
	$a \equiv b \pmod{m}$	a is congruent to b modulo m	161
	$(a_k a_{k-1} \cdots a_1 a_0)_b$	base b representation	169
MATRICES	$[a_{ij}]$	matrix with entries a_{ij}	197
	$\mathbf{A} + \mathbf{B}$	matrix sum of $\mathbf{A}$ and $\mathbf{B}$	197
	$\mathbf{AB}$	matrix product of $\mathbf{A}$ and $\mathbf{B}$	198
	$\mathbf{I}_n$	identity matrix of order n	201
	$\mathbf{A}^t$	transpose of $\mathbf{A}$	201
	$\mathbf{A} \vee \mathbf{B}$	join of $\mathbf{A}$ and $\mathbf{B}$	202
	$\mathbf{A} \wedge \mathbf{B}$	the meet of $\mathbf{A}$ and $\mathbf{B}$	202
	$\mathbf{A} \odot \mathbf{B}$	Boolean product of $\mathbf{A}$ and $\mathbf{B}$	202
	$\mathbf{A}^{[n]}$	nth Boolean power of $\mathbf{A}$	203

(List of Symbols continued at back of book)

IMPORTANT:

HERE IS YOUR REGISTRATION CODE TO ACCESS
YOUR PREMIUM McGRAW-HILL ONLINE RESOURCES.

For key premium online resources you need THIS CODE to gain access. Once the code is entered, you will be able to use the Web resources for the length of your course.

If your course is using **WebCT** or **Blackboard**, you'll be able to use this code to access the McGraw-Hill content within your instructor's online course.

Access is provided if you have purchased a new book. If the registration code is missing from this book, the registration screen on our Website, and within your WebCT or Blackboard course, will tell you how to obtain your new code.

Registering for McGraw-Hill Online Resources

To gain access to your McGraw-Hill web resources simply follow the steps below:

1. USE YOUR WEB BROWSER TO GO TO: **http://www.mhhe.com/rosen/**

2. CLICK ON **FIRST TIME USER**.

3. ENTER THE REGISTRATION CODE* PRINTED ON THE TEAR-OFF BOOKMARK ON THE RIGHT.

4. AFTER YOU HAVE ENTERED YOUR REGISTRATION CODE, CLICK **REGISTER**.

5. FOLLOW THE INSTRUCTIONS TO SET-UP YOUR PERSONAL UserID AND PASSWORD.

6. WRITE YOUR UserID AND PASSWORD DOWN FOR FUTURE REFERENCE.
   KEEP IT IN A SAFE PLACE.

TO GAIN ACCESS to the McGraw-Hill content in your instructor's **WebCT** or **Blackboard** course simply log in to the course with the UserID and Password provided by your instructor. Enter the registration code exactly as it appears in the box to the right when prompted by the system. You will only need to use the code the first time you click on McGraw-Hill content.

Thank you, and welcome to your McGraw-Hill online Resources!

0-07-242434-6 ROSEN: DISCRETE MATHEMATICS AND ITS APPLICATIONS, 5/E

MCGRAW-HILL
ONLINE RESOURCES

REGISTRATION CODE

unoccupied-51671142

Mc Graw Hill Higher Education

Discrete Mathematics and Its Applications

Fifth Edition

Kenneth H. Rosen

AT&T Laboratories

Boston   Burr Ridge, IL   Dubuque, IA   Madison, WI   New York   San Francisco   St. Louis
Bangkok   Bogotá   Caracas   Kuala Lumpur   Lisbon   London   Madrid   Mexico City
Milan   Montreal   New Delhi   Santiago   Seoul   Singapore   Sydney   Taipei   Toronto

McGraw-Hill Higher Education

A Division of The **McGraw-Hill** Companies

DISCRETE MATHEMATICS AND ITS APPLICATIONS, FIFTH EDITION

Published by McGraw-Hill, a business unit of The McGraw-Hill Companies, Inc., 1221 Avenue of the Americas, New York, NY 10020. Copyright © 2003, 1999, 1995, 1991, 1988 by American Telephone and Telegraph Company. All rights reserved. No part of this publication may be reproduced or distributed in any form or by any means, or stored in a database or retrieval system, without the prior written consent of The McGraw-Hill Companies, Inc., including, but not limited to, in any network or other electronic storage or transmission, or broadcast for distance learning.

Some ancillaries, including electronic and print components, may not be available to customers outside the United States.

This book is printed on acid-free paper.

International   1 2 3 4 5 6 7 8 9 0 VNH/VNH 0 9 8 7 6 5 4 3 2
Domestic   1 2 3 4 5 6 7 8 9 0 VNH/VNH 0 9 8 7 6 5 4 3 2

ISBN 0–07–242434–6
ISBN 0–07–119881–4 (ISE)

Publisher: *William K. Barter*
Executive editor: *Robert E. Ross*
Senior developmental editor: *Michelle D. Munn*
Executive marketing manager: *Marianne C. P. Rutter*
Senior project manager: *Joyce M. Berendes*
Senior production supervisor: *Laura Fuller*
Senior media project manager: *Stacy A. Patch*
Media technology producer: *Jeff Huettman*
Designer: *K. Wayne Harms*
Interior designer: *Andrew Ogus*
Lead photo research coordinator: *Carrie K. Burger*
Photo research: *Toni Michaels/PhotoFind LLC*
Supplement producer: *Brenda A. Ernzen*
Compositor: *Techsetters*
Typeface: *10/12 Times Roman*
Printer: *Von Hoffmann Press, Inc.*

Cover image: *Jasper Johns'* Between the Clock and the Bed, *1982–83.* © *Jasper Johns/Licensed by VAGA, New York, NY*

The credits section for this book begins on page C-1 and is considered an extension of the copyright page.

Library of Congress Cataloging-in-Publication Data

Rosen, Kenneth H.
    Discrete mathematics and its applications / Kenneth H. Rosen.—5th ed.
        p.   cm.
    Includes bibliographical references and index.
    ISBN 0-07-242434-6
    1. Mathematics.   2. Computer science—Mathematics.   I. Title.

QA39.3 .R67 2003
511—dc21   2002070890
                                                                        CIP

INTERNATIONAL EDITION ISBN 0–07–119881–4
Copyright © 2003. Exclusive rights by The McGraw-Hill Companies, Inc., for manufacture and export. This book cannot be re-exported from the country to which it is sold by McGraw-Hill. The International Edition is not available in North America.

www.mhhe.com

Contents

Preface vii

The Companion Website xvii

To the Student xix

1 The Foundations: Logic and Proof, Sets, and Functions 1

1.1	Logic	1
1.2	Propositional Equivalences	20
1.3	Predicates and Quantifiers	28
1.4	Nested Quantifiers	44
1.5	Methods of Proof	56
1.6	Sets	77
1.7	Set Operations	86
1.8	Functions	97
	End-of-Chapter Material	111

2 The Fundamentals: Algorithms, the Integers, and Matrices 119

2.1	Algorithms	120
2.2	The Growth of Functions	131
2.3	Complexity of Algorithms	144
2.4	The Integers and Division	153
2.5	Integers and Algorithms	169
2.6	Applications of Number Theory	181
2.7	Matrices	196
	End-of-Chapter Material	206

3 Mathematical Reasoning, Induction, and Recursion 213

3.1	Proof Strategy	214
3.2	Sequences and Summations	225
3.3	Mathematical Induction	238
3.4	Recursive Definitions and Structural Induction	256
3.5	Recursive Algorithms	274
3.6	Program Correctness	284
	End-of-Chapter Material	290

4   Counting 301

4.1	The Basics of Counting	301
4.2	The Pigeonhole Principle	313
4.3	Permutations and Combinations	320
4.4	Binomial Coefficients	327
4.5	Generalized Permutations and Combinations	335
4.6	Generating Permutations and Combinations	344
	End-of-Chapter Material	349

5   Discrete Probability 355

5.1	An Introduction to Discrete Probability	355
5.2	Probability Theory	362
5.3	Expected Value and Variance	379
	End-of-Chapter Material	394

6   Advanced Counting Techniques 401

6.1	Recurrence Relations	401
6.2	Solving Recurrence Relations	413
6.3	Divide-and-Conquer Algorithms and Recurrence Relations	425
6.4	Generating Functions	435
6.5	Inclusion–Exclusion	451
6.6	Applications of Inclusion–Exclusion	457
	End-of-Chapter Material	465

7   Relations 471

7.1	Relations and Their Properties	471
7.2	n-ary Relations and Their Applications	482
7.3	Representing Relations	489
7.4	Closures of Relations	496
7.5	Equivalence Relations	507
7.6	Partial Orderings	516
	End-of-Chapter Material	530

8   Graphs 537

8.1	Introduction to Graphs	537
8.2	Graph Terminology	545
8.3	Representing Graphs and Graph Isomorphism	557
8.4	Connectivity	567
8.5	Euler and Hamilton Paths	577
8.6	Shortest-Path Problems	593

8.7   Planar Graphs   603
8.8   Graph Coloring   613
       End-of-Chapter Material   622

9   Trees 631

9.1   Introduction to Trees   631
9.2   Applications of Trees   644
9.3   Tree Traversal   660
9.4   Spanning Trees   674
9.5   Minimum Spanning Trees   688
       End-of-Chapter Material   694

10   Boolean Algebra 701

10.1   Boolean Functions   701
10.2   Representing Boolean Functions   709
10.3   Logic Gates   712
10.4   Minimization of Circuits   719
        End-of-Chapter Material   734

11   Modeling Computation 739

11.1   Languages and Grammars   739
11.2   Finite-State Machines with Output   751
11.3   Finite-State Machines with No Output   758
11.4   Language Recognition   765
11.5   Turing Machines   775
        End-of-Chapter Material   783

Appendixes

A.1   Exponential and Logarithmic Functions   A–1
A.2   Pseudocode   A–4

Suggested Readings B–1
Answers to Odd-Numbered Exercises S–1
Photo Credits C–1
Index of Biographies I–1
Index I–2

About the Author

Kenneth H. Rosen is a Distinguished Member of the Technical Staff in AT&T Laboratories in Middletown, New Jersey.

Dr. Rosen received his B.S. in Mathematics from the University of Michigan, Ann Arbor (1972), and his Ph.D. in Mathematics from M.I.T. (1976), where he wrote his thesis in the area of number theory under the direction of Harold Stark. Before joining Bell Laboratories in 1982, he held positions at the University of Colorado, Boulder; the Ohio State University, Columbus; and the University of Maine, Orono, where he was an associate professor of mathematics. While working at AT&T Labs, Ken has taught at Monmouth University, teaching courses in discrete mathematics, coding theory, and data security.

Dr. Rosen has published numerous articles in professional journals in the areas of number theory and mathematical modeling. He is the author of the textbooks *Elementary Number Theory and Its Applications,* currently in its fourth edition, published by Addison-Wesley, and *Discrete Mathematics and Its Applications,* in its fifth edition, published by McGraw-Hill. Both books have been used extensively at hundreds of universities. He is coauthor of *UNIX: The Complete Reference, UNIX System V Release 4: An Introduction,* and *Best UNIX Tips Ever*, published by Osborne McGraw-Hill. These books have sold more than 100,000 copies, with translations into Chinese, German, Spanish, and Italian. Ken is also the editor of the *Handbook of Discrete and Combinatorial Mathematics,* published in 2000 by CRC Press, and he is the advisory editor of the CRC series of books in discrete mathematics. Ken is also interested in integrating mathematical software into the educational and professional environments and has worked on projects with Waterloo MAPLE software in both these areas.

At Bell Laboratories, and now AT&T Laboratories, Dr. Rosen has worked on a wide range of projects, including operations research studies and product line planning for computers and data communications equipment. He has helped plan AT&T's future products and services in the area of multimedia, including video communications, speech recognition and synthesis, and image networking. He has evaluated new technology for use by AT&T. He has also invented many new services, and holds or has submitted more than 65 patents. One of his more interesting projects involved helping evaluate technology for the AT&T attraction at EPCOT Center.

Preface

In writing this book, I was guided by my long-standing experience and interest in teaching discrete mathematics. For the student, my purpose was to present material in a precise, readable manner, with the concepts and techniques of discrete mathematics clearly presented and demonstrated. My goal was to show the relevance and practicality of discrete mathematics to students, who are often skeptical. I wanted to give students studying computer science all the mathematical foundations they need for their future studies; I wanted to give mathematics students an understanding of important mathematical concepts together with a sense of why these concepts are important for applications. And I wanted to accomplish these goals without watering down the material.

For the instructor, my purpose was to design a flexible, comprehensive teaching tool using proven pedagogical techniques in mathematics. I wanted to provide instructors with a package of materials that they could use to teach discrete mathematics effectively and efficiently in the most appropriate manner for their particular set of students. I hope that I have achieved these goals.

I have been extremely gratified by the tremendous success of this text. The many improvements in the fifth edition have been made possible by the feedback and suggestions of a large number of instructors and students at many of the more than 500 schools where this book has been successfully used. There are many enhancements in this edition. The ancillary package has been enriched, and a companion website provides helpful material, making it easier for students and instructors to achieve their goals.

This text is designed for a one- or two-term introductory discrete mathematics course to be taken by students in a wide variety of majors, including mathematics, computer science, and engineering. College algebra is the only explicit prerequisite.

Goals of a Discrete Mathematics Course

A discrete mathematics course has more than one purpose. Students should learn a particular set of mathematical facts and how to apply them; more importantly, such a course should teach students how to think mathematically. To achieve these goals, this text stresses mathematical reasoning and the different ways problems are solved. Five important themes are interwoven in this text: mathematical reasoning, combinatorial analysis, discrete structures, algorithmic thinking, and applications and modeling. A successful discrete mathematics course should carefully blend and balance all five themes.

1. *Mathematical Reasoning:* Students must understand mathematical reasoning in order to read, comprehend, and construct mathematical arguments. This text starts with a discussion of mathematical logic, which serves as the foundation for the subsequent discussions of methods of proof. The technique of mathematical induction is stressed through many different types of examples of such proofs and a careful explanation of why mathematical induction is a valid proof technique.

2. *Combinatorial Analysis:* An important problem-solving skill is the ability to count or enumerate objects. The discussion of enumeration in this book begins with the basic techniques of counting. The stress is on performing combinatorial analysis to solve counting problems, not on applying formulae.

3. *Discrete Structures:* A course in discrete mathematics should teach students how to work with discrete structures, which are the abstract mathematical structures used to represent discrete objects and relationships between these objects. These discrete structures include sets, permutations, relations, graphs, trees, and finite-state machines.

4. *Algorithmic Thinking:* Certain classes of problems are solved by the specification of an algorithm. After an algorithm has been described, a computer program can be constructed implementing it. The mathematical portions of this activity, which include the specification of the algorithm, the verification that it works properly, and the analysis of the computer memory and time required to perform it, are all covered in this text. Algorithms are described using both English and an easily understood form of pseudocode.

5. *Applications and Modeling:* Discrete mathematics has applications to almost every conceivable area of study. There are many applications to computer science and data networking in this text, as well as applications to such diverse areas as chemistry, botany, zoology, linguistics, geography, business, and the Internet. These applications are natural and important uses of discrete mathematics and are not contrived. Modeling with discrete mathematics is an extremely important problem-solving skill, which students have the opportunity to develop by constructing their own models in some of the exercises.

Changes in the Fifth Edition

The fourth edition of this book has been used successfully at over 500 schools in the United States, dozens of Canadian universities, and at universities throughout Europe, Asia, and Oceania. Although the fourth edition has been an extremely effective text, many instructors, including longtime users, have requested changes designed to make this book more effective. I have devoted a significant amount of time and energy to satisfy these requests.

The result is a fifth edition that offers both instructors and students much more than the fourth edition did. Most significantly, an improved organization of topics has been implemented in this fifth edition, making the book a more effective teaching tool. Substantial enhancements to the material devoted to logic, method of proof, and proof strategies are designed to help students master mathematical reasoning. Additional explanations and examples have been added to clarify material where students often have difficulty. New exercises, both routine and challenging, have been inserted into the exercise sets. Highly relevant applications, including many related to the Web and computer science, have been added. The companion website has benefited from extensive development activity and now provides tools students can use to master key concepts and explore the world of discrete mathematics.

Improved Organization

- Coverage of mathematical reasoning is concentrated in Chapter 1, flowing from propositional and predicate logic to rules of inference and basic proof techniques.

- Big-*O* and related notation is discussed immediately before the complexity of algorithms.

- Sequences and sums are treated immediately before the section on mathematical induction.

- Binomial coefficients are covered in a separate section.
- Probability theory is covered in its own chapter.

- Sorting is introduced in Chapter 2, with more sorting algorithms addressed.

Logic

- Implications receive in-depth coverage, with additional treatment of the converse, inverse, and contrapositive of an implication.
- A subsection on logic puzzles has been added.
- Quantifiers are now covered in two sections.
- More explanation is provided on how to translate from English and mathematical statements to logical expressions and vice versa.

- Negations of quantifications are covered in greater depth.
- Resolution and the use of predicate logic in Prolog are treated.
- The application of logic to system specifications—a topic of interest to system, hardware, and software engineers, is described.

Writing and Understanding Proofs

- Methods of proof are now introduced in Chapter 1, allowing this material to be used explicitly throughout early chapters.
- Uniqueness proofs are now explicitly discussed.
- Section 3.1 further explores proof strategies, expanding on the introductory treatment of proof strategies in Chapter 1.

- Coverage of mathematical induction and strong induction have been enhanced with additional explanations and new examples.
- Structural induction is covered explicitly.
- Proving the correctness of a recursive algorithm is now treated.

Algorithms

- Greedy algorithms are introduced in Chapter 2.
- The coverage of recursive algorithms has been expanded.
- Coverage of depth-first search and breadth-first search algorithms has been increased.
- The complexity of more algorithms is analyzed or discussed.

- Treatment of divide-and-conquer algorithms and recurrence relations used to study their complexity has been expanded.
- Algorithms for fast modular exponentiation, solving the closest-pair problem, Huffman coding, and the greedy algorithm for making change have been added.

Applications

- Graph models added include the Web graph, telephone call graphs, the Hollywood graph, acquaintanceship graphs, and collaboration graphs.
- Search techniques used by Web spiders are now discussed.

- The use of Gray codes in K-maps is now described.
- Coverage of Backus–Naur form has been increased.
- The treatment of n-ary relations and relational databases has been expanded.

Number Theory, Combinatorics, and Probability Theory

- Number theory relevant to public key cryptography, including pseudoprimes, Carmichael numbers, and probabilistic primality testing, is covered in greater depth.
- The material on conversions between different base expansions has been expanded.

- Coverage of binomial coefficients and the binomial theorem is now in a separate section.
- Coverage of probability theory has been expanded and is now in a separate chapter with expanded treatments of many key topics.

Graphs and Trees

- An explanation of how to inductively construct n-cubes has been added.
- Sufficient conditions for the existence of Hamilton circuits are now covered in greater detail.

- Game trees and minmax strategies are discussed.
- Huffman coding is now introduced.
- Depth-first and breadth-first search now receive expanded coverage.

Exercise Sets

- Over 600 new exercises have been added, ranging from routine to challenging, with a special emphasis on new exercises related to logic and proof, including mathematical induction.

- Exercises have been added to balance odd-numbered exercises with answers and even-numbered exercises.

Additional Biographies and Historical Notes

- Biographies have been added for Aristotle, Sheffer, Smullyan, Rivest, Shamir, Adleman, Carmichael, McCarthy, and Huffman.
- Images of people described in the biographies are now included, except in the few cases where no image is available.

- Many biographies found in the previous edition have been enhanced.
- New historical notes have been added.

The Website (www.mhhe.com/rosen)

- Expanded annotated links to hundreds of Web resources have been developed.
- Extra examples in key areas are hosted.
- More detailed explanation of certain examples and proofs can be accessed.
- Tools for self-assessment of key topics including implications, quantifiers, proof methods, functions, big-O notation, mathematical induction, and counting problems are available.

- Icons are placed in the book indicating the type of associated content on the Web.
- Interactive demos of key algorithms have been developed for integrated use with the text.

Special Features

ACCESSIBILITY   This text has proven to be easily read and understood by beginning students. There are no mathematical prerequisites beyond college algebra for almost all of this text. The few places in the book where calculus is referred to are explicitly noted. Most students should easily understand the pseudocode used in the text to express algorithms, regardless of whether they have formally studied programming languages. There is no formal computer science prerequisite.

Each chapter begins at an easily understood and accessible level. Once basic mathematical concepts have been carefully developed, more difficult material and applications to other areas of study are presented.

FLEXIBILITY   This text has been carefully designed for flexible use. The dependence of chapters on previous material has been minimized. Each chapter is divided into sections of approximately the same length, and each section is divided into subsections that form natural blocks of material for teaching. Instructors can easily pace their lectures using these blocks.

WRITING STYLE   The writing style in this book is direct and pragmatic. Precise mathematical language is used without excessive formalism and abstraction. Care has been taken to balance the mix of notation and words in mathematical statements.

EXTENSIVE CLASSROOM USE   This book has been used at over 500 schools, and more than 400 have used it more than once. The feedback from instructors and students at many of the schools has helped make this fifth edition an even more successful teaching tool than previous editions.

MATHEMATICAL RIGOR AND PRECISION   All definitions and theorems in this text are stated extremely carefully so that students will appreciate the precision of language and rigor needed in mathematics. Proofs are motivated and developed slowly; their steps are all carefully justified. Recursive definitions are explained and used extensively.

WORKED EXAMPLES   Over 700 examples are used to illustrate concepts, relate different topics, and introduce applications. In most examples, a question is first posed, then its solution is presented with the appropriate amount of detail.

APPLICATIONS   The applications included in this text demonstrate the utility of discrete mathematics in the solution of real-world problems. This text includes applications to a wide variety of areas, including computer science, data networking, psychology, chemistry, engineering, linguistics, biology, business, and the Internet.

ALGORITHMS   Results in discrete mathematics are often expressed in terms of algorithms; hence, key algorithms are introduced in each chapter of the book. These algorithms are expressed in words and in an easily understood form of structured pseudocode, which is described and specified in Appendix A.2. The computational complexity of the algorithms in the text is also analyzed at an elementary level.

HISTORICAL INFORMATION   The background of many topics is succinctly described in the text. Brief biographies of more than 60 mathematicians and computer scientists, accompanied by photos or images, are included as footnotes. These biographies include information about the lives, careers, and accomplishments of these important contributors to discrete mathematics and images of these contributors are displayed. In addition, numerous historical footnotes are included that supplement the historical information in the main body of the text.

KEY TERMS AND RESULTS   A list of key terms and results follows each chapter. The key terms include only the most important that students should learn, not every term defined in the chapter.

EXERCISES   There are over 3500 exercises in the text. There are many different types of questions posed. There is an ample supply of straightforward exercises that develop basic skills, a large number of intermediate exercises, and many challenging exercises. Exercises are stated clearly and unambiguously, and all are carefully graded for level of difficulty. Exercise sets contain special discussions, with exercises, that develop new concepts not covered in the text, enabling students to discover new ideas through their own work.

Exercises that are somewhat more difficult than average are marked with a single star; those that are much more challenging are marked with two stars. Exercises whose solutions require calculus are explicitly noted. Exercises that develop results used in the text are clearly identified with the symbol ☞. Answers or outlined solutions to all odd-numbered exercises are provided at the back of the text. The solutions include proofs in which most of the steps are clearly spelled out.

REVIEW QUESTIONS   A set of review questions is provided at the end of each chapter. These questions are designed to help students focus their study on the most important concepts and techniques of that chapter. To answer these questions students need to write long answers, rather than just perform calculations or give short replies.

SUPPLEMENTARY EXERCISE SETS   Each chapter is followed by a rich and varied set of supplementary exercises. These exercises are generally more difficult than those in the exercise sets following the sections. The supplementary exercises reinforce the concepts of the chapter and integrate different topics more effectively.

COMPUTER PROJECTS   Each chapter is followed by a set of computer projects. The approximately 150 computer projects tie together what students may have learned in computing and in discrete mathematics. Computer projects that are more difficult than average, from both a mathematical and a programming point of view, are marked with a star, and those that are extremely challenging are marked with two stars.

COMPUTATIONS AND EXPLORATIONS   A set of computations and explorations is included at the conclusion of each chapter. These exercises (approximately 100 in total) are designed to be completed using existing software tools, such as programs that students or instructors have written or mathematical computation packages such as MAPLE or Mathematica. Many of these exercises give students the opportunity to uncover new facts and ideas through computation. (Some of these exercises are discussed in the book, *Exploring Discrete Mathematics with MAPLE*.)

WRITING PROJECTS   Each chapter is followed by a set of writing projects. To do these projects students need to consult the mathematical literature. Some of these projects are historical in nature and may involve looking up original sources. Others are designed to serve as gateways to new topics and ideas. All are designed to expose students to ideas not covered in depth in the text. These projects tie together mathematical concepts and the writing process and help expose students to possible areas for future study. (Suggested references for these projects can be found in the *Student Solutions Guide*.)

APPENDIXES   There are two appendixes to the text. The first covers exponential and logarithmic functions, reviewing some basic material used heavily in the course; the second specifies the pseudocode used to describe algorithms in this text.

SUGGESTED READINGS   A list of suggested readings for each chapter is provided in a section at the end of the text. These suggested readings include books at or below the level of this text, more difficult books, expository articles, and articles in which discoveries in discrete mathematics were originally published.

How to Use This Book

This text has been carefully written and constructed to support discrete mathematics courses at several levels and with differing foci. The following table identifies the core and optional sections. An introductory one-term course in discrete mathematics at the sophomore level can be based on the core sections of the text, with other sections covered at the discretion of the instructor. A two-term introductory course could include all the optional mathematics sections in addition to the core sections. A course with a strong computer science emphasis can be taught by covering some or all of the optional computer science sections.

Chapter	Core Sections	Optional Computer Science Sections	Optional Mathematics Sections
1	1.1–1.8 (as needed)		
2	2.1–2.4, 2.7 (as needed)	2.5	2.6
3	3.1–3.4	3.5, 3.6	
4	4.1–4.3	4.6	4.4, 4.5
5	5.1	5.3	5.2
6	6.1, 6.5	6.3	6.2, 6.4, 6.6
7	7.1, 7.3, 7.5	7.2	7.4, 7.6
8	8.1–8.5		8.6–8.8
9	9.1	9.2, 9.3	9.4, 9.5
10		10.1–10.4	
11		11.1–11.5	

Instructors using this book can adjust the level of difficulty of their course by choosing either to cover or to omit the more challenging examples at the end of sections, as well as the more challenging exercises. The dependence of chapters on earlier chapters is shown in the following chart.

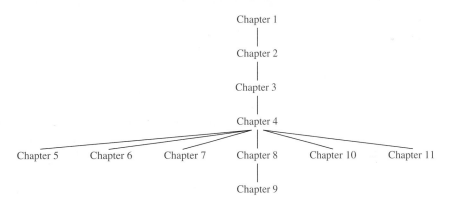

Ancillaries

STUDENT SOLUTIONS GUIDE   This student manual, available separately, contains *full* solutions to all the odd-numbered problems in the exercise sets. These solutions explain why a particular method is used and why it works. For some exercises, one or two other possible approaches are described to show that a problem can be solved in several different ways. Suggested references for the writing projects found at the end of each chapter are also included in this volume. The guide contains a guide to writing proofs and an extensive description of common mistakes students make in discrete mathematics. It also includes sample tests and a sample crib sheet for each chapter, both designed to help students prepare for exams. Students find this guide extremely useful.

INSTRUCTOR'S RESOURCE GUIDE   This manual contains full solutions to even-numbered exercises in the text. It also provides suggestions on how to teach the material in each chapter of the book, including the points to stress in each section and how to put the material into perspective. Furthermore, the manual contains a test bank of sample examination questions for each chapter, including some sample tests as well as the solutions to the sample questions. Finally, sample syllabi are presented.

APPLICATIONS OF DISCRETE MATHEMATICS   This ancillary, available in print format or downloadable from the website, can be used either in conjunction with the text or independently. It contains more than 20 chapters (each with its own set of exercises) written by instructors who have used the text. Following a common format similar to that of the text, the chapters in this book can be used as a text for a separate course, for a student seminar, or for a student doing independent study.

TEST BANK   An extensive test bank of more than 1600 questions is available for use on Windows or Macintosh systems. Instructors can use this software to create their own tests by selecting questions of their choice or by random selection. Instructors can add their own headings and instructions, print scrambled versions of the same test, and edit the existing questions or add their own. A printed version of this test bank, including the questions and their answers, is included in the Instructor's Resource Guide.

EXPLORING DISCRETE MATHEMATICS AND ITS APPLICATIONS WITH MAPLE   This ancillary is designed to help students use the MAPLE computer algebra system to do a wide range of computations in discrete mathematics. For each chapter of this text, this ancillary includes the following: a description of relevant MAPLE functions and how they are used, MAPLE programs that carry out relevant computations, suggestions and examples showing how MAPLE can be used for the computations and explorations at the end of each chapter, and exercises that can be worked using MAPLE.

Acknowledgments

I would like to thank the many instructors and students at a variety of schools who have used this book and provided me with their valuable feedback and helpful suggestions. Their input has made this a much better book than it would have been otherwise. I especially want to thank Jerrold Grossman and John Michaels for their technical reviews of the fifth edition and their "eagle eyes," which have helped ensure the accuracy of this book. I also appreciate the help provided by all those who have submitted comments via the website.

I thank the reviewers of this fifth and the four previous editions. These reviewers have provided much helpful criticism and encouragement to me. I hope this edition lives up to their high expectations.

Reviewers for the Fifth Edition

Kendall Atkinson,
  University of Iowa, Iowa City

Zhaojun Bai,
  University of California, Davis

Klaus Bichteler,
  University of Texas, Austin

Scott Buffett,
  University of New Brunswick

E. Rodney Canfield,
  University of Georgia

George J. Davis,
  Georgia State University

Bruce S. Elenbogen,
  University of Michigan, Dearborn

Jonathan Goldstine,
  Pennsylvania State University

Brian Gray,
  Howard Community College

Jonathan Gross,
  Columbia University

Jerrold Grossman,
  Oakland University

David F. Hayes,
  San Jose State University

Akihiro Kanamori,
  Boston University

Takashi Kimura,
  Boston University

Shui F. Lam,
  California State University, Long Beach

Harbir Lamba,
  George Mason University

Sheau Dong Lang,
  University of Central Florida

Cary Lee,
  *Grossmont Community
  College*

Stephen C. Locke,
  Florida Atlantic University

George F. Luger,
  University of New Mexico

John G. Michaels,
  SUNY Brockport

Thomas D. Morley,
  Georgia Tech

Timothy S. Norfolk, Sr.
  University of Akron

Truc T. Nguyen,
  *Bowling Green State
  University*

George Novacky,
  University of Pittsburgh

Jaroslav Opatrny,
  Concordia University

Jonathan Pakianathan,
  University of Rochester

Halina Przymusinska,
  *California State Polytechnic
  University, Pomona*

Don Reichman,
  *Mercer County Community
  College*

Robert Rodman,
  *North Carolina State
  University*

Matthew J. Saltzman,
  Clemson University

Michael J. Schlosser,
  The Ohio State University

Alistair Sinclair,
  *University of California,
  Berkeley*

Carl H. Smith,
  University of Maryland

Patrick Tantalo,
  *University of California, Santa
  Cruz*

Reviewers of Previous Editions

Eric Allender,
  Rutgers University

Stephen Andrilli,
  La Salle University

Jack R. Barone,
  Baruch College

Alfred E. Borm,
  *Southwest Texas State
  University*

Ken W. Bosworth,
  University of Maryland

Lois Brady,
  *California Polytechnic State
  University, San Luis Obispo*

Russell Campbell,
  University of Northern Iowa

Kevin Carolan,
  Marist College

Tim Carroll,
  Bloomsburg University

Kit C. Chan,
  *Bowling Green State
  University*

Allan C. Cochran,
  University of Arkansas

Peter Collinge,
  Monroe Community College

Ron Davis,
  Millersville University

Nachum Dershowitz,
  *University of Illinois,
  Urbana–Champaign*

Thomas Dowling,
  The Ohio State University

Patrick C. Fischer,
  Vanderbilt University

Jane Fritz,
  University of New Brunswick

Ladnor Geissinger,
  University of North Carolina

Paul Gormley,
  Villanova University

Laxmi N. Gupta,
  *Rochester Institute of
  Technology*

Daniel Gusfield,
  *University of California at
  Davis*

Xin He,
  SUNY at Buffalo

Arthur M. Hobbs,
  Texas A&M University

Donald Hutchison,
  *Clackamas Community
  College*

Kenneth Johnson,
  North Dakota State University

David Jonah,
  Wayne State University

W. Thomas Kiley,
  George Mason University

Nancy Kinnersley,
  University of Kansas

Gary Klatt,
  University of Wisconsin

Nicholas Krier,
  Colorado State University

Lawrence S. Kroll,
  San Francisco State University

Robert Lavelle,
  Iona College

Yi-Hsin Liu,
  *University of Nebraska at
  Omaha*

George Luger,
  University of New Mexico

David S. McAllister,
  *North Carolina State
  University*

Robert McGuigan,
  Westfield State College

Michael Maller,
  Queens College

Ernie Manes,
  University of Massachusetts

Francis Masat,
  Glassboro State College

J. M. Metzger,
  University of North Dakota

D. R. Morrison,
  University of New Mexico

Ho Kuen Ng,
  San Jose State University

Jeffrey Nunemacher,
  Ohio Wesleyan University

Charles Parry,
  *Virginia Polytechnic Institute
  and State University*

Thomas W. Parsons,
  Hofstra University

Mary K. Prisco,
  *University of Wisconsin–Green
  Bay*

Harold Reiter,
  University of North Carolina

Adrian Riskin,
  *Northern Arizona State
  University*

Amy L. Rocha,
  San Jose State University

Janet Roll,
  University of Findlay

Alyssa Sankey,
  Slippery Rock University

Dinesh Sarvate,
  College of Charleston

Steven R. Seidel,
  Michigan Technological University

Douglas Shier,
  Clemson University

Hunter S. Snevily,
  University of Idaho

Daniel Somerville,
  University of Massachusetts, Boston

Bharti Temkin,
  Texas Tech University

Wallace Terwilligen,
  *Bowling Green State
  University*

Philip D. Tiu,
  Oglethorpe University

Lisa Townsley-Kulich,
  Illinois Benedictine College

George Trapp,
  West Virginia University

David S. Tucker,
  Midwestern State University

Thomas Upson,
  *Rochester Institute of
  Technology*

Roman Voronka,
  *New Jersey Institute of
  Technology*

James Walker,
  University of South Carolina

Anthony S. Wojcik,
  Michigan State University

I would like to thank the staff of McGraw-Hill for their strong support of this project. In particular, thanks go to Bill Barter, Publisher, for his backing; to Robert Ross, Executive Editor, for his advocacy and enthusiasm; and to Michelle Munn, Developmental Editor, for her dedication and attention. I would also like to thank the original editor, Wayne Yuhasz, whose insights and skills helped ensure the book's success, as well as Maggie Rogers, the original Sponsoring Editor of this edition, who helped plan and initiate work on this edition and with whom I worked for many years.

To the staff involved in the production of this fifth edition, I offer my thanks to Joyce Berendes, Senior Project Manager; K. Wayne Harms, Designer; Jeff Huettman, Web Developer; Laura Fuller, Supplements Coordinator; and Mary Kittell, Marketing Manager. I am also grateful to Georgia Kamvosoulis Mederer who checked the entire manuscript for accuracy, Mary Ellen Oliver, the technical proofreader, and Paul Lorczak who checked the accuracy of the solutions manual and answer sections.

As always, I am grateful for the support given to me by my management at AT&T Laboratories. They have provided an environment that has helped me develop professionally and generously given me the resources needed to make this book a success.

Kenneth H. Rosen

The Companion Website

An extensive companion website has been developed and will be maintained and improved on a continuing basis. You can use this site in many ways to enhance your experience studying discrete mathematics. The address of this site is:

www.mhhe.com/rosen

Following this address takes you to the home page for the site. On this site you will find links for:

- Information Center
- Student Center
- Instructor Center

THE INFORMATION CENTER   The Information Center contains some basic information about the book and its ancillaries. Instructors can find out about the *Page Out*^{TM} system they can use to build their own course Web pages. They can also learn about custom publishing options. An overview of the website for this book can be reached by following the appropriate link from the Information Center.

THE STUDENT CENTER   The Student Center contains a wealth of resources available for student use, including the following resources tied into the text and for which a special icon is shown in the text.

- **The Web Resources Guide**   This guide provides annotated links to hundreds of external websites containing relevant material. You can browse these links or access them by page number in the text or by keywords. These links will take you to websites containing historical and biographical information, puzzles and problems, discussions, applets, programs, and other types of resources.

- **Extra Examples**   You can find a large number of additional examples on the site. These examples are concentrated in areas where students often ask for additional material. Although most of these examples amplify the basic concepts, some more challenging examples can also be found here.

- **Additional Steps**   Further explanations are provided to help you understand some troublesome points in the text, especially in some proofs and examples.

- **Assessments**   You can assess your understanding of seven key concepts. Each assessment provides a question bank where each question includes a brief tutorial, followed by a multiple-choice question. If you select an incorrect answer, advice is provided to help you understand your error. Using these assessments, you should be able to diagnose your problems and focus on available remedies.

- **Interactive Demonstrations**   We have developed eight interactive demonstrations which you can use to explore how important algorithms work. These demonstrations are keyed to material in the text.

From the Student Center you can access *Net Tutor*™, which provides online tutorial help. You can ask questions relating to this text and receive answers in real-time during regular, scheduled hours or ask questions and later receive answers.

The Student Center also supports a *Bulletin Board* on which you can post messages. You can post questions and respond to messages from other students using this facility.

Additional resources in the Student Center include:

- A Guide to Writing Proofs

- Common Mistakes in Discrete Mathematics

- Advice on Writing Projects

- MAPLE software

THE INSTRUCTOR CENTER   This part of the website provides links to the resources in the Student Center and the Information Center, as well as:

- Sample syllabi

- Teaching suggestions

- Transparency masters

- Several chapters from the book *Applications of Discrete Mathematics*

To the Student

What is discrete mathematics? Discrete mathematics is the part of mathematics devoted to the study of discrete objects. (Here *discrete* means consisting of distinct or unconnected elements.) The kind of problems solved using discrete mathematics include:

- How many ways are there to choose a valid password on a computer system?

- What is the probability of winning a lottery?

- Is there a link between two computers in a network?

- What is the shortest path between two cities using a transportation system?

- How can a list of integers be sorted so that the integers are in increasing order?

- How many steps are required to do such a sorting?

- How can it be proved that a sorting algorithm correctly sorts a list?

- How can a circuit that adds two integers be designed?

- How many valid Internet addresses are there?

You will learn the discrete structures and techniques needed to solve problems such as these.

More generally, discrete mathematics is used whenever objects are counted, when relationships between finite (or countable) sets are studied, and when processes involving a finite number of steps are analyzed. A key reason for the growth in the importance of discrete mathematics is that information is stored and manipulated by computing machines in a discrete fashion.

WHY STUDY DISCRETE MATHEMATICS?   There are several important reasons for studying discrete mathematics. First, through this course you can develop your mathematical maturity, that is, your ability to understand and create mathematical arguments. You will not get very far in your studies in the mathematical sciences without these skills.

Second, discrete mathematics is the gateway to more advanced courses in all parts of the mathematical sciences. Discrete mathematics provides the mathematical foundations for many computer science courses, including data structures, algorithms, database theory, automata theory, formal languages, compiler theory, computer security, and operating systems. Students find these courses much more difficult when they have not had the appropriate mathematical foundations from discrete math. One student has sent me an electronic mail message to tell me that she used the contents of this book in every computer science course she took!

Math courses based on the material studied in discrete mathematics include logic, set theory, number theory, linear algebra, abstract algebra, combinatorics, graph theory, and probability theory (the discrete part of the subject).

Also, discrete mathematics contains the necessary mathematical background for solving problems in operations research (including many discrete optimization techniques), chemistry, engineering, biology, and so on. In the text, we will study applications to some of these areas.

Many students find their introductory discrete mathematics course to be significantly more challenging than courses they have previously taken. One reason for this is that one of the primary goals of this course is to teach mathematical reasoning and problem solving, rather than a discrete set of skills. The exercises in this book are designed to reflect this goal. Although there are plenty of exercises in this text similar to those addressed in the examples, a large percentage of the exercises require original thought. This is intentional. The material discussed in the text provides the tools needed to solve these exercises, but your job is to successfully apply these tools using your own creativity. One of the primary goals of this course is to learn how to attack problems that may be somewhat different from any you may have previously seen. Unfortunately, learning how to solve only particular types of exercises is not sufficient for success in developing the problem-solving skills needed in subsequent courses and professional work. This text addresses many different topics, but discrete mathematics is an extremely diverse and large area of study. One of my goals as an author is to help you develop the skills needed to master the additional material you will need in your own future pursuits.

THE EXERCISES I would like to offer some advice about how you can best learn discrete mathematics (and other subjects in the mathematical and computing sciences). You will learn the most by actively working exercises. I suggest that you solve as many as you possibly can. After working the exercises your instructor has assigned, I encourage you to solve additional exercises such as those in the exercise sets following each section of the text and in the supplementary exercises at the end of each chapter. (Note the key explaining the markings preceding exercises.)

Key to the Exercises

No marking	A routine exercise
*	A difficult exercise
**	An extremely challenging exercise
☞	An exercise containing a result used in the book
(Calculus required)	An exercise whose solution requires the use of limits or concepts from differential or integral calculus

The best approach is to try exercises yourself before you consult the answer section at the end of this book or the complete solutions provided for you in the *Student Solutions Guide*. The *answers* to all odd-numbered exercises in this text are provided at the end of this book; note that these are answers only and they are not full solutions. In particular, the reasoning required to obtain answers is omitted in these answers. The *Student Solutions Guide* provides complete, worked solutions to all odd-numbered exercises in this text. When you hit an impasse trying to solve an odd-numbered exercise, I suggest you consult the *Student Solutions Guide* and look for some guidance as to how to solve the problem. The more work you do yourself rather than passively reading or copying solutions, the more you will learn. The answers and solutions to the even-numbered exercises are intentionally not available from the publisher; ask your instructor if you have trouble with these.

WEB RESOURCES You are *strongly* encouraged to take advantage of additional resources available on the Web, especially those on the site for this book found at www.mhhe.com/rosen. You will find many extra examples, additional steps designed to clarify places where students often have trouble, assessments for gauging how well you understand key concepts, animated algorithms for exploring key algorithms, and a potpourri of links to external sites that you can use to explore the world of discrete mathematics. You will also find a bulletin board for discussions with other students. You can use this bulletin board to ask other students for help understanding concepts or for guidance attacking exercises. Students who can help might find that answering questions helps them better master the material. You will even have access to an online tutorial service that you can use to receive help from tutors either via real-time chat or via messages. (Please follow your instructor's wishes concerning your use of work done by others.) For more details on the website, see the description of the companion website immediately preceding this "To the Student" message.

THE VALUE OF THIS BOOK Finally, I understand that this book is costly. My intention is to make your investment in this text an excellent value. The book, the associated ancillaries, and companion website have taken many years of effort to develop and refine. I am confident that most of you will find that the text and associated materials will help you master discrete mathematics. Even though it is likely that you will not cover some chapters in your current course, you should find it helpful, as many other students have, to read the relevant sections of the book as you take additional courses. Most of you will return to this book as a useful tool throughout your future studies, especially for those of you who continue in computer science, mathematics, and engineering.

Kenneth H. Rosen

1 The Foundations: Logic and Proof, Sets, and Functions

This chapter reviews the foundations of discrete mathematics. Three important topics are covered: logic, sets, and functions. The rules of logic specify the meaning of mathematical statements. For instance, these rules help us understand and reason with statements such as "There exists an integer that is not the sum of two squares," and "For every positive integer n the sum of the positive integers not exceeding n is $n(n + 1)/2$." Logic is the basis of all mathematical reasoning, and it has practical applications to the design of computing machines, to system specifications, to artificial intelligence, to computer programming, to programming languages, and to other areas of computer science, as well as to many other fields of study.

To understand mathematics, we must understand what makes up a correct mathematical argument, that is, a proof. Moreover, to learn mathematics, a person needs to actively construct mathematical arguments and not just read exposition. In this chapter, we explain what makes up a correct mathematical argument and introduce tools to construct these arguments. Proofs are important not only in mathematics, but also in many parts of computer science, including program verification, algorithm correctness, and system security. Furthermore, automated reasoning systems have been constructed that allow computers to construct their own proofs.

Much of discrete mathematics is devoted to the study of discrete structures, which are used to represent discrete objects. Many important discrete structures are built using sets, which are collections of objects. Among the discrete structures built from sets are combinations, which are unordered collections of objects used extensively in counting; relations, which are sets of ordered pairs that represent relationships between objects; graphs, which consist of sets of vertices and of edges that connect vertices; and finite state machines, which are used to model computing machines.

The concept of a function is extremely important in discrete mathematics. A function assigns to each element of a set precisely one element of a set. Useful structures such as sequences and strings are special types of functions. Functions play important roles throughout discrete mathematics. They are used to represent the computational complexity of algorithms, to study the size of sets, to count objects of different kinds, and in a myriad of other ways.

1.1 Logic

INTRODUCTION

The rules of logic give precise meaning to mathematical statements. These rules are used to distinguish between valid and invalid mathematical arguments. Since a major goal of this book is to teach the reader how to understand and how to construct correct mathematical arguments, we begin our study of discrete mathematics with an introduction to logic.

In addition to its importance in understanding mathematical reasoning, logic has numerous applications in computer science. These rules are used in the design of computer circuits, the construction of computer programs, the verification of the correctness of programs, and in many other ways. We will discuss each of these applications in the upcoming chapters.

PROPOSITIONS

Our discussion begins with an introduction to the basic building blocks of logic—propositions. A **proposition** is a declarative sentence that is either true or false, but not both.

EXAMPLE 1   All the following declarative sentences are propositions.

1. Washington, D.C., is the capital of the United States of America.
2. Toronto is the capital of Canada.
3. $1 + 1 = 2$.
4. $2 + 2 = 3$.

Propositions 1 and 3 are true, whereas 2 and 4 are false. ◀

Some sentences that are not propositions are given in the next example.

EXAMPLE 2   Consider the following sentences.

1. What time is it?
2. Read this carefully.

ARISTOTLE (384 B.C.E.–322 B.C.E.) Aristotle was born in Stargirus in northern Greece. His father was the personal physician of the King of Macedonia. Because his father died when Aristotle was young, Aristotle could not follow the custom of following his father's profession. Aristotle became an orphan at a young age when his mother also died. His guardian who raised him taught him poetry, rhetoric, and Greek. At the age of 17, his guardian sent him to Athens to further his education. Aristotle joined Plato's Academy where for 20 years he attended Plato's lectures, later presenting his own lectures on rhetoric. When Plato died in 347 B.C.E., Aristotle was not chosen to succeed him because his views differed too much from those of Plato. Instead, Aristotle joined the court of King Hermeas where he remained for three years, and married the niece of the King. When the Persians defeated Hermeas, Aristotle moved to Mytilene and, at the invitation of King Philip of Macedonia, he tutored Alexander, Philip's son, who later became Alexander the Great. Aristotle tutored Alexander for five years and after the death of King Philip, he returned to Athens and set up his own school, called the Lyceum.

Aristotle's followers were called the peripatetics, which means "to walk about," because Aristotle often walked around as he discussed philosophical questions. Aristotle taught at the Lyceum for 13 years where he lectured to his advanced students in the morning and gave popular lectures to a broad audience in the evening. When Alexander the Great died in 323 B.C.E., a backlash against anything related to Alexander led to trumped-up charges of impiety against Aristotle. Aristotle fled to Chalcis to avoid prosecution. He only lived one year in Chalcis, dying of a stomach ailment in 322 B.C.E.

Aristotle wrote three types of works: those written for a popular audience, compilations of scientific facts, and systematic treatises. The systematic treatises included works on logic, philosophy, psychology, physics, and natural history. Aristotle's writings were preserved by a student and were hidden in a vault where a wealthy book collector discovered them about 200 years later. They were taken to Rome, where they were studied by scholars and issued in new editions, preserving them for posterity.

    3. $x + 1 = 2$.

    4. $x + y = z$.

Sentences 1 and 2 are not propositions because they are not declarative sentences. Sentences 3 and 4 are not propositions because they are neither true nor false, since the variables in these sentences have not been assigned values. Various ways to form propositions from sentences of this type will be discussed in Section 1.3. ◄

    Letters are used to denote propositions, just as letters are used to denote variables. The conventional letters used for this purpose are $p, q, r, s, \ldots$. The **truth value** of a proposition is true, denoted by T, if it is a true proposition and false, denoted by F, if it is a false proposition.

    The area of logic that deals with propositions is called the **propositional calculus** or **propositional logic.** It was first developed systematically by the Greek philosopher Aristotle more than 2300 years ago.

Links

    We now turn our attention to methods for producing new propositions from those that we already have. These methods were discussed by the English mathematician George Boole in 1854 in his book *The Laws of Thought.* Many mathematical statements are constructed by combining one or more propositions. New propositions, called **compound propositions,** are formed from existing propositions using logical operators.

DEFINITION 1

Let p be a proposition. The statement

    "It is not the case that p"

is another proposition, called the *negation* of p. The negation of p is denoted by $\neg p$. The proposition $\neg p$ is read "not p."

EXAMPLE 3

Find the negation of the proposition

    "Today is Friday."

Extra Examples

and express this in simple English.

Solution: The negation is

    "It is not the case that today is Friday."

This negation can be more simply expressed by

    "Today is not Friday,"

or

    "It is not Friday today." ◄

TABLE 1 The Truth Table for the Negation of a Proposition.

p	$\neg p$
T	F
F	T

Remark: Strictly speaking, sentences involving variable times such as those in Example 3 are not propositions unless a fixed time is assumed. The same holds for variable places unless a fixed place is assumed and for pronouns unless a particular person is assumed.

    A **truth table** displays the relationships between the truth values of propositions. Truth tables are especially valuable in the determination of the truth values of propositions constructed from simpler propositions. Table 1 displays the two possible truth values of a proposition p and the corresponding truth values of its negation $\neg p$.

The negation of a proposition can also be considered the result of the operation of the **negation operator** on a proposition. The negation operator constructs a new proposition from a single existing proposition. We will now introduce the logical operators that are used to form new propositions from two or more existing propositions. These logical operators are also called **connectives.**

DEFINITION 2

Let p and q be propositions. The proposition "p and q," denoted $p \wedge q$, is the proposition that is true when both p and q are true and is false otherwise. The proposition $p \wedge q$ is called the *conjunction* of p and q.

The truth table for $p \wedge q$ is shown in Table 2. Note that there are four rows in this truth table, one row for each possible combination of truth values for the propositions p and q.

EXAMPLE 4

Find the conjunction of the propositions p and q where p is the proposition "Today is Friday" and q is the proposition "It is raining today."

Solution: The conjunction of these propositions, $p \wedge q$, is the proposition "Today is Friday and it is raining today." This proposition is true on rainy Fridays and is false on any day that is not a Friday and on Fridays when it does not rain. ◄

DEFINITION 3

Let p and q be propositions. The proposition "p or q," denoted $p \vee q$, is the proposition that is false when p and q are both false and true otherwise. The proposition $p \vee q$ is called the *disjunction* of p and q.

The truth table for $p \vee q$ is shown in Table 3.

The use of the connective *or* in a disjunction corresponds to one of the two ways the word *or* is used in English, namely, in an inclusive way. A disjunction is true when at least one of the two propositions is true. For instance, the inclusive or is being used in the statement

"Students who have taken calculus or computer science can take this class."

GEORGE BOOLE (1815–1864) George Boole, the son of a cobbler, was born in Lincoln, England, in November 1815. Because of his family's difficult financial situation, Boole had to struggle to educate himself while supporting his family. Nevertheless, he became one of the most important mathematicians of the 1800s. Although he considered a career as a clergyman, he decided instead to go into teaching and soon afterward opened a school of his own. In his preparation for teaching mathematics, Boole—unsatisfied with textbooks of his day—decided to read the works of the great mathematicians. While reading papers of the great French mathematician Lagrange, Boole made discoveries in the calculus of variations, the branch of analysis dealing with finding curves and surfaces optimizing certain parameters.

In 1848 Boole published *The Mathematical Analysis of Logic,* the first of his contributions to symbolic logic. In 1849 he was appointed professor of mathematics at Queen's College in Cork, Ireland. In 1854 he published *The Laws of Thought,* his most famous work. In this book Boole introduced what is now called *Boolean algebra* in his honor. Boole wrote textbooks on differential equations and on difference equations that were used in Great Britain until the end of the nineteenth century. Boole married in 1855; his wife was the niece of the professor of Greek at Queen's College. In 1864 Boole died from pneumonia, which he contracted as a result of keeping a lecture engagement even though he was soaking wet from a rainstorm.

TABLE 2 The Truth Table for the Conjunction of Two Propositions.		
p	q	$p \wedge q$
T	T	T
T	F	F
F	T	F
F	F	F

TABLE 3 The Truth Table for the Disjunction of Two Propositions.		
p	q	$p \vee q$
T	T	T
T	F	T
F	T	T
F	F	F

Here, we mean that students who have taken both calculus and computer science can take the class, as well as the students who have taken only one of the two subjects. On the other hand, we are using the exclusive or when we say

> "Students who have taken calculus or computer science, but not both, can enroll in this class."

Here, we mean that students who have taken both calculus and a computer science course cannot take the class. Only those who have taken exactly one of the two courses can take the class.

Similarly, when a menu at a restaurant states, "Soup or salad comes with an entrée," the restaurant almost always means that customers can have either soup or salad, but not both. Hence, this is an exclusive, rather than an inclusive, or.

EXAMPLE 5   What is the disjunction of the propositions p and q where p and q are the same propositions as in Example 4?

Solution: The disjunction of p and q, $p \vee q$, is the proposition

> "Today is Friday or it is raining today."

This proposition is true on any day that is either a Friday or a rainy day (including rainy Fridays). It is only false on days that are not Fridays when it also does not rain.   ◀

As was previously remarked, the use of the connective *or* in a disjunction corresponds to one of the two ways the word *or* is used in English, namely, in an inclusive way. Thus, a disjunction is true when at least one of the two propositions in it is true. Sometimes, we use *or* in an exclusive sense. When the exclusive or is used to connect the propositions p and q, the proposition "p or q (but not both)" is obtained. This proposition is true when p is true and q is false, and when p is false and q is true. It is false when both p and q are false and when both are true.

Extra Examples

DEFINITION 4   Let p and q be propositions. The *exclusive or* of p and q, denoted by $p \oplus q$, is the proposition that is true when exactly one of p and q is true and is false otherwise.

The truth table for the exclusive or of two propositions is displayed in Table 4.

TABLE 4 The Truth Table for the Exclusive Or of Two Propositions.		
p	*q*	*p* ⊕ *q*
T	T	F
T	F	T
F	T	T
F	F	F

TABLE 5 The Truth Table for the Implication $p \rightarrow q$.		
p	*q*	$p \rightarrow q$
T	T	T
T	F	F
F	T	T
F	F	T

IMPLICATIONS

We will discuss several other important ways in which propositions can be combined.

DEFINITION 5

Assessment

Let p and q be propositions. The *implication* $p \rightarrow q$ is the proposition that is false when p is true and q is false, and true otherwise. In this implication p is called the *hypothesis* (or *antecedent* or *premise*) and q is called the *conclusion* (or *consequence*).

The truth table for the implication $p \rightarrow q$ is shown in Table 5. An implication is sometimes called a **conditional statement.**

Because implications play such an essential role in mathematical reasoning, a variety of terminology is used to express $p \rightarrow q$. You will encounter most if not all of the following ways to express this implication:

Extra Examples

"if p, then q"   "p implies q"
"if p, q"   "p only if q"
"p is sufficient for q"   "a sufficient condition for q is p"
"q if p"   "q whenever p"
"q when p"   "q is necessary for p"
"a necessary condition for p is q"   "q follows from p"

The implication $p \rightarrow q$ is false only in the case that p is true, but q is false. It is true when both p and q are true, and when p is false (no matter what truth value q has).

A useful way to understand the truth value of an implication is to think of an obligation or a contract. For example, the pledge many politicians make when running for office is:

"If I am elected, then I will lower taxes."

If the politician is elected, voters would expect this politician to lower taxes. Furthermore, if the politician is not elected, then voters will not have any expectation that this person will lower taxes, although the person may have sufficient influence to cause those in power to lower taxes. It is only when the politician is elected but does not lower taxes that voters can say that the politician has broken the campaign pledge. This last scenario corresponds to the case when p is true, but q is false in $p \rightarrow q$.

Similarly, consider a statement that a professor might make:

"If you get 100% on the final, then you will get an A."

If you manage to get a 100% on the final, then you would expect to receive an A. If you do not get 100% you may or may not receive an A depending on other factors. However, if you do get 100%, but the professor does not give you an A, you will feel cheated.

Many people find it confusing that "p only if q" expresses the same thing as "if p then q." To remember this, note that "p only if q" says that p cannot be true when q is not true. That is, the statement is false if p is true, but q is false. When p is false, q may be either true or false, because the statement says nothing about the truth value of q. A common error is for people to think that "q only if p" is a way of expressing $p \rightarrow q$. However, these statements have different truth values when p and q have different truth values.

The way we have defined implications is more general than the meaning attached to implications in the English language. For instance, the implication

"If it is sunny today, then we will go to the beach."

is an implication used in normal language, since there is a relationship between the hypothesis and the conclusion. Further, this implication is considered valid unless it is indeed sunny today, but we do not go to the beach. On the other hand, the implication

"If today is Friday, then $2 + 3 = 5$."

is true from the definition of implication, since its conclusion is true. (The truth value of the hypothesis does not matter then.) The implication

"If today is Friday, then $2 + 3 = 6$."

is true every day except Friday, even though $2 + 3 = 6$ is false.

We would not use these last two implications in natural language (except perhaps in sarcasm), since there is no relationship between the hypothesis and the conclusion in either implication. In mathematical reasoning we consider implications of a more general sort than we use in English. The mathematical concept of an implication is independent of a cause-and-effect relationship between hypothesis and conclusion. Our definition of an implication specifies its truth values; it is not based on English usage.

The if-then construction used in many programming languages is different from that used in logic. Most programming languages contain statements such as **if** p **then** S, where p is a proposition and S is a program segment (one or more statements to be executed). When execution of a program encounters such a statement, S is executed if p is true, but S is not executed if p is false, as illustrated in Example 6.

EXAMPLE 6   What is the value of the variable x after the statement

if $2 + 2 = 4$ **then** $x := x + 1$

if $x = 0$ before this statement is encountered? (The symbol $:=$ stands for assignment. The statement $x := x + 1$ means the assignment of the value of $x + 1$ to x.)

Solution: Since $2 + 2 = 4$ is true, the assignment statement $x := x + 1$ is executed. Hence, x has the value $0 + 1 = 1$ after this statement is encountered.   ◀

CONVERSE, CONTRAPOSITIVE, AND INVERSE There are some related implications that can be formed from $p \rightarrow q$. The proposition $q \rightarrow p$ is called the **converse** of $p \rightarrow q$. The **contrapositive** of $p \rightarrow q$ is the proposition $\neg q \rightarrow \neg p$. The proposition $\neg p \rightarrow \neg q$ is called the **inverse** of $p \rightarrow q$.

The contrapositive, $\neg q \rightarrow \neg p$, of an implication $p \rightarrow q$ has the same truth value as $p \rightarrow q$. To see this, note that the contrapositive is false only when $\neg p$ is false and $\neg q$ is true, that is, only when p is true and q is false. On the other hand, neither the converse, $q \rightarrow p$, nor the inverse, $\neg p \rightarrow \neg q$, has the same truth value as $p \rightarrow q$ for all possible truth values of p and q. To see this, note that when p is true and q is false, the original implication is false, but the converse and the inverse are both true. When two compound propositions always have the same truth value we call them **equivalent,** so that an implication and its contrapositive are equivalent. The converse and the inverse of an implication are also equivalent, as the reader can verify. (We will study equivalent propositions in Section 1.2.) One of the most common logical errors is to assume that the converse or the inverse of an implication is equivalent to this implication.

We illustrate the use of implications in Example 7.

EXAMPLE 7   What are the contrapositive, the converse, and the inverse of the implication

Extra Examples

"The home team wins whenever it is raining."?

Solution: Because "q whenever p" is one of the ways to express the implication $p \rightarrow q$, the original statement can be rewritten as

"If it is raining, then the home team wins."

Consequently, the contrapositive of this implication is

"If the home team does not win, then it is not raining."

The converse is

"If the home team wins, then it is raining."

The inverse is

"If it is not raining, then the home team does not win."

Only the contrapositive is equivalent to the original statement.   ◀

We now introduce another way to combine propositions.

DEFINITION 6

> Let p and q be propositions. The *biconditional* $p \leftrightarrow q$ is the proposition that is true when p and q have the same truth values, and is false otherwise.

The truth table for $p \leftrightarrow q$ is shown in Table 6. Note that the biconditional $p \leftrightarrow q$ is true precisely when both the implications $p \rightarrow q$ and $q \rightarrow p$ are true. Because of this, the terminology

"p if and only if q"

is used for this biconditional and it is symbolically written by combining the symbols $\rightarrow$ and $\leftarrow$. There are some other common ways to express $p \leftrightarrow q$:

TABLE 6 The Truth Table for the Biconditional $p \leftrightarrow q$.		
p	q	$p \leftrightarrow q$
T	T	T
T	F	F
F	T	F
F	F	T

"p is necessary and sufficient for q"
"if p then q, and conversely"
"p if q".

The last way of expressing the biconditional uses the abbreviation "iff" for "if and only if." Note that $p \leftrightarrow q$ has exactly the same truth value as $(p \rightarrow q) \wedge (q \rightarrow p)$.

EXAMPLE 8

Extra Examples

Let p be the statement "You can take the flight" and let q be the statement "You buy a ticket." Then $p \leftrightarrow q$ is the statement

"You can take the flight if and only if you buy a ticket."

This statement is true if p and q are either both true or both false, that is, if you buy a ticket and can take the flight or if you do not buy a ticket and you cannot take the flight. It is false when p and q have opposite truth values, that is, when you do not buy a ticket, but you can take the flight (such as when you get a free trip) and when you buy a ticket and cannot take the flight (such as when the airline bumps you). ◀

The "if and only if" construction used in biconditionals is rarely used in common language. Instead, biconditionals are often expressed using an "if, then" or an "only if" construction. The other part of the "if and only if" is implicit. For example, consider the statement in English "If you finish your meal, then you can have dessert." What is really meant is "You can have dessert if and only if you finish your meal." This last statement is logically equivalent to the two statements "If you finish your meal, then you can have dessert" and "You can have dessert, only if you finish your meal." Because of this imprecision in natural language, we need to make an assumption whether a conditional statement in natural language implicitly includes its converse. Because precision is essential in mathematics and in logic, we will always distinguish between the conditional statement $p \rightarrow q$ and the biconditional statement $p \leftrightarrow q$.

PRECEDENCE OF LOGICAL OPERATORS

We can construct compound propositions using the negation operator and the logical operators defined so far. We will generally use parentheses to specify the order in which logical operators in a compound proposition are to be applied. For instance, $(p \vee q) \wedge (\neg r)$ is the conjunction of $p \vee q$ and $\neg r$. However, to reduce the number of parentheses, we specify that the negation operator is applied before all other logical operators. This means that $\neg p \wedge q$ is the conjunction of $\neg p$ and q, namely, $(\neg p) \wedge q$, not the negation of the conjunction of p and q, namely $\neg(p \wedge q)$.

TABLE 7 Precedence of Logical Operators.	
Operator	**Precedence**
¬	1
∧	2
∨	3
→	4
↔	5

Another general rule of precedence is that the conjunction operator takes precedence over the disjunction operator, so that $p \land q \lor r$ means $(p \land q) \lor r$ rather than $p \land (q \lor r)$. Because this rule may be difficult to remember, we will continue to use parentheses so that the order of the disjunction and conjunction operators is clear.

Finally, it is an accepted rule that the conditional and biconditional operators → and ↔ have lower precedence than the conjunction and disjunction operators, ∧ and ∨. Consequently, $p \lor q \to r$ is the same as $(p \lor q) \to r$. We will use parentheses when the order of the conditional operator and biconditional operator is at issue, although the conditional operator has precedence over the biconditional operator. Table 7 displays the precedence levels of the logical operators, ¬, ∧, ∨, →, and ↔.

TRANSLATING ENGLISH SENTENCES

There are many reasons to translate English sentences into expressions involving propositional variables and logical connectives. In particular, English (and every other human language) is often ambiguous. Translating sentences into logical expressions removes the ambiguity. Note that this may involve making a set of reasonable assumptions based on the intended meaning of the sentence. Moreover, once we have translated sentences from English into logical expressions we can analyze these logical expressions to determine their truth values, we can manipulate them, and we can use rules of inference (which are discussed in Section 1.5) to reason about them.

To illustrate the process of translating an English sentence into a logical expression, consider Examples 9 and 10.

EXAMPLE 9 How can this English sentence be translated into a logical expression?

"You can access the Internet from campus only if you are a computer science major or you are not a freshman."

Solution: There are many ways to translate this sentence into a logical expression. Although it is possible to represent the sentence by a single propositional variable, such as p, this would not be useful when analyzing its meaning or reasoning with it. Instead, we will use propositional variables to represent each sentence part and determine the appropriate logical connectives between them. In particular, we let a, c, and f represent "You can access the Internet from campus," "You are a computer science major," and "You are a freshman," respectively. Noting that "only if" is one way an implication can be expressed, this sentence can be represented as

$$a \to (c \lor \neg f).$$ ◄

EXAMPLE 10 How can this English sentence be translated into a logical expression?

"You cannot ride the roller coaster if you are under 4 feet tall unless you are older than 16 years old."

Solution: There are many ways to translate this sentence into a logical expression. The simplest but least useful way is simply to represent the sentence by a single propositional variable, say, p. Although this is not wrong, doing this would not assist us when we try to analyze the sentence or reason using it. More appropriately, what we can do is to use propositional variables to represent each of the sentence parts and to decide on the appropriate logical connectives between them. In particular, we let q, r, and s represent

"You can ride the roller coaster," "You are under 4 feet tall," and "You are older than 16 years old," respectively. Then the sentence can be translated to

$$(r \wedge \neg s) \rightarrow \neg q.$$

Of course, there are other ways to represent the original sentence as a logical expression, but the one we have used should meet our needs. ◀

SYSTEM SPECIFICATIONS

Translating sentences in natural language (such as English) into logical expressions is an essential part of specifying both hardware and software systems. System and software engineers take requirements in natural language and produce precise and unambiguous specifications that can be used as the basis for system development. Example 11 shows how propositional expressions can be used in this process.

EXAMPLE 11

Extra Examples

Express the specification "The automated reply cannot be sent when the file system is full" using logical connectives.

Solution: One way to translate this is to let p denote "The automated reply can be sent" and q denote "The file system is full." Then $\neg p$ represents "It is not the case that the automated reply can be sent," which can also be expressed as "The automated reply cannot be sent." Consequently, our specification can be represented by the implication $q \rightarrow \neg p$. ◀

System specifications should not contain conflicting requirements. If they did there would be no way to develop a system that satisfies all specifications. Consequently, propositional expressions representing these specifications need to be **consistent.** That is, there must be an assignment of truth values to the variables in the expressions that makes all the expressions true.

EXAMPLE 12

Determine whether these system specifications are consistent:

"The diagnostic message is stored in the buffer or it is retransmitted."
"The diagnostic message is not stored in the buffer."
"If the diagnostic message is stored in the buffer, then it is retransmitted."

Extra Examples

Solution: To determine whether these specifications are consistent, we first express them using logical expressions. Let p denote "The diagnostic message is stored in the buffer" and let q denote "The diagnostic message is retransmitted." The specifications can then be written as $p \vee q, \neg p,$ and $p \rightarrow q$. An assignment of truth values that makes all three specifications true must have p false to make $\neg p$ true. Since we want $p \vee q$ to be true but p must be false, q must be true. Because $p \rightarrow q$ is true when p is false and q is true, we conclude that these specifications are consistent since they are all true when p is false and q is true. We could come to the same conclusion by use of a truth table to examine the four possible assignments of truth values to p and q. ◀

EXAMPLE 13

Do the system specifications in Example 12 remain consistent if the specification "The diagnostic message is not retransmitted" is added?

Solution: By the reasoning in Example 12, the three specifications from that example are true only in the case when *p* is false and *q* is true. However, this new specification is ¬*q*, which is false when *q* is true. Consequently, these four specifications are inconsistent. ◄

BOOLEAN SEARCHES

Logical connectives are used extensively in searches of large collections of information, such as indexes of Web pages. Because these searches employ techniques from propositional logic, they are called **Boolean searches.**

In Boolean searches, the connective *AND* is used to match records that contain both of two search terms, the connective *OR* is used to match one or both of two search terms, and the connective *NOT* (sometimes written as *AND NOT*) is used to exclude a particular search term. Careful planning of how logical connectives are used is often required when Boolean searches are used to locate information of potential interest. Example 14 illustrates how Boolean searches are carried out.

EXAMPLE 14   **Web Page Searching.** Most Web search engines support Boolean searching techniques, which usually can help find Web pages about particular subjects. For instance, using Boolean searching to find Web pages about universities in New Mexico, we can look for pages matching NEW *AND* MEXICO *AND* UNIVERSITIES. The results of this search will include those pages that contain the three words NEW, MEXICO, and UNIVERSITIES. This will include all of the pages of interest, together with others such as a page about new universities in Mexico. Next, to find pages that deal with universities in New Mexico or Arizona, we can search for pages matching (NEW *AND* MEXICO *OR* ARIZONA) *AND* UNIVERSITIES. (*Note:* Here the *AND* operator takes precedence over the *OR* operator.) The results of this search will include all pages that contain the word UNIVERSITIES and either both the words NEW and MEXICO or the word ARIZONA. Again, pages besides those of interest will be listed. Finally, to find Web pages that deal with universities in Mexico (and not New Mexico), we might first look for pages matching MEXICO *AND* UNIVERSITIES, but since the results of this search will include pages about universities in New Mexico, as well as universities in Mexico, it might be better to search for pages matching (MEXICO *AND* UNIVERSITIES) *NOT* NEW. The results of this search include pages that contain both the words MEXICO and UNIVERSITIES but do not contain the word NEW. ◄

LOGIC PUZZLES

Puzzles that can be solved using logical reasoning are known as **logic puzzles.** Solving logic puzzles is an excellent way to practice working with the rules of logic. Also, computer programs designed to carry out logical reasoning often use well-known logic puzzles to illustrate their capabilities. Many people enjoy solving logic puzzles, which are published in books and periodicals as a recreational activity.

We will discuss two logic puzzles here. We begin with a puzzle that was originally posed by Raymond Smullyan, a master of logic puzzles, who has published more than a dozen books containing challenging puzzles that involve logical reasoning.

EXAMPLE 15   In [Sm78] Smullyan posed many puzzles about an island that has two kinds of inhabitants, knights, who always tell the truth, and their opposites, knaves, who always lie. You encounter two people *A* and *B*. What are *A* and *B* if *A* says "*B* is a knight" and *B* says "The two of us are opposite types"?

Solution: Let p and q be the statements that A is a knight and B is a knight, respectively, so that $\neg p$ and $\neg q$ are the statements that A is a knave and that B is a knave, respectively.

We first consider the possibility that A is a knight; this is the statement that p is true. If A is a knight, then he is telling the truth when he says that B is a knight, so that q is true, and A and B are the same type. However, if B is a knight, then B's statement that A and B are of opposite types, the statement $(p \wedge \neg q) \vee (\neg p \wedge q)$, would have to be true, which it is not, because A and B are both knights. Consequently, we can conclude that A is not a knight, that is, that p is false.

If A is a knave, then because everything a knave says is false, A's statement that B is a knight, that is, that q is true, is a lie, which means that q is false and B is also a knave. Furthermore, if B is a knave, then B's statement that A and B are opposite types is a lie, which is consistent with both A and B being knaves. We can conclude that both A and B are knaves.   ◄

We pose more of Smullyan's puzzles about knights and knaves in Exercises 51–55 at the end of this section. Next, we pose a puzzle known as the **muddy children puzzle** for the case of two children.

EXAMPLE 16   A father tells his two children, a boy and a girl, to play in their backyard without getting dirty. However, while playing, both children get mud on their foreheads. When the children stop playing, the father says "At least one of you has a muddy forehead," and then asks the children to answer "Yes" or "No" to the question: "Do you know whether you have a muddy forehead?" The father asks this question twice. What will the children answer each time this question is asked, assuming that a child can see whether his or her sibling has a muddy forehead, but cannot see his or her own forehead? Assume that both children are honest and that the children answer each question simultaneously.

Solution: Let s be the statement that the son has a muddy forehead and let d be the statement that the daughter has a muddy forehead. When the father says that at least

Links

RAYMOND SMULLYAN (BORN 1919) Raymond Smullyan dropped out of high school. He wanted to study what he was really interested in and not standard high school material. After jumping from one university to the next, he earned an undergraduate degree in mathematics at the University of Chicago in 1955. He paid his college expenses by performing magic tricks at parties and clubs. He obtained a Ph.D. in logic in 1959 at Princeton, studying under Alonzo Church. After graduating from Princeton, he taught mathematics and logic at Dartmouth College, Princeton University, Yeshiva University, and the City University of New York. He joined the philosophy department at Indiana University in 1981 where he is now an emeritus professor.

Smullyan has written many books on recreational logic and mathematics, including *Satan, Cantor, and Infinity; What Is the Name of This Book?; The Lady or the Tiger?; Alice in Puzzleland; To Mock a Mockingbird; Forever Undecided;* and *The Riddle of Scheherazade: Amazing Logic Puzzles, Ancient and Modern.* Because his logic puzzles are challenging, entertaining, and thought-provoking, he is considered to be a modern-day Lewis Carroll. Smullyan has also written several books about the application of deductive logic to chess, three collections of philosophical essays and aphorisms, and several advanced books on mathematical logic and set theory. He is particularly interested in self-reference and has worked on extending some of Godel's results that show that it is impossible to write a computer program that can solve all mathematical problems. He is also particularly interested in explaining ideas from mathematical logic to the public.

Smullyan is a talented musician and often plays piano with his wife, who is a concert-level pianist. Making telescopes is one of his hobbies. He is also interested in optics and stereo photography. He states "I've never had a conflict between teaching and research as some people do because when I'm teaching, I'm doing research."

TABLE 8 Table for the Bit Operators *OR*, *AND*, and *XOR*.

x	y	$x \vee y$	$x \wedge y$	$x \oplus y$
0	0	0	0	0
0	1	1	0	1
1	0	1	0	1
1	1	1	1	0

one of the two children has a muddy forehead he is stating that the disjunction $s \vee d$ is true. Both children will answer "No" the first time the question is asked because each sees mud on the other child's forehead. That is, the son knows that d is true, but does not know whether s is true, and the daughter knows that s is true, but does not know whether d is true.

After the son has answered "No" to the first question, the daughter can determine that d must be true. This follows because when the first question is asked, the son knows that $s \vee d$ is true, but cannot determine whether s is true. Using this information, the daughter can conclude that d must be true, for if d were false, the son could have reasoned that because $s \vee d$ is true, then s must be true, and he would have answered "Yes" to the first question. The son can reason in a similar way to determine that s must be true. It follows that both children answer "Yes" the second time the question is asked. ◀

LOGIC AND BIT OPERATIONS

Truth Value	Bit
T	1
F	0

Links

Computers represent information using bits. A **bit** has two possible values, namely, 0 (zero) and 1 (one). This meaning of the word bit comes from *binary dig*it, since zeros and ones are the digits used in binary representations of numbers. The well-known statistician John Tukey introduced this terminology in 1946. A bit can be used to represent a truth value, since there are two truth values, namely, *true* and *false.* As is customarily done, we will use a 1 bit to represent true and a 0 bit to represent false. That is, 1 represents T (true), 0 represents F (false). A variable is called a **Boolean variable** if its value is either true or false. Consequently, a Boolean variable can be represented using a bit.

Computer **bit operations** correspond to the logical connectives. By replacing true by a one and false by a zero in the truth tables for the operators $\wedge$, $\vee$, and $\oplus$, the tables shown in Table 8 for the corresponding bit operations are obtained. We will also use the notation *OR, AND,* and *XOR* for the operators $\vee$, $\wedge$, and $\oplus$, as is done in various programming languages.

Information is often represented using bit strings, which are sequences of zeros and ones. When this is done, operations on the bit strings can be used to manipulate this information.

DEFINITION 7

A *bit string* is a sequence of zero or more bits. The *length* of this string is the number of bits in the string.

EXAMPLE 17   101010011 is a bit string of length nine. ◀

We can extend bit operations to bit strings. We define the **bitwise *OR*, bitwise *AND*,** and **bitwise *XOR*** of two strings of the same length to be the strings that have as their bits the *OR, AND,* and *XOR* of the corresponding bits in the two strings, respectively. We use the symbols ∨, ∧, and ⊕ to represent the bitwise *OR*, bitwise *AND*, and bitwise *XOR* operations, respectively. We illustrate bitwise operations on bit strings with Example 18.

EXAMPLE 18   Find the bitwise *OR*, bitwise *AND*, and bitwise *XOR* of the bit strings 01 1011 0110 and 11 0001 1101. (Here, and throughout this book, bit strings will be split into blocks of four bits to make them easier to read.)

Solution: The bitwise *OR*, bitwise *AND*, and bitwise *XOR* of these strings are obtained by taking the *OR, AND,* and *XOR* of the corresponding bits, respectively. This gives us

$$
\begin{array}{ll}
01\ 1011\ 0110 & \\
11\ 0001\ 1101 & \\
\hline
11\ 1011\ 1111 & \text{bitwise } OR \\
01\ 0001\ 0100 & \text{bitwise } AND \\
10\ 1010\ 1011 & \text{bitwise } XOR
\end{array}
$$

◀

Exercises

1. Which of these sentences are propositions? What are the truth values of those that are propositions?
   a) Boston is the capital of Massachusetts.
   b) Miami is the capital of Florida.
   c) $2 + 3 = 5$.   d) $5 + 7 = 10$.
   e) $x + 2 = 11$.   f) Answer this question.
   g) $x + y = y + x$ for every pair of real numbers x and y.

2. Which of these are propositions? What are the truth values of those that are propositions?
   a) Do not pass go.   b) What time is it?
   c) There are no black flies in Maine.
   d) $4 + x = 5$.
   e) $x + 1 = 5$ if $x = 1$.
   f) $x + y = y + z$ if $x = z$.

Links

JOHN WILDER TUKEY (1915–2000) Tukey, born in New Bedford, Massachusetts, was an only child. His parents, both teachers, decided home schooling would best develop his potential. His formal education began at Brown University, where he studied mathematics and chemistry. He received a master's degree in chemistry from Brown and continued his studies at Princeton University, changing his field of study from chemistry to mathematics. He received his Ph.D. from Princeton in 1939 for work in topology, when he was appointed an instructor in mathematics at Princeton. With the start of World War II, he joined the Fire Control Research Office, where he began working in statistics. Tukey found statistical research to his liking and impressed several leading statisticians with his skills. In 1945, at the conclusion of the war, Tukey returned to the mathematics department at Princeton as a professor of statistics, and he also took a position at AT&T Bell Laboratories. Tukey founded the Statistics Department at Princeton in 1966 and was its first chairman. Tukey made significant contributions to many areas of statistics, including the analysis of variance, the estimation of spectra of time series, inferences about the values of a set of parameters from a single experiment, and the philosophy of statistics. However, he is best known for his invention, with J. W. Cooley, of the fast Fourier transform.

Tukey contributed his insight and expertise by serving on the President's Science Advisory Committee. He chaired several important committees dealing with the environment, education, and chemicals and health. He also served on committees working on nuclear disarmament. Tukey received many awards, including the National Medal of Science.

HISTORICAL NOTE There were several other suggested words for a binary digit, including *binit* and *bigit*, that never were widely accepted. The adoption of the word *bit* may be due to its meaning as a common English word. For an account of Tukey's coining of the word *bit*, see the April 1984 issue of *Annals of the History of Computing*.

3. What is the negation of each of these propositions?

  a) Today is Thursday.
  b) There is no pollution in New Jersey.
  c) $2 + 1 = 3$.
  d) The summer in Maine is hot and sunny.

4. Let p and q be the propositions

  p : I bought a lottery ticket this week.
  q : I won the million dollar jackpot on Friday.

Express each of these propositions as an English sentence.

  a) $\neg p$ **b)** $p \vee q$ **c)** $p \rightarrow q$
  d) $p \wedge q$ **e)** $p \leftrightarrow q$ **f)** $\neg p \rightarrow \neg q$
  g) $\neg p \wedge \neg q$ **h)** $\neg p \vee (p \wedge q)$

5. Let p and q be the propositions "Swimming at the New Jersey shore is allowed" and "Sharks have been spotted near the shore," respectively. Express each of these compound propositions as an English sentence.

  a) $\neg q$ **b)** $p \wedge q$ **c)** $\neg p \vee q$
  d) $p \rightarrow \neg q$ **e)** $\neg q \rightarrow p$ **f)** $\neg p \rightarrow \neg q$
  g) $p \leftrightarrow \neg q$ **h)** $\neg p \wedge (p \vee \neg q)$

6. Let p and q be the propositions "The election is decided" and "The votes have been counted," respectively. Express each of these compound propositions as an English sentence.

  a) $\neg p$ **b)** $p \vee q$ **c)** $\neg p \wedge q$
  d) $q \rightarrow p$ **e)** $\neg q \rightarrow \neg p$ **f)** $\neg p \rightarrow \neg q$
  g) $p \leftrightarrow q$ **h)** $\neg q \vee (\neg p \wedge q)$

7. Let p and q be the propositions

  p : It is below freezing.
  q : It is snowing.

Write these propositions using p and q and logical connectives.

  a) It is below freezing and snowing.
  b) It is below freezing but not snowing.
  c) It is not below freezing and it is not snowing.
  d) It is either snowing or below freezing (or both).
  e) If it is below freezing, it is also snowing.
  f) It is either below freezing or it is snowing, but it is not snowing if it is below freezing.
  g) That it is below freezing is necessary and sufficient for it to be snowing.

8. Let p, q, and r be the propositions

  p : You have the flu.
  q : You miss the final examination.
  r : You pass the course.

Express each of these propositions as an English sentence.

  a) $p \rightarrow q$ **b)** $\neg q \leftrightarrow r$
  c) $q \rightarrow \neg r$ **d)** $p \vee q \vee r$
  e) $(p \rightarrow \neg r) \vee (q \rightarrow \neg r)$
  f) $(p \wedge q) \vee (\neg q \wedge r)$

9. Let p and q be the propositions

  p : You drive over 65 miles per hour.
  q : You get a speeding ticket.

Write these propositions using p and q and logical connectives.

  a) You do not drive over 65 miles per hour.
  b) You drive over 65 miles per hour, but you do not get a speeding ticket.
  c) You will get a speeding ticket if you drive over 65 miles per hour.
  d) If you do not drive over 65 miles per hour, then you will not get a speeding ticket.
  e) Driving over 65 miles per hour is sufficient for getting a speeding ticket.
  f) You get a speeding ticket, but you do not drive over 65 miles per hour.
  g) Whenever you get a speeding ticket, you are driving over 65 miles per hour.

10. Let p, q, and r be the propositions

  p : You get an A on the final exam.
  q : You do every exercise in this book.
  r : You get an A in this class.

Write these propositions using p, q, and r and logical connectives.

  a) You get an A in this class, but you do not do every exercise in this book.
  b) You get an A on the final, you do every exercise in this book, and you get an A in this class.
  c) To get an A in this class, it is necessary for you to get an A on the final.
  d) You get an A on the final, but you don't do every exercise in this book; nevertheless, you get an A in this class.
  e) Getting an A on the final and doing every exercise in this book is sufficient for getting an A in this class.
  f) You will get an A in this class if and only if you either do every exercise in this book or you get an A on the final.

11. Let p, q, and r be the propositions

  p : Grizzly bears have been seen in the area.
  q : Hiking is safe on the trail.
  r : Berries are ripe along the trail.

Write these propositions using p, q, and r and logical connectives.

  a) Berries are ripe along the trail, but grizzly bears have not been seen in the area.
  b) Grizzly bears have not been seen in the area and hiking on the trail is safe, but berries are ripe along the trail.
  c) If berries are ripe along the trail, hiking is safe if and only if grizzly bears have not been seen in the area.

d) It is not safe to hike on the trail, but grizzly bears have not been seen in the area and the berries along the trail are ripe.

e) For hiking on the trail to be safe, it is necessary but not sufficient that berries not be ripe along the trail and for grizzly bears not to have been seen in the area.

f) Hiking is not safe on the trail whenever grizzly bears have been seen in the area and berries are ripe along the trail.

12. Determine whether these biconditionals are true or false.

a) $2 + 2 = 4$ if and only if $1 + 1 = 2$.

b) $1 + 1 = 2$ if and only if $2 + 3 = 4$.

c) It is winter if and only if it is not spring, summer, or fall.

d) $1 + 1 = 3$ if and only if pigs can fly.

e) $0 > 1$ if and only if $2 > 1$.

13. Determine whether each of these implications is true or false.

a) If $1 + 1 = 2$, then $2 + 2 = 5$.

b) If $1 + 1 = 3$, then $2 + 2 = 4$.

c) If $1 + 1 = 3$, then $2 + 2 = 5$.

d) If pigs can fly, then $1 + 1 = 3$.

e) If $1 + 1 = 3$, then God exists.

f) If $1 + 1 = 3$, then pigs can fly.

g) If $1 + 1 = 2$, then pigs can fly.

h) If $2 + 2 = 4$, then $1 + 2 = 3$.

14. For each of these sentences, determine whether an inclusive or an exclusive or is intended. Explain your answer.

a) Experience with C++ or Java is required.

b) Lunch includes soup or salad.

c) To enter the country you need a passport or a voter registration card.

d) Publish or perish.

15. For each of these sentences, state what the sentence means if the or is an inclusive or (that is, a disjunction) versus an exclusive or. Which of these meanings of or do you think is intended?

a) To take discrete mathematics, you must have taken calculus or a course in computer science.

b) When you buy a new car from Acme Motor Company, you get $2000 back in cash or a 2% car loan.

c) Dinner for two includes two items from column A or three items from column B.

d) School is closed if more than 2 feet of snow falls or if the wind chill is below −100.

16. Write each of these statements in the form "if p, then q" in English. (*Hint:* Refer to the list of common ways to express implications provided in this section.)

a) It is necessary to wash the boss's car to get promoted.

b) Winds from the south imply a spring thaw.

c) A sufficient condition for the warranty to be good

is that you bought the computer less than a year ago.

d) Willy gets caught whenever he cheats.

e) You can access the website only if you pay a subscription fee.

f) Getting elected follows from knowing the right people.

g) Carol gets seasick whenever she is on a boat.

17. Write each of these statements in the form "if p, then q" in English. (*Hint:* Refer to the list of common ways to express implications provided in this section.)

a) It snows whenever the wind blows from the northeast.

b) The apple trees will bloom if it stays warm for a week.

c) That the Pistons win the championship implies that they beat the Lakers.

d) It is necessary to walk 8 miles to get to the top of Long's Peak.

e) To get tenure as a professor, it is sufficient to be world-famous.

f) If you drive more than 400 miles, you will need to buy gasoline.

g) Your guarantee is good only if you bought your CD player less than 90 days ago.

18. Write each of these statements in the form "if p, then q" in English. (*Hint:* Refer to the list of common ways to express implications provided in this section.)

a) I will remember to send you the address only if you send me an e-mail message.

b) To be a citizen of this country, it is sufficient that you were born in the United States.

c) If you keep your textbook, it will be a useful reference in your future courses.

d) The Red Wings will win the Stanley Cup if their goalie plays well.

e) That you get the job implies that you had the best credentials.

f) The beach erodes whenever there is a storm.

g) It is necessary to have a valid password to log on to the server.

19. Write each of these propositions in the form "p if and only if q" in English.

a) If it is hot outside you buy an ice cream cone, and if you buy an ice cream cone it is hot outside.

b) For you to win the contest it is necessary and sufficient that you have the only winning ticket.

c) You get promoted only if you have connections, and you have connections only if you get promoted.

d) If you watch television your mind will decay, and conversely.

e) The trains run late on exactly those days when I take it.

20. Write each of these propositions in the form "*p* if and only if *q*" in English.

   a) For you to get an A in this course, it is necessary and sufficient that you learn how to solve discrete mathematics problems.

   b) If you read the newspaper every day, you will be informed, and conversely.

   c) It rains if it is a weekend day, and it is a weekend day if it rains.

   d) You can see the wizard only if the wizard is not in, and the wizard is not in only if you can see him.

21. State the converse, contrapositive, and inverse of each of these implications.

   a) If it snows today, I will ski tomorrow.

   b) I come to class whenever there is going to be a quiz.

   c) A positive integer is a prime only if it has no divisors other than 1 and itself.

22. State the converse, contrapositive, and inverse of each of these implications.

   a) If it snows tonight, then I will stay at home.

   b) I go to the beach whenever it is a sunny summer day.

   c) When I stay up late, it is necessary that I sleep until noon.

23. Construct a truth table for each of these compound propositions.

   a) $p \wedge \neg p$ **b)** $p \vee \neg p$

   c) $(p \vee \neg q) \to q$ **d)** $(p \vee q) \to (p \wedge q)$

   e) $(p \to q) \leftrightarrow (\neg q \to \neg p)$

   f) $(p \to q) \to (q \to p)$

24. Construct a truth table for each of these compound propositions.

   a) $p \to \neg p$ **b)** $p \leftrightarrow \neg p$

   c) $p \oplus (p \vee q)$ **d)** $(p \wedge q) \to (p \vee q)$

   e) $(q \to \neg p) \leftrightarrow (p \leftrightarrow q)$

   f) $(p \leftrightarrow q) \oplus (p \leftrightarrow \neg q)$

25. Construct a truth table for each of these compound propositions.

   a) $(p \vee q) \to (p \oplus q)$

   b) $(p \oplus q) \to (p \wedge q)$

   c) $(p \vee q) \oplus (p \wedge q)$

   d) $(p \leftrightarrow q) \oplus (\neg p \leftrightarrow q)$

   e) $(p \leftrightarrow q) \oplus (\neg p \leftrightarrow \neg r)$

   f) $(p \oplus q) \to (p \oplus \neg q)$

26. Construct a truth table for each of these compound propositions.

   a) $p \oplus p$ **b)** $p \oplus \neg p$

   c) $p \oplus \neg q$ **d)** $\neg p \oplus \neg q$

   e) $(p \oplus q) \vee (p \oplus \neg q)$

   f) $(p \oplus q) \wedge (p \oplus \neg q)$

27. Construct a truth table for each of these compound propositions.

   a) $p \to \neg q$ **b)** $\neg p \leftrightarrow q$

   c) $(p \to q) \vee (\neg p \to q)$

   d) $(p \to q) \wedge (\neg p \to q)$

   e) $(p \leftrightarrow q) \vee (\neg p \leftrightarrow q)$

   f) $(\neg p \leftrightarrow \neg q) \leftrightarrow (p \leftrightarrow q)$

28. Construct a truth table for each of these compound propositions.

   a) $(p \vee q) \vee r$ **b)** $(p \vee q) \wedge r$

   c) $(p \wedge q) \vee r$ **d)** $(p \wedge q) \wedge r$

   e) $(p \vee q) \wedge \neg r$ **f)** $(p \wedge q) \vee \neg r$

29. Construct a truth table for each of these compound propositions.

   a) $p \to (\neg q \vee r)$

   b) $\neg p \to (q \to r)$

   c) $(p \to q) \vee (\neg p \to r)$

   d) $(p \to q) \wedge (\neg p \to r)$

   e) $(p \leftrightarrow q) \vee (\neg q \leftrightarrow r)$

   f) $(\neg p \leftrightarrow \neg q) \leftrightarrow (q \leftrightarrow r)$

30. Construct a truth table for $((p \to q) \to r) \to s$.

31. Construct a truth table for $(p \leftrightarrow q) \leftrightarrow (r \leftrightarrow s)$.

32. What is the value of *x* after each of these statements is encountered in a computer program, if $x = 1$ before the statement is reached?

   a) **if** $1 + 2 = 3$ **then** $x := x + 1$

   b) **if** $(1 + 1 = 3)$ *OR* $(2 + 2 = 3)$ **then** $x := x + 1$

   c) **if** $(2 + 3 = 5)$ *AND* $(3 + 4 = 7)$ **then** $x := x + 1$

   d) **if** $(1 + 1 = 2)$ *XOR* $(1 + 2 = 3)$ **then** $x := x + 1$

   e) **if** $x < 2$ **then** $x := x + 1$

33. Find the bitwise *OR*, bitwise *AND*, and bitwise *XOR* of each of these pairs of bit strings.

   a) 101 1110, 010 0001

   b) 1111 0000, 1010 1010

   c) 00 0111 0001, 10 0100 1000

   d) 11 1111 1111, 00 0000 0000

34. Evaluate each of these expressions.

   a) 1 1000 $\wedge$ (0 1011 $\vee$ 1 1011)

   b) (0 1111 $\wedge$ 1 0101) $\vee$ 0 1000

   c) (0 1010 $\oplus$ 1 1011) $\oplus$ 0 1000

   d) (1 1011 $\vee$ 0 1010) $\wedge$ (1 0001 $\vee$ 1 1011)

Fuzzy logic is used in artificial intelligence. In fuzzy logic, a proposition has a truth value that is a number between 0 and 1, inclusive. A proposition with a truth value of 0 is false and one with a truth value of 1 is true. Truth values that are between 0 and 1 indicate varying degrees of truth. For instance, the truth value 0.8 can be assigned to the statement "Fred is happy," since Fred is happy most of the time, and the truth value 0.4 can be assigned to the statement "John is happy," since John is happy slightly less than half the time.

35. The truth value of the negation of a proposition in fuzzy logic is 1 minus the truth value of the proposition. What are the truth values of the statements "Fred is not happy" and "John is not happy"?

36. The truth value of the conjunction of two propositions in fuzzy logic is the minimum of the truth values of the two propositions. What are the truth values of the statements "Fred and John are happy" and "Neither Fred nor John is happy"?

37. The truth value of the disjunction of two propositions in fuzzy logic is the maximum of the truth values of the two propositions. What are the truth values of the statements "Fred is happy, or John is happy" and "Fred is not happy, or John is not happy"?

*38. Is the assertion "This statement is false" a proposition?

*39. The nth statement in a list of 100 statements is "Exactly n of the statements in this list are false."

    a) What conclusions can you draw from these statements?

    b) Answer part (a) if the nth statement is "At least n of the statements in this list are false."

    c) Answer part (b) assuming that the list contains 99 statements.

40. An ancient Sicilian legend says that the barber in a remote town who can be reached only by traveling a dangerous mountain road shaves those people, and only those people, who do not shave themselves. Can there be such a barber?

41. Each inhabitant of a remote village always tells the truth or always lies. A villager will only give a "Yes" or a "No" response to a question a tourist asks. Suppose you are a tourist visiting this area and come to a fork in the road. One branch leads to the ruins you want to visit; the other branch leads deep into the jungle. A villager is standing at the fork in the road. What one question can you ask the villager to determine which branch to take?

42. An explorer is captured by a group of cannibals. There are two types of cannibals—those who always tell the truth and those who always lie. The cannibals will barbecue the explorer unless he can determine whether a particular cannibal always lies or always tells the truth. He is allowed to ask the cannibal exactly one question.

    a) Explain why the question "Are you a liar?" does not work.

    b) Find a question that the explorer can use to determine whether the cannibal always lies or always tells the truth.

43. Express these system specifications using the propositions p "The message is scanned for viruses" and q "The message was sent from an unknown system" together with logical connectives.

    a) "The message is scanned for viruses whenever the message was sent from an unknown system."

    b) "The message was sent from an unknown system but it was not scanned for viruses."

    c) "It is necessary to scan the message for viruses whenever it was sent from an unknown system."

    d) "When a message is not sent from an unknown system it is not scanned for viruses."

44. Express these system specifications using the propositions p "The user enters a valid password," q "Access is granted," and r "The user has paid the subscription fee" and logical connectives.

    a) "The user has paid the subscription fee, but does not enter a valid password."

    b) "Access is granted whenever the user has paid the subscription fee and enters a valid password."

    c) "Access is denied if the user has not paid the subscription fee."

    d) "If the user has not entered a valid password but has paid the subscription fee, then access is granted."

45. Are these system specifications consistent? "The system is in multiuser state if and only if it is operating normally. If the system is operating normally, the kernel is functioning. The kernel is not functioning or the system is in interrupt mode. If the system is not in multiuser state, then it is in interrupt mode. The system is not in interrupt mode."

46. Are these system specifications consistent? "Whenever the system software is being upgraded, users cannot access the file system. If users can access the file system, then they can save new files. If users cannot save new files, then the system software is not being upgraded."

47. Are these system specifications consistent? "The router can send packets to the edge system only if it supports the new address space, For the router to support the new address space it is necessary that the latest software release be installed, The router can send packets to the edge system if the latest software release is installed, The router does not support the new address space."

48. Are these system specifications consistent? "If the file system is not locked, then new messages will be queued. If the file system is not locked, then the system is functioning normally, and conversely. If new messages are not queued, then they will be sent to the message buffer. If the file system is not locked, then new messages will be sent to the message buffer. New messages will not be sent to the message buffer."

49. What Boolean search would you use to look for Web pages about beaches in New Jersey? What if you wanted to find Web pages about beaches on the isle of Jersey (in the English Channel)?

50. What Boolean search would you use to look for Web pages about hiking in West Virginia? What if you wanted to find Web pages about hiking in Virginia, but not in West Virginia?

Exercises 51–55 relate to inhabitants of the island of knights and knaves created by Smullyan, where knights always tell the truth and knaves always lie. You encounter two people, *A* and *B*. Determine, if possible, what *A* and *B* are if they address you in the ways described. If you cannot determine what these two people are, can you draw any conclusions?

51. *A* says "At least one of us is a knave" and *B* says nothing.

52. *A* says "The two of us are both knights" and *B* says "*A* is a knave."

53. *A* says "I am a knave or *B* is a knight" and *B* says nothing.

54. Both *A* and *B* say "I am a knight."

55. *A* says "We are both knaves" and *B* says nothing.

Exercises 56–61 are puzzles that can be solved by translating statements into logical expressions and reasoning from these expressions using truth tables.

56. The police have three suspects for the murder of Mr. Cooper: Mr. Smith, Mr. Jones, and Mr. Williams. Smith, Jones, and Williams each declare that they did not kill Cooper. Smith also states that Cooper was a friend of Jones and that Williams disliked him. Jones also states that he did not know Cooper and that he was out of town the day Cooper was killed. Williams also states that he saw both Smith and Jones with Cooper the day of the killing and that either Smith or Jones must have killed him. Can you determine who the murderer was if

   a) one of the three men is guilty, the two innocent men are telling the truth, but the statements of the guilty man may or may not be true?

   b) innocent men do not lie?

57. Steve would like to determine the relative salaries of three coworkers using two facts. First, he knows that if Fred is not the highest paid of the three, then Janice is. Second, he knows that if Janice is not the lowest paid, then Maggie is paid the most. Is it possible to determine the relative salaries of Fred, Maggie, and Janice from what Steve knows? If so, who is paid the most and who the least? Explain your reasoning.

58. Five friends have access to a chat room. Is it possible to determine who is chatting if the following information is known? Either Kevin or Heather, or both, are chatting. Either Randy or Vijay, but not both, are chatting. If Abby is chatting, so is Randy.

Vijay and Kevin are either both chatting or neither is. If Heather is chatting, then so are Abby and Kevin. Explain your reasoning.

59. A detective has interviewed four witnesses to a crime. From the stories of the witnesses the detective has concluded that if the butler is telling the truth then so is the cook; the cook and the gardener cannot both be telling the truth; the gardener and the handyman are not both lying; and if the handyman is telling the truth then the cook is lying. For each of the four witnesses, can the detective determine whether that person is telling the truth or lying? Explain your reasoning.

60. Four friends have been identified as suspects for an unauthorized access into a computer system. They have made statements to the investigating authorities. Alice said "Carlos did it." John said "I did not do it." Carlos said "Diana did it." Diana said "Carlos lied when he said that I did it."

   a) If the authorities also know that exactly one of the four suspects is telling the truth, who did it? Explain your reasoning.

   b) If the authorities also know that exactly one is lying, who did it? Explain your reasoning.

*61. Solve this famous logic puzzle, attributed to Albert Einstein, and known as the **zebra puzzle.** Five men with different nationalities and with different jobs live in consecutive houses on a street. These houses are painted different colors. The men have different pets and have different favorite drinks. Determine who owns a zebra and whose favorite drink is mineral water (which is one of the favorite drinks) given these clues: The Englishman lives in the red house. The Spaniard owns a dog. The Japanese man is a painter. The Italian drinks tea. The Norwegian lives in the first house on the left. The green house is on the right of the white one. The photographer breeds snails. The diplomat lives in the yellow house. Milk is drunk in the middle house. The owner of the green house drinks coffee. The Norwegian's house is next to the blue one. The violinist drinks orange juice. The fox is in a house next to that of the physician. The horse is in a house next to that of the diplomat. (*Hint:* Make a table where the rows represent the men and columns represent the color of their houses, their jobs, their pets, and their favorite drinks and use logical reasoning to determine the correct entries in the table.)

1.2 Propositional Equivalences

INTRODUCTION

An important type of step used in a mathematical argument is the replacement of a statement with another statement with the same truth value. Because of this, methods that produce propositions with the same truth value as a given compound proposition are used extensively in the construction of mathematical arguments.

    We begin our discussion with a classification of compound propositions according to their possible truth values.

DEFINITION 1

> A compound proposition that is always true, no matter what the truth values of the propositions that occur in it, is called a *tautology*. A compound proposition that is always false is called a *contradiction*. Finally, a proposition that is neither a tautology nor a contradiction is called a *contingency*.

Tautologies and contradictions are often important in mathematical reasoning. The following example illustrates these types of propositions.

EXAMPLE 1 We can construct examples of tautologies and contradictions using just one proposition. Consider the truth tables of $p \lor \neg p$ and $p \land \neg p$, shown in Table 1. Since $p \lor \neg p$ is always true, it is a tautology. Since $p \land \neg p$ is always false, it is a contradiction. ◀

LOGICAL EQUIVALENCES

Compound propositions that have the same truth values in all possible cases are called **logically equivalent.** We can also define this notion as follows.

DEFINITION 2

> The propositions p and q are called *logically equivalent* if $p \leftrightarrow q$ is a tautology. The notation $p \equiv q$ denotes that p and q are logically equivalent.

Remark: The symbol $\equiv$ is not a logical connective since $p \equiv q$ is not a compound proposition, but rather is the statement that $p \leftrightarrow q$ is a tautology. The symbol $\Leftrightarrow$ is sometimes used instead of $\equiv$ to denote logical equivalence.

    One way to determine whether two propositions are equivalent is to use a truth table. In particular, the propositions p and q are equivalent if and only if the columns giving their truth values agree. The following example illustrates this method.

EXAMPLE 2 Show that $\neg(p \lor q)$ and $\neg p \land \neg q$ are logically equivalent. This equivalence is one of *De Morgan's laws* for propositions, named after the English mathematician Augustus De Morgan, of the mid-nineteenth century.

TABLE 1 Examples of a Tautology and a Contradiction.

p	$\neg p$	$p \lor \neg p$	$p \land \neg p$
T	F	T	F
F	T	T	F

TABLE 2 Truth Tables for $\neg(p \vee q)$ and $\neg p \wedge \neg q$.

p	q	$p \vee q$	$\neg(p \vee q)$	$\neg p$	$\neg q$	$\neg p \wedge \neg q$
T	T	T	F	F	F	F
T	F	T	F	F	T	F
F	T	T	F	T	F	F
F	F	F	T	T	T	T

TABLE 3 Truth Tables for $\neg p \vee q$ and $p \rightarrow q$.

p	q	$\neg p$	$\neg p \vee q$	$p \rightarrow q$
T	T	F	T	T
T	F	F	F	F
F	T	T	T	T
F	F	T	T	T

Solution: The truth tables for these propositions are displayed in Table 2. Since the truth values of the propositions $\neg(p \vee q)$ and $\neg p \wedge \neg q$ agree for all possible combinations of the truth values of p and q, it follows that $\neg(p \vee q) \leftrightarrow (\neg p \wedge \neg q)$ is a tautology and that these propositions are logically equivalent. ◄

EXAMPLE 3 Show that the propositions $p \rightarrow q$ and $\neg p \vee q$ are logically equivalent.

Solution: We construct the truth table for these propositions in Table 3. Since the truth values of $\neg p \vee q$ and $p \rightarrow q$ agree, these propositions are logically equivalent. ◄

EXAMPLE 4 Show that the propositions $p \vee (q \wedge r)$ and $(p \vee q) \wedge (p \vee r)$ are logically equivalent. This is the *distributive law* of disjunction over conjunction.

Solution: We construct the truth table for these propositions in Table 4. Since the truth values of $p \vee (q \wedge r)$ and $(p \vee q) \wedge (p \vee r)$ agree, these propositions are logically equivalent. ◄

Remark: A truth table of a compound proposition involving three different propositions requires eight rows, one for each possible combination of truth values of the three propositions. In general, 2^n rows are required if a compound proposition involves n propositions.

Table 5 contains some important equivalences.* In these equivalences, **T** denotes any proposition that is always true and **F** denotes any proposition that is always false. We also display some useful equivalences for compound propositions involving implications and biconditionals in Tables 6 and 7, respectively. The reader is asked to verify the equivalences in Tables 5–7 in the exercises at the end of the section.

---

*These identities are a special case of identities that hold for any Boolean algebra. Compare them with set identities in Table 1 in Section 1.7 and with Boolean identities in Table 5 in Section 10.1.

TABLE 4 A Demonstration That $p \vee (q \wedge r)$ and $(p \vee q) \wedge (p \vee r)$ Are Logically Equivalent.							
p	q	r	$q \wedge r$	$p \vee (q \wedge r)$	$p \vee q$	$p \vee r$	$(p \vee q) \wedge (p \vee r)$
T	T	T	T	T	T	T	T
T	T	F	F	T	T	T	T
T	F	T	F	T	T	T	T
T	F	F	F	T	T	T	T
F	T	T	T	T	T	T	T
F	T	F	F	F	T	F	F
F	F	T	F	F	F	T	F
F	F	F	F	F	F	F	F

The associative law for disjunction shows that the expression $p \vee q \vee r$ is well defined, in the sense that it does not matter whether we first take the disjunction of p and q and then the disjunction of $p \vee q$ with r, or if we first take the disjunction of q and r and then take the disjunction of p and $q \vee r$. Similarly, the expression $p \wedge q \wedge r$ is well defined. By extending this reasoning, it follows that $p_1 \vee p_2 \vee \cdots \vee p_n$ and $p_1 \wedge p_2 \wedge \cdots \wedge p_n$ are well defined whenever $p_1, p_2, \ldots, p_n$ are propositions. Furthermore, note that De Morgan's laws extend to

$$\neg(p_1 \vee p_2 \vee \cdots \vee p_n) \equiv (\neg p_1 \wedge \neg p_2 \wedge \cdots \wedge \neg p_n)$$

and

$$\neg(p_1 \wedge p_2 \wedge \cdots \wedge p_n) \equiv (\neg p_1 \vee \neg p_2 \vee \cdots \vee \neg p_n).$$

(Methods for proving these identities will be given in Section 3.3.)

Links

AUGUSTUS DE MORGAN (1806–1871) Augustus De Morgan was born in India, where his father was a colonel in the Indian army. De Morgan's family moved to England when he was 7 months old. He attended private schools, where he developed a strong interest in mathematics in his early teens. De Morgan studied at Trinity College, Cambridge, graduating in 1827. Although he considered entering medicine or law, he decided on a career in mathematics. He won a position at University College, London, in 1828, but resigned when the college dismissed a fellow professor without giving reasons. However, he resumed this position in 1836 when his successor died, staying there until 1866.

De Morgan was a noted teacher who stressed principles over techniques. His students included many famous mathematicians, including Ada Augusta, Countess of Lovelace, who was Charles Babbage's collaborator in his work on computing machines (see page 25 for biographical notes on Ada Augusta). (De Morgan cautioned the countess against studying too much mathematics, since it might interfere with her childbearing abilities!)

De Morgan was an extremely prolific writer. He wrote more than 1000 articles for more than 15 periodicals. De Morgan also wrote textbooks on many subjects, including logic, probability, calculus, and algebra. In 1838 he presented what was perhaps the first clear explanation of an important proof technique known as *mathematical induction* (discussed in Section 3.3 of this text), a term he coined. In the 1840s De Morgan made fundamental contributions to the development of symbolic logic. He invented notations that helped him prove propositional equivalences, such as the laws that are named after him. In 1842 De Morgan presented what was perhaps the first precise definition of a limit and developed some tests for convergence of infinite series. De Morgan was also interested in the history of mathematics and wrote biographies of Newton and Halley.

In 1837 De Morgan married Sophia Frend, who wrote his biography in 1882. De Morgan's research, writing, and teaching left little time for his family or social life. Nevertheless, he was noted for his kindness, humor, and wide range of knowledge.

TABLE 5 Logical Equivalences.	
Equivalence	*Name*
$p \wedge \mathbf{T} \equiv p$ $p \vee \mathbf{F} \equiv p$	Identity laws
$p \vee \mathbf{T} \equiv \mathbf{T}$ $p \wedge \mathbf{F} \equiv \mathbf{F}$	Domination laws
$p \vee p \equiv p$ $p \wedge p \equiv p$	Idempotent laws
$\neg(\neg p) \equiv p$	Double negation law
$p \vee q \equiv q \vee p$ $p \wedge q \equiv q \wedge p$	Commutative laws
$(p \vee q) \vee r \equiv p \vee (q \vee r)$ $(p \wedge q) \wedge r \equiv p \wedge (q \wedge r)$	Associative laws
$p \vee (q \wedge r) \equiv (p \vee q) \wedge (p \vee r)$ $p \wedge (q \vee r) \equiv (p \wedge q) \vee (p \wedge r)$	Distributive laws
$\neg(p \wedge q) \equiv \neg p \vee \neg q$ $\neg(p \vee q) \equiv \neg p \wedge \neg q$	De Morgan's laws
$p \vee (p \wedge q) \equiv p$ $p \wedge (p \vee q) \equiv p$	Absorption laws
$p \vee \neg p \equiv \mathbf{T}$ $p \wedge \neg p \equiv \mathbf{F}$	Negation laws

TABLE 6 Logical Equivalences Involving Implications.
$p \rightarrow q \equiv \neg p \vee q$
$p \rightarrow q \equiv \neg q \rightarrow \neg p$
$p \vee q \equiv \neg p \rightarrow q$
$p \wedge q \equiv \neg(p \rightarrow \neg q)$
$\neg(p \rightarrow q) \equiv p \wedge \neg q$
$(p \rightarrow q) \wedge (p \rightarrow r) \equiv p \rightarrow (q \wedge r)$
$(p \rightarrow r) \wedge (q \rightarrow r) \equiv (p \vee q) \rightarrow r$
$(p \rightarrow q) \vee (p \rightarrow r) \equiv p \rightarrow (q \vee r)$
$(p \rightarrow r) \vee (q \rightarrow r) \equiv (p \wedge q) \rightarrow r$

TABLE 7 Logical Equivalences Involving Biconditionals.
$p \leftrightarrow q \equiv (p \rightarrow q) \wedge (q \rightarrow p)$
$p \leftrightarrow q \equiv \neg p \leftrightarrow \neg q$
$p \leftrightarrow q \equiv (p \wedge q) \vee (\neg p \wedge \neg q)$
$\neg(p \leftrightarrow q) \equiv p \leftrightarrow \neg q$

Extra Examples

The logical equivalences in Table 5, as well as any others that have been established (such as those shown in Tables 6 and 7), can be used to construct additional logical equivalences. The reason for this is that a proposition in a compound proposition can be replaced by one that is logically equivalent to it without changing the truth value of the compound proposition. This technique is illustrated in Examples 5 and 6, where we also use the fact that if p and q are logically equivalent and q and r are logically equivalent, then p and r are logically equivalent (see Exercise 50).

EXAMPLE 5   Show that $\neg(p \vee (\neg p \wedge q))$ and $\neg p \wedge \neg q$ are logically equivalent.

Solution: We could use a truth table to show that these compound propositions are equivalent. Instead, we will establish this equivalence by developing a series of logical equivalences, using one of the equivalences in Table 5 at a time, starting with $\neg(p \vee (\neg p \wedge q))$ and ending with $\neg p \wedge \neg q$. We have the following equivalences.

$$\begin{aligned}
\neg(p \vee (\neg p \wedge q)) &\equiv \neg p \wedge \neg(\neg p \wedge q) && \text{from the second De Morgan's law} \\
&\equiv \neg p \wedge [\neg(\neg p) \vee \neg q] && \text{from the first De Morgan's law} \\
&\equiv \neg p \wedge (p \vee \neg q) && \text{from the double negation law} \\
&\equiv (\neg p \wedge p) \vee (\neg p \wedge \neg q) && \text{from the second distributive law} \\
&\equiv \mathbf{F} \vee (\neg p \wedge \neg q) && \text{since } \neg p \wedge p \equiv \mathbf{F} \\
&\equiv (\neg p \wedge \neg q) \vee \mathbf{F} && \text{from the commutative law} \\
& && \qquad \text{for disjunction} \\
&\equiv \neg p \wedge \neg q && \text{from the identity law for } \mathbf{F}
\end{aligned}$$

Consequently $\neg(p \vee (\neg p \wedge q))$ and $\neg p \wedge \neg q$ are logically equivalent.   ◄

EXAMPLE 6   Show that $(p \wedge q) \rightarrow (p \vee q)$ is a tautology.

Solution: To show that this statement is a tautology, we will use logical equivalences to demonstrate that it is logically equivalent to $\mathbf{T}$. (*Note:* This could also be done using a truth table.)

$$\begin{aligned}
(p \wedge q) \rightarrow (p \vee q) &\equiv \neg(p \wedge q) \vee (p \vee q) && \text{by Example 3} \\
&\equiv (\neg p \vee \neg q) \vee (p \vee q) && \text{by the first De Morgan's law} \\
&\equiv (\neg p \vee p) \vee (\neg q \vee q) && \text{by the associative and commutative} \\
& && \qquad \text{laws for disjunction} \\
&\equiv \mathbf{T} \vee \mathbf{T} && \text{by Example 1 and the commutative} \\
& && \qquad \text{law for disjunction} \\
&\equiv \mathbf{T} && \text{by the domination law} \qquad ◄
\end{aligned}$$

A truth table can be used to determine whether a compound proposition is a tautology. This can be done by hand for a proposition with a small number of variables, but when the number of variables grows, this becomes impractical. For instance, there are $2^{20} = 1{,}048{,}576$ rows in the truth value table for a proposition with 20 variables. Clearly, you need a computer to help you determine, in this way, whether a compound proposition in

ADA AUGUSTA, COUNTESS OF LOVELACE (1815–1852)   Ada Augusta was the only child from the marriage of the famous poet Lord Byron and Annabella Millbanke, who separated when Ada was 1 month old. She was raised by her mother, who encouraged her intellectual talents. She was taught by the mathematicians William Frend and Augustus De Morgan. In 1838 she married Lord King, later elevated to Earl of Lovelace. Together they had three children.

Ada Augusta continued her mathematical studies after her marriage, assisting Charles Babbage in his work on an early computing machine, called the Analytic Engine. The most complete accounts of this machine are found in her writings. After 1845 she and Babbage worked toward the development of a system to predict horse races. Unfortunately, their system did not work well, leaving Ada heavily in debt at the time of her death. The programming language Ada is named in honor of the Countess of Lovelace.

20 variables is a tautology. But when there are 1000 variables, can even a computer determine in a reasonable amount of time whether a compound proposition is a tautology? Checking every one of the 2^{1000} (a number with more than 300 decimal digits) possible combinations of truth values simply cannot be done by a computer in even trillions of years. Furthermore, no other procedures are known that a computer can follow to determine in a reasonable amount of time whether a compound proposition in such a large number of variables is a tautology. We will study questions such as this in Chapter 2, when we study the complexity of algorithms.

Exercises

1. Use truth tables to verify these equivalences.

   a) $p \land \mathbf{T} \equiv p$ **b)** $p \lor \mathbf{F} \equiv p$

   c) $p \land \mathbf{F} \equiv \mathbf{F}$ **d)** $p \lor \mathbf{T} \equiv \mathbf{T}$

   e) $p \lor p \equiv p$ **f)** $p \land p \equiv p$

2. Show that $\neg(\neg p)$ and p are logically equivalent.

3. Use truth tables to verify the commutative laws

   a) $p \lor q \equiv q \lor p$

   b) $p \land q \equiv q \land p$

4. Use truth tables to verify the associative laws

   a) $(p \lor q) \lor r \equiv p \lor (q \lor r)$

   b) $(p \land q) \land r \equiv p \land (q \land r)$

5. Use a truth table to verify the distributive law
$p \land (q \lor r) \equiv (p \land q) \lor (p \land r)$.

6. Use a truth table to verify the equivalence
$\neg(p \land q) \equiv \neg p \lor \neg q$.

7. Show that each of these implications is a tautology by using truth tables.

   a) $(p \land q) \to p$ **b)** $p \to (p \lor q)$

   c) $\neg p \to (p \to q)$ **d)** $(p \land q) \to (p \to q)$

   e) $\neg(p \to q) \to p$ **f)** $\neg(p \to q) \to \neg q$

8. Show that each of these implications is a tautology by using truth tables.

   a) $[\neg p \land (p \lor q)] \to q$

   b) $[(p \to q) \land (q \to r)] \to (p \to r)$

   c) $[p \land (p \to q)] \to q$

   d) $[(p \lor q) \land (p \to r) \land (q \to r)] \to r$

9. Show that each implication in Exercise 7 is a tautology without using truth tables.

10. Show that each implication in Exercise 8 is a tautology without using truth tables.

11. Use truth tables to verify the absorption laws.

   a) $p \lor (p \land q) \equiv p$ **b)** $p \land (p \lor q) \equiv p$

12. Determine whether $(\neg p \land (p \to q)) \to \neg q$ is a tautology.

13. Determine whether $(\neg q \land (p \to q)) \to \neg p$ is a tautology.

Links

HENRY MAURICE SHEFFER (1883–1964) Henry Maurice Sheffer, born to Jewish parents in the western Ukraine, emigrated to the United States in 1892 with his parents and six siblings. He studied at the Boston Latin School before entering Harvard, where he completed his undergraduate degree in 1905, his master's in 1907, and his Ph.D. in philosophy in 1908. After holding a postdoctoral position at Harvard, Henry traveled to Europe on a fellowship. Upon returning to the United States, he became an academic nomad, spending one year each at the University of Washington, Cornell, the University of Minnesota, the University of Missouri, and City College in New York. In 1916 he returned to Harvard as a faculty member in the philosophy department. He remained at Harvard until his retirement in 1952.

   Sheffer introduced what is now known as the Sheffer stroke in 1913; it became well known only after its use in the 1925 edition of Whitehead and Russell's *Principia Mathematica*. In this same edition Russell wrote that Sheffer had invented a powerful method that could be used to simplify the *Principia*. Because of this comment, Sheffer was something of a mystery man to logicians, especially because Sheffer, who published little in his career, never published the details of this method, only describing it in mimeographed notes and in a brief published abstract.

   Sheffer was a dedicated teacher of mathematical logic. He liked his classes to be small and did not like auditors. When strangers appeared in his classroom, Sheffer would order them to leave, even his colleagues or distinguished guests visiting Harvard. Sheffer was barely five feet tall; he was noted for his wit and vigor, as well as for his nervousness and irritability. Although widely liked, he was quite lonely. He is noted for a quip he spoke at his retirement: "Old professors never die, they just become emeriti." Sheffer is also credited with coining the term "Boolean algebra" (the subject of Chapter 10 of this text). Sheffer was briefly married and lived most of his later life in small rooms at a hotel packed with his logic books and vast files of slips of paper he used to jot down his ideas. Unfortunately, Sheffer suffered from severe depression during the last two decades of his life.

14. Show that $p \leftrightarrow q$ and $(p \wedge q) \vee (\neg p \wedge \neg q)$ are equivalent.

15. Show that $(p \rightarrow q) \rightarrow r$ and $p \rightarrow (q \rightarrow r)$ are not equivalent.

16. Show that $p \rightarrow q$ and $\neg q \rightarrow \neg p$ are logically equivalent.

17. Show that $\neg p \leftrightarrow q$ and $p \leftrightarrow \neg q$ are logically equivalent.

18. Show that $\neg(p \oplus q)$ and $p \leftrightarrow q$ are logically equivalent.

19. Show that $\neg(p \leftrightarrow q)$ and $\neg p \leftrightarrow q$ are logically equivalent.

20. Show that $(p \rightarrow q) \wedge (p \rightarrow r)$ and $p \rightarrow (q \wedge r)$ are logically equivalent.

21. Show that $(p \rightarrow r) \wedge (q \rightarrow r)$ and $(p \vee q) \rightarrow r$ are logically equivalent.

22. Show that $(p \rightarrow q) \vee (p \rightarrow r)$ and $p \rightarrow (q \vee r)$ are logically equivalent.

23. Show that $(p \rightarrow r) \vee (q \rightarrow r)$ and $(p \wedge q) \rightarrow r$ are logically equivalent.

24. Show that $\neg p \rightarrow (q \rightarrow r)$ and $q \rightarrow (p \vee r)$ are logically equivalent.

25. Show that $p \leftrightarrow q$ and $(p \rightarrow q) \wedge (q \rightarrow p)$ are logically equivalent.

26. Show that $p \leftrightarrow q$ and $\neg p \leftrightarrow \neg q$ are logically equivalent.

27. Show that $\neg(p \leftrightarrow q)$ and $p \leftrightarrow \neg q$ are logically equivalent.

28. Show that $(p \vee q) \wedge (\neg p \vee r) \rightarrow (q \vee r)$ is a tautology.

29. Show that $(p \rightarrow q) \wedge (q \rightarrow r) \rightarrow (p \rightarrow r)$ is a tautology.

The **dual** of a compound proposition that contains only the logical operators $\vee$, $\wedge$, and $\neg$ is the proposition obtained by replacing each $\vee$ by $\wedge$, each $\wedge$ by $\vee$, each **T** by **F**, and each **F** by **T**. The dual of proposition s is denoted by s^*.

30. Find the dual of each of these propositions.
   a) $p \wedge \neg q \wedge \neg r$ **b)** $(p \wedge q \wedge r) \vee s$
   c) $(p \vee \mathbf{F}) \wedge (q \vee \mathbf{T})$

31. Show that $(s^*)^* = s$.

32. Show that the logical equivalences in Table 5, except for the double negation law, come in pairs, where each pair contains propositions that are duals of each other.

**33. Why are the duals of two equivalent compound propositions also equivalent, where these compound propositions contain only the operators $\wedge$, $\vee$, and $\neg$?

34. Find a compound proposition involving the propositions p, q, and r that is true when p and q are true and r is false, but is false otherwise. (*Hint:* Use a conjunction of each proposition or its negation.)

35. Find a compound proposition involving the propositions p, q, and r that is true when exactly two of p, q,

and r are true and is false otherwise. (*Hint:* Form a disjunction of conjunctions. Include a conjunction for each combination of values for which the proposition is true. Each conjunction should include each of the three propositions or their negations.)

36. Suppose that a truth table in n propositional variables is specified. Show that a compound proposition with this truth table can be formed by taking the disjunction of conjunctions of the variables or their negations, with one conjunction included for each combination of values for which the compound proposition is true. The resulting compound proposition is said to be in **disjunctive normal form.**

A collection of logical operators is called **functionally complete** if every compound proposition is logically equivalent to a compound proposition involving only these logical operators.

37. Show that $\neg$, $\wedge$, and $\vee$ form a functionally complete collection of logical operators. (*Hint:* Use the fact that every proposition is logically equivalent to one in disjunctive normal form, as shown in Exercise 36.)

*38. Show that $\neg$ and $\wedge$ form a functionally complete collection of logical operators. (*Hint:* First use De Morgan's law to show that $p \vee q$ is equivalent to $\neg(\neg p \wedge \neg q)$.)

*39. Show that $\neg$ and $\vee$ form a functionally complete collection of logical operators.

The following exercises involve the logical operators *NAND* and *NOR*. The proposition p *NAND* q is true when either p or q, or both, are false; and it is false when both p and q are true. The proposition p *NOR* q is true when both p and q are false, and it is false otherwise. The propositions p *NAND* q and p *NOR* q are denoted by $p \mid q$ and $p \downarrow q$, respectively. (The operators $\mid$ and $\downarrow$ are called the **Sheffer stroke** and the **Peirce arrow** after H. M. Sheffer and C. S. Peirce, respectively.)

40. Construct a truth table for the logical operator *NAND*.

41. Show that $p \mid q$ is logically equivalent to $\neg(p \wedge q)$.

42. Construct a truth table for the logical operator *NOR*.

43. Show that $p \downarrow q$ is logically equivalent to $\neg(p \vee q)$.

44. In this exercise we will show that $\{\downarrow\}$ is a functionally complete collection of logical operators.
   a) Show that $p \downarrow p$ is logically equivalent to $\neg p$.
   b) Show that $(p \downarrow q) \downarrow (p \downarrow q)$ is logically equivalent to $p \vee q$.
   c) Conclude from parts (a) and (b), and Exercise 39, that $\{\downarrow\}$ is a functionally complete collection of logical operators.

*45. Find a proposition equivalent to $p \rightarrow q$ using only the logical operator $\downarrow$.

46. Show that $\{\mid\}$ is a functionally complete collection of logical operators.

47. Show that $p \mid q$ and $q \mid p$ are equivalent.

48. Show that $p \mid (q \mid r)$ and $(p \mid q) \mid r$ are not equivalent, so that the logical operator $\mid$ is not associative.

*49. How many different truth tables of compound propositions are there that involve the propositions p and q?

50. Show that if p, q, and r are compound propositions such that p and q are logically equivalent and q and r are logically equivalent, then p and r are logically equivalent.

51. The following sentence is taken from the specification of a telephone system: "If the directory database is opened, then the monitor is put in a closed state, if the system is not in its initial state." This specification is hard to understand since it involves two implications. Find an equivalent, easier-to-understand specification that involves disjunctions and negations but not implications.

52. How many of the disjunctions $p \vee \neg q$, $\neg p \vee q$, $q \vee r$, $q \vee \neg r$, $\neg q \vee \neg r$ can be made simultaneously true by an assignment of truth values to p, q, and r?

53. How many of the disjunctions $p \vee \neg q \vee s$, $\neg p \vee \neg r \vee s$, $\neg p \vee \neg r \vee \neg s$, $\neg p \vee q \vee \neg s$, $q \vee r \vee \neg s$, $q \vee \neg r \vee \neg s$, $\neg p \vee \neg q \vee \neg s$, $p \vee r \vee s$, $p \vee r \vee \neg s$ can be made simultaneously true by an assignment of truth values to p, q, r, and s?

A compound proposition is **satisfiable** if there is an assignment of truth values to the variables in the proposition that makes the compound proposition true.

54. Which of these compound propositions are satisfiable?

   a) $(p \vee q \vee \neg r) \wedge (p \vee \neg q \vee \neg s) \wedge (p \vee \neg r \vee \neg s) \wedge (\neg p \vee \neg q \vee \neg s) \wedge (p \vee q \vee \neg s)$

   b) $(\neg p \vee \neg q \vee r) \wedge (\neg p \vee q \vee \neg s) \wedge (p \vee \neg q \vee \neg s) \wedge (\neg p \vee \neg r \vee \neg s) \wedge (p \vee q \vee \neg r) \wedge (p \vee \neg r \vee \neg s)$

   c) $(p \vee q \vee r) \wedge (p \vee \neg q \vee \neg s) \wedge (q \vee \neg r \vee s) \wedge (\neg p \vee r \vee s) \wedge (\neg p \vee q \vee \neg s) \wedge (p \vee \neg q \vee \neg r) \wedge (\neg p \vee \neg q \vee s) \wedge (\neg p \vee \neg r \vee \neg s)$

55. Explain how an algorithm for determining whether a compound proposition is satisfiable can be used to determine whether a compound proposition is a tautology. (*Hint:* Look at $\neg p$, where p is the proposition that is being examined.)

1.3   Predicates and Quantifiers

INTRODUCTION

Statements involving variables, such as

    "$x > 3$,"   "$x = y + 3$,"   and   "$x + y = z$,"

are often found in mathematical assertions and in computer programs. These statements are neither true nor false when the values of the variables are not specified. In this section we will discuss the ways that propositions can be produced from such statements.

    The statement "x is greater than 3" has two parts. The first part, the variable x, is the subject of the statement. The second part—the **predicate**, "is greater than 3"—refers to a property that the subject of the statement can have. We can denote the statement "x is greater than 3" by $P(x)$, where P denotes the predicate "is greater than 3" and x is the variable. The statement $P(x)$ is also said to be the value of the **propositional function** P at x. Once a value has been assigned to the variable x, the statement $P(x)$ becomes a proposition and has a truth value. Consider Example 1.

EXAMPLE 1   Let $P(x)$ denote the statement "$x > 3$." What are the truth values of $P(4)$ and $P(2)$?

    Solution: We obtain the statement $P(4)$ by setting $x = 4$ in the statement "$x > 3$." Hence, $P(4)$, which is the statement "$4 > 3$," is true. However, $P(2)$, which is the statement "$2 > 3$," is false.   ◀

    We can also have statements that involve more than one variable. For instance, consider the statement "$x = y + 3$." We can denote this statement by $Q(x, y)$, where x and y are variables and Q is the predicate. When values are assigned to the variables x and y, the statement $Q(x, y)$ has a truth value.

EXAMPLE 2   Let $Q(x, y)$ denote the statement "$x = y + 3$." What are the truth values of the propositions $Q(1, 2)$ and $Q(3, 0)$?

Solution: To obtain $Q(1, 2)$, set $x = 1$ and $y = 2$ in the statement $Q(x, y)$. Hence, $Q(1, 2)$ is the statement "$1 = 2 + 3$," which is false. The statement $Q(3, 0)$ is the proposition "$3 = 0 + 3$," which is true.   ◄

Similarly, we can let $R(x, y, z)$ denote the statement "$x + y = z$." When values are assigned to the variables x, y, and z, this statement has a truth value.

EXAMPLE 3   What are the truth values of the propositions $R(1, 2, 3)$ and $R(0, 0, 1)$?

Solution: The proposition $R(1, 2, 3)$ is obtained by setting $x = 1$, $y = 2$, and $z = 3$ in the statement $R(x, y, z)$. We see that $R(1, 2, 3)$ is the statement "$1 + 2 = 3$," which is true. Also note that $R(0, 0, 1)$, which is the statement "$0 + 0 = 1$," is false.   ◄

In general, a statement involving the n variables $x_1, x_2, \ldots, x_n$ can be denoted by

$$P(x_1, x_2, \ldots, x_n).$$

A statement of the form $P(x_1, x_2, \ldots, x_n)$ is the value of the **propositional function** P at the n-tuple $(x_1, x_2, \ldots, x_n)$, and P is also called a **predicate.**

CHARLES SANDERS PEIRCE (1839–1914)   Many consider Charles Peirce the most original and versatile intellect from the United States; he was born in Cambridge, Massachusetts. His father, Benjamin Peirce, was a professor of mathematics and natural philosophy at Harvard. Peirce attended Harvard (1855–1859) and received a Harvard master of arts degree (1862) and an advanced degree in chemistry from the Lawrence Scientific School (1863). His father encouraged him to pursue a career in science, but instead he chose to study logic and scientific methodology.

In 1861, Peirce became an aide in the United States Coast Survey, with the goal of better understanding scientific methodology. His service for the Survey exempted him from military service during the Civil War. While working for the Survey, Peirce carried out astronomical and geodesic work. He made fundamental contributions to the design of pendulums and to map projections, applying new mathematical developments in the theory of elliptic functions. He was the first person to use the wavelength of light as a unit of measurement. Peirce rose to the position of Assistant for the Survey, a position he held until he was forced to resign in 1891 when he disagreed with the direction taken by the Survey's new administration.

Although making his living from work in the physical sciences, Peirce developed a hierarchy of sciences, with mathematics at the top rung, in which the methods of one science could be adapted for use by those sciences under it in the hierarchy. He was also the founder of the American philosophical theory of pragmatism.

The only academic position Peirce ever held was as a lecturer in logic at Johns Hopkins University in Baltimore from 1879 to 1884. His mathematical work during this time included contributions to logic, set theory, abstract algebra, and the philosophy of mathematics. His work is still relevant today; some of his work on logic has been recently applied to artificial intelligence. Peirce believed that the study of mathematics could develop the mind's powers of imagination, abstraction, and generalization. His diverse activities after retiring from the Survey included writing for newspapers and journals, contributing to scholarly dictionaries, translating scientific papers, guest lecturing, and textbook writing. Unfortunately, the income from these pursuits was insufficient to protect him and his second wife from abject poverty. He was supported in his later years by a fund created by his many admirers and administered by the philosopher William James, his lifelong friend. Although Peirce wrote and published voluminously in a vast range of subjects, he left more than 100,000 pages of unpublished manuscripts. Because of the difficulty of studying his unpublished writings, scholars have only recently started to understand some of his varied contributions. A group of people is devoted to making his work available over the Internet to bring a better appreciation of Peirce's accomplishments to the world.

Propositional functions occur in computer programs, as Example 4 demonstrates.

EXAMPLE 4   Consider the statement

$$\textbf{if } x > 0 \textbf{ then } x := x + 1.$$

When this statement is encountered in a program, the value of the variable x at that point in the execution of the program is inserted into $P(x)$, which is "$x > 0$." If $P(x)$ is true for this value of x, the assignment statement $x := x + 1$ is executed, so the value of x is increased by 1. If $P(x)$ is false for this value of x, the assignment statement is not executed, so the value of x is not changed.   ◀

QUANTIFIERS

When all the variables in a propositional function are assigned values, the resulting statement becomes a proposition with a certain truth value. However, there is another important way, called **quantification,** to create a proposition from a propositional function. Two types of quantification will be discussed here, namely, universal quantification and existential quantification. The area of logic that deals with predicates and quantifiers is called the **predicate calculus.**

THE UNIVERSAL QUANTIFIER Many mathematical statements assert that a property is true for all values of a variable in a particular domain, called the **universe of discourse** or the **domain.** Such a statement is expressed using a universal quantification. The universal quantification of a propositional function is the proposition that asserts that $P(x)$ is true for all values of x in the universe of discourse. The universe of discourse specifies the possible values of the variable x.

DEFINITION 1

> The *universal quantification* of $P(x)$ is the proposition
>
> "$P(x)$ is true for all values of x in the universe of discourse."

The notation

$$\forall x\, P(x)$$

denotes the universal quantification of $P(x)$. Here $\forall$ is called the **universal quantifier.** The proposition $\forall x\, P(x)$ is read as

"for all $x\, P(x)$"   or   "for every $x\, P(x)$."

Remark: It is best to avoid using "for any x" since it is often ambiguous as to whether "any" means "every" or "some." In some cases, "any" is unambiguous, such as when it is used in negatives, for example, "there is not any reason to avoid studying."

We illustrate the use of the universal quantifier in Examples 5–10.

EXAMPLE 5   Let $P(x)$ be the statement "$x + 1 > x$." What is the truth value of the quantification $\forall x\, P(x)$, where the universe of discourse consists of all real numbers?

Solution: Since $P(x)$ is true for all real numbers x, the quantification

$$\forall x \, P(x)$$

is true.   ◄

EXAMPLE 6   Let $Q(x)$ be the statement "$x < 2$." What is the truth value of the quantification $\forall x \, Q(x)$, where the universe of discourse consists of all real numbers?

Solution: $Q(x)$ is not true for every real number x, since, for instance, $Q(3)$ is false. Thus

$$\forall x \, Q(x)$$

is false.   ◄

When all the elements in the universe of discourse can be listed—say, $x_1, x_2, \ldots,$ x_n—it follows that the universal quantification $\forall x \, P(x)$ is the same as the conjunction

$$P(x_1) \wedge P(x_2) \wedge \cdots \wedge P(x_n),$$

since this conjunction is true if and only if $P(x_1), P(x_2), \ldots, P(x_n)$ are all true.

EXAMPLE 7   What is the truth value of $\forall x \, P(x)$, where $P(x)$ is the statement "$x^2 < 10$" and the universe of discourse consists of the positive integers not exceeding 4?

Solution: The statement $\forall x \, P(x)$ is the same as the conjunction

$$P(1) \wedge P(2) \wedge P(3) \wedge P(4),$$

since the universe of discourse consists of the integers $1, 2, 3,$ and 4. Since $P(4)$, which is the statement "$4^2 < 10$," is false, it follows that $\forall x \, P(x)$ is false.   ◄

EXAMPLE 8   What does the statement $\forall x \, T(x)$ mean if $T(x)$ is "x has two parents" and the universe of discourse consists of all people?

Solution: The statement $\forall x \, T(x)$ means that for every person x, that person has two parents. This statement can be expressed in English as "Every person has two parents." This statement is true (except for clones).   ◄

Specifying the universe of discourse is important when quantifiers are used. The truth value of a quantified statement often depends on which elements are in this universe of discourse, as Example 9 shows.

EXAMPLE 9   What is the truth value of $\forall x (x^2 \geq x)$ if the universe of discourse consists of all real numbers and what is its truth value if the universe of discourse consists of all integers?

Solution: Note that $x^2 \geq x$ if and only if $x^2 - x = x(x - 1) \geq 0$. Consequently, $x^2 \geq x$ if and only if $x \leq 0$ or $x \geq 1$. It follows that $\forall x (x^2 \geq x)$ is false if the universe of discourse consists of all real numbers (since the inequality is false for all real numbers x with $0 < x < 1$). However, if the universe of discourse consists of the integers, $\forall x (x^2 \geq x)$ is true, since there are no integers x with $0 < x < 1$.   ◄

To show that a statement of the form $\forall x \, P(x)$ is false, where $P(x)$ is a propositional function, we need only find one value of x in the universe of discourse for which $P(x)$ is false. Such a value of x is called a **counterexample** to the statement $\forall x \, P(x)$.

EXAMPLE 10   Suppose that $P(x)$ is "$x^2 > 0$." To show the statement $\forall x\, P(x)$ is false where the universe of discourse consists of all integers, we give a counterexample. We see that $x = 0$ is a counterexample since $x^2 = 0$ when $x = 0$ so that x^2 is not greater than 0 when $x = 0$. ◀

Looking for counterexamples to universally quantified statements is an important activity in the study of mathematics, as we will see in subsequent sections of this book.

THE EXISTENTIAL QUANTIFIER Many mathematical statements assert that there is an element with a certain property. Such statements are expressed using existential quantification. With existential quantification, we form a proposition that is true if and only if $P(x)$ is true for at least one value of x in the universe of discourse.

DEFINITION 2

> The *existential quantification* of $P(x)$ is the proposition
>
> "There exists an element x in the universe of discourse such that $P(x)$ is true."

We use the notation

$$\exists x\, P(x)$$

for the existential quantification of $P(x)$. Here $\exists$ is called the **existential quantifier.** The existential quantification $\exists x\, P(x)$ is read as

"There is an x such that $P(x)$,"
"There is at least one x such that $P(x)$,"

or

"For some $x\, P(x)$."

We illustrate the use of the existential quantifier in Examples 11–13.

EXAMPLE 11   Let $P(x)$ denote the statement "$x > 3$." What is the truth value of the quantification $\exists x\, P(x)$, where the universe of discourse consists of all real numbers?

Solution: Since "$x > 3$" is true—for instance, when $x = 4$—the existential quantification of $P(x)$, which is $\exists x\, P(x)$, is true. ◀

EXAMPLE 12   Let $Q(x)$ denote the statement "$x = x + 1$." What is the truth value of the quantification $\exists x\, Q(x)$, where the universe of discourse consists of all real numbers?

Solution: Since $Q(x)$ is false for every real number x, the existential quantification of $Q(x)$, which is $\exists x\, Q(x)$, is false. ◀

When all elements in the universe of discourse can be listed—say, $x_1, x_2, \ldots, x_n$— the existential quantification $\exists x\, P(x)$ is the same as the disjunction

$$P(x_1) \vee P(x_2) \vee \cdots \vee P(x_n),$$

since this disjunction is true if and only if at least one of $P(x_1), P(x_2), \ldots, P(x_n)$ is true.

TABLE 1 Quantifiers.		
Statement	*When True?*	*When False?*
$\forall x\, P(x)$	$P(x)$ is true for every x.	There is an x for which $P(x)$ is false.
$\exists x\, P(x)$	There is an x for which $P(x)$ is true.	$P(x)$ is false for every x.

EXAMPLE 13   What is the truth value of $\exists x\, P(x)$ where $P(x)$ is the statement "$x^2 > 10$" and the universe of discourse consists of the positive integers not exceeding 4?

Solution: Since the universe of discourse is $\{1, 2, 3, 4\}$, the proposition $\exists x\, P(x)$ is the same as the disjunction

$$P(1) \vee P(2) \vee P(3) \vee P(4).$$

Since $P(4)$, which is the statement "$4^2 > 10$," is true, it follows that $\exists x\, P(x)$ is true.   ◀

Table 1 summarizes the meaning of the universal and the existential quantifiers.

It is sometimes helpful to think in terms of looping and searching when determining the truth value of a quantification. Suppose that there are n objects in the universe of discourse for the variable x. To determine whether $\forall x\, P(x)$ is true, we can loop through all n values of x to see if $P(x)$ is always true. If we encounter a value x for which $P(x)$ is false, then we have shown that $\forall x\, P(x)$ is false. Otherwise, $\forall x\, P(x)$ is true. To see whether $\exists x\, P(x)$ is true, we loop through the n values of x searching for a value for which $P(x)$ is true. If we find one, then $\exists x\, P(x)$ is true. If we never find such an x, we have determined that $\exists x\, P(x)$ is false. (Note that this searching procedure does not apply if there are infinitely many values in the universe of discourse. However, it is still a useful way of thinking about the truth values of quantifications.)

BINDING VARIABLES

When a quantifier is used on the variable x or when we assign a value to this variable, we say that this occurrence of the variable is **bound.** An occurrence of a variable that is not bound by a quantifier or set equal to a particular value is said to be **free.** All the variables that occur in a propositional function must be bound to turn it into a proposition. This can be done using a combination of universal quantifiers, existential quantifiers, and value assignments.

The part of a logical expression to which a quantifier is applied is called the **scope** of this quantifier. Consequently, a variable is free if it is outside the scope of all quantifiers in the formula that specifies this variable.

EXAMPLE 14   In the statement $\exists x\, Q(x, y)$, the variable x is bound by the existential quantification $\exists x$, but the variable y is free because it is not bound by a quantifier and no value is assigned to this variable.

In the statement $\exists x(P(x) \wedge Q(x)) \vee \forall x\, R(x)$, all variables are bound. The scope of the first quantifier, $\exists x$, is the expression $P(x) \wedge Q(x)$ because $\exists x$ is applied only to $P(x) \wedge Q(x)$, and not to the rest of the statement. Similarly, the scope of the second quantifier, $\forall x$, is the expression $R(x)$. That is, the existential quantifier binds the variable x in $P(x) \wedge Q(x)$ and the universal quantifier $\forall x$ binds the variable x in $R(x)$. Observe that we could have written our statement using two different variables x and y, as $\exists x (P(x) \wedge Q(x)) \vee \forall y R(y)$, because the scopes of the two quantifiers do not overlap. The reader should be aware that in common usage, the same letter is often used to represent variables bound by different quantifiers with scopes that do not overlap.   ◀

NEGATIONS

We will often want to consider the negation of a quantified expression. For instance, consider the negation of the statement

"Every student in the class has taken a course in calculus."

This statement is a universal quantification, namely,

$$\forall x P(x),$$

where $P(x)$ is the statement "x has taken a course in calculus." The negation of this statement is "It is not the case that every student in the class has taken a course in calculus." This is equivalent to "There is a student in the class who has not taken a course in calculus." And this is simply the existential quantification of the negation of the original propositional function, namely,

$$\exists x \neg P(x).$$

This example illustrates the following equivalence:

$$\neg \forall x P(x) \equiv \exists x \neg P(x).$$

Extra Examples

Suppose we wish to negate an existential quantification. For instance, consider the proposition "There is a student in this class who has taken a course in calculus." This is the existential quantification

$$\exists x Q(x),$$

where $Q(x)$ is the statement "x has taken a course in calculus." The negation of this statement is the proposition "It is not the case that there is a student in this class who has taken a course in calculus." This is equivalent to "Every student in this class has not taken calculus," which is just the universal quantification of the negation of the original propositional function, or, phrased in the language of quantifiers,

$$\forall x \neg Q(x).$$

This example illustrates the equivalence

$$\neg \exists x Q(x) \equiv \forall x \neg Q(x).$$

Negations of quantifiers are summarized in Table 2.

Remark: When the universe of discourse of a predicate $P(x)$ consists of n elements, where n is a positive integer, the rules for negating quantified statements are exactly the same as De Morgan's laws discussed in Section 1.2. This follows because $\neg \forall x P(x)$ is the same as $\neg (P(x_1) \wedge P(x_2) \wedge \cdots \wedge P(x_n))$, which is equivalent to $\neg P(x_1) \vee \neg P(x_2) \vee \cdots \vee \neg P(x_n)$ by De Morgan's laws, and this is the same as $\exists x \neg P(x)$. Similarly, $\neg \exists x P(x)$ is the same as $\neg (P(x_1) \vee P(x_2) \vee \cdots \vee P(x_n))$, which by De Morgan's laws is equivalent to $\neg P(x_1) \wedge \neg P(x_2) \wedge \cdots \wedge \neg P(x_n)$, and this is the same as $\forall x \neg P(x)$.

We illustrate the negation of quantified statements in Examples 15 and 16.

	TABLE 2	**Negating Quantifiers.**		
Negation	*Equivalent Statement*	*When Is Negation True?*	*When False?*
$\neg \exists x\, P(x)$	$\forall x \neg P(x)$	For every x, $P(x)$ is false.	There is an x for which $P(x)$ is true.
$\neg \forall x\, P(x)$	$\exists x \neg P(x)$	There is an x for which $P(x)$ is false.	$P(x)$ is true for every x.

EXAMPLE 15 What are the negations of the statements "There is an honest politician" and "All Americans eat cheeseburgers"?

Solution: Let $H(x)$ denote "x is honest." Then the statement "There is an honest politician" is represented by $\exists x\, H(x)$, where the universe of discourse consists of all politicians. The negation of this statement is $\neg \exists x\, H(x)$, which is equivalent to $\forall x \neg H(x)$. This negation can be expressed as "Every politician is dishonest." (*Note:* In English the statement "All politicians are not honest" is ambiguous. In common usage this statement often means "Not all politicians are honest." Consequently, we do not use this statement to express this negation.)

Let $C(x)$ denote "x eats cheeseburgers." Then the statement "All Americans eat cheeseburgers" is represented by $\forall x\, C(x)$, where the universe of discourse consists of all Americans. The negation of this statement is $\neg \forall x\, C(x)$, which is equivalent to $\exists x \neg C(x)$. This negation can be expressed in several different ways, including "Some American does not eat cheeseburgers" and "There is an American who does not eat cheeseburgers." ◀

EXAMPLE 16 What are the negations of the statements $\forall x(x^2 > x)$ and $\exists x(x^2 = 2)$?

Solution: The negation of $\forall x(x^2 > x)$ is the statement $\neg \forall x(x^2 > x)$, which is equivalent to $\exists x \neg(x^2 > x)$. This can be rewritten as $\exists x(x^2 \leq x)$. The negation of $\exists x(x^2 = 2)$ is the statement $\neg \exists x(x^2 = 2)$, which is equivalent to $\forall x \neg(x^2 = 2)$. This can be rewritten as $\forall x(x^2 \neq 2)$. The truth values of these statements depend on the universe of discourse. ◀

TRANSLATING FROM ENGLISH INTO LOGICAL EXPRESSIONS

Translating sentences in English (or other natural languages) into logical expressions is a crucial task in mathematics, logic programming, artificial intelligence, software engineering, and many other disciplines. We began studying this topic in Section 1.1, where we used propositions to express sentences in logical expressions. In that discussion, we purposely avoided sentences whose translations required predicates and quantifiers. Translating from English to logical expressions becomes even more complex when quantifiers are needed. Furthermore, there can be many ways to translate a particular sentence. (As a consequence, there is no "cookbook" approach that can be followed step by step.) We will use some examples to illustrate how to translate sentences from English into logical expressions. The goal in this translation is to produce simple and useful logical expressions. In this section, we restrict ourselves to sentences that can be translated into logical expressions using a single quantifier; in the next section, we will look at more complicated sentences that require multiple quantifiers.

EXAMPLE 17

Extra
Examples

Express the statement "Every student in this class has studied calculus" using predicates and quantifiers.

Solution: First, we rewrite the statement so that we can clearly identify the appropriate quantifiers to use. Doing so, we obtain:

"For every student in this class, that student has studied calculus."

Next, we introduce a variable x so that our statement becomes

"For every student x in this class, x has studied calculus."

Continuing, we introduce the predicate $C(x)$, which is the statement "x has studied calculus." Consequently, if the universe of discourse for x consists of the students in the class, we can translate our statement as $\forall x\, C(x)$.

However, there are other correct approaches; different universes of discourse and other predicates can be used. The approach we select depends on the subsequent reasoning we want to carry out. For example, we may be interested in a wider group of people than only those in this class. If we change the universe of discourse to consist of all people, we will need to express our statement as

"For every person x, if person x is a student in this class then x has studied calculus."

If $S(x)$ represents the statement that person x is in this class, we see that our statement can be expressed as $\forall x(S(x) \to C(x))$. [*Caution!* Our statement *cannot* be expressed as $\forall x(S(x) \land C(x))$ since this statement says that all people are students in this class and have studied calculus!]

Finally, when we are interested in the background of people in subjects besides calculus, we may prefer to use the two-variable quantifier $Q(x, y)$ for the statement "student x has studied subject y." Then we would replace $C(x)$ by $Q(x, \text{calculus})$ in both approaches we have followed to obtain $\forall x\, Q(x, \text{calculus})$ or $\forall x(S(x) \to Q(x, \text{calculus}))$. ◀

In Example 17 we displayed different approaches for expressing the same statement using predicates and quantifiers. However, we should always adopt the simplest approach that is adequate for use in subsequent reasoning.

EXAMPLE 18

Express the statements "Some student in this class has visited Mexico" and "Every student in this class has visited either Canada or Mexico" using predicates and quantifiers.

Solution: The statement "Some student in this class has visited Mexico" means that

"There is a student in this class with the property that the student has visited Mexico."

We can introduce a variable x, so that our statement becomes

"There is a student x in this class having the property x has visited Mexico."

We introduce the predicate $M(x)$, which is the statement "x has visited Mexico." If the universe of discourse for x consists of the students in this class, we can translate this first statement as $\exists x\, M(x)$.

However, if we are interested in people other than those in this class, we look at the statement a little differently. Our statement can be expressed as

"There is a person x having the properties that x is a student in this class and x has visited Mexico."

In this case, the universe of discourse for the variable x consists of all people. We introduce the predicate $S(x)$, "x is a student in this class." Our solution becomes $\exists x(S(x) \wedge M(x))$ since the statement is that there is a person x who is a student in this class and who has visited Mexico. [*Caution!* Our statement cannot be expressed as $\exists x(S(x) \rightarrow M(x))$, which is true when there is someone not in the class.]

Similarly, the second statement can be expressed as

"For every x in this class, x has the property that x has visited Mexico or x has visited Canada."

(Note that we are assuming the inclusive, rather than the exclusive, *or* here.) We let $C(x)$ be the statement "x has visited Canada." Following our earlier reasoning, we see that if the universe of discourse for x consists of the students in this class, this second statement can be expressed as $\forall x(C(x) \vee M(x))$. However, if the universe of discourse for x consists of all people, our statement can be expressed as

"For every person x, if x is a student in this class, then x has visited Mexico or x has visited Canada."

In this case, the statement can be expressed as $\forall x(S(x) \rightarrow (C(x) \vee M(x)))$.

Instead of using the predicates $M(x)$ and $C(x)$ to represent that x has visited Mexico and x has visited Canada, respectively, we could use a two-place predicate $V(x, y)$ to represent "x has visited country y." In this case, $V(x, \text{Mexico})$ and $V(x, \text{Canada})$ would have the same meaning as $M(x)$ and $C(x)$ and could replace them in our answers. If we are working with many statements that involve people visiting different countries, we might prefer to use this two-variable approach. Otherwise, for simplicity, we would stick with the one-variable predicates $M(x)$ and $C(x)$. ◀

EXAMPLES FROM LEWIS CARROLL

Lewis Carroll (really C. L. Dodgson writing under a pseudonym), the author of *Alice in Wonderland,* is also the author of several works on symbolic logic. His books contain many examples of reasoning using quantifiers. Examples 19 and 20 come from his book

Links

CHARLES LUTWIDGE DODGSON (1832–1898) We know Charles Dodgson as *Lewis Carroll—* the pseudonym he used in his writings on logic. Dodgson, the son of a clergyman, was the third of 11 children, all of whom stuttered. He was uncomfortable in the company of adults and is said to have spoken without stuttering only to young girls, many of whom he entertained, corresponded with, and photographed (often in the nude). Although attracted to young girls, he was extremely puritanical and religious. His friendship with the three young daughters of Dean Liddell led to his writing *Alice in Wonderland,* which brought him money and fame.

Dodgson graduated from Oxford in 1854 and obtained his master of arts degree in 1857. He was appointed lecturer in mathematics at Christ Church College, Oxford, in 1855. He was ordained in the Church of England in 1861 but never practiced his ministry. His writings include articles and books on geometry, determinants, and the mathematics of tournaments and elections. (He also used the pseudonym Lewis Carroll for his many works on recreational logic.)

Symbolic Logic; other examples from that book are given in the exercise set at the end of this section. These examples illustrate how quantifiers are used to express various types of statements.

EXAMPLE 19   Consider these statements. The first two are called *premises* and the third is called the *conclusion.* The entire set is called an *argument.*

> "All lions are fierce."
> "Some lions do not drink coffee."
> "Some fierce creatures do not drink coffee."

(In Section 1.5 we will discuss the issue of determining whether the conclusion is a valid consequence of the premises. In this example, it is.) Let $P(x)$, $Q(x)$, and $R(x)$ be the statements "x is a lion," "x is fierce," and "x drinks coffee," respectively. Assuming that the universe of discourse is the set of all creatures, express the statements in the argument using quantifiers and $P(x)$, $Q(x)$, and $R(x)$.

Solution: We can express these statements as:

> $\forall x(P(x) \rightarrow Q(x))$.
> $\exists x(P(x) \wedge \neg R(x))$.
> $\exists x(Q(x) \wedge \neg R(x))$.

Notice that the second statement cannot be written as $\exists x(P(x) \rightarrow \neg R(x))$. The reason is that $P(x) \rightarrow \neg R(x)$ is true whenever x is not a lion, so that $\exists x(P(x) \rightarrow \neg R(x))$ is true as long as there is at least one creature that is not a lion, even if every lion drinks coffee. Similarly, the third statement cannot be written as

> $\exists x(Q(x) \rightarrow \neg R(x))$.   ◄

EXAMPLE 20   Consider these statements, of which the first three are premises and the fourth is a valid conclusion.

> "All hummingbirds are richly colored."
> "No large birds live on honey."
> "Birds that do not live on honey are dull in color."
> "Hummingbirds are small."

Let $P(x)$, $Q(x)$, $R(x)$, and $S(x)$ be the statements "x is a hummingbird," "x is large," "x lives on honey," and "x is richly colored," respectively. Assuming that the universe of discourse is the set of all birds, express the statements in the argument using quantifiers and $P(x)$, $Q(x)$, $R(x)$, and $S(x)$.

Solution: We can express the statements in the argument as

> $\forall x(P(x) \rightarrow S(x))$.
> $\neg \exists x(Q(x) \wedge R(x))$.
> $\forall x(\neg R(x) \rightarrow \neg S(x))$.
> $\forall x(P(x) \rightarrow \neg Q(x))$.

(Note we have assumed that "small" is the same as "not large" and that "dull in color" is the same as "not richly colored." To show that the fourth statement is a valid conclusion of the first three, we need to use rules of inference that will be discussed in Section 1.5.)   ◄

LOGIC PROGRAMMING

An important type of programming language is designed to reason using the rules of predicate logic. Prolog (from *Pro*gramming in *Log*ic), developed in the 1970s by computer scientists working in the area of artificial intelligence, is an example of such a language. Prolog programs include a set of declarations consisting of two types of statements, **Prolog facts** and **Prolog rules.** Prolog facts define predicates by specifying the elements that satisfy these predicates. Prolog rules are used to define new predicates using those already defined by Prolog facts. Example 21 illustrates these notions.

EXAMPLE 21   Consider a Prolog program given facts telling it the instructor of each class and in which classes students are enrolled. The program uses these facts to answer queries concerning the professors who teach particular students. Such a program could use the predicates *instructor*(p, c) and *enrolled*(s, c) to represent that professor p is the instructor of course c and that student s is enrolled in course c, respectively. For example, the Prolog facts in such a program might include:

```
instructor(chan,math273)
instructor(patel,ee222)
instructor(grossman,cs301)
enrolled(kevin,math273)
enrolled(juana,ee222)
enrolled(juana,cs301)
enrolled(kiko,math273)
enrolled(kiko,cs301)
```

(Lowercase letters have been used for entries because Prolog considers names beginning with an uppercase letter to be variables.)

A new predicate *teaches*(p, s), representing that professor p teaches student s, can be defined using the Prolog rule

```
teaches(P,S) :- instructor(P,C), enrolled(S,C)
```

which means that *teaches*(p, s) is true if there exists a class c such that professor p is the instructor of class c and student s is enrolled in class c. (Note that a comma is used to represent a conjunction of predicates in Prolog. Similarly, a semicolon is used to represent a disjunction of predicates.)

Prolog answers queries using the facts and rules it is given. For example, using the facts and rules listed, the query

```
?enrolled(kevin,math273)
```

produces the response

```
yes
```

since the fact *enrolled*(kevin, math273) was provided as input. The query

```
?enrolled(X,math273)
```

produces the response

```
kevin
kiko
```

To produce this response, Prolog determines all possible values of X for which *enrolled*$(X, \text{math273})$ has been included as a Prolog fact. Similarly, to find all the professors who are instructors in classes being taken by Juana, we use the query

```
?teaches(X,juana)
```

This query returns

```
patel
grossman
```

◀

Exercises

1. Let $P(x)$ denote the statement "$x \le 4$." What are the truth values?

   a) $P(0)$ **b)** $P(4)$ **c)** $P(6)$

2. Let $P(x)$ be the statement "the word x contains the letter a." What are the truth values?

   a) $P(\text{orange})$ **b)** $P(\text{lemon})$
   c) $P(\text{true})$ **d)** $P(\text{false})$

3. Let $Q(x, y)$ denote the statement "x is the capital of y." What are these truth values?

   a) $Q(\text{Denver, Colorado})$
   b) $Q(\text{Detroit, Michigan})$
   c) $Q(\text{Massachusetts, Boston})$
   d) $Q(\text{New York, New York})$

4. State the value of x after the statement **if** $P(x)$ **then** $x := 1$ is executed, where $P(x)$ is the statement "$x > 1$," if the value of x when this statement is reached is

   a) $x = 0$.   **b)** $x = 1$.   **c)** $x = 2$.

5. Let $P(x)$ be the statement "x spends more than five hours every weekday in class," where the universe of discourse for x consists of all students. Express each of these quantifications in English.

   a) $\exists x\, P(x)$ **b)** $\forall x\, P(x)$
   c) $\exists x\, \neg P(x)$ **d)** $\forall x\, \neg P(x)$

6. Let $N(x)$ be the statement "x has visited North Dakota," where the universe of discourse consists of the students in your school. Express each of these quantifications in English.

   a) $\exists x\, N(x)$ **b)** $\forall x\, N(x)$ **c)** $\neg \exists x\, N(x)$
   d) $\exists x\, \neg N(x)$ **e)** $\neg \forall x\, N(x)$ **f)** $\forall x\, \neg N(x)$

7. Translate these statements into English, where $C(x)$ is "x is a comedian" and $F(x)$ is "x is funny" and the universe of discourse consists of all people.

   a) $\forall x(C(x) \rightarrow F(x))$ **b)** $\forall x(C(x) \wedge F(x))$
   c) $\exists x(C(x) \rightarrow F(x))$ **d)** $\exists x(C(x) \wedge F(x))$

8. Translate these statements into English, where $R(x)$ is "x is a rabbit" and $H(x)$ is "x hops" and the universe of discourse consists of all animals.

   a) $\forall x(R(x) \rightarrow H(x))$ **b)** $\forall x(R(x) \wedge H(x))$
   c) $\exists x(R(x) \rightarrow H(x))$ **d)** $\exists x(R(x) \wedge H(x))$

9. Let $P(x)$ be the statement "x can speak Russian" and let $Q(x)$ be the statement "x knows the computer language C++." Express each of these sentences in terms of $P(x)$, $Q(x)$, quantifiers, and logical connectives. The universe of discourse for quantifiers consists of all students at your school.

   a) There is a student at your school who can speak Russian and who knows C++.
   b) There is a student at your school who can speak Russian but who doesn't know C++.
   c) Every student at your school either can speak Russian or knows C++.
   d) No student at your school can speak Russian or knows C++.

10. Let $C(x)$ be the statement "x has a cat," let $D(x)$ be the statement "x has a dog," and let $F(x)$ be the statement "x has a ferret." Express each of these statements in terms of $C(x)$, $D(x)$, $F(x)$, quantifiers, and logical connectives. Let the universe of discourse consist of all students in your class.

   a) A student in your class has a cat, a dog, and a ferret.
   b) All students in your class have a cat, a dog, or a ferret.
   c) Some student in your class has a cat and a ferret, but not a dog.
   d) No student in your class has a cat, a dog, and a ferret.
   e) For each of the three animals, cats, dogs, and ferrets, there is a student in your class who has one of these animals as a pet.

11. Let $P(x)$ be the statement "$x = x^2$." If the universe of discourse consists of the integers, what are the truth values?

   a) $P(0)$ **b)** $P(1)$ **c)** $P(2)$
   d) $P(-1)$ **e)** $\exists x\, P(x)$ **f)** $\forall x\, P(x)$

12. Let $Q(x)$ be the statement "$x + 1 > 2x$." If the universe of discourse consists of all integers, what are these truth values?

a) $Q(0)$ **b)** $Q(-1)$ **c)** $Q(1)$
d) $\exists x \, Q(x)$ **e)** $\forall x \, Q(x)$ **f)** $\exists x \neg Q(x)$
g) $\forall x \neg Q(x)$

13. Determine the truth value of each of these statements if the universe of discourse consists of all integers.

a) $\forall n (n + 1 > n)$ **b)** $\exists n (2n = 3n)$
c) $\exists n (n = -n)$ **d)** $\forall n (n^2 \geq n)$

14. Determine the truth value of each of these statements if the universe of discourse consists of all real numbers.

a) $\exists x (x^3 = -1)$ **b)** $\exists x (x^4 < x^2)$
c) $\forall x ((-x)^2 = x^2)$ **d)** $\forall x (2x > x)$

15. Determine the truth value of each of these statements if the universe of discourse for all variables consists of all integers.

a) $\forall n (n^2 \geq 0)$ **b)** $\exists n (n^2 = 2)$
c) $\forall n (n^2 \geq n)$ **d)** $\exists n (n^2 < 0)$

16. Determine the truth value of each of these statements if the universe of discourse of each variable consists of all real numbers.

a) $\exists x (x^2 = 2)$ **b)** $\exists x (x^2 = -1)$
c) $\forall x (x^2 + 2 \geq 1)$ **d)** $\forall x (x^2 \neq x)$

17. Suppose that the universe of discourse of the propositional function $P(x)$ consists of the integers 0, 1, 2, 3, and 4. Write out each of these propositions using disjunctions, conjunctions, and negations.

a) $\exists x \, P(x)$ **b)** $\forall x \, P(x)$ **c)** $\exists x \neg P(x)$
d) $\forall x \neg P(x)$ **e)** $\neg \exists x \, P(x)$ **f)** $\neg \forall x \, P(x)$

18. Suppose that the universe of discourse of the propositional function $Q(x)$ consists of the integers $-2, -1,$ 0, 1, and 2. Write out each of these propositions using disjunctions, conjunctions, and negations.

a) $\exists x \, P(x)$ **b)** $\forall x \, P(x)$ **c)** $\exists x \neg P(x)$
d) $\forall x \neg P(x)$ **e)** $\neg \exists x \, P(x)$ **f)** $\neg \forall x \, P(x)$

19. Suppose that the universe of discourse of the propositional function $P(x)$ consists of the integers $1, 2, 3, 4,$ and 5. Express these statements without using quantifiers, instead using only negations, disjunctions, and conjunctions.

a) $\exists x \, P(x)$ **b)** $\forall x \, P(x)$
c) $\neg \exists x \, P(x)$ **d)** $\neg \forall x \, P(x)$
e) $\forall x ((x \neq 3) \to P(x)) \vee \exists x \neg P(x)$

20. Suppose that the universe of discourse of the propositional function $P(x)$ consists of $-5, -3, -1, 1, 3,$ and 5. Express these statements without using quantifiers, instead using only negations, disjunctions, and conjunctions.

a) $\exists x \, P(x)$ **b)** $\forall x \, P(x)$
c) $\forall x ((x \neq 1) \to P(x))$

d) $\exists x ((x \geq 0) \wedge P(x))$
e) $\exists x (\neg P(x)) \wedge \forall x ((x < 0) \to P(x))$

21. Translate in two ways each of these statements into logical expressions using predicates, quantifiers, and logical connectives. First, let the universe of discourse consist of the students in your class and second, let it consist of all people.

a) Someone in your class can speak Hindi.
b) Everyone in your class is friendly.
c) There is a person in your class who was not born in California.
d) A student in your class has been in a movie.
e) No student in your class has taken a course in logic programming.

22. Translate in two ways each of these statements into logical expressions using predicates, quantifiers, and logical connectives. First, let the universe of discourse consist of the students in your class and second, let it consist of all people.

a) Everyone in your class has a cellular phone.
b) Somebody in your class has seen a foreign movie.
c) There is a person in your class who cannot swim.
d) All students in your class can solve quadratic equations.
e) Some student in your class does not want to be rich.

23. Translate each of these statements into logical expressions using predicates, quantifiers, and logical connectives.

a) No one is perfect.
b) Not everyone is perfect.
c) All your friends are perfect.
d) One of your friends is perfect.
e) Everyone is your friend and is perfect.
f) Not everybody is your friend or someone is not perfect.

24. Translate each of these statements into logical expressions in three different ways by varying the universe of discourse and by using predicates with one and with two variables.

a) Someone in your school has visited Uzbekistan.
b) Everyone in your class has studied calculus and C++.
c) No one in your school owns both a bicycle and a motorcycle.
d) There is a person in your school who is not happy.
e) Everyone in your school was born in the twentieth century.

25. Translate each of these statements into logical expressions in three different ways by varying the universe of discourse and by using predicates with one and with two variables.

a) A student in your school has lived in Vietnam.
b) There is a student in your school who cannot speak Hindi.

c) A student in your school knows Java, Prolog, and C++.

d) Everyone in your class enjoys Thai food.

e) Someone in your class does not play hockey.

26. Translate each of these statements into logical expressions using predicates, quantifiers, and logical connectives.

a) Something is not in the correct place.

b) All tools are in the correct place and are in excellent condition.

c) Everything is in the correct place and in excellent condition.

d) Nothing is in the correct place and is in excellent condition.

e) One of your tools is not in the correct place, but it is in excellent condition.

27. Express each of these statements using logical operators, predicates, and quantifiers.

a) Some propositions are tautologies.

b) The negation of a contradiction is a tautology.

c) The disjunction of two contingencies can be a tautology.

d) The conjunction of two tautologies is a tautology.

28. Suppose the universe of discourse of the propositional function $P(x, y)$ consists of pairs x and y, where x is 1, 2, or 3 and y is 1, 2, or 3. Write out these propositions using disjunctions and conjunctions.

a) $\exists x\, P(x, 3)$ **b)** $\forall y\, P(1, y)$

c) $\exists y \neg P(2, y)$ **d)** $\forall x \neg P(x, 2)$

29. Suppose that the universe of discourse of $Q(x, y, z)$ consists of triples x, y, z, where $x = 0, 1,$ or $2, y = 0$ or 1, and $z = 0$ or 1. Write out these propositions using disjunctions and conjunctions.

a) $\forall y\, Q(0, y, 0)$

b) $\exists x\, Q(x, 1, 1)$

c) $\exists z \neg Q(0, 0, z)$

d) $\exists x \neg Q(x, 0, 1)$

30. Express each of these statements using quantifiers. Then form the negation of the statement so that no negation is to the left of a quantifier. Next, express the negation in simple English. (Do not simply use the words "It is not the case that.")

a) All dogs have fleas.

b) There is a horse that can add.

c) Every koala can climb.

d) No monkey can speak French.

e) There exists a pig that can swim and catch fish.

31. Express each of these statements using quantifiers. Then form the negation of the statement, so that no negation is to the left of a quantifier. Next, express the negation in simple English. (Do not simply use the words "It is not the case that.")

a) Some old dogs can learn new tricks.

b) No rabbit knows calculus.

c) Every bird can fly.

d) There is no dog that can talk.

e) There is no one in this class who knows French and Russian.

32. Express the negation of these propositions using quantifiers, and then express the negation in English.

a) Some drivers do not obey the speed limit.

b) All Swedish movies are serious.

c) No one can keep a secret.

d) There is someone in this class who does not have a good attitude.

33. Find a counterexample, if possible, to these universally quantified statements, where the universe of discourse for all variables consists of all integers.

a) $\forall x(x^2 \geq x)$ **b)** $\forall x(x > 0 \vee x < 0)$

c) $\forall x(x = 1)$

34. Find a counterexample, if possible, to these universally quantified statements, where the universe of discourse for all variables consists of all real numbers.

a) $\forall x(x^2 \neq x)$ **b)** $\forall x(x^2 \neq 2)$

c) $\forall x(|x| > 0)$

35. Express each of these statements using predicates and quantifiers.

a) A passenger on an airline qualifies as an elite flyer if the passenger flies more than 25,000 miles in a year or takes more than 25 flights during that year.

b) A man qualifies for the marathon if his best previous time is less than 3 hours and a woman qualifies for the marathon if her best previous time is less than 3.5 hours.

c) A student must take at least 60 course hours, or at least 45 course hours and write a master's thesis, and receive a grade no lower than a B in all required courses, to receive a master's degree.

d) There is a student who has taken more than 21 credit hours in a semester and received all A's.

Exercises 36–40 deal with the translation between system specification and logical expressions involving quantifiers.

36. Translate these system specifications into English where the predicate $S(x, y)$ is "x is in state y" and where the universe of discourse for x and y consists of all systems and all possible states, respectively.

a) $\exists x\, S(x, \text{open})$

b) $\forall x(S(x, \text{malfunctioning}) \vee S(x, \text{diagnostic}))$

c) $\exists x\, S(x, \text{open}) \vee \exists x\, S(x, \text{diagnostic})$

d) $\exists x \neg S(x, \text{available})$

e) $\forall x \neg S(x, \text{working})$

37. Translate these specifications into English where $F(p)$ is "Printer p is out of service," $B(p)$ is "Printer p is busy," $L(j)$ is "Print job j is lost," and $Q(j)$ is "Print job j is queued."

a) $\exists p(F(p) \wedge B(p)) \rightarrow \exists j\, L(j)$

b) $\forall p\, B(p) \rightarrow \exists j\, Q(j)$

c) $\exists j\, (Q(j) \wedge L(j)) \rightarrow \exists p\, F(p)$

d) $(\forall p\, B(p) \wedge \forall j\, Q(j)) \rightarrow \exists j\, L(j)$

38. Express each of these system specifications using predicates, quantifiers, and logical connectives.

  a) When there is less than 30 megabytes free on the hard disk, a warning message is sent to all users.

  b) No directories in the file system can be opened and no files can be closed when system errors have been detected.

  c) The file system cannot be backed up if there is a user currently logged on.

  d) Video on demand can be delivered when there are at least 8 megabytes of memory available and the connection speed is at least 56 kilobits per second.

39. Express each of these system specifications using predicates, quantifiers, and logical connectives.

  a) At least one mail message can be saved if there is a disk with more than 10 kilobytes of free space.

  b) Whenever there is an active alert, all queued messages are transmitted.

  c) The diagnostic monitor tracks the status of all systems except the main console.

  d) Each participant on the conference call whom the host of the call did not put on a special list was billed.

40. Express each of these system specifications using predicates, quantifiers, and logical connectives.

  a) Every user has access to an electronic mailbox.

  b) The system mailbox can be accessed by everyone in the group if the file system is locked.

  c) The firewall is in a diagnostic state only if the proxy server is in a diagnostic state.

  d) At least one router is functioning normally if the throughput is between 100 kbps and 500 kbps and the proxy server is not in diagnostic mode.

41. Determine whether $\forall x(P(x) \rightarrow Q(x))$ and $\forall x\, P(x) \rightarrow \forall x\, Q(x)$ have the same truth value.

42. Show that $\forall x(P(x) \wedge Q(x))$ and $\forall x\, P(x) \wedge \forall x\, Q(x)$ have the same truth value.

43. Show that $\exists x(P(x) \vee Q(x))$ and $\exists x\, P(x) \vee \exists x\, Q(x)$ have the same truth value.

44. Establish these logical equivalences, where A is a proposition not involving any quantifiers.

  a) $(\forall x\, P(x)) \vee A \equiv \forall x(P(x) \vee A)$

  b) $(\exists x\, P(x)) \vee A \equiv \exists x(P(x) \vee A)$

45. Establish these logical equivalences, where A is a proposition not involving any quantifiers.

  a) $(\forall x\, P(x)) \wedge A \equiv \forall x(P(x) \wedge A)$

  b) $(\exists x\, P(x)) \wedge A \equiv \exists x(P(x) \wedge A)$

46. Show that $\forall x\, P(x) \vee \forall x\, Q(x)$ and $\forall x(P(x) \vee Q(x))$ are not logically equivalent.

47. Show that $\exists x\, P(x) \wedge \exists x\, Q(x)$ and $\exists x(P(x) \wedge Q(x))$ are not logically equivalent.

48. The notation $\exists! x\, P(x)$ denotes the proposition

    "There exists a unique x such that $P(x)$ is true."

If the universe of discourse consists of all integers, what are the truth values?

  a) $\exists! x(x > 1)$ **b)** $\exists! x(x^2 = 1)$

  c) $\exists! x(x + 3 = 2x)$ **d)** $\exists! x(x = x + 1)$

49. What are the truth values of these statements?

  a) $\exists! x\, P(x) \rightarrow \exists x\, P(x)$

  b) $\forall x\, P(x) \rightarrow \exists! x\, P(x)$

  c) $\exists! x\, \neg P(x) \rightarrow \neg \forall x\, P(x)$

50. Write out $\exists! x\, P(x)$, where the universe of discourse consists of the integers 1, 2, and 3, in terms of negations, conjunctions, and disjunctions.

51. Given the Prolog facts in Example 21, what would Prolog return given these queries?

  a) `?instructor(chan,math273)`

  b) `?instructor(patel,cs301)`

  c) `?enrolled(X,cs301)`

  d) `?enrolled(kiko,Y)`

  e) `?teaches(grossman,Y)`

52. Given the Prolog facts in Example 21, what would Prolog return when given these queries?

  a) `?enrolled(kevin,ee222)`

  b) `?enrolled(kiko,math273)`

  c) `?instructor(grossman,X)`

  d) `?instructor(X,cs301)`

  e) `?teaches(X,kevin)`

53. Suppose that Prolog facts are used to define the predicates *mother*(M, Y) and *father*(F, X), which represent that M is the mother of Y and F is the father of X, respectively. Give a Prolog rule to define the predicate *sibling*(X, Y), which represents that X and Y are siblings (that is, have the same mother and the same father).

54. Suppose that Prolog facts are used to define the predicates *mother*(M, Y) and *father*(F, X), which represent that M is the mother of Y and F is the father of X, respectively. Give a Prolog rule to define the predicate *grandfather*(X, Y), which represents that X is the grandfather of Y. (*Hint:* You can write a disjunction in Prolog either by using a semicolon to separate predicates or by putting these predicates on separate lines.)

Exercises 55–58 are based on questions found in the book *Symbolic Logic* by Lewis Carroll.

55. Let $P(x), Q(x)$, and $R(x)$ be the statements "x is a professor," "x is ignorant," and "x is vain," respectively. Express each of these statements using quantifiers; logical connectives; and $P(x), Q(x)$, and $R(x)$, where the universe of discourse consists of all people.

  a) No professors are ignorant.

  b) All ignorant people are vain.

  c) No professors are vain.

d) Does (c) follow from (a) and (b)? If not, is there a correct conclusion?

56. Let $P(x)$, $Q(x)$, and $R(x)$ be the statements "x is a clear explanation," "x is satisfactory," and "x is an excuse," respectively. Suppose that the universe of discourse for x consists of all English text. Express each of these statements using quantifiers, logical connectives, and $P(x)$, $Q(x)$, and $R(x)$.

  a) All clear explanations are satisfactory.
  b) Some excuses are unsatisfactory.
  c) Some excuses are not clear explanations.
 * d) Does (c) follow from (a) and (b)? If not, is there a correct conclusion?

57. Let $P(x)$, $Q(x)$, $R(x)$, and $S(x)$ be the statements "x is a baby," "x is logical," "x is able to manage a crocodile," and "x is despised," respectively. Suppose that the universe of discourse consists of all people. Express each of these statements using quantifiers;

logical connectives; and $P(x)$, $Q(x)$, $R(x)$, and $S(x)$.

  a) Babies are illogical.
  b) Nobody is despised who can manage a crocodile.
  c) Illogical persons are despised.
  d) Babies cannot manage crocodiles.
 * e) Does (d) follow from (a), (b), and (c)? If not, is there a correct conclusion?

58. Let $P(x)$, $Q(x)$, $R(x)$, and $S(x)$ be the statements "x is a duck," "x is one of my poultry," "x is an officer," and "x is willing to waltz," respectively. Express each of these statements using quantifiers; logical connectives; and $P(x)$, $Q(x)$, $R(x)$, and $S(x)$.

  a) No ducks are willing to waltz.
  b) No officers ever decline to waltz.
  c) All my poultry are ducks.
  d) My poultry are not officers.
 * e) Does (d) follow from (a), (b), and (c)? If not, is there a correct conclusion?

1.4   Nested Quantifiers

INTRODUCTION

In Section 1.3 we defined the existential and universal quantifiers and showed how they can be used to represent mathematical statements. We also explained how they can be used to translate English sentences into logical expressions. In this section we will study **nested quantifiers,** which are quantifiers that occur within the scope of other quantifiers, such as in the statement $\forall x \exists y(x + y = 0)$. Nested quantifiers commonly occur in mathematics and computer science. Although nested quantifiers can sometimes be difficult to understand, the rules we have already studied in Section 1.3 can help us use them.

TRANSLATING STATEMENTS INVOLVING NESTED QUANTIFIERS

Complicated expressions involving quantifiers arise in many contexts. To understand these statements involving many quantifiers, we need to unravel what the quantifiers and predicates that appear mean. This is illustrated in Example 1.

EXAMPLE 1

Additional Steps

Assume that the universe of discourse for the variables x and y consists of all real numbers. The statement

$$\forall x \forall y(x + y = y + x)$$

says that $x + y = y + x$ for all real numbers x and y. This is the commutative law for addition of real numbers. Likewise, the statement

$$\forall x \exists y(x + y = 0)$$

says that for every real number x there is a real number y such that $x + y = 0$. This states that every real number has an additive inverse. Similarly, the statement

Extra Examples

$$\forall x \forall y \forall z(x + (y + z) = (x + y) + z)$$

is the associative law for addition of real numbers.

◀

EXAMPLE 2   Translate into English the statement

$$\forall x \forall y ((x > 0) \land (y < 0) \rightarrow (xy < 0)),$$

where the universe of discourse for both variables consists of all real numbers.

Solution: This statement says that for every real number x and for every real number y, if $x > 0$ and $y < 0$, then $xy < 0$. That is, this statement says that for real numbers x and y, if x is positive and y is negative, then xy is negative. This can be stated more succinctly as "The product of a positive real number and a negative real number is a negative real number." ◄

Expressions with nested quantifiers expressing statements in English can be quite complicated. The first step in translating such an expression is to write out what the quantifiers and predicates in the expression mean. The next step is to express this meaning in a simpler sentence. This process is illustrated in Examples 3 and 4.

EXAMPLE 3   Translate the statement

$$\forall x (C(x) \lor \exists y (C(y) \land F(x, y)))$$

into English, where $C(x)$ is "x has a computer," $F(x, y)$ is "x and y are friends," and the universe of discourse for both x and y consists of all students in your school.

Solution: The statement says that for every student x in your school x has a computer or there is a student y such that y has a computer and x and y are friends. In other words, every student in your school has a computer or has a friend who has a computer. ◄

EXAMPLE 4   Translate the statement

$$\exists x \forall y \forall z ((F(x, y) \land F(x, z) \land (y \neq z)) \rightarrow \neg F(y, z))$$

into English, where $F(a, b)$ means a and b are friends and the universe of discourse for $x, y,$ and z consists of all students in your school.

Solution: We first examine the expression $(F(x, y) \land F(x, z) \land (y \neq z)) \rightarrow \neg F(y, z)$. This expression says that if students x and y are friends, and students x and z are friends, and furthermore, if y and z are not the same student, then y and z are not friends. It follows that the original statement, which is triply quantified, says that there is a student x such that for all students y and all students z other than y, if x and y are friends and x and z are friends, then y and z are not friends. In other words, there is a student none of whose friends are also friends with each other. ◄

TRANSLATING SENTENCES INTO LOGICAL EXPRESSIONS

In Section 1.3 we showed how quantifiers can be used to translate sentences into logical expressions. However, we avoided sentences whose translation into logical expressions required the use of nested quantifiers. We now address the translation of such sentences.

EXAMPLE 5   Express the statement "If a person is female and is a parent, then this person is someone's mother" as a logical expression involving predicates, quantifiers with a universe of discourse consisting of all people, and logical connectives.

Solution: The statement "If a person is female and is a parent, then this person is some-one's mother" can be expressed as "For every person x, if person x is female and person x is a parent, then there exists a person y such that person x is the mother of person y." We introduce the predicates $F(x)$ to represent "x is female," $P(x)$ to represent "x is a parent," and $M(x, y)$ to represent "x is the mother of y." The original statement can be represented as

$$\forall x((F(x) \land P(x)) \rightarrow \exists y M(x, y)).$$

We can move $\exists y$ all the way to the left, because y does not appear in $F(x) \land P(x)$, to obtain an equivalent expression

$$\forall x \exists y((F(x) \land P(x)) \rightarrow M(x, y)). \qquad \blacktriangleleft$$

EXAMPLE 6 Express the statement "Everyone has exactly one best friend" as a logical expression involving predicates, quantifiers with a universe of discourse consisting of all people, and logical connectives.

Solution: The statement "Everyone has exactly one best friend" can be expressed as "For every person x, person x has exactly one best friend." Introducing the universal quantifier, we see that this statement is the same as "$\forall x$ (person x has exactly one best friend)" where the universe of discourse consists of all people.

To say that x has exactly one best friend means that there is a person y who is the best friend of x, and, furthermore, that for every person z, if person z is not person y, then z is not the best friend of x. When we introduce the predicate $B(x, y)$ to be the statement "y is the best friend of x," the statement that x has exactly one best friend can be represented as

$$\exists y(B(x, y) \land \forall z((z \neq y) \rightarrow \neg B(x, z))).$$

Consequently, our original statement can be expressed as

$$\forall x \exists y(B(x, y) \land \forall z((z \neq y) \rightarrow \neg B(x, z))).$$

(Note that we can write this statement as $\forall x \exists! y B(x, y)$, where $\exists!$ is the "uniqueness quantifier" defined in Exercise 48 of Section 1.3. However, the "uniqueness quantifier" is not really a quantifier; rather, it is a shorthand for expressing certain statements that can be expressed using the quantifiers $\forall$ and $\exists$. The "uniqueness quantifier" $\exists!$ can be thought of as a macro.) $\qquad \blacktriangleleft$

EXAMPLE 7 Use quantifiers to express the statement "There is a woman who has taken a flight on every airline in the world."

Solution: Let $P(w, f)$ be "w has taken f" and $Q(f, a)$ be "f is a flight on a." We can express the statement as

$$\exists w \forall a \exists f(P(w, f) \land Q(f, a)),$$

where the universes of discourse for w, f, and a consist of all the women in the world, all airplane flights, and all airlines, respectively.

The statement could also be expressed as

$$\exists w \forall a \exists f R(w, f, a),$$

where $R(w, f, a)$ is "w has taken f on a." Although this is more compact, it somewhat obscures the relationships among the variables. Consequently, the first solution is usually preferable. $\qquad \blacktriangleleft$

Mathematical statements expressed in English can be translated into logical expressions as Examples 8–10 show.

EXAMPLE 8   Translate the statement "The sum of two positive integers is positive" into a logical expression.

Additional Steps

Solution: To translate this statement into a logical expression, we first rewrite it so that the implied quantifiers are shown: "For every two positive integers, the sum of these integers is positive." Next, we introduce the variables x and y to obtain "For all positive integers x and y, $x + y$ is positive." Consequently, we can express this statement as

$$\forall x \forall y ((x > 0) \wedge (y > 0) \rightarrow (x + y > 0)),$$

where the universe of discourse for both variables consists of all integers.   ◀

EXAMPLE 9   Translate the statement "Every real number except zero has a multiplicative inverse."

Extra Examples

Solution: We first rewrite this as "For every real number x except zero, x has a multiplicative inverse." We can rewrite this as "For every real number x, if $x \neq 0$, then there exists a real number y such that $xy = 1$." This can be rewritten as

$$\forall x ((x \neq 0) \rightarrow \exists y (xy = 1)).$$ ◀

One example that you may be familiar with is the concept of limit, which is important in calculus.

EXAMPLE 10   **(Calculus required)** Express the definition of a limit using quantifiers.

Solution: Recall that the definition of the statement

$$\lim_{x \to a} f(x) = L$$

is: For every real number $\epsilon > 0$ there exists a real number $\delta > 0$ such that $|f(x) - L| < \epsilon$ whenever $0 < |x - a| < \delta$. This definition of a limit can be phrased in terms of quantifiers by

$$\forall \epsilon \exists \delta \forall x (0 < |x - a| < \delta \rightarrow |f(x) - L| < \epsilon),$$

where the universe of discourse for the variables δ and ϵ consists of all positive real numbers and for x consists of all real numbers.

This definition can also be expressed as

$$\forall \epsilon {>} 0 \; \exists \delta {>} 0 \; \forall x (0 < |x - a| < \delta \rightarrow |f(x) - L| < \epsilon)$$

when the universe of discourse for the variables ϵ and δ consists of all real numbers, rather than just the positive real numbers.   ◀

NEGATING NESTED QUANTIFIERS

Statements involving nested quantifiers can be negated by successively applying the rules for negating statements involving a single quantifier. This is illustrated in Examples 11–13.

EXAMPLE 11

Extra
Examples

Express the negation of the statement $\forall x \exists y(xy = 1)$ so that no negation precedes a quantifier.

Solution: By successively applying the rules for negating quantified statements given in Table 2 of Section 1.3, we can move the negation in $\neg\forall x \exists y(xy = 1)$ inside all the quantifiers. We find that $\neg\forall x \exists y(xy = 1)$ is equivalent to $\exists x \neg \exists y(xy = 1)$, which is equivalent to $\exists x \forall y \neg(xy = 1)$. Since $\neg(xy = 1)$ can be expressed more simply as $xy \neq 1$, we conclude that our negated statement can be expressed as $\exists x \forall y(xy \neq 1)$. ◄

EXAMPLE 12

Use quantifiers to express the statement that "There does not exist a woman who has taken a flight on every airline in the world."

Solution: This statement is the negation of the statement "There is a woman who has taken a flight on every airline in the world" from Example 7. By Example 7, our statement can be expressed as $\neg\exists w \forall a \exists f(P(w, f) \wedge Q(f, a))$, where $P(w, f)$ is "w has taken f" and $Q(f, a)$ is "f is a flight on a." By successively applying the rules for negating quantified statements from Table 2 of Section 1.3 to move the negation inside successive quantifiers and by applying De Morgan's law in the last step, we find that our statement is equivalent to each of this sequence of statements:

$$\forall w \neg \forall a \exists f(P(w, f) \wedge Q(f, a)) \equiv \forall w \exists a \neg \exists f(P(w, f) \wedge Q(f, a))$$

$$\equiv \forall w \exists a \forall f \neg(P(w, f) \wedge Q(f, a))$$

$$\equiv \forall w \exists a \forall f(\neg P(w, f) \vee \neg Q(f, a)).$$

This last statement states "For every woman there is an airline such that for all flights, this woman has not taken that flight or that flight is not on this airline." ◄

EXAMPLE 13

Use quantifiers and predicates to express the fact that $\lim_{x \to a} f(x)$ does not exist.

Solution: To say that $\lim_{x \to a} f(x)$ does not exist means that for all real numbers L, $\lim_{x \to a} f(x) \neq L$. By using Example 10, the statement $\lim_{x \to a} f(x) \neq L$ can be expressed as

$$\neg\forall \epsilon > 0 \, \exists \delta > 0 \, \forall x(0 < |x - a| < \delta \to |f(x) - L| < \epsilon).$$

Successively applying the rules for negating quantified expressions, we construct this sequence of equivalent statements

$$\neg\forall \epsilon > 0 \, \exists \delta > 0 \, \forall x(0 < |x - a| < \delta \to |f(x) - L| < \epsilon)$$

$$\equiv \exists \epsilon > 0 \, \neg \exists \delta > 0 \, \forall x(0 < |x - a| < \delta \to |f(x) - L| < \epsilon)$$

$$\equiv \exists \epsilon > 0 \, \forall \delta > 0 \, \neg \forall x(0 < |x - a| < \delta \to |f(x) - L| < \epsilon)$$

$$\equiv \exists \epsilon > 0 \, \forall \delta > 0 \, \exists x \, \neg(0 < |x - a| < \delta \to |f(x) - L| < \epsilon)$$

$$\equiv \exists \epsilon > 0 \, \forall \delta > 0 \, \exists x(0 < |x - a| < \delta \wedge |f(x) - L| \geq \epsilon).$$

We use the equivalence $\neg(p \to q) \equiv p \wedge \neg q$, in the last step.

Because the statement $\lim_{x \to a} f(x)$ does not exist means for all real numbers L, $\lim_{x \to a} f(x) \neq L$. This statement can be expressed as

$$\forall L \exists \epsilon > 0 \, \forall \delta > 0 \, \exists x(0 < |x - a| < \delta \wedge |f(x) - L| \geq \epsilon).$$

This last statement says that for every real number L there is a real number $\epsilon > 0$ such that for every real number $\delta > 0$, there exists a real number x such that $0 < |x - a| < \delta$ and $|f(x) - L| \geq \epsilon$. ◄

THE ORDER OF QUANTIFIERS

Many mathematical statements involve multiple quantifications of propositional functions involving more than one variable. It is important to note that the order of the quantifiers is important, unless all the quantifiers are universal quantifiers or all are existential quantifiers. These remarks are illustrated by Examples 14–16. In each of these examples the universe of discourse for each variable consists of all real numbers.

EXAMPLE 14   Let $P(x, y)$ be the statement "$x + y = y + x$." What is the truth value of the quantification $\forall x \forall y P(x, y)$?

Solution: The quantification

$$\forall x \forall y P(x, y)$$

denotes the proposition

"For all real numbers x and for all real numbers $y, x + y = y + x$."

Since $P(x, y)$ is true for all real numbers x and y, the proposition $\forall x \forall y P(x, y)$ is true.   ◄

EXAMPLE 15   Let $Q(x, y)$ denote "$x + y = 0$." What are the truth values of the quantifications $\exists y \forall x Q(x, y)$ and $\forall x \exists y Q(x, y)$?

Solution: The quantification

$$\exists y \forall x Q(x, y)$$

denotes the proposition

"There is a real number y such that for every real number $x, Q(x, y)$."

No matter what value of y is chosen, there is only one value of x for which $x + y = 0$. Since there is no real number y such that $x + y = 0$ for all real numbers x, the statement $\exists y \forall x Q(x, y)$ is false.

The quantification

$$\forall x \exists y Q(x, y)$$

denotes the proposition

"For every real number x there is a real number y such that $Q(x, y)$."

Given a real number x, there is a real number y such that $x + y = 0$; namely, $y = -x$. Hence, the statement $\forall x \exists y Q(x, y)$ is true.   ◄

Example 15 illustrates that the order in which quantifiers appear makes a difference. The statements $\exists y \forall x P(x, y)$ and $\forall x \exists y P(x, y)$ are not logically equivalent. The statement $\exists y \forall x P(x, y)$ is true if and only if there is a y that makes $P(x, y)$ true for every x. So, for this statement to be true, there must be a particular value of y for which $P(x, y)$ is true regardless of the choice of x. On the other hand, $\forall x \exists y P(x, y)$ is true if and only if for every value of x there is a value of y for which $P(x, y)$ is true. So, for this statement to be true, no matter which x you choose, there must be a value of y (possibly depending on the x you choose) for which $P(x, y)$ is true. In other words, in the second case y can depend on x, whereas in the first case y is a constant independent of x.

TABLE 1 Quantifications of Two Variables.		
Statement	*When True?*	*When False?*
$\forall x \forall y P(x, y)$ $\forall y \forall x P(x, y)$	$P(x, y)$ is true for every pair x, y.	There is a pair x, y for which $P(x, y)$ is false.
$\forall x \exists y P(x, y)$	For every x there is a y for which $P(x, y)$ is true.	There is an x such that $P(x, y)$ is false for every y.
$\exists x \forall y P(x, y)$	There is an x for which $P(x, y)$ is true for every y.	For every x there is a y for which $P(x, y)$ is false.
$\exists x \exists y P(x, y)$ $\exists y \exists x P(x, y)$	There is a pair x, y for which $P(x, y)$ is true.	$P(x, y)$ is false for every pair x, y.

From these observations, it follows that if $\exists y \forall x P(x, y)$ is true, then $\forall x \exists y P(x, y)$ must also be true. However, if $\forall x \exists y P(x, y)$ is true, it is not necessary for $\exists y \forall x P(x, y)$ to be true. (See Supplementary Exercises 14 and 16 at the end of this chapter.)

THINKING OF QUANTIFICATION AS LOOPS In working with quantifications of more than one variable, it is sometimes helpful to think in terms of nested loops. (Of course, if there are infinitely many elements in the universe of discourse of some variable, we cannot actually loop through all values. Nevertheless, this way of thinking is helpful in understanding nested quantifiers.) For example, to see whether $\forall x \forall y P(x, y)$ is true, we loop through the values for x, and for each x we loop through the values for y. If we find that $P(x, y)$ is true for all values for x and y, we have determined that $\forall x \forall y P(x, y)$ is true. If we ever hit a value x for which we hit a value y for which $P(x, y)$ is false, we have shown that $\forall x \forall y P(x, y)$ is false.

Similarly, to determine whether $\forall x \exists y P(x, y)$ is true, we loop through the values for x. For each x we loop through the values for y until we find a y for which $P(x, y)$ is true. If for all x we hit such a y, then $\forall x \exists y P(x, y)$ is true; if for some x we never hit such a y, then $\forall x \exists y P(x, y)$ is false.

To see whether $\exists x \forall y P(x, y)$ is true, we loop through the values for x until we find an x for which $P(x, y)$ is always true when we loop through all values for y. Once we find such an x, we know that $\exists x \forall y P(x, y)$ is true. If we never hit such an x, then we know that $\exists x \forall y P(x, y)$ is false.

Finally, to see whether $\exists x \exists y P(x, y)$ is true, we loop through the values for x, where for each x we loop through the values for y until we hit an x for which we hit a y for which $P(x, y)$ is true. The statement $\exists x \exists y P(x, y)$ is false only if we never hit an x for which we hit a y such that $P(x, y)$ is true.

Table 1 summarizes the meanings of the different possible quantifications involving two variables.

Quantifications of more than two variables are also common, as Example 16 illustrates.

EXAMPLE 16 Let $Q(x, y, z)$ be the statement "$x + y = z$." What are the truth values of the statements $\forall x \forall y \exists z Q(x, y, z)$ and $\exists z \forall x \forall y Q(x, y, z)$?

Solution: Suppose that x and y are assigned values. Then, there exists a real number z such that $x + y = z$. Consequently, the quantification

$$\forall x \forall y \exists z\, Q(x, y, z),$$

which is the statement

"For all real numbers x and for all real numbers y there is a real number z such that $x + y = z$,"

is true. The order of the quantification here is important, since the quantification

$$\exists z \forall x \forall y\, Q(x, y, z),$$

which is the statement

"There is a real number z such that for all real numbers x and for all real numbers y it is true that $x + y = z$,"

is false, since there is no value of z that satisfies the equation $x + y = z$ for all values of x and y.   ◀

Exercises

1. Translate these statements into English, where the universe of discourse for each variable consists of all real numbers.

   a) $\forall x \exists y (x < y)$
   b) $\forall x \forall y (((x \geq 0) \wedge (y \geq 0)) \rightarrow (xy \geq 0))$
   c) $\forall x \forall y \exists z (xy = z)$

2. Translate these statements into English, where the universe of discourse for each variable consists of all real numbers.

   a) $\exists x \forall y (xy = y)$
   b) $\forall x \forall y (((x \geq 0) \wedge (y < 0)) \rightarrow (x - y > 0))$
   c) $\forall x \forall y \exists z (x = y + z)$

3. Let $Q(x, y)$ be the statement "x has sent an e-mail message to y," where the universe of discourse for both x and y consists of all students in your class. Express each of these quantifications in English.

   a) $\exists x \exists y\, Q(x, y)$ **b)** $\exists x \forall y\, Q(x, y)$
   c) $\forall x \exists y\, Q(x, y)$ **d)** $\exists y \forall x\, Q(x, y)$
   e) $\forall y \exists x\, Q(x, y)$ **f)** $\forall x \forall y\, Q(x, y)$

4. Let $P(x, y)$ be the statement "student x has taken class y," where the universe of discourse for x consists of all students in your class and for y consists of all computer science courses at your school. Express each of these quantifications in English.

   a) $\exists x \exists y\, P(x, y)$ **b)** $\exists x \forall y\, P(x, y)$
   c) $\forall x \exists y\, P(x, y)$ **d)** $\exists y \forall x\, P(x, y)$
   e) $\forall y \exists x\, P(x, y)$ **f)** $\forall x \forall y\, P(x, y)$

5. Let $W(x, y)$ mean that student x has visited website y, where the universe of discourse for x consists of all students in your school and the universe of discourse

for y consists of all websites. Express each of these statements by a simple English sentence.

   a) $W(\text{Sarah Smith, www.att.com})$
   b) $\exists x\, W(x, \text{www.imdb.org})$
   c) $\exists y\, W(\text{Jose Orez}, y)$
   d) $\exists y (W(\text{Ashok Puri}, y) \wedge W(\text{Cindy Yoon}, y))$
   e) $\exists y \forall z (y \neq (\text{David Belcher}) \wedge (W(\text{David Belcher}, z) \rightarrow W(y, z)))$
   f) $\exists x \exists y \forall z ((x \neq y) \wedge (W(x, z) \leftrightarrow W(y, z)))$

6. Let $C(x, y)$ mean that student x is enrolled in class y, where the universe of discourse for x consists of all students in your school and the universe of discourse for y is the set of all classes being given at your school. Express each of these statements by a simple English sentence.

   a) $C(\text{Randy Goldberg, CS 252})$
   b) $\exists x\, C(x, \text{Math 695})$
   c) $\exists y\, C(\text{Carol Sitea}, y)$
   d) $\exists x (C(x, \text{Math 222}) \wedge C(x, \text{CS 252}))$
   e) $\exists x \exists y \forall z ((x \neq y) \wedge (C(x, z) \rightarrow C(y, z)))$
   f) $\exists x \exists y \forall z ((x \neq y) \wedge (C(x, z) \leftrightarrow C(y, z)))$

7. Let $T(x, y)$ mean that student x likes cuisine y, where the universe of discourse for x consists of all students at your school and the universe of discourse for y consists of all cuisines. Express each of these statements by a simple English sentence.

   a) $\neg T(\text{Abdallah Hussein, Japanese})$
   b) $\exists x\, T(x, \text{Korean}) \wedge \forall x\, T(x, \text{Mexican})$
   c) $\exists y (T(\text{Monique Arsenault}, y) \vee T(\text{Jay Johnson}, y))$
   d) $\forall x \forall z \exists y ((x \neq z) \rightarrow \neg(T(x, y) \wedge T(z, y)))$

   e) $\exists x \exists z \forall y (T(x, y) \leftrightarrow T(z, y))$

   f) $\forall x \forall z \exists y (T(x, y) \leftrightarrow T(z, y))$

8. Let $Q(x, y)$ be the statement "student x has been a contestant on quiz show y." Express each of these sentences in terms of $Q(x, y)$, quantifiers, and logical connectives, where the universe of discourse for x consists of all students at your school and for y consists of all quiz shows on television.

   a) There is a student at your school who has been a contestant on a television quiz show.

   b) No student at your school has ever been a contestant on a television quiz show.

   c) There is a student at your school who has been a contestant on *Jeopardy* and on *Wheel of Fortune*.

   d) Every television quiz show has had a student from your school as a contestant.

   e) At least two students from your school have been contestants on *Jeopardy*.

9. Let $L(x, y)$ be the statement "x loves y," where the universe of discourse for both x and y consists of all people in the world. Use quantifiers to express each of these statements.

   a) Everybody loves Jerry.

   b) Everybody loves somebody.

   c) There is somebody whom everybody loves.

   d) Nobody loves everybody.

   e) There is somebody whom Lydia does not love.

   f) There is somebody whom no one loves.

   g) There is exactly one person whom everybody loves.

   h) There are exactly two people whom Lynn loves.

   i) Everyone loves himself or herself.

   j) There is someone who loves no one besides himself or herself.

10. Let $F(x, y)$ be the statement "x can fool y," where the universe of discourse consists of all people in the world. Use quantifiers to express each of these statements.

   a) Everybody can fool Fred.

   b) Evelyn can fool everybody.

   c) Everybody can fool somebody.

   d) There is no one who can fool everybody.

   e) Everyone can be fooled by somebody.

   f) No one can fool both Fred and Jerry.

   g) Nancy can fool exactly two people.

   h) There is exactly one person whom everybody can fool.

   i) No one can fool himself or herself.

   j) There is someone who can fool exactly one person besides himself or herself.

11. Let $S(x)$ be the predicate "x is a student," $F(x)$ the predicate "x is a faculty member," and $A(x, y)$ the predicate "x has asked y a question," where the universe of discourse consists of all people associated with your school. Use quantifiers to express each of these statements.

   a) Lois has asked Professor Michaels a question.

   b) Every student has asked Professor Gross a question.

   c) Every faculty member has either asked Professor Miller a question or been asked a question by Professor Miller.

   d) Some student has not asked any faculty member a question.

   e) There is a faculty member who has never been asked a question by a student.

   f) Some student has asked every faculty member a question.

   g) There is a faculty member who has asked every other faculty member a question.

   h) Some student has never been asked a question by a faculty member.

12. Let $I(x)$ be the statement "x has an Internet connection" and $C(x, y)$ be the statement "x and y have chatted over the Internet," where the universe of discourse for the variables x and y consists of all students in your class. Use quantifiers to express each of these statements.

   a) Jerry does not have an Internet connection.

   b) Rachel has not chatted over the Internet with Chelsea.

   c) Jan and Sharon have never chatted over the Internet.

   d) No one in the class has chatted with Bob.

   e) Sanjay has chatted with everyone except Joseph.

   f) Someone in your class does not have an Internet connection.

   g) Not everyone in your class has an Internet connection.

   h) Exactly one student in your class has an Internet connection.

   i) Everyone except one student in your class has an Internet connection.

   j) Everyone in your class with an Internet connection has chatted over the Internet with at least one other student in your class.

   k) Someone in your class has an Internet connection but has not chatted with anyone else in your class.

   l) There are two students in your class who have not chatted with each other over the Internet.

   m) There is a student in your class who has chatted with everyone in your class over the Internet.

   n) There are at least two students in your class who have not chatted with the same person in your class.

   o) There are two students in the class who between them have chatted with everyone else in the class.

13. Let $M(x, y)$ be "x has sent y an e-mail message" and $T(x, y)$ be "x has telephoned y," where the universe of discourse consists of all students in your class. Use quantifiers to express each of these statements. (Assume that all e-mail messages that were sent are received, which is not the way things often work.)

a) Chou has never sent an e-mail message to Koko.

b) Arlene has never sent an e-mail message to or telephoned Sarah.

c) Jose has never received an e-mail message from Deborah.

d) Every student in your class has sent an e-mail message to Ken.

e) No one in your class has telephoned Nina.

f) Everyone in your class has either telephoned Avi or sent him an e-mail message.

g) There is a student in your class who has sent everyone else in your class an e-mail message.

h) There is someone in your class who has either sent an e-mail message or telephoned everyone else in your class.

i) There are two students in your class who have sent each other e-mail messages.

j) There is a student who has sent himself or herself an e-mail message.

k) There is a student in your class who has not received an e-mail message from anyone else in the class and who has not been called by any other student in the class.

l) Every student in the class has either received an e-mail message or received a telephone call from another student in the class.

m) There are at least two students in your class such that one student has sent the other e-mail and the second student has telephoned the first student.

n) There are two students in your class who between them have sent an e-mail message to or telephoned everyone else in the class.

14. Use quantifiers and predicates with more than one variable to express these statements.

a) There is a student in this class who can speak Hindi.

b) Every student in this class plays some sport.

c) Some student in this class has visited Alaska but has not visited Hawaii.

d) All students in this class have learned at least one programming language.

e) There is a student in this class who has taken every course offered by one of the departments in this school.

f) Some student in this class grew up in the same town as exactly one other student in this class.

g) Every student in this class has chatted with at least one other student in at least one chat group.

15. Use quantifiers and predicates with more than one variable to express these statements.

a) Every computer science student needs a course in discrete mathematics.

b) There is a student in this class who owns a personal computer.

c) Every student in this class has taken at least one computer science course.

d) There is a student in this class who has taken at least one course in computer science.

e) Every student in this class has been in every building on campus.

f) There is a student in this class who has been in every room of at least one building on campus.

g) Every student in this class has been in at least one room of every building on campus.

16. A discrete mathematics class contains 1 mathematics major who is a freshman, 12 mathematics majors who are sophomores, 15 computer science majors who are sophomores, 2 mathematics majors who are juniors, 2 computer science majors who are juniors, and 1 computer science major who is a senior. Express each of these statements in terms of quantifiers and then determine its truth value.

a) There is a student in the class who is a junior.

b) Every student in the class is a computer science major.

c) There is a student in the class who is neither a mathematics major nor a junior.

d) Every student in the class is either a sophomore or a computer science major.

e) There is a major such that there is a student in the class in every year of study with that major.

17. Express each of these system specifications using predicates, quantifiers, and logical connectives, if necessary.

a) Every user has access to exactly one mailbox.

b) There is a process that continues to run during all error conditions only if the kernel is working correctly.

c) All users on the campus network can access all websites whose url has a .edu extension.

* d) There are exactly two systems that monitor every remote server.

18. Express each of these system specifications using predicates, quantifiers, and logical connectives, if necessary.

a) At least one console must be accessible during every fault condition.

b) The e-mail address of every user can be retrieved whenever the archive contains at least one message sent by every user on the system.

c) For every security breach there is at least one mechanism that can detect that breach if and only if there is a process that has not been compromised.

d) There are at least two paths connecting every two endpoints on the network.

e) No one knows the password of every user on the system except for the system administrator.

19. Express each of these statements using mathematical and logical operators, predicates, and quantifiers, where the universe of discourse consists of all integers.

   a) The sum of two negative integers is negative.
   b) The difference of two positive integers is not necessarily positive.
   c) The sum of the squares of two integers is greater than or equal to the square of their sum.
   d) The absolute value of the product of two integers is the product of their absolute values.

20. Express each of these statements using predicates, quantifiers, logical connectives, and mathematical operators where the universe of discourse consists of all integers.

   a) The product of two negative integers is positive.
   b) The average of two positive integers is positive.
   c) The difference of two negative integers is not necessarily negative.
   d) The absolute value of the sum of two integers does not exceed the sum of the absolute values of these integers.

21. Use predicates, quantifiers, logical connectives, and mathematical operators to express the statement that every positive integer is the sum of the squares of four integers.

22. Use predicates, quantifiers, logical connectives, and mathematical operators to express the statement that there is a positive integer that is not the sum of three squares.

23. Express each of these mathematical statements using predicates, quantifiers, logical connectives, and mathematical operators.

   a) The product of two negative real numbers is positive.
   b) The difference of a real number and itself is zero.
   c) Every positive real number has exactly two square roots.
   d) A negative real number does not have a square root that is a real number.

24. Translate each of these nested quantifications into an English statement that expresses a mathematical fact. The universe of discourse in each case consists of all real numbers.

   a) $\exists x \forall y (x + y = y)$
   b) $\forall x \forall y (((x \geq 0) \wedge (y < 0)) \rightarrow (x - y > 0))$
   c) $\exists x \exists y (((x \leq 0) \wedge (y \leq 0)) \wedge (x - y > 0))$
   d) $\forall x \forall y ((x \neq 0) \wedge (y \neq 0) \leftrightarrow (xy \neq 0))$

25. Translate each of these nested quantifications into an English statement that expresses a mathematical fact. The universe of discourse in each case consists of all real numbers.

   a) $\exists x \forall y (xy = y)$
   b) $\forall x \forall y (((x < 0) \wedge (y < 0)) \rightarrow (xy > 0))$

   c) $\exists x \exists y ((x^2 > y) \wedge (x < y))$
   d) $\forall x \forall y \exists z (x + y = z)$

26. Let $Q(x, y)$ be the statement "$x + y = x - y$." If the universe of discourse for both variables consists of all integers, what are the truth values?

   a) $Q(1, 1)$ **b)** $Q(2, 0)$ **c)** $\forall y Q(1, y)$
   d) $\exists x Q(x, 2)$ **e)** $\exists x \exists y Q(x, y)$ **f)** $\forall x \exists y Q(x, y)$
   g) $\exists y \forall x Q(x, y)$
   h) $\forall y \exists x Q(x, y)$
   i) $\forall x \forall y Q(x, y)$

27. Determine the truth value of each of these statements if the universe of discourse for all variables consists of all integers.

   a) $\forall n \exists m (n^2 < m)$ **b)** $\exists n \forall m (n < m^2)$
   c) $\forall n \exists m (n + m = 0)$ **d)** $\exists n \forall m (nm = m)$
   e) $\exists n \exists m (n^2 + m^2 = 5)$ **f)** $\exists n \exists m (n^2 + m^2 = 6)$
   g) $\exists n \exists m (n + m = 4 \wedge n - m = 1)$
   h) $\exists n \exists m (n + m = 4 \wedge n - m = 2)$
   i) $\forall n \forall m \exists p (p = (m + n)/2)$

28. Determine the truth value of each of these statements if the universe of discourse of each variable consists of all real numbers.

   a) $\forall x \exists y (x^2 = y)$ **b)** $\forall x \exists y (x = y^2)$
   c) $\exists x \forall y (xy = 0)$ **d)** $\exists x \exists y (x + y \neq y + x)$
   e) $\forall x (x \neq 0 \rightarrow \exists y (xy = 1))$
   f) $\exists x \forall y (y \neq 0 \rightarrow xy = 1)$
   g) $\forall x \exists y (x + y = 1)$
   h) $\exists x \exists y (x + 2y = 2 \wedge 2x + 4y = 5)$
   i) $\forall x \exists y (x + y = 2 \wedge 2x - y = 1)$
   j) $\forall x \forall y \exists z (z = (x + y)/2)$

29. Suppose the universe of discourse of the propositional function $P(x, y)$ consists of pairs x and y, where x is 1, 2, or 3 and y is 1, 2, or 3. Write out these propositions using disjunctions and conjunctions.

   a) $\forall x \forall y P(x, y)$ **b)** $\exists x \exists y P(x, y)$
   c) $\exists x \forall y P(x, y)$ **d)** $\forall y \exists x P(x, y)$

30. Rewrite each of these statements so that negations appear only within predicates (that is, so that no negation is outside a quantifier or an expression involving logical connectives).

   a) $\neg \exists y \exists x P(x, y)$ **b)** $\neg \forall x \exists y P(x, y)$
   c) $\neg \exists y (Q(y) \wedge \forall x \neg R(x, y))$
   d) $\neg \exists y (\exists x R(x, y) \vee \forall x S(x, y))$
   e) $\neg \exists y (\forall x \exists z T(x, y, z) \vee \exists x \forall z U(x, y, z))$

31. Express the negations of each of these statements so that all negation symbols immediately precede predicates.

   a) $\forall x \exists y \forall z T(x, y, z)$
   b) $\forall x \exists y P(x, y) \vee \forall x \exists y Q(x, y)$
   c) $\forall x \exists y (P(x, y) \wedge \exists z R(x, y, z))$
   d) $\forall x \exists y (P(x, y) \rightarrow Q(x, y))$

32. Express the negations of each of these statements so that all negation symbols immediately precede predicates.

a) $\exists z \forall y \forall x T(x, y, z)$
b) $\exists x \exists y P(x, y) \wedge \forall x \forall y Q(x, y)$
c) $\exists x \exists y (Q(x, y) \leftrightarrow Q(y, x))$
d) $\forall y \exists x \exists z (T(x, y, z) \vee Q(x, y))$

33. Rewrite each of these statements so that negations appear only within predicates (that is, so that no negation is outside a quantifier or an expression involving logical connectives).

a) $\neg \forall x \forall y P(x, y)$ **b)** $\neg \forall y \exists x P(x, y)$
c) $\neg \forall y \forall x (P(x, y) \vee Q(x, y))$
d) $\neg(\exists x \exists y \neg P(x, y) \wedge \forall x \forall y Q(x, y))$
e) $\neg \forall x (\exists y \forall z P(x, y, z) \wedge \exists z \forall y P(x, y, z))$

34. Express each of these statements using quantifiers. Then form the negation of the statement so that no negation is to the left of a quantifier. Next, express the negation in simple English. (Do not simply use the words "It is not the case that.")

a) No one has lost more than one thousand dollars playing the lottery.
b) There is a student in this class who has chatted with exactly one other student.
c) No student in this class has sent e-mail to exactly two other students in this class.
d) Some student has solved every exercise in this book.
e) No student has solved at least one exercise in every section of this book.

35. Express each of these statements using quantifiers. Then form the negation of the statement so that no negation is to the left of a quantifier. Next, express the negation in simple English. (Do not simply use the words "It is not the case that.")

a) Every student in this class has taken exactly two mathematics classes at this school.
b) Someone has visited every country in the world except Libya.
c) No one has climbed every mountain in the Himalayas.
d) Every movie actor has either been in a movie with Kevin Bacon or has been in a movie with someone who has been in a movie with Kevin Bacon.

36. Express the negations of these propositions using quantifiers, and in English.

a) Every student in this class likes mathematics.
b) There is a student in this class who has never seen a computer.
c) There is a student in this class who has taken every mathematics course offered at this school.
d) There is a student in this class who has been in at least one room of every building on campus.

37. Find a counterexample, if possible, to these universally quantified statements, where the universe of discourse for all variables consists of all integers.

a) $\forall x \forall y (x^2 = y^2 \rightarrow x = y)$

b) $\forall x \exists y (y^2 = x)$ **c)** $\forall x \forall y (xy \geq x)$

38. Find a counterexample, if possible, to these universally quantified statements, where the universe of discourse for all variables consists of all integers.

a) $\forall x \exists y (x = 1/y)$ **b)** $\forall x \exists y (y^2 - x < 100)$
c) $\forall x \forall y (x^2 \neq y^3)$

39. Use quantifiers to express the associative law for multiplication of real numbers.

40. Use quantifiers to express the distributive laws of multiplication over addition for real numbers.

41. Determine the truth value of the statement $\forall x \exists y (xy = 1)$ if the universe of discourse for the variables consists of

a) the nonzero real numbers.
b) the nonzero integers.
c) the positive real numbers.

42. Determine the truth value of the statement $\exists x \forall y (x \leq y^2)$ if the universe of discourse for the variables consists of

a) the positive real numbers.
b) the integers.
c) the nonzero real numbers.

43. Show that the two statements $\neg \exists x \forall y P(x, y)$ and $\forall x \exists y \neg P(x, y)$ have the same truth value.

*44. Show that $\forall x P(x) \vee \forall x Q(x)$ and $\forall x \forall y (P(x) \vee Q(y))$ are logically equivalent. (The new variable y is used to combine the quantifications correctly.)

*45. **a)** Show that $\forall x P(x) \wedge \exists x Q(x)$ is equivalent to $\forall x \exists y (P(x) \wedge Q(y))$.
b) Show that $\forall x P(x) \vee \exists x Q(x)$ is equivalent to $\forall x \exists y (P(x) \vee Q(y))$.

A statement is in **prenex normal form (PNF)** if and only if it is of the form

$$Q_1 x_1 Q_2 x_2 \cdots Q_k x_k P(x_1, x_2, \ldots, x_k),$$

where each Q_i, $i = 1, 2, \ldots, k$, is either the existential quantifier or the universal quantifier, and $P(x_1, \ldots, x_k)$ is a predicate involving no quantifiers. For example, $\exists x \forall y (P(x, y) \wedge Q(y))$ is in prenex normal form, whereas $\exists x P(x) \vee \forall x Q(x)$ is not (since the quantifiers do not all occur first).

Every statement formed from propositional variables, predicates, **T**, and **F** using logical connectives and quantifiers is equivalent to a statement in prenex normal form. Exercise 47 asks for a proof of this fact.

*46. Put these statements in prenex normal form. (*Hint:* Use logical equivalence from Tables 5 and 6 in Section 1.2, Table 2 in Section 1.3, Exercises 42–45 in Section 1.3, and Exercises 44 and 45 in this section.)

a) $\exists x P(x) \vee \exists x Q(x) \vee A$, where A is a proposition not involving any quantifiers.
b) $\neg(\forall x P(x) \vee \forall x Q(x))$
c) $\exists x P(x) \rightarrow \exists x Q(x)$

**47. Show how to transform an arbitrary statement to a statement in prenex normal form that is equivalent to the given statement.

48. A real number x is called an **upper bound** of a set S of real numbers if x is greater than or equal to every member of S. The real number x is called the **least upper bound** of a set S of real numbers if x is an upper bound of S and x is less than or equal to every upper bound of S; if the least upper bound of a set S exists, it is unique.

   a) Using quantifiers, express the fact that x is an upper bound of S.

   b) Using quantifiers, express the fact that x is the least upper bound of S.

*49. Express the quantification $\exists! x\, P(x)$ using universal quantifications, existential quantifications, and logical operators.

The statement $\lim_{n \to \infty} a_n = L$ means that for every positive real number ϵ there is a positive integer N such that $|a_n - L| < \epsilon$ whenever $n > N$.

50. (Calculus required) Use quantifiers to express the statement that $\lim_{n \to \infty} a_n = L$.

51. (Calculus required) Use quantifiers to express the statement that $\lim_{n \to \infty} a_n$ does not exist.

52. (Calculus required) Use quantifiers to express this definition: A sequence $\{a_n\}$ is a Cauchy sequence if for every real number $\epsilon > 0$ there exists a positive integer N such that $|a_m - a_n| < \epsilon$ for every pair of positive integers m and n with $m > N$ and $n > N$.

53. (Calculus required) Use quantifiers and logical connectives to express this definition: A number L is the **limit superior** of a sequence $\{a_n\}$ if for every real number $\epsilon > 0$, $a_n > L - \epsilon$ for infinitely many n and $a_n > L + \epsilon$ for only finitely many n.

1.5   Methods of Proof

INTRODUCTION

Two important questions that arise in the study of mathematics are: (1) When is a mathematical argument correct? (2) What methods can be used to construct mathematical arguments? This section helps answer these questions by describing various forms of correct and incorrect mathematical arguments.

A **theorem** is a statement that can be shown to be true. (Theorems are sometimes called *propositions, facts,* or *results.*) We demonstrate that a theorem is true with a sequence of statements that form an argument, called a **proof.** To construct proofs, methods are needed to derive new statements from old ones. The statements used in a proof can include **axioms** or **postulates,** which are the underlying assumptions about mathematical structures, the hypotheses of the theorem to be proved, and previously proved theorems. The **rules of inference,** which are the means used to draw conclusions from other assertions, tie together the steps of a proof.

In this section rules of inference will be discussed. This will help clarify what makes up a correct proof. Some common forms of incorrect reasoning, called **fallacies,** will also be described. Then various methods commonly used to prove theorems will be introduced.

The terms *lemma* and *corollary* are used for certain types of theorems. A **lemma** (plural **lemmas** or **lemmata**) is a simple theorem used in the proof of other theorems. Complicated proofs are usually easier to understand when they are proved using a series of lemmas, where each lemma is proved individually. A **corollary** is a proposition that can be established directly from a theorem that has been proved. A **conjecture** is a statement whose truth value is unknown. When a proof of a conjecture is found, the conjecture becomes a theorem. Many times conjectures are shown to be false, so they are not theorems.

The methods of proof discussed in this chapter are important not only because they are used to prove mathematical theorems, but also for their many applications to computer science. These applications include verifying that computer programs are correct, establishing that operating systems are secure, making inferences in the area of artificial intelligence, showing that system specifications are consistent, and so on. Consequently, understanding the techniques used in proofs is essential both in mathematics and in computer science.

RULES OF INFERENCE

We will now introduce rules of inference for propositional logic. These rules provide the justification of the steps used to show that a conclusion follows logically from a set of hypotheses. The tautology $(p \land (p \rightarrow q)) \rightarrow q$ is the basis of the rule of inference called **modus ponens,** or the **law of detachment.** This tautology is written in the following way:

$$\frac{\begin{array}{c} p \\ p \rightarrow q \end{array}}{\therefore q}$$

Using this notation, the hypotheses are written in a column and the conclusion below a bar. (The symbol $\therefore$ denotes "therefore.") Modus ponens states that if both an implication and its hypothesis are known to be true, then the conclusion of this implication is true.

Extra Examples

EXAMPLE 1   Suppose that the implication "if it snows today, then we will go skiing" and its hypothesis, "it is snowing today," are true. Then, by modus ponens, it follows that the conclusion of the implication, "we will go skiing," is true. ◀

EXAMPLE 2   Assume that the implication "if n is greater than 3, then n^2 is greater than 9" is true. Consequently, if n is greater than 3, then, by modus ponens, it follows that n^2 is greater than 9. ◀

Table 1 lists some important rules of inference. The verifications of these rules of inference can be found as exercises in Section 1.2. Here are some examples of arguments using these rules of inference.

EXAMPLE 3   State which rule of inference is the basis of the following argument: "It is below freezing now. Therefore, it is either below freezing or raining now."

Solution: Let p be the proposition "It is below freezing now" and q the proposition "It is raining now." Then this argument is of the form

$$\frac{p}{\therefore p \lor q}$$

This is an argument that uses the addition rule. ◀

EXAMPLE 4   State which rule of inference is the basis of the following argument: "It is below freezing and raining now. Therefore, it is below freezing now."

Solution: Let p be the proposition "It is below freezing now," and let q be the proposition "It is raining now." This argument is of the form

$$\frac{p \land q}{\therefore p}$$

This argument uses the simplification rule. ◀

EXAMPLE 5   State which rule of inference is used in the argument:

If it rains today, then we will not have a barbecue today. If we do not have a barbecue today, then we will have a barbecue tomorrow. Therefore, if it rains today, then we will have a barbecue tomorrow.

TABLE 1 Rules of Inference.		
Rule of Inference	*Tautology*	*Name*
$\dfrac{p}{\therefore\ p \vee q}$	$p \to (p \vee q)$	Addition
$\dfrac{p \wedge q}{\therefore\ p}$	$(p \wedge q) \to p$	Simplification
$\begin{array}{c} p \\ \underline{q} \\ \therefore\ p \wedge q \end{array}$	$((p) \wedge (q)) \to (p \wedge q)$	Conjunction
$\begin{array}{c} p \\ \underline{p \to q} \\ \therefore\ q \end{array}$	$[p \wedge (p \to q)] \to q$	Modus ponens
$\begin{array}{c} \neg q \\ \underline{p \to q} \\ \therefore\ \neg p \end{array}$	$[\neg q \wedge (p \to q)] \to \neg p$	Modus tollens
$\begin{array}{c} p \to q \\ \underline{q \to r} \\ \therefore\ p \to r \end{array}$	$[(p \to q) \wedge (q \to r)] \to (p \to r)$	Hypothetical syllogism
$\begin{array}{c} p \vee q \\ \underline{\neg p} \\ \therefore\ q \end{array}$	$[(p \vee q) \wedge \neg p] \to q$	Disjunctive syllogism
$\begin{array}{c} p \vee q \\ \underline{\neg p \vee r} \\ \therefore\ q \vee r \end{array}$	$[(p \vee q) \wedge (\neg p \vee r)] \to (q \vee r)$	Resolution

Solution: Let p be the proposition "It is raining today," let q be the proposition "We will not have a barbecue today," and let r be the proposition "We will have a barbecue tomorrow." Then this argument is of the form

$$\begin{array}{c} p \to q \\ \underline{q \to r} \\ \therefore\ p \to r \end{array}$$

Hence, this argument is a hypothetical syllogism. ◀

VALID ARGUMENTS

An argument form is called **valid** if whenever all the hypotheses are true, the conclusion is also true. Consequently, showing that q logically follows from the hypotheses $p_1, p_2, \ldots, p_n$ is the same as showing that the implication

$$(p_1 \wedge p_2 \wedge \cdots \wedge p_n) \to q$$

is true. When all propositions used in a valid argument are true, it leads to a correct conclusion. However, a valid argument can lead to an incorrect conclusion if one or more false propositions are used within the argument. For example,

"If $\sqrt{2} > \frac{1}{2}$, then $\left(\sqrt{2}\right)^2 > \left(\frac{3}{2}\right)^2$. We know that $\sqrt{2} > \frac{3}{2}$. Consequently, $\left(\sqrt{2}\right)^2 = 2 > \left(\frac{3}{2}\right)^2 = \frac{9}{4}$."

is a valid argument form based on modus ponens. However, the conclusion of this argument is false, because $2 < \frac{9}{4}$. The false proposition "$\sqrt{2} > \frac{3}{2}$" has been used in the argument, which means that the conclusion of the argument may be false.

When there are many premises, several rules of inference are often needed to show that an argument is valid. This is illustrated by the following examples, where the steps of arguments are displayed step by step, with the reason for each step explicitly stated. These examples also show how arguments in English can be analyzed using rules of inference.

Extra Examples

EXAMPLE 6 Show that the hypotheses "It is not sunny this afternoon and it is colder than yesterday," "We will go swimming only if it is sunny," "If we do not go swimming, then we will take a canoe trip," and "If we take a canoe trip, then we will be home by sunset" lead to the conclusion "We will be home by sunset."

Solution: Let p be the proposition "It is sunny this afternoon," q the proposition "It is colder than yesterday," r the proposition "We will go swimming," s the proposition "We will take a canoe trip," and t the proposition "We will be home by sunset." Then the hypotheses become $\neg p \wedge q$, $r \rightarrow p$, $\neg r \rightarrow s$, and $s \rightarrow t$. The conclusion is simply t.

We construct an argument to show that our hypotheses lead to the desired conclusion as follows.

Step	Reason
1. $\neg p \wedge q$	Hypothesis
2. $\neg p$	Simplification using Step 1
3. $r \rightarrow p$	Hypothesis
4. $\neg r$	Modus tollens using Steps 2 and 3
5. $\neg r \rightarrow s$	Hypothesis
6. s	Modus ponens using Steps 4 and 5
7. $s \rightarrow t$	Hypothesis
8. t	Modus ponens using Steps 6 and 7

◀

EXAMPLE 7 Show that the hypotheses "If you send me an e-mail message, then I will finish writing the program," "If you do not send me an e-mail message, then I will go to sleep early," and "If I go to sleep early, then I will wake up feeling refreshed" lead to the conclusion "If I do not finish writing the program, then I will wake up feeling refreshed."

Solution: Let p be the proposition "You send me an e-mail message," q the proposition "I will finish writing the program," r the proposition "I will go to sleep early," and s the proposition "I will wake up feeling refreshed." Then the hypotheses are $p \rightarrow q$, $\neg p \rightarrow r$, and $r \rightarrow s$. The desired conclusion is $\neg q \rightarrow s$.

This argument form shows that our hypotheses lead to the desired conclusion.

Step	Reason
1. $p \rightarrow q$	Hypothesis
2. $\neg q \rightarrow \neg p$	Contrapositive of Step 1
3. $\neg p \rightarrow r$	Hypothesis
4. $\neg q \rightarrow r$	Hypothetical syllogism using Steps 2 and 3
5. $r \rightarrow s$	Hypothesis
6. $\neg q \rightarrow s$	Hypothetical syllogism using Steps 4 and 5

◀

RESOLUTION

Computer programs have been developed to automate the task of reasoning and proving theorems. Many of these programs make use of a rule of inference known as **resolution.** This rule of inference is based on the tautology

$$((p \lor q) \land (\neg p \lor r)) \rightarrow (q \lor r).$$

(The verification that this is a tautology was addressed in Exercise 28 in Section 1.2.) The final disjunction in the resolution rule, $q \lor r$, is called the **resolvent.** When we let $q = r$ in this tautology, we obtain $(p \lor q) \land (\neg p \lor q) \rightarrow q$. Furthermore, when we let $r = \mathbf{F}$, we obtain $(p \lor q) \land (\neg p) \rightarrow q$ (because $q \lor \mathbf{F} \equiv q$), which is the tautology on which the rule of disjunctive syllogism is based.

EXAMPLE 8

Use resolution to show that the hypotheses "Jasmine is skiing or it is not snowing" and "It is snowing or Bart is playing hockey" imply that "Jasmine is skiing or Bart is playing hockey."

Solution: Let p be the proposition "It is snowing," q the proposition "Jasmine is skiing," and r the proposition "Bart is playing hockey." We can represent the hypotheses as $\neg p \lor q$ and $p \lor r$, respectively. Using resolution, the proposition $q \lor r$, "Jasmine is skiing or Bart is playing hockey," follows. ◀

Resolution plays an important role in programming languages based on the rules of logic, such as Prolog (where resolution rules for quantified statements are applied). Furthermore, it can be used to build automatic theorem proving systems. To construct proofs in propositional logic using resolution as the only rule of inference, the hypotheses and the conclusion must be expressed as **clauses,** where a clause is a disjunction of variables or negations of these variables. We can replace a statement in propositional logic that is not a clause by one or more equivalent statements that are clauses. For example, suppose we have a statement of the form $p \lor (q \land r)$. Because $p \lor (q \land r) \equiv (p \lor q) \land (p \lor r)$, we can replace the single statement $p \lor (q \land r)$ by two statements $p \lor q$ and $p \lor r$, each of which is a clause. We can replace a statement of the form $\neg(p \lor q)$ by the two statements $\neg p$ and $\neg q$ because De Morgan's law tells us that $\neg(p \lor q) \equiv \neg p \land \neg q$. We can also replace an implication $p \rightarrow q$ with the equivalent disjunction $\neg p \lor q$.

EXAMPLE 9

Show that the hypotheses $(p \land q) \lor r$ and $r \rightarrow s$ imply the conclusion $p \lor s$.

Solution: We can rewrite the hypothesis $(p \land q) \lor r$ as two clauses, $p \lor r$ and $q \lor r$. We can also replace $r \rightarrow s$ by the equivalent clause $\neg r \lor s$. Using the two clauses $p \lor r$ and $\neg r \lor s$, we can use resolution to conclude $p \lor s$. ◀

FALLACIES

Several common fallacies arise in incorrect arguments. These fallacies resemble rules of inference but are based on contingencies rather than tautologies. These are discussed here to show the distinction between correct and incorrect reasoning.

The proposition $[(p \rightarrow q) \land q] \rightarrow p$ is not a tautology, since it is false when p is false and q is true. However, there are many incorrect arguments that treat this as a tautology. This type of incorrect reasoning is called the **fallacy of affirming the conclusion.**

EXAMPLE 10   Is the following argument valid?

If you do every problem in this book, then you will learn discrete mathematics. You learned discrete mathematics.

Therefore, you did every problem in this book.

Solution: Let p be the proposition "You did every problem in this book." Let q be the proposition "You learned discrete mathematics." Then this argument is of the form: if $p \rightarrow q$ and q, then p. This is an example of an incorrect argument using the fallacy of affirming the conclusion. Indeed, it is possible for you to learn discrete mathematics in some way other than by doing every problem in this book. (You may learn discrete mathematics by reading, listening to lectures, doing some but not all the problems in this book, and so on.) ◀

The proposition $[(p \rightarrow q) \wedge \neg p] \rightarrow \neg q$ is not a tautology, since it is false when p is false and q is true. Many incorrect arguments use this incorrectly as a rule of inference. This type of incorrect reasoning is called the **fallacy of denying the hypothesis.**

EXAMPLE 11   Let p and q be as in Example 10. If the implication $p \rightarrow q$ is true, and $\neg p$ is true, is it correct to conclude that $\neg q$ is true? In other words, is it correct to assume that you did not learn discrete mathematics if you did not do every problem in the book, assuming that if you do every problem in this book, then you will learn discrete mathematics?

Solution: It is possible that you learned discrete mathematics even if you did not do every problem in this book. This incorrect argument is of the form $p \rightarrow q$ and $\neg p$ imply $\neg q$, which is an example of the fallacy of denying the hypothesis. ◀

RULES OF INFERENCE FOR QUANTIFIED STATEMENTS

We discussed rules of inference for propositions. We will now describe some important rules of inference for statements involving quantifiers. These rules of inference are used extensively in mathematical arguments, often without being explicitly mentioned.

Universal instantiation is the rule of inference used to conclude that $P(c)$ is true, where c is a particular member of the universe of discourse, given the premise $\forall x P(x)$. Universal instantiation is used when we conclude from the statement "All women are wise" that "Lisa is wise," where Lisa is a member of the universe of discourse of all women.

Universal generalization is the rule of inference that states that $\forall x P(x)$ is true, given the premise that $P(c)$ is true for all elements c in the universe of discourse. Universal generalization is used when we show that $\forall x P(x)$ is true by taking an arbitrary element c from the universe of discourse and showing that $P(c)$ is true. The element c that we select must be an arbitrary, and not a specific, element of the universe of discourse. Universal generalization is used implicitly in many proofs in mathematics and is seldom mentioned explicitly.

Existential instantiation is the rule that allows us to conclude that there is an element c in the universe of discourse for which $P(c)$ is true if we know that $\exists x P(x)$ is true. We cannot select an arbitrary value of c here, but rather it must be a c for which $P(c)$ is true. Usually we have no knowledge of what c is, only that it exists. Since it exists, we may give it a name (c) and continue our argument.

TABLE 2 Rules of Inference for Quantified Statements.	
Rule of Inference	**Name**
$\forall x\, P(x)$ $\therefore\ P(c)$	Universal instantiation
$P(c)$ for an arbitrary c $\therefore\ \forall x\, P(x)$	Universal generalization
$\exists x\, P(x)$ $\therefore\ P(c)$ for some element c	Existential instantiation
$P(c)$ for some element c $\therefore\ \exists x\, P(x)$	Existential generalization

Existential generalization is the rule of inference that is used to conclude that $\exists x\, P(x)$ is true when a particular element c with $P(c)$ true is known. That is, if we know one element c in the universe of discourse for which $P(c)$ is true, then we know that $\exists x\, P(x)$ is true.

We summarize these rules of inference in Table 2. We will illustrate how one of these rules of inference for quantified statements is used in Example 12.

EXAMPLE 12   Show that the premises "Everyone in this discrete mathematics class has taken a course in computer science" and "Marla is a student in this class" imply the conclusion "Marla has taken a course in computer science."

Solution: Let $D(x)$ denote "x is in this discrete mathematics class," and let $C(x)$ denote "x has taken a course in computer science." Then the premises are $\forall x(D(x) \rightarrow C(x))$ and $D(\text{Marla})$. The conclusion is $C(\text{Marla})$.

The following steps can be used to establish the conclusion from the premises.

Step	Reason
1. $\forall x(D(x) \rightarrow C(x))$	Premise
2. $D(\text{Marla}) \rightarrow C(\text{Marla})$	Universal instantiation from (1)
3. $D(\text{Marla})$	Premise
4. $C(\text{Marla})$	Modus ponens from (2) and (3) ◀

EXAMPLE 13   Show that the premises "A student in this class has not read the book," and "Everyone in this class passed the first exam" imply the conclusion "Someone who passed the first exam has not read the book."

Solution: Let $C(x)$ be "x is in this class," $B(x)$ be "x has read the book," and $P(x)$ be "x passed the first exam." The premises are $\exists x(C(x) \wedge \neg B(x))$ and $\forall x(C(x) \rightarrow P(x))$. The conclusion is $\exists x(P(x) \wedge \neg B(x))$. These steps can be used to establish the conclusion from the premises.

Step	Reason
1. $\exists x(C(x) \wedge \neg B(x))$	Premise
2. $C(a) \wedge \neg B(a)$	Existential instantiation from (1)
3. $C(a)$	Simplification from (2)
4. $\forall x(C(x) \rightarrow P(x))$	Premise

Step	Reason
5. $C(a) \rightarrow P(a)$	Universal instantiation from (4)
6. $P(a)$	Modus ponens from (3) and (5)
7. $\neg B(a)$	Simplification from (2)
8. $P(a) \wedge \neg B(a)$	Conjunction from (6) and (7)
9. $\exists x(P(x) \wedge \neg B(x))$	Existential generalization from (8) ◀

Remark: Mathematical arguments often include steps where both a rule of inference for propositions and a rule of inference for quantifiers are used. For example, universal instantiation and modus ponens are often used together. When these rules of inference are combined, the hypothesis $\forall x(P(x) \rightarrow Q(x))$ and $P(c)$, where c is a member of the universe of discourse, show that the conclusion $Q(c)$ is true.

Remark: Many theorems in mathematics state that a property holds for all elements in a particular set, such as the set of integers or the set of real numbers. Although the precise statement of such theorems needs to include a universal quantifier, the standard convention in mathematics is to omit it. For example, the statement "If $x > y$, where x and y are positive real numbers, then $x^2 > y^2$" really means "For all positive real numbers x and y, if $x > y$, then $x^2 > y^2$." Furthermore, when theorems of this type are proved, the law of universal generalization is often used without explicit mention. The first step of the proof usually involves selecting a general element of the universe of discourse. Subsequent steps show that this element has the property in question. Universal generalization implies that the theorem holds for all members of the universe of discourse.

In our subsequent discussions, we will follow the usual conventions and not explicitly mention the use of universal quantification and universal generalization. However, you should always understand when this rule of inference is being implicitly applied.

METHODS OF PROVING THEOREMS

Proving theorems can be difficult. We need all the ammunition that is available to help us prove different results. We now introduce a battery of different proof methods. These methods should become part of your repertoire for proving theorems. Because many theorems are implications, the techniques for proving implications are important. Recall that $p \rightarrow q$ is true unless p is true but q is false. Note that when the statement $p \rightarrow q$ is proved, it need only be shown that q is true if p is true; it is *not* usually the case that q is proved to be true. The following discussion will give the most common techniques for proving implications.

DIRECT PROOFS The implication $p \rightarrow q$ can be proved by showing that if p is true, then q must also be true. This shows that the combination p true and q false never occurs. A proof of this kind is called a **direct proof.** To carry out such a proof, assume that p is true and use rules of inference and theorems already proved to show that q must also be true.

Before we give an example of a direct proof, we need a definition.

DEFINITION 1

The integer n is *even* if there exists an integer k such that $n = 2k$ and it is *odd* if there exists an integer k such that $n = 2k + 1$. (Note that an integer is either even or odd.)

EXAMPLE 14   Give a direct proof of the theorem "If n is an odd integer, then n^2 is an odd integer."

Solution: Assume that the hypothesis of this implication is true, namely, suppose that n is odd. Then $n = 2k + 1$, where k is an integer. It follows that $n^2 = (2k + 1)^2 = 4k^2 + 4k + 1 = 2(2k^2 + 2k) + 1$. Therefore, n^2 is an odd integer (it is one more than twice an integer).   ◀

INDIRECT PROOFS Since the implication $p \rightarrow q$ is equivalent to its contrapositive, $\neg q \rightarrow \neg p$, the implication $p \rightarrow q$ can be proved by showing that its contrapositive, $\neg q \rightarrow \neg p$, is true. This related implication is usually proved directly, but any proof technique can be used. An argument of this type is called an **indirect proof.**

EXAMPLE 15   Give an indirect proof of the theorem "If $3n + 2$ is odd, then n is odd."

Solution: Assume that the conclusion of this implication is false; namely, assume that n is even. Then $n = 2k$ for some integer k. It follows that $3n + 2 = 3(2k) + 2 = 6k + 2 = 2(3k + 1)$, so $3n + 2$ is even (since it is a multiple of 2) and therefore not odd. Because the negation of the conclusion of the implication implies that the hypothesis is false, the original implication is true.   ◀

VACUOUS AND TRIVIAL PROOFS Suppose that the hypothesis p of an implication $p \rightarrow q$ is false. Then the implication $p \rightarrow q$ is true, because the statement has the form $\mathbf{F} \rightarrow \mathbf{T}$ or $\mathbf{F} \rightarrow \mathbf{F}$, and hence is true. Consequently, if it can be shown that p is false, then a proof, called a **vacuous proof,** of the implication $p \rightarrow q$ can be given. Vacuous proofs are often used to establish special cases of theorems that state that an implication is true for all positive integers [i.e., a theorem of the kind $\forall n\, P(n)$ where $P(n)$ is a propositional function]. Proof techniques for theorems of this kind will be discussed in Section 3.3.

EXAMPLE 16   Show that the proposition $P(0)$ is true where $P(n)$ is the propositional function "If $n > 1$, then $n^2 > n$."

Solution: Note that the proposition $P(0)$ is the implication "If $0 > 1$, then $0^2 > 0$." Since the hypothesis $0 > 1$ is false, the implication $P(0)$ is automatically true.   ◀

Remark: The fact that the conclusion of this implication, $0^2 > 0$, is false is irrelevant to the truth value of the implication, because an implication with a false hypothesis is guaranteed to be true.

Suppose that the conclusion q of an implication $p \rightarrow q$ is true. Then $p \rightarrow q$ is true, since the statement has the form $\mathbf{T} \rightarrow \mathbf{T}$ or $\mathbf{F} \rightarrow \mathbf{T}$, which are true. Hence, if it can be shown that q is true, then a proof, called a **trivial proof,** of $p \rightarrow q$ can be given. Trivial proofs are often important when special cases of theorems are proved (see the discussion of proof by cases) and in mathematical induction, which is a proof technique discussed in Section 3.3.

EXAMPLE 17   Let $P(n)$ be "If a and b are positive integers with $a \geq b$, then $a^n \geq b^n$." Show that the proposition $P(0)$ is true.

Solution: The proposition $P(0)$ is "If $a \geq b$, then $a^0 \geq b^0$." Since $a^0 = b^0 = 1$, the conclusion of $P(0)$ is true. Hence, $P(0)$ is true. This is an example of a trivial proof. Note that the hypothesis, which is the statement "$a \geq b$," was not needed in this proof. ◄

A LITTLE PROOF STRATEGY We have described both direct and indirect proofs and we have provided an example of how they are used. However, when confronted with an implication to prove, which method should you use? First, quickly evaluate whether a direct proof looks promising. Begin by expanding the definitions in the hypotheses. Then begin to reason using them, together with axioms and available theorems. If a direct proof does not seem to go anywhere, try the same thing with an indirect proof. Recall that in an indirect proof you assume that the conclusion of the implication is false and use a direct proof to show this implies that the hypothesis must be false. Sometimes when there is no obvious way to approach a direct proof, an indirect proof works nicely. We illustrate this strategy in Examples 18 and 19.

Extra
Examples

Before we present our next example, we need a definition.

DEFINITION 2

> The real number r is *rational* if there exist integers p and q with $q \neq 0$ such that $r = p/q$. A real number that is not rational is called *irrational*.

EXAMPLE 18 Prove that the sum of two rational numbers is rational.

Solution: We first attempt a direct proof. To begin, suppose that r and s are rational numbers. From the definition of a rational number, it follows that there are integers p and q, with $q \neq 0$, such that $r = p/q$, and integers t and u, with $u \neq 0$, such that $s = t/u$. Can we use this information to show that $r + s$ is rational? The obvious next step is to add $r = p/q$ and $s = t/u$, to obtain

$$r + s = \frac{p}{q} + \frac{t}{u} = \frac{pu + qt}{qu}.$$

Because $q \neq 0$ and $u \neq 0$, it follows that $qu \neq 0$. Consequently, we have expressed $r + s$ as the ratio of two integers, $pu + qt$ and qu, where $qu \neq 0$. This means that $r + s$ is rational. Our attempt to find a direct proof succeeded. ◄

EXAMPLE 19 Prove that if n is an integer and n^2 is odd, then n is odd.

Solution: We first attempt a direct proof. Suppose that n is an integer and n^2 is odd. Then, there exists an integer k such that $n^2 = 2k + 1$. Can we use this information to show that n is odd? There seems to be no obvious approach to show that n is odd because solving for n produces the equation $n = \pm\sqrt{2k + 1}$, which is not terribly useful.

Because this attempt to use a direct proof did not bear fruit, we next attempt an indirect proof. We take as our hypothesis the statement that n is not odd. Because every integer is odd or even, this means that n is even. This implies that there exists an integer k such that $n = 2k$. To prove the theorem, we need to show that this hypothesis implies the conclusion that n^2 is not odd, that is, that n^2 is even. Can we use the equation $n = 2k$ to achieve this? By squaring both sides of this equation, we obtain $n^2 = 4k^2 = 2(2k^2)$, which implies that n^2 is also even since $n^2 = 2t$, where $t = 2k^2$. Our attempt to find an indirect proof succeeded. ◄

PROOFS BY CONTRADICTION There are other approaches we can use when neither a direct proof nor an indirect proof succeeds. We now introduce several additional proof techniques.

Suppose that a contradiction q can be found so that $\neg p \to q$ is true, that is, $\neg p \to \mathbf{F}$ is true. Then the proposition $\neg p$ must be false. Consequently, p must be true. This technique can be used when a contradiction, such as $r \wedge \neg r$, can be found so that it is possible to show that the implication $\neg p \to (r \wedge \neg r)$ is true. An argument of this type is called a **proof by contradiction.**

We provide three examples of proof by contradiction. The first is an example of an application of the pigeonhole principle, a combinatorial technique which we will cover in depth in Section 4.2.

EXAMPLE 20 Show that at least four of any 22 days must fall on the same day of the week.

Solution: Let p be the proposition "At least four of the 22 chosen days are the same day of the week." Suppose that $\neg p$ is true. Then at most three of the 22 days are the same day of the week. Because there are seven days of the week, this implies that at most 21 days could have been chosen since three is the most days chosen that could be a particular day of the week. This is a contradiction. ◀

EXAMPLE 21 Prove that $\sqrt{2}$ is irrational by giving a proof by contradiction.

Solution: Let p be the proposition "$\sqrt{2}$ is irrational." Suppose that $\neg p$ is true. Then $\sqrt{2}$ is rational. We will show that this leads to a contradiction. Under the assumption that $\sqrt{2}$ is rational, there exist integers a and b with $\sqrt{2} = a/b$, where a and b have no common factors (so that the fraction a/b is in lowest terms). Since $\sqrt{2} = a/b$, when both sides of this equation are squared, it follows that

$$2 = a^2/b^2.$$

Hence,

$$2b^2 = a^2.$$

This means that a^2 is even, implying that a is even. Furthermore, since a is even, $a = 2c$ for some integer c. Thus

$$2b^2 = 4c^2,$$

so

$$b^2 = 2c^2.$$

This means that b^2 is even. Hence, b must be even as well.

It has been shown that $\neg p$ implies that $\sqrt{2} = a/b$, where a and b have no common factors, and 2 divides a and b. This is a contradiction since we have shown that $\neg p$ implies both r and $\neg r$ where r is the statement that a and b are integers with no common factors. Hence, $\neg p$ is false, so that p: "$\sqrt{2}$ is irrational" is true. ◀

An indirect proof of an implication can be rewritten as a proof by contradiction. In an indirect proof we show that $p \to q$ is true by using a direct proof to show that $\neg q \to \neg p$ is true. That is, in an indirect proof of $p \to q$ we assume that $\neg q$ is true and show that $\neg p$ must also be true. To rewrite an indirect proof of $p \to q$ as a proof by contradiction, we suppose that both p and $\neg q$ are true. Then we use the steps from the direct proof of $\neg q \to \neg p$ to show that $\neg p$ must also be true. This leads to the contradiction $p \wedge \neg p$, completing the proof by contradiction. Example 22 illustrates how an indirect proof of an implication can be rewritten as a proof by contradiction.

EXAMPLE 22   Give a proof by contradiction of the theorem "If $3n + 2$ is odd, then n is odd."

Solution: We assume that $3n + 2$ is odd and that n is not odd, so that n is even. Following the same steps as in the solution of Example 15 (an indirect proof of this theorem), we can show that if n is even, then $3n + 2$ is even. This contradicts the assumption that $3n + 2$ is odd, completing the proof.   ◀

PROOF BY CASES   To prove an implication of the form

$$(p_1 \vee p_2 \vee \cdots \vee p_n) \to q$$

the tautology

$$[(p_1 \vee p_2 \vee \cdots \vee p_n) \to q] \leftrightarrow [(p_1 \to q) \wedge (p_2 \to q) \wedge \cdots \wedge (p_n \to q)]$$

can be used as a rule of inference. This shows that the original implication with a hypothesis made up of a disjunction of the propositions $p_1, p_2, \ldots, p_n$ can be proved by proving each of the n implications $p_i \to q, i = 1, 2, \ldots, n$, individually. Such an argument is called a **proof by cases.** Sometimes to prove that an implication $p \to q$ is true, it is convenient to use a disjunction $p_1 \vee p_2 \vee \cdots \vee p_n$ instead of p as the hypothesis of the implication, where p and $p_1 \vee p_2 \vee \cdots \vee p_n$ are equivalent. Consider Example 23.

Extra Examples

EXAMPLE 23   Use a proof by cases to show that $|xy| = |x||y|$, where x and y are real numbers. (Recall that $|x|$, the absolute value of x, equals x when $x \geq 0$ and equals $-x$ when $x \leq 0$.)

Solution: Let p be "x and y are real numbers" and let q be "$|xy| = |x||y|$." Note that p is equivalent to $p_1 \vee p_2 \vee p_3 \vee p_4$, where p_1 is "$x \geq 0 \wedge y \geq 0$," p_2 is "$x \geq 0 \wedge y < 0$," p_3 is "$x < 0 \wedge y \geq 0$," and p_4 is "$x < 0 \wedge y < 0$." Hence, to show that $p \to q$, we can show that $p_1 \to q$, $p_2 \to q$, $p_3 \to q$, and $p_4 \to q$. (We have used these four cases because we can remove the absolute value signs by making the appropriate choice of signs within each case.)

We see that $p_1 \to q$ because $xy \geq 0$ when $x \geq 0$ and $y \geq 0$, so that $|xy| = xy = |x||y|$.

To see that $p_2 \to q$, note that if $x \geq 0$ and $y < 0$, then $xy \leq 0$, so that $|xy| = -xy = x(-y) = |x||y|$. (Here, because $y < 0$, we have $|y| = -y$.)

To see that $p_3 \to q$, we follow the same reasoning as the previous case with the roles of x and y reversed.

To see that $p_4 \to q$, note that when $x < 0$ and $y < 0$, it follows that $xy > 0$. Hence $|xy| = xy = (-x)(-y) = |x||y|$. This completes the proof.   ◀

PROOFS OF EQUIVALENCE   To prove a theorem that is a biconditional, that is, one that is a statement of the form $p \leftrightarrow q$ where p and q are propositions, the tautology

$$(p \leftrightarrow q) \leftrightarrow [(p \to q) \wedge (q \to p)]$$

can be used. That is, the proposition "p if and only if q" can be proved if both the implications "if p, then q" and "if q, then p" are proved.

EXAMPLE 24   Prove the theorem "The integer n is odd if and only if n^2 is odd."

Solution: This theorem has the form "p if and only if q," where p is "n is odd" and q is "n^2 is odd." To prove this theorem, we need to show that $p \rightarrow q$ and $q \rightarrow p$ are true.

We have already shown (in Example 14) that $p \rightarrow q$ is true and (in Example 19) that $q \rightarrow p$ is true.

Since we have shown that both $p \rightarrow q$ and $q \rightarrow p$ are true, we have shown that the theorem is true.   ◀

Sometimes a theorem states that several propositions are equivalent. Such a theorem states that propositions p_1, p_2, p_3, ..., p_n are equivalent. This can be written as

$$p_1 \leftrightarrow p_2 \leftrightarrow \cdots \leftrightarrow p_n,$$

which states that all n propositions have the same truth values, and consequently, that for all i and j with $1 \le i \le n$ and $1 \le j \le n$, p_i and p_j are equivalent. One way to prove these mutually equivalent is to use the tautology

$$[p_1 \leftrightarrow p_2 \leftrightarrow \cdots \leftrightarrow p_n] \leftrightarrow [(p_1 \rightarrow p_2) \wedge (p_2 \rightarrow p_3) \wedge \cdots \wedge (p_n \rightarrow p_1)].$$

This shows that if the implications $p_1 \rightarrow p_2, p_2 \rightarrow p_3, \ldots, p_n \rightarrow p_1$ can be shown to be true, then the propositions p_1, p_2, ..., p_n are all equivalent.

This is much more efficient than proving that $p_i \rightarrow p_j$ for all $i \ne j$ with $1 \le i \le n$ and $1 \le j \le n$.

When we prove that a group of statements are equivalent, we can establish any chain of implications we choose as long as it is possible to work through the chain to go from any one of these statements to any other statement. For example, we can show that p_1, p_2, and p_3 are equivalent by showing that $p_1 \rightarrow p_3, p_3 \rightarrow p_2,$ and $p_2 \rightarrow p_1$.

EXAMPLE 25   Show that these statements are equivalent:

p_1: n is an even integer.
p_2: $n - 1$ is an odd integer.
p_3: n^2 is an even integer.

Solution: We will show that these three statements are equivalent by showing that the implications $p_1 \rightarrow p_2, p_2 \rightarrow p_3,$ and $p_3 \rightarrow p_1$ are true.

We use a direct proof to show that $p_1 \rightarrow p_2$. Suppose that n is even. Then $n = 2k$ for some integer k. Consequently, $n - 1 = 2k - 1 = 2(k - 1) + 1$. This means that $n - 1$ is odd since it is of the form $2m + 1$, where m is the integer $k - 1$.

We also use a direct proof to show that $p_2 \rightarrow p_3$. Now suppose $n - 1$ is odd. Then $n - 1 = 2k + 1$ for some integer k. Hence, $n = 2k + 2$ so that $n^2 = (2k + 2)^2 = 4k^2 + 8k + 4 = 2(2k^2 + 4k + 2)$. This means that n^2 is twice the integer $2k^2 + 4k + 2$, and hence is even.

To prove $p_3 \rightarrow p_1$, we use an indirect proof. That is, we prove that if n is not even, then n^2 is not even. This is the same as proving that if n is odd, then n^2 is odd, which we have already done in Example 14. This completes the proof.   ◀

THEOREMS AND QUANTIFIERS

Many theorems are stated as propositions that involve quantifiers. A variety of methods are used to prove theorems that are quantifications. We will describe some of the most important of these here.

EXISTENCE PROOFS Many theorems are assertions that objects of a particular type exist. A theorem of this type is a proposition of the form $\exists x \, P(x)$, where P is a predicate. A proof of a proposition of the form $\exists x \, P(x)$ is called an **existence proof.** There are several ways to prove a theorem of this type. Sometimes an existence proof of $\exists x \, P(x)$ can be given by finding an element a such that $P(a)$ is true. Such an existence proof is called **constructive.** It is also possible to give an existence proof that is **nonconstructive;** that is, we do not find an element a such that $P(a)$ is true, but rather prove that $\exists x \, P(x)$ is true in some other way. One common method of giving a nonconstructive existence proof is to use proof by contradiction and show that the negation of the existential quantification implies a contradiction. The concept of a constructive existence proof is illustrated by Example 26.

Extra
Examples

EXAMPLE 26 **A Constructive Existence Proof.** Show that there is a positive integer that can be written as the sum of cubes of positive integers in two different ways.

Solution: After considerable computation (such as a computer search) we find that

$$1729 = 10^3 + 9^3 = 12^3 + 1^3.$$

Because we have displayed a positive integer that can be written as the sum of cubes in two different ways, we are done. ◀

EXAMPLE 27 **A Nonconstructive Existence Proof.** Show that there exist irrational numbers x and y such that x^y is rational.

Solution: By Example 21 we know that $\sqrt{2}$ is irrational. Consider the number $\sqrt{2}^{\sqrt{2}}$. If it is rational, we have two irrational numbers x and y with x^y rational, namely, $x = \sqrt{2}$ and $y = \sqrt{2}$. On the other hand if $\sqrt{2}^{\sqrt{2}}$ is irrational, then we can let $x = \sqrt{2}^{\sqrt{2}}$ and $y = \sqrt{2}$ so that $x^y = (\sqrt{2}^{\sqrt{2}})^{\sqrt{2}} = \sqrt{2}^{(\sqrt{2} \cdot \sqrt{2})} = \sqrt{2}^2 = 2$.

This proof is an example of a nonconstructive existence proof because we have not found irrational numbers x and y such that x^y is rational. Rather, we have shown that either the pair $x = \sqrt{2}, y = \sqrt{2}$ or the pair $x = \sqrt{2}^{\sqrt{2}}, y = \sqrt{2}$ have the desired property, but we do not know which of these two pairs works! ◀

UNIQUENESS PROOFS Some theorems assert the existence of a unique element with a particular property. In other words, these theorems assert that there is exactly one element with this property. To prove a statement of this type we need to show that an element with this property exists and that no other element has this property. The two parts of a uniqueness proof are:

Existence: We show that an element x with the desired property exists.
Uniqueness: We show that if $y \neq x$, then y does not have the desired property.

HISTORICAL NOTE The English mathematician G. H. Hardy, when visiting the ailing Indian prodigy Ramanujan in the hospital, remarked that 1729, the number of the cab he took, was rather dull. Ramanujan replied "No, it is a very interesting number; it is the smallest number expressible as the sum of cubes in two different ways." (See the Supplementary Exercises in Chapter 3 for biographies of Hardy and Ramanujan.)

Remark: Showing that there is a unique element x such that $P(x)$ is the same as proving the statement $\exists x (P(x) \wedge \forall y (y \neq x \rightarrow \neg P(y)))$.

EXAMPLE 28

Show that every integer has a unique additive inverse. That is, show that if p is an integer, then there exists a unique integer q such that $p + q = 0$.

Solution: If p is an integer, we find that $p + q = 0$ when $q = -p$ and q is also an integer. Consequently, there exists an integer q such that $p + q = 0$.

To show that given the integer p, the integer q with $p + q = 0$ is unique, suppose that r is an integer with $r \neq q$ such that $p + r = 0$. Then $p + q = p + r$. By subtracting p from both sides of the equation, it follows that $q = r$, which contradicts our assumption that $q \neq r$. Consequently, there is a unique integer q such that $p + q = 0$. ◀

COUNTEREXAMPLES In Section 1.3 we mentioned that we can show that a statement of the form $\forall x P(x)$ is false if we can find a counterexample, that is, an example x for which $P(x)$ is false. When we are presented with a statement of the form $\forall x P(x)$, either which we believe to be false or which has resisted all attempts to find a proof, we look for a counterexample. We illustrate the hunt for a counterexample in Example 29.

EXAMPLE 29

Show that the statement "Every positive integer is the sum of the squares of three integers" is false.

Solution: We can show that this statement is false if we can find a counterexample. That is, the statement is false if we can show that there is a particular integer that is not the sum of the squares of three integers. To look for a counterexample, we try to write successive positive integers as a sum of three squares. We find that $1 = 0^2 + 0^2 + 1^2$, $2 = 0^2 + 1^2 + 1^2$, $3 = 1^2 + 1^2 + 1^2$, $4 = 0^2 + 0^2 + 2^2$, $5 = 0^2 + 1^2 + 2^2$, $6 = 1^2 + 1^2 + 2^2$, but we cannot find a way to write 7 as the sum of three squares. To show that there are not three squares that add up to 7, we note that the only possible squares we can use are those not exceeding 7, namely, 0, 1, and 4. Since no three terms where each term is 0, 1, or 4 add up to 7, it follows that 7 is a counterexample. We conclude that the statement "Every positive integer is the sum of the squares of three integers" is false. ◀

A common error is to assume that one or more examples establish the truth of a statement. No matter how many examples there are where $P(x)$ is true, the universal quantification $\forall x P(x)$ may still be false. Consider Example 30.

EXAMPLE 30

Is it true that every positive integer is the sum of 18 fourth powers of integers? That is, is the statement $\forall n P(n)$ a theorem where $P(n)$ is the statement "n can be written as the sum of 18 fourth powers of integers" and the universe of discourse consists of all positive integers?

Solution: To determine whether n can be written as the sum of 18 fourth powers of integers, we might begin by examining whether n is the sum of 18 fourth powers of integers for the smallest positive integers. Because the fourth powers of integers are 0, 1, 16, 81, ..., if we can select 18 terms from these numbers that add up to n, then n is the sum of 18 fourth powers. We can show that all positive integers up to 78 can be written as the sum of 18 fourth powers. (The details are left to the reader.) However, if we decided this was enough checking, we would come to the wrong conclusion. It is not true that every positive integer is the sum of 18 fourth powers because 79 is not the sum of 18 fourth powers (as the reader can verify). ◀

MISTAKES IN PROOFS

There are many common errors made in constructing mathematical proofs. We will briefly describe some of these here. Among the most common errors are mistakes in arithmetic and basic algebra. Even professional mathematicians make such errors, especially when working with complicated formulas. Whenever you use such computations you should check them as carefully as possible. (You should also review any troublesome aspects of basic algebra, especially before you study Section 3.3.)

Each step of a mathematical proof needs to be correct and the conclusion needs to logically follow from the steps that precede it. Many mistakes result from the introduction of steps that do not logically follow from those that precede it. This is illustrated in Examples 31–33.

EXAMPLE 31   What is wrong with this famous supposed "proof" that $1 = 2$?

"Proof:" We use these steps, where a and b are two equal positive integers.

Step	Reason
1. $a = b$	Given
2. $a^2 = ab$	Multiply both sides of (1) by a
3. $a^2 - b^2 = ab - b^2$	Subtract b^2 from both sides of (2)
4. $(a - b)(a + b) = b(a - b)$	Factor both sides of (3)
5. $a + b = b$	Divide both sides of (4) by $a - b$
6. $2b = b$	Replace a by b in (5) because $a = b$ and simplify
7. $2 = 1$	Divide both sides of (6) by b

Solution: Every step is valid except for one, step 5 where we divided both sides by $a - b$. The error is that $a - b$ equals zero; division of both sides of an equation by the same quantity is valid as long as this quantity is not zero. ◀

EXAMPLE 32   What is wrong with this "proof"?

"Theorem:" If n^2 is positive, then n is positive.

"Proof:" Suppose that n^2 is positive. Because the implication "If n is positive, then n^2 is positive" is true, we can conclude that n is positive.

Solution: Let $P(n)$ be "n is positive" and $Q(n)$ be "n^2 is positive." Then our hypothesis is $Q(n)$. The statement "If n is positive, then n^2 is positive" is the statement $\forall n(P(n) \to Q(n))$. From the hypothesis $Q(n)$ and the statement $\forall n(P(n) \to Q(n))$ we cannot conclude $P(n)$, because we are not using a valid rule of inference. Instead, this is an example of the fallacy of affirming the conclusion. A counterexample is supplied by $n = -1$ for which $n^2 = 1$ is positive, but n is negative. ◀

EXAMPLE 33   What is wrong with this "proof"?

"Theorem:" If n is not positive, then n^2 is not positive. (This is the contrapositive of the "theorem" in Example 32.)

"Proof:"   Suppose that n is not positive. Because the implication "If n is positive, then n^2 is positive" is true, we can conclude that n^2 is not positive.

Solution: Let $P(n)$ and $Q(n)$ be as in the solution of Example 32. Then our hypothesis is $\neg P(n)$ and the statement "If n is positive, then n^2 is positive" is the statement $\forall n(P(n) \rightarrow Q(n))$. From the hypothesis $\neg P(n)$ and the statement $\forall n(P(n) \rightarrow Q(n))$ we cannot conclude $\neg Q(n)$, because we are not using a valid rule of inference. Instead, this is an example of the fallacy of denying the hypothesis. A counterexample is supplied by $n = -1$, as in Example 32.   ◀

A common error in making unwarranted assumptions occurs in proofs by cases, where not all cases are considered. This is illustrated in Example 34.

EXAMPLE 34   What is wrong with this "proof"?

"Theorem:" If x is a real number, then x^2 is a positive real number.

"Proof:"   Let p_1 be "x is positive," let p_2 be "x is negative," and let q be "x^2 is positive." To show that $p_1 \rightarrow q$, note that when x is positive, x^2 is positive since it is the product of two positive numbers, x and x. To show that $p_2 \rightarrow q$, note that when x is negative, x^2 is positive since it is the product of two negative numbers, x and x. This completes the proof.

Solution: The problem with the proof we have given is that we missed the case $x = 0$. When $x = 0$, $x^2 = 0$ is not positive, so the supposed theorem is false. If p is "x is a real number," then we can prove results where p is the hypothesis with three cases, p_1, p_2, and p_3, where p_1 is "x is positive," p_2 is "x is negative," and p_3 is "$x = 0$" because of the equivalence $p \leftrightarrow p_1 \vee p_2 \vee p_3$.   ◀

Finally, we briefly discuss a particularly nasty type of error. Many incorrect arguments are based on a fallacy called **begging the question.** This fallacy occurs when one or more steps of a proof are based on the truth of the statement being proved. In other words, this fallacy arises when a statement is proved using itself, or a statement equivalent to it. That is why this fallacy is also called **circular reasoning.**

EXAMPLE 35   Is the following argument correct? It supposedly shows that n is an even integer whenever n^2 is an even integer.

Suppose that n^2 is even. Then $n^2 = 2k$ for some integer k. Let $n = 2l$ for some integer l. This shows that n is even.

Solution: This argument is incorrect. The statement "let $n = 2l$ for some integer l" occurs in the proof. No argument has been given to show that n can be written as $2l$ for some integer l. This is circular reasoning because this statement is equivalent to the statement being proved, namely, "n is even." Of course, the result itself is correct; only the method of proof is wrong.   ◀

Making mistakes in proofs is part of the learning process. When you make a mistake that someone else finds, you should carefully analyze where you went wrong and make sure that you do not make the same mistake again. Even professional mathematicians make mistakes in proofs. More than a few incorrect proofs of important results have fooled people for many years before subtle errors in them were found.

JUST A BEGINNING

We have introduced a variety of methods for proving theorems. Observe that no algorithm for proving theorems has been given here or even mentioned. It is a deep result that no such procedure exists.

There are many theorems whose proofs are easy to find by directly working through the hypotheses and definitions of the terms of the theorem. However, it is often difficult to prove a theorem without resorting to a clever use of an indirect proof or a proof by contradiction, or some other proof technique. Constructing proofs is an art that can be learned only through experience, including writing proofs, having your proofs critiqued, reading and analyzing other proofs, and so on.

We will present a variety of proofs in the rest of this chapter and in Chapter 2 before we return to the subject of proofs. In Chapter 3 we will address some of the art and the strategy in proving theorems and in working with conjectures. We will also introduce several important proof techniques in Chapter 3, including mathematical induction, which can be used to prove results that hold for all positive integers. In Chapter 4 we will introduce the notion of combinatorial proofs.

Exercises

1. What rule of inference is used in each of these arguments?

   a) Alice is a mathematics major. Therefore, Alice is either a mathematics major or a computer science major.

   b) Jerry is a mathematics major and a computer science major. Therefore, Jerry is a mathematics major.

   c) If it is rainy, then the pool will be closed. It is rainy. Therefore, the pool is closed.

   d) If it snows today, the university will close. The university is not closed today. Therefore, it did not snow today.

   e) If I go swimming, then I will stay in the sun too long. If I stay in the sun too long, then I will sunburn. Therefore, if I go swimming, then I will sunburn.

2. What rule of inference is used in each of these arguments?

   a) Kangaroos live in Australia and are marsupials. Therefore, kangaroos are marsupials.

   b) It is either hotter than 100 degrees today or the pollution is dangerous. It is less than 100 degrees outside today. Therefore, the pollution is dangerous.

   c) Linda is an excellent swimmer. If Linda is an excellent swimmer, then she can work as a lifeguard. Therefore, Linda can work as a lifeguard.

   d) Steve will work at a computer company this summer. Therefore, this summer Steve will work at a computer company or he will be a beach bum.

   e) If I work all night on this homework, then I can answer all the exercises. If I answer all the exercises, I will understand the material. Therefore, if I work all night on this homework, then I will understand the material.

3. Construct an argument using rules of inference to show that the hypotheses "Randy works hard," "If Randy works hard, then he is a dull boy," and "If Randy is a dull boy, then he will not get the job" imply the conclusion "Randy will not get the job."

4. Construct an argument using rules of inference to show that the hypotheses "If it does not rain or if it is not foggy, then the sailing race will be held and the lifesaving demonstration will go on," "If the sailing race is held, then the trophy will be awarded," and "The trophy was not awarded" imply the conclusion "It rained."

5. What rules of inference are used in this famous argument? "All men are mortal. Socrates is a man. Therefore, Socrates is mortal."

6. What rules of inference are used in this argument? "No man is an island. Manhattan is an island. Therefore, Manhattan is not a man."

7. For each of these sets of premises, what relevant conclusion or conclusions can be drawn? Explain the rules of inference used to obtain each conclusion from the premises.

a) "If I take the day off, it either rains or snows." "I took Tuesday off or I took Thursday off." "It was sunny on Tuesday." "It did not snow on Thursday."

b) "If I eat spicy foods, then I have strange dreams." "I have strange dreams if there is thunder while I sleep." "I did not have strange dreams."

c) "I am either clever or lucky." "I am not lucky." "If I am lucky, then I will win the lottery."

d) "Every computer science major has a personal computer." "Ralph does not have a personal computer." "Ann has a personal computer."

e) "What is good for corporations is good for the United States." "What is good for the United States is good for you." "What is good for corporations is for you to buy lots of stuff."

f) "All rodents gnaw their food." "Mice are rodents." "Rabbits do not gnaw their food." "Bats are not rodents."

8. For each of these sets of premises, what relevant conclusion or conclusions can be drawn? Explain the rules of inference used to obtain each conclusion from the premises.

a) "If I play hockey, then I am sore the next day." "I use the whirlpool if I am sore." "I did not use the whirlpool."

b) "If I work, it is either sunny or partly sunny." "I worked last Monday or I worked last Friday." "It was not sunny on Tuesday." "It was not partly sunny on Friday."

c) "All insects have six legs." "Dragonflies are insects." "Spiders do not have six legs." "Spiders eat dragonflies."

d) "Every student has an Internet account." "Homer does not have an Internet account." "Maggie has an Internet account."

e) "All foods that are healthy to eat do not taste good." "Tofu is healthy to eat." "You only eat what tastes good." "You do not eat tofu." "Cheeseburgers are not healthy to eat."

f) "I am either dreaming or hallucinating." "I am not dreaming." "If I am hallucinating, I see elephants running down the road."

9. For each of these arguments, explain which rules of inference are used for each step.

a) "Doug, a student in this class, knows how to write programs in JAVA. Everyone who knows how to write programs in JAVA can get a high-paying job. Therefore, someone in this class can get a high-paying job."

b) "Somebody in this class enjoys whale watching. Every person who enjoys whale watching cares about ocean pollution. Therefore, there is a person in this class who cares about ocean pollution."

c) "Each of the 93 students in this class owns a personal computer. Everyone who owns a personal computer can use a word processing program. Therefore, Zeke, a student in this class, can use a word processing program."

d) "Everyone in New Jersey lives within 50 miles of the ocean. Someone in New Jersey has never seen the ocean. Therefore, someone who lives within 50 miles of the ocean has never seen the ocean."

10. For each of these arguments, explain which rules of inference are used for each step.

a) "Linda, a student in this class, owns a red convertible. Everyone who owns a red convertible has gotten at least one speeding ticket. Therefore, someone in this class has gotten a speeding ticket."

b) "Each of five roommates, Melissa, Aaron, Ralph, Veneesha, and Keeshawn, has taken a course in discrete mathematics. Every student who has taken a course in discrete mathematics can take a course in algorithms. Therefore, all five roommates can take a course in algorithms next year."

c) "All movies produced by John Sayles are wonderful. John Sayles produced a movie about coal miners. Therefore, there is a wonderful movie about coal miners."

d) "There is someone in this class who has been to France. Everyone who goes to France visits the Louvre. Therefore, someone in this class has visited the Louvre."

11. For each of these arguments determine whether the argument is correct or incorrect and explain why.

a) All students in this class understand logic. Xavier is a student in this class. Therefore, Xavier understands logic.

b) Every computer science major takes discrete mathematics. Natasha is taking discrete mathematics. Therefore, Natasha is a computer science major.

c) All parrots like fruit. My pet bird is not a parrot. Therefore, my pet bird does not like fruit.

d) Everyone who eats granola every day is healthy. Linda is not healthy. Therefore, Linda does not eat granola every day.

12. For each of these arguments determine whether the argument is correct or incorrect and explain why.

a) Everyone enrolled in the university has lived in a dormitory. Mia has never lived in a dormitory. Therefore, Mia is not enrolled in the university.

b) A convertible car is fun to drive. Isaac's car is not a convertible. Therefore, Isaac's car is not fun to drive.

c) Quincy likes all action movies. Quincy likes the movie *Eight Men Out*. Therefore, *Eight Men Out* is an action movie.

d) All lobstermen set at least a dozen traps. Hamilton is a lobsterman. Therefore, Hamilton sets at least a dozen traps.

13. Determine whether each of these arguments is valid. If an argument is correct, what rule of inference is being used? If it is not, what logical error occurs?

a) If n is a real number such that $n > 1$, then $n^2 > 1$. Suppose that $n^2 > 1$. Then $n > 1$.

b) The number $\log_2 3$ is irrational if it is not the ratio of two integers. Therefore, since $\log_2 3$ cannot be written in the form a/b where a and b are integers, it is irrational.

c) If n is a real number with $n > 3$, then $n^2 > 9$. Suppose that $n^2 \le 9$. Then $n \le 3$.

d) If n is a real number with $n > 2$, then $n^2 > 4$. Suppose that $n \le 2$. Then $n^2 \le 4$.

14. Determine whether these are valid arguments.

a) "If x^2 is irrational, then x is irrational. Therefore, if x is irrational, it follows that x^2 is irrational."

b) "If x^2 is irrational, then x is irrational. The number $x = \pi^2$ is irrational. Therefore, the number $x = \pi$ is irrational."

15. What is wrong with this argument? Let $H(x)$ be "x is happy." Given the premise $\exists x H(x)$, we conclude that $H(\text{Lola})$. Therefore, Lola is happy.

16. What is wrong with this argument? Let $S(x, y)$ be "x is shorter than y." Given the premise $\exists s S(s, \text{Max})$, it follows that $S(\text{Max}, \text{Max})$. Then by existential generalization it follows that $\exists x S(x, x)$, so that someone is shorter than himself.

17. Prove the proposition $P(0)$, where $P(n)$ is the proposition "If n is a positive integer greater than 1, then $n^2 > n$." What kind of proof did you use?

18. Prove the proposition $P(1)$, where $P(n)$ is the proposition "If n is a positive integer, then $n^2 \ge n$." What kind of proof did you use?

19. Let $P(n)$ be the proposition "If a and b are positive real numbers, then $(a + b)^n \ge a^n + b^n$." Prove that $P(1)$ is true. What kind of proof did you use?

20. Prove that the square of an even number is an even number using

a) a direct proof.   **b)** an indirect proof.
c) a proof by contradiction.

21. Prove that if n is an integer and $n^3 + 5$ is odd, then n is even using

a) an indirect proof.
b) a proof by contradiction.

22. Prove that if n is an integer and $3n + 2$ is even, then n is even using

a) an indirect proof.
b) a proof by contradiction.

23. Prove that the sum of two odd integers is even.

24. Prove that the product of two odd numbers is odd.

25. Prove that the sum of an irrational number and a rational number is irrational using a proof by contradiction.

26. Prove that the product of two rational numbers is rational.

27. Prove or disprove that the product of two irrational numbers is irrational.

28. Prove or disprove that the product of a nonzero rational number and an irrational number is irrational.

29. Prove that if x is irrational, then $1/x$ is irrational.

30. Prove that if x is rational and $x \ne 0$, then $1/x$ is rational.

31. Show that at least 10 of any 64 days chosen must fall on the same day of the week.

32. Show that at least 3 of any 25 days chosen must fall in the same month of the year.

33. Prove that if x and y are real numbers, then $\max(x, y) + \min(x, y) = x + y$. (*Hint:* Use a proof by cases, with the two cases corresponding to $x \ge y$ and $x < y$, respectively.)

34. Use a proof by cases to show that $\min(a, \min(b, c)) = \min(\min(a, b), c)$ whenever a, b, and c are real numbers.

35. Prove the **triangle inequality,** which states that if x and y are real numbers, then $|x| + |y| \ge |x + y|$ (where $|x|$ represents the absolute value of x, which equals x if $x \ge 0$ and equals $-x$ if $x < 0$).

36. Prove that a square of an integer ends with a 0, 1, 4, 5, 6, or 9. (*Hint:* Let $n = 10k + l$ where $l = 0, 1, \ldots, 9$.)

37. Prove that a fourth power of an integer ends with a 0, 1, 5, or 6.

38. Prove that if n is a positive integer, then n is even if and only if $7n + 4$ is even.

39. Prove that if n is a positive integer, then n is odd if and only if $5n + 6$ is odd.

40. Prove that $m^2 = n^2$ if and only if $m = n$ or $m = -n$.

41. Prove or disprove that if m and n are integers such that $mn = 1$, then either $m = 1$ and $n = 1$, or else $m = -1$ and $n = -1$.

42. Show that these three statements are equivalent where a and b are real numbers: (*i*) a is less than b, (*ii*) the average of a and b is greater than a, and (*iii*) the average of a and b is less than b.

43. Show that these statements are equivalent: (*i*) $3x + 2$ is an even integer, (*ii*) $x + 5$ is an odd integer, (*iii*) x^2 is an even integer.

44. Show that these statements are equivalent: (*i*) x is rational, (*ii*) $x/2$ is rational, and (*iii*) $3x - 1$ is rational.

45. Show that these statements are equivalent: (*i*) x is irrational, (*ii*) $3x + 2$ is irrational, (*iii*) $x/2$ is irrational.

46. Is this reasoning for finding the solutions of the equation $\sqrt{2x^2 - 1} = x$ correct? (*1*) $\sqrt{2x^2 - 1} = x$ is given; (*2*) $2x^2 - 1 = x^2$, obtained by squaring both sides of (1); (*3*) $x^2 - 1 = 0$, obtained by subtracting x^2 from both sides of (2); (*4*) $(x - 1)(x + 1) = 0$, obtained by factoring the left-hand side of $x^2 - 1$; (*5*) $x = 1$

or $x = -1$, which follows since $ab = 0$ implies that $a = 0$ or $b = 0$.

47. Are these steps for finding the solutions of $\sqrt{x+3} = 3 - x$ correct? (*1*) $\sqrt{x+3} = 3 - x$ is given; (*2*) $x + 3 = x^2 - 6x + 9$, obtained by squaring both sides of (*1*); (*3*) $0 = x^2 - 7x + 6$, obtained by subtracting $x + 3$ from both sides of (*2*); (*4*) $0 = (x-1)(x-6)$, obtained by factoring the right-hand side of (*3*); (*5*) $x = 1$ or $x = 6$, which follows from (*4*) since $ab = 0$ implies that $a = 0$ or $b = 0$.

48. Prove that there is a positive integer that equals the sum of the positive integers not exceeding it. Is your proof constructive or nonconstructive?

49. Prove that there are 100 consecutive positive integers that are not perfect squares. Is your proof constructive or nonconstructive?

50. Prove that either $2 \cdot 10^{500} + 15$ or $2 \cdot 10^{500} + 16$ is not a perfect square. Is your proof constructive or nonconstructive?

51. Prove that there exists a pair of consecutive integers such that one of these integers is a perfect square and the other is a perfect cube.

52. Show that the product of two of the numbers $65^{1000} - 8^{2001} + 3^{177}$, $79^{1212} - 9^{2399} + 2^{2001}$, and $24^{4493} - 5^{8192} + 7^{1777}$ is nonnegative. Is your proof constructive or nonconstructive? (*Hint:* Do not try to evaluate these numbers!)

53. Show that each of these statements can be used to express the fact that there is a unique element x such that $P(x)$ is true. [Note that by Exercise 48 in Section 1.3, this is the statement $\exists! P(x)$.]
   a) $\exists x \forall y (P(y) \leftrightarrow x = y)$
   b) $\exists x P(x) \land \forall x \forall y (P(x) \land P(y) \rightarrow x = y)$
   c) $\exists x (P(x) \land \forall y (P(y) \rightarrow x = y))$

54. Show that if a, b, and c are real numbers and $a \neq 0$, then there is a unique solution of the equation $ax + b = c$.

55. Suppose that a and b are odd integers with $a \neq b$. Show there is a unique integer c such that $|a - c| = |b - c|$.

56. Show that if r is an irrational number, there is a unique integer n such that the distance between r and n is less than $1/2$.

57. Show that if n is an odd integer, then there is a unique integer k such that n is the sum of $k - 2$ and $k + 3$.

58. Prove that given a real number x there exist unique numbers n and ϵ such that $x = n + \epsilon$, n is an integer, and $0 \leq \epsilon < 1$.

59. Prove that given a real number x there exist unique numbers n and ϵ such that $x = n - \epsilon$, n is an integer, and $0 \leq \epsilon < 1$.

60. Use resolution to show the hypotheses "Allen is a bad boy or Hillary is a good girl" and "Allen is a good boy or David is happy" imply the conclusion "Hillary is a good girl or David is happy."

61. Use resolution to show that the hypotheses "It is not raining or Yvette has her umbrella," "Yvette does not have her umbrella or she does not get wet," and "It is raining or Yvette does not get wet" imply that "Yvette does not get wet."

62. Show that the equivalence $p \land \neg p \equiv \mathbf{F}$ can be derived using resolution together with the fact that an implication with a false hypothesis is true. (*Hint:* Let $q = r = \mathbf{F}$ in resolution.)

63. Use resolution to show that the compound proposition $(p \lor q) \land (\neg p \lor q) \land (p \lor \neg q) \land (\neg p \lor \neg q)$ is not satisfiable.

64. Prove or disprove that if a and b are rational numbers, then a^b is also rational.

65. Prove or disprove that there is a rational number x and an irrational number y such that x^y is irrational.

66. Show that the propositions p_1, p_2, p_3, and p_4 can be shown to be equivalent by showing that $p_1 \leftrightarrow p_4$, $p_2 \leftrightarrow p_3$, and $p_1 \leftrightarrow p_3$.

67. Show that the propositions p_1, p_2, p_3, p_4, and p_5 can be shown to be equivalent by proving that the implications $p_1 \rightarrow p_4$, $p_3 \rightarrow p_1$, $p_4 \rightarrow p_2$, $p_2 \rightarrow p_5$, and $p_5 \rightarrow p_3$ are true.

68. Prove that an 8×8 chessboard can be completely covered using dominos (1×2 pieces).

*69. Prove that it is impossible to cover completely with dominos the 8×8 chessboard with two squares at opposite corners of the board removed.

*70. The Logic Problem, taken from *WFF'N PROOF, The Game of Logic*, has these two assumptions:
   1. "Logic is difficult or not many students like logic."
   2. "If mathematics is easy, then logic is not difficult."

   By translating these assumptions into statements involving propositional variables and logical connectives, determine whether each of the following are valid conclusions of these assumptions:
   a) That mathematics is not easy, if many students like logic.
   b) That not many students like logic, if mathematics is not easy.
   c) That mathematics is not easy or logic is difficult.
   d) That logic is not difficult or mathematics is not easy.
   e) That if not many students like logic, then either mathematics is not easy or logic is not difficult.

71. Prove that at least one of the real numbers a_1, $a_2, \ldots, a_n$ is greater than or equal to the average of these numbers. What kind of proof did you use?

72. Use Exercise 71 to show that if the first 10 positive integers are placed around a circle, in any order, there exist three integers in consecutive locations around the circle that have a sum greater than or equal to 17.

73. Prove that if n is an integer, these four statements are equivalent: (*i*) n is even, (*ii*) $n + 1$ is odd, (*iii*) $3n + 1$ is odd, (*iv*) $3n$ is even.

74. Prove that these four statements are equivalent: (*i*) n^2 is odd, (*ii*) $1 - n$ is even, (*iii*) n^3 is odd, (*iv*) $n^2 + 1$ is even.

75. Which rules of inference are used to establish the conclusion of Lewis Carroll's argument described in Example 19 of Section 1.3?

76. Which rules of inference are used to establish the conclusion of Lewis Carroll's argument described in Example 20 of Section 1.3?

*77. Determine whether this argument, taken from Backhouse [Ba86], is valid.

> If Superman were able and willing to prevent evil, he would do so. If Superman were unable to prevent evil, he would be impotent; if he were unwilling to prevent evil, he would be malevolent. Superman does not prevent evil. If Superman exists, he is neither impotent nor malevolent. Therefore, Superman does not exist.

1.6   Sets

INTRODUCTION

We will study a wide variety of discrete structures in this book. These include relations, which consist of ordered pairs of elements; combinations, which are unordered collections of elements; and graphs, which are sets of vertices and edges connecting vertices. Moreover, we will illustrate how these and other discrete structures are used in modeling and problem solving. In particular, many examples of the use of discrete structures in the storage, communication, and manipulation of data will be described. In this section we study the fundamental discrete structure upon which all other discrete structures are built, namely, the set.

Sets are used to group objects together. Often, the objects in a set have similar properties. For instance, all the students who are currently enrolled in your school make up a set. Likewise, all the students currently taking a course in discrete mathematics at any school make up a set. In addition, those students enrolled in your school who are taking a course in discrete mathematics form a set that can be obtained by taking the elements common to the first two collections. The language of sets is a means to study such collections in an organized fashion. We now provide a definition of a set.

DEFINITION 1   A *set* is an unordered collection of objects.

Note that the term *object* has been used without specifying what an object is. This description of a set as a collection of objects, based on the intuitive notion of an object, was first stated by the German mathematician Georg Cantor in 1895. The theory that results from this intuitive definition of a set leads to **paradoxes,** or logical inconsistencies, as the English philosopher Bertrand Russell showed in 1902 (see Exercise 30 for a description of one of these paradoxes). These logical inconsistencies can be avoided by building set theory starting with basic assumptions, called **axioms.** We will use Cantor's original version of set theory, known as **naive set theory,** without developing an axiomatic version of set theory, since all sets considered in this book can be treated consistently using Cantor's original theory.

We now proceed with our discussion of sets.

DEFINITION 2   The objects in a set are also called the *elements*, or *members*, of the set. A set is said to *contain* its elements.

There are several ways to describe a set. One way is to list all the members of a set, when this is possible. We use a notation where all members of the set are listed between braces. For example, the notation $\{a, b, c, d\}$ represents the set with the four elements $a, b, c,$ and d.

EXAMPLE 1   The set V of all vowels in the English alphabet can be written as $V = \{a, e, i, o, u\}$. ◀

EXAMPLE 2   The set O of odd positive integers less than 10 can be expressed by $O = \{1, 3, 5, 7, 9\}$. ◀

EXAMPLE 3   Although sets are usually used to group together elements with common properties, there is nothing that prevents a set from having seemingly unrelated elements. For instance, $\{a, 2, \text{Fred}, \text{New Jersey}\}$ is the set containing the four elements $a, 2,$ Fred, and New Jersey. ◀

Sometimes the brace notation is used to describe a set without listing all its members. Some members of the set are listed, and then *ellipses* (. . .) are used when the general pattern of the elements is obvious.

EXAMPLE 4   The set of positive integers less than 100 can be denoted by $\{1, 2, 3, \ldots, 99\}$. ◀

These sets, each denoted using a boldface letter, play an important role in discrete mathematics:

$\mathbf{N} = \{0, 1, 2, 3, \ldots\}$, the set of **natural numbers**
$\mathbf{Z} = \{\ldots, -2, -1, 0, 1, 2, \ldots\}$, the set of **integers**
$\mathbf{Z}^{+} = \{1, 2, 3, \ldots\}$, the set of **positive integers**
$\mathbf{Q} = \{p/q \mid p \in \mathbf{Z}, q \in \mathbf{Z}, q \neq 0\}$, the set of **rational numbers**
$\mathbf{R}$, the set of **real numbers**

Links

GEORG CANTOR (1845–1918) Georg Cantor was born in St. Petersburg, Russia, where his father was a successful merchant. Cantor developed his interest in mathematics in his teens. He began his university studies in Zurich in 1862, but when his father died he left Zurich. He continued his university studies at the University of Berlin in 1863, where he studied under the eminent mathematicians Weierstrass, Kummer, and Kronecker. He received his doctor's degree in 1867 after having written a dissertation on number theory. Cantor assumed a position at the University of Halle in 1869, where he continued working until his death.

Cantor is considered the founder of set theory. His contributions in this area include the discovery that the set of real numbers is uncountable. He is also noted for his many important contributions to analysis. Cantor also was interested in philosophy and wrote papers relating his theory of sets with metaphysics.

Cantor married in 1874 and had five children. His melancholy temperament was balanced by his wife's happy disposition. Although he received a large inheritance from his father, he was poorly paid as a professor. To mitigate this, he tried to obtain a better-paying position at the University of Berlin. His appointment there was blocked by Kronecker, who did not agree with Cantor's views on set theory. Cantor suffered from mental illness throughout the later years of his life. He died in 1918 in a psychiatric clinic.

(Note that some people do not consider 0 a natural number, so be careful to check how the term *natural numbers* is used when you read other books.)

Since many mathematical statements assert that two differently specified collections of objects are really the same set, we need to understand what it means for two sets to be equal.

DEFINITION 3   Two sets are *equal* if and only if they have the same elements.

EXAMPLE 5   The sets {1, 3, 5} and {3, 5, 1} are equal, since they have the same elements. Note that the order in which the elements of a set are listed does not matter. Note also that it does not matter if an element of a set is listed more than once, so that {1, 3, 3, 3, 5, 5, 5, 5} is the same as the set {1, 3, 5} since they have the same elements. ◀

Extra Examples

Another way to describe a set is to use **set builder** notation. We characterize all those elements in the set by stating the property or properties they must have to be members. For instance, the set O of all odd positive integers less than 10 can be written as

$O = \{x \mid x$ is an odd positive integer less than 10$\}$.

We often use this type of notation to describe sets when it is impossible to list all the elements of the set. For instance, the set of all real numbers can be written as

$\mathbf{R} = \{x \mid x$ is a real number$\}$.

Sets can also be represented graphically using Venn diagrams, named after the English mathematician John Venn, who introduced their use in 1881. In Venn diagrams the **universal set** U, which contains all the objects under consideration, is represented by a rectangle. Inside this rectangle, circles or other geometrical figures are used to represent sets. Sometimes points are used to represent the particular elements of the set. Venn diagrams are often used to indicate the relationships between sets. We show how a Venn diagram can be used in the following example.

EXAMPLE 6   Draw a Venn diagram that represents V, the set of vowels in the English alphabet.

Solution: We draw a rectangle to indicate the universal set U, which is the set of the 26 letters of the English alphabet. Inside this rectangle we draw a circle to represent V. Inside this circle we indicate the elements of V with points (see Figure 1). ◀

Links

BERTRAND RUSSELL (1872–1970)   Bertrand Russell was born into a prominent English family active in the progressive movement and having a strong commitment to liberty. He became an orphan at an early age and was placed in the care of his father's parents, who had him educated at home. He entered Trinity College, Cambridge, in 1890, where he excelled in mathematics and in moral science. He won a fellowship on the basis of his work on the foundations of geometry. In 1910 Trinity College appointed him to a lectureship in logic and the philosophy of mathematics.

Russell fought for progressive causes throughout his life. He held strong pacifist views, and his protests against World War I led to dismissal from his position at Trinity College. He was imprisoned for 6 months in 1918 because of an article he wrote that was branded as seditious. Russell fought for women's suffrage in Great Britain. In 1961, at the age of 89, he was imprisoned for the second time for his protests advocating nuclear disarmament.

Russell's greatest work was in his development of principles that could be used as a foundation for all of mathematics. His most famous work is *Principia Mathematica,* written with Alfred North Whitehead, which attempts to deduce all of mathematics using a set of primitive axioms. He wrote many books on philosophy, physics, and his political ideas. Russell won the Nobel Prize for literature in 1950.

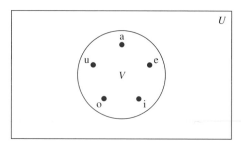

FIGURE 1   Venn Diagram for the Set of Vowels.

We will now introduce notation used to describe membership in sets. We write $a \in A$ to denote that a is an element of the set A. The notation $a \notin A$ denotes that a is not a member of the set A. Note that lowercase letters are usually used to denote elements of sets.

There is a special set that has no elements. This set is called the **empty set,** or **null set,** and is denoted by $\emptyset$. The empty set can also be denoted by { } (that is, we represent the empty set with a pair of braces that encloses all the elements in this set). Often, a set of elements with certain properties turns out to be the null set. For instance, the set of all positive integers that are greater than their squares is the null set.

A common error is to mistake the empty set $\emptyset$ with the set $\{\emptyset\}$, which is a **singleton set**, that is, a set with one element. The single element of the set $\{\emptyset\}$ is the empty set itself!

DEFINITION 4

The set A is said to be a *subset* of B if and only if every element of A is also an element of B. We use the notation $A \subseteq B$ to indicate that A is a subset of the set B.

We see that $A \subseteq B$ if and only if the quantification

$$\forall x (x \in A \rightarrow x \in B)$$

is true. For instance, the set of all odd positive integers less than 10 is a subset of the set of all positive integers less than 10. The set of all computer science majors at your school is a subset of the set of all students at your school.

Theorem 1 shows that every nonempty set S is guaranteed to have at least two subsets, the empty set and the set S itself, that is, $\emptyset \subseteq S$ and $S \subseteq S$.

JOHN VENN (1834–1923)   John Venn was born into a London suburban family noted for its philanthropy. He attended London schools and got his mathematics degree from Caius College, Cambridge, in 1857. He was elected a fellow of this college and held his fellowship there until his death. He took holy orders in 1859 and, after a brief stint of religious work, returned to Cambridge, where he developed programs in the moral sciences. Besides his mathematical work, Venn had an interest in history and wrote extensively about his college and family.

Venn's book *Symbolic Logic* clarifies ideas originally presented by Boole. In this book, Venn presents a systematic development of a method that uses geometric figures, known now as *Venn diagrams*. Today these diagrams are primarily used to analyze logical arguments and to illustrate relationships between sets. In addition to his work on symbolic logic, Venn made contributions to probability theory described in his widely used textbook on that subject.

Links

THEOREM 1

For any set S,

(i) $\emptyset \subseteq S$ and (ii) $S \subseteq S$.

Proof: We will prove (i) and leave the proof of (ii) as an exercise.

Let S be a set. To show that $\emptyset \subseteq S$, we must show that $\forall x (x \in \emptyset \rightarrow x \in S)$ is true. Since the empty set contains no elements, it follows that $x \in \emptyset$ is always false. It follows that the implication $x \in \emptyset \rightarrow x \in S$ is always true, since its hypothesis is always false and an implication with a false hypothesis is true. That is, $\forall x (x \in \emptyset \rightarrow x \in S)$ is true. This completes the proof of (i). Note that this is an example of a vacuous proof. ◁

When we wish to emphasize that a set A is a subset of the set B but that $A \neq B$, we write $A \subset B$ and say that A is a **proper subset** of B. Venn diagrams can be used to show that a set A is a subset of a set B. We draw the universal set U as a rectangle. Within this rectangle we draw a circle for B. Since A is a subset of B, we draw the circle for A within the circle for B. This relationship is shown in Figure 2.

One way to show that two sets have the same elements is to show that each set is a subset of the other. In other words, we can show that if A and B are sets with $A \subseteq B$ and $B \subseteq A$, then $A = B$. This turns out to be a useful way to show that two sets are equal. That is, $A = B$, where A and B are sets, if and only if $\forall x (x \in A \rightarrow x \in B)$ and $\forall x (x \in B \rightarrow x \in A)$, or equivalently if and only if $\forall x (x \in A \leftrightarrow x \in B)$.

Sets may have other sets as members. For instance, we have the sets

$$\{\emptyset, \{a\}, \{b\}, \{a, b\}\} \quad \text{and} \quad \{x \mid x \text{ is a subset of the set } \{a, b\}\}.$$

Note that these two sets are equal.

Sets are used extensively in counting problems, and for such applications we need to discuss the size of sets.

DEFINITION 5

Let S be a set. If there are exactly n distinct elements in S where n is a nonnegative integer, we say that S is a *finite set* and that n is the *cardinality* of S. The cardinality of S is denoted by $|S|$.

EXAMPLE 7

Let A be the set of odd positive integers less than 10. Then $|A| = 5$. ◀

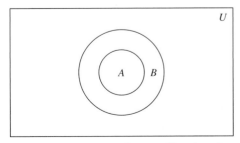

FIGURE 2   Venn Diagram Showing that
A Is a Subset of B.

EXAMPLE 8   Let S be the set of letters in the English alphabet. Then $|S| = 26$.   ◀

EXAMPLE 9   Since the null set has no elements, it follows that $|\emptyset| = 0$.   ◀

We will also be interested in sets that are not finite.

DEFINITION 6   A set is said to be *infinite* if it is not finite.

EXAMPLE 10   The set of positive integers is infinite.   ◀

Extra Examples

The cardinality of infinite sets will be discussed in Section 3.2. In that section, we will discuss what it means for a set to be countable and show that certain sets are countable while others are not.

THE POWER SET

Many problems involve testing all combinations of elements of a set to see if they satisfy some property. To consider all such combinations of elements of a set S, we build a new set that has as its members all the subsets of S.

DEFINITION 7   Given a set S, the *power set* of S is the set of all subsets of the set S. The power set of S is denoted by $P(S)$.

EXAMPLE 11   What is the power set of the set $\{0, 1, 2\}$?

Solution: The power set $P(\{0, 1, 2\})$ is the set of all subsets of $\{0, 1, 2\}$. Hence,

$$P(\{0, 1, 2\}) = \{\emptyset, \{0\}, \{1\}, \{2\}, \{0, 1\}, \{0, 2\}, \{1, 2\}, \{0, 1, 2\}\}.$$

Note that the empty set and the set itself are members of this set of subsets.   ◀

EXAMPLE 12   What is the power set of the empty set? What is the power set of the set $\{\emptyset\}$?

Solution: The empty set has exactly one subset, namely, itself. Consequently,

$$P(\emptyset) = \{\emptyset\}.$$

The set $\{\emptyset\}$ has exactly two subsets, namely, $\emptyset$ and the set $\{\emptyset\}$ itself. Therefore,

$$P(\{\emptyset\}) = \{\emptyset, \{\emptyset\}\}.$$

◀

If a set has n elements, then its power set has 2^n elements. We will demonstrate this fact in several ways in subsequent sections of the text.

CARTESIAN PRODUCTS

The order of elements in a collection is often important. Since sets are unordered, a different structure is needed to represent ordered collections. This is provided by **ordered n-tuples.**

DEFINITION 8

> The *ordered n-tuple* $(a_1, a_2, \ldots, a_n)$ is the ordered collection that has a_1 as its first element, a_2 as its second element, $\ldots$, and a_n as its nth element.

We say that two ordered n-tuples are equal if and only if each corresponding pair of their elements is equal. In other words, $(a_1, a_2, \ldots, a_n) = (b_1, b_2, \ldots, b_n)$ if and only if $a_i = b_i$, for $i = 1, 2, \ldots, n$. In particular, 2-tuples are called **ordered pairs.** The ordered pairs (a, b) and (c, d) are equal if and only if $a = c$ and $b = d$. Note that (a, b) and (b, a) are not equal unless $a = b$.

Many of the discrete structures we will study in later chapters are based on the notion of the *Cartesian product* of sets (named after René Descartes). We first define the Cartesian product of two sets.

DEFINITION 9

> Let A and B be sets. The *Cartesian product* of A and B, denoted by $A \times B$, is the set of all ordered pairs (a, b) where $a \in A$ and $b \in B$. Hence,
>
> $$A \times B = \{(a, b) \mid a \in A \land b \in B\}.$$

EXAMPLE 13

Let A represent the set of all students at a university, and let B represent the set of all courses offered at the university. What is the Cartesian product $A \times B$?

Solution: The Cartesian product $A \times B$ consists of all the ordered pairs of the form (a, b), where a is a student at the university and b is a course offered at the university. The set $A \times B$ can be used to represent all possible enrollments of students in courses at the university. ◀

RENÉ DESCARTES (1596–1650) René Descartes was born into a noble family near Tours, France, about 200 miles southwest of Paris. He was the third child of his father's first wife; she died several days after his birth. Because of René's poor health, his father, a provincial judge, let his son's formal lessons slide until, at the age of 8, René entered the Jesuit college at La Flèche. The rector of the school took a liking to him and permitted him to stay in bed until late in the morning because of his frail health. From then on, Descartes spent his mornings in bed; he considered these times his most productive hours for thinking.

Descartes left school in 1612, moving to Paris, where he spent 2 years studying mathematics. He earned a law degree in 1616 from the University of Poitiers. At 18 Descartes became disgusted with studying and decided to see the world. He moved to Paris and became a successful gambler. However, he grew tired of bawdy living and moved to the suburb of Saint-Germain, where he devoted himself to mathematical study. When his gambling friends found him, he decided to leave France and undertake a military career. However, he never did any fighting. One day, while escaping the cold in an overheated room at a military encampment, he had several feverish dreams, which revealed his future career as a mathematician and philosopher.

After ending his military career, he traveled throughout Europe. He then spent several years in Paris, where he studied mathematics and philosophy and constructed optical instruments. Descartes decided to move to Holland, where he spent 20 years wandering around the country, accomplishing his most important work. During this time he wrote several books, including the *Discours,* which contains his contributions to analytic geometry, for which he is best known. He also made fundamental contributions to philosophy.

In 1649 Descartes was invited by Queen Christina to visit her court in Sweden to tutor her in philosophy. Although he was reluctant to live in what he called "the land of bears amongst rocks and ice," he finally accepted the invitation and moved to Sweden. Unfortunately, the winter of 1649–1650 was extremely bitter. Descartes caught pneumonia and died in mid-February.

EXAMPLE 14   What is the Cartesian product of $A = \{1, 2\}$ and $B = \{a, b, c\}$?

Solution: The Cartesian product $A \times B$ is

$$A \times B = \{(1, a), (1, b), (1, c), (2, a), (2, b), (2, c)\}.$$ ◄

A subset R of the Cartesian product $A \times B$ is called a **relation** from the set A to the set B. The elements of R are ordered pairs, where the first element belongs to A and the second to B. For example, $R = \{(a, 0), (a, 1), (a, 3), (b, 1), (b, 2), (c, 0), (c, 3)\}$ is a relation from the set $\{a, b, c\}$ to the set $\{0, 1, 2, 3\}$. We will study relations at length in Chapter 7.

The Cartesian products $A \times B$ and $B \times A$ are not equal, unless $A = \emptyset$ or $B = \emptyset$ (so that $A \times B = \emptyset$) or unless $A = B$ (see Exercise 26, at the end of this section). This is illustrated in Example 15.

EXAMPLE 15   Show that the Cartesian product $B \times A$ is not equal to the Cartesian product $A \times B$, where A and B are as in Example 14.

Solution: The Cartesian product $B \times A$ is

$$B \times A = \{(a, 1), (a, 2), (b, 1), (b, 2), (c, 1), (c, 2)\}.$$

This is not equal to $A \times B$, which was found in Example 14.   ◄

The Cartesian product of more than two sets can also be defined.

DEFINITION 10

The *Cartesian product* of the sets $A_1, A_2, \ldots, A_n$, denoted by $A_1 \times A_2 \times \cdots \times A_n$, is the set of ordered n-tuples $(a_1, a_2, \ldots, a_n)$, where a_i belongs to A_i for $i = 1, 2, \ldots, n$. In other words

$$A_1 \times A_2 \times \cdots \times A_n = \{(a_1, a_2, \ldots, a_n) \mid a_i \in A_i \text{ for } i = 1, 2, \ldots, n\}.$$

EXAMPLE 16   What is the Cartesian product $A \times B \times C$, where $A = \{0, 1\}$, $B = \{1, 2\}$, and $C = \{0, 1, 2\}$?

Solution: The Cartesian product $A \times B \times C$ consists of all ordered triples (a, b, c), where $a \in A$, $b \in B$, and $c \in C$. Hence,

$$A \times B \times C = \{(0,1,0), (0,1,1), (0,1,2), (0,2,0), (0,2,1), (0,2,2), (1,1,0), (1,1,1),$$
$$(1,1,2), (1,2,0), (1,2,1), (1,2,2)\}.$$ ◄

USING SET NOTATION WITH QUANTIFIERS

Sometimes we specify the universe of discourse of a statement explicitly in the notation. In particular, $\forall x \in S\, P(x)$ denotes the universal quantification of $P(x)$, where the universe of discourse is the set S. Similarly, $\exists x \in S\, P(x)$ denotes the existential quantification of $P(x)$, where the universe of discourse is S.

EXAMPLE 17   What do the statements $\forall x \in \mathbf{R}\ (x^2 \geq 0)$ and $\exists x \in \mathbf{Z}\ (x^2 = 1)$ mean?

Solution: The statement $\forall x \in \mathbf{R}(x^2 \geq 0)$ states that for every real number x, $x^2 \geq 0$. This statement can be expressed as "The square of every real number is nonnegative." This is a true statement.

The statement $\exists x \in \mathbf{Z}(x^2 = 1)$ states that there exists an integer x such that $x^2 = 1$. This statement can be expressed as "There is an integer whose square is 1." This is also a true statement since $x = 1$ is such an integer (as is -1). ◀

Exercises

1. List the members of these sets.

   a) $\{x \mid x$ is a real number such that $x^2 = 1\}$
   b) $\{x \mid x$ is a positive integer less than 12$\}$
   c) $\{x \mid x$ is the square of an integer and $x < 100\}$
   d) $\{x \mid x$ is an integer such that $x^2 = 2\}$

2. Use set builder notation to give a description of each of these sets.

   a) $\{0, 3, 6, 9, 12\}$
   b) $\{-3, -2, -1, 0, 1, 2, 3\}$
   c) $\{m, n, o, p\}$

3. Determine whether each of these pairs of sets are equal.

   a) $\{1, 3, 3, 3, 5, 5, 5, 5, 5\}$, $\{5, 3, 1\}$
   b) $\{\{1\}\}$, $\{1, \{1\}\}$
   c) $\emptyset$, $\{\emptyset\}$

4. Suppose that $A = \{2, 4, 6\}$, $B = \{2, 6\}$, $C = \{4, 6\}$, and $D = \{4, 6, 8\}$. Determine which of these sets are subsets of which other of these sets.

5. For each of the following sets, determine whether 2 is an element of that set.

   a) $\{x \in \mathbf{R} \mid x$ is an integer greater than 1$\}$
   b) $\{x \in \mathbf{R} \mid x$ is the square of an integer$\}$
   c) $\{2, \{2\}\}$
   d) $\{\{2\}, \{\{2\}\}\}$
   e) $\{\{2\}, \{2, \{2\}\}\}$
   f) $\{\{\{2\}\}\}$

6. For each of the sets in Exercise 5, determine whether $\{2\}$ is an element of that set.

7. Determine whether each of these statements is true or false.

   a) $0 \in \emptyset$ **b)** $\emptyset \in \{0\}$
   c) $\{0\} \subset \emptyset$ **d)** $\emptyset \subset \{0\}$
   e) $\{0\} \in \{0\}$ **f)** $\{0\} \subset \{0\}$
   g) $\{\emptyset\} \subseteq \{\emptyset\}$

8. Determine whether these statements are true or false.

   a) $\emptyset \in \{\emptyset\}$ **b)** $\emptyset \in \{\emptyset, \{\emptyset\}\}$
   c) $\{\emptyset\} \in \{\emptyset\}$ **d)** $\{\emptyset\} \in \{\{\emptyset\}\}$
   e) $\{\emptyset\} \subset \{\emptyset, \{\emptyset\}\}$ **f)** $\{\{\emptyset\}\} \subset \{\emptyset, \{\emptyset\}\}$
   g) $\{\{\emptyset\}\} \subset \{\{\emptyset\}, \{\emptyset\}\}$

9. Determine whether each of these statements is true or false.

   a) $x \in \{x\}$ **b)** $\{x\} \subseteq \{x\}$ **c)** $\{x\} \in \{x\}$
   d) $\{x\} \in \{\{x\}\}$ **e)** $\emptyset \subseteq \{x\}$ **f)** $\emptyset \in \{x\}$

10. Use a Venn diagram to illustrate the relationship $A \subseteq B$ and $B \subseteq C$.

11. Suppose that A, B, and C are sets such that $A \subseteq B$ and $B \subseteq C$. Show that $A \subseteq C$.

12. Find two sets A and B such that $A \in B$ and $A \subseteq B$.

13. What is the cardinality of each of these sets?

   a) $\{a\}$ **b)** $\{\{a\}\}$
   c) $\{a, \{a\}\}$ **d)** $\{a, \{a\}, \{a, \{a\}\}\}$

14. What is the cardinality of each of these sets?

   a) $\emptyset$ **b)** $\{\emptyset\}$
   c) $\{\emptyset, \{\emptyset\}\}$ **d)** $\{\emptyset, \{\emptyset\}, \{\emptyset, \{\emptyset\}\}\}$

15. Find the power set of each of these sets.

   a) $\{a\}$ **b)** $\{a, b\}$ **c)** $\{\emptyset, \{\emptyset\}\}$

16. Can you conclude that $A = B$ if A and B are two sets with the same power set?

17. How many elements does each of these sets have?

   a) $P(\{a, b, \{a, b\}\})$
   b) $P(\{\emptyset, a, \{a\}, \{\{a\}\}\})$
   c) $P(P(\emptyset))$

18. Determine whether each of these sets is the power set of a set.

   a) $\emptyset$ **b)** $\{\emptyset, \{a\}\}$
   c) $\{\emptyset, \{a\}, \{\emptyset, a\}\}$ **d)** $\{\emptyset, \{a\}, \{b\}, \{a, b\}\}$

19. Let $A = \{a, b, c, d\}$ and $B = \{y, z\}$. Find

   a) $A \times B$ **b)** $B \times A$

20. What is the Cartesian product $A \times B$, where A is the set of courses offered by the mathematics department at a university and B is the set of mathematics professors at this university?

21. What is the Cartesian product $A \times B \times C$, where A is the set of all airlines and B and C are both the set of all cities in the United States?

22. Suppose that $A \times B = \emptyset$, where A and B are sets. What can you conclude?

23. Let A be a set. Show that $\emptyset \times A = A \times \emptyset = \emptyset$.

24. Let $A = \{a, b, c\}$, $B = \{x, y\}$, and $C = \{0, 1\}$. Find

   a) $A \times B \times C$ **b)** $C \times B \times A$
   c) $C \times A \times B$ **d)** $B \times B \times B$

25. How many different elements does $A \times B$ have if A has m elements and B has n elements?

26. Show that $A \times B \neq B \times A$, when A and B are nonempty, unless $A = B$.

27. Translate each of these quantifications into English and determine its truth value.

    a) $\forall x \in \mathbf{R} \, (x^2 \neq -1)$ **b)** $\exists x \in \mathbf{Z} \, (x^2 = 2)$
    c) $\forall x \in \mathbf{Z} \, (x^2 > 0)$ **d)** $\exists x \in \mathbf{R} \, (x^2 = x)$

28. Translate each of these quantifications into English and determine its truth value.

    a) $\exists x \in \mathbf{R} \, (x^3 = -1)$
    b) $\exists x \in \mathbf{Z} \, (x + 1 > x)$
    c) $\forall x \in \mathbf{Z} \, (x - 1 \in \mathbf{Z})$
    d) $\forall x \in \mathbf{Z} \, (x^2 \in \mathbf{Z})$

*29. Show that the ordered pair (a, b) can be defined in terms of sets as $\{\{a\}, \{a, b\}\}$. (*Hint:* First show that $\{\{a\}, \{a, b\}\} = \{\{c\}, \{c, d\}\}$ if and only if $a = c$ and $b = d$.)

*30. In this exercise **Russell's paradox** is presented. Let S be the set that contains a set x if the set x does not belong to itself, so that $S = \{x \mid x \notin x\}$.

    a) Show that the assumption that S is a member of S leads to a contradiction.

    b) Show that the assumption that S is not a member of S leads to a contradiction.

From parts (a) and (b) it follows that the set S cannot be defined as it was. This paradox can be avoided by restricting the types of elements that sets can have.

*31. Describe a procedure for listing all the subsets of a finite set.

1.7   Set Operations

INTRODUCTION

Two sets can be combined in many different ways. For instance, starting with the set of mathematics majors and the set of computer science majors at your school, we can form the set of students who are mathematics majors or computer science majors, the set of students who are joint majors in mathematics and computer science, the set of all students not majoring in mathematics, and so on.

DEFINITION 1

> Let A and B be sets. The *union* of the sets A and B, denoted by $A \cup B$, is the set that contains those elements that are either in A or in B, or in both.

An element x belongs to the union of the sets A and B if and only if x belongs to A or x belongs to B. This tells us that

$$A \cup B = \{x \mid x \in A \vee x \in B\}.$$

The Venn diagram shown in Figure 1 represents the union of two sets A and B. The shaded area within the circle representing A or the circle representing B is the area that represents the union of A and B.

We will give some examples of the union of sets.

EXAMPLE 1   The union of the sets $\{1, 3, 5\}$ and $\{1, 2, 3\}$ is the set $\{1, 2, 3, 5\}$; that is, $\{1, 3, 5\} \cup \{1, 2, 3\} = \{1, 2, 3, 5\}$.   ◄

EXAMPLE 2   The union of the set of all computer science majors at your school and the set of all mathematics majors at your school is the set of students at your school who are majoring either in mathematics or in computer science (or in both).   ◄

DEFINITION 2

> Let A and B be sets. The *intersection* of the sets A and B, denoted by $A \cap B$, is the set containing those elements in both A and B.

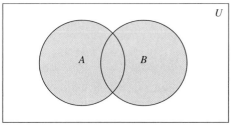

$A \cup B$ is shaded.

FIGURE 1 **Venn Diagram Representing the Union of *A* and *B*.**

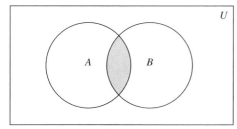

$A \cap B$ is shaded.

FIGURE 2 **Venn Diagram Representing the Intersection of *A* and *B*.**

An element x belongs to the intersection of the sets A and B if and only if x belongs to A and x belongs to B. This tells us that

$$A \cap B = \{x \mid x \in A \land x \in B\}.$$

The Venn diagram shown in Figure 2 represents the intersection of two sets A and B. The shaded area that is within both the circles representing the sets A and B is the area that represents the intersection of A and B.

We give some examples of the intersection of sets.

EXAMPLE 3 The intersection of the sets $\{1, 3, 5\}$ and $\{1, 2, 3\}$ is the set $\{1, 3\}$; that is, $\{1, 3, 5\} \cap \{1, 2, 3\} = \{1, 3\}$. ◀

EXAMPLE 4 The intersection of the set of all computer science majors at your school and the set of all mathematics majors is the set of all students who are joint majors in mathematics and computer science. ◀

DEFINITION 3 Two sets are called *disjoint* if their intersection is the empty set.

EXAMPLE 5 Let $A = \{1, 3, 5, 7, 9\}$ and $B = \{2, 4, 6, 8, 10\}$. Since $A \cap B = \emptyset$, A and B are disjoint. ◀

We often are interested in finding the cardinality of the union of sets. To find the number of elements in the union of two finite sets A and B, note that $|A| + |B|$ counts each element that is in A but not in B or in B but not in A exactly once, and each element that is in both A and B exactly twice. Thus, if the number of elements that are in both A and B is subtracted from $|A| + |B|$, elements in $A \cap B$ will be counted only once. Hence,

$$|A \cup B| = |A| + |B| - |A \cap B|.$$

The generalization of this result to unions of an arbitrary number of sets is called the **principle of inclusion–exclusion.** The principle of inclusion–exclusion is an important technique used in the art of enumeration. We will discuss this principle and other counting techniques in detail in Chapters 4 and 6.

There are other important ways to combine sets.

DEFINITION 4 Let A and B be sets. The *difference* of A and B, denoted by $A - B$, is the set containing those elements that are in A but not in B. The difference of A and B is also called the *complement of B with respect to A.*

An element x belongs to the difference of A and B if and only if $x \in A$ and $x \notin B$. This tells us that

$$A - B = \{x \mid x \in A \wedge x \notin B\}.$$

The Venn diagram shown in Figure 3 represents the difference of the sets A and B. The shaded area inside the circle that represents A and outside the circle that represents B is the area that represents $A - B$.

We give some examples of differences of sets.

EXAMPLE 6 The difference of $\{1, 3, 5\}$ and $\{1, 2, 3\}$ is the set $\{5\}$; that is, $\{1, 3, 5\} - \{1, 2, 3\} = \{5\}$. This is different from the difference of $\{1, 2, 3\}$ and $\{1, 3, 5\}$, which is the set $\{2\}$. ◀

EXAMPLE 7 The difference of the set of computer science majors at your school and the set of mathematics majors at your school is the set of all computer science majors at your school who are not also mathematics majors. ◀

Once the universal set U has been specified, the **complement** of a set can be defined.

DEFINITION 5 Let U be the universal set. The *complement* of the set A, denoted by $\overline{A}$, is the complement of A with respect to U. In other words, the complement of the set A is $U - A$.

An element belongs to $\overline{A}$ if and only if $x \notin A$. This tells us that

$$\overline{A} = \{x \mid x \notin A\}.$$

In Figure 4 the shaded area outside the circle representing A is the area representing $\overline{A}$.

We give some examples of the complement of a set.

EXAMPLE 8 Let $A = \{a, e, i, o, u\}$ (where the universal set is the set of letters of the English alphabet). Then $\overline{A} = \{b, c, d, f, g, h, j, k, l, m, n, p, q, r, s, t, v, w, x, y, z\}$. ◀

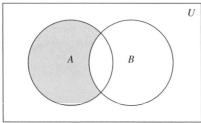

$A - B$ is shaded.

FIGURE 3   Venn Diagram for the Difference of A and B.

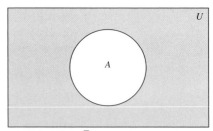

$\overline{A}$ is shaded.

FIGURE 4   Venn Diagram for the Complement of the Set A.

TABLE 1 Set Identities.	
Identity	**Name**
$A \cup \emptyset = A$ $A \cap U = A$	Identity laws
$A \cup U = U$ $A \cap \emptyset = \emptyset$	Domination laws
$A \cup A = A$ $A \cap A = A$	Idempotent laws
$\overline{(\overline{A})} = A$	Complementation law
$A \cup B = B \cup A$ $A \cap B = B \cap A$	Commutative laws
$A \cup (B \cup C) = (A \cup B) \cup C$ $A \cap (B \cap C) = (A \cap B) \cap C$	Associative laws
$A \cap (B \cup C) = (A \cap B) \cup (A \cap C)$ $A \cup (B \cap C) = (A \cup B) \cap (A \cup C)$	Distributive laws
$\overline{A \cup B} = \overline{A} \cap \overline{B}$ $\overline{A \cap B} = \overline{A} \cup \overline{B}$	De Morgan's laws
$A \cup (A \cap B) = A$ $A \cap (A \cup B) = A$	Absorption laws
$A \cup \overline{A} = U$ $A \cap \overline{A} = \emptyset$	Complement laws

EXAMPLE 9   Let A be the set of positive integers greater than 10 (with universal set the set of all positive integers). Then $\overline{A} = \{1, 2, 3, 4, 5, 6, 7, 8, 9, 10\}$. ◀

SET IDENTITIES

Table 1 lists the most important set identities. We will prove several of these identities here, using three different methods. These methods are presented to illustrate that there are often many different approaches to the solution of a problem. The proofs of the remaining identities will be left as exercises. The reader should note the similarity between these set identities and the logical equivalences discussed in Section 1.2. In fact, the set identities given can be proved directly from the corresponding logical equivalences. Furthermore, both are special cases of identities that hold for Boolean algebra (discussed in Chapter 10).

Extra Examples

One way to prove that two sets are equal is to show that one of the sets is a subset of the other and vice versa. We illustrate this type of proof by establishing the second of De Morgan's laws.

EXAMPLE 10   Prove that $\overline{A \cap B} = \overline{A} \cup \overline{B}$.

Solution: We will prove that these two sets are equal by showing that each is a subset of the other.

First, suppose that $x \in \overline{A \cap B}$. By the definition of complement, $x \notin A \cap B$. By the definition of the intersection, $\neg((x \in A) \wedge (x \in B))$ is true. Applying De Morgan's law (from logic), we see that $\neg(x \in A)$ or $\neg(x \in B)$. Hence, by the definition of negation, $x \notin A$ or $x \notin B$. By the definition of complement, $x \in \overline{A}$ or $x \in \overline{B}$. It follows by the definition of union that $x \in \overline{A} \cup \overline{B}$. This shows that $\overline{A \cap B} \subseteq \overline{A} \cup \overline{B}$.

Now suppose that $x \in \overline{A} \cup \overline{B}$. By the definition of union, $x \in \overline{A}$ or $x \in \overline{B}$. Using the definition of complement, we see that $x \notin A$ or $x \notin B$. Consequently, $\neg(x \in A) \vee \neg(x \in B)$ is true. By De Morgan's law (from logic), we conclude that $\neg((x \in A) \wedge (x \in B))$ is true. By the definition of intersection, it follows that $\neg(x \in A \cap B)$ holds. We use the definition of complement to conclude that $x \in \overline{A \cap B}$. This shows that $\overline{A} \cup \overline{B} \subseteq \overline{A \cap B}$. Since we have shown that each set is a subset of the other, the two sets are equal, and the identity is proved. ◀

We can more succinctly express the reasoning used in Example 10 using set builder notation, as Example 11 illustrates.

EXAMPLE 11 Use set builder notation and logical equivalences to show that $\overline{A \cap B} = \overline{A} \cup \overline{B}$.

Solution: The following chain of equalities provides a demonstration of this identity:

$$
\begin{aligned}
\overline{A \cap B} &= \{x \mid x \notin A \cap B\} \\
&= \{x \mid \neg(x \in (A \cap B))\} \\
&= \{x \mid \neg(x \in A \wedge x \in B)\} \\
&= \{x \mid x \notin A \vee x \notin B\} \\
&= \{x \mid x \in \overline{A} \vee x \in \overline{B}\} \\
&= \{x \mid x \in \overline{A} \cup \overline{B}\}. \\
&= \overline{A} \cup \overline{B}
\end{aligned}
$$

Note that the second De Morgan's law for logical equivalences was used in the fourth equality of this chain. ◀

Proving a set identity involving more than two sets by showing each side of the identity is a subset of the other often requires that we keep track of different cases, as illustrated by the proof in Example 12 of one of the distributive laws for sets.

EXAMPLE 12 Prove that $A \cap (B \cup C) = (A \cap B) \cup (A \cap C)$ for all sets A, B, and C.

Solution: We will prove this identity by showing that each side is a subset of the other side.

Suppose that $x \in A \cap (B \cup C)$. Then $x \in A$ and $x \in B \cup C$. By the definition of union, it follows that $x \in A$, and $x \in B$ or $x \in C$ (or both). Consequently, we know that $x \in A$ and $x \in B$ or that $x \in A$ and $x \in C$. By the definition of intersection, it follows that $x \in A \cap B$ or $x \in A \cap C$. Using the definition of union, we conclude that $x \in (A \cap B) \cup (A \cap C)$. We conclude that $A \cap (B \cup C) \subseteq (A \cap B) \cup (A \cap C)$.

Now suppose that $x \in (A \cap B) \cup (A \cap C)$. Then, by the definition of union, $x \in A \cap B$ or $x \in A \cap C$. By the definition of intersection, it follows that $x \in A$ and $x \in B$ or that $x \in A$ and $x \in C$. From this we see that $x \in A$, and $x \in B$ or $x \in C$. Consequently, by the definition of union we see that $x \in A$ and $x \in B \cup C$.

TABLE 2 A Membership Table for the Distributive Property.

A	B	C	$B \cup C$	$A \cap (B \cup C)$	$A \cap B$	$A \cap C$	$(A \cap B) \cup (A \cap C)$
1	1	1	1	1	1	1	1
1	1	0	1	1	1	0	1
1	0	1	1	1	0	1	1
1	0	0	0	0	0	0	0
0	1	1	1	0	0	0	0
0	1	0	1	0	0	0	0
0	0	1	1	0	0	0	0
0	0	0	0	0	0	0	0

Furthermore, by the definition of intersection, it follows that $x \in A \cap (B \cup C)$. We conclude that $(A \cap B) \cup (A \cap C) \subseteq A \cap (B \cup C)$. This completes the proof of the identity. ◄

Set identities can also be proved using **membership tables.** We consider each combination of sets that an element can belong to and verify that elements in the same combinations of sets belong to both the sets in the identity. To indicate that an element is in a set, a 1 is used; to indicate that an element is not in a set, a 0 is used. (The reader should note the similarity between membership tables and truth tables.)

EXAMPLE 13   Use a membership table to show that $A \cap (B \cup C) = (A \cap B) \cup (A \cap C)$.

Solution: The membership table for these combinations of sets is shown in Table 2. This table has eight rows. Since the columns for $A \cap (B \cup C)$ and $(A \cap B) \cup (A \cap C)$ are the same, the identity is valid. ◄

Additional set identities can be established using those that we have already proved. Consider Example 14.

EXAMPLE 14   Let A, B, and C be sets. Show that

$$\overline{A \cup (B \cap C)} = (\overline{C} \cup \overline{B}) \cap \overline{A}.$$

Solution: We have

$$\overline{A \cup (B \cap C)} = \overline{A} \cap \overline{(B \cap C)} \quad \text{by the first De Morgan's law}$$
$$= \overline{A} \cap (\overline{B} \cup \overline{C}) \quad \text{by the second De Morgan's law}$$
$$= (\overline{B} \cup \overline{C}) \cap \overline{A} \quad \text{by the commutative law for intersections}$$
$$= (\overline{C} \cup \overline{B}) \cap \overline{A} \quad \text{by the commutative law for unions.} \quad ◄$$

GENERALIZED UNIONS AND INTERSECTIONS

Since unions and intersections of sets satisfy associative laws, the sets $A \cup B \cup C$ and $A \cap B \cap C$ are well defined when A, B, and C are sets. Note that $A \cup B \cup C$ contains those elements that are in at least one of the sets A, B, and C, and that $A \cap B \cap C$

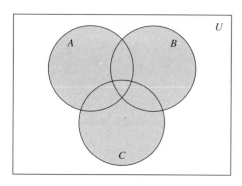
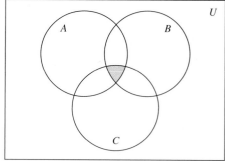

(a) $A \cup B \cup C$ is shaded.   (b) $A \cap B \cap C$ is shaded.

FIGURE 5   The Union and Intersection of A, B, and C.

contains those elements that are in all of A, B, and C. These combinations of the three sets, A, B, and C, are shown in Figure 5.

EXAMPLE 15   Let $A = \{0, 2, 4, 6, 8\}$, $B = \{0, 1, 2, 3, 4\}$, and $C = \{0, 3, 6, 9\}$. What are $A \cup B \cup C$ and $A \cap B \cap C$?

Solution: The set $A \cup B \cup C$ contains those elements in at least one of A, B, and C. Hence,

$$A \cup B \cup C = \{0, 1, 2, 3, 4, 6, 8, 9\}.$$

The set $A \cap B \cap C$ contains those elements in all three of A, B, and C. Thus,

$$A \cap B \cap C = \{0\}. \qquad\qquad \blacktriangleleft$$

We can also consider unions and intersections of an arbitrary number of sets. We use these definitions.

DEFINITION 6   The *union* of a collection of sets is the set that contains those elements that are members of at least one set in the collection.

We use the notation

$$A_1 \cup A_2 \cup \cdots \cup A_n = \bigcup_{i=1}^{n} A_i$$

to denote the union of the sets $A_1, A_2, \ldots, A_n$.

DEFINITION 7   The *intersection* of a collection of sets is the set that contains those elements that are members of all the sets in the collection.

We use the notation

$$A_1 \cap A_2 \cap \cdots \cap A_n = \bigcap_{i=1}^{n} A_i$$

to denote the intersection of the sets $A_1, A_2, \ldots, A_n$. We illustrate generalized unions and intersections with Example 16.

EXAMPLE 16   Let $A_i = \{i, i + 1, i + 2, \ldots\}$. Then

$$\bigcup_{i=1}^{n} A_i = \bigcup_{i=1}^{n} \{i, i + 1, i + 2, \ldots\} = \{1, 2, 3, \ldots\},$$

and

$$\bigcap_{i=1}^{n} A_i = \bigcap_{i=1}^{n} \{i, i + 1, i + 2, \ldots\} = \{n, n + 1, n + 2, \ldots\}. \qquad \blacktriangleleft$$

COMPUTER REPRESENTATION OF SETS

There are various ways to represent sets using a computer. One method is to store the elements of the set in an unordered fashion. However, if this is done, the operations of computing the union, intersection, or difference of two sets would be time-consuming, since each of these operations would require a large amount of searching for elements. We will present a method for storing elements using an arbitrary ordering of the elements of the universal set. This method of representing sets makes computing combinations of sets easy.

Assume that the universal set U is finite (and of reasonable size so that the number of elements of U is not larger than the memory size of the computer being used). First, specify an arbitrary ordering of the elements of U, for instance $a_1, a_2, \ldots, a_n$. Represent a subset A of U with the bit string of length n, where the ith bit in this string is 1 if a_i belongs to A and is 0 if a_i does not belong to A. Example 17 illustrates this technique.

EXAMPLE 17   Let $U = \{1, 2, 3, 4, 5, 6, 7, 8, 9, 10\}$, and the ordering of elements of U has the elements in increasing order; i.e., $a_i = i$. What bit strings represent the subset of all odd integers in U, the subset of all even integers in U, and the subset of integers not exceeding 5 in U?

Solution: The bit string that represents the set of odd integers in U, namely, $\{1, 3, 5, 7, 9\}$, has a one bit in the first, third, fifth, seventh, and ninth positions, and a zero elsewhere. It is

    10 1010 1010.

(We have split this bit string of length ten into blocks of length four for easy reading since long bit strings are difficult to read.) Similarly, we represent the subset of all even integers in U, namely, $\{2, 4, 6, 8, 10\}$, by the string

    01 0101 0101.

The set of all integers in U that do not exceed 5, namely, $\{1, 2, 3, 4, 5\}$, is represented by the string

    11 1110 0000. $\qquad \blacktriangleleft$

Using bit strings to represent sets, it is easy to find complements of sets and unions, intersections, and differences of sets. To find the bit string for the complement of a set from the bit string for that set, we simply change each 1 to a 0 and each 0 to 1, since $x \in A$ if and only if $x \notin \overline{A}$. Note that this operation corresponds to taking the negation of each bit when we associate a bit with a truth value—with 1 representing true and 0, false.

EXAMPLE 18   We have seen that the bit string for the set $\{1, 3, 5, 7, 9\}$ (with universal set $\{1, 2, 3, 4, 5, 6, 7, 8, 9, 10\}$) is

10 1010 1010.

What is the bit string for the complement of this set?

Solution: The bit string for the complement of this set is obtained by replacing 0s with 1s and vice versa. This yields the string

01 0101 0101,

which corresponds to the set $\{2, 4, 6, 8, 10\}$.   ◄

To obtain the bit string for the union and intersection of two sets we perform bitwise Boolean operations on the bit strings representing the two sets. The bit in the ith position of the bit string of the union is 1 if either of the bits in the ith position in the two strings is 1 (or both are 1), and is 0 when both bits are 0. Hence, the bit string for the union is the bitwise *OR* of the bit strings for the two sets. The bit in the ith position of the bit string of the intersection is 1 when the bits in the corresponding position in the two strings are both 1, and is 0 when either of the two bits is 0 (or both are). Hence, the bit string for the intersection is the bitwise *AND* of the bit strings for the two sets.

EXAMPLE 19   The bit strings for the sets $\{1, 2, 3, 4, 5\}$ and $\{1, 3, 5, 7, 9\}$ are 11 1110 0000 and 10 1010 1010, respectively. Use bit strings to find the union and intersection of these sets.

Solution: The bit string for the union of these sets is

11 1110 0000 $\vee$ 10 1010 1010 = 11 1110 1010,

which corresponds to the set $\{1, 2, 3, 4, 5, 7, 9\}$. The bit string for the intersection of these sets is

11 1110 0000 $\wedge$ 10 1010 1010 = 10 1010 0000,

which corresponds to the set $\{1, 3, 5\}$.   ◄

Exercises

1. Let A be the set of students who live within one mile of school and let B be the set of students who walk to classes. Describe the students in each of these sets.

a) $A \cap B$ **b)** $A \cup B$
c) $A - B$ **d)** $B - A$

2. Suppose that A is the set of sophomores at your school and B is the set of students in discrete mathematics

at your school. Express each of these sets in terms of A and B.

a) the set of sophomores taking discrete mathematics in your school

b) the set of sophomores at your school who are not taking discrete mathematics

c) the set of students at your school who either are sophomores or are taking discrete mathematics

d) the set of students at your school who either are not sophomores or are not taking discrete mathematics

3. Let $A = \{1, 2, 3, 4, 5\}$ and $B = \{0, 3, 6\}$. Find

a) $A \cup B$.   **b)** $A \cap B$.

c) $A - B$.   **d)** $B - A$.

4. Let $A = \{a, b, c, d, e\}$ and $B = \{a, b, c, d, e, f, g, h\}$. Find

a) $A \cup B$.   **b)** $A \cap B$.

c) $A - B$.   **d)** $B - A$.

5. Let A be a set. Show that $\overline{\overline{A}} = A$.

6. Let A be a set. Show that

a) $A \cup \emptyset = A$.   **b)** $A \cap \emptyset = \emptyset$.

c) $A \cup A = A$.   **d)** $A \cap A = A$.

e) $A - \emptyset = A$.   **f)** $A \cup U = U$.

g) $A \cap U = A$.   **h)** $\emptyset - A = \emptyset$.

7. Let A and B be sets. Show that

a) $A \cup B = B \cup A$.   **b)** $A \cap B = B \cap A$.

8. Let A and B be sets. Show that $A \cup (A \cap B) = A$.

9. Let A and B be sets. Show that $A \cap (A \cup B) = A$.

10. Find the sets A and B if $A - B = \{1, 5, 7, 8\}$, $B - A = \{2, 10\}$, and $A \cap B = \{3, 6, 9\}$.

11. Show that if A and B are sets, then $\overline{A \cup B} = \overline{A} \cap \overline{B}$

a) by showing each side is a subset of the other side.

b) using a membership table.

12. Let A and B be sets. Show that

a) $(A \cap B) \subseteq A$.   **b)** $A \subseteq (A \cup B)$.

c) $A - B \subseteq A$.   **d)** $A \cap (B - A) = \emptyset$.

e) $A \cup (B - A) = A \cup B$.

13. Show that if A, B, and C are sets, then $\overline{A \cap B \cap C} = \overline{A} \cup \overline{B} \cup \overline{C}$

a) by showing each side is a subset of the other side.

b) using a membership table.

14. Let A, B, and C be sets. Show that

a) $(A \cup B) \subseteq (A \cup B \cup C)$.

b) $(A \cap B \cap C) \subseteq (A \cap B)$.

c) $(A - B) - C \subseteq A - C$.

d) $(A - C) \cap (C - B) = \emptyset$.

e) $(B - A) \cup (C - A) = (B \cup C) - A$.

15. Show that if A and B are sets, then $A - B = A \cap \overline{B}$.

16. Show that if A and B are sets, then $(A \cap B) \cup (A \cap \overline{B}) = A$.

17. Let A, B, and C be sets. Show that

a) $A \cup (B \cup C) = (A \cup B) \cup C$.

b) $A \cap (B \cap C) = (A \cap B) \cap C$.

c) $A \cup (B \cap C) = (A \cup B) \cap (A \cup C)$.

18. Let A, B, and C be sets. Show that $(A - B) - C = (A - C) - (B - C)$.

19. Let $A = \{0, 2, 4, 6, 8, 10\}$, $B = \{0, 1, 2, 3, 4, 5, 6\}$, and $C = \{4, 5, 6, 7, 8, 9, 10\}$. Find

a) $A \cap B \cap C$.   **b)** $A \cup B \cup C$.

c) $(A \cup B) \cap C$.   **d)** $(A \cap B) \cup C$.

20. Draw the Venn diagrams for each of these combinations of the sets A, B, and C.

a) $A \cap (B \cup C)$ **b)** $\overline{A} \cap \overline{B} \cap \overline{C}$

c) $(A - B) \cup (A - C) \cup (B - C)$

21. What can you say about the sets A and B if we know that

a) $A \cup B = A$?   **b)** $A \cap B = A$?

c) $A - B = A$?   **d)** $A \cap B = B \cap A$?

e) $A - B = B - A$?

22. Can you conclude that $A = B$ if A, B, and C are sets such that

a) $A \cup C = B \cup C$?   **b)** $A \cap C = B \cap C$?

23. Let A and B be subsets of a universal set U. Show that $A \subseteq B$ if and only if $\overline{B} \subseteq \overline{A}$.

The **symmetric difference** of A and B, denoted by $A \oplus B$, is the set containing those elements in either A or B, but not in both A and B.

24. Find the symmetric difference of $\{1, 3, 5\}$ and $\{1, 2, 3\}$.

25. Find the symmetric difference of the set of computer science majors at a school and the set of mathematics majors at this school.

26. Draw a Venn diagram for the symmetric difference of the sets A and B.

27. Show that $A \oplus B = (A \cup B) - (A \cap B)$.

28. Show that $A \oplus B = (A - B) \cup (B - A)$.

29. Show that if A is a subset of a universal set U, then

a) $A \oplus A = \emptyset$.   **b)** $A \oplus \emptyset = A$.

c) $A \oplus U = \overline{A}$.   **d)** $A \oplus \overline{A} = U$.

30. Show that if A and B are sets, then

a) $A \oplus B = B \oplus A$.   **b)** $(A \oplus B) \oplus B = A$.

31. What can you say about the sets A and B if $A \oplus B = A$?

*32. Determine whether the symmetric difference is associative; that is, if A, B, and C are sets, does it follow that $A \oplus (B \oplus C) = (A \oplus B) \oplus C$?

*33. Suppose that A, B, and C are sets such that $A \oplus C = B \oplus C$. Must it be the case that $A = B$?

34. If A, B, C, and D are sets, does it follow that $(A \oplus B) \oplus (C \oplus D) = (A \oplus C) \oplus (B \oplus D)$?

35. If A, B, C, and D are sets, does it follow that $(A \oplus B) \oplus (C \oplus D) = (A \oplus D) \oplus (B \oplus C)$?

*36. Show that if A, B, and C are sets, then

$$|A \cup B \cup C| = |A| + |B| + |C| - |A \cap B|$$
$$- |A \cap C| - |B \cap C| + |A \cap B \cap C|.$$

(This is a special case of the inclusion–exclusion principle, which will be studied in Chapter 6.)

*37. Let $A_i = \{1, 2, 3, \ldots, i\}$ for $i = 1, 2, 3, \ldots$. Find

a) $\displaystyle\bigcup_{i=1}^{n} A_i$.   **b)** $\displaystyle\bigcap_{i=1}^{n} A_i$.

*38. Let $A_i = \{\ldots, -2, -1, 0, 1, \ldots, i\}$. Find

a) $\displaystyle\bigcup_{i=1}^{n} A_i$.   **b)** $\displaystyle\bigcap_{i=1}^{n} A_i$.

39. Let A_i be the set of all nonempty bit strings (that is, bit strings of length at least one) of length not exceeding i. Find

a) $\displaystyle\bigcup_{i=1}^{n} A_i.$ **b)** $\displaystyle\bigcap_{i=1}^{n} A_i.$

40. Suppose that the universal set is $U = \{1, 2, 3, 4, 5, 6, 7, 8, 9, 10\}$. Express each of these sets with bit strings where the ith bit in the string is 1 if i is in the set and 0 otherwise.

a) $\{3, 4, 5\}$ **b)** $\{1, 3, 6, 10\}$ **c)** $\{2, 3, 4, 7, 8, 9\}$

41. Using the same universal set as in the last problem, find the set specified by each of these bit strings.

a) 11 1100 1111 **b)** 01 0111 1000
c) 10 0000 0001

42. What subsets of a finite universal set do these bit strings represent?

a) the string with all zeros
b) the string with all ones

43. What is the bit string corresponding to the difference of two sets?

44. What is the bit string corresponding to the symmetric difference of two sets?

45. Show how bitwise operations on bit strings can be used to find these combinations of $A = \{a, b, c, d, e\}$, $B = \{b, c, d, g, p, t, v\}$, $C = \{c, e, i, o, u, x, y, z\}$, and $D = \{d, e, h, i, n, o, t, u, x, y\}$.

a) $A \cup B$ **b)** $A \cap B$
c) $(A \cup D) \cap (B \cup C)$
d) $A \cup B \cup C \cup D$

46. How can the union and intersection of n sets that all are subsets of the universal set U be found using bit strings?

47. The **successor** of the set A is the set $A \cup \{A\}$. Find the successors of the following sets.

a) $\{1, 2, 3\}$ **b)** $\emptyset$
c) $\{\emptyset\}$ **d)** $\{\emptyset, \{\emptyset\}\}$

48. How many elements does the successor of a set with n elements have?

Sometimes the number of times that an element occurs in an unordered collection matters. **Multisets** are unordered collections of elements where an element can occur as a member more than once. The notation $\{m_1 \cdot a_1, m_2 \cdot a_2, \dots, m_r \cdot a_r\}$ denotes the multiset with element a_1 occurring m_1 times, element a_2 occurring m_2 times, and so on. The numbers $m_i, i = 1, 2, \dots, r$ are called the **multiplicities** of the elements $a_i, i = 1, 2, \dots, r$.

Let P and Q be multisets. The **union** of the multisets P and Q is the multiset where the multiplicity of an element is the maximum of its multiplicities in P and Q. The **intersection** of P and Q is the multiset where the multiplicity of an element is the minimum of its multi-

plicities in P and Q. The **difference** of P and Q is the multiset where the multiplicity of an element is the multiplicity of the element in P less its multiplicity in Q unless this difference is negative, in which case the multiplicity is 0. The **sum** of P and Q is the multiset where the multiplicity of an element is the sum of multiplicities in P and Q. The union, intersection, and difference of P and Q are denoted by $P \cup Q, P \cap Q,$ and $P - Q$, respectively (where these operations should not be confused with the analogous operations for sets). The sum of P and Q is denoted by $P + Q$.

49. Let A and B be the multisets $\{3 \cdot a, 2 \cdot b, 1 \cdot c\}$ and $\{2 \cdot a, 3 \cdot b, 4 \cdot d\}$, respectively. Find

a) $A \cup B$. **b)** $A \cap B$. **c)** $A - B$.
d) $B - A$. **e)** $A + B$.

50. Suppose that A is the multiset that has as its elements the types of computer equipment needed by one department of a university where the multiplicities are the number of pieces of each type needed, and B is the analogous multiset for a second department of the university. For instance, A could be the multiset $\{107 \cdot \text{personal computers}, 44 \cdot \text{routers}, 6 \cdot \text{servers}\}$ and B could be the multiset $\{14 \cdot \text{personal computers}, 6 \cdot \text{routers}, 2 \cdot \text{mainframes}\}$.

a) What combination of A and B represents the equipment the university should buy assuming both departments use the same equipment?
b) What combination of A and B represents the equipment that will be used by both departments if both departments use the same equipment?
c) What combination of A and B represents the equipment that the second department uses, but the first department does not, if both departments use the same equipment?
d) What combination of A and B represents the equipment that the university should purchase if the departments do not share equipment?

Fuzzy sets are used in artificial intelligence. Each element in the universal set U has a **degree of membership,** which is a real number between 0 and 1 (including 0 and 1), in a fuzzy set S. The fuzzy set S is denoted by listing the elements with their degrees of membership (elements with 0 degree of membership are not listed). For instance, we write $\{0.6 \text{ Alice}, 0.9 \text{ Brian}, 0.4 \text{ Fred}, 0.1 \text{ Oscar}, 0.5 \text{ Rita}\}$ for the set F (of famous people) to indicate that Alice has a 0.6 degree of membership in F, Brian has a 0.9 degree of membership in F, Fred has a 0.4 degree of membership in F, Oscar has a 0.1 degree of membership in F, and Rita has a 0.5 degree of membership in F (so that Brian is the most famous and Oscar is the least famous of these people). Also suppose that R is the set of rich people with $R = \{0.4 \text{ Alice}, 0.8 \text{ Brian}, 0.2 \text{ Fred}, 0.9 \text{ Oscar}, 0.7 \text{ Rita}\}$.

51. The **complement** of a fuzzy set S is the set $\overline{S}$, with the degree of the membership of an element in $\overline{S}$ equal to 1 minus the degree of membership of this element in S. Find $\overline{F}$ (the fuzzy set of people who are not famous) and $\overline{R}$ (the fuzzy set of people who are not rich).

52. The **union** of two fuzzy sets S and T is the fuzzy set $S \cup T$, where the degree of membership of an element in $S \cup T$ is the maximum of the degrees of member-

ship of this element in S and in T. Find the fuzzy set $F \cup R$ of rich or famous people.

53. The **intersection** of two fuzzy sets S and T is the fuzzy set $S \cap T$, where the degree of membership of an element in $S \cap T$ is the minimum of the degrees of membership of this element in S and in T. Find the fuzzy set $F \cap R$ of rich and famous people.

1.8 Functions

INTRODUCTION

In many instances we assign to each element of a set a particular element of a second set (which may be the same as the first). For example, suppose that each student in a discrete mathematics class is assigned a letter grade from the set $\{A, B, C, D, F\}$. And suppose that the grades are A for Adams, C for Chou, B for Goodfriend, A for Rodriguez, and F for Stevens. This assignment of grades is illustrated in Figure 1.

This assignment is an example of a function. The concept of a function is extremely important in discrete mathematics. Functions are used in the definition of such discrete structures as sequences and strings. Functions are also used to represent how long it takes a computer to solve problems of a given size. Recursive functions, which are functions defined in terms of themselves, are used throughout computer science; they will be studied in Chapter 3. This section reviews the basic concepts involving functions needed in discrete mathematics.

DEFINITION 1

Let A and B be sets. A *function* f from A to B is an assignment of exactly one element of B to each element of A. We write $f(a) = b$ if b is the unique element of B assigned by the function f to the element a of A. If f is a function from A to B, we write $f : A \rightarrow B$.

Functions are specified in many different ways. Sometimes we explicitly state the assignments, as in Figure 1. Often we give a formula, such as $f(x) = x + 1$, to define a function. Other times we use a computer program to specify a function.

Remark: A function $f : A \rightarrow B$ is sometimes defined in terms of a relation from A to B, defined in Section 1.6. We will discuss this approach in Chapter 7.

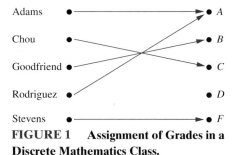

FIGURE 1   Assignment of Grades in a Discrete Mathematics Class.

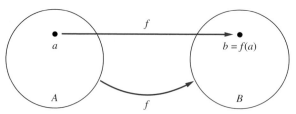

FIGURE 2   The Function f Maps A to B.

DEFINITION 2

> If f is a function from A to B, we say that A is the *domain* of f and B is the *codomain* of f. If $f(a) = b$, we say that b is the *image* of a and a is a *pre-image* of b. The *range* of f is the set of all images of elements of A. Also, if f is a function from A to B, we say that f *maps* A to B.

Figure 2 represents a function f from A to B.

Consider the example that began this section. Let G be the function that assigns a grade to a student in our discrete mathematics class. Note that $G(\text{Adams}) = A$, for instance. The domain of G is the set {Adams, Chou, Goodfriend, Rodriguez, Stevens}, and the codomain is the set $\{A, B, C, D, F\}$. The range of G is the set $\{A, B, C, F\}$, because each grade except D is assigned to some student. Also consider Examples 1 and 2.

EXAMPLE 1

Let f be the function that assigns the last two bits of a bit string of length 2 or greater to that string. Then, the domain of f is the set of all bit strings of length 2 or greater, and both the codomain and range are the set {00, 01, 10, 11}. ◀

EXAMPLE 2

Let $f: \mathbf{Z} \to \mathbf{Z}$ assign the square of an integer to this integer. Then, $f(x) = x^2$, where the domain of f is the set of all integers, the codomain of f can be chosen to be the set of all integers, and the range of f is the set of all nonnegative integers that are perfect squares, namely, $\{0, 1, 4, 9, \dots\}$. ◀

EXAMPLE 3

The domain and codomain of functions are often specified in programming languages. For instance, the Java statement

   int **floor**(float real){. . .}

and the Pascal statement

   function *floor*(x: **real**): **integer**

both state that the domain of the floor function is the set of real numbers and its codomain is the set of integers. ◀

Two real-valued functions with the same domain can be added and multiplied.

DEFINITION 3

> Let f_1 and f_2 be functions from A to $\mathbf{R}$. Then $f_1 + f_2$ and $f_1 f_2$ are also functions from A to $\mathbf{R}$ defined by
> $$(f_1 + f_2)(x) = f_1(x) + f_2(x),$$
> $$(f_1 f_2)(x) = f_1(x) f_2(x).$$

Note that the functions $f_1 + f_2$ and $f_1 f_2$ have been defined by specifying their values at x in terms of the values of f_1 and f_2 at x.

EXAMPLE 4   Let f_1 and f_2 be functions from **R** to **R** such that $f_1(x) = x^2$ and $f_2(x) = x - x^2$. What are the functions $f_1 + f_2$ and $f_1 f_2$?

Solution: From the definition of the sum and product of functions, it follows that

$$(f_1 + f_2)(x) = f_1(x) + f_2(x) = x^2 + (x - x^2) = x$$

and

$$(f_1 f_2)(x) = x^2(x - x^2) = x^3 - x^4.$$ ◄

When f is a function from a set A to a set B, the image of a subset of A can also be defined.

DEFINITION 4

> Let f be a function from the set A to the set B and let S be a subset of A. The *image* of S is the subset of B that consists of the images of the elements of S. We denote the image of S by $f(S)$, so that
>
> $$f(S) = \{f(s) \mid s \in S\}.$$

EXAMPLE 5   Let $A = \{a, b, c, d, e\}$ and $B = \{1, 2, 3, 4\}$ with $f(a) = 2$, $f(b) = 1$, $f(c) = 4$, $f(d) = 1$, and $f(e) = 1$. The image of the subset $S = \{b, c, d\}$ is the set $f(S) = \{1, 4\}$.   ◄

ONE-TO-ONE AND ONTO FUNCTIONS

Some functions have distinct images at distinct members of their domain. These functions are said to be **one-to-one.**

DEFINITION 5

> A function f is said to be *one-to-one*, or *injective*, if and only if $f(x) = f(y)$ implies that $x = y$ for all x and y in the domain of f. A function is said to be an *injection* if it is one-to-one.

Remark: A function f is one-to-one if and only if $f(x) \neq f(y)$ whenever $x \neq y$. This way of expressing that f is one-to-one is obtained by taking the contrapositive of the implication in the definition. Note that we can express that f is one-to-one using quantifiers as $\forall x \forall y (f(x) = f(y) \rightarrow x = y)$ or equivalently $\forall x \forall y (x \neq y \rightarrow f(x) \neq f(y))$, where the universe of discourse is the domain of the function.

We illustrate this concept by giving examples of functions that are one-to-one and other functions that are not one-to-one.

EXAMPLE 6   Determine whether the function f from $\{a, b, c, d\}$ to $\{1, 2, 3, 4, 5\}$ with $f(a) = 4$, $f(b) = 5$, $f(c) = 1$, and $f(d) = 3$ is one-to-one.

Solution: The function f is one-to-one since f takes on different values at the four elements of its domain. This is illustrated in Figure 3. ◀

EXAMPLE 7 Determine whether the function $f(x) = x^2$ from the set of integers to the set of integers is one-to-one.

Solution: The function $f(x) = x^2$ is not one-to-one because, for instance, $f(1) = f(-1) = 1$, but $1 \neq -1$. Note that the function f is one-to-one if its domain is restricted to $\mathbf{Z}^+$. ◀

EXAMPLE 8 Determine whether the function $f(x) = x + 1$ is one-to-one.

Solution: The function $f(x) = x + 1$ is a one-to-one function. To demonstrate this, note that $x + 1 \neq y + 1$ when $x \neq y$. ◀

We now give some conditions that guarantee that a function is one-to-one.

DEFINITION 6
> A function f whose domain and codomain are subsets of the set of real numbers is called *strictly increasing* if $f(x) < f(y)$ whenever $x < y$ and x and y are in the domain of f. Similarly, f is called *strictly decreasing* if $f(x) > f(y)$ whenever $x < y$ and x and y are in the domain of f.

Remark: A function f is strictly increasing if $\forall x \forall y((x < y) \rightarrow (f(x) < f(y)))$ and is strictly decreasing if $\forall x \forall y((x < y) \rightarrow (f(x) > f(y)))$, where the universe of discourse is the domain of f.

From these definitions, we see that a function that is either strictly increasing or strictly decreasing must be one-to-one.

For some functions the range and the codomain are equal. That is, every member of the codomain is the image of some element of the domain. Functions with this property are called **onto** functions.

DEFINITION 7
> A function f from A to B is called *onto,* or *surjective,* if and only if for every element $b \in B$ there is an element $a \in A$ with $f(a) = b$. A function f is called a *surjection* if it is onto.

Remark: A function f is onto if $\forall y \exists x(f(x) = y)$, where the universe of discourse for x is the domain of the function and the universe of discourse for y is the codomain of the function.

We now give examples of onto functions and functions that are not onto.

EXAMPLE 9 Let f be the function from $\{a, b, c, d\}$ to $\{1, 2, 3\}$ defined by $f(a) = 3$, $f(b) = 2$, $f(c) = 1$, and $f(d) = 3$. Is f an onto function?

Solution: Since all three elements of the codomain are images of elements in the domain, we see that f is onto. This is illustrated in Figure 4. Note that if the codomain were $\{1, 2, 3, 4\}$ then f would not be onto. ◀

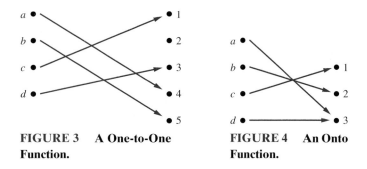

FIGURE 3   A One-to-One Function.   **FIGURE 4   An Onto Function.**

EXAMPLE 10   Is the function $f(x) = x^2$ from the set of integers to the set of integers onto?

Solution: The function f is not onto since there is no integer x with $x^2 = -1$, for instance.   ◄

EXAMPLE 11   Is the function $f(x) = x + 1$ from the set of integers to the set of integers onto?

Solution: This function is onto, since for every integer y there is an integer x such that $f(x) = y$. To see this, note that $f(x) = y$ if and only if $x + 1 = y$, which holds if and only if $x = y - 1$.   ◄

DEFINITION 8   The function f is a *one-to-one correspondence*, or a *bijection*, if it is both one-to-one and onto.

Examples 12 and 13 illustrate the concept of a bijection.

EXAMPLE 12   Let f be the function from $\{a, b, c, d\}$ to $\{1, 2, 3, 4\}$ with $f(a) = 4, f(b) = 2, f(c) = 1$, and $f(d) = 3$. Is f a bijection?

Solution: The function f is one-to-one and onto. It is one-to-one since the function takes on distinct values. It is onto since all four elements of the codomain are images of elements in the domain. Hence, f is a bijection.   ◄

Figure 5 displays four functions where the first is one-to-one but not onto, the second is onto but not one-to-one, the third is both one-to-one and onto, and the fourth is neither one-to-one nor onto. The fifth correspondence in Figure 5 is not a function, since it sends an element to two different elements.

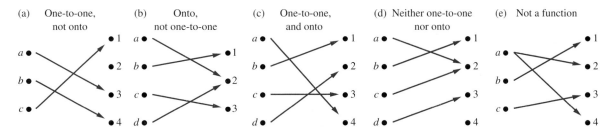

FIGURE 5   Examples of Different Types of Correspondences.

Suppose that f is a function from a set A to itself. If A is finite, then f is one-to-one if and only if it is onto. (This follows from the result in Exercise 64 at the end of this section.) This is not necessarily the case if A is infinite (as will be shown in Section 3.2).

EXAMPLE 13   Let A be a set. The *identity function* on A is the function $\iota_A : A \to A$ where

$$\iota_A(x) = x$$

where $x \in A$. In other words, the identity function ι_A is the function that assigns each element to itself. The function ι_A is one-to-one and onto, so that it is a bijection.   ◄

INVERSE FUNCTIONS AND COMPOSITIONS OF FUNCTIONS

Now consider a one-to-one correspondence f from the set A to the set B. Since f is an onto function, every element of B is the image of some element in A. Furthermore, because f is also a one-to-one function, every element of B is the image of a *unique* element of A. Consequently, we can define a new function from B to A that reverses the correspondence given by f. This leads to Definition 9.

DEFINITION 9

Let f be a one-to-one correspondence from the set A to the set B. The *inverse function* of f is the function that assigns to an element b belonging to B the unique element a in A such that $f(a) = b$. The inverse function of f is denoted by f^{-1}. Hence, $f^{-1}(b) = a$ when $f(a) = b$.

Figure 6 illustrates the concept of an inverse function.

If a function f is not a one-to-one correspondence, we cannot define an inverse function of f. When f is not a one-to-one correspondence, either it is not one-to-one or it is not onto. If f is not one-to-one, some element b in the codomain is the image of more than one element in the domain. If f is not onto, for some element b in the codomain, no element a in the domain exists for which $f(a) = b$. Consequently, if f is not a one-to-one correspondence, we cannot assign to each element b in the codomain a unique element a in the domain such that $f(a) = b$ (because for some b there is either more than one such a or no such a).

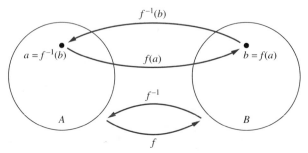

FIGURE 6   The Function f^{-1} Is the Inverse of Function f.

A one-to-one correspondence is called **invertible** since we can define an inverse of this function. A function is **not invertible** if it is not a one-to-one correspondence, since the inverse of such a function does not exist.

EXAMPLE 14 Let f be the function from $\{a, b, c\}$ to $\{1, 2, 3\}$ such that $f(a) = 2$, $f(b) = 3$, and $f(c) = 1$. Is f invertible, and if it is, what is its inverse?

Solution: The function f is invertible since it is a one-to-one correspondence. The inverse function f^{-1} reverses the correspondence given by f, so that $f^{-1}(1) = c$, $f^{-1}(2) = a$, and $f^{-1}(3) = b$. ◄

EXAMPLE 15 Let f be the function from the set of integers to the set of integers such that $f(x) = x+1$. Is f invertible, and if it is, what is its inverse?

Solution: The function f has an inverse since it is a one-to-one correspondence, as we have shown. To reverse the correspondence, suppose that y is the image of x, so that $y = x + 1$. Then $x = y - 1$. This means that $y - 1$ is the unique element of $\mathbf{Z}$ that is sent to y by f. Consequently, $f^{-1}(y) = y - 1$. ◄

EXAMPLE 16 Let f be the function from $\mathbf{Z}$ to $\mathbf{Z}$ with $f(x) = x^2$. Is f invertible?

Solution: Since $f(-1) = f(1) = 1$, f is not one-to-one. If an inverse function were defined, it would have to assign two elements to 1. Hence, f is not invertible. ◄

DEFINITION 10 Let g be a function from the set A to the set B and let f be a function from the set B to the set C. The *composition* of the functions f and g, denoted by $f \circ g$, is defined by

$$(f \circ g)(a) = f(g(a)).$$

In other words, $f \circ g$ is the function that assigns to the element a of A the element assigned by f to $g(a)$. Note that the composition $f \circ g$ cannot be defined unless the range of g is a subset of the domain of f. In Figure 7 the composition of functions is shown.

EXAMPLE 17 Let g be the function from the set $\{a, b, c\}$ to itself such that $g(a) = b$, $g(b) = c$, and $g(c) = a$. Let f be the function from the set $\{a, b, c\}$ to the set $\{1, 2, 3\}$ such that $f(a) = 3$, $f(b) = 2$, and $f(c) = 1$. What is the composition of f and g, and what is the composition of g and f?

Solution: The composition $f \circ g$ is defined by $(f \circ g)(a) = f(g(a)) = f(b) = 2$, $(f \circ g)(b) = f(g(b)) = f(c) = 1$, and $(f \circ g)(c) = f(g(c)) = f(a) = 3$.
    Note that $g \circ f$ is not defined, because the range of f is not a subset of the domain of g. ◄

EXAMPLE 18 Let f and g be the functions from the set of integers to the set of integers defined by $f(x) = 2x + 3$ and $g(x) = 3x + 2$. What is the composition of f and g? What is the composition of g and f?

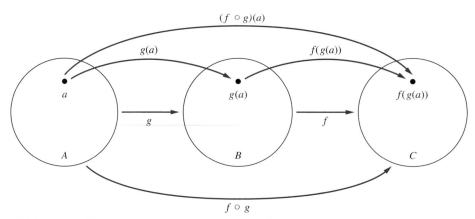

FIGURE 7   **The Composition of the Functions f and g.**

Solution: Both the compositions $f \circ g$ and $g \circ f$ are defined. Moreover,

$$(f \circ g)(x) = f(g(x)) = f(3x + 2) = 2(3x + 2) + 3 = 6x + 7$$

and

$$(g \circ f)(x) = g(f(x)) = g(2x + 3) = 3(2x + 3) + 2 = 6x + 11. \qquad \blacktriangleleft$$

Remark: Note that even though $f \circ g$ and $g \circ f$ are defined for the functions f and g in Example 18, $f \circ g$ and $g \circ f$ are not equal. In other words, the commutative law does not hold for the composition of functions.

When the composition of a function and its inverse is formed, in either order, an identity function is obtained. To see this, suppose that f is a one-to-one correspondence from the set A to the set B. Then the inverse function f^{-1} exists and is a one-to-one correspondence from B to A. The inverse function reverses the correspondence of the original function, so that $f^{-1}(b) = a$ when $f(a) = b$, and $f(a) = b$ when $f^{-1}(b) = a$. Hence,

$$(f^{-1} \circ f)(a) = f^{-1}(f(a)) = f^{-1}(b) = a,$$

and

$$(f \circ f^{-1})(b) = f(f^{-1}(b)) = f(a) = b.$$

Consequently $f^{-1} \circ f = \iota_A$ and $f \circ f^{-1} = \iota_B$, where ι_A and ι_B are the identity functions on the sets A and B, respectively. That is, $(f^{-1})^{-1} = f$.

THE GRAPHS OF FUNCTIONS

We can associate a set of pairs in $A \times B$ to each function from A to B. This set of pairs is called the **graph** of the function and is often displayed pictorially to aid in understanding the behavior of the function.

DEFINITION 11

Let f be a function from the set A to the set B. The *graph* of the function f is the set of ordered pairs $\{(a, b) \mid a \in A \text{ and } f(a) = b\}$.

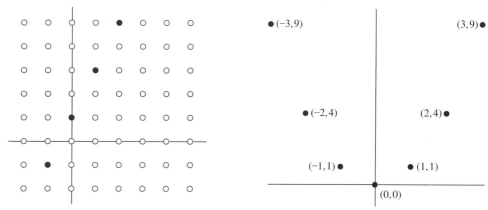

FIGURE 8 **The Graph of the** **FIGURE 9** **The Graph of $f(x) = x^2$**
Function $f(n) = 2n + 1$ from Z to Z. **from Z to Z.**

From the definition, the graph of a function f from A to B is the subset of $A \times B$ containing the ordered pairs with the second entry equal to the element of B assigned by f to the first entry.

EXAMPLE 19 Display the graph of the function $f(n) = 2n + 1$ from the set of integers to the set of integers.

Solution: The graph of f is the set of ordered pairs of the form $(n, 2n + 1)$ where n is an integer. This graph is displayed in Figure 8. ◄

EXAMPLE 20 Display the graph of the function $f(x) = x^2$ from the set of integers to the set of integers.

Solution: The graph of f is the set of ordered pairs of the form $(x, f(x)) = (x, x^2)$ where x is an integer. This graph is displayed in Figure 9. ◄

SOME IMPORTANT FUNCTIONS

Next, we introduce two important functions in discrete mathematics, namely, the floor and ceiling functions. Let x be a real number. The floor function rounds x down to the closest integer less than or equal to x, and the ceiling function rounds x up to the closest integer greater than or equal to x. These functions are often used when objects are counted. They play an important role in the analysis of the number of steps used by procedures to solve problems of a particular size.

DEFINITION 12

The *floor function* assigns to the real number x the largest integer that is less than or equal to x. The value of the floor function at x is denoted by $\lfloor x \rfloor$. The *ceiling function* assigns to the real number x the smallest integer that is greater than or equal to x. The value of the ceiling function at x is denoted by $\lceil x \rceil$.

Remark: The floor function is often also called the *greatest integer function*. It is often denoted by $[x]$.

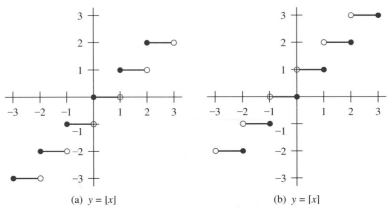

(a) $y = \lfloor x \rfloor$ (b) $y = \lceil x \rceil$

FIGURE 10   **Graphs of the (a) Floor and (b) Ceiling Functions.**

EXAMPLE 21

These are some values of the floor and ceiling functions:

$$\lfloor \tfrac{1}{2} \rfloor = 0, \qquad \lceil \tfrac{1}{2} \rceil = 1, \qquad \lfloor -\tfrac{1}{2} \rfloor = -1, \qquad \lceil -\tfrac{1}{2} \rceil = 0,$$
$$\lfloor 3.1 \rfloor = 3, \qquad \lceil 3.1 \rceil = 4, \qquad \lfloor 7 \rfloor = 7, \qquad \lceil 7 \rceil = 7. \qquad \blacktriangleleft$$

We display the graphs of the floor and ceiling functions in Figure 10. In Figure 10(a) we display the graph of the floor function $\lfloor x \rfloor$. Note that this function has the same value throughout the interval $[n, n + 1)$, namely n, and then it jumps up to $n + 1$ when $x = n + 1$. In Figure 10(b) we display the graph of the ceiling function $\lceil x \rceil$. Note that this function has the same value throughout the interval $(n, n + 1]$, namely $n + 1$, and then jumps to $n + 2$ when x is a little larger than $n + 1$.

The floor and ceiling functions are useful in a wide variety of applications, including those involving data storage and data transmission. Consider Examples 22 and 23, typical of basic calculations done when database and data communications problems are studied.

EXAMPLE 22

Data stored on a computer disk or transmitted over a data network are usually represented as a string of bytes. Each byte is made up of 8 bits. How many bytes are required to encode 100 bits of data?

Solution: To determine the number of bytes needed, we determine the smallest integer that is at least as large as the quotient when 100 is divided by 8, the number of bits in a byte. Consequently, $\lceil 100/8 \rceil = \lceil 12.5 \rceil = 13$ bytes are required. $\qquad \blacktriangleleft$

EXAMPLE 23

In asynchronous transfer mode (ATM) (a communications protocol used on backbone networks), data are organized into cells of 53 bytes. How many ATM cells can be transmitted in 1 minute over a connection that transmits data at the rate of 500 kilobits per second?

Solution: In 1 minute, this connection can transmit $500,000 \cdot 60 = 30,000,000$ bits. Each ATM cell is 53 bytes long, which means that it is $53 \cdot 8 = 424$ bits long. To determine the number of cells that can be transmitted in 1 minute, we determine the largest integer not exceeding the quotient when $30,000,000$ is divided by 424. Consequently, $\lfloor 30,000,000/424 \rfloor = 70,754$ ATM cells can be transmitted in 1 minute over a 500 kilobit per second connection. $\qquad \blacktriangleleft$

> **TABLE 1 Useful Properties of the Floor and Ceiling Functions.** (n is an integer)
>
> ---
>
> (1a) $\lfloor x \rfloor = n$ if and only if $n \le x < n + 1$
> (1b) $\lceil x \rceil = n$ if and only if $n - 1 < x \le n$
> (1c) $\lfloor x \rfloor = n$ if and only if $x - 1 < n \le x$
> (1d) $\lceil x \rceil = n$ if and only if $x \le n < x + 1$
>
> ---
>
> (2) $x - 1 < \lfloor x \rfloor \le x \le \lceil x \rceil < x + 1$
>
> ---
>
> (3a) $\lfloor -x \rfloor = -\lceil x \rceil$
> (3b) $\lceil -x \rceil = -\lfloor x \rfloor$
>
> ---
>
> (4a) $\lfloor x + n \rfloor = \lfloor x \rfloor + n$
> (4b) $\lceil x + n \rceil = \lceil x \rceil + n$

Table 1, with x denoting a real number, displays some simple but important properties of the floor and ceiling functions. Since these functions appear so frequently in discrete mathematics, it is useful to look over these identities. Each property in this table can be established using the definitions of the floor and ceiling functions. Properties (1a), (1b), (1c), and (1d) follow directly from these definitions. For example, (1a) states that $\lfloor x \rfloor = n$ if and only if the integer n is less than or equal to x and $n + 1$ is larger than x. This is precisely what it means for n to be the greatest integer not exceeding x, which is the definition of $\lfloor x \rfloor = n$. Properties (1b), (1c), and (1d) can be established similarly. We will prove property (4a) using a direct proof.

Proof: Suppose that $\lfloor x \rfloor = m$, where m is a positive integer. By property (1a), it follows that $m \le x < m + 1$. Adding n to both sides of this inequality shows that $m + n \le x + n < m + n + 1$. Using property (1a) again, we see that $\lfloor x + n \rfloor = m + n = \lfloor x \rfloor + n$. This completes the proof. We defer establishing the other properties to the exercises. ◁

The floor and ceiling functions enjoy many other useful properties besides those displayed in Table 1. There are also many statements about these functions that may appear to be correct, but actually are not. We will consider statements about the floor and ceiling functions in Examples 24 and 25.

A useful approach for considering statements about the floor function is to let $x = n + \epsilon$, where $n = \lfloor x \rfloor$ is an integer and ϵ, the fractional part of x, satisfies the inequality $0 \le \epsilon < 1$. Similarly, when considering statements about the ceiling function, it is useful to write $x = n - \epsilon$, where $n = \lceil x \rceil$ is an integer and $0 \le \epsilon < 1$.

EXAMPLE 24   Prove that if x is a real number, then $\lfloor 2x \rfloor = \lfloor x \rfloor + \lfloor x + \frac{1}{2} \rfloor$.

Solution: To prove this statement we let $x = n + \epsilon$, where n is a positive integer and $0 \le \epsilon < 1$. There are two cases to consider depending whether ϵ is less than $\frac{1}{2}$. (The reason we choose these two cases will be made clear in the proof.)

We first consider the case when $0 \le \epsilon < \frac{1}{2}$. In this case, $2x = 2n + 2\epsilon$ and $\lfloor 2x \rfloor = 2n$ because $0 \le 2\epsilon < 1$. Similarly, $x + \frac{1}{2} = n + (\frac{1}{2} + \epsilon)$, so that $\lfloor x + \frac{1}{2} \rfloor = n$, because $0 < \frac{1}{2} + \epsilon < 1$. Consequently, $\lfloor 2x \rfloor = 2n$ and $\lfloor x \rfloor + \lfloor x + \frac{1}{2} \rfloor = n + n = 2n$.

Next, we consider the case when $\frac{1}{2} \le \epsilon < 1$. In this case, $2x = 2n + 2\epsilon = (2n + 1) + (2\epsilon - 1)$. Since $0 \le 2\epsilon - 1 < 1$, it follows that $\lfloor 2x \rfloor = 2n + 1$. Because

$\lfloor x + \frac{1}{2} \rfloor = \lfloor n + (\frac{1}{2} + \epsilon) \rfloor = \lfloor n + 1 + (\epsilon - \frac{1}{2}) \rfloor$ and $0 \leq \epsilon - \frac{1}{2} < 1$, it follows that $\lfloor x + \frac{1}{2} \rfloor = n + 1$. Consequently, $\lfloor 2x \rfloor = 2n + 1$ and $\lfloor x \rfloor + \lfloor x + \frac{1}{2} \rfloor = n + (n + 1) = 2n + 1$. This concludes the proof. ◀

EXAMPLE 25   Prove or disprove that $\lceil x + y \rceil = \lceil x \rceil + \lceil y \rceil$ for all real numbers x and y.

Solution: Although this statement may appear reasonable, it is false. A counterexample is supplied by $x = \frac{1}{2}$ and $y = \frac{1}{2}$. With these values we find that $\lceil x + y \rceil = \lceil \frac{1}{2} + \frac{1}{2} \rceil = \lceil 1 \rceil = 1$, but $\lceil x \rceil + \lceil y \rceil = \lceil \frac{1}{2} \rceil + \lceil \frac{1}{2} \rceil = 1 + 1 = 2$. ◀

There are certain types of functions that will be used throughout the text. These include polynomial, logarithmic, and exponential functions. A brief review of the properties of these functions needed in this text is given in Appendix 1. In this book the notation $\log x$ will be used to denote the logarithm to the base 2 of x, since 2 is the base that we will usually use for logarithms. We will denote logarithms to the base b, where b is any real number greater than 1, by $\log_b x$, and the natural logarithm by $\ln x$.

Another function we will use throughout this text is the **factorial function** $f : \mathbf{N} \rightarrow \mathbf{Z}^+$, denoted by $f(n) = n!$. The value of $f(n) = n!$ is the product of the first n positive integers, so that $f(n) = 1 \cdot 2 \cdots (n - 1) \cdot n$ [and $f(0) = 0! = 1$].

EXAMPLE 26   We have $f(6) = 6! = 1 \cdot 2 \cdot 3 \cdot 4 \cdot 5 \cdot 6 = 720$. ◀

Exercises

1. Why is f not a function from $\mathbf{R}$ to $\mathbf{R}$ if
  a) $f(x) = 1/x$?
  b) $f(x) = \sqrt{x}$?
  c) $f(x) = \pm\sqrt{(x^2 + 1)}$?

2. Determine whether f is a function from $\mathbf{Z}$ to $\mathbf{R}$ if
  a) $f(n) = \pm n$.
  b) $f(n) = \sqrt{n^2 + 1}$.
  c) $f(n) = 1/(n^2 - 4)$.

3. Determine whether f is a function from the set of all bit strings to the set of integers if
  a) $f(S)$ is the position of a 0 bit in S.
  b) $f(S)$ is the number of 1 bits in S.
  c) $f(S)$ is the smallest integer i such that the ith bit of S is 1 and $f(S) = 0$ when S is the empty string, the string with no bits.

4. Find the domain and range of these functions.
  a) the function that assigns to each nonnegative integer its last digit
  b) the function that assigns the next largest integer to a positive integer
  c) the function that assigns to a bit string the number of one bits in the string
  d) the function that assigns to a bit string the number of bits in the string

5. Find the domain and range of these functions.
  a) the function that assigns to each bit string the difference between the number of ones and the number of zeros
  b) the function that assigns to each bit string twice the number of zeros in that string
  c) the function that assigns the number of bits left over when a bit string is split into bytes (which are blocks of 8 bits)
  d) the function that assigns to each positive integer the largest perfect square not exceeding this integer

6. Find the domain and range of these functions.
  a) the function that assigns to each pair of positive integers the first integer of the pair
  b) the function that assigns to each positive integer its largest decimal digit
  c) the function that assigns to a bit string the difference between the number of ones and the number of zeros in the string
  d) the function that assigns to each positive integer the largest integer not exceeding the square root of the integer
  e) the function that assigns to a bit string the longest string of ones in the string

7. Find the domain and range of these functions.

   a) the function that assigns to each pair of positive integers the maximum of these two integers

   b) the function that assigns to each positive integer the number of the digits $0, 1, 2, 3, 4, 5, 6, 7, 8, 9$ that do not appear as decimal digits of the integer

   c) the function that assigns to a bit string the number of times the block 11 appears

   d) the function that assigns to a bit string the numerical position of the first 1 in the string and that assigns the value 0 to a bit string consisting of all 0s

8. Find these values.

   a) $\lfloor 1.1 \rfloor$ **b)** $\lceil 1.1 \rceil$ **c)** $\lfloor -0.1 \rfloor$
   d) $\lceil -0.1 \rceil$ **e)** $\lceil 2.99 \rceil$ **f)** $\lceil -2.99 \rceil$
   g) $\lfloor \frac{1}{2} + \lceil \frac{1}{2} \rceil \rfloor$ **h)** $\lceil \lfloor \frac{1}{2} \rfloor + \lceil \frac{1}{2} \rceil + \frac{1}{2} \rceil$

9. Find these values.

   a) $\lceil \frac{3}{4} \rceil$ **b)** $\lfloor \frac{7}{8} \rfloor$ **c)** $\lceil -\frac{3}{4} \rceil$ **d)** $\lfloor -\frac{7}{8} \rfloor$
   e) $\lceil 3 \rceil$ **f)** $\lfloor -1 \rfloor$ **g)** $\lfloor \frac{1}{2} + \lceil \frac{3}{2} \rceil \rfloor$ **h)** $\lfloor \frac{1}{2} \cdot \lfloor \frac{5}{2} \rfloor \rfloor$

10. Determine whether each of these functions from $\{a, b, c, d\}$ to itself is one-to-one.

   a) $f(a) = b, f(b) = a, f(c) = c, f(d) = d$
   b) $f(a) = b, f(b) = b, f(c) = d, f(d) = c$
   c) $f(a) = d, f(b) = b, f(c) = c, f(d) = d$

11. Which functions in Exercise 10 are onto?

12. Determine whether each of these functions from **Z** to **Z** is one-to-one.

   a) $f(n) = n - 1$ **b)** $f(n) = n^2 + 1$
   c) $f(n) = n^3$ **d)** $f(n) = \lceil n/2 \rceil$

13. Which functions in Exercise 12 are onto?

14. Determine whether $f: \mathbf{Z} \times \mathbf{Z} \to \mathbf{Z}$ is onto if

   a) $f(m, n) = 2m - n$.
   b) $f(m, n) = m^2 - n^2$.
   c) $f(m, n) = m + n + 1$.
   d) $f(m, n) = |m| - |n|$.
   e) $f(m, n) = m^2 - 4$.

15. Determine whether the function $f: \mathbf{Z} \times \mathbf{Z} \to \mathbf{Z}$ is onto if

   a) $f(m, n) = m + n$.
   b) $f(m, n) = m^2 + n^2$.
   c) $f(m, n) = m$.
   d) $f(m, n) = |n|$.
   e) $f(m, n) = m - n$.

16. Give an example of a function from **N** to **N** that is

   a) one-to-one but not onto.
   b) onto but not one-to-one.
   c) both onto and one-to-one (but different from the identity function).
   d) neither one-to-one nor onto.

17. Give an explicit formula for a function from the set of integers to the set of positive integers that is

   a) one-to-one, but not onto.
   b) onto, but not one-to-one.

   c) one-to-one and onto.
   d) neither one-to-one nor onto.

18. Determine whether each of these functions is a bijection from **R** to **R**.

   a) $f(x) = -3x + 4$
   b) $f(x) = -3x^2 + 7$
   c) $f(x) = (x + 1)/(x + 2)$
   d) $f(x) = x^5 + 1$

19. Determine whether each of these functions is a bijection from **R** to **R**.

   a) $f(x) = 2x + 1$
   b) $f(x) = x^2 + 1$
   c) $f(x) = x^3$
   d) $f(x) = (x^2 + 1)/(x^2 + 2)$

20. Let $f: \mathbf{R} \to \mathbf{R}$ and let $f(x) > 0$ for all $x \in \mathbf{R}$. Show that $f(x)$ is strictly increasing if and only if the function $g(x) = 1/f(x)$ is strictly decreasing.

21. Let $f: \mathbf{R} \to \mathbf{R}$ and let $f(x) > 0$. Show that $f(x)$ is strictly decreasing if and only if the function $g(x) = 1/f(x)$ is strictly increasing.

22. Let $S = \{-1, 0, 2, 4, 7\}$. Find $f(S)$ if

   a) $f(x) = 1$. **b)** $f(x) = 2x + 1$.
   c) $f(x) = \lceil x/5 \rceil$. **d)** $f(x) = \lfloor (x^2 + 1)/3 \rfloor$.

23. Let $f(x) = \lfloor x^2/3 \rfloor$. Find $f(S)$ if

   a) $S = \{-2, -1, 0, 1, 2, 3\}$.
   b) $S = \{0, 1, 2, 3, 4, 5\}$.
   c) $S = \{1, 5, 7, 11\}$.
   d) $S = \{2, 6, 10, 14\}$.

24. Let $f(x) = 2x$. What is

   a) $f(\mathbf{Z})$? **b)** $f(\mathbf{N})$? **c)** $f(\mathbf{R})$?

25. Suppose that g is a function from A to B and f is a function from B to C.

   a) Show that if both f and g are one-to-one functions, then $f \circ g$ is also one-to-one.
   b) Show that if both f and g are onto functions, then $f \circ g$ is also onto.

*26. If f and $f \circ g$ are one-to-one, does it follow that g is one-to-one? Justify your answer.

*27. If f and $f \circ g$ are onto, does it follow that g is onto? Justify your answer.

28. Find $f \circ g$ and $g \circ f$ where $f(x) = x^2 + 1$ and $g(x) = x + 2$ are functions from **R** to **R**.

29. Find $f + g$ and fg for the functions f and g given in Exercise 28.

30. Let $f(x) = ax + b$ and $g(x) = cx + d$ where $a, b, c,$ and d are constants. Determine for which constants $a, b, c,$ and d it is true that $f \circ g = g \circ f$.

31. Show that the function $f(x) = ax + b$ from **R** to **R** is invertible, where a and b are constants, with $a \neq 0$, and find the inverse of f.

32. Let f be a function from the set A to the set B. Let S and T be subsets of A. Show that

   a) $f(S \cup T) = f(S) \cup f(T)$.
   b) $f(S \cap T) \subseteq f(S) \cap f(T)$.

33. Give an example to show that the inclusion in part (b) in Exercise 32 may be proper.

Let f be a function from the set A to the set B. Let S be a subset of B. We define the **inverse image** of S to be the subset of A containing all pre-images of all elements of S. We denote the inverse image of S by $f^{-1}(S)$, so that $f^{-1}(S) = \{a \in A \mid f(a) \in S\}$.

34. Let f be the function from **R** to **R** defined by $f(x) = x^2$. Find

  a) $f^{-1}(\{1\})$.   **b)** $f^{-1}(\{x \mid 0 < x < 1\})$.
  c) $f^{-1}(\{x \mid x > 4\})$.

35. Let $g(x) = \lfloor x \rfloor$. Find

  a) $g^{-1}(\{0\})$.
  b) $g^{-1}(\{-1, 0, 1\})$.
  c) $g^{-1}(\{x \mid 0 < x < 1\})$.

36. Let f be a function from A to B. Let S and T be subsets of B. Show that

  a) $f^{-1}(S \cup T) = f^{-1}(S) \cup f^{-1}(T)$.
  b) $f^{-1}(S \cap T) = f^{-1}(S) \cap f^{-1}(T)$.

37. Let f be a function from A to B. Let S be a subset of B. Show that $f^{-1}(\overline{S}) = \overline{f^{-1}(S)}$.

38. Show that $\lfloor x + \frac{1}{2} \rfloor$ is the closest integer to the number x, except when x is midway between two integers, when it is the larger of these two integers.

39. Show that $\lceil x - \frac{1}{2} \rceil$ is the closest integer to the number x, except when x is midway between two integers, when it is the smaller of these two integers.

40. Show that if x is a real number, then $\lceil x \rceil - \lfloor x \rfloor = 1$ if x is not an integer and $\lceil x \rceil - \lfloor x \rfloor = 0$ if x is an integer.

41. Show that if x is a real number, then $x - 1 < \lfloor x \rfloor \le x \le \lceil x \rceil < x + 1$.

42. Show that if x is a real number and m is an integer, then $\lceil x + m \rceil = \lceil x \rceil + m$.

43. Show that if x is a real number and n is an integer, then

  a) $x < n$ if and only if $\lfloor x \rfloor < n$.
  b) $n < x$ if and only if $n < \lceil x \rceil$.

44. Show that if x is a real number and n is an integer, then

  a) $x \le n$ if and only if $\lceil x \rceil \le n$.
  b) $n \le x$ if and only if $n \le \lfloor x \rfloor$.

45. Prove that if n is an integer, then $\lfloor n/2 \rfloor = n/2$ if n is even and $(n-1)/2$ if n is odd.

46. Prove that if x is a real number, then $\lfloor -x \rfloor = -\lceil x \rceil$ and $\lceil -x \rceil = -\lfloor x \rfloor$.

47. The function INT is found on some calculators, where $\text{INT}(x) = \lfloor x \rfloor$ when x is a nonnegative real number and $\text{INT}(x) = \lceil x \rceil$ when x is a negative real number. Show that this INT function satisfies the identity $\text{INT}(-x) = -\text{INT}(x)$.

48. Let a and b be real numbers with $a < b$. Use the floor and/or ceiling functions to express the number of integers n that satisfy the inequality $a \le n \le b$.

49. Let a and b be real numbers with $a < b$. Use the floor and/or ceiling functions to express the number of integers n that satisfy the inequality $a < n < b$.

50. How many bytes are required to encode n bits of data where n equals

  a) 4?   **b)** 10?   **c)** 500?   **d)** 3000?

51. How many bytes are required to encode n bits of data where n equals

  a) 7?   **b)** 17?   **c)** 1001?   **d)** 28,800?

52. How many ATM cells (described in Example 23) can be transmitted in 10 seconds over a link operating at the following rates?

  a) 128 kilobits per second (1 kilobit = 1000 bits)
  b) 300 kilobits per second
  c) 1 megabit per second (1 megabit = 1,000,000 bits)

53. Data are transmitted over a particular Ethernet network in blocks of 1500 octets (blocks of 8 bits). How many blocks are required to transmit the following amounts of data over this Ethernet network? (Note that a byte is a synonym for an octet, a kilobyte is 1000 bytes, and a megabyte is 1,000,000 bytes.)

  a) 150 kilobytes of data
  b) 384 kilobytes of data
  c) 1.544 megabytes of data
  d) 45.3 megabytes of data

54. Draw the graph of the function $f(n) = 1 - n^2$ from **Z** to **Z**.

55. Draw the graph of the function $f(x) = \lfloor 2x \rfloor$ from **R** to **R**.

56. Draw the graph of the function $f(x) = \lfloor x/2 \rfloor$ from **R** to **R**.

57. Draw the graph of the function $f(x) = \lfloor x \rfloor + \lfloor x/2 \rfloor$ from **R** to **R**.

58. Draw the graph of the function $f(x) = \lceil x \rceil + \lfloor x/2 \rfloor$ from **R** to **R**.

59. Draw graphs of each of these functions.

  a) $f(x) = \lfloor x + \frac{1}{2} \rfloor$ **b)** $f(x) = \lfloor 2x + 1 \rfloor$
  c) $f(x) = \lceil x/3 \rceil$ **d)** $f(x) = \lceil 1/x \rceil$
  e) $f(x) = \lceil x - 2 \rceil + \lfloor x + 2 \rfloor$
  f) $f(x) = \lfloor 2x \rfloor \lceil x/2 \rceil$
  g) $f(x) = \lceil \lfloor x - \frac{1}{2} \rfloor + \frac{1}{2} \rceil$

60. Draw graphs of each of these functions.

  a) $f(x) = \lceil 3x - 2 \rceil$ **b)** $f(x) = \lceil 0.2x \rceil$
  c) $f(x) = \lfloor -1/x \rfloor$ **d)** $f(x) = \lfloor x^2 \rfloor$
  e) $f(x) = \lceil x/2 \rceil \lfloor x/2 \rfloor$
  f) $f(x) = \lfloor x/2 \rfloor + \lceil x/2 \rceil$
  g) $f(x) = \lfloor 2 \lceil x/2 \rceil + \frac{1}{2} \rfloor$

61. Find the inverse function of $f(x) = x^3 + 1$.

62. Suppose that f is an invertible function from Y to Z and g is an invertible function from X to Y. Show that the inverse of the composition $f \circ g$ is given by $(f \circ g)^{-1} = g^{-1} \circ f^{-1}$.

63. Let S be a subset of a universal set U. The **characteristic function** f_S of S is the function from U to the set $\{0, 1\}$ such that $f_S(x) = 1$ if x belongs to S and $f_S(x) = 0$ if x does not belong to S. Let A and B be sets. Show that for all x

a) $f_{A \cap B}(x) = f_A(x) \cdot f_B(x)$
b) $f_{A \cup B}(x) = f_A(x) + f_B(x) - f_A(x) \cdot f_B(x)$
c) $f_{\overline{A}}(x) = 1 - f_A(x)$
d) $f_{A \oplus B}(x) = f_A(x) + f_B(x) - 2f_A(x)f_B(x)$

64. Suppose that f is a function from A to B, where A and B are finite sets with $|A| = |B|$. Show that f is one-to-one if and only if it is onto.

65. Prove or disprove each of these statements about the floor and ceiling functions.

a) $\lceil \lfloor x \rfloor \rceil = \lfloor x \rfloor$ for all real numbers x.
b) $\lfloor 2x \rfloor = 2\lfloor x \rfloor$ whenever x is a real number.
c) $\lceil x \rceil + \lceil y \rceil - \lceil x + y \rceil = 0$ or 1 whenever x and y are real numbers.
d) $\lceil xy \rceil = \lceil x \rceil \lceil y \rceil$ for all real numbers x and y.
e) $\left\lceil \dfrac{x}{2} \right\rceil = \left\lfloor \dfrac{x + 1}{2} \right\rfloor$ for all real numbers x.

66. Prove or disprove each of these statements about the floor and ceiling functions.

a) $\lfloor \lceil x \rceil \rfloor = \lceil x \rceil$ for all real numbers x.
b) $\lfloor x + y \rfloor = \lfloor x \rfloor + \lfloor y \rfloor$ for all real numbers x and y.
c) $\lceil \lceil x/2 \rceil /2 \rceil = \lceil x/4 \rceil$ for all real numbers x.
d) $\lfloor \sqrt{\lceil x \rceil} \rfloor = \lfloor \sqrt{x} \rfloor$ for all real numbers x.
e) $\lfloor x \rfloor + \lfloor y \rfloor + \lfloor x + y \rfloor \le \lfloor 2x \rfloor + \lfloor 2y \rfloor$ for all real numbers x and y.

67. Prove that if x is a positive real number, then

a) $\lfloor \sqrt{\lfloor x \rfloor} \rfloor = \lfloor \sqrt{x} \rfloor$.
b) $\lceil \sqrt{\lceil x \rceil} \rceil = \lceil \sqrt{x} \rceil$.

68. Let x be a real number. Show that $\lfloor 3x \rfloor = \lfloor x \rfloor + \lfloor x + \frac{1}{3} \rfloor + \lfloor x + \frac{2}{3} \rfloor$.

A program designed to evaluate a function may not produce the correct value of the function for all elements in the domain of this function. For example, a program may not produce a correct value because evaluating the function may lead to an infinite loop or an overflow.

To study such situations, we use the concept of a partial function. A **partial function** f from a set A to a set B is an assignment to each element a in a subset of A, called the **domain of definition** of f, of a unique element b in B. The sets A and B are called the **domain** and **codomain** of f, respectively. We say that f is **undefined** for elements in A that are not in the domain of definition of f. We write $f : A \to B$ to denote that f is a partial function from A to B. (This is the same notation as is used for functions. The context in which the notation is used determines whether f is a partial function or a total function.) When the domain of definition of f equals A, we say that f is a **total function.**

69. For each of these partial functions, determine its domain, codomain, domain of definition, and the set of values for which it is undefined. Also, determine whether it is a total function.

a) $f : \mathbf{Z} \to \mathbf{R}$, $f(n) = 1/n$
b) $f : \mathbf{Z} \to \mathbf{Z}$, $f(n) = \lceil n/2 \rceil$
c) $f : \mathbf{Z} \times \mathbf{Z} \to \mathbf{Q}$, $f(m, n) = m/n$
d) $f : \mathbf{Z} \times \mathbf{Z} \to \mathbf{Z}$, $f(m, n) = mn$
e) $f : \mathbf{Z} \times \mathbf{Z} \to \mathbf{Z}$, $f(m, n) = m - n$ if $m > n$

70. a) Show that a partial function from A to B can be viewed as a function f^* from A to $B \cup \{u\}$ where u is not an element of B and

$$f^*(a) = \begin{cases} f(a) & \text{if } a \text{ belongs to the domain} \\ & \text{of definition of } f \\ u & \text{if } f \text{ is undefined at } a. \end{cases}$$

b) Using the construction in (a), find the function f^* corresponding to each partial function in Exercise 69.

*71. Show that the function $f : \mathbf{Z}^+ \times \mathbf{Z}^+ \to \mathbf{Z}^+$ with $f(m, n) = (m + n - 2)(m + n - 1)/2 + m$ is one-to-one and onto.

Key Terms and Results

LOGIC (SECTIONS 1–4):

TERMS

proposition: a statement that is true or false

truth value: true or false

$\neg p$ (negation of p): the proposition with truth value opposite to the truth value of p

logical operators: operators used to combine propositions

compound proposition: a proposition constructed by combining propositions using logical operators

truth table: a table displaying the truth values of propositions

$p \vee q$ (disjunction of p and q): the proposition that is true unless both p and q are false

$p \wedge q$ (conjunction of p and q): the proposition that is true only when both p and q are true

$p \oplus q$ (exclusive or of p and q): the proposition that is true when exactly one of p and q is true

$p \rightarrow q$ (p implies q): the proposition that is false only when p is true and q is false

converse of $p \rightarrow q$: the implication $q \rightarrow p$

contrapositive of $p \rightarrow q$: the implication $\neg q \rightarrow \neg p$

inverse of $p \rightarrow q$: the implication $\neg p \rightarrow \neg q$

$p \leftrightarrow q$ (biconditional): the proposition that is true only when p and q have the same truth value

bit: either a 0 or a 1

Boolean variable: a variable that has a value of 0 or 1

bit operation: an operation on a bit or bits

bit string: a list of bits

bitwise operations: operations on bit strings that operate on each bit in one string and the corresponding bit in the other string

tautology: a compound proposition that is always true

contradiction: a compound proposition that is always false

contingency: a compound proposition that is sometimes true and sometimes false

consistent compound propositions: compound propositions for which there is an assignment of truth values to the variables that makes all these propositions true

METHODS OF PROOF (SECTION 5):

TERMS

theorem: a mathematical assertion that can be shown to be true

conjecture: a mathematical assertion whose truth value is unknown

proof: a demonstration that a theorem is true

lemma: a simple theorem used to prove other theorems

corollary: a proposition that can be proved as a consequence of a theorem that has just been proved

rule of inference: an implication that is a tautology that is then used to draw conclusions from known assertions

fallacy: an implication that is a contingency that is often incorrectly used to draw conclusions

circular reasoning or **begging the question:** reasoning where one or more steps are based on the truth of the statement being proved

vacuous proof: a proof that the implication $p \rightarrow q$ is true based on the fact that p is false

trivial proof: a proof that the implication $p \rightarrow q$ is true based on the fact that q is true

direct proof: a proof that the implication $p \rightarrow q$ is true

SETS (SECTIONS 6–7):

TERMS

set: a collection of distinct objects

axiom: a basic assumption of a theory

paradox: a logical inconsistency

element, member of a set: an object in a set

logical equivalence: compound propositions are logically equivalent if they always have the same truth values

propositional function: the combination of a variable and a predicate

universe of discourse: the domain of the variable in a propositional function

$\exists x \, P(x)$ (existential quantification of $P(x)$): the proposition that is true if and only if there exists an x in the universe of discourse such that $P(x)$ is true

$\forall x \, P(x)$ (universal quantification of $P(x)$): the proposition that is true if and only if $P(x)$ is true for all x in the universe of discourse

free variable: a variable not bound in a propositional function

RESULTS

The logical equivalences given in Tables 5, 6, and 7 in Section 1.2.

that proceeds by showing that q must be true when p is true

indirect proof: a proof that the implication $p \rightarrow q$ is true that proceeds by showing that p must be false when q is false

proof by contradiction: a proof that a proposition p is true based on the truth of the implication $\neg p \rightarrow q$ where q is a contradiction

proof by cases: a proof of an implication where the hypothesis is a disjunction of propositions that shows that each hypothesis separately implies the conclusion

counterexample: an element x such that $P(x)$ is false

constructive existence proof: a proof that an element with a specified property exists that explicitly finds such an element

nonconstructive existence proof: a proof that an element with a specified property exists that does not explicitly find such an element

rational number: a number that can be expressed as the ratio of two integers p and q such that $q \neq 0$

uniqueness proof: a proof that there is exactly one element satisfying a specified property

$\emptyset$ (empty set, null set): the set with no members

universal set: the set containing all objects under consideration

Venn diagram: a graphical representation of a set or sets

$S = T$ (set equality): S and T have the same elements

$S \subseteq T$ (**S is a subset of T**): every element of S is also an element of T

$S \subset T$ (**S is a proper subset of T**): S is a subset of T and $S \neq T$

finite set: a set with n elements where n is a nonnegative integer

infinite set: a set that is not finite

$|S|$ (**the cardinality of S**): the number of elements in S

$P(S)$ (**the power set of S**): the set of all subsets of S

$A \cup B$ (**the union of A and B**): the set containing those elements that are in at least one of A and B

$A \cap B$ (**the intersection of A and B**): the set containing

those elements that are in both A and B.

$A - B$ (**the difference of A and B**): the set containing those elements that are in A but not in B

$\overline{A}$ (**the complement of A**): the set of elements in the universal set that are not in A

$A \oplus B$ (**the symmetric difference of A and B**): the set containing those elements in exactly one of A and B

membership table: a table displaying the membership of elements in sets

RESULTS

The set identities given in Table 1 in Section 1.7

FUNCTIONS (SECTION 8):

TERMS

function from A to B: an assignment of exactly one element of B to each element of A

domain of f: the set A where f is a function from A to B

codomain of f: the set B where f is a function from A to B

b is the image of a under f: $b = f(a)$

a is a pre-image of b under f: $f(a) = b$

range of f: the set of images of f

onto function, surjection: a function from A to B such that every element of B is the image of some element in A

one-to-one function, injection: a function such that the images of elements in its domain are all different

one-to-one correspondence, bijection: a function that is both one-to-one and onto

inverse of f: the function that reverses the correspondence given by f (when f is a bijection)

$f \circ g$ (composition of f and g): the function that assigns $f(g(x))$ to x

$\lfloor x \rfloor$ (**floor function**): the largest integer not exceeding x

$\lceil x \rceil$ (**ceiling function**): the smallest integer greater than or equal to x

Review Questions

1. **a)** Define the negation of a proposition.
   b) What is the negation of "This is a boring course"?
2. **a)** Define (using truth tables) the disjunction, conjunction, exclusive or, conditional, and biconditional of the propositions p and q.
   b) What are the disjunction, conjunction, exclusive or, implication, and biconditional of the propositions "I'll go to the movies tonight" and "I'll finish my discrete mathematics homework"?
3. **a)** Describe at least five different ways to write the implication $p \rightarrow q$ in English.
   b) Define the converse and contrapositive of an implication.
   c) State the converse and the contrapositive of the implication "If it is sunny tomorrow, then I will go for a walk in the woods."
4. **a)** What does it mean for two propositions to be logically equivalent?
   b) Describe the different ways to show that two compound propositions are logically equivalent.

   c) Show in at least two different ways that the compound propositions $\neg p \vee (r \rightarrow \neg q)$ and $\neg p \vee \neg q \vee \neg r$ are equivalent.

5. (*Depends on the Exercise Set in Section 1.2*)

   a) Given a truth table, explain how to use disjunctive normal form to construct a compound proposition with this truth table.
   b) Explain why part (a) shows that the operators $\wedge$, $\vee$, and $\neg$ are functionally complete.
   c) Is there an operator such that the set containing just this operator is functionally complete?

6. What are the universal and existential quantifications of a predicate $P(x)$? What are their negations?

7. **a)** What is the difference between the quantification $\exists x \forall y P(x, y)$ and $\forall y \exists x P(x, y)$, where $P(x, y)$ is a predicate?
   b) Give an example of a predicate $P(x, y)$ such that $\exists x \forall y P(x, y)$ and $\forall y \exists x P(x, y)$ have different truth values.

8. a) Describe what is meant by a direct proof, an indirect proof, and a proof by contradiction of an implication $p \rightarrow q$.

b) Give a direct proof, an indirect proof, and a proof by contradiction of the statement: "If n is even, then $n + 4$ is even."

9. a) Describe one way to prove the biconditional $p \leftrightarrow q$.

b) Prove the statement: "The integer $3n + 2$ is odd if and only if the integer $9n + 5$ is even, where n is an integer."

10. To prove that the statements p_1, p_2, p_3, and p_4 are equivalent, is it sufficient to show that the implications $p_4 \rightarrow p_2$, $p_3 \rightarrow p_1$, and $p_1 \rightarrow p_2$ are valid? If not, provide another set of implications that can be used to show that the four statements are equivalent.

11. a) Suppose that a statement of the form $\forall x\, P(x)$ is false. How can this be proved?

b) Show that the statement "For every positive integer n, $n^2 \geq 2n$" is false.

12. What is the difference between a constructive and nonconstructive existence proof? Give an example of each.

13. What are the elements of a proof that there is a unique element x such that $P(x)$, where $P(x)$ is a propositional function?

14. a) Define the union, intersection, difference, and symmetric difference of two sets.

b) What are the union, intersection, difference, and symmetric difference of the set of positive integers and the set of odd integers?

15. a) Define what it means for two sets to be equal.

b) Describe the ways to show that two sets are equal.

c) Show in at least two different ways that the sets $A - (B \cap C)$ and $(A - B) \cup (A - C)$ are equal.

16. Explain the relationship between logical equivalences and set identities.

17. a) Define $|S|$, the cardinality of the set S.

b) Give a formula for $|A \cup B|$ where A and B are sets.

18. a) Define the power set of a set S.

b) When is the empty set in the power set of a set S?

c) How many elements does the power set of a set S with n elements have?

19. a) Define the domain, codomain, and the range of a function.

b) Let $f(n)$ be the function from the set of integers to the set of integers such that $f(n) = n^2 + 1$. What are the domain, codomain, and range of this function?

20. a) Define what it means for a function from the set of positive integers to the set of positive integers to be one-to-one.

b) Define what it means for a function from the set of positive integers to the set of positive integers to be onto.

c) Give an example of a function from the set of positive integers to the set of positive integers that is both one-to-one and onto.

d) Give an example of a function from the set of positive integers to the set of positive integers that is one-to-one but not onto.

e) Give an example of a function from the set of positive integers to the set of positive integers that is not one-to-one but is onto.

f) Give an example of a function from the set of positive integers to the set of positive integers that is neither one-to-one nor onto.

21. a) Define the inverse of a function.

b) When does a function have an inverse?

c) Does the function $f(n) = 10 - n$ from the set of integers to the set of integers have an inverse? If so, what is it?

22. a) Define the floor and ceiling functions from the set of real numbers to the set of integers.

b) For which real numbers x is it true that $\lfloor x \rfloor = \lceil x \rceil$?

Supplementary Exercises

1. Let p be the proposition "I will do every exercise in this book" and q be the proposition "I will get an 'A' in this course." Express each of these as a combination of p and q.

a) I will get an 'A' in this course only if I do every exercise in this book.

b) I will get an 'A' in this course and I will do every exercise in this book.

c) Either I will not get an 'A' in this course or I will not do every exercise in this book.

d) For me to get an 'A' in this course it is necessary and sufficient that I do every exercise in this book.

2. Find the truth table of the compound proposition $(p \vee q) \rightarrow (p \wedge \neg r)$.

3. Show that these propositions are tautologies.

a) $(\neg q \wedge (p \rightarrow q)) \rightarrow \neg p$

b) $((p \vee q) \wedge \neg p) \rightarrow q$

4. Give the converse, the contrapositive, and the inverse of these implications.

a) If it rains today, then I will drive to work.

b) If $|x| = x$, then $x \geq 0$.

c) If n is greater than 3, then n^2 is greater than 9.

5. Find a compound proposition involving the propositional variables p, q, r, and s that is true when exactly three of these propositional variables are true and is false otherwise.

6. Show that these statements are inconsistent: "If Sergei takes the job offer then he will get a signing bonus." "If Sergei takes the job offer, then he will receive a higher salary." "If Sergei gets a signing bonus, then he will not receive a higher salary." "Sergei takes the job offer."

7. Show that these statements are inconsistent: "If Miranda does not take a course in discrete mathematics, then she will not graduate." "If Miranda does not graduate, then she is not qualified for the job." "If Miranda reads this book, then she is qualified for the job." "Miranda does not take a course in discrete mathematics but she reads this book."

8. Suppose that you meet three people, A, B, and C, on the island of knights and knaves described in Example 15 in Section 1.1. What are A, B, and C if A says "I am a knave and B is a knight" and B says "Exactly one of the three of us is a knight."

9. (Adapted from [Sm78]) Suppose that on an island there are three types of people, knights, knaves, and normals. Knights always tell the truth, knaves always lie, and normals sometimes lie and sometimes tell the truth. Detectives questioned three inhabitants of the island—Amy, Brenda, and Claire—as part of the investigation of a crime. The detectives knew that one of the three committed the crime, but not which one. They also knew that the criminal was a knight, and that the other two were not. Additionally, the detectives recorded these statements: Amy: "I am innocent." Brenda: "What Amy says is true." Claire: "Brenda is not a normal." After analyzing their information, the detectives positively identified the guilty party. Who was it?

10. Let $P(x)$ be the statement "student x knows calculus" and let $Q(y)$ be the statement "class y contains a student who knows calculus." Express each of these as quantifications of $P(x)$ and $Q(y)$.

   a) Some students know calculus.
   b) Not every student knows calculus.
   c) Every class has a student in it who knows calculus.
   d) Every student in every class knows calculus.
   e) There is at least one class with no students who know calculus.

11. Let $P(m, n)$ be the statement "m divides n," where the universe of discourse for both variables is the set of positive integers. Determine the truth values of each of these propositions.

   a) $P(4, 5)$ **b)** $P(2, 4)$
   c) $\forall m \, \forall n \, P(m, n)$ **d)** $\exists m \, \forall n \, P(m, n)$
   e) $\exists n \, \forall m \, P(m, n)$ **f)** $\forall n \, P(1, n)$

12. Use quantifiers to express the statement "No one has more than three grandmothers" using the propositional function $G(x, y)$, which represents "x is the grandmother of y."

13. Use quantifiers to express the statement "Everyone has exactly two biological parents" using the propositional function $P(x, y)$, which represents "x is the biological parent of y."

14. Let $P(x, y)$ be a propositional function. Show that the implication $\exists x \, \forall y \, P(x, y) \rightarrow \forall y \, \exists x \, P(x, y)$ is a tautology.

15. Let $P(x)$ and $Q(x)$ be propositional functions. Show that $\exists x \, (P(x) \rightarrow Q(x))$ and $\forall x \, P(x) \rightarrow \exists x \, Q(x)$ always have the same truth value.

16. If $\forall y \, \exists x \, P(x, y)$ is true, does it necessarily follow that $\exists x \, \forall y \, P(x, y)$ is true?

17. If $\forall x \, \exists y \, P(x, y)$ is true, does it necessarily follow that $\exists x \, \forall y \, P(x, y)$ is true?

18. Find the negations of these statements.

   a) If it snows today, then I will go skiing tomorrow.
   b) Every person in this class understands mathematical induction.
   c) Some students in this class do not like discrete mathematics.
   d) In every mathematics class there is some student who falls asleep during lectures.

19. Express this statement using quantifiers: "Every student in this class has taken some course in every department in the school of mathematical sciences."

20. Express this statement using quantifiers: "There is a building on the campus of some college in the United States in which every room is painted white."

21. Let A be the set of English words that contain the letter x, and let B be the set of English words that contain the letter q. Express each of these sets as a combination of A and B.

   a) The set of English words that do not contain the letter x.
   b) The set of English words that contain both an x and a q.
   c) The set of English words that contain an x but not a q.
   d) The set of English words that do not contain either an x or a q.
   e) The set of English words that contain an x or a q, but not both.

22. Describe a rule of inference that can be used to prove that there are exactly two elements x and y in a universe of discourse such that $P(x)$ and $P(y)$ are true. Express this rule of inference as a statement in English.

23. What is wrong with this argument? Given the premise $\exists x \, P(x) \wedge \exists x \, Q(x)$, use simplification to obtain $\exists x \, P(x)$; use existential instantiation to obtain $P(c)$ for some element c; use simplification again to obtain $\exists x\, Q(x)$; use existential instantiation to obtain $Q(c)$ for some element c; use conjunction to conclude that $P(c) \wedge Q(c)$; and finally, use existential generalization to conclude that $\exists x (P(x) \wedge Q(x))$.

24. Use rules of inference to show that the premises $\forall x (P(x) \rightarrow Q(x))$, $\forall x (Q(x) \rightarrow R(x))$ and $\neg R(a)$, where a is in the universe of discourse, imply the conclusion $\neg P(a)$.

25. Use rules of inference to show that the premises $\forall x (P(x) \wedge Q(x))$ and $\forall x (\neg P(x) \wedge Q(x) \rightarrow R(x))$ imply the conclusion $\forall x (\neg R(x) \rightarrow P(x))$.

26. Prove that if x^3 is irrational, then x is irrational.

27. Prove that if x is irrational and $x \geq 0$, then $\sqrt{x}$ is irrational.

28. Prove that given a nonnegative integer n, there is a unique nonnegative integer m such that $m^2 \leq n < (m+1)^2$.

29. Prove that there exists an integer m such that $m^2 > 10^{1000}$. Is your proof constructive or nonconstructive?

30. Prove that there is a positive integer that can be written as the sum of squares of positive integers in two different ways. (Use a computer or calculator to speed up your work.)

31. Disprove the statement that every positive integer is the sum of the cubes of eight nonnegative integers.

32. Disprove the statement that every positive integer is the sum of at most two squares and a cube of nonnegative integers.

33. Disprove the statement that every positive integer is the sum of 36 fifth powers of nonnegative integers.

34. Show that if A is a subset of B, then the power set of A is a subset of the power set of B.

35. Suppose that A and B are sets such that the power set of A is a subset of the power set of B. Does it follow that A is a subset of B?

36. Let **E** denote the set of even integers and **O** denote the set of odd integers. As usual, let **Z** denote the set of all integers. Determine each of these sets.
a) $\mathbf{E} \cup \mathbf{O}$ b) $\mathbf{E} \cap \mathbf{O}$ c) $\mathbf{Z} - \mathbf{E}$ d) $\mathbf{Z} - \mathbf{O}$

37. Show that if A is a set and U is the universal set, then
a) $A \cap \overline{A} = \emptyset$.   b) $A \cup \overline{A} = U$.

38. Show that if A and B are sets, then
a) $A = A \cap (A \cup B)$.
b) $A = A \cup (A \cap B)$.

39. Show that if A and B are sets, then $A - (A - B) = A \cap B$.

40. Let A and B be sets. Show that $A \subseteq B$ if and only if $A \cap B = A$.

41. Let A, B, and C be sets. Show that $(A - B) - C$ is not necessarily equal to $A - (B - C)$.

42. Suppose that A, B, and C are sets. Prove or disprove that $(A - B) - C = (A - C) - B$.

43. Suppose that A, B, C, and D are sets. Prove or disprove that $(A - B) - (C - D) = (A - C) - (B - D)$.

44. Show that if A and B are finite sets, then $|A \cap B| \leq |A \cup B|$. Determine when this relationship is an equality.

45. Let A and B be sets in a finite universal set U. List the following in order of increasing size.
a) $|A|$, $|A \cup B|$, $|A \cap B|$, $|U|$, $|\emptyset|$
b) $|A - B|$, $|A \oplus B|$, $|A| + |B|$, $|A \cup B|$, $|\emptyset|$

46. Let A and B be subsets of the finite universal set U. Show that $|\overline{A} \cap \overline{B}| = |U| - |A| - |B| + |A \cap B|$.

47. Let f and g be functions from $\{1, 2, 3, 4\}$ to $\{a, b, c, d\}$ and from $\{a, b, c, d\}$ to $\{1, 2, 3, 4\}$, respectively, such that $f(1) = d$, $f(2) = c$, $f(3) = a$, $f(4) = b$, and $g(a) = 2, g(b) = 1, g(c) = 3, g(d) = 2$.
a) Is f one-to-one? Is g one-to-one?
b) Is f onto? Is g onto?
c) Does either f or g have an inverse? If so, find this inverse.

48. Let f be a one-to-one function from the set A to the set B. Let S and T be subsets of A. Show that $f(S \cap T) = f(S) \cap f(T)$.

49. Give an example to show that the equality in Exercise 48 may not hold if f is not one-to-one.

50. Show that if n is an integer, then $n = \lceil n/2 \rceil + \lfloor n/2 \rfloor$.

51. For which real numbers x and y is it true that $\lfloor x + y \rfloor = \lfloor x \rfloor + \lfloor y \rfloor$?

52. For which real numbers x and y is it true that $\lceil x + y \rceil = \lceil x \rceil + \lceil y \rceil$?

Computer Projects

WRITE PROGRAMS WITH THE SPECIFIED INPUT AND OUTPUT.

1. Given the truth values of the propositions p and q, find the truth values of the conjunction, disjunction, exclusive or, implication, and biconditional of these propositions.

2. Given two bit strings of length n, find the bitwise *AND,* bitwise *OR,* and bitwise *XOR* of these strings.

3. Given the truth values of the propositions p and q in fuzzy logic, find the truth value of the disjunction and the conjunction of p and q (see Exercises 35–37 of Section 1.1).

4. Given subsets A and B of a set with n elements, use bit strings to find $\overline{A}$, $A \cup B$, $A \cap B$, $A - B$, and $A \oplus B$.

5. Given multisets A and B from the same universal set, find $A \cup B, A \cap B, A - B$, and $A + B$ (see preamble to Exercise 49 of Section 1.7).

6. Given fuzzy sets A and B, find $\overline{A}, A \cup B$, and $A \cap B$ (see preamble to Exercise 51 of Section 1.7).

7. Given a function f from $\{1, 2, \ldots, n\}$ to the set of integers, determine whether f is one-to-one.

8. Given a function f from $\{1, 2, \ldots, n\}$ to itself, determine whether f is onto.

9. Given a bijection f from the set $\{1, 2, \ldots, n\}$ to itself, find f^{-1}.

Computations and Explorations

USE A COMPUTATIONAL PROGRAM OR PROGRAMS YOU HAVE WRITTEN TO DO THESE EXERCISES.

1. Look for positive integers that are not the sum of the cubes of nine different positive integers.

2. Look for positive integers greater than 79 that are not the sum of the fourth powers of 18 positive integers.

3. Find as many positive integers as you can that can be written as the sum of cubes of positive integers in two different ways, sharing this property with 1729.

4. Calculate the number of one-to-one functions from a set S to a set T, where S and T are finite sets of various sizes. Can you determine a formula for the number of such functions? (We will find such a formula in Chapter 4.)

5. Calculate the number of onto functions from a set S to a set T where S and T are finite sets of various sizes. Can you determine a formula for the number of such functions? (We will find such a formula in Chapter 6.)

Writing Projects

RESPOND TO THESE WITH ESSAYS USING OUTSIDE SOURCES.

1. Discuss logical paradoxes including the paradox of Epimenides the Cretan, Jourdain's card paradox, and the barber paradox and how they are resolved.

2. Describe how fuzzy logic is being applied to practical applications. Consult one or more of the recent books on fuzzy logic written for general audiences.

3. Describe the basic rules of *WFF'N PROOF, The Game of Modern Logic*, developed by Layman Allen. Give examples of some of the games included in *WFF'N PROOF*.

4. Read some of the writings of Lewis Carroll on symbolic logic. Describe in detail some of the models he used to represent logical arguments and the rules of inference he used in these arguments.

5. Extend the discussion of Prolog given in Section 1.3, explaining in more depth how Prolog employs resolution.

6. Discuss some of the techniques used in computational logic, including Skolem's rule.

7. "Automated theorem proving" is the task of using computers to mechanically prove theorems. Discuss the goals and applications of automated theorem proving and the progress made in developing automated theorem provers.

8. Discuss how an axiomatic set theory can be developed to avoid Russell's paradox. (See Exercise 30 of Section 1.6.)

9. Research where the concept of a function first arose, and describe how this concept was first used.

10. Describe how DNA computing has been used to solve instances of the satisfiability problem.

2

The Fundamentals: Algorithms, the Integers, and Matrices

Many problems can be solved by considering them as special cases of general problems. For instance, consider the problem of locating the largest integer in the sequence 101, 12, 144, 212, 98. This is a specific case of the problem of locating the largest integer in a sequence of integers. To solve this general problem we must give an algorithm, which specifies a sequence of steps used to solve this general problem. We will study algorithms for solving many different types of problems in this book. For instance, algorithms will be developed for finding the greatest common divisor of two integers, for generating all the orderings of a finite set, for searching a list, for sorting the elements of a sequence, and for finding the shortest path between nodes in a network.

One important consideration concerning an algorithm is its computational complexity. That is, what are the computer resources needed to use this algorithm to solve a problem of a specified size? To measure the complexity of algorithms we use big-O and big-Theta notation, which we develop in this chapter. We will illustrate the analysis of the complexity of algorithms in this chapter. Furthermore, we will discuss what the complexity of an algorithm means in practical and theoretical terms.

The set of integers plays a fundamental role in discrete mathematics. In particular, the concept of division of integers is fundamental to computer arithmetic. We will briefly review some of the important concepts of number theory, the study of integers and their properties. Some important algorithms involving integers will be studied, including the Euclidean algorithm for computing greatest common divisors, which was first described thousands of years ago. Integers can be represented using any positive integer greater than 1 as a base. Binary expansions, which are used throughout computer science, are representations with 2 as the base. In this chapter we discuss base b representations of integers and give an algorithm for finding them. Algorithms for integer arithmetic, which were the first procedures called algorithms, will also be discussed. This chapter also introduces several important applications of number theory. For example, in this chapter we will use number theory to make messages secret, to generate pseudorandom numbers, and to assign memory locations to computer files. Number theory, once considered the purest of subjects, has become an essential tool in providing computer and Internet security.

Matrices are used in discrete mathematics to represent a variety of discrete structures. We review the basic material about matrices and matrix arithmetic needed to represent relations and graphs. Matrix arithmetic will be used in numerous algorithms involving these structures.

2.1 Algorithms

INTRODUCTION

There are many general classes of problems that arise in discrete mathematics. For instance: given a sequence of integers, find the largest one; given a set, list all of its subsets; given a set of integers, put them in increasing order; given a network, find the shortest path between two vertices. When presented with such a problem, the first thing to do is to construct a model that translates the problem into a mathematical context. Discrete structures used in such models include sets, sequences, and functions—structures discussed in Chapter 1—as well as such other structures as permutations, relations, graphs, trees, networks, and finite state machines—concepts that will be discussed in later chapters.

Setting up the appropriate mathematical model is only part of the solution. To complete the solution, a method is needed that will solve the general problem using the model. Ideally, what is required is a procedure that follows a sequence of steps that leads to the desired answer. Such a sequence of steps is called an **algorithm.**

DEFINITION 1   An *algorithm* is a finite set of precise instructions for performing a computation or for solving a problem.

The term *algorithm* is a corruption of the name *al-Khowarizmi*, a mathematician of the ninth century, whose book on Hindu numerals is the basis of modern decimal notation. Originally, the word *algorism* was used for the rules for performing arithmetic using decimal notation. *Algorism* evolved into the word *algorithm* by the eighteenth century. With the growing interest in computing machines, the concept of an algorithm was given a more general meaning, to include all definite procedures for solving problems, not just the procedures for performing arithmetic. (We will discuss algorithms for performing arithmetic with integers in Section 2.5.)

In this book, we will discuss algorithms that solve a wide variety of problems. In this section we will use the problem of finding the largest integer in a finite sequence of integers to illustrate the concept of an algorithm and the properties algorithms have. Also, we will describe algorithms for locating a particular element in a finite set. In subsequent sections, procedures for finding the greatest common divisor of two integers,

ABU JA'FAR MOHAMMED IBN MUSA AL-KHOWARIZMI (C. 780–C. 850) al-Khowarizmi, an astronomer and mathematician, was a member of the House of Wisdom, an academy of scientists in Baghdad. The name al-Khowarizmi means "from the town of Kowarzizm," which is now called *Khiva* and is part of Uzbekistan. al-Khowarizmi wrote books on mathematics, astronomy, and geography. Western Europeans first learned about algebra from his works. The word *algebra* comes from al-jabr, part of the title of his book *Kitab al-jabr w'al muquabala*. This book was translated into Latin and was a widely used textbook. His book on the use of Hindu numerals describes procedures for arithmetic operations using these numerals. European authors used a Latin corruption of his name, which later evolved to the word *algorithm,* to describe the subject of arithmetic with Hindu numerals.

for finding the shortest path between two points in a network, for multiplying matrices, and so on, will be discussed.

EXAMPLE 1

Describe an algorithm for finding the maximum (largest) value in a finite sequence of integers.

Even though the problem of finding the maximum element in a sequence is relatively trivial, it provides a good illustration of the concept of an algorithm. Also, there are many instances where the largest integer in a finite sequence of integers is required. For instance, a university may need to find the highest score on a competitive exam taken by thousands of students. Or a sports organization may want to identify the member with the highest rating each month. We want to develop an algorithm that can be used whenever the problem of finding the largest element in a finite sequence of integers arises.

We can specify a procedure for solving this problem in several ways. One method is simply to use the English language to describe the sequence of steps used. We now provide such a solution.

Solution of Example 1: We perform the following steps.

1. Set the temporary maximum equal to the first integer in the sequence. (The temporary maximum will be the largest integer examined at any stage of the procedure.)

2. Compare the next integer in the sequence to the temporary maximum, and if it is larger than the temporary maximum, set the temporary maximum equal to this integer.

3. Repeat the previous step if there are more integers in the sequence.

4. Stop when there are no integers left in the sequence. The temporary maximum at this point is the largest integer in the sequence. ◀

An algorithm can also be described using a computer language. However, when that is done, only those instructions permitted in the language can be used. This often leads to a description of the algorithm that is complicated and difficult to understand. Furthermore, since many programming languages are in common use, it would be undesirable to choose one particular language. So, instead of using a particular computer language to specify algorithms, a form of **pseudocode** will be used in this book. (All algorithms will also be described using the English language.) Pseudocode provides an intermediate step between an English language description of an algorithm and an implementation of this algorithm in a programming language. The steps of the algorithm are specified using instructions resembling those used in programming languages. However, in pseudocode, the instructions used can include any well-defined operations or statements. A computer program can be produced in any computer language using the pseudocode description as a starting point.

The pseudocode used in this book is loosely based on the programming language Pascal. However, the syntax of Pascal, or that of other programming languages, will not be followed. Furthermore, any well-defined instruction can be used in this pseudocode. The details of the pseudocode used in the text are given in Appendix 2. The reader should refer to this appendix whenever the need arises.

A pseudocode description of the algorithm for finding the maximum element in a finite sequence follows.

ALGORITHM 1 Finding the Maximum Element in a Finite Sequence.

---

procedure $max(a_1, a_2, \ldots, a_n$: integers)
$max := a_1$
for $i := 2$ **to** n
    if $max < a_i$ **then** $max := a_i$
$\{max$ is the largest element$\}$

This algorithm first assigns the initial term of the sequence, a_1, to the variable *max*. The "for" loop is used to successively examine terms of the sequence. If a term is greater than the current value of *max*, it is assigned to be the new value of *max*.

There are several properties that algorithms generally share. They are useful to keep in mind when algorithms are described. These properties are:

- *Input.* An algorithm has input values from a specified set.

- *Output.* From each set of input values an algorithm produces output values from a specified set. The output values are the solution to the problem.

- *Definiteness.* The steps of an algorithm must be defined precisely.

- *Correctness.* An algorithm should produce the correct output values for each set of input values.

- *Finiteness.* An algorithm should produce the desired output after a finite (but perhaps large) number of steps for any input in the set.

- *Effectiveness.* It must be possible to perform each step of an algorithm exactly and in a finite amount of time.

- *Generality.* The procedure should be applicable for all problems of the desired form, not just for a particular set of input values.

EXAMPLE 2 Show that Algorithm 1 for finding the maximum element in a finite sequence of integers has all the properties listed.

Solution: The input to Algorithm 1 is a sequence of integers. The output is the largest integer in the sequence. Each step of the algorithm is precisely defined, since only assignments, a finite loop, and conditional statements occur. To show that the algorithm is correct, we must show that when the algorithm terminates, the value of the variable *max* equals the maximum of the terms of the sequence. To see this, note that the initial value of *max* is the first term of the sequence; as successive terms of the sequence are examined, *max* is updated to the value of a term if the term exceeds the maximum of the terms previously examined. This (informal) argument shows that when all the terms have been examined, *max* equals the value of the largest term. (A rigorous proof of this requires techniques developed in Section 3.3.) The algorithm uses a finite number of steps, since it terminates after all the integers in the sequence have been examined. The algorithm can be carried out in a finite amount of time since each step is either a comparison or an assignment. Finally, Algorithm 1 is general, since it can be used to find the maximum of any finite sequence of integers. ◀

SEARCHING ALGORITHMS

The problem of locating an element in an ordered list occurs in many contexts. For instance, a program that checks the spelling of words searches for them in a dictionary, which is just an ordered list of words. Problems of this kind are called **searching problems.** We will discuss several algorithms for searching in this section. We will study the number of steps used by each of these algorithms in Section 2.3.

The general searching problem can be described as follows: Locate an element x in a list of distinct elements $a_1, a_2, \ldots, a_n$, or determine that it is not in the list. The solution to this search problem is the location of the term in the list that equals x (that is, i is the solution if $x = a_i$) and is 0 if x is not in the list.

THE LINEAR SEARCH The first algorithm that we will present is called the **linear search,** or **sequential search,** algorithm. The linear search algorithm begins by comparing x and a_1. When $x = a_1$, the solution is the location of a_1, namely, 1. When $x \neq a_1$, compare x with a_2. If $x = a_2$, the solution is the location of a_2, namely, 2. When $x \neq a_2$, compare x with a_3. Continue this process, comparing x successively with each term of the list until a match is found, where the solution is the location of that term, unless no match occurs. If the entire list has been searched without locating x, the solution is 0. The pseudocode for the linear search algorithm is displayed as Algorithm 2.

ALGORITHM 2 The Linear Search Algorithm.

---

procedure *linear search*(x: integer, $a_1, a_2, \ldots, a_n$: distinct integers)
$i := 1$
while ($i \leq n$ and $x \neq a_i$)
    $i := i + 1$
if $i \leq n$ **then** *location* := i
else *location* := 0
{*location* is the subscript of the term that equals x, or is 0 if x is not
    found}

THE BINARY SEARCH We will now consider another searching algorithm. This algorithm can be used when the list has terms occurring in order of increasing size (for instance: if the terms are numbers, they are listed from smallest to largest; if they are words, they are listed in lexicographic, or alphabetic, order). This second algorithm is called the **binary search algorithm.** It proceeds by comparing the element to be located to the middle term of the list. The list is then split into two smaller sublists of the same size, or where one of these smaller lists has one fewer term than the other. The search continues by restricting the search to the appropriate sublist based on the comparison of the element to be located and the middle term. In Section 2.3, it will be shown that the binary search algorithm is much more efficient than the linear search algorithm. Example 3 demonstrates how a binary search works.

EXAMPLE 3 To search for 19 in the list

       1 2 3 5 6 7 8 10 12 13 15 16 18 19 20 22,

first split this list, which has 16 terms, into two smaller lists with eight terms each, namely,

1 2 3 5 6 7 8 10   12 13 15 16 18 19 20 22.

Then, compare 19 and the largest term in the first list. Since $10 < 19$, the search for 19 can be restricted to the list containing the 9th through the 16th terms of the original list. Next, split this list, which has eight terms, into the two smaller lists of four terms each, namely,

12 13 15 16   18 19 20 22.

Since $16 < 19$ (comparing 19 with the largest term of the first list) the search is restricted to the second of these lists, which contains the 13th through the 16th terms of the original list. The list 18 19 20 22 is split into two lists, namely,

18 19   20 22.

Since 19 is not greater than the largest term of the first of these two lists, which is also 19, the search is restricted to the first list: 18 19, which contains the 13th and 14th terms of the original list. Next, this list of two terms is split into two lists of one term each: 18 and 19. Since $18 < 19$, the search is restricted to the second list: the list containing the 14th term of the list, which is 19. Now that the search has been narrowed down to one term, a comparison is made, and 19 is located as the 14th term in the original list. ◀

We now specify the steps of the binary search algorithm. To search for the integer x in the list $a_1, a_2, \ldots, a_n$, where $a_1 < a_2 < \cdots < a_n$, begin by comparing x with the middle term of the sequence, a_m, where $m = \lfloor (n+1)/2 \rfloor$. (Recall that $\lfloor x \rfloor$ is the greatest integer not exceeding x.) If $x > a_m$, the search for x can be restricted to the second half of the sequence, which is $a_{m+1}, a_{m+2}, \ldots, a_n$. If x is not greater than a_m, the search for x can be restricted to the first half of the sequence, which is $a_1, a_2, \ldots, a_m$.

The search has now been restricted to a list with no more than $\lceil n/2 \rceil$ elements. (Recall that $\lceil x \rceil$ is the smallest integer greater than or equal to x.) Using the same procedure, compare x to the middle term of the restricted list. Then restrict the search to the first or second half of the list. Repeat this process until a list with one term is obtained. Then determine whether this term is x. Pseudocode for the binary search algorithm is displayed as Algorithm 3.

ALGORITHM 3 The Binary Search Algorithm.

---

procedure *binary search*(x: integer, $a_1, a_2, \ldots, a_n$: increasing integers)
$i := 1$ {i is left endpoint of search interval}
$j := n$ {j is right endpoint of search interval}
while $i < j$
begin
　　$m := \lfloor (i+j)/2 \rfloor$
　　if $x > a_m$ **then** $i := m + 1$
　　else $j := m$
end
if $x = a_i$ **then** *location* $:= i$
else *location* $:= 0$
{*location* is the subscript of the term equal to x, or 0 if x is not found}

Algorithm 3 proceeds by successively narrowing down the part of the sequence being searched. At any given stage only the terms beginning with a_i and ending with a_j are under consideration. In other words, i and j are the smallest and largest subscripts of the remaining terms, respectively. Algorithm 3 continues narrowing the part of the sequence being searched until only one term of the sequence remains. When this is done, a comparison is made to see whether this term equals x.

SORTING

Demo

Ordering the elements of a list is a problem that occurs in many contexts. For example, to produce a telephone directory it is necessary to alphabetize the names of subscribers. Putting addresses in order in an e-mail mailing list can determine whether there are duplicated addresses. Creating a useful dictionary requires that words be put in alphabetical order. Similarly, generating a parts list requires that we order them according to increasing part number.

Suppose that we have a list of elements of a set. Furthermore, suppose that we have a way to order elements of the set. (The notion of ordering elements of sets will be discussed in detail in Section 7.6.) A **sorting** is putting these elements into a list in which the elements are in increasing order. For instance, sorting the list $7, 2, 1, 4, 5, 9$ produces the list $1, 2, 4, 5, 7, 9$. Sorting the list d, h, c, a, f (using alphabetical order) produces the list a, c, d, f, h.

An amazingly large percentage of computing resources is devoted to sorting one thing or another. Hence, much effort has been devoted to the development of sorting algorithms. A surprisingly large number of sorting algorithms have been devised using distinct strategies. In his fundamental work, *The Art of Computer Programming,* Donald Knuth devotes close to 400 pages to sorting, covering around 15 different sorting algorithms in depth! There are many reasons why sorting algorithms interest computer scientists and mathematicians. Among these reasons are that some algorithms are easier to implement, some algorithms are more efficient (either in general, or when given input with certain characteristics, such as lists slightly out of order), some algorithms take advantage of particular computer architectures, and some algorithms are particularly clever. In this section we will introduce two sorting algorithms, the bubble sort and the insertion sort. Two other sorting algorithms, the selection sort and the binary insertion sort, are introduced in the exercises at the end of this section, and the shaker sort is introduced in the Supplementary Exercises at the end of this chapter. In Section 3.5 we will discuss the merge sort and introduce the quick sort in the exercises in that section; the tournament sort is introduced in the exercise set in Section 9.2. We cover sorting algorithms both because sorting is an important problem and because these algorithms can serve as examples of many important concepts.

Links

THE BUBBLE SORT The **bubble sort** is one of the simplest sorting algorithms, but not one of the most efficient. It puts a list into increasing order by successively comparing adjacent elements, interchanging them if they are in the wrong order. To carry out the bubble sort, we perform the basic operation, that is, interchanging a larger element with a smaller one following it, starting at the beginning of the list, for a full pass. We iterate this procedure until the sort is complete. Pseudocode for the bubble sort is given as Algorithm 4. We can imagine the elements in the list placed in a column. In the bubble sort, the smaller elements "bubble" to the top as they are interchanged with larger elements. The larger elements "sink" to the bottom. This is illustrated in Example 4.

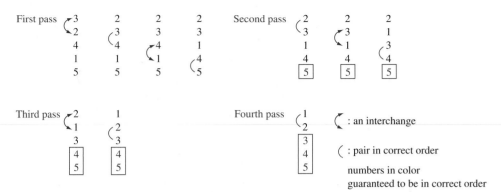

FIGURE 1   The Steps of a Bubble Sort.

EXAMPLE 4   Use the bubble sort to put $3, 2, 4, 1, 5$ into increasing order.

Solution: The steps of this algorithm are illustrated in Figure 1. Begin by comparing the first two elements, 3 and 2. Since $3 > 2$, interchange 3 and 2, producing the list $2, 3, 4, 1, 5$. Since $3 < 4$, continue by comparing 4 and 1. Since $4 > 1$, interchange 1 and 4, producing the list $2, 3, 1, 4, 5$. Since $4 < 5$, the first pass is complete. The first pass guarantees that the largest element, 5, is in the correct position.

The second pass begins by comparing 2 and 3. Since these are in the correct order, 3 and 1 are compared. Since $3 > 1$, these numbers are interchanged, producing $2, 1, 3, 4, 5$. Since $3 < 4$, these numbers are in the correct order. It is not necessary to do any more comparisons for this pass because 5 is already in the correct position. The second pass guarantees that the two largest elements, 4 and 5, are in their correct positions.

The third pass begins by comparing 2 and 1. These are interchanged since $2 > 1$, producing $1, 2, 3, 4, 5$. Because $2 < 3$, these two elements are in the correct order. It is not necessary to do any more comparisons for this pass because 4 and 5 are already in the correct positions. The third pass guarantees that the three largest elements, $3, 4$, and 5, are in their correct positions.

The fourth pass consists of one comparison, namely, the comparison of 1 and 2. Since $1 < 2$, these elements are in the correct order. This completes the bubble sort.   ◀

ALGORITHM 4 The Bubble Sort.

procedure *bubblesort*$(a_1, \ldots, a_n)$
for $i := 1$ **to** $n - 1$
    for $j := 1$ **to** $n - i$
        if $a_j > a_{j+1}$ **then** interchange a_j and a_{j+1}
$\{a_1, \ldots, a_n$ is in increasing order$\}$

THE INSERTION SORT The **insertion sort** is a simple sorting algorithm, but it is usually not the most efficient. To sort a list with n elements, the insertion sort begins with the second element. The insertion sort compares this second element with the first element and inserts it before the first element if it does not exceed the first element and after the first element if it exceeds the first element. At this point, the first two elements are in the correct order. The third element is then compared with the first element, and if it is larger than the first element, it is compared with the second element; it is inserted into the correct position among the first three elements.

In general, in the jth step of the insertion sort, the jth element of the list is inserted into the correct position in the list of the previously sorted $j - 1$ elements. To insert the jth element in the list, a linear search technique is used (see Exercise 43); the jth element is successively compared with the already sorted $j - 1$ elements at the start of the list until the first element that is not less than this element is found or until it has been compared with all $j - 1$ elements; the jth element is inserted in the correct position so that the first j elements are sorted. The algorithm continues until the last element is placed in the correct position relative to the already sorted list of the first $n - 1$ elements. The insertion sort is described in pseudocode in Algorithm 5.

EXAMPLE 5   Use the insertion sort to put the elements of the list 3, 2, 4, 1, 5 in increasing order.

Solution: The insertion sort first compares 2 and 3. Since $3 > 2$, it places 2 in the first position, producing the list 2, 3, 4, 1, 5 (the sorted part of the list is shown in color). At this point, 2 and 3 are in the correct order. Next, it inserts the third element, 4, into the already sorted part of the list by making the comparisons $4 > 2$ and $4 > 3$. Since $4 > 3$, 4 is placed in the third position. At this point, the list is 2, 3, 4, 1, 5 and we know that the ordering of the first three elements is correct. Next, we find the correct place for the fourth element, 1, among the already sorted elements, 2, 3, 4. Since $1 < 2$, we obtain the list 1, 2, 3, 4, 5. Finally, we insert 5 into the correct position by successively comparing it to 1, 2, 3, and 4. Since $5 > 4$, it goes at the end of the list, producing the correct order for the entire list.   ◄

ALGORITHM 5 The Insertion Sort.

---

procedure *insertion sort*($a_1, a_2, \ldots, a_n$: real numbers with $n \geq 2$)
for $j := 2$ **to** n
begin
$\quad i := 1$
$\quad$ **while** $a_j > a_i$
$\quad\quad i := i + 1$
$\quad m := a_j$
$\quad$ **for** $k := 0$ **to** $j - i - 1$
$\quad\quad a_{j-k} := a_{j-k-1}$
$\quad a_i := m$
end $\{a_1, a_2, \ldots, a_n$ are sorted$\}$

GREEDY ALGORITHMS

Many algorithms we will study in this book are designed to solve **optimization problems.** The goal of such problems is to find a solution to the given problem that either minimizes or maximizes the value of some parameter. Optimization problems studied later in this text include finding a route between two cities with smallest total mileage, determining a way to encode messages using the fewest bits possible, and finding a set of fiber links between network nodes using the least amount of fiber.

Surprisingly, one of the simplest approaches often leads to a solution of an optimization problem. This approach selects the best choice at each step, instead of considering all sequences of steps that may lead to an optimal solution. Algorithms that make what seems to be the "best" choice at each step are called **greedy algorithms.** Once we know that a greedy algorithm finds a feasible solution, we need to determine whether it has found an optimal solution. To do this, we either prove that the solution is optimal or we show that there is a counterexample where the algorithm yields a nonoptimal solution. To make these concepts more concrete, we will consider an algorithm that makes change using coins.

EXAMPLE 6   Consider the problem of making n cents change with quarters, dimes, nickels, and pennies, and using the least total number of coins. We can devise a greedy algorithm for making change for n cents by making a locally optimal choice at each step; that is, at each step we choose the coin of the largest denomination possible to add to the pile of change without exceeding n cents. For example, to make change for 67 cents, we first select a quarter (leaving 42 cents). We next select a second quarter (leaving 17 cents), followed by a dime (leaving 7 cents), followed by a nickel (leaving 2 cents), followed by a penny (leaving 1 cent), followed by a penny. ◀

We display a greedy change-making algorithm for n cents, using any set of denominations of coins, as Algorithm 6.

ALGORITHM 6 Greedy Change-Making Algorithm.

---

procedure $change(c_1, c_2, \ldots, c_r$: values of denominations of coins, where
$\qquad c_1 > c_2 > \cdots > c_r$; n: a positive integer)
for $i := 1$ **to** r
$\qquad$ **while** $n \geq c_i$
$\qquad$ **begin**
$\qquad\qquad$ add a coin with value c_i to the change
$\qquad\qquad$ $n := n - c_i$
$\qquad$ **end**

We have described a greedy algorithm for making change using quarters, dimes, nickels, and pennies. We will show that this algorithm leads to an optimal solution in the sense that it uses the fewest coins possible. Before we embark on our proof, we show that there are sets of coins for which the greedy algorithm (Algorithm 6) does not necessarily produce change using the fewest coins possible. For example, if we have only quarters, dimes, and pennies (and no nickels) to use, the greedy algorithm would make change for 30 cents using six coins—a quarter and five pennies—whereas we could have used three coins, namely, three dimes.

LEMMA 1   If n is a positive integer, then n cents in change using quarters, dimes, nickels, and pennies using the fewest coins possible has at most two dimes, at most one nickel, at most four pennies, and cannot have two dimes and a nickel. The amount of change in dimes, nickels, and pennies cannot exceed 24 cents.

Proof: We use a proof by contradiction. We will show that if we had more than the specified number of coins of each type, we could replace them using fewer coins that have the same value. We note that if we had three dimes we could replace them with a quarter and a nickel, if we had two nickels we could replace them with a dime, if we had five pennies we could replace them with a nickel, and if we had two dimes and a nickel we could replace them with a quarter. Because we can have at most two dimes, one nickel, and four pennies, but we cannot have two dimes and a nickel, it follows that 24 cents is the most money we can have in dimes, nickels, and pennies when we make change using the fewest number of coins for n cents. ◁

THEOREM 1   The greedy algorithm (Algorithm 6) produces change using the fewest coins possible.

Proof: We will use a proof by contradiction. Suppose that there is a positive integer n such that there is a way to make change for n cents using quarters, dimes, nickels, and pennies that uses fewer coins than the greedy algorithm finds. We first note that q', the number of quarters used in this optimal way to make change for n cents, must be the same as q, the number of quarters used by the greedy algorithm. To show this, first note that the greedy algorithm uses the most quarters possible, so $q' \leq q$. However, it is also the case that q' cannot be less than q. If it were, we would need to make up at least 25 cents from dimes, nickels, and pennies in this optimal way to make change. But this is impossible by Lemma 1.

Since there must be the same number of quarters in the two ways to make change, the value of the dimes, nickels, and pennies in these two ways must be the same, and these coins are worth no more than 24 cents. There must be the same number of dimes, because the greedy algorithm used the most dimes possible and by Lemma 1, when change is made using the fewest coins possible, at most one nickel and at most four pennies are used, so that the most dimes possible are also used in the optimal way to make change. Similarly, we have the same number of nickels and, finally, the same number of pennies. ◁

Exercises

1. List all the steps used by Algorithm 1 to find the maximum of the list 1, 8, 12, 9, 11, 2, 14, 5, 10, 4.

2. Determine which characteristics of an algorithm the following procedures have and which they lack.

  a) procedure *double*(n: positive integer)
    while $n > 0$
      $n := 2n$

  b) procedure *divide*(n: positive integer)
    while $n \geq 0$
    begin
      $m := 1/n$
      $n := n - 1$
    end

  c) procedure *sum*(n: positive integer)
    $sum := 0$
    while $i < 10$
      $sum := sum + i$

  d) procedure *choose*(a,b: integers)
    $x :=$ either a or b

3. Devise an algorithm that finds the sum of all the integers in a list.

4. Describe an algorithm that takes as input a list of n integers and produces as output the largest difference between consecutive integers in the list.

5. Describe an algorithm that takes as input a list of n integers in nondecreasing order and produces the list of all values that occur more than once.

6. Describe an algorithm that takes as input a list of n integers and finds the number of negative integers in the list.

7. Describe an algorithm that takes as input a list of n integers and finds the location of the last even integer in the list or returns 0 if there are no even integers in the list.

8. Describe an algorithm that takes as input a list of n distinct integers and finds the location of the largest even integer in the list or returns 0 if there are no even integers in the list.

9. A **palindrome** is a string that reads the same forward and backward. Describe an algorithm for determining whether a string of n characters is a palindrome.

10. Devise an algorithm to compute x^n, where x is a real number and n is an integer. (*Hint:* First give a procedure for computing x^n when n is nonnegative by successive multiplication by x, starting with 1. Then extend this procedure, and use the fact that $x^{-n} = 1/x^n$ to compute x^n when n is negative.)

11. Describe an algorithm that interchanges the values of the variables x and y, using only assignments. What is the minimum number of assignment statements needed to do this?

12. Describe an algorithm that uses only assignment statements that replaces the triple (x, y, z) with (y, z, x). What is the minimum number of assignment statements needed?

13. List all the steps used to search for 9 in the sequence $1, 3, 4, 5, 6, 8, 9, 11$ using
   a) a linear search.   **b)** a binary search.

14. List all the steps used to search for 7 in the sequence given in Exercise 13.

15. Describe an algorithm that inserts an integer x in the appropriate position into the list $a_1, a_2, \ldots, a_n$ of integers that are in increasing order.

16. Describe an algorithm for finding the smallest integer in a finite sequence of natural numbers.

17. Describe an algorithm that locates the first occurrence of the largest element in a finite list of integers, where the integers in the list are not necessarily distinct.

18. Describe an algorithm that locates the last occurrence of the smallest element in a finite list of integers, where the integers in the list are not necessarily distinct.

19. Describe an algorithm that produces the maximum, median, mean, and minimum of a set of three integers. (The **median** of a set of integers is the middle element in the list when these integers are listed in order of increasing size. The **mean** of a set of integers is the sum of the integers divided by the number of integers in the set.)

20. Describe an algorithm for finding both the largest and the smallest integers in a finite sequence of integers.

21. Describe an algorithm that puts the first three terms of a sequence of integers of arbitrary length in increasing order.

22. Describe an algorithm to find the longest word in an English sentence (where a word is a string of letters and a sentence is a list of words, separated by blanks).

23. Describe an algorithm that determines whether a function from a finite set to another finite set is onto.

24. Describe an algorithm that determines whether a function from a finite set to another finite set is one-to-one.

25. Describe an algorithm that will count the number of 1s in a bit string by examining each bit of the string to determine whether it is a 1 bit.

26. Change Algorithm 3 so that the binary search procedure compares x to a_m at each stage of the algorithm, with the algorithm terminating if $x = a_m$. What advantage does this version of the algorithm have?

27. The **ternary search algorithm** locates an element in a list of increasing integers by successively splitting the list into three sublists of equal (or as close to equal as possible) size, and restricting the search to the appropriate piece. Specify the steps of this algorithm.

28. Specify the steps of an algorithm that locates an element in a list of increasing integers by successively splitting the list into four sublists of equal (or as close to equal as possible) size, and restricting the search to the appropriate piece.

29. A **mode** of a list of integers is an element that occurs at least as often as each of the other elements. Devise an algorithm that finds a mode in a list of nondecreasing integers.

30. Devise an algorithm that finds all modes (defined in Exercise 29) in a list of nondecreasing integers.

31. Devise an algorithm that finds the first term of a sequence of integers that equals some previous term in the sequence.

32. Devise an algorithm that finds all terms of a finite sequence of integers that are greater than the sum of all previous terms of the sequence.

33. Devise an algorithm that finds the first term of a sequence of positive integers that is less than the immediately preceding term of the sequence.

34. Use the bubble sort to sort $6, 2, 3, 1, 5, 4$, showing the lists obtained at each step.

35. Use the bubble sort to sort $3, 1, 5, 7, 4$, showing the lists obtained at each step.

36. Use the bubble sort to sort d, f, k, m, a, b, showing the lists obtained at each step.

*37. Adapt the bubble sort algorithm so that it stops when no interchanges are required. Express this more efficient version of the algorithm in pseudocode.

38. Use the insertion sort to sort the list in Exercise 34, showing the lists obtained at each step.

39. Use the insertion sort to sort the list in Exercise 35, showing the lists obtained at each step.

40. Use the insertion sort to sort the list in Exercise 36, showing the lists obtained at each step.

The **selection sort** begins by finding the least element in the list. This element is moved to the front. Then the least element among the remaining elements is found and put into the second position. This procedure is repeated until the entire list has been sorted.

41. Sort these lists using the selection sort.
   a) $3, 5, 4, 1, 2$ **b)** $5, 4, 3, 2, 1$ **c)** $1, 2, 3, 4, 5$
42. Write the selection sort algorithm in pseudocode.
43. Describe an algorithm based on the linear search for determining the correct position in which to insert a new element in an already sorted list.
44. Describe an algorithm based on the binary search for determining the correct position in which to insert a new element in an already sorted list.
45. How many comparisons does the insertion sort use to sort the list $1, 2, \ldots, n$?
46. How many comparisons does the insertion sort use to sort the list $n, n - 1, \ldots, 2, 1$?

The **binary insertion sort** is a variation of the insertion sort that uses a binary search technique (see Exercise 44) rather than a linear search technique to insert the ith element in the correct place among the previously sorted elements.

47. Show all the steps used by the binary insertion sort to sort the list $3, 2, 4, 5, 1, 6$.
48. Compare the number of comparisons used by the insertion sort and the binary insertion sort to sort the list $7, 4, 3, 8, 1, 5, 4, 2$.
***49.** Express the binary insertion sort in pseudocode.
50. a) Devise a variation of the insertion sort that uses a linear search technique that inserts the jth element in the correct place by first comparing it with the $(j - 1)$th element, then the $(j - 2)$th element if necessary, and so on.
   b) Use your algorithm to sort $3, 2, 4, 5, 1, 6$.
   c) Answer Exercise 45 using this algorithm.
   d) Answer Exercise 46 using this algorithm.
51. When a list of elements is in close to the correct order, would it be better to use an insertion sort or its variation described in Exercise 50?
52. Use the greedy algorithm to make change using quarters, dimes, nickels, and pennies for
   a) 87 cents.   **b)** 49 cents.   **c)** 99 cents.   **d)** 33 cents.

53. Use the greedy algorithm to make change using quarters, dimes, nickels, and pennies for
   a) 51 cents.   **b)** 69 cents.
   c) 76 cents.   **d)** 60 cents.
54. Use the greedy algorithm to make change using quarters, dimes, and pennies (but no nickels) for each of the amounts given in Exercise 52. For which of these amounts does the greedy algorithm use the fewest coins of these denominations possible?
55. Use the greedy algorithm to make change using quarters, dimes, and pennies (but no nickels) for each of the amounts given in Exercise 53. For which of these amounts does the greedy algorithm use the fewest coins of these denominations possible?
56. Show that if there were a coin worth 12 cents, the greedy algorithm using quarters, 12-cent coins, dimes, nickels, and pennies would not always produce change using the fewest coins possible.
57. a) Describe a greedy algorithm that solves the problem of scheduling a subset of a collection of proposed talks in a lecture hall by selecting at each step the proposed talk that has the earliest finish time among all those that do not conflict with talks already scheduled. (In Section 3.3 we will show that this greedy algorithm always produces a schedule that includes the largest number of talks possible.)
   b) Use your algorithm from part (a) to schedule the largest number of talks in a lecture hall from a proposed set of talks, if the starting and finishing times of the talks are 9:00 A.M. and 9:45 A.M.; 9:30 A.M. and 10:00 A.M.; 9:50 A.M. and 10:15 A.M.; 10:00 A.M. and 10:30 A.M.; 10:10 A.M. and 10:25 A.M.; 10:30 A.M. and 10:55 A.M.; 10:15 A.M. and 10:45 A.M.; 10:30 A.M. and 11:00 A.M.; 10:45 A.M. and 11:30 A.M.; 10:55 A.M. and 11:25 A.M.; 11:00 A.M. and 11:15 A.M.

2.2 The Growth of Functions

INTRODUCTION

In Section 2.1 we discussed the concept of an algorithm. We introduced algorithms that solve a variety of problems, including searching for an element in a list and sorting a list. In Section 2.3 we will study the number of operations used by these algorithms. In particular, we will estimate the number of comparisons used by the linear and binary search algorithms to find an element in a sequence of n elements. We will also estimate the number of comparisons used by the bubble sort and by the insertion sort to sort a list of n elements. The time required to solve a problem depends on more than only the number of operations it uses. The time also depends on the hardware and software used to run the program that implements the algorithm. However, when we change the hardware and software used to implement an algorithm, we can closely approximate the time required to solve a problem of size n by multiplying the previous time required by a constant. For example, on a supercomputer we might be able to solve a problem of size n a million times faster than we can on a PC. However, this factor of one million will not depend on n (except perhaps in some minor ways). One of the advantages of using **big-O notation,** which we introduce in this section, is that we can estimate the growth of a function without worrying about constant multipliers or smaller order terms. This means that, using big-O notation, we do not have to worry about the hardware and software used to implement an algorithm. Furthermore, using big-O notation, we can assume that the different operations used in an algorithm take the same time, which simplifies the analysis considerably.

Big-O notation is used extensively to estimate the number of operations an algorithm uses as its input grows. With the help of this notation, we can determine whether it is practical to use a particular algorithm to solve a problem as the size of the input increases. Furthermore, using big-O notation, we can compare two algorithms to determine which is more efficient as the size of the input grows. For instance, if we have two algorithms for solving a problem, one using $100n^2 + 17n + 4$ operations and the other using n^3 operations, big-O notation can help us see that the first algorithm uses far fewer operations when n is large, even though it uses more operations for small values of n, such as $n = 10$.

This section introduces big-O notation and the related big-Omega and big-Theta notations. We will explain how big-O, big-Omega, and big-Theta estimates are constructed and establish estimates for some important functions that are used in the analysis of algorithms.

BIG-O NOTATION

The growth of functions is often described using a special notation. Definition 1 describes this notation.

DEFINITION 1

Let f and g be functions from the set of integers or the set of real numbers to the set of real numbers. We say that $f(x)$ is $O(g(x))$ if there are constants C and k such that

$$|f(x)| \leq C|g(x)|$$

whenever $x > k$. (This is read as "$f(x)$ is big-oh of $g(x)$.")

The constants C and k in the definition of big-O notation are called **witnesses** to the relationship $f(x)$ is $O(g(x))$. To establish that $f(x)$ is $O(g(x))$ we need only one pair of witnesses to this relationship. That is, to show that $f(x)$ is $O(g(x))$, we need find only *one* pair of constants C and k, the witnesses, such that $|f(x)| \leq C|g(x)|$ whenever $x > k$.

Note that when there is one pair of witnesses to the relationship $f(x)$ is $O(g(x))$, there are *infinitely many* pairs of witnesses. To see this, note that if C and k are one pair of witnesses, then any pair C' and k', where $C < C'$ and $k < k'$, is also a pair of witnesses, since $|f(x)| \leq C|g(x)| \leq C'|g(x)|$ whenever $x > k' > k$.

A useful approach for finding a pair of witnesses is to first select a value of k for which the size of $|f(x)|$ can be readily estimated when $x > k$ and to see whether we can use this estimate to find a value of C for which $|f(x)| < C|g(x)|$ for $x > k$. This approach is illustrated in Example 1.

EXAMPLE 1   Show that $f(x) = x^2 + 2x + 1$ is $O(x^2)$.

Solution: We observe that we can readily estimate the size of $f(x)$ when $x > 1$ because $x < x^2$ and $1 < x^2$ when $x > 1$. It follows that

$$0 \le x^2 + 2x + 1 \le x^2 + 2x^2 + x^2 = 4x^2$$

whenever $x > 1$, as shown in Figure 1. Consequently, we can take $C = 4$ and $k = 1$ as witnesses to show that $f(x)$ is $O(x^2)$. That is, $f(x) = x^2 + 2x + 1 < 4x^2$ whenever $x > 1$. (Note that it is not necessary to use absolute values here because all functions in these equalities are positive when x is positive.)

Alternatively, we can estimate the size of $f(x)$ when $x > 2$. When $x > 2$, we have $2x \le x^2$ and $1 \le x^2$. Consequently, if $x > 2$, we have

$$0 \le x^2 + 2x + 1 \le x^2 + x^2 + x^2 = 3x^2.$$

It follows that $C = 3$ and $k = 2$ are also witnesses to the relation $f(x)$ is $O(x^2)$.

Observe that in the relationship $f(x)$ is $O(x^2)$, x^2 can be replaced by any function with larger values than x^2. For example, $f(x)$ is $O(x^3)$, $f(x)$ is $O(x^2 + x + 7)$, and so on.

It is also true that x^2 is $O(x^2 + 2x + 1)$, because $x^2 < x^2 + 2x + 1$ whenever $x > 1$. This means that $C = 1$ and $k = 1$ are witnesses to the relationship x^2 is $O(x^2 + 2x + 1)$.

◀

Note that in Example 1 we have two functions, $f(x) = x^2 + 2x + 1$ and $g(x) = x^2$, such that $f(x)$ is $O(g(x))$ and $g(x)$ is $O(f(x))$—the latter fact following from the inequality $x^2 \le x^2 + 2x + 1$, which holds for all nonnegative real numbers x. We say that two functions $f(x)$ and $g(x)$ that satisfy both of these big-O relationships are of the **same order.** We will return to this notion later in this section.

Remark: The fact that $f(x)$ is $O(g(x))$ is sometimes written $f(x) = O(g(x))$. However, the equals sign in this notation does *not* represent a genuine equality. Rather, this notation tells us that an inequality holds relating the values of the functions f and g for sufficiently large numbers in the domains of these functions.

Big-O notation has been used in mathematics for more than a century. In computer science it is widely used in the analysis of algorithms, as will be seen in Section 2.3. The German mathematician Paul Bachmann first introduced big-O notation in 1892 in an

PAUL GUSTAV HEINRICH BACHMANN (1837–1920) Paul Bachmann, the son of a Lutheran pastor, shared his father's pious lifestyle and love of music. His mathematical talent was discovered by one of his teachers, even though he had difficulties with some of his early mathematical studies. After recuperating from tuberculosis in Switzerland, Bachmann studied mathematics, first at the University of Berlin and later at Göttingen, where he attended lectures presented by the famous number theorist Dirichlet. He received his doctorate under the German number theorist Kummer in 1862; his thesis was on group theory. Bachmann was a professor at Breslau and later at Münster. After he retired from his professorship, he continued his mathematical writing, played the piano, and served as a music critic for newspapers. Bachmann's mathematical writings include a five-volume survey of results and methods in number theory, a two-volume work on elementary number theory, a book on irrational numbers, and a book on the famous conjecture known as Fermat's Last Theorem. He introduced big-O notation in his 1892 book *Analytische Zahlentheorie.*

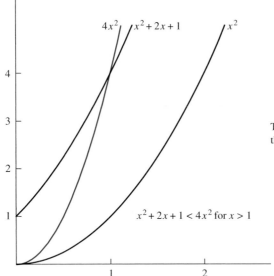

$4x^2$ $x^2 + 2x + 1$ x^2

The part of the graph of $f(x) = x^2 + 2x + 1$
that satisfies $f(x) < 4x^2$ is shown in color.

$x^2 + 2x + 1 < 4x^2$ for $x > 1$

FIGURE 1   **The Function $x^2 + 2x + 1$ is $O(x^2)$.**

important book on number theory. The big-O symbol is sometimes called a **Landau symbol** after the German mathematician Edmund Landau, who used this notation throughout his work. The use of big-O notation in computer science was popularized by Donald Knuth, who also introduced the big-Omega and big-Theta notations defined later in this section.

When $f(x)$ is $O(g(x))$, and $h(x)$ is a function that has larger absolute values than $g(x)$ does for sufficiently large values of x, it follows that $f(x)$ is $O(h(x))$. In other words, the function $g(x)$ in the relationship $f(x)$ is $O(g(x))$ can be replaced by a function with larger absolute values. To see this, note that if

$$|f(x)| \leq C|g(x)| \qquad \text{if } x > k,$$

and if $|h(x)| > |g(x)|$ for all $x > k$, then

$$|f(x)| \leq C|h(x)| \qquad \text{if } x > k.$$

Hence, $f(x)$ is $O(h(x))$.

When big-O notation is used, the function g in the relationship $f(x)$ is $O(g(x))$ is chosen to be as small as possible (sometimes from a set of reference functions, such as functions of the form x^n, where n is a positive integer).

In subsequent discussions, we will almost always deal with functions that take on only positive values. All references to absolute values can be dropped when working with big-O estimates for such functions. Figure 2 illustrates the relationship $f(x)$ is $O(g(x))$.

Example 2 illustrates how big-O notation is used to estimate the growth of functions.

EDMUND LANDAU (1877–1938) Edmund Landau, the son of a Berlin gynecologist, attended high school and university in Berlin. He received his doctorate in 1899, under the direction of Frobenius. Landau first taught at the University of Berlin and then moved to Göttingen, where he was a full professor until the Nazis forced him to stop teaching. Landau's main contributions to mathematics were in the field of analytic number theory. In particular, he established several important results concerning the distribution of primes. He authored a three-volume exposition on number theory as well as other books on number theory and mathematical analysis.

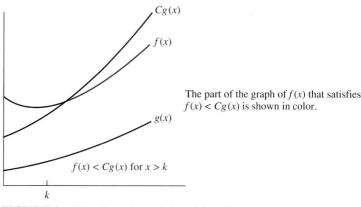

The part of the graph of $f(x)$ that satisfies $f(x) < Cg(x)$ is shown in color.

$f(x) < Cg(x)$ for $x > k$

FIGURE 2   The Function $f(x)$ is $O(g(x))$.

EXAMPLE 2   Show that $7x^2$ is $O(x^3)$.

Solution: Note that when $x > 7$, we have $7x^2 < x^3$. (We can obtain this inequality by multiplying both sides of $x > 7$ by x^2.) Consequently, we can take $C = 1$ and $k = 7$

Links

DONALD E. KNUTH (BORN 1938)   Knuth grew up in Milwaukee, where his father taught book-keeping at a Lutheran high school and owned a small printing business. He was an excellent student, earning academic achievement awards. He applied his intelligence in unconventional ways, winning a contest when he was in the eighth grade by finding over 4,500 words that could be formed from the letters in "Ziegler's Giant Bar." This won a television set for his school and a candy bar for everyone in his class.

Knuth had a difficult time choosing physics over music as his major at the Case Institute of Technology. He then switched from physics to mathematics, and in 1960 he received his bachelor of science degree, simultaneously receiving a master of science degree by a special award of the faculty who considered his work outstanding. At Case, he managed the basketball team and applied his talents by constructing a formula for the value of each player. This novel approach was covered by *Newsweek* and by Walter Cronkite on the CBS television network. Knuth began graduate work at the California Institute of Technology in 1960 and received his Ph.D. there in 1963. During this time he worked as a consultant, writing compilers for different computers.

Knuth joined the staff of the California Institute of Technology in 1963, where he remained until 1968, when he took a job as a full professor at Stanford University. He retired as Professor Emeritus in 1992 to concentrate on writing. He is especially interested in updating and completing new volumes of his series *The Art of Computer Programming,* a work that has had a profound influence on the development of computer science, which he began writing as a graduate student in 1962, focusing on compilers. In common jargon, "Knuth," referring to *The Art of Computer Programming,* has come to mean the reference that answers all questions about such topics as data structures and algorithms.

Knuth is the founder of the modern study of computational complexity. He has made fundamental contributions to the subject of compilers. His dissatisfaction with mathematics typography sparked him to invent the now widely used TeX and Metafont systems. TeX has become a standard language for computer typography. Two of the many awards Knuth has received are the 1974 Turing Award and the 1979 National Medal of Technology, awarded to him by President Carter.

Knuth has written for a wide range of professional journals in computer science and in mathematics. However, his first publication, in 1957, when he was a college freshman, was a parody of the metric system called "The Potrzebie Systems of Weights and Measures," which appeared in *MAD Magazine* and has been in reprint several times. He is a church organist, as his father was. He is also a composer of music for the organ. Knuth believes that writing computer programs can be an aesthetic experience, much like writing poetry or composing music.

Knuth pays $2.56 for the first person to find each error in his books and $0.32 for significant suggestions. If you send him a letter with an error (you will need to use regular mail, since he has given up reading e-mail), he will eventually inform you whether you were the first person to tell him about this error. Be prepared for a long wait, since he receives an overwhelming amount of mail. (The author received a letter years after sending an error report to Knuth, noting that this report arrived several months after the first report of this error.)

as witnesses to establish the relationship $7x^2$ is $O(x^3)$. Alternatively, when $x > 1$, we have $7x^2 < 7x^3$, so that $C = 7$ and $k = 1$ are also witnesses to the relationship $7x^2$ is $O(x^3)$. ◀

EXAMPLE 3 Example 2 shows that $7x^2$ is $O(x^3)$. Is it also true that x^3 is $O(7x^2)$?

Solution: To determine whether x^3 is $O(7x^2)$, we need to determine whether there are constants C and k such that $x^3 \leq C(7x^2)$ whenever $x > k$. The inequality $x^3 \leq C(7x^2)$ is equivalent to the inequality $x \leq 7C$, which follows by dividing the original inequality by the positive quantity x^2. Note that no C exists for which $x \leq 7C$ for all $x > k$ no matter what k is, because x can be made arbitrarily large. It follows that no witnesses C and k exist for this proposed big-O relationship. Hence, x^3 *is not* $O(7x^2)$. ◀

SOME IMPORTANT BIG-*O* RESULTS

Polynomials can often be used to estimate the growth of functions. Instead of analyzing the growth of polynomials each time they occur, we would like a result that can always be used to estimate the growth of a polynomial. The following theorem does this. It shows that the leading term of a polynomial dominates its growth by asserting that a polynomial of degree n or less is $O(x^n)$.

THEOREM 1 Let $f(x) = a_nx^n + a_{n-1}x^{n-1} + \cdots + a_1x + a_0$, where $a_0, a_1, \ldots, a_{n-1}, a_n$ are real numbers. Then $f(x)$ is $O(x^n)$.

Proof: Using the triangle inequality (see Exercise 35 in Section 1.5), if $x > 1$ we have

$$|f(x)| = |a_nx^n + a_{n-1}x^{n-1} + \cdots + a_1x + a_0|$$
$$\leq |a_n|x^n + |a_{n-1}|x^{n-1} + \cdots + |a_1|x + |a_0|$$
$$= x^n \left(|a_n| + |a_{n-1}|/x + \cdots + |a_1|/x^{n-1} + |a_0|/x^n\right)$$
$$\leq x^n \left(|a_n| + |a_{n-1}| + \cdots + |a_1| + |a_0|\right).$$

This shows that

$$|f(x)| \leq Cx^n$$

where $C = |a_n| + |a_{n-1}| + \cdots + |a_0|$ whenever $x > 1$. Hence, the witnesses $C = |a_n| + |a_{n-1}| + \cdots + |a_0|$ and $k = 1$ show that $f(x)$ is $O(x^n)$. ◁

We now give some examples involving functions that have the set of positive integers as their domains.

EXAMPLE 4 How can big-O notation be used to estimate the sum of the first n positive integers?

Solution: Since each of the integers in the sum of the first n positive integers does not exceed n, it follows that

$$1 + 2 + \cdots + n \leq n + n + \cdots + n = n^2.$$

From this inequality it follows that $1 + 2 + 3 + \cdots + n$ is $O(n^2)$, taking $C = 1$ and $k = 1$ as witnesses. (In this example the domains of the functions in the big-O relationship are the set of positive integers.) ◀

In Example 5 big-O estimates will be developed for the factorial function and its logarithm. These estimates will be important in the analysis of the number of steps used in sorting procedures.

EXAMPLE 5   Give big-O estimates for the factorial function and the logarithm of the factorial function, where the factorial function $f(n) = n!$ is defined by

$$n! = 1 \cdot 2 \cdot 3 \cdot \ \cdots \ \cdot n$$

whenever n is a positive integer, and $0! = 1$. For example,

$$1! = 1, \quad 2! = 1 \cdot 2 = 2, \quad 3! = 1 \cdot 2 \cdot 3 = 6, \quad 4! = 1 \cdot 2 \cdot 3 \cdot 4 = 24.$$

Note that the function $n!$ grows rapidly. For instance,

$$20! = 2,432,902,008,176,640,000.$$

Solution: A big-O estimate for $n!$ can be obtained by noting that each term in the product does not exceed n. Hence,

$$
\begin{aligned}
n! &= 1 \cdot 2 \cdot 3 \cdot \ \cdots \ \cdot n \\
&\leq n \cdot n \cdot n \cdot \ \cdots \ \cdot n \\
&= n^n.
\end{aligned}
$$

This inequality shows that $n!$ is $O(n^n)$, taking $C = 1$ and $k = 1$ as witnesses. Taking logarithms of both sides of the inequality established for $n!$, we obtain

$$\log n! \leq \log n^n = n \log n.$$

This implies that $\log n!$ is $O(n \log n)$, again taking $C = 1$ and $k = 1$ as witnesses. ◀

EXAMPLE 6   In Section 3.3 we will show that

$$n < 2^n$$

whenever n is a positive integer. Using this inequality we can conclude that n is $O(2^n)$. (Take $k = C = 1$ as witnesses.) Since the logarithm function is increasing, taking logarithms (base 2) of both sides of this inequality shows that

$$\log n < n.$$

It follows that

$$\log n \text{ is } O(n).$$

(Again we take $C = k = 1$ as witnesses.)

If we have logarithms to a base b, where b is different from 2, we still have $\log_b n$ is $O(n)$ since

$$\log_b n = \frac{\log n}{\log b} < \frac{n}{\log b}$$

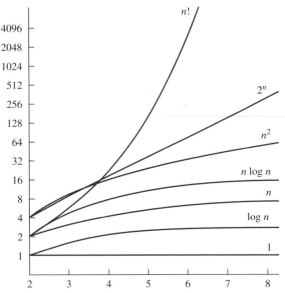

FIGURE 3   A Display of the Growth of Functions Commonly Used in Big-O Estimates.

whenever n is a positive integer. We take $C = 1/\log b$ and $k = 1$ as witnesses. (We have used Theorem 3 in Appendix 1 to see that $\log_b n = \log n / \log b$.) ◀

As mentioned before, big-O notation is used to estimate the number of operations needed to solve a problem using a specified procedure or algorithm. The functions used in these estimates often include the following:

$$1,\ \log n,\ n,\ n \log n,\ n^2,\ 2^n,\ n!$$

Using calculus it can be shown that each function in the list is smaller than the succeeding function, in the sense that the ratio of a function and the succeeding function tends to zero as n grows without bound. Figure 3 displays the graphs of these functions, using a scale for the values of the functions that doubles for each successive marking on the graph.

THE GROWTH OF COMBINATIONS OF FUNCTIONS

Many algorithms are made up of two or more separate subprocedures. The number of steps used by a computer to solve a problem with input of a specified size using such an algorithm is the sum of the number of steps used by these subprocedures. To give a big-O estimate for the number of steps needed, it is necessary to find big-O estimates for the number of steps used by each subprocedure and then combine these estimates.

Big-O estimates of combinations of functions can be provided if care is taken when different big-O estimates are combined. In particular, it is often necessary to estimate the growth of the sum and the product of two functions. What can be said if big-O estimates for each of two functions are known? To see what sort of estimates hold for the sum and the product of two functions, suppose that $f_1(x)$ is $O(g_1(x))$ and $f_2(x)$ is $O(g_2(x))$.

From the definition of big-O notation, there are constants C_1, C_2, k_1, and k_2 such that

$$|f_1(x)| \leq C_1 |g_1(x)|$$

when $x > k_1$, and

$$|f_2(x)| \leq C_2|g_2(x)|$$

when $x > k_2$. To estimate the sum of $f_1(x)$ and $f_2(x)$, note that

$$|(f_1 + f_2)(x)| = |f_1(x) + f_2(x)|$$

$$\leq |f_1(x)| + |f_2(x)| \quad \text{using triangle inequality } |a + b| \leq |a| + |b|.$$

When x is greater than both k_1 and k_2, it follows from the inequalities for $|f_1(x)|$ and $|f_2(x)|$ that

$$|f_1(x)| + |f_2(x)| \leq C_1|g_1(x)| + C_2|g_2(x)|$$

$$\leq C_1|g(x)| + C_2|g(x)|$$

$$= (C_1 + C_2)|g(x)|$$

$$= C|g(x)|,$$

where $C = C_1 + C_2$ and $g(x) = \max(|g_1(x)|, |g_2(x)|)$. [Here $\max(a, b)$ denotes the maximum, or larger, of a and b.]

This inequality shows that $|(f_1 + f_2)(x)| \leq C|g(x)|$ whenever $x > k$, where $k = \max(k_1, k_2)$. We state this useful result as Theorem 2.

THEOREM 2   Suppose that $f_1(x)$ is $O(g_1(x))$ and $f_2(x)$ is $O(g_2(x))$. Then $(f_1 + f_2)(x)$ is $O(\max(|g_1(x)|, |g_2(x)|))$.

We often have big-O estimates for f_1 and f_2 in terms of the same function g. In this situation, Theorem 2 can be used to show that $(f_1 + f_2)(x)$ is also $O(g(x))$, since $\max(g(x), g(x)) = g(x)$. This result is stated in Corollary 1.

COROLLARY 1   Suppose that $f_1(x)$ and $f_2(x)$ are both $O(g(x))$. Then $(f_1 + f_2)(x)$ is $O(g(x))$.

In a similar way big-O estimates can be derived for the product of the functions f_1 and f_2. When x is greater than $\max(k_1, k_2)$ it follows that

$$|(f_1 f_2)(x)| = |f_1(x)||f_2(x)|$$

$$\leq C_1|g_1(x)|C_2|g_2(x)|$$

$$\leq C_1 C_2|(g_1 g_2)(x)|$$

$$\leq C|(g_1 g_2)(x)|,$$

where $C = C_1 C_2$. From this inequality, it follows that $f_1(x)f_2(x)$ is $O(g_1 g_2)$, since there are constants C and k, namely, $C = C_1 C_2$ and $k = \max(k_1, k_2)$, such that $|(f_1 f_2)(x)| \leq C|g_1(x)g_2(x)|$ whenever $x > k$. This result is stated in Theorem 3.

THEOREM 3   Suppose that $f_1(x)$ is $O(g_1(x))$ and $f_2(x)$ is $O(g_2(x))$. Then $(f_1 f_2)(x)$ is $O(g_1(x)g_2(x))$.

The goal in using big-O notation to estimate functions is to choose a function $g(x)$ that grows relatively slowly so that $f(x)$ is $O(g(x))$. The following examples illustrate how to use Theorems 2 and 3 to do this. The type of analysis given in these examples is often used in the analysis of the time used to solve problems using computer programs.

EXAMPLE 7   Give a big-O estimate for $f(n) = 3n \log(n!) + (n^2 + 3) \log n$, where n is a positive integer.

Solution: First, the product $3n \log(n!)$ will be estimated. From Example 5 we know that $\log(n!)$ is $O(n \log n)$. Using this estimate and the fact that $3n$ is $O(n)$, Theorem 3 gives the estimate that $3n \log(n!)$ is $O(n^2 \log n)$.

Next, the product $(n^2 + 3) \log n$ will be estimated. Since $(n^2 + 3) < 2n^2$ when $n > 2$, it follows that $n^2 + 3$ is $O(n^2)$. Thus, from Theorem 3 it follows that $(n^2 + 3) \log n$ is $O(n^2 \log n)$. Using Theorem 2 to combine the two big-O estimates for the products shows that $f(n) = 3n \log(n!) + (n^2 + 3) \log n$ is $O(n^2 \log n)$.   ◀

EXAMPLE 8   Give a big-O estimate for $f(x) = (x + 1) \log(x^2 + 1) + 3x^2$.

Solution: First, a big-O estimate for $(x + 1) \log(x^2 + 1)$ will be found. Note that $(x + 1)$ is $O(x)$. Furthermore, $x^2 + 1 \leq 2x^2$ when $x > 1$. Hence,

$$\log(x^2 + 1) \leq \log(2x^2) = \log 2 + \log x^2 = \log 2 + 2 \log x \leq 3 \log x,$$

if $x > 2$. This shows that $\log(x^2 + 1)$ is $O(\log x)$.

From Theorem 3 it follows that $(x + 1) \log(x^2 + 1)$ is $O(x \log x)$. Since $3x^2$ is $O(x^2)$, Theorem 2 tells us that $f(x)$ is $O(\max(x \log x, x^2))$. Since $x \log x \leq x^2$, for $x > 1$, it follows that $f(x)$ is $O(x^2)$.   ◀

BIG-OMEGA AND BIG-THETA NOTATION

Big-O notation is used extensively to describe the growth of functions, but it has limitations. In particular, when $f(x)$ is $O(g(x))$, we have an upper bound, in terms of $g(x)$, for the size of $f(x)$ for large values of x. However, big-O notation does not provide a lower bound for the size of $f(x)$ for large x. For this, we use **big-Omega notation.** When we want to give both an upper and a lower bound on the size of a function $f(x)$, relative to a reference function $g(x)$, we use **big-Theta notation.** Both big-Omega and big-Theta notation were introduced by Donald Knuth in the 1970s. His motivation for introducing these notations was the common misuse of big-O notation when both an upper and a lower bound on the size of a function are needed.

We now define big-Omega notation and illustrate its use. After doing so, we will do the same for big-Theta notation.

There is a strong connection between big-O and big-Omega notation. In particular, $f(x)$ is $\Omega(g(x))$ if and only if $g(x)$ is $O(f(x))$. We leave the verification of this fact as a straightforward exercise for the reader.

DEFINITION 2   Let f and g be functions from the set of integers or the set of real numbers to the set of real numbers. We say that $f(x)$ is $\Omega(g(x))$ if there are positive constants C and k such that

$$|f(x)| \geq C|g(x)|$$

whenever $x > k$. (This is read as "$f(x)$ is big-Omega of $g(x)$".)

EXAMPLE 9   The function $f(x) = 8x^3 + 5x^2 + 7$ is $\Omega(g(x))$, where $g(x)$ is the function $g(x) = x^3$. This is easy to see since $f(x) = 8x^3 + 5x^2 + 7 \geq 8x^3$ for all positive real numbers x. This is equivalent to saying that $g(x) = x^3$ is $O(8x^3 + 5x^2 + 7)$, which can be established directly by turning the inequality around.   ◄

Often, it is important to know the order of growth of a function in terms of some relatively simple reference function such as x^n when n is a positive integer or c^x, where $c > 1$. Knowing the order of growth requires that we have both an upper bound and a lower bound for the size of the function. That is, given a function $f(x)$, we want a reference function $g(x)$ such that $f(x)$ is $O(g(x))$ and $f(x)$ is $\Omega(g(x))$. Big-Theta notation, defined as follows, is used to express both of these relationships, providing both an upper and a lower bound on the size of a function.

DEFINITION 3   Let f and g be functions from the set of integers or the set of real numbers to the set of real numbers. We say that $f(x)$ is $\Theta(g(x))$ if $f(x)$ is $O(g(x))$ and $f(x)$ is $\Omega(g(x))$. When $f(x)$ is $\Theta(g(x))$, we say that "f is big-Theta of $g(x)$" and we also say that $f(x)$ is of *order* $g(x)$.

When $f(x)$ is $\Theta(g(x))$, it is also the case that $g(x)$ is $\Theta(f(x))$. Also note that $f(x)$ is $\Theta(g(x))$ if and only if $f(x)$ is $O(g(x))$ and $g(x)$ is $O(f(x))$ (see Exercise 25).

Usually, when big-Theta notation is used, the function $g(x)$ in $\Theta(g(x))$ is a relatively simple reference function, such as x^n, c^x, $\log x$, and so on, while $f(x)$ can be relatively complicated.

EXAMPLE 10   We showed (in Example 4) that the sum of the first n positive integers is $O(n^2)$. Is this sum of order n^2?

Solution: Let $f(n) = 1 + 2 + 3 + \cdots + n$. Since we already know that $f(n) = O(n^2)$, to show that $f(n)$ is of order n^2 we need to find a positive constant C such that $f(n) > Cn^2$ for sufficiently large integers n. To obtain a lower bound for this sum, we can ignore the first half of the terms. Summing only the terms greater than $\lceil n/2 \rceil$, we find that

$$1 + 2 + \cdots + n \geq \lceil n/2 \rceil + (\lceil n/2 \rceil + 1) + \cdots + n$$
$$\geq \lceil n/2 \rceil + \lceil n/2 \rceil + \cdots + \lceil n/2 \rceil$$
$$= (n - \lceil n/2 \rceil + 1) \lceil n/2 \rceil$$
$$\geq (n/2)(n/2)$$
$$= n^2/4.$$

This shows that $f(n)$ is $\Omega(n^2)$. We conclude that $f(n)$ is of order n^2, or in symbols, $f(n)$ is $\Theta(n^2)$.   ◄

We can show that $f(x)$ is $\Theta(g(x))$ if we can find positive real numbers C_1 and C_2 and a positive real number k such that

$$C_1|g(x)| \leq |f(x)| \leq C_2|g(x)|$$

whenever $x > k$. This shows that $f(x)$ is $O(g(x))$ and $f(x)$ is $\Omega(g(x))$.

EXAMPLE 11   Show that $3x^2 + 8x \log x$ is $\Theta(x^2)$.

Solution: Since $0 \le 8x \log x \le 8x^2$, it follows that $3x^2 + 8x \log x \le 11x^2$ for $x > 1$. Consequently, $3x^2 + 8x \log x$ is $O(x^2)$. Clearly, x^2 is $O(3x^2 + 8x \log x)$. Consequently, $3x^2 + 8x \log x$ is $\Theta(x^2)$. ◀

One useful fact is that the leading term of a polynomial determines its order. For example, if $f(x) = 3x^5 + x^4 + 17x^3 + 2$, then $f(x)$ is of order x^5. This is stated in the following theorem, whose proof is left as an exercise at the end of this section.

THEOREM 4   Let $f(x) = a_n x^n + a_{n-1} x^{n-1} + \cdots + a_1 x + a_0$, where $a_0, a_1, \ldots, a_n$ are real numbers with $a_n \ne 0$. Then $f(x)$ is of order x^n.

EXAMPLE 12   The polynomials $3x^8 + 10x^7 + 221x^2 + 1444$, $x^{19} - 18x^4 - 10,112$, and $-x^{99} + 40001x^{98} + 100003x$ are of orders x^8, x^{19}, and x^{99}, respectively. ◀

Unfortunately, as Knuth observed, big-O notation is often used by careless writers and speakers as if it had the same meaning as big-Theta notation. Keep this in mind when you see big-O notation used. The recent trend has been to use big-Theta notation whenever both upper and lower bounds on the size of a function are needed.

Exercises

In Exercises 1–14, to establish a big-O relationship, find witnesses C and k such that $|f(x)| \le C|g(x)|$ whenever $x > k$.

1. Determine whether each of these functions is $O(x)$.

a) $f(x) = 10$ **b)** $f(x) = 3x + 7$
c) $f(x) = x^2 + x + 1$ **d)** $f(x) = 5 \log x$
e) $f(x) = \lfloor x \rfloor$ **f)** $f(x) = \lceil x/2 \rceil$

2. Determine whether each of these functions is $O(x^2)$.

a) $f(x) = 17x + 11$ **b)** $f(x) = x^2 + 1000$
c) $f(x) = x \log x$ **d)** $f(x) = x^4/2$
e) $f(x) = 2^x$ **f)** $f(x) = \lfloor x \rfloor \cdot \lceil x \rceil$

3. Use the definition of the fact that $f(x)$ is $O(g(x))$ to show that $x^4 + 9x^3 + 4x + 7$ is $O(x^4)$.

4. Use the definition of the fact that $f(x)$ is $O(g(x))$ to show that $2^x + 17$ is $O(3^x)$.

5. Show that $(x^2 + 1)/(x + 1)$ is $O(x)$.

6. Show that $(x^3 + 2x)/(2x + 1)$ is $O(x^2)$.

7. Find the least integer n such that $f(x)$ is $O(x^n)$ for each of these functions.

a) $f(x) = 2x^3 + x^2 \log x$
b) $f(x) = 3x^3 + (\log x)^4$
c) $f(x) = (x^4 + x^2 + 1)/(x^3 + 1)$
d) $f(x) = (x^4 + 5 \log x)/(x^4 + 1)$

8. Find the least integer n such that $f(x)$ is $O(x^n)$ for each of these functions.

a) $f(x) = 2x^2 + x^3 \log x$

b) $f(x) = 3x^5 + (\log x)^4$
c) $f(x) = (x^4 + x^2 + 1)/(x^4 + 1)$
d) $f(x) = (x^3 + 5 \log x)/(x^4 + 1)$

9. Show that $x^2 + 4x + 17$ is $O(x^3)$ but that x^3 is not $O(x^2 + 4x + 17)$.

10. Show that x^3 is $O(x^4)$ but that x^4 is not $O(x^3)$.

11. Show that $3x^4 + 1$ is $O(x^4/2)$ and $x^4/2$ is $O(3x^4 + 1)$.

12. Show that $x \log x$ is $O(x^2)$ but that x^2 is not $O(x \log x)$.

13. Show that 2^n is $O(3^n)$ but that 3^n is not $O(2^n)$.

14. Is it true that x^3 is $O(g(x))$, if g is the given function? [For example, if $g(x) = x + 1$, this question asks whether x^3 is $O(x + 1)$.]

a) $g(x) = x^2$ **b)** $g(x) = x^3$
c) $g(x) = x^2 + x^3$ **d)** $g(x) = x^2 + x^4$
e) $g(x) = 3^x$ **f)** $g(x) = x^3/2$

15. Explain what it means for a function to be $O(1)$.

16. Show that if $f(x)$ is $O(x)$, then $f(x)$ is $O(x^2)$.

17. Suppose that $f(x), g(x)$, and $h(x)$ are functions such that $f(x)$ is $O(g(x))$ and $g(x)$ is $O(h(x))$. Show that $f(x)$ is $O(h(x))$.

18. Let k be a positive integer. Show that $1^k + 2^k + \cdots + n^k$ is $O(n^{k+1})$.

19. Give as good a big-O estimate as possible for each of these functions.

a) $(n^2 + 8)(n + 1)$
b) $(n \log n + n^2)(n^3 + 2)$
c) $(n! + 2^n)(n^3 + \log(n^2 + 1))$

20. Give a big-O estimate for each of these functions. For the function g in your estimate $f(x)$ is $O(g)$, use a simple function g of smallest order.

a) $(n^3 + n^2 \log n)(\log n + 1) + (17 \log n + 19)(n^3 + 2)$

b) $(2^n + n^2)(n^3 + 3^n)$

c) $(n^n + n2^n + 5^n)(n! + 5^n)$

21. Give a big-O estimate for each of these functions. For the function g in your estimate that $f(x)$ is $O(g(x))$ use a simple function g of the smallest order.

a) $n \log(n^2 + 1) + n^2 \log n$

b) $(n \log n + 1)^2 + (\log n + 1)(n^2 + 1)$

c) $n^{2^n} + n^{n^2}$

22. For each function in Exercise 1, determine whether that function is $\Omega(x)$ and whether it is $\Theta(x)$.

23. For each function in Exercise 2, determine whether that function is $\Omega(x^2)$ and whether it is $\Theta(x^2)$.

24. a) Show that $3x + 7$ is $\Theta(x)$.

b) Show that $2x^2 + x - 7$ is $\Theta(x^2)$.

c) Show that $\lfloor x + 1/2 \rfloor$ is $\Theta(x)$.

d) Show that $\log(x^2 + 1)$ is $\Theta(\log_2 x)$.

e) Show that $\log_{10} x$ is $\Theta(\log_2 x)$.

25. Show that $f(x)$ is $\Theta(g(x))$ if and only if $f(x)$ is $O(g(x))$ and $g(x)$ is $O(f(x))$.

26. Show that if $f(x)$ and $g(x)$ are functions from the set of real numbers to the set of real numbers, then $f(x)$ is $O(g(x))$ if and only if $g(x)$ is $\Omega(f(x))$.

27. Show that if $f(x)$ and $g(x)$ are functions from the set of real numbers to the set of real numbers, then $f(x)$ is $\Theta(g(x))$ if and only if there are positive constants k, C_1, and C_2 such that $C_1|g(x)| \leq |f(x)| \leq C_2|g(x)|$ whenever $x > k$.

28. a) Show that $3x^2 + x + 1$ is $\Theta(3x^2)$ by directly finding the constants k, C_1, C_2 in Exercise 27.

b) Express the relationship in part (a) using a picture showing the functions $3x^2 + x + 1$, $C_1 \cdot 3x^2$, and $C_2 \cdot 3x^2$, and the constant k on the x-axis, where C_1, C_2, and k are the constants you found in part (a) to show that $3x^2 + x + 1$ is $\Theta(3x^2)$.

29. Express the relationship $f(x)$ is $\Theta(g(x))$ using a picture. Show the graphs of the functions $f(x)$, $C_1|g(x)|$, and $C_2|g(x)|$, as well as the constant k on the x-axis.

30. Explain what it means for a function to be $\Omega(1)$.

31. Explain what it means for a function to be $\Theta(1)$.

32. Give a big-O estimate of the product of the first n odd positive integers.

33. Show that if f and g are real-valued functions such that $f(x)$ is $O(g(x))$, then $f^k(x)$ is $O(g^k(x))$. [Note that $f^k(x) = f(x)^k$.]

34. Show that if $f(x)$ is $O(\log_b x)$ where $b > 1$, then $f(x)$ is $O(\log_a x)$ where $a > 1$.

35. Suppose that $f(x)$ is $O(g(x))$ where f and g are increasing and unbounded functions. Show that $\log |f(x)|$ is $O(\log |g(x)|)$.

36. Suppose that $f(x)$ is $O(g(x))$. Does it follow that $2^{f(x)}$ is $O(2^{g(x)})$?

37. Let $f_1(x)$ and $f_2(x)$ be functions from the set of real numbers to the set of positive real numbers. Show that if $f_1(x)$ and $f_2(x)$ are both $\Theta(g(x))$, where $g(x)$ is a function from the set of real numbers to the set of positive real numbers, then $f_1(x) + f_2(x)$ is $\Theta(g(x))$. Is this still true if $f_1(x)$ and $f_2(x)$ can take negative values?

38. Suppose that $f(x)$, $g(x)$, and $h(x)$ are functions such that $f(x)$ is $\Theta(g(x))$ and $g(x)$ is $\Theta(h(x))$. Show that $f(x)$ is $\Theta(h(x))$.

39. If $f_1(x)$ and $f_2(x)$ are functions from the set of positive integers to the set of positive real numbers and $f_1(x)$ and $f_2(x)$ are both $\Theta(g(x))$, is $(f_1 - f_2)(x)$ also $\Theta(g(x))$? Either prove that it is or give a counterexample.

40. Show that if $f_1(x)$ and $f_2(x)$ are functions from the set of positive integers to the set of real numbers and $f_1(x)$ is $\Theta(g_1(x))$ and $f_2(x)$ is $\Theta(g_2(x))$, then $(f_1 f_2)(x)$ is $\Theta(g_1 g_2(x))$.

41. Find functions f and g from the set of positive integers to the set of real numbers such that $f(n)$ is not $O(g(n))$ and $g(n)$ is not $O(f(n))$ simultaneously.

42. Express the relationship $f(x)$ is $\Omega(g(x))$ using a picture. Show the graphs of the functions $f(x)$ and $Cg(x)$, as well as the constant k on the real axis.

43. Show that if $f_1(x)$ is $\Theta(g_1(x))$, $f_2(x)$ is $\Theta(g_2(x))$, and $f_2(x) \neq 0$ and $g_2(x) \neq 0$ for all real numbers $x > 0$, then $(f_1/f_2)(x)$ is $\Theta((g_1/g_2)(x))$.

44. Show that if $f(x) = a_n x^n + a_{n-1} x^{n-1} + \cdots + a_1 x + a_0$, where $a_0, a_1, \ldots, a_{n-1}, a_n$ are real numbers and $a_n \neq 0$, then $f(x)$ is $\Theta(x^n)$.

Big-O, big-Theta, and big-Omega notation can be extended to functions in more than one variable. For example, the statement $f(x, y)$ is $O(g(x, y))$ means that there exist constants C, k_1, and k_2 such that $|f(x, y)| \leq C|g(x, y)|$ whenever $x > k_1$ and $y > k_2$.

45. Define the statement $f(x, y)$ is $\Theta(g(x, y))$.

46. Define the statement $f(x, y)$ is $\Omega(g(x, y))$.

47. Show that $(x^2 + xy + x \log y)^3$ is $O(x^6 y^3)$.

48. Show that $x^5 y^3 + x^4 y^4 + x^3 y^5$ is $\Omega(x^3 y^3)$.

49. Show that $\lfloor xy \rfloor$ is $O(xy)$.

50. Show that $\lceil xy \rceil$ is $\Omega(xy)$.

The following problems deal with another type of asymptotic notation, called **little-o** notation. Because little-o notation is based on the concept of limits, a knowledge of calculus is needed for these problems. We say that $f(x)$ is $o(g(x))$ [read $f(x)$ is "little-oh" of $g(x)$], when

$$\lim_{x \to \infty} \frac{f(x)}{g(x)} = 0.$$

51. (Calculus required) Show that

a) x^2 is $o(x^3)$.

b) $x \log x$ is $o(x^2)$.

c) x^2 is $o(2^x)$.

d) $x^2 + x + 1$ is not $o(x^2)$.

52. (Calculus required)

   a) Show that if $f(x)$ and $g(x)$ are functions such that $f(x)$ is $o(g(x))$ and c is a constant, then $cf(x)$ is $o(g(x))$ where $(cf)(x) = cf(x)$.

   b) Show that if $f_1(x)$, $f_2(x)$, and $g(x)$ are functions such that $f_1(x)$ is $o(g(x))$ and $f_2(x)$ is $o(g(x))$, then $(f_1 + f_2)(x)$ is $o(g(x))$, where $(f_1 + f_2)(x) = f_1(x) + f_2(x)$.

53. (Calculus required) Represent pictorially that $x \log x$ is $o(x^2)$ by graphing $x \log x$, x^2, and $x \log x / x^2$. Explain how this picture shows that $x \log x$ is $o(x^2)$.

54. (Calculus required) Express the relationship $f(x)$ is $o(g(x))$ using a picture. Show the graphs of $f(x)$, $g(x)$, and $f(x)/g(x)$.

*55. (Calculus required) Suppose that $f(x)$ is $o(g(x))$. Does it follow that $2^{f(x)}$ is $o(2^{g(x)})$?

*56. (Calculus required) Suppose that $f(x)$ is $o(g(x))$. Does it follow that $\log|f(x)|$ is $o(\log|g(x)|)$?

57. (Calculus required) The two parts of this exercise describe the relationship between little-o and big-O notation.

   a) Show that if $f(x)$ and $g(x)$ are functions such that $f(x)$ is $o(g(x))$, then $f(x)$ is $O(g(x))$.

   b) Show that if $f(x)$ and $g(x)$ are functions such that $f(x)$ is $O(g(x))$, then it does not necessarily follow that $f(x)$ is $o(g(x))$.

58. (Calculus required) Show that if $f(x)$ is a polynomial of degree n and $g(x)$ is a polynomial of degree m where $m > n$, then $f(x)$ is $o(g(x))$.

59. (Calculus required) Show that if $f_1(x)$ is $O(g(x))$ and $f_2(x)$ is $o(g(x))$, then $f_1(x) + f_2(x)$ is $O(g(x))$.

60. (Calculus required) Let H_n be the nth **harmonic number**

$$H_n = 1 + \frac{1}{2} + \frac{1}{3} + \cdots + \frac{1}{n}.$$

Show that H_n is $O(\log n)$. (*Hint:* First establish the inequality

$$\sum_{j=2}^{n} \frac{1}{j} < \int_1^n \frac{1}{x} dx$$

by showing that the sum of the areas of the rectangles of height $1/j$ with base from $j-1$ to j, for $j = 2, 3, \ldots, n$, is less than the area under the curve $y = 1/x$ from 2 to n.)

*61. Show that $n \log n$ is $O(\log n!)$.

62. Determine whether $\log(n!)$ is $\Theta(n \log n)$. Justify your answer.

Let $f(x)$ and $g(x)$ be functions from the set of real numbers to the set of real numbers. We say that the functions f and g are **asymptotic** and write $f(x) \sim g(x)$ if $\lim_{x \to \infty} f(x)/g(x) = 1$.

63. (Calculus required) For each of these pairs of functions, determine whether f and g are asymptotic.

   a) $f(x) = x^2 + 3x + 7$, $g(x) = x^2 + 10$

   b) $f(x) = x^2 \log x$, $g(x) = x^3$

   c) $f(x) = x^4 + \log(3x^8 + 7)$, $g(x) = (x^2 + 17x + 3)^2$

   d) $f(x) = (x^3 + x^2 + x + 1)^4$, $g(x) = (x^4 + x^3 + x^2 + x + 1)^3$.

   e) $f(x) = \log(x^2 + 1)$, $g(x) = \log x$

   f) $f(x) = 2^{x+3}$, $g(x) = 2^{x+7}$

   g) $f(x) = 2^{2^x}$, $g(x) = 2^{x^2}$

2.3   Complexity of Algorithms

INTRODUCTION

When does an algorithm provide a satisfactory solution to a problem? First, it must always produce the correct answer. How this can be demonstrated will be discussed in Chapter 3. Second, it should be efficient. The efficiency of algorithms will be discussed in this section.

How can the efficiency of an algorithm be analyzed? One measure of efficiency is the time used by a computer to solve a problem using the algorithm, when input values are of a specified size. A second measure is the amount of computer memory required to implement the algorithm when input values are of a specified size.

Questions such as these involve the **computational complexity** of the algorithm. An analysis of the time required to solve a problem of a particular size involves the **time complexity** of the algorithm. An analysis of the computer memory required involves the **space complexity** of the algorithm. Considerations of the time and space complexity of an algorithm are essential when algorithms are implemented. It is obviously important to know whether an algorithm will produce an answer in a microsecond, a minute, or a billion years. Likewise, the required memory must be available to solve a problem, so that space complexity must be taken into account.

Considerations of space complexity are tied in with the particular data structures used to implement the algorithm. Because data structures are not dealt with in detail in this book, space complexity will not be considered. We will restrict our attention to time complexity.

TIME COMPLEXITY

The time complexity of an algorithm can be expressed in terms of the number of operations used by the algorithm when the input has a particular size. The operations used to measure time complexity can be the comparison of integers, the addition of integers, the multiplication of integers, the division of integers, or any other basic operation.

Time complexity is described in terms of the number of operations required instead of actual computer time because of the difference in time needed for different computers to perform basic operations. Moreover, it is quite complicated to break all operations down to the basic bit operations that a computer uses. Furthermore, the fastest computers in existence can perform basic bit operations (for instance, adding, multiplying, comparing, or exchanging two bits) in 10^{-9} second (1 nanosecond), but personal computers may require 10^{-6} second (1 microsecond), which is 1000 times as long, to do the same operations.

We illustrate how to analyze the time complexity of an algorithm by considering Algorithm 1 of Section 2.1, which finds the maximum of a finite set of integers.

EXAMPLE 1   Describe the time complexity of Algorithm 1 of Section 2.1 for finding the maximum element in a set.

Solution: The number of comparisons will be used as the measure of the time complexity of the algorithm, since comparisons are the basic operations used.

To find the maximum element of a set with n elements, listed in an arbitrary order, the temporary maximum is first set equal to the initial term in the list. Then, after a comparison has been done to determine that the end of the list has not yet been reached, the temporary maximum and second term are compared, updating the temporary maximum to the value of the second term if it is larger. This procedure is continued, using two additional comparisons for each term of the list—one to determine that the end of the list has not been reached and another to determine whether to update the temporary maximum. Since two comparisons are used for each of the second through the nth elements and one more comparison is used to exit the loop when $i = n + 1$, exactly $2(n - 1) + 1 = 2n - 1$ comparisons are used whenever this algorithm is applied. Hence, the algorithm for finding the maximum of a set of n elements has time complexity $\Theta(n)$, measured in terms of the number of comparisons used.   ◀

Next, we will analyze the time complexity of searching algorithms.

EXAMPLE 2   Describe the time complexity of the linear search algorithm.

Solution: The number of comparisons used by the algorithm will be taken as the measure of the time complexity. At each step of the loop in the algorithm, two comparisons are performed—one to see whether the end of the list has been reached and one to compare the element x with a term of the list. Finally, one more comparison is made outside the loop. Consequently, if $x = a_i$, $2i + 1$ comparisons are used. The most comparisons, $2n + 2$, are required when the element is not in the list. In this case, $2n$ comparisons are used to determine that x is not a_i, for $i = 1, 2, \ldots, n$, an additional comparison is used to exit the loop, and one comparison is made outside the loop. So when x is not in the list, a total of $2n + 2$ comparisons are used. Hence, a linear search requires at most $O(n)$ comparisons. ◄

WORST-CASE COMPLEXITY The type of complexity analysis done in Example 2 is a **worst-case** analysis. By the worst-case performance of an algorithm, we mean the largest number of operations needed to solve the given problem using this algorithm on input of specified size. Worst-case analysis tells us how many operations an algorithm requires to guarantee that it will produce a solution.

EXAMPLE 3   Describe the time complexity of the binary search algorithm.

Solution: For simplicity, assume there are $n = 2^k$ elements in the list $a_1, a_2, \ldots, a_n$, where k is a nonnegative integer. Note that $k = \log n$. (If n, the number of elements in the list, is not a power of 2, the list can be considered part of a larger list with 2^{k+1} elements, where $2^k < n < 2^{k+1}$. Here 2^{k+1} is the smallest power of 2 larger than n.)

At each stage of the algorithm, i and j, the locations of the first term and the last term of the restricted list at that stage, are compared to see whether the restricted list has more than one term. If $i < j$, a comparison is done to determine whether x is greater than the middle term of the restricted list.

At the first stage the search is restricted to a list with 2^{k-1} terms. So far, two comparisons have been used. This procedure is continued, using two comparisons at each stage to restrict the search to a list with half as many terms. In other words, two comparisons are used at the first stage of the algorithm when the list has 2^k elements, two more when the search has been reduced to a list with 2^{k-1} elements, two more when the search has been reduced to a list with 2^{k-2} elements, and so on, until two comparisons are used when the search has been reduced to a list with $2^1 = 2$ elements. Finally, when one term is left in the list, one comparison tells us that there are no additional terms left, and one more comparison is used to determine if this term is x.

Hence, at most $2k + 2 = 2 \log n + 2$ comparisons are required to perform a binary search when the list being searched has 2^k elements. (If n is not a power of 2, the original list is expanded to a list with 2^{k+1} terms, where $k = \lfloor \log n \rfloor$, and the search requires at most $2 \lceil \log n \rceil + 2$ comparisons.) Consequently, a binary search requires at most $\Theta(\log n)$ comparisons. From this analysis it follows that the binary search algorithm is more efficient, in the worst case, than a linear search. ◄

AVERAGE-CASE COMPLEXITY Another important type of complexity analysis, besides worst-case analysis, is called **average-case** analysis. The average number of operations used to solve the problem over all inputs of a given size is found in this type of analysis. Average-case time complexity analysis is usually much more complicated than worst-case analysis. However, the average-case analysis for the linear search algorithm can be done without difficulty, as shown in Example 4.

EXAMPLE 4   Describe the average-case performance of the linear search algorithm, assuming that the element x is in the list.

Solution: There are n types of possible inputs when x is known to be in the list. If x is the first term of the list, three comparisons are needed, one to determine whether the end of the list has been reached, one to compare x and the first term, and one outside the loop.

If x is the second term of the list, two more comparisons are needed, so that a total of five comparisons are used. In general, if x is the ith term of the list, two comparisons will be used at each of the i steps of the loop, and one outside the loop, so that a total of $2i + 1$ comparisons are needed. Hence, the average number of comparisons used equals

$$\frac{3 + 5 + 7 + \cdots + (2n + 1)}{n} = \frac{2(1 + 2 + 3 + \cdots + n) + n}{n}.$$

In Section 3.3 we will show that

$$1 + 2 + 3 + \cdots + n = \frac{n(n + 1)}{2}.$$

Hence, the average number of comparisons used by the linear search algorithm (when x is known to be in the list) is

$$\frac{2[n(n + 1)/2]}{n} + 1 = n + 2,$$

which is $\Theta(n)$. ◀

Remark: In the analysis in Chapter 4 it has been assumed that x is in the list being searched and it is equally likely that x is in any position. It is also possible to do an average-case analysis of this algorithm when x may not be in the list (see Exercise 13 at the end of this section).

Remark: Although we have counted the comparisons needed to determine whether we have reached the end of a loop, these comparisons are often not counted. From this point on we will ignore such comparisons.

WORST-CASE COMPLEXITY OF TWO SORTING ALGORITHMS We analyze the worst-case complexity of the bubble sort and the insertion sort in Examples 5 and 6.

EXAMPLE 5   What is the worst-case complexity of the bubble sort in terms of the number of comparisons made?

Solution: The bubble sort (described in Example 4 in Section 2.1) sorts a list by performing a sequence of passes through the list. During each pass the bubble sort successively compares adjacent elements, interchanging them if necessary. When the ith pass begins, the $i - 1$ largest elements are guaranteed to be in the correct positions. During this pass, $n - i$ comparisons are used. Consequently, the total number of comparisons used by the bubble sort to order a list of n elements is

$$(n - 1) + (n - 2) + \cdots + 2 + 1 = \frac{(n - 1)n}{2}$$

using a summation formula that we will prove in Section 3.3. Note that the bubble sort always uses this many comparisons, because it continues even if the list becomes completely sorted at some intermediate step. Consequently, the bubble sort uses $(n - 1)n/2$ comparisons, so it has $\Theta(n^2)$ worst-case complexity in terms of the number of comparisons used. ◀

EXAMPLE 6   What is the worst-case complexity of the insertion sort in terms of the number of comparisons made?

Solution: The insertion sort (described in Example 5 in Section 2.1) inserts the jth element into the correct position among the first $j - 1$ elements that have already been put into the correct order. It does this by using a linear search technique, successively comparing the jth element with successive terms until a term that is greater than or equal to it is found or it compares a_j with itself and stops because a_j is not less than itself. Consequently, in the worst case, j comparisons are required to insert the jth element into the correct position. Therefore, the total number of comparisons used by the insertion sort to sort a list of n elements is

$$2 + 3 + \cdots + n = \frac{n(n + 1)}{2} - 1$$

using the summation formula for sum of consecutive integers that we will prove in Section 3.3 and noting that the first term, 1, is missing in this sum. Note that the insertion sort may use considerably fewer comparisons if the smaller elements started out at the end of the list. We conclude that the insertion sort has worst-case complexity $\Theta(n^2)$. ◀

UNDERSTANDING THE COMPLEXITY OF ALGORITHMS

Table 1 displays some common terminology used to describe the time complexity of algorithms. For example, an algorithm that finds the largest of the first 100 terms of a list of n elements by applying Algorithm 1 to the sequence of the first 100 terms, where n is an integer with $n \geq 100$, has **constant complexity** since it uses 99 comparisons no matter what n is (as the reader can verify). The linear search algorithm has **linear** (worst-case or average-case) **complexity** and the binary search algorithm has **logarithmic** (worst-case) **complexity.** Many important algorithms have $n \log n$ complexity, such as the merge sort, which we will introduce in Chapter 3.

An algorithm has **polynomial complexity** if it has complexity $O(n^b)$, where b is an integer with $b \geq 1$. For example, the bubble sort algorithm is a polynomial-time algorithm because it uses $O(n^2)$ comparisons in the worst case. An algorithm has **exponential complexity** if it has time complexity $O(b^n)$, where $b > 1$. The algorithm that determines whether a compound proposition in n variables is satisfiable by checking all possible assignments of truth variables is an algorithm with exponential complexity because it uses $O(2^n)$ operations. Finally, an algorithm has **factorial complexity** if it has $O(n!)$ time com-

---

TABLE 1 Commonly Used Terminology for the Complexity of Algorithms.

Complexity	*Terminology*
$O(1)$	Constant complexity
$O(\log n)$	Logarithmic complexity
$O(n)$	Linear complexity
$O(n \log n)$	$n \log n$ complexity
$O(n^b)$	Polynomial complexity
$O(b^n)$, where $b > 1$	Exponential complexity
$O(n!)$	Factorial complexity

plexity. The algorithm that finds all orders that a travelling salesman could use to visit *n* cities has factorial complexity; we will discuss this algorithm in Chapter 8.

A problem that is solvable using an algorithm with polynomial worst-case complexity is called **tractable,** since the expectation is that the algorithm will produce the solution to the problem for reasonably sized input in a relatively short time. However, if the polynomial in the big-*O* estimate has high degree (such as degree 100) or if the coefficients are extremely large, the algorithm may take an extremely long time to solve the problem. Consequently, that a problem can be solved using an algorithm with polynomial worst-case time complexity is no guarantee that the problem can be solved in a reasonable amount of time for even relatively small input values. Fortunately, in practice, the degree and coefficients of polynomials in such estimates are small.

The situation is much worse for problems that cannot be solved using an algorithm with worst-case polynomial time complexity. Such problems are called **intractable.** Usually, but not always, an extremely large amount of time is required to solve the problem for the worst cases of even small input values. In practice, however, there are situations where an algorithm with worst-case time complexity may be able to solve a problem much more quickly for most cases than for its worst case. When we are willing to allow that some, perhaps small, number of cases may not be solved in a reasonable amount of time, the average-case time complexity is a better measure of how long an algorithm takes to solve a problem. Many problems important in industry are thought to be intractable but can be practically solved for essentially all sets of input that arise in daily life. Another way that intractable problems are handled when they arise in practical applications is that instead of looking for exact solutions of a problem, approximate solutions are sought. It may be the case that fast algorithms exist for finding such approximate solutions, perhaps even with a guarantee that they do not differ by very much from an exact solution.

Some problems even exist for which it can be shown that no algorithm exists for solving them. Such problems are called **unsolvable** (as opposed to **solvable** problems that can be solved using an algorithm). The first proof that there are unsolvable problems was provided by the great English mathematician and computer scientist Alan Turing. The problem he showed unsolvable is the **halting problem.** This problem takes as its input a program together with input to this program. The problem asks whether the program will halt when executed with the input to the program. We will study the halting problem in Section 3.1. (A biography of Alan Turing and a description of some of his other work can be found in Chapter 11.)

The study of the complexity of algorithms goes far beyond what we can describe here. Note, however, that many solvable problems are believed to have the property that no algorithm with polynomial worst-case time complexity solves them, but that a solution, if known, can be checked in polynomial time. Problems for which a solution can be checked in polynomial time are said to belong to the **class NP** (tractable problems are said to belong to **class P**).* There is also an important class of problems, called **NP-complete problems,** with the property that if any of these problems can be solved by a polynomial worst-case time algorithm, then all can be solved by polynomial worst-case time algorithms.

The satisfiability problem is an example of an NP-complete problem—we can quickly verify that an assignment of truth values to the variables of a compound proposition makes it true, but no polynomial time algorithm has been discovered for finding such an assignment of truth values. [For example, an exhaustive search of all possible truth

Links

---

*NP stands for *nondeterministic polynomial* time.

values requires $\Theta(2^n)$ bit operations where n is the number of variables in the compound proposition.] Furthermore, if a polynomial time algorithm for solving the satisfiability problem were known, there would be polynomial time algorithms for all problems known to be in this class of problems (and there are many important problems in this class).

Despite extensive research, no polynomial worst-case time algorithm has been found for any problem in this class. It is generally accepted, although not proven, that no NP-complete problem can be solved in polynomial time. For more information about the complexity of algorithms, consult the references, including [CoLeRi 90], for this section listed at the end of this book.

Note that a big-O estimate of the time complexity of an algorithm expresses how the time required to solve the problem changes as the input grows in size. In practice, the best estimate (that is, with the smallest reference function) that can be shown is used. However, big-O estimates of time complexity cannot be directly translated into the actual amount of computer time used. One reason is that a big-O estimate $f(n)$ is $O(g(n))$, where $f(n)$ is the time complexity of an algorithm and $g(n)$ is a reference function, means that $f(n) \leq Cg(n)$ when $n > k$, where C and k are constants. So without knowing the constants C and k in the inequality, this estimate cannot be used to determine an upper bound on the number of operations used. Moreover, as remarked before, the time required for an operation depends on the type of operation and the computer being used. (Also note that a big-O estimate on the time complexity of an algorithm provides an upper, but not a lower, bound, on the worst-case time required for the algorithm as a function of the input size. To provide a lower bound, a big-Theta estimate should be used. However, for simplicity, we will use big-O estimates when describing the time complexity of algorithms, with the understanding that big-Theta estimates would provide more information.)

However, the time required for an algorithm to solve a problem of a specified size can be determined if all operations can be reduced to the bit operations used by the computer. Table 2 displays the time needed to solve problems of various sizes with an algorithm using the indicated number of bit operations. Times of more than 10^{100} years are indicated with an asterisk. (In Section 2.5 the number of bit operations used to add and multiply two integers will be discussed.) In the construction of this table, each bit operation is assumed to take 10^{-9} second, which is the time required for a bit operation using the fastest computers today. In the future, these times will decrease as faster computers are developed.

It is important to know how long a computer will need to solve a problem. For instance, if an algorithm requires 10 hours, it may be worthwhile to spend the computer time (and money) required to solve this problem. But, if an algorithm requires 10 billion

TABLE 2 The Computer Time Used by Algorithms.

Problem Size	*Bit Operations Used*					
n	$\log n$	n	$n \log n$	n^2	2^n	$n!$
10	3×10^{-9} s	10^{-8} s	3×10^{-8} s	10^{-7} s	10^{-6} s	3×10^{-3} s
10^2	7×10^{-9} s	10^{-7} s	7×10^{-7} s	10^{-5} s	4×10^{13} yr	*
10^3	1.0×10^{-8} s	10^{-6} s	1×10^{-5} s	10^{-3} s	*	*
10^4	1.3×10^{-8} s	10^{-5} s	1×10^{-4} s	10^{-1} s	*	*
10^5	1.7×10^{-8} s	10^{-4} s	2×10^{-3} s	10 s	*	*
10^6	2×10^{-8} s	10^{-3} s	2×10^{-2} s	17 min	*	*

years to solve a problem, it would be unreasonable to use resources to implement this algorithm. One of the most interesting phenomena of modern technology is the tremendous increase in the speed and memory space of computers. Another important factor that decreases the time needed to solve problems on computers is **parallel processing,** which is the technique of performing sequences of operations simultaneously.

Efficient algorithms, including most algorithms with polynomial time complexity, benefit most from significant technology improvements. However, these technology improvements offer little help in overcoming the complexity of algorithms of exponential or factorial time complexity. Because of the increased speed of computation, increases in computer memory, and the use of algorithms that take advantage of parallel processing, many problems that were considered impossible to solve 5 years ago are now routinely solved, and certainly 5 years from now this statement will still be true.

Exercises

1. How many comparisons are used by the algorithm given in Exercise 16 of Section 2.1 to find the smallest natural number in a sequence of n natural numbers?

2. Write the algorithm that puts the first four terms of a list of arbitrary length in increasing order. Show that this algorithm has time complexity $O(1)$ in terms of the number of comparisons used.

3. Suppose that an element is known to be among the first four elements in a list of 32 elements. Would a linear search or a binary search locate this element more rapidly?

4. Determine the number of multiplications used to find x^{2^k} starting with x and successively squaring (to find x^2, x^4, and so on). Is this a more efficient way to find x^{2^k} than by multiplying x by itself the appropriate number of times?

5. Give a big-O estimate for the number of comparisons used by the algorithm that determines the number of 1s in a bit string by examining each bit of the string to determine whether it is a 1 bit (see Exercise 25 of Section 2.1).

* 6. a) Show that this algorithm determines the number of 1 bits in the bit string S:

> **procedure** *bit count*(*S*: bit string)
> *count* := 0
> **while** $S \neq 0$
> **begin**
>   *count* := *count* + 1
> $S := S \wedge (S - 1)$
> **end** {*count* is the number of 1s in S}

Here $S - 1$ is the bit string obtained by changing the rightmost 1 bit of S to a 0 and all the 0 bits to the right of this to 1s. [Recall that $S \wedge (S - 1)$ is the bitwise *AND* of S and $S - 1$.]

b) How many bitwise *AND* operations are needed to find the number of 1 bits in a string S?

7. The conventional algorithm for evaluating a polynomial $a_nx^n + a_{n-1}x^{n-1} + \cdots + a_1x + a_0$ at $x = c$ can be expressed in pseudocode by

> **procedure** *polynomial*(*c*, $a_0, a_1, \ldots, a_n$: real numbers)
> *power* := 1
> $y := a_0$
> **for** $i := 1$ **to** n
> **begin**
>   *power* := *power* $* c$
> $y := y + a_i * power$
> **end** {$y = a_nc^n + a_{n-1}c^{n-1} + \cdots + a_1c + a_0$}

where the final value of y is the value of the polynomial at $x = c$.

a) Evaluate $3x^2 + x + 1$ at $x = 2$ by working through each step of the algorithm.

b) Exactly how many multiplications and additions are used to evaluate a polynomial of degree n at $x = c$? (Do not count additions used to increment the loop variable.)

8. There is a more efficient algorithm (in terms of the number of multiplications and additions used) for evaluating polynomials than the conventional algorithm described in the previous exercise. It is called **Horner's method.** This pseudocode shows how to use this method to find the value of $a_nx^n + a_{n-1}x^{n-1} + \cdots + a_1x + a_0$ at $x = c$.

> **procedure** *Horner*(*c*, $a_0, a_1, a_2, \ldots, a_n$: real numbers)
> $y := a_n$
> **for** $i := 1$ **to** n
> $y := y * c + a_{n-i}$
> {$y = a_nc^n + a_{n-1}c^{n-1} + \cdots + a_1c + a_0$}

a) Evaluate $3x^2 + x + 1$ at $x = 2$ by working through each step of the algorithm.

b) Exactly how many multiplications and additions are used by this algorithm to evaluate a polynomial of degree n at $x = c$? (Do not count additions used to increment the loop variable.)

9. How large a problem can be solved in 1 second using an algorithm that requires $f(n)$ bit operations, where each bit operation is carried out in 10^{-9} second, with these values for $f(n)$?

    a) $\log n$ **b)** n **c)** $n \log n$
    d) n^2 **e)** 2^n **f)** $n!$

10. How much time does an algorithm take to solve a problem of size n if this algorithm uses $2n^2 + 2^n$ bit operations, each requiring 10^{-9} second, with these values of n?

    a) 10   **b)** 20
    c) 50   **d)** 100

11. How much time does an algorithm using 2^{50} bit operations need if each bit operation takes these amounts of time?

    a) 10^{-6} second
    b) 10^{-9} second
    c) 10^{-12} second

12. Determine the least number of comparisons, or best-case performance,

    a) required to find the maximum of a sequence of n integers, using Algorithm 1 of Section 2.1.
    b) used to locate an element in a list of n terms with a linear search.
    c) used to locate an element in a list of n terms using a binary search.

13. Analyze the average-case performance of the linear search algorithm, if exactly half the time element x is not in the list and if x is in the list it is equally likely to be in any position.

14. An algorithm is called **optimal** for the solution of a problem with respect to a specified operation if there is no algorithm for solving this problem using fewer operations.

    a) Show that Algorithm 1 in Section 2.1 is an optimal algorithm with respect to the number of comparisons of integers. (*Note:* Comparisons used for bookkeeping in the loop are not of concern here.)
    b) Is the linear search algorithm optimal with respect to the number of comparisons of integers (not including comparisons used for bookkeeping in the loop)?

15. Describe the worst-case time complexity, measured in terms of comparisons, of the ternary search algorithm described in Exercise 27 of Section 2.1.

16. Describe the worst-case time complexity, measured in terms of comparisons, of the search algorithm described in Exercise 28 of Section 2.1.

17. Analyze the worst-case time complexity of the algo-

rithm you devised in Exercise 29 of Section 2.1 for locating a mode in a list of nondecreasing integers.

18. Analyze the worst-case time complexity of the algorithm you devised in Exercise 30 of Section 2.1 for locating all modes in a list of nondecreasing integers.

19. Analyze the worst-case time complexity of the algorithm you devised in Exercise 31 of Section 2.1 for finding the first term of a sequence of integers equal to some previous term.

20. Analyze the worst-case time complexity of the algorithm you devised in Exercise 32 of Section 2.1 for finding all terms of a sequence that are greater than the sum of all previous terms.

21. Analyze the worst-case time complexity of the algorithm you devised in Exercise 33 of Section 2.1 for finding the first term of a sequence less than the immediately preceding term.

22. Determine the worst-case complexity in terms of comparisons of the algorithm from Exercise 5 in Section 2.1 for determining all values that occur more than once in a sorted list of integers.

23. Determine the worst-case complexity in terms of comparisons of the algorithm from Exercise 9 in Section 2.1 for determining whether a string is a palindrome.

24. How many comparisons does the selection sort (see preamble to Exercise 41 in Section 2.1) use to sort n items? Use your answer to give a big-O estimate of the complexity of the selection sort in terms of number of comparisons for the selection sort.

25. Find a big-O estimate for the worst-case complexity in terms of number of comparisons used and the number of terms swapped by the binary insertion sort described in the preamble to Exercise 47 in Section 2.1.

26. Show that the greedy algorithm for making change for n cents using quarters, dimes, nickels, and pennies has $O(n)$ complexity measured in terms of comparisons needed.

27. Describe how the number of comparisons used in the worst case changes when these algorithms are used to search for an element of a list when the size of the list doubles from n to $2n$, where n is a positive integer.

    a) linear search
    b) binary search

28. Describe how the number of comparisons used in the worst case changes when the size of the list to be sorted doubles from n to $2n$, where n is a positive integer when these sorting algorithms are used.
    a) bubble sort
    b) insertion sort
    c) selection sort (described in the preamble to Exercise 41 in Section 2.1)
    d) binary insertion sort (described in the preamble to Exercise 47 in Section 2.1)

2.4 The Integers and Division

INTRODUCTION

The part of discrete mathematics involving the integers and their properties belongs to the branch of mathematics called **number theory.** This section is the beginning of a three-section introduction to number theory. In this section we will review some basic concepts of number theory, including divisibility, greatest common divisors, and modular arithmetic. In Section 2.5 we will describe several important algorithms from number theory, tying together the material in Sections 2.1 and 2.3 on algorithms and their complexity with the notions introduced in this section. For example, we will introduce algorithms for finding the greatest common divisor of two positive integers and for performing computer arithmetic using binary expansions. Finally, in Section 2.6, we will continue our study of number theory by introducing some important results and their applications to computer arithmetic and cryptology, the study of secret messages.

The ideas that we will develop in this section are based on the notion of divisibility. One important concept based on divisibility is that of a prime number. A prime is an integer greater than 1 that is divisible only by 1 and by itself. Determining whether an integer is prime is important in applications to cryptology. An important theorem from number theory, the Fundamental Theorem of Arithmetic, asserts that every positive integer can be written uniquely as the product of prime numbers. Factoring integers into their prime factors is important in cryptology. Division of an integer by a positive integer produces a quotient and a remainder. Working with these remainders leads to modular arithmetic, which is used throughout computer science. We will discuss three applications of modular arithmetic in this section: generating pseudorandom numbers, assigning computer memory locations to files, and encrypting and decrypting messages.

DIVISION

When one integer is divided by a second, nonzero integer, the quotient may or may not be an integer. For example, $12/3 = 4$ is an integer, whereas $11/4 = 2.75$ is not. This leads to the following definition.

DEFINITION 1   If a and b are integers with $a \neq 0$, we say that a *divides* b if there is an integer c such that $b = ac$. When a divides b we say that a is a *factor* of b and that b is a *multiple* of a. The notation $a \mid b$ denotes that a divides b. We write $a \nmid b$ when a does not divide b.

Remark: We can express $a \mid b$ using quantifiers as $\exists c (ac = b)$, where the universe of discourse is the set of integers.

In Figure 1 a number line indicates which integers are divisible by the positive integer d.

EXAMPLE 1   Determine whether $3 \mid 7$ and whether $3 \mid 12$.

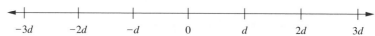

FIGURE 1   **Integers Divisible by the Positive Integer *d*.**

Solution: It follows that $3 \nmid 7$, since 7/3 is not an integer. On the other hand, $3 \mid 12$ since 12/3 = 4. ◀

EXAMPLE 2

Extra
Examples

Let *n* and *d* be positive integers. How many positive integers not exceeding *n* are divisible by *d*?

Solution: The positive integers divisible by *d* are all the integers of the form *dk*, where *k* is a positive integer. Hence, the number of positive integers divisible by *d* that do not exceed *n* equals the number of integers *k* with $0 < dk \leq n$, or with $0 < k \leq n/d$. Therefore, there are $\lfloor n/d \rfloor$ positive integers not exceeding *n* that are divisible by *d*. ◀

Some of the basic properties of divisibility of integers are given in Theorem 1.

THEOREM 1

Let *a*, *b*, and *c* be integers. Then

1. if $a \mid b$ and $a \mid c$, then $a \mid (b + c)$;
2. if $a \mid b$, then $a \mid bc$ for all integers *c*;
3. if $a \mid b$ and $b \mid c$, then $a \mid c$.

Proof: To prove (*1*) suppose that $a \mid b$ and $a \mid c$. Then, from the definition of divisibility, it follows that there are integers *s* and *t* with *b* = *as* and *c* = *at*. Hence,

$$b + c = as + at = a(s + t).$$

Therefore, *a* divides *b* + *c*. This establishes part (*1*) of the theorem. The proofs of parts (*2*) and (*3*) are left as exercises for the reader. ◁

Theorem 1 has this useful consequence.

COROLLARY 1

If *a*, *b*, and *c* are integers such that $a \mid b$ and $a \mid c$, then $a \mid mb + nc$ whenever *m* and *n* are integers.

Proof: By part (*2*) of Theorem 1 it follows that $a \mid mb$ and $a \mid nc$ whenever *m* and *n* are integers. By part (*1*) of Theorem 1 it follows that $a \mid mb + nc$. ◁

PRIMES

Every positive integer greater than 1 is divisible by at least two integers, since a positive integer is divisible by 1 and by itself. Integers that have exactly two different positive integer factors are called **primes.**

DEFINITION 2   A positive integer p greater than 1 is called *prime* if the only positive factors of p are 1 and p. A positive integer that is greater than 1 and is not prime is called *composite.*

Remark: The integer n is composite if and only if there exists an integer a such that $a \mid n$ and $1 < a < n$.

EXAMPLE 3   The integer 7 is prime since its only positive factors are 1 and 7, whereas the integer 9 is composite since it is divisible by 3.   ◀

The primes less than 100 are $2, 3, 5, 7, 11, 13, 17, 19, 23, 29, 31, 37, 41, 43, 47, 53, 59, 61,$ $67, 71, 73, 79, 83, 89,$ and 97. In Section 6.6 we introduce a procedure, known as the **sieve of Eratosthenes,** which can be used to find all the primes not exceeding an integer n.

The primes are the building blocks of positive integers, as the Fundamental Theorem of Arithmetic shows. The proof will be given in Section 3.3.

THEOREM 2   **THE FUNDAMENTAL THEOREM OF ARITHMETIC** Every positive integer greater than 1 can be written uniquely as a prime or as the product of two or more primes where the prime factors are written in order of nondecreasing size.

Example 4 gives some prime factorizations of integers.

EXAMPLE 4   The prime factorizations of $100, 641, 999,$ and 1024 are given by

Extra
Examples

$$100 = 2 \cdot 2 \cdot 5 \cdot 5 = 2^2 5^2,$$
$$641 = 641,$$
$$999 = 3 \cdot 3 \cdot 3 \cdot 37 = 3^3 \cdot 37,$$
$$1024 = 2 \cdot 2 \cdot 2 \cdot 2 \cdot 2 \cdot 2 \cdot 2 \cdot 2 \cdot 2 \cdot 2 = 2^{10}.$$ ◀

It is often important to show that a given integer is prime. For instance, in cryptology large primes are used in some methods for making messages secret. One procedure for showing that an integer is prime is based on the following observation.

THEOREM 3   If n is a composite integer, then n has a prime divisor less than or equal to $\sqrt{n}$.

Proof: If n is composite, it has a factor a with $1 < a < n$. Hence, $n = ab$, where both a and b are positive integers greater than 1. We see that $a \le \sqrt{n}$ or $b \le \sqrt{n}$, since otherwise $ab > \sqrt{n} \cdot \sqrt{n} = n$. Hence, n has a positive divisor not exceeding $\sqrt{n}$. This divisor is either prime or, by the Fundamental Theorem of Arithmetic, has a prime divisor. In either case, n has a prime divisor less than or equal to $\sqrt{n}$.   ◁

From Theorem 3, it follows that an integer is prime if it is not divisible by any prime less than or equal to its square root. In the following example this observation is used to show that 101 is prime.

EXAMPLE 5   Show that 101 is prime.

Solution: The only primes not exceeding $\sqrt{101}$ are $2, 3, 5,$ and 7. Since 101 is not divisible by $2, 3, 5,$ or 7 (the quotient of 101 and each of these integers is not an integer), it follows that 101 is prime.   ◀

Since every integer has a prime factorization, it would be useful to have a procedure for finding this prime factorization. Consider the problem of finding the prime factorization of n. Begin by dividing n by successive primes, starting with the smallest prime, 2. If n has a prime factor, then by Theorem 3 a prime factor p not exceeding $\sqrt{n}$ will be found. So, if no prime factor not exceeding $\sqrt{n}$ is found, then n is prime. Otherwise, if a prime factor p is found, continue by factoring n/p. Note that n/p has no prime factors less than p. Again, if n/p has no prime factor greater than or equal to p and not exceeding its square root, then it is prime. Otherwise, if it has a prime factor q, continue by factoring $n/(pq)$. This procedure is continued until the factorization has been reduced to a prime. This procedure is illustrated in Example 6.

EXAMPLE 6   Find the prime factorization of 7007.

Solution: To find the prime factorization of 7007, first perform divisions of 7007 by successive primes, beginning with 2. None of the primes 2, 3, and 5 divides 7007. However, 7 divides 7007, with $7007/7 = 1001$. Next, divide 1001 by successive primes, beginning with 7. It is immediately seen that 7 also divides 1001, since $1001/7 = 143$. Continue by dividing 143 by successive primes, beginning with 7. Although 7 does not divide 143, 11 does divide 143, and $143/11 = 13$. Since 13 is prime, the procedure is completed. It follows that the prime factorization of 7007 is $7 \cdot 7 \cdot 11 \cdot 13 = 7^2 \cdot 11 \cdot 13$.   ◀

Prime numbers were studied in ancient times for philosophical reasons. Today, there are highly practical reasons for their study. In particular, large primes play a crucial role in cryptography, as we will see in Section 2.6.

THE INFINITUDE OF PRIMES   It has long been known that there are infinitely many primes. We will prove this fact using a proof given by Euclid in his famous mathematics text, the *Elements*.

THEOREM 4   There are infinitely many primes.

Proof: We will prove this theorem using a proof by contradiction. We assume that there are only finitely many primes, $p_1, p_2, \ldots, p_n$. Let

$$Q = p_1 p_2 \cdots p_n + 1.$$

By the Fundamental Theorem of Arithmetic, Q is prime or else it can be written as the product of two or more primes. However, none of the primes p_j divides Q, for if $p_j \mid Q$, then p_j divides $Q - p_1 p_2 \cdots p_n = 1$. This is a contradiction because we assumed that we have listed all the primes. Consequently, there are infinitely many primes. (Note that in this proof we do *not* state that Q is prime!)   ◁

Since there are infinitely many primes, given any positive integer there are primes greater than this integer. There is an ongoing quest to discover larger and larger prime numbers; for almost all the last 300 years, the largest prime known has been an integer of the special form $2^p - 1$, where p is also prime. Such primes are called **Mersenne primes,** after the French monk Marin Mersenne, who studied them in the seventeenth century. The reason that the largest known prime has usually been a Mersenne prime is that there is an extremely efficient test, known as the Lucas–Lehmer test, for determining whether $2^p - 1$ is prime. Furthermore, it is not currently possible to test numbers not of certain special forms anywhere near as quickly to determine whether they are prime.

EXAMPLE 7   The numbers $2^2 - 1 = 3$, $2^3 - 1 = 7$, and $2^5 - 1 = 31$ are Mersenne primes, while $2^{11} - 1 = 2047$ is not a Mersenne prime since $2047 = 23 \cdot 89$.   ◀

Progress in finding Mersenne primes has been steady since computers were invented. As of mid-2002, 39 Mersenne primes were known, with eight found since 1990. The largest Mersenne prime known (as of mid-2002) is $2^{13,466,917} - 1$, a number with over four million digits, which was shown to be prime in late 2001. A communal effort, the Great Internet Mersenne Prime Search (GIMPS), has been organized to look for new Mersenne primes. By the way, even the search for Mersenne primes has practical implications. One quality control test for supercomputers has been to replicate the Lucas–Lehmer test that establishes the primality of a large Mersenne prime.

THE DISTRIBUTION OF PRIMES Theorem 4 tells us that there are infinitely many primes. However, how many primes are less than a positive number x? This question interested mathematicians for many years; in the late eighteenth century mathematicians produced large tables of prime numbers to gather evidence concerning the distribution of primes. Using this evidence, the great mathematicians of the day, including Gauss and Legendre, conjectured, but did not prove, Theorem 5.

THEOREM 5   **THE PRIME NUMBER THEOREM**   The ratio of the number of primes not exceeding x and $x/\ln x$ approaches 1 as x grows without bound. (Here $\ln x$ is the natural logarithm of x.)

MARIN MERSENNE (1588–1648)   Mersenne was born in Maine, France, into a family of laborers and attended the College of Mans and the Jesuit College at La Flèche. He continued his education at the Sorbonne, studying theology from 1609 to 1611. He joined the religious order of the Minims in 1611, a group whose name comes from the word *minimi* (the members of this group considered themselves the least religious order). Besides prayer, the members of this group devoted their energy to scholarship and study. In 1612 he became a priest at the Place Royale in Paris; between 1614 and 1618 he taught philosophy at the Minim Convent at Nevers. He returned to Paris in 1619, where his cell in the Minims de l'Annociade became a place for meetings of French scientists, philosophers, and mathematicians, including Fermat and Pascal. Mersenne corresponded extensively with scholars throughout Europe, serving as a clearinghouse for mathematical and scientific knowledge, a function later served by mathematical journals (and today also by the Internet). Mersenne wrote books covering mechanics, mathematical physics, mathematics, music, and acoustics. He studied prime numbers and tried unsuccessfully to construct a formula representing all primes. In 1644 Mersenne claimed that $2^p - 1$ is prime for $p = 2, 3, 5, 7, 13, 17, 19, 31, 67, 127, 257$ but is composite for all other primes less than 257. It took over 300 years to determine that Mersenne's claim was wrong five times. Specifically, $2^p - 1$ is not prime for $p = 67$ and $p = 257$ but is prime for $p = 61$, $p = 87$, and $p = 107$. It is also noteworthy that Mersenne defended two of the most famous men of his time, Descartes and Galileo, from religious critics. He also helped expose alchemists and astrologers as frauds.

Links

The Prime Number Theorem was first proved in 1896 by the French mathematician Jacques Hadamard and the Belgian mathematician Charles-Jean-Gustave-Nicholas de la Valleé-Poussin using the theory of complex variables. Although proofs not using complex variables have been found, all known proofs of the Prime Number Theorem are quite complicated.

We can use the Prime Number Theorem to estimate the odds that a randomly chosen number of a certain size is prime. The Prime Number Theorem tells us that the number of primes not exceeding x can be approximated by $x/\ln x$. Consequently, the odds that a randomly selected positive integer x is prime are approximately $(x/\ln x)/x = 1/\ln x$. For example, the odds that an integer near 10^{1000} is prime are approximately $1/\ln 10^{1000}$, which is approximately $1/2300$. (Of course, by choosing only odd numbers, we double our chances of finding a prime.)

Using trial division with Theorem 3 gives procedures for factoring and for primality testing. However, these procedures are not efficient algorithms; many much more practical and efficient algorithms for these tasks have been developed. Factoring and primality testing have become important in the applications of number theory to cryptography. This has led to a great interest in developing efficient algorithms for both tasks. Clever procedures have been devised in the last 30 years for efficiently generating large primes. However, even though powerful new factorization methods have been developed in the same time frame, factoring large numbers remains extraordinarily more time consuming. Nevertheless, the challenge of factoring large numbers interests many people. There is a communal effort on the Internet to factor large numbers, especially those of the special form $k^n \pm 1$, where k is a small positive integer and n is a large positive integer (such numbers are called *Cunningham numbers*). At any given time, there is a list of the "Ten Most Wanted" large numbers of this type awaiting factorization.

Links

THE DIVISION ALGORITHM

When an integer is divided by a positive integer, there is a quotient and a remainder, as the division algorithm shows.

THEOREM 6

> **THE DIVISION ALGORITHM** Let a be an integer and d a positive integer. Then there are unique integers q and r, with $0 \le r < d$, such that $a = dq + r$.

Remark: Theorem 6 is not really an algorithm. (Why not?) Nevertheless, we use its traditional name.

DEFINITION 3

> In the equality given in the division algorithm, d is called the *divisor*, a is called the *dividend*, q is called the *quotient*, and r is called the *remainder*. This notation is used to express the quotient and remainder:
>
> $$q = a \textbf{ div } d, \quad r = a \textbf{ mod } d.$$

Examples 8 and 9 illustrate the division algorithm.

EXAMPLE 8   What are the quotient and remainder when 101 is divided by 11?

Solution: We have

$$101 = 11 \cdot 9 + 2.$$

Hence, the quotient when 101 is divided by 11 is $9 = 101$ **div** 11, and the remainder is $2 = 101$ **mod** 11. ◀

EXAMPLE 9   What are the quotient and remainder when -11 is divided by 3?

Solution: We have

$$-11 = 3(-4) + 1.$$

Hence, the quotient when -11 is divided by 3 is $-4 = -11$ **div** 3, and the remainder is $1 = -11$ **mod** 3.

Note that the remainder cannot be negative. Consequently, the remainder is *not* -2, even though

$$-11 = 3(-3) - 2,$$

since $r = -2$ does not satisfy $0 \le r < 3$. ◀

Note that the integer a is divisible by the integer d if and only if the remainder is zero when a is divided by d.

GREATEST COMMON DIVISORS AND LEAST COMMON MULTIPLES

The largest integer that divides both of two integers is called the **greatest common divisor** of these integers.

DEFINITION 4   Let a and b be integers, not both zero. The largest integer d such that $d \mid a$ and $d \mid b$ is called the *greatest common divisor* of a and b. The greatest common divisor of a and b is denoted by $\gcd(a, b)$.

The greatest common divisor of two integers, not both zero, exists because the set of common divisors of these integers is finite. One way to find the greatest common divisor of two integers is to find all the positive common divisors of both integers and then take the largest divisor. This is done in the following examples. Later, a more efficient method of finding greatest common divisors will be given.

EXAMPLE 10   What is the greatest common divisor of 24 and 36?

Solution: The positive common divisors of 24 and 36 are 1, 2, 3, 4, 6, and 12. Hence, $\gcd(24, 36) = 12$. ◀

EXAMPLE 11   What is the greatest common divisor of 17 and 22?

Solution: The integers 17 and 22 have no positive common divisors other than 1, so that $\gcd(17, 22) = 1$. ◀

Since it is often important to specify that two integers have no common positive divisor other than 1, we have the following definition.

DEFINITION 5 The integers a and b are *relatively prime* if their greatest common divisor is 1.

EXAMPLE 12 From Example 11 it follows that the integers 17 and 22 are relatively prime, since $\gcd(17, 22) = 1$. ◀

Since we often need to specify that no two integers in a set of integers have a common positive divisor greater than 1, we make Definition 6.

DEFINITION 6 The integers $a_1, a_2, \ldots, a_n$ are *pairwise relatively prime* if $\gcd(a_i, a_j) = 1$ whenever $1 \le i < j \le n$.

EXAMPLE 13 Determine whether the integers 10, 17, and 21 are pairwise relatively prime and whether the integers 10, 19, and 24 are pairwise relatively prime.

Solution: Since $\gcd(10, 17) = 1$, $\gcd(10, 21) = 1$, and $\gcd(17, 21) = 1$, we conclude that 10, 17, and 21 are pairwise relatively prime.

Since $\gcd(10, 24) = 2 > 1$, we see that 10, 19, and 24 are not pairwise relatively prime. ◀

Another way to find the greatest common divisor of two integers is to use the prime factorizations of these integers. Suppose that the prime factorizations of the integers a and b, neither equal to zero, are

$$a = p_1^{a_1} p_2^{a_2} \cdots p_n^{a_n}, \ b = p_1^{b_1} p_2^{b_2} \cdots p_n^{b_n},$$

where each exponent is a nonnegative integer, and where all primes occurring in the prime factorization of either a or b are included in both factorizations, with zero exponents if necessary. Then $\gcd(a, b)$ is given by

$$\gcd(a, b) = p_1^{\min(a_1, b_1)} p_2^{\min(a_2, b_2)} \cdots p_n^{\min(a_n, b_n)},$$

where $\min(x, y)$ represents the minimum of the two numbers x and y. To show that this formula for $\gcd(a, b)$ is valid, we must show that the integer on the right-hand side divides both a and b, and that no larger integer also does. This integer does divide both a and b, since the power of each prime in the factorization does not exceed the power of this prime in either the factorization of a or that of b. Further, no larger integer can divide both a and b, because the exponents of the primes in this factorization cannot be increased, and no other primes can be included.

EXAMPLE 14 Since the prime factorizations of 120 and 500 are $120 = 2^3 \cdot 3 \cdot 5$ and $500 = 2^2 \cdot 5^3$, the greatest common divisor is

$$\gcd(120, 500) = 2^{\min(3, 2)} 3^{\min(1, 0)} 5^{\min(1, 3)} = 2^2 3^0 5^1 = 20.$$ ◀

Prime factorizations can also be used to find the **least common multiple** of two integers.

DEFINITION 7

The *least common multiple* of the positive integers a and b is the smallest positive integer that is divisible by both a and b. The least common multiple of a and b is denoted by $\text{lcm}(a, b)$.

The least common multiple exists because the set of integers divisible by both a and b is nonempty, and every nonempty set of positive integers has a least element (by the well-ordering property, which will be discussed in Section 3.3). Suppose that the prime factorizations of a and b are as before. Then the least common multiple of a and b is given by

$$\text{lcm}(a, b) = p_1^{\max(a_1, b_1)} p_2^{\max(a_2, b_2)} \cdots p_n^{\max(a_n, b_n)}$$

where $\max(x, y)$ denotes the maximum of the two numbers x and y. This formula is valid since a common multiple of a and b has at least $\max(a_i, b_i)$ factors of p_i in its prime factorization, and the least common multiple has no other prime factors besides those in a and b.

EXAMPLE 15

What is the least common multiple of $2^3 3^5 7^2$ and $2^4 3^3$?

Solution: We have

$$\text{lcm}(2^3 3^5 7^2, 2^4 3^3) = 2^{\max(3, 4)} 3^{\max(5, 3)} 7^{\max(2, 0)} = 2^4 3^5 7^2.$$ ◄

The following theorem gives the relationship between the greatest common divisor and least common multiple of two integers. It can be proved using the formulae we have derived for these quantities. The proof of this theorem is left as an exercise for the reader.

THEOREM 7

Let a and b be positive integers. Then

$$ab = \gcd(a, b) \cdot \text{lcm}(a, b).$$

MODULAR ARITHMETIC

In some situations we care only about the remainder of an integer when it is divided by some specified positive integer. For instance, when we ask what time it will be (on a 24-hour clock) 50 hours from now, we care only about the remainder when 50 plus the current hour is divided by 24. Since we are often interested only in remainders, we have special notations for them.

We have a notation to indicate that two integers have the same remainder when they are divided by the positive integer m.

DEFINITION 8

If a and b are integers and m is a positive integer, then a is *congruent to b modulo m* if m divides $a - b$. We use the notation $a \equiv b \pmod{m}$ to indicate that a is congruent to b modulo m. If a and b are not congruent modulo m, we write $a \not\equiv b \pmod{m}$.

The connection between the notations used when working with remainders is made clear in Theorem 8.

THEOREM 8   Let a and b be integers, and let m be a positive integer. Then $a \equiv b \pmod{m}$ if and only if $a \bmod m = b \bmod m$.

The proof of Theorem 8 is left as Exercises 21 and 22 at the end of this section.

EXAMPLE 16   Determine whether 17 is congruent to 5 modulo 6 and whether 24 and 14 are congruent modulo 6.

Solution: Since 6 divides $17 - 5 = 12$, we see that $17 \equiv 5 \pmod{6}$. However, since $24 - 14 = 10$ is not divisible by 6, we see that $24 \not\equiv 14 \pmod{6}$.   ◄

The great German mathematician Karl Friedrich Gauss developed the concept of congruences at the end of the eighteenth century.

The notion of congruences has played an important role in the development of number theory. Theorem 9 provides a useful way to work with congruences.

THEOREM 9   Let m be a positive integer. The integers a and b are congruent modulo m if and only if there is an integer k such that $a = b + km$.

Proof: If $a \equiv b \pmod{m}$, then $m \mid (a - b)$. This means that there is an integer k such that $a - b = km$, so that $a = b + km$. Conversely, if there is an integer k such that $a = b + km$, then $km = a - b$. Hence, m divides $a - b$, so that $a \equiv b \pmod{m}$.   ◁

The set of all integers congruent to an integer a modulo m is called the **congruence class** of a modulo m. In Chapter 7 we will show that there are m pairwise disjoint equivalence classes modulo m and that the union of these equivalence classes is the set of integers.

Theorem 10 shows how congruences work with respect to addition and multiplication.

Links

KARL FRIEDRICH GAUSS (1777–1855) Karl Friedrich Gauss, the son of a bricklayer, was a child prodigy. He demonstrated his potential at the age of 10, when he quickly solved a problem assigned by a teacher to keep the class busy. The teacher asked the students to find the sum of the first 100 positive integers. Gauss realized that this sum could be found by forming 50 pairs, each with the sum 101: $1 + 100, 2 + 99, \ldots, 50 + 51$. This brilliance attracted the sponsorship of patrons, including Duke Ferdinand of Brunswick, who made it possible for Gauss to attend Caroline College and the University of Göttingen. While a student, he invented the method of least squares, which is used to estimate the most likely value of a variable from experimental results. In 1796 Gauss made a fundamental discovery in geometry, advancing a subject that had not advanced since ancient times. He showed that a 17-sided regular polygon could be drawn using just a ruler and compass.

In 1799 Gauss presented the first rigorous proof of the Fundamental Theorem of Algebra, which states that a polynomial of degree n has exactly n roots (counting multiplicities). Gauss achieved worldwide fame when he successfully calculated the orbit of the first asteroid discovered, Ceres, using scanty data.

Gauss was called the Prince of Mathematics by his contemporary mathematicians. Although Gauss is noted for his many discoveries in geometry, algebra, analysis, astronomy, and physics, he had a special interest in number theory, which can be seen from his statement "Mathematics is the queen of the sciences, and the theory of numbers is the queen of mathematics." Gauss laid the foundations for modern number theory with the publication of his book *Disquisitiones Arithmeticae* in 1801.

THEOREM 10   Let m be a positive integer. If $a \equiv b \pmod{m}$ and $c \equiv d \pmod{m}$, then

$$a + c \equiv b + d \pmod{m} \qquad \text{and} \qquad ac \equiv bd \pmod{m}.$$

Proof: Since $a \equiv b \pmod{m}$ and $c \equiv d \pmod{m}$, there are integers s and t with $b = a + sm$ and $d = c + tm$. Hence,

$$b + d = (a + sm) + (c + tm) = (a + c) + m(s + t)$$

and

$$bd = (a + sm)(c + tm) = ac + m(at + cs + stm).$$

Hence,

$$a + c \equiv b + d \pmod{m} \qquad \text{and} \qquad ac \equiv bd \pmod{m}. \qquad \triangleleft$$

EXAMPLE 17   Since $7 \equiv 2 \pmod 5$ and $11 \equiv 1 \pmod 5$, it follows from Theorem 10 that

$$18 = 7 + 11 \equiv 2 + 1 = 3 \pmod 5$$

and that

$$77 = 7 \cdot 11 \equiv 2 \cdot 1 = 2 \pmod 5. \qquad \blacktriangleleft$$

APPLICATIONS OF CONGRUENCES

Number theory has applications to a wide range of areas. We will introduce three applications in this section: the use of congruences to assign memory locations to computer files, the generation of pseudorandom numbers, and cryptosystems based on modular arithmetic.

EXAMPLE 18   **Hashing Functions** The central computer at your school maintains records for each student. How can memory locations be assigned so that student records can be retrieved quickly? The solution to this problem is to use a suitably chosen **hashing function.** Records are identified using a **key,** which uniquely identifies each student's records. For instance, student records are often identified using the Social Security number of the student as the key. A hashing function h assigns memory location $h(k)$ to the record that has k as its key.

In practice, many different hashing functions are used. One of the most common is the function

$$h(k) = k \bmod m$$

where m is the number of available memory locations.

Hashing functions should be easily evaluated so that files can be quickly located. The hashing function $h(k) = k \bmod m$ meets this requirement; to find $h(k)$, we need only compute the remainder when k is divided by m. Furthermore, the hashing function should be onto, so that all memory locations are possible. The function $h(k) = k \bmod m$ also satisfies this property.

For example, when $m = 111$, the record of the student with Social Security number 064212848 is assigned to memory location 14, since

$$h(064212848) = 064212848 \bmod 111 = 14.$$

Similarly, since

$$h(037149212) = 037149212 \bmod 111 = 65,$$

the record of the student with Social Security number 037149212 is assigned to memory location 65.

Since a hashing function is not one-to-one (since there are more possible keys than memory locations), more than one file may be assigned to a memory location. When this happens, we say that a **collision** occurs. One way to resolve a collision is to assign the first free location following the occupied memory location assigned by the hashing function. For example, after making the two earlier assignments, we assign location 15 to the record of the student with the Social Security number 107405723. To see this, first note that $h(k)$ maps this Social Security number to location 14, since

$$h(107405723) = 107405723 \bmod 111 = 14,$$

but this location is already occupied (by the file of the student with Social Security number 064212848). However, memory location 15, the first location following memory location 14, is free.

There are many more sophisticated ways to resolve collisions that are more efficient than the simple method we have described. These are discussed in the references on hashing functions given at the end of the book. ◄

EXAMPLE 19

Links

Pseudorandom Numbers Randomly chosen numbers are often needed for computer simulations. Different methods have been devised for generating numbers that have properties of randomly chosen numbers. Because numbers generated by systematic methods are not truly random, they are called **pseudorandom numbers.**

The most commonly used procedure for generating pseudorandom numbers is the **linear congruential method.** We choose four integers: the **modulus** m, **multiplier** a, **increment** c, and **seed** x_0, with $2 \leq a < m, 0 \leq c < m$, and $0 \leq x_0 < m$. We generate a sequence of pseudorandom numbers $\{x_n\}$, with $0 \leq x_n < m$ for all n, by successively using the congruence

$$x_{n+1} = (ax_n + c) \bmod m.$$

(This is an example of a recursive definition, discussed in Section 3.4. In that section we will show that such sequences are well defined.)

Many computer experiments require the generation of pseudorandom numbers between 0 and 1. To generate such numbers, we divide numbers generated with a linear congruential generator by the modulus: that is, we use the numbers x_n/m.

For instance, the sequence of pseudorandom numbers generated by choosing $m = 9, a = 7, c = 4$, and $x_0 = 3$, can be found as follows:

$$x_1 = 7x_0 + 4 = 7 \cdot 3 + 4 = 25 \bmod 9 = 7,$$
$$x_2 = 7x_1 + 4 = 7 \cdot 7 + 4 = 53 \bmod 9 = 8,$$
$$x_3 = 7x_2 + 4 = 7 \cdot 8 + 4 = 60 \bmod 9 = 6,$$
$$x_4 = 7x_3 + 4 = 7 \cdot 6 + 4 = 46 \bmod 9 = 1,$$

$$x_5 = 7x_4 + 4 = 7 \cdot 1 + 4 = 11 \text{ mod } 9 = 2,$$
$$x_6 = 7x_5 + 4 = 7 \cdot 2 + 4 = 18 \text{ mod } 9 = 0,$$
$$x_7 = 7x_6 + 4 = 7 \cdot 0 + 4 = 4 \text{ mod } 9 = 4,$$
$$x_8 = 7x_7 + 4 = 7 \cdot 4 + 4 = 32 \text{ mod } 9 = 5,$$
$$x_9 = 7x_8 + 4 = 7 \cdot 5 + 4 = 39 \text{ mod } 9 = 3.$$

Since $x_9 = x_0$ and since each term depends only on the previous term, this sequence is generated:

$$3, 7, 8, 6, 1, 2, 0, 4, 5, 3, 7, 8, 6, 1, 2, 0, 4, 5, 3, \dots.$$

This sequence contains nine different numbers before repeating.

Most computers do use linear congruential generators to generate pseudorandom numbers. Often, a linear congruential generator with increment $c = 0$ is used. Such a generator is called a **pure multiplicative generator.** For example, the pure multiplicative generator with modulus $2^{31} - 1$ and multiplier $7^5 = 16,807$ is widely used. With these values, it can be shown that $2^{31} - 2$ numbers are generated before repetition begins. ◀

CRYPTOLOGY

Congruences have many applications to discrete mathematics and computer science. Discussions of these applications can be found in the suggested readings given at the end of the book. One of the most important applications of congruences involves **cryptology,** which is the study of secret messages. One of the earliest known uses of cryptology was by Julius Caesar. He made messages secret by shifting each letter three letters forward in the alphabet (sending the last three letters of the alphabet to the first three). For instance, using this scheme the letter B is sent to E and the letter X is sent to A. This is an example of **encryption,** that is, the process of making a message secret.

To express Caesar's encryption process mathematically, first replace each letter by an integer from 0 to 25, based on its position in the alphabet. For example, replace A by 0, K by 10, and Z by 25. Caesar's encryption method can be represented by the function f that assigns to the nonnegative integer p, $p \le 25$, the integer $f(p)$ in the set $\{0, 1, 2, \dots, 25\}$ with

$$f(p) = (p + 3) \text{ mod } 26.$$

In the encrypted version of the message, the letter represented by p is replaced with the letter represented by $(p + 3) \text{ mod } 26$.

EXAMPLE 20 What is the secret message produced from the message "MEET YOU IN THE PARK" using the Caesar cipher?

Solution: First replace the letters in the message with numbers. This produces

12 4 4 19   24 14 20   8 13   19 7 4   15 0 17 10.

Now replace each of these numbers p by $f(p) = (p + 3) \text{ mod } 26$. This gives

15 7 7 22   1 17 23   11 16   22 10 7   18 3 20 13.

Translating this back to letters produces the encrypted message "PHHW BRX LQ WKH SDUN." ◀

To recover the original message from a secret message encrypted by the Caesar cipher, the function f^{-1}, the inverse of f, is used. Note that the function f^{-1} sends an integer p from $\{0, 1, 2, \ldots, 25\}$ to $f^{-1}(p) = (p - 3) \bmod 26$. In other words, to find the original message, each letter is shifted back three letters in the alphabet, with the first three letters sent to the last three letters of the alphabet. The process of determining the original message from the encrypted message is called **decryption.**

There are various ways to generalize the Caesar cipher. For example, instead of shifting each letter by 3, we can shift each letter by k, so that

$$f(p) = (p + k) \bmod 26.$$

Such a cipher is called a **shift cipher.** Note that decryption can be carried out using

$$f^{-1}(p) = (p - k) \bmod 26.$$

Obviously, Caesar's method and shift ciphers do not provide a high level of security. There are various ways to enhance this method. One approach that slightly enhances the security is to use a function of the form

$$f(p) = (ap + b) \bmod 26,$$

where a and b are integers, chosen such that f is a bijection. (Such a mapping is called an *affine transformation*.) This provides a number of possible encryption systems. The use of one of these systems is illustrated in the following example.

EXAMPLE 21   What letter replaces the letter K when the function $f(p) = (7p + 3) \bmod 26$ is used for encryption?

Solution: First, note that 10 represents K. Then, using the encryption function specified, it follows that $f(10) = (7 \cdot 10 + 3) \bmod 26 = 21$. Since 21 represents V, K is replaced by V in the encrypted message.   ◀

Caesar's encryption method, and the generalization of this method, proceed by replacing each letter of the alphabet by another letter in the alphabet. Encryption methods of this kind are vulnerable to attacks based on the frequency of occurrence of letters in the message. More sophisticated encryption methods are based on replacing blocks of letters with other blocks of letters. There are a number of techniques based on modular arithmetic for encrypting blocks of letters. A discussion of these can be found in the suggested readings listed at the end of the book.

Exercises

1. Does 17 divide each of these numbers?
   a) 68   **b)** 84   **c)** 357   **d)** 1001
2. Show that if a is an integer other than 0, then
   a) 1 divides a.   **b)** a divides 0.
3. Show that part (2) of Theorem 1 is true.
4. Show that part (3) of Theorem 1 is true.
5. Show that if $a \mid b$ and $b \mid a$, where a and b are integers, then $a = b$ or $a = -b$.

6. Show that if a, b, c, and d are integers such that $a \mid c$ and $b \mid d$, then $ab \mid cd$.
7. Show that if a, b, and c are integers such that $ac \mid bc$, then $a \mid b$.
8. Are these integers primes?
   a) 19   **b)** 27
   c) 93   **d)** 101
   e) 107   **f)** 113

9. What are the quotient and remainder when
   a) 19 is divided by 7?
   b) −111 is divided by 11?
   c) 789 is divided by 23?
   d) 1001 is divided by 13?
   e) 0 is divided by 19?
   f) 3 is divided by 5?
   g) −1 is divided by 3?
   h) 4 is divided by 1?

10. What are the quotient and remainder when
    a) 44 is divided by 8?
    b) 777 is divided by 21?
    c) −123 is divided by 19?
    d) −1 is divided by 23?
    e) −2002 is divided by 87?
    f) 0 is divided by 17?
    g) 1,234,567 is divided by 1001?
    h) −100 is divided by 101?

11. Find the prime factorization of each of these integers.
    a) 88   b) 126   c) 729
    d) 1001   e) 1111   f) 909,090

12. Find the prime factorization of each of these integers.
    a) 39   b) 81   c) 101
    d) 143   e) 289   f) 899

13. Find the prime factorization of 10!.

*14. How many zeros are there at the end of 100!?

*15. Show that $\log_2 3$ is an irrational number. Recall that an irrational number is a real number x that cannot be written as the ratio of two integers.

16. Which positive integers less than 12 are relatively prime to 12?

17. Which positive integers less than 30 are relatively prime to 30?

18. Determine whether the integers in each of these sets are pairwise relatively prime.
    a) 21, 34, 55   b) 14, 17, 85
    c) 25, 41, 49, 64   d) 17, 18, 19, 23

19. Determine whether the integers in each of these sets are pairwise relatively prime.
    a) 11, 15, 19   b) 14, 15, 21
    c) 12, 17, 31, 37   d) 7, 8, 9, 11

20. We call a positive integer **perfect** if it equals the sum of its positive divisors other than itself.
    a) Show that 6 and 28 are perfect.
    b) Show that $2^{p-1}(2^p - 1)$ is a perfect number when $2^p - 1$ is prime.

21. Let m be a positive integer. Show that $a \equiv b \pmod{m}$ if $a \bmod m = b \bmod m$.

22. Let m be a positive integer. Show that $a \bmod m = b \bmod m$ if $a \equiv b \pmod{m}$.

23. Show that if $2^n - 1$ is prime, then n is prime. [*Hint:* Use the identity $2^{ab} - 1 = (2^a - 1) \cdot (2^{a(b-1)} + 2^{a(b-2)} + \cdots + 2^a + 1)$.]

24. Determine whether each of these integers is prime, verifying some of Mersenne's claims.
    a) $2^7 - 1$ b) $2^9 - 1$
    c) $2^{11} - 1$ d) $2^{13} - 1$

25. The value of the **Euler ϕ-function** at the positive integer n is defined to be the number of positive integers less than or equal to n that are relatively prime to n. (*Note:* ϕ is the Greek letter phi.) Find
    a) $\phi(4)$.   b) $\phi(10)$.   c) $\phi(13)$.

26. Show that n is prime if and only if $\phi(n) = n - 1$.

27. What is the value of $\phi(p^k)$ when p is prime and k is a positive integer?

28. What are the greatest common divisors of these pairs of integers?
    a) $2^2 \cdot 3^3 \cdot 5^5$, $2^5 \cdot 3^3 \cdot 5^2$
    b) $2 \cdot 3 \cdot 5 \cdot 7 \cdot 11 \cdot 13$, $2^{11} \cdot 3^9 \cdot 11 \cdot 17^{14}$
    c) 17, 17^{17}
    d) $2^2 \cdot 7$, $5^3 \cdot 13$
    e) 0, 5
    f) $2 \cdot 3 \cdot 5 \cdot 7$, $2 \cdot 3 \cdot 5 \cdot 7$

29. What are the greatest common divisors of these pairs of integers?
    a) $3^7 \cdot 5^3 \cdot 7^3$, $2^{11} \cdot 3^5 \cdot 5^9$
    b) $11 \cdot 13 \cdot 17$, $2^9 \cdot 3^7 \cdot 5^5 \cdot 7^3$
    c) 23^{31}, 23^{17}
    d) $41 \cdot 43 \cdot 53$, $41 \cdot 43 \cdot 53$
    e) $3^{13} \cdot 5^{17}$, $2^{12} \cdot 7^{21}$
    f) 1111, 0

30. What is the least common multiple of each pair in Exercise 28?

31. What is the least common multiple of each pair in Exercise 29?

32. Find $\gcd(1000, 625)$ and $\text{lcm}(1000, 625)$ and verify that $\gcd(1000, 625) \cdot \text{lcm}(1000, 625) = 1000 \cdot 625$.

*33. Show that if n and k are positive integers, then $\lceil n/k \rceil = \lfloor (n-1)/k \rfloor + 1$.

34. Show that if a is an integer and d is a positive integer greater than 1, then the quotient and remainder obtained when a is divided by d are $\lfloor a/d \rfloor$ and $a - d\lfloor a/d \rfloor$, respectively.

35. Find a formula for the integer with smallest absolute value that is congruent to an integer a modulo m, where m is a positive integer.

36. Evaluate these quantities.
    a) $-17 \bmod 2$ b) $144 \bmod 7$
    c) $-101 \bmod 13$ d) $199 \bmod 19$

37. Evaluate these quantities.
    a) $13 \bmod 3$ b) $-97 \bmod 11$
    c) $155 \bmod 19$ d) $-221 \bmod 23$

38. List five integers that are congruent to 4 modulo 12.

39. Decide whether each of these integers is congruent to 5 modulo 17.

a) 80   **b)** 103
c) −29   **d)** −122

40. If the product of two integers is $2^7 3^8 5^2 7^{11}$ and their greatest common divisor is $2^3 3^4 5$, what is their least common multiple?

41. Show that if a and b are positive integers then $ab = \gcd(a, b) \cdot \text{lcm}(a, b)$. [*Hint:* Use the prime factorizations of a and b and the formulae for $\gcd(a, b)$ and $\text{lcm}(a, b)$ in terms of these factorizations.]

42. Show that if $a \equiv b \pmod{m}$ and $c \equiv d \pmod{m}$, where $a, b, c, d,$ and m are integers with $m \geq 2$, then $a - c \equiv b - d \pmod{m}$.

43. Show that if $n \mid m$, where n and m are positive integers greater than 1, and if $a \equiv b \pmod{m}$, where a and b are integers, then $a \equiv b \pmod{n}$.

44. Show that if $a, b, c,$ and m are integers such that $m \geq 2$, $c > 0$, and $a \equiv b \pmod{m}$, then $ac \equiv bc \pmod{mc}$.

45. Show that $ac \equiv bc \pmod{m}$, where $a, b, c,$ and m are integers with $m \geq 2$, does not necessarily imply that $a \equiv b \pmod{m}$.

46. Show that if $a, b,$ and m are integers such that $m \geq 2$ and $a \equiv b \pmod{m}$, then $\gcd(a, m) = \gcd(b, m)$.

47. Show that if $a, b, k,$ and m are integers such that $k \geq 1$, $m \geq 2$, and $a \equiv b \pmod{m}$, then $a^k \equiv b^k \pmod{m}$ whenever k is a positive integer.

48. Which memory locations are assigned by the hashing function $h(k) = k \bmod 101$ to the records of students with these Social Security numbers?

a) 104578690   **b)** 432222187
c) 372201919   **d)** 501338753

49. A parking lot has 31 visitor spaces, numbered from 0 to 30. Visitors are assigned parking spaces using the hashing function $h(k) = k \bmod 31$, where k is the number formed from the first three digits on a visitor's license plate.

  a) Which spaces are assigned by the hashing function to cars that have these first three digits on their license plates?

$$317, 918, 007, 100, 111, 310$$

  b) Describe a procedure visitors should follow to find a free parking space, when the space they are assigned is occupied.

50. What sequence of pseudorandom numbers is generated using the linear congruential generator $x_{n+1} = (4x_n + 1) \bmod 7$ with seed $x_0 = 3$?

51. What sequence of pseudorandom numbers is generated using the pure multiplicative generator $x_{n+1} = 3x_n \bmod 11$ with seed $x_0 = 2$?

52. Write an algorithm in pseudocode for generating a sequence of pseudorandom numbers using a linear congruential generator.

53. Encrypt the message "DO NOT PASS GO" by translating the letters into numbers, applying the encryption function given, and then translating the numbers back into letters.

  a) $f(p) = (p + 3) \bmod 26$ (the Caesar cipher)
  b) $f(p) = (p + 13) \bmod 26$
  c) $f(p) = (3p + 7) \bmod 26$

54. Decrypt these messages encrypted using the Caesar cipher.
  a) EOXH MHDQV
  b) WHVW WRGDB
  c) HDW GLP VXP

Books are identified by an **International Standard Book Number (ISBN)**, a 10-digit code $x_1 x_2 \ldots x_{10}$, assigned by the publisher. These 10 digits consist of blocks identifying the language, the publisher, the number assigned to the book by its publishing company, and finally, a 1-digit check digit that is either a digit or the letter X (used to represent 10). This check digit is selected so that $\sum_{i=1}^{10} i x_i \equiv 0 \pmod{11}$ and is used to detect errors in individual digits and transposition of digits.

55. The first nine digits of the ISBN of the third edition of this book are 0-07-053965. What is the check digit for this book?

56. The ISBN of *Elementary Number Theory and Its Applications,* 3rd ed., is 0-201-57Q89-1, where Q is a digit. Find the value of Q.

57. Determine whether the check digit of the ISBN for this textbook was computed correctly by the publisher.

58. Find the smallest positive integer with exactly n different factors when n is

  a) 3.   **b)** 4.   **c)** 5.
  d) 6.   **e)** 10.

59. Can you find a formula or rule for the nth term of a sequence related to the prime numbers or prime factorizations so that the initial terms of the sequence have these values?

  a) $0, 1, 1, 0, 1, 0, 1, 0, 0, 0, 1, 0, 1, \ldots$
  b) $1, 2, 3, 2, 5, 2, 7, 2, 3, 2, 11, 2, 13, 2, \ldots$
  c) $1, 2, 2, 3, 2, 4, 2, 4, 3, 4, 2, 6, 2, 4, \ldots$
  d) $1, 1, 1, 0, 1, 1, 1, 0, 0, 1, 1, 0, 1, 1, \ldots$
  e) $1, 2, 3, 3, 5, 5, 7, 7, 7, 7, 11, 11, 13, 13, \ldots$
  f) $1, 2, 6, 30, 210, 2310, 30030, 510510, 9699690, 223092870, \ldots$

60. Can you find a formula or rule for the nth term of a sequence related to the prime numbers or prime factorizations so that the initial terms of the sequence have these values?

  a) $2, 2, 3, 5, 5, 7, 7, 11, 11, 11, 11, 13, 13, \ldots$
  b) $0, 1, 2, 2, 3, 3, 4, 4, 4, 4, 5, 5, 6, 6, \ldots$
  c) $1, 0, 0, 1, 0, 1, 0, 1, 1, 1, 0, 1, 0, 1, \ldots$
  d) $1, -1, -1, 0, -1, 1, -1, 0, 0, 1, -1, 0, -1, 1, 1, \ldots$
  e) $1, 1, 1, 1, 1, 0, 1, 1, 1, 0, 1, 0, 1, 0, 0, \ldots$
  f) $4, 9, 25, 49, 121, 169, 289, 361, 529, 841, 961, 1369, \ldots$

2.5 Integers and Algorithms

INTRODUCTION

As mentioned in Section 2.1, the term *algorithm* originally referred to procedures for performing arithmetic operations using the decimal representations of integers. These algorithms, adapted for use with binary representations, are the basis for computer arithmetic. They provide good illustrations of the concept of an algorithm and the complexity of algorithms. For these reasons, they will be discussed in this section.

There are many important algorithms involving integers besides those used in arithmetic, including the Euclidean algorithm, which is one of the most useful algorithms, and perhaps the oldest algorithm, in mathematics. We will also describe an algorithm for finding the base b expansion of a positive integer for any base b and for modular exponentiation, an algorithm important in cryptography.

REPRESENTATIONS OF INTEGERS

In everyday life we use decimal notation to express integers. For example, 965 is used to denote $9 \cdot 10^2 + 6 \cdot 10 + 5$. However, it is often convenient to use bases other than 10. In particular, computers usually use binary notation (with 2 as the base) when carrying out arithmetic, and octal (base 8) or hexadecimal (base 16) notation when expressing characters, such as letters or digits. In fact, we can use any positive integer greater than 1 as the base when expressing integers. This is stated in Theorem 1.

THEOREM 1

Let b be a positive integer greater than 1. Then if n is a positive integer, it can be expressed uniquely in the form

$$n = a_k b^k + a_{k-1} b^{k-1} + \cdots + a_1 b + a_0,$$

where k is a nonnegative integer, $a_0, a_1, \ldots, a_k$ are nonnegative integers less than b, and $a_k \neq 0$.

The proof of this theorem can be found in [Ro99]. The representation of n given in Theorem 1 is called the **base b expansion of n.** The base b expansion of n is denoted by $(a_k a_{k-1} \cdots a_1 a_0)_b$. For instance, $(245)_8$ represents $2 \cdot 8^2 + 4 \cdot 8 + 5 = 165$.

BINARY EXPANSIONS Choosing 2 as the base gives **binary expansions** of integers. In binary notation each digit is either a 0 or a 1. In other words, the binary expansion of an integer is just a bit string. Binary expansions (and related expansions that are variants of binary expansions) are used by computers to represent and do arithmetic with integers.

EXAMPLE 1 What is the decimal expansion of the integer that has $(1\ 0101\ 1111)_2$ as its binary expansion?

Solution: We have

$$(1\ 0101\ 1111)_2 = 1 \cdot 2^8 + 0 \cdot 2^7 + 1 \cdot 2^6 + 0 \cdot 2^5 + 1 \cdot 2^4$$
$$+ 1 \cdot 2^3 + 1 \cdot 2^2 + 1 \cdot 2^1 + 1 \cdot 2^0 = 351.$$ ◀

HEXADECIMAL EXPANSIONS Sixteen is another base used in computer science. The base 16 expansion of an integer is called its **hexadecimal** expansion. Sixteen different digits are required for such expansions. Usually, the hexadecimal digits used are 0, 1, 2, 3, 4, 5, 6, 7, 8, 9, A, B, C, D, E, and F, where the letters A through F represent the digits corresponding to the numbers 10 through 15 (in decimal notation).

EXAMPLE 2   What is the decimal expansion of the hexadecimal expansion of $(2AE0B)_{16}$?

Solution: We have

$$(2AE0B)_{16} = 2 \cdot 16^4 + 10 \cdot 16^3 + 14 \cdot 16^2 + 0 \cdot 16 + 11 = (175627)_{10}. \quad \blacktriangleleft$$

Each hexadecimal digit can be represented using four bits. For instance, we see that $(1110\,0101)_2 = (E5)_{16}$ since $(1110)_2 = (E)_{16}$ and $(0101)_2 = (5)_{16}$. **Bytes,** which are bit strings of length eight, can be represented by two hexadecimal digits.

BASE CONVERSION We will now describe an algorithm for constructing the base b expansion of an integer n. First, divide n by b to obtain a quotient and remainder, that is,

$$n = bq_0 + a_0, \qquad 0 \le a_0 < b.$$

The remainder, a_0, is the rightmost digit in the base b expansion of n. Next, divide q_0 by b to obtain

$$q_0 = bq_1 + a_1, \qquad 0 \le a_1 < b.$$

We see that a_1 is the second digit from the right in the base b expansion of n. Continue this process, successively dividing the quotients by b, obtaining additional base b digits as the remainders. This process terminates when we obtain a quotient equal to zero.

EXAMPLE 3   Find the base 8, or **octal,** expansion of $(12345)_{10}$.

Solution: First, divide 12345 by 8 to obtain

$$12345 = 8 \cdot 1543 + 1.$$

Successively dividing quotients by 8 gives

$$1543 = 8 \cdot 192 + 7,$$
$$192 = 8 \cdot 24 + 0,$$
$$24 = 8 \cdot 3 + 0,$$
$$3 = 8 \cdot 0 + 3.$$

Since the remainders are the digits of the base 8 expansion of 12345, it follows that

$$(12345)_{10} = (30071)_8. \quad \blacktriangleleft$$

EXAMPLE 4   Find the hexadecimal expansion of $(177130)_{10}$.

Solution: First divide 177130 by 16 to obtain

$$177130 = 16 \cdot 11070 + 10.$$

Successively dividing quotients by 16 gives

$$11070 = 16 \cdot 691 + 14,$$
$$691 = 16 \cdot 43 + 3,$$
$$43 = 16 \cdot 2 + 11,$$
$$2 = 16 \cdot 0 + 2.$$

Since the remainders are the digits of the hexadecimal (base 16) expansion of $(177130)_{10}$, it follows that

$$(177130)_{10} = (2B3EA)_{16}.$$

(Recall that the integers $10, 11$, and 14 correspond to the hexadecimal digits A, B, and E, respectively.) ◀

EXAMPLE 5   Find the binary expansion of $(241)_{10}$.

Solution: First divide 241 by 2 to obtain

$$241 = 2 \cdot 120 + 1.$$

Successively dividing quotients by 2 gives

$$120 = 2 \cdot 60 + 0,$$
$$60 = 2 \cdot 30 + 0,$$
$$30 = 2 \cdot 15 + 0,$$
$$15 = 2 \cdot 7 + 1,$$
$$7 = 2 \cdot 3 + 1,$$
$$3 = 2 \cdot 1 + 1,$$
$$1 = 2 \cdot 0 + 1.$$

Since the remainders are the digits of the binary (base 2) expansion of $(241)_{10}$, it follows that

$$(241)_{10} = (1111\ 0001)_2.$$ ◀

The pseudocode given in Algorithm 1 finds the base b expansion $(a_{k-1} \cdots a_1 a_0)_b$ of the integer n.

ALGORITHM 1 Constructing Base b Expansions.

---

procedure *base b expansion*(n: positive integer)
$q := n$
$k := 0$
while $q \neq 0$
begin
    $a_k := q \bmod b$
    $q := \lfloor q/b \rfloor$
    $k := k + 1$
end {the base b expansion of n is $(a_{k-1} \cdots a_1 a_0)_b$}

TABLE 1 Hexadecimal, Octal, and Binary Representation of the Integers 0 through 15.															
Decimal	0	1	2	3	4	5	6	7	8	9	10	11	12	13	14	15
Hexadecimal	0	1	2	3	4	5	6	7	8	9	A	B	C	D	E	F
Octal	0	1	2	3	4	5	6	7	10	11	12	13	14	15	16	17
Binary	0	1	10	11	100	101	110	111	1000	1001	1010	1011	1100	1101	1110	1111

In Algorithm 1, q represents the quotient obtained by successive divisions by b, starting with $q = n$. The digits in the base b expansion are the remainders of these divisions and are given by q **mod** b. The algorithm terminates when a quotient $q = 0$ is reached.

Remark: Note that Algorithm 1 can be thought of as a greedy algorithm.

Conversion between binary and hexadecimal expansions is extremely easy because each hexadecimal digit corresponds to a block of four binary digits, with this correspondence shown in Table 1 without initial 0s shown. (We leave it as Exercises 11 and 12 at the end of this section to show that this is the case.) This conversion is illustrated in Example 6.

EXAMPLE 6   Find the hexadecimal expansion of $(11\ 1110\ 1011\ 1100)_2$ and the binary expansion of $(A8D)_{16}$.

Solution: To convert $(11\ 1110\ 1011\ 1100)_2$ into hexadecimal notation we group the binary digits into blocks of four, adding initial zeros at the start of the leftmost block if necessary. These blocks are 0011, 1110, 1011, and 1100, which correspond to the hexadecimal digits 3, E, B, and C, respectively. Consequently, $(11\ 1110\ 1011\ 1100)_2 = (3EBC)_{16}$.

To convert $(A8D)_{16}$ into binary notation, we replace each hexadecimal digit by a block of four binary digits. These blocks are 1010, 1000, 1101. Consequently, $(A8D)_{16} = (1010\ 1000\ 1101)_2$. ◄

ALGORITHMS FOR INTEGER OPERATIONS

The algorithms for performing operations with integers using their binary expansions are extremely important in computer arithmetic. We will describe algorithms for the addition and the multiplication of two integers expressed in binary notation. We will also analyze the computational complexity of these algorithms, in terms of the actual number of bit operations used. Throughout this discussion, suppose that the binary expansions of a and b are

$$a = (a_{n-1}a_{n-2} \cdots a_1a_0)_2, \quad b = (b_{n-1}b_{n-2} \cdots b_1b_0)_2,$$

so that a and b each have n bits (putting bits equal to 0 at the beginning of one of these expansions if necessary).

We will measure the complexity of algorithms for integer arithmetic in terms of the number of bits in these numbers.

Consider the problem of adding two integers in binary notation. A procedure to perform addition can be based on the usual method for adding numbers with pencil and paper. This method proceeds by adding pairs of binary digits together with carries, when they occur, to compute the sum of two integers. This procedure will now be specified in detail.

To add a and b, first add their rightmost bits. This gives

$$a_0 + b_0 = c_0 \cdot 2 + s_0,$$

where s_0 is the rightmost bit in the binary expansion of $a + b$ and c_0 is the **carry**, which is either 0 or 1. Then add the next pair of bits and the carry,

$$a_1 + b_1 + c_0 = c_1 \cdot 2 + s_1,$$

where s_1 is the next bit (from the right) in the binary expansion of $a + b$, and c_1 is the carry. Continue this process, adding the corresponding bits in the two binary expansions and the carry, to determine the next bit from the right in the binary expansion of $a + b$. At the last stage, add a_{n-1}, b_{n-1}, and c_{n-2} to obtain $c_{n-1} \cdot 2 + s_{n-1}$. The leading bit of the sum is $s_n = c_{n-1}$. This procedure produces the binary expansion of the sum, namely, $a + b = (s_n s_{n-1} s_{n-2} \cdots s_1 s_0)_2$.

EXAMPLE 7 Add $a = (1110)_2$ and $b = (1011)_2$.

Solution: Following the procedure specified in the algorithm, first note that

$$a_0 + b_0 = 0 + 1 = 0 \cdot 2 + 1,$$

so that $c_0 = 0$ and $s_0 = 1$. Then, since

$$a_1 + b_1 + c_0 = 1 + 1 + 0 = 1 \cdot 2 + 0,$$

it follows that $c_1 = 1$ and $s_1 = 0$. Continuing,

$$a_2 + b_2 + c_1 = 1 + 0 + 1 = 1 \cdot 2 + 0,$$

so that $c_2 = 1$ and $s_2 = 0$. Finally, since

$$a_3 + b_3 + c_2 = 1 + 1 + 1 = 1 \cdot 2 + 1,$$

it follows that $c_3 = 1$ and $s_3 = 1$. This means that $s_4 = c_3 = 1$. Therefore, $s = a + b = (1\ 1001)_2$. This addition is displayed in Figure 1. ◀

```
  1 1 1
    1 1 1 0
    1 0 1 1
  ─────────
  1 1 0 0 1
```

FIGURE 1 Adding $(1110)_2$ and $(1011)_2$.

The algorithm for addition can be described using pseudocode as follows.

ALGORITHM 2 Addition of Integers.

---

procedure $add(a, b$: positive integers)
{the binary expansions of a and b are $(a_{n-1}a_{n-2} \cdots a_1 a_0)_2$
     and $(b_{n-1}b_{n-2} \cdots b_1 b_0)_2$, respectively}
$c := 0$
for $j := 0$ **to** $n - 1$
begin
    $d := \lfloor (a_j + b_j + c)/2 \rfloor$
    $s_j := a_j + b_j + c - 2d$
    $c := d$
end
$s_n := c$
{the binary expansion of the sum is $(s_n s_{n-1} \cdots s_0)_2$}

Next, the number of additions of bits used by Algorithm 2 will be analyzed.

EXAMPLE 8 How many additions of bits are required to use Algorithm 2 to add two integers with n bits (or less) in their binary representations?

Solution: Two integers are added by successively adding pairs of bits and, when it occurs, a carry. Adding each pair of bits and the carry requires three or fewer additions of bits. Thus, the total number of additions of bits used is less than three times the number of bits in the expansion. Hence, the number of additions of bits used by Algorithm 2 to add two n-bit integers is $O(n)$. ◄

Next, consider the multiplication of two n-bit integers a and b. The conventional algorithm (used when multiplying with pencil and paper) works as follows. Using the distributive law, we see that

$$ab = a(b_0 2^0 + b_1 2^1 + \cdots + b_{n-1} 2^{n-1})$$
$$= a(b_0 2^0) + a(b_1 2^1) + \cdots + a(b_{n-1} 2^{n-1}).$$

We can compute ab using this equation. We first note that $ab_j = a$ if $b_j = 1$ and $ab_j = 0$ if $b_j = 0$. Each time we multiply a term by 2, we shift its binary expansion one place to the left and add a zero at the tail end of the expansion. Consequently, we can obtain $(ab_j)2^j$ by **shifting** the binary expansion of ab_j j places to the left, adding j zero bits at the tail end of this binary expansion. Finally, we obtain ab by adding the n integers $ab_j 2^j$, $j = 0, 1, 2, \ldots, n - 1$.

This procedure for multiplication can be described using the pseudocode shown as Algorithm 3.

ALGORITHM 3 Multiplying Integers.

procedure *multiply*(a, b: positive integers)
{the binary expansions of a and b are $(a_{n-1}a_{n-2} \cdots a_1 a_0)_2$
        and $(b_{n-1}b_{n-2} \cdots b_1 b_0)_2$, respectively}
for $j := 0$ **to** $n - 1$
begin
    if $b_j = 1$ **then** $c_j := a$ shifted j places
    else $c_j := 0$
end
{$c_0, c_1, \ldots, c_{n-1}$ are the partial products}
$p := 0$
for $j := 0$ **to** $n - 1$
    $p := p + c_j$
{p is the value of ab}

Example 9 illustrates the use of this algorithm.

EXAMPLE 9 Find the product of $a = (110)_2$ and $b = (101)_2$.

Solution: First note that

$$ab_0 \cdot 2^0 = (110)_2 \cdot 1 \cdot 2^0 = (110)_2,$$

$$ab_1 \cdot 2^1 = (110)_2 \cdot 0 \cdot 2^1 = (0000)_2,$$

```
  1 1 0
  1 0 1
  -----
  1 1 0
0 0 0
1 1 0
-------
1 1 1 1 0
```

FIGURE 2
Multiplying
(110)$_2$ and (101)$_2$.

and

$$ab_2 \cdot 2^2 = (110)_2 \cdot 1 \cdot 2^2 = (11000)_2.$$

To find the product, add $(110)_2$, $(0000)_2$, and $(11000)_2$. Carrying out these additions (using Algorithm 2, including initial zero bits when necessary) shows that $ab = (1\ 1110)_2$. This multiplication is displayed in Figure 2. ◄

Next, we determine the number of additions of bits and shifts of bits used by Algorithm 3 to multiply two integers.

EXAMPLE 10   How many additions of bits and shifts of bits are used to multiply a and b using Algorithm 3?

Solution: Algorithm 3 computes the products of a and b by adding the partial products $c_0, c_1, c_2, \ldots,$ and c_{n-1}. When $b_j = 1$, we compute the partial product c_j by shifting the binary expansion of a j bits. When $b_j = 0$, no shifts are required since $c_j = 0$. Hence, to find all n of the integers $ab_j 2^j$, $j = 0, 1, \ldots, n-1$, requires at most

$$0 + 1 + 2 + \cdots + n - 1$$

shifts. Hence, by Example 4 in Section 2.2 the number of shifts required is $O(n^2)$.

To add the integers ab_j from $j = 0$ to $j = n - 1$ requires the addition of an n-bit integer, an $(n+1)$-bit integer, $\ldots,$ and a $(2n)$-bit integer. We know from Example 8 that each of these additions requires $O(n)$ additions of bits. Consequently, a total of $O(n^2)$ additions of bits are required for all n additions. ◄

Surprisingly, there are more efficient algorithms than the conventional algorithm for multiplying integers. One such algorithm, which uses $O(n^{1.585})$ bit operations to multiply n-bit numbers, will be described in Section 6.3.

Given integers a and $d, d > 0$, we can find $q = a$ **div** d and $r = a$ **mod** d using Algorithm 4. In this algorithm, when a is positive we subtract d from a as many times as necessary until what is left is less than d. The number of times we perform this subtraction is the quotient and what is left over after all these subtractions is the remainder. Algorithm 4 also covers the case where a is negative. It finds the quotient q, and remainder r when $|a|$ is divided by d and when $r > 0$, uses these to find the quotient $-(q + 1)$ and remainder $d - r$ when a is divided by d. We leave it to the reader (Exercise 55) to show that, assuming that $a > d$, this algorithm uses $O(q \log a)$ bit operations.

There are more efficient algorithms than Algorithm 4 for determining the quotient $q = a$ **div** d and the remainder $r = a$ **mod** d when a positive integer a is divided by a positive integer d (see [Kn98] for details). These algorithms require $O(\log a \cdot \log d)$ bit operations. If both of the binary expansions of a and d contain n or fewer bits, then we can replace $\log a \cdot \log d$ by n^2. This means that we need $O(n^2)$ bit operations to find the quotient and remainder when a is divided by d.

MODULAR EXPONENTIATION

In cryptography it is important to be able to efficiently find b^n **mod** m, where $b, n,$ and m are large integers. It is impractical to first compute b^n and then find its remainder when divided by m because b^n will be a huge number. Instead, we can use an algorithm that employs the binary expansion of the exponent n, say $n = (a_{k-1} \cdots a_1 a_0)_2$. The algorithm

---

ALGORITHM 4 Computing div and mod.

procedure *division algorithm*(a: integer, d: positive integer)
$q := 0$
$r := |a|$
while $r \geq d$
begin
    $r := r - d$
    $q := q + 1$
end
if $a < 0$ and $r > 0$ **then**
begin
    $r := d - r$
    $q := -(q + 1)$
end
$\{q = a$ **div** d is the quotient, $r = a$ **mod** d is the remainder$\}$

ALGORITHM 5 Modular Exponentiation.

procedure *modular exponentiation*(b: integer, $n = (a_{k-1}a_{k-2}\cdots a_1 a_0)_2$,
    m: positive integers)
$x := 1$
$power := b$ **mod** m
for $i := 0$ **to** $k - 1$
begin
    if $a_i = 1$ **then** $x := (x \cdot power)$ **mod** m
    $power := (power \cdot power)$ **mod** m
end
$\{x$ equals b^n **mod** $m\}$

---

successively finds b **mod** m, b^2 **mod** m, b^4 **mod** $m, \ldots, b^{2^{k-1}}$ **mod** m and multiplies together those terms b^{2^j} **mod** m where $a_j = 1$, finding the remainder of the product when divided by m after each multiplication. Pseudocode for this algorithm is shown in Algorithm 5.

We illustrate how Algorithm 5 works in Example 11.

EXAMPLE 11   Use Algorithm 5 to find 2^{644} **mod** 645.

Solution: Algorithm 5 initially sets $x = 1$ and $power = 2$ **mod** $645 = 2$. In the computation of 2^{644} **mod** 645, this algorithm determines 2^{2^j} **mod** 645 for $j = 1, 2, \ldots, 9$ by successively squaring and reducing modulo 645. If $a_j = 1$ (where a_j is the bit in the jth position in the binary expansion of 644), it multiplies the current value of x by 2^{2^j} **mod** 645 and reduces the result modulo 645. Here are the steps used:

$i = 0$: Because $a_0 = 0$, we have $x = 1$ and $power = 2^2 = 4 \bmod 645 = 4$;

$i = 1$: Because $a_1 = 0$, we have $x = 1$ and $power = 4^2 = 16 \bmod 645 = 16$;

$i = 2$: Because $a_2 = 1$, we have $x = 1 \cdot 16 \bmod 645 = 16$ and $power = 16^2 = 256 \bmod 645 = 256$;

$i = 3$: Because $a_3 = 0$, we have $x = 16$ and $power = 256^2 = 65{,}536 \bmod 645 = 391$;

$i = 4$: Because $a_4 = 0$, we have $x = 16$ and $power = 391^2 = 152{,}881 \bmod 645 = 16$;

$i = 5$: Because $a_5 = 0$, we have $x = 16$ and $power = 16^2 = 256 \bmod 645 = 256$;

$i = 6$: Because $a_6 = 0$, we have $x = 16$ and $power = 256^2 = 65{,}536 \bmod 645 = 391$;

$i = 7$: Because $a_7 = 1$, we find that $x = (16 \cdot 391) \bmod 645 = 451$ and $power = 391^2 = 152{,}881 \bmod 645 = 16$;

$i = 8$: Because $a_8 = 0$, we have $x = 451$ and $power = 16^2 = 256 \bmod 645 = 256$;

$i = 9$: Because $a_9 = 1$, we find that $x = (451 \cdot 256) \bmod 645 = 1$.

This shows that following the steps of Algorithm 5 produces the result $2^{644} \bmod 645 = 1$.

◀

Algorithm 5 is quite efficient; it uses $O((\log m)^2 \log n)$ bit operations to find $b^n \bmod m$ (see Exercise 54).

THE EUCLIDEAN ALGORITHM

The method described in Section 2.4 for computing the greatest common divisor of two integers, using the prime factorizations of these integers, is inefficient. The reason is that it is time-consuming to find prime factorizations. We will give a more efficient method of finding the greatest common divisor, called the **Euclidean algorithm.** This algorithm has been known since ancient times. It is named after the ancient Greek mathematician Euclid, who included a description of this algorithm in his *Elements*.

Before describing the Euclidean algorithm, we will show how it is used to find $\gcd(91, 287)$. First, divide 287, the larger of the two integers, by 91, the smaller, to obtain

$$287 = 91 \cdot 3 + 14.$$

Any divisor of 91 and 287 must also be a divisor of $287 - 91 \cdot 3 = 14$. Also, any divisor of 91 and 14 must also be a divisor of $287 = 91 \cdot 3 + 14$. Hence, the greatest common divisor of 91 and 287 is the same as the greatest common divisor of 91 and 14. This means that the problem of finding $\gcd(91, 287)$ has been reduced to the problem of finding $\gcd(91, 14)$.

Next, divide 91 by 14 to obtain

$$91 = 14 \cdot 6 + 7.$$

Since any common divisor of 91 and 14 also divides $91 - 14 \cdot 6 = 7$ and any common divisor of 14 and 7 divides 91, it follows that $\gcd(91, 14) = \gcd(14, 7)$.

EUCLID (C. 325–265 B.C.E.) Euclid was the author of the most successful mathematics book ever written, the *Elements,* which appeared in over 1000 different editions from ancient to modern times. Little is known about Euclid's life, other than that he taught at the famous academy at Alexandria. Apparently, Euclid did not stress applications. When a student asked what he would get by learning geometry, Euclid explained that knowledge was worth acquiring for its own sake and told his servant to give the student a coin "since he must make a profit from what he learns."

Continue by dividing 14 by 7, to obtain

$$14 = 7 \cdot 2.$$

Since 7 divides 14, it follows that $\gcd(14, 7) = 7$. Furthermore, because $\gcd(287, 91) = \gcd(91, 14) = \gcd(14, 7) = 7$, the original problem has been solved.

We now describe how the Euclidean algorithm works in generality. We will use successive divisions to reduce the problem of finding the greatest common divisor of two positive integers to the same problem with smaller integers, until one of the integers is zero.

The Euclidean algorithm is based on the following result about greatest common divisors and the division algorithm.

LEMMA 1   Let $a = bq + r$, where $a, b, q,$ and r are integers. Then $\gcd(a, b) = \gcd(b, r)$.

Proof: If we can show that the common divisors of a and b are the same as the common divisors of b and r, we will have shown that $\gcd(a, b) = \gcd(b, r)$, since both pairs must have the same *greatest* common divisor.

So suppose that d divides both a and b. Then it follows that d also divides $a - bq = r$ (from Theorem 1 of Section 2.4). Hence, any common divisor of a and b is also a common divisor of b and r.

Likewise, suppose that d divides both b and r. Then d also divides $bq + r = a$. Hence, any common divisor of b and r is also a common divisor of a and b.

Consequently, $\gcd(a, b) = \gcd(b, r)$.   ◁

Suppose that a and b are positive integers with $a \geq b$. Let $r_0 = a$ and $r_1 = b$. When we successively apply the division algorithm, we obtain

$$r_0 = r_1 q_1 + r_2 \qquad 0 \leq r_2 < r_1,$$
$$r_1 = r_2 q_2 + r_3 \qquad 0 \leq r_3 < r_2,$$
$$\vdots$$
$$r_{n-2} = r_{n-1} q_{n-1} + r_n \quad 0 \leq r_n < r_{n-1},$$
$$r_{n-1} = r_n q_n.$$

Eventually a remainder of zero occurs in this sequence of successive divisions, since the sequence of remainders $a = r_0 > r_1 > r_2 > \cdots \geq 0$ cannot contain more than a terms. Furthermore, it follows from Lemma 1 that

$$\gcd(a, b) = \gcd(r_0, r_1) = \gcd(r_1, r_2) = \cdots = \gcd(r_{n-2}, r_{n-1})$$
$$= \gcd(r_{n-1}, r_n) = \gcd(r_n, 0) = r_n.$$

Hence, the greatest common divisor is the last nonzero remainder in the sequence of divisions.

EXAMPLE 12   Find the greatest common divisor of 414 and 662 using the Euclidean algorithm.

Solution: Successive uses of the division algorithm give:

$$662 = 414 \cdot 1 + 248$$

$$414 = 248 \cdot 1 + 166$$

$$248 = 166 \cdot 1 + 82$$

$$166 = 82 \cdot 2 + 2$$

$$82 = 2 \cdot 41.$$

Hence, gcd(414, 662) = 2, since 2 is the last nonzero remainder. ◄

The Euclidean algorithm is expressed in pseudocode in Algorithm 6.

ALGORITHM 6 The Euclidean Algorithm.

procedure *gcd*(*a*, *b*: positive integers)
x := *a*
y := *b*
while *y* ≠ 0
begin
    r := *x* **mod** *y*
    x := *y*
    y := *r*
end {gcd(*a*, *b*) is *x*}

In Algorithm 6, the initial values of x and y are a and b, respectively. At each stage of the procedure, x is replaced by y, and y is replaced by x **mod** y, which is the remainder when x is divided by y. This process is repeated as long as $y \neq 0$. The algorithm terminates when $y = 0$, and the value of x at that point, the last nonzero remainder in the procedure, is the greatest common divisor of a and b.

We will study the time complexity of the Euclidean algorithm in Section 3.4, where we will show that the number of divisions required to find the greatest common divisor of a and b, where $a \geq b$, is $O(\log b)$.

Exercises

1. Convert these integers from decimal notation to binary notation.

  a) 231   **b)** 4532   **c)** 97644

2. Convert these integers from decimal notation to binary notation.

  a) 321   **b)** 1023   **c)** 100632

3. Convert these integers from binary notation to decimal notation.

  a) 1 1111   **b)** 10 0000 0001
  c) 1 0101 0101   **d)** 110 1001 0001 0000

4. Convert these integers from binary notation to decimal notation.

  a) 1 1011   **b)** 10 1011 0101
  c) 11 1011 1110 **d)** 111 1100 0001 1111

5. Convert these integers from hexadecimal notation to binary notation.

  a) 80E   **b)** 135AB
  c) ABBA   **d)** DEFACED

6. Convert $(\text{BADFACED})_{16}$ from its hexadecimal expansion to its binary expansion.

7. Convert $(ABCDEF)_{16}$ from its hexadecimal expansion to its binary expansion.

8. Convert each of these integers from binary notation to hexadecimal notation.
   a) 1111 0111
   b) 1010 1010 1010
   c) 111 0111 0111 0111

9. Convert $(1011\ 0111\ 1011)_2$ from its binary expansion to its hexadecimal expansion.

10. Convert $(1\ 1000\ 0110\ 0011)_2$ from its binary expansion to its hexadecimal expansion.

11. Show that the hexadecimal expansion of a positive integer can be obtained from its binary expansion by grouping together blocks of four binary digits, adding initial digits if necessary, and translating each block of four binary digits into a single hexadecimal digit.

12. Show that the binary expansion of a positive integer can be obtained from its hexadecimal expansion by translating each hexadecimal digit into a block of four binary digits.

13. Give a simple procedure for converting from the binary expansion of an integer to its octal expansion.

14. Give a simple procedure for converting from the octal expansion of an integer to its binary expansion.

15. Convert $(7345321)_8$ to its binary expansion and $(10\ 1011\ 1011)_2$ to its octal expansion.

16. Give a procedure for converting from the hexadecimal expansion of an integer to its octal expansion using binary notation as an intermediate step.

17. Give a procedure for converting from the octal expansion of an integer to its hexadecimal expansion using binary notation as an intermediate step.

18. Convert $(12345670)_8$ to its hexadecimal expansion and $(ABB093BABBA)_{16}$ to its octal expansion.

19. Use Algorithm 5 to find $3^{2003} \bmod 99$.

20. Use Algorithm 5 to find $123^{1001} \bmod 101$.

21. Use the Euclidean algorithm to find
   a) $\gcd(12, 18)$.   b) $\gcd(111, 201)$.
   c) $\gcd(1001, 1331)$.   d) $\gcd(12345, 54321)$.
   e) $\gcd(1000, 5040)$.   f) $\gcd(9888, 6060)$.

22. Use the Euclidean algorithm to find
   a) $\gcd(1, 5)$.   b) $\gcd(100, 101)$.
   c) $\gcd(123, 277)$.   d) $\gcd(1529, 14039)$.
   e) $\gcd(1529, 14038)$.   f) $\gcd(11111, 111111)$.

23. How many divisions are required to find $\gcd(21, 34)$ using the Euclidean algorithm?

24. How many divisions are required to find $\gcd(34, 55)$ using the Euclidean algorithm?

25. Show that every positive integer can be represented uniquely as the sum of distinct powers of 2. (*Hint:* Consider binary expansions of integers.)

26. It can be shown that every integer can be uniquely represented in the form

$$e_k 3^k + e_{k-1} 3^{k-1} + \cdots + e_1 3 + e_0,$$

where $e_j = -1, 0,$ or 1 for $j = 0, 1, 2, \ldots, k$. Expansions of this type are called **balanced ternary expansions.** Find the balanced ternary expansions of
   a) 5.   b) 13.   c) 37.   d) 79.

27. Show that a positive integer is divisible by 3 if and only if the sum of its decimal digits is divisible by 3.

28. Show that a positive integer is divisible by 11 if and only if the difference of the sum of its decimal digits in even-numbered positions and the sum of its decimal digits in odd-numbered positions is divisible by 11.

29. Show that a positive integer is divisible by 3 if and only if the difference of the sum of its binary digits in even-numbered positions and the sum of its binary digits in odd-numbered positions is divisible by 3.

One's complement representations of integers are used to simplify computer arithmetic. To represent positive and negative integers with absolute value less than 2^{n-1}, a total of n bits is used. The leftmost bit is used to represent the sign. A 0 bit in this position is used for positive integers, and a 1 bit in this position is used for negative integers. For positive integers the remaining bits are identical to the binary expansion of the integer. For negative integers, the remaining bits are obtained by first finding the binary expansion of the absolute value of the integer, and then taking the complement of each of these bits, where the complement of a 1 is a 0 and the complement of a 0 is a 1.

30. Find the one's complement representations, using bit strings of length six, of the following integers.
   a) 22   b) 31   c) -7 d) -19

31. What integer does each of the following one's complement representations of length five represent?
   a) 11001   b) 01101   c) 10001   d) 11111

32. If m is a positive integer less than 2^{n-1}, how is the one's complement representation of $-m$ obtained from the one's complement of m, when bit strings of length n are used?

33. How is the one's complement representation of the sum of two integers obtained from the one's complement representations of these integers?

34. How is the one's complement representation of the difference of two integers obtained from the one's complement representations of these integers?

35. Show that the integer m with one's complement representation $(a_{n-1} a_{n-2} \cdots a_1 a_0)$ can be found using the equation $m = -a_{n-1}(2^{n-1} - 1) + a_{n-2} 2^{n-2} + \cdots + a_1 \cdot 2 + a_0$.

Two's complement representations of integers are also used to simplify computer arithmetic and are used more commonly than one's complement representations. To represent an integer x with $-2^{n-1} \le x \le 2^{n-1} - 1$ for a specified positive integer n, a total of n bits is used. The leftmost bit is used to represent the sign. A 0 bit in this position is used for positive integers, and a 1 bit in this position is used for negative integers, just as in one's complement expansions. For a positive integer, the remaining bits are identical to the binary expansion of the integer. For a negative integer, the remaining bits are the bits of the binary expansion of $2^{n-1} - |x|$. Two's complement expansions of integers are often used by computers because addition and subtraction of integers can be performed easily using these expansions, where these integers can be either positive or negative.

36. Answer Exercise 30, but this time find the two's complement expansion using bit strings of length six.

37. Answer Exercise 31 if each expansion is a two's complement expansion of length five.

38. Answer Exercise 32 for two's complement expansions.

39. Answer Exercise 33 for two's complement expansions.

40. Answer Exercise 34 for two's complement expansions.

41. Show that the integer m with two's complement representation $(a_{n-1}a_{n-2}\cdots a_1a_0)$ can be found using the equation $m = -a_{n-1} \cdot 2^{n-1} + a_{n-2}2^{n-2} + \cdots + a_1 \cdot 2 + a_0$.

42. Give a simple algorithm for forming the two's complement representation of an integer from its one's complement representation.

43. Sometimes integers are encoded by using four-digit binary expansions to represent each decimal digit. This produces the **binary coded decimal** form of the integer. For instance, 791 is encoded in this way by 011110010001. How many bits are required to represent a number with n decimal digits using this type of encoding?

A **Cantor expansion** is a sum of the form

$$a_n n! + a_{n-1}(n-1)! + \cdots + a_2 2! + a_1 1!,$$

where a_i is an integer with $0 \le a_i \le i$ for $i = 1, 2, \ldots, n$.

44. Find the Cantor expansions of

    a) 2.   **b)** 7.   **c)** 19.

    d) 87.   **e)** 1000.   **f)** 1,000,000.

*45. Describe an algorithm that finds the Cantor expansion of an integer.

*46. Describe an algorithm to add two integers from their Cantor expansions.

47. Add $(10111)_2$ and $(11010)_2$ by working through each step of the algorithm for addition given in the text.

48. Multiply $(1110)_2$ and $(1010)_2$ by working through each step of the algorithm for multiplication given in the text.

49. Describe an algorithm for finding the difference of two binary expansions.

50. Estimate the number of bit operations used to subtract two binary expansions.

51. Devise an algorithm that, given the binary expansions of the integers a and b, determines whether $a > b$, $a = b$, or $a < b$.

52. How many bit operations does the comparison algorithm from Exercise 51 use when the larger of a and b has n bits in its binary expansion?

53. Estimate the complexity of Algorithm 1 for finding the base b expansion of an integer n in terms of the number of divisions used.

*54. Show that Algorithm 5 uses $O((\log m)^2 \log n)$ bit operations to find $b^n \bmod m$.

55. Show that Algorithm 4 uses $O(q \log |a|)$ bit operations, assuming that $a > d$.

2.6 **Applications of Number Theory**

INTRODUCTION

Number theory has many applications, especially to computer science. In Section 2.4 we described several of these applications, including hashing functions, the generation of pseudorandom numbers, and shift ciphers. This section continues our introduction to number theory, developing some key results and presenting two important applications: a method for performing arithmetic with large integers and a recently invented type of cryptosystem, called a *public key system*. In such a cryptosystem, we do not have to keep encryption keys secret, since knowledge of an encryption key does not help someone decrypt messages in a realistic amount of time. Privately held decryption keys are used to decrypt messages.

Before developing these applications, we will introduce some key results that play a central role in number theory and its applications. For example, we will show how to solve systems of linear congruences modulo pairwise relatively prime integers using the Chinese Remainder Theorem, and then show how to use this result as a basis for performing arithmetic with large integers. We will introduce Fermat's Little Theorem and the concept of a pseudoprime and will show how to use these concepts to develop a public key cryptosystem.

SOME USEFUL RESULTS

An important result we will use throughout this section is that the greatest common divisor of two integers a and b can be expressed in the form

$$sa + tb,$$

where s and t are integers. In other words, $\gcd(a, b)$ can be expressed as a **linear combination** with integer coefficients of a and b. For example, $\gcd(6, 14) = 2$, and $2 = (-2) \cdot 6 + 1 \cdot 14$. We state this fact as Theorem 1.

THEOREM 1   If a and b are positive integers, then there exist integers s and t such that $\gcd(a, b) = sa + tb$.

We will not give a formal proof of Theorem 1 here (see Exercise 66 in Section 3.3 and [Ro00] for proofs), but we will provide an example of a method for finding a linear combination of two integers equal to their greatest common divisor. (In this section, we will assume that a linear combination has integer coefficients.) The method proceeds by working backward through the divisions of the Euclidean algorithm. (We also describe an algorithm called the **extended Euclidean algorithm** that can be used to express $\gcd(a, b)$ as a linear combination of a and b in the preamble to Exercise 48.)

EXAMPLE 1   Express $\gcd(252, 198) = 18$ as a linear combination of 252 and 198.

Solution: To show that $\gcd(252, 198) = 18$, the Euclidean algorithm uses these divisions:

$$252 = 1 \cdot 198 + 54$$
$$198 = 3 \cdot 54 + 36$$
$$54 = 1 \cdot 36 + 18$$
$$36 = 2 \cdot 18.$$

Using the next-to-last division (the third division), we can express $\gcd(252, 198) = 18$ as a linear combination of 54 and 36. We find that

$$18 = 54 - 1 \cdot 36.$$

The second division tells us that

$$36 = 198 - 3 \cdot 54.$$

Substituting this expression for 36 into the previous equation, we can express 18 as a linear combination of 54 and 198. We have

$$18 = 54 - 1 \cdot 36 = 54 - 1 \cdot (198 - 3 \cdot 54) = 4 \cdot 54 - 1 \cdot 198.$$

The first division tells us that

$$54 = 252 - 1 \cdot 198.$$

Substituting this expression for 54 into the previous equation, we can express 18 as a linear combination of 252 and 198. We conclude that

$$18 = 4 \cdot (252 - 1 \cdot 198) - 1 \cdot 198 = 4 \cdot 252 - 5 \cdot 198,$$

completing the solution.   ◄

We will use Theorem 1 to develop several useful results. One of our goals will be to prove the part of the Fundamental Theorem of Arithmetic asserting that a positive integer has at most one prime factorization. We will show that if a positive integer has a factorization into primes, where the primes are written in nondecreasing order, then this factorization is unique.

First, we need to develop some results about divisibility.

LEMMA 1   If $a, b,$ and c are positive integers such that $\gcd(a, b) = 1$ and $a \mid bc,$ then $a \mid c.$

Proof: Since $\gcd(a, b) = 1,$ by Theorem 1 there are integers s and t such that

$$sa + tb = 1.$$

Multiplying both sides of this equation by $c,$ we obtain

$$sac + tbc = c.$$

Using Theorem 1 of Section 2.4, we can use this last equation to show that $a \mid c.$ By part 2 of that theorem, $a \mid tbc.$ Since $a \mid sac$ and $a \mid tbc,$ by part 1 of that theorem, we conclude that a divides $sac + tbc,$ and hence $a \mid c.$ This finishes the proof.   ◁

We will use the following generalization of Lemma 1 in the proof of uniqueness of prime factorizations. (The proof of Lemma 2 is left as an exercise in Section 3.3, since it can be most easily carried out using the method of mathematical induction, which will be covered in that section.)

LEMMA 2   If p is a prime and $p \mid a_1 a_2 \cdots a_n$ where each a_i is an integer, then $p \mid a_i$ for some $i.$

We can now show that a factorization of an integer into primes is unique. That is, we will show that every integer can be written as the product of primes in nondecreasing order in at most one way. This is part of the Fundamental Theorem of Arithmetic. We will prove the other part, that every integer has a factorization into primes, in Section 3.3.

Proof (of the uniqueness of the prime factorization of a positive integer): We will use a proof by contradiction. Suppose that the positive integer n can be written as the product of primes in two different ways, say, $n = p_1 p_2 \cdots p_s$ and $n = q_1 q_2 \cdots q_t,$ each p_i and q_j are primes such that $p_1 \le p_2 \le \cdots \le p_s$ and $q_1 \le q_2 \le \cdots \le q_t.$

When we remove all common primes from the two factorizations, we have

$$p_{i_1} p_{i_2} \cdots p_{i_u} = q_{j_1} q_{j_2} \cdots q_{j_v},$$

where no prime occurs on both sides of this equation and u and v are positive integers. By Lemma 1 it follows that p_{i_1} divides q_{j_k} for some k. Since no prime divides another prime, this is impossible. Consequently, there can be at most one factorization of n into primes in nondecreasing order. ◁

Lemma 1 can also be used to prove a result about dividing both sides of a congruence by the same integer. We have shown (Theorem 10 in Section 2.4) that we can multiply both sides of a congruence by the same integer. However, dividing both sides of a congruence by an integer does not always produce a valid congruence, as Example 2 shows.

EXAMPLE 2   The congruence $14 \equiv 8 \pmod 6$ holds, but both sides of this congruence cannot be divided by 2 since $14/2 = 7$ and $8/2 = 4$, but $7 \not\equiv 4 \pmod 6$. ◀

However, using Lemma 1, we can show that we can divide both sides of a congruence by an integer relatively prime to the modulus. This is stated as Theorem 2.

THEOREM 2

Let m be a positive integer and let a, b, and c be integers. If $ac \equiv bc \pmod m$ and $\gcd(c, m) = 1$, then $a \equiv b \pmod m$.

Proof: Since $ac \equiv bc \pmod m$, $m \mid ac - bc = c(a - b)$. By Lemma 1, since $\gcd(c, m) = 1$, it follows that $m \mid a - b$. We conclude that $a \equiv b \pmod m$. ◁

LINEAR CONGRUENCES

A congruence of the form

$$ax \equiv b \pmod m,$$

where m is a positive integer, a and b are integers, and x is a variable, is called a **linear congruence.** Such congruences arise throughout number theory and its applications.

How can we solve the linear congruence $ax \equiv b \pmod m$, that is, find all integers x that satisfy this congruence? One method that we will describe uses an integer $\bar{a}$ such that $\bar{a}a \equiv 1 \pmod m$, if such an integer exists. Such an integer $\bar{a}$ is said to be an **inverse** of a modulo m. Theorem 3 guarantees that an inverse of a modulo m exists whenever a and m are relatively prime.

THEOREM 3

If a and m are relatively prime integers and $m > 1$, then an inverse of a modulo m exists. Furthermore, this inverse is unique modulo m. (That is, there is a unique positive integer $\bar{a}$ less than m that is an inverse of a modulo m and every other inverse of a modulo m is congruent to $\bar{a}$ modulo m.)

Proof: By Theorem 1, since $\gcd(a, m) = 1$, there are integers s and t such that

$$sa + tm = 1.$$

This implies that

$$sa + tm \equiv 1 \pmod m.$$

Since $tm \equiv 0 \pmod{m}$, it follows that

$$sa \equiv 1 \pmod{m}.$$

Consequently, s is an inverse of a modulo m. That this inverse is unique modulo m is left as Exercise 9 at the end of this section. ◁

The proof of Theorem 3 describes a method for finding the inverse of a modulo m when a and m are relatively prime: find a linear combination of a and m that equals 1 (which can be done by working backward through the steps of the Euclidean algorithm); the coefficient of a in this linear combination is an inverse of a modulo m. We illustrate this procedure in Example 3.

EXAMPLE 3   Find an inverse of 3 modulo 7.

Solution: Since $\gcd(3, 7) = 1$, Theorem 3 tells us that an inverse of 3 modulo 7 exists. The Euclidean algorithm ends quickly when used to find the greatest common divisor of 3 and 7:

$$7 = 2 \cdot 3 + 1.$$

From this equation we see that

$$-2 \cdot 3 + 1 \cdot 7 = 1.$$

This shows that -2 is an inverse of 3 modulo 7. (Note that every integer congruent to -2 modulo 7 is also an inverse of 3, such as $5, -9, 12$, and so on.) ◀

When we have an inverse $\overline{a}$ of a modulo m, we can easily solve the congruence $ax \equiv b \pmod{m}$ by multiplying both sides of the linear congruence by $\overline{a}$, as Example 4 illustrates.

EXAMPLE 4   What are the solutions of the linear congruence $3x \equiv 4 \pmod{7}$?

Solution: By Example 3 we know that -2 is an inverse of 3 modulo 7. Multiplying both sides of the congruence by -2 shows that

$$-2 \cdot 3x \equiv -2 \cdot 4 \pmod{7}.$$

Since $-6 \equiv 1 \pmod{7}$ and $-8 \equiv 6 \pmod{7}$, it follows that if x is a solution, then $x \equiv -8 \equiv 6 \pmod{7}$.

We need to determine whether every x with $x \equiv 6 \pmod{7}$ is a solution. Assume that $x \equiv 6 \pmod{7}$. Then, by Theorem 10 of Section 2.4, it follows that

$$3x \equiv 3 \cdot 6 = 18 \equiv 4 \pmod{7},$$

which shows that all such x satisfy the congruence. We conclude that the solutions to the congruence are the integers x such that $x \equiv 6 \pmod{7}$, namely, $6, 13, 20, \ldots$ and $-1, -8, -15, \ldots$. ◀

THE CHINESE REMAINDER THEOREM

Links

Systems of linear congruences arise in many contexts. For example, as we will see later, they are the basis for a method that can be used to perform arithmetic with large integers.

Such systems can even be found as word puzzles in the writings of ancient Chinese and Hindu mathematicians, such as that given in Example 5.

EXAMPLE 5   In the first century, the Chinese mathematician Sun-Tsu asked:

> There are certain things whose number is unknown. When divided by 3, the remainder is 2; when divided by 5, the remainder is 3; and when divided by 7, the remainder is 2. What will be the number of things?

This puzzle can be translated into the following question: What are the solutions of the systems of congruences

$$x \equiv 2 \ (\text{mod } 3),$$
$$x \equiv 3 \ (\text{mod } 5),$$
$$x \equiv 2 \ (\text{mod } 7)?$$

We will solve this system, and with it Sun-Tsu's puzzle, later in this section.   ◀

The *Chinese Remainder Theorem*, named after the Chinese heritage of problems involving systems of linear congruences, states that when the moduli of a system of linear congruences are pairwise relatively prime, there is a unique solution of the system modulo the product of the moduli.

THEOREM 4

THE CHINESE REMAINDER THEOREM Let $m_1, m_2, \ldots, m_n$ be pairwise relatively prime positive integers. The system

$$x \equiv a_1 \ (\text{mod } m_1),$$
$$x \equiv a_2 \ (\text{mod } m_2),$$

$$\cdot$$
$$\cdot$$
$$\cdot$$

$$x \equiv a_n \ (\text{mod } m_n)$$

has a unique solution modulo $m = m_1 m_2 \cdots m_n$. (That is, there is a solution x with $0 \leq x < m$, and all other solutions are congruent modulo m to this solution.)

Proof: To establish this theorem, we need to show that a solution exists and that it is unique modulo m. We will show that a solution exists by describing a way to construct this solution; showing that the solution is unique modulo m is Exercise 24 at the end of this section.

To construct a simultaneous solution, first let

$$M_k = m/m_k$$

for $k = 1, 2, \ldots, n$. That is, M_k is the product of the moduli except for m_k. Since m_i and m_k have no common factors greater than 1 when $i \neq k$, it follows that $\gcd(m_k, M_k) = 1$. Consequently, by Theorem 3, we know that there is an integer y_k, an inverse of M_k modulo m_k, such that

$$M_k y_k \equiv 1 \ (\text{mod } m_k).$$

To construct a simultaneous solution, form the sum

$$x = a_1 M_1 y_1 + a_2 M_2 y_2 + \cdots + a_n M_n y_n.$$

We will now show that x is a simultaneous solution. First, note that since $M_j \equiv 0 \pmod{m_k}$ whenever $j \neq k$, all terms except the kth term in this sum are congruent to 0 modulo m_k. Since $M_k y_k \equiv 1 \pmod{m_k}$ we see that

$$x \equiv a_k M_k y_k \equiv a_k \pmod{m_k},$$

for $k = 1, 2, \ldots, n$. We have shown that x is a simultaneous solution to the n congruences.   ◁

The following example illustrates how to use the construction given in the proof of Theorem 4 to solve a system of congruences. We will solve the system given in Example 5, arising in Sun-Tsu's puzzle.

EXAMPLE 6   To solve the system of congruences in Example 5, first let $m = 3 \cdot 5 \cdot 7 = 105$, $M_1 = m/3 = 35$, $M_2 = m/5 = 21$, and $M_3 = m/7 = 15$. We see that 2 is an inverse of $M_1 = 35$ modulo 3, since $35 \equiv 2 \pmod 3$; 1 is an inverse of $M_2 = 21$ modulo 5, since $21 \equiv 1 \pmod 5$; and 1 is an inverse of $M_3 = 15 \pmod 7$, since $15 \equiv 1 \pmod 7$. The solutions to this system are those x such that

$$\begin{aligned}
x &\equiv a_1 M_1 y_1 + a_2 M_2 y_2 + a_3 M_3 y_3 = 2 \cdot 35 \cdot 2 + 3 \cdot 21 \cdot 1 + 2 \cdot 15 \cdot 1 \\
&= 233 \equiv 23 \pmod{105}.
\end{aligned}$$

It follows that 23 is the smallest positive integer that is a simultaneous solution. We conclude that 23 is the smallest positive integer that leaves a remainder of 2 when divided by 3, a remainder of 3 when divided by 5, and a remainder of 2 when divided by 7.   ◀

COMPUTER ARITHMETIC WITH LARGE INTEGERS

Suppose that $m_1, m_2, \ldots, m_n$ are pairwise relatively prime integers greater than or equal to 2 and let m be their product. By the Chinese Remainder Theorem, we can show (see Exercise 22) that an integer a with $0 \leq a < m$ can be uniquely represented by the n-tuple consisting of its remainders upon division by m_i, $i = 1, 2, \ldots, n$. That is, we can uniquely represent a by

$$(a \bmod m_1, a \bmod m_2, \ldots, a \bmod m_n).$$

EXAMPLE 7   What are the pairs used to represent the nonnegative integers less than 12 when they are represented by the ordered pair where the first component is the remainder of the integer upon division by 3 and the second component is the remainder of the integer upon division by 4?

Solution: We have the following representations, obtained by finding the remainder of each integer when it is divided by 3 and by 4:

$$\begin{array}{lll}
0 = (0, 0) & 4 = (1, 0) & 8 = (2, 0) \\
1 = (1, 1) & 5 = (2, 1) & 9 = (0, 1) \\
2 = (2, 2) & 6 = (0, 2) & 10 = (1, 2) \\
3 = (0, 3) & 7 = (1, 3) & 11 = (2, 3).
\end{array}$$

◀

To perform arithmetic with large integers, we select moduli $m_1, m_2, \ldots, m_n$, where each m_i is an integer greater than 2, $\gcd(m_i, m_j) = 1$ whenever $i \neq j$, and $m = m_1 m_2 \cdots m_n$ is greater than the result of the arithmetic operations we want to carry out.

Once we have selected our moduli, we carry out arithmetic operations with large integers by performing componentwise operations on the n-tuples representing these integers using their remainders upon division by m_i, $i = 1, 2, \ldots, n$. Once we have computed the value of each component in the result, we recover its value by solving a system of n congruences modulo m_i, $i = 1, 2, \ldots, n$. This method of performing arithmetic with large integers has several valuable features. First, it can be used to perform arithmetic with integers larger than can ordinarily be carried out on a computer. Second, computations with respect to the different moduli can be done in parallel, speeding up the arithmetic.

EXAMPLE 8   Suppose that performing arithmetic with integers less than 100 on a certain processor is much quicker than doing arithmetic with larger integers. We can restrict almost all our computations to integers less than 100 if we represent integers using their remainders modulo pairwise relatively prime integers less than 100. For example, we can use the moduli of 99, 98, 97, and 95. (These integers are relatively prime pairwise, since no two have a common factor greater than 1.)

By the Chinese Remainder Theorem, every nonnegative integer less than $99 \cdot 98 \cdot 97 \cdot 95 = 89{,}403{,}930$ can be represented uniquely by its remainders when divided by these four moduli. For example, we represent 123,684 as $(33, 8, 9, 89)$, since 123,684 **mod** 99 $= 33$, 123,684 **mod** 98 $= 8$; 123,684 **mod** 97 $= 9$; and 123,684 **mod** 95 $= 89$. Similarly, we represent 413,456 as $(32, 92, 42, 16)$.

To find the sum of 123,684 and 413,456, we work with these 4-tuples instead of these two integers directly. We add the 4-tuples componentwise and reduce each component with respect to the appropriate modulus. This yields

$$(33, 8, 9, 89) + (32, 92, 42, 16)$$
$$= (65 \textbf{ mod } 99, 100 \textbf{ mod } 98, 51 \textbf{ mod } 97, 105 \textbf{ mod } 95)$$
$$= (65, 2, 51, 10).$$

To find the sum, that is, the integer represented by $(65, 2, 51, 10)$, we need to solve the system of congruences

$$x \equiv 65 \ (\text{mod } 99)$$
$$x \equiv \ 2 \ (\text{mod } 98)$$
$$x \equiv 51 \ (\text{mod } 97)$$
$$x \equiv 10 \ (\text{mod } 95)$$

It can be shown (see Exercise 39) that 537,140 is the unique nonnegative solution of this system less than 89,403,930. Consequently, 537,140 is the sum. Note that it is only when we have to recover the integer represented by $(65, 2, 51, 10)$ that we have to do arithmetic with integers larger than 100.   ◄

Particularly good choices for moduli for arithmetic with large integers are sets of integers of the form $2^k - 1$, where k is a positive integer, since it is easy to do binary arithmetic modulo such integers, and since it is easy to find sets of such integers that are pairwise relatively prime. [The second reason is a consequence of the fact that $\gcd(2^a - 1, 2^b - 1) = 2^{\gcd(a, b)} - 1$, as Exercise 41 shows.] Suppose, for instance, that we can do arithmetic with integers less than 2^{35} easily on our computer, but that working with larger integers requires special procedures. We can use pairwise relatively prime moduli less than 2^{35} to perform arithmetic with integers as large as their product. For example, as Exercise 42 shows, the integers $2^{35} - 1, 2^{34} - 1, 2^{33} - 1, 2^{31} - 1, 2^{29} - 1$, and $2^{23} - 1$ are pairwise relatively prime. Since the product of these six moduli exceeds 2^{184}, we can perform arithmetic with integers as large as 2^{184} (as long as the results do not exceed this number) by doing arithmetic modulo for each of these six moduli, none of which exceeds 2^{35}.

PSEUDOPRIMES

In Section 2.4 we showed that an integer n is prime when it is not divisible by any prime p with $p \leq \sqrt{n}$. Unfortunately, using this criterion to show that a given integer is prime is inefficient. It requires that we find all primes not exceeding $\sqrt{n}$ and that we carry out trial division by each such prime to see whether it divides n.

Are there more efficient ways to determine whether an integer is prime? According to some sources, ancient Chinese mathematicians believed that n was prime if and only if

$$2^{n-1} \equiv 1 \pmod{n}.$$

If this were true, it would provide an efficient primality test. Why did they believe this congruence could be used to determine whether an integer is prime? First, they observed that the congruence holds whenever n is prime. For example, 5 is prime and

$$2^{5-1} = 2^4 = 16 \equiv 1 \pmod{5}.$$

Second, they never found a composite integer n for which the congruence holds. The ancient Chinese were only partially correct. They were correct in thinking that the congruence holds whenever n is prime, but they were incorrect in concluding that n is necessarily prime if the congruence holds.

The great French mathematician Fermat showed that the congruence holds when n is prime. He proved the following, more general result.

THEOREM 5

FERMAT'S LITTLE THEOREM If p is prime and a is an integer not divisible by p, then
$$a^{p-1} \equiv 1 \pmod{p}.$$
Furthermore, for every integer a we have
$$a^p \equiv a \pmod{p}.$$

The proof of Theorem 5 is outlined in Exercise 17 at the end of this section.

PIERRE DE FERMAT (1601–1665) Pierre de Fermat, one of the most important mathematicians of the seventeenth century, was a lawyer by profession. He is the most famous amateur mathematician in history. Fermat published little of his mathematical discoveries. It is through his correspondence with other mathematicians that we know of his work. Fermat was one of the inventors of analytic geometry and developed some of the fundamental ideas of calculus. Fermat, along with Pascal, gave probability theory a mathematical basis. Fermat formulated what was the most famous unsolved problem in mathematics. He asserted that the equation $x^n + y^n = z^n$ has no nontrivial positive integer solutions when n is an integer greater than 2. For more than 300 years, no proof (or counterexample) was found. In his copy of the works of the ancient Greek mathematician Diophantus, Fermat wrote that he had a proof but that it would not fit in the margin. Because the first proof, found by Andrew Wiles in 1994, relies on sophisticated, modern mathematics, most people think that Fermat thought he had a proof, but it was incorrect. However, he may have been tempting others to look for a proof, not being able to find one himself.

Unfortunately, there are composite integers n such that $2^{n-1} \equiv 1 \pmod{n}$. Such integers are called **pseudoprimes** to the base 2.

EXAMPLE 9   The integer 341 is a pseudoprime to the base 2 since it is composite ($341 = 11 \cdot 31$) and as Exercise 27 shows

$$2^{340} \equiv 1 \pmod{341}.$$ ◀

We can use an integer other than 2 as the base when we study pseudoprimes.

DEFINITION 1   Let b be a positive integer. If n is a composite positive integer, and $b^{n-1} \equiv 1 \pmod{n}$, then n is called a *pseudoprime to the base b*.

Given a positive integer n, determining whether $2^{n-1} \equiv 1 \pmod{n}$ is a useful test that provides some evidence concerning whether n is prime. In particular, if n satisfies this congruence, then it is either prime or a pseudoprime to the base 2; if n does not satisfy this congruence, it is composite. We can perform similar tests using bases b other than 2 and obtain more evidence whether n is prime. If n passes all such tests, it is either prime or a pseudoprime to all the bases b we have chosen. Furthermore, among the positive integers not exceeding x, where x is a positive real number, compared to primes there are relatively few pseudoprimes to the base b, where b is a positive integer. For example, less than 10^{10} there are 455,052,512 primes, but only 14,884 pseudoprimes to the base 2. Unfortunately, we cannot distinguish between primes and pseudoprimes just by choosing sufficiently many bases, because there are composite integers n that pass all tests with bases with $\gcd(b, n) = 1$. This leads to Definition 2.

DEFINITION 2   A composite integer n that satisfies the congruence $b^{n-1} \equiv 1 \pmod{n}$ for all positive integers b with $\gcd(b, n) = 1$ is called a *Carmichael number*. (These numbers are named after Robert Carmichael, who studied them in the early twentieth century.)

EXAMPLE 10   The integer 561 is a Carmichael number. To see this, first note that 561 is composite since $561 = 3 \cdot 11 \cdot 17$. Next, note that if $\gcd(b, 561) = 1$, then $\gcd(b, 3) = \gcd(b, 11) = \gcd(b, 17) = 1$.
Using Fermat's Little Theorem we find that

$$b^2 \equiv 1 \pmod{3}, b^{10} \equiv 1 \pmod{11}, \text{ and } b^{16} \equiv 1 \pmod{17}.$$

ROBERT DANIEL CARMICHAEL (1879–1967)   Robert Daniel Carmichael was born in Alabama. He received his undergraduate degree from Lineville College in 1898 and his Ph.D. in 1911 from Princeton. Carmichael held positions at Indiana University from 1911 until 1915 and at the University of Illinois from 1915 until 1947. Carmichael was an active researcher in a wide variety of areas, including number theory, real analysis, differential equations, mathematical physics, and group theory. His Ph.D. thesis, written under the direction of G. D. Birkhoff, is considered the first significant American contribution to the subject of differential equations.

It follows that

$$b^{560} = (b^2)^{280} \equiv 1 \pmod{3},$$
$$b^{560} = (b^{10})^{56} \equiv 1 \pmod{11},$$
$$b^{560} = (b^{16})^{35} \equiv 1 \pmod{17}.$$

By Exercise 23 at the end of this section, it follows that $b^{560} \equiv 1 \pmod{561}$ for all positive integers b with $\gcd(b, 561) = 1$. Hence 561 is a Carmichael number.   ◄

Although there are infinitely many Carmichael numbers, more delicate tests, described in the exercise set at the end of this section, can be devised that can be used as the basis for efficient probabilistic primality tests. Such tests can be used to quickly show that it is almost certainly the case that a given integer is prime. More precisely, if an integer is not prime, then the probability that it passes a series of tests is close to 0. We will describe such a test in Chapter 5 and discuss the notions from probability theory that this test relies on. These probabilistic primality tests can be used, and are used, to find large primes extremely rapidly on computers.

PUBLIC KEY CRYPTOGRAPHY

In Section 2.4 we introduced methods for encrypting messages based on congruences. When these encryption methods are used, messages, which are strings of characters, are translated into numbers. Then the number for each character is transformed into another number, either using a shift or an affine transformation modulo 26. These methods are examples of **private key cryptosystems.** Knowing the encryption key lets you quickly find the decryption key. For example, when a shift cipher is used with encryption key k, a number p representing a letter is sent to

$$c = (p + k) \bmod 26.$$

Decryption is carried out by shifting by $-k$; that is,

$$p = (c - k) \bmod 26.$$

When a private key cryptosystem is used, a pair of people who wish to communicate in secret must have a separate key. Since anyone knowing this key can both encrypt and decrypt messages easily, these two people need to securely exchange the key.

In the mid-1970s, cryptologists introduced the concept of **public key cryptosystems.** When such cryptosystems are used, knowing how to send someone a message does not help you decrypt messages sent to this person. In such a system, every person can have a publicly known encryption key. Only the decryption keys are kept secret, and only the intended recipient of a message can decrypt it, since the encryption key does not let someone find the decryption key without an extraordinary amount of work (such as more than 2 billion years of computer time).

In 1976, three researchers at M.I.T.—Ronald Rivest, Adi Shamir, and Leonard Adleman—introduced a public key cryptosystem, known as the **RSA system,** from the initials of its inventors. The RSA cryptosystem is based on modular exponentiation modulo, the product of two large primes, which can be done rapidly using Algorithm 5 in Section 2.5. Each individual has an encryption key consisting of a modulus $n = pq$, where p and q are large primes, say, with 200 digits each, and an exponent e that is relatively prime to $(p-1)(q-1)$. To produce a usable key, two large primes must be found. This can be done quickly on a computer using probabilistic primality tests, referred to earlier in this section. However, the product of these primes $n = pq$, with approximately 400 digits, cannot be factored in a reasonable length of time. As we will see, this is an important reason why decryption cannot be done quickly without a separate decryption key.

RSA ENCRYPTION

In the RSA encryption method, messages are translated into sequences of integers. This can be done by translating each letter into an integer, as is done with the Caesar cipher. These integers are grouped together to form larger integers, each representing a block of letters. The encryption proceeds by transforming the integer M, representing the plaintext (the original message), to an integer C, representing the ciphertext (the encrypted message), using the function

$$C = M^e \bmod n.$$

(To perform the encryption, we use an algorithm for fast modular exponentiation, such as Algorithm 5 in Section 2.5.) We leave the encrypted message as blocks of numbers and send these to the intended recipient.

Example 11 illustrates how RSA encryption is performed. For practical reasons we use small primes p and q in this example, rather than primes with 100 or more digits. Although the cipher described in this example is not secure, it does illustrate the techniques used in the RSA cipher.

EXAMPLE 11 Encrypt the message STOP using the RSA cryptosystem with $p = 43$ and $q = 59$, so that $n = 43 \cdot 59 = 2537$, and with $e = 13$. Note that

$$\gcd(e, (p - 1)(q - 1)) = \gcd(13, 42 \cdot 58) = 1.$$

RONALD RIVEST (BORN 1948) Ronald Rivest received a B.A. from Yale in 1969 and his Ph.D. in computer science from Stanford in 1974. Rivest is a computer science professor at M.I.T. and was a cofounder of RSA Data Security, which held the patent on the RSA cryptosystem that he invented together with Adi Shamir and Leonard Adleman. Areas that Rivest has worked in besides cryptography include machine learning, VLSI design, and computer algorithms. He is a coauthor of a popular text on algorithms ([CoLeRiSt01]).

ADI SHAMIR (BORN 1952) Adi Shamir was born in Tel Aviv, Israel. His undergraduate degree is from Tel Aviv University (1972) and his Ph.D. is from the Weizmann Institute of Science (1977). Shamir was a research assistant at the University of Warwick and an assistant professor at M.I.T. He is currently a professor in the Applied Mathematics Department at the Weizmann Institute and leads a group studying computer security. Shamir's contributions to cryptography, besides the RSA cryptosystem, include cracking knapsack cryptosystems, cryptanalysis of the Data Encryption Standard (DES), and the design of many cryptographic protocols.

LEONARD ADLEMAN (BORN 1945) Leonard Adleman was born in San Francisco, California. He received a B.S. in mathematics (1968) and his Ph.D. in computer science (1976) from the University of California, Berkeley. Adleman was a member of the mathematics faculty at M.I.T. from 1976 until 1980, where he was a co-inventor of the RSA cryptosystem, and in 1980 he took a position in the computer science department at the University of Southern California (USC). He was appointed to a chaired position at USC in 1985. Adleman has worked on computer security, computational complexity, immunology, and molecular biology. He invented the term "computer virus." Adleman's recent work on DNA computing has sparked great interest. He was a technical adviser for the movie *Sneakers*, in which computer security played an important role.

Solution: We translate the letters in STOP into their numerical equivalents and then group the numbers into blocks of four. We obtain

1819   1415.

We encrypt each block using the mapping

$$C = M^{13} \bmod 2537.$$

Computations using fast modular multiplication show that $1819^{13} \bmod 2537 = 2081$ and $1415^{13} \bmod 2537 = 2182$. The encrypted message is 2081 2182.   ◄

RSA DECRYPTION

The plaintext message can be quickly recovered when the decryption key d, an inverse of e modulo $(p-1)(q-1)$, is known. [Such an inverse exists since $\gcd(e, (p-1)(q-1)) = 1$.] To see this, note that if $de \equiv 1 \pmod{(p-1)(q-1)}$, there is an integer k such that $de = 1 + k(p-1)(q-1)$. It follows that

$$C^d \equiv (M^e)^d = M^{de} = M^{1+k(p-1)(q-1)} \pmod{n}.$$

By Fermat's Little Theorem [assuming that $\gcd(M, p) = \gcd(M, q) = 1$, which holds except in rare cases], it follows that $M^{p-1} \equiv 1 \pmod{p}$ and $M^{q-1} \equiv 1 \pmod{q}$. Consequently,

$$C^d \equiv M \cdot (M^{p-1})^{k(q-1)} \equiv M \cdot 1 \equiv M \pmod{p}$$

and

$$C^d \equiv M \cdot (M^{q-1})^{k(p-1)} \equiv M \cdot 1 \equiv M \pmod{q}.$$

Since $\gcd(p, q) = 1$, it follows by the Chinese Remainder Theorem that

$$C^d \equiv M \pmod{pq}.$$

Example 12 illustrates how to decrypt messages sent using the RSA cryptosystem.

EXAMPLE 12   We receive the encrypted message 0981 0461. What is the decrypted message if it was encrypted using the RSA cipher from Example 11?

Solution: The message was encrypted using the RSA cryptosystem with $n = 43 \cdot 59$ and exponent 13. As Exercise 4 shows, $d = 937$ is an inverse of 13 modulo $42 \cdot 58 = 2436$. We use 937 as our decryption exponent. Consequently, to decrypt a block C, we compute

$$P = C^{937} \bmod 2537.$$

To decrypt the message, we use the fast modular exponentiation algorithm to compute $0981^{937} \bmod 2537 = 0704$ and $0461^{937} \bmod 2537 = 1115$. Consequently, the numerical version of the original message is 0704 1115. Translating this back to English letters, we see that the message is HELP.   ◄

RSA AS A PUBLIC KEY SYSTEM

Why is the RSA cryptosystem suitable for public key cryptography? When we know the factorization of the modulus n, that is, when we know p and q, we can use the Euclidean algorithm to quickly find an exponent d inverse to e modulo $(p-1)(q-1)$. This lets us decrypt messages sent using our key. However, no method is known to decrypt messages that is not based on finding a factorization of n, or that does not also lead to the factorization of n. Factorization is believed to be a difficult problem, as opposed to finding large primes p and q, which can be done quickly. The most efficient factorization methods known (as of 2002) require billions of years to factor 400-digit integers. Consequently, when p and q are 200-digit primes, messages encrypted using $n = pq$ as the modulus cannot be found in a reasonable time unless the primes p and q are known.

Active research is under way to find new ways to efficiently factor integers. Integers that were thought, as recently as several years ago, to be far too large to be factored in a reasonable amount of time can now be factored routinely. Integers with more than 100 digits, as well as some with more than 150 digits, have been factored using team efforts. When new factorization techniques are found, it will be necessary to use larger primes to ensure secrecy of messages. Unfortunately, messages that were considered secure earlier can be saved and subsequently decrypted by unintended recipients when it becomes feasible to factor the $n = pq$ in the key used for RSA encryption.

The RSA method is now widely used. However, the most commonly used cryptosystems are private key cryptosystems. The use of public key cryptography, via the RSA system, is growing. However, there are applications that use both private key and public key systems. For example, a public key cryptosystem, such as RSA, can be used to distribute private keys to pairs of individuals when they wish to communicate. These people then use a private key system for encryption and decryption of messages.

Exercises

1. Express the greatest common divisor of each of these pairs of integers as a linear combination of these integers.

   a) $10, 11$ **b)** $21, 44$ **c)** $36, 48$
   d) $34, 55$ **e)** $117, 213$ **f)** $0, 223$
   g) $123, 2347$ **h)** $3454, 4666$ **i)** $9999, 11111$

2. Express the greatest common divisor of each of these pairs of integers as a linear combination of these integers.

   a) $9, 11$ **b)** $33, 44$ **c)** $35, 78$
   d) $21, 55$ **e)** $101, 203$ **f)** $124, 323$
   g) $2002, 2339$ **h)** $3457, 4669$ **i)** $10001, 13422$

3. Show that 15 is an inverse of 7 modulo 26.

4. Show that 937 is an inverse of 13 modulo 2436.

5. Find an inverse of 4 modulo 9.

6. Find an inverse of 2 modulo 17.

7. Find an inverse of 19 modulo 141.

8. Find an inverse of 144 modulo 233.

*9. Show that if a and m are relatively prime positive integers, then the inverse of a modulo m is unique modulo m. [*Hint:* Assume that there are two solutions b and c of the congruence $ax \equiv 1 \pmod{m}$. Use Theorem 2 to show that $b \equiv c \pmod{m}$.]

10. Show that an inverse of a modulo m does not exist if $\gcd(a, m) > 1$.

11. Solve the congruence $4x \equiv 5 \pmod{9}$.

12. Solve the congruence $2x \equiv 7 \pmod{17}$.

*13. Show that if m is a positive integer greater than 1 and $ac \equiv bc \pmod{m}$, then $a \equiv b \bmod m / \gcd(c, m)$.

14. a) Show that the positive integers less than 11, except 1 and 10, can be split into pairs of integers such that each pair consists of integers that are inverses of each other modulo 11.

    b) Use part (a) to show that $10! \equiv -1 \pmod{11}$.

15. Show that if p is prime, the only solutions of $x^2 \equiv 1 \pmod{p}$ are integers x such that $x \equiv 1 \pmod{p}$ and $x \equiv -1 \pmod{p}$.

*16. a) Generalize the result in part (a) of Exercise 14; that is, show that if p is a prime, the positive integers less than p, except 1 and $p - 1$, can be split into $(p-3)/2$ pairs of integers such that each pair consists of integers that are inverses of each other. (*Hint:* Use the result of Exercise 15.)

    b) From part (a) conclude that $(p-1)! \equiv -1 \pmod{p}$ whenever p is prime. This result is known as **Wilson's Theorem.**

    c) What can we conclude if n is a positive integer such that $(n-1)! \not\equiv -1 \pmod{n}$?

*17. This exercise outlines a proof of Fermat's Little Theorem.

a) Suppose that a is not divisible by the prime p. Show that no two of the integers $1 \cdot a, 2 \cdot a, \ldots,$ $(p-1)a$ are congruent modulo p.

b) Conclude from part (a) that the product of $1, 2, \ldots, p-1$ is congruent modulo p to the product of $a, 2a, \ldots, (p-1)a$. Use this to show that

$$(p-1)! \equiv a^{p-1}(p-1)! \pmod{p}.$$

c) Use Wilson's theorem (proved in Exercise 16) to show that $a^{p-1} \equiv 1 \pmod{p}$ if $p \nmid a$.

d) Use part (c) to show that $a^p \equiv a \pmod{p}$ for all integers a.

18. Find all solutions to the system of congruences.
$x \equiv 2 \pmod{3}$
$x \equiv 1 \pmod{4}$
$x \equiv 3 \pmod{5}$

19. Find all solutions to the system of congruences.
$x \equiv 1 \pmod{2}$
$x \equiv 2 \pmod{3}$
$x \equiv 3 \pmod{5}$
$x \equiv 4 \pmod{11}$

*20. Find all solutions, if any, to the system of congruences.
$x \equiv 5 \pmod{6}$
$x \equiv 3 \pmod{10}$
$x \equiv 8 \pmod{15}$

*21. Find all solutions, if any, to the system of congruences.
$x \equiv 7 \pmod{9}$
$x \equiv 4 \pmod{12}$
$x \equiv 16 \pmod{21}$

22. Use the Chinese Remainder Theorem to show that an integer a, with $0 \le a < m = m_1 m_2 \cdots m_n$, where the integers $m_1, m_2, \ldots, m_n$ are pairwise relatively prime, can be represented uniquely by the n-tuple $(a \bmod m_1, a \bmod m_2, \ldots, a \bmod m_n)$.

*23. Let $m_1, m_2, \ldots, m_n$ be pairwise relatively prime integers greater than or equal to 2. Show that if $a \equiv b \pmod{m_i}$ for $i = 1, 2, \ldots, n$, then $a \equiv b \pmod{m}$, where $m = m_1 m_2 \cdots m_n$.

*24. Complete the proof of the Chinese Remainder Theorem by showing that the simultaneous solution of a system of linear congruences modulo pairwise relatively prime integers is unique modulo the product of these moduli. (*Hint:* Assume that x and y are two simultaneous solutions. Show that $m_i \mid x - y$ for all i. Using Exercise 23, conclude that $m = m_1 m_2 \cdots m_n \mid x - y$.)

25. Which integers leave a remainder of 1 when divided by 2 and also leave a remainder of 1 when divided by 3?

26. Which integers are divisible by 5 but leave a remainder of 1 when divided by 3?

27. a) Show that $2^{340} \equiv 1 \pmod{11}$ by Fermat's Little Theorem and noting that $2^{340} = (2^{10})^{34}$.

b) Show that $2^{340} \equiv 1 \pmod{31}$ using the fact that $2^{340} = (2^5)^{68} = 32^{68}$.

c) Conclude from parts (a) and (b) that $2^{340} \equiv 1 \pmod{341}$.

28. a) Use Fermat's Little Theorem to compute $3^{302} \bmod 5$, $3^{302} \bmod 7$, and $3^{302} \bmod 11$.

b) Use your results from part (a) and the Chinese Remainder Theorem to find $3^{302} \bmod 385$. (Note that $385 = 5 \cdot 7 \cdot 11$.)

29. a) Use Fermat's Little Theorem to compute $5^{2003} \bmod 7$, $5^{2003} \bmod 11$, and $5^{2003} \bmod 13$.

b) Use your results from part (a) and the Chinese Remainder Theorem to find $5^{2003} \bmod 1001$. (Note that $1001 = 7 \cdot 11 \cdot 13$.)

Let n be a positive integer and let $n - 1 = 2^s t$, where s is a nonnegative integer and t is an odd positive integer. We say that n passes **Miller's test for the base** b if either $b^t \equiv 1 \pmod{n}$ or $b^{2^j t} \equiv -1 \pmod{n}$ for some j with $0 \le j \le s - 1$. It can be shown (see [Ro00]) that a composite integer n passes Miller's test for fewer than $n/4$ bases b with $1 < b < n$.

30. Show that if n is prime and b is a positive integer with $n \nmid b$, then n passes Miller's test to the base b.

31. Show that 2047 passes Miller's test to the base 2, but that it is composite. A composite positive integer n that passes Miller's test to the base b is called a **strong pseudoprime to the base** b. It follows that 2047 is a strong pseudoprime to the base 2.

32. Show that 1729 is a Carmichael number.

33. Show that 2821 is a Carmichael number.

*34. Show that if $n = p_1 p_2 \cdots p_k$, where $p_1, p_2, \ldots, p_k$ are distinct primes that satisfy $p_j - 1 \mid n - 1$ for $j = 1, 2, \ldots, k$, then n is a Carmichael number.

35. a) Use Exercise 34 to show that every integer of the form $(6m + 1)(12m + 1)(18m + 1)$, where m is a positive integer and $6m + 1, 12m + 1$, and $18m + 1$ are all primes, is a Carmichael number.

b) Use part (a) to show that 172,947,529 is a Carmichael number.

36. Find the nonnegative integer a less than 28 represented by each of these pairs, where each pair represents $(a \bmod 4, a \bmod 7)$.

a) $(0, 0)$	**b)** $(1, 0)$	**c)** $(1, 1)$
d) $(2, 1)$	**e)** $(2, 2)$	**f)** $(0, 3)$
g) $(2, 0)$	**h)** $(3, 5)$	**i)** $(3, 6)$

37. Express each nonnegative integer a less than 15 using the pair $(a \bmod 3, a \bmod 5)$.

38. Explain how to use the pairs found in Exercise 37 to add 4 and 7.

39. Solve the system of congruences that arises in Example 8.

*40. Show that if a and b are positive integers, then $(2^a - 1) \bmod (2^b - 1) = 2^{a \bmod b} - 1$.

**41. Use Exercise 40 to show that if a and b are positive integers, then $\gcd(2^a - 1, 2^b - 1) = 2^{\gcd(a, b)} - 1$. [*Hint:* Show that the remainders obtained when the Euclidean algorithm is used to compute $\gcd(2^a - 1,$

$2^b - 1$) are of the form $2^r - 1$, where r is a remainder arising when the Euclidean algorithm is used to find $\gcd(a, b)$.]

42. Use Exercise 41 to show that the integers $2^{35} - 1$, $2^{34} - 1, 2^{33} - 1, 2^{31} - 1, 2^{29} - 1$, and $2^{23} - 1$ are pairwise relatively prime.

43. Show that if p is an odd prime, then every divisor of the Mersenne number $2^p - 1$ is of the form $2kp + 1$, where k is a nonnegative integer. (*Hint:* Use Fermat's Little Theorem and Exercise 41.)

44. Use Exercise 43 to determine whether $M_{13} = 2^{13} - 1 = 8191$ and $M_{23} = 2^{23} - 1 = 8,388,607$ are prime.

*45. Show that we can easily factor n when we know that n is the product of two primes, p and q, and we know the value of $(p - 1)(q - 1)$.

46. Encrypt the message ATTACK using the RSA system with $n = 43 \cdot 59$ and $e = 13$, translating each letter into integers and grouping together pairs of integers, as done in Example 11.

47. What is the original message encrypted using the RSA system with $n = 43 \cdot 59$ and $e = 13$ if the encrypted message is 0667 1947 0671? (*Note:* Some computational aid is needed to do this in a realistic amount of time.)

The **extended Euclidean algorithm** can be used to express $\gcd(a, b)$ as a linear combination with integer coefficients of the integers a and b. We set $s_0 = 1, s_1 = 0, t_0 = 0$, and $t_1 = 1$ and let $s_j = s_{j-2} - q_{j-1}s_{j-1}$ and $t_j = t_{j-2} - q_{j-1}t_{j-1}$ for $j = 2, 3, \ldots, n$, where the q_j are the quotients in the divisions used when the Euclidean algorithm finds $\gcd(a, b)$ (see page 178). It can be shown (see [Ro00]) that $\gcd(a, b) = s_n a + t_n b$.

48. Use the extended Euclidean algorithm to express $\gcd(252, 356)$ as a linear combination of 252 and 356.

49. Use the extended Euclidean algorithm to express $\gcd(144, 89)$ as a linear combination of 144 and 89.

50. Use the extended Euclidean algorithm to express $\gcd(1001, 100001)$ as a linear combination of 1001 and 100001.

51. Describe the extended Euclidean algorithm using pseudocode.

If m is a positive integer, the integer a is a **quadratic residue** of m if $\gcd(a, m) = 1$ and the congruence

$x^2 \equiv a \pmod{m}$ has a solution. In other words, a quadratic residue of m is an integer relatively prime to m that is a perfect square modulo m. For example, 2 is a quadratic residue of 7 since $\gcd(2, 7) = 1$ and $3^2 \equiv 2 \pmod 7$ and 3 is a quadratic nonresidue of 7 since $\gcd(3, 7) = 1$ and $x^2 \equiv 3 \pmod 7$ has no solution.

52. Which integers are quadratic residues of 11?

53. Show that if p is an odd prime and a is an integer not divisible by p, then the congruence $x^2 \equiv a \pmod p$ has either no solutions or exactly two incongruent solutions modulo p.

54. Show that if p is an odd prime, then there are exactly $(p - 1)/2$ quadratic residues of p among the integers $1, 2, \ldots, p - 1$.

If p is an odd prime and a is an integer not divisible by p, the **Legendre symbol** $\left(\dfrac{a}{p}\right)$ is defined to be 1 if a is a quadratic residue of p and -1 otherwise.

55. Show that if p is an odd prime and a and b are integers with $a \equiv b \pmod p$, then

$$\left(\frac{a}{p}\right) = \left(\frac{b}{p}\right).$$

56. Prove that if p is an odd prime and a is a positive integer not divisible by p, then

$$\left(\frac{a}{p}\right) \equiv a^{(p-1)/2} \pmod p.$$

57. Use Exercise 56 to show that if p is an odd prime and a and b are integers not divisible by p, then

$$\left(\frac{ab}{p}\right) = \left(\frac{a}{p}\right)\left(\frac{b}{p}\right).$$

58. Show that if p is an odd prime, then -1 is a quadratic residue of p if $p \equiv 1 \pmod 4$ and 1 is not a quadratic residue of p if $p \equiv 3 \pmod 4$. (*Hint:* Use Exercise 56.)

59. Find all solutions of the congruence $x^2 \equiv 29 \pmod{35}$. (*Hint:* Find the solutions of this congruence modulo 5 and modulo 7, and then use the Chinese Remainder Theorem.)

60. Find all solutions of the congruence $x^2 \equiv 16 \pmod{105}$. (*Hint:* Find the solutions of this congruence modulo 3, modulo 5, and modulo 7, and then use the Chinese Remainder Theorem.)

2.7   Matrices

INTRODUCTION

Links

Matrices are used throughout discrete mathematics to express relationships between elements in sets. In subsequent chapters we will use matrices in a wide variety of models. For instance, matrices will be used in models of communications networks and transportation systems. Many algorithms will be developed that use these matrix models. This section reviews matrix arithmetic that will be used in these algorithms.

DEFINITION 1

A *matrix* is a rectangular array of numbers. A matrix with m rows and n columns is called an $m \times n$ matrix. The plural of matrix is *matrices*. A matrix with the same number of rows as columns is called *square*. Two matrices are *equal* if they have the same number of rows and the same number of columns and the corresponding entries in every position are equal.

EXAMPLE 1 The matrix $\begin{bmatrix} 1 & 1 \\ 0 & 2 \\ 1 & 3 \end{bmatrix}$ is a 3×2 matrix. ◀

We now introduce some terminology about matrices. Boldface uppercase letters will be used to represent matrices.

DEFINITION 2

Let

$$\mathbf{A} = \begin{bmatrix} a_{11} & a_{12} & \cdots & a_{1n} \\ a_{21} & a_{22} & \cdots & a_{2n} \\ \cdot & \cdot & & \cdot \\ \cdot & \cdot & & \cdot \\ \cdot & \cdot & & \cdot \\ a_{n1} & a_{n2} & \cdots & a_{nn} \end{bmatrix}.$$

The ith *row* of $\mathbf{A}$ is the $1 \times n$ matrix $[a_{i1}, a_{i2}, \ldots, a_{in}]$. The jth *column* of $\mathbf{A}$ is the $n \times 1$ matrix

$$\begin{bmatrix} a_{1j} \\ a_{2j} \\ \cdot \\ \cdot \\ \cdot \\ a_{nj} \end{bmatrix}.$$

The (i, j)th *element* or *entry* of $\mathbf{A}$ is the element a_{ij}, that is, the number in the ith row and jth column of $\mathbf{A}$. A convenient shorthand notation for expressing the matrix $\mathbf{A}$ is to write $\mathbf{A} = [a_{ij}]$, which indicates that $\mathbf{A}$ is the matrix with its (i, j)th element equal to a_{ij}.

MATRIX ARITHMETIC

The basic operations of matrix arithmetic will now be discussed, beginning with a definition of matrix addition.

DEFINITION 3

Let $\mathbf{A} = [a_{ij}]$ and $\mathbf{B} = [b_{ij}]$ be $m \times n$ matrices. The *sum* of $\mathbf{A}$ and $\mathbf{B}$, denoted by $\mathbf{A} + \mathbf{B}$, is the $m \times n$ matrix that has $a_{ij} + b_{ij}$ as its (i, j)th element. In other words, $\mathbf{A} + \mathbf{B} = [a_{ij} + b_{ij}]$.

The sum of two matrices of the same size is obtained by adding elements in the corresponding positions. Matrices of different sizes cannot be added, since the sum of two matrices is defined only when both matrices have the same number of rows and the same number of columns.

EXAMPLE 2 We have $\begin{bmatrix} 1 & 0 & -1 \\ 2 & 2 & -3 \\ 3 & 4 & 0 \end{bmatrix} + \begin{bmatrix} 3 & 4 & -1 \\ 1 & -3 & 0 \\ -1 & 1 & 2 \end{bmatrix} = \begin{bmatrix} 4 & 4 & -2 \\ 3 & -1 & -3 \\ 2 & 5 & 2 \end{bmatrix}.$ ◀

We now discuss matrix products. A product of two matrices is defined only when the number of columns in the first matrix equals the number of rows of the second matrix.

DEFINITION 4

Let **A** be an $m \times k$ matrix and **B** be a $k \times n$ matrix. The *product* of **A** and **B**, denoted by **AB**, is the $m \times n$ matrix with its (i, j)th entry equal to the sum of the products of the corresponding elements from the ith row of **A** and the jth column of **B**. In other words, if $\mathbf{AB} = [c_{ij}]$, then

$$c_{ij} = a_{i1}b_{1j} + a_{i2}b_{2j} + \cdots + a_{ik}b_{kj}.$$

In Figure 1 the colored row of **A** and the colored column of **B** are used to compute the element c_{ij} of **AB**. The product of two matrices is not defined when the number of columns in the first matrix and the number of rows in the second matrix are not the same. We now give some examples of matrix products.

EXAMPLE 3 Let

$$\mathbf{A} = \begin{bmatrix} 1 & 0 & 4 \\ 2 & 1 & 1 \\ 3 & 1 & 0 \\ 0 & 2 & 2 \end{bmatrix} \quad \text{and} \quad \mathbf{B} = \begin{bmatrix} 2 & 4 \\ 1 & 1 \\ 3 & 0 \end{bmatrix}.$$

Find **AB** if it is defined.

Solution: Since **A** is a 4×3 matrix and **B** is a 3×2 matrix, the product **AB** is defined and is a 4×2 matrix. To find the elements of **AB**, the corresponding elements of the rows of **A** and the columns of **B** are first multiplied and then these products are added. For instance, the element in the $(3, 1)$th position of **AB** is the sum of the products of the corresponding elements of the third row of **A** and the first column of **B**; namely,

$$\begin{bmatrix} a_{11} & a_{12} & \cdots & a_{1k} \\ a_{21} & a_{22} & \cdots & a_{2k} \\ \vdots & \vdots & & \vdots \\ a_{i1} & a_{i2} & \cdots & a_{ik} \\ \vdots & \vdots & & \vdots \\ a_{m1} & a_{m2} & \cdots & a_{mk} \end{bmatrix} \begin{bmatrix} b_{11} & b_{12} & \cdots & b_{1j} & \cdots & b_{1n} \\ b_{21} & b_{22} & \cdots & b_{2j} & \cdots & b_{2n} \\ \vdots & & & \vdots & & \vdots \\ b_{k1} & b_{k2} & \cdots & b_{kj} & \cdots & b_{kn} \end{bmatrix} = \begin{bmatrix} c_{11} & c_{12} & \cdots & c_{1n} \\ c_{21} & c_{22} & \cdots & c_{2n} \\ \vdots & & c_{ij} & \vdots \\ c_{m1} & c_{m2} & \cdots & c_{mn} \end{bmatrix}$$

FIGURE 1   **The Product of $\mathbf{A} = [a_{ij}]$ and $\mathbf{B} = [b_{ij}]$.**

$3 \cdot 2 + 1 \cdot 1 + 0 \cdot 3 = 7$. When all the elements of **AB** are computed, we see that

$$\mathbf{AB} = \begin{bmatrix} 14 & 4 \\ 8 & 9 \\ 7 & 13 \\ 8 & 2 \end{bmatrix}.$$

◀

Matrix multiplication is *not* commutative. That is, if **A** and **B** are two matrices, it is not necessarily true that **AB** and **BA** are the same. In fact, it may be that only one of these two products is defined. For instance, if **A** is 2×3 and **B** is 3×4, then **AB** is defined and is 2×4; however, **BA** is not defined, since it is impossible to multiply a 3×4 matrix and a 2×3 matrix.

In general, suppose that **A** is an $m \times n$ matrix and **B** is an $r \times s$ matrix. Then **AB** is defined only when $n = r$ and **BA** is defined only when $s = m$. Moreover, even when **AB** and **BA** are both defined, they will not be the same size unless $m = n = r = s$. Hence, if both **AB** and **BA** are defined and are the same size, then both **A** and **B** must be square and of the same size. Furthermore, even with **A** and **B** both $n \times n$ matrices, **AB** and **BA** are not necessarily equal, as the following example demonstrates.

EXAMPLE 4 Let

$$\mathbf{A} = \begin{bmatrix} 1 & 1 \\ 2 & 1 \end{bmatrix} \quad \text{and} \quad \mathbf{B} = \begin{bmatrix} 2 & 1 \\ 1 & 1 \end{bmatrix}.$$

Does **AB** = **BA**?

Solution: We find that

$$\mathbf{AB} = \begin{bmatrix} 3 & 2 \\ 5 & 3 \end{bmatrix} \quad \text{and} \quad \mathbf{BA} = \begin{bmatrix} 4 & 3 \\ 3 & 2 \end{bmatrix}.$$

Hence, **AB** ≠ **BA**.

◀

ALGORITHMS FOR MATRIX MULTIPLICATION

The definition of the product of two matrices leads to an algorithm that computes the product of two matrices. Suppose that $\mathbf{C} = [c_{ij}]$ is the $m \times n$ matrix that is the product of the $m \times k$ matrix $\mathbf{A} = [a_{ij}]$ and the $k \times n$ matrix $\mathbf{B} = [b_{ij}]$. The algorithm based on the definition of the matrix product is expressed in pseudocode in Algorithm 1.

ALGORITHM 1 Matrix Multiplication.

---

procedure *matrix multiplication*(**A**, **B**: matrices)
for $i := 1$ **to** m
    for $j := 1$ **to** n
    begin
        $c_{ij} := 0$
        for $q := 1$ **to** k
            $c_{ij} := c_{ij} + a_{iq}b_{qj}$
    end
{$\mathbf{C} = [c_{ij}]$ is the product of **A** and **B**}

We can determine the complexity of this algorithm in terms of the number of additions and multiplications used.

EXAMPLE 5 How many additions of integers and multiplications of integers are used by Algorithm 1 to multiply two $n \times n$ matrices with integer entries?

Solution: There are n^2 entries in the product of **A** and **B**. To find each entry requires a total of n multiplications and $n - 1$ additions. Hence, a total of n^3 multiplications and $n^2(n - 1)$ additions are used. ◄

Surprisingly, there are more efficient algorithms for matrix multiplication than that given in Algorithm 1. As Example 5 shows, multiplying two $n \times n$ matrices directly from the definition requires $O(n^3)$ multiplications and additions. Using other algorithms, two $n \times n$ matrices can be multiplied using $O(n^{\sqrt{7}})$ multiplications and additions. (Details of such algorithms can be found in [CoLeRiSt01].)

MATRIX-CHAIN MULTIPLICATION There is another important problem involving the complexity of the multiplication of matrices. How should the **matrix-chain** $\mathbf{A}_1\mathbf{A}_2 \cdots \mathbf{A}_n$ be computed using the fewest multiplications of integers, where $\mathbf{A}_1, \mathbf{A}_2, \ldots, \mathbf{A}_n$ are $m_1 \times m_2, m_2 \times m_3, \ldots, m_n \times m_{n+1}$ matrices, respectively, and each has integers as entries? (Since matrix multiplication is associative, as shown in Exercise 13 at the end of this section, the order of the multiplication used does not matter.) Before studying this problem, note that $m_1 m_2 m_3$ multiplications of integers are performed to multiply an $m_1 \times m_2$ matrix and an $m_2 \times m_3$ matrix using Algorithm 1 (see Exercise 23 at the end of this section). Example 6 illustrates this problem.

EXAMPLE 6 In which order should the matrices $\mathbf{A}_1, \mathbf{A}_2$, and $\mathbf{A}_3$—where $\mathbf{A}_1$ is 30×20, $\mathbf{A}_2$ is 20×40, and $\mathbf{A}_3$ is 40×10, all with integer entries—be multiplied to use the least number of multiplications of integers?

Solution: There are two possible ways to compute $\mathbf{A}_1\mathbf{A}_2\mathbf{A}_3$. These are $\mathbf{A}_1(\mathbf{A}_2\mathbf{A}_3)$ and $(\mathbf{A}_1\mathbf{A}_2)\mathbf{A}_3$.

If $\mathbf{A}_2$ and $\mathbf{A}_3$ are first multiplied, a total of $20 \cdot 40 \cdot 10 = 8000$ multiplications of integers are used to obtain the 20×10 matrix $\mathbf{A}_2\mathbf{A}_3$. Then, to multiply $\mathbf{A}_1$ and $\mathbf{A}_2\mathbf{A}_3$ requires $30 \cdot 20 \cdot 10 = 6000$ multiplications. Hence, a total of

$$8000 + 6000 = 14{,}000$$

multiplications are used. On the other hand, if $\mathbf{A}_1$ and $\mathbf{A}_2$ are first multiplied, then $30 \cdot 20 \cdot 40 = 24{,}000$ multiplications are used to obtain the 30×40 matrix $\mathbf{A}_1\mathbf{A}_2$. Then, to multiply $\mathbf{A}_1\mathbf{A}_2$ and $\mathbf{A}_3$ requires $30 \times 40 \times 10 = 12{,}000$ multiplications. Hence, a total of

$$24{,}000 + 12{,}000 = 36{,}000$$

multiplications are used.

Clearly, the first method is more efficient. ◄

Algorithms for determining the most efficient way to carry out matrix-chain multiplication are discussed in [CoLeRiSt01].

TRANSPOSES AND POWERS OF MATRICES

We now introduce an important matrix with entries that are zeros and ones.

DEFINITION 5

The *identity matrix of order n* is the $n \times n$ matrix $\mathbf{I}_n = [\delta_{ij}]$, where $\delta_{ij} = 1$ if $i = j$ and $\delta_{ij} = 0$ if $i \neq j$. Hence

$$\mathbf{I}_n = \begin{bmatrix} 1 & 0 & \cdots & 0 \\ 0 & 1 & \cdots & 0 \\ \cdot & \cdot & & \cdot \\ \cdot & \cdot & & \cdot \\ \cdot & \cdot & & \cdot \\ 0 & 0 & \cdots & 1 \end{bmatrix}.$$

Multiplying a matrix by an appropriately sized identity matrix does not change this matrix. In other words, when $\mathbf{A}$ is an $m \times n$ matrix, we have

$$\mathbf{AI}_n = \mathbf{I}_m \mathbf{A} = \mathbf{A}.$$

Powers of square matrices can be defined. When $\mathbf{A}$ is an $n \times n$ matrix, we have

$$\mathbf{A}^0 = \mathbf{I}_n, \qquad \mathbf{A}^r = \underbrace{\mathbf{AAA} \cdots \mathbf{A}}_{r \text{ times}}.$$

The operation of interchanging the rows and columns of a square matrix is used in many algorithms.

DEFINITION 6

Let $\mathbf{A} = [a_{ij}]$ be an $m \times n$ matrix. The *transpose* of $\mathbf{A}$, denoted by $\mathbf{A}^t$, is the $n \times m$ matrix obtained by interchanging the rows and columns of $\mathbf{A}$. In other words, if $\mathbf{A}^t = [b_{ij}]$, then $b_{ij} = a_{ji}$ for $i = 1, 2, \ldots, n$ and $j = 1, 2, \ldots, m$.

EXAMPLE 7 The transpose of the matrix $\begin{bmatrix} 1 & 2 & 3 \\ 4 & 5 & 6 \end{bmatrix}$ is the matrix $\begin{bmatrix} 1 & 4 \\ 2 & 5 \\ 3 & 6 \end{bmatrix}$. ◄

Matrices that do not change when their rows and columns are interchanged are often important.

DEFINITION 7

A square matrix $\mathbf{A}$ is called *symmetric* if $\mathbf{A} = \mathbf{A}^t$. Thus $\mathbf{A} = [a_{ij}]$ is symmetric if $a_{ij} = a_{ji}$ for all i and j with $1 \leq i \leq n$ and $1 \leq j \leq n$.

Note that a matrix is symmetric if and only if it is square and it is symmetric with respect to its main diagonal (which consists of entries that are in the ith row and ith column for some i). This symmetry is displayed in Figure 2.

EXAMPLE 8 The matrix $\begin{bmatrix} 1 & 1 & 0 \\ 1 & 0 & 1 \\ 0 & 1 & 0 \end{bmatrix}$ is symmetric. ◄

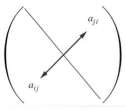

FIGURE 2   A Symmetric Matrix.

ZERO–ONE MATRICES

A matrix with entries that are either 0 or 1 is called a **zero–one matrix.** Zero–one matrices are often used to represent discrete structures, as we will see in Chapters 7 and 8. Algorithms using these structures are based on Boolean arithmetic with zero–one matrices. This arithmetic is based on the Boolean operations $\vee$ and $\wedge$, which operate on pairs of bits, defined by

$$b_1 \wedge b_2 = \begin{cases} 1 & \text{if } b_1 = b_2 = 1 \\ 0 & \text{otherwise,} \end{cases}$$

$$b_1 \vee b_2 = \begin{cases} 1 & \text{if } b_1 = 1 \text{ or } b_2 = 1 \\ 0 & \text{otherwise.} \end{cases}$$

DEFINITION 8

Let $\mathbf{A} = [a_{ij}]$ and $\mathbf{B} = [b_{ij}]$ be $m \times n$ zero–one matrices. Then the *join* of $\mathbf{A}$ and $\mathbf{B}$ is the zero–one matrix with (i, j)th entry $a_{ij} \vee b_{ij}$. The join of $\mathbf{A}$ and $\mathbf{B}$ is denoted by $\mathbf{A} \vee \mathbf{B}.$ The *meet* of $\mathbf{A}$ and $\mathbf{B}$ is the zero–one matrix with (i, j)th entry $a_{ij} \wedge b_{ij}$. The meet of $\mathbf{A}$ and $\mathbf{B}$ is denoted by $\mathbf{A} \wedge \mathbf{B}.$

EXAMPLE 9   Find the join and meet of the zero–one matrices

$$\mathbf{A} = \begin{bmatrix} 1 & 0 & 1 \\ 0 & 1 & 0 \end{bmatrix}, \qquad \mathbf{B} = \begin{bmatrix} 0 & 1 & 0 \\ 1 & 1 & 0 \end{bmatrix}.$$

Solution: We find that the join of $\mathbf{A}$ and $\mathbf{B}$ is

$$\mathbf{A} \vee \mathbf{B} = \begin{bmatrix} 1 \vee 0 & 0 \vee 1 & 1 \vee 0 \\ 0 \vee 1 & 1 \vee 1 & 0 \vee 0 \end{bmatrix} = \begin{bmatrix} 1 & 1 & 1 \\ 1 & 1 & 0 \end{bmatrix}.$$

The meet of $\mathbf{A}$ and $\mathbf{B}$ is

$$\mathbf{A} \wedge \mathbf{B} = \begin{bmatrix} 1 \wedge 0 & 0 \wedge 1 & 1 \wedge 0 \\ 0 \wedge 1 & 1 \wedge 1 & 0 \wedge 0 \end{bmatrix} = \begin{bmatrix} 0 & 0 & 0 \\ 0 & 1 & 0 \end{bmatrix}. \qquad \blacktriangleleft$$

We now define the **Boolean product** of two matrices.

DEFINITION 9

Let $\mathbf{A} = [a_{ij}]$ be an $m \times k$ zero–one matrix and $\mathbf{B} = [b_{ij}]$ be a $k \times n$ zero–one matrix. Then the *Boolean product* of $\mathbf{A}$ and $\mathbf{B},$ denoted by $\mathbf{A} \odot \mathbf{B},$ is the $m \times n$ matrix with (i, j)th entry $[c_{ij}]$ where

$$c_{ij} = (a_{i1} \wedge b_{1j}) \vee (a_{i2} \wedge b_{2j}) \vee \cdots \vee (a_{ik} \wedge b_{kj}).$$

Note that the Boolean product of $\mathbf{A}$ and $\mathbf{B}$ is obtained in an analogous way to the ordinary product of these matrices, but with addition replaced with the operation $\vee$ and with multiplication replaced with the operation $\wedge$. We give an example of the Boolean products of matrices.

EXAMPLE 10   Find the Boolean product of $\mathbf{A}$ and $\mathbf{B},$ where

$$\mathbf{A} = \begin{bmatrix} 1 & 0 \\ 0 & 1 \\ 1 & 0 \end{bmatrix}, \qquad \mathbf{B} = \begin{bmatrix} 1 & 1 & 0 \\ 0 & 1 & 1 \end{bmatrix}.$$

Solution: The Boolean product $\mathbf{A} \odot \mathbf{B}$ is given by

$$\mathbf{A} \odot \mathbf{B} = \begin{bmatrix} (1 \wedge 1) \vee (0 \wedge 0) & (1 \wedge 1) \vee (0 \wedge 1) & (1 \wedge 0) \vee (0 \wedge 1) \\ (0 \wedge 1) \vee (1 \wedge 0) & (0 \wedge 1) \vee (1 \wedge 1) & (0 \wedge 0) \vee (1 \wedge 1) \\ (1 \wedge 1) \vee (0 \wedge 0) & (1 \wedge 1) \vee (0 \wedge 1) & (1 \wedge 0) \vee (0 \wedge 1) \end{bmatrix}$$

$$= \begin{bmatrix} 1 \vee 0 & 1 \vee 0 & 0 \vee 0 \\ 0 \vee 0 & 0 \vee 1 & 0 \vee 1 \\ 1 \vee 0 & 1 \vee 0 & 0 \vee 0 \end{bmatrix}$$

$$= \begin{bmatrix} 1 & 1 & 0 \\ 0 & 1 & 1 \\ 1 & 1 & 0 \end{bmatrix}.$$ ◄

Algorithm 2 displays pseudocode for computing the Boolean product of two matrices.

ALGORITHM 2 The Boolean Product.

---

procedure *Boolean product* ($\mathbf{A}, \mathbf{B}$: zero–one matrices)
for $i := 1$ **to** m
    for $j := 1$ **to** n
    begin
        $c_{ij} := 0$
        for $q := 1$ **to** k
            $c_{ij} := c_{ij} \vee (a_{iq} \wedge b_{qj})$
    end
$\{\mathbf{C} = [c_{ij}]$ is the Boolean product of $\mathbf{A}$ and $\mathbf{B}\}$

We can also define the Boolean powers of a square zero–one matrix. These powers will be used in our subsequent studies of paths in graphs, which are used to model such things as communications paths in computer networks.

DEFINITION 10 Let $\mathbf{A}$ be a square zero–one matrix and let r be a positive integer. The rth *Boolean power* of $\mathbf{A}$ is the Boolean product of r factors of $\mathbf{A}$. The rth Boolean product of $\mathbf{A}$ is denoted by $\mathbf{A}^{[r]}$. Hence

$$\mathbf{A}^{[r]} = \underbrace{\mathbf{A} \odot \mathbf{A} \odot \mathbf{A} \odot \cdots \odot \mathbf{A}}_{r \text{ times}}.$$

(This is well defined since the Boolean product of matrices is associative.) We also define $\mathbf{A}^{[0]}$ to be $\mathbf{I}_n$.

EXAMPLE 11 Let $\mathbf{A} = \begin{bmatrix} 0 & 0 & 1 \\ 1 & 0 & 0 \\ 1 & 1 & 0 \end{bmatrix}$. Find $\mathbf{A}^{[n]}$ for all positive integers n.

Solution: We find that

$$\mathbf{A}^{[2]} = \mathbf{A} \odot \mathbf{A} = \begin{bmatrix} 1 & 1 & 0 \\ 0 & 0 & 1 \\ 1 & 0 & 1 \end{bmatrix}.$$

We also find that

$$\mathbf{A}^{[3]} = \mathbf{A}^{[2]} \odot \mathbf{A} = \begin{bmatrix} 1 & 0 & 1 \\ 1 & 1 & 0 \\ 1 & 1 & 1 \end{bmatrix}, \qquad \mathbf{A}^{[4]} = \mathbf{A}^{[3]} \odot \mathbf{A} = \begin{bmatrix} 1 & 1 & 1 \\ 1 & 0 & 1 \\ 1 & 1 & 1 \end{bmatrix}.$$

Additional computation shows that

$$\mathbf{A}^{[5]} = \begin{bmatrix} 1 & 1 & 1 \\ 1 & 1 & 1 \\ 1 & 1 & 1 \end{bmatrix}.$$

The reader can now see that $\mathbf{A}^{[n]} = \mathbf{A}^{[5]}$ for all positive integers n with $n \geq 5$. ◀

The number of bit operations used to find the Boolean product of two $n \times n$ matrices can be easily determined.

EXAMPLE 12   How many bit operations are used to find $\mathbf{A} \odot \mathbf{B}$, where $\mathbf{A}$ and $\mathbf{B}$ are $n \times n$ zero–one matrices?

Solution: There are n^2 entries in $\mathbf{A} \odot \mathbf{B}$. Using Algorithm 2, a total of n *OR*s and n *AND*s are used to find an entry of $\mathbf{A} \odot \mathbf{B}$. Hence, $2n$ bit operations are used to find each entry. Therefore, $2n^3$ bit operations are required to compute $\mathbf{A} \odot \mathbf{B}$ using Algorithm 2. ◀

Exercises

1. Let $\mathbf{A} = \begin{bmatrix} 1 & 1 & 1 & 3 \\ 2 & 0 & 4 & 6 \\ 1 & 1 & 3 & 7 \end{bmatrix}$.

a) What size is $\mathbf{A}$?
b) What is the third column of $\mathbf{A}$?
c) What is the second row of $\mathbf{A}$?
d) What is the element of $\mathbf{A}$ in the $(3, 2)$th position?
e) What is $\mathbf{A}^t$?

2. Find $\mathbf{A} + \mathbf{B}$, where

a) $\mathbf{A} = \begin{bmatrix} 1 & 0 & 4 \\ -1 & 2 & 2 \\ 0 & -2 & -3 \end{bmatrix}$,

$\mathbf{B} = \begin{bmatrix} -1 & 3 & 5 \\ 2 & 2 & -3 \\ 2 & -3 & 0 \end{bmatrix}$.

b) $\mathbf{A} = \begin{bmatrix} -1 & 0 & 5 & 6 \\ -4 & -3 & 5 & -2 \end{bmatrix}$,

$\mathbf{B} = \begin{bmatrix} -3 & 9 & -3 & 4 \\ 0 & -2 & -1 & 2 \end{bmatrix}$.

3. Find $\mathbf{AB}$ if

a) $\mathbf{A} = \begin{bmatrix} 2 & 1 \\ 3 & 2 \end{bmatrix}, \mathbf{B} = \begin{bmatrix} 0 & 4 \\ 1 & 3 \end{bmatrix}$.

b) $\mathbf{A} = \begin{bmatrix} 1 & -1 \\ 0 & 1 \\ 2 & 3 \end{bmatrix}, \mathbf{B} = \begin{bmatrix} 3 & -2 & -1 \\ 1 & 0 & 2 \end{bmatrix}$.

c) $\mathbf{A} = \begin{bmatrix} 4 & -3 \\ 3 & -1 \\ 0 & -2 \\ -1 & 5 \end{bmatrix}, \mathbf{B} = \begin{bmatrix} -1 & 3 & 2 & -2 \\ 0 & -1 & 4 & -3 \end{bmatrix}$.

4. Find the product $\mathbf{AB}$, where

a) $\mathbf{A} = \begin{bmatrix} 1 & 0 & 1 \\ 0 & -1 & -1 \\ -1 & 1 & 0 \end{bmatrix}, \mathbf{B} = \begin{bmatrix} 0 & 1 & -1 \\ 1 & -1 & 0 \\ -1 & 0 & 1 \end{bmatrix}$.

b) $\mathbf{A} = \begin{bmatrix} 1 & -3 & 0 \\ 1 & 2 & 2 \\ 2 & 1 & -1 \end{bmatrix}$,

$\mathbf{B} = \begin{bmatrix} 1 & -1 & 2 & 3 \\ -1 & 0 & 3 & -1 \\ -3 & -2 & 0 & 2 \end{bmatrix}$.

c) $\mathbf{A} = \begin{bmatrix} 0 & -1 \\ 7 & 2 \\ -4 & -3 \end{bmatrix}, \mathbf{B} = \begin{bmatrix} 4 & -1 & 2 & 3 & 0 \\ -2 & 0 & 3 & 4 & 1 \end{bmatrix}$.

5. Find a matrix $\mathbf{A}$ such that

$$\begin{bmatrix} 2 & 3 \\ 1 & 4 \end{bmatrix} \mathbf{A} = \begin{bmatrix} 3 & 0 \\ 1 & 2 \end{bmatrix}.$$

(*Hint:* Finding $\mathbf{A}$ requires that you solve systems of linear equations.)

6. Find a matrix $\mathbf{A}$ such that

$$\begin{bmatrix} 1 & 3 & 2 \\ 2 & 1 & 1 \\ 4 & 0 & 3 \end{bmatrix} \mathbf{A} = \begin{bmatrix} 7 & 1 & 3 \\ 1 & 0 & 3 \\ -1 & -3 & 7 \end{bmatrix}$$

7. Let $\mathbf{A}$ be an $m \times n$ matrix and let $\mathbf{0}$ be the $m \times n$ matrix that has all entries equal to zero. Show that $\mathbf{A} = \mathbf{0} + \mathbf{A} = \mathbf{A} + \mathbf{0}$.

8. Show that matrix addition is commutative; that is, show that if $\mathbf{A}$ and $\mathbf{B}$ are both $m \times n$ matrices, then $\mathbf{A} + \mathbf{B} = \mathbf{B} + \mathbf{A}$.

9. Show that matrix addition is associative; that is, show that if $\mathbf{A}$, $\mathbf{B}$, and $\mathbf{C}$ are all $m \times n$ matrices, then $\mathbf{A} + (\mathbf{B} + \mathbf{C}) = (\mathbf{A} + \mathbf{B}) + \mathbf{C}$.

10. Let $\mathbf{A}$ be a 3×4 matrix, $\mathbf{B}$ be a 4×5 matrix, and $\mathbf{C}$ be a 4×4 matrix. Determine which of the following products are defined and find the size of those that are defined.

a) **AB** b) **BA** c) **AC**
d) **CA** e) **BC** f) **CB**

11. What do we know about the sizes of the matrices $\mathbf{A}$ and $\mathbf{B}$ if both of the products $\mathbf{AB}$ and $\mathbf{BA}$ are defined?

12. In this exercise we show that matrix multiplication is distributive over matrix addition.

 a) Suppose that $\mathbf{A}$ and $\mathbf{B}$ are $m \times k$ matrices and that $\mathbf{C}$ is a $k \times n$ matrix. Show that $(\mathbf{A} + \mathbf{B})\mathbf{C} = \mathbf{AC} + \mathbf{BC}$.

 b) Suppose that $\mathbf{C}$ is an $m \times k$ matrix and that $\mathbf{A}$ and $\mathbf{B}$ are $k \times n$ matrices. Show that $\mathbf{C}(\mathbf{A} + \mathbf{B}) = \mathbf{CA} + \mathbf{CB}$.

13. In this exercise we show that matrix multiplication is associative. Suppose that $\mathbf{A}$ is an $m \times p$ matrix, $\mathbf{B}$ is a $p \times k$ matrix, and $\mathbf{C}$ is a $k \times n$ matrix. Show that $\mathbf{A}(\mathbf{BC}) = (\mathbf{AB})\mathbf{C}$.

14. The $n \times n$ matrix $\mathbf{A} = [a_{ij}]$ is called a **diagonal matrix** if $a_{ij} = 0$ when $i \neq j$. Show that the product of two $n \times n$ diagonal matrices is again a diagonal matrix. Give a simple rule for determining this product.

15. Let

$$\mathbf{A} = \begin{bmatrix} 1 & 1 \\ 0 & 1 \end{bmatrix}.$$

 Find a formula for $\mathbf{A}^n$, whenever n is a positive integer.

16. Show that $(\mathbf{A}^t)^t = \mathbf{A}$.

17. Let $\mathbf{A}$ and $\mathbf{B}$ be two $n \times n$ matrices. Show that

a) $(\mathbf{A} + \mathbf{B})^t = \mathbf{A}^t + \mathbf{B}^t$. b) $(\mathbf{AB})^t = \mathbf{B}^t\mathbf{A}^t$.

If $\mathbf{A}$ and $\mathbf{B}$ are $n \times n$ matrices with $\mathbf{AB} = \mathbf{BA} = \mathbf{I}_n$, then $\mathbf{B}$ is called the **inverse** of $\mathbf{A}$ (this terminology is appropriate

since such a matrix $\mathbf{B}$ is unique) and $\mathbf{A}$ is said to be **invertible**. The notation $\mathbf{B} = \mathbf{A}^{-1}$ denotes that $\mathbf{B}$ is the inverse of $\mathbf{A}$.

18. Show that

$$\begin{bmatrix} 2 & 3 & -1 \\ 1 & 2 & 1 \\ -1 & -1 & 3 \end{bmatrix}$$

 is the inverse of

$$\begin{bmatrix} 7 & -8 & 5 \\ -4 & 5 & -3 \\ 1 & -1 & 1 \end{bmatrix}.$$

19. Let $\mathbf{A}$ be a 2×2 matrix with

$$\mathbf{A} = \begin{bmatrix} a & b \\ c & d \end{bmatrix}.$$

 Show that if $ad - bc \neq 0$, then

$$\mathbf{A}^{-1} = \begin{bmatrix} \dfrac{d}{ad - bc} & \dfrac{-b}{ad - bc} \\[2ex] \dfrac{-c}{ad - bc} & \dfrac{a}{ad - bc} \end{bmatrix}.$$

20. Let

$$\mathbf{A} = \begin{bmatrix} -1 & 2 \\ 1 & 3 \end{bmatrix}.$$

 a) Find $\mathbf{A}^{-1}$. (*Hint:* Use Exercise 19.)
 b) Find $\mathbf{A}^3$.
 c) Find $(\mathbf{A}^{-1})^3$.
 d) Use your answers to (b) and (c) to show that $(\mathbf{A}^{-1})^3$ is the inverse of $\mathbf{A}^3$.

21. Let $\mathbf{A}$ be an invertible matrix. Show that $(\mathbf{A}^n)^{-1} = (\mathbf{A}^{-1})^n$ whenever n is a positive integer.

22. Let $\mathbf{A}$ be a matrix. Show that the matrix $\mathbf{AA}^t$ is symmetric. (*Hint:* Show that this matrix equals its transpose with the help of Exercise 17b.)

23. Show that the conventional algorithm uses $m_1 m_2 m_3$ multiplications to compute the product of the $m_1 \times m_2$ matrix $\mathbf{A}$ and the $m_2 \times m_3$ matrix $\mathbf{B}$.

24. What is the most efficient way to multiply the matrices $\mathbf{A}_1, \mathbf{A}_2$, and $\mathbf{A}_3$ with sizes

 a) $20 \times 50, 50 \times 10, 10 \times 40$?
 b) $10 \times 5, 5 \times 50, 50 \times 1$?

25. What is the most efficient way to multiply the matrices $\mathbf{A}_1, \mathbf{A}_2, \mathbf{A}_3$, and $\mathbf{A}_4$ if the dimensions of these matrices are $10 \times 2, 2 \times 5, 5 \times 20$, and 20×3, respectively?

26. **a)** Show that the system of simultaneous linear equations

$$a_{11}x_1 + a_{12}x_2 + \cdots + a_{1n}x_n = b_1$$

$$a_{21}x_1 + a_{22}x_2 + \cdots + a_{2n}x_n = b_2$$

$$\vdots$$

$$a_{n1}x_1 + a_{n2}x_2 + \cdots + a_{nn}x_n = b_n$$

in the variables $x_1, x_2, \ldots, x_n$ can be expressed as $\mathbf{AX} = \mathbf{B}$, where $\mathbf{A} = [a_{ij}]$, $\mathbf{X}$ is an $n \times 1$ matrix with x_i the entry in its ith row, and $\mathbf{B}$ is an $n \times 1$ matrix with b_i the entry in its ith row.

b) Show that if the matrix $\mathbf{A} = [a_{ij}]$ is invertible (as defined in the preamble to Exercise 18), then the solution of the system in part (a) can be found using the equation $\mathbf{X} = \mathbf{A}^{-1}\mathbf{B}$.

27. Use Exercises 18 and 26 to solve the system

$$7x_1 - 8x_2 + 5x_3 = 5$$

$$-4x_1 + 5x_2 - 3x_3 = -3$$

$$x_1 - x_2 + x_3 = 0$$

28. Let

$$\mathbf{A} = \begin{bmatrix} 1 & 1 \\ 0 & 1 \end{bmatrix} \quad \text{and} \quad \mathbf{B} = \begin{bmatrix} 0 & 1 \\ 1 & 0 \end{bmatrix}.$$

Find

a) $\mathbf{A} \vee \mathbf{B}$.   **b)** $\mathbf{A} \wedge \mathbf{B}$.
c) $\mathbf{A} \odot \mathbf{B}$.

29. Let

$$\mathbf{A} = \begin{bmatrix} 1 & 0 & 1 \\ 1 & 1 & 0 \\ 0 & 0 & 1 \end{bmatrix} \quad \text{and} \quad \mathbf{B} = \begin{bmatrix} 0 & 1 & 1 \\ 1 & 0 & 1 \\ 1 & 0 & 1 \end{bmatrix}.$$

Find

a) $\mathbf{A} \vee \mathbf{B}$.
b) $\mathbf{A} \wedge \mathbf{B}$.
c) $\mathbf{A} \odot \mathbf{B}$.

30. Find the Boolean product of $\mathbf{A}$ and $\mathbf{B}$, where

$$\mathbf{A} = \begin{bmatrix} 1 & 0 & 0 & 1 \\ 0 & 1 & 0 & 1 \\ 1 & 1 & 1 & 1 \end{bmatrix} \quad \text{and} \quad \mathbf{B} = \begin{bmatrix} 1 & 0 \\ 0 & 1 \\ 1 & 1 \\ 1 & 0 \end{bmatrix}.$$

31. Let

$$\mathbf{A} = \begin{bmatrix} 1 & 0 & 0 \\ 1 & 0 & 1 \\ 0 & 1 & 0 \end{bmatrix}.$$

Find

a) $\mathbf{A}^{[2]}$.   **b)** $\mathbf{A}^{[3]}$.   **c)** $\mathbf{A} \vee \mathbf{A}^{[2]} \vee \mathbf{A}^{[3]}$.

32. Let $\mathbf{A}$ be a zero–one matrix. Show that

a) $\mathbf{A} \vee \mathbf{A} = \mathbf{A}$.
b) $\mathbf{A} \wedge \mathbf{A} = \mathbf{A}$.

33. In this exercise we show that the meet and join operations are commutative. Let $\mathbf{A}$ and $\mathbf{B}$ be $m \times n$ zero–one matrices. Show that

a) $\mathbf{A} \vee \mathbf{B} = \mathbf{B} \vee \mathbf{A}$.
b) $\mathbf{B} \wedge \mathbf{A} = \mathbf{A} \wedge \mathbf{B}$.

34. In this exercise we show that the meet and join operations are associative. Let $\mathbf{A}$, $\mathbf{B}$, and $\mathbf{C}$ be $m \times n$ zero–one matrices. Show that

a) $(\mathbf{A} \vee \mathbf{B}) \vee \mathbf{C} = \mathbf{A} \vee (\mathbf{B} \vee \mathbf{C})$.
b) $(\mathbf{A} \wedge \mathbf{B}) \wedge \mathbf{C} = \mathbf{A} \wedge (\mathbf{B} \wedge \mathbf{C})$.

35. We will establish distributive laws of the meet over the join operation in this exercise. Let $\mathbf{A}, \mathbf{B}$, and $\mathbf{C}$ be $m \times n$ zero–one matrices. Show that

a) $\mathbf{A} \vee (\mathbf{B} \wedge \mathbf{C}) = (\mathbf{A} \vee \mathbf{B}) \wedge (\mathbf{A} \vee \mathbf{C})$.
b) $\mathbf{A} \wedge (\mathbf{B} \vee \mathbf{C}) = (\mathbf{A} \wedge \mathbf{B}) \vee (\mathbf{A} \wedge \mathbf{C})$.

36. Let $\mathbf{A}$ be an $n \times n$ zero–one matrix. Let $\mathbf{I}$ be the $n \times n$ identity matrix. Show that $\mathbf{A} \odot \mathbf{I} = \mathbf{I} \odot \mathbf{A} = \mathbf{A}$.

37. In this exercise we will show that the Boolean product of zero–one matrices is associative. Assume that $\mathbf{A}$ is an $m \times p$ zero–one matrix, $\mathbf{B}$ is a $p \times k$ zero–one matrix, and $\mathbf{C}$ is a $k \times n$ zero–one matrix. Show that $\mathbf{A} \odot (\mathbf{B} \odot \mathbf{C}) = (\mathbf{A} \odot \mathbf{B}) \odot \mathbf{C}$.

Key Terms and Results

TERMS

algorithm: a finite set of precise instructions for performing a computation or solving a problem

searching algorithm: the problem of locating an element in a list

linear search algorithm: a procedure for searching a list element by element

binary search algorithm: a procedure for searching an ordered list by successively splitting the list in half

sorting: the reordering of the elements of a list into nondecreasing order

greedy algorithm: an algorithm that makes the best choice at each step

$f(x)$ is $O(g(x))$: the fact that $|f(x)| \leq C|g(x)|$ for all $x > k$ for some constants C and k

witness to the relationship $f(x)$ is $O(g(x))$: a pair C and k such that $|f(x)| \leq C|g(x)|$ whenever $x > k$

$f(x)$ is $\Omega(g(x))$: the fact that $|f(x)| \geq C|g(x)|$ for all $x > k$ for some positive constants C and k

$f(x)$ is $\Theta(g(x))$: the fact that $f(x)$ is $O(g(x))$ and $f(x)$ is $\Omega(g(x))$

time complexity: the amount of time required for an algorithm to solve a problem

space complexity: the amount of storage space required for an algorithm to solve a problem

worst-case time complexity: the greatest amount of time required for an algorithm to solve a problem of a given size

average-case time complexity: the average amount of time required for an algorithm to solve a problem of a given size

a | *b* (*a* **divides** *b*): there is an integer c such that $b = ac$

prime: a positive integer greater than 1 with exactly two positive integer divisors

composite: a positive integer greater than 1 that is not prime

Mersenne prime: a prime of the form $2^p - 1$, where p is prime

gcd(*a*, *b*) (greatest common divisor of *a* and *b*): the largest integer that divides both a and b

relatively prime integers: integers a and b such that $\gcd(a, b) = 1$

pairwise relatively prime integers: a set of integers with the property that every pair of these integers is relatively prime

lcm(*a*, *b*) (least common multiple of *a* and *b*): the smallest positive integer that is divisible by both a and b

a **mod** *b*: the remainder when the integer a is divided by the positive integer b

$a \equiv b \pmod{m}$ (*a* **is congruent to** *b* **modulo** *m*): $a - b$ is divisible by m

encryption: the process of making a message secret

decryption: the process of returning a secret message to its original form

$n = (a_k a_{k-1} \cdots a_1 a_0)_b$: the base b representation of n

binary representation: the base 2 representation of an integer

hexadecimal representation: the base 16 representation of an integer

octal representation: the base 8 representation of an integer

linear combination of *a* and *b* with integer coefficients: a number of the form $sa + tb$ where s and t are integers

inverse of *a* modulo *m*: an integer $\bar{a}$ such that $\bar{a}a \equiv 1 \pmod{m}$

linear congruence: a congruence of the form $ax \equiv b \pmod{m}$ where x is a variable

pseudoprime to the base 2: a composite integer n such that $2^{n-1} \equiv 1 \pmod{n}$

pseudoprime to the base *b*: a composite integer n such that $b^{n-1} \equiv 1 \pmod{n}$

Carmichael number: a composite integer n such that n is a pseudoprime to the base b for all positive integers b with $\gcd(b, n) = 1$

private key encryption: encryption where both encryption keys and decryption keys must be kept secret

public key encryption: encryption where encryption keys are public knowledge, but decryption keys are kept secret

matrix: a rectangular array of numbers

matrix addition: see page 197

matrix multiplication: see page 198

I_n (**identity matrix of order *n*):** the $n \times n$ matrix that has entries equal to 1 on its diagonal and 0s elsewhere

$\mathbf{A}^t$ (**transpose of A**): the matrix obtained from $\mathbf{A}$ by interchanging the rows and columns

symmetric: a matrix is symmetric if it equals its transpose

zero–one matrix: a matrix with each entry equal to either 0 or 1

A ∨ B (the join of A and B): see page 202

A ∧ B (the meet of A and B): see page 202

A ⊙ B (the Boolean product of A and B): see page 202

RESULTS

linear and binary search algorithms: (given in Section 2.1)

bubble sort: a sorting that uses passes where successive items are interchanged if they are out of order

insertion sort: a sorting that at the jth step inserts the jth element into the correct position in the list of the sorted first $j - 1$ elements

The linear search has $O(n)$ complexity.

The binary search has $O(\log n)$ complexity.

The bubble and insertion sorts have $O(n^2)$ complexity.

$\log n!$ is $O(n \log n)$.

If $f_1(x)$ is $O(g_1(x))$ and $f_2(x)$ is $O(g_2(x))$, then $(f_1 + f_2)(x)$ is $O(\max(g_1(x), g_2(x)))$ and $(f_1 f_2)(x)$ is $O(g_1(x) g_2(x))$.

If $a_0, a_1, \ldots, a_n$ are real numbers with $a_n \neq 0$, then $a_n x^n + a_{n-1} x^{n-1} + \cdots + a_1 x + a_0$ is $O(x^n)$ and $\Theta(x^n)$.

Fundamental Theorem of Arithmetic: Every positive integer can be written uniquely as the product of primes, where the prime factors are written in order of increasing size.

division algorithm: Let a and d be integers with d positive. Then there are unique integers q and r with $0 \le r < d$ such that $a = dq + r$.

If a and b are positive integers, then $ab = \gcd(a, b) \cdot \text{lcm}(a, b)$.

Euclidean algorithm: for finding greatest common divisors (see Algorithm 6 in Section 2.5)

Let b be a positive integer greater than 1. Then if n is a positive integer, it can be expressed uniquely in the form $n = a_k b^k + a_{k-1} b^{k-1} + \cdots + a_1 b + a_0$.

The algorithm for finding the base b expansion of an integer (see Algorithm 1 in Section 2.5)

The conventional algorithms for addition and multiplication of integers (given in Section 2.5)

The modular exponentiation algorithm (see Algorithm 5 in Section 2.5)

The greatest common divisor of two integers can be expressed as a linear combination with integer coefficients of these integers.

If m is a positive integer and $\gcd(a, m) = 1$, then a has a unique inverse modulo m.

Chinese Remainder Theorem: A system of linear congruences modulo pairwise relatively prime integers has a unique solution modulo the product of these moduli.

Fermat's Little Theorem: If p is prime and $p \nmid a$), then $a^{p-1} \equiv 1 \pmod{p}$.

Review Questions

1. a) Define the term *algorithm*.

b) What are the different ways to describe algorithms?

c) What is the difference between an algorithm for solving a problem and a computer program that solves this problem?

2. a) Describe, using English, an algorithm for finding the largest integer in a list of n integers.

b) Express this algorithm in pseudocode.

c) How many comparisons does the algorithm use?

3. a) State the definition of the fact that $f(n)$ is $O(g(n))$, where $f(n)$ and $g(n)$ are functions from the set of positive integers to the set of real numbers.

b) Use the definition of the fact that $f(n)$ is $O(g(n))$ directly to prove or disprove that $n^2 + 18n + 107$ is $O(n^3)$.

c) Use the definition of the fact that $f(n)$ is $O(g(n))$ directly to prove or disprove that n^3 is $O(n^2 + 18n + 107)$.

4. a) How can you produce a big-O estimate for a function that is the sum of different terms where each term is the product of several functions?

b) Give a big-O estimate for the function $f(n) = (n! + 1)(2^n + 1) + (n^{n-2} + 8n^{n-3})(n^3 + 2^n)$. For the function g in your estimate $f(x)$ is $O(g(x))$ use a simple function of smallest possible order.

5. a) Define what the worst-case time complexity, average-case time complexity, and best-case time complexity (in terms of comparisons) mean for an algorithm that finds the smallest integer in a list of n integers.

b) What are the worst-case, average-case, and best-case time complexities, in terms of comparisons, of the algorithm that finds the smallest integer in a list of n integers by comparing each of the integers with the smallest integer found so far?

6. a) Describe the linear search and binary search algorithm for finding an integer in a list of integers in increasing order.

b) Compare the worst-case time complexities of these two algorithms.

c) Is one of these algorithms always faster than the other (measured in terms of comparisons)?

7. a) Describe the bubble sort algorithm.

b) Use the bubble sort algorithm to sort the list 5, 2, 4, 1, 3.

c) Give a big-O estimate for the number of comparisons used by the bubble sort.

8. a) Describe the insertion sort algorithm.

b) Use the insertion sort algorithm to sort the list 2, 5, 1, 4, 3.

c) Give a big-O estimate for the number of comparisons used by the insertion sort.

9. a) Explain the concept of a greedy algorithm.

b) Provide an example of a greedy algorithm that produces an optimal solution and explain why it produces an optimal solution.

c) Provide an example of a greedy algorithm that does not always produce an optimal solution and explain why it fails to do so.

10. State the Fundamental theorem of arithmetic.

11. a) Describe a procedure for finding the prime factorization of an integer.

b) Use this procedure to find the prime factorization of 80,707.

12. a) Define the greatest common divisor of two integers.

b) Describe at least three different ways to find the greatest common divisor of two integers. When does each method work best?

c) Find the greatest common divisor of 1,234,567 and 7,654,321.

d) Find the greatest common divisor of $2^3 3^5 5^7 7^9 11$ and $2^9 3^7 5^5 7^3 13$.

13. a) Define what it means for a and b to be congruent modulo 7.

b) Which pairs of the integers $-11, -8, -7, -1, 0, 3$, and 17 are congruent modulo 7?

c) Show that if a and b are congruent modulo 7, then $10a + 13$ and $-4b + 20$ are also congruent modulo 7.

14. Describe a procedure for converting decimal (base 10) expansions of integers into hexadecimal expansions.

15. a) How can you find a linear combination (with integer coefficients) of two integers that equals their greatest common divisor?

b) Express $\gcd(84, 119)$ as a linear combination of 84 and 119.

16. a) What does it mean for $\bar{a}$ to be an inverse of a modulo m?

b) How can you find an inverse of a modulo m when m is a positive integer and $\gcd(a, m) = 1$?

c) Find an inverse of 7 modulo 19.

17. a) How can an inverse of a modulo m be used to solve the linear congruence $ax \equiv b \pmod{m}$ when $\gcd(a, m) = 1$?

b) Solve the linear congruence $7x \equiv 13 \pmod{19}$.

18. a) State the Chinese Remainder Theorem.

b) Find the solutions to the system $x \equiv 1 \pmod 4$, $x \equiv 2 \pmod 5$, and $x \equiv 3 \pmod 7$.

19. Suppose that $2^{n-1} \equiv 1 \pmod n$. Is n necessarily prime?

20. a) What is the difference between a public key and a private key cryptosystem?

b) Explain why using shift ciphers is a private key system.

c) Explain why the RSA cipher system is a public key system.

21. Define the product of two matrices **A** and **B**. When is this product defined?

22. a) How many different ways are there to evaluate the product $\mathbf{A_1 A_2 A_3 A_4}$ by successively multiplying pairs of matrices, when this product is defined?

b) Suppose that $\mathbf{A_1, A_2, A_3,}$ and $\mathbf{A_4}$ are $10 \times 20, 20 \times 5,$ $5 \times 10,$ and 10×5 matrices, respectively. How should $\mathbf{A_1 A_2 A_3 A_4}$ be computed to use the least number of multiplications of entries?

Supplementary Exercises

1. a) Describe an algorithm for locating the last occurrence of the largest number in a list of integers.

b) Estimate the number of comparisons used.

2. a) Describe an algorithm for finding the first and second largest elements in a list of integers.

b) Estimate the number of comparisons used.

3. a) Give an algorithm to determine whether a bit string contains a pair of consecutive zeros.

b) How many comparisons does the algorithm use?

4. a) Suppose that a list contains integers that are in order of largest to smallest and an integer can appear repeatedly in this list. Devise an algorithm that locates all occurrences of an integer x in the list.

b) Estimate the number of comparisons used.

5. a) Adapt Algorithm 1 in Section 2.1 to find the maximum and the minimum of a sequence of n elements by employing a temporary maximum and a temporary minimum that is updated as each successive element is examined.

b) Describe the algorithm from part (a) in pseudocode.

c) How many comparisons of elements in the sequence are carried out by this algorithm? (Do not count comparisons used to determine whether the end of the sequence has been reached.)

6. a) Describe in detail the steps of an algorithm that finds the maximum and minimum of a sequence of n elements by examining pairs of successive elements, keeping track of a temporary maximum and a temporary minimum. If n is odd, both the temporary maximum and temporary minimum should initially equal the first term, and if n is even, the temporary minimum and temporary maximum should be found by comparing the initial two elements. The temporary maximum and temporary minimum should be updated by comparing them with the maximum and minimum of the pair of elements being examined.

b) Express the algorithm described in part (a) in pseudocode.

c) How many comparisons of elements of the sequence are carried out by this algorithm? (Do not count comparisons used to determine whether the end of the sequence has been reached.) How does this compare to the number of comparisons used by the algorithm in Exercise 5?

* 7. Show that the worst-case complexity in terms of comparisons of an algorithm that finds the maximum and minimum of n elements is at least $\lceil 3n/2 \rceil - 2$.

8. Devise an efficient algorithm for finding the second largest element in a sequence of n elements and determine the worst-case complexity of your algorithm.

The **shaker sort** (or **bidirectional bubble sort**) successively compares adjacent pairs of elements, exchanging them if they are out of order, and alternately passing through the list from the beginning to the end and then from the end to the beginning until no exchanges are needed.

9. Show the steps used by the shaker sort to sort the list 3, 5, 1, 4, 6, 2.

10. Express the shaker sort in pseudocode.

11. Show that the shaker sort has $O(n^2)$ complexity measured in terms of the number of comparisons it uses.

12. Explain why the shaker sort is efficient for sorting lists that are already in close to the correct order.

13. Show that $(n \log n + n^2)^3$ is $O(n^6)$.

14. Show that $8x^3 + 12x + 100 \log x$ is $O(x^3)$.

15. Give a big-O estimate for $(x^2 + x(\log x)^3) \cdot (2^x + x^3)$.

16. Find a big-O estimate for $\sum_{j=1}^{n} j(j+1)$.

*17. Show that $n!$ is not $O(2^n)$.

*18. Show that n^n is not $O(n!)$.

19. Find four numbers congruent to 5 modulo 17.

20. Show that if a and d are positive integers, then there are integers q and r such that $a = dq + r$ where $-d/2 < r \le d/2$.

*21. Show that if $ac \equiv bc \pmod{m}$, then $a \equiv b \pmod{m/d}$ where $d = \gcd(m, c)$.

*22. How many zeros are at the end of the binary expansion of $100_{10}!$?

23. Use the Euclidean algorithm to find the greatest common divisor of 10,223 and 33,341.

24. How many divisions are required to find gcd(144, 233) using the Euclidean algorithm?

25. Find $\gcd(2n + 1, 3n + 2)$, where n is a positive integer. (*Hint:* Use the Euclidean algorithm.)

26. a) Show that if a and b are positive integers with $a \ge b$, then $\gcd(a, b) = a$ if $a = b$, $\gcd(a, b) = 2 \gcd(a/2, b/2)$ if a and b are even, $\gcd(a, b) =$

$\gcd(a/2, b)$ if a is even and b is odd, and $\gcd(a, b) = \gcd(a - b, b)$ if both a and b are odd.

b) Explain how to use (a) to construct an algorithm for computing the greatest common divisor of two positive integers that uses only comparisons, subtractions, and shifts of binary expansions, without using any divisions.

c) Find $\gcd(1202, 4848)$ using this algorithm.

27. Show that an integer is divisible by 9 if and only if the sum of its decimal digits is divisible by 9.

28. a) Explain why n **div** 7 equals the number of weeks in n days.

b) Explain why n **div** 24 equals the number of days in n hours.

A set of integers is called **mutually relatively prime** if the greatest common divisor of these integers is 1.

29. Determine whether these sets of integers are mutually relatively prime.

a) $8, 10, 12$ **b)** $12, 15, 25$

c) $15, 21, 28$ **d)** $21, 24, 28, 32$

30. Find a set of four mutually relatively prime integers such that no two of them are relatively prime.

31. a) Suppose that messages are encrypted using the function $f(p) = (ap + b)$ **mod** 26 such that $\gcd(a, 26) = 1$. Determine a function that can be used to decrypt messages.

b) The encrypted version of a message is LJMKG MGMXF QEXMW. If it was encrypted using the function $f(p) = (7p + 10)$ **mod** 26, what was the original message?

32. Show that the system of congruences $x \equiv 2 \pmod{6}$ and $x \equiv 3 \pmod{9}$ has no solutions.

33. Find all solutions of the system of congruences $x \equiv 4 \pmod{6}$ and $x \equiv 13 \pmod{15}$.

***34. a)** Show that the system of congruences $x \equiv a_1 \pmod{m_1}$ and $x \equiv a_2 \pmod{m_2}$ has a solution if and only if $\gcd(m_1, m_2) \mid a_1 - a_2$.

b) Show that the solution in part (a) is unique modulo $\operatorname{lcm}(m_1, m_2)$.

35. Find $\mathbf{A}^n$ if $\mathbf{A}$ is

$$\begin{bmatrix} 0 & 1 \\ -1 & 0 \end{bmatrix}.$$

36. Show that if $\mathbf{A} = c\mathbf{I}$, where c is a real number and $\mathbf{I}$ is the $n \times n$ identity matrix, then $\mathbf{AB} = \mathbf{BA}$ whenever $\mathbf{B}$ is an $n \times n$ matrix.

37. Show that if $\mathbf{A}$ is a 2×2 matrix such that $\mathbf{AB} = \mathbf{BA}$ whenever $\mathbf{B}$ is a 2×2 matrix, then $\mathbf{A} = c\mathbf{I}$, where c is a real number and $\mathbf{I}$ is the 2×2 identity matrix.

An $n \times n$ matrix is called **upper triangular** if $a_{ij} = 0$ whenever $i > j$.

38. From the definition of the matrix product, devise an algorithm for computing the product of two upper triangular matrices that ignores those products in the computation that are automatically equal to zero.

39. Give a pseudocode description of the algorithm in Exercise 38 for multiplying two upper triangular matrices.

40. How many multiplications of entries are used by the algorithm found in Exercise 38 for multiplying two $n \times n$ upper triangular matrices?

41. Show that if $\mathbf{A}$ and $\mathbf{B}$ are invertible matrices and $\mathbf{AB}$ exists, then $(\mathbf{AB})^{-1} = \mathbf{B}^{-1}\mathbf{A}^{-1}$.

42. What is the best order to form the product $\mathbf{ABCD}$ if $\mathbf{A}, \mathbf{B}, \mathbf{C}$, and $\mathbf{D}$ are matrices with dimensions 30×10, 10×40, 40×50, and 50×30, respectively? Assume that the number of multiplications of entries used to multiply a $p \times q$ matrix and a $q \times r$ matrix is pqr.

43. Let $\mathbf{A}$ be an $n \times n$ matrix and let $\mathbf{0}$ be the $n \times n$ matrix all of whose entries are zero. Show that the following are true.

a) $\mathbf{A} \odot \mathbf{0} = \mathbf{0} \odot \mathbf{A} = \mathbf{0}$ **b)** $\mathbf{A} \vee \mathbf{0} = \mathbf{0} \vee \mathbf{A} = \mathbf{A}$

c) $\mathbf{A} \wedge \mathbf{0} = \mathbf{0} \wedge \mathbf{A} = \mathbf{0}$

44. Show that if the denominations of coins are $c^0, c^1, \ldots, c^k$, where k is a positive integer and c is a positive integer, $c > 1$, the greedy algorithm always produces change using the fewest coins possible.

45. Devise an algorithm for guessing a number between 1 and $2^n - 1$ by successively guessing each bit in its binary expansion.

46. Determine the complexity, in terms of the number of guesses, needed to determine a number between 1 and $2^n - 1$ by successively guessing the bits in its binary expansion.

Computer Projects

WRITE PROGRAMS WITH THESE INPUTS AND OUTPUTS.

1. Given a list of n integers, find the largest integer in the list.

2. Given a list of n integers, find the first and last occurrences of the largest integer in the list.

3. Given a list of n distinct integers, determine the position of an integer in the list using a linear search.

4. Given an ordered list of n distinct integers, determine the position of an integer in the list using a binary search.

5. Given a list of n integers, sort them using a bubble sort.

6. Given a list of n integers, sort them using an insertion sort.

7. Given an integer n, use the greedy algorithm to find the change for n cents using quarters, dimes, nickels, and pennies.

8. Given an ordered list of n integers and an integer x, find the number of comparisons used to determine the position of an integer in the list using a linear search and using a binary search.

9. Given a list of integers, determine the number of comparisons used by the bubble sort and by the insertion sort to sort this list.

10. Given a positive integer, determine whether it is prime.

11. Given a message, encrypt this message using the Caesar cipher; and given a message encrypted using the Caesar cipher, decrypt this message.

12. Given two positive integers, find their greatest common divisor using the Euclidean algorithm.

13. Given two positive integers, find their least common multiple.

*14. Given a positive integer, find the prime factorization of this integer.

15. Given a positive integer and a positive integer b greater than 1, find the base b expansion of this integer.

16. Given the positive integers a, b, and $m > 1$, find $a^b \bmod m$.

17. Given a positive integer, find the Cantor expansion of this integer (see the preamble to Exercise 44 of Section 2.5).

18. Given a positive integer n, a modulus m, multiplier a, increment c, and seed x_0, with $0 \le a < m, 0 \le c < m$, and $0 \le x_0 < m$, generate the sequence of n pseudorandom numbers using the linear congruential generator $x_{n+1} = (ax_n + c) \bmod m$.

19. Given positive integers a and b, find integers s and t such that $sa + tb = \gcd(a, b)$.

20. Given n linear congruences modulo pairwise relatively prime moduli, find the simultaneous solution of these congruences modulo the product of these moduli.

21. Given an $m \times k$ matrix **A** and a $k \times n$ matrix **B**, find **AB**.

22. Given a square matrix **A** and a positive integer n, find **A**^{n}.

23. Given a square matrix, determine whether it is symmetric.

24. Given an $n_1 \times n_2$ matrix **A**, an $n_2 \times n_3$ matrix **B**, an $n_3 \times n_4$ matrix **C**, and an $n_4 \times n_5$ matrix **D**, all with integer entries, determine the most efficient order to multiply these matrices (in terms of the number of multiplications and additions of integers).

25. Given two $m \times n$ Boolean matrices, find their meet and join.

26. Given an $m \times k$ Boolean matrix **A** and a $k \times n$ Boolean matrix **B**, find the Boolean product of **A** and **B**.

27. Given a square Boolean matrix **A** and a positive integer n, find **A**$^{[n]}$.

Computations and Explorations

USE A COMPUTATIONAL PROGRAM OR PROGRAMS YOU HAVE WRITTEN TO DO THESE EXERCISES.

1. We know that n^b is $O(d^n)$ when b and d are positive numbers with $d \ge 2$. Give values of the constants C and k such that $n^b \le Cd^n$ whenever $x > k$ for each of these sets of values: $b = 10, d = 2; b = 20, d = 3; b = 1000, d = 7$.

2. Compute the change for different values of n with coins of different denominations using the greedy algorithm and determine whether the smallest number of coins was used. Can you find conditions so that the greedy algorithm is guaranteed to use the fewest coins possible?

3. Using a generator of random orderings of the integers $1, 2, \ldots, n$, find the number of comparisons used by the bubble sort, insertion sort, binary insertion sort, and selection sort to sort these integers.

4. Determine whether $2^p - 1$ is prime for each of the primes not exceeding 100.

5. Test a range of large Mersenne numbers $2^p - 1$ to determine whether they are prime. (You may want to use software from the GIMPS project.)

6. Show that $n^2 + n + 41$ is prime for all integers n with $0 \le n \le 39$, but is not prime when $n = 40$. Is there a polynomial in n with integer coefficients and degree greater than zero that always takes on a prime value when n is a positive integer?

7. Find as many primes of the form $n^2 + 1$ where n is a positive integer as you can. It is not known whether there are infinitely many such primes.

8. Find 10 different primes each with 100 digits.

9. How many primes are there less than 1,000,000, less than 10,000,000, and less than 100,000,000? Can you propose an estimate for the number of primes less than x where x is a positive integer?

10. Find a prime factor of each of 10 different 20-digit odd integers, selected at random. Keep track of how long it takes to find a factor of each of these integers.

Do the same thing for 10 different 30-digit odd integers, 10 different 40-digit odd integers, and so on, continuing as long as possible.

11. Find all pseudoprimes to the base 2, that is, composite integers n such that $2^{n-1} \equiv 1 \pmod{n}$, where n does not exceed 10,000.

Writing Projects

RESPOND TO THESE QUESTIONS WITH ESSAYS USING OUTSIDE SOURCES.

1. Examine the history of the word *algorithm* and describe the use of this word in early writings.

2. Look up Bachmann's original introduction of big-O notation. Explain how he and others have used this notation.

3. Explain how sorting algorithms can be classified into a taxonomy based on the underlying principle on which they are based.

4. Describe the radix sort algorithm.

5. Describe what is meant by a parallel algorithm. Explain how the pseudocode used in this book can be extended to handle parallel algorithms.

6. Explain how the complexity of parallel algorithms can be measured. Give some examples to illustrate this concept, showing how a parallel algorithm can work more quickly than one that does not operate in parallel.

7. Describe the Lucas–Lehmer test for determining whether a Mersenne number is prime. Discuss the progress of the GIMPS project in finding Mersenne primes using this test.

8. Explain how probabilistic primality tests are used in practice to produce extremely large numbers that are almost certainly prime. Do such tests have any potential drawbacks?

9. The question of whether there are infinitely many Carmichael numbers was solved recently after being open for more than 75 years. Describe the ingredients that went into the proof that there are infinitely many such numbers.

10. Summarize the current status of factoring algorithms in terms of their complexity and the size of numbers that can currently be factored. When do you think

that it will be feasible to factor 200-digit numbers?

11. Describe the algorithms that are actually used in modern computers to add, subtract, multiply, and divide positive integers.

12. Describe the history of the Chinese Remainder Theorem. Describe some of the relevant problems posed in Chinese and Hindu writings and how the Chinese Remainder Theorem applies to them.

13. When are the numbers of a sequence truly random numbers, and not pseudorandom? What shortcomings have been observed in simulations and experiments in which pseudorandom numbers have been used? What are the properties that pseudorandom numbers can have that random numbers should not have?

14. Describe how public key cryptography is being applied. Are the ways it is applied secure given the status of factoring algorithms? Will information kept secure using public key cryptography become insecure in the future?

15. Describe how public key cryptography can be used to send signed secret messages so that the recipient is relatively sure the message was sent by the person claiming to have sent it.

16. Show how a congruence can be used to tell the day of the week for any given date.

17. Describe some of the algorithms used to efficiently multiply large integers.

18. Describe some of the algorithms used to efficiently multiply large matrices.

19. Describe the Rabin public key cryptosystem, explaining how to encrypt and how to decrypt messages and why it is suitable for use as a public key cryptosystem.

3

Mathematical Reasoning, Induction, and Recursion

I n Chapter 1 we introduced rules of inference and methods of proof. In the later sections of Chapter 1 and in Chapter 2 we used these proof methods to prove a variety of results. However, we did not discuss the process of formulating conjectures and then attempting to determine whether these conjectures are correct. In this chapter we will briefly address this process and provide guidance for selecting techniques from our arsenal of proof methods and for constructing proofs using these methods. We will also discuss the role of counterexamples in this process.

We will discuss some important properties of sequences and strings in this chapter. We will define the notion of a countable set and prove that the set of rational numbers is countable. We will prove that the set of real numbers is not countable using a proof technique called the diagonalization argument.

Many mathematical statements assert that a property is true for all positive integers. Examples of such statements are that for every positive integer n: $n! \leq n^n$, $n^3 - n$ is divisible by 3, and the sum of the first n positive integers is $n(n + 1)/2$. A major goal of this chapter, and the book, is to give the student a thorough understanding of mathematical induction, which is used to prove results of this kind.

In Chapters 1 and 2 we explicitly defined sets and functions. That is, we described sets by listing their elements or by giving some property that characterizes these elements. We gave formulae for the values of functions. There is another important way to define such objects, based on mathematical induction. To define functions, some initial terms are specified, and a rule is given for finding subsequent values from values already known. Sets can be defined by listing some of their elements and giving rules for constructing elements from those already known to be in the set. Such definitions, called *recursive definitions,* are used throughout discrete mathematics and computer science. Once we have defined a set recursively, we can use a proof method called structural induction to prove results about this set.

When a procedure is specified for solving a problem, this procedure *always* solves the problem correctly. Just testing to see that the correct result is obtained for a set of input values does not show that the procedure always works correctly. The correctness of a procedure can be guaranteed only by proving that it always yields the correct result. The final section of this chapter contains an introduction to the techniques of program verification. This is a formal technique to verify that procedures are correct. Program verification serves as the basis for attempts under way to prove in a mechanical fashion that programs are correct.

3.1 Proof Strategy

INTRODUCTION

In Section 1.5 we introduced some of the most important methods of proof and illustrated how each method is used. In Sections 1.6–1.8 and in Chapter 2 we used these methods to prove many different theorems. However, we discussed only briefly the strategy behind constructing proofs. This strategy includes selecting a proof method and then successfully constructing an argument step by step, based on this method. In this section we will study some additional aspects of the art and science of proofs. We will provide advice on how to find a proof of a theorem. Because we already have a portfolio of proof methods, we can investigate more closely how proofs are actually constructed. In particular, we will describe some tricks of the trade, including how proofs can be found by working backward, by adapting existing proofs, and by taking advantage of the method of proof by cases.

When mathematicians work, they formulate conjectures and attempt to prove or disprove them. We will briefly describe this process here. Besides providing useful advice about constructing proofs and formulating conjectures, we will also supply some words of advice about counterexamples. We will describe the role counterexamples play in resolving some long-standing conjectures. Further, we discuss some interesting problems that remain unsolved.

Finally, we will present a proof by contradiction of an extremely important result in computer science. We will prove that no procedure exists that can determine, given a computer program and its input, whether this program will eventually stop when provided with this input.

PROOF STRATEGIES

Finding proofs can be a challenging business. When you are confronted with a statement to prove, you should first replace terms by their definitions and then carefully analyze what the hypotheses and the conclusion mean. After doing so, you can attempt to prove the result using one of the available methods of proof. Generally, if the statement is an implication, you should first try a direct proof; if this fails, you can try an indirect proof. If neither of these approaches works, you might try a proof by contradiction.

FORWARD AND BACKWARD REASONING Whichever method you choose, you need a starting point for your proof. To begin a direct proof of an implication, you can start with the hypotheses. Using these hypotheses, together with axioms and known theorems, you can construct a proof using a sequence of steps that leads to the conclusion. This type of reasoning, called *forward reasoning,* is the most common type of reasoning used to prove relatively simple results. Similarly, with indirect reasoning you can start with the negation of the conclusion and, using a sequence of steps, obtain the negation of the hypotheses.

Unfortunately, forward reasoning is often difficult to use to prove more complicated results, because the reasoning needed to reach the desired conclusion may be far from obvious. In such cases it may be helpful to use *backward reasoning*. To reason backward to prove a statement q, we find a statement p that we can prove with the property that $p \rightarrow q$. (Note that it is not helpful to find a statement r that you can prove such that

$q \rightarrow r$, because it is the fallacy of begging the question to conclude from $q \rightarrow r$ and r that q is true.) Backward reasoning is illustrated in Examples 1 and 2.

EXAMPLE 1

Extra
Examples

Given two positive real numbers a and b, their **arithmetic mean** is $(a + b)/2$ and their **geometric mean** is $\sqrt{ab}$. When we compare the arithmetic and geometric means of pairs of distinct positive real numbers, we find that the arithmetic mean is always greater than the geometric mean. (For example, when $a = 4$ and $b = 6$, we have $5 = (4 + 6)/2 > \sqrt{4 \cdot 6} = \sqrt{24}$.) Can we prove that this inequality is always true?

Solution: To prove that $(a + b)/2 > \sqrt{ab}$ when a and b are distinct positive real numbers, we can work backwards. We construct a sequence of equivalent inequalities. (We leave it to the reader to show that each successive pair of these inequalities is equivalent, using the fact that a and b are positive real numbers.) The equivalent inequalities are

$$(a + b)/2 > \sqrt{ab},$$
$$(a + b)^2/4 > ab,$$
$$(a + b)^2 > 4\,ab,$$
$$a^2 + 2\,ab + b^2 > 4\,ab,$$
$$a^2 - 2\,ab + b^2 > 0,$$
$$(a - b)^2 > 0.$$

Because $(a - b)^2 > 0$ when $a \neq b$, it follows that the final inequality is true. Since all these inequalities are equivalent, it follows that $(a + b)/2 > \sqrt{ab}$ when $a \neq b$. Once we have constructed this proof, we can easily construct a proof using forward reasoning. We leave this to the reader. ◀

EXAMPLE 2

Suppose that two people play a game taking turns removing 1, 2, or 3 stones at a time from a pile that begins with 15 stones. The person who removes the last stone wins the game. Show that the first player can win the game no matter what the second player does.

Solution: To prove that the first player can always win the game, we work backwards. At the last step, the first player can win if this player is left with a pile containing 1, 2, or 3 stones. The second player will be forced to leave 1, 2, or 3 stones if this player has to remove stones from a pile containing 4 stones. Consequently, the next-to-last move of the first player should leave 4 stones for the second player. This can be done when there are 5, 6, or 7 stones left, which happens when the second player has to remove stones from a pile with 8 stones. Consequently, the first player should leave 8 stones for the second player at the second-to-last move for the first player. This means that there are 9, 10, or 11 stones when the first player makes this move. Similarly, the first player should leave 12 stones when this player makes the first move. We can reverse this argument to show that the first player can always make moves so this player wins the game no matter what the second player does. These moves successively leave 12, 8, and 4 stones for the second player. ◀

LEVERAGING PROOF BY CASES When it is not possible to consider all cases of a proof at the same time, a proof by cases (introduced in Section 1.5) should be considered. When should you use such a proof? Generally, look for a proof by cases when there is no obvious way to begin a proof, but when extra information in each case helps move the proof forward. For example, to prove results about the integers, it may be advantageous to consider odd and even integers in separate cases or positive and negative integers in separate cases (such as when absolute values of variables are involved). Similarly, you may want to separately consider cases when an integer is divisible by 5 and when it is not, or the five cases when an integer is of the form $5k, 5k + 1, 5k + 2, 5k + 3$, or $5k + 4$ for some integer k. Beware of a common error in proofs by cases: many people forget to consider all cases so that a purported proof does not establish the result for all possible values. Example 3 illustrates a situation when a proof by cases is called for.

EXAMPLE 3   Prove that if n is an integer not divisible by 2 or 3, then $n^2 - 1$ is divisible by 24.

Solution: To attempt a proof without using cases, we first observe that $n^2 - 1 = (n - 1) \cdot (n + 1)$. Unfortunately, there seems to be no clear way to show this number is divisible by 24 when n is not divisible by either 2 or 3 without looking at different cases. A proof by cases seems to be the next reasonable approach, but which cases should be used? Because we want to prove a result about integers not divisible by either 2 or 3, looking separately at the cases where n is of the form $6k + j$ for $j = 0, 1, 2, 3, 4, 5$ may help. (By the division algorithm, every integer is of one of these forms.) Because we are interested only in integers not divisible by either 2 or 3, we can immediately eliminate the cases where $n = 6k, 6k + 2, 6k + 3$, or $6k + 4$ for some integer k. This leaves only two cases, where n is of the form $6k + 1$ and where it is of the form $6k + 5$. We now consider these two cases separately.

Suppose that $n = 6k + 1$ for some integer k. Then $n^2 - 1 = (n+1)(n-1) = (6k+2)6k = 12(3k + 1)k$. Note that $(3k + 1)k$ is an even integer; when k is even, this follows immediately, and when k is odd, $3k + 1$ is even, so that $(3k + 1)k$ is also even. Because $(3k + 1)k$ is even, there is an integer q such that $(3k + 1)k = 2q$. Hence $n^2 - 1 = 12(3k + 1)k = 24q$. It follows that 24 divides $n^2 - 1$.

Next, suppose that $n = 6k + 5$ for some integer k. Then, $n^2 - 1 = (n + 1)(n - 1) = (6k + 6)(6k + 4) = 12(k + 1)(3k + 2)$. Note that $(k + 1)(3k + 2)$ is even because when k is even, $3k + 2$ is even, and when k is odd, $k + 1$ is even. Hence, there is an integer q such that $(k + 1)(3k + 2) = 2q$. It follows that $n^2 - 1 = 12 \cdot 2q = 24q$. We conclude that 24 divides $n^2 - 1$. This completes the proof.   ◄

A common error of reasoning is to draw incorrect conclusions from examples. (For instance, see Example 8.) No matter how many separate examples are considered, a theorem is not proved by considering examples unless every possible case is covered. The problem of proving a theorem is analogous to showing that a computer program always produces the output desired. No matter how many input values are tested, unless all input values are tested, we cannot conclude that the program always produces the correct output. Sometimes, however, we are lucky and we can prove a theorem by considering only a few special cases, as Example 4 demonstrates.

EXAMPLE 4   Show that there are no solutions in integers x and y of $x^2 + 3y^2 = 8$.

Solution: We can quickly reduce this problem to checking just a few cases since $x^2 > 8$ when $|x| \geq 3$ and $3y^2 > 8$ when $|y| \geq 2$. This leaves the cases when x takes on one of the values $-2, -1, 0, 1$, or 2 and y takes on one of the values $-1, 0$, or 1. To dispense with these cases, we note that possible values for x^2 are 0, 1, and 4, and possible values for $3y^2$ are 0 and 3, and the largest sum of possible values for x^2 and $3y^2$ is 7. Consequently, it is impossible for $x^2 + 3y^2 = 8$ to hold when x and y are integers.   ◄

ADAPTING EXISTING PROOFS An excellent way to look for possible approaches that can be used to prove a statement is to take advantage of existing proofs. Often an existing proof can be adapted to prove a new result. Even when this is not the case, some of the ideas used in existing proofs may be helpful. Because existing proofs provide clues for new proofs, you should read and understand the proofs you encounter in your studies. Example 5 illustrates how to adapt a proof you have already seen to prove a new result.

EXAMPLE 5

Extra
Examples

Prove that there are infinitely many primes of the form $4k + 3$, where k is a nonnegative integer.

Solution: A proof we might be able to adapt is the proof given in Section 2.4 that there are infinitely many primes. Recall that this proof assumes that there are only finitely many primes $p_1, p_2, \ldots, p_n$, and forms the number $p_1 p_2 \cdots p_n + 1$. This number is either prime or has a prime factor different from each of the primes $p_1, p_2, \ldots, p_n$, proving that there are infinitely many primes. So, let us suppose that there are only finitely many primes of the form $4k + 3$, namely, $q_1, q_2, \ldots, q_n$, where $q_1 = 3, q_2 = 7$, and so on.

What number can we form that is not divisible by any of these primes, but that must be divisible by a prime of the form $4k+3$? We might consider the number $4q_1 q_2 \cdots q_n + 3$. Unfortunately, this number is not prime and it is divisible by 3. (This follows since $q_1 = 3$.) Instead, we consider the number

$$Q = 4q_1 q_2 \cdots q_n - 1.$$

Note that Q is of the form $4k + 3$ (where $k = q_1 q_2 \cdots q_n - 1$. If Q is prime, we have found a prime of the desired form different from all those listed. If Q is not prime, Q has at least one prime factor not in the list $q_1, q_2, q_3, \ldots$ because the remainder when Q is divided by q_j is $q_j - 1$, and $q_j - 1 \neq 0$, so $q_j \nmid Q$ for $j = 1, 2, \ldots, n$. Because all odd primes are either of the form $4k + 1$ or of the form $4k + 3$, and the product of primes of the form $4k + 1$ is also of this form (as the reader can verify), there must be a factor of Q of the form $4k + 3$ different from the primes we listed. This completes the proof. We were able to adapt a proof we already had, making some minor modifications to prove a new result. ◄

A famous theorem from number theory, which has only been proved by advanced methods, states that there are infinitely many primes of the form $ak + b$ whenever a and b are relatively prime positive integers. We leave it to the reader to explore whether the proof we have constructed in Example 5 can be adapted to show that there are infinitely many primes of the form $ak + b$ for other pairs a, b besides $a = 4$ and $b = 3$. (See Exercises 29 and 30, for example.)

CONJECTURE AND PROOF

Mathematics is generally taught as if mathematical facts were carved in stone. Mathematics texts (including the bulk of this book) formally present theorems and their proofs. Such presentations do not convey the discovery process in mathematics. This process begins with the exploration of concepts and examples, the formulation of conjectures, and attempts to settle these conjectures either by proof or by counterexample. These are the day-to-day activities of mathematicians. Believe it or not, the material taught in textbooks was originally developed in this way.

People formulate conjectures on the basis of many types of possible evidence. The examination of special cases can lead to a conjecture, as can the identification of possible patterns. Altering the hypotheses and conclusions of known theorems also can lead to plausible conjectures. At other times, conjectures are made based on intuition or a belief that a result holds. No matter how a conjecture was made, once it has been formulated, the goal is to prove or disprove it. When mathematicians believe a conjecture may be true, they try to find a proof. If they cannot find a proof, they may look for a counterexample. When they cannot find a counterexample, they may switch gears and once again try to prove the conjecture. Although many conjectures are quickly settled, a few conjectures resist attack for hundreds of years and lead to the development of new parts of mathematics. We illustrate the process of formulating conjectures and attempting to prove them in Examples 6 and 7.

EXAMPLE 6   In Chapter 2 we mentioned that the largest primes known are Mersenne primes. Recall that these are primes of the form $2^p - 1$, where p is prime; that is, Mersenne primes are 1 less than a prime power of 2. Do other primes of the special form $a^n - 1$, where a and n are positive integers, exist? After some numerical explorations (such as noting that none of $2^6 - 1 = 63$, $2^8 - 1 = 255$, $3^4 - 1 = 80$, and $4^5 - 1 = 1023$ are prime) and finding no other primes of this form besides Mersenne primes, we conjecture that $a^n - 1$ is composite when $a > 2$ or when $a = 2$ and n is composite. Can we prove this conjecture?

   If we can find a proper factor of $a^n - 1$ when $a > 2$ or $a = 2$ and n is composite, then we have proved this conjecture. A useful approach is to use the factorization $x^n - 1 = (x - 1)(x^{n-1} + x^{n-2} + \cdots + x + 1)$. (You may recall this factorization from your earlier studies. If you do not, you can look at reference books such as [Zw02] that list factorizations of special types of polynomials.) We let $x = a$ in this factorization, to conclude that $a - 1$ is a factor of $a^n - 1$. When $a = 2$ we have $a - 1 = 2 - 1 = 1$, so this factorization does not yield a proper factor of $a^n - 1$. However, when $a > 2$, the factor $a - 1$ satisfies $1 < a - 1 < a^n - 1$, which tells us that $a^n - 1$ is not prime.

   The case when $a = 2$ remains. When n is not prime, there are positive integers r and s with $1 < r < n$ and $1 < s < n$ such that $n = rs$. Using the factorization $x^n - 1 = x^{rs} - 1 = (x^r - 1)(x^{r(s-1)} + \cdots + x^r + 1)$, setting $x = a$ shows that $a^r - 1$ is a factor of $a^n - 1$. When $a^r > 2$, $a^r - 1$ is a positive integer greater than 1. This produces a nontrivial factor of $2^n - 1$ when n is composite.   ◄

   Putting everything together, we can present a succinct direct proof of the conjecture we made in Example 6. We use a proof by cases, since the case when $a > 2$ needs to be considered separately from the case when $a = 2$.

THEOREM 1   The integer $a^n - 1$ is composite when $a > 2$ or when $a = 2$ and n is composite.

Proof: We use a proof by cases.

Case (i): When $a > 2$, the integer $a - 1$ is a factor of $a^n - 1$ because $a^n - 1 = (a - 1) \cdot (a^{n-1} + \cdots + a + 1)$ and $1 < a - 1 < a^n - 1$. Consequently, $a^n - 1$ is composite.

Case (ii): When $a = 2$ and n is composite, there exist integers r and s such that $n = rs$ with $1 < r \le s < n$. In this case, $2^r - 1$ is a factor of $2^n - 1$ because $2^n - 1 = (2^r - 1) \cdot (2^{r(s-1)} + \cdots + 2^r + 1)$ and $1 < 2^r - 1 < 2^n - 1$. This shows that $2^n - 1$ is composite.   ◁

EXAMPLE 7   We can easily find long strings of consecutive composite integers, such as $24, 25, 26, 27, 28$ and $90, 91, 92, 93, 94, 95, 96$. From such evidence, we formulate the conjecture that given a positive integer n, there are n consecutive composite positive integers. How might we prove this conjecture?

Solution: One possible approach to proving this conjecture is to work backward. We reason that if there is an integer X such that $X, X + 1, X + 2, \ldots, X + (n - 1)$ are all composite, then we have proved the result. (This is simply what it means for these n integers to be consecutive.) Given n, can we find an integer X such that each of the integers $X + j, j = 0, 1, \ldots, n - 1$, has a nontrivial factor? One possibility is to let $X = n!$. Then $X = n!$ is composite, and, for $j = 2, 3, \ldots, n - 1$, we immediately see that each of the integers $X + j = n! + j$ is composite because j divides $n! + j$. However, we do not know whether $X + 1 = n! + 1$ is composite.

To get around this problem, we instead take $X = (n + 1)! + 1$. We see that the n consecutive integers $X + 1, X + 2, \ldots, X + n$, which are the integers

$$(n + 1)! + 2, (n + 1)! + 3, \ldots, (n + 1)! + (n + 1),$$

are all composite, because j divides $(n + 1)! + j$ for $j = 2, \ldots, n + 1$. Consequently, we have found n consecutive composite integers. It is a short task, which we will not do here, to rewrite the proof without showing our backward reasoning. ◄

CONJECTURE AND COUNTEREXAMPLES

Not all the plausible conjectures we make turn out to be true. To dispose of those that turn out to be false, we need to find counterexamples. Examples 8 and 9 illustrate this process.

EXAMPLE 8

Extra Examples

It would be useful to have a function $f(n)$ such that $f(n)$ is prime for all positive integers n. If we had such a function, we could find large primes for use in cryptography and other applications. Looking for such a function, we might check out different polynomial functions, as some mathematicians did several hundred years ago. After a lot of computation we may encounter the polynomial $f(n) = n^2 - n + 41$. This polynomial has the interesting property that $f(n)$ is prime for all positive integers n not exceeding 40. [We have $f(1) = 41, f(2) = 43, f(3) = 47, f(4) = 53$, and so on.] This can lead us to the conjecture that $f(n)$ is prime for all positive integers n. Can we settle this conjecture?

Solution: Perhaps not surprisingly, this conjecture turns out to be false; we do not have to look far to find a positive integer n for which $f(n)$ is composite, because $f(41) = 41^2 - 41 + 41 = 41^2$. Because $f(n) = n^2 - n + 41$ is prime for all positive integers n with $1 \leq n \leq 40$, we might be tempted to find a different polynomial with the property that $f(n)$ is prime for *all* positive integers n. However, there is no such polynomial. It can be shown that for every polynomial $f(n)$ with integer coefficients there is a positive integer y such that $f(y)$ is composite. (See Exercise 37.) ◄

Example 9 illustrates that surprisingly large counterexamples are sometimes needed. This reiterates the important point that even a vast amount of numerical calculation does not establish a theorem.

EXAMPLE 9

In the eighteenth century, the great Swiss mathematician Leonhard Euler conjectured that for every integer n with $n \geq 3$, the sum of $n - 1$ nth powers of positive integers cannot be an nth power. The numerical evidence found for close to two hundred years supported this conjecture. However, no proof could be found, even for a single value of n. Then in 1966, L. J. Lander and T. R. Parkin found a surprisingly small counterexample for $n = 5$. They showed that $27^5 + 84^5 + 110^5 + 133^5 = 144^5$.

This counterexample left open the possibility that the conjecture is true for other values of n with $n \geq 3$. But in 1988, Noam Elkies found a counterexample for $n = 4$. All known counterexamples are much larger than that found for the $n = 5$ case. The smallest counterexample known for $n = 4$ is $95{,}800^4 + 217{,}519^4 + 414{,}560^4 = 422{,}481^4$ (where by smallest we mean the smallest in terms of the sum of the integers whose fourth powers are being computed). All cases with $n \geq 6$ are still unsettled. Looking hard for a counterexample for the $n = 6$ case may pay off for someone, but it is possible (though considered unlikely by many) that the conjecture is true for $n = 6$.   ◄

THE ROLE OF OPEN PROBLEMS Many advances in mathematics have been made by people trying to solve famous unsolved problems. For instance, one famous problem unsolved for approximately three hundred years led to the development of an entire branch of number theory. This problem asked whether the statement known as **Fermat's Last Theorem** is true.

THEOREM 2

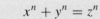
Links

FERMAT'S LAST THEOREM The equation

$$x^n + y^n = z^n$$

has no solutions in integers x, y, and z with $xyz \neq 0$ whenever n is an integer with $n > 2$.

Remark: The equation $x^2 + y^2 = z^2$ has infinitely many solutions in integers x, y, and z; these solutions are called Pythagorean triples and correspond to the lengths of the sides of right triangles with integer lengths. See Exercise 10.

This problem has a fascinating history. In the seventeenth century, Fermat jotted in the margin of his copy of the works of Diophantus that he had a "wondrous proof" that there are no nontrivial integer solutions of $x^n + y^n = z^n$ when n is an integer greater than 2. However, he never published a proof (Fermat published almost nothing) and no proof could be found in the papers he left when he died. Mathematicians looked for a proof for three centuries without success, although many people were convinced that a relatively simple proof could be found. (Proofs of special cases were found, such as the proof of the case when $n = 3$ by Euler and the proof of the $n = 4$ case by Fermat himself.) Over the years, several established mathematicians thought that they had proved this theorem. In the nineteenth century, one of these failed attempts led to the development of the part of number theory called algebraic number theory. A correct proof, requiring hundreds of pages of advanced mathematics, was not found until the 1990s, when Andrew Wiles used recently developed ideas from a sophisticated area of number theory called the theory of elliptic curves to prove Fermat's Last Theorem. Wiles's quest to find a proof of Fermat's Last Theorem using this powerful theory, documented in a program in the *Nova* series on public television, took close to ten years! (The interested reader should consult [Ro99] for more information about Fermat's Last Theorem and for additional references concerning this problem and its resolution.)

Many famous problems still await ultimate resolution by clever people. We describe a few of the most accessible and better known of these open problems in Examples 10–13.

EXAMPLE 10   **Goldbach's Conjecture** In 1742, Christian Goldbach, in a letter to Leonhard Euler, conjectured that every odd integer n, $n > 5$, is the sum of three primes. Euler replied that this conjecture is equivalent to the conjecture that every even integer n, $n > 2$, is the sum of two primes (see Exercise 39). The conjecture that every even integer $n, n > 2,$ is the sum of two primes is now called **Goldbach's conjecture.** We can check this conjecture for small even numbers. For example, $4 = 2 + 2, 6 = 3 + 3, 8 = 5 + 3, 10 = 7 + 3, 12 = 7 + 5,$ and so on. Goldbach's conjecture was verified by hand calculations for numbers up to the millions prior to the advent of computers. With computers it can be checked for extremely large numbers. As of mid-2002, the conjecture has been checked for all positive even integers up to $4 \cdot 10^{14}$.

Although no proof of Goldbach's conjecture has been found, most mathematicians believe it is true. Several theorems have been proved, using complicated methods from analytic number theory, establishing results weaker than Goldbach's conjecture. Among these are the result that every even positive integer greater than 2 is the sum of at most six primes (proved in 1995 by O. Ramaré) and that every sufficiently large positive integer is the sum of a prime and a number that is either prime or the product of two primes (proved in 1966 by J. R. Chen). Perhaps Goldbach's conjecture will be settled in the not too distant future. ◄

EXAMPLE 11

There are many conjectures asserting that there are infinitely many primes of certain forms. For example, there is a conjecture that there are infinitely many Mersenne primes. (Recall from Section 2.4 that these are primes of the form $2^p - 1$, where p is prime.) Another conjecture of this sort is the conjecture that there are infinitely many primes of the form $n^2 + 1$, where n is a positive integer. For example, $5 = 2^2 + 1, 17 = 4^2 + 1, 37 = 6^2 + 1$, and so on. The best result currently known is that there are infinitely many positive integers n such that $n^2 + 1$ is prime or the product of at most two primes (proved by Henryk Iwaniec in 1973). ◄

EXAMPLE 12

The Twin Prime Conjecture **Twin primes** are primes that differ by 2, such as 3 and 5, 5 and 7, 11 and 13, 17 and 19, and 4967 and 4969. The twin prime conjecture asserts that there are infinitely many twin primes. The strongest result proved concerning twin primes is that there are infinitely many pairs p and $p + 2$, where p is prime and $p + 2$ is prime or the product of two primes (proved by J. R. Chen in 1966). The world's record for twin primes, as of mid-2002, consists of the numbers $318,032,361 \cdot 2^{107,001} \pm 1$, numbers with 32,220 digits. ◄

EXAMPLE 13

The $3x + 1$ Conjecture Let $f(x)$ be the function that maps an even integer x to $x/2$ and an odd integer x to $3x + 1$. There is a famous conjecture, sometimes known as the **$3x + 1$ conjecture,** that states that for all positive integers x, when we repeatedly apply the function f we will eventually reach the integer 1. For example, starting with $x = 13$, we find $f(13) = 3 \cdot 13 + 1 = 40, f(40) = 40/2 = 20, f(20) = 20/2 = 10, f(10) = 10/2 = 5, f(5) = 3 \cdot 5 + 1 = 16, f(16) = 8, f(8) = 4, f(4) = 2,$ and $f(2) = 1$. The $3x + 1$ conjecture has been verified for all integers x up to $5.6 \cdot 10^{13}$.

The $3x + 1$ conjecture has an interesting history and has attracted the attention of mathematicians since the 1950s. The conjecture has been raised many times and goes by many other names, including the Collatz problem, Hasse's algorithm, Ulam's problem, the

CHRISTIAN GOLDBACH (1690–1764) Christian Goldbach was born in Königsberg, Prussia, the city noted for its famous bridge problem (which will be studied in Section 8.5). He became professor of mathematics at the Academy in St. Petersburg in 1725. In 1728 Goldbach went to Moscow to tutor the son of the Tsar. He entered the world of politics when, in 1742, he became a staff member in the Russian Ministry of Foreign Affairs. Goldbach is best known for his correspondence with eminent mathematicians, including Euler and Bernoulli, for his famous conjectures in number theory, and for several contributions to analysis.

Syracuse problem, and Kakutani's problem. Many mathematicians have been diverted from their work to spend time attacking this conjecture. This led to the joke that this problem was part of a conspiracy to slow down American mathematical research. See the article by Jeffrey Lagarias [La85] for a fascinating discussion of this problem and the results that have been found by mathematicians attacking it. ◀

THE HALTING PROBLEM

We will now describe a proof of one of the most famous theorems in computer science. We will show that there is a problem that cannot be solved using any procedure. That is, we will show there are unsolvable problems, as was mentioned in Section 2.3. The problem we will study is the **halting problem.** It asks whether there is a procedure that does this: It takes as input a computer program and input to the program and determines whether the program will eventually stop when run with this input. It would be convenient to have such a procedure, if it existed. Certainly being able to test whether a program entered into an infinite loop would be helpful when writing and debugging programs. However, in 1936 Alan Turing showed that no such procedure exists (see his biography in Section 11.4).

Before we present a proof that the halting problem is unsolvable, first note that we cannot simply run a program and observe what it does to determine whether it terminates when run with the given input. If the program halts, we have our answer, but if it is still running after any fixed length of time has elapsed, we do not know whether it will never halt or we just did not wait long enough for it to terminate. After all, it is not hard to design a program that will stop only after more than a billion years has elapsed.

We will describe Turing's proof that the halting problem is unsolvable; it is a proof by contradiction. (The reader should note that our proof is not completely rigorous, since we have not explicitly defined what a procedure is. To remedy this, the concept of a Turing machine is needed. This concept is introduced in Section 11.5.)

Proof: Assume there is a solution to the halting problem, a procedure called $H(P, I)$. The procedure $H(P, I)$ takes two inputs, one a program P and the other I, an input to the program P. $H(P, I)$ generates the string "halt" as output if H determines that P stops when given I as input. Otherwise, $H(P, I)$ generates the string "loops forever" as output. We will now derive a contradiction.

When a procedure is coded, it is expressed as a string of characters; this string can be interpreted as a sequence of bits. This means that a program itself can be used as data. Therefore a program can be thought of as input to another program, or even itself. Hence, H can take a program P as both of its inputs, which are a program and input to this program. H should be able to determine if P will halt when it is given a copy of itself as input.

To show that no procedure H exists that solves the halting problem, we construct a simple procedure $K(P)$, which works as follows, making use of the output $H(P, P)$. If the output of $H(P, P)$ is "loops forever," which means that P loops forever when given a copy of itself as input, then $K(P)$ halts. If the output of $H(P, P)$ is "halt," which means that P halts when given a copy of itself as input, then $K(P)$ loops forever. That is, $K(P)$ does the opposite of what the output of $H(P, P)$ specifies. (See Figure 1.)

Now suppose we provide K as input to K. We note that if the output of $H(K, K)$ is "loops forever," then by the definition of K we see that $K(K)$ halts. Otherwise, if the output of $H(K, K)$ is "halt," then by the definition of K we see that $K(K)$ loops forever, in violation of what H tells us. In both cases, we have a contradiction.

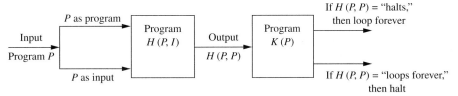

FIGURE 1

Thus, *H* cannot always give the correct answers. Consequently, there is no procedure that solves the halting problem.   ◁

ADDITIONAL PROOF METHODS

In Chapter 1 we introduced the basic methods used in proofs and in this section we described how to leverage these methods to prove a variety of results. However, there are many important proof methods besides those we have covered. We will introduce some of these methods later in this book. In particular, in Section 3.3 we will discuss mathematical induction, which is an extremely useful method for proving statements of the form $\forall n\, P(n)$, where the universe of discourse is the set of positive integers. In Section 3.4 we will introduce structural induction, which can be used to prove results about recursively defined sets. We will use the Cantor diagonalization method, which can be used to prove results about the size of infinite sets, in Section 3.2. Finally, in Chapter 4 we will introduce the notion of combinatorial proofs, which can be used to prove results by counting arguments. The reader should note that entire books have been devoted to the activities discussed in this section, including many excellent works by George Pólya ([Po61], [Po71], [Po90]).

Exercises

1. Prove that the product of any three consecutive integers is divisible by 6.

2. Prove that if *n* is an odd positive integer, then $n^2 \equiv 1 \pmod{8}$.

3. Prove that if *n* is an odd positive integer, then $n^4 \equiv 1 \pmod{16}$.

4. Prove that there are no solutions in integers *x* and *y* to the equation $2x^2 + 5y^2 = 14$.

5. Prove that there are no solutions in positive integers *x* and *y* to the equation $x^4 + y^4 = 625$.

6. Prove that there are no integers *x* and *y* such that $x^2 - 4y = 3$.

7. Prove that there are no integers *x* and *y* such that $3x^2 - 8y = 1$.

8. Prove that there are no solutions in integers *x* and *y* to the equation $x^2 - 5y^2 = 2$. (*Hint:* Consider this equation modulo 5.)

9. Prove that there are no solutions in integers *x* and *y* to the equation $x^4 - 16y^4 = 3$. (*Hint:* Look at the congruence obtained by considering this equation modulo 16 and use Exercise 3.)

10. Prove that there are infinitely many solutions in positive integers *x*, *y*, and *z* to the equation $x^2 + y^2 = z^2$. (*Hint:* Let $x = m^2 - n^2$, $y = 2mn$, and $z = m^2 + n^2$, where *m* and *n* are integers.)

11. The **harmonic mean** of two integers *a* and *b* equals $\frac{2ab}{a+b}$. By computing the harmonic and geometric means of different pairs of positive integers, formulate a conjecture about their relative sizes and prove your conjecture.

12. The **quadratic mean** of two integers *a* and *b* equals $\sqrt{\frac{a^2+b^2}{2}}$. By computing the arithmetic and quadratic means of different pairs of positive integers, formulate a conjecture about their relative sizes and prove your conjecture.

*13. Write the numbers $1, 2, \ldots, 2n$ on a blackboard, where n is an odd integer. Pick any two of the numbers, j and k, write $|j - k|$ on the board and erase j and k. Continue this process until only one integer is written on the board. Prove that this integer must be odd.

*14. Suppose that five ones and four zeros are arranged around a circle. Between any two equal bits you insert a 0 and between any two unequal bits you insert a 1 to produce nine new bits. Then you erase the nine original bits. Show that when you iterate this procedure, you can never get nine zeros. (*Hint:* Work backwards, assuming that you did end up with nine zeros.)

*15. Prove or disprove that $n^2 - 79n + 1601$ is prime whenever n is a positive integer.

16. Prove or disprove that $2^n + 1$ is prime for all non-negative integers n.

17. Prove or disprove that $\sqrt[3]{3}$ is irrational.

*18. Show that $\sqrt{n}$ is irrational if n is a positive integer that is not a perfect square.

*19. Prove or disprove that there is no rational number x such that $x^3 + x + 1 = 0$.

20. Prove that the square of an integer not divisible by 5 leaves a remainder of 1 or 4 when divided by 5. (*Hint:* Use a proof by cases.)

*21. Let p be prime. Prove that $a^2 \equiv b^2 \pmod{p}$ if and only if $a \equiv b \pmod{p}$ or $a \equiv -b \pmod{p}$.

22. Prove or disprove that $n^2 - 1$ is composite whenever n is a positive integer greater than 1.

☞ 23. Prove that $(a \bmod m)(b \bmod m) \bmod m = ab \bmod m$ for all integers a and b whenever m is a positive integer.

24. Prove or disprove that $a \bmod m + b \bmod m = (a + b) \bmod m$ for all integers a and b whenever m is a positive integer.

25. Prove or disprove that there are three consecutive odd positive integers that are primes, that is, odd primes of the form p, $p + 2$, and $p + 4$.

26. Prove or disprove that given a positive integer n, there are n consecutive odd positive integers that are primes.

27. Give a constructive proof of the proposition: "For every positive integer n there is an integer divisible by more than n primes."

28. Find a counterexample to the proposition: "For every prime number n, $n + 2$ is prime."

29. Adapt the proof given in Example 5 to show that there are infinitely many primes of the form $6k + 5$.

30. What goes wrong when you try to adapt the proof that there are infinitely many primes to show that there are infinitely many primes of the form $4k + 1$?

*31. Prove that if x is a real number and n is a positive integer, then $\lfloor \frac{x}{3} \rfloor + \lfloor \frac{x+n}{3} \rfloor + \lfloor \frac{x+2n}{3} \rfloor = \gcd(3, n) \lfloor \frac{x}{\gcd(3, n)} \rfloor + (n - 1) + \frac{\gcd(3, n) - 1}{2}$.

32. Prove that $\lfloor n/2 \rfloor \lceil n/2 \rceil = \lfloor n^2/4 \rfloor$ for all integers n.

*33. Let $S = x_1 y_1 + x_2 y_2 + \cdots + x_n y_n$, where $x_1, x_2, \ldots, x_n$ and $y_1, y_2, \ldots, y_n$ are orderings of two different sequences of positive real numbers, each containing n elements.

a) Show that S takes its maximum value over all orderings of the two sequences when both sequences are sorted (so that the elements in each sequence are in nondecreasing order).

b) Show that S takes its minimum value over all orderings of the two sequences when one sequence is sorted into nondecreasing order and the other is sorted into nonincreasing order.

34. Prove that if m and n are positive integers and x is a real number, then

$$\left\lfloor \frac{\lfloor x \rfloor + n}{m} \right\rfloor = \left\lfloor \frac{x + n}{m} \right\rfloor.$$

*35. Prove that if m is a positive integer and x is a real number, then

$$\lfloor mx \rfloor = \lfloor x \rfloor + \left\lfloor x + \frac{1}{m} \right\rfloor + \left\lfloor x + \frac{2}{m} \right\rfloor + \cdots$$
$$+ \left\lfloor x + \frac{m - 1}{m} \right\rfloor.$$

**36. Show that if a and b are positive irrational numbers such that $1/a + 1/b = 1$, then every positive integer can be uniquely expressed as either $\lfloor ka \rfloor$ or $\lfloor kb \rfloor$ for some positive integer k.

*37. Prove that if $f(x)$ is a nonconstant polynomial with integer coefficients, then there is an integer y such that $f(y)$ is composite. [*Hint:* Assume that $f(x_0) = p$ is prime. Show that p divides $f(x_0 + kp)$ for all integers k. Obtain a contradiction of the fact that a polynomial of degree n, where $n > 1$, takes on each value at most n times.]

38. Prove that if n is a positive integer such that the sum of its divisors is $n + 1$, then n is prime.

39. Show that Goldbach's conjecture, which states that every even integer greater than 2 is the sum of two primes, is equivalent to the statement that every integer greater than 5 is the sum of three primes.

40. A conjecture known as **de Polignac's conjecture** asserts that every positive even integer can be written as the difference of two consecutive primes in infinitely many ways. Show that de Polignac's conjecture implies the twin prime conjecture.

41. Verify the $3x + 1$ conjecture for these integers.
   a) 6   **b)** 7   **c)** 17   **d)** 21

42. Verify the $3x + 1$ conjecture for these integers.
   a) 16   **b)** 11   **c)** 35   **d)** 113

A **perfect number** is a positive integer that equals the sum of its proper divisors (that is, divisors other than itself).

43. Show that 6, 28, and 496 are perfect.

44. Prove that if $\gcd(s, t) = 1$, where s and t are positive integers, then the sum of the divisors of st is the prod-

uct of the sum of the divisors of s and the sum of the divisors of t.

45. Prove that the integer $2^{p-1}(2^p - 1)$ is perfect when $2^p - 1$ is a Mersenne prime.

**46. Prove that if n is an even integer that is perfect, then $n = 2^p(2^p - 1)$, where $2^p - 1$ is a Mersenne prime.

47. Prove or disprove that if you have an eight-gallon jug of water and two empty jugs with capacities of five gallons and three gallons, respectively, then you can measure four gallons by successively pouring some of or all of the water in a jug into another jug.

*48. Prove or disprove that if n is a positive integer, then $\lfloor \sqrt{n} + \sqrt{n+1} \rfloor = \lfloor \sqrt{4n+2} \rfloor$.

49. Show that the problem of determining whether a program with a given input ever prints the digit 1 is unsolvable.

50. Show that the problem of deciding whether a specific program with a specific input halts is solvable.

51. Show that the following problem is solvable. Given two programs with their inputs and the knowledge that exactly one of them halts, determine which halts.

3.2 Sequences and Summations

INTRODUCTION

Sequences are used to represent ordered lists of elements. Sequences are used in discrete mathematics in many ways. They can be used to represent solutions to certain counting problems, as we will see in Chapter 6. They are also an important data structure in computer science. This section contains a review of the concept of a function, as well as the notation used to represent sequences and sums of terms of sequences.

When the elements of an infinite set can be listed, the set is called countable. We will conclude this section with a discussion of both countable and uncountable sets. We will prove that the set of rational numbers is countable, but the set of real numbers is not.

SEQUENCES

A sequence is a discrete structure used to represent an ordered list.

DEFINITION 1

A *sequence* is a function from a subset of the set of integers (usually either the set $\{0, 1, 2, \ldots\}$ or the set $\{1, 2, 3, \ldots\}$) to a set S. We use the notation a_n to denote the image of the integer n. We call a_n a *term* of the sequence.

We use the notation $\{a_n\}$ to describe the sequence. (Note that a_n represents an individual term of the sequence $\{a_n\}$. Also note that the notation $\{a_n\}$ for a sequence conflicts with the notation for a set. However, the context in which we use this notation will always make it clear when we are dealing with sets and when we are dealing with sequences. Note also that the choice of the letter a is arbitrary.)

We describe sequences by listing the terms of the sequence in order of increasing subscripts.

EXAMPLE 1   Consider the sequence $\{a_n\}$, where

$$a_n = 1/n.$$

The list of the terms of this sequence, beginning with a_1, namely,

$$a_1, a_2, a_3, a_4, \ldots,$$

starts with

$$1, \frac{1}{2}, \frac{1}{3}, \frac{1}{4}, \ldots.$$ ◄

DEFINITION 2

A *geometric progression* is a sequence of the form

$$a, ar, ar^2, \ldots, ar^n,$$

where the *initial term a* and the *common ratio r* are real numbers.

Remark: A geometric progression is a discrete analogue of the exponential function $f(x) = ar^x$.

EXAMPLE 2
The sequences $\{b_n\}$ with $b_n = (-1)^n$, $\{c_n\}$ with $c_n = 2 \cdot 5^n$, and $\{d_n\}$ with $d_n = 6 \cdot (1/3)^n$ are geometric progressions with initial term and common ratio equal to -1 and -1; 10 and 5; and 2 and 1/3, respectively. The list of terms $b_1, b_2, b_3, b_4, \ldots$ begins with

$$-1, 1, -1, 1, \ldots;$$

the list of terms $c_1, c_2, c_3, c_4, \ldots$ begins with

$$10, 50, 250, 1250, \ldots;$$

and the list of terms $d_1, d_2, d_3, d_4, \ldots$ begins with

$$2, 2/3, 2/9, 2/27, \ldots.$$ ◄

DEFINITION 3

An *arithmetic progression* is a sequence of the form

$$a, a + d, a + 2d, \ldots, a + nd,$$

where the *initial term a* and the *common difference d* are real numbers.

Remark: An arithmetic progression is a discrete analogue of the linear function $f(x) = dx + a$.

EXAMPLE 3
The sequences $\{s_n\}$ with $s_n = -1 + 4n$ and $\{t_n\}$ with $t_n = 7 - 3n$ are both arithmetic progressions with initial terms and common differences equal to -1 and 4, and 7 and -3, respectively. The list of terms, starting with the term with $n = 0$, $s_0, s_1, s_2, s_3, \ldots$ begins with

$$-1, 3, 7, 11, \ldots,$$

and the list of terms $t_0, t_1, t_2, t_3, \ldots$ begins with

$$7, 4, 1, -2, \ldots.$$ ◄

Sequences of the form $a_1, a_2, \ldots, a_n$ are often used in computer science. These finite sequences are also called **strings.** This string is also denoted by $a_1 a_2 \cdots a_n$. (Recall that bit strings, which are finite sequences of bits, were introduced in Section 1.1.) The **length** of the string S is the number of terms in this string. The **empty string,** denoted by λ, is the string that has no terms. The empty string has length zero.

EXAMPLE 4
The string *abcd* is a string of length four. ◄

SPECIAL INTEGER SEQUENCES

A common problem in discrete mathematics is finding a formula or a general rule for constructing the terms of a sequence. Sometimes only a few terms of a sequence solving a problem are known; the goal is to identify the sequence. Even though the initial terms of a sequence do not determine the entire sequence (after all, there are infinitely many different sequences that start with any finite set of initial terms), knowing the first few terms may help you make an educated conjecture about the identity of your sequence. Once you have made this conjecture, you can try to verify that you have the correct sequence.

When trying to deduce a possible formula or rule for the terms of a sequence from the initial terms, try to find a pattern in these terms. You might also see whether you can determine how a term might have been produced from those preceding it. There are many questions you could ask, but some of the more useful are:

- Are there runs of the same value?
- Are terms obtained from previous terms by adding the same amount or an amount that depends on the position in the sequence?
- Are terms obtained from previous terms by multiplying by a particular amount?
- Are terms obtained by combining previous terms in a certain way?
- Are there cycles among the terms?

EXAMPLE 5   Find formulas for the sequences with the following first five terms: (a) $1, 1/2, 1/4, 1/8, 1/16$ (b) $1, 3, 5, 7, 9$ (c) $1, -1, 1, -1, 1$.

Solution: (a) We recognize that the denominators are powers of 2. The sequence with $a_n = 1/2^{n-1}$ is a possible match. This proposed sequence is a geometric progression with $a = 1$ and $r = 1/2$.

(b) We note that each term is obtained by adding 2 to the previous term. The sequence with $a_n = 2n - 1$ is a possible match. This proposed sequence is an arithmetic progression with $a = 1$ and $d = 2$.

(c) The terms alternate between 1 and -1. The sequence with $a_n = (-1)^{n+1}$ is a possible match. This proposed sequence is a geometric progression with $a = 1$ and $r = -1$.   ◀

Examples 6 and 7 illustrate how we can analyze sequences to find how the terms are constructed.

EXAMPLE 6   How can we produce the terms of a sequence if the first 10 terms are $1, 2, 2, 3, 3, 3, 4, 4, 4, 4$?

Solution: Note that the integer 1 appears once, the integer 2 appears twice, the integer 3 appears three times, and the integer 4 appears four times. A reasonable rule for generating this sequence is that the integer n appears exactly n times, so the next five terms of the sequence would all be 5, the following six terms would all be 6, and so on. The sequence generated this way is a possible match.   ◀

EXAMPLE 7   How can we produce the terms of a sequence if the first 10 terms are $5, 11, 17, 23, 29, 35, 41, 47, 53, 59$?

Solution: Note that each of the first 10 terms of this sequence after the first is obtained by adding 6 to the previous term. (We could see this by noticing that the difference between consecutive terms is 6.) Consequently, the nth term could be produced by starting with 5 and adding 6 a total of $n - 1$ times; that is, a reasonable guess is that the nth term is $5 + 6(n - 1) = 6n - 1$. (This is an arithmetic progression with $a = 5$ and $d = 6$.) ◄

Another useful technique for finding a rule for generating the terms of a sequence is to compare the terms of a sequence of interest with the terms of a well-known integer sequence, such as terms of an arithmetic progression, terms of a geometric progression, perfect squares, perfect cubes, and so on. The first 10 terms of some sequences you may want to keep in mind are displayed in Table 1.

EXAMPLE 8   Conjecture a simple formula for a_n if the first 10 terms of the sequence $\{a_n\}$ are $1, 7, 25,$ $79, 241, 727, 2185, 6559, 19681, 59047$.

Solution: To attack this problem, we begin by looking at the difference of consecutive terms, but we do not see a pattern. When we form the ratio of consecutive terms to see whether each term is a multiple of the previous term, we find that this ratio, although not a constant, is close to 3. So it is reasonable to suspect that the terms of this sequence are generated by a formula involving 3^n. Comparing these terms with the corresponding terms of the sequence $\{3^n\}$, we notice that the nth term is 2 less than the corresponding power of 3. We see that $a_n = 3^n - 2$ for $1 \leq n \leq 10$ and conjecture that this formula holds for all n. ◄

We will see throughout this text that integer sequences appear in a wide range of contexts in discrete mathematics. Sequences we have or will encounter include the sequence of prime numbers (Chapter 2), the number of ways to order n discrete objects (Chapter 4), the number of the moves required to solve the famous Tower of Hanoi puzzle with n disks (Chapter 6), and the number of rabbits on an island after n months (Chapter 6).

Integer sequences appear in an amazingly wide range of subject areas besides discrete mathematics, including biology, engineering, chemistry, and physics, as well as in puzzles. A wonderfully diverse collection of over 8000 different integer sequences has been constructed over the past 20 years by the mathematician Neil Sloane, who has teamed up with Simon Plouffe, to produce *The Encyclopedia of Integer Sequences* ([SlPl95]). An extended list of the sequences is available on the Web, with new sequences added regularly. There is also a program accessible via the Web that you can use to find sequences from the encyclopedia that match initial terms you provide.

Links

TABLE 1 Some Useful Sequences.	
nth Term	*First 10 Terms*
n^2	$1, 4, 9, 16, 25, 36, 49, 64, 81, 100, \ldots$
n^3	$1, 8, 27, 64, 125, 216, 343, 512, 729, 1000, \ldots$
n^4	$1, 16, 81, 256, 625, 1296, 2401, 4096, 6561, 10000, \ldots$
2^n	$2, 4, 8, 16, 32, 64, 128, 256, 512, 1024, \ldots$
3^n	$3, 9, 27, 81, 243, 729, 2187, 6561, 19683, 59049, \ldots$
$n!$	$1, 2, 6, 24, 120, 720, 5040, 40320, 362880, 3628800, \ldots$

SUMMATIONS

Next, we introduce **summation notation.** We begin by describing the notation used to express the sum of the terms

$$a_m, a_{m+1}, \ldots, a_n$$

from the sequence $\{a_n\}$. We use the notation

$$\sum_{j=m}^{n} a_j \quad \text{or} \quad \sum_{j=m}^{n} a_j$$

to represent

$$a_m + a_{m+1} + \cdots + a_n.$$

Here, the variable j is called the **index of summation,** and the choice of the letter j as the variable is arbitrary; that is, we could have used any other letter, such as i or k. Or, in notation,

$$\sum_{j=m}^{n} a_j = \sum_{i=m}^{n} a_i = \sum_{k=m}^{n} a_k.$$

Here, the index of summation runs through all integers starting with its **lower limit** m and ending with its **upper limit** n. The uppercase Greek letter sigma, Σ, is used to denote summation. We give some examples of summation notation.

NEIL SLOANE (BORN 1939)   Neil Sloane studied mathematics and electrical engineering at the University of Melbourne on a scholarship from the Australian state telephone company. He mastered many telephone-related jobs, such as erecting telephone poles, in his summer work. After graduating, he designed minimal cost telephone networks in Australia. In 1962 he came to the United States and studied electrical engineering at Cornell University. His Ph.D. thesis was on what are now called neural networks. He took a job at Bell Labs in 1969, working in many areas, including network design, coding theory, and sphere packing. He now works for AT&T Labs, moving there from Bell Labs when AT&T split up in 1996. One of his favorite problems is the **kissing problem** (a name he coined), which asks how many spheres can be arranged in n dimensions so that they all touch a central sphere of the same size. (In two dimensions the answer is 6, since 6 pennies can be placed so that they touch a central penny. In three dimensions, 12 billiard balls can be placed so that they touch a central billiard ball. Two billiard balls that just touch are said to "kiss," giving rise to the terminology "kissing problem" and "kissing number.") Sloane, together with Andrew Odlyzko, showed that in 8 and 24 dimensions the optimal kissing numbers are, respectively, 240 and 196,560. The kissing number is known in dimensions 1, 2, 3, 8, and 24, but not in any other dimensions. Sloane's books include *Sphere Packings, Lattices and Groups,* 3d ed., with John Conway; *The Theory of Error-Correcting Codes* with Jessie MacWilliams; *The Encyclopedia of Integer Sequences* with Simon Plouffe; and *The Rock-Climbing Guide to New Jersey Crags* with Paul Nick. The last book demonstrates his interest in rock climbing; it includes more than 50 climbing sites in New Jersey.

EXAMPLE 9

Express the sum of the first 100 terms of the sequence $\{a_n\}$, where $a_n = 1/n$ for $n = 1, 2, 3, \ldots$.

Solution: The lower limit for the index of summation is 1, and the upper limit is 100. We write this sum as

$$\sum_{j=1}^{100} \frac{1}{j}.$$ ◀

EXAMPLE 10

What is the value of $\sum_{j=1}^{5} j^2$?

Solution: We have

$$\sum_{j=1}^{5} j^2 = 1^2 + 2^2 + 3^2 + 4^2 + 5^2$$
$$= 1 + 4 + 9 + 16 + 25$$
$$= 55.$$ ◀

EXAMPLE 11

What is the value of $\sum_{k=4}^{8} (-1)^k$?

Solution: We have

$$\sum_{k=4}^{8} (-1)^k = (-1)^4 + (-1)^5 + (-1)^6 + (-1)^7 + (-1)^8$$
$$= 1 + (-1) + 1 + (-1) + 1$$
$$= 1.$$ ◀

Sometimes it is useful to shift the index of summation in a sum. This is often done when two sums need to be added but their indices of summation do not match. When shifting an index of summation, it is important to make the appropriate changes in the corresponding summand. This is illustrated by the following example.

EXAMPLE 12

Suppose we have the sum

$$\sum_{j=1}^{5} j^2$$

but want the index of summation to run between 0 and 4 rather than from 1 to 5. To do this, we let $k = j - 1$. Then the new summation index runs from 0 to 4, and the term j^2 becomes $(k + 1)^2$. Hence

$$\sum_{j=1}^{5} j^2 = \sum_{k=0}^{4} (k + 1)^2.$$

It is easily checked that both sums are $1 + 4 + 9 + 16 + 25 = 55$.   ◀

Sums of terms of geometric progressions commonly arise (such sums are called **geometric series**). Theorem 1 gives us a formula for the sum of terms of a geometric progression.

THEOREM 1   If a and r are real numbers and $r \neq 0$, then

$$\sum_{j=0}^{n} ar^j = \begin{cases} \dfrac{ar^{n+1} - a}{r - 1} & \text{if } r \neq 1 \\[2ex] (n + 1)a & \text{if } r = 1. \end{cases}$$

Proof: Let

$$S = \sum_{j=0}^{n} ar^j.$$

To compute S, first multiply both sides of the equality by r and then manipulate the resulting sum as follows:

$$
\begin{aligned}
rS &= r \sum_{j=0}^{n} ar^j \\
&= \sum_{j=0}^{n} ar^{j+1} \\
&= \sum_{k=1}^{n+1} ar^k \\
&= \sum_{k=0}^{n} ar^k + (ar^{n+1} - a) \qquad \text{We obtain this equality by shifting the index of} \\
&\qquad\qquad\qquad\qquad\qquad\qquad \text{summation, setting } k = j + 1. \\
&= S + (ar^{n+1} - a).
\end{aligned}
$$

From these equalities, we see that

$$rS = S + (ar^{n+1} - a).$$

Solving for S shows that if $r \neq 1$

$$S = \frac{ar^{n+1} - a}{r - 1}.$$

If $r = 1$, then clearly the sum equals $(n + 1)a$.   ◁

EXAMPLE 13   Double summations arise in many contexts (as in the analysis of nested loops in computer programs). An example of a double summation is

$$\sum_{i=1}^{4} \sum_{j=1}^{3} ij.$$

To evaluate the double sum, first expand the inner summation and then continue by computing the outer summation:

$$
\begin{aligned}
\sum_{i=1}^{4} \sum_{j=1}^{3} ij &= \sum_{i=1}^{4} (i + 2i + 3i) \\
&= \sum_{i=1}^{4} 6i \\
&= 6 + 12 + 18 + 24 = 60.
\end{aligned}
$$

◀

We can also use summation notation to add all values of a function, or terms of an indexed set, where the index of summation runs over all values in a set. That is, we write

$$\sum_{s \in S} f(s)$$

to represent the sum of the values $f(s)$, for all members s of S.

EXAMPLE 14   What is the value of $\sum_{s \in \{0,2,4\}} s$?

Solution: Since $\sum_{s \in \{0,2,4\}} s$ represents the sum of the values of s for all the members of the set $\{0, 2, 4\}$, it follows that

$$\sum_{s \in \{0,2,4\}} s = 0 + 2 + 4 = 6.$$ ◀

Certain sums arise repeatedly throughout discrete mathematics. Having a collection of formulae for such sums can be useful, so Table 2 provides a small table of formulae for commonly occurring sums.

We derived the first formula in this table in Theorem 1. The next three formulae give us the sum of the first n positive integers, the sum of their squares, and the sum of their cubes. These three formulae can be derived in many different ways (for example, see Exercises 21 and 22 at the end of this section). Also note that each of these formulae, once known, can easily be proved using mathematical induction, the subject of Section 3.3. The last two formulae in the table involve infinite series and will be discussed shortly.

Example 15 illustrates how the formulae in Table 2 can be useful.

EXAMPLE 15   Find $\sum_{k=50}^{100} k^2$.

TABLE 2 Some Useful Summation Formulae.	
Sum	***Closed Form***
$\sum_{k=0}^{n} ar^k \ (r \neq 0)$	$\dfrac{ar^{n+1} - a}{r - 1}, r \neq 1$
$\sum_{k=1}^{n} k$	$\dfrac{n(n + 1)}{2}$
$\sum_{k=1}^{n} k^2$	$\dfrac{n(n + 1)(2n + 1)}{6}$
$\sum_{k=1}^{n} k^3$	$\dfrac{n^2(n + 1)^2}{4}$
$\sum_{k=0}^{\infty} x^k, |x| < 1$	$\dfrac{1}{1 - x}$
$\sum_{k=1}^{\infty} kx^{k-1}, |x| < 1$	$\dfrac{1}{(1 - x)^2}$

Solution: First note that since $\sum_{k=1}^{100} k^2 = \sum_{k=1}^{49} k^2 + \sum_{k=50}^{100} k^2$, we have

$$\sum_{k=50}^{100} k^2 = \sum_{k=1}^{100} k^2 - \sum_{k=1}^{49} k^2.$$

Using the formula $\sum_{k=1}^{n} k^2 = n(n+1)(2n+1)/6$ from Table 2, we see that

$$\sum_{k=50}^{100} k^2 = \frac{100 \cdot 101 \cdot 201}{6} - \frac{49 \cdot 50 \cdot 99}{6} = 338{,}350 - 40{,}425 = 297{,}925. \quad \blacktriangleleft$$

SOME INFINITE SERIES Although most of the summations in this book are finite sums, infinite series are important in some parts of discrete mathematics. The closed forms for the infinite series in Examples 16 and 17 are quite useful.

EXAMPLE 16 (Requires calculus) Let x be a real number with $|x| < 1$. Find $\sum_{n=0}^{\infty} x^n$.

Solution: By Theorem 1 with $a = 1$ and $r = x$ we see that $\sum_{n=0}^{k} x^n = \frac{x^{k+1} - 1}{x - 1}$. Because $|x| < 1$, x^{k+1} approaches 0 as k approaches infinity. It follows that

$$\sum_{n=0}^{\infty} x^n = \lim_{k \to \infty} \frac{x^{k+1} - 1}{x - 1} = \frac{-1}{x - 1} = \frac{1}{1 - x}. \quad \blacktriangleleft$$

We can produce new summation formulae by differentiating or integrating existing formulae.

EXAMPLE 17 (Requires calculus) Differentiating both sides of the equation

$$\sum_{k=0}^{\infty} x^k = \frac{1}{1 - x},$$

from Example 16, we find that

$$\sum_{k=1}^{\infty} k x^{k-1} = \frac{1}{(1 - x)^2}.$$

(This differentiation is valid for $|x| < 1$ by a theorem about infinite series.) $\quad \blacktriangleleft$

CARDINALITY

Recall that in Section 1.6, the cardinality of a finite set was defined to be the number of elements in the set. It is possible to extend the concept of cardinality to all sets, both finite and infinite, with Definition 4.

DEFINITION 4

The sets A and B have the same *cardinality* if and only if there is a one-to-one correspondence from A to B.

To see that this definition agrees with the previous definition of the cardinality of a finite set as the number of elements in that set, note that there is a one-to-one correspondence between any two finite sets with n elements, where n is a nonnegative integer.

We will now split infinite sets into two groups, those with the same cardinality as the set of natural numbers and those with different cardinality.

DEFINITION 5

A set that is either finite or has the same cardinality as the set of positive integers is called *countable*. A set that is not countable is called *uncountable*.

We now give examples of countable and uncountable sets.

EXAMPLE 18

Show that the set of odd positive integers is a countable set.

Solution: To show that the set of odd positive integers is countable, we will exhibit a one-to-one correspondence between this set and the set of positive integers. Consider the function

$$f(n) = 2n - 1$$

from $\mathbf{Z}^+$ to the set of odd positive integers. We show that f is a one-to-one correspondence by showing that it is both one-to-one and onto. To see that it is one-to-one, suppose that $f(n) = f(m)$. Then $2n - 1 = 2m - 1$, so that $n = m$. To see that it is onto, suppose that t is an odd positive integer. Then t is 1 less than an even integer $2k$, where k is a natural number. Hence $t = 2k - 1 = f(k)$. We display this one-to-one correspondence in Figure 1. ◀

An infinite set is countable if and only if it is possible to list the elements of the set in a sequence (indexed by the positive integers). The reason for this is that a one-to-one correspondence f from the set of positive integers to a set S can be expressed in terms of a sequence $a_1, a_2, \ldots, a_n, \ldots$, where $a_1 = f(1), a_2 = f(2), \ldots, a_n = f(n), \ldots$. For instance, the set of odd integers can be listed in a sequence $a_1, a_2, \ldots, a_n, \ldots$, where $a_n = 2n - 1$.

EXAMPLE 19

Show that the set of positive rational numbers is countable.

Solution: It may seem surprising that the set of positive rational numbers is countable, but we will show how we can list the positive rational numbers as a sequence $r_1, r_2, \ldots, r_n, \ldots$. First, note that every positive rational number is the quotient p/q of two positive integers. We can arrange the positive rational numbers by listing those with denominator $q = 1$ in the first row, those with denominator $q = 2$ in the second row, and so on, as displayed in Figure 2.

The key to listing the rational numbers in a sequence is to first list the positive rational numbers p/q with $p + q = 2$, followed by those with $p + q = 3$, followed by those with $p + q = 4$, and so on, following the path shown in Figure 2. Whenever we encounter a number p/q that is already listed, we do not list it again. For example, when we come to $2/2 = 1$ we do not list it since we have already listed $1/1 = 1$. The initial terms in the list of positive rational numbers we have constructed are 1, 1/2, 2, 3, 1/3, 1/4, 2/3, 3/2, 4, 5, and so on. Because all rational numbers are listed once, as the reader can verify, we have shown that the set of rational numbers is countable. ◀

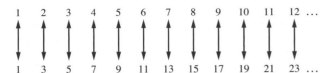

FIGURE 1   **A One-to-One Correspondence Between $\mathbf{Z}^+$ and the Set of Odd Positive Integers.**

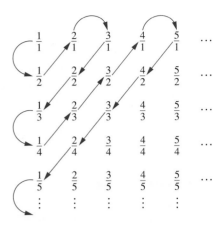

FIGURE 2   **The Positive Rational Numbers Are Countable.**

Example 20 shows that the set of real numbers is uncountable. Georg Cantor discovered this fact in 1879. We use an important proof method, known as the **Cantor diagonalization argument,** to prove that the set of real numbers is not countable. This proof method is used extensively in mathematical logic and in the theory of computation.

EXAMPLE 20   Show that the set of real numbers is an uncountable set.

Solution: To show that the set of real numbers is uncountable, we suppose that the set of real numbers is countable and arrive at a contradiction. Then, the subset of all real numbers that fall between 0 and 1 would also be countable (since any subset of a countable set is also countable; see Exercise 34 at the end of the section). Under this assumption, the real numbers between 0 and 1 can be listed in some order, say, $r_1, r_2, r_3, \ldots$. Let the decimal representation of these real numbers be

$$r_1 = 0.d_{11}d_{12}d_{13}d_{14} \ldots$$
$$r_2 = 0.d_{21}d_{22}d_{23}d_{24} \ldots$$
$$r_3 = 0.d_{31}d_{32}d_{33}d_{34} \ldots$$
$$r_4 = 0.d_{41}d_{42}d_{43}d_{44} \ldots$$
$$\vdots$$

where $d_{ij} \in \{0, 1, 2, 3, 4, 5, 6, 7, 8, 9\}$. (For example, if $r_1 = 0.23794102\ldots$, we have $d_{11} = 2, d_{12} = 3, d_{13} = 7$, and so on.) Then, form a new real number with decimal expansion $r = 0.d_1d_2d_3d_4 \ldots$, where the decimal digits are determined by the following rule:

$$d_i = \begin{cases} 4 & \text{if } d_{ii} \neq 4 \\ 5 & \text{if } d_{ii} = 4. \end{cases}$$

(As an example, suppose that $r_1 = 0.23794102\ldots$, $r_2 = 0.44590138\ldots$, $r_3 = 0.09118764\ldots$, $r_4 = 0.80553900\ldots$, and so on. Then we have $r = 0.d_1d_2d_3d_4 \ldots = 0.4544\ldots$, where $d_1 = 4$ since $d_{11} \neq 4, d_2 = 5$ since $d_{22} = 4, d_3 = 4$ since $d_{33} \neq 4$, $d_4 = 4$ since $d_{44} \neq 4$, and so on.)

Every real number has a unique decimal expansion (when the possibility that the expansion has a tail end that consists entirely of the digit 9 is excluded). Then, the real number r is not equal to any of $r_1, r_2, \ldots$, since the decimal expansion of r differs from the decimal expansion of r_i in the ith place to the right of the decimal point, for each i.

Since there is a real number r between 0 and 1 that is not in the list, the assumption that all the real numbers between 0 and 1 could be listed must be false. Therefore, all the real numbers between 0 and 1 cannot be listed, so that the set of real numbers between 0 and 1 is uncountable. Any set with an uncountable subset is uncountable (see Exercise 35 at the end of this section). Hence, the set of real numbers is uncountable. ◀

Exercises

1. Find these terms of the sequence $\{a_n\}$ where $a_n = 2 \cdot (-3)^n + 5^n$.

   a) a_0 **b)** a_1 **c)** a_4 **d)** a_5

2. What is the term a_8 of the sequence $\{a_n\}$ if a_n equals

   a) 2^{n-1}?   **b)** 7?

   c) $1 + (-1)^n$?   **d)** $-(-2)^n$?

3. What are the terms a_0, a_1, a_2, and a_3 of the sequence $\{a_n\}$, where a_n equals

   a) $2^n + 1$?   **b)** $(n+1)^{n+1}$?

   c) $\lfloor n/2 \rfloor$?   **d)** $\lfloor n/2 \rfloor + \lceil n/2 \rceil$?

4. What are the terms a_0, a_1, a_2, and a_3 of the sequence $\{a_n\}$, where a_n equals

   a) $(-2)^n$?   **b)** 3?

   c) $7 + 4^n$?   **d)** $2^n + (-2)^n$?

5. List the first 10 terms of each of these sequences.

   a) the sequence that begins with 2 and in which each successive term is 3 more than the preceding term

   b) the sequence that lists each positive integer three times, in increasing order

   c) the sequence that lists the odd positive integers in increasing order, listing each odd integer twice

   d) the sequence whose nth term is $n! - 2^n$

   e) the sequence that begins with 3, where each succeeding term is twice the preceding term

   f) the sequence whose first two terms are 1 and each succeeding term is the sum of the two preceding terms (This is the famous Fibonacci sequence, which we will study later in this text.)

   g) the sequence whose nth term is the number of bits in the binary expansion of the number n (defined in Section 2.5)

   h) the sequence where the nth term is the number of letters in the English word for the index n

6. List the first 10 terms of each of these sequences.

   a) the sequence obtained by starting with 10 and obtaining each term by subtracting 3 from the previous term

   b) the sequence whose nth term is the sum of the first n positive integers

   c) the sequence whose nth term is $3^n - 2^n$

   d) the sequence whose nth term is $\lfloor \sqrt{n} \rfloor$

   e) the sequence whose first two terms are 1 and 2 and each succeeding term is the sum of the two previous terms

   f) the sequence whose nth term is the largest integer whose binary expansion (defined in Section 2.5) has n bits (Write your answer in decimal notation.)

   g) the sequence whose terms are constructed sequentially as follows: start with 1, then add 1, then multiply by 1, then add 2, then multiply by 2, and so on

   h) the sequence whose nth term is the largest integer k such that $k! \le n$

7. Find at least three different sequences beginning with the terms 1, 2, 4 whose terms are generated by a simple formula or rule.

8. Find at least three different sequences beginning with the terms 3, 5, 7 whose terms are generated by a simple formula or rule.

9. For each of these lists of integers, provide a simple formula or rule that generates the terms of an integer sequence that begins with the given list.

   a) 1, 0, 1, 1, 0, 0, 1, 1, 1, 0, 0, 0, 1, ...

   b) 1, 2, 2, 3, 4, 4, 5, 6, 6, 7, 8, 8, ...

   c) 1, 0, 2, 0, 4, 0, 8, 0, 16, 0, ...

   d) 3, 6, 12, 24, 48, 96, 192, ...

   e) 15, 8, 1, −6, −13, −20, −27, ...

   f) 3, 5, 8, 12, 17, 23, 30, 38, 47, ...

   g) 2, 16, 54, 128, 250, 432, 686, ...

   h) 2, 3, 7, 25, 121, 721, 5041, 40321, ...

10. For each of these lists of integers, provide a simple formula or rule that generates the terms of an integer sequence that begins with the given list.

   a) 3, 6, 11, 18, 27, 38, 51, 66, 83, 102, ...

   b) 7, 11, 15, 19, 23, 27, 31, 35, 39, 43, ...

   c) 1, 10, 11, 100, 101, 110, 111, 1000, 1001, 1010, 1011, ...

   d) 1, 2, 2, 2, 3, 3, 3, 3, 3, 5, 5, 5, 5, 5, 5, 5, ...

   e) 0, 2, 8, 26, 80, 242, 728, 2186, 6560, 19682, ...

f) 1, 3, 15, 105, 945, 10395, 135135, 2027025, 34459425, . . .

g) 1, 0, 0, 1, 1, 1, 0, 0, 0, 0, 1, 1, 1, 1, 1, . . .

h) 2, 4, 16, 256, 65536, 4294967296, . . .

*11. Show that if a_n denotes the nth positive integer that is not a perfect square, then $a_n = n + \{\sqrt{n}\}$, where $\{x\}$ denotes the integer closest to the real number x.

*12. Let a_n be the nth term of the sequence 1, 2, 2, 3, 3, 3, 4, 4, 4, 4, 5, 5, 5, 5, 5, 6, 6, 6, 6, 6, 6, . . . , constructed by including the integer k exactly k times. Show that $a_n = \lfloor \sqrt{2n} + \frac{1}{2} \rfloor$.

13. What are the values of these sums?

a) $\displaystyle\sum_{k=1}^{5} (k+1)$ **b)** $\displaystyle\sum_{j=0}^{4} (-2)^j$

c) $\displaystyle\sum_{i=1}^{10} 3$ **d)** $\displaystyle\sum_{j=0}^{8} (2^{j+1} - 2^j)$

14. What are the values of these sums, where $S = \{1, 3, 5, 7\}$?

a) $\displaystyle\sum_{j \in S} j$ **b)** $\displaystyle\sum_{j \in S} j^2$

c) $\displaystyle\sum_{j \in S} (1/j)$ **d)** $\displaystyle\sum_{j \in S} 1$

15. What is the value of each of these sums of terms of a geometric progression?

a) $\displaystyle\sum_{j=0}^{8} 3 \cdot 2^j$ **b)** $\displaystyle\sum_{j=1}^{8} 2^j$

c) $\displaystyle\sum_{j=2}^{8} (-3)^j$ **d)** $\displaystyle\sum_{j=0}^{8} 2 \cdot (-3)^j$

16. Find the value of each of these sums.

a) $\displaystyle\sum_{j=0}^{8} (1 + (-1)^j)$ **b)** $\displaystyle\sum_{j=0}^{8} (3^j - 2^j)$

c) $\displaystyle\sum_{j=0}^{8} (2 \cdot 3^j + 3 \cdot 2^j)$ **d)** $\displaystyle\sum_{j=0}^{8} (2^{j+1} - 2^j)$

17. Compute each of these double sums.

a) $\displaystyle\sum_{i=1}^{2} \sum_{j=1}^{3} (i + j)$ **b)** $\displaystyle\sum_{i=0}^{2} \sum_{j=0}^{3} (2i + 3j)$

c) $\displaystyle\sum_{i=1}^{3} \sum_{j=0}^{2} i$ **d)** $\displaystyle\sum_{i=0}^{2} \sum_{j=1}^{3} ij$

18. Compute each of these double sums.

a) $\displaystyle\sum_{i=1}^{3} \sum_{j=1}^{2} (i - j)$ **b)** $\displaystyle\sum_{i=0}^{3} \sum_{j=0}^{2} (3i + 2j)$

c) $\displaystyle\sum_{i=1}^{3} \sum_{j=0}^{2} j$ **d)** $\displaystyle\sum_{i=0}^{2} \sum_{j=0}^{3} i^2 j^3$

19. Show that $\sum_{j=1}^{n} (a_j - a_{j-1}) = a_n - a_0$ where $a_0, a_1, \ldots, a_n$ is a sequence of real numbers. This type of sum is called **telescoping.**

20. Use the identity $1/(k(k+1)) = 1/k - 1/(k+1)$ and Exercise 19 to compute $\sum_{k=1}^{n} 1/(k(k+1))$.

21. Sum both sides of the identity $k^2 - (k-1)^2 = 2k - 1$ from $k = 1$ to $k = n$ and use Exercise 19 to find

a) a formula for $\sum_{k=1}^{n} (2k - 1)$ (the sum of the first n odd natural numbers).

b) a formula for $\sum_{k=1}^{n} k$.

*22. Use the technique given in Exercise 19, together with the result of Exercise 21b, to find a formula for $\sum_{k=1}^{n} k^2$.

23. Find $\sum_{k=100}^{200} k$. (Use Table 2.)

24. Find $\sum_{k=99}^{200} k^3$. (Use Table 2.)

*25. Find a formula for $\sum_{k=0}^{m} \lfloor \sqrt{k} \rfloor$, when m is a positive integer. (*Hint:* Use the formula for $\sum_{k=1}^{n} k^2$.)

*26. Find a formula for $\sum_{k=0}^{m} \lfloor \sqrt[3]{k} \rfloor$, when m is a positive integer. (*Hint:* Use the formula for $\sum_{k=1}^{n} k^3$.)

There is also a special notation for products. The product of $a_m, a_{m+1}, \ldots, a_n$ is represented by

$$\prod_{j=m}^{n} a_j.$$

27. What are the values of the following products?

a) $\displaystyle\prod_{i=0}^{10} i$ **b)** $\displaystyle\prod_{i=5}^{8} i$

c) $\displaystyle\prod_{i=1}^{100} (-1)^i$ **d)** $\displaystyle\prod_{i=1}^{10} 2$

Recall that the value of the factorial function at a positive integer n, denoted by $n!$, is the product of the positive integers from 1 to n, inclusive. Also, we specify that $0! = 1$.

28. Express $n!$ using product notation.

29. Find $\sum_{j=0}^{4} j!$.

30. Find $\prod_{j=0}^{4} j!$.

31. Determine whether each of these sets is countable or uncountable. For those that are countable, exhibit a one-to-one correspondence between the set of natural numbers and that set.

a) the negative integers

b) the even integers

c) the real numbers between 0 and $\frac{1}{2}$

d) integers that are multiples of 7

*32. Determine whether each of these sets is countable or uncountable. For those that are countable, exhibit a one-to-one correspondence between the set of natural numbers and that set.

a) integers not divisible by 3

b) integers divisible by 5 but not by 7

c) the real numbers with decimal representations consisting of all 1s

d) the real numbers with decimal representations of all 1s or 9s

33. If A is an uncountable set and B is a countable set, must $A - B$ be uncountable?

34. Show that a subset of a countable set is also countable.

35. Show that if A is an uncountable set and $A \subseteq B$, then B is uncountable.

36. Show that the union of two countable sets is countable.

**37. Show that the union of a countable number of countable sets is countable.

38. Show that the set $\mathbf{Z}^+ \times \mathbf{Z}^+$ is countable.

*39. Show that the set of all bit strings is countable.

*40. Show that the set of real numbers that are solutions of quadratic equations $ax^2 + bx + c = 0$, where a, b, and c are integers, is countable.

*41. Show that the set of all computer programs in a particular programming language is countable. (*Hint:* A computer program written in a programming language can be thought of as a string of symbols from a finite alphabet.)

*42. Show that the set of functions from the positive integers to the set $\{0, 1, 2, 3, 4, 5, 6, 7, 8, 9\}$ is uncountable. [*Hint:* First set up a one-to-one correspondence between the set of real numbers between 0 and 1 and a subset of these functions. Do this by associating to the real number $0.d_1 d_2 \ldots d_n \ldots$ the function f with $f(n) = d_n$.]

*43. We say that a function is **computable** if there is a computer program that finds the values of this function. Use Exercises 41 and 42 to show that there are functions that are not computable.

*44. Prove that the set of positive rational numbers is countable by setting up a function that assigns to a rational number p/q with $\gcd(p, q) = 1$ the base 11 number formed from the decimal representation of p followed by the base 11 digit A, which corresponds to the decimal number 10, followed by the decimal representation of q.

*45. Prove that the set of positive rational numbers is countable by showing that the function K is a one-to-one correspondence between the set of positive rational numbers and the set of positive integers if $K(m/n) = p_1^{2a_1} p_2^{2a_2} \cdots p_s^{2a_s} q_1^{2b_1 - 1} q_2^{2b_2 - 1} \cdots q_t^{2b_t - 1}$, where $\gcd(m, n) = 1$ and the prime-power factorizations of m and n are $m = p_1^{a_1} p_2^{a_2} \cdots p_s^{a_s}$ and $n = q_1^{b_1} q_2^{b_2} \cdots q_t^{b_t}$.

3.3   Mathematical Induction

INTRODUCTION

Links

What is a formula for the sum of the first n positive odd integers? The sums of the first n positive odd integers for $n = 1, 2, 3, 4, 5$ are

$$1 = 1, \qquad\qquad 1 + 3 = 4, \qquad\qquad 1 + 3 + 5 = 9,$$
$$1 + 3 + 5 + 7 = 16, \quad 1 + 3 + 5 + 7 + 9 = 25.$$

From these values it is reasonable to guess that the sum of the first n positive odd integers is n^2. We need a method to *prove* that this *guess* is correct, if in fact it is.

Mathematical induction is an extremely important proof technique that can be used to prove assertions of this type. As we will see in this section and in subsequent chapters, mathematical induction is used extensively to prove results about a large variety of discrete objects. For example, it is used to prove results about the complexity of algorithms, the correctness of certain types of computer programs, theorems about graphs and trees, as well as a wide range of identities and inequalities.

In this section we will describe how mathematical induction can be used and why it is a valid proof technique. It is extremely important to note that mathematical induction can be used only to prove results obtained in some other way. It is *not* a tool for discovering formulae or theorems.

There are several useful illustrations of mathematical induction that can help you remember how this principle works. One of these involves a line of people, person one, person two, and so on. A secret is told to person one, and each person tells the secret to the next person in line, if the former person hears it. Let $P(n)$ be the proposition that person n knows the secret. Then $P(1)$ is true, since the secret is told to person one; $P(2)$ is true, since person one tells person two the secret; $P(3)$ is true, since person two tells person three the secret; and so on. By the principle of mathematical induction, every person in line learns the secret. This is illustrated in Figure 1. (Of course, it has been assumed that each person relays the secret in an unchanged manner to the next person, which is usually not true in real life.)

FIGURE 1   People Telling Secrets.

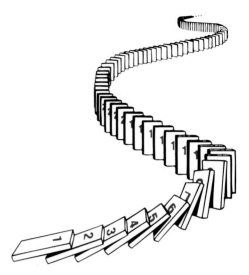

FIGURE 2   Illustrating How Mathematical Induction Works Using Dominos.

Another way to illustrate the principle of mathematical induction is to consider an infinite row of dominos, labeled $1, 2, 3, \ldots, n$, where each domino is standing up. Let $P(n)$ be the proposition that domino n is knocked over. If the first domino is knocked over—i.e., if $P(1)$ is true—and if, whenever the nth domino is knocked over, it also knocks the $(n + 1)$th domino over—i.e., if $P(n) \to P(n + 1)$ is true—then all the dominos are knocked over. This is illustrated in Figure 2.

MATHEMATICAL INDUCTION

Many theorems state that $P(n)$ is true for all positive integers n, where $P(n)$ is a propositional function, such as the statement that $1 + 2 + \cdots + n = n(n + 1)/2$ or the statement that $n \leq 2^n$. Mathematical induction is a technique for proving theorems of this kind. In other words, mathematical induction is used to prove propositions of the form $\forall n \, P(n)$, where the universe of discourse is the set of positive integers.

A proof by mathematical induction that $P(n)$ is true for every positive integer n consists of two steps:

BASIS STEP: The proposition $P(1)$ is shown to be true.

INDUCTIVE STEP: The implication $P(k) \to P(k + 1)$ is shown to be true for every positive integer k.

Here, the statement $P(k)$ for a fixed positive integer k is called the **inductive hypothesis.** When we complete both steps of a proof by mathematical induction, we have proved that $P(n)$ is true for all positive integers n; that is, we have shown that $\forall n \, P(n)$ is true.

Expressed as a rule of inference, this proof technique can be stated as

$$[P(1) \land \forall k \, (P(k) \to P(k + 1))] \to \forall n \, P(n).$$

Since mathematical induction is such an important technique, it is worthwhile to explain in detail the steps of a proof using this technique. The first thing we do to prove that $P(n)$

is true for all positive integers n is to show that $P(1)$ is true. This amounts to showing that the particular statement obtained when n is replaced by 1 in $P(n)$ is true. Then we must show that $P(k) \rightarrow P(k+1)$ is true for every positive integer k. To prove that this implication is true for every positive integer k, we need to show that $P(k+1)$ cannot be false when $P(k)$ is true. This can be accomplished by assuming that $P(k)$ is true and showing that *under this hypothesis* $P(k+1)$ must also be true.

Remark: In a proof by mathematical induction it is *not* assumed that $P(k)$ is true for all positive integers! It is only shown that *if it is assumed* that $P(k)$ is true, then $P(k+1)$ is also true. Thus, a proof by mathematical induction is not a case of begging the question, or circular reasoning.

When we use mathematical induction to prove a theorem, we first show that $P(1)$ is true. Then we know that $P(2)$ is true, since $P(1)$ implies $P(2)$. Further, we know that $P(3)$ is true, since $P(2)$ implies $P(3)$. Continuing along these lines, we see that $P(n)$ is true, for every positive integer n.

EXAMPLES OF PROOFS BY MATHEMATICAL INDUCTION

Links

Extra Examples

We will use a variety of examples to illustrate how theorems are proved using mathematical induction. We begin by proving a formula for the sum of the first n odd positive integers. (Many theorems proved in this section via mathematical induction can be proved using different methods. However, it is worthwhile to try to prove a theorem in more than one way, since one method of attack may succeed whereas another approach may not.)

EXAMPLE 1   Use mathematical induction to prove that the sum of the first n odd positive integers is n^2.

Solution: Let $P(n)$ denote the proposition that the sum of the first n odd positive integers is n^2. We must first complete the basis step; that is, we must show that $P(1)$ is true. Then we must carry out the inductive step; that is, we must show that $P(k+1)$ is true when $P(k)$ is assumed to be true.

BASIS STEP: $P(1)$ states that the sum of the first one odd positive integer is 1^2. This is true since the sum of the first odd positive integer is 1.

INDUCTIVE STEP: To complete the inductive step we must show that the proposition $P(k) \rightarrow P(k+1)$ is true for every positive integer k. To do this, suppose that $P(k)$ is true for a positive integer k; that is,

$$1 + 3 + 5 + \cdots + (2k - 1) = k^2.$$

Links

HISTORICAL NOTE   The first known use of mathematical induction is in the work of the sixteenth-century mathematician Francesco Maurolico (1494–1575). Maurolico wrote extensively on the works of classical mathematics and made many contributions to geometry and optics. In his book *Arithmeticorum Libri Duo*, Maurolico presented a variety of properties of the integers together with proofs of these properties. To prove some of these properties he devised the method of mathematical induction. His first use of mathematical induction in this book was to prove that the sum of the first n odd positive integers equals n^2.

[Note that the kth odd positive integer is $(2k - 1)$, since this integer is obtained by adding 2 a total of $k - 1$ times to 1.] We must show that $P(k + 1)$ is true, assuming that $P(k)$ is true. Note that $P(k + 1)$ is the statement that

$$1 + 3 + 5 + \cdots + (2k - 1) + (2k + 1) = (k + 1)^2.$$

So, assuming that $P(k)$ is true, it follows that

$$1 + 3 + 5 + \cdots + (2k - 1) + (2k + 1) = [1 + 3 + \cdots + (2k - 1)] + (2k + 1)$$
$$= k^2 + (2k + 1)$$
$$= k^2 + 2k + 1$$
$$= (k + 1)^2.$$

This shows that $P(k + 1)$ follows from $P(k)$. Note that we used the inductive hypothesis $P(k)$ in the second equality to replace the sum of the first k odd positive integers by k^2.

Since $P(1)$ is true and the implication $P(k) \rightarrow P(k + 1)$ is true for all positive integers k, the principle of mathematical induction shows that $P(n)$ is true for all positive integers n. ◀

Example 2 uses the principle of mathematical induction to prove an inequality.

EXAMPLE 2 Use mathematical induction to prove the inequality

$$n < 2^n$$

for all positive integers n.

Solution: Let $P(n)$ be the proposition "$n < 2^n$."

BASIS STEP: $P(1)$ is true, since $1 < 2^1 = 2$.

INDUCTIVE STEP: Assume that $P(k)$ is true for the positive integer k. That is, assume that $k < 2^k$. We need to show that $P(k + 1)$ is true. That is, we need to show that $k + 1 < 2^{k+1}$. Adding 1 to both sides of $k < 2^k$, and then noting that $1 \leq 2^k$, gives

$$k + 1 < 2^k + 1 \leq 2^k + 2^k = 2^{k+1}.$$

We have shown that $P(k + 1)$ is true, namely, that $k + 1 < 2^{k+1}$, based on the assumption that $P(k)$ is true. The induction step is complete.

Therefore, by the principle of mathematical induction, it has been shown that $n < 2^n$ is true for all positive integers n. ◀

We will now use mathematical induction to prove a theorem involving the divisibility of integers.

EXAMPLE 3 Use mathematical induction to prove that $n^3 - n$ is divisible by 3 whenever n is a positive integer.

Solution: To construct the proof, let $P(n)$ denote the proposition: "$n^3 - n$ is divisible by 3."

BASIS STEP: $P(1)$ is true, since $1^3 - 1 = 0$ is divisible by 3.

INDUCTIVE STEP: Assume that $P(k)$ is true; that is, $k^3 - k$ is divisible by 3. We must show that $P(k+1)$ is true. That is, we must show that $(k+1)^3 - (k+1)$ is divisible by 3. Note that

$$(k+1)^3 - (k+1) = (k^3 + 3k^2 + 3k + 1) - (k+1)$$
$$= (k^3 - k) + 3(k^2 + k).$$

Since both terms in this sum are divisible by 3 (the first by the assumption of the inductive step, and the second because it is 3 times an integer), it follows that $(k+1)^3 - (k+1)$ is also divisible by 3. This completes the induction step. Thus, by the principle of mathematical induction, $n^3 - n$ is divisible by 3 whenever n is a positive integer. ◀

Sometimes we need to show that $P(n)$ is true for $n = b, b+1, b+2, \ldots$, where b is an integer other than 1. We can use mathematical induction to accomplish this as long as we change the basis step. For instance, consider Example 4, which proves that a summation formula is valid for all nonnegative integers, so that we need to prove that $P(n)$ is true for $n = 0, 1, 2, \ldots$.

EXAMPLE 4   Use mathematical induction to show that

$$1 + 2 + 2^2 + \cdots + 2^n = 2^{n+1} - 1$$

for all nonnegative integers n.

Solution: Let $P(n)$ be the proposition that this formula is correct for the integer n.

BASIS STEP: $P(0)$ is true since $2^0 = 1 = 2^1 - 1$.

INDUCTIVE STEP: Assume that $P(k)$ is true. To carry out the inductive step using this assumption, it must be shown that $P(k+1)$ is true, namely,

$$1 + 2 + 2^2 + \cdots + 2^k + 2^{k+1} = 2^{(k+1)+1} - 1 = 2^{k+2} - 1.$$

Using the inductive hypothesis $P(k)$, it follows that

$$1 + 2 + 2^2 + \cdots + 2^k + 2^{k+1} = (1 + 2 + 2^2 + \cdots + 2^k) + 2^{k+1}$$
$$= (2^{k+1} - 1) + 2^{k+1}$$
$$= 2 \cdot 2^{k+1} - 1$$
$$= 2^{k+2} - 1.$$

This finishes the inductive step, which completes the proof. ◀

As Example 4 demonstrates, to use mathematical induction to show that $P(n)$ is true for $n = b, b+1, b+2, \ldots$, where b is an integer other than 1, we show that $P(b)$ is true (the basis step) and then show that the implication $P(k) \rightarrow P(k+1)$ is true for $k = b, b+1, b+2, \ldots$ (the inductive step). Note that b can be negative, zero, or positive. Following the domino analogy we used earlier, imagine that we begin by knocking down the bth domino (the basis step), and as each domino falls, it knocks down the next domino (the inductive step). We leave it to the reader to show that this form of induction is valid (see Exercise 76).

The formula given in Example 4 is a special case of a general result for the sum of terms of a geometric progression (Theorem 1 in Section 3.2). We will use mathematical induction to provide an alternate proof of this formula.

EXAMPLE 5   **Sums of Geometric Progressions**   Use mathematical induction to prove this formula for the sum of a finite number of terms of a geometric progression:

$$\sum_{j=0}^{n} ar^j = a + ar + ar^2 + \cdots + ar^n = \frac{ar^{n+1} - a}{r - 1}, \qquad \text{when } r \neq 1.$$

Solution: To prove this formula using mathematical induction, let $P(n)$ be the proposition that the sum of the first $n + 1$ terms of a geometric progression in this formula is correct.

BASIS STEP: $P(0)$ is true, since

$$a = \frac{ar - a}{r - 1}.$$

INDUCTIVE STEP: Assume that $P(k)$ is true. That is, assume

$$a + ar + ar^2 + \cdots + ar^k = \frac{ar^{k+1} - a}{r - 1}.$$

To show that this implies that $P(k + 1)$ is true, add ar^{k+1} to both sides of this equation to obtain

$$a + ar + ar^2 + \cdots + ar^k + ar^{k+1} = \frac{ar^{k+1} - a}{r - 1} + ar^{k+1}.$$

Rewriting the right-hand side of this equation shows that

$$\frac{ar^{k+1} - a}{r - 1} + ar^{k+1} = \frac{ar^{k+1} - a}{r - 1} + \frac{ar^{k+2} - ar^{k+1}}{r - 1}$$

$$= \frac{ar^{k+2} - a}{r - 1}.$$

Combining these last two equations gives

$$a + ar + ar^2 + \cdots + ar^k + ar^{k+1} = \frac{ar^{k+2} - a}{r - 1}.$$

This shows that if $P(k)$ is true, then $P(k + 1)$ must also be true. This completes the inductive argument and shows that the formula for the sum of the terms of a geometric series is correct.   ◀

As previously mentioned, the formula in Example 4 is the case of the formula in Example 5 with $a = 1$ and $r = 2$. The reader should verify that putting these values for a and r in the general formula gives the same formula as in Example 4.

An important inequality for the sum of the reciprocals of a set of positive integers will be proved in the next example.

EXAMPLE 6   **An Inequality for Harmonic Numbers**   The **harmonic numbers** H_j, $j = 1, 2, 3, \ldots$, are defined by

$$H_j = 1 + \frac{1}{2} + \frac{1}{3} + \cdots + \frac{1}{j}.$$

For instance,

$$H_4 = 1 + \frac{1}{2} + \frac{1}{3} + \frac{1}{4} = \frac{25}{12}.$$

Use mathematical induction to show that

$$H_{2^n} \geq 1 + \frac{n}{2},$$

whenever n is a nonnegative integer.

Solution: To carry out the proof, let $P(n)$ be the proposition that $H_{2^n} \geq 1 + n/2$.

BASIS STEP: $P(0)$ is true, since $H_{2^0} = H_1 = 1 \geq 1 + 0/2$.

INDUCTIVE STEP: Assume that $P(k)$ is true, so that $H_{2^k} \geq 1 + k/2$. It must be shown that $P(k + 1)$, which states that $H_{2^{k+1}} \geq 1 + (k + 1)/2$, must also be true under this assumption. This can be done since

$$
\begin{aligned}
H_{2^{k+1}} &= 1 + \frac{1}{2} + \frac{1}{3} + \cdots + \frac{1}{2^k} + \frac{1}{2^k + 1} + \cdots + \frac{1}{2^{k+1}} \quad \text{definition of harmonic number} \\
&= H_{2^k} + \frac{1}{2^k + 1} + \cdots + \frac{1}{2^{k+1}} \quad \text{definition of harmonic number} \\
&\geq \left(1 + \frac{k}{2}\right) + \frac{1}{2^k + 1} + \cdots + \frac{1}{2^{k+1}} \quad \text{by the inductive hypothesis} \\
&\geq \left(1 + \frac{k}{2}\right) + 2^k \cdot \frac{1}{2^{k+1}} \quad \text{since there are } 2^k \text{ terms each not less than } 1/2^{k+1} \\
&\geq \left(1 + \frac{k}{2}\right) + \frac{1}{2} \\
&= 1 + \frac{k + 1}{2}.
\end{aligned}
$$

This establishes the inductive step of the proof. Thus, the inequality for the harmonic numbers is valid for all nonnegative integers n.   ◀

Remark: The inequality established here shows that the **harmonic series**

$$1 + \frac{1}{2} + \frac{1}{3} + \cdots + \frac{1}{n} + \cdots$$

is a divergent infinite series. This is an important example in the study of infinite series.

Example 7 shows how mathematical induction can be used to verify a formula for the number of subsets of a finite set.

EXAMPLE 7   **The Number of Subsets of a Finite Set**   Use mathematical induction to show that if S is a finite set with n elements, then S has 2^n subsets. (We will prove this result directly in several ways in Chapter 4.)

Solution: Let $P(n)$ be the proposition that a set with n elements has 2^n subsets.

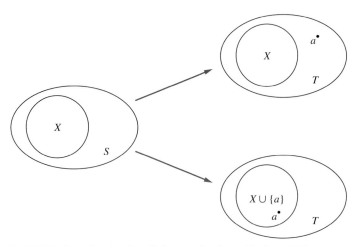

FIGURE 3   Generating Subsets of a Set with $k + 1$ Elements. Here $T = S \cup \{a\}$.

BASIS STEP: $P(0)$ is true, since a set with zero elements, the empty set, has exactly $2^0 = 1$ subsets, since it has one subset, namely, itself.

INDUCTIVE STEP: Assume that $P(k)$ is true, that is, that every set with k elements has 2^k subsets. It must be shown that under this assumption $P(k + 1)$, which is the statement that every set with $k + 1$ elements has 2^{k+1} subsets, must also be true. To show this, let T be a set with $k + 1$ elements. Then, it is possible to write $T = S \cup \{a\}$ where a is one of the elements of T and $S = T - \{a\}$. The subsets of T can be obtained in the following way. For each subset X of S there are exactly two subsets of T, namely, X and $X \cup \{a\}$. (This is illustrated in Figure 3.) These constitute all the subsets of T and are all distinct. Since there are 2^k subsets of S, there are $2 \cdot 2^k = 2^{k+1}$ subsets of T. This finishes the induction argument.   ◀

EXAMPLE 8   Show that if n is a positive integer,

$$1 + 2 + \cdots + n = n(n + 1)/2.$$

Solution: Let $P(n)$ be the proposition that the sum of the first n positive integers is $n(n + 1)/2$. We must do two things to prove that $P(n)$ is true for $n = 1, 2, 3, \ldots$. Namely, we must show that $P(1)$ is true and that the implication $P(k)$ implies $P(k + 1)$ is true for $k = 1, 2, 3, \ldots$.

BASIS STEP: $P(1)$ is true, since $1 = 1(1 + 1)/2$.

INDUCTIVE STEP: Assume that $P(k)$ holds so that

$$1 + 2 + \cdots + k = k(k + 1)/2.$$

Under this assumption, it must be shown that $P(k + 1)$ is true, namely, that

$$1 + 2 + \cdots + k + (k + 1) = (k + 1)[(k + 1) + 1]/2 = (k + 1)(k + 2)/2$$

is also true. Add $k + 1$ to both sides of the equation in $P(k)$ to obtain

$$1 + 2 + \cdots + k + (k + 1) = k(k + 1)/2 + (k + 1)$$
$$= [(k/2) + 1](k + 1)$$
$$= (k + 1)(k + 2)/2.$$

This last equation shows that $P(k + 1)$ is true. This completes the inductive step and completes the proof. ◀

EXAMPLE 9 Use mathematical induction to prove that $2^n < n!$ for every positive integer n with $n \geq 4$.

Solution: Let $P(n)$ be the proposition that $2^n < n!$.

BASIS STEP: To prove the inequality for $n \geq 4$ requires that the basis step be $P(4)$. Note that $P(4)$ is true, since $2^4 = 16 < 4! = 24$.

INDUCTIVE STEP: Assume that $P(k)$ is true. That is, assume that $2^k < k!$. We must show that $P(k + 1)$ is true. That is, we must show that $2^{k+1} < (k + 1)!$. Multiplying both sides of the inequality $2^k < k!$ by 2, it follows that

$$2 \cdot 2^k < 2 \cdot k!$$
$$< (k + 1) \cdot k!$$
$$= (k + 1)!.$$

This shows that $P(k + 1)$ is true when $P(k)$ is true. This completes the inductive step of the proof. Hence, it follows that $2^n < n!$ is true for all integers n with $n \geq 4$. ◀

EXAMPLE 10 Use mathematical induction to prove the following generalization of one of De Morgan's laws:

$$\overline{\bigcap_{j=1}^{n} A_j} = \bigcup_{j=1}^{n} \overline{A_j},$$

whenever $A_1, A_2, \ldots, A_n$ are subsets of a universal set U and $n \geq 2$.

Solution: Let $P(n)$ be the identity for n sets.

BASIS STEP: The statement $P(2)$ asserts that $\overline{A_1 \cap A_2} = \overline{A_1} \cup \overline{A_2}$. This is one of De Morgan's laws; it was proved in Section 1.7.

INDUCTIVE STEP: Assume that $P(k)$ is true, that is,

$$\overline{\bigcap_{j=1}^{k} A_j} = \bigcup_{j=1}^{k} \overline{A_j}$$

whenever $A_1, A_2, \ldots, A_k$ are subsets of the universal set U. To carry out the inductive step it must be shown that if this equality holds for any k subsets of U, it must also be valid for any $k + 1$ subsets of U. Suppose that $A_1, A_2, \ldots, A_k, A_{k+1}$ are subsets of U. When the inductive hypothesis is assumed to hold, it follows that

$$\overline{\bigcap_{j=1}^{k+1} A_j} = \overline{\left(\bigcap_{j=1}^{k} A_j\right) \cap A_{k+1}}$$

$$= \overline{\left(\bigcap_{j=1}^{k} A_j\right)} \cup \overline{A_{k+1}} \quad \text{by De Morgan's law}$$

$$= \left(\bigcup_{j=1}^{k} \overline{A_j}\right) \cup \overline{A_{k+1}} \quad \text{by the inductive hypothesis}$$

$$= \bigcup_{j=1}^{k+1} \overline{A_j}.$$

This completes the proof by induction.   ◀

Example 11 illustrates how mathematical induction can be used to prove a result about covering chessboards with pieces shaped like the letter "L."

EXAMPLE 11   Let n be a positive integer. Show that any $2^n \times 2^n$ chessboard with one square removed can be tiled using L-shaped pieces, where these pieces cover three squares at a time, as shown in Figure 4.

Solution: Let $P(n)$ be the proposition that any $2^n \times 2^n$ chessboard with one square removed can be tiled using L-shaped pieces. We can use mathematical induction to prove that $P(n)$ is true for all positive integers n.

BASIS STEP: $P(1)$ is true, since any of the four 2×2 chessboards with one square removed can be tiled using one L-shaped piece, as shown in Figure 5.

FIGURE 4   An L-Shaped Piece.

INDUCTIVE STEP: Assume that $P(k)$ is true; that is, assume that any $2^k \times 2^k$ chessboard with one square removed can be tiled using L-shaped pieces. It must be shown that under this assumption $P(k + 1)$ must also be true; that is, any $2^{k+1} \times 2^{k+1}$ chessboard with one square removed can be tiled using L-shaped pieces.

To see this, consider a $2^{k+1} \times 2^{k+1}$ chessboard with one square removed. Split this chessboard into four chessboards of size $2^k \times 2^k$, by dividing it in half in both directions. This is illustrated in Figure 6. No square has been removed from three of these four

FIGURE 5   Tiling 2 × 2 Chessboards with One Square Removed.

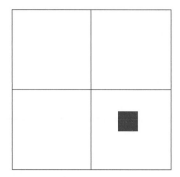

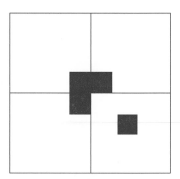

FIGURE 6   Dividing a $2^{k+1} \times 2^{k+1}$ Chessboard into Four $2^k \times 2^k$ Chessboards.

FIGURE 7   Tiling the $2^{k+1} \times 2^{k+1}$ Chessboard with One Square Removed.

chessboards. The fourth $2^k \times 2^k$ chessboard has one square removed, so by the inductive hypothesis, it can be covered by L-shaped pieces. Now temporarily remove the square from each of the other three $2^k \times 2^k$ chessboards that has the center of the original, larger chessboard as one of its corners, as shown in Figure 7. By the inductive hypothesis, each of these three $2^k \times 2^k$ chessboards with a square removed can be tiled by L-shaped pieces. Furthermore, the three squares that were temporarily removed can be covered by one L-shaped piece. Hence, the entire $2^{k+1} \times 2^{k+1}$ chessboard can be tiled with L-shaped pieces. This completes the proof. ◀

Next, we provide an example that illustrates one of many ways mathematical induction is used in the study of algorithms. We will show how mathematical induction can be used to prove that a greedy algorithm yields an optimal solution. (For an introduction to greedy algorithms, see Section 2.1.)

EXAMPLE 12   We can use a greedy algorithm to schedule a subset of m proposed talks $t_1, t_2, \ldots, t_m$ in a single lecture hall. Suppose that talk t_j begins at time b_j and ends at time e_j. (No two lectures can proceed at the same time and a lecture can begin at the same time one ends.) We assume that the talks are listed in order of nondecreasing ending time, so that $e_1 \leq e_2 \leq \cdots \leq e_m$. The greedy algorithm proceeds by selecting at each stage a talk with the earliest ending time among all those talks that begin after all talks already scheduled end. (A lecture with an earliest end time is always added first by the algorithm.) We will show that this greedy algorithm is optimal in the sense that it always schedules the most talks possible. To prove the optimality of this algorithm we use mathematical induction on the variable n, the number of talks scheduled by the algorithm. We let $P(n)$ be the proposition that if the greedy algorithm schedules n talks, then it is not possible to schedule more than n talks.

BASIS STEP: Suppose that the greedy algorithm managed to schedule just one talk, t_1. This means that every other talk cannot start after e_1, the ending time of t_1. Otherwise, the first such talk we come to as we go through the talks in order of nondecreasing end time could be added. Hence, at time e_1 each of the remaining talks needs to use the lecture hall since they all start at or before e_1 and end after e_1. It follows that no two talks can be scheduled since both need to use the lecture hall at time e_1. This shows that $P(1)$ is true and completes the basis step.

INDUCTIVE STEP: Assume that $P(k)$ is true, that is, that the greedy algorithm always schedules the most possible talks when it selects k talks, given any set of talks (no matter how large). Now assume that the algorithm has selected $k + 1$ talks. We must show that the greedy algorithm has selected the largest number of talks possible, given the assumption that it always produces an optimal solution when it schedules k talks. That is, we need to show that $P(k + 1)$ is true, assuming that $P(k)$ is true.

To complete the inductive step, we first show there is a schedule including the most talks possible that contains talk t_1, a talk with the earliest end time. This is easy to see since a schedule that begins with the talk t_i in the list, where $i > 1$, can be changed so that talk t_1 replaces talk t_i. To see this, note that since $e_1 \leq e_i$, all talks that were scheduled to follow talk t_i can still be scheduled.

Once we included talk t_1, scheduling the talks so that as many as possible are scheduled is reduced to scheduling as many talks as possible that begin at or after time e_1. So, if we have scheduled as many talks as possible, the schedule of talks other than talk t_1 is an optimal schedule of the original talks that begin once talk t_1 has ended. Since the greedy algorithm schedules k talks when it creates this schedule, we can apply the induction hypothesis to conclude that it has scheduled the most possible talks. It follows that the greedy algorithm has scheduled the most possible talks, $k + 1$, when it produced a schedule with $k + 1$ talks, so that $P(k + 1)$ is true. This completes the induction step, finishing the proof that $P(n)$ is true for all positive integers n, and completes the proof of optimality.   ◄

STRONG INDUCTION

There is another form of mathematical induction that is often useful in proofs. With this form we use the same basis step as before, but we use a different inductive step. We assume that $P(j)$ is true for $j = 1, \ldots, k$ and show that $P(k + 1)$ must also be true based on this assumption. This is called **strong induction** (and is sometimes also known as the **second principle of mathematical induction**).

We summarize the two steps used to show that $P(n)$ is true for all positive integers n:

BASIS STEP: The proposition $P(1)$ is shown to be true.

INDUCTIVE STEP: It is shown that $[P(1) \land P(2) \land \cdots \land P(k)] \to P(k + 1)$ is true for every positive integer k.

The two forms of mathematical induction are equivalent; that is, each can be shown to be a valid proof technique assuming the other. We leave it as an exercise for the reader to show this. We now give three examples that show how the strong induction is used.

EXAMPLE 13   Consider a game in which two players take turns removing any number of matches they want from one of two piles of matches. The player who removes the last match wins the game. Show that if the two piles contain the same number of matches initially, the second player can always guarantee a win.

Solution: Let n be the number of matches in each pile. We will use strong induction to prove $P(n)$, the statement that the second player can win when there are initially n matches in each pile.

BASIS STEP: When $n = 1$, the first player has only one choice, removing one match from one of the piles, leaving a single pile with a single match, which the second player can remove to win the game.

INDUCTIVE STEP: Suppose that $P(j)$ is true for all j with $1 \leq j \leq k$, that is, that the second player can always win whenever there are j matches where $1 \leq j \leq k$ in each of the two piles at the start of the game. Now suppose that there are $k + 1$ matches in each of the two piles at the start of the game and suppose that the first player removes j matches $(1 \leq j \leq k)$ from one of the piles, leaving $k + 1 - j$ matches in this pile. By removing the same number of matches from the other pile, player two creates the situation where there are two piles each with $k + 1 - j$ matches. Because $1 \leq k + 1 - j \leq k$ the second player can always win by the induction hypothesis. We complete the proof by noting that if the first player removes all $k + 1$ matches from one of the piles, the second player can win by removing all the remaining matches. ◄

EXAMPLE 14   Show that if n is an integer greater than 1, then n can be written as the product of primes.

Solution: Let $P(n)$ be the proposition that n can be written as the product of primes.

BASIS STEP: $P(2)$ is true, since 2 can be written as the product of one prime, itself. [Note that $P(2)$ is the first case we need to establish.]

INDUCTIVE STEP: Assume that $P(j)$ is true for all positive integers j with $j \leq k$. To complete the inductive step, it must be shown that $P(k + 1)$ is true under this assumption.

There are two cases to consider, namely, when $k + 1$ is prime and when $k + 1$ is composite. If $k + 1$ is prime, we immediately see that $P(k + 1)$ is true. Otherwise, $k + 1$ is composite and can be written as the product of two positive integers a and b with $2 \leq a \leq b < k + 1$. By the induction hypothesis, both a and b can be written as the product of primes. Thus, if $k + 1$ is composite, it can be written as the product of primes, namely, those primes in the factorization of a and those in the factorization of b. ◄

Remark: Since 1 is a product of primes, namely, the *empty* product of no primes, we could have started the proof in Example 14 with $P(1)$ as the basis step. We chose not to do this because many people find this confusing.

Note that Example 14 completes the proof of the Fundamental Theorem of Arithmetic, which asserts that every nonnegative integer can be written uniquely as the product of primes in nondecreasing order. We showed in Section 2.6 (see page 183) that an integer has at most one such factorization into primes. Example 14 shows there is at least one such factorization.

Using the principle of mathematical induction, instead of strong induction, to prove the result in Example 14 is difficult. However, as Example 15 shows, some results can be readily proved using either the principle of mathematical induction or strong induction.

EXAMPLE 15   Prove that every amount of postage of 12 cents or more can be formed using just 4-cent and 5-cent stamps.

Solution: We will prove this result using the principle of mathematical induction. Then we will present a proof using strong induction. Let $P(n)$ be the statement that postage of n cents can be formed using 4-cent and 5-cent stamps.

We begin by using the principle of mathematical induction.

BASIS STEP: Postage of 12 cents can be formed using three 4-cent stamps.

INDUCTIVE STEP: Assume that $P(k)$ is true, so that postage of k cents can be formed using 4-cent and 5-cent stamps. If at least one 4-cent stamp was used, replace it with a 5-cent stamp to form postage of $k + 1$ cents. If no 4-cent stamps were used, postage of k cents was formed using just 5-cent stamps. Since $k \geq 12$, at least three 5-cent stamps were used. So, replace three 5-cent stamps with four 4-cent stamps to form postage of $k + 1$ cents. This completes the inductive step, as well as the proof by the principle of mathematical induction.

Next, we will use strong induction. We will show that postage of 12, 13, 14, and 15 cents can be formed and then show how to get postage of $k + 1$ cents for $k \geq 15$ from postage of $k - 3$ cents.

BASIS STEP: We can form postage of 12, 13, 14, and 15 cents using three 4-cent stamps, two 4-cent stamps and one 5-cent stamp, one 4-cent stamp and two 5-cent stamps, and three 5-cent stamps, respectively.

INDUCTIVE STEP: Let $k \geq 15$. Assume that we can form postage of j cents, where $12 \leq j \leq k$. To form postage of $k + 1$ cents, use the stamps that form postage of $k - 3$ cents together with a 4-cent stamp. This completes the inductive step, as well as the proof by strong induction.

(There are other ways to approach this problem besides those described here. Can you find a solution that does not use mathematical induction?)   ◀

Remark: Example 15 shows how we can adapt strong induction to handle cases where the inductive step is valid only for sufficiently large values of k. In particular, to prove that $P(n)$ is true for $n = j, j + 1, j + 2, \ldots$, where j is an integer, we first show that $P(j), P(j + 1), P(j + 2), \ldots, P(l)$ are true (the basis step), and then we show that $[P(j) \wedge P(j + 1) \wedge P(j + 2) \wedge \cdots \wedge P(k)] \rightarrow P(k + 1)$ is true for every integer $k \geq l$ (the inductive step). For example, the basis step of the second proof in the solution of Example 15 shows that $P(12), P(13), P(14)$, and $P(15)$ are true. We need to prove these cases separately since the inductive step, which shows that $[P(12) \wedge P(13) \wedge \cdots \wedge P(k)] \rightarrow P(k + 1)$, holds only when $k \geq 15$.

THE WELL-ORDERING PROPERTY

The validity of mathematical induction follows from the following fundamental axiom about the set of integers.

THE WELL-ORDERING PROPERTY Every nonempty set of nonnegative integers has a least element.

The well-ordering property can often be used directly in proofs.

EXAMPLE 16   Use the well-ordering property to prove the division algorithm. Recall that the division algorithm states that if a is an integer and d is a positive integer, then there are unique integers q and r with $0 \leq r < d$ and $a = dq + r$.

Solution: Let S be the set of nonnegative integers of the form $a - dq$ where q is an integer. This set is nonempty since $-dq$ can be made as large as desired (taking q to be a negative integer with large absolute value). By the well-ordering property S has a least element $r = a - dq_0$.

The integer r is nonnegative. It is also the case that $r < d$. If it were not, then there would be a smaller nonnegative element in S, namely, $a - d(q_0 + 1)$. To see this, suppose that $r \geq d$. Since $a = dq_0 + r$, it follows that $a - d(q_0 + 1) = (a - dq_0) - d = r - d \geq 0$.

Consequently, there are integers q and r with $0 \le r < d$. The proof that q and r are unique is left as an exercise for the reader. ◄

EXAMPLE 17   In a round-robin tournament every player plays every other player exactly once and each match has a winner and loser. We say that the players $p_1, p_2, \ldots, p_m$ form a *cycle* if p_1 beats p_2, p_2 beats p_3, $\ldots$, p_{m-1} beats p_m, and p_m beats p_1. Use the well-ordering principle to show that if there is a cycle of length m ($m \ge 3$) among the players in a round-robin tournament, there must be a cycle of three of these players.

Solution: We assume that there is no cycle of three players. Since there is at least one cycle in the round-robin tournament, the set of all positive integers n for which there is a cycle of length n is nonempty. By the well-ordering property, this set of positive integers has a least element k, which by assumption must be greater than three. Consequently, there exists a cycle of players $p_1, p_2, p_3, \ldots, p_k$ and no shorter cycle exists.

Now suppose that there is no cycle of three of these players, so that $k > 3$. Consider the first three elements of this cycle, p_1, p_2, p_3. There are two possible outcomes of the match between p_1 and p_3. If p_3 beats p_1, it follows that p_1, p_2, p_3 is a cycle of length three, contradicting our assumption that there is no cycle of three players. Consequently, it must be the case that p_1 beats p_3. This means that we can omit p_2 from the cycle $p_1, p_2, p_3, \ldots, p_k$ to obtain the cycle $p_1, p_3, p_4, \ldots, p_k$ of length $k - 1$, contradicting the assumption that the smallest cycle has length k. We conclude that there must be a cycle of length three. ◄

INFINITE DESCENT We will now describe a proof method, the **method of infinite descent,** introduced by Pierre de Fermat in the 1600s. The method of infinite descent is often used to show that for a propositional function $P(n)$, $P(k)$ is false for all positive integers k. The method is based on the observation that if $P(k)$ is true for at least one integer k, then the well-ordering property implies that there is a least positive integer s such that $P(s)$ is true. The method proceeds by finding a positive integer s' with $s' < s$ for which $P(s')$ is true. It follows that $P(n)$ must be false for all positive integers. (This technique is called the method of *infinite* descent since the procedure of finding smaller integers for which the propositional function is true could be continued indefinitely, producing an infinite sequence of decreasing positive integers, which is impossible by the well-ordering property.) The method of infinite descent is often used to show that there are no solutions in integers to certain equations. In particular, Fermat used it to prove the $n = 4$ case of Fermat's Last Theorem, which states that the equation $x^4 + y^4 = z^4$ has no solutions in positive integers. We illustrate the use of infinite descent in Example 18.

EXAMPLE 18   In Example 21 in Section 1.5 we showed that $\sqrt{2}$ is irrational. Here we will provide a different proof of this fact using infinite descent. First, suppose that $\sqrt{2}$ is rational. Then there exist positive integers m and n such that $\sqrt{2} = m/n$. By the well-ordering property, there is a least positive integer N such that $\sqrt{2} = M/N$ for some positive integer M. (This would make N the smallest possible denominator of ratios of two positive integers that equal $\sqrt{2}$.)

To carry out the proof by infinite descent, we will show that $\sqrt{2} = (2N-M)/(M-N)$ and $0 < M - N < N$. This contradicts the choice of N as the least positive integer such that $\sqrt{2} = M/N$ for some positive integer M. To show that $\sqrt{2} = (2N - M)/(M - N)$ we need only show that $(2N - M)/(M - N) = M/N$. To show this, first note that because $(M/N)^2 = 2$, it follows that $M^2 = 2N^2$. Consequently,

$$\frac{2N - M}{M - N} = \frac{(2N - M)N}{(M - N)N} = \frac{2N^2 - MN}{(M - N)N} = \frac{M^2 - MN}{(M - N)N} = \frac{(M - N)M}{(M - N)N} = \frac{M}{N}.$$

To finish the proof, we need only show that the denominator, $M - N$, is positive and smaller than N. To see this, note that because $1 < \sqrt{2} < 2$ and $\sqrt{2} = M/N$, it follows that $1 < M/N < 2$, and hence, that $N < M < 2N$. Subtracting N, we conclude that $0 < M - N < N$. ◀

WHY MATHEMATICAL INDUCTION IS VALID

Why is mathematical induction a valid proof technique? The reason comes from the well-ordering property. Suppose we know that $P(1)$ is true and that the proposition $P(k) \rightarrow P(k + 1)$ is true for all positive integers k. To show that $P(n)$ must be true for all positive integers, assume that there is at least one positive integer for which $P(n)$ is false. Then the set S of positive integers for which $P(n)$ is false is nonempty. Thus, by the well-ordering property, S has a least element, which will be denoted by m. We know that m cannot be 1, since $P(1)$ is true. Since m is positive and greater than 1, $m - 1$ is a positive integer. Furthermore, since $m - 1$ is less than m, it is not in S, so $P(m - 1)$ must be true. Since the implication $P(m - 1) \rightarrow P(m)$ is also true, it must be the case that $P(m)$ is true. This contradicts the choice of m. Hence, $P(n)$ must be true for every positive integer n.

Exercises

1. Find a formula for the sum of the first n even positive integers.
2. Use mathematical induction to prove the formula that you found in Exercise 1.
3. Use mathematical induction to prove that $3 + 3 \cdot 5 + 3 \cdot 5^2 + \cdots + 3 \cdot 5^n = 3(5^{n+1} - 1)/4$ whenever n is a nonnegative integer.
4. Use mathematical induction to prove that $2 - 2 \cdot 7 + 2 \cdot 7^2 - \cdots + 2(-7)^n = (1 - (-7)^{n+1})/4$ whenever n is a nonnegative integer.
5. Find a formula for
$$\frac{1}{2} + \frac{1}{4} + \frac{1}{8} + \cdots + \frac{1}{2^n}$$
by examining the values of this expression for small values of n. Use mathematical induction to prove your result.
6. Find a formula for
$$\frac{1}{1 \cdot 2} + \frac{1}{2 \cdot 3} + \cdots + \frac{1}{n(n + 1)}$$
by examining the values of this expression for small values of n. Use mathematical induction to prove your result.
7. Show that $1^2 + 2^2 + \cdots + n^2 = n(n + 1)(2n + 1)/6$ whenever n is a positive integer.
8. Show that $1^3 + 2^3 + \cdots + n^3 = [n(n + 1)/2]^2$ whenever n is a positive integer.
9. Prove that $1^2 + 3^2 + 5^2 + \cdots + (2n + 1)^2 = (n + 1)(2n + 1)(2n + 3)/3$ whenever n is a nonnegative integer.

10. Prove that $1 \cdot 1! + 2 \cdot 2! + \cdots + n \cdot n! = (n + 1)! - 1$ whenever n is a positive integer.
*11. Show by mathematical induction that if $h > -1$, then $1 + nh \leq (1 + h)^n$ for all nonnegative integers n. This is called **Bernoulli's inequality.**
12. Prove that $3^n < n!$ whenever n is a positive integer greater than 6.
13. Show that $2^n > n^2$ whenever n is an integer greater than 4.
14. Use mathematical induction to prove that $n! < n^n$ whenever n is a positive integer greater than 1.
15. Prove using mathematical induction that
$$1 \cdot 2 + 2 \cdot 3 + \cdots + n(n + 1) = n(n + 1)(n + 2)/3$$
whenever n is a positive integer.
16. Use mathematical induction to prove that
$$1 \cdot 2 \cdot 3 + 2 \cdot 3 \cdot 4 + \cdots + n(n + 1)(n + 2)$$
$$= n(n + 1)(n + 2)(n + 3)/4.$$
17. Show that $1^2 - 2^2 + 3^2 - \cdots + (-1)^{n-1}n^2 = (-1)^{n-1}n(n + 1)/2$ whenever n is a positive integer.
18. Prove that
$$1 + \frac{1}{4} + \frac{1}{9} + \cdots + \frac{1}{n^2} < 2 - \frac{1}{n}$$
whenever n is a positive integer greater than 1.
19. Show that any postage that is a positive integer number of cents greater than 7 cents can be formed using just 3-cent stamps and 5-cent stamps.

20. Use mathematical induction to show that 3 divides $n^3 + 2n$ whenever n is a nonnegative integer.

21. Use mathematical induction to show that 5 divides $n^5 - n$ whenever n is a nonnegative integer.

22. Use mathematical induction to show that 6 divides $n^3 - n$ whenever n is a nonnegative integer.

*23. Use mathematical induction to show that $n^2 - 1$ is divisible by 8 whenever n is an odd positive integer.

24. Use mathematical induction to show that $n^2 - 7n + 12$ is nonnegative if n is an integer greater than 3.

25. Use mathematical induction to prove that a set with n elements has $n(n-1)/2$ subsets containing exactly two elements whenever n is an integer greater than or equal to 2.

*26. Use mathematical induction to prove that a set with n elements has $n(n-1)(n-2)/6$ subsets containing exactly three elements whenever n is an integer greater than or equal to 3.

27. Use mathematical induction to prove that $\sum_{j=1}^{n} j^4 = n(n+1)(2n+1)(3n^2 + 3n - 1)/30$ whenever n is a positive integer.

28. For which nonnegative integers n is $n^2 \le n!$? Prove your answer using mathematical induction.

29. For which nonnegative integers n is $2n + 3 \le 2^n$? Prove your answer using mathematical induction.

30. Use mathematical induction to show that $1/(2n) \le [1 \cdot 3 \cdot 5 \cdots (2n-1)]/(2 \cdot 4 \cdots 2n)$ whenever n is a positive integer.

31. a) Determine which amounts of postage can be formed using just 5-cent and 6-cent stamps.

b) Prove your answer to (a) using the principle of mathematical induction.

c) Prove your answer to (a) using the second principle of mathematical induction.

32. Which amounts of money can be formed using just dimes and quarters? Prove your answer using a form of mathematical induction.

33. An automatic teller machine has only \$20 bills and \$50 bills. Which amounts of money can the machine dispense, assuming the machine has a limitless supply of these two denominations of bills? Prove your answer using a form of mathematical induction.

34. Assume that a chocolate bar consists of n squares arranged in a rectangular pattern. The bar or a smaller rectangular piece of the bar can be broken along a vertical or a horizontal line separating the squares. Assuming that only one piece can be broken at a time, determine how many breaks you must successively make to break the bar into n separate squares. Use strong induction to prove your answer.

35. Consider this variation of the game of Nim. The game begins with n matches. Two players take turns removing matches, one, two, or three at a time. The player removing the last match loses. Use strong induction to show that if each player plays the best strategy pos-

sible, the first player wins if $n = 4j, 4j + 2,$ or $4j + 3$ for some nonnegative integer j and the second player wins in the remaining case when $n = 4j + 1$ for some nonnegative integer j.

36. Prove that $\sum_{k=1}^{n} k2^k = (n-1)2^{n+1} + 2$ using mathematical induction.

37. Show that if n is a positive integer, then

$$\sum_{\{a_1,\ldots,a_k\} \subseteq \{1,2,\ldots,n\}} \frac{1}{a_1 a_2 \cdots a_k} = n.$$

(Here the sum is over all nonempty subsets of the set of the n smallest positive integers.)

38. Use mathematical induction to show that given a set of $n + 1$ positive integers, none exceeding $2n$, there is at least one integer in this set that divides another integer in the set.

*39. A knight on a chessboard can move one space horizontally (in either direction) and two spaces vertically (in either direction) or two spaces horizontally (in either direction) and one space vertically (in either direction). Use mathematical induction to show that for every square a knight starting at $(0, 0)$, the corner of an infinite chessboard made up of all squares (m, n), where m and n are nonnegative integers, can visit this square using a finite sequence of moves. (*Hint:* Use induction on the variable $s = m + n$.)

40. Suppose you begin with a pile of n stones and split this pile into n piles of one stone each by successively splitting a pile of stones into two smaller piles. Each time you split a pile you multiply the number of stones in each of the two smaller piles you form, so that if these piles have r and s stones in them, respectively, you compute rs. Show that no matter how you split the piles, the sum of the products computed at each step equals $n(n-1)/2$.

41. (Calculus required) Use mathematical induction to prove that the derivative of $f(x) = x^n$ equals nx^{n-1} whenever n is a positive integer. (For the inductive step, use the product rule for derivatives.)

42. Suppose that

$$\mathbf{A} = \begin{bmatrix} a & 0 \\ 0 & b \end{bmatrix}$$

where a and b are real numbers. Show that

$$\mathbf{A}^n = \begin{bmatrix} a^n & 0 \\ 0 & b^n \end{bmatrix}$$

for every positive integer n.

43. Suppose that $\mathbf{A}$ and $\mathbf{B}$ are square matrices with the property $\mathbf{AB} = \mathbf{BA}$. Show that $\mathbf{AB}^n = \mathbf{B}^n \mathbf{A}$ for every positive integer n.

44. Suppose that m is a positive integer. Use mathematical induction to prove that if a and b are integers with $a \equiv b \pmod{m}$, then $a^k \equiv b^k \pmod{m}$ whenever k is a nonnegative integer.

45. Use mathematical induction to show that if A_1, $A_2, \ldots, A_n$ and B are sets, then

$$(A_1 \cup A_2 \cup \cdots \cup A_n) \cap B$$
$$= (A_1 \cap B) \cup (A_2 \cap B) \cup \cdots \cup (A_n \cap B).$$

46. Prove that if $A_1, A_2, \ldots, A_n$ and $B_1, B_2, \ldots, B_n$ are sets such that $A_k \subseteq B_k$ for $k = 1, 2, \ldots, n$, then

a) $\displaystyle\bigcup_{k=1}^{n} A_k \subseteq \bigcup_{k=1}^{n} B_k.$ **b)** $\displaystyle\bigcap_{k=1}^{n} A_k \subseteq \bigcap_{k=1}^{n} B_k.$

47. Use mathematical induction to prove that if A_1, $A_2, \ldots, A_n$ are subsets of a universal set U, then

$$\overline{\bigcup_{k=1}^{n} A_k} = \bigcap_{k=1}^{n} \overline{A_k}.$$

48. Use mathematical induction to show that $\neg(p_1 \vee p_2 \vee \cdots \vee p_n)$ is equivalent to $\neg p_1 \wedge \neg p_2 \wedge \cdots \wedge \neg p_n$ whenever $p_1, p_2, \ldots, p_n$ are propositions.

*49. Show that

$$[(p_1 \rightarrow p_2) \wedge (p_2 \rightarrow p_3) \wedge \cdots \wedge (p_{n-1} \rightarrow p_n)]$$
$$\rightarrow [(p_1 \wedge p_2 \wedge \cdots \wedge p_{n-1}) \rightarrow p_n]$$

is a tautology whenever $p_1, p_2, \ldots, p_n$ are propositions.

50. What is wrong with this "proof"?

"Theorem" For every positive integer n, $\sum_{i=1}^{n} i = (n + \frac{1}{2})^2/2$.

Basis Step: The formula is true for $n = 1$.

Inductive Step: Suppose that $\sum_{i=1}^{n} i = (n + \frac{1}{2})^2/2$. Then $\sum_{i=1}^{n+1} i = (\sum_{i=1}^{n} i) + (n + 1)$. By the inductive hypothesis, $\sum_{i=1}^{n+1} i = (n + \frac{1}{2})^2/2 + n + 1 = (n^2 + n + \frac{1}{4})/2 + n + 1 = (n^2 + 3n + \frac{9}{4})/2 = (n + \frac{3}{2})^2/2 = [(n + 1) + \frac{1}{2}]^2/2$, completing the inductive step.

51. What is wrong with this "proof" that all horses are the same color?

Let $P(n)$ be the proposition that all the horses in a set of n horses are the same color.

Basis Step: Clearly, $P(1)$ is true.

Inductive Step: Assume that $P(k)$ is true, so that all the horses in any set of k horses are the same color. Consider any $k + 1$ horses; number these as horses $1, 2, 3, \ldots, k, k + 1$. Now the first k of these horses all must have the same color, and the last k of these must also have the same color. Since the set of the first k horses and the set of the last k horses overlap, all $k + 1$ must be the same color. This shows that $P(k + 1)$ is true and finishes the proof by induction.

52. What is wrong with this "proof"?

"Theorem" For every positive integer n, if x and y are positive integers with $\max(x, y) = n$, then $x = y$.

Basis Step: Suppose that $n = 1$. If $\max(x, y) = 1$ and x and y are positive integers, we have $x = 1$ and $y = 1$.

Inductive Step: Let k be a positive integer. Assume that whenever $\max(x, y) = k$ and x and y are positive integers, then $x = y$. Now let $\max(x, y) = k + 1$, where x and y are positive integers. Then $\max(x - 1, y - 1) = k$, so by the inductive hypothesis, $x - 1 = y - 1$. It follows that $x = y$, completing the inductive step.

53. What is wrong with this "proof" by strong induction?

"Theorem" For every nonnegative integer n, $5n = 0$.

Basis Step: $5 \cdot 0 = 0$.

Inductive Step: Suppose that $5j = 0$ for all nonnegative integers j with $0 \leq j \leq k$. Write $k + 1 = i + j$, where i and j are natural numbers less than $k + 1$. By the induction hypothesis, $5(k + 1) = 5(i + j) = 5i + 5j = 0 + 0 = 0$.

*54. Find the flaw with the following "proof" that $a^n = 1$ for all nonnegative integers n, whenever a is a nonzero real number.

Basis Step: $a^0 = 1$ is true by the definition of a^0.

Inductive Step: Assume that $a^j = 1$ for all nonnegative integers j with $j \leq k$. Then note that

$$a^{k+1} = \frac{a^k \cdot a^k}{a^{k-1}} = \frac{1 \cdot 1}{1} = 1.$$

*55. Show that strong induction is a valid method of proof by showing that it follows from the well-ordering property.

*56. Show that the following form of mathematical induction is a valid method to prove that $P(n)$ is true for all positive integers n.

Basis Step: $P(1)$ and $P(2)$ are true.

Inductive Step: For each positive integer k, if $P(k)$ and $P(k + 1)$ are both true, then $P(k + 2)$ is true.

In Exercises 57 and 58, H_n denotes the nth harmonic number.

*57. Use mathematical induction to show that $H_{2^n} \leq 1 + n$ whenever n is a nonnegative integer.

*58. Use mathematical induction to prove that

$$H_1 + H_2 + \cdots + H_n = (n + 1)H_n - n.$$

*59. Prove that

$$1 + \frac{1}{\sqrt{2}} + \frac{1}{\sqrt{3}} + \cdots + \frac{1}{\sqrt{n}} > 2(\sqrt{n+1} - 1).$$

*60. Show that n lines separate the plane into $(n^2 + n + 2)/2$ regions if no two of these lines are parallel and no three pass through a common point.

61. Let $a_1, a_2, \ldots, a_n$ be positive real numbers. The **arithmetic mean of these numbers is defined by

$$A = (a_1 + a_2 + \cdots + a_n)/n,$$

and the **geometric mean** of these numbers is defined by

$$G = (a_1 a_2 \cdots a_n)^{1/n}.$$

Use mathematical induction to prove that $A \geq G$.

*62. Use mathematical induction to show that 21 divides $4^{n+1} + 5^{2n-1}$ whenever n is a positive integer.

63. Use mathematical induction to prove Lemma 2 of Section 2.6, which states that if p is a prime and $p \mid a_1 a_2 \cdots a_n$, where a_i is an integer for $i = 1, 2, 3, \ldots, n$, then $p \mid a_i$ for some integer i.

64. Use infinite descent to show that the equation $8x^4 + 4y^4 + 2z^4 = w^4$ has no solutions in positive integers x, y, z, and w.

65. Use infinite descent to show that there are no solutions in positive integers w, x, y, and z to $w^2 + x^2 + y^2 + z^2 = 2wxyz$. (*Hint:* First show that if this equation holds, then all of w, x, y, and z must be even. Then show that all four of these integers must be divisible by 4, by 8, and so on.)

*66. The well-ordering property can be used to show that there is a unique greatest common divisor of two positive integers. Let a and b be positive integers, and let S be the set of positive integers of the form $as + bt$, where s and t are integers.

a) Show that S is nonempty.

b) Use the well-ordering property to show that S has a smallest element c.

c) Show that if d is a common divisor of a and b, then d is a divisor of c.

d) Show that $c \mid a$ and $c \mid b$. (*Hint:* First, assume that $c \nmid a$. Then $a = qc + r$, where $0 < r < c$. Show that $r \in S$, contradicting the choice of c.)

e) Conclude from (c) and (d) that the greatest common divisor of a and b exists. Finish the proof by showing that this greatest common divisor of two positive integers is unique.

*67. Show that if $a_1, a_2, \ldots, a_n$ are n distinct real numbers, exactly $n - 1$ multiplications are used to compute the product of these n numbers no matter how parenthe-

ses are inserted into their product. (*Hint:* Use strong induction and consider the last multiplication.)

68. Construct a tiling using L-shaped pieces of the 4×4 chessboard with the square in the upper left corner removed.

69. Construct a tiling using L-shaped pieces of the 8×8 chessboard with the square in the upper left corner removed.

70. Prove or disprove that all chessboards of these shapes can be completely covered using L-shaped pieces whenever n is a positive integer.

a) 3×2^n b) 6×2^n
c) $3^n \times 3^n$ d) $6^n \times 6^n$

*71. Show that a three-dimensional $2^n \times 2^n \times 2^n$ chessboard with one $1 \times 1 \times 1$ cube missing can be completely covered by $2 \times 2 \times 2$ cubes with one $1 \times 1 \times 1$ cube removed.

*72. Show that an $n \times n$ chessboard with one square removed can be completely covered using L-shaped pieces if $n > 5, n$ is odd, and $3 \nmid n$.

73. Show that a 5×5 chessboard with a corner square removed can be tiled using L-shaped pieces.

*74. Find a 5×5 chessboard with a square removed that cannot be tiled using L-shaped pieces. Prove that such a tiling does not exist for this board.

75. Let a be an integer and d be a positive integer. Show that the integers q and r with $a = dq + r$ and $0 \leq r < d$, which were shown to exist in Example 16, are unique.

76. Use the principle of mathematical induction to show that $P(n)$ is true for $n = b, b + 1, b + 2, \ldots$, where b is an integer, if $P(b)$ is true and the implication $P(k) \rightarrow P(k + 1)$ is true for all positive integers k with $k \geq b$.

**77. Can you use the well-ordering property to prove this statement? "Every positive integer can be described using no more than 15 English words"?

78. Use the well-ordering principle to show that if x and y are real numbers with $x < y$, then there is a rational number r with $x < r < y$. [*Hint:* Show that there exists a positive integer A with $A > 1/(y - x)$. Then show that there is a rational number r with denominator A between x and y by looking at the numbers $\lfloor x \rfloor + j/A$, where j is a positive integer.]

3.4   Recursive Definitions and Structural Induction

INTRODUCTION

Sometimes it is difficult to define an object explicitly. However, it may be easy to define this object in terms of itself. This process is called **recursion.** For instance, the picture shown in Figure 1 is produced recursively. First, an original picture is given. Then a process of

FIGURE 1   **A Recursively Defined Picture.**

successively superimposing centered smaller pictures on top of the previous pictures is carried out.

We can use recursion to define sequences, functions, and sets. In previous discussions, we specified the terms of a sequence using an explicit formula. For instance, the sequence of powers of 2 is given by $a_n = 2^n$ for $n = 0, 1, 2, \ldots$. However, this sequence can also be defined by giving the first term of the sequence, namely, $a_0 = 1$, and a rule for finding a term of the sequence from the previous one, namely, $a_{n+1} = 2a_n$ for $n = 0, 1, 2, \ldots$. When we define a sequence *recursively* by specifying how terms of the sequence are found from previous terms, we can use induction to prove results about the sequence.

When we define sets recursively, we specify some initial elements in a basis step and provide a rule for constructing new elements from those we already have in the recursive step. To prove results about recursively defined sets we use a method called *structural induction.*

RECURSIVELY DEFINED FUNCTIONS

We use two steps to define a function with the set of nonnegative integers as its domain:

BASIS STEP: Specify the value of the function at zero.

RECURSIVE STEP: Give a rule for finding its value at an integer from its values at smaller integers.

Such a definition is called a **recursive** or **inductive definition.**

EXAMPLE 1

Suppose that f is defined recursively by

$$f(0) = 3,$$
$$f(n + 1) = 2f(n) + 3.$$

Find $f(1), f(2), f(3)$, and $f(4)$.

Solution: From the recursive definition it follows that

$$f(1) = 2f(0) + 3 = 2 \cdot 3 + 3 = 9,$$
$$f(2) = 2f(1) + 3 = 2 \cdot 9 + 3 = 21,$$
$$f(3) = 2f(2) + 3 = 2 \cdot 21 + 3 = 45,$$
$$f(4) = 2f(3) + 3 = 2 \cdot 45 + 3 = 93.$$ ◀

Many functions can be studied using their recursive definitions. The factorial function is one such example.

EXAMPLE 2

Give an inductive definition of the factorial function $F(n) = n!$.

Solution: We can define the factorial function by specifying the initial value of this function, namely, $F(0) = 1$, and giving a rule for finding $F(n+1)$ from $F(n)$. This is obtained by noting that $(n + 1)!$ is computed from $n!$ by multiplying by $n + 1$. Hence, the desired rule is

$$F(n + 1) = (n + 1)F(n).$$ ◀

To determine a value of the factorial function, such as $F(5) = 5!$, from the recursive definition found in Example 2, it is necessary to use the rule that shows how to express $F(n + 1)$ in terms of $F(n)$ several times:

$$F(5) = 5F(4) = 5 \cdot 4F(3) = 5 \cdot 4 \cdot 3F(2) = 5 \cdot 4 \cdot 3 \cdot 2F(1)$$
$$= 5 \cdot 4 \cdot 3 \cdot 2 \cdot 1 \cdot F(0) = 5 \cdot 4 \cdot 3 \cdot 2 \cdot 1 \cdot 1 = 120.$$

Once $F(0)$ is the only value of the function that occurs, no more reductions are necessary. The only thing left to do is to insert the value of $F(0)$ into the formula.

Recursively defined functions are well defined. This is a consequence of the principle of mathematical induction. (See Exercise 56 at the end of this section.) Additional examples of recursive definitions are given in the following examples.

EXAMPLE 3

Give a recursive definition of a^n, where a is a nonzero real number and n is a nonnegative integer.

Solution: The recursive definition contains two parts. First a^0 is specified, namely, $a^0 = 1$. Then the rule for finding a^{n+1} from a^n, namely, $a^{n+1} = a \cdot a^n$, for $n = 0, 1, 2, 3, \ldots$, is given. These two equations uniquely define a^n for all nonnegative integers n. ◀

EXAMPLE 4

Give a recursive definition of

$$\sum_{k=0}^{n} a_k.$$

Solution: The first part of the recursive definition is

$$\sum_{k=0}^{0} a_k = a_0.$$

The second part is

$$\sum_{k=0}^{n+1} a_k = \left(\sum_{k=0}^{n} a_k\right) + a_{n+1}. \qquad \blacktriangleleft$$

In some recursive definitions of functions, the values of the function at the first k positive integers are specified, and a rule is given for determining the value of the function at larger integers from its values at some or all of the preceding k integers. That such definitions produce well-defined functions follows from strong induction (see Exercise 57 at the end of this section).

DEFINITION 1

The *Fibonacci numbers*, $f_0, f_1, f_2, \ldots$, are defined by the equations $f_0 = 0$, $f_1 = 1$, and

$$f_n = f_{n-1} + f_{n-2}$$

for $n = 2, 3, 4, \ldots$.

EXAMPLE 5   Find the Fibonacci numbers f_2, f_3, f_4, f_5, and f_6.

Solution: Since the first part of the definition states that $f_0 = 0$ and $f_1 = 1$, it follows from the second part of the definition that

$$f_2 = f_1 + f_0 = 1 + 0 = 1,$$
$$f_3 = f_2 + f_1 = 1 + 1 = 2,$$
$$f_4 = f_3 + f_2 = 2 + 1 = 3,$$
$$f_5 = f_4 + f_3 = 3 + 2 = 5,$$
$$f_6 = f_5 + f_4 = 5 + 3 = 8.$$
$$\blacktriangleleft$$

We can use the recursive definition of the Fibonacci numbers to prove many properties of these numbers. We give one such property in Example 6.

FIBONACCI (1170–1250)   Fibonacci (short for *filius Bonacci,* or "son of Bonacci") was also known as Leonardo of Pisa. He was born in the Italian commercial center of Pisa. Fibonacci was a merchant who traveled extensively throughout the Mideast, where he came into contact with Arabian mathematics. In his book *Liber Abaci,* Fibonacci introduced the European world to Arabic notation for numerals and algorithms for arithmetic. It was in this book that his famous rabbit problem (described in Section 6.1) appeared. Fibonacci also wrote books on geometry and trigonometry and on Diophantine equations, which involve finding integer solutions to equations.

EXAMPLE 6   Show that whenever $n \geq 3$, $f_n > \alpha^{n-2}$, where $\alpha = (1 + \sqrt{5})/2$.

Solution: We can use strong induction to prove this inequality. Let $P(n)$ be the statement $f_n > \alpha^{n-2}$. We want to show that $P(n)$ is true whenever n is an integer greater than or equal to 3.

BASIS STEP: First, note that

$$\alpha < 2 = f_3, \qquad \alpha^2 = (3 + \sqrt{5})/2 < 3 = f_4,$$

so that $P(3)$ and $P(4)$ are true.

INDUCTIVE STEP: Assume that $P(j)$ is true, namely, that $f_j > \alpha^{j-2}$, for all integers j with $3 \leq j \leq k$, where $k \geq 4$. We must show that $P(k + 1)$ is true, that is, that $f_{k+1} > \alpha^{k-1}$. Since α is a solution of $x^2 - x - 1 = 0$ (as the quadratic formula shows), it follows that $\alpha^2 = \alpha + 1$. Therefore,

$$\alpha^{k-1} = \alpha^2 \cdot \alpha^{k-3} = (\alpha + 1)\alpha^{k-3} = \alpha \cdot \alpha^{k-3} + 1 \cdot \alpha^{k-3} = \alpha^{k-2} + \alpha^{k-3}.$$

By the inductive hypothesis, if $k \geq 4$, it follows that

$$f_{k-1} > \alpha^{k-3}, \qquad f_k > \alpha^{k-2}.$$

Therefore, we have

$$f_{k+1} = f_k + f_{k-1} > \alpha^{k-2} + \alpha^{k-3} = \alpha^{k-1}.$$

It follows that $P(k + 1)$ is true. This completes the proof.   ◄

Remark: The inductive step shows that whenever $k \geq 4$, $P(k + 1)$ follows from the assumption that $P(j)$ is true for $3 \leq j \leq k$. Hence, the inductive step does *not* show that $P(3) \rightarrow P(4)$. Therefore, we had to show that $P(4)$ is true separately.

We can now show that the Euclidean algorithm uses $O(\log b)$ divisions to find the greatest common divisor of the positive integers a and b, where $a \geq b$.

THEOREM 1

> **LAMÉ'S THEOREM**   Let a and b be positive integers with $a \geq b$. Then the number of divisions used by the Euclidean algorithm to find $\gcd(a, b)$ is less than or equal to five times the number of decimal digits in b.

GABRIEL LAMÉ (1795–1870)   Gabriel Lamé entered the École Polytechnique in 1813, graduating in 1817. He continued his education at the École des Mines, graduating in 1820.

In 1820 Lamé went to Russia, where he was appointed director of the Schools of Highways and Transportation in St. Petersburg. Not only did he teach, but he also planned roads and bridges while in Russia. He returned to Paris in 1832, where he helped found an engineering firm. However, he soon left the firm, accepting the chair of physics at the École Polytechnique, which he held until 1844. While holding this position, he was active outside academia as an engineering consultant, serving as chief engineer of mines and participating in the building of railways.

Lamé contributed original work to number theory, applied mathematics, and thermodynamics. His best-known work involves the introduction of curvilinear coordinates. His work on number theory includes proving Fermat's Last Theorem for $n = 7$, as well as providing the upper bound for the number of divisions used by the Euclidean algorithm given in this text.

In the opinion of Gauss, one of the most important mathematicians of all time, Lamé was the foremost French mathematician of his time. However, French mathematicians considered him too practical, whereas French scientists considered him too theoretical.

Proof: Recall that when the Euclidean algorithm is applied to find $\gcd(a, b)$ with $a \geq b$, this sequence of equations (where $a = r_0$ and $b = r_1$) is obtained.

$$r_0 = r_1 q_1 + r_2 \qquad\qquad 0 \leq r_2 < r_1$$

$$r_1 = r_2 q_2 + r_3 \qquad\qquad 0 \leq r_3 < r_2$$

$$\cdot$$
$$\cdot$$
$$\cdot$$

$$r_{n-2} = r_{n-1} q_{n-1} + r_n \qquad 0 \leq r_n < r_{n-1}$$

$$r_{n-1} = r_n q_n.$$

Here n divisions have been used to find $r_n = \gcd(a, b)$. Note that the quotients $q_1, q_2, \ldots, q_{n-1}$ are all at least 1. Moreover, $q_n \geq 2$, since $r_n < r_{n-1}$. This implies that

$$r_n \geq 1 = f_2,$$

$$r_{n-1} \geq 2r_n \geq 2f_2 = f_3,$$

$$r_{n-2} \geq r_{n-1} + r_n \geq f_3 + f_2 = f_4,$$

$$\cdot$$
$$\cdot$$
$$\cdot$$

$$r_2 \geq r_3 + r_4 \geq f_{n-1} + f_{n-2} = f_n,$$

$$b = r_1 \geq r_2 + r_3 \geq f_n + f_{n-1} = f_{n+1}.$$

It follows that if n divisions are used by the Euclidean algorithm to find $\gcd(a, b)$ with $a \geq b$, then $b \geq f_{n+1}$. From Example 6 we know that $f_{n+1} > \alpha^{n-1}$ for $n > 2$, where $\alpha = (1 + \sqrt{5})/2$. Therefore, it follows that $b > \alpha^{n-1}$. Furthermore, since $\log_{10} \alpha \sim 0.208 > 1/5$, we see that

$$\log_{10} b > (n - 1) \log_{10} \alpha > (n - 1)/5.$$

Hence, $n - 1 < 5 \cdot \log_{10} b$. Now suppose that b has k decimal digits. Then $b < 10^k$ and $\log_{10} b < k$. It follows that $n - 1 < 5k$, and since k is an integer, it follows that $n \leq 5k$. This finishes the proof. ◁

Since the number of decimal digits in b, which equals $\lfloor \log_{10} b \rfloor + 1$, is less than or equal to $\log_{10} b + 1$, Theorem 1 tells us that the number of divisions required to find $\gcd(a, b)$ with $a > b$ is less than or equal to $5(\log_{10} b + 1)$. Since $5(\log_{10} b + 1)$ is $O(\log b)$, we see that $O(\log b)$ divisions are used by the Euclidean algorithm to find $\gcd(a, b)$ whenever $a > b$.

RECURSIVELY DEFINED SETS AND STRUCTURES

We have explored how functions can be defined recursively. We now turn our attention to how sets can be defined recursively. Just as in the recursive definition of functions, recursive definitions of sets have two parts, a **basis step** and a **recursive step.** In the basis step, an initial collection of elements is specified. In the recursive step, rules for forming new elements in the set from those already known to be in the set are provided. Recursive definitions may also include an **exclusion rule,** which specifies that a recursively defined set contains nothing other than those elements specified in the basis step or generated by applications of the recursive step. In our discussions, we will always tacitly assume that the exclusion rule holds and no element belongs to a recursively defined set unless it is in the initial collection specified in the basis step or can be generated using the recursive step one or more times. Later we will see how we can use a technique known as structural induction to prove results about recursively defined sets.

Examples 7, 8, 10, and 11 illustrate the recursive definition of sets. In each example, we show those elements generated by the first few applications of the recursive step.

EXAMPLE 7   Consider the subset S of the set of integers defined by

BASIS STEP: $3 \in S$.

Extra Examples

RECURSIVE STEP: If $x \in S$ and $y \in S$, then $x + y \in S$.

The new elements found to be in S are 3 by the basis step, $3 + 3 = 6$ at the first application of the recursive step, $3 + 6 = 6 + 3 = 9$ and $6 + 6 = 12$ at the second application of the recursive step, and so on.   ◀

Recursive definitions play an important role in the study of strings. (See Chapter 11 for an introduction to the theory of formal languages, for example.) Recall from Section 3.2 that a string over an alphabet Σ is a finite sequence of symbols from Σ. We can define Σ^*, the set of strings over Σ, recursively, as Definition 2 shows.

DEFINITION 2   The set Σ^* of *strings* over the alphabet Σ can be defined recursively by

BASIS STEP: $\lambda \in \Sigma^*$ (where λ is the empty string containing no symbols).

RECURSIVE STEP: If $w \in \Sigma^*$ and $x \in \Sigma$, then $wx \in \Sigma^*$.

The basis step of the recursive definition of strings says that the empty string belongs to Σ^*. The recursive step states that new strings are produced by adding a symbol from Σ to the end of strings in Σ^*. At each application of the recursive step, strings containing one additional symbol are generated.

EXAMPLE 8   If $\Sigma = \{0, 1\}$, the strings found to be in Σ^*, the set of all bit strings, are λ, specified to be in Σ^* in the basis step, 0 and 1 formed during the first application of the recursive step, 00, 01, 10, and 11 formed during the second application of the recursive step, and so on.   ◀

Recursive definitions can be used to define operations or functions on the elements of recursively defined sets. This is illustrated in Definition 3 of the concatenation of two strings and Example 9 concerning the length of a string.

DEFINITION 3

> Two strings can be combined via the operation of *concatenation*. Let Σ be a set of symbols and Σ^* the set of strings formed from symbols in Σ. We can define the concatenation of two strings, denoted by $\cdot$, recursively as follows.
>
> *BASIS STEP:* If $w \in \Sigma^*$, then $w \cdot \lambda = w$, where λ is the empty string.
>
> *RECURSIVE STEP:* If $w_1 \in \Sigma^*$ and $w_2 \in \Sigma^*$ and $x \in \Sigma$, then $w_1 \cdot (w_2 x) = (w_1 \cdot w_2)x$.

The concatenation of the strings w_1 and w_2 is often written as $w_1 w_2$ rather than $w_1 \cdot w_2$. By repeated application of the recursive definition, it follows that the concatenation of two strings w_1 and w_2 consists of the symbols in w_1 followed by the symbols in w_2. For instance, the concatenation of $w_1 = abra$ and $w_2 = cadabra$ is $w_1 w_2 = abracadabra$.

EXAMPLE 9   **Length of a String** Give a recursive definition of $l(w)$, the length of the string w.

Solution: The length of a string can be defined by

$$l(\lambda) = 0;$$
$$l(wx) = l(w) + 1 \text{ if } w \in \Sigma^* \text{ and } x \in \Sigma. \qquad \blacktriangleleft$$

Another important use of recursive definitions is to define **well-formed formulae** of various types. This is illustrated in Examples 10 and 11.

EXAMPLE 10   **Well-Formed Formulae for Compound Propositions** We can define the set of well-formed formulae for compound propositions involving $\mathbf{T}, \mathbf{F}$, propositional variables, and operators from the set $\{\neg, \wedge, \vee, \rightarrow, \leftrightarrow\}$.

BASIS STEP: $\mathbf{T}$, $\mathbf{F}$, and p, where p is a propositional variable, are well-formed formulae.

RECURSIVE STEP: If E and F are well-formed formulae, then $(\neg E), (E \wedge F), (E \vee F), (E \rightarrow F)$, and $(E \leftrightarrow F)$ are well-formed formulae.

For example, by the basis step we know that $\mathbf{T}, \mathbf{F}, p$, and q are well-formed formulae, where p and q are propositional variables. From an initial application of the recursive step, we know that $(p \vee q), (p \rightarrow \mathbf{F}), (\mathbf{F} \rightarrow q)$, and $(q \wedge \mathbf{F})$ are well-formed formulae. A second application of the recursive step shows that $((p \vee q) \rightarrow (q \wedge \mathbf{F})), (q \vee (p \vee q))$, and $((p \rightarrow \mathbf{F}) \rightarrow \mathbf{T})$ are well-formed formulae. $\qquad \blacktriangleleft$

EXAMPLE 11   **Well-Formed Formulae of Operators and Operands** We can define the set of well-formed formulae consisting of variables, numerals, and operators from the set $\{+, -, *, /, \uparrow\}$ (where $*$ denotes multiplication and $\uparrow$ denotes exponentiation) recursively.

BASIS STEP: x is a well-formed formula if x is a numeral or variable.

RECURSIVE STEP: If F and G are well-formed formulae, then $(F + G)$, $(F - G)$, $(F * G)$, (F/G), and $(F \uparrow G)$ are well-formed formulae.

For example, by the basis step we see that $x, y, 0,$ and 3 are well-formed formulae (as is any variable or numeral). Well-formed formulae generated by applying the recursive step once include $(x + 3)$, $(3 + y)$, $(x - y)$, $(3 - 0)$, $(x * 3)$, $(3 * y)$, $(3/0)$, (x/y), $(3 \uparrow x)$, and $(0 \uparrow 3)$. Applying the recursive step twice shows that formulae such as $((x + 3) + 3)$ and $(x - (3 * y))$ are well-formed formulae. [Note that $(3/0)$ is a well-formed formula since we are concerned only with syntax matters here.] ◀

We will study trees extensively in Chapter 9. A tree is a special type of a graph; a graph is made up of vertices and edges connecting some pairs of vertices. We will study graphs in Chapter 8. We will briefly introduce them here to illustrate how they can be defined recursively.

DEFINITION 4

The set of *rooted trees,* where a rooted tree consists of a set of vertices containing a distinguished vertex called the *root,* and edges connecting these vertices, can be defined recursively by these steps:

BASIS STEP: A single vertex r is a rooted tree.

RECURSIVE STEP: Suppose that $T_1, T_2, \ldots, T_n$ are rooted trees with roots $r_1, r_2, \ldots, r_n$, respectively. Then the graph formed by starting with a root r, which is not in any of the rooted trees $T_1, T_2, \ldots, T_n$, and adding an edge from r to each of the vertices $r_1, r_2, \ldots, r_n$, is also a rooted tree.

In Figure 2 we illustrate some of the rooted trees formed starting with the basis step and applying the recursive step one time and two times. Note that infinitely many rooted trees are formed at each application of the recursive definition.

Rooted trees are a special type of binary trees. We will provide recursive definitions of two types of binary trees, full binary trees and extended binary trees. In the recursive step

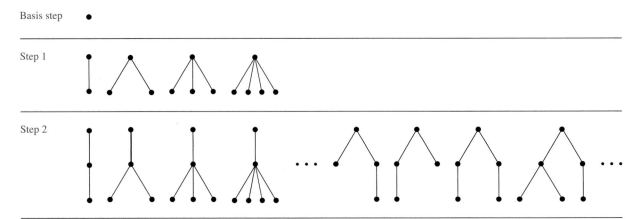

FIGURE 2   **Building Up Rooted Trees.**

of the definition of each type of binary tree, two binary trees are combined to form a new tree with one of these trees designated the left subtree and the other the right subtree. In extended binary trees, the left subtree or the right subtree can be empty, but in full binary trees this is not possible. Binary trees are one of the most important types of structures in computer science. In Chapter 9 we will see how they can be used in searching and sorting algorithms, in algorithms for compressing data, and in many other applications. We first define extended binary trees.

DEFINITION 5

The set of *extended binary trees* can be defined recursively by these steps:

BASIS STEP: The empty set is an extended binary tree.

RECURSIVE STEP: If T_1 and T_2 are extended binary trees, there is an extended binary tree, denoted by $T_1 \cdot T_2$, consisting of a root r together with edges connecting the root to each of the roots of the left subtree T_1 and the right subtree T_2 when these trees are nonempty.

Figure 3 shows how extended binary trees are built up by applying the recursive step from one to three times.

We now show how to define the set of full binary trees. Note that the difference between this recursive definition and that of extended binary trees lies entirely in the basis step.

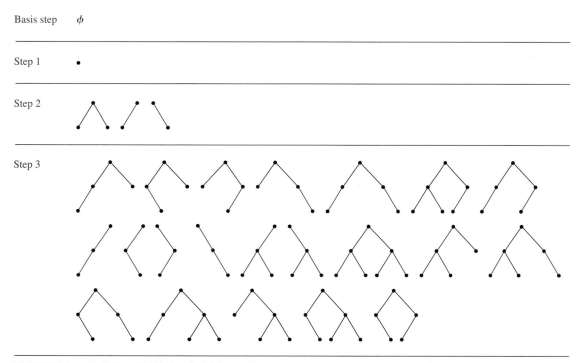

FIGURE 3   Building Up Extended Binary Trees.

Basis step

Step 1

Step 2

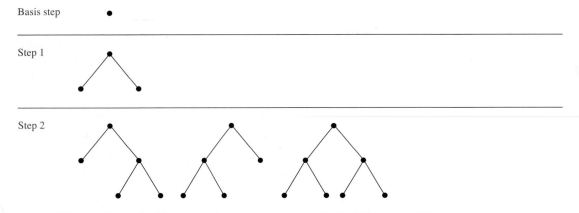

FIGURE 4   Building Up Full Binary Trees.

DEFINITION 6

The set of full binary trees can be defined recursively by these steps:

BASIS STEP: There is a full binary tree consisting only of a single vertex r.

RECURSIVE STEP: If T_1 and T_2 are full binary trees, there is a full binary tree, denoted by $T_1 \cdot T_2$, consisting of a root r together with edges connecting the root to each of the roots of the left subtree T_1 and the right subtree T_2.

Figure 4 shows how full binary trees are built up by applying the recursive step one and two times.

STRUCTURAL INDUCTION

To prove results about recursively defined sets we generally use some form of mathematical induction. Example 12 illustrates the connection between recursively defined sets and mathematical induction.

EXAMPLE 12   Show that the set S defined in Example 7 is the set of all positive integers that are multiples of 3.

Solution: Let A be the set of all positive integers divisible by 3. To prove that $A = S$, we must show that A is a subset of S and that S is a subset of A. To prove that A is a subset of S, we must show that every positive integer divisible by 3 is in S. We will use mathematical induction to prove this.

   Let $P(n)$ be the statement that $3n$ belongs to S. The basis step holds since by the first part of the recursive definition of S, $3 \cdot 1 = 3$ is in S. To establish the inductive step, assume that $P(k)$ is true, namely, that $3k$ is in S. Since $3k$ is in S and since 3 is in S, it follows from the second part of the recursive definition of S that $3k + 3 = 3(k + 1)$ is also in S.

   To prove that S is a subset of A, we use the recursive definition of S. First, the basis step of the definition specifies that 3 is in S. Since $3 = 3 \cdot 1$, all elements specified to be in S in this step are divisible by 3. To finish the proof, we must show that all integers in S generated using the second part of the recursive definition are in A. This consists of showing that $x + y$ is in A whenever x and y are elements of S also assumed to be in A. Now if x and y are both in A, it follows that $3 \mid x$ and $3 \mid y$. By Theorem 1 of Section 2.4, it follows that $3 \mid x + y$, completing the proof. ◀

In Example 12 we used mathematical induction over the set of positive integers and a recursive definition to prove a result about a recursively defined set. However, instead of using mathematical induction directly to prove results about recursively defined sets, we can use a more convenient form of induction known as **structural induction.** A proof by structural induction consists of two parts. These parts are

BASIS STEP: Show that the result holds for all elements specified in the basis step of the recursive definition to be in the set.

RECURSIVE STEP: Show that if the statement is true for each of the elements used to construct new elements in the recursive step of the definition, the result holds for these new elements.

The validity of structural induction follows from the principle of mathematical induction for the nonnegative integers. To see this, let $P(n)$ state that the claim is true for all elements of the set that are generated by n or fewer applications of the rules in the recursive step of a recursive definition. We will have established that the principle of mathematical induction implies the principle of structural induction if we can show that $P(n)$ is true whenever n is a positive integer. In the basis step of a proof by structural induction we show that $P(0)$ is true. That is, we show that the result is true of all elements specified to be in the set in the basis step of the definition. A consequence of the inductive step is that if we assume $P(k)$ is true, it follows that $P(k + 1)$ is true. When we have completed a proof using structural induction, we have shown that $P(0)$ is true and that $P(k)$ implies $P(k + 1)$. By mathematical induction it follows that $P(n)$ is true for all nonnegative integers n. This also shows that the result is true for all elements generated by the recursive definition, and shows that structural induction is a valid proof technique.

EXAMPLES OF PROOFS USING STRUCTURAL INDUCTION To use structural induction to prove a result about the set of well-formed expressions defined in Example 10, we need to complete this basis step and this recursive step.

BASIS STEP: Show that the result is true for $\mathbf{T}, \mathbf{F}$, and p whenever p is a propositional variable.

RECURSIVE STEP: Show that if the result is true for the compound propositions p and q, it is also true for $(\neg p), (p \lor q), (p \land q), (p \rightarrow q)$, and $(p \leftrightarrow q)$.

Example 13 illustrates how we can prove results about well-formed formulae using structural induction.

EXAMPLE 13 Show that every well-formed formula for compound propositions, as defined in Example 10, contains an equal number of left and right parentheses. ◀

Proof:

BASIS STEP: Each of the formulae $\mathbf{T}, \mathbf{F}$, and p contains no parentheses, so clearly they contain an equal number of left and right parentheses.

RECURSIVE STEP: Assume p and q are well-formed formulae each containing an equal number of left and right parentheses. That is, if l_p and l_q are the number of left parentheses in p and q, respectively, and r_p and r_q are the number of right parentheses in p and q, respectively, then $l_p = r_p$ and $l_q = r_q$. To complete the inductive step, we need to show that each of $(\neg p)$, $(p \vee q)$, $(p \wedge q)$, $(p \rightarrow q)$, and $(p \leftrightarrow q)$ also contains an equal number of left and right parentheses. The number of left parentheses in the first of these compound propositions equals $l_p + 1$ and in each of the other compound propositions equals $l_p + l_q + 1$. Similarly, the number of right parentheses in the first of these compound propositions equals $r_p + 1$ and in each of the other compound propositions equals $r_p + r_q + 1$. Since $l_p = r_p$ and $l_q = r_q$, it follows that each of these compound expressions contains the same number of left and right parentheses. This completes the inductive proof.   ◁

Suppose that $P(w)$ is a propositional function over the set of strings $w \in \Sigma^*$. To use structural induction to prove that $P(w)$ holds for all strings $w \in \Sigma^*$, we need to complete both a basis step and a recursive step. These steps are:

BASIS STEP: Show that $P(\lambda)$ is true.

RECURSIVE STEP: Assume that $P(w)$ is true, where $w \in \Sigma^*$. Show that if $x \in \Sigma$, then $P(wx)$ must also be true.

Example 14 illustrates how structural induction can be used in proofs about strings.

EXAMPLE 14   Use structural induction to prove that $l(xy) = l(x) + l(y)$, where x and y belong to Σ^*, the set of strings over the alphabet Σ.

Solution: We will base our proof on the recursive definition of the set Σ^* given in Definition 2 and the definition of the length of a string in Example 9. Let $P(y)$ be the statement that $l(xy) = l(x) + l(y)$ whenever x belongs to Σ^*.

BASIS STEP: To complete the basis step, we must show that $P(\lambda)$ is true. That is, we must show that $l(x\lambda) = l(x) + l(\lambda)$ for all $x \in \Sigma^*$. Since $l(x\lambda) = l(x) = l(x) + 0 = l(x) + l(\lambda)$ for every string x, it follows that $P(\lambda)$ is true.

RECURSIVE STEP: To complete the inductive step, we assume that $P(y)$ is true and show that this implies that $P(ya)$ is true whenever $a \in \Sigma$. What we need to show is that $l(xya) = l(x) + l(ya)$ for every $a \in \Sigma$. To show this, note that by the recursive definition of $l(w)$ (given in Example 9), we have $l(xya) = l(xy) + 1$ and $l(ya) = l(y) + 1$. And, by the inductive hypothesis, $l(xy) = l(x) + l(y)$. We conclude that $l(xya) = l(x) + l(y) + 1 = l(x) + l(ya)$.   ◀

We can prove results about trees or special classes of trees using structural induction. For example, to prove a result about full binary trees using structural induction we need to complete this basis step and this recursive step.

BASIS STEP: Show that the result is true for the tree consisting of a single vertex.

RECURSIVE STEP: Show that if the result is true for the trees T_1 and T_2, then it is true for tree $T_1 \cdot T_2$ consisting of a root r, which has T_1 as its left subtree and T_2 as its right subtree.

Before we provide an example showing how structural induction can be used to prove a result about full binary trees, we need some definitions. We will recursively define the height $h(T)$ and the number of vertices $n(T)$ of a full binary tree T. We begin by defining the height of a full binary tree.

DEFINITION 7

We define the height $h(T)$ of a full binary tree T recursively.

BASIS STEP: The height of the full binary tree T consisting of only a root r is $h(T) = 0$.

RECURSIVE STEP: If T_1 and T_2 are full binary trees, then the full binary tree $T = T_1 \cdot T_2$ has height $h(T) = 1 + \max(h(T_1), h(T_2))$.

If we let $n(T)$ denote the number of vertices in a full binary tree, we observe that $n(T)$ satisfies the following recursive formula:

BASIS STEP: The number of vertices $n(T)$ of the full binary tree T consisting of only a root r is $n(T) = 1$.

RECURSIVE STEP: If T_1 and T_2 are full binary trees, then the number of vertices of the full binary tree $T = T_1 \cdot T_2$ is $n(T) = 1 + n(T_1) + n(T_2)$.

We now show how structural induction can be used to prove a result about full binary trees.

THEOREM 2

If T is a full binary tree T, then $n(T) \leq 2^{h(T)+1} - 1$.

Proof: We prove this inequality using structural induction.

BASIS STEP: For the full binary tree consisting of just the root r the result is true since $n(T) = 1$ and $h(T) = 0$, so that $n(T) = 1 \leq 2^{0+1} - 1 = 1$.

INDUCTIVE STEP: For the inductive hypothesis we assume that $n(T_1) \leq 2^{h(T_1)+1} - 1$ and $n(T_2) \leq 2^{h(T_2)+1} - 1$ whenever T_1 and T_2 are full binary trees. By the recursive formulae for $n(T)$ and $h(T)$ we have $n(T) = 1 + n(T_1) + n(T_2)$ and $h(T) = 1 + \max(h(T_1), h(T_2))$.

We find that

$$n(T) = 1 + n(T_1) + n(T_2) \quad \text{by the recursive formula for } n(T)$$
$$\leq 1 + (2^{h(T_1)+1} - 1) + (2^{h(T_2)+1} - 1) \quad \text{by the inductive hypothesis}$$
$$= 2 \cdot \max(2^{h(T_1)+1}, 2^{h(T_2)+1}) - 1 \quad \text{since the sum of two terms is at most}$$
$$\text{2 times the larger}$$
$$= 2 \cdot 2^{\max(h(T_1), h(T_2))+1} - 1$$
$$= 2 \cdot 2^{h(T)} - 1 \quad \text{by the recursive definition of } h(T)$$
$$= 2^{h(T)+1} - 1.$$

This completes the inductive step.   ◁

GENERALIZED INDUCTION

We can extend mathematical induction to prove results about other sets that have the well-ordering property besides the set of integers. Although we will discuss this concept in detail in Section 7.6, we provide an example here to illustrate the usefulness of such an approach.

As an example, note that we can define an ordering on $\mathbf{N} \times \mathbf{N}$, the ordered pairs of nonnegative integers, by specifying that (x_1, y_1) is less than or equal to (x_2, y_2) if either $x_1 < x_2$, or $x_1 = x_2$ and $y_1 < y_2$; this is called the **lexicographic ordering.** The set $\mathbf{N} \times \mathbf{N}$ with this ordering has the property that every subset of $\mathbf{N} \times \mathbf{N}$ has a least element (see Supplementary Exercise 47 in Chapter 7). This implies that we can recursively define the terms $a_{m,n}$, with $m \in \mathbf{N}$ and $n \in \mathbf{N}$, and prove results about them using a variant of mathematical induction, as illustrated in Example 15.

EXAMPLE 15   Suppose that $a_{m,n}$ is defined recursively for $(m, n) \in \mathbf{N} \times \mathbf{N}$ by $a_{0,0} = 0$ and

$$a_{m,n} = \begin{cases} a_{m-1,n} + 1 & \text{if } n = 0 \text{ and } m > 0 \\ a_{m,n-1} + n & \text{if } n > 0. \end{cases}$$

Show that $a_{m,n} = m + n(n + 1)/2$ for all $(m, n) \in \mathbf{N} \times \mathbf{N}$, that is, for all pairs of nonnegative integers.

Solution: We can prove that $a_{m,n} = m + n(n+1)/2$ using a generalized version of mathematical induction. The basis step requires that we show that this formula is valid when $(m, n) = (0, 0)$. The induction step requires that we show that if the formula holds for all pairs smaller than (m, n) in the lexicographic ordering of $\mathbf{N} \times \mathbf{N}$, then it also holds for (m, n).

BASIS STEP: Let $(m, n) = (0, 0)$. Then by the basis case of the recursive definition of $a_{m,n}$ we have $a_{0,0} = 0$. Furthermore, when $m = n = 0$, $m + n(n + 1)/2 = 0 + (0 \cdot 1)/2 = 0$. This completes the basis step.

INDUCTIVE STEP: Suppose that $a_{m',n'} = m' + n'(n' + 1)/2$ whenever (m', n') is less than (m, n) in the lexicographic ordering of $\mathbf{N} \times \mathbf{N}$. By the recursive definition, if $n = 0$, then $a_{m,n} = a_{m-1,n} + 1$. Because $(m - 1, n)$ is smaller than (m, n), the induction hypothesis tells us that $a_{m-1,n} = m - 1 + n(n + 1)/2$, so that $a_{m,n} = m - 1 + n(n + 1)/2 + 1 = m + n(n + 1)/2$, giving us the desired equality. Now suppose that $n > 0$, so $a_{m,n} = a_{m,n-1} + n$. Since $(m, n - 1)$ is smaller than (m, n), the induction hypothesis tells us that $a_{m,n-1} = m + (n - 1)n/2$, so $a_{m,n} = m + (n - 1)n/2 + n = m + (n^2 - n + 2n)/2 = m + n(n + 1)/2$. This finishes the inductive step.   ◄

As mentioned, we will justify this proof technique in Section 7.6.

Exercises

1. Find $f(1)$, $f(2)$, $f(3)$, and $f(4)$ if $f(n)$ is defined recursively by $f(0) = 1$ and for $n = 0, 1, 2, \ldots$

   a) $f(n + 1) = f(n) + 2$.
   b) $f(n + 1) = 3f(n)$.
   c) $f(n + 1) = 2^{f(n)}$.
   d) $f(n + 1) = f(n)^2 + f(n) + 1$.

2. Find $f(1)$, $f(2)$, $f(3)$, $f(4)$, and $f(5)$ if $f(n)$ is defined recursively by $f(0) = 3$ and for $n = 0, 1, 2, \ldots$

   a) $f(n + 1) = -2f(n)$.
   b) $f(n + 1) = 3f(n) + 7$.
   c) $f(n + 1) = f(n)^2 - 2f(n) - 2$.
   d) $f(n + 1) = 3^{f(n)/3}$.

3. Find $f(2)$, $f(3)$, $f(4)$, and $f(5)$ if f is defined recursively by $f(0) = -1$, $f(1) = 2$ and for $n = 1, 2, \ldots$

   a) $f(n + 1) = f(n) + 3f(n - 1)$.
   b) $f(n + 1) = f(n)^2 f(n - 1)$.

c) $f(n+1) = 3f(n)^2 - 4f(n-1)^2$.
d) $f(n+1) = f(n-1)/f(n)$.

4. Find $f(2)$, $f(3)$, $f(4)$, and $f(5)$ if f is defined recursively by $f(0) = f(1) = 1$ and for $n = 1, 2, \ldots$

a) $f(n+1) = f(n) - f(n-1)$.
b) $f(n+1) = f(n)f(n-1)$.
c) $f(n+1) = f(n)^2 + f(n-1)^3$.
d) $f(n+1) = f(n)/f(n-1)$.

5. Determine whether each of these proposed definitions is a valid recursive definition of a function f from the set of nonnegative integers to the set of integers. If f is well defined, find a formula for $f(n)$ when n is a nonnegative integer and prove that your formula is valid.

a) $f(0) = 0$, $f(n) = 2f(n-2)$ for $n \geq 1$
b) $f(0) = 1$, $f(n) = f(n-1) - 1$ for $n \geq 1$
c) $f(0) = 2$, $f(1) = 3$, $f(n) = f(n-1) - 1$ for $n \geq 2$
d) $f(0) = 1$, $f(1) = 2$, $f(n) = 2f(n-2)$ for $n \geq 2$
e) $f(0) = 1$, $f(n) = 3f(n-1)$ if n is odd and $n \geq 1$ and $f(n) = 9f(n-2)$ if n is even and $n \geq 2$

6. Determine whether each of these proposed definitions is a valid recursive definition of a function f from the set of nonnegative integers to the set of integers. If f is well defined, find a formula for $f(n)$ when n is a nonnegative integer and prove that your formula is valid.

a) $f(0) = 1$, $f(n) = -f(n-1)$ for $n \geq 1$
b) $f(0) = 1$, $f(1) = 0$, $f(2) = 2$, $f(n) = 2f(n-3)$ for $n \geq 3$
c) $f(0) = 0$, $f(1) = 1$, $f(n) = 2f(n+1)$ for $n \geq 2$
d) $f(0) = 0$, $f(1) = 1$, $f(n) = 2f(n-1)$ for $n \geq 1$
e) $f(0) = 2$, $f(n) = f(n-1)$ if n is odd and $n \geq 1$ and $f(n) = 2f(n-2)$ if $n \geq 2$

7. Give a recursive definition of the sequence $\{a_n\}$, $n = 1, 2, 3, \ldots$ if

a) $a_n = 6n$.   **b)** $a_n = 2n + 1$.
c) $a_n = 10^n$.   **d)** $a_n = 5$.

8. Give a recursive definition of the sequence $\{a_n\}$, $n = 1, 2, 3, \ldots$ if

a) $a_n = 4n - 2$.   **b)** $a_n = 1 + (-1)^n$.
c) $a_n = n(n+1)$.   **d)** $a_n = n^2$.

9. Let F be the function such that $F(n)$ is the sum of the first n positive integers. Give a recursive definition of $F(n)$.

10. Give a recursive definition of $S_m(n)$, the sum of the integer m and the nonnegative integer n.

11. Give a recursive definition of $P_m(n)$, the product of the integer m and the nonnegative integer n.

In Exercises 12–19 f_n is the nth Fibonacci number.

12. Prove that $f_1^2 + f_2^2 + \cdots + f_n^2 = f_n f_{n+1}$ whenever n is a positive integer.

13. Prove that $f_1 + f_3 + \cdots + f_{2n-1} = f_{2n}$ whenever n is a positive integer.

*14. Show that $f_{n+1} f_{n-1} - f_n^2 = (-1)^n$ whenever n is a positive integer.

*15. Show that $f_0 f_1 + f_1 f_2 + \cdots + f_{2n-1} f_{2n} = f_{2n}^2$ whenever n is a positive integer.

*16. Show that $f_0 - f_1 + f_2 - \cdots - f_{2n-1} + f_{2n} = f_{2n-1} - 1$ whenever n is a positive integer.

17. Determine the number of divisions used by the Euclidean algorithm to find the greatest common divisor of the Fibonacci numbers f_n and f_{n+1} where n is a nonnegative integer. Verify your answer using mathematical induction.

18. Let

$$\mathbf{A} = \begin{bmatrix} 1 & 1 \\ 1 & 0 \end{bmatrix}$$

Show that

$$\mathbf{A}^n = \begin{bmatrix} f_{n+1} & f_n \\ f_n & f_{n-1} \end{bmatrix}$$

whenever n is a positive integer.

19. By taking determinants of both sides of the equation in Exercise 18, prove the identity given in Exercise 14. (This exercise depends on the notion of the determinant of a 2×2 matrix.)

*20. Give a recursive definition of the functions max and min so that $\max(a_1, a_2, \ldots, a_n)$ and $\min(a_1, a_2, \ldots, a_n)$ are the maximum and minimum of the n numbers $a_1, a_2, \ldots, a_n$, respectively.

*21. Let $a_1, a_2 \ldots, a_n$, and $b_1, b_2, \ldots, b_n$ be real numbers. Use the recursive definitions that you gave in Exercise 20 to prove these.

a) $\max(-a_1, -a_2, \ldots, -a_n) = -\min(a_1, a_2, \ldots, a_n)$
b) $\max(a_1 + b_1, a_2 + b_2, \ldots, a_n + b_n)$
$\leq \max(a_1, a_2, \ldots, a_n) + \max(b_1, b_2, \ldots, b_n)$
c) $\min(a_1 + b_1, a_2 + b_2, \ldots, a_n + b_n)$
$\geq \min(a_1, a_2, \ldots, a_n) + \min(b_1, b_2, \ldots, b_n)$

22. Show that the set S defined by $1 \in S$ and $s + t \in S$ whenever $s \in S$ and $t \in S$ is the set of positive integers.

23. Give a recursive definition of the set of positive integers that are multiples of 5.

24. Give a recursive definition of

a) the set of odd positive integers.
b) the set of positive integer powers of 3.
c) the set of polynomials with integer coefficients.

25. Give a recursive definition of

a) the set of even integers.
b) the set of positive integers congruent to 2 modulo 3.
c) the set of positive integers not divisible by 5.

26. Let S be the subset of the set of ordered pairs of integers defined recursively by

Basis step: $(0, 0) \in S$.

Recursive step: If $(a, b) \in S$, then $(a + 2, b + 3) \in S$ and $(a + 3, b + 2) \in S$.

a) List the elements of S produced by the first five applications of the recursive definition.

b) Use strong induction on the number of applications of the recursive step of the definition to show that $5 \mid a + b$ when $(a, b) \in S$.

c) Use structural induction to show that $5 \mid a + b$ when $(a, b) \in S$.

27. Let S be the subset of the set of ordered pairs of integers defined recursively by

Basis step: $(0, 0) \in S$.

Recursive step: If $(a, b) \in S$, then $(a, b + 1) \in S$, $(a + 1, b + 1) \in S$, and $(a + 2, b + 1) \in S$.

a) List the elements of S produced by the first four applications of the recursive definition.

b) Use strong induction on the number of applications of the recursive step of the definition to show that $a \leq 2b$ whenever $(a, b) \in S$.

c) Use structural induction to show that $a \leq 2b$ whenever $(a, b) \in S$.

28. Give a recursive definition of each of these sets of ordered pairs of positive integers. (*Hint:* Plot the points in the set in the plane and look for lines containing points in the set.)

a) $S = \{(a, b) \mid a \in \mathbf{Z}^+, b \in \mathbf{Z}^+, \text{ and } a + b \text{ is odd}\}$

b) $S = \{(a, b) \mid a \in \mathbf{Z}^+, b \in \mathbf{Z}^+, \text{ and } a \mid b\}$

c) $S = \{(a, b) \mid a \in \mathbf{Z}^+, b \in \mathbf{Z}^+, \text{ and } 3 \mid a + b\}$

29. Give a recursive definition of each of these sets of ordered pairs of positive integers. Use structural induction to prove that the recursive definition you found is correct. (*Hint:* To find a recursive definition plot the points in the set in the plane and look for patterns.)

a) $S = \{(a, b) \mid a \in \mathbf{Z}^+, b \in \mathbf{Z}^+, \text{ and } a + b \text{ is even}\}$

b) $S = \{(a, b) \mid a \in \mathbf{Z}^+, b \in \mathbf{Z}^+, \text{ and } a \text{ or } b \text{ is odd}\}$

c) $S = \{(a, b) \mid a \in \mathbf{Z}^+, b \in \mathbf{Z}^+, \text{ and } a + b \text{ is odd and } 3 \mid b\}$

30. Prove that in a bit string, the string 01 occurs at most one more time than the string 10.

31. Define well-formed formulae of sets, variables representing sets, and operators from $\{\overline{}, \cup, \cap, -\}$.

32. a) Give a recursive definition of the function $ones(s)$, which counts the number of ones in a bit string s.

b) Use structural induction to prove that $ones(st) = ones(s) + ones(t)$.

33. a) Give a recursive definition of the function $m(s)$, which equals the smallest digit in a nonempty string of decimal digits.

b) Use structural induction to prove that $m(st) = \min(m(s), m(t))$.

The **reversal** of a string is the string consisting of the symbols of the string in reverse order. The reversal of the string w is denoted by w^R.

34. Find the reversal of the following bit strings.

a) 0101   **b)** 1 1011   **c)** 1000 1001 0111

35. Give a recursive definition of the reversal of a string. (*Hint:* First define the reversal of the empty string. Then write a string w of length $n + 1$ as xy, where x is a string of length n, and express the reversal of w in terms of x^R and y.)

*36. Use structural induction to prove that $(w_1 w_2)^R = w_2^R w_1^R$.

37. Give a recursive definition of w^i where w is a string and i is a nonnegative integer. (Here w^i represents the concatenation of i copies of the string w.)

*38. Give a recursive definition of the set of bit strings that are palindromes.

39. When does a string belong to the set A of bit strings defined recursively by

$$\lambda \in A$$
$$0x1 \in A \text{ if } x \in A,$$

where λ is the empty string?

*40. Recursively define the set of bit strings that have more zeros than ones.

41. Use Exercise 37 and mathematical induction to show that $l(w^i) = i \cdot l(w)$, where w is a string and i is a nonnegative integer.

*42. Show that $(w^R)^i = (w^i)^R$ whenever w is a string and i is a nonnegative integer; that is, show that the ith power of the reversal of a string is the reversal of the ith power of the string.

43. Use structural induction to show that $n(T) \geq 2h(T) + 1$, where T is a full binary tree, $n(T)$ equals the number of vertices of T, and $h(T)$ is the height of T.

The set of leaves and the set of internal vertices of a full binary tree can be defined recursively.

Basis step: The root r is a leaf of the full binary tree with exactly one vertex r. This tree has no internal vertices.

Recursive step: The set of leaves of the tree $T = T_1 \cdot T_2$ is the union of the set of leaves of T_1 and the set of leaves of T_2. The internal vertices of T are the root r of T and the union of the set of internal vertices of T_1 and the set of internal vertices of T_2.

44. Use structural induction to show that $l(T)$, the number of leaves of a full binary tree T, is 1 more than $i(T)$, the number of internal vertices of T.

45. Use generalized induction as was done in Example 15 to show that if $a_{m,n}$ is defined recursively by $a_{0,0} = 0$ and

$$a_{m,n} = \begin{cases} a_{m-1,n} + 1 & \text{if } n = 0 \text{ and } m > 0 \\ a_{m,n-1} + 1 & \text{if } n > 0, \end{cases}$$

then $a_{m,n} = m + n$ for all $(m, n) \in \mathbf{N} \times \mathbf{N}$.

46. Use generalized induction as was done in Example 15 to show that if $a_{m,n}$ is defined recursively by $a_{1,1} = 5$ and

$$a_{m,n} = \begin{cases} a_{m-1,n} + 2 & \text{if } n = 1 \text{ and } m > 1 \\ a_{m,n-1} + 2 & \text{if } n > 1, \end{cases}$$

then $a_{m,n} = 2(m+n) + 1$ for all $(m, n) \in \mathbf{Z}^+ \times \mathbf{Z}^+$.

*47. A **partition** of a positive integer n is a way to write n as a sum of positive integers. For instance, $7 = 3 + 2 + 1 + 1$ is a partition of 7. Let P_m equal the number of different partitions of m, where the order of terms in the sum does not matter, and let $P_{m,n}$ be the number of different ways to express m as the sum of positive integers not exceeding n.

a) Show that $P_{m,m} = P_m$.

b) Show that the following recursive definition for $P_{m,n}$ is correct:

$$P_{m,n} = \begin{cases} 1 & \text{if } m = 1 \\ 1 & \text{if } n = 1 \\ P_{m,m} & \text{if } m < n \\ 1 + P_{m,m-1} & \text{if } m = n > 1 \\ P_{m,n-1} + P_{m-n,n} & \text{if } m > n > 1. \end{cases}$$

c) Find the number of partitions of 5 and of 6 using this recursive definition.

Consider an inductive definition of a version of **Ackermann's function.** This function was named after Wilhelm Ackermann, a German mathematician who was a student of the great mathematician David Hilbert. Ackermann's function plays an important role in the theory of recursive functions and in the study of the complexity of certain algorithms involving set unions. (There are several different variants of this function. All are called Ackermann's function and have similar properties even though their values do not always agree.)

$$A(m, n) = \begin{cases} 2n & \text{if } m = 0 \\ 0 & \text{if } m \geq 1 \text{ and } n = 0 \\ 2 & \text{if } m \geq 1 \text{ and } n = 1 \\ A(m-1, A(m, n-1)) \\ & \text{if } m \geq 1 \text{ and } n \geq 2 \end{cases}$$

Exercises 48–55 involve this version of Ackermann's function.

48. Find these values of Ackermann's function.

a) $A(1, 0)$ **b)** $A(0, 1)$
c) $A(1, 1)$ **d)** $A(2, 2)$

49. Show that $A(m, 2) = 4$ whenever $m \geq 1$.
50. Show that $A(1, n) = 2^n$ whenever $n \geq 1$.
51. Find these values of Ackermann's function.

a) $A(2, 3)$ ***b)** $A(3, 3)$

*52. Find $A(3, 4)$.
**53. Prove that $A(m, n+1) > A(m, n)$ whenever m and n are nonnegative integers.
*54. Prove that $A(m+1, n) \geq A(m, n)$ whenever m and n are nonnegative integers.

55. Prove that $A(i, j) \geq j$ whenever i and j are nonnegative integers.
56. Use mathematical induction to prove that a function F defined by specifying $F(0)$ and a rule for obtaining $F(n+1)$ from $F(n)$ is well defined.
57. Use strong induction to prove that a function F defined by specifying $F(0)$ and a rule for obtaining $F(n+1)$ from the values $F(k)$ for $k = 0, 1, 2, \ldots, n$ is well defined.
58. Show that each of these proposed recursive definitions of a function on the set of positive integers does not produce a well-defined function.

a) $F(n) = 1 + F(\lfloor n/2 \rfloor)$ for $n \geq 1$ and $F(1) = 1$.
b) $F(n) = 1 + F(n - 3)$ for $n \geq 2$, $F(1) = 2$, and $F(2) = 3$.
c) $F(n) = 1 + F(n/2)$ for $n \geq 2$, $F(1) = 1$, and $F(2) = 2$.
d) $F(n) = 1 + F(n/2)$ if n is even and $n \geq 2$, $F(n) = 1 - F(n - 1)$ if n is odd, and $F(1) = 1$.
e) $F(n) = 1 + F(n/2)$ if n is even and $n \geq 2$, $F(n) = F(3n - 1)$ if n is odd and $n \geq 3$, and $F(1) = 1$.

59. Show that each of these proposed recursive definitions of a function on the set of positive integers does not produce a well-defined function.

a) $F(n) = 1 + F(\lfloor (n+1)/2 \rfloor)$ for $n \geq 1$ and $F(1) = 1$.
b) $F(n) = 1 + F(n - 2)$ for $n \geq 2$ and $F(1) = 0$.
c) $F(n) = 1 + F(n/3)$ for $n \geq 3$, $F(1) = 1$, $F(2) = 2$, and $F(3) = 3$.
d) $F(n) = 1 + F(n/2)$ if n is even and $n \geq 2$, $F(n) = 1 + F(n - 2)$ if n is odd, and $F(1) = 1$.
e) $F(n) = 1 + F(F(n - 1))$ if $n \geq 2$ and $F(1) = 2$.

Exercises 60–62 deal with iterations of the logarithm function. Let $\log n$ denote the logarithm of n to the base 2, as usual. The function $\log^{(k)} n$ is defined recursively by

$$\log^{(k)} n = \begin{cases} n & \text{if } k = 0 \\ \log(\log^{(k-1)} n) & \text{if } \log^{(k-1)} n \text{ is defined} \\ & \text{and positive} \\ \text{undefined} & \text{otherwise.} \end{cases}$$

The **iterated logarithm** is the function $\log^* n$ whose value at n is the smallest nonnegative integer k such that $\log^{(k)} n \leq 1$.

60. Find each of these values:

a) $\log^{(2)} 16$
b) $\log^{(3)} 256$
c) $\log^{(3)} 2^{65536}$
d) $\log^{(4)} 2^{2^{65536}}$

61. Find the value of $\log^* n$ for each of these values of n:

a) 2   **b)** 4   **c)** 8   **d)** 16
e) 256   **f)** 65536   **g)** 2^{2048}

62. Find the largest integer n such that $\log^* n = 5$. Determine the number of decimal digits in this number.

Exercises 63–65 deal with values of iterated functions. Suppose that $f(n)$ is a function from the set of real numbers, or positive real numbers, or some other set of real numbers, to the set of real numbers such that $f(n)$ is monotonically increasing [that is, $f(n) < f(m)$ when $n < m$] and $f(n) < n$ for all n in the domain of f.] The function $f^{(k)}(n)$ is defined recursively by

$$f^{(k)}(n) = \begin{cases} n & \text{if } k = 0 \\ f(f^{(k-1)}(n)) & \text{if } k > 0. \end{cases}$$

Furthermore, let c be a positive real number. The **iterated** function f_c^* is the number of iterations of f required to reduce its argument to c or less, so that $f_c^*(n)$ is the smallest nonnegative integer k such that $f^k(n) \leq c$.

63. Let $f(n) = n - a$, where a is a positive integer. Find a formula for $f^{(k)}(n)$. What is the value of $f_0^*(n)$ when n is a positive integer?

64. Let $f(n) = n/2$. Find a formula for $f^{(k)}(n)$. What is the value of $f_1^*(n)$ when n is a positive integer?

65. Let $f(n) = \sqrt{n}$. Find a formula for $f^{(k)}(n)$. What is the value of $f_2^*(n)$ when n is a positive integer?

3.5   Recursive Algorithms

INTRODUCTION

Sometimes we can reduce the solution to a problem with a particular set of input to the solution of the same problem with smaller input values. For instance, the problem of finding the greatest common divisor of two positive integers a and b where $b > a$ can be reduced to finding the greatest common divisor of a pair of smaller integers, namely, $b \bmod a$ and a, since $\gcd(b \bmod a, a) = \gcd(a, b)$. When such a reduction can be done, the solution to the original problem can be found with a sequence of reductions, until the problem has been reduced to some initial case for which the solution is known. For instance, for finding the greatest common divisor, the reduction continues until the smaller of the two numbers is zero, since $\gcd(a, 0) = a$ when $a > 0$.

We will see that algorithms that successively reduce a problem to the same problem with smaller input are used to solve a wide variety of problems.

DEFINITION 1   An algorithm is called *recursive* if it solves a problem by reducing it to an instance of the same problem with smaller input.

We will describe several different recursive algorithms in Examples 1, 2, 4, 5, and 6. The first example shows how a recursive algorithm can be constructed to evaluate a function from its recursive definition.

EXAMPLE 1   Give a recursive algorithm for computing a^n where a is a nonzero real number and n is a nonnegative integer.

Solution: We can base a recursive algorithm on the recursive definition of a^n. This definition states that $a^{n+1} = a \cdot a^n$ for $n > 0$ and the initial condition $a^0 = 1$. To find a^n, successively use the recursive condition to reduce the exponent until it becomes zero. We give this procedure in Algorithm 1.   ◀

ALGORITHM 1 A Recursive Algorithm for Computing a^n.

procedure *power*(a: nonzero real number, n: nonnegative integer)
if $n = 0$ **then** *power*(a, n) := 1
else *power*(a, n) := $a \cdot$ *power*($a, n - 1$)

EXAMPLE 2   Devise a recursive algorithm for computing b^n **mod** m, where b, n, and m are integers with $m \geq 2, n \geq 0$, and $1 \leq b < m$.

Solution: We can base a recursive algorithm on the fact that b^n **mod** $m = (b \cdot (b^{n-1} \bmod m))$ **mod** m, which follows by Exercise 23 in Section 3.1, and the initial condition b^0 **mod** $m = 1$. We leave this as Exercise 6 for the reader at the end of the section.

However, we can devise a much more efficient recursive algorithm, which we describe in pseudocode as Algorithm 2. ◀

ALGORITHM 2 Recursive Modular Exponentiation.

procedure *mpower*(b, n, m: integers with $m \geq 2, n \geq 0; 1 \leq b < m$)
if $n = 0$ **then**
    mpower(b, n, m) = 1
else if n is even **then**
    mpower(b, n, m) = *mpower*($b, n/2, m$)^2 **mod** m
else
    mpower(b, n, m) = (*mpower*($b, \lfloor n/2 \rfloor, m$)^2 **mod** $m \cdot b$ **mod** m) **mod** m
{*mpower*(b, n, m) = b^n **mod** m}

Strong induction can be used to prove that a recursive algorithm is correct, that is, that it produces the desired output for all possible input values. We illustrate how this is done in Example 3 by proving that Algorithm 2 is correct.

EXAMPLE 3   Prove that Algorithm 2, which computes modular powers, is correct.

Solution: We use strong induction on the exponent n.

BASIS STEP: When $n = 0$, *mpower*(b, n, m) = 1. Because b^0 **mod** $m = 1$ whenever b is an integer, m is an integer with $m \geq 2$, and $1 \leq b < m$, the basis step is complete.

INDUCTIVE STEP: The inductive hypothesis is that *mpower*(b, j, m) = b^j **mod** m for all integers $0 \leq j < k$ whenever b is a positive integer, m is an integer with $m \geq 2$, and $b < m$.

When k is even, we have *mpower*(b, k, m) = *mpower*($b, k/2, m$)^2 **mod** $m = (b^{k/2}$ **mod** $m)^2$ **mod** $m = b^k$ **mod** m, where we have used the inductive hypothesis to replace *mpower*($b, k/2, m$) by $b^{k/2}$ **mod** m.

When k is odd, we have *mpower*(b, k, m) = ((*mpower*($b, \lfloor k/2 \rfloor, m$))^2 **mod** $m \cdot b$ **mod** m) **mod** $m = ((b^{\lfloor k/2 \rfloor}$ **mod** $m)^2$ **mod** $m \cdot b$ **mod** m) **mod** $m = b^{2\lfloor k/2 \rfloor + 1}$ **mod** $m = b^k$ **mod** m, using Exercise 23 in Section 3.1, because $2\lfloor k/2 \rfloor + 1 = 2(k - 1)/2 + 1 = k$

when k is odd. Here we have used the inductive hypothesis to replace $mpower(b, \lfloor k/2 \rfloor, m)$ by $b^{\lfloor k/2 \rfloor}$ **mod** m. This completes the inductive step and shows that Algorithm 2 is correct. ◀

Next we give a recursive algorithm for finding greatest common divisors.

EXAMPLE 4 Give a recursive algorithm for computing the greatest common divisor of two nonnegative integers a and b with $a < b$.

Solution: We can base a recursive algorithm on the reduction $\gcd(a, b) = \gcd(b \bmod a, a)$ and the condition $\gcd(0, b) = b$ when $b > 0$. This produces the procedure in Algorithm 3. ◀

ALGORITHM 3 A Recursive Algorithm for Computing gcd(a, b).

---

procedure $gcd(a, b$: nonnegative integers with $a < b)$
if $a = 0$ **then** $gcd(a, b) := b$
else $gcd(a, b) := gcd(b \bmod a, a)$

We will now give recursive versions of searching algorithms which were introduced in Section 2.1.

EXAMPLE 5 Express the linear search algorithm as a recursive procedure.

Solution: To *search* for x in the search sequence $a_1, a_2, \ldots, a_n$, at the ith step of the algorithm x and a_i are compared. If x equals a_i then i is the location of x. Otherwise, the search for x is reduced to a search in a sequence with one fewer element, namely, the sequence $a_{i+1}, \ldots, a_n$. We can now give a recursive procedure, which is displayed as pseudocode in Algorithm 4.

Let *search*(i, j, x) be the procedure that searches for x in the sequence $a_i, a_{i+1}, \ldots, a_j$. The input to the procedure consists of the triple $(1, n, x)$. The procedure terminates at a step if the first term of the remaining sequence is x or if there is only one term of the sequence and this is not x. If x is not the first term and there are additional terms, the same procedure is carried out but with a search sequence of one fewer term, obtained by deleting the first term of the search sequence. ◀

ALGORITHM 4 A Recursive Linear Search Algorithm.

---

procedure $search(i, j, x)$
if $a_i = x$ **then**
    $location := i$
else if $i = j$ **then**
    $location := 0$
else
    $search(i + 1, j, x)$

EXAMPLE 6   Construct a recursive version of a binary search algorithm.

Solution: Suppose we want to locate x in the sequence $a_1, a_2, \ldots, a_n$. To perform a binary search, we begin by comparing x with the middle term, $a_{\lfloor (n+1)/2 \rfloor}$. Our algorithm will terminate if x equals this term. Otherwise, we reduce the search to a smaller search sequence, namely, the first half of the sequence if x is smaller than the middle term of the original sequence, and the second half otherwise. We have reduced the solution of the search problem to the solution of the same problem with a sequence approximately half as long. We express this recursive version of a binary search algorithm as Algorithm 5. ◄

ALGORITHM 5 A Recursive Binary Search Algorithm.

---

procedure *binary search*(x, i, j)
$m := \lfloor (i + j)/2 \rfloor$
if $x = a_m$ **then**
    location := m
else if $(x < a_m$ and $i < m)$ **then**
    binary search$(x, i, m - 1)$
else if $(x > a_m$ and $j > m)$ **then**
    binary search$(x, m + 1, j)$
else *location* := 0

RECURSION AND ITERATION

A recursive definition expresses the value of a function at a positive integer in terms of the values of the function at smaller integers. This means that we can devise a recursive algorithm to evaluate a recursively defined function at a positive integer.

EXAMPLE 7   The recursive procedure in Algorithm 6 gives the value of $n!$ when the input is a positive integer n.   ◄

ALGORITHM 6 A Recursive Procedure for Factorials.

---

procedure *factorial*$(n$: positive integer$)$
if $n = 1$ **then**
    factorial(n) := 1
else
    factorial(n) := $n \cdot$ *factorial*$(n - 1)$

There is another way to evaluate the factorial function at an integer from its recursive definition. Instead of successively reducing the computation to the evaluation of the function at smaller integers, we can start with the value of the function at 1 and successively apply the recursive definition to find the values of the function at successive larger integers. Such a procedure is called **iterative.** In other words, to find $n!$ using an iterative procedure, we start with 1, the value of the factorial function at 1, and multiply successively by each positive integer less than or equal to n. This procedure is shown in Algorithm 7.

ALGORITHM 7 An Iterative Procedure for Factorials.

procedure *iterative factorial*(n: positive integer)
$x := 1$
for $i := 1$ **to** n
    $x := i \cdot x$
$\{x$ is $n!\}$

After this code has been executed, the value of the variable x is $n!$. For instance, going through the loop six times gives $6! = 1 \cdot 2 \cdot 3 \cdot 4 \cdot 5 \cdot 6 = 720$.

Often an iterative approach for the evaluation of a recursively defined sequence requires much less computation than a procedure using recursion (unless special-purpose recursive machines are used). This is illustrated by the iterative and recursive procedures for finding the nth Fibonacci number. The recursive procedure is given first.

ALGORITHM 8 A Recursive Algorithm for Fibonacci Numbers.

procedure *fibonacci*(n: nonnegative integer)
if $n = 0$ **then** *fibonacci*(0) := 0
else if $n = 1$ **then** *fibonacci*(1) := 1
else *fibonacci*(n) := *fibonacci*($n - 1$) + *fibonacci*($n - 2$)

When we use a recursive procedure to find f_n, we first express f_n as $f_{n-1} + f_{n-2}$. Then we replace both of these Fibonacci numbers by the sum of two previous Fibonacci numbers, and so on. When f_1 or f_0 arises, it is replaced by its value.

Note that at each stage of the recursion, until f_1 or f_0 is obtained, the number of Fibonacci numbers to be evaluated has doubled. For instance, when we find f_4 using this recursive algorithm, we must carry out all the computations illustrated in the tree diagram in Figure 1. This tree consists of a root labeled with f_4, and branches from the root to vertices labeled with the two Fibonacci numbers f_3 and f_2 that occur in the reduction of the computation of f_4. Each subsequent reduction produces two branches in the tree. This branching ends when f_0 and f_1 are reached. The reader can verify that this algorithm requires $f_{n+1} - 1$ additions to find f_n.

Now consider the amount of computation required to find f_n using the iterative approach in Algorithm 9.

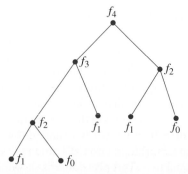

FIGURE 1   Evaluating f_4 Recursively.

ALGORITHM 9 An Iterative Algorithm for Computing Fibonacci Numbers.

procedure *iterative fibonacci*(n: nonnegative integer)
if $n = 0$ **then** $y := 0$
else
begin
    $x := 0$
    $y := 1$
    for $i := 1$ to $n - 1$
    begin
        $z := x + y$
        $x := y$
        $y := z$
    end
end
{y is the nth Fibonacci number}

This procedure initializes x as $f_0 = 0$ and y as $f_1 = 1$. When the loop is traversed, the sum of x and y is assigned to the auxiliary variable z. Then x is assigned the value of y and y is assigned the value of the auxiliary variable z. Therefore, after going through the loop the first time, it follows that x equals f_1 and y equals $f_0 + f_1 = f_2$. Furthermore, after going through the loop $n - 1$ times, x equals f_{n-1} and y equals f_n (the reader should verify this statement). Only $n - 1$ additions have been used to find f_n with this iterative approach when $n > 1$. Consequently, this algorithm requires far less computation than does the recursive algorithm.

We have shown that a recursive algorithm may require far more computation than an iterative one when a recursively defined function is evaluated. It is sometimes preferable to use a recursive procedure even if it is less efficient than the iterative procedure. In particular, this is true when the recursive approach is easily implemented and the iterative approach is not. (Also, machines designed to handle recursion may be available that eliminate the advantage of using iteration.)

THE MERGE SORT

We now describe a recursive sorting algorithm called the **merge sort** algorithm. We will demonstrate how the merge sort algorithm works with an example before describing it in generality.

EXAMPLE 8 We will sort the list 8, 2, 4, 6, 9, 7, 10, 1, 5, 3 using the merge sort. A merge sort begins by splitting the list into individual elements by successively splitting lists in two. The progression of sublists for this example is represented with the balanced binary tree of height 4 shown in the upper half of Figure 2.

Sorting is done by successively merging pairs of lists. At the first stage, pairs of individual elements are merged into lists of length two in increasing order. Then successive merges of pairs of lists are performed until the entire list is put into increasing order. The succession of merged lists in increasing order is represented by the balanced binary tree of height 4 shown in the lower half of Figure 2 (note that this tree is displayed "upside down"). ◄

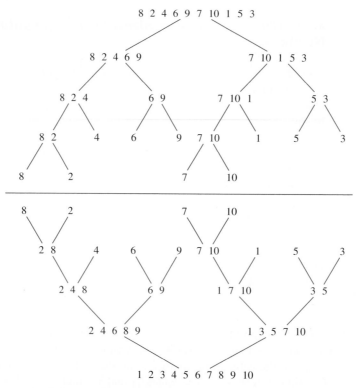

FIGURE 2 **The Merge Sort of 8, 2, 4, 6, 9, 7, 10, 1, 5, 3.**

In general, a merge sort proceeds by iteratively splitting lists into two sublists of equal length (or where one sublist has one more element than the other) until each sublist contains one element. This succession of sublists can be represented by a balanced binary tree. The procedure continues by successively merging pairs of lists, where both lists are in increasing order, into a larger list with elements in increasing order, until the original list is put into increasing order. The succession of merged lists can be represented by a balanced binary tree.

We can also describe the merge sort recursively. To do a merge sort, we split a list into two sublists of equal, or approximately equal, size, sorting each sublist using the merge sort algorithm, and then merging the two lists. The recursive version of the merge sort is given in Algorithm 10. This algorithm uses the subroutine *merge*, which is described in Algorithm 11.

ALGORITHM 10 A Recursive Merge Sort.

procedure *mergesort*($L = a_1, \ldots, a_n$)
if $n > 1$ **then**
    $m := \lfloor n/2 \rfloor$
    $L_1 := a_1, a_2, \ldots, a_m$
    $L_2 := a_{m+1}, a_{m+2}, \ldots, a_n$
    $L := merge(mergesort(L_1), \ mergesort(L_2))$
{L is now sorted into elements in nondecreasing order}

An efficient algorithm for merging two ordered lists into a larger ordered list is needed to implement the merge sort. We will now describe such a procedure.

EXAMPLE 9 We will describe how to merge the two lists $2, 3, 5, 6$ and $1, 4$. Table 1 illustrates the steps we use.

First, compare the smallest elements in the two lists, 2 and 1, respectively. Since 1 is the smaller, put it at the beginning of the merged list and remove it from the second list. At this stage, the first list is $2, 3, 5, 6$, the second is 4, and the combined list is 1.

Next, compare 2 and 4, the smallest elements of the two lists. Since 2 is the smaller, add it to the combined list and remove it from the first list. At this stage the first list is $3, 5, 6$, the second is 4, and the combined list is $1, 2$.

Continue by comparing 3 and 4, the smallest elements of their respective lists. Since 3 is the smaller of these two elements, add it to the combined list and remove it from the first list. At this stage the first list is $5, 6$, and the second is 4. The combined list is $1, 2, 3$.

Then compare 5 and 4, the smallest elements in the two lists. Since 4 is the smaller of these two elements, add it to the combined list and remove it from the second list. At this stage the first list is $5, 6$, the second list is empty, and the combined list is $1, 2, 3, 4$.

Finally, since the second list is empty, all elements of the first list can be appended to the end of the combined list in the order they occur in the first list. This produces the ordered list $1, 2, 3, 4, 5, 6$. ◀

We will now consider the general problem of merging two ordered lists L_1 and L_2, into an ordered list L. We will describe an algorithm for solving this problem. Start with an empty list L. Compare the smallest elements of the two lists. Put the smaller of these two elements at the left end of L, and remove it from the list it was in. Next, if one of L_1 and L_2 is empty, append the other (nonempty) list to L, which completes the merging. If neither L_1 nor L_2 is empty, repeat this process. Algorithm 11 gives a pseudocode description of this procedure.

We will need estimates for the number of comparisons used to merge two ordered lists in the analysis of the merge sort. We can easily obtain such an estimate for Algorithm 11. Each time a comparison of an element from L_1 and an element from L_2 is made, an additional element is added to the merged list L. However, when either L_1 or L_2 is empty, no more comparisons are needed. Hence, Algorithm 11 is least efficient when $m + n - 2$ comparisons are carried out, where m and n are the number of elements in L_1 and L_2, respectively, leaving one element in each of L_1 and L_2. The next comparison will be the last one needed, because it will make one of these lists empty. Hence, Algorithm 11 uses no more than $m + n - 1$ comparisons. Lemma 1 summarizes this estimate.

TABLE 1 **Merging the Two Sorted Lists 2, 3, 5, 6 and 1, 4.**			
First List	*Second List*	*Merged List*	*Comparison*
2 3 5 6	1 4		$1 < 2$
2 3 5 6	4	1	$2 < 4$
3 5 6	4	1 2	$3 < 4$
5 6	4	1 2 3	$4 < 5$
5 6		1 2 3 4	
		1 2 3 4 5 6	

ALGORITHM 11 Merging Two Lists.

---

procedure $merge(L_1, L_2 :$ lists)
$L :=$ empty list
while L_1 and L_2 are both nonempty
begin
    remove smaller of first element of L_1 and L_2 from the list it is
        in and put it at the left end of L
    if removal of this element makes one list empty **then** remove
        all elements from the other list and append them to L
end {L is the merged list with elements in increasing order}

LEMMA 1 Two sorted lists with m elements and n elements can be merged into a sorted list using no more than $m + n - 1$ comparisons.

Sometimes two sorted lists of length m and n can be merged using far fewer than $m + n - 1$ comparisons. For instance, when $m = 1$, a binary search procedure can be applied to put the one element in the first list into the second list. This requires only $\lceil \log n \rceil$ comparisons, which is much smaller than $m + n - 1 = n$, for $m = 1$. On the other hand, for some values of m and n, Lemma 1 gives the best possible bound. That is, there are lists with m and n elements that cannot be merged using fewer than $m + n - 1$ comparisons. (See Exercise 35 at the end of this section.)

We can now analyze the complexity of the merge sort. Instead of studying the general problem, we will assume that n, the number of elements in the list, is a power of 2, say 2^m. This will make the analysis less complicated, but when this is not the case, various modifications can be applied that will yield the same estimate.

At the first stage of the splitting procedure, the list is split into two sublists, of 2^{m-1} elements each, at level 1 of the tree generated by the splitting. This process continues, splitting the two sublists with 2^{m-1} elements into four sublists of 2^{m-2} elements each at level 2, and so on. In general, there are 2^{k-1} lists at level $k - 1$, each with 2^{m-k+1} elements. These lists at level $k - 1$ are split into 2^k lists at level k, each with 2^{m-k} elements. At the end of this process, we have 2^m lists each with one element at level m.

We start merging by combining pairs of the 2^m lists of one element into 2^{m-1} lists, at level $m - 1$, each with two elements. To do this, 2^{m-1} pairs of lists with one element each are merged. The merger of each pair requires exactly one comparison.

The procedure continues, so that at level k ($k = m, m - 1, m - 2, \ldots, 3, 2, 1$), 2^k lists each with 2^{m-k} elements are merged into 2^{k-1} lists, each with 2^{m-k+1} elements, at level $k - 1$. To do this a total of 2^{k-1} mergers of two lists, each with 2^{m-k} elements, are needed. But, by Lemma 1, each of these mergers can be carried out using at most $2^{m-k} + 2^{m-k} - 1 = 2^{m-k+1} - 1$ comparisons. Hence, going from level k to $k - 1$ can be accomplished using at most $2^{k-1}(2^{m-k+1} - 1)$ comparisons. Summing all these estimates shows that the number of comparisons required for the merge sort is at most

$$\sum_{k=1}^{m} 2^{k-1}(2^{m-k+1} - 1) = \sum_{k=1}^{m} 2^m - \sum_{k=1}^{m} 2^{k-1} = m2^m - (2^m - 1) = n \log n - n + 1,$$

since $m = \log n$ and $n = 2^m$. (We evaluated $\sum_{k=1}^{m} 2^m$ by noting that it is the sum of m identical terms, each equal to 2^m. We evaluated $\sum_{k=1}^{m} 2^{k-1}$ using the formula for the sum of the terms of a geometric progression from Theorem 1 of Section 3.2.)

This analysis shows that the merge sort achieves the best possible big-O estimate for the number of comparisons needed by sorting algorithms, as stated in the following theorem.

THEOREM 1 The number of comparisons needed to merge sort a list with n elements is $O(n \log n)$.

We describe another efficient algorithm, the quick sort, in the exercises.

Exercises

1. Give a recursive algorithm for computing nx whenever n is a positive integer and x is an integer.
2. Give a recursive algorithm for finding the sum of the first n positive integers.
3. Give a recursive algorithm for finding the sum of the first n odd positive integers.
4. Give a recursive algorithm for finding the maximum of a finite set of integers.
5. Give a recursive algorithm for finding the minimum of a finite set of integers.
6. Devise a recursive algorithm for finding $x^n \bmod m$ whenever n, x, and m are positive integers based on the fact that $x^n \bmod m = (x^{n-1} \bmod m \cdot x \bmod m) \bmod m$.
7. Give a recursive algorithm for finding $n! \bmod m$ whenever n and m are positive integers.
8. Give a recursive algorithm for finding a **mode** of a list of integers. (A **mode** is an element in the list that occurs at least as often as every other element.)
9. Devise a recursive algorithm for computing the greatest common divisor of two nonnegative integers a and b with $a < b$ if $\gcd(a, b) = \gcd(a, b - a)$.
10. Prove that the recursive algorithm for finding the sum of the first n positive integers you found in Exercise 2 is correct.
11. Describe a recursive algorithm for multiplying two nonnegative integers x and y based on the fact that $xy = 2(x \cdot (y/2))$ when y is even and $xy = 2(x \cdot \lfloor y/2 \rfloor) + x$ when y is odd, together with the initial condition $xy = 0$ when $y = 0$.
12. Prove that the algorithm you devised in Exercise 11 is correct.
13. Prove that the recursive algorithm that you found in Exercise 1 is correct.
14. Prove that the recursive algorithm that you found in Exercise 4 is correct.

15. Prove that Algorithm 1 is correct.
16. Devise a recursive algorithm to find a^{2^n} where a is a real number and n is a positive integer. [*Hint:* Use the equality $a^{2^{n+1}} = (a^{2^n})^2$.]
17. How does the number of multiplications used by the algorithm in Exercise 16 compare to the number of multiplications used by Algorithm 1 to evaluate a^{2^n}?
*18. Use the algorithm in Exercise 16 to devise an algorithm for evaluating a^n when n is a nonnegative integer. (*Hint:* Use the binary expansion of n.)
*19. How does the number of multiplications used by the algorithm in Exercise 18 compare to the number of multiplications used by Algorithm 1 to evaluate a^n?
20. How many additions are used by the recursive and iterative algorithms given in Algorithms 8 and 9, respectively, to find the Fibonacci number f_7?
21. Devise a recursive algorithm to find the nth term of the sequence defined by $a_0 = 1$, $a_1 = 2$, and $a_n = a_{n-1} \cdot a_{n-2}$, for $n = 2, 3, 4, \ldots$.
22. Devise an iterative algorithm to find the nth term of the sequence defined in Exercise 21.
23. Is the recursive or the iterative algorithm for finding the sequence in Exercise 21 more efficient?
24. Devise a recursive algorithm to find the nth term of the sequence defined by $a_0 = 1$, $a_1 = 2$, $a_2 = 3$, and $a_n = a_{n-1} + a_{n-2} + a_{n-3}$, for $n = 3, 4, 5, \ldots$.
25. Devise an iterative algorithm to find the nth term of the sequence defined in Exercise 24.
26. Is the recursive or the iterative algorithm for finding the sequence in Exercise 24 more efficient?
27. Give iterative and recursive algorithms for finding the nth term of the sequence defined by $a_0 = 1$, $a_1 = 3$, $a_2 = 5$, and $a_n = a_{n-1} \cdot a_{n-2}^2 \cdot a_{n-3}^3$. Which is more efficient?

28. Give a recursive algorithm to find the number of partitions of a positive integer based on the recursive definition given in Exercise 47 in Section 3.4.
29. Give a recursive algorithm for finding the reversal of a bit string. (See the definition of the reversal of a bit string in the preamble of Exercise 34 in Section 3.4.)
30. Give a recursive algorithm for finding the string w^i, the concatenation of i copies of w, when w is a bit string.
31. Give a recursive algorithm for computing values of the Ackermann function. (*Hint:* See the preamble to Exercise 48 in Section 3.4.)
32. Use a merge sort to sort $4, 3, 2, 5, 1, 8, 7, 6$. Show all the steps used by the algorithm.
33. Use a merge sort to sort $b, d, a, f, g, h, z, p, o, k$. Show all the steps used by the algorithm.
34. How many comparisons are required to merge these pairs of lists using Algorithm 11?

    a) $1, 3, 5, 7, 9; 2, 4, 6, 8, 10$
    b) $1, 2, 3, 4, 5; 6, 7, 8, 9, 10$
    c) $1, 5, 6, 7, 8; 2, 3, 4, 9, 10$

35. Show that there are lists with m elements and n elements such that they cannot be merged into one sorted list using Algorithm 11 with fewer than $m + n - 1$ comparisons.
*36. What is the least number of comparisons needed to merge any two lists in increasing order into one list in increasing order when the number of elements in the two lists are

    a) $1, 4$?   b) $2, 4$?   c) $3, 4$?   d) $4, 4$?

*37. Prove that the merge sort algorithm is correct.

The **quick sort** is an efficient algorithm. To sort $a_1, a_2, \ldots, a_n$, this algorithm begins by taking the first element a_1 and forming two sublists, the first containing those elements that are less than a_1, in the order they arise, and the second containing those elements greater than a_1, in the order they arise. Then a_1 is put at the end of the first sublist. This procedure is repeated recursively for each sublist, until all sublists contain one item. The ordered list of n items is obtained by combining the sublists of one item in the order they occur.

38. Sort $3, 5, 7, 8, 1, 9, 2, 4, 6$ using the quick sort.
39. Let $a_1, a_2, \ldots, a_n$ be a list of n distinct real numbers. How many comparisons are needed to form two sublists from this list, the first containing elements less than a_1 and the second containing elements greater than a_1?
40. Describe the quick sort algorithm using pseudocode.
41. What is the largest number of comparisons needed to order a list of four elements using the quick sort algorithm?
42. What is the least number of comparisons needed to order a list of four elements using the quick sort algorithm?
43. Determine the worst-case complexity of the quick sort algorithm in terms of the number of comparisons used.
*44. Show that $\log n!$ is greater than $(n \log n)/4$ for $n > 4$. [*Hint:* Begin with the inequality $n! > n(n-1)(n-2) \cdots \lceil n/2 \rceil$.]

3.6   Program Correctness

INTRODUCTION

Suppose that we have designed an algorithm to solve a problem and have written a program to implement it. How can we be sure that the program always produces the correct answer? After all the bugs have been removed so that the syntax is correct, we can test the program with sample input. It is not correct if an incorrect result is produced for any sample input. But even if the program gives the correct answer for all sample input, it may not always produce the correct answer (unless all possible input has been tested). We need a proof to show that the program *always* gives the correct output.

Program verification, the proof of correctness of programs, uses the rules of inference and proof techniques described in this chapter, including mathematical induction. Since an incorrect program can lead to disastrous effects, a large amount of methodology has been constructed for verifying programs. Efforts have been devoted to automating program verification so that it can be carried out using a computer. However, only limited progress has been made toward this goal. Indeed, some mathematicians and theoretical computer scientists argue that it will never be realistic to mechanize the proof of correctness of complex programs.

Some of the concepts and methods used to prove that programs are correct will be introduced in this section. However, a complete methodology for program verification will not be developed in this book. This section is meant to be a brief introduction to the area of program verification, which ties together the rules of logic, proof techniques, and the concept of an algorithm.

PROGRAM VERIFICATION

A program is said to be **correct** if it produces the correct output for every possible input. A proof that a program is correct consists of two parts. The first part shows that the correct answer is obtained if the program terminates. This part of the proof establishes the **partial correctness** of the program. The second part of the proof shows that the program always terminates.

To specify what it means for a program to produce the correct output, two propositions are used. The first is the **initial assertion,** which gives the properties that the input values must have. The second is the **final assertion,** which gives the properties that the output of the program should have, if the program did what was intended. The appropriate initial and final assertions must be provided when a program is checked.

DEFINITION 1

> A program, or program segment, S is said to be *partially correct with respect to* the initial assertion p and the final assertion q if whenever p is true for the input values of S and S terminates, then q is true for the output values of S. The notation $p\{S\}q$ indicates that the program, or program segment, S is partially correct with respect to the initial assertion p and the final assertion q.

Note: The notation $p\{S\}q$ is known as a *Hoare triple* after Tony Hoare, who introduced the concept of partial correctness.

Note that the notion of partial correctness has nothing to do with whether a program terminates; it focuses only on whether the program does what it is expected to do if it terminates.

A simple example illustrates the concepts of initial and final assertions.

EXAMPLE 1

Show that the program segment

$$y := 2$$
$$z := x + y$$

is correct with respect to the initial assertion $p: x = 1$ and the final assertion $q: z = 3$.

Solution: Suppose that p is true, so that $x = 1$ as the program begins. Then y is assigned the value 2, and z is assigned the sum of the values of x and y, which is 3. Hence, S is correct with respect to the initial assertion p and the final assertion q. Thus, $p\{S\}q$ is true. ◀

RULES OF INFERENCE

A useful rule of inference proves that a program is correct by splitting the program into a series of subprograms and then showing that each subprogram is correct.

Suppose that the program S is split into subprograms S_1 and S_2. Write $S = S_1; S_2$ to indicate that S is made up of S_1 followed by S_2. Suppose that the correctness of S_1 with respect to the initial assertion p and final assertion q, and the correctness of S_2 with respect to the initial assertion q and the final assertion r, have been established. It follows that if p is true and S_1 is executed and terminates, then q is true; and if q is true, and S_2 executes and terminates, then r is true. Thus, if p is true and $S = S_1; S_2$ is executed and terminates, then r is true. This rule of inference, called the **composition rule,** can be stated as

$$p\{S_1\}q$$
$$q\{S_2\}r$$
$$\overline{\therefore\ p\{S_1; S_2\}r.}$$

This rule of inference will be used later in this section.

Next, some rules of inference for program segments involving conditional statements and loops will be given. Since programs can be split into segments for proofs of correctness, this will let us verify many different programs.

CONDITIONAL STATEMENTS

First, rules of inference for conditional statements will be given. Suppose that a program segment has the form

> **if** *condition* **then**
> $\quad S$

where S is a block of statements. Then S is executed if *condition* is true, and it is not executed when *condition* is false. To verify that this segment is correct with respect to the initial assertion p and final assertion q, two things must be done. First, it must be shown that when p is true and *condition* is also true, then q is true after S terminates. Second, it must be shown that when p is true and *condition* is false, then q is true (since in this case S does not execute).

This leads to the following rule of inference:

$$(p \wedge condition)\{S\}q$$
$$(p \wedge \neg condition) \rightarrow q$$
$$\overline{\therefore\ p\{\textbf{if } condition \textbf{ then } S\}q.}$$

Example 2 illustrates how this rule of inference is used.

C. ANTHONY R. HOARE (BORN 1934) Tony Hoare is currently Professor of Computer Science at Oxford University, England, and is a Fellow of the Royal Society. Hoare has made many important contributions to the theory of programming languages and to programming methodology. He was the first person to define a programming language based on how programs could be proved to be correct with respect to their specifications. Hoare is also the creator of the quick sort, one of the most commonly used and studied sorting algorithms (see the exercise set in Section 3.5). Hoare is a noted writer in the technical and social aspects of computer science.

EXAMPLE 2   Verify that the program segment

> **if** $x > y$ **then**
> > $y := x$

is correct with respect to the initial assertion **T** and the final assertion $y \geq x$.

Solution: When the initial assertion is true and $x > y$, the assignment $y := x$ is carried out. Hence, the final assertion, which asserts that $y \geq x$, is true in this case. Moreover, when the initial assertion is true and $x > y$ is false, so that $x \leq y$, the final assertion is again true. Hence, using the rule of inference for program segments of this type, this program is correct with respect to the given initial and final assertions.   ◀

Similarly, suppose that a program has a statement of the form

> **if** *condition* **then**
> > S_1
> **else**
> > S_2

If *condition* is true, then S_1 executes; if *condition* is false, then S_2 executes. To verify that this program segment is correct with respect to the initial assertion p and the final assertion q, two things must be done. First, it must be shown that when p is true and *condition* is true, then q is true after S_1 terminates. Second, it must be shown that when p is true and *condition* is false, then q is true after S_2 terminates. This leads to the following rule of inference:

$$(p \wedge condition)\{S_1\}q$$
$$(p \wedge \neg condition)\{S_2\}q$$

$$\therefore p\{\textbf{if } condition \textbf{ then } S_1 \textbf{ else } S_2\}q.$$

The following example illustrates how this rule of inference is used.

EXAMPLE 3   Verify that the program segment

> **if** $x < 0$ **then**
> > $abs := -x$
> **else**
> > $abs := x$

is correct with respect to the initial assertion **T** and the final assertion $abs = |x|$.

Solution: Two things must be demonstrated. First, it must be shown that if the initial assertion is true and $x < 0$, then $abs = |x|$. This is correct, since when $x < 0$ the assignment statement $abs := -x$ sets $abs = -x$, which is $|x|$ by definition when $x < 0$. Second, it must be shown that if the initial assertion is true and $x < 0$ is false, so that $x \geq 0$, then $abs = |x|$.

This is also correct, since in this case the program uses the assignment statement $abs := x$, and x is $|x|$ by definition when $x \geq 0$, so that $abs := x$. Hence, using the rule of inference for program segments of this type, this segment is correct with respect to the given initial and final assertions. ◀

LOOP INVARIANTS

Next, proofs of correctness of **while** loops will be described. To develop a rule of inference for program segments of the type

> **while** *condition*
>   *S*

note that S is repeatedly executed until *condition* becomes false. An assertion that remains true each time S is executed must be chosen. Such an assertion is called a **loop invariant.** In other words, p is a loop invariant if $(p \wedge condition)\{S\}p$ is true.

Suppose that p is a loop invariant. It follows that if p is true before the program segment is executed, p and $\neg condition$ are true after termination, if it occurs. This rule of inference is

$$\frac{(p \wedge condition)\{S\}p}{\therefore p\{\textbf{while } condition \ \ S\}(\neg condition \wedge p).}$$

The use of a loop invariant is illustrated in Example 4.

EXAMPLE 4   A loop invariant is needed to verify that the program segment

> $i := 1$
> $factorial := 1$
> **while** $i < n$
> **begin**
> $i := i + 1$
> $factorial := factorial \cdot i$
> **end**

terminates with $factorial = n!$ when n is a positive integer.

Let p be the assertion "$factorial = i!$ and $i \leq n$." We first prove that p is a loop invariant. Suppose that, at the beginning of one execution of the **while** loop, p is true and the condition of the **while** loop holds; in other words, assume that $factorial = i!$ and that $i < n$. The new values i_{new} and $factorial_{new}$ of i and $factorial$ are $i_{new} = i + 1$ and $factorial_{new} = factorial \cdot (i + 1) = (i + 1)! = i_{new}!$. Since $i < n$, we also have $i_{new} = i + 1 \leq n$. Thus p is true at the end of the execution of the loop. This shows that p is a loop invariant.

Now we consider the program segment. Just before entering the loop, $i = 1 \leq n$ and $factorial = 1 = 1! = i!$ both hold, so p is true. Since p is a loop invariant, the rule of inference just introduced implied that if the **while** loop terminates, it terminates with p true and with $i < n$ false. In this case, at the end, *factorial* $= i!$ and $i \leq n$ are true, but $i < n$ is false; in other words, $i = n$ and *factorial* $= i! = n!$, as desired.

Finally, we need to check that the **while** loop actually terminates. At the beginning of the program i is assigned the value 1, so after $n - 1$ traversals of the loop, the new value of i will be n, and the loop terminates at that point. ◄

A final example will be given to show how the various rules of inference can be used to verify the correctness of a longer program.

EXAMPLE 5   We will outline how to verify the correctness of the program S for computing the product of two integers.

---

procedure *multiply*(m, n: integers)

S_1 $\begin{cases} \textbf{if } n < 0 \textbf{ then } a := -n \\ \textbf{else } a := n \end{cases}$

S_2 $\begin{cases} k := 0 \\ x := 0 \end{cases}$

S_3 $\begin{cases} \textbf{while } k < a \\ \textbf{begin} \\ \quad x := x + m \\ \quad k := k + 1 \\ \textbf{end} \end{cases}$

S_4 $\begin{cases} \textbf{if } n < 0 \textbf{ then } product := -x \\ \textbf{else } product := x \end{cases}$

The goal is to prove that after S is executed, *product* has the value mn. The proof of correctness can be carried out by splitting S into four segments, with $S = S_1; S_2; S_3; S_4$, as shown in the listing of S. The rule of composition can be used to build the correctness proof. Here is how the argument proceeds. The details will be left as an exercise for the reader.

Let p be the initial assertion that m and n are integers. Then, it can be shown that $p\{S_1\}q$ is true, when q is the proposition $p \wedge (a = |n|)$. Next, let r be the proposition $q \wedge (k = 0) \wedge (x = 0)$. It is easily verified that $q\{S_2\}r$ is true. It can be shown that "$x = mk$ and $k \leq a$" is an invariant for the loop in S_3. Furthermore, it is easy to see that the loop terminates after a iterations, with $k = a$, so that $x = ma$ at this point. Since r implies that $x = m \cdot 0$ and $0 \leq a$, the loop invariant is true before the loop is entered. Since the loop terminates with $k = a$, it follows that $r\{S_3\}s$ is true where s is the proposition "$x = ma$ and $a = |n|$." Finally, it can be shown that S_4 is correct with respect to the initial assertion s and final assertion t, where t is the proposition "*product* $= mn$."

Putting all this together, since $p\{S_1\}q$, $q\{S_2\}r$, $r\{S_3\}s$, and $s\{S_4\}t$ are all true, it follows from the rule of composition that $p\{S\}t$ is true. Furthermore, since all four segments terminate, S does terminate. This verifies the correctness of the program. ◄

Exercises

1. Prove that the program segment

$$y := 1$$
$$z := x + y$$

is correct with respect to the initial assertion $x = 0$ and the final assertion $z = 1$.

2. Verify that the program segment

if $x < 0$ **then** $x := 0$

is correct with respect to the initial assertion **T** and the final assertion $x \geq 0$.

3. Verify that the program segment

$$x := 2$$
$$z := x + y$$
if $y > 0$ **then**
$$z := z + 1$$
else
$$z := 0$$

is correct with respect to the initial assertion $y = 3$ and the final assertion $z = 6$.

4. Verify that the program segment

if $x < y$ **then**
$$min := x$$
else
$$min := y$$

is correct with respect to the initial assertion **T** and the final assertion
$(x \leq y \wedge min = x) \vee (x > y \wedge min = y)$.

*5. Devise a rule of inference for verification of partial correctness of statements of the form

if *condition* 1 **then**
$$S_1$$
else if *condition* 2 **then**
$$S_2$$
$$\vdots$$
else
$$S_n$$

where $S_1, S_2, \ldots, S_n$ are blocks.

6. Use the rule of inference developed in Exercise 5 to verify that the program

if $x < 0$ **then**
$$y := -2|x|/x$$
else if $x > 0$ **then**

$$y := 2|x|/x$$
else if $x = 0$ **then**
$$y := 2$$

is correct with respect to the initial assertion **T** and the final assertion $y = 2$.

7. Use a loop invariant to prove that the following program segment for computing the nth power, where n is a positive integer, of a real number x is correct.

$$power := 1$$
$$i := 1$$
while $i \leq n$
begin
$$power := power * x$$
$$i := i + 1$$
end

*8. Prove that the iterative program for finding f_n given in Section 3.5 is correct.

9. Provide all the details in the proof of correctness given in Example 5.

10. Suppose that both the implication $p_0 \to p_1$ and the program assertion $p_1\{S\}q$ are true. Show that $p_0\{S\}q$ also must be true.

11. Suppose that both the program assertion $p\{S\}q_0$ and the implication $q_0 \to q_1$ are true. Show that $p\{S\}q_1$ also must be true.

12. This program computes quotients and remainders.

$$r := a$$
$$q := 0$$
while $r \geq d$
begin
$$r := r - d$$
$$q := q + 1$$
end

Verify that it is partially correct with respect to the initial assertion "a and d are positive integers" and the final assertion "q and r are integers such that $a = dq + r$ and $0 \leq r < d$."

13. Use a loop invariant to verify that the Euclidean algorithm (Algorithm 6 in Section 2.5) is partially correct with respect to the initial assertion "a and b are positive integers" and the final assertion "$x = \gcd(a, b)$."

Key Terms and Results

TERMS

sequence: a function with domain that is a subset of the set of integers

geometric progression: a sequence of the form a, ar, $ar^2, \ldots$ where a and r are real numbers
arithmetic progression: a sequence of the form $a, a + d$, $a + 2d, \ldots$ where a and d are real numbers

string: a finite sequence

empty string: a string of length zero

$\sum_{i=1}^{n}$: the sum $a_1 + a_2 + \cdots + a_n$

$\prod_{i=1}^{n}$: the product $a_1 a_2 \cdots a_n$

countable set: a set that is either finite or can be placed in one-to-one correspondence with the set of positive integers

uncountable set: a set that is not countable

Cantor diagonalization argument: a proof technique that can be used to show that the set of rational numbers is countable

the principle of mathematical induction: the statement $\forall n\, P(n)$ is true if $P(1)$ is true and $\forall k[P(k) \to P(k+1)]$ is true.

basis step: the proof of $P(1)$ in a proof by mathematical induction of $\forall n\, P(n)$

inductive step: the proof of $P(k) \to P(k+1)$ in a proof by mathematical induction of $\forall n\, P(n)$

strong induction: The statement $\forall n\, P(n)$ is true if $P(1)$ is true and $\forall k[(P(1) \wedge \cdots \wedge P(k)) \to P(k+1)]$ is true.

well-ordering property: Every nonempty set of nonnegative integers has a least element.

recursive definition of a function: a definition of a function that specifies an initial set of values and a rule for obtaining values of this function at integers from its values at smaller integers

recursive definition of a set: a definition of a set that specifies an initial set of elements in the set and a rule for obtaining other elements from those in the set

structural induction: a technique for proving results about recursively defined sets

recursive algorithm: an algorithm that proceeds by reducing a problem to the same problem with smaller input

merge sort: a sorting algorithm that sorts a list by splitting it in two, sorting each of the two resulting lists, and merging the results into a sorted list

iteration: a procedure based on the repeated use of operations in a loop

program correctness: verification that a procedure always produces the correct result

loop invariant: a property that remains true during every traversal of a loop

initial assertion: the statement specifying the properties of the input values of a program

final assertion: the statement specifying the properties the output values should have if the program worked correctly

RESULTS

halting problem: No procedure exists that, when given a program and input to this program, determines whether the program terminates when given this input.

the sum of terms of a geometric series: $\sum_{j=0}^{n} ar^j = (ar^{n+1} - a)/(r - 1)$ if $r \neq 1$.

The set of rational numbers is countable.

The set of real numbers is uncountable.

Review Questions

1. Describe the process of formulating and resolving conjecture and demonstrate this process with the example of finding primes of the form $a^n - 1$ where a and n are positive integers.

2. Show that the set of odd integers is countable.

3. Give an example of an uncountable set.

4. **a)** Can you use the principle of mathematical induction to find a formula for the sum of the first n terms of a sequence?

   b) Can you use the principle of mathematical induction to determine whether a given formula for the sum of the first n terms of a sequence is correct?

   c) Find a formula for the sum of the first n even positive integers, and prove it using mathematical induction.

5. **a)** For which positive integers n is it true that $11n + 17 \leq 2^n$?

   b) Prove the conjecture you made in part (a) using mathematical induction.

6. **a)** Which amounts of postage can be formed using only 5-cent and 9-cent stamps?

   b) Prove the conjecture you made using mathematical induction.

   c) Prove the conjecture you made using the second principle of mathematical induction.

   d) Find a proof of your conjecture different from the ones you gave in (b) and (c).

7. Give two different examples of proofs that use strong induction.

8. **a)** State the well-ordering property for the set of positive integers.

   b) Use this property to show that every positive integer can be written as the product of primes.

9. **a)** Explain why a function is well-defined if it is defined recursively by specifying $f(1)$ and a rule for finding $f(n)$ from $f(n - 1)$.

   b) Provide a recursive definition of the function $f(n) = (n + 1)!$.

10. **a)** Give a recursive definition of the Fibonacci numbers.

b) Show that $f_n > \alpha^{n-2}$ whenever $n \geq 3$ where f_n is the nth term of the Fibonacci sequence and $\alpha = (1 + \sqrt{5})/2$.

11. a) Explain why a sequence a_n is well-defined if it is defined recursively by specifying a_1 and a_2 and a rule for finding a_n from $a_1, a_2, \ldots, a_{n-1}$ for $n = 3, 4, 5, \ldots$.

b) Find the value of a_n if $a_1 = 1$, $a_2 = 2$, and $a_n = a_{n-1} + a_{n-2} + \cdots + a_1$, for $n = 3, 4, 5, \ldots$.

12. Give two examples of how well-formed formulae are defined recursively for different sets of elements and operators.

13. a) Give a recursive definition of the length of a string.

b) Use the recursive definition from part (a) and structural induction to prove that $l(xy) = l(x) + l(y)$.

14. a) What is a recursive algorithm?

b) Describe a recursive algorithm for computing the sum of n numbers in a sequence.

15. Describe a recursive algorithm for computing the greatest common divisor of two positive integers.

16. a) Describe the merge sort algorithm.

b) Use the merge sort algorithm to put the list 4, 10, 1, 5, 3, 8, 7, 2, 6, 9 in increasing order.

c) Give a big-O estimate for the number of comparisons used by the merge sort.

17. a) Does testing a computer program to see whether it produces the correct output for certain input values verify that the program always produces the correct output?

b) Does showing that a computer program is partially correct with respect to an initial assertion and a final assertion verify that the program always produces the correct output? If not, what else is needed?

18. What techniques can you use to show that a long computer program is partially correct with respect to an initial assertion and a final assertion?

19. What is a loop invariant? How is a loop invariant used?

Supplementary Exercises

1. Prove that $2^{2^5} + 1$ is composite.

2. a) Formulate a conjecture concerning which integers of the form $a^n + 1$, where $a \in \mathbf{Z}^+$ and $n \in \mathbf{Z}^+$, are necessarily composite.

b) Prove your conjecture.

3. Prove that $\sqrt[3]{7}$ is irrational.

4. Prove that $\sqrt{5}$ is irrational.

5. Show that there are no integer solutions of $x^4 + y^4 = 1000$.

6. Prove that $n^4 - 1$ is divisible by 5 when n is not divisible by 5. Use a proof by cases, with four different cases—one for each of the nonzero remainders that an integer not divisible by 5 can have when you divide it by 5.

7. Prove or disprove that $|x - y| \geq |x| - |y|$ whenever x and y are real numbers.

* 8. We define the **Ulam numbers** by setting $u_1 = 1$ and $u_2 = 2$. Furthermore, after determining whether the integers less than n are Ulam numbers, we set n equal to the next Ulam number if it can be written uniquely as the sum of two different Ulam numbers. Note that $u_3 = 3, u_4 = 4, u_5 = 6$, and $u_6 = 8$.

a) Find the first 20 Ulam numbers.

b) Prove that there are infinitely many Ulam numbers.

9. Give a constructive proof that there is a polynomial $P(x)$ such that $P(x_1) = y_1, P(x_2) = y_2, \ldots, P(x_n) = y_n$, where $x_1, \ldots, x_n, y_1, \ldots, y_n$ are real numbers and the x_i are distinct. [*Hint:* Let

$$P(x) = \sum_{i=1}^{n} \left(\prod_{i \neq j} \frac{x - x_j}{x_i - x_j} \right) y_i.]$$

10. Show that $1^3 + 3^3 + 5^3 + \cdots + (2n + 1)^3 = (n + 1)^2 (2n^2 + 4n + 1)$ whenever n is a positive integer.

11. Show that $1 \cdot 2^0 + 2 \cdot 2^1 + 3 \cdot 2^2 + \cdots + n \cdot 2^{n-1} = (n - 1) \cdot 2^n + 1$ whenever n is a positive integer.

12. Show that

$$\frac{1}{1 \cdot 3} + \frac{1}{3 \cdot 5} + \cdots + \frac{1}{(2n - 1)(2n + 1)} = \frac{n}{2n + 1}$$

whenever n is a positive integer.

13. Show that

$$\frac{1}{1 \cdot 4} + \frac{1}{4 \cdot 7} + \cdots + \frac{1}{(3n - 2)(3n + 1)} = \frac{n}{3n + 1}$$

whenever n is a positive integer.

14. Use mathematical induction to show that $2^n > n^2 + n$ whenever n is an integer greater than 4.

15. Use mathematical induction to show that $2^n > n^3$ whenever n is an integer greater than 9.

16. Find an integer N such that $2^n > n^4$ whenever n is greater than N. Prove that your result is correct using mathematical induction.

17. Use mathematical induction to prove that $a - b$ is a factor of $a^n - b^n$ whenever n is a positive integer.

18. Use mathematical induction to prove that 9 divides $n^3 + (n + 1)^3 + (n + 2)^3$ whenever n is a nonnegative integer.

19. Use mathematical induction to prove this formula for the sum of the terms of an arithmetic progression.

$$a + (a + d) + \cdots + (a + nd) = (n + 1)(2a + nd)/2$$

20. Suppose that $a_j \equiv b_j \pmod{m}$ for $j = 1, 2, \ldots, n$. Use mathematical induction to prove that

a) $\displaystyle\sum_{j=1}^{n} a_j \equiv \sum_{j=1}^{n} b_j \pmod{m}$.

b) $\displaystyle\prod_{j=1}^{n} a_j \equiv \prod_{j=1}^{n} b_j \pmod{m}$.

21. Show that if n is a positive integer, then

$$\sum_{k=1}^{n} \frac{k+4}{k(k+1)(k+2)} = \frac{n(3n+7)}{2(n+1)(n+2)}.$$

22. For which positive integers n is $n + 6 < (n^2 - 8n)/16$? Prove your answer using mathematical induction.

23. (Requires calculus) Suppose that $f(x) = e^x$ and $g(x) = xe^x$. Use mathematical induction together with the product rule and the fact that $f'(x) = e^x$ to prove that $g^{(n)}(x) = (x + n)e^x$ whenever n is a positive integer.

24. (Requires calculus) Suppose that $f(x) = e^x$ and $g(x) = e^{cx}$, where c is a constant. Use mathematical induction together with the chain rule and the fact that $f'(x) = e^x$ to prove that $g^{(n)} = c^n e^{cx}$ whenever n is a positive integer.

*25. Determine which Fibonacci numbers are even, and use a form of mathematical induction to prove your conjecture.

*26. Determine which Fibonacci numbers are divisible by 3. Use a form of mathematical induction to prove your conjecture.

*27. Prove that $f_k f_n + f_{k+1} f_{n+1} = f_{n+k+1}$ for all nonnegative integers n, where k is a nonnegative integer and f_i denotes the ith Fibonacci number.

The sequence of **Lucas numbers** is defined by $l_0 = 2$, $l_1 = 1$, and $l_n = l_{n-1} + l_{n-2}$ for $n = 2, 3, 4, \ldots$.

28. Show that $f_n + f_{n+2} = l_{n+1}$ whenever n is a positive integer, where f_i and l_i are the ith Fibonacci number and ith Lucas number, respectively.

29. Show that $l_0^2 + l_1^2 + \cdots + l_n^2 = l_n l_{n+1} + 2$ whenever n is a nonnegative integer and l_i is the ith Lucas number.

*30. Use mathematical induction to show that the product of any n consecutive positive integers is divisible by $n!$. [*Hint:* Use the identity $m(m + 1) \cdots (m + n - 1)/n! = (m-1)m(m + 1) \cdots (m + n - 2)/n! + m(m + 1) \cdots (m + n - 2)/(n - 1)!.$]

31. Use mathematical induction to show that $(\cos x + i \sin x)^n = \cos nx + i \sin nx$ whenever n is a positive integer. [*Hint:* Use the identities $\cos(a + b) = \cos a \cos b - \sin a \sin b$ and $\sin(a + b) = \sin a \cos b + \cos a \sin b$.]

*32. Use mathematical induction to show that $\sum_{j=1}^{n} \cos jx = \cos[(n + 1)x/2] \sin(nx/2)/\sin(x/2)$ whenever n is a positive integer and $\sin(x/2) \neq 0$.

33. Use mathematical induction to prove that $\sum_{j=1}^{n} j^2 2^j = n^2 2^{n+1} - n2^{n+2} + 3 \cdot 2^{n+1} - 6$ for every positive integer n.

34. (Requires calculus) Suppose that the sequence $x_1, x_2, \ldots, x_n, \ldots$ is recursively defined by $x_1 = 0$ and $x_{n+1} = \sqrt{x_n + 6}$.

a) Use mathematical induction to show that $x_1 < x_2 < \cdots < x_n < \cdots$, that is, the sequence $\{x_n\}$ is monotonically increasing.

b) Use mathematical induction to prove that $x_n < 3$ for $n = 1, 2, \ldots$.

c) Show that $\lim_{n \to \infty} x_n = 3$.

35. Use mathematical induction to prove that if n people stand in a line, where n is a positive integer, and if the first person in the line is a woman and the last person in line is a man, then somewhere in the line there is a woman directly in front of a man.

*36. Suppose that in a country there is a direct one-way road connecting every pair of cities. Use mathematical induction to show that there is a city that can be reached from every other city either directly or via exactly one other city.

37. Use mathematical induction to show that when n circles divide the plane into regions, these regions can be colored with two different colors such that no regions with a common boundary are colored the same.

*38. Suppose that among a group of cars on a circular track there is enough fuel for one car to complete a lap. Use mathematical induction to show that there is a car in the group that can complete a lap by obtaining gas from other cars as it travels around the track.

39. Show that if n is a positive integer, then

$$\sum_{j=1}^{n} (2j - 1) \left(\sum_{k=j}^{n} 1/k \right) = n(n + 1)/2.$$

40. A **unit** or **Egyptian fraction** is a fraction of the form $1/n$, where n is a positive integer. In this exercise, we will use strong induction to show that a greedy algorithm can be used to express every rational number p/q with $0 < p/q < 1$ as the sum of distinct unit fractions. At each step of the algorithm, we find the smallest positive integer n such that $1/n$ can be added to the sum without exceeding p/q. For example, to express $5/7$ we first start the sum with $1/2$. Since $5/7 - 1/2 = 3/14$ we add $1/5$ to the sum since 5 is the smallest positive integer k such that $1/k < 3/14$. Since $3/14 - 1/5 = 1/70$, the algorithm terminates, showing that $5/7 = 1/2 + 1/5 + 1/70$. Let $T(p)$ be the statement that this algorithm terminates for all rational numbers p/q with $0 < p/q < 1$. We will prove that the algorithm always terminates by showing that $T(p)$ holds for all positive integers p.

a) Show that the basis step $T(1)$ holds.

b) Suppose that $T(k)$ holds for positive integers k with $k < p$. That is, assume that the algorithm terminates for all rational numbers k/r, where $1 \le k < p$. Show that if we start with p/q and the fraction $1/n$ is selected in the first step of the algorithm, then $p/q = p'/q' + 1/n$, where $p' = np - q$ and $q' = nq$. After considering the case where $p/q = 1/n$, use the induction hypothesis to show that the greedy algorithm terminates when it begins with p'/q' and complete the inductive step.

The **McCarthy 91 function** (defined by John McCarthy, one of the founders of artificial intelligence) is defined using the rule

$$M(n) = \begin{cases} n - 10 & \text{if } n > 100 \\ M(M(n + 11)) & \text{if } n \le 100, \end{cases}$$

for all positive integers n.

41. By successively using the defining rule for $M(n)$, find

  a) $M(102)$. **b)** $M(101)$. **c)** $M(99)$.
  d) $M(97)$. **e)** $M(87)$. **f)** $M(76)$.

**42. Show that the function $M(n)$ is a well-defined function from the set of positive integers to the set of positive integers. [*Hint:* Prove that $M(n) = 91$ for all positive integers n with $n \le 101$.]

43. Is this proof that

$$\frac{1}{1 \cdot 2} + \frac{1}{2 \cdot 3} + \cdots + \frac{1}{(n-1)n} = \frac{3}{2} - \frac{1}{n},$$

whenever n is a positive integer, correct? Justify your answer.

Basis step: The result is true when $n = 1$ since

$$\frac{1}{1 \cdot 2} = \frac{3}{2} - \frac{1}{1}.$$

Inductive step: Assume that the result is true for n. Then

$$\frac{1}{1 \cdot 2} + \frac{1}{2 \cdot 3} + \cdots + \frac{1}{(n-1)n} + \frac{1}{n(n+1)}$$
$$= \frac{3}{2} - \frac{1}{n} + \left(\frac{1}{n} - \frac{1}{n+1}\right)$$
$$= \frac{3}{2} - \frac{1}{n+1}.$$

Hence, the result is true for $n + 1$ if it is true for n. This completes the proof.

*44. A jigsaw puzzle is put together by successively joining pieces that fit together into blocks. A move is made each time a piece is added to a block, or when two blocks are joined. Use the second form of mathematical induction to prove that no matter how the moves are carried out, exactly $n - 1$ moves are required to assemble a puzzle with n pieces.

*45. Show that n circles divide the plane into $n^2 - n + 2$ regions if every two circles intersect in exactly two points and no three circles contain a common point.

*46. Show that n planes divide three-dimensional space into $(n^3 + 5n + 6)/6$ regions if any three of these planes have a point in common and no four contain a common point.

*47. Use the well-ordering property to show that $\sqrt{2}$ is irrational. (*Hint:* Assume that $\sqrt{2}$ is rational. Show that the set of positive integers of the form $b\sqrt{2}$ has a least element a. Then show that $a\sqrt{2} - a$ is a smaller positive integer of this form.)

Links

JOHN McCARTHY (BORN 1927) John McCarthy was born in Boston. He grew up in Boston and in Los Angeles. He studied mathematics both as an undergraduate and a graduate student, receiving his B.S. in 1948 from the California Institute of Technology and his Ph.D. in 1951 from Princeton. After graduating from Princeton, McCarthy held positions at Princeton, Stanford, Dartmouth, and M.I.T. He held a position at Stanford from 1962 until 1994, and is now an emeritus professor there. At Stanford, he was the director of the Artificial Intelligence Laboratory, held a named chair in the School of Engineering, and was a senior fellow in the Hoover Institution.

    McCarthy was a pioneer in the study of artificial intelligence, a term he coined in 1955. He worked on problems related to the reasoning and information needs required for intelligent computer behavior. McCarthy was among the first computer scientists to design time-sharing computer systems. He developed LISP, a programming language for computing using symbolic expressions. He played an important role in using logic to verify the correctness of computer programs. McCarthy has also worked on the social implications of computer technology. He is currently working on the problem of how people and computers make conjectures through assumptions that complications are absent from situations. McCarthy is an advocate of the sustainability of human progress and is an optimist about the future of humanity. He has also begun writing science fiction stories. Some of his recent writing explores the possibility that the world is a computer program written by some higher force.

    Among the awards McCarthy has won are the Turing Award from the Association for Computing Machinery, the Research Excellence Award of the International Conference on Artificial Intelligence, the Kyoto Prize, and the National Medal of Science.

48. A set is **well-ordered** if every nonempty subset of this set has a least element. Determine whether each of the following sets is well-ordered.

   a) the set of integers
   b) the set of integers greater than -100
   c) the set of positive rationals
   d) the set of positive rationals with denominator less than 100

***49.** Show that the well-ordering property can be proved when the principle of mathematical induction is taken as an axiom.

***50.** Show that the principle of mathematical induction and strong induction are equivalent; that is, each can be shown to be valid from the other.

51. a) Show that if $a_1, a_2, \ldots, a_n$ are positive integers, then $\gcd(a_1, a_2, \ldots, a_{n-1}, a_n) = \gcd(a_1, a_2, \ldots, a_{n-2}, \gcd(a_{n-1}, a_n))$.
   b) Use part (a), together with the Euclidean algorithm, to develop a recursive algorithm for computing the greatest common divisor of a set of n positive integers.

***52.** Describe a recursive algorithm for writing the greatest common divisor of n positive integers as a linear combination of these integers.

53. Find an explicit formula for $f(n)$ if $f(1) = 1$ and $f(n) = f(n-1) + 2n - 1$ for $n \geq 2$. Prove your result using mathematical induction.

****54.** Give a recursive definition of the set of bit strings that contain twice as many 0s as 1s.

55. Let S be the set of bit strings defined recursively by $\lambda \in S$ and $0x \in S$, $x1 \in S$ if $x \in S$, where λ is the empty string.

   a) Find all strings in S of length not exceeding five.
   b) Give an explicit description of the elements of S.

56. Let S be the set of strings defined recursively by $abc \in S$, $bac \in S$, $acb \in S$, and $abcx \in S$; $abxc \in S$, $axbc \in S$, $xabc \in S$ if $x \in S$.

   a) Find all elements of S of length eight or less.
   b) Show that every element of S has a length divisible by three.

The set B of all **balanced strings of parentheses** is defined recursively by $\lambda \in B$, where λ is the empty string; $(x) \in B$, $xy \in B$ if $x, y \in B$.

57. Show that $(()())$ is a balanced string of parentheses and $((()))$ is not a balanced string of parentheses.

58. Find all balanced strings of parentheses with exactly six symbols.

59. Find all balanced strings of parentheses with four or fewer symbols.

60. Use induction to show that if x is a balanced string of parentheses, then the number of left parentheses equals the number of right parentheses in x.

Define the function N on the set of strings of parentheses by

$$N(\lambda) = 0, N\left((\right) = 1, N\left( ) \right) = -1,$$
$$N(uv) = N(u) + N(v),$$

where λ is the empty string, and u and v are strings. It can be shown that N is well-defined.

61. Find

   a) $N\left(() \right)$.
   b) $N\left( )))()((\right)$.
   c) $N\left((()(() \right)$.
   d) $N\left(()(((()))()()) \right)$.

****62.** Show that a string w of parentheses is balanced if and only if $N(w) = 0$ and $N(u) \geq 0$ whenever u is a prefix of w, that is, $w = uv$.

***63.** Give a recursive algorithm for finding all balanced strings of parentheses containing n or fewer symbols.

64. Give a recursive algorithm for finding $\gcd(a, b)$, where a and b are nonnegative integers, based on these facts: $\gcd(a, b) = \gcd(b, a)$ if $a > b$, $\gcd(0, b) = b$, $\gcd(a, b) = 2 \gcd(a/2, b/2)$ if a and b are even, $\gcd(a, b) = \gcd(a/2, b)$ if a is even and b is odd, and $\gcd(a, b) = \gcd(a, b - a)$.

65. Verify the program segment

   if $x > y$ **then**
       $x := y$

with respect to the initial assertion **T** and the final assertion $x \leq y$.

***66.** Develop a rule of inference for verifying recursive programs and use it to verify the recursive program for computing factorials given in Section 3.5.

67. Devise a recursive algorithm that counts the number of times the integer 0 occurs in a list of integers.

Exercises 68–75 deal with some unusual sequences, informally called **self-generating sequences**, produced by simple recurrence relations or rules. In particular, Exercises 68–75 deal with the sequence $\{a(n)\}$ defined by $a(n) = n - a(a(n - 1))$ for $n \geq 1$ and $a(0) = 0$. (This sequence, as well as those in Exercises 72 and 73, are defined in Douglas Hofstader's fascinating book *Gödel, Escher, Bach* ([Ho99]).

68. Find the first 10 terms of the sequence $\{a(n)\}$ defined in the preamble to this exercise.

***69.** Prove that this sequence is well defined. That is, show that $a(n)$ is uniquely defined for all nonnegative integers n.

****70.** Prove that $a(n) = \lfloor (n + 1)\mu \rfloor$ where $\mu = (-1 + \sqrt{5})/2$. [*Hint:* First show for all $n > 0$ that $(\mu n - \lfloor \mu n \rfloor) + (\mu^2 n - \lfloor \mu^2 n \rfloor) = 1$. Then show for all real numbers α with $0 \leq \alpha < 1$ and $\alpha \neq 1 - \mu$ that $\lfloor (1 + \mu)(1 - \alpha) \rfloor + \lfloor \alpha + \mu \rfloor = 1$, considering the cases $0 \leq \alpha < 1 - \mu$ and $1 - \mu < \alpha < 1$ separately.]

***71.** Use the formula from Exercise 70 to show that $a(n) = a(n-1)$ if $\mu n - \lfloor \mu n \rfloor < 1 - \mu$ and $a(n) = a(n-1) + 1$ otherwise.

72. Find the first 10 terms of each of the following self-generating sequences:

a) $a(n) = n - a(a(a(n-1)))$ for $n \geq 1$ with $a(0) = 0$.

b) $a(n) = n - a(a(a(a(n-1))))$ for $n \geq 1$ with $a(0) = 0$.

c) $a(n) = a(n - a(n-1)) + a(n - a(n-2))$ for $n \geq 3$ with $a(1) = 1$ and $a(2) = 1$.

73. Find the first 10 terms of both the sequences $m(n)$ and $f(n)$ defined by the following pair of interwoven recurrence relations: $m(n) = n - f(m(n-1))$, $f(n) = n - m(f(n-1))$ for $n \geq 1$ with $f(0) = 1$ and $m(0) = 0$.

Golomb's self-generating sequence is the unique nondecreasing sequence of positive integers $a_1, a_2, a_3, \ldots$, which has the property that it contains exactly a_k occurrences of k for each positive integer k.

74. Find the first 20 terms of Golomb's self-generating sequence.

*75. Show that if $f(n)$ is the largest integer m such that $a_m = n$, where a_m is the mth term of Golomb's self-generating sequence, then $f(n) = \sum_{k=1}^{n} a_k$ and $f(f(n)) = \sum_{k=1}^{n} k a_k$.

The set $\mathcal{L}$ of **logarithmico-exponential functions,** introduced by the famous British mathematician G. H. Hardy, is the smallest set of functions such that:

- the function $f(n) = \alpha$ belongs to $\mathcal{L}$, whenever α is a real number;

- the function $f(n) = n$ belongs to $\mathcal{L}$;

- if the functions $f(n)$ and $g(n)$ belong to $\mathcal{L}$, then $f(n) - g(n)$ belongs to $\mathcal{L}$;

- if the function $f(n)$ belongs to $\mathcal{L}$, then $e^{f(n)}$ belongs to $\mathcal{L}$;

Links

GODFREY HAROLD HARDY (1877–1947) Hardy, born in Cranleigh, Surrey, England, was the older of two children of Isaac Hardy and Sophia Hall Hardy. His father was the geography and drawing master at the Cranleigh School and also gave singing lessons and played soccer. His mother gave piano lessons and helped run a boardinghouse for young students. Hardy's parents were devoted to their children's education. Hardy demonstrated his numerical ability at the early age of two when he began writing down numbers into the millions. He had a private mathematics tutor rather than attending regular classes at the Cranleigh School. He moved to Winchester College, a private high school, when he was 13 and was awarded a scholarship. He excelled in his studies and demonstrated a strong interest in mathematics. He entered Trinity College, Cambridge, in 1896 on a scholarship and won several prizes during his time there, graduating in 1899.

Hardy held the position of lecturer in mathematics at Trinity College at Cambridge University from 1906 to 1919, when he was appointed to the Sullivan chair of geometry at Oxford. He had become unhappy at Cambridge with the dismissal of the famous philosopher and mathematician Bertrand Russell from Trinity for antiwar activities and did not like a heavy load of administrative duties. In 1931 he returned to Cambridge as the Sadleirian professor of pure mathematics, where he remained until his retirement in 1942. He was a pure mathematician and held an elitist view of mathematics, hoping his research could never be applied. Ironically, he is perhaps best known as one of the developers of the Hardy–Weinberg law, which predicts patterns of inheritance. His work in this area appeared as a letter to the journal *Science* in which he used simple algebraic ideas to demonstrate errors in an article on genetics. Hardy worked primarily in number theory and function theory, working on such topics as the Riemann zeta function, Fourier series, and the distribution of primes. He made many important contributions to many important problems, such as Waring's problem about representing positive integers as the sum of kth powers and the problem of representing odd integers as the sum of three primes. Hardy is also remembered for his collaborations with John E. Littlewood, a colleague at Cambridge, with whom he wrote more than 100 papers and the famous Indian mathematical prodigy Srinivasa Ramanujan. His collaboration with Littlewood led to the joke that there were only three important English mathematicians at that time, Hardy, Littlewood, and Hardy–Littlewood, although some people thought that Hardy had invented a fictitious person, Littlewood, since Littlewood was seldom seen outside Cambridge. Hardy had the wisdom of recognizing Ramanujan's genius from unconventional but extremely creative writings Ramanujan sent him, while other mathematicians failed to see the genius. Hardy brought Ramanujan to Cambridge and collaborated on important joint papers, establishing new results on the number of partitions of an integer. Hardy was interested in mathematics education, and his book *A Course in Pure Mathematics* had a profound effect on undergraduate instruction in mathematics in the first half of the twentieth century. Hardy also wrote *A Mathematician's Apology* in which he gives his answer to the question whether it is worthwhile to devote one's life to the study of mathematics. It presents Hardy's view of what mathematics is and what a mathematician does.

Hardy had a strong interest in sports. He was an avid cricket fan and followed scores closely. One peculiar trait he had was that he did not like his picture taken (only five snapshots are known) and disliked mirrors, covering them with towels immediately on entering a hotel room.

■ if $f(n)$ belongs to $\mathcal{L}$ and there exists an integer N such that $f(n) > 0$ for $n \geq N$ (this means that f is called **eventually positive**), then $\ln f(n)$ belongs to $\mathcal{L}$, where $\ln x$ denotes the natural logarithm of x, as usual.

Hardy showed for every logarithmico-exponential function not identically zero that $f(n)$ is either eventually positive or eventually negative. He proved that if $f(n)$ and $g(n)$ belong to $\mathcal{L}$, then either $f(n)$ is $o(g(n))$, $g(n)$ is $o(f(n))$, or $f(n)$ and $g(n)$ are of the same order.

76. Show that if $f(n)$ and $g(n)$ belong to $\mathcal{L}$, then $f(n) + g(n)$ belongs to $\mathcal{L}$.

77. Show that if $f(n)$ and $g(n)$ belong to $\mathcal{L}$ and are eventually positive, then $f(n)g(n)$ and $f(n)/g(n)$ belong to $\mathcal{L}$, and that this implies, using the fact that every logarithmico-exponential function is eventually positive, eventually negative, or identically zero, that the product and quotient of any two functions, not identically zero, in $\mathcal{L}$ are in $\mathcal{L}$.

78. Use Exercises 76 and 77 to show that every polynomial $f(n) = a_m n^m + a_{m-1} n^{m-1} + \cdots + a_0$ with real coefficients belongs to $\mathcal{L}$.

79. Show that if $f(n)$ belongs to $\mathcal{L}$ and is eventually positive, then $\sqrt{f(n)}$ belongs to $\mathcal{L}$.

80. Show that the function $e^{\sqrt{n}\sqrt{\ln n \ln \ln \ln n}}$ belongs to $\mathcal{L}$.

Links

SRINIVASA RAMANUJAN (1887–1920) The famous mathematical prodigy Ramanujan was born and raised in southern India near the city of Madras. His father was a clerk in a cloth shop. His mother contributed to the family income by singing at a local temple. Ramanujan studied at the local English language school, displaying his talent and interest for mathematics. At 13 he mastered a textbook used by college students. When he was 15, a university student lent him a copy of *Synopsis of Pure Mathematics*. Ramanujan decided to work out the over 6000 results in this book, stated without proof or explanation, writing on sheets later collected to form notebooks. He graduated from high school in 1904, winning a scholarship to the University of Madras. Enrolling in a fine arts curriculum, he neglected his subjects other than mathematics and lost his scholarship. He failed to pass examinations at the university four times from 1904 to 1907, doing well only in mathematics. During this time he filled his notebooks with original writings, sometimes rediscovering already published work and at other times making new discoveries.

Without a university degree, it was difficult for Ramanujan to find a decent job. To survive, he had to depend on the goodwill of his friends. He tutored students in mathematics, but his unconventional ways of thinking and failure to stick to the syllabus caused problems. He was married in 1909 in an arranged marriage to a young woman nine years his junior. Needing to support himself and his wife, he moved to Madras and sought a job. He showed his notebooks of mathematical writings to his potential employers, but the books bewildered them. However, a professor at the Presidency College recognized his genius and supported him for a while, and in 1912 he found work as an accounts clerk, earning a small salary.

Ramanujan continued his mathematical work during this time and published his first paper in 1910 in an Indian journal. He realized that his work was beyond that of Indian mathematicians and decided to write to leading English mathematicians. The first mathematicians he wrote to turned down his request for help. But in January 1913 he wrote to G. H. Hardy, who was inclined to turn Ramanujan down, but the mathematical statements in the letter, although stated without proof, puzzled Hardy. He decided to examine them closely with the help of his colleague and collaborator J. E. Littlewood. They decided, after careful study, that Ramanujan was probably a genius, since his statements "could only be written down by a mathematician of the highest class; they must be true, because if they were not true, no one would have the imagination to invent them."

Hardy arranged a scholarship for Ramanujan, bringing him to England in 1914. Hardy personally tutored him in mathematical analysis, and they collaborated for five years, proving significant theorems about the number of partitions of integers. During this time, Ramanujan made important contributions to number theory and also worked on continued fractions, infinite series, and elliptic functions. Ramanujan had amazing insight involving certain types of functions and series, but his purported theorems on prime numbers were often wrong, illustrating his vague idea of what constitutes a correct proof. He was one of the youngest members ever appointed a Fellow of the Royal Society. Unfortunately, in 1917 Ramanujan became extremely ill. At the time, it was thought that he had trouble with the English climate and had contracted tuberculosis. It is now thought that he suffered from a vitamin deficiency, brought on by Ramanujan's strict vegetarianism and shortages in wartime England. He returned to India in 1919, continuing to do mathematics even when confined to his bed. He was religious and thought his mathematical talent came from his family deity, Namagiri. He considered mathematics and religion to be linked. He said that "an equation for me has no meaning unless it expresses a thought of God." His short life came to an end in April 1920, when he was 32 years old. Ramanujan left several notebooks of unpublished results. The writings in these notebooks illustrate Ramanujan's insights but are quite sketchy. Several mathematicians have devoted many years of study to explaining and justifying the results in these notebooks.

Computer Projects

WRITE PROGRAMS WITH THESE INPUT AND OUTPUT.

1. Given the terms of a sequence $a_1, a_2, \ldots, a_n$, find $\sum_{j=1}^{n} a_j$ and $\prod_{j=1}^{n} a_j$.

2. Given a geometric progression $a, ar, ar^2, \ldots, ar^n$, find the sum of its terms.

3. Given a nonnegative integer n, find the sum of the n smallest positive integers.

** 4. Given a $2^n \times 2^n$ chessboard with one square missing, construct a tiling of this chessboard using L-shaped pieces.

** 5. Generate all well-formed formulae for expressions involving the variables x, y, and z and the operators $\{+, *, /, -\}$ with n or fewer symbols.

** 6. Generate all well-formed formulae for propositions with n or fewer symbols where each symbol is **T**, **F**, one of the propositional variables p and q, or an operator from $\{\neg, \vee, \wedge, \rightarrow, \leftrightarrow\}$.

7. Given a string, find its reversal.

8. Given a real number a and a nonnegative integer n, find a^n using recursion.

9. Given a real number a and a nonnegative integer n, find a^{2^n} using recursion.

*10. Given a real number a and a nonnegative integer n, find a^n using the binary expansion of n and a recursive algorithm for computing a^{2^k}.

11. Given two integers not both zero, find their greatest common divisor using recursion.

12. Given a list of integers and an element x, locate x in this list using a recursive implementation of a linear search.

13. Given a list of integers and an element x, locate x in this list using a recursive implementation of a binary search.

14. Given a nonnegative integer n, find the nth Fibonacci number using iteration.

15. Given a nonnegative integer n, find the nth Fibonacci number using recursion.

16. Given a positive integer, find the number of partitions of this integer. (See Exercise 47 of Section 3.4.)

17. Given positive integers m and n, find $A(m, n)$, the value of Ackermann's function at the pair (m, n). (See the preamble to Exercise 48 of Section 3.4.)

18. Given a list of n integers, sort these integers using the merge sort.

Computations and Explorations

USE A COMPUTATIONAL PROGRAM OR PROGRAMS YOU HAVE WRITTEN TO DO THESE EXERCISES.

1. Verify Goldbach's conjecture, which states that every even positive integer n is the sum of two primes, for $n \leq 10,000$.

2. Find the smallest prime factor of $n! + 1$ for all positive integers n with $n \leq 20$.

3. Find the smallest set of n consecutive composite integers for each positive integer n with $n \leq 10$.

4. Find as many twin primes as you can.

5. Verify the $3x + 1$ conjecture for as many positive integers as possible.

6. What are the largest values of n for which $n!$ has fewer than 100 decimal digits and fewer than 1000 decimal digits?

7. Determine which Fibonacci numbers are divisible by 5, which are divisible by 7, and which are divisible by 11. Prove that your conjectures are correct.

8. Construct tilings using L-shaped pieces of various 16×16, 32×32, and 64×64 chessboards with one square missing.

9. Explore which $m \times n$ chessboards can be completely covered by L-shaped pieces. Can you make a conjecture that answers this question?

10. Which values of Ackermann's function are small enough that you are able to compute them?

11. Compare either the number of operations or the time needed to compute Fibonacci numbers recursively versus that needed to compute them iteratively.

Writing Projects

RESPOND TO THESE WITH ESSAYS USING OUTSIDE SOURCES.

1. Provide a brief explanation of the types of methods used to prove what is currently known about Goldbach's conjecture.

2. Describe the conjectures that have been made about the number of twin primes not exceeding a real number x. On what basis were these conjectures made?

3. Describe some open questions in number theory other than the ones mentioned in Section 3.1.

4. Explain the different ways in which the *Encyclopedia of Integer Sequences* has been found useful. Also, describe a few of the more unusual sequences in this encyclopedia and how they arise.

5. Define the recently invented EKG sequence and describe some of its properties and open questions about it.

6. Look up the definition of a transcendental number. Explain how to show that such numbers exist and how such numbers can be constructed. Which famous numbers can be shown to be transcendental and for which famous numbers is it still unknown whether they are transcendental?

7. Describe the origins of mathematical induction. Who were the first people to use it and to which problems did they apply it?

8. In the past 25 years, several important theorems have been proved based on extensive computer computations. Discuss the validity of such proofs and describe the controversy surrounding proofs based on computer calculations.

9. The L-shaped pieces used in the exercises of Section 3.3 are examples of *polyominoes*, introduced by Golomb in 1954. Describe some of the problems and associated results concerning tiling chessboards with polyominoes.

10. Discuss the uses of Ackermann's function both in the theory of recursive definitions and in the analysis of the complexity of algorithms for set unions.

11. Discuss some of the various methodologies used to establish the correctness of programs and compare them to Hoare's methods described in Section 3.6.

12. Explain how the ideas and concepts of program correctness can be extended to prove that operating systems are secure.

Counting

Combinatorics, the study of arrangements of objects, is an important part of discrete mathematics. This subject was studied as long ago as the seventeenth century, when combinatorial questions arose in the study of gambling games. Enumeration, the counting of objects with certain properties, is an important part of combinatorics. We must count objects to solve many different types of problems. For instance, counting is used to determine the complexity of algorithms. Counting is also required to determine whether there are enough telephone numbers or Internet protocol addresses to meet demand. Furthermore, counting techniques are used extensively when probabilities of events are computed.

The basic rules of counting, which we will study in Section 4.1, can solve a tremendous variety of problems. For instance, we can use these rules to enumerate the different phone numbers possible in the United States, the allowable passwords on a computer system, and the different orders in which the runners in a race can finish. Another important combinatorial tool is the pigeonhole principle, which we will study in Section 4.2. This states that when objects are placed in boxes and there are more objects than boxes, then there is a box containing at least two objects. For instance, we can use this principle to show that among a set of 15 or more students, at least 3 were born on the same day of the week.

We can phrase many counting problems in terms of ordered or unordered arrangements of the objects of a set. These arrangements, called permutations and combinations, are used in many counting problems. For instance, suppose the 100 top finishers on a competitive exam taken by 2000 students are invited to a banquet. We can enumerate the possible sets of 100 students that will be invited, as well as the ways the top 10 prizes can be awarded.

Another problem in combinatorics involves generating all the arrangements of a specified kind. This is often important in computer simulations. We will devise algorithms to generate arrangements of various types.

4.1 The Basics of Counting

INTRODUCTION

A password on a computer system consists of six, seven, or eight characters. Each of these characters must be a digit or a letter of the alphabet. Each password must contain at least one digit. How many such passwords are there? The techniques needed to answer this question and a wide variety of other counting problems will be introduced in this section.

Counting problems arise throughout mathematics and computer science. For example, we must count the successful outcomes of experiments and all the possible outcomes of these experiments to determine probabilities of discrete events. We need to count the number of operations used by an algorithm to study its time complexity.

We will introduce the basic techniques of counting in this section. These methods serve as the foundation for almost all counting techniques.

BASIC COUNTING PRINCIPLES

Assessment

We will present two basic counting principles, the **product rule** and the **sum rule.** Then we will show how they can be used to solve many different counting problems.

The product rule applies when a procedure is made up of separate tasks.

> **THE PRODUCT RULE** Suppose that a procedure can be broken down into a sequence of two tasks. If there are n_1 ways to do the first task and n_2 ways to do the second task after the first task has been done, then there are $n_1 n_2$ ways to do the procedure.

Extra Examples

Examples 1–9 show how the product rule is used.

EXAMPLE 1   The chairs of an auditorium are to be labeled with a letter and a positive integer not exceeding 100. What is the largest number of chairs that can be labeled differently?

Solution: The procedure of labeling a chair consists of two tasks, namely, assigning one of the 26 letters and then assigning one of the 100 possible integers to the seat. The product rule shows that there are $26 \cdot 100 = 2600$ different ways that a chair can be labeled. Therefore, the largest number of chairs that can be labeled differently is 2600. ◄

EXAMPLE 2   There are 32 microcomputers in a computer center. Each microcomputer has 24 ports. How many different ports to a microcomputer in the center are there?

Solution: The procedure of choosing a port consists of two tasks, first picking a microcomputer and then picking a port on this microcomputer. Since there are 32 ways to choose the microcomputer and 24 ways to choose the port no matter which microcomputer has been selected, the product rule shows that there are $32 \cdot 24 = 768$ ports. ◄

An extended version of the product rule is often useful. Suppose that a procedure is carried out by performing the tasks $T_1, T_2, \ldots, T_m$ in sequence. If task T_i can be done in n_i ways after tasks $T_1, T_2, \ldots,$ and T_{i-1} have been done, then there are $n_1 \cdot n_2 \cdot \cdots \cdot n_m$ ways to carry out the procedure. This version of the product rule can be proved by mathematical induction from the product rule for two tasks (see Exercise 56 at the end of the section).

EXAMPLE 3   How many different bit strings are there of length seven?

Solution: Each of the seven bits can be chosen in two ways, since each bit is either 0 or 1. Therefore, the product rule shows there are a total of $2^7 = 128$ different bit strings of length seven. ◄

‒ ‒ ‒ ‒ ‒ ‒ ‒
2 choices for each bit

EXAMPLE 4   How many different license plates are available if each plate contains a sequence of three letters followed by three digits (and no sequences of letters are prohibited, even if they are obscene)?

$\underbrace{\overbrace{\text{— — —}}}_{\substack{\text{26 choices} \\ \text{for each} \\ \text{letter}}} \underbrace{\overbrace{\text{— — —}}}_{\substack{\text{10 choices} \\ \text{for each} \\ \text{digit}}}$

Solution: There are 26 choices for each of the three letters and ten choices for each of the three digits. Hence, by the product rule there are a total of $26 \cdot 26 \cdot 26 \cdot 10 \cdot 10 \cdot 10 = 17,576,000$ possible license plates. ◄

EXAMPLE 5   **Counting Functions** How many functions are there from a set with m elements to one with n elements?

Solution: A function corresponds to a choice of one of the n elements in the codomain for each of the m elements in the domain. Hence, by the product rule there are $n \cdot n \cdot \cdots \cdot n = n^m$ functions from a set with m elements to one with n elements. For example, there are 5^3 different functions from a set with three elements to a set with five elements. ◄

EXAMPLE 6   **Counting One-to-One Functions** How many one-to-one functions are there from a set with m elements to one with n elements?

Solution: First note when $m > n$ there are no one-to-one functions from a set with m elements to a set with n elements. Now let $m \leq n$. Suppose the elements in the domain are $a_1, a_2, \ldots, a_m$. There are n ways to choose the value of the function at a_1. Since the function is one-to-one, the value of the function at a_2 can be picked in $n - 1$ ways (since the value used for a_1 cannot be used again). In general, the value of the function at a_k can be chosen in $n - k + 1$ ways. By the product rule, there are $n(n-1)(n-2) \cdots (n - m + 1)$ one-to-one functions from a set with m elements to one with n elements. For example, there are $5 \cdot 4 \cdot 3 = 60$ one-to-one functions from a set with three elements to a set with five elements. ◄

EXAMPLE 7   **The Telephone Numbering Plan** The format of telephone numbers in North America is specified by a *numbering plan*. A telephone number consists of ten digits, which are split into a three-digit area code, a three-digit office code, and a four-digit station code. Because of signaling considerations, there are certain restrictions on some of these digits. To specify the allowable format, let X denote a digit that can take any of the values 0 through 9, let N denote a digit that can take any of the values 2 through 9, and let Y denote a digit that must be a 0 or a 1. Two numbering plans, which will be called the old plan and the new plan, will be discussed. (The old plan, in use in the 1960s, has been replaced by the new plan, but the recent rapid growth in demand for new numbers will make even this new plan obsolete.) As will be shown, the new plan allows the use of more numbers.

In the old plan, the formats of the area code, office code, and station code are *NYX, NNX,* and *XXXX,* respectively, so that telephone numbers had the form *NYX-NNX-XXXX.* In the new plan, the formats of these codes are *NXX, NXX,* and *XXXX,* respectively, so that telephone numbers have the form *NXX-NXX-XXXX.* How many different North American telephone numbers are possible under the old plan and under the new plan?

Solution: By the product rule, there are $8 \cdot 2 \cdot 10 = 160$ area codes with format *NYX* and $8 \cdot 10 \cdot 10 = 800$ area codes with format *NXX.* Similarly, by the product rule, there are $8 \cdot 8 \cdot 10 = 640$ office codes with format *NNX*. The product rule also shows that there are $10 \cdot 10 \cdot 10 \cdot 10 = 10,000$ station codes with format *XXXX*.

Consequently, applying the product rule again, it follows that under the old plan there are

$$160 \cdot 640 \cdot 10,000 = 1,024,000,000$$

different numbers available in North America. Under the new plan there are

$$800 \cdot 800 \cdot 10,000 = 6,400,000,000$$

different numbers available. ◄

EXAMPLE 8   What is the value of k after the following code has been executed?

```
k := 0
for i₁ := 1 to n₁
    for i₂ := 1 to n₂
        .
        .
        .
        for iₘ := 1 to nₘ
            k := k + 1
```

Solution: The initial value of k is zero. Each time the nested loop is traversed, 1 is added to k. Let T_i be the task of traversing the ith loop. Then the number of times the loop is traversed is the number of ways to do the tasks $T_1, T_2, \ldots, T_m$. The number of ways to carry out the task T_j, $j = 1, 2, \ldots, m$, is n_j, since the jth loop is traversed once for each integer i_j with $1 \le i_j \le n_j$. By the product rule, it follows that the nested loop is traversed $n_1 n_2 \cdots n_m$ times. Hence, the final value of k is $n_1 n_2 \cdots n_m$. ◄

EXAMPLE 9   **Counting Subsets of a Finite Set**   Use the product rule to show that the number of different subsets of a finite set S is $2^{|S|}$.

Solution: Let S be a finite set. List the elements of S in arbitrary order. Recall that there is a one-to-one correspondence between subsets of S and bit strings of length $|S|$. Namely, a subset of S is associated with the bit string with a 1 in the ith position if the ith element in the list is in the subset, and a 0 in this position otherwise. By the product rule, there are $2^{|S|}$ bit strings of length $|S|$. Hence, $|P(S)| = 2^{|S|}$. ◄

The product rule is often phrased in terms of sets in this way: If $A_1, A_2, \ldots, A_m$ are finite sets, then the number of elements in the Cartesian product of these sets is the product of the number of elements in each set. To relate this to the product rule, note that the task of choosing an element in the Cartesian product $A_1 \times A_2 \times \cdots \times A_m$ is done by choosing an element in A_1, an element in A_2, $\ldots$, and an element in A_m. By the product rule it follows that

$$|A_1 \times A_2 \times \cdots \times A_m| = |A_1| \cdot |A_2| \cdot \cdots \cdot |A_m|.$$

We now introduce the sum rule.

> **THE SUM RULE** If a first task can be done in n_1 ways and a second task in n_2 ways, and if these tasks cannot be done at the same time, then there are $n_1 + n_2$ ways to do one of these tasks.

Example 10 illustrates how the sum rule is used.

EXAMPLE 10 Suppose that either a member of the mathematics faculty or a student who is a mathematics major is chosen as a representative to a university committee. How many different choices are there for this representative if there are 37 members of the mathematics faculty and 83 mathematics majors?

Solution: The first task, choosing a member of the mathematics faculty, can be done in 37 ways. The second task, choosing a mathematics major, can be done in 83 ways. From the sum rule it follows that there are $37 + 83 = 120$ possible ways to pick this representative. ◄

We can extend the sum rule to more than two tasks. Suppose that the tasks T_1, $T_2, \ldots, T_m$ can be done in $n_1, n_2, \ldots, n_m$ ways, respectively, and no two of these tasks can be done at the same time. Then the number of ways to do one of these tasks is $n_1 + n_2 + \cdots + n_m$. This extended version of the sum rule is often useful in counting problems, as Examples 11 and 12 show. This version of the sum rule can be proved using mathematical induction from the sum rule for two sets. (This is Exercise 55 at the end of the section.)

EXAMPLE 11 A student can choose a computer project from one of three lists. The three lists contain 23, 15, and 19 possible projects, respectively. How many possible projects are there to choose from?

Solution: The student can choose a project from the first list in 23 ways, from the second list in 15 ways, and from the third list in 19 ways. Hence, there are $23 + 15 + 19 = 57$ projects to choose from. ◄

EXAMPLE 12 What is the value of k after the following code has been executed?

```
k := 0
for i₁ := 1 to n₁
    k := k + 1
for i₂ := 1 to n₂
    k := k + 1

    .
    .
    .

for iₘ := 1 to nₘ
    k := k + 1
```

Solution: The initial value of k is zero. This block of code is made up of m different loops. Each time a loop is traversed, 1 is added to k. Let T_i be the task of traversing the ith loop. The task T_i can be done in n_i ways, since the ith loop is traversed n_i times. Since no two of these tasks can be done at the same time, the sum rule shows that the final value of k, which is the number of ways to do one of the tasks $T_i, i = 1, 2, \ldots, m$, is $n_1 + n_2 + \cdots + n_m$. ◀

The sum rule can be phrased in terms of sets as: If $A_1, A_2, \ldots, A_m$ are disjoint sets, then the number of elements in the union of these sets is the sum of the numbers of elements in them. To relate this to our statement of the sum rule, let T_i be the task of choosing an element from A_i for $i = 1, 2, \ldots, m$. There are $|A_i|$ ways to do T_i. From the sum rule, since no two of the tasks can be done at the same time, the number of ways to choose an element from one of the sets, which is the number of elements in the union, is

$$|A_1 \cup A_2 \cup \cdots \cup A_m| = |A_1| + |A_2| + \cdots + |A_m|.$$

This equality applies only when the sets in question are disjoint. The situation is much more complicated when these sets have elements in common. That situation will be briefly discussed later in this section and discussed in more depth in Chapter 6.

MORE COMPLEX COUNTING PROBLEMS

Many counting problems cannot be solved using just the sum rule or just the product rule. However, many complicated counting problems can be solved using both of these rules.

EXAMPLE 13

Extra
Examples

In a version of the computer language BASIC, the name of a variable is a string of one or two alphanumeric characters, where uppercase and lowercase letters are not distinguished. (An *alphanumeric* character is either one of the 26 English letters or one of the 10 digits.) Moreover, a variable name must begin with a letter and must be different from the five strings of two characters that are reserved for programming use. How many different variable names are there in this version of BASIC?

Solution: Let V equal the number of different variable names in this version of BASIC. Let V_1 be the number of these that are one character long and V_2 be the number of these that are two characters long. Then by the sum rule, $V = V_1 + V_2$. Note that $V_1 = 26$, since a one-character variable name must be a letter. Furthermore, by the product rule there are $26 \cdot 36$ strings of length two that begin with a letter and end with an alphanumeric character. However, five of these are excluded, so that $V_2 = 26 \cdot 36 - 5 = 931$. Hence, there are $V = V_1 + V_2 = 26 + 931 = 957$ different names for variables in this version of BASIC. ◀

EXAMPLE 14

Each user on a computer system has a password, which is six to eight characters long, where each character is an uppercase letter or a digit. Each password must contain at least one digit. How many possible passwords are there?

Solution: Let P be the total number of possible passwords, and let P_6, P_7, and P_8 denote the number of possible passwords of length 6, 7, and 8, respectively. By the sum rule, $P = P_6 + P_7 + P_8$. We will now find P_6, P_7, and P_8. Finding P_6 directly is difficult. To find P_6 it is easier to find the number of strings of uppercase letters and digits that are six characters long, including those with no digits, and subtract from this the number of strings with no digits. By the product rule, the number of strings of six characters is 36^6, and the number of strings with no digits is 26^6. Hence,

$$P_6 = 36^6 - 26^6 = 2{,}176{,}782{,}336 - 308{,}915{,}776 = 1{,}867{,}866{,}560.$$

Similarly, it can be shown that

$$P_7 = 36^7 - 26^7 = 78{,}364{,}164{,}096 - 8{,}031{,}810{,}176 = 70{,}332{,}353{,}920$$

and

$$P_8 = 36^8 - 26^8 = 2{,}821{,}109{,}907{,}456 - 208{,}827{,}064{,}576$$
$$= 2{,}612{,}282{,}842{,}880.$$

Consequently,

$$P = P_6 + P_7 + P_8 = 2{,}684{,}483{,}063{,}360. \quad \blacktriangleleft$$

EXAMPLE 15

Links

Counting Internet Addresses In the Internet, which is made up of interconnected physical networks of computers, each computer (or more precisely, each network connection of a computer) is assigned an *Internet address*. In Version 4 of the Internet Protocol (IPv4), now in use, an address is a string of 32 bits. It begins with a *network number* (*netid*). The netid is followed by a *host number* (*hostid*), which identifies a computer as a member of a particular network.

Three forms of addresses are used, with different numbers of bits used for netids and hostids. **Class A addresses,** used for the largest networks, consist of 0, followed by a 7-bit netid and a 24-bit hostid. **Class B addresses,** used for medium-sized networks, consist of 10, followed by a 14-bit netid and a 16-bit hostid. **Class C addresses,** used for the smallest networks, consist of 110, followed by a 21-bit netid and an 8-bit hostid. There are several restrictions on addresses because of special uses: 1111111 is not available as the netid of a Class A network, and the hostids consisting of all 0s and all 1s are not available for use in any network. A computer on the Internet has either a Class A, a Class B, or a Class C address. (Besides Class A, B, and C addresses, there are also Class D addresses, reserved for use in multicasting when multiple computers are addressed at a single time, consisting of 1110 followed by 28 bits, and Class E addresses, reserved for future use, consisting of 11110 followed by 27 bits. Neither Class D nor Class E addresses are assigned as the IP address of a computer on the Internet.) Figure 1 illustrates IPv4 addressing. (Limitations on the number of Class A and Class B netids have made IPv4 addressing inadequate; IPv6, which will replace IPv4, uses 128-bit addresses to solve this problem.)

How many different IPv4 addresses are available for computers on the Internet?

Bit Number	0	1	2	3	4		8	16	24	31
Class A	0		netid					hostid		
Class B	1	0			netid				hostid	
Class C	1	1	0			netid				hostid
Class D	1	1	1	0			Multicast Address			
Class E	1	1	1	1	0		Address			

FIGURE 1   **Internet Addresses (IPv4).**

Solution: Let x be the number of available addresses for computers on the Internet, and let $x_A, x_B,$ and x_C denote the number of Class A, Class B, and Class C addresses available, respectively. By the sum rule, $x = x_A + x_B + x_C$.

To find x_A, note that there are $2^7 - 1 = 127$ Class A netids, recalling that the netid 1111111 is unavailable. For each netid, there are $2^{24} - 2 = 16,777,214$ hostids, recalling that the hostids consisting of all 0s and all 1s are unavailable. Consequently, $x_A = 127 \cdot 16,777,214 = 2,130,706,178$.

To find x_B and x_C, note that there are $2^{14} = 16,384$ Class B netids and $2^{21} = 2,097,152$ Class C netids. For each Class B netid, there are $2^{16} - 2 = 65,534$ hostids, and for each Class C netid, there are $2^8 - 2 = 254$ hostids, recalling that in each network the hostids consisting of all 0s and all 1s are unavailable. Consequently, $x_B = 1,073,709,056$ and $x_C = 532,676,608$.

We conclude that the total number of IPv4 addresses available is $x = x_A + x_B + x_C = 2,130,706,178 + 1,073,709,056 + 532,676,608 = 3,737,091,842$. ◀

THE INCLUSION–EXCLUSION PRINCIPLE

When two tasks can be done at the same time, we cannot use the sum rule to count the number of ways to do one of the two tasks. Adding the number of ways to do each task leads to an overcount, since the ways to do both tasks are counted twice. To correctly count the number of ways to do one of the two tasks, we add the number of ways to do each of the two tasks and then subtract the number of ways to do both tasks. This technique is called the **principle of inclusion–exclusion.** Example 16 illustrates how we can solve counting problems using this principle.

EXAMPLE 16   How many bit strings of length eight either start with a 1 bit or end with the two bits 00?

Extra
Examples

Solution: The first task, constructing a bit string of length eight beginning with a 1 bit, can be done in $2^7 = 128$ ways. This follows by the product rule, since the first bit can be chosen in only one way and each of the other seven bits can be chosen in two ways.

The second task, constructing a bit string of length eight ending with the two bits 00, can be done in $2^6 = 64$ ways. This follows by the product rule, since each of the first six bits can be chosen in two ways and the last two bits can be chosen in only one way.

Both tasks, constructing a bit string of length eight that begins with a 1 and ends with 00, can be done in $2^5 = 32$ ways. This follows by the product rule, since the first bit can be chosen in only one way, each of the second through the sixth bits can be chosen in two ways, and the last two bits can be chosen in one way. Consequently, the number of bit strings of length eight that begin with a 1 or end with a 00, which equals the number of ways to do either the first task or the second task, equals $128 + 64 - 32 = 160$. ◀

We can phrase this counting principle in terms of sets. Let A_1 and A_2 be sets and let T_1 be the task of choosing an element from A_1 and T_2 the task of choosing an element from A_2. There are $|A_1|$ ways to do T_1 and $|A_2|$ ways to do T_2. The number of ways to do either T_1 or T_2 is the sum of the number of ways to do T_1 and the number of ways to do T_2, minus the number of ways to do both T_1 and T_2. Since there are $|A_1 \cup A_2|$ ways to do either T_1 or T_2 and $|A_1 \cap A_2|$ ways to do both T_1 and T_2, we have

$$|A_1 \cup A_2| = |A_1| + |A_2| - |A_1 \cap A_2|.$$

This is the formula given in Section 1.7 for the number of elements in the union of two sets.

The principle of inclusion–exclusion can be generalized to find the number of ways to do one of n different tasks or, equivalently, to find the number of elements in the union of n sets, whenever n is a positive integer. We will study the inclusion–exclusion principle and some of its many applications in Chapter 6.

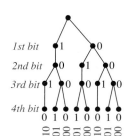

FIGURE 2
Bit Strings of Length Four without Consecutive 1s.

TREE DIAGRAMS

Counting problems can be solved using **tree diagrams.** A tree consists of a root, a number of branches leaving the root, and possible additional branches leaving the endpoints of other branches. (We will study trees in detail in Chapter 9.) To use trees in counting, we use a branch to represent each possible choice. We represent the possible outcomes by the leaves, which are the endpoints of branches not having other branches starting at them.

Note that when a tree diagram is used to solve a counting problem, the number of choices required to reach a leaf can vary (see Example 18, for example).

EXAMPLE 17 How many bit strings of length four do not have two consecutive 1s?

Solution: The tree diagram in Figure 2 displays all bit strings of length four without two consecutive 1s. We see that there are eight bit strings of length four without two consecutive 1s. ◄

EXAMPLE 18 A playoff between two teams consists of at most five games. The first team that wins three games wins the playoff. In how many different ways can the playoff occur?

Solution: The tree diagram in Figure 3 displays all the ways the playoff can proceed, with the winner of each game shown. We see that there are 20 different ways for the playoff to occur. ◄

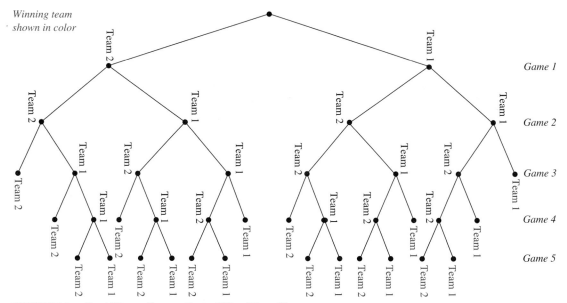

FIGURE 3 **Best Three Games Out of Five Playoffs.**

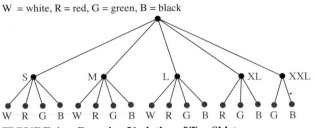

W = white, R = red, G = green, B = black

FIGURE 4   Counting Varieties of Tee Shirts.

EXAMPLE 19   Suppose that "I Love New Jersey" tee shirts come in five different sizes: S, M, L, XL, and XXL. Further suppose that each size comes in four colors, white, red, green, and black, except for XL, which comes only in red, green, and black, and XXL, which comes only in green and black. How many different shirts does a souvenir shop have to stock to have at least one of each available size and color of the tee shirt?

Solution: The tree diagram in Figure 4 displays all possible size and color pairs. It follows that the souvenir shop owner needs to stock 17 different tee shirts.   ◀

Exercises

1. There are 18 mathematics majors and 325 computer science majors at a college.

   a) How many ways are there to pick two representatives, so that one is a mathematics major and the other is a computer science major?

   b) How many ways are there to pick one representative who is either a mathematics major or a computer science major?

2. An office building contains 27 floors and has 37 offices on each floor. How many offices are in the building?

3. A multiple-choice test contains ten questions. There are four possible answers for each question.

   a) How many ways can a student answer the questions on the test if every question is answered?

   b) How many ways can a student answer the questions on the test if the student can leave answers blank?

4. A particular brand of shirt comes in 12 colors, has a male version and a female version, and comes in three sizes for each sex. How many different types of this shirt are made?

5. There are six different airlines that fly from New York to Denver and seven that fly from Denver to San Francisco. How many different possibilities are there for a trip from New York to San Francisco via Denver, when an airline is picked for the flight to Denver and an airline is picked for the continuation flight to San Francisco?

6. There are four major auto routes from Boston to Detroit and six from Detroit to Los Angeles. How many major auto routes are there from Boston to Los Angeles via Detroit?

7. How many different three-letter initials can people have?

8. How many different three-letter initials with none of the letters repeated can people have?

9. How many different three-letter initials are there that begin with an *A*?

10. How many bit strings are there of length eight?

11. How many bit strings of length ten begin and end with a 1?

12. How many bit strings are there of length six or less?

13. How many bit strings with length not exceeding n, where n is a positive integer, consist entirely of 1s?

14. How many bit strings of length n, where n is a positive integer, start and end with 1s?

15. How many strings are there of lowercase letters of length four or less?

16. How many strings are there of four lowercase letters that have the letter x in them?

17. How many strings of five ASCII characters contain the character @ (at sign) at least once? (*Note:* There are 128 different ASCII characters.)

18. How many positive integers less than 1000

   a) are divisible by 7?

   b) are divisible by 7 but not by 11?

   c) are divisible by both 7 and 11?

d) are divisible by either 7 or 11?

e) are divisible by exactly one of 7 and 11?

f) are divisible by neither 7 nor 11?

g) have distinct digits?

h) have distinct digits and are even?

19. How many positive integers between 100 and 999 inclusive

a) are divisible by 7?

b) are odd?

c) have the same three decimal digits?

d) are not divisible by 4?

e) are divisible by 3 or 4?

f) are not divisible by either 3 or 4?

g) are divisible by 3 but not by 4?

h) are divisible by 3 and 4?

20. How many positive integers between 1000 and 9999 inclusive

a) are divisible by 9?

b) are even?

c) have distinct digits?

d) are not divisible by 3?

e) are divisible by 5 or 7?

f) are not divisible by either 5 or 7?

g) are divisible by 5 but not by 7?

h) are divisible by 5 and 7?

21. How many strings of three decimal digits

a) do not contain the same digit three times?

b) begin with an odd digit?

c) have exactly two digits that are 4s?

22. How many strings of four decimal digits

a) do not contain the same digit twice?

b) end with an even digit?

c) have exactly three digits that are 9s?

23. A committee is formed containing either the governor or one of the two senators of each of the 50 states. How many ways are there to form this committee?

24. How many license plates can be made using either three digits followed by three letters or three letters followed by three digits?

25. How many license plates can be made using either two letters followed by four digits or two digits followed by four letters?

26. How many license plates can be made using either three letters followed by three digits or four letters followed by two digits?

27. How many license plates can be made using either two or three letters followed by either two or three digits?

28. How many strings of eight English letters are there

a) if letters can be repeated?

b) if no letter can be repeated?

c) that start with X, if letters can be repeated?

d) that start with X, if no letter can be repeated?

e) that start and end with X, if letters can be repeated?

f) that start with the letters BO (in that order), if letters can be repeated?

g) that start and end with the letters BO (in that order), if letters can be repeated?

h) that start or end with the letters BO (in that order), if letters can be repeated?

29. How many strings of eight English letters are there

a) that contain no vowels, if letters can be repeated?

b) that contain no vowels, if letters cannot be repeated?

c) that start with a vowel, if letters can be repeated?

d) that start with a vowel, if letters cannot be repeated?

e) that contain at least one vowel, if letters can be repeated?

f) that contain exactly one vowel, if letters can be repeated?

g) that start with X and contain at least one vowel, if letters can be repeated?

h) that start and end with X and contain at least one vowel, if letters can be repeated?

30. How many different functions are there from a set with 10 elements to sets with the following numbers of elements?

a) 2   b) 3   c) 4   d) 5

31. How many one-to-one functions are there from a set with five elements to sets with the following number of elements?

a) 4   b) 5   c) 6   d) 7

32. How many functions are there from the set $\{1, 2, \ldots, n\}$, where n is a positive integer, to the set $\{0, 1\}$?

33. How many functions are there from the set $\{1, 2, \ldots, n\}$, where n is a positive integer, to the set $\{0, 1\}$

a) that are one-to-one?

b) that assign 0 to both 1 and n?

c) that assign 1 to exactly one of the positive integers less than n?

34. How many partial functions (see the exercises in Section 1.8) are there from a set with five elements to sets with each of these number of elements?

a) 1   b) 2   c) 5   d) 9

35. How many partial functions (see the exercises in Section 1.8) are there from a set with m elements to a set with n elements, where m and n are positive integers?

36. How many subsets of a set with 100 elements have more than one element?

37. A **palindrome** is a string whose reversal is identical to the string. How many bit strings of length n are palindromes?

38. In how many ways can a photographer at a wedding arrange 6 people in a row from a group of 10 people, where the bride and the groom are among these 10 people, if

   a) the bride must be in the picture?
   b) both the bride and groom must be in the picture?
   c) exactly one of the bride and the groom is in the picture?

39. In how many ways can a photographer at a wedding arrange six people in a row, including the bride and groom, if

   a) the bride must be next to the groom?
   b) the bride is not next to the groom?
   c) the bride is positioned somewhere to the left of the groom?

40. How many bit strings of length seven either begin with two 0s or end with three 1s?

41. How many bit strings of length 10 either begin with three 0s or end with two 0s?

*42. How many bit strings of length 10 contain either five consecutive 0s or five consecutive 1s?

**43. How many bit strings of length eight contain either three consecutive 0s or four consecutive 1s?

44. Every student in a discrete mathematics class is either a computer science or a mathematics major or is a joint major in these two subjects. How many students are in the class if there are 38 computer science majors (including joint majors), 23 mathematics majors (including joint majors), and 7 joint majors?

45. How many positive integers not exceeding 100 are divisible either by 4 or by 6?

46. The name of a variable in the C programming language is a string that can contain uppercase letters, lowercase letters, digits, or underscores. Further, the first character in the string must be a letter, either uppercase or lowercase, or an underscore. If the name of a variable is determined by its first eight characters, how many different variables can be named in C? (Note that the name of a variable may contain fewer than eight characters.)

47. Suppose that at some future time every telephone in the world is assigned a number that contains a country code 1 to 3 digits long, that is, of the form X, XX, or XXX, followed by a ten-digit telephone number of the form NXX-NXX-XXXX (as described in Example 7). How many different telephone numbers would be available worldwide under this numbering plan?

48. Use a tree diagram to find the number of bit strings of length four with no three consecutive 0s.

49. How many ways are there to arrange the letters $a, b, c,$ and d such that a is not followed immediately by b?

50. Use a tree diagram to find the number of ways that the World Series can occur, where the first team that wins four games out of seven wins the series.

51. Use a tree diagram to determine the number of subsets of {3, 7, 9, 11, 24} with the property that the sum of the elements in the subset is less than 28.

52. a) Suppose that a store sells six varieties of soft drinks: cola, ginger ale, orange, root beer, lemonade, and cream soda. Use a tree diagram to determine the number of different types of bottles the store must stock to have all varieties available in all size bottles if all varieties are available in 12-ounce bottles, all but lemonade are available in 20-ounce bottles, only cola and ginger ale are available in 32-ounce bottles, and all but lemonade and cream soda are available in 64-ounce bottles?

   b) Answer the question in part (a) using counting rules.

53. a) Suppose that a popular style of running shoe is available for both men and women. The woman's shoe comes in sizes 6, 7, 8, and 9, and the man's shoe comes in sizes 8, 9, 10, 11, and 12. The man's shoe comes in white and black, while the woman's shoe comes in white, red, and black. Use a tree diagram to determine the number of different shoes that a store has to stock to have at least one pair of this type of running shoe for all available sizes and colors for both men and women.

   b) Answer the question in part (a) using counting rules.

*54. Use the product rule to show that there are 2^{2^n} different truth tables for propositions in n variables.

55. Use mathematical induction to prove the sum rule for m tasks from the sum rule for two tasks.

56. Use mathematical induction to prove the product rule for m tasks from the product rule for two tasks.

57. How many diagonals does a convex polygon with n sides have? (A polygon is convex if every line segment connecting two points in the interior or boundary of the polygon lies entirely within this set.)

58. Data are transmitted over the Internet in **datagrams,** which are structured blocks of bits. Each datagram contains header information organized into a maximum of 14 different fields (specifying many things, including the source and destination addresses) and a data area that contains the actual data that are transmitted. One of the 14 header fields is the **header length field** (denoted by HLEN), which is specified by the protocol to be 4 bits long and that specifies the header length in terms of 32-bit blocks of bits. For example, if HLEN = 0110, the header is made up of six 32-bit blocks. Another of the 14 header fields is the 16-bit-long **total length field** (denoted by TOTAL LENGTH), which specifies the length in bits of the entire datagram, including both the header fields and the data area. The length of the data area is the total length of the datagram minus the length of the header.

a) The largest possible value of TOTAL LENGTH (which is 16 bits long) determines the maximum total length in octets (blocks of 8 bits) of an Internet datagram. What is this value?

b) The largest possible value of HLEN (which is 4 bits long) determines the maximum total header length in 32-bit blocks. What is this value? What is the maximum total header length in octets?

c) The minimum (and most common) header length is 20 octets. What is the maximum total length in octets of the data area of an Internet datagram?

d) How many different strings of octets in the data area can be transmitted if the header length is 20 octets and the total length is as long as possible?

4.2   The Pigeonhole Principle

INTRODUCTION

Suppose that a flock of pigeons flies into a set of pigeonholes to roost. The **pigeonhole principle** states that if there are more pigeons than pigeonholes, then there must be at least one pigeonhole with at least two pigeons in it (see Figure 1). Of course, this principle applies to other objects besides pigeons and pigeonholes.

THEOREM 1

> **THE PIGEONHOLE PRINCIPLE** If $k + 1$ or more objects are placed into k boxes, then there is at least one box containing two or more of the objects.

Proof: Suppose that none of the k boxes contains more than one object. Then the total number of objects would be at most k. This is a contradiction, since there are at least $k + 1$ objects.   ◁

The pigeonhole principle is also called the **Dirichlet drawer principle,** after the nineteenth-century German mathematician Dirichlet, who often used this principle in his work. The following examples show how the pigeonhole principle is used.

EXAMPLE 1   Among any group of 367 people, there must be at least two with the same birthday, because there are only 366 possible birthdays.   ◀

EXAMPLE 2   In any group of 27 English words, there must be at least two that begin with the same letter, since there are 26 letters in the English alphabet.   ◀

EXAMPLE 3   How many students must be in a class to guarantee that at least two students receive the same score on the final exam, if the exam is graded on a scale from 0 to 100 points?

G. LEJEUNE DIRICHLET (1805–1859)   G. Lejeune Dirichlet was born into a French family living near Cologne, Germany. He studied at the University of Paris and held positions at the University of Breslau and the University of Berlin. In 1855 he was chosen to succeed Gauss at the University of Göttingen. Dirichlet is said to be the first person to master Gauss's *Disquisitiones Arithmeticae*, which appeared 20 years earlier. He is said to have kept a copy at his side even when he traveled. Dirichlet made many important discoveries in number theory, including the theorem that there are infinitely many primes in arithmetical progressions $an + b$ when a and b are relatively prime. He proved the $n = 5$ case of Fermat's Last Theorem, that there are no nontrivial solutions in integers to $x^5 + y^5 = z^5$. Dirichlet also made many contributions to analysis.

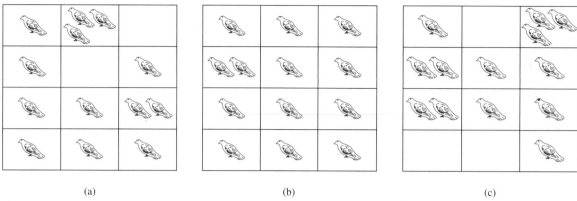

(a)   (b)   (c)

FIGURE 1   There Are More Pigeons Than Pigeonholes.

Solution: There are 101 possible scores on the final. The pigeonhole principle shows that among any 102 students there must be at least 2 students with the same score.   ◀

The pigeonhole principle is a useful tool in many proofs, including proofs of surprising results such as that given in Example 4.

EXAMPLE 4   Show that for every integer n there is a multiple of n that has only 0s and 1s in its decimal expansion.

Solution: Let n be a positive integer. Consider the n integers $1, 11, 111, \ldots, 11 \cdots 1$ (where the last integer in this list is the integer with $n + 1$ 1s in its decimal expansion). Note that there are n possible remainders when an integer is divided by n. Since there are $n + 1$ integers in this list, by the pigeonhole principle there must be two with the same remainder when divided by n. The difference of these two integers has a decimal expansion consisting entirely of 0s and 1s and is divisible by n.   ◀

THE GENERALIZED PIGEONHOLE PRINCIPLE

The pigeonhole principle states that there must be at least two objects in the same box when there are more objects than boxes. However, even more can be said when the number of objects exceeds a multiple of the number of boxes. For instance, among any set of 21 decimal digits there must be 3 that are the same. This follows because when 21 objects are distributed into 10 boxes, one box must have more than 2 objects.

THEOREM 2   **THE GENERALIZED PIGEONHOLE PRINCIPLE**   If N objects are placed into k boxes, then there is at least one box containing at least $\lceil N/k \rceil$ objects.

Proof: Suppose that none of the boxes contains more than $\lceil N/k \rceil - 1$ objects. Then, the total number of objects is at most

$$k \left(\left\lceil \frac{N}{k} \right\rceil - 1 \right) < k \left(\left(\frac{N}{k} + 1 \right) - 1 \right) = N,$$

where the inequality $\lceil N/k \rceil < (N/k) + 1$ has been used. This is a contradiction since there are a total of N objects.   ◁

A common type of problem asks for the minimum number of objects so that at least r of these objects must be in one of k boxes when these objects are distributed among the boxes. When we have N objects, the generalized pigeonhole principle tells us there must be at least r objects in one of the boxes as long as $\lceil N/k \rceil \geq r$. The smallest integer N with $N/k > r - 1$, namely, $N = k(r - 1) + 1$, is the smallest integer satisfying the inequality $\lceil N/k \rceil \geq r$. Could a smaller value of N suffice? The answer is no, because if we have $k(r - 1)$ objects, we could put $r - 1$ of them in each of the k boxes and no box would have at least r objects.

When thinking about problems of this type, it is useful to consider how you can avoid having at least r objects in one of the boxes as you add successive objects. To avoid adding a rth object to any box, you eventually end up with $r - 1$ objects in each box. There is no way to add the next object without putting a rth object in that box.

Examples 5–8 illustrate how the generalized pigeonhole principle is applied.

EXAMPLE 5   Among 100 people there are at least $\lceil 100/12 \rceil = 9$ who were born in the same month. ◀

EXAMPLE 6   What is the minimum number of students required in a discrete mathematics class to be sure that at least six will receive the same grade, if there are five possible grades, A, B, C, D, and F?

Extra Examples

Solution: The minimum number of students needed to ensure that at least six students receive the same grade is the smallest integer N such that $\lceil N/5 \rceil = 6$. The smallest such integer is $N = 5 \cdot 5 + 1 = 26$. If you have only 25 students, it is possible for there to be five who have received each grade so that no six students have received the same grade. Thus, 26 is the minimum number of students needed to ensure that at least six students will receive the same grade.   ◀

EXAMPLE 7   **a)** How many cards must be selected from a standard deck of 52 cards to guarantee that at least three cards of the same suit are chosen?
b) How many must be selected to guarantee that at least three hearts are selected?

Solution:
a) Suppose there are four boxes, and as cards are selected they are placed in the box reserved for cards of that suit. Using the generalized pigeonhole principle, we see that if N cards are selected, there is at least one box containing at least $\lceil N/4 \rceil$ cards. Consequently, we know that at least three cards of one suit are selected if $\lceil N/4 \rceil \geq 3$. The smallest integer N such that $\lceil N/4 \rceil \geq 3$ is $N = 2 \cdot 4 + 1 = 9$, so nine cards suffice. Note that if eight cards are selected, it is possible to have two cards of each suit, so that more than eight cards are needed. Consequently, nine cards must be selected to guarantee that at least three cards of one suit are chosen. One good way to think about this is to note that after the eighth card is chosen, there is no way to avoid having a third card of some suit.
b) We do not use the generalized pigeonhole principle to answer this question, since we want to make sure that there are three hearts, not just three cards of one suit. Note that in the worst case, we can select all the clubs, diamonds, and spades, 39 cards in all, before we select a single heart. The next three cards will be all hearts, so we may need to select 42 cards to get three hearts.   ◀

EXAMPLE 8   What is the least number of area codes needed to guarantee that the 25 million phones in a state have distinct ten-digit telephone numbers? (Assume that telephone numbers are of the form *NXX-NXX-XXXX*, where the first three digits form the area code, *N* represents a digit from 2 to 9 inclusive, and *X* represents any digit.)

Solution: There are eight million different phone numbers of the form *NXX-XXXX* (as shown in Example 7 of Section 4.1). Hence, by the generalized pigeonhole principle, among 25 million telephones, at least $\lceil 25,000,000/8,000,000 \rceil$ of them must have identical phone numbers. Hence, at least four area codes are required to ensure that all ten-digit numbers are different. ◄

Example 9, although not an application of the generalized pigeonhole principle, makes use of similar principles.

EXAMPLE 9 Suppose that a computer science laboratory has 15 workstations and ten servers. A cable can be used to directly connect a workstation to a server. Only one direct connection to a server can be active at any time. We want to guarantee that at any time any set of ten or fewer workstations can simultaneously access different servers via direct connections. Although we could do this by connecting every workstation directly to every server (using 150 connections), what is the minimum number of direct connections needed to achieve this goal?

Solution: Suppose that we label the workstations $W_1, W_2, \ldots, W_{15}$ and the servers $S_1, S_2, \ldots, S_{10}$. Furthermore, suppose that we connect W_k to S_k for $k = 1, 2, \ldots, 10$ and each of $W_{11}, W_{12}, W_{13}, W_{14}$, and W_{15} to all ten servers. We have a total of 60 direct connections. Clearly any set of ten or fewer workstations can simultaneously access different servers. We see this by noting that if workstation W_j is included with $1 \leq j \leq 10$, it can access server S_j, and for each workstation W_k with $k \geq 11$ included, there must be a corresponding workstation W_j with $1 \leq j \leq 10$ not included, so that W_k can access server S_j. (This follows since there are at least as many available servers S_j as there are workstations W_j with $1 \leq j \leq 10$ not included.)

Now suppose there are fewer than 60 direct connections between workstations and servers. Then some server would be connected to at most $\lfloor 59/10 \rfloor = 5$ workstations. (If all servers were connected to at least six workstations, there would be at least $6 \cdot 10 = 60$ direct connections.) This means that the remaining nine servers are not enough to allow the other ten workstations to simultaneously access different servers. Consequently, at least 60 direct connections are needed. It follows that 60 is the answer. ◄

SOME ELEGANT APPLICATIONS OF THE PIGEONHOLE PRINCIPLE

In many interesting applications of the pigeonhole principle, the objects to be placed in boxes must be chosen in a clever way. A few such applications will be described here.

EXAMPLE 10 During a month with 30 days a baseball team plays at least one game a day, but no more than 45 games. Show that there must be a period of some number of consecutive days during which the team must play exactly 14 games.

Solution: Let a_j be the number of games played on or before the jth day of the month. Then $a_1, a_2, \ldots, a_{30}$ is an increasing sequence of distinct positive integers, with $1 \leq a_j \leq 45$. Moreover, $a_1 + 14, a_2 + 14, \ldots, a_{30} + 14$ is also an increasing sequence of distinct positive integers, with $15 \leq a_j + 14 \leq 59$.

The 60 positive integers $a_1, a_2, \ldots, a_{30}, a_1 + 14, a_2 + 14, \ldots, a_{30} + 14$ are all less than or equal to 59. Hence, by the pigeonhole principle two of these integers are equal. Since the integers a_j, $j = 1, 2, \ldots, 30$ are all distinct and the integers $a_j + 14$, $j = 1, 2, \ldots, 30$ are all distinct, there must be indices i and j with $a_i = a_j + 14$. This means that exactly 14 games were played from day $j + 1$ to day i. ◀

EXAMPLE 11 Show that among any $n + 1$ positive integers not exceeding $2n$ there must be an integer that divides one of the other integers.

Solution: Write each of the $n + 1$ integers $a_1, a_2, \ldots, a_{n+1}$ as a power of 2 times an odd integer. In other words, let $a_j = 2^{k_j} q_j$ for $j = 1, 2, \ldots, n + 1$, where k_j is a nonnegative integer and q_j is odd. The integers $q_1, q_2, \ldots, q_{n+1}$ are all odd positive integers less than $2n$. Since there are only n odd positive integers less than $2n$, it follows from the pigeonhole principle that two of the integers $q_1, q_2, \ldots, q_{n+1}$ must be equal. Therefore, there are integers i and j such that $q_i = q_j$. Let q be the common value of q_i and q_j. Then, $a_i = 2^{k_i} q$ and $a_j = 2^{k_j} q$. It follows that if $k_i < k_j$, then a_i divides a_j; while if $k_i > k_j$, then a_j divides a_i. ◀

A clever application of the pigeonhole principle shows the existence of an increasing or a decreasing subsequence of a certain length in a sequence of distinct integers. Some definitions will be reviewed before this application is presented. Suppose that $a_1, a_2, \ldots, a_N$ is a sequence of real numbers. A **subsequence** of this sequence is a sequence of the form $a_{i_1}, a_{i_2}, \ldots, a_{i_m}$, where $1 \leq i_1 < i_2 < \cdots < i_m \leq N$. Hence, a subsequence is a sequence obtained from the original sequence by including some of the terms of the original sequence in their original order, and perhaps not including other terms. A sequence is called **strictly increasing** if each term is larger than the one that precedes it, and it is called **strictly decreasing** if each term is smaller than the one that precedes it.

THEOREM 3 Every sequence of $n^2 + 1$ distinct real numbers contains a subsequence of length $n + 1$ that is either strictly increasing or strictly decreasing.

We give an example before presenting the proof of Theorem 3.

EXAMPLE 12 The sequence $8, 11, 9, 1, 4, 6, 12, 10, 5, 7$ contains ten terms. Note that $10 = 3^2 + 1$. There are four increasing subsequences of length four, namely, $1, 4, 6, 12$; $1, 4, 6, 7$; $1, 4, 6, 10$; and $1, 4, 5, 7$. There is also a decreasing subsequence of length four, namely, $11, 9, 6, 5$. ◀

The proof of the theorem will now be given.

Proof: Let $a_1, a_2, \ldots, a_{n^2+1}$ be a sequence of $n^2 + 1$ distinct real numbers. Associate an ordered pair with each term of the sequence, namely, associate (i_k, d_k) to the term a_k, where i_k is the length of the longest increasing subsequence starting at a_k, and d_k is the length of the longest decreasing subsequence starting at a_k.

Suppose that there are no increasing or decreasing subsequences of length $n + 1$. Then i_k and d_k are both positive integers less than or equal to n, for $k = 1, 2, \ldots, n^2 + 1$. Hence, by the product rule there are n^2 possible ordered pairs for (i_k, d_k). By the pigeonhole principle, two of these $n^2 + 1$ ordered pairs are equal. In other words, there exist terms a_s and a_t, with $s < t$ such that $i_s = i_t$ and $d_s = d_t$. We will show that this is impossible. Because the terms of the sequence are distinct, either $a_s < a_t$ or $a_s > a_t$. If $a_s < a_t$, then, since $i_s = i_t$, an increasing subsequence of length $i_t + 1$ can be built starting at a_s, by taking a_s followed by an increasing subsequence of length i_t beginning at a_t. This is a contradiction. Similarly, if $a_s > a_t$, it can be shown that d_s must be greater than d_t, which is a contradiction. ◁

Links

The final example shows how the generalized pigeonhole principle can be applied to an important part of combinatorics called **Ramsey theory,** after the English mathematician F. P. Ramsey. In general, Ramsey theory deals with the distribution of subsets of elements of sets.

EXAMPLE 13   Assume that in a group of six people, each pair of individuals consists of two friends or two enemies. Show that there are either three mutual friends or three mutual enemies in the group.

Solution: Let A be one of the six people. Of the five other people in the group, there are either three or more who are friends of A, or three or more who are enemies of A. This follows from the generalized pigeonhole principle, since when five objects are divided into two sets, one of the sets has at least $\lceil 5/2 \rceil = 3$ elements. In the former case, suppose that B, C, and D are friends of A. If any two of these three individuals are friends, then these two and A form a group of three mutual friends. Otherwise, B, C, and D form a set of three mutual enemies. The proof in the latter case, when there are three or more enemies of A, proceeds in a similar manner. ◀

The **Ramsey number** $R(m, n)$, where m and n are positive integers greater than or equal to 2, denotes the minimum number of people at a party so that there are either m mutual friends or n mutual enemies, assuming that every pair of people at the party are friends or enemies. Example 13 shows that $R(3, 3) \le 6$. We conclude that $R(3, 3) = 6$ since in a group of five people where every two people are friends or enemies, there may not be three mutual friends or three mutual enemies (see Exercise 24).

It is possible to prove some useful properties about Ramsey numbers, but for the most part it is difficult to find their exact values. Note that by symmetry it can be shown that $R(m, n) = R(n, m)$ (see Exercise 28). We also have $R(2, n) = n$ for every positive integer $n \ge 2$ (see Exercise 27). The exact values of only nine Ramsey numbers $R(m, n)$ with $3 \le m \le n$ are known, including $R(4, 4) = 18$. Only bounds are known for many other Ramsey numbers, including $R(5, 5)$, which is known to satisfy $43 \le R(5, 5) \le 49$. The reader interested in learning more about Ramsey numbers should consult [MiRo91] or [GrRoSp90].

Links

FRANK PLUMPTON RAMSEY (1903–1930)   Frank Plumpton Ramsey, son of the president of Magdalene College, Cambridge, was educated at Winchester and Trinity Colleges. After graduating in 1923, he was elected a fellow of King's College, Cambridge, where he spent the remainder of his life. Ramsey made important contributions to mathematical logic. What we now call Ramsey theory began with his clever combinatorial arguments, published in the paper "On a Problem of Formal Logic." Ramsey also made contributions to the mathematical theory of economics. He was noted as an excellent lecturer on the foundations of mathematics. His death at the age of 26 deprived the mathematical community and Cambridge University of a brilliant young scholar.

Exercises

1. Show that in any set of six classes there must be two that meet on the same day, assuming that no classes are held on weekends.

2. Show that if there are 30 students in a class, then at least two have last names that begin with the same letter.

3. A drawer contains a dozen brown socks and a dozen black socks, all unmatched. A man takes socks out at random in the dark.
   a) How many socks must he take out to be sure that he has at least two socks of the same color?
   b) How many socks must he take out to be sure that he has at least two black socks?

4. A bowl contains ten red balls and ten blue balls. A woman selects balls at random without looking at them.
   a) How many balls must she select to be sure of having at least three balls of the same color?
   b) How many balls must she select to be sure of having at least three blue balls?

5. Show that among any group of five (not necessarily consecutive) integers, there are two with the same remainder when divided by 4.

6. Let d be a positive integer. Show that among any group of $d + 1$ (not necessarily consecutive) integers there are two with exactly the same remainder when they are divided by d.

7. Let n be a positive integer. Show that in any set of n consecutive integers there is exactly one divisible by n.

8. Show that if f is a function from S to T where S and T are finite sets with $|S| > |T|$, then there are elements s_1 and s_2 in S such that $f(s_1) = f(s_2)$, or in other words, f is not one-to-one.

9. What is the minimum number of students, each of whom comes from one of the 50 states, enrolled in a university to guarantee that there are at least 100 who come from the same state?

*10. Let (x_i, y_i), $i = 1, 2, 3, 4, 5$, be a set of five distinct points with integer coordinates in the xy plane. Show that the midpoint of the line joining at least one pair of these points has integer coordinates.

*11. Let (x_i, y_i, z_i), $i = 1, 2, 3, 4, 5, 6, 7, 8, 9$, be a set of nine distinct points with integer coordinates in xyz space. Show that the midpoint of at least one pair of these points has integer coordinates.

12. How many ordered pairs of integers (a, b) are needed to guarantee that there are two ordered pairs (a_1, b_1) and (a_2, b_2) such that $a_1 \bmod 5 = a_2 \bmod 5$ and $b_1 \bmod 5 = b_2 \bmod 5$?

13. a) Show that if five integers are selected from the first eight positive integers, there must be a pair of these integers with a sum equal to 9.
   b) Is the conclusion in part (a) true if four integers are selected rather than five?

14. a) Show that if seven integers are selected from the first 10 positive integers, there must be at least two pairs of these integers with the sum 11.
   b) Is the conclusion in part (a) true if six integers are selected rather than seven?

15. How many numbers must be selected from the set $\{1, 2, 3, 4, 5, 6\}$ to guarantee that at least one pair of these numbers add up to 7?

16. How many numbers must be selected from the set $\{1, 3, 5, 7, 9, 11, 13, 15\}$ to guarantee that at least one pair of these numbers add up to 16?

17. A company stores products in a warehouse. Storage bins in this warehouse are specified by their aisle, location in the aisle, and shelf. There are 50 aisles, 85 horizontal locations in each aisle, and 5 shelves throughout the warehouse. What is the least number of products the company can have so that at least two products must be stored in the same bin?

18. Suppose that there are nine students in a discrete mathematics class at a small college.
   a) Show that the class must have at least five male students or at least five female students.
   b) Show that the class must have at least three male students or at least seven female students.

19. Suppose that every student in a discrete mathematics class of 25 students is a freshman, a sophomore, or a junior.
   a) Show that there are at least nine freshmen, at least nine sophomores, or at least nine juniors in the class.
   b) Show that there are either at least three freshmen, at least 19 sophomores, or at least five juniors in the class.

20. Find an increasing subsequence of maximal length and a decreasing subsequence of maximal length in the sequence 22, 5, 7, 2, 23, 10, 15, 21, 3, 17.

21. Construct a sequence of 16 positive integers that has no increasing or decreasing subsequence of five terms.

22. Show that if there are 101 people of different heights standing in a line, it is possible to find 11 people in the order they are standing in the line with heights that are either increasing or decreasing.

*23. Describe an algorithm in pseudocode for producing the largest increasing or decreasing subsequence of a sequence of distinct integers.

24. Show that in a group of five people (where any two people are either friends or enemies), there are not necessarily three mutual friends or three mutual enemies.

25. Show that in a group of ten people (where any two people are either friends or enemies), there are either three mutual friends or four mutual enemies, and there are either three mutual enemies or four mutual friends.

26. Use Exercise 25 to show that among any group of 20 people (where any two people are either friends or enemies), there are either four mutual friends or four mutual enemies.

27. Show that if n is a positive integer with $n \geq 2$, then the Ramsey number $R(2, n)$ equals n.

28. Show that if m and n are positive integers with $m \geq 2$ and $n \geq 2$, then the Ramsey numbers $R(m, n)$ and $R(n, m)$ are equal.

29. Show that there are at least six people in California (population: 34 million) with the same three initials who were born on the same day of the year (but not necessarily in the same year).

30. Show that if there are 100,000,000 wage earners in the United States who earn less than 1,000,000 dollars, then there are two who earned exactly the same amount of money, to the penny, last year.

31. There are 38 different time periods during which classes at a university can be scheduled. If there are 677 different classes, how many different rooms will be needed?

32. A computer network consists of six computers. Each computer is directly connected to at least one of the other computers. Show that there are at least two computers in the network that are directly connected to the same number of other computers.

33. A computer network consists of six computers. Each computer is directly connected to zero or more of the other computers. Show that there are at least two computers in the network that are directly connected to the same number of other computers.

34. Find the least number of cables required to connect eight computers to four printers to guarantee that four computers can directly access four different printers. Justify your answer.

35. Find the least number of cables required to connect 100 computers to 20 printers to guarantee that 20 computers can directly access 20 different printers. Justify your answer.

*36. Prove that at a party where there are at least two people, there are two people who know the same number of other people there.

37. An arm wrestler is the champion for a period of 75 hours. The arm wrestler had at least one match an hour, but no more than 125 total matches. Show that there is a period of consecutive hours during which the arm wrestler had exactly 24 matches.

*38. Is the statement in Exercise 37 true if 24 is replaced by

a) 2?   **b)** 23?   **c)** 25?   **d)** 30?

39. Show that if f is a function from S to T where S and T are finite sets and $m = \lceil |S|/|T| \rceil$, then there are at least m elements of S mapped to the same value of T. That is, show that there are elements $s_1, s_2, \ldots, s_m$ of S such that $f(s_1) = f(s_2) = \cdots = f(s_m)$.

40. There are 51 houses on a street. Each house has an address between 1000 and 1099, inclusive. Show that at least two houses have addresses that are consecutive integers.

*41. Let x be an irrational number. Show that for some positive integer j not exceeding n, the absolute value of the difference between jx and the nearest integer to jx is less than $1/n$.

42. Let $n_1, n_2, \ldots, n_t$ be positive integers. Show that if $n_1 + n_2 + \cdots + n_t - t + 1$ objects are placed into t boxes, then for some $i, i = 1, 2, \ldots, t$, the ith box contains at least n_i objects.

*43. A proof of Theorem 3 based on the generalized pigeonhole principle is outlined in this problem. The notation used is the same as that used in the proof in the text.

a) Assume that $i_k \leq n$ for $k = 1, 2, \ldots, n^2 + 1$. Use the generalized pigeonhole principle to show that there are $n + 1$ terms $a_{k_1}, a_{k_2}, \ldots, a_{k_{n+1}}$ with $i_{k_1} = i_{k_2} = \cdots = i_{k_{n+1}}$, where $1 \leq k_1 < k_2 < \cdots < k_{n+1}$.

b) Show that $a_{k_j} > a_{k_{j+1}}$ for $j = 1, 2, \ldots, n$. (*Hint:* Assume that $a_{k_j} < a_{k_{j+1}}$, and show that this implies that $i_{k_j} > i_{k_{j+1}}$, which is a contradiction.)

c) Use parts (a) and (b) to show that if there is no increasing subsequence of length $n + 1$, then there must be a decreasing subsequence of this length.

4.3   Permutations and Combinations

INTRODUCTION

Suppose that a tennis team has ten members. The coach has to select five players to make the trip to a match at another school. In addition, the coach has to prepare an ordered list of four players to play the four singles matches. In this section, methods will be developed to count the different unordered collections of the five players that are selected to make the trip and the different lists of four players to play the four singles matches. More generally, techniques will be introduced for counting the unordered selections of distinct objects and the ordered arrangements of objects of a finite set.

PERMUTATIONS

A **permutation** of a set of distinct objects is an ordered arrangement of these objects. We also are interested in ordered arrangements of some of the elements of a set. An ordered arrangement of r elements of a set is called an **r-permutation.**

EXAMPLE 1   Let $S = \{1, 2, 3\}$. The arrangement 3, 1, 2 is a permutation of S. The arrangement 3, 2 is a 2-permutation of S.   ◀

The number of r-permutations of a set with n elements is denoted by $P(n, r)$. We can find $P(n, r)$ using the product rule.

THEOREM 1   The number of r-permutations of a set with n distinct elements is

$$P(n, r) = n(n - 1)(n - 2) \cdots (n - r + 1).$$

Proof: The first element of the permutation can be chosen in n ways, since there are n elements in the set. There are $n - 1$ ways to choose the second element of the permutation, since there are $n - 1$ elements left in the set after using the element picked for the first position. Similarly, there are $n - 2$ ways to choose the third element, and so on, until there are exactly $n - (r - 1) = n - r + 1$ ways to choose the rth element. Consequently, by the product rule, there are

$$n(n - 1)(n - 2) \cdots (n - r + 1)$$

r-permutations of the set.   ◁

From Theorem 1 it follows that

$$P(n, r) = n(n - 1)(n - 2) \cdots (n - r + 1) = \frac{n!}{(n - r)!}$$

In particular, note that $P(n, n) = n!$. We will illustrate this result with some examples.

EXAMPLE 2   How many ways are there to select a first-prize winner, a second-prize winner, and a third-prize winner from 100 different people who have entered a contest?

Solution: Because it matters which person wins which prize, the number of ways to pick the three prize winners is the number of ordered selections of three elements from a set of 100 elements, that is, the number of 3-permutations of a set of 100 elements. Consequently, the answer is

$$P(100, 3) = 100 \cdot 99 \cdot 98 = 970,200.$$ ◀

EXAMPLE 3   Suppose that there are eight runners in a race. The winner receives a gold medal the second-place finisher receives a silver medal, and the third-place finisher receives a bronze medal. How many different ways are there to award these medals, if all possible outcomes of the race can occur and there are no ties?

Solution: The number of different ways to award the medals is the number of 3-permutations of a set with eight elements. Hence, there are $P(8, 3) = 8 \cdot 7 \cdot 6 = 336$ possible ways to award the medals. ◀

EXAMPLE 4 Suppose that a saleswoman has to visit eight different cities. She must begin her trip in a specified city, but she can visit the other seven cities in any order she wishes. How many possible orders can the saleswoman use when visiting these cities?

Solution: The number of possible paths between the cities is the number of permutations of seven elements, since the first city is determined, but the remaining seven can be ordered arbitrarily. Consequently, there are $7! = 7 \cdot 6 \cdot 5 \cdot 4 \cdot 3 \cdot 2 \cdot 1 = 5040$ ways for the saleswoman to choose her tour. If, for instance, the saleswoman wishes to find the path between the cities with minimum distance, and she computes the total distance for each possible path, she must consider a total of 5040 paths! ◀

EXAMPLE 5 How many permutations of the letters $ABCDEFGH$ contain the string ABC?

Solution: Because the letters ABC must occur as a block, we can find the answer by finding the number of permutations of six objects, namely, the block ABC and the individual letters D, E, F, G, and H. Because these six objects can occur in any order, there are $6! = 720$ permutations of the letters $ABCDEFGH$ in which ABC occurs as a block. ◀

COMBINATIONS

An *r-combination* of elements of a set is an unordered selection of r elements from the set. Thus, an r-combination is simply a subset of the set with r elements.

EXAMPLE 6 Let S be the set $\{1, 2, 3, 4\}$. Then $\{1, 3, 4\}$ is a 3-combination from S. ◀

The number of r-combinations of a set with n distinct elements is denoted by $C(n, r)$. Note that $C(n, r)$ is also denoted by $\binom{n}{r}$ and is called a **binomial coefficient.** We will learn where this terminology arises in Section 4.4.

EXAMPLE 7 We see that $C(4, 2) = 6$, since the 2-combinations of $\{a, b, c, d\}$ are the six subsets $\{a, b\}$, $\{a, c\}$, $\{a, d\}$, $\{b, c\}$, $\{b, d\}$, and $\{c, d\}$. ◀

We can determine the number of r-combinations of a set with n elements using the formula for the number of r-permutations of a set. To do this, note that the r-permutations of a set can be obtained by first forming r-combinations and then ordering the elements in these combinations. The proof of the following theorem, which gives the value of $C(n, r)$, is based on this observation.

THEOREM 2 The number of r-combinations of a set with n elements, where n is a nonnegative integer and r is an integer with $0 \leq r \leq n$, equals

$$C(n, r) = \frac{n!}{r! \, (n - r)!}.$$

Proof: The r-permutations of the set can be obtained by forming the $C(n, r)$ r-combinations of the set, and then ordering the elements in each r-combination, which can be done in $P(r, r)$ ways. Consequently,

$$P(n, r) = C(n, r) \cdot P(r, r).$$

This implies that

$$C(n, r) = \frac{P(n, r)}{P(r, r)} = \frac{n!/(n-r)!}{r!/(r-r)!} = \frac{n!}{r!\,(n-r)!}.$$ ◁

Remark: The formula in Theorem 2, although explicit, is not helpful when $C(n, r)$ is computed for large values of n and r. The reasons are that it is practical to compute exact values of factorials exactly only for small integer values, and when floating point arithmetic is used, the formula in Theorem 2 may produce a value that is not an integer. In practice, to compute $C(n, r)$ you cancel out all the terms in the larger factorial in the denominator from the numerator and denominator, multiply all the terms that do not cancel in the numerator and then divide by the smaller factorial in the denominator. Many calculators have a built-in function for $C(n, r)$.

Corollary 1 displays a useful identity for the number of r-combinations of a set.

COROLLARY 1   Let n and r be nonnegative integers with $r \leq n$. Then $C(n, r) = C(n, n - r)$.

Proof: From Theorem 2 it follows that

$$C(n, r) = \frac{n!}{r!\,(n-r)!}$$

and

$$C(n, n - r) = \frac{n!}{(n-r)!\,[n-(n-r)]!} = \frac{n!}{(n-r)!\,r!}.$$

Hence, $C(n, r) = C(n, n - r)$. ◁

We can also prove Corollary 1 using a proof that shows that both sides of the equation in Corollary 1 count the same objects using different reasoning. We describe this important type of proof in Definition 1.

DEFINITION 1   A *combinatorial proof* is a proof that uses counting arguments to prove a theorem, rather than some other method such as algebraic techniques.

Many identities involving binomial coefficients can be proved using combinatorial proofs. An identity can be proved using a combinatorial proof if it can be shown that the two sides of the identity count the same elements, but in different ways. We now provide a combinatorial proof of Corollary 1.

Proof: Suppose that S is a set with n elements. Every subset A of S with r elements corresponds to a subset of S with $n - r$ elements, namely $\overline{A}$. Consequently, $C(n, r) = C(n, n - r)$.   ◁

EXAMPLE 8

Extra Examples

How many ways are there to select five players from a 10-member tennis team to make a trip to a match at another school?

Solution: The answer is given by the number of 5-combinations of a set with ten elements. By Theorem 2, the number of such combinations is

$$C(10, 5) = \frac{10!}{5!\,5!} = 252.$$ ◀

EXAMPLE 9 A group of 30 people have been trained as astronauts to go on the first mission to Mars. How many ways are there to select a crew of six people to go on this mission (assuming that all crew members have the same job)?

Solution: The number of ways to select a crew of six from the pool of 30 people is the number of 6-combinations of a set with 30 elements, because the order in which these people are chosen does not matter. By Theorem 2, the number of such combinations is

$$C(30, 6) = \frac{30!}{6!\,24!} = \frac{30 \cdot 29 \cdot 28 \cdot 27 \cdot 26 \cdot 25}{6 \cdot 5 \cdot 4 \cdot 3 \cdot 2 \cdot 1} = 593,775.$$ ◀

EXAMPLE 10 How many bit strings of length n contain exactly r 1s?

Solution: The positions of r 1s in a bit string of length n form an r-combination of the set $\{1, 2, 3, \ldots, n\}$. Hence, there are $C(n, r)$ bit strings of length n that contain exactly r 1s.   ◀

EXAMPLE 11 How many ways are there to select a committee to develop a discrete mathematics course at a school if the committee is to consist of three faculty members from the mathematics department and four from the computer science department, if there are nine faculty members of the mathematics department and 11 of the computer science department?

Solution: By the product rule, the answer is the product of the number of 3-combinations of a set with nine elements and the number of 4-combinations of a set with 11 elements. By Theorem 2, the number of ways to select the committee is

$$C(9, 3) \cdot C(11, 4) = \frac{9!}{3!\,6!} \cdot \frac{11!}{4!\,7!} = 84 \cdot 330 = 27,720.$$ ◀

Exercises

1. List all the permutations of $\{a, b, c\}$.
2. How many permutations are there of the set $\{a, b, c, d, e, f, g\}$?
3. How many permutations of $\{a, b, c, d, e, f, g\}$ end with a?
4. Let $S = \{1, 2, 3, 4, 5\}$.
   a) List all the 3-permutations of S.
   b) List all the 3-combinations of S.

5. Find the value of each of these quantities.
   a) $P(6, 3)$ b) $P(6, 5)$ c) $P(8, 1)$
   d) $P(8, 5)$ e) $P(8, 8)$ f) $P(10, 9)$

6. Find the value of each of these quantities.
   a) $C(5, 1)$ b) $C(5, 3)$ c) $C(8, 4)$
   d) $C(8, 8)$ e) $C(8, 0)$ f) $C(12, 6)$

7. Find the number of 5-permutations of a set with nine elements.

8. In how many different orders can five runners finish a race if no ties are allowed?

9. How many possibilities are there for the win, place, and show (first, second, and third) positions in a horse race with 12 horses if all orders of finish are possible?

10. There are six different candidates for governor of a state. In how many different orders can the names of the candidates be printed on a ballot?

11. How many bit strings of length ten contain

   a) exactly four 1s?
   b) at most four 1s?
   c) at least four 1s?
   d) an equal number of 0s and 1s?

12. How many bit strings of length 12 contain

   a) exactly three 1s?
   b) at most three 1s?
   c) at least three 1s?
   d) an equal number of 0s and 1s?

13. A group contains n men and n women. How many ways are there to arrange these people in a row if the men and women alternate?

14. In how many ways can a set of two positive integers less than 100 be chosen?

15. In how many ways can a set of five letters be selected from the English alphabet?

16. How many subsets with an odd number of elements does a set with ten elements have?

17. How many subsets with more than two elements does a set with 100 elements have?

18. A coin is flipped eight times where each flip comes up either heads or tails. How many possible outcomes

   a) are there in total?
   b) contain exactly three heads?
   c) contain at least three heads?
   d) contain the same number of heads and tails?

19. A coin is flipped ten times where each flip comes up either heads or tails. How many possible outcomes

   a) are there in total?
   b) contain exactly two heads?
   c) contain at most three tails?
   d) contain the same number of heads and tails?

20. How many bit strings of length ten have

   a) exactly three 0s?
   b) more 0s than 1s?
   c) at least seven 1s?
   d) at least three 1s?

21. How many permutations of the letters *ABCDEFG* contain

   a) the string *BCD*?

   b) the string *CFGA*?
   c) the strings *BA* and *GF*?
   d) the strings *ABC* and *DE*?
   e) the strings *ABC* and *CDE*?
   f) the strings *CBA* and *BED*?

22. How many permutations of the letters *ABCDEFGH* contain

   a) the string *ED*?
   b) the string *CDE*?
   c) the strings *BA* and *FGH*?
   d) the strings *AB*, *DE*, and *GH*?
   e) the strings *CAB* and *BED*?
   f) the strings *BCA* and *ABF*?

23. How many ways are there for eight men and five women to stand in a line so that no two women stand next to each other? (*Hint:* First position the men and then consider possible positions for the women.)

24. How many ways are there for ten women and six men to stand in a line so that no two men stand next to each other? (*Hint:* First position the women and then consider possible positions for the men.)

25. One hundred tickets, numbered 1, 2, 3, . . . , 100, are sold to 100 different people for a drawing. Four different prizes are awarded, including a grand prize (a trip to Tahiti). How many ways are there to award the prizes if

   a) there are no restrictions?
   b) the person holding ticket 47 wins the grand prize?
   c) the person holding ticket 47 wins one of the prizes?
   d) the person holding ticket 47 does not win a prize?
   e) the people holding tickets 19 and 47 both win prizes?
   f) the people holding tickets 19, 47, and 73 all win prizes?
   g) the people holding tickets 19, 47, 73, and 97 all win prizes?
   h) none of the people holding tickets 19, 47, 73, and 97 wins a prize?
   i) the grand prize winner is a person holding ticket 19, 47, 73, or 97?
   j) the people holding tickets 19 and 47 win prizes, but the people holding tickets 73 and 97 do not win prizes?

26. Thirteen people on a softball team show up for a game.

   a) How many ways are there to choose ten players to take the field?
   b) How many ways are there to assign the ten positions by selecting players from the 13 people who show up?
   c) Of the 13 people who show up, three are women. How many ways are there to choose ten players

to take the field if at least one of these players must be a woman?

27. A club has 25 members.

   a) How many ways are there to choose four members of the club to serve on an executive committee?

   b) How many ways are there to choose a president, vice president, secretary, and treasurer of the club?

28. A professor writes 40 discrete mathematics true/false questions. Of the statements in these questions, 17 are true. If the questions can be positioned in any order, how many different answer keys are possible?

29. How many 4-permutations of the positive integers not exceeding 100 contain three consecutive integers in the correct order

   a) where consecutive means in the usual order of the integers and where these consecutive integers can perhaps be separated by other integers in the permutation?

   b) where consecutive means both that the numbers be consecutive integers and that they be in consecutive positions in the permutation?

30. Seven women and nine men are on the faculty in the mathematics department at a school.

   a) How many ways are there to select a committee of five members of the department if at least one woman must be on the committee?

   b) How many ways are there to select a committee of five members of the department if at least one woman and at least one man must be on the committee?

31. The English alphabet contains 21 consonants and five vowels. How many strings of six lowercase letters of the English alphabet contain

   a) exactly one vowel?   b) exactly two vowels?
   c) at least one vowel?   d) at least two vowels?

32. How many strings of six lowercase letters from the English alphabet contain

   a) the letter a?
   b) the letters a and b?
   c) the letters a and b in consecutive positions with a preceding b, with all the letters distinct?
   d) the letters a and b, where a is somewhere to the left of b in the string, with all the letters distinct?

33. Suppose that a department contains ten men and 15 women. How many ways are there to form a committee with six members if it must have the same number of men and women?

34. Suppose that a department contains ten men and 15 women. How many ways are there to form a committee with six members if it must have more women than men?

35. How many bit strings contain exactly eight 0s and ten 1s if every 0 must be immediately followed by a 1?

36. How many bit strings contain exactly five 0s and 14 1s if every 0 must be immediately followed by two 1s?

37. How many bit strings of length ten contain at least three 1s and at least three 0s?

38. How many ways are there to select 12 countries in the United Nations to serve on a council if 3 are selected from a block of 45, 4 are selected from a block of 57, and the others are selected from the remaining 69 countries?

39. How many license plates consisting of three letters followed by three digits contain no letter or digit twice?

40. How many ways are there to seat six people around a circular table, where seatings are considered to be the same if they can be obtained from each other by rotating the table?

41. How many ways are there for a horse race with three horses to finish if ties are possible? (*Note:* Two or three horses may tie.)

*42. How many ways are there for a horse race with four horses to finish if ties are possible? (*Note:* Any number of the four horses may tie.)

*43. There are six runners in the 100-yard dash. How many ways are there for three medals to be awarded if ties are possible? (The runner or runners who finish with the fastest time receive gold medals, the runner or runners who finish with exactly one runner ahead receive silver medals, and the runner or runners who finish with exactly two runners ahead receive bronze medals.)

*44. This procedure is used to break ties in games in the championship round of the World Cup soccer tournament. Each team selects five players in a prescribed order. Each of these players takes a penalty kick, with a player from the first team followed by a player from the second team and so on, following the order of players specified. If the score is still tied at the end of the ten penalty kicks, this procedure is repeated. If the score is till tied after 20 penalty kicks, a sudden-death shootout occurs, with the first team scoring an unanswered goal victorious.

   a) How many different scoring scenarios are possible if the game is settled in the first round of ten penalty kicks, where the round ends once it is impossible for a team to equal the number of goals scored by the other team?

   b) How many different scoring scenarios for the first and second groups of penalty kicks are possible if the game is settled in the second round of ten penalty kicks?

   c) How many scoring scenarios are possible for the full set of penalty kicks if the game is settled with no more than ten total additional kicks after the two rounds of five kicks for each team?

4.4 Binomial Coefficients

As we remarked in Section 4.3, the number of r-combinations from a set with n elements is often denoted by $\binom{n}{r}$. This number is also called a **binomial coefficient** because these numbers occur as coefficients in the expansion of powers of binomial expressions such as $(a + b)^n$. We will discuss the **Binomial Theorem,** which gives a power of a binomial expression as a sum of terms involving binomial coefficients. We will prove this theorem using a combinatorial proof. We will also show how combinatorial proofs can be used to establish some of the many different identities that express relationships among binomial coefficients.

THE BINOMIAL THEOREM

Links

The binomial theorem gives the coefficients of the expansion of powers of binomial expressions. A **binomial** expression is simply the sum of two terms, such as $x + y$. (The terms can be products of constants and variables, but that does not concern us here.) Example 1 illustrates why this theorem holds.

EXAMPLE 1 The expansion of $(x + y)^3$ can be found using combinatorial reasoning instead of multiplying the three terms out. When $(x + y)^3 = (x + y)(x + y)(x + y)$ is expanded, all products of a term in the first sum, a term in the second sum, and a term in the third sum are added. Terms of the form x^3, x^2y, xy^2, and y^3 arise. To obtain a term of the form x^3, an x must be chosen in each of the sums, and this can be done in only one way. Thus, the x^3 term in the product has a coefficient of 1. To obtain a term of the form x^2y, an x must be chosen in two of the three sums (and consequently a y in the other sum). Hence, the number of such terms is the number of 2-combinations of three objects, namely $\binom{3}{2}$. Similarly, the number of terms of the form xy^2 is the number of ways to pick one of the three sums to obtain an x (and consequently take a y from each of the other two sums). This can be done in $\binom{3}{1}$ ways. Finally, the only way to obtain a y^3 term is to choose the y for each of the three sums in the product, and this can be done in exactly one way. Consequently, it follows that

$$(x + y)^3 = (x + y)(x + y)(x + y) = (xx + xy + yx + yy)(x + y)$$
$$= xxx + xxy + xyx + xyy + yxx + yxy + yyx + yyy$$
$$= x^3 + 3x^2y + 3xy^2 + y^3. \qquad \blacktriangleleft$$

The Binomial Theorem will now be stated.

THEOREM 1 **THE BINOMIAL THEOREM** Let x and y be variables, and let n be a nonnegative integer. Then

$$(x + y)^n = \sum_{j=0}^{n} \binom{n}{j} x^{n-j} y^j$$

$$= \binom{n}{0} x^n + \binom{n}{1} x^{n-1}y + \binom{n}{2} x^{n-2}y^2 + \cdots + \binom{n}{n-1} xy^{n-1} + \binom{n}{n} y^n.$$

Proof: A combinatorial proof of the theorem will be given. The terms in the product when it is expanded are of the form $x^{n-j}y^j$ for $j = 0, 1, 2, \ldots, n$. To count the number of terms of the form $x^{n-j}y^j$, note that to obtain such a term it is necessary to choose $n - j$ xs from the n sums (so that the other j terms in the product are ys). Therefore, the coefficient of $x^{n-j}y^j$ is $\binom{n}{n-j}$, which is equal to $\binom{n}{j}$. This proves the theorem. ◁

The use of the Binomial Theorem is illustrated by Examples 2–4.

EXAMPLE 2   What is the expansion of $(x + y)^4$?

Solution: From the Binomial Theorem it follows that

$$(x + y)^4 = \sum_{j=0}^{4} \binom{4}{j} x^{4-j} y^j$$

$$= \binom{4}{0} x^4 + \binom{4}{1} x^3 y + \binom{4}{2} x^2 y^2 + \binom{4}{3} x y^3 + \binom{4}{4} y^4$$

$$= x^4 + 4x^3 y + 6x^2 y^2 + 4x y^3 + y^4.$$

◀

EXAMPLE 3   What is the coefficient of $x^{12}y^{13}$ in the expansion of $(x + y)^{25}$?

Solution: From the Binomial Theorem it follows that this coefficient is

$$\binom{25}{13} = \frac{25!}{13!\,12!} = 5,200,300.$$

◀

EXAMPLE 4   What is the coefficient of $x^{12}y^{13}$ in the expansion of $(2x - 3y)^{25}$?

Solution: First, note that this expression equals $(2x + (-3y))^{25}$. By the Binomial Theorem, we have

$$(2x + (-3y))^{25} = \sum_{j=0}^{25} \binom{25}{j} (2x)^{25-j} (-3y)^j.$$

Consequently, the coefficient of $x^{12}y^{13}$ in the expansion is obtained when $j = 13$, namely,

$$\binom{25}{13} 2^{12}(-3)^{13} = -\frac{25!}{13!\,12!} 2^{12} 3^{13}.$$

◀

We can prove some useful identities using the Binomial Theorem, as Corollaries 1, 2, and 3 demonstrate.

COROLLARY 1   Let n be a nonnegative integer. Then

$$\sum_{k=0}^{n} \binom{n}{k} = 2^n.$$

Proof: Using the Binomial Theorem with $x = 1$ and $y = 1$, we see that

$$2^n = (1 + 1)^n = \sum_{k=0}^{n} \binom{n}{k} 1^k 1^{n-k} = \sum_{k=0}^{n} \binom{n}{k}.$$

This is the desired result.   ◁

There is also a nice combinatorial proof of Corollary 1, which we now present.

Proof: A set with n elements has a total of 2^n different subsets. Each subset has zero elements, one element, two elements, ..., or n elements in it. There are $\binom{n}{0}$ subsets with zero elements, $\binom{n}{1}$ subsets with one element, $\binom{n}{2}$ subsets with two elements, ..., and $\binom{n}{n}$ subsets with n elements. Therefore,

$$\sum_{k=0}^{n} \binom{n}{k}$$

counts the total number of subsets of a set with n elements. This shows that

$$\sum_{k=0}^{n} \binom{n}{k} = 2^n.$$ ◁

COROLLARY 2   Let n be a positive integer. Then

$$\sum_{k=0}^{n} (-1)^k \binom{n}{k} = 0.$$

Proof: By the Binomial Theorem it follows that

$$0 = 0^n = ((-1) + 1)^n = \sum_{k=0}^{n} \binom{n}{k} (-1)^k 1^{n-k} = \sum_{k=0}^{n} \binom{n}{k} (-1)^k.$$

This proves the corollary.   ◁

Remark: Corollary 2 implies that

$$\binom{n}{0} + \binom{n}{2} + \binom{n}{4} + \cdots = \binom{n}{1} + \binom{n}{3} + \binom{n}{5} + \cdots.$$

COROLLARY 3   Let n be a nonnegative integer. Then

$$\sum_{k=0}^{n} 2^k \binom{n}{k} = 3^n.$$

Proof: We recognize that the left-hand side of this formula is the expansion of $(1 + 2)^n$ provided by the Binomial Theorem. Therefore, by the Binomial Theorem, we see that

$$(1 + 2)^n = \sum_{k=0}^{n} \binom{n}{k} 1^{n-k} 2^k = \sum_{k=0}^{n} \binom{n}{k} 2^k.$$

Hence

$$\sum_{k=0}^{n} 2^k \binom{n}{k} = 3^n.$$

◁

PASCAL'S IDENTITY AND TRIANGLE

The binomial coefficients satisfy many different identities. We introduce one of the most important of these now.

THEOREM 2

PASCAL'S IDENTITY Let n and k be positive integers with $n \geq k$. Then

$$\binom{n+1}{k} = \binom{n}{k-1} + \binom{n}{k}.$$

Proof: Suppose that T is a set containing $n + 1$ elements. Let a be an element in T, and let $S = T - \{a\}$. Note that there are $\binom{n+1}{k}$ subsets of T containing k elements. However, a subset of T with k elements either contains a together with $k - 1$ elements of S, or contains k elements of S and does not contain a. Since there are $\binom{n}{k-1}$ subsets of $k - 1$ elements of S, there are $\binom{n}{k-1}$ subsets of k elements of T that contain a. And there are $\binom{n}{k}$ subsets of k elements of T that do not contain a, since there are $\binom{n}{k}$ subsets of k elements of S. Consequently,

$$\binom{n+1}{k} = \binom{n}{k-1} + \binom{n}{k}.$$

◁

Remark: A combinatorial proof of Pascal's Identity has been given. It is also possible to prove this identity by algebraic manipulation from the formula for $\binom{n}{r}$ (see Exercise 19 at the end of this section).

Remark: Pascal's Identity, together with the initial conditions $\binom{n}{0} = \binom{n}{n} = 1$ for all integers n, can be used to recursively define binomial coefficients. This recursive definition

BLAISE PASCAL (1623–1662) Blaise Pascal exhibited his talents at an early age, although his father, who had made discoveries in analytic geometry, kept mathematics books away from him to encourage other interests. At 16 Pascal discovered an important result concerning conic sections. At 18 he designed a calculating machine, which he built and sold. Pascal, along with Fermat, laid the foundations for the modern theory of probability. In this work he made new discoveries concerning what is now called Pascal's triangle. In 1654, Pascal abandoned his mathematical pursuits to devote himself to theology. After this, he returned to mathematics only once. One night, distracted by a severe toothache, he sought comfort by studying the mathematical properties of the cycloid. Miraculously, his pain subsided, which he took as a sign of divine approval of the study of mathematics.

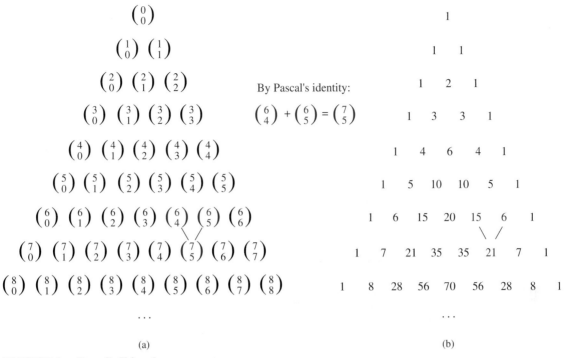

By Pascal's identity:

$$\binom{6}{4} + \binom{6}{5} = \binom{7}{5}$$

(a)

(b)

FIGURE 1   Pascal's Triangle.

is useful in the computation of binomial coefficients since only the addition of integers is needed to use this recursive definition.

Pascal's Identity is the basis for a geometric arrangement of the binomial coefficients in a triangle, as shown in Figure 1.

The nth row in the triangle consists of the binomial coefficients

$$\binom{n}{k}, \; k = 0, 1, \ldots, n.$$

This triangle is known as **Pascal's triangle.** Pascal's Identity shows that when two adjacent binomial coefficients in this triangle are added, the binomial coefficient in the next row between these two coefficients is produced.

SOME OTHER IDENTITIES OF THE BINOMIAL COEFFICIENTS

We conclude this section with combinatorial proofs of two of the many identities enjoyed by the binomial coefficients.

THEOREM 3

VANDERMONDE'S IDENTITY Let $m, n,$ and r be nonnegative integers with r not exceeding either m or n. Then

$$\binom{m + n}{r} = \sum_{k=0}^{r} \binom{m}{r - k} \binom{n}{k}.$$

Remark: This identity was discovered by mathematician Alexandre-Théophile Vandermonde in the eighteenth century.

Proof: Suppose that there are m items in one set and n items in a second set. Then the total number of ways to pick r elements from the union of these sets is $\binom{m+n}{r}$. Another way to pick r elements from the union is to pick k elements from the first set and then $r - k$ elements from the second set, where k is an integer with $0 \leq k \leq r$. This can be done in $\binom{m}{k}\binom{n}{r-k}$ ways, using the product rule. Hence, the total number of ways to pick r elements from the union also equals

$$\binom{m+n}{r} = \sum_{k=0}^{r} \binom{m}{r-k}\binom{n}{k}.$$

This proves Vandermonde's Identity. ◁

Corollary 4 follows from Vandermonde's Identity.

COROLLARY 4
If n is a nonnegative integer, then

$$\binom{2n}{n} = \sum_{k=0}^{n} \binom{n}{k}^2.$$

Proof: We use Vandermonde's Identity with $m = r = n$ to obtain

$$\binom{2n}{n} = \sum_{k=0}^{n} \binom{n}{n-k}\binom{n}{k} = \sum_{k=0}^{n} \binom{n}{k}^2.$$

The last equality was obtained using the identity $\binom{n}{k} = \binom{n}{n-k}$. ◁

We can prove combinatorial identities by counting bit strings with different properties, as the proof of Theorem 4 will demonstrate.

THEOREM 4
Let n and r be nonnegative integers with $r \leq n$. Then

$$\binom{n+1}{r+1} = \sum_{j=r}^{n} \binom{j}{r}.$$

Proof: We use a combinatorial proof. By Example 10 in Section 4.3, the left-hand side, $\binom{n+1}{r+1}$, counts the bit strings of length $n + 1$ containing $r + 1$ ones.

ALEXANDRE-THÉOPHILE VANDERMONDE (1735–1796) Because Alexandre-Théophile Vandermonde was a sickly child, his physician father directed him to a career in music. However, he later developed an interest in mathematics. His complete mathematical work consists of four papers published in 1771–1772. These papers include fundamental contributions on the roots of equations, on the theory of determinants, and on the knight's tour problem (introduced in the exercises in Section 8.5). Vandermonde's interest in mathematics lasted for only 2 years. Afterward, he published papers on harmony, experiments with cold, and the manufacture of steel. He also became interested in politics, joining the cause of the French revolution and holding several different positions in government.

We show that the right-hand side counts the same objects by considering the cases corresponding to the possible locations of the final 1 in a string with $r + 1$ ones. This final one must occur at position $r + 1, r + 2, \ldots,$ or $n + 1$. Furthermore, if the last one is the kth bit there must be r ones among the first $k - 1$ positions. Consequently, by Example 10 in Section 4.3, there are $\binom{k-1}{r}$ such bit strings. Summing over k with $r + 1 \leq k \leq n + 1$, we find that there are

$$\sum_{k=r+1}^{n+1} \binom{k-1}{r} = \sum_{j=r}^{n} \binom{j}{r}$$

bit strings of length n containing exactly $r + 1$ ones. (Note that the last step follows from the change of variables $j = k - 1$.) Since the left-hand side and the right-hand side count the same objects, they are equal. This completes the proof. ◁

Exercises

1. Find the expansion of $(x + y)^4$
   a) using combinatorial reasoning, as in Example 1.
   b) using the Binomial Theorem.
2. Find the expansion of $(x + y)^5$
   a) using combinatorial reasoning, as in Example 1.
   b) using the Binomial Theorem.
3. Find the expansion of $(x + y)^6$.
4. Find the coefficient of $x^5 y^8$ in $(x + y)^{13}$.
5. How many terms are there in the expansion of $(x + y)^{100}$?
6. What is the coefficient of x^7 in $(1 + x)^{11}$?
7. What is the coefficient of x^9 in $(2 - x)^{19}$?
8. What is the coefficient of $x^8 y^9$ in the expansion of $(3x + 2y)^{17}$?
9. What is the coefficient of $x^{101} y^{99}$ in the expansion of $(2x - 3y)^{200}$?
*10. Give a formula for the coefficient of x^k in the expansion of $(x + 1/x)^{100}$, where k is an integer.
*11. Give a formula for the coefficient of x^k in the expansion of $(x^2 - 1/x)^{100}$, where k is an integer.
12. The row of Pascal's triangle containing the binomial coefficients $\binom{10}{k}, 0 \leq k \leq 10$, is:

   1 10 45 120 210 252 210 120 45 10 1

   Use Pascal's identity to produce the row immediately following this row in Pascal's triangle.
13. What is the row of Pascal's triangle containing the binomial coefficients $\binom{9}{k}, 0 \leq k \leq 9$?
14. Show that if n is a positive integer, then $1 = \binom{n}{0} < \binom{n}{1} < \cdots < \binom{n}{\lfloor n/2 \rfloor} = \binom{n}{\lceil n/2 \rceil} > \cdots > \binom{n}{n-1} > \binom{n}{n} = 1$.
15. Show that $\binom{n}{k} \leq 2^n$ for all positive integers n and k with $0 \leq k \leq n$.
16. a) Use Exercise 14 and Corollary 1 to show that if n is an integer greater than 1, then $\binom{n}{\lfloor n/2 \rfloor} \geq 2^n/n$.

   b) Conclude from part (a) that if n is a positive integer, then $\binom{2n}{n} \geq 4^n/2n$.
☞17. Show that if n and k are integers with $1 \leq k \leq n$, then $\binom{n}{k} \leq n^k/2^{k-1}$.
18. Suppose that b is an integer with $b \geq 7$. Use the Binomial Theorem and the appropriate row of Pascal's triangle to find the base-b expansion of $(11)_b^4$ [that is, the fourth power of the number $(11)_b$ in base-b notation].
19. Prove Pascal's Identity, using the formula for $\binom{n}{r}$.
20. Suppose that k and n are integers with $1 \leq k < n$. Prove the **hexagon identity**
$$\binom{n-1}{k-1}\binom{n}{k+1}\binom{n+1}{k} = \binom{n-1}{k}\binom{n}{k-1}\binom{n+1}{k+1},$$
which relates terms in Pascal's triangle that form a hexagon.
21. Prove that if n and k are integers with $1 \leq k \leq n$, then $k\binom{n}{k} = n\binom{n-1}{k-1}$
   a) using a combinatorial proof. (*Hint:* Show that the two sides of the identity count the number of ways to select a subset with k elements from a set with n elements and then an element of this subset.)
   b) using an algebraic proof based on the formula for $\binom{n}{r}$ given in Theorem 2 in Section 4.3.
22. Prove the identity $\binom{n}{r}\binom{r}{k} = \binom{n}{k}\binom{n-k}{r-k}$, whenever $n, r,$ and k are nonnegative integers with $r \leq n$ and $k \leq r$,
   a) using a combinatorial argument.
   b) using an argument based on the formula for the number of r-combinations of a set with n elements.
23. Show that if n and k are positive integers, then
$$\binom{n+1}{k} = (n+1)\binom{n}{k-1}/k.$$

Use this identity to construct an inductive definition of the binomial coefficients.

24. Show that if p is a prime and k is an integer such that $1 \le k \le p - 1$, then p divides $\binom{p}{k}$.

25. Let n be a positive integer. Show that $\binom{2n}{n+1} + \binom{2n}{n} = \binom{2n+2}{n+1}/2$.

*26. Let n and k be integers with $1 \le k \le n$. Show that $\sum_{k=1}^{n} \binom{n}{k}\binom{n}{k-1} = \binom{2n+2}{n+1}/2 - \binom{2n}{n}$.

*27. Prove that

$$\sum_{k=0}^{r} \binom{n+k}{k} = \binom{n+r+1}{r}$$

whenever n and r are positive integers,

a) using a combinatorial argument.

b) using Pascal's identity.

28. Show that if n is a positive integer, then $\binom{2n}{2} = 2\binom{n}{2} + n^2$

a) using a combinatorial argument.

b) by algebraic manipulation.

*29. Give a combinatorial proof that $\sum_{k=1}^{n} k\binom{n}{k} = n2^{n-1}$. (*Hint:* Count in two ways the number of ways to select a committee and to then select a leader of the committee.)

*30. Give a combinatorial proof that $\sum_{k=1}^{n} k\binom{n}{k}^2 = n\binom{2n-1}{n-1}$. (*Hint:* Count in two ways the number of ways to select a committee, with n members from a group of n mathematics professors and n computer science professors, such that the chairperson of the committee is a mathematics professor.)

31. Show that a nonempty set has the same number of subsets with an odd number of elements as it does subsets with an even number of elements.

*32. Prove the Binomial Theorem using mathematical induction.

33. In this exercise we will count the number of paths in the xy plane between the origin $(0, 0)$ and point (m, n) such that each path is made up of a series of steps, where each step is a move one unit to the right or a move one unit upward. (No moves to the left or downward are allowed.) Two such paths from $(0, 0)$ to $(5, 3)$ are illustrated here.

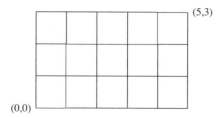

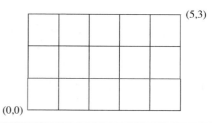

a) Show that each path of the type described can be represented by a bit string consisting of m 0s and n 1s, where a 0 represents a move one unit to the right and a 1 represents a move one unit upward.

b) Conclude from part (a) that there are $\binom{m+n}{n}$ paths of the desired type.

34. Use Exercise 33 to prove that $\binom{n}{k} = \binom{n}{n-k}$ whenever k is an integer with $0 \le k \le n$. [*Hint:* Consider the number of paths of the type described in Exercise 33 from $(0, 0)$ to $(n - k, k)$ and from $(0, 0)$ to $(k, n - k)$.]

35. Use Exercise 33 to prove Theorem 4. [*Hint:* Count the number of paths with n steps of the type described in Exercise 33. Every such path must end at one of the points $(n - k, k)$ for $k = 0, 1, 2, \ldots, n$.]

36. Use Exercise 33 to prove Pascal's Identity. [*Hint:* Show that a path of the type described in Exercise 33 from $(0, 0)$ to $(n + 1 - k, k)$ passes through either $(n + 1 - k, k - 1)$ or $(n - k, k)$, but not through both.]

37. Prove the identity in Exercise 27 using Exercise 33. [*Hint:* First, note that the number of paths from $(0, 0)$ to $(n+1, r)$ equals $\binom{n+1+r}{r}$. Second, count the number of paths by summing the number of these paths that start by going k units upward for $k = 0, 1, 2, \ldots, r$.]

38. Give a combinatorial proof that if n is a positive integer then $\sum_{k=0}^{n} k^2 \binom{n}{k} = n(n + 1)2^{n-2}$. [*Hint:* Show that both sides count the ways to select a subset of a set of n elements together with two not necessarily distinct elements from this subset. Furthermore, express the right-hand side as $n(n - 1)2^{n-2} + n2^{n-1}$.]

*39. Determine a formula involving binomial coefficients for the nth term of a sequence if its initial terms are those listed. [*Hint:* Looking at Pascal's triangle will be helpful. Although infinitely many sequences start with a specified set of terms, each of the following lists is the start of a sequence of the type desired.]

a) $1, 3, 6, 10, 15, 21, 28, 36, 45, 55, 66, \ldots$

b) $1, 4, 10, 20, 35, 56, 84, 120, 165, 220, \ldots$

c) $1, 2, 6, 20, 70, 252, 924, 3432, 12870, 48620, \ldots$

d) $1, 1, 2, 3, 6, 10, 20, 35, 70, 126, \ldots$

e) $1, 1, 1, 3, 1, 5, 15, 35, 1, 9, \ldots$

f) $1, 3, 15, 84, 495, 3003, 18564, 116280, 735471, 4686825, \ldots$

4.5 Generalized Permutations and Combinations

INTRODUCTION

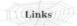

In many counting problems, elements may be used repeatedly. For instance, a letter or digit may be used more than once on a license plate. When a dozen donuts are selected, each variety can be chosen repeatedly. This contrasts with the counting problems discussed earlier in the chapter where we only considered permutations and combinations in which each item could be used at most once. In this section we will show how to solve counting problems where elements may be used more than once.

Also, some counting problems involve indistinguishable elements. For instance, to count the number of ways the letters of the word *SUCCESS* can be rearranged, the placement of identical letters must be considered. This contrasts with the counting problems discussed earlier where all elements were considered distinguishable. In this section we will describe how to solve counting problems in which some elements are indistinguishable.

Moreover, in this section we will explain how to solve another important class of counting problems, problems involving counting the ways to place distinguishable elements in boxes. An example of this type of problem is the number of different ways poker hands can be dealt to four players.

Taken together, the methods described earlier in this chapter and the methods introduced in this section form a useful toolbox for solving a wide range of counting problems. When the additional methods discussed in Chapter 6 are added to this arsenal, you will be able to solve a large percentage of the counting problems that arise in a wide range of areas of study.

PERMUTATIONS WITH REPETITION

Counting permutations when repetition of elements is allowed can easily be done using the product rule, as Example 1 shows.

EXAMPLE 1 How many strings of length n can be formed from the English alphabet?

Solution: By the product rule, since there are 26 letters, and since each letter can be used repeatedly, we see that there are 26^n strings of length n. ◀

The number of r-permutations of a set with n elements when repetition is allowed is given in Theorem 1.

THEOREM 1   The number of r-permutations of a set of n objects with repetition allowed is n^r.

Proof: There are n ways to select an element of the set for each of the r positions in the r-permutation when repetition is allowed, since for each choice all n objects are available. Hence, by the product rule there are n^r r-permutations when repetition is allowed. ◁

COMBINATIONS WITH REPETITION

Consider these examples of combinations with repetition of elements allowed.

EXAMPLE 2   How many ways are there to select four pieces of fruit from a bowl containing apples, oranges, and pears if the order in which the pieces are selected does not matter, only the type of fruit and not the individual piece matters, and there are at least four pieces of each type of fruit in the bowl?

Solution: To solve this problem we list all the ways possible to select the fruit. There are 15 ways:

4 apples	4 oranges	4 pears
3 apples, 1 orange	3 apples, 1 pear	3 oranges, 1 apple
3 oranges, 1 pear	3 pears, 1 apple	3 pears, 1 orange
2 apples, 2 oranges	2 apples, 2 pears	2 oranges, 2 pears
2 apples, 1 orange, 1 pear	2 oranges, 1 apple, 1 pear	2 pears, 1 apple, 1 orange

The solution is the number of 4-combinations with repetition allowed from a three-element set, {*apple, orange, pear*}.   ◄

To solve more complex counting problems of this type, we need a general method for counting the r-combinations of an n-element set. In Example 3 we will illustrate such a method.

EXAMPLE 3   How many ways are there to select five bills from a cash box containing $1 bills, $2 bills, $5 bills, $10 bills, $20 bills, $50 bills, and $100 bills? Assume that the order in which the bills are chosen does not matter, that the bills of each denomination are indistinguishable, and that there are at least five bills of each type.

Solution: Since the order in which the bills are selected does not matter and seven different types of bills can be selected as many as five times, this problem involves counting 5-combinations with repetition allowed from a set with seven elements. Listing all possibilities would be tedious, since there are a large number of solutions. Instead, we will illustrate the use of a technique for counting combinations with repetition allowed.

Suppose that a cash box has seven compartments, one to hold each type of bill, as illustrated in Figure 1. These bins are separated by six dividers, as shown in the picture. The choice of five bills corresponds to placing five markers in the compartments holding different types of bills. Figure 2 illustrates this correspondence for three different ways to select five bills, where the six dividers are represented by bars and the five bills by stars.

The number of ways to select five bills corresponds to the number of ways to arrange six bars and five stars. Consequently, the number of ways to select the five bills is the number of ways to select the positions of the five stars, from 11 possible positions. This corresponds to the number of unordered selections of 5 objects from a set of 11 objects, which can be done in $C(11, 5)$ ways. Consequently, there are

$$C(11, 5) = \frac{11!}{5! \, 6!} = 462$$

ways to choose five bills from the cash box with seven types of bills.   ◄

Theorem 2 generalizes this discussion.

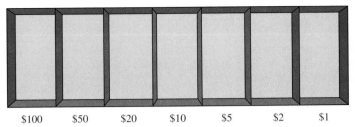

$100 $50 $20 $10 $5 $2 $1
FIGURE 1   **Cash Box with Seven Types of Bills.**

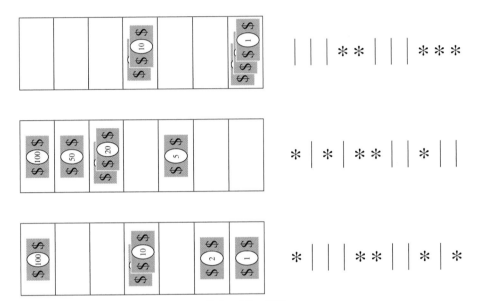

FIGURE 2   **Examples of Ways to Select Five Bills.**

THEOREM 2   There are $C(n + r - 1, r)$ r-combinations from a set with n elements when repetition of elements is allowed.

Proof: Each r-combination of a set with n elements when repetition is allowed can be represented by a list of $n - 1$ bars and r stars. The $n - 1$ bars are used to mark off n different cells, with the ith cell containing a star for each time the ith element of the set occurs in the combination. For instance, a 6-combination of a set with four elements is represented with three bars and six stars. Here

$$** \mid * \mid \mid ***$$

represents the combination containing exactly two of the first element, one of the second element, none of the third element, and three of the fourth element of the set.

As we have seen, each different list containing $n - 1$ bars and r stars corresponds to an r-combination of the set with n elements, when repetition is allowed. The number of such lists is $C(n - 1 + r, r)$, since each list corresponds to a choice of the r positions to place the r stars from the $n - 1 + r$ positions that contain r stars and $n - 1$ bars.   ◁

Examples 4–7 show how Theorem 2 is applied.

EXAMPLE 4

Extra
Examples

Suppose that a cookie shop has four different kinds of cookies. How many different ways can six cookies be chosen? Assume that only the type of cookie, and not the individual cookies or the order in which they are chosen, matters.

Solution: The number of ways to choose six cookies is the number of 6-combinations of a set with four elements. From Theorem 2 this equals $C(4 + 6 - 1, 6) = C(9, 6)$. Since

$$C(9, 6) = C(9, 3) = \frac{9 \cdot 8 \cdot 7}{1 \cdot 2 \cdot 3} = 84,$$

there are 84 different ways to choose the six cookies. ◄

Theorem 2 can also be used to find the number of solutions of certain linear equations where the variables are integers subject to constraints. This is illustrated by Example 5.

EXAMPLE 5

How many solutions does the equation

$$x_1 + x_2 + x_3 = 11$$

have, where x_1, x_2, and x_3 are nonnegative integers?

Solution: To count the number of solutions, we note that a solution corresponds to a way of selecting 11 items from a set with three elements, so that x_1 items of type one, x_2 items of type two, and x_3 items of type three are chosen. Hence, the number of solutions is equal to the number of 11-combinations with repetition allowed from a set with three elements. From Theorem 2 it follows that there are

$$C(3 + 11 - 1, 11) = C(13, 11) = C(13, 2) = \frac{13 \cdot 12}{1 \cdot 2} = 78$$

solutions.

The number of solutions of this equation can also be found when the variables are subject to constraints. For instance, we can find the number of solutions where the variables are integers with $x_1 \geq 1, x_2 \geq 2$, and $x_3 \geq 3$. A solution to the equation subject to these constraints corresponds to a selection of 11 items with x_1 items of type one, x_2 items of type two, and x_3 items of type three, where, in addition, there is at least one item of type one, two items of type two, and three items of type three. So, choose one item of type one, two of type two, and three of type three. Then select five additional items. By Theorem 2 this can be done in

$$C(3 + 5 - 1, 5) = C(7, 5) = C(7, 2) = \frac{7 \cdot 6}{1 \cdot 2} = 21$$

ways. Thus, there are 21 solutions of the equation subject to the given constraints. ◄

There is a one-to-one correspondence between r-combinations from a set with n elements when repetition is allowed and the ways to place r indistinguishable balls into n distinguishable bins. To set up this correspondence, we put a ball in the ith bin each time the ith element of the set is included in the r-combination.

EXAMPLE 6

How many ways are there to place ten indistinguishable balls into eight distinguishable bins?

Solution: The number of ways to place ten indistinguishable balls into eight bins equals the number of 10-combinations from a set with eight elements when repetition is allowed. Consequently, there are

$$C(8 + 10 - 1, 10) = C(17, 10) = \frac{17!}{10!7!} = 19,448.$$ ◀

Example 7 shows how counting the number of combinations with repetition allowed arises in determining the value of a variable that is incremented each time a certain type of nested loop is traversed.

EXAMPLE 7   What is the value of k after the following pseudocode has been executed?

```
k := 0
for i₁ := 1 to n
    for i₂ := 1 to i₁
        .
        .
        .
            for iₘ := 1 to iₘ₋₁
                k := k + 1
```

Solution: Note that the initial value of k is 0 and that 1 is added to k each time the nested loop is traversed with a sequence of integers $i_1, i_2, \ldots, i_m$ such that

$$1 \le i_m \le i_{m-1} \le \cdots \le i_1 \le n.$$

The number of such sequences of integers is the number of ways to choose m integers from $\{1, 2, \ldots, n\}$, with repetition allowed. (To see this, note that once such a sequence has been selected, if we order the integers in the sequence in nondecreasing order, this uniquely defines an assignment of $i_m, i_{m-1}, \ldots, i_1$. Conversely, every such assignment corresponds to a unique unordered set.) Hence, from Theorem 2, it follows that $k = C(n + m - 1, m)$ after this code has been executed. ◀

The formulae for the numbers of ordered and unordered selections of r elements, chosen with and without repetition allowed from a set with n elements, are shown in Table 1.

PERMUTATIONS WITH INDISTINGUISHABLE OBJECTS

Some elements may be indistinguishable in counting problems. When this is the case, care must be taken to avoid counting things more than once. Consider Example 8.

EXAMPLE 8   How many different strings can be made by reordering the letters of the word

SUCCESS?

Solution: Because some of the letters of *SUCCESS* are the same, the answer is *not* given by the number of permutations of seven letters. This word contains three Ss, two Cs,

TABLE 1 Combinations and Permutations with and without Repetition.

Type	Repetition Allowed?	Formula
r-permutations	No	$\dfrac{n!}{(n-r)!}$
r-combinations	No	$\dfrac{n!}{r!\,(n-r)!}$
r-permutations	Yes	n^r
r-combinations	Yes	$\dfrac{(n+r-1)!}{r!\,(n-1)!}$

one U, and one E. To determine the number of different strings that can be made by reordering the letters, first note that the three Ss can be placed among the seven positions in $C(7, 3)$ different ways, leaving four positions free. Then the two Cs can be placed in $C(4, 2)$ ways, leaving two free positions. The U can be placed in $C(2, 1)$ ways, leaving just one position free. Hence E can be placed in $C(1, 1)$ way. Consequently, from the product rule, the number of different strings that can be made is

$$C(7, 3)C(4, 2)C(2, 1)C(1, 1) = \frac{7!}{3!\,4!} \cdot \frac{4!}{2!\,2!} \cdot \frac{2!}{1!\,1!} \cdot \frac{1!}{1!\,0!}$$

$$= \frac{7!}{3!\,2!\,1!\,1!}$$

$$= 420 \,. \qquad \blacktriangleleft$$

We can prove Theorem 3 using the same sort of reasoning as in Example 8.

THEOREM 3

The number of different permutations of n objects, where there are n_1 indistinguishable objects of type 1, n_2 indistinguishable objects of type 2, ..., and n_k indistinguishable objects of type k, is

$$\frac{n!}{n_1!\,n_2!\cdots n_k!}.$$

Proof: To determine the number of permutations, first note that the n_1 objects of type one can be placed among the n positions in $C(n, n_1)$ ways, leaving $n - n_1$ positions free. Then the objects of type two can be placed in $C(n - n_1, n_2)$ ways, leaving $n - n_1 - n_2$ positions free. Continue placing the objects of type three, ..., type $k - 1$, until at the last stage n_k objects of type k can be placed in $C(n - n_1 - n_2 - \cdots - n_{k-1}, n_k)$ ways. Hence, by the product rule, the total number of different permutations is

$$C(n, n_1)C(n - n_1, n_2) \cdots C(n - n_1 - \cdots - n_{k-1}, n_k)$$

$$= \frac{n!}{n_1!\,(n - n_1)!} \frac{(n - n_1)!}{n_2!\,(n - n_1 - n_2)!} \cdots \frac{(n - n_1 - \cdots - n_{k-1})!}{n_k!\,0!}$$

$$= \frac{n!}{n_1!\,n_2!\cdots n_k!}. \qquad \triangleleft$$

DISTRIBUTING OBJECTS INTO BOXES

Some counting problems can be solved by enumerating the ways distinguishable objects can be placed into distinguishable boxes. Consider the following example in which the objects are cards and the "boxes" are hands of players.

EXAMPLE 9   How many ways are there to distribute hands of 5 cards to each of four players from the standard deck of 52 cards?

Solution: We will use the product rule to solve this problem. To begin, note that the first player can be dealt 5 cards in $C(52, 5)$ ways. The second player can be dealt 5 cards in $C(47, 5)$ ways, since only 47 cards are left. The third player can be dealt 5 cards in $C(42, 5)$ ways. Finally, the fourth player can be dealt 5 cards in $C(37, 5)$ ways. Hence, the total number of ways to deal four players 5 cards each is

$$
C(52, 5)C(47, 5)C(42, 5)C(37, 5) = \frac{52!}{47!\,5!} \cdot \frac{47!}{42!\,5!} \cdot \frac{42!}{37!\,5!} \cdot \frac{37!}{32!\,5!}
$$

$$
= \frac{52!}{5!\,5!\,5!\,5!\,32!}. \qquad \blacktriangleleft
$$

Remark: The solution to Example 9 equals the number of permutations of 52 objects, with 5 indistinguishable objects of each of four different types, and 32 objects of a fifth type. This equality can be seen by defining a one-to-one correspondence between permutations of this type and distributions of cards to the players. To define this correspondence, first order the cards from 1 to 52. Then cards dealt to the first player correspond to the cards in the positions assigned to objects of the first type in the permutation. Similarly, cards dealt to the second, third, and fourth players, respectively, correspond to cards in the positions assigned to objects of the second, third, and fourth type, respectively. The cards not dealt to any player correspond to cards in the positions assigned to objects of the fifth type. The reader should verify that this is a one-to-one correspondence.

Example 9 is a typical problem that involves distributing distinguishable objects into distinguishable boxes. The distinguishable objects are the 52 cards, and the five distinguishable boxes are the hands of the four players and the rest of the deck. Counting problems that involve distributing distinguishable objects into boxes can be solved using the following theorem.

THEOREM 4   The number of ways to distribute n distinguishable objects into k distinguishable boxes so that n_i objects are placed into box $i, i = 1, 2, \ldots, k$, equals

$$
\frac{n!}{n_1!\,n_2!\cdots n_k!}.
$$

Theorem 4 can be proved using the product rule. We leave the details as Exercise 47. It can also be proved (see Exercise 48) by setting up a one-to-one correspondence between the permutations counted by Theorem 3 and the ways to distribute objects counted by Theorem 4.

Exercises

1. In how many different ways can five elements be selected in order from a set with three elements when repetition is allowed?

2. In how many different ways can five elements be selected in order from a set with five elements when repetition is allowed?

3. How many strings of six letters are there?

4. Every day a student randomly chooses a sandwich for lunch from a pile of wrapped sandwiches. If there are six kinds of sandwiches, how many different ways are there for the student to choose sandwiches for the seven days of a week if the order in which the sandwiches are chosen matters?

5. How many ways are there to assign three jobs to five employees if each employee can be given more than one job?

6. How many ways are there to select five unordered elements from a set with three elements when repetition is allowed?

7. How many ways are there to select three unordered elements from a set with five elements when repetition is allowed?

8. How many different ways are there to choose a dozen donuts from the 21 varieties at a donut shop?

9. A bagel shop has onion bagels, poppy seed bagels, egg bagels, salty bagels, pumpernickel bagels, sesame seed bagels, raisin bagels, and plain bagels. How many ways are there to choose

   a) six bagels?
   b) a dozen bagels?
   c) two dozen bagels?
   d) a dozen bagels with at least one of each kind?
   e) a dozen bagels with at least three egg bagels and no more than two salty bagels?

10. A croissant shop has plain croissants, cherry croissants, chocolate croissants, almond croissants, apple croissants, and broccoli croissants. How many ways are there to choose

   a) a dozen croissants?
   b) three dozen croissants?
   c) two dozen croissants with at least two of each kind?
   d) two dozen croissants with no more than two broccoli croissants?
   e) two dozen croissants with at least five chocolate croissants and at least three almond croissants?
   f) two dozen croissants with at least one plain croissant, at least two cherry croissants, at least three chocolate croissants, at least one almond croissant, at least two apple croissants, and no more than three broccoli croissants?

11. How many ways are there to choose eight coins from a piggy bank containing 100 identical pennies and 80 identical nickels?

12. How many different combinations of pennies, nickels, dimes, quarters, and half dollars can a piggy bank contain if it has 20 coins in it?

13. A book publisher has 3000 copies of a discrete mathematics book. How many ways are there to store these books in their three warehouses if the copies of the book are indistinguishable?

14. How many solutions are there to the equation

$$x_1 + x_2 + x_3 + x_4 = 17,$$

where x_1, x_2, x_3, and x_4 are nonnegative integers?

15. How many solutions are there to the equation

$$x_1 + x_2 + x_3 + x_4 + x_5 = 21,$$

where x_i, $i = 1, 2, 3, 4, 5$, is a nonnegative integer such that

   a) $x_1 \geq 1$?
   b) $x_i \geq 2$ for $i = 1, 2, 3, 4, 5$?
   c) $0 \leq x_1 \leq 10$?
   d) $0 \leq x_1 \leq 3, 1 \leq x_2 < 4$, and $x_3 \geq 15$?

16. How many solutions are there to the equation

$$x_1 + x_2 + x_3 + x_4 + x_5 + x_6 = 29,$$

where x_i, $i = 1, 2, 3, 4, 5, 6$, is a nonnegative integer such that

   a) $x_i > 1$ for $i = 1, 2, 3, 4, 5, 6$?
   b) $x_1 \geq 1, x_2 \geq 2, x_3 \geq 3, x_4 \geq 4, x_5 > 5$, and $x_6 \geq 6$?
   c) $x_1 \leq 5$?
   d) $x_1 < 8$ and $x_2 > 8$?

17. How many strings of 10 ternary digits (0, 1, or 2) are there that contain exactly two 0s, three 1s, and five 2s?

18. How many strings of 20 decimal digits are there that contain two 0s, four 1s, three 2s, one 3, two 4s, three 5s, two 7s, and three 9s?

19. Suppose that a large family has 14 children, including two sets of identical triplets, three sets of identical twins, and two individual children. How many ways are there to seat these children in a row of chairs if the identical triplets or twins cannot be distinguished from one another?

20. How many solutions are there to the inequality

$$x_1 + x_2 + x_3 \leq 11,$$

where x_1, x_2, and x_3 are nonnegative integers? (*Hint:* Introduce an auxiliary variable x_4 so that $x_1 + x_2 + x_3 + x_4 = 11$.)

21. How many ways are there to distribute six indistinguishable balls into nine distinguishable bins?

22. How many ways are there to distribute 12 indistinguishable balls into six distinguishable bins?

23. How many ways are there to distribute 12 distinguishable objects into six distinguishable boxes so that two objects are placed in each box?

24. How many ways are there to distribute 15 distinguishable objects into five distinguishable boxes so that the boxes have one, two, three, four, and five objects in them, respectively.

25. How many positive integers less than 1,000,000 have the sum of their digits equal to 19?

26. How many positive integers less than 1,000,000 have exactly one digit equal to 9 and have a sum of digits equal to 13?

27. There are 10 questions on a discrete mathematics final exam. How many ways are there to assign scores to the problems if the sum of the scores is 100 and each question is worth at least 5 points?

28. Show that there are $C(n + r - q_1 - q_2 - \cdots - q_r - 1,$ $n - q_1 - q_2 - \cdots - q_r)$ different unordered selections of n objects of r different types that include at least q_1 objects of type one, q_2 objects of type two, ..., and q_r objects of type r.

29. How many different bit strings can be transmitted if the string must begin with a 1 bit, must include three additional 1 bits (so that a total of four 1 bits is sent), must include a total of twelve 0 bits, and must have at least two 0 bits following each 1 bit?

30. How many different strings can be made from the letters in *MISSISSIPPI*, using all the letters?

31. How many different strings can be made from the letters in *ABRACADABRA*, using all the letters?

32. How many different strings can be made from the letters in *AARDVARK*, using all the letters, if all three *A*s must be consecutive?

33. How many different strings can be made from the letters in *ORONO*, using some or all of the letters?

34. How many strings with five or more characters can be formed from the letters in *SEERESS*?

35. How many strings with seven or more characters can be formed from the letters in *EVERGREEN*?

36. How many different bit strings can be formed using six 1s and eight 0s?

37. A student has three mangos, two papayas, and two kiwi fruits. If the student eats one piece of fruit each day, and only the type of fruit matters, in how many different ways can these fruits be consumed?

38. A professor packs her collection of 40 issues of a mathematics journal in four boxes with 10 issues per box. How many ways can she distribute the journals if

    a) each box is numbered, so that they are distinguishable?

    b) the boxes are identical, so that they cannot be distinguished?

39. How many ways are there to travel in xyz space from the origin $(0, 0, 0)$ to the point $(4, 3, 5)$ by taking steps one unit in the positive x direction, one unit in the positive y direction, or one unit in the positive z direction? (Moving in the negative x, y, or z direction is prohibited, so that no backtracking is allowed.)

40. How many ways are there to travel in $xyzw$ space from the origin $(0, 0, 0, 0)$ to the point $(4, 3, 5, 4)$ by taking steps one unit in the positive x, positive y, positive z, or positive w direction?

41. How many ways are there to deal hands of seven cards to each of five players from a standard deck of 52 cards?

42. In bridge, the 52 cards of a standard deck are dealt to four players. How many different ways are there to deal bridge hands to four players?

43. What is the probability that each player has a hand containing an ace when the 52 cards of a standard deck are dealt to four players?

44. In how many ways can a dozen books be placed on four distinguishable shelves

    a) if the books are indistinguishable copies of the same title?

    b) if no two books are the same, and the positions of the books on the shelves matter? (*Hint:* Break this into 12 tasks, placing each book separately. Start with the sequence 1, 2, 3, 4 to represent the shelves. Represent the books by b_i, $i = 1, 2, \ldots, 12$. Place b_1 to the right of one of the terms in 1, 2, 3, 4. Then successively place $b_2, b_3, \ldots,$ and b_{12}.)

45. How many ways can n books be placed on k distinguishable shelves

    a) if the books are indistinguishable copies of the same title?

    b) if no two books are the same, and the positions of the books on the shelves matter?

46. A shelf holds 12 books in a row. How many ways are there to choose five books so that no two adjacent books are chosen? (*Hint:* Represent the books that are chosen by bars and the books not chosen by stars. Count the number of sequences of five bars and seven stars so that no two bars are adjacent.)

*47. Use the product rule to prove Theorem 4, by first placing objects in the first box, then placing objects in the second box, and so on.

*48. Prove Theorem 4 by first setting up a one-to-one correspondence between permutations of n objects with n_i indistinguishable objects of type i, $i = 1, 2, 3, \ldots, k$, and the distributions of n objects in k boxes such that n_i objects are placed in box i, $i = 1, 2, 3, \ldots, k$ and then applying Theorem 3.

*49. In this exercise we will prove Theorem 2 by setting up a one-to-one correspondence between the set of r-combinations with repetition allowed of $S = \{1, 2, 3, \ldots, n\}$ and the set of r-combinations of the set $T = \{1, 2, 3, \ldots, n + r - 1\}$.

a) Arrange the elements in an r-combination, with repetition allowed, of S into an increasing sequence $x_1 \leq x_2 \leq \cdots \leq x_r$. Show that the sequence formed by adding $k - 1$ to the kth term is strictly increasing. Conclude that this sequence is made up of r distinct elements from T.

b) Show that the procedure described in (a) defines a one-to-one correspondence between the set of r-combinations, with repetition allowed, of S and the r-combinations of T. (*Hint:* Show the correspondence can be reversed by associating to the r-combination $\{x_1, x_2, \ldots, x_r\}$ of T, with $1 \leq x_1 < x_2 < \cdots < x_r \leq n + r - 1$, the r-combination with repetition allowed from S, formed by subtracting $k - 1$ from the kth element.)

c) Conclude that there are $C(n + r - 1, r)$ r-combinations with repetition allowed from a set with n elements.

50. How many ways are there to distribute five distin-

guishable objects into three indistinguishable boxes?

51. How many ways are there to distribute five indistinguishable objects into three indistinguishable boxes?

52. How many different terms are there in the expansion of $(x_1 + x_2 + \cdots + x_m)^n$ after all terms with identical sets of exponents are added?

*53. Prove the **Multinomial Theorem:** If n is a positive integer, then

$$(x_1 + x_2 + \cdots + x_m)^n$$

$$= \sum_{n_1 + n_2 + \cdots + n_m = n} C(n; n_1, n_2, \ldots, n_m) x_1^{n_1} x_2^{n_2} \cdots x_n^{n_m},$$

where

$$C(n; n_1, n_2, \ldots, n_m) = \frac{n!}{n_1! \, n_2! \cdots n_m!}$$

is a **multinomial coefficient.**

54. Find the expansion of $(x + y + z)^4$.

55. Find the coefficient of $x^3 y^2 z^5$ in $(x + y + z)^{10}$.

56. How many terms are there in the expansion of $(x + y + z)^{100}$?

4.6   Generating Permutations and Combinations

INTRODUCTION

Methods for counting various types of permutations and combinations were described in the previous sections of this chapter, but sometimes permutations or combinations need to be generated, not just counted. Consider the following three problems. First, suppose that a salesman must visit six different cities. In which order should these cities be visited to minimize total travel time? One way to determine the best order is to determine the travel time for each of the $6! = 720$ different orders in which the cities can be visited and choose the one with the smallest travel time. Second, suppose some numbers from a set of six numbers have 100 as their sum. One way to find these numbers is to generate all the $2^6 = 64$ subsets and check the sum of their terms. Third, suppose a laboratory has 95 employees. A group of 12 of these employees with a particular set of 25 skills is needed for a project. (Each employee can have one or more of these skills.) One way to find such a set of employees is to generate all sets of 12 of these employees and check whether they have the desired skills. These examples show that it is often necessary to generate permutations and combinations to solve problems.

GENERATING PERMUTATIONS

Any set with n elements can be placed in one-to-one correspondence with the set $\{1, 2, 3, \ldots, n\}$. We can list the permutations of any set of n elements by generating the permutations of the n smallest positive integers and then replacing these integers with the corresponding elements. Many different algorithms have been developed to generate the $n!$ permutations of this set. We will describe one of these that is based on the **lexicographic ordering** of the set of permutations of $\{1, 2, 3, \ldots, n\}$. In this ordering, the permutation $a_1a_2 \cdots a_n$ precedes the permutation of $b_1b_2 \cdots b_n$, if for some k, with $1 \leq k \leq n$, $a_1 = b_1, a_2 = b_2, \ldots, a_{k-1} = b_{k-1}$, and $a_k < b_k$. In other words, a permutation of the set of the n smallest positive integers precedes (in lexicographic order) a second permutation if the number in this permutation in the first position where the two permutations disagree is smaller than the number in that position in the second permutation.

EXAMPLE 1   The permutation 23415 of the set $\{1, 2, 3, 4, 5\}$ precedes the permutation 23514, since these permutations agree in the first two positions, but the number in the third position in the first permutation, 4, is smaller than the number in the third position in the second permutation, 5. Similarly, the permutation 41532 precedes 52143.   ◀

An algorithm for generating the permutations of $\{1, 2, \ldots, n\}$ can be based on a procedure that constructs the next permutation in lexicographic order following a given permutation $a_1a_2 \cdots a_n$. We will show how this can be done. First, suppose that $a_{n-1} < a_n$. Interchange a_{n-1} and a_n to obtain a larger permutation. No other permutation is both larger than the original permutation and smaller than the permutation obtained by interchanging a_{n-1} and a_n. For instance, the next largest permutation after 234156 is 234165. On the other hand, if $a_{n-1} > a_n$, then a larger permutation cannot be obtained by interchanging these last two terms in the permutation. Look at the last three integers in the permutation. If $a_{n-2} < a_{n-1}$, then the last three integers in the permutation can be rearranged to obtain the next largest permutation. Put the smaller of the two integers a_{n-1} and a_n that is greater than a_{n-2} in position $n - 2$. Then, place the remaining integer and a_{n-2} into the last two positions in increasing order. For instance, the next largest permutation after 234165 is 234516.

On the other hand, if $a_{n-2} > a_{n-1}$ (and $a_{n-1} > a_n$), then a larger permutation cannot be obtained by permuting the last three terms in the permutation. Based on these observations, a general method can be described for producing the next largest permutation in increasing order following a given permutation $a_1a_2 \cdots a_n$. First, find the integers a_j and a_{j+1} with $a_j < a_{j+1}$ and

$$a_{j+1} > a_{j+2} > \cdots > a_n,$$

that is, the last pair of adjacent integers in the permutation where the first integer in the pair is smaller than the second. Then, the next largest permutation in lexicographic order is obtained by putting in the jth position the least integer among $a_{j+1}, a_{j+2}, \ldots$, and a_n that is greater than a_j and listing in increasing order the rest of the integers $a_j, a_{j+1}, \ldots, a_n$ in positions $j + 1$ to n. It is easy to see that there is no other permutation larger than the permutation $a_1a_2 \cdots a_n$ but smaller than the new permutation produced. (The verification of this fact is left as an exercise for the reader.)

EXAMPLE 2   What is the next largest permutation in lexicographic order after 362541?

Solution: The last pair of integers a_j and a_{j+1} where $a_j < a_{j+1}$ is $a_3 = 2$ and $a_4 = 5$. The least integer to the right of 2 that is greater than 2 in the permutation is $a_5 = 4$. Hence, 4 is placed in the third position. Then the integers 2, 5, and 1 are placed in order in the last three positions, giving 125 as the last three positions of the permutation. Hence, the next permutation is 364125.   ◀

To produce the $n!$ permutations of the integers $1, 2, 3, \ldots, n$, begin with the smallest permutation in lexicographic order, namely, $123 \cdots n$, and successively apply the procedure described for producing the next largest permutation of $n! - 1$ times. This yields all the permutations of the n smallest integers in lexicographic order.

EXAMPLE 3   Generate the permutations of the integers 1, 2, 3 in lexicographic order.

Solution: Begin with 123. The next permutation is obtained by interchanging 3 and 2 to obtain 132. Next, since 3 > 2 and 1 < 3, permute the three integers in 132. Put the smaller of 3 and 2 in the first position, and then put 1 and 3 in increasing order in positions 2 and 3 to obtain 213. This is followed by 231, obtained by interchanging 1 and 3, since 1 < 3. The next largest permutation has 3 in the first position, followed by 1 and 2 in increasing order, namely, 312. Finally, interchange 1 and 2 to obtain the last permutation, 321.   ◀

Algorithm 1 displays the procedure for finding the next largest permutation in lexicographic order after a permutation that is not $n\ n-1\ n-2\ \cdots\ 2\ 1$, which is the largest permutation.

ALGORITHM 1 Generating the Next Largest Permutation in Lexicographic Order.

---

procedure *next permutation*$(a_1 a_2 \cdots a_n$: permutation of
$\{1, 2, \ldots, n\}$ not equal to $n\ n-1\ \cdots\ 2\ 1)$
$j := n - 1$
while $a_j > a_{j+1}$
$\quad j := j - 1$
$\{j$ is the largest subscript with $a_j < a_{j+1}\}$
$k := n$
while $a_j > a_k$
$\quad k := k - 1$
$\{a_k$ is the smallest integer greater than a_j to the right of $a_j\}$
interchange a_j and a_k
$r := n$
$s := j + 1$
while $r > s$
begin
$\quad$ interchange a_r and a_s
$\quad r := r - 1$
$\quad s := s + 1$
end
$\{$this puts the tail end of the permutation after the jth position in
$\quad$ increasing order$\}$

GENERATING COMBINATIONS

How can we generate all the combinations of the elements of a finite set? Since a combination is just a subset, we can use the correspondence between subsets of $\{a_1, a_2, \ldots, a_n\}$ and bit strings of length n.

Recall that the bit string corresponding to a subset has a 1 in position k if a_k is in the subset, and has a 0 in this position if a_k is not in the subset. If all the bit strings of length n can be listed, then by the correspondence between subsets and bit strings, a list of all the subsets is obtained.

Recall that a bit string of length n is also the binary expansion of an integer between 0 and $2^n - 1$. The 2^n bit strings can be listed in order of their increasing size as integers in their binary expansions. To produce all binary expansions of length n, start with the bit string $000\ldots00$, with n zeros. Then, successively find the next largest expansion until the bit string $111\ldots11$ is obtained. At each stage the next largest binary expansion is found by locating the first position from the right that is not a 1, then changing all the 1s to the right of this position to 0s and making this first 0 (from the right) a 1.

EXAMPLE 4   Find the next largest bit string after 10 0010 0111.

Solution: The first bit from the right that is not a 1 is the fourth bit from the right. Change this bit to a 1 and change all the following bits to 0s. This produces the next largest bit string, 10 0010 1000.   ◀

The procedure for producing the next largest bit string after $b_{n-1}b_{n-2}\ldots b_1b_0$ is given as Algorithm 2.

ALGORITHM 2 Generating the Next Largest Bit String.

---

procedure *next bit string*$(b_{n-1}\, b_{n-2}\ \ldots\ b_1b_0$: bit string not equal to $11\ \ldots\ 11)$
$i := 0$
while $b_i = 1$
begin
   $b_i := 0$
   $i := i + 1$
end
$b_i := 1$

Next, an algorithm for generating the r-combinations of the set $\{1, 2, 3, \ldots, n\}$ will be given. An r-combination can be represented by a sequence containing the elements in the subset in increasing order. The r-combinations can be listed using lexicographic order on these sequences. The next combinations after $a_1a_2 \cdots a_r$ can be obtained in the following way: First, locate the last element a_i in the sequence such that $a_i \neq n - r + i$. Then, replace a_i with $a_i + 1$ and a_j with $a_i + j - i + 1$, for $j = i + 1, i + 2, \ldots, r$. It is left for the reader to show that this produces the next largest combination in lexicographic order. This procedure is illustrated with the following example.

EXAMPLE 5   Find the next largest 4-combination of the set $\{1, 2, 3, 4, 5, 6\}$ after $\{1, 2, 5, 6\}$.

Solution: The last term among the terms a_i with $a_1 = 1, a_2 = 2, a_3 = 5$, and $a_4 = 6$ such that $a_i \neq 6 - 4 + i$ is $a_2 = 2$. To obtain the next largest 4-combination, increment a_2 by 1 to obtain $a_2 = 3$. Then set $a_3 = 3 + 1 = 4$ and $a_4 = 3 + 2 = 5$. Hence the next largest 4-combination is $\{1, 3, 4, 5\}$.   ◀

Algorithm 3 displays pseudocode for this procedure.

**ALGORITHM 3 Generating the Next *r*-Combination
in Lexicographic Order.**

procedure *next r-combination*($\{a_1, a_2, \ldots, a_r\}$: proper subset of
$\quad \{1, 2, \ldots, n\}$ not equal to $\{n - r + 1, \ldots, n\}$ with
$\quad\quad a_1 < a_2 < \cdots < a_r$)
$i := r$
while $a_i = n - r + i$
$\quad i := i - 1$
$a_i := a_i + 1$
for $j := i + 1$ **to** r
$\quad a_j := a_i + j - i$

Exercises

1. Find the next largest permutation in lexicographic order after each of these permutations.

   a) 1432   **b)** 54123
   c) 12453   **d)** 45231
   e) 6714235   **f)** 31528764

2. Find the next largest permutation in lexicographic order after each of these permutations.

   a) 1342   **b)** 45321
   c) 13245   **d)** 612345
   e) 1623547   **f)** 23587416

3. Place these permutations of $\{1, 2, 3, 4, 5\}$ in lexicographic order: 43521, 15432, 45321, 23451, 23514, 14532, 21345, 45213, 31452, 31542.

4. Place these permutations of $\{1,2,3,4,5,6\}$ in lexicographic order: 234561, 231456, 165432, 156423, 543216, 541236, 231465, 314562, 432561, 654321, 654312, 435612.

5. Use Algorithm 1 to generate the 24 permutations of the first four positive integers in lexicographic order.

6. Use Algorithm 2 to list all the subsets of the set $\{1, 2, 3, 4\}$.

7. Use Algorithm 3 to list all the 3-combinations of $\{1, 2, 3, 4, 5\}$.

8. Show that Algorithm 1 produces the next largest permutation in lexicographic order.

9. Show that Algorithm 3 produces the next largest *r*-combination in lexicographic order after a given *r*-combination.

10. Develop an algorithm for generating the *r*-permutations of a set of *n* elements.

11. List all 3-permutations of $\{1, 2, 3, 4, 5\}$.

The remaining exercises in this section develop another algorithm for generating the permutations of $\{1, 2, 3, \ldots, n\}$. This algorithm is based on Cantor expansions of integers. Every nonnegative integer less than $n!$ has a unique Cantor expansion

$$a_1 1! + a_2 2! + \cdots + a_{n-1}(n - 1)!$$

where a_i is a nonnegative integer not exceeding i, for $i = 1, 2, \ldots, n - 1$. The integers $a_1, a_2, \ldots, a_{n-1}$ are called the **Cantor digits** of this integer.

   Given a permutation of $\{1, 2, \ldots, n\}$, let $a_{k-1}, k = 2, 3, \ldots, n$, be the number of integers less than k that follow k in the permutation. For instance, in the permutation 43215, a_1 is the number of integers less than 2 that follow 2, so that $a_1 = 1$. Similarly, for this example $a_2 = 2$, $a_3 = 3$, and $a_4 = 0$. Consider the function from the set of permutations $\{1, 2, 3, \ldots, n\}$ to the set of nonnegative integers less than $n!$ that sends a permutation to the integer that has $a_1, a_2, \ldots, a_{n-1}$, defined in this way, as its Cantor digits.

12. Find the integers that correspond to these permutations.

   a) 246531   **b)** 12345   **c)** 654321

*13. Show that the correspondence described here is a bijection between the set of permutations of $\{1, 2, 3, \ldots, n\}$ and the nonnegative integers less than $n!$.

14. Find the permutations of $\{1, 2, 3, 4, 5\}$ that correspond to these integers with respect to the correspondence between Cantor expansions and permutations as described in the preamble to Exercise 12.

   a) 3   **b)** 89   **c)** 111

15. Develop an algorithm for producing all permutations of a set of n elements based on the correspondence described in the preamble to Exercise 12.

*16. The following method can be used to generate a random permutation of a sequence of n terms. First, interchange the nth term and the $r(n)$th term where $r(n)$ is a randomly selected integer with $1 \le r(n) \le n$. Next, interchange the $(n-1)$th term of the resulting sequence with its $r(n-1)$th term where $r(n-1)$ is a randomly selected integer with $1 \le r(n-1) \le n-1$. Continue this process until $j = n$, where at the jth step you interchange the $(n-j+1)$th term of the resulting sequence with its $r(n-j+1)$th term, where $r(n-j+1)$ is a randomly selected integer with $1 \le r(n-j+1) \le n-j+1$. Show that when this method is followed each of the $n!$ different permutations of the terms of the sequence is equally likely to be generated. [*Hint:* Use mathematical induction, assuming that the probability that each of the permutations of $n-1$ terms produced by this procedure for a sequence of $n-1$ terms is $1/(n-1)!$.]

Key Terms and Results

TERMS

combinatorics: the study of arrangements of objects

enumeration: the counting of arrangements of objects

tree diagram: a diagram made up of a root, branches leaving the root, and other branches leaving some of the endpoints of branches

permutation: an ordered arrangement of the elements of a set

r-permutation: an ordered arrangement of r elements of a set

$P(n, r)$: the number of r-permutations of a set with n elements

r-combination: an unordered selection of r elements of a set

$C(n, r)$: the number of r-combinations of a set with n elements

$\binom{n}{r}$ **(binomial coefficient):** also the number of r-combinations of a set with n elements

combinatorial proof: a proof based on counting arguments

Pascal's triangle: a representation of the binomial coefficients where the ith row of the triangle contains $\binom{i}{j}$ for $j = 0, 1, 2, \ldots, i$

RESULTS

the product rule: a basic counting technique, which states that the number of ways to do a procedure that consists of two subtasks is the product of the number of ways to do the first task and the number of ways to do the second task after the first task has been done

the sum rule: a basic counting technique, which states that the number of ways to do a task in one of two ways is the sum of the number of ways to do these tasks if they cannot be done simultaneously

the pigeonhole principle: When more than k objects are placed in k boxes, there must be a box containing more than one object.

the generalized pigeonhole principle: When N objects are placed in k boxes, there must be a box containing at least $\lceil N/k \rceil$ objects.

$$P(n, r) = \frac{n!}{(n-r)!}$$

$$C(n, r) = \binom{n}{r} = \frac{n!}{r!(n-r)!}$$

Pascal's Identity: $\binom{n+1}{k} = \binom{n}{k-1} + \binom{n}{k}$

the Binomial Theorem: $(x+y)^n = \sum_{k=0}^{n} \binom{n}{k} x^{n-k} y^k$

There are n^r r-permutations of a set with n elements when repetition is allowed.

There are $C(n+r-1, r)$ r-combinations of a set with n elements when repetition is allowed.

There are $n!/(n_1! n_2! \cdots n_k!)$ permutations of n objects where there are n_i indistinguishable objects of type i for $i = 1, 2, 3, \ldots, k$.

The algorithm for generating the permutations of the set $\{1, 2, \ldots, n\}$.

Review Questions

1. Explain how the sum and product rules can be used to find the number of bit strings with a length not exceeding 10.

2. Explain how to find the number of bit strings of length not exceeding 10 that have at least one 0 bit.

3. **a)** How can the product rule be used to find the number of functions from a set with m elements to a set with n elements?
   b) How many functions are there from a set with five elements to a set with ten elements?
   c) How can the product rule be used to find the number of one-to-one functions from a set with m elements to a set with n elements?
   d) How many one-to-one functions are there from a set with five elements to a set with ten elements?
   e) How many onto functions are there from a set with five elements to a set with ten elements?

4. How can you find the number of possible outcomes of a playoff between two teams where the first team that wins four games wins the playoff?

5. How can you find the number of bit strings of length ten that either begin with 101 or end with 010?

6. **a)** State the pigeonhole principle.
   b) Explain how the pigeonhole principle can be used to show that among any 11 integers, at least two must have the same last digit.

7. **a)** State the generalized pigeonhole principle.
   b) Explain how the generalized pigeonhole principle can be used to show that among any 91 integers, there are at least ten that end with the same digit.

8. **a)** What is the difference between an r-combination and an r-permutation of a set with n elements?
   b) Derive an equation that relates the number of r-combinations and the number of r-permutations of a set with n elements.
   c) How many ways are there to select six students from a class of 25 to serve on a committee?
   d) How many ways are there to select six students from a class of 25 to hold six different executive positions on a committee?

9. **a)** What is Pascal's triangle?
   b) How can a row of Pascal's triangle be produced from the one above it?

10. What is meant by a combinatorial proof of an identity? How is such a proof different from an algebraic one?

11. Explain how to prove Pascal's Identity using a combinatorial argument.

12. **a)** State the Binomial Theorem.
    b) Explain how to prove the Binomial Theorem using a combinatorial argument.

c) Find the coefficient of $x^{100}y^{101}$ in the expansion of $(2x + 5y)^{201}$.

13. **a)** Explain how to find a formula for the number of ways to select r objects from n objects when repetition is allowed and order does not matter.
    b) How many ways are there to select a dozen objects from among objects of five different types if objects of the same type are indistinguishable?
    c) How many ways are there to select a dozen objects from these five different types if there must be at least three objects of the first type?
    d) How many ways are there to select a dozen objects from these five different types if there cannot be more than four objects of the first type?
    e) How many ways are there to select a dozen objects from these five different types if there must be at least two objects of the first type but no more than three objects of the second type?

14. **a)** Let n and r be positive integers. Explain why the number of solutions of the equation $x_1 + x_2 + \cdots + x_n = r$, where x_i is a nonnegative integer for $i = 1, 2, 3, \ldots, n$, equals the number of r-combinations of a set with n elements.
    b) How many solutions in nonnegative integers are there to the equation $x_1 + x_2 + x_3 + x_4 = 17$?
    c) How many solutions in positive integers are there to the equation in part (b)?

15. **a)** Derive a formula for the number of permutations of n objects of k different types where there are n_1 indistinguishable objects of type one, n_2 indistinguishable objects of type two, ..., and n_k indistinguishable objects of type k.
    b) How many ways are there to order the letters of the word *INDISCREETNESS*?

16. Describe an algorithm for generating all the permutations of the set of the n smallest positive integers.

17. **a)** How many ways are there to deal hands of five cards to six players from a standard 52-card deck?
    b) How many ways are there to distribute n distinguishable objects into k distinguishable boxes so that n_i objects are placed in box i?

18. Describe an algorithm for generating all the combinations of the set of the n smallest positive integers.

Supplementary Exercises

1. How many ways are there to choose 6 items from 10 distinct items when
   a) the items in the choices are ordered and repetition is not allowed?
   b) the items in the choices are ordered and repetition is allowed?
   c) the items in the choices are unordered and repetition is not allowed?

d) the items in the choices are unordered and repetition is allowed?

2. How many ways are there to choose 10 items from 6 distinct items when

   a) the items in the choices are ordered and repetition is not allowed?

   b) the items in the choices are ordered and repetition is allowed?

   c) the items in the choices are unordered and repetition is not allowed?

   d) the items in the choices are unordered and repetition is allowed?

3. A test contains 100 true/false questions. How many different ways can a student answer the questions on the test, if answers may be left blank?

4. How many bit strings of length 10 either start with 000 or end with 1111?

5. How many bit strings of length 10 over the alphabet $\{a, b, c\}$ have either exactly three as or exactly four bs?

6. The internal telephone numbers in the phone system on a campus consist of five digits, with the first digit not equal to zero. How many different numbers can be assigned in this system?

7. An ice cream parlor has 28 different flavors, eight different kinds of sauce, and 12 toppings.

   a) In how many different ways can a dish of three scoops of ice cream be made where each flavor can be used more than once and the order of the scoops does not matter?

   b) How many different kinds of small sundaes are there if a small sundae contains one scoop of ice cream, a sauce, and a topping?

   c) How many different kinds of large sundaes are there if a large sundae contains three scoops of ice cream, where each flavor can be used more than once and the order of the scoops does not matter; two kinds of sauce, where each sauce can be used only once and the order of the sauces does not matter; and three toppings, where each topping can be used only once and the order of the toppings does not matter?

8. How many positive integers less than 1000

   a) have exactly three decimal digits?

   b) have an odd number of decimal digits?

   c) have at least one decimal digit equal to 9?

   d) have no odd decimal digits?

   e) have two consecutive decimal digits equal to 5?

   f) are palindromes (that is, read the same forward and backward)?

9. When the numbers from 1 to 1000 are written out in decimal notation, how many of each of these digits are used?

   a) 0   **b)** 1   **c)** 2   **d)** 9

10. There are 12 signs of the zodiac. How many people are needed to guarantee that at least six of these people have the same sign?

11. A fortune cookie company makes 213 different fortunes. A student eats at a restaurant that uses fortunes from this company. What is the largest possible number of times that the student can eat at the restaurant without getting the same fortune four times?

12. How many people are needed to guarantee that at least two were born on the same day of the week and in the same month (perhaps in different years)?

13. Show that there are at least two different five-element subsets of a set of 10 positive integers not exceeding 50 that have the same sum.

14. A package of baseball cards contains 20 cards. How many packages must be purchased to ensure that two cards in these packages are identical if there are a total of 550 different cards?

15. a) How many cards must be chosen from a deck to guarantee that at least two aces are chosen?

   b) How many cards must be chosen from a deck to guarantee that at least two aces and two kinds are chosen?

   c) How many cards must be chosen from a deck to guarantee that there are at least two cards of the same kind?

   d) How many cards must be chosen from a deck to guarantee that there are at least two cards of two different kinds?

*16. Show that in any set of $n + 1$ positive integers not exceeding $2n$ there must be two that are relatively prime.

*17. Show that in a sequence of m integers there exists one or more consecutive terms with a sum divisible by m.

18. Show that if five points are picked in the interior of a square with a side length of 2, then at least two of these points are no farther than $\sqrt{2}$ apart.

19. Show that the decimal expansion of a rational number must repeat itself from some point onward.

20. How many diagonals does a regular polygon with n sides have, where n is a positive integer with $n \geq 3$?

21. How many ways are there to choose a dozen donuts from 20 varieties

   a) if there are no two donuts of the same variety?

   b) if all donuts are of the same variety?

   c) if there are no restrictions?

   d) if there are at least two varieties?

   e) if there must be at least six blueberry-filled donuts?

   f) if there can be no more than six blueberry-filled donuts?

22. Find n if

   a) $P(n, 2) = 110$.   **b)** $P(n, n) = 5040$.

   c) $P(n, 4) = 12P(n, 2)$.

23. Find n if

a) $C(n, 2) = 45$.   **b)** $C(n, 3) = P(n, 2)$.
c) $C(n, 5) = C(n, 2)$.

24. Show that if n and r are nonnegative integers and $n \geq r$, then

$$P(n + 1, r) = P(n, r)(n + 1)/(n + 1 - r).$$

*25. Suppose that S is a set with n elements. How many ordered pairs (A, B) are there such that A and B are subsets of S with $A \subseteq B$? (*Hint:* Show that each element of S belongs to $A, B - A$, or $S - B$.)

26. Give a combinatorial proof of Corollary 2 of Section 4.4 by setting up a correspondence between the subsets of a set with an even number of elements and the subsets of this set with an odd number of elements. (*Hint:* Take an element a in the set. Set up the correspondence by putting a in the subset if it is not already in it and taking it out if it is in the subset.)

27. Let n and r be nonnegative integers with $r < n$. Show that

$$C(n, r - 1) = C(n + 2, r + 1)$$
$$- 2C(n + 1, r + 1) + C(n, r + 1).$$

28. Prove using mathematical induction that $\sum_{j=2}^{n} C(j, 2) = C(n+1, 3)$ whenever n is an integer greater than 1.

29. Show that if n is an integer then

$$\sum_{k=0}^{n} 3^k \binom{n}{k} = 4^n.$$

30. In this exercise we will derive a formula for the sum of the squares of the n smallest positive integers. We will count the number of triples (i, j, k) such that i, j, and k are integers such that $0 \leq i < k, 0 \leq j < k$, and $1 \leq k \leq n$ in two ways.

a) Show that there are k^2 such triples with a fixed k. Conclude that there are $\sum_{k=1}^{n} k^2$ such triples.
b) Show that the number of such triples with $0 \leq i < j < k$ and the number of such triples with $0 \leq j < i < k$ both equal $C(n + 1, 3)$.
c) Show that the number of such triples with $0 \leq i = j < k$ equals $C(n + 1, 2)$.
d) Combining part (a) with parts (b) and (c), conclude that

$$\sum_{k=1}^{n} k^2 = 2C(n + 1, 3) + C(n + 1, 2)$$

$$= n(n + 1)(2n + 1)/6.$$

*31. How many bit strings of length n, where $n \geq 4$, contain exactly two occurrences of 01?

32. Let S be a set. We say that a collection of subsets $A_1, A_2, \ldots, A_n$ each containing d elements, where $d \geq 2$, is *2-colorable* if it is possible to assign to each

element of S one of two different colors such that in every subset A_i there are elements that have been assigned each color. Let $m(d)$ be the largest integer such that every collection of fewer than $m(d)$ sets each containing d elements is 2-colorable.

a) Show that the collection of all subsets with d elements of a set S with $2d - 1$ elements is not 2-colorable.
b) Show that $m(2) = 3$.
** c) Show that $m(3) = 7$. (*Hint:* Show that the collection $\{1, 3, 5\}, \{1, 2, 6\}, \{1, 4, 7\}, \{2, 3, 4\}, \{2, 5, 7\}, \{3, 6, 7\}, \{4, 5, 6\}$ is not 2-colorable. Then show that all collections of six sets with three elements each are 2-colorable.)

33. A professor writes 20 multiple-choice questions, each with the possible answer a, b, c, or d, for a discrete mathematics test. If the number of questions with a, b, c, and d as their answer is $8, 3, 4$, and 5, respectively, how many different answer keys are possible, if the questions can be placed in any order?

34. How many different arrangements are there of eight people seated at a round table, where two arrangements are considered the same if one can be obtained from the other by a rotation?

35. How many ways are there to assign 24 students to five faculty advisors?

36. How many ways are there to choose a dozen apples from a bushel containing 20 indistinguishable Delicious apples, 20 indistinguishable Macintosh apples, and 20 indistinguishable Granny Smith apples, if at least three of each kind must be chosen?

37. How many solutions are there to the equation $x_1 + x_2 + x_3 = 17$, where x_1, x_2, and x_3 are nonnegative integers with

a) $x_1 > 1, x_2 > 2$, and $x_3 > 3$?
b) $x_1 < 6$ and $x_3 > 5$?
c) $x_1 < 4, x_2 < 3$, and $x_3 > 5$?

38. a) How many different strings can be made from the word *PEPPERCORN* when all the letters are used?
b) How many of these strings start and end with the letter P?
c) In how many of these strings are the three letter Ps consecutive?

39. How many subsets of a set with ten elements

a) have fewer than five elements?
b) have more than seven elements?
c) have an odd number of elements?

40. A witness to a hit-and-run accident tells the police that the license plate of the car in the accident, which contains three letters followed by three digits, starts with the letters AS and contains both the digits 1 and 2. How many different license plates can fit this description?

41. How many ways are there to put n identical objects into m distinct containers so that no container is empty?

42. How many ways are there to seat six boys and eight girls in a row of chairs so that no two boys are seated next to each other?

43. Devise an algorithm for generating all the r-permutations of a finite set when repetition is allowed.

44. Devise an algorithm for generating all the r-combinations of a finite set when repetition is allowed.

*45. Show that if m and n are integers with $m \geq 3$ and $n \geq 3$, then $R(m, n) \leq R(m, n - 1) + R(m - 1, n)$.

*46. Show that $R(3, 4) \geq 7$ by showing that in a group of six people, where any two people are friends or enemies, there are not necessarily three mutual friends or four mutual enemies.

Computer Projects

WRITE PROGRAMS WITH THESE INPUT AND OUTPUT.

1. Given a positive integer n and a nonnegative integer not exceeding n, find the number of r-permutations and r-combinations of a set with n elements.

2. Given positive integers n and r, find the number of r-permutations when repetition is allowed and r-combinations when repetition is allowed of a set with n elements.

3. Given a sequence of positive integers, find the longest increasing and the longest decreasing subsequence of the sequence.

*4. Given an equation $x_1 + x_2 + \cdots + x_n = C$, where C is a constant, and $x_1, x_2, \ldots, x_n$ are nonnegative integers, list all the solutions.

5. Given a positive integer n, list all the permutations of the set $\{1, 2, 3, \ldots, n\}$ in lexicographic order.

6. Given a positive integer n and a nonnegative integer r not exceeding n, list all the r-combinations of the set $\{1, 2, 3, \ldots, n\}$ in lexicographic order.

7. Given a positive integer n and a nonnegative integer r not exceeding n, list all the r-permutations of the set $\{1, 2, 3, \ldots, n\}$ in lexicographic order.

8. Given a positive integer n, list all the combinations of the set $\{1, 2, 3, \ldots, n\}$.

9. Given positive integers n and r, list all the r-permutations, with repetition allowed, of the set $\{1, 2, 3, \ldots, n\}$.

10. Given positive integers n and r, list all the r-combinations, with repetition allowed, of the set $\{1, 2, 3, \ldots, n\}$.

Computations and Explorations

USE A COMPUTATIONAL PROGRAM OR PROGRAMS YOU HAVE WRITTEN TO DO THESE EXERCISES.

1. Find the number of possible outcomes in a two-team playoff when the winner is the first team to win 5 out of 9, 6 out of 11, 7 out of 13, and 8 out of 15.

2. Which binomial coefficients are odd? Can you formulate a conjecture based on numerical evidence?

3. It is not known whether the binomial coefficient $C(2n, n)$ must be divisible by the square of a prime or whether the largest exponent in the prime factorization of $C(2n, n)$ grows without bound as n grows. Explore these questions by finding the smallest and largest powers of primes in the factorization of $C(2n, n)$ for as many positive integers n as feasible.

4. Generate all the permutations of a set with eight elements.

5. Generate all the 6-permutations of a set with nine elements.

6. Generate all combinations of a set with eight elements.

7. Generate all 5-combinations with repetition allowed of a set with seven elements.

Writing Projects

RESPOND TO THESE WITH ESSAYS USING OUTSIDE SOURCES.

1. Describe some of the earliest uses of the pigeonhole principle by Dirichlet and other mathematicians.

2. Discuss ways in which the current telephone numbering plan can be extended to accommodate the rapid demand for more telephone numbers. (See if you can find some of the proposals coming from the telecommunications industry.) For each new numbering plan you discuss, show how to find the number of different telephone numbers it supports.

3. Many combinatorial identities are described in this book. Find some sources of such identities and describe important combinatorial identities besides those already introduced in this book. Give some representative proofs, including combinatorial ones, of some of these identities.

4. Describe the different models used to model the distribution of particles in statistical mechanics, including Maxwell–Boltzmann, Bose–Einstein, and Fermi–Dirac statistics. In each case, describe the counting techniques used in the model.

5. Define the Stirling numbers of the first kind and describe some of their properties and the identities they satisfy.

6. Define the Stirling numbers of the second kind and describe some of their properties and the identities they satisfy.

7. Describe the latest discoveries of values and bounds for Ramsey numbers.

8. Describe additional ways to generate all the permutations of a set with n elements besides those found in Section 4.6. Compare these algorithms and the algorithms described in the text and exercises of Section 4.6 in terms of their computational complexity.

9. Describe at least one way to generate all the partitions of a positive integer n. (See Exercise 47 in Section 3.4.)

Discrete Probability

Combinatorics and probability theory share common origins. The theory of probability was first developed more than three hundred years ago, when certain gambling games were analyzed by Blaise Pascal. It was in these studies that Pascal discovered various properties of the binomial coefficients. Later, the French mathematician Laplace, who also studied gambling, defined the probability of an event, which led to the development of much of basic probability theory.

Although probability theory was originally invented to study gambling, it now plays an essential role in a wide variety of disciplines. For example, probability theory is extensively applied in the study of genetics, where it can be used to help understand the inheritance of traits. Of course, probability still remains an extremely popular part of mathematics because of its applicability to gambling, which continues to be an extremely popular human endeavor.

In computer science, probability theory plays an important role in the study of the complexity of algorithms. In particular, ideas and techniques from probability theory are used to determine the average-case complexity of algorithms. Probabilistic algorithms can be used to solve many problems that cannot be easily or practically solved by deterministic algorithms. In a probabilistic algorithm, instead of always following the same steps when given the same input, as a deterministic algorithm does, the algorithm makes one or more random choices, which may lead to different output. In combinatorics, probability theory can even be used to show that objects with certain properties exist. The probabilistic method, a technique in combinatorics introduced by Paul Erdős and Alfréd Rényi, shows that an object with a specified property exists by showing that there is a positive probability that there is an object with this property.

5.1 An Introduction to Discrete Probability

INTRODUCTION

Probability theory dates back to the seventeenth century when the French mathematician Blaise Pascal determined the odds of winning some popular bets based on the outcome when a pair of dice is repeatedly rolled. In the eighteenth century, the French mathematician Laplace, who also studied gambling, defined the probability of an event as the number of successful outcomes divided by the number of possible outcomes. For instance, the probability a die comes up an odd number when it is rolled is the number of successful outcomes—namely, the number of ways it can come up odd—divided by the number of possible outcomes—namely, the number of different ways the die can come up. There are a total of six possible outcomes—namely, 1, 2, 3, 4, 5, and 6—and exactly three of these are successful outcomes—namely, 1, 3, and 5. Hence, the probability that the die comes up an odd number is $3/6 = 1/2$. (Note that it has been assumed that all possible outcomes are equally likely, or, in other words, that the die is fair.)

In this section we will restrict ourselves to experiments that have finitely many, equally likely, outcomes. This permits us to use Laplace's definition of the probability of an event. We will continue our study of probability in Section 5.2, where we will study experiments with finitely many outcomes that are not necessarily equally likely. In Section 5.2 we will also introduce some key concepts in probability theory, including conditional probability, independence of events, and random variables. In Section 5.3 we will introduce the concepts of the expectation and variance of a random variable.

FINITE PROBABILITY

An **experiment** is a procedure that yields one of a given set of possible outcomes. The **sample space** of the experiment is the set of possible outcomes. An **event** is a subset of the sample space. Laplace's definition of the probability of an event with finitely many possible outcomes will now be stated.

DEFINITION 1

The *probability* of an event E, which is a subset of a finite sample space S of equally likely outcomes, is $p(E) = \dfrac{|E|}{|S|}$.

Examples 1–8 illustrate how the probability of an event is found.

EXAMPLE 1
An urn contains four blue balls and five red balls. What is the probability that a ball chosen from the urn is blue?

Extra Examples

Solution: To calculate the probability, note that there are nine possible outcomes, and four of these possible outcomes produce a blue ball. Hence, the probability that a blue ball is chosen is $4/9$. ◀

EXAMPLE 2
What is the probability that when two dice are rolled, the sum of the numbers on the two dice is 7?

Links

PIERRE-SIMON LAPLACE (1749–1827) Pierre-Simon Laplace came from humble origins in Normandy. In his childhood he was educated in a school run by the Benedictines. At 16 he entered the University of Caen intending to study theology. However, he soon realized his true interests were in mathematics. After completing his studies he was named a provisional professor at Caen, and in 1769 he became professor of mathematics at the Paris Military School.

Laplace is best known for his contributions to celestial mechanics, the study of the motions of heavenly bodies. His *Traité du Mécanique Céleste* is considered one of the greatest scientific works of the early nineteenth century. Laplace was one of the founders of probability theory and made many contributions to mathematical statistics. His work in this area is documented in his book *Thèorie Analytique des Probabilitées*, in which he defined the probability of an event as the ratio of the number of favorable outcomes to the total number of outcomes of an experiment.

Laplace was famous for his political flexibility. He was loyal, in succession, to the French Republic, Napoleon, and King Louis XVIII. This permitted him to be productive before, during, and after the French Revolution.

Solution: There are a total of 36 equally likely possible outcomes when two dice are rolled. (The product rule can be used to see this; since each die has six possible outcomes, the total number of outcomes when two dice are rolled is $6^2 = 36$.) There are six successful outcomes, namely, $(1, 6)$, $(2, 5)$, $(3, 4)$, $(4, 3)$, $(5, 2)$, and $(6, 1)$, where the values of the first and second dice are represented by an ordered pair. Hence, the probability that a seven comes up when two fair dice are rolled is $6/36 = 1/6$). ◄

Lotteries have become extremely popular recently. We can easily compute the odds of winning different types of lotteries.

EXAMPLE 3   In a lottery, players win a large prize when they pick four digits that match, in the correct order, four digits selected by a random mechanical process. A smaller prize is won if only three digits are matched. What is the probability that a player wins the large prize? What is the probability that a player wins the small prize?

Solution: There is only one way to choose all four digits correctly. By the product rule, there are $10^4 = 10,000$ ways to choose four digits. Hence, the probability that a player wins the large prize is $1/10,000 = 0.0001$.

Players win the smaller prize when they correctly choose exactly three of the four digits. Exactly one digit must be wrong to get three digits correct, but not all four correct. By the sum rule, the number of ways to choose exactly three digits correctly can be obtained by adding the number of ways to choose four digits matching the digits picked in all but the ith position, for $i = 1, 2, 3, 4$. To count the number of successes with the first digit incorrect, note that there are nine possible choices for the first digit (all but the one correct digit), and one choice for each of the other digits, namely, the correct digits for these slots. Hence, there are nine ways to choose four digits where the first digit is incorrect, but the last three are correct. Similarly, there are nine ways to choose four digits where the second digit is incorrect, nine with the third digit incorrect, and nine with the fourth digit incorrect. Hence, there is a total of 36 ways to choose four digits with exactly three of the four digits correct. Thus, the probability that a player wins the smaller prize is $36/10,000 = 9/2500 = 0.0036$. ◄

EXAMPLE 4   There are many lotteries now that award enormous prizes to people who correctly choose a set of six numbers out of the first n positive integers, where n is usually between 30 and 60. What is the probability that a person picks the correct six numbers out of 40?

Solution: There is only one winning combination. The total number of ways to choose six numbers out of 40 is

$$C(40, 6) = \frac{40!}{34!\,6!} = 3,838,380.$$

Consequently, the probability of picking a winning combination is $1/3,838,380 \approx 0.00000026$. (Here the symbol $\approx$ means approximately equal to.) ◄

We can find the probability of hands in card games using the techniques developed so far. A deck of cards contains 52 cards. There are 13 different kinds of cards, with four cards of each kind. These kinds are twos, threes, fours, fives, sixes, sevens, eights, nines, tens, jacks, queens, kings, and aces. There are also four suits, spades, clubs, hearts, and diamonds, each containing 13 cards, with one card of each kind in a suit.

EXAMPLE 5 How many different hands of five cards from the deck of 52 are there?

Solution: There are $C(52, 5) = 2{,}598{,}960$ different hands with five cards. ◄

EXAMPLE 6 Find the probability that a hand of five cards in poker contains four cards of one kind.

Solution: By the product rule, the number of hands of five cards with four cards of one kind is the product of the number of ways to pick one kind, the number of ways to pick the four of this kind out of the four in the deck of this kind, and the number of ways to pick the fifth card. This is

$$C(13, 1)C(4, 4)C(48, 1).$$

Since there is a total of $C(52, 5)$ different hands of five cards, the probability that a hand contains four cards of one kind is

$$\frac{C(13, 1)C(4, 4)C(48, 1)}{C(52, 5)} = \frac{13 \cdot 1 \cdot 48}{2{,}598{,}960} \approx 0.00024.$$ ◄

EXAMPLE 7 What is the probability that a poker hand contains a full house, that is, three of one kind and two of another kind?

Solution: By the product rule, the number of hands containing a full house is the product of the number of ways to pick two kinds in order, the number of ways to pick three out of four for the first kind, and the number of ways to pick two out of four for the second kind. (Note that the order of the two kinds matters, since, for instance, three queens and two aces is different than three aces and two queens.) We see that the number of hands containing a full house is

$$P(13, 2)C(4, 3)C(4, 2) = 13 \cdot 12 \cdot 4 \cdot 6 = 3744.$$

Since there are $2{,}598{,}960$ poker hands, the probability of a full house is

$$\frac{3744}{2{,}598{,}960} \approx 0.0014.$$ ◄

EXAMPLE 8 What is the probability that the numbers $11, 4, 17, 39, 23$ are drawn in that order from a bin containing 50 balls labeled with the numbers $1, 2, \ldots, 50$ if (a) the ball selected is not returned to the bin before the next ball is selected and (b) the ball selected is returned to the bin before the next ball is selected?

Solution:
(a) By the product rule, there are $50 \cdot 49 \cdot 48 \cdot 47 \cdot 46 = 254{,}251{,}200$ ways to select the balls because each time a ball is drawn there is one fewer ball to choose from. Consequently, the probability that $11, 4, 17, 39, 23$ are drawn is $1/254{,}251{,}200$. This is an example of **sampling without replacement.**
(b) By the product rule, there are $50^5 = 312{,}500{,}000$ ways to select the balls because there are 50 possible balls to choose from each time a ball is drawn. Consequently, the probability that $11, 4, 17, 39, 23$ are drawn is $1/312{,}500{,}000$. This is an example of **sampling with replacement.** ◄

THE PROBABILITY OF COMBINATIONS OF EVENTS

We can use counting techniques to find the probability of events derived from other events.

THEOREM 1 Let E be an event in a sample space S. The probability of the event $\overline{E}$, the complementary event of E, is given by
$$p(\overline{E}) = 1 - p(E).$$

Proof: To find the probability of the event $\overline{E}$, note that $|\overline{E}| = |S| - |E|$. Hence,
$$p(\overline{E}) = \frac{|S| - |E|}{|S|} = 1 - \frac{|E|}{|S|} = 1 - p(E). \qquad \triangleleft$$

There is an alternative strategy for finding the probability of an event when a direct approach does not work well. Instead of determining the probability of the event, the probability of its complement can be found. This is often easier to do, as Example 9 shows.

EXAMPLE 9 A sequence of ten bits is randomly generated. What is the probability that at least one of these bits is 0?

Solution: Let E be the event that at least one of the ten bits is 0. Then $\overline{E}$ is the event that all the bits are 1s. Since the sample space S is the set of all bit strings of length ten, it follows that
$$p(E) = 1 - p(\overline{E}) = 1 - \frac{|\overline{E}|}{|S|} = 1 - \frac{1}{2^{10}}$$
$$= 1 - \frac{1}{1024} = \frac{1023}{1024}.$$

Hence, the probability that the bit string will contain at least one 0 bit is $1023/1024$. It is quite difficult to find this probability directly without using Theorem 1. ◀

We can also find the probability of the union of two events.

THEOREM 2 Let E_1 and E_2 be events in the sample space S. Then
$$p(E_1 \cup E_2) = p(E_1) + p(E_2) - p(E_1 \cap E_2).$$

Proof: Using the formula given in Section 1.7 for the number of elements in the union of two sets, it follows that
$$|E_1 \cup E_2| = |E_1| + |E_2| - |E_1 \cap E_2|.$$
Hence,
$$p(E_1 \cup E_2) = \frac{|E_1 \cup E_2|}{|S|}$$
$$= \frac{|E_1| + |E_2| - |E_1 \cap E_2|}{|S|}$$
$$= \frac{|E_1|}{|S|} + \frac{|E_2|}{|S|} - \frac{|E_1 \cap E_2|}{|S|}$$
$$= p(E_1) + p(E_2) - p(E_1 \cap E_2). \qquad \triangleleft$$

EXAMPLE 10   What is the probability that a positive integer selected at random from the set of positive integers not exceeding 100 is divisible by either 2 or 5?

Solution: Let E_1 be the event that the integer selected is divisible by 2, and let E_2 be the event that it is divisible by 5. Then $E_1 \cup E_2$ is the event that it is divisible by either 2 or 5. Also, $E_1 \cap E_2$ is the event that it is divisible by both 2 and 5, or equivalently, that it is divisible by 10. Since $|E_1| = 50, |E_2| = 20,$ and $|E_1 \cap E_2| = 10,$ it follows that

$$p(E_1 \cup E_2) = p(E_1) + p(E_2) - p(E_1 \cap E_2)$$

$$= \frac{50}{100} + \frac{20}{100} - \frac{10}{100} = \frac{3}{5}. \blacktriangleleft$$

PROBABILISTIC REASONING

A common problem is to determine which of two events is more likely. Analyzing the probabilities of such events can be tricky. The following example describes a problem of this type. It discusses a famous problem originating with the television game show "Let's Make a Deal."

EXAMPLE 11

Links

The Monty Hall Three Door Puzzle Suppose you are a game show contestant. You have a chance to win a large prize. You are asked to select one of three doors to open; the large prize is behind one of the three doors and the other two doors are losers. Once you select a door, the game show host, who knows what is behind each door, does the following. First, whether or not you selected the winning door, he opens one of the other two doors that he knows is a losing door (selecting at random if both are losing doors). Then he asks you whether you would like to switch doors. Which strategy should you use? Should you change doors or keep your original selection, or does it not matter?

Additional Steps

Solution: The probability you select the correct door (before the host opens a door and asks you whether you want to change) is $1/3$, since the three doors are equally likely to be the correct door. The probability this is the correct door does not change once the game show host opens one of the other doors, since he will always open a door that the prize is not behind.

The probability that you selected incorrectly is the probability the prize is behind one of the two doors you did not select. Consequently, the probability that you selected incorrectly is $2/3$. If you selected incorrectly, when the game show host opens a door to show you that the prize is not behind it, the prize is behind the other door. You will always win if your initial choice was incorrect and you change doors. So, by changing doors, the probability you win is $2/3$. In other words, you should always change doors when given the chance to do so by the game show host. This doubles the probability that you will win. $\blacktriangleleft$

Exercises

1. What is the probability that a card selected from a deck is an ace?

2. What is the probability that a die comes up six when it is rolled?

3. What is the probability that a randomly selected integer chosen from the first 100 positive integers is odd?

4. What is the probability that a randomly selected day of the year (from the 366 possible days) is in April?

5. What is the probability that the sum of the numbers on two dice is even when they are rolled?

6. What is the probability that a card selected from a deck is an ace or a heart?

7. What is the probability that a coin lands heads up six times in a row?

8. What is the probability that a five-card poker hand contains the ace of hearts?

9. What is the probability that a five-card poker hand does not contain the queen of hearts?

10. What is the probability that a five-card poker hand contains the two of diamonds and the three of spades?

11. What is the probability that a five-card poker hand contains the two of diamonds, the three of spades, the six of hearts, the ten of clubs, and the king of hearts?

12. What is the probability that a five-card poker hand contains exactly one ace?

13. What is the probability that a five-card poker hand contains at least one ace?

14. What is the probability that a five-card poker hand contains cards of five different kinds?

15. What is the probability that a five-card poker hand contains two pairs (that is, two of each of two different kinds and a fifth card of a third kind)?

16. What is the probability that a five-card poker hand contains a flush, that is, five cards of the same suit?

17. What is the probability that a five-card poker hand contains a straight, that is, five cards that have consecutive kinds? (Note that an ace can be considered either the lowest card of an A-2-3-4-5 straight or the highest card of a 10-J-Q-K-A straight.)

18. What is the probability that a five-card poker hand contains a straight flush, that is, five cards of the same suit of consecutive kinds?

*19. What is the probability that a five-card poker hand contains cards of five different kinds and does not contain a flush or a straight?

20. What is the probability that a five-card poker hand contains a royal flush, that is, the 10, jack, queen, king, and ace of one suit?

21. What is the probability that a die never comes up an even number when it is rolled six times?

22. What is the probability that a positive integer not exceeding 100 selected at random is divisible by 3?

23. What is the probability that a positive integer not exceeding 100 selected at random is divisible by 5 or 7?

24. Find the probability of winning the lottery by selecting the correct six integers, where the order in which these integers are selected does not matter, from the positive integers not exceeding

a) 30. **b)** 36. **c)** 42. **d)** 48.

25. Find the probability of winning the lottery by selecting the correct six integers, where the order in which

these integers are selected does not matter, from the positive integers not exceeding

a) 50. **b)** 52. **c)** 56. **d)** 60.

26. Find the probability of selecting none of the correct six integers, where the order in which these integers are selected does not matter, from the positive integers not exceeding

a) 40. **b)** 48. **c)** 56. **d)** 64.

27. Find the probability of selecting exactly one of the correct six integers, where the order in which these integers are selected does not matter, from the positive integers not exceeding

a) 40. **b)** 48. **c)** 56. **d)** 64.

28. To play the Pennsylvania superlottery, a player selects 7 numbers out of the first 80 positive integers. What is the probability that a person wins the grand prize by picking 7 numbers that are among the 11 numbers selected by the Pennsylvania lottery commission?

29. In a superlottery, players win a fortune if they choose the eight numbers selected by a computer from the positive integers not exceeding 100. What is the probability that a player wins this superlottery?

30. What is the probability that a player wins the prize offered for correctly choosing five (but not six) numbers out of six integers chosen between 1 and 40, inclusive, by a computer?

31. Suppose that 100 people enter a contest and that different winners are selected at random for first, second, and third prizes. What is the probability that Michelle wins one of these prizes if she is one of the contestants?

32. Suppose that 100 people enter a contest and that different winners are selected at random for first, second, and third prizes. What is the probability that Kumar, Janice, and Pedro each win a prize if each has entered the contest?

33. What is the probability that Abby, Barry, and Sylvia win the first, second, and third prizes, respectively, in a drawing if 200 people enter a contest and

a) no one can win more than one prize.

b) winning more than one prize is allowed.

34. What is the probability that Bo, Colleen, Jeff, and Rohini win the first, second, third, and fourth prizes, respectively, in a drawing if 50 people enter a contest and

a) no one can win more than one prize.

b) winning more than one prize is allowed.

35. In roulette, a wheel with 38 numbers is spun. Of these, 18 are red, and 18 are black. The other two numbers, which are neither black nor red, are 0 and 00. The probability that when the wheel is spun it lands on any particular number is 1/38.

a) What is the probability that the wheel lands on a red number?

b) What is the probability that the wheel lands on a black number twice in a row?

c) What is the probability that the wheel lands on 0 or 00?

d) What is the probability that in five spins the wheel never lands on either 0 or 00?

e) What is the probability that the wheel lands on a number between 1 and 6, inclusive, on one spin, but does not land between them on the next spin?

36. Which is more likely: rolling a total of 8 when two dice are rolled or rolling a total of 8 when three dice are rolled?

37. Which is more likely: rolling a total of 9 when two dice are rolled or rolling a total of 9 when three dice are rolled?

38. Two events E_1 and E_2 are called **independent** if $p(E_1 \cap E_2) = p(E_1)p(E_2)$. For each of the following pairs of events, which are subsets of the set of all possible outcomes when a coin is tossed three times, determine whether or not they are independent.

a) E_1: the first coin comes up tails; E_2: the second coin comes up heads.

b) E_1: the first coin comes up tails; E_2: two, and not three, heads come up in a row.

c) E_1: the second coin comes up tails; E_2: two, and not three, heads come up in a row.

(We will study independence of events in more depth in Section 5.2.)

39. Explain what is wrong with the statement that in the Monty Hall Three Door Puzzle the probability that the prize is behind the first door you select and the probability that the prize is behind the other of the two doors that Monty does not open are both 1/2, since there are two doors left.

40. Suppose that instead of three doors, there are four doors in the Monty Hall puzzle. What is the probability that you win by not changing once the host, who knows what is behind each door, opens a losing door and gives you the chance to change doors? What is the probability that you win by changing the door you select to one of the two remaining doors among the three that you did not select?

41. This problem was posed by the Chevalier de Méré and was solved by Blaise Pascal and Pierre de Fermat.

a) Find the probability of rolling a six when a die is rolled four times.

b) Find the probability that a double six comes up when a pair of dice is rolled 24 times. Answer the query the Chevalier de Méré made to Pascal asking whether this probability was greater than 1/2.

c) Is it more likely that a six comes up when a die is rolled four times or that a double six comes up when a pair of dice is rolled 24 times?

5.2 Probability Theory

INTRODUCTION

In Section 5.1 we introduced the notion of the probability of an event. (Recall that an event is a subset of the possible outcomes of an experiment.) We defined the probability of an event E as Laplace did, that is,

$$p(E) = \frac{|E|}{|S|},$$

the number of outcomes in E divided by the total number of outcomes. This definition assumes that all outcomes are equally likely. However, many experiments have outcomes that are not equally likely. For instance, a coin may be biased so that it comes up heads

HISTORICAL NOTE The Chevalier de Méré was a French nobleman, a famous gambler, and a bon vivant. He was successful at making bets with odds slightly greater than 1/2 (such as having at least one six come up in four tosses of a die). His correspondence with Pascal asking about the probability of having at least one double six come up when a pair of dice is rolled 24 times led to the development of probability theory. According to one account, Pascal wrote to Fermat about the Chevalier saying something like "He's a good guy but, alas, he's no mathematician."

twice as often as tails. Similarly, the likelihood that the input of a linear search is a particular element in a list, or is not in the list, depends on how the input is generated. How can we model the likelihood of events in such situations? In this section we will show how to define probabilities of outcomes to study probabilities of experiments where outcomes may not be equally likely.

Suppose that a fair coin is flipped four times, and the first time it comes up heads. Given this information, what is the probability that heads comes up three times? To answer this and similar questions, we will introduce the concept of *conditional probability*. Does knowing that the first flip comes up heads change the probability that heads comes up three times? If not, these two events are called *independent*, a concept studied later in this section.

Many questions address a particular numerical value associated with the outcome of an experiment. For instance, when we flip a coin 100 times, what is the probability that exactly 40 heads appear? How many heads should we expect to appear? In this section we will introduce *random variables*, which are functions that associate numerical values to the outcomes of experiments.

ASSIGNING PROBABILITIES

Let S be the sample space of an experiment with a finite or countable number of outcomes. We assign a probability $p(s)$ to each outcome s. We require that two conditions be met:

(*i*) $0 \leq p(s) \leq 1$ for each $s \in S$

and

(*ii*) $\displaystyle\sum_{s \in S} p(s) = 1.$

Condition (*i*) states that the probability of each outcome is a nonnegative real number no greater than 1. Condition (*ii*) states that the sum of the probabilities of all possible outcomes should be 1; that is, when we do the experiment, it is a certainty that one of these outcomes occurs. This is a generalization of Laplace's definition in which each of n outcomes is assigned a probability of $1/n$. Indeed, conditions (*i*) and (*ii*) are met when Laplace's definition of probabilities of equally likely outcomes is used and S is finite. (See Exercise 4.)

Note when there are n possible outcomes, $x_1, x_2, \ldots, x_n$, the two conditions to be met are

(*i*) $0 \leq p(x_i) \leq 1$ for $i = 1, 2, \ldots, n$

and

(*ii*) $\displaystyle\sum_{i=1}^{n} p(x_i) = 1.$

The function p from the set of all events of the sample space S is called a **probability distribution.**

To model an experiment, the probability $p(s)$ assigned to an outcome s should equal the limit of the number of times s occurs divided by the number of times the experiment is performed, as this number grows without bound. (We will assume that all experiments discussed have outcomes that are predictable on the average, so that this limit exists. We also assume that the outcomes of successive trials of an experiment do not depend on past results.)

Remark: We will not discuss probabilities of events when the set of outcomes is not discrete, such as when the outcome of an experiment can be any real number. In such cases integral calculus is usually required for the study of the probabilities of events.

We can model experiments in which outcomes are either equally likely or not equally likely by choosing the appropriate function $p(s)$, as Example 1 illustrates.

EXAMPLE 1   What probabilities should we assign to the outcomes H (heads) and T (tails) when a fair coin is flipped? What probabilities should be assigned to these events when the coin is biased so that heads comes up twice as often as tails?

Solution: For a fair coin, the probability that heads comes up when the coin is flipped equals the probability that tails comes up, so that the events are equally likely. Consequently, we assign the probability $1/2$ to each of the two possible outcomes, that is, $p(H) = p(T) = 1/2$.

For the biased coin we have

$$p(H) = 2p(T).$$

Since

$$p(H) + p(T) = 1,$$

it follows that

$$2p(T) + p(T) = 3p(T) = 1.$$

We conclude that $p(T) = 1/3$ and $p(H) = 2/3$.   ◀

DEFINITION 1   Suppose that S is a set with n elements. The *uniform distribution* assigns the probability $1/n$ to each element of S.

Note that the uniform distribution assigns the same probability to an event that Laplace's original definition of probability assigns to this event. The experiment of selecting an element from a sample space with a uniform distribution is called selecting an element of S **at random.**

We now define the probability of an event as the sum of the probabilities of the outcomes in this event.

DEFINITION 2   The *probability* of the event E is the sum of the probabilities of the outcomes in E. That is,

$$p(E) = \sum_{s \in E} p(s).$$

Note that when there are n outcomes in the event E, that is, if $E = \{a_1, a_2, \ldots, a_n\}$, then $p(E) = \sum_{i=1}^{n} p(a_i)$.

EXAMPLE 2 Suppose that a die is biased (or loaded) so that 3 appears twice as often as each other number but that the other five outcomes are equally likely. What is the probability that an odd number appears when we roll this die?

Solution: We want to find the probability of the event $E = \{1, 3, 5\}$. By Exercise 2 at the end of this section, we have

$$p(1) = p(2) = p(4) = p(5) = p(6) = 1/7; \ p(3) = 2/7.$$

It follows that

$$p(E) = p(1) + p(3) + p(5) = 1/7 + 2/7 + 1/7 = 4/7. \qquad \blacktriangleleft$$

When events are equally likely and there are a finite number of possible outcomes, the definition of the probability of an event given in this section (Definition 2) agrees with Laplace's definition (Definition 1 of Section 5.1). To see this, suppose that there are n equally likely outcomes; each possible outcome has probability $1/n$, since the sum of their probabilities is 1. Suppose the event E contains m outcomes. According to Definition 2,

$$p(E) = \sum_{i=1}^{m} \frac{1}{n} = \frac{m}{n}.$$

Since $|E| = m$ and $|S| = n$, it follows that

$$p(E) = \frac{m}{n} = \frac{|E|}{|S|}.$$

This is Laplace's definition of the probability of the event E.

COMBINATIONS OF EVENTS

The formulae for probabilities of combinations of events in Section 5.1 continue to hold when we use Definition 2 to define the probability of an event. For example, Theorem 1 of Section 5.1 asserts that

$$p(\overline{E}) = 1 - p(E)$$

where $\overline{E}$ is the complementary event of the event E. This equality also holds when Definition 2 is used. To see this, note that since the sum of the probabilities of the n possible outcomes is 1, and each outcome is either in E or in $\overline{E}$, but not in both, we have

$$\sum_{s \in S} p(s) = 1 = p(E) + p(\overline{E}).$$

Hence, $p(\overline{E}) = 1 - p(E)$.

Under Laplace's definition, by Theorem 2 in Section 5.1, we have

$$p(E_1 \cup E_2) = p(E_1) + p(E_2) - p(E_1 \cap E_2)$$

whenever E_1 and E_2 are events in a sample space S. This also holds when we define the probability of an event as we do in this section. To see this, note that $p(E_1 \cup E_2)$ is the sum of the probabilities of the outcomes in $E_1 \cup E_2$. When an outcome x is in one, but not both, of E_1 and E_2, $p(x)$ occurs in exactly one of the sums for $p(E_1)$ and $p(E_2)$. When an outcome x is in both E_1 and E_2, $p(x)$ occurs in the sum for $p(E_1)$, in the sum for $p(E_2)$, and in the sum for $p(E_1 \cap E_2)$, so that it occurs $1 + 1 - 1 = 1$ time on the right-hand side. Consequently, the left-hand side and right-hand side are equal.

The probability of the union of pairwise disjoint events can be found using Theorem 1.

THEOREM 1   If $E_1, E_2, \ldots$ is a sequence of pairwise disjoint events in a sample space S, then

$$p\left(\bigcup_i E_i\right) = \sum_i p(E_i).$$

(Note that this theorem applies when the sequence $E_1, E_2, \ldots$ consists of a finite number or a countably infinite number of pairwise disjoint events.)

We leave the proof of Theorem 1 to the reader (see Exercises 36 and 37).

CONDITIONAL PROBABILITY

Suppose that we flip a coin three times, and all eight possibilities are equally likely. Moreover, suppose we know that the event F, that the first flip comes up tails, occurs. Given this information, what is the probability of the event E, that an odd number of tails appears? Since the first flip comes up tails, there are only four possible outcomes: TTT, TTH, THT, and THH, where H and T represent heads and tails, respectively. An odd number of tails appears only for the outcomes TTT and THH. Since the eight outcomes have equal probability, each of the four possible outcomes, given that F occurs, should also have an equal probability of $1/4$. This suggests that we should assign the probability of $2/4 = 1/2$ to E, given that F occurs. This probability is called the **conditional probability** of E given F.

In general, to find the conditional probability of E given F, we use F as the sample space. For an outcome from E to occur, this outcome must also belong to $E \cap F$. With this motivation, we make the following definition.

DEFINITION 3   Let E and F be events with $p(F) > 0$. The *conditional probability* of E given F, denoted by $p(E \mid F)$, is defined as

$$p(E \mid F) = \frac{p(E \cap F)}{p(F)}.$$

EXAMPLE 3

Extra Examples

A bit string of length four is generated at random so that each of the 16 bit strings of length four is equally likely. What is the probability that it contains at least two consecutive 0s, given that its first bit is a 0? (We assume that 0 bits and 1 bits are equally likely.)

Solution: Let E be the event that a bit string of length four contains at least two consecutive 0s, and let F be the event that the first bit of a bit string of length four is a 0. The probability that a bit string of length four has at least two consecutive 0s, given that its first bit is a 0, equals

$$p(E \mid F) = \frac{p(E \cap F)}{p(F)}.$$

Since $E \cap F = \{0000, 0001, 0010, 0011, 0100\}$, we see that $p(E \cap F) = 5/16$. Since there are eight bit strings of length four that start with a 0, we have $p(F) = 8/16 = 1/2$. Consequently,

$$p(E \mid F) = \frac{5/16}{1/2} = \frac{5}{8}.$$ ◀

EXAMPLE 4 What is the conditional probability that a family with two children has two boys, given they have at least one boy? Assume that each of the possibilities BB, BG, GB, and GG is equally likely, where B represents a boy and G represents a girl.

Solution: Let E be the event that a family with two children has two boys, and let F be the event that a family with two children has at least one boy. It follows that $E = \{BB\}$, $F = \{BB, BG, GB\}$, and $E \cap F = \{BB\}$. Since the four possibilities are equally likely, it follows that $p(F) = 3/4$ and $p(E \cap F) = 1/4$. We conclude that

$$p(E \mid F) = \frac{p(E \cap F)}{p(F)} = \frac{1/4}{3/4} = \frac{1}{3}.$$ ◀

INDEPENDENCE

Suppose a coin is flipped three times, as described in the introduction to our discussion of conditional probability. Does knowing that the first flip comes up tails (event F) alter the probability that tails comes up an odd number of times (event E)? In other words, is it the case that $p(E \mid F) = p(E)$? This equality is valid for the events E and F, since $p(E \mid F) = 1/2$ and $p(E) = 1/2$. Because this equality holds, we say that E and F are **independent events.**

Since $p(E \mid F) = p(E \cap F)/p(F)$, asking whether $p(E \mid F) = p(E)$ is the same as asking whether $p(E \cap F) = p(E)p(F)$. This leads to Definition 4.

DEFINITION 4 The events E and F are *independent* if and only if $p(E \cap F) = p(E)p(F)$.

EXAMPLE 5 Suppose E is the event that a randomly generated bit string of length four begins with a 1 and F is the event that this bit string contains an even number of ones. Are E and F independent, if the 16 bit strings of length four are equally likely?

Solution: There are eight bit strings of length four that begin with a one: 1000, 1001, 1010, 1011, 1100, 1101, 1110, and 1111. There are also eight bit strings of length four that contain an even number of ones: 0000, 0011, 0101, 0110, 1001, 1010, 1100, 1111. Since there are 16 bit strings of length four, it follows that

$$p(E) = p(F) = 8/16 = 1/2.$$

Because $E \cap F = \{1111, 1100, 1010, 1001\}$, we see that

$$p(E \cap F) = 4/16 = 1/4.$$

Since

$$p(E \cap F) = 1/4 = (1/2)(1/2) = p(E)p(F),$$

we conclude that E and F are independent. ◀

Probability has many applications to genetics, as Examples 6 and 7 illustrate.

EXAMPLE 6   Assume, as in Example 4, that each of the four ways a family can have two children is equally likely. Are the events E, that a family with two children has two boys, and F, that a family with two children has at least one boy, independent?

Solution: Since $E = \{BB\}$, we have $p(E) = 1/4$. In Example 4 we showed that $p(F) = 3/4$ and that $p(E \cap F) = 1/4$. Since $p(E \cap F) = 1/4 \neq 3/16 = (1/4)(3/4) = p(E)p(F)$, events E and F are not independent.   ◀

EXAMPLE 7   Are the events E, that a family with three children has children of both sexes, and F, that a family with three children has at most one boy, independent? Assume that the eight ways a family can have three children are equally likely.

Solution: By assumption, each of the eight ways a family can have three children, $BBB, BBG, BGB, BGG, GBB, GBG, GGB,$ and GGG, has a probability of $1/8$. Since $E = \{BBG, BGB, BGG, GBB, GBG, GGB\}$, $F = \{BGG, GBG, GGB, GGG\}$, and $E \cap F = \{BGG, GBG, GGB\}$, it follows that $p(E) = 6/8 = 3/4$, $p(F) = 4/8 = 1/2$, and $p(E \cap F) = 3/8$. Since

$$p(E \cap F) = \frac{3}{8} = \frac{3}{4} \cdot \frac{1}{2} = p(E)p(F),$$

we conclude that E and F are independent. (This conclusion may seem surprising. Indeed, if we change the number of children, the conclusion may no longer hold. See Exercise 27 at the end of this section.)   ◀

BERNOULLI TRIALS AND THE BINOMIAL DISTRIBUTION

Suppose that an experiment can have only two possible outcomes. For instance, when a bit is generated at random, the possible outcomes are 0 and 1. When a coin is flipped, the possible outcomes are heads and tails. Each performance of an experiment with two possible outcomes is called a **Bernoulli trial,** after James Bernoulli, who made important contributions to probability theory. In general, a possible outcome of a Bernoulli trial is called a **success** or a **failure.** If p is the probability of a success and q is the probability of a failure, it follows that $p + q = 1$.

Many problems can be solved by determining the probability of k successes when an experiment consists of n independent Bernoulli trials. Consider the following example.

EXAMPLE 8   A coin is biased so that the probability of heads is $2/3$. What is the probability that exactly four heads come up when the coin is flipped seven times, assuming that the flips are independent?

Solution: There are $2^7 = 128$ possible outcomes when a coin is flipped seven times. The number of ways four of the seven flips can be heads is $C(7, 4)$. Since the seven flips are independent, the probability of each of these outcomes is $(2/3)^4(1/3)^3$. Consequently, the probability that exactly four heads appear is

$$C(7, 4)(2/3)^4(1/3)^3 = \frac{35 \cdot 16}{3^7} = \frac{560}{2187}.$$ ◀

Following the same reasoning as was used in Example 8, we can find the probability of k successes in n independent Bernoulli trials.

THEOREM 2   The probability of exactly k successes in n independent Bernoulli trials, with probability of success p and probability of failure $q = 1 - p$, is

$$C(n, k) p^k q^{n-k}.$$

Proof: When n Bernoulli trials are carried out, the outcome is an n-tuple $(t_1, t_2, \ldots, t_n)$, where $t_i = S$ (for success) or $t_i = F$ (for failure) for $i = 1, 2, \ldots, n$. Since the n trials are independent, the probability of each outcome of n trials consisting of k successes and $n - k$ failures (in any order) is $p^k q^{n-k}$. Since there are $C(n, k)$ n-tuples of Ss and Fs that contain k Ss, the probability of k successes is

$$C(n, k) p^k q^{n-k}. \qquad \qquad \qquad \triangleleft$$

We denote by $b(k; n, p)$ the probability of k successes in n independent Bernoulli trials with probability of success p and probability of failure $q = 1 - p$. Considered as a function of k, we call this function the **binomial distribution.** Theorem 2 tells us that $b(k; n, p) = C(n, k) p^k q^{n-k}$.

EXAMPLE 9   What is the probability that exactly eight 0 bits are generated when 10 bits are generated with the probability that a 0 bit is generated is 0.9, the probability that a 1 bit is generated is 0.1, and the bits are generated independently?

Extra Examples

Solution: By Theorem 2, the probability that exactly eight 0 bits are generated is

$$b(8; 10, 0.9) = C(10, 8)(0.9)^8 (0.1)^2 = 0.1937102445. \qquad \blacktriangleleft$$

Note that the sum of the probabilities that there are k successes when n independent Bernoulli trials are carried out, for $k = 0, 1, 2, \ldots, n$, equals

$$\sum_{k=0}^{n} C(n, k) p^k q^{n-k} = (p + q)^n = 1,$$

as should be the case. The first equality in this string of equalities is a consequence of the binomial theorem. The second equality follows since $q = 1 - p$.

JAMES BERNOULLI (1654–1705)   James Bernoulli (also known as Jacob I), was born in Basel, Switzerland. He is one of the eight prominent mathematicians in the Bernoulli family (see Section 9.1 for the Bernoulli family tree of mathematicians). Following his father's wish, James studied theology and entered the ministry. But contrary to the desires of his parents, he also studied mathematics and astronomy. He traveled throughout Europe from 1676 to 1682, learning about the latest discoveries in mathematics and the sciences. Upon returning to Basel in 1682, he founded a school for mathematics and the sciences. He was appointed professor of mathematics at the University of Basel in 1687, remaining in this position for the rest of his life.

Links

James Bernoulli is best known for the work *Ars Conjectandi*, published eight years after his death. In this work, he described the known results in probability theory and in enumeration, often providing alternative proofs of known results. This work also includes the application of probability theory to games of chance and his introduction of the theorem known as the **law of large numbers.** This law states that if $\epsilon > 0$, as n becomes arbitrarily large the probability approaches 1 that the number of times an event E occurs during n trials is within ϵ of $p(E)$.

RANDOM VARIABLES

Many problems are concerned with a numerical value associated with the outcome of an experiment. For instance, we may want to know the probability that there are nine one bits generated when ten bits are randomly generated, or we may want to know the probability that a coin comes up tails 11 times when it is flipped 20 times. To study problems of this type we introduce the concept of a random variable.

DEFINITION 5

A *random variable* is a function from the sample space of an experiment to the set of real numbers. That is, a random variable assigns a real number to each possible outcome.

Remark: Note that a random variable is a function. It is not a variable, and it is not random!

EXAMPLE 10   Suppose that a coin is flipped three times. Let $X(t)$ be the random variable that equals the number of heads that appear when t is the outcome. Then $X(t)$ takes on the following values:

$$X(HHH) = 3,$$
$$X(HHT) = X(HTH) = X(THH) = 2,$$
$$X(TTH) = X(THT) = X(HTT) = 1,$$
$$X(TTT) = 0.$$
◀

DEFINITION 6

The *distribution* of a random variable X on a sample space S is the set of pairs $(r, p(X = r))$ for all $r \in X(S)$, where $p(X = r)$ is the probability that X takes the value r. A distribution is usually described by specifying $p(X = r)$ for each $r \in X(S)$.

EXAMPLE 11   Because each of the eight possible outcomes when three coins are flipped has probability $1/8$, the distribution of the random variable $X(t)$ in Example 10 is given by $P(X = 3) = 1/8, P(X = 2) = 3/8, P(X = 1) = 3/8, P(X = 0) = 1/8$.   ◀

EXAMPLE 12   Let X be the sum of the numbers that appear when a pair of dice is rolled. What are the values of this random variable for the 36 possible outcomes (i, j), where i and j are the numbers that appear on the first die and the second die, respectively, when these two dice are rolled?

Solution: The random variable X takes on the following values:

$$X((1, 1)) = 2,$$
$$X((1, 2)) = X((2, 1)) = 3,$$
$$X((1, 3)) = X((2, 2)) = X((3, 1)) = 4,$$
$$X((1, 4)) = X((2, 3)) = X((3, 2)) = X((4, 1)) = 5,$$
$$X((1, 5)) = X((2, 4)) = X((3, 3)) = X((4, 2)) = X((5, 1)) = 6,$$
$$X((1, 6)) = X((2, 5)) = X((3, 4)) = X((4, 3)) = X((5, 2)) = X((6, 1)) = 7,$$
$$X((2, 6)) = X((3, 5)) = X((4, 4)) = X((5, 3)) = X((6, 2)) = 8,$$

$$X((3, 6)) = X((4, 5)) = X((5, 4)) = X((6, 3)) = 9,$$
$$X((4, 6)) = X((5, 5)) = X((6, 4)) = 10,$$
$$X((5, 6)) = X((6, 5)) = 11,$$
$$X((6, 6)) = 12. \qquad \blacktriangleleft$$

We will continue our study of random variables in Section 5.3, where we will show how they can be used in a variety of applications.

THE BIRTHDAY PROBLEM

A famous puzzle asks for the smallest number of people needed in a room so that it is more likely than not that at least two of them have the same day of the year as their birthday. Most people find the answer, which we determine in Example 13, to be surprisingly small. After we solve this famous problem, we will show how similar reasoning can be adapted to solve a question about hashing functions.

EXAMPLE 13

Links

The Birthday Problem What is the minimum number of people who need to be in a room so that the probability that at least two of them have the same birthday is greater than 1/2?

Solution: First, we state some assumptions. We assume that the birthdays of the people in the room are independent. Furthermore, we assume that each birthday is equally likely and that there are 366 days in the year. (In reality, more people are born on some days of the year than others, such as days nine months after some holidays including New Year's Eve, and only leap years have 366 days.)

To find the probability that at least two of n people in a room have the same birthday, we first calculate the probability p_n that these people all have different birthdays. Then, the probability that at least two people have the same birthday is $1 - p_n$. To compute p_n, we consider the birthdays of the n people in some fixed order. Imagine them entering the room one at a time; we will compute the probability that each successive person entering the room has a birthday different from those of the people already in the room.

The birthday of the first person certainly does not match the birthday of someone already in the room. The probability that the birthday of the second person is different from that of the first person is $365/366$ because the second person has a different birthday when he or she was born on one of the 365 days of the year other than the day the first person was born. (The assumption that it is equally likely for someone to be born on any of the 366 days of the year enters into this and subsequent steps.) The probability that the third person has a birthday different from both the birthdays of the first and second people given that these two people have different birthdays is $364/366$. In general, the probability that the jth person, with $2 \leq j \leq 366$, has a birthday different from the birthdays of the $j - 1$ people already in the room given that these $j - 1$ people have different birthdays is

$$\frac{366 - (j - 1)}{366} = \frac{367 - j}{366}.$$

Because we have assumed that the birthdays of the people in the room are independent, we can conclude that the probability that the n people in the room have different birthdays is

$$p_n = \frac{365}{366} \frac{364}{366} \frac{363}{366} \cdots \frac{367 - n}{366}.$$

It follows that the probability that among n people there are at least two people with the same birthday is

$$1 - p_n = 1 - \frac{365}{366} \frac{364}{366} \frac{363}{366} \cdots \frac{367 - n}{366}.$$

To determine the minimum number of people in the room so that the probability that at least two of them have the same birthday is greater than $1/2$, we use the formula we have found for $1 - p_n$ to compute it for increasing values of n until it becomes greater than $1/2$. (There are more sophisticated approaches using calculus that can eliminate this computation, but we will not use them here.) After considerable computation we find that for $n = 22, 1 - p_n \approx 0.475$, while for $n = 23, 1 - p_n \approx 0.506$. Consequently, the minimum number of people needed so that the probability that at least two people have the same birthday is greater than $1/2$ is 23. ◄

The solution to the birthday problem leads to the solution of the question in Example 14 about hashing functions.

EXAMPLE 14 **Probability of a Collision in Hashing Functions** Recall from Section 2.4 that a hashing function $h(k)$ is a mapping of the keys (of the records that are to be stored in a database) to storage locations. Hashing functions map a large universe of keys (such as the approximately 300 million Social Security numbers in the United States) to a much smaller set of storage locations. A good hashing function yields few **collisions,** which are mappings of two different keys to the same memory location, when relatively few of the records are in play in a given application. What is the probability that no two keys are mapped to the same location by a hashing function, or, in other words, that there are no collisions?

Solution: To calculate this probability, we assume that the probability that a randomly selected key is mapped to a location is $1/m$, where m is the number of available locations, that is, the hashing function distributes keys uniformly. (In practice, hashing functions may not satisfy this assumption. However, for a good hashing function, this assumption should be close to correct.) Furthermore, we assume that the keys of the records selected have an equal probability to be any of the elements of the key universe and that these keys are independently selected.

Suppose that the keys are $k_1, k_2, \ldots, k_n$. When we add the second record, the probability that it is mapped to a location different from the location of the first record, that $h(k_2) \neq h(k_1)$, is $(m - 1)/m$ because there are $m - 1$ free locations after the first record has been placed. The probability that the third record is mapped to a free location after the first and second records have been placed without a collision is $(m - 2)/m$. In general, the probability that the jth record is mapped to a free location after the first $j - 1$ records have been mapped to locations $h(k_1), h(k_2), \ldots, h(k_{j-1})$ without collisions is $(m - (j - 1))/m$ because $j - 1$ of the m locations are taken.

Since the keys are independent, the probability that all n keys are mapped to different locations is

$$p_n = \frac{m - 1}{m} \cdot \frac{m - 2}{m} \cdot \ldots \cdot \frac{m - n + 1}{m}.$$

It follows that the probability that there is at least one collision, that is, at least two keys are mapped to the same location, is

$$1 - p_n = 1 - \frac{m - 1}{m} \cdot \frac{m - 2}{m} \cdot \ldots \cdot \frac{m - n + 1}{m}.$$

We can use techniques from calculus to find the smallest value of n given a value of m such that the probability of a collision is greater than a particular threshold. It can be shown that the smallest integer n such that the probability of a collision is greater than $1/2$ is approximately $n = 1.177\sqrt{m}$. For example, when $m = 1,000,000$, the smallest integer n such that the probability of a collision is greater than $1/2$ is 1178. ◄

MONTE CARLO ALGORITHMS

The algorithms discussed so far in this book are all deterministic. That is, each algorithm always proceeds in the same way whenever given the same input. However, there are many situations where we would like an algorithm to make a random choice at one or more steps. Such a situation arises when a deterministic algorithm would have to go through a huge number, or even an unknown number, of possible cases. Algorithms that make random choices at one or more steps are called **probabilistic algorithms.** We will discuss a particular class of probabilistic algorithms in this section, namely, **Monte Carlo algorithms,** for decision problems. Monte Carlo algorithms always produce answers to problems, but a small probability remains that these answers may be incorrect. However, the probability that the answer is incorrect decreases rapidly when the algorithm carries out sufficient computation. Decision problems have either "true" or "false" as their answer.

A Monte Carlo algorithm for a decision problem uses a sequence of tests. The probability that the algorithm answers the decision problem correctly increases as more tests are carried out. At each step of the algorithm, possible responses are "true," which means that the answer is "true" and no additional iterations are needed, or "unknown," which means that the answer could be either "true" or "false." After running all the iterations in such an algorithm, the final answer produced is "true" if any iteration yields the answer "true," and the answer is "false" if every iteration yields the answer "unknown." If the correct answer is "false" the algorithm answers "false," since every iteration will yield "unknown." However, if the correct answer is "true," the algorithm could answer either "true" or "false," since it may be possible that each iteration produced the response "unknown" even though the correct response was "true." We will show that this possibility becomes extremely unlikely as the number of tests increases.

Suppose that p is the probability that the response of a test is "true," given that the answer is "true." It follows that $1 - p$ is the probability that the response is "unknown," given that the answer is "true." Because the algorithm answers "false" when all n iterations yield the answer "unknown" and the iterations perform independent tests, the probability of error is $(1 - p)^n$. When $p \neq 0$, this probability approaches 0 as the number of tests increases. Consequently, the probability that the algorithm answers "true" when the answer is "true" approaches 1.

EXAMPLE 15 **Quality Control** (This example is adapted from [AhUl95].) Suppose that a manufacturer orders processor chips in batches of size n, where n is a positive integer. The chip maker has tested only some of these batches to make sure all the chips in the batch are good. In previously untested batches the probability that a particular chip is bad has been observed to be 0.1 when random testing is done. The PC manufacturer wants to decide whether all the chips in a batch are good. To do this, the PC manufacturer can test each chip in a batch to see whether it is good. However, this requires n tests. Assuming that each test can be carried out in constant time, these tests require $O(n)$ seconds. Can the PC manufacturer test whether a batch of chips has been tested by the chip maker using less time?

Solution: We can use a Monte Carlo algorithm to determine whether a batch of chips has been tested by the chip maker as long as we are willing to accept some probability of error. The algorithm is set up to answer the question: "Has this batch of chips not been tested by the chip maker?" It proceeds by successively selecting chips at random from the batch and testing them one by one. When a bad chip is encountered, the algorithm answers "true" and stops. If a tested chip is good, the algorithm answers "unknown" and goes on to the next chip. After the algorithm has tested a specified number of chips, say k chips, without getting an answer of "true," the algorithm terminates with the answer "false"; that is, the algorithm concludes that the batch is good, that is, that the chip maker has tested all the chips in the batch.

The only way for this algorithm to answer incorrectly is for it to conclude that an untested batch of chips has been tested by the chip maker. The probability that a chip is good, but that it came from an untested batch, is $1 - 0.1 = 0.9$. Because the events of testing different chips from a batch are independent, the probability that all k steps of the algorithm produce the answer "unknown," given that the batch of chips is untested, is 0.9^k.

By taking k large enough, we can make this probability as small as we like. For example, by testing 66 chips, the probability that the algorithm decides a batch has been tested by the chip maker is 0.9^{66}, which is less than 0.001. That is, the probability is less than 1 in 1000 that the algorithm has answered incorrectly. Note that this probability is independent of n, the number of chips in a batch. That is, the Monte Carlo algorithm uses a constant number, or $O(1)$, tests and requires $O(1)$ seconds, no matter how many chips are in a batch. As long as the PC manufacturer can live with an error rate of less than 1 in 1000, the Monte Carlo algorithm will save the PC manufacturer a lot of testing. If a smaller error rate is needed, the PC manufacturer can test more chips in each batch; the reader can verify that 132 tests lower the error rate to less than 1 in 1,000,000.   ◄

EXAMPLE 16   **Probabilistic Primality Testing** In Chapter 2 we remarked that a composite integer, that is, an integer that is not prime, passes Miller's test (see Exercise 30 in Section 2.6) for fewer than $n/4$ bases b with $1 < b < n$. This observation is the basis for a Monte Carlo algorithm to determine whether a positive integer is prime. Since large primes play an essential role in public-key cryptography (see Section 2.6), being able to generate large primes quickly has become extremely important.

The goal of the algorithm is to decide the question "Is n composite?" Given an integer n, we select an integer b at random with $1 < b < n$ and determine whether n passes Miller's test to the base b. If n fails the test, the answer is "true" since n must be composite, and the algorithm ends. Otherwise, we repeat the test k times, where k is an integer. Each time we select a random integer b and determine whether n passes Miller's test to the base b. If the answer is "unknown" at each step, the algorithm answers "false," that is, it says that n is not composite, so that it is prime. The only possibility for the algorithm to return an incorrect answer occurs when n is composite, and the answer "unknown" is the output at each of the k iterations. The probability that a composite integer n passes Miller's test for a randomly selected base b is less than $1/4$. Since the integer b with $1 < b < n$ is selected at random at each iteration and these iterations are independent, the probability that n is composite but the algorithm responds that n is prime is less than $(1/4)^k$. By taking k to be sufficiently large, we can make this probability extremely small. For example, with 10 iterations, the probability that the algorithm decides that n is prime when it really is composite is less than 1 in 1,000,000. With 30 iterations, this probability drops to less than 1 in 10^{18}, an extremely unlikely event.

To generate large primes, say with 200 digits, we randomly choose an integer n with 200 digits and run this algorithm, with 30 iterations. If the algorithm decides that n

is prime, we can use it as one of the two primes used in an encryption key for the RSA cryptosystem. If n is actually composite and is used as part of the key, the procedures used to decrypt messages will not produce the original encrypted message. The key is then discarded and two new possible primes are used. ◀

THE PROBABILISTIC METHOD

We discussed existence proofs in Chapter 1 and illustrated the difference between constructive existence proofs and nonconstructive existence proofs. The probabilistic method, introduced by Paul Erdős and Alfréd Rényi, is a powerful technique that can be used to create nonconstructive existence proofs. To use the probabilistic method to prove results about a set S, such as the existence of an element in S with a specified property, we assign probabilities to the elements of S. We then use methods from probability theory to prove results about the elements of S. In particular, we can show that an element with a specified property exists by showing that the probability an element $x \in S$ has this property is positive. The probabilistic method is based on the equivalent statement in Theorem 3.

THEOREM 3

> **THE PROBABILISTIC METHOD** If the probability that an element of a set S does not have a particular property is less than 1, there exists an element in S with this property.

An existence proof based on the probabilistic method is nonconstructive since it does not find a particular element with the desired property.

We illustrate the power of the probabilistic method by finding a lower bound for the Ramsey number $R(k, k)$. Recall from Section 4.2 that $R(k, k)$ equals the minimum number of people at a party to ensure that there are at least k mutual friends or k mutual enemies (assuming that any two people are friends or enemies).

THEOREM 4

> If k is an integer with $k \geq 2$, then $R(k, k) \geq 2^{k/2}$.

Proof: We note that the theorem holds for $k = 2$ and $k = 3$ because $R(2, 2) = 2$ and $R(3, 3) = 6$, as shown in Section 4.2. Now suppose that $k \geq 4$. We will use the probabilistic method to show that if there are fewer than $2^{k/2}$ people at a party, it is possible that no k of them are mutual friends or mutual enemies. This will show that $R(k, k)$ is at least $2^{k/2}$.

To use the probabilistic method, we assume that it is equally likely for two people to be friends or enemies. (Note that this assumption does not have to be realistic.) Suppose there are n people at the party. It follows that there are $\binom{n}{k}$ different sets of k people at this party, which we list as $S_1, S_2, \ldots, S_{\binom{n}{k}}$. Let E_i be the event that all k people in S_i are either mutual friends or mutual enemies. The probability that there are either k mutual friends or k mutual enemies among the n people equals $p(\bigcup_{i=1}^{\binom{n}{k}} E_i)$.

According to our assumption it is equally likely for two people to be friends or enemies. The probability that two people are friends equals the probability that they are enemies; both probabilities equal $1/2$. Furthermore, there are $\binom{k}{2} = k(k - 1)/2$ pairs of people in S_i because there are k people in S_i. Hence, the probability that all k people in

S_i are mutual friends and the probability that all k people in S_i are mutual enemies both equal $(1/2)^{k(k-1)/2}$. It follows that $p(E_i) = 2(1/2)^{k(k-1)/2}$.

The probability that there are either k mutual friends or k mutual enemies in the group of n people equals $p(\bigcup_{i=1}^{\binom{n}{k}} E_i)$. Using Boole's Inequality (Exercise 15), it follows that

$$p\left(\bigcup_{i=1}^{\binom{n}{k}} E_i\right) \leq \sum_{i=1}^{\binom{n}{k}} p(E_i) = \binom{n}{k} \cdot 2\left(\frac{1}{2}\right)^{k(k-1)/2}.$$

By Exercise 17 in Section 4.4, we have $\binom{n}{k} \leq n^k / 2^{k-1}$. Hence,

$$\binom{n}{k} 2\left(\frac{1}{2}\right)^{k(k-1)/2} \leq \frac{n^k}{2^{k-1}} 2\left(\frac{1}{2}\right)^{k(k-1)/2}.$$

Now if $n < 2^{k/2}$, we have

$$\frac{n^k}{2^{k-1}} 2\left(\frac{1}{2}\right)^{k(k-1)/2} < \frac{2^{k(k/2)}}{2^{k-1}} 2\left(\frac{1}{2}\right)^{k(k-1)/2} = 2^{2-(k/2)} \leq 1$$

where the last step follows because $k \geq 4$.

We can now conclude that $p(\bigcup_{i=1}^{\binom{n}{k}} E_i) < 1$ when $k \geq 4$. Hence, the probability of the complementary event, that there is no set of either k mutual friends or mutual enemies at the party, is greater than 0. It follows that if $n < 2^{k/2}$, there is at least one set such that no subset of k people are mutual friends or mutual enemies. ◁

Exercises

1. What probability should be assigned to the outcome of heads when a biased coin is tossed, if heads is three times as likely to come up as tails? What probability should be assigned to the outcome of tails?

2. Find the probability of each outcome when a loaded die is rolled, if a 3 is twice as likely to appear as each of the other five numbers on the die.

3. Find the probability of each outcome when a biased die is rolled, if rolling a 2 or rolling a 4 is three times as likely as rolling each of the other four numbers on the die and it is equally likely to roll a 2 or a 4.

4. Show that conditions (*i*) and (*ii*) are met under Laplace's definition of probability, when outcomes are equally likely.

5. A pair of dice is loaded. The probability that a 4 appears on the first die is 2/7, and the probability that a 3 appears on the second die is 2/7. Other outcomes for each die appear with probability 1/7. What is the probability of 7 appearing as the sum of the numbers when the two dice are rolled?

6. What is the probability of these events when we randomly select a permutation of $\{1, 2, 3\}$?

   a) 1 precedes 3.

   b) 3 precedes 1.

   c) 3 precedes 1 and 3 precedes 2.

7. What is the probability of these events when we randomly select a permutation of $\{1, 2, 3, 4\}$?

   a) 1 precedes 4.

   b) 4 precedes 1.

   c) 4 precedes 1 and 4 precedes 2.

   d) 4 precedes 1, 4 precedes 2, and 4 precedes 3.

   e) 4 precedes 3 and 2 precedes 1.

8. What is the probability of these events when we randomly select a permutation of $\{1, 2, \ldots, n\}$ where $n \geq 4$?

   a) 1 precedes 2.

   b) 2 precedes 1.

   c) 1 immediately precedes 2.

   d) n precedes 1 and $n-1$ precedes 2.

   e) n precedes 1 and n precedes 2.

9. What is the probability of these events when we randomly select a permutation of the 26 lowercase letters of the English alphabet?

   a) The permutation consists of the letters in reverse alphabetic order.

b) z is the first letter of the permutation.

c) z precedes a in the permutation.

d) a immediately precedes z in the permutation.

e) a immediately precedes m, which immediately precedes z in the permutation.

f) $m, n,$ and o are in their original places in the permutation.

10. What is the probability of these events when we randomly select a permutation of the 26 lowercase letters of the English alphabet?

a) The first 13 letters of the permutation are in alphabetic order.

b) a is the first letter of the permutation and z is the last letter.

c) a and z are next to each other in the permutation.

d) a and b are not next to each other in the permutation.

e) a and z are separated by at least 23 letters in the permutation.

f) z precedes both a and b in the permutation.

11. Suppose that E and F are events such that $p(E) = 0.7$ and $p(F) = 0.5$. Show that $p(E \cup F) \geq 0.7$ and $p(E \cap F) \geq 0.2$.

12. Suppose that E and F are events such that $p(E) = 0.8$ and $p(F) = 0.6$. Show that $p(E \cup F) \geq 0.8$ and $p(E \cap F) \geq 0.4$.

13. Show that if E and F are events, then $p(E \cap F) \geq p(E) + p(F) - 1$. This is known as **Bonferroni's Inequality.**

14. Use mathematical induction to prove the following generalization of Bonferroni's Inequality:

$$p(E_1 \cap E_2 \cap \cdots \cap E_n)$$
$$\geq p(E_1) + p(E_2) + \cdots + p(E_n) - (n - 1),$$

where $E_1, E_2, \ldots, E_n$ are n events.

☞ **15.** Show that if $E_1, E_2, \ldots, E_n$ are events from a finite sample space, then

$$p(E_1 \cup E_2 \cup \cdots \cup E_n)$$
$$\leq p(E_1) + p(E_2) + \cdots + p(E_n).$$

This is known as **Boole's Inequality.**

16. Show that if E and F are independent events, then $\overline{E}$ and $\overline{F}$ are also independent events.

☞ **17.** If E and F are independent events, prove or disprove that $\overline{E}$ and F are necessarily independent events.

18. a) What is the probability that two people were born on the same day of the week?

b) What is the probability that in a group of n people, there are at least two born on the same day of the week?

c) How many people are needed to make the probability greater than 1/2 that there are at least two people born on the same day of the week?

19. a) What is the probability that two people were born during the same month of the year?

b) What is the probability that in a group of n people, there are at least two born in the same month of the year?

c) How many people are needed to make the probability greater than 1/2 that there are at least two people born in the same month of the year?

20. Find the smallest number of people in a room so that the probability that someone has a birthday today exceeds 1/2. (Assume all birthdays are equally likely and that the year has 366 days.)

21. Find the smallest number of people in a room so that the probability that two people in the room were both born on April 1 exceeds 1/2. (Assume all birthdays are equally likely and that the year has 366 days.)

*22. February 29 occurs only in leap years. Years divisible by 4, but not by 100, are always leap years. Years divisible by 100, but not by 400, are not leap years, but years divisible by 400 are leap years.

a) What probability distribution for birthdays should be used to reflect how often February 29 occurs?

b) Using the probability distribution from part (a), what is the probability that in a group of n people at least two have the same birthday?

23. What is the conditional probability that exactly four heads appear when a fair coin is flipped five times, given that the first flip came up heads?

24. What is the conditional probability that exactly four heads appear when a fair coin is flipped five times, given that the first flip came up tails?

25. What is the conditional probability that a randomly generated bit string of length four contains at least two consecutive 0s, given that the first bit is a 1? (Assume the probabilities of a 0 and a 1 are the same.)

26. Let E be the event that a randomly generated bit string of length three contains an odd number of 1s, and let F be the event that the string starts with 1. Are E and F independent?

27. Let E and F be the events that a family of n children has children of both sexes and has at most one boy, respectively. Are E and F independent if

a) $n = 2$? **b)** $n = 4$? **c)** $n = 5$?

28. Assume that the probability a child is a boy is 0.51 and that the sexes of children born into a family are independent. What is the probability that a family of five children has

a) exactly three boys?

b) at least one boy?

c) at least one girl?

d) all children of the same sex?

29. A group of six people play the game of "odd person out" to determine who will buy refreshments. Each person flips a fair coin. If there is a person whose outcome is not the same as that of any other member of the group, this person has to buy the refreshments. What is the probability that there is an odd person out after the coins are flipped once?

30. Find the probability that a randomly generated bit string of length 10 does not contain a 0 if bits are independent and if

    a) a 0 bit and a 1 bit are equally likely.

    b) the probability that a bit is a 1 is 0.6.

    c) the probability that the ith bit is a 1 is $1/2^i$ for $i = 1, 2, 3, \ldots, 10$.

31. Find the probability that a family with five children does not have a boy, if the sexes of children are independent and if

    a) a boy and a girl are equally likely.

    b) the probability of a boy is 0.51.

    c) the probability that the ith child is a boy is $0.51 - (i/100)$.

32. Find the probability that a randomly generated bit string of length 10 begins with a 1 or ends with a 00 for the same conditions as in parts (a), (b), and (c) of Exercise 30, if bits are generated independently.

33. Find the probability that the first child of a family with five children is a boy or that the last two children of the family are girls, for the same conditions as in parts (a), (b), and (c) of Exercise 31.

34. Find each of the following probabilities when n independent Bernoulli trials are carried out with probability of success p.

    a) the probability of no successes

    b) the probability of at least one success

    c) the probability of at most one success

    d) the probability of at least two successes

35. Find each of the following probabilities when n independent Bernoulli trials are carried out with probability of success p.

    a) the probability of no failures

    b) the probability of at least one failure

    c) the probability of at most one failure

    d) the probability of at least two failures

36. Use mathematical induction to prove that if $E_1, E_2, \ldots, E_n$ is a sequence of n pairwise disjoint events in a sample space S, where n is a positive integer, then $p(\bigcup_{i=1}^{n} E_i) = \sum_{i=1}^{n} p(E_i)$.

*37. (Requires calculus) Show that if $E_1, E_2, \ldots$ is an infinite sequence of pairwise disjoint events in a sample space S, then $p(\bigcup_{i=1}^{\infty} E_i) = \sum_{i=1}^{\infty} p(E_i)$. [*Hint:* Use Exercise 36 and take limits.]

38. A pair of dice is rolled in a remote location and an honest observer informs us that at least one of the dice came up six.

    a) What is the probability that the sum of the numbers that came up on the two dice is seven,

given the information provided by the honest observer?

    b) Suppose that the honest observer tells us that at least one die came up five. What is the probability the sum of the numbers that came up on the dice is seven, given this information?

39. This exercise employs the probabilistic method to prove a result about round-robin tournaments. In a **round-robin tournament with m players, every two players play one game in which one player wins and the other loses.

We want to find conditions on positive integers m and k with $k < m$ so that it is possible for the outcomes of the tournament to have the property that for every set of k players, there is a player who beats every member in this set. So that we can use probabilistic reasoning to draw conclusions about round-robin tournaments, we assume that when two players compete it is equally likely that either player wins the game and we assume that the outcomes of different games are independent. Let E be the event that for every set S with k players, where k is a positive integer less than m, there is a player who has beaten all k players in S.

    a) Show that $p(\overline{E}) \le \sum_{j=1}^{\binom{m}{k}} p(F_j)$, where F_j is the event that there is no player who beats all k players from the jth set in a list of the $\binom{m}{k}$ sets of k players.

    b) Show that the probability of F_j is $(1 - 2^{-k})^{m-k}$.

    c) Conclude from parts (a) and (b) that $p(\overline{E}) \le \binom{m}{k}(1 - 2^{-k})^{m-k}$ and therefore, that there must be a tournament with the described property if $\binom{m}{k}(1 - 2^{-k})^{m-k} < 1$.

    d) Use part (c) to find values of m such that there is a tournament with m players such that for every set S of two players, there is a player who has beaten both players in S. Repeat for sets of three players.

*40. Devise a Monte Carlo algorithm that determines whether a permutation of the integers 1 through n has already been sorted (that is, it is in increasing order), or instead, is a random permutation. A step of the algorithm should answer "true" if it determines the list is not sorted and "unknown" otherwise. After k steps, the algorithm decides that the integers are sorted if the answer is "unknown" in each step. Show that as the number of steps increases, the probability that the algorithm produces an incorrect answer is extremely small. (*Hint:* For each step, test whether certain elements are in the correct order. Make sure these tests are independent.)

41. Use pseudocode to write out the probabilistic primality test described in Example 16.

5.3   **Expected Value and Variance**

INTRODUCTION

The **expected value** of a random variable is the sum over all elements in a sample space of the product of the probability of the element and the value of the random variable at this element. Consequently, the expected value is a weighted average of the values of a random variable. The expected value of a random variable provides a central point for the distribution of values of this random variable. We can solve many problems using the notion of the expected value of a random variable, such as determining who has an advantage in gambling games and computing the average-case complexity of algorithms. Another useful measure of a random variable is its **variance,** which tells us how spread out the values of this random variable are. We can use the variance of a random variable to help us estimate the probability that a random variable takes values far removed from its expected value.

EXPECTED VALUES

Many questions can be formulated in terms of the value we expect a random variable to take, or more precisely, the average value of a random variable when an experiment is performed a large number of times. Questions of this kind include: How many heads are expected to appear when a coin is flipped 100 times? What is the expected number of comparisons used to find an element in a list using a linear search? To study such questions we introduce the concept of the expected value of a random variable.

DEFINITION 1

The *expected value* (or *expectation*) of the random variable $X(s)$ on the sample space S is equal to

$$E(X) = \sum_{s \in S} p(s) X(s).$$

Note that when the sample space S has n elements $S = \{x_1, x_2, \ldots, x_n\}$, $E(X) = \sum_{i=1}^{n} p(x_i) X(x_i)$.

Remark: We are concerned only with random variables with finite expected values here.

EXAMPLE 1   **Expected Value of a Die** Let X be the number that comes up when a die is rolled. What is the expected value of X?

Solution: The random variable X takes the values 1, 2, 3, 4, 5, or 6, each with probability $1/6$. It follows that

$$E(X) = \frac{1}{6} \cdot 1 + \frac{1}{6} \cdot 2 + \frac{1}{6} \cdot 3 + \frac{1}{6} \cdot 4 + \frac{1}{6} \cdot 5 + \frac{1}{6} \cdot 6 = \frac{21}{6} = \frac{7}{2}. \qquad \blacktriangleleft$$

EXAMPLE 2   A fair coin is flipped three times. Let S be the sample space of the eight possible outcomes, and let X be the random variable that assigns to an outcome the number of heads in this outcome. What is the expected value of X?

Solution: In Example 10 of Section 5.2 we listed the values of X for the eight possible outcomes when a coin is flipped three times. Since the coin is fair and the flips are independent, the probability of each outcome is $1/8$. Consequently,

$$E(X) = \frac{1}{8}[X(HHH) + X(HHT) + X(HTH) + X(THH) + X(TTH)$$

$$+ X(THT) + X(HTT) + X(TTT)]$$

$$= \frac{1}{8}(3 + 2 + 2 + 2 + 1 + 1 + 1 + 0) = \frac{12}{8}$$

$$= \frac{3}{2}.$$

Consequently, the average number of heads that come up when a fair coin is flipped three times is $3/2$. ◀

When an experiment has relatively few outcomes, we can compute the expected value of a random variable directly from its definition, as was done in Example 2. However, when an experiment has a large number of outcomes, it may be inconvenient to compute the expected value of a random variable directly from its definition. Instead, we can find the expected value of a random variable by grouping together all outcomes assigned the same value by the random variable, as Theorem 1 shows.

THEOREM 1   If X is a random variable and $p(X = r)$ is the probability that $X = r$, so that $p(X = r) = \sum_{s \in S, X(s)=r} p(s)$, then

$$E(X) = \sum_{r \in X(S)} p(X = r)r.$$

Proof: Suppose that X is a random variable with range $X(S)$, and let $p(X = r)$ be the probability that the random variable X takes the value r. Consequently, $p(X = r)$ is the sum of the probabilities of the outcomes s such that $X(s) = r$. It follows that

$$E(X) = \sum_{r \in X(S)} p(X = r)r.$$ ◁

Example 3 and the proof of Theorem 2 illustrate the use of this formula. In Example 3 we will find the expected value of the sum of the numbers that appear on two fair dice when they are rolled. In Theorem 2 we will find the expected value of the number of successes when n Bernoulli trials are performed.

EXAMPLE 3   What is the expected value of the sum of the numbers that appear when a pair of fair dice is rolled?

Solution: Let X be the random variable equal to the sum of the numbers that appear when a pair of dice is rolled. In Example 12 of Section 5.2 we listed the value of X for the 36 outcomes of this experiment. The range of X is $\{2, 3, 4, 5, 6, 7, 8, 9, 10, 11, 12\}$. By Example 12 of Section 5.2 we see that

$$p(X = 2) = p(X = 12) = 1/36,$$

$$p(X = 3) = p(X = 11) = 2/36 = 1/18,$$

$$p(X = 4) = p(X = 10) = 3/36 = 1/12,$$

$$p(X = 5) = p(X = 9) = 4/36 = 1/9,$$

$$p(X = 6) = p(X = 8) = 5/36,$$

$$p(X = 7) = 6/36 = 1/6.$$

Substituting these values in the formula, we have

$$E(X) = 2 \cdot \frac{1}{36} + 3 \cdot \frac{1}{18} + 4 \cdot \frac{1}{12} + 5 \cdot \frac{1}{9} + 6 \cdot \frac{5}{36} + 7 \cdot \frac{1}{6}$$

$$+ 8 \cdot \frac{5}{36} + 9 \cdot \frac{1}{9} + 10 \cdot \frac{1}{12} + 11 \cdot \frac{1}{18} + 12 \cdot \frac{1}{36}$$

$$= 7.$$ ◀

THEOREM 2   The expected number of successes when n Bernoulli trials are performed, where p is the probability of success on each trial, is np.

Proof: Let X be the random variable equal to the number of successes in n trials. By Theorem 2 of Section 5.2 we see that $p(X = k) = C(n, k)p^k q^{n-k}$. Hence, we have

$$E(X) = \sum_{k=1}^{n} kp(X = k) \qquad \text{by Theorem 1}$$

$$= \sum_{k=1}^{n} kC(n, k)p^k q^{n-k} \qquad \text{by Theorem 2 in Section 5.2}$$

$$= \sum_{k=1}^{n} nC(n - 1, k - 1)p^k q^{n-k} \qquad \text{by Exercise 21 in Section 4.4}$$

$$= np \sum_{k=1}^{n} C(n - 1, k - 1)p^{k-1} q^{n-k} \qquad \text{factoring } np \text{ from each term}$$

$$= np \sum_{j=0}^{n-1} C(n - 1, j)p^j q^{n-1-j} \qquad \text{shifting index of summation with } j = k - 1$$

$$= np(p + q)^{n-1} \qquad \text{by the binomial theorem}$$

$$= np. \qquad \text{because } p + q = 1$$

This completes the proof because it shows that the expected number of successes in n Bernoulli trials is np.   ◁

LINEARITY OF EXPECTATIONS

Theorem 3 tells us that expected values are linear. For example, the expected value of the sum of random variables is the sum of their expected values. We will find this property exceedingly useful.

THEOREM 3

If $X_i, i = 1, 2, \ldots, n$ with n a positive integer, are random variables on S, and if a and b are real numbers, then

(*i*) $E(X_1 + X_2 + \cdots + X_n) = E(X_1) + E(X_2) + \cdots + E(X_n)$
(*ii*) $E(aX + b) = aE(X) + b.$

Proof: The first result for $n = 2$ follows directly from the definition of expected value, since

$$E(X_1 + X_2) = \sum_{s \in S} p(s)(X_1(s) + X_2(s))$$

$$= \sum_{s \in S} p(s)X_1(s) + \sum_{s \in S} p(s)X_2(s)$$

$$= E(X_1) + E(X_2).$$

The case with n random variables follows easily using mathematical induction from the case of two random variables. Finally, note that $E(aX + b) = \sum_{s \in S} p(s)(aX(s) + b) = a \sum_{s \in S} p(s)X(s) + b \sum_{s \in S} p(s) = aE(X) + b$ since $\sum_{s \in S} p(s) = 1$. ◁

Examples 4 and 5 illustrate how to use Theorem 3.

EXAMPLE 4

Use Theorem 3 to find the expected value of the sum of the numbers that appear when a pair of fair dice is rolled. (This was done in Example 3 without the benefit of this theorem.)

Solution: Let X_1 and X_2 be the random variables with $X_1((i, j)) = i$ and $X_2((i, j)) = j$, so that X_1 is the number appearing on the first die and X_2 is the number appearing on the second die. It is easy to see that $E(X_1) = E(X_2) = 7/2$ since both equal $(1 + 2 + 3 + 4 + 5 + 6)/6 = 21/6 = 7/2$. The sum of the two numbers that appear when the two dice are rolled is the sum $X_1 + X_2$. By Theorem 3, the expected value of the sum is $E(X_1 + X_2) = E(X_1) + E(X_2) = 7/2 + 7/2 = 7$. ◀

EXAMPLE 5

In the proof of Theorem 2 we found the expected value of the number of successes when n Bernoulli trials are performed, where p is the probability of success on each trial by direct computation. Show how this result can be found using Theorem 3.

Solution: Let X_i be the random variable with $X_i((t_1, t_2, \ldots, t_n)) = 1$ if t_i is a success and $X_i((t_1, t_2, \ldots, t_n)) = 0$ if t_i is a failure. The expected value of X_i is $E(X_i) = 1 \cdot p + 0 \cdot (1 - p) = p$ for $i = 1, 2, \ldots, n$. Let $X = X_1 + X_2 + \cdots + X_n$ so that X counts the number of successes when these n Bernoulli trials are performed. Theorem 3, applied to the sum of n random variables, shows that $E(X) = E(X_1) + E(X_2) + \cdots + E(X_n) = np$. ◀

We can take advantage of the linearity of expectations to find the solutions of many seemingly difficult problems. The key step is to express a random variable whose expectation we wish to find as the sum of random variables whose expectations are easy to find. Examples 6 and 7 illustrate this technique.

EXAMPLE 6   **Expected Value in the Hatcheck Problem** A new employee checks the hats of n people at a restaurant, forgetting to put claim check numbers on the hats. When customers return for their hats, the checker gives them back hats chosen at random from the remaining hats. What is the expected number of hats that are returned correctly?

Solution: Let X be the random variable that equals the number of people who receive the correct hat from the checker. Let X_i be the random variable with $X_i = 1$ if the ith person receives the correct hat and $X_i = 0$ otherwise. It follows that

$$X = X_1 + X_2 + \cdots + X_n.$$

Because it is equally likely that the checker returns any of the hats to this person, it follows that the probability that the ith person receives the correct hat is $1/n$. Consequently, by Theorem 1, for all i we have

$$E(X_i) = 1 \cdot P(X_i = 1) + 0 \cdot P(X_i = 0) = 1 \cdot 1/n + 0 = 1/n.$$

By the linearity of expectations (Theorem 3), it follows that

$$E(X) = E(X_1) + E(X_2) + \cdots + E(X_n) = n \cdot 1/n = 1.$$

Consequently, the average number of people who receive the correct hat is exactly 1. Note that this answer is independent of the number of people who have checked their hats! (We will find an explicit formula for the probability that no one receives the correct hat in Section 6.5.)   ◀

EXAMPLE 7   **Expected Number of Inversions in a Permutation** The ordered pair (i, j) is called an **inversion** in a permutation of the first n positive integers if $i < j$ but j precedes i in the permutation. For instance, there are six inversions in the permutation $3, 5, 1, 4, 2$; these inversions are

$$(1, 3), (1, 5), (2, 3), (2, 4), (2, 5), (4, 5).$$

To find the expected number of inversions in a random permutation of the first n positive integers, we let $I_{i,j}$ be the random variable on the set of all permutations of the first n positive integers with $I_{i,j} = 1$ if (i, j) is an inversion of the permutation and $I_{i,j} = 0$ otherwise. It follows that if X is the random variable equal to the number of inversions in the permutation, then

$$X = \sum_{1 \le i < j \le n} I_{i,j}.$$

Note that it is equally likely for i to precede j in a randomly chosen permutation as it is for j to precede i. (To see this, note that there are an equal number of permutations with each of these properties.) Consequently, for all pairs i and j we have

$$E(I_{i,j}) = 1 \cdot p(I_{i,j} = 1) + 0 \cdot p(I_{i,j} = 0) = 1 \cdot 1/2 + 0 = 1/2.$$

Because there are $\binom{n}{2}$ pairs i and j with $1 \le i < j \le n$ and by the linearity of expectations (Theorem 3), we have

$$E(X) = \sum_{1 \le i < j \le n} I_{i,j} = \binom{n}{2} \cdot \frac{1}{2} = \frac{n(n - 1)}{4}.$$

It follows that there are an average of $n(n-1)/4$ inversions in a permutation of the first n positive integers. ◀

AVERAGE-CASE COMPUTATIONAL COMPLEXITY

Computing the average-case computational complexity of an algorithm can be interpreted as computing the expected value of a random variable. Let the sample space of an experiment be the set of possible inputs a_j, $j = 1, 2, \ldots, n$, and let X be the random variable that assigns to a_j the number of operations used by the algorithm when given a_j as input. Based on our knowledge of the input, we assign a probability $p(a_j)$ to each possible input value a_j. Then, the average-case complexity of the algorithm is

$$E(X) = \sum_{j=1}^{n} p(a_j)X(a_j).$$

This is the expected value of X.

Finding the average-case computational complexity of an algorithm is usually much more difficult than finding its worst-case computational complexity, and often involves the use of sophisticated methods. However, there are some algorithms for which the analysis required to find the average-case computational complexity is not difficult. For instance, in Example 8 we will illustrate how to find the average-case computational complexity of the linear search algorithm under different assumptions concerning the probability that the element for which we search is an element of the list.

EXAMPLE 8 **Average-Case Complexity of the Linear Search Algorithm** We are given an element x and a list of n distinct real numbers. The linear search algorithm, described in Section 2.1, locates x by successively comparing it to each element in the list, terminating when x is located or when all the elements have been examined and it has been determined that x is not in the list. What is the average-case computational complexity of the linear search algorithm if the probability that x is in the list is p and it is equally likely that x is any of the n elements in the list? (There are $n + 1$ possible types of input: the n numbers in the list and a number not in the list, which we treat as a single input.)

Solution: In Example 4 of Section 2.3 we showed that $2i + 1$ comparisons are used if x equals the ith element of the list and, in Example 2 of Section 2.3, that $2n+2$ comparisons are used if x is not in the list. The probability that x equals a_i, the ith element in the list, is p/n, and the probability that x is not in the list is $q = 1 - p$. It follows that the average-case computational complexity of the linear search algorithm is

$$E = 3p/n + 5p/n + \cdots + (2n+1)p/n + (2n+2)q$$

$$= \frac{p}{n}(3 + 5 + \cdots + (2n+1)) + (2n+2)q$$

$$= \frac{p}{n}((n+1)^2 - 1) + (2n+2)q$$

$$= p(n+2) + (2n+2)q.$$

(The third equality follows from Example 21 of Section 3.2.) For instance, when x is guaranteed to be in the list, we have $p = 1$ (so that the probability that $x = a_i$ is $1/n$ for each i) and $q = 0$. Then $E = n + 2$, as we showed in Example 4 in Section 2.3.

When p, the probability that x is in the list, is $1/2$, it follows that $q = 1 - p = 1/2$, so that $E = (n+2)/2 + n + 1 = (3n+4)/2$. Similarly, if the probability that x is in the list is $3/4$, we have $p = 3/4$ and $q = 1/4$, so that $E = 3(n+2)/4 + (n+1)/2 = (5n+8)/4$.

Finally, when x is guaranteed not to be in the list, we have $p = 0$ and $q = 1$. It follows that $E = 2n + 2$, which is not surprising since we have to search the entire list.   ◄

Example 9 illustrates how the linearity of expectations can help us find the average-case complexity of a sorting algorithm, the insertion sort.

EXAMPLE 9   **Average-Case Complexity of the Insertion Sort** What is the average number of comparisons used by the insertion sort to sort n distinct elements?

Solution: We first suppose that X is the random variable equal to the number of comparisons used by the insertion sort (described in Section 2.1) to sort a list $a_1, a_2, \ldots, a_n$ of n distinct elements. Then $E(X)$ is the average number of comparisons used. (Recall that at step i for $i = 2, \ldots, n$, the insertion sort inserts the ith element in the original list into the correct position in the sorted list of the first $i - 1$ elements of the original list.)

We let X_i be the random variable equal to the number of comparisons used to insert a_i into the proper position after the first $i - 1$ elements $a_1, a_2, \ldots, a_{i-1}$ have been sorted. Because

$$X = X_2 + X_3 + \cdots + X_n,$$

we can use the linearity of expectations to conclude that

$$E(X) = E(X_2 + X_3 + \cdots + X_n) = E(X_2) + E(X_3) + \cdots + E(X_n).$$

To find $E(X_i)$ for $i = 2, 3, \ldots, n$, let $p_j(k)$ denote the probability that the largest of the first j elements in the list occurs at the kth position, that is, that $\max(a_1, a_2, \ldots, a_j) = a_k$, where $1 \leq k \leq j$. Because the elements of the list are randomly distributed, it is equally likely for the largest element among the first j elements to occur at any position. Consequently, $p_j(k) = 1/j$. If $X_i(k)$ equals the number of comparisons used by the insertion sort if a_i is inserted into the kth position in the list once $a_1, a_2, \ldots, a_{i-1}$ have been sorted, it follows that $X_i(k) = k$. Because it is possible that a_i is inserted in any of the first i positions, we find that

$$E(X_i) = \sum_{k=1}^{i} p_i(k) \cdot X_i(k) = \sum_{k=1}^{i} \frac{1}{i} \cdot k = \frac{1}{i} \cdot \sum_{k=1}^{i} k = \frac{1}{i} \cdot \frac{i(i+1)}{2} = \frac{i+1}{2}.$$

It follows that

$$E(X) = \sum_{i=2}^{n} E(X_i) = \sum_{i=2}^{n} \frac{i+1}{2} = \frac{1}{2} \sum_{j=3}^{n+1} j$$

$$= \frac{1}{2} \frac{(n+1)(n+2)}{2} - \frac{1}{2}(1+2) = \frac{n^2 + 3n - 4}{4}.$$

To obtain the third of these equalities we shifted the index of summation, setting $j = i + 1$. To obtain the fourth equality, we used the formula $\sum_{k=1}^{m} k = m(m+1)/2$ (from Table 1 in Section 3.2) with $m = n + 1$, subtracting off the missing terms with $j = 1$ and $j = 2$. We conclude that the average number of comparisons used by the insertion sort to sort n elements equals $(n^2 + 3n - 4)/4$, which is $\Theta(n^2)$.   ◄

THE GEOMETRIC DISTRIBUTION

Consider the problem in Example 10.

EXAMPLE 10   Suppose that the probability that a coin comes up tails is p. This coin is flipped repeatedly until it comes up tails. What is the expected number of flips until this coin comes up tails?

Solution: We first note that the sample space consists of all sequences that begin with any number of heads, denoted by H, followed by a tail, denoted by T. Therefore, the sample space is the set $\{T, HT, HHT, HHHT, HHHHT, \ldots\}$. Note that this is an infinite sample space. We can determine the probability of an element of the sample space by noting that the coin flips are independent and that the probability of a head is $1 - p$. Therefore, $p(T) = p$, $p(HT) = (1 - p)p$, $p(HHT) = (1 - p)^2 p$, and in general the probability that the coin is flipped n times before a tail comes up, that is, that $n - 1$ heads come up followed by a tail, is $(1 - p)^{n-1} p$. (Exercise 14 asks for a verification that the sum of the probabilities of the points in the sample space is 1.)

Now let X be the random variable equal to the number of flips in an element in the sample space. That is, $X(T) = 1$, $X(HT) = 2$, $X(HHT) = 3$, and so on. Note that $p(X = j) = (1 - p)^{j-1} p$. The expected number of flips until the coin comes up tails equals $E(X)$.

Using Theorem 1, we find that

$$E(X) = \sum_{j=1}^{\infty} j \cdot p(X = j) = \sum_{j=1}^{\infty} j(1 - p)^{j-1} p = p \sum_{j=1}^{\infty} j(1 - p)^{j-1} = p \cdot \frac{1}{p^2} = \frac{1}{p}.$$

(The third equality in this chain follows from Table 2 in Section 3.2, which tells us that $\sum_{j=1}^{\infty} j(1 - p)^{j-1} = 1/(1 - (1 - p))^2 = 1/p^2$.) It follows that the expected number of times the coin is flipped until tails comes up is $1/p$. Note that when the coin is fair we have $p = 1/2$, so that the expected number of flips until it comes up tails is $1/(1/2) = 2$.   ◀

The random variable X that equals the number of flips expected before a coin comes up tails is an example of a random variable with a **geometric distribution.**

DEFINITION 2   A random variable X has a *geometric distribution with parameter p* if $p(X = k) = (1 - p)^{k-1} p$ for $k = 1, 2, 3, \ldots$.

Geometric distributions arise in many applications since they are used to study the time required before a particular event happens, such as the time required before we find an object with a certain property, the number of attempts before an experiment succeeds, the number of times a product can be used before it fails, and so on.

When we computed the expected value of the number of flips required before a coin comes up tails, we proved Theorem 4.

THEOREM 4   If the random variable X has the geometric distribution with parameter p, then $E(X) = 1/p$.

INDEPENDENT RANDOM VARIABLES

We have already discussed independent events. We will now define what it means for two random variables to be independent.

DEFINITION 3

The random variables X and Y on a sample space S are *independent* if

$$p(X(s) = r_1 \text{ and } Y(s) = r_2) = p(X(s) = r_1) \cdot p(Y(s) = r_2),$$

or in words, if the probability that $X(s) = r_1$ and $Y(s) = r_2$ equals the product of the probabilities that $X(s) = r_1$ and $Y(s) = r_2$, for all real numbers r_1 and r_2.

EXAMPLE 11 Are the random variables X_1 and X_2 from Example 4 independent?

Solution: Let $S = \{1, 2, 3, 4, 5, 6\}$, and let $i \in S$ and $j \in S$. Since there are 36 possible outcomes when the pair of dice is rolled and each is equally likely, we have

$$p(X_1 = i \text{ and } X_2 = j) = 1/36.$$

Furthermore, $p(X_1 = i) = 1/6$ and $p(X_2 = j) = 1/6$, since the probability that i appears on the first die and the probability that j appears on the second die are both $1/6$. It follows that

$$p(X_1 = i \text{ and } X_2 = j) = 1/36 = (1/6)(1/6) = p(X_1 = i)p(X_2 = j),$$

so X_1 and X_2 are independent. ◀

EXAMPLE 12 Show that the random variables X_1 and $X = X_1 + X_2$, where X_1 and X_2 are as defined in Example 4, are not independent.

Solution: Note that $p(X_1 = 1 \text{ and } X = 12) = 0$, since $X_1 = 1$ means the number appearing on the first die is 1, which implies that the sum of the numbers appearing on the two dice cannot equal 12. On the other hand, $p(X_1 = 1) = 1/6$ and $p(X = 12) = 1/36$. Hence $p(X_1 = 1 \text{ and } X = 12) \neq p(X_1 = 1) \cdot p(X = 12)$. This counterexample shows that X_1 and X are not independent. ◀

The expected value of the product of two independent random variables is the product of their expected values, as Theorem 5 shows.

THEOREM 5

If X and Y are independent random variables on a space S, then $E(XY) = E(X)E(Y)$.

Proof: From the definition of expected value and since X and Y are independent random variables, it follows that

$$E(XY) = \sum_{s \in S} X(s)Y(s)p(s)$$

$$= \sum_{r_1 \in X(S), r_2 \in Y(S)} r_1 r_2 \cdot p(X(s) = r_1 \text{ and } Y(s) = r_2)$$

$$= \sum_{r_1 \in X(S), r_2 \in Y(S)} r_1 r_2 \cdot p(X(s) = r_1) \cdot p(Y(s) = r_2)$$

$$= \left(\sum_{r_1 \in X(S)} r_1 p(X(s) = r_1) \right) \cdot \left(\sum_{r_2 \in Y(S)} r_2 p(Y(s) = r_2) \right)$$

$$= E(X)E(Y).$$

This completes the proof. ◁

Note that when X and Y are random variables that are not independent, we cannot conclude that $E(XY) = E(X)E(Y)$, as Example 13 shows.

EXAMPLE 13   Let X and Y be random variables that count the number of heads and the number of tails when a coin is flipped twice. Since $p(X = 2) = 1/4$, $p(X = 1) = 1/2$, and $p(X = 0) = 1/4$, by Theorem 1 we have

$$E(X) = 2 \cdot \frac{1}{4} + 1 \cdot \frac{1}{2} + 0 \cdot \frac{1}{4} = 1.$$

A similar computation shows that $E(Y) = 1$. We note that $XY = 0$ when either two heads and no tails or two tails and no heads come up and that $XY = 1$ when one head and one tail come up. Hence

$$E(XY) = 1 \cdot \frac{1}{2} + 0 \cdot \frac{1}{2} = \frac{1}{2}.$$

It follows that

$$E(XY) \neq E(X)E(Y).$$

This does not contradict Theorem 5 since X and Y are not independent, as the reader should verify (see Exercise 16). ◀

VARIANCE

The expected value of a random variable tells us its average value, but nothing about how widely its values are distributed. For example, if X and Y are the random variables on the set $S = \{1, 2, 3, 4, 5, 6\}$, with $X(s) = 0$ for all $s \in S$ and $Y(s) = -1$ if $s \in \{1, 2, 3\}$ and $Y(s) = 1$ if $s \in \{4, 5, 6\}$, then the expected values of X and Y are both zero. However, the random variable X never varies from 0, while the random variable Y always differs from 0 by 1. The variance of a random variable helps us characterize how widely a random variable is distributed.

DEFINITION 4   Let X be a random variable on a sample space S. The *variance* of X, denoted by $V(X)$, is

$$V(X) = \sum_{s \in S} (X(s) - E(X))^2 p(s).$$

The *standard deviation* of X, denoted $\sigma(X)$, is defined to be $\sqrt{V(X)}$.

Theorem 6 provides a useful simple expression for the variance of a random variable.

THEOREM 6   If X is a random variable on a sample space S, then $V(X) = E(X^2) - E(X)^2$.

Proof: Note that

$$V(X) = \sum_{s \in S} (X(s) - E(X))^2 p(s)$$

$$= \sum_{s \in S} X(s)^2 p(s) - 2E(X) \sum_{s \in S} X(s) p(s) + E(X)^2 \sum_{s \in S} p(s)$$

$$= E(X^2) - 2E(X)E(X) + E(X)^2$$

$$= E(X^2) - E(X)^2.$$

We have used the fact that $\sum_{s \in S} p(s) = 1$ in the next-to-last step.   ◁

EXAMPLE 14   What is the variance of the random variable X with $X(t) = 1$ if a Bernoulli trial is a success and $X(t) = 0$ if it is a failure, where p is the probability of success?

Solution: Since X takes only the values 0 and 1, it follows that $X^2(t) = X(t)$. Hence,

$$V(X) = E(X^2) - E(X)^2 = p - p^2 = p(1 - p) = pq.$$ ◀

EXAMPLE 15   **Variance of the Value of a Die** What is the variance of the random variable X, where X is the number that comes up when a die is rolled?

Solution: We have $V(X) = E(X^2) - E(X)^2$. By Example 1 we know that $E(X) = 7/2$. To find $E(X^2)$ note that X^2 takes the values $i^2, i = 1, 2, \ldots, 6$, each with probability $1/6$. It follows that

$$E(X^2) = \frac{1}{6}(1^2 + 2^2 + 3^2 + 4^2 + 5^2 + 6^2) = \frac{91}{6}.$$

We conclude that

$$V(X) = \frac{91}{6} - \left(\frac{7}{2}\right)^2 = \frac{35}{12}.$$ ◀

EXAMPLE 16   What is the variance of the random variable $X((i, j)) = 2i$, where i is the number appearing on the first die and j is the number appearing on the second die, when two dice are rolled?

Solution: We will use Theorem 6 to find the variance of X. To do so, we need to find the expected values of X and X^2. Note that since $p(X = k)$ is $1/6$ for $k = 2, 4, 6, 8, 10, 12$ and is 0 otherwise,

$$E(X) = (2 + 4 + 6 + 8 + 10 + 12)/6 = 7,$$

and

$$E(X^2) = (2^2 + 4^2 + 6^2 + 8^2 + 10^2 + 12^2)/6 = 182/3.$$

It follows from Theorem 6 that

$$V(X) = E(X^2) - E(X)^2 = 182/3 - 49 = 35/3.$$ ◀

Another useful fact about variances is that the variance of the sum of two independent random variables is the sum of their variances. This result is useful for computing the variance of the result of n independent Bernoulli trials, for instance.

THEOREM 7

> If X and Y are two independent random variables on a sample space S, then $V(X + Y) = V(X) + V(Y)$. Furthermore, if X_i, $i = 1, 2, \ldots, n$, with n a positive integer, are pairwise independent random variables on S, then $V(X_1 + X_2 + \cdots + X_n) = V(X_1) + V(X_2) + \cdots + V(X_n)$.

Proof: From Theorem 6, we have

$$V(X + Y) = E((X + Y)^2) - E(X + Y)^2.$$

It follows that

$$V(X + Y) = E(X^2 + 2XY + Y^2) - (E(X) + E(Y))^2$$
$$= E(X^2) + 2E(XY) + E(Y^2) - E(X)^2 - 2E(X)E(Y) - E(Y)^2.$$

Since X and Y are independent, by Theorem 5 we have $E(XY) = E(X)E(Y)$. It follows that

$$V(X + Y) = (E(X^2) - E(X)^2) + (E(Y^2) - E(Y)^2)$$
$$= V(X) + V(Y).$$

The case with n pairwise independent random variables can be proved using mathematical induction; the proof is left to the reader. ◁

EXAMPLE 17

Find the variance and standard deviation of the random variable X whose value when two dice are rolled is $X((i, j)) = i + j$, where i is the number appearing on the first die and j is the number appearing on the second die.

Solution: Let X_1 and X_2 be the random variables defined by $X_1((i, j)) = i$ and $X_2((i, j)) = j$ for a roll of the dice. Then $X = X_1 + X_2$ and X_1 and X_2 are independent, as Example 11 showed. From Theorem 7 it follows that $V(X) = V(X_1) + V(X_2)$. A simple computation as in Example 16 together with Exercise 27 in the Supplementary Exercises at the end of the chapter tell us that $V(X_1) = V(X_2) = 35/12$. Hence, $V(X) = 35/12 + 35/12 = 35/6$ and $\sigma(X) = \sqrt{35/6}$. ◀

We will now find the variance of the random variable that counts the number of successes when n independent Bernoulli trials are carried out.

EXAMPLE 18

What is the variance of the number of successes when n independent Bernoulli trials are performed, where p is the probability of success on each trial?

Solution: Let X_i be the random variable with $X_i((t_1, t_2, \ldots, t_n)) = 1$ if t_i is a success and $X_i((t_1, t_2, \ldots, t_n)) = 0$ if t_i is a failure. Let $X = X_1 + X_2 + \cdots + X_n$. Then X counts the number of successes in the n trials. From Theorem 7 it follows that $V(X) = V(X_1) + V(X_2) + \cdots + V(X_n)$. Using Example 14 we have that $V(X_i) = pq$ for $i = 1, 2, \ldots, n$. It follows that $V(X) = npq$. ◀

CHEBYSHEV'S INEQUALITY

How likely is it that a random variable takes a value far from its expected value? The following theorem, called Chebyshev's Inequality, helps answer this question by providing an upper bound on the probability that the value of a random variable differs from the expected value of the random variable by more than a specified amount.

THEOREM 8 **CHEBYSHEV'S INEQUALITY**   Let X be a random variable on a sample space S with probability function p. If r is a positive real number, then

$$p(|X(s) - E(X)| \geq r) \leq V(X)/r^2.$$

Proof: Let A be the event

$$A = \{s \in S \,|\, |X(s) - E(X)| \geq r\}.$$

What we want to prove is that $p(A) \leq V(X)/r^2$. Note that

$$V(X) = \sum_{s \in S} (X(s) - E(X))^2 p(s)$$

$$= \sum_{s \in A} (X(s) - E(X))^2 p(s) + \sum_{s \notin A} (X(s) - E(X))^2 p(s).$$

The second sum in this expression is nonnegative, since each of its summands is nonnegative. Also, since for each element s in A, $(X(s) - E(X))^2 \geq r^2$, the first sum in this expression is at least $\sum_{s \in A} r^2 p(s)$. Hence, $V(x) \geq \sum_{s \in A} r^2 p(s) = r^2 p(A)$. This is what we needed to prove. $\triangleleft$

EXAMPLE 19 **Deviation from the Mean when Counting Tails**   Suppose that X is the random variable that counts the number of tails when a fair coin is tossed n times. Note that X is the number of successes when n independent Bernoulli trials, each with probability of success $1/2$, are performed. It follows that $E(X) = n/2$ (by Theorem 2) and $V(X) = n/4$ (by Example 18). Applying Chebyshev's Theorem with $r = \sqrt{n}$ shows that

$$p(|X(s) - n/2| \geq \sqrt{n}) \leq (n/4)/(\sqrt{n})^2 = 1/4.$$

Consequently, the probability is no more than $1/4$ that the number of tails that come up when a fair coin is tossed n times deviates from the mean by more than $\sqrt{n}$.   ◀

Chebyshev's Inequality, although applicable to any random variable, often fails to provide a practical estimate for the probability that the value of a random variable exceeds its mean by a large amount. This is illustrated by the following example.

Links

PAFNUTY LVOVICH CHEBYSHEV (1821–1894)   Chebyshev was born into the gentry in Okatovo, Russia. His father was a retired army officer who fought against Napoleon. In 1832 the family, with its nine children, moved to Moscow, where Pafnuty completed his high school education at home. He entered the Department of Physics and Mathematics at Moscow University. As a student, he developed a new method for approximating the roots of equations. He graduated from Moscow University in 1841 with a degree in mathematics, and he continued his studies, passing his master's exam in 1843 and completing his master's thesis in 1846.

Chebyshev was appointed in 1847 to a position as an assistant at the University of St. Petersburg. He wrote and defended a thesis in 1847. He became a professor at St. Petersburg in 1860, a position he held until 1882. His book on the theory of congruences written in 1849 was influential in the development of number theory. His work on the distribution of prime numbers was seminal. He proved Bertrand's conjecture that for every integer $n > 3$, there is a prime between n and $2n - 2$. Chebyshev helped develop ideas that were later used to prove the Prime Number Theorem. Chebyshev's work on the approximation of functions using polynomials is used extensively when computers are used to find values of functions. Chebyshev was also interested in mechanics. He studied the conversion of rotary motion into rectilinear motion by mechanical coupling. The Chebyshev parallel motion is three linked bars approximating rectilinear motion.

EXAMPLE 20 Let X be the random variable whose value is the number appearing when a fair die is rolled. We have $E(X) = 7/2$ (see Example 1) and $V(X) = 35/12$ (see Example 15). Since the only possible values of X are $1, 2, 3, 4, 5$, and 6, X cannot take a value more than $5/2$ from its mean, $E(X) = 7/2$. Hence, $p(|X - 7/2| \geq r) = 0$ if $r > 5/2$. By Chebyshev's Inequality we know that $p(|X - 7/2| \geq r) \leq (35/12)/r^2$. For example, when $r = 3$, Chebyshev's Inequality tells us that $p(|X - 7/2| \geq 3) \leq (35/12)/9 = 35/108$, which is a poor estimate, since $p(|X - 7/2| \geq 3) = 0$. ◀

Exercises

1. What is the expected number of heads that come up when a fair coin is flipped five times?

2. What is the expected number of heads that come up when a fair coin is flipped ten times?

3. What is the expected number of times a 6 appears when a fair die is rolled ten times?

4. A coin is biased so that the probability a head comes up when it is flipped is 0.6. What is the expected number of heads that come up when it is flipped ten times?

5. What is the expected sum of the numbers that appear on two dice, each biased so that a 3 comes up twice as often as each other number?

6. What is the expected value when a $1 lottery ticket is bought in which the purchaser wins exactly $10 million if the ticket contains the six winning numbers chosen from the set $\{1, 2, 3, \ldots, 50\}$ and the purchaser wins nothing otherwise?

7. The final exam of a discrete mathematics course consists of 50 true/false questions, each worth two points, and 25 multiple-choice questions, each worth four points. The probability that Linda answers a true/false question correctly is 0.9, and the probability that she answers a multiple-choice question correctly is 0.8. What is her expected score on the final?

8. What is the expected sum of the numbers that appear when three fair dice are rolled?

9. Suppose that the probability that x is in a list of n distinct integers is $2/3$ and that it is equally likely that x equals any element in the list. Find the average number of comparisons used by the linear search algorithm to find x or to determine that it is not in the list.

10. Suppose that we flip a coin until either it comes up tails twice or we have flipped it six times. What is the expected number of times we flip the coin?

11. Suppose that we roll a die until a 6 comes up or we have rolled it ten times. What is the expected number of times we roll the die?

12. Suppose that we roll a die until a 6 comes up.

   a) What is the probability that we roll the die n times?

   b) What is the expected number of times we roll the die?

13. Suppose that we roll a pair of dice until the sum of the numbers on the dice is seven. What is the expected number of times we roll the dice?

14. Show that the sum of the probabilities of a random variable with geometric distribution with parameter p, where $0 < p \leq 1$, equals 1.

15. Show that if the random variable X has the geometric distribution with parameter p, and j is a positive integer, then $p(X \geq j) = (1 - p)^{j-1}$.

16. Let X and Y be the random variables that count the number of heads and the number of tails that come up when two coins are flipped. Show that X and Y are not independent.

17. Estimate the expected number of integers with 1000 digits that need to be selected at random to find a prime, if the probability a number with 1000 digits is prime is approximately $1/2302$.

18. Suppose that X and Y are random variables and that X and Y are nonnegative for all points in a sample space S. Let Z be the random variable defined by $Z(s) = \max(X(s), Y(s))$ for all elements $s \in S$. Show that $E(Z) \leq E(X) + E(Y)$.

19. Let X be the number appearing on the first die when two dice are rolled and let Y be the sum of the numbers appearing on the two dice. Show that $E(X)E(Y) \neq E(XY)$.

20. Let A be an event. Then I_A, the **indicator random variable** of A, equals 1 if A occurs and equals 0 otherwise. Show that the expectation of the indicator random variable of A equals the probability of A, that is, $E(I_A) = p(A)$.

21. A **run** is a maximal sequence of successes in a sequence of Bernoulli trials. For example, in the sequence $S, S, S, F, S, S, F, F, S$, where S represents success and F represents failure, there are three runs consisting of three successes, two successes, and one success, respectively. Let R denote the random variable on the set of sequences of n independent Bernoulli trials that counts the number of runs in this

sequence. Find $E(R)$. [*Hint:* Show that $R = \sum_{j=1}^{n} I_j$, where $I_j = 1$ if a run begins at the jth Bernoulli trial and $I_j = 0$ otherwise. Find $E(I_1)$ and then find $E(I_j)$, where $1 < j \leq n$.]

22. Let $X(s)$ be a random variable, where $X(s)$ is a non-negative integer for all $s \in S$, and let A_k be the event that $X(s) \geq k$. Show that $E(X) = \sum_{k=1}^{\infty} p(A_k)$.

23. What is the variance of the number of heads that come up when a fair coin is flipped 10 times?

24. What is the variance of the number of times a 6 appears when a fair die is rolled 10 times?

25. Let X_n be the random variable that counts the difference in the number of tails and the number of heads when n coins are flipped.

   a) What is the expected value of X_n?
   b) What is the variance of X_n?

26. Provide an example that shows that the variance of the sum of two random variables is not necessarily equal to the sum of their variances when the random variables are not independent.

27. Use Chebyshev's Inequality to find an upper bound on the probability that the number of tails that come up when a fair coin is tossed n times deviates from the mean by more than $5\sqrt{n}$.

28. Use Chebyshev's Inequality to find an upper bound on the probability that the number of tails that come up when a biased coin with probability of heads equal to 0.6 is tossed n times deviates from the mean by more than $\sqrt{n}$.

29. Let X be a random variable on a sample space S such that $X(s) \geq 0$ for all $s \in S$. Show that $p(X(s) \geq a) \leq E(X)/a$ for every positive real number a. This inequality is called **Markov's Inequality.**

30. Suppose that the number of cans of soda pop filled in a day at a bottling plant is a random variable with an expected value of 10,000 and a variance of 1000.

   a) Use Markov's Inequality (Exercise 29) to obtain an upper bound on the probability that the plant will fill more than 11,000 cans on a particular day.
   b) Use Chebyshev's Inequality to obtain a lower bound on the probability that the plant will fill between 9000 and 11,000 cans on a particular day.

31. Suppose that the number of tin cans recycled in a day at a recycling center is a random variable with an expected value of 50,000 and a variance of 2500.

   a) Use Markov's Inequality (Exercise 29) to find an upper bound on the probability that the center will recycle more than 55,000 cans on a particular day.
   b) Use Chebyshev's Inequality to provide a lower bound on the probability that the center will recycle 40,000 to 60,000 cans on a certain day.

*32. Suppose the probability that x is the ith element in a list of n distinct integers is $i/[n(n+1)]$. Find the aver-

age number of comparisons used by the linear search algorithm to find x or to determine that it is not in the list.

*33. In this exercise we derive an estimate of the average-case complexity of the variant of the bubble sort algorithm that terminates once a pass has been made with no interchanges. Let X be the random variable on the set of permutations of a set of n distinct integers $\{a_1, a_2, \ldots, a_n\}$ with $a_1 < a_2 < \cdots < a_n$ such that $X(P)$ equals the number of comparisons used by the bubble sort to put these integers into increasing order.

   a) Show that, under the assumption that the input is equally likely to be any of the $n!$ permutations of these integers, the average number of comparisons used by the bubble sort equals $E(X)$.
   b) Use Example 5 in Section 2.3 to show that $E(X) \leq n(n-1)/2$.
   c) Show that the sort makes at least one comparison for every inversion of two integers in the input.
   d) Let $I(P)$ be the random variable that equals the number of inversions in the permutation P. Show that $E(X) \geq E(I)$.
   e) Let $I_{j,k}$ be the random variable with $I_{j,k}(P) = 1$ if a_k precedes a_j in P and $I_{j,k} = 0$ otherwise. Show that $I(P) = \sum_k \sum_{j<k} I_{j,k}(P)$.
   f) Show that $E(I) = \sum_k \sum_{j<k} E(I_{j,k})$.
   g) Show that $E(I_{j,k}) = 1/2$. [*Hint:* Show that $E(I_{j,k}) =$ probability that a_k precedes a_j in a permutation P. Then show it is equally likely for a_k to precede a_j as it is for a_j to precede a_k in a permutation.]
   h) Use parts (f) and (g) to show that $E(I) = n(n-1)/4$.
   i) Conclude from parts (a), (b), and (h) that the average number of comparisons used to sort n integers is $\Theta(n^2)$.

*34. In this exercise we find the average-case complexity of the quick sort algorithm, described in the preamble to Exercise 38 in Section 3.5, assuming a uniform distribution on the set of permutations.

   a) Let X be the number of comparisons used by the quick sort algorithm to sort a list of n distinct integers. Show that the average number of comparisons used by the quick sort algorithm is $E(X)$ (where the sample space is the set of all $n!$ permutations of n integers).
   b) Let $I_{j,k}$ denote the random variable that equals 1 if the jth smallest element and the kth smallest element of the initial list are ever compared as the quick sort algorithm sorts the list and equals 0 otherwise. Show that $X = \sum_{k=2}^{n} \sum_{j=1}^{k-1} I_{j,k}$.
   c) Show that $E(X) = \sum_{k=2}^{n} \sum_{j=1}^{k-1} p$ (the jth smallest element and the kth smallest element are compared).

d) Show that p (the jth smallest element and the kth smallest element are compared) where $k > j$ equals $2/(k - j + 1)$.

e) Use parts (c) and (d) to show that $E(X) = 2(n + 1)(\sum_{i=2}^{n} 1/i) - 2(n - 1)$.

f) Conclude from part (e) and the fact that $\sum_{j=1}^{n} 1/j \approx \ln n + \gamma$, where $\gamma = 0.57721\ldots$ is Euler's constant, that the average number of comparisons used by the quick sort algorithm is $\Theta(n \log n)$.

*35. What is the variance of the number of **fixed elements,** that is, elements left in the same position, of a randomly selected permutation of n elements? (*Hint:* Let X denote the number of fixed points of a random permutation. Write $X = X_1 + X_2 + \cdots + X_n$, where $X_i = 1$ if the permutation fixes the ith element and $X_i = 0$ otherwise.)

The **covariance** of two random variables X and Y on a sample space S, denoted by $\text{Cov}(X, Y)$, is defined to be the expected value of the random variable $(X - E(X))(Y - E(Y))$. That is, $\text{Cov}(X, Y) = E((X - E(X))(Y - E(Y)))$.

36. Show that $\text{Cov}(X, Y) = E(XY) - E(X)E(Y)$, and use this result to conclude that $\text{Cov}(X, Y) = 0$ if X and Y are independent random variables.

37. Show that $V(X + Y) = V(X) + V(Y) + 2\,\text{Cov}(X, Y)$.

38. Find $\text{Cov}(X, Y)$ if X and Y are the random variables with $X((i, j)) = 2i$ and $Y((i, j)) = i + j$, where i and j are the numbers that appear on the first and second of two dice when they are rolled.

39. When m balls are distributed into n bins uniformly at random, what is the probability that the first bin remains empty?

40. What is the expected number of balls that fall into the first bin when m balls are distributed into n bins uniformly at random?

41. What is the expected number of bins that remain empty when m balls are distributed into n bins uniformly at random?

Key Terms and Results

TERMS

probability of an event: the number of successful outcomes of this event divided by the number of possible outcomes

$p(E \mid F)$ (conditional probability of E given F): the ratio $p(E \cap F)/p(F)$

independent events: events E and F such that $p(E \cap F) = p(E)p(F)$

random variable: a function that assigns a real number to each outcome of an experiment

distribution of a random variable X: the set of pairs $(r, p(X = r))$ for $r \in X(S)$

uniform distribution: the assignment of equal probabilities to the elements of a finite set

expected value of a random variable: the weighted average of a random variable, with values of the random variable weighted by the probability of outcomes, that is, $E(X) = \sum_{s \in S} p(s)X(s)$

geometric distribution: the distribution of a random variable X such that $p(X = k) = (1 - p)^{k-1}p$ for $k = 1, 2, \ldots$ for some p

independent random variables: random variables X and Y such that $p(X(s) = r_1$ and $Y(s) = r_2) = p(X(s) = r_1)p(Y(s) = r_2)$ for all real numbers r_1 and r_2

variance of a random variable: the weighted average of the square of the difference between the value of the random variable and its expected value, with weights given by the probability of outcomes, that is, $V(X) = \sum_{s \in S}(X(s) - E(X))^2 p(s)$

Bernoulli trial: an experiment with two possible outcomes

probabilistic algorithm: an algorithm in which random choices are made at one or more steps

probabilistic method: a technique for proving results about objects in a set with certain properties that proceeds by assigning probabilities to objects and showing that the probability that an object has these properties is positive

RESULTS

The probability of k successes when n independent Bernoulli trials are carried out equals $C(n, k)p^k q^{n-k}$, where p is the probability of success and $q = 1 - p$ is the probability of failure.

$E(X) = \sum_{r \in X(S)} p(X = r)r$.

linearity of expectations: $E(X_1 + X_2 + \cdots + X_n) = E(X_1) + E(X_2) + \cdots + E(X_n)$ if $X_1, X_2, \ldots, X_n$ are random variables

If X and Y are independent random variables, then $E(XY) = E(X)E(Y)$.

If $X_1, X_2, \ldots, X_n$ are independent random variables, then $V(X_1 + X_2 + \cdots + X_n) = V(X_1) + V(X_2) + \cdots + V(X_n)$.

Chebyshev's Inequality: $p(|X(s) - E(X)| \geq r) \leq V(X)/r^2$, where X is a random variable with probability function p and r is a positive real number.

Review Questions

1. a) Define the probability of an event when all outcomes are equally likely.

b) What is the probability that you select the six winning numbers in a lottery if the six different winning numbers are selected from the first 50 positive integers?

2. a) What conditions should be met by the probabilities assigned to the outcomes from a finite sample space?

b) What probabilities should be assigned to the outcome of heads and the outcome of tails if heads comes up three times as often as tails?

3. a) Define the conditional probability of an event E given an event F.

b) Suppose E is the event that when a die is rolled it comes up an even number, and F is the event that when a die is rolled it comes up 1, 2, or 3. What is the probability of F given E?

4. a) When are two events E and F independent?

b) Suppose E is the event that an even number appears when a fair die is rolled, and F is the event that a 5 or 6 comes up. Are E and F independent?

5. a) What is a random variable?

b) What are the values assigned by the random variable X that assigns to a roll of two dice the larger number that appears on the two dice?

6. a) Define the expected value of a random variable X.

b) What is the expected value of the random variable X that assigns to a roll of two dice the larger number that appears on the two dice?

7. a) Explain how the average-case computational complexity of an algorithm, with finitely many possible input values, can be interpreted as an expected value.

b) What is the average-case computational complexity of the linear search algorithm, if the probability that the element for which we search is in the list is 1/3, and it is equally likely that this element is any of the n elements in the list?

8. a) What is meant by a Bernoulli trial?

b) What is the probability of k successes in n independent Bernoulli trials?

c) What is the expected value of the number of successes in n independent Bernoulli trials?

9. a) What does the linearity of expectations of random variables mean?

b) How can the linearity of expectations help us find the expected number of people who receive the correct hat when a hatcheck person returns hats at random?

10. a) How can probability be used to solve a decision problem, if a small probability of error is acceptable?

b) How can we quickly determine whether a positive integer is prime, if we are willing to accept a small probability of making an error?

11. a) What does it mean to say that a random variable has a geometric distribution with probability p?

b) What is the mean of a geometric distribution with probability p?

12. a) What is the variance of a random variable?

b) What is the variance of a Bernoulli trial with probability p of success?

13. a) What is the variance of the sum of n independent random variables?

b) What is the variance of the number of successes when n independent Bernoulli trials, each with probability p of success, are carried out?

14. What does Chebyshev's Inequality tell us about the probability that a random variable deviates from its mean by more than a specified amount?

Supplementary Exercises

1. What is the probability that six consecutive numbers will be chosen as the winning numbers in a lottery where each number chosen is between 1 and 40 (inclusive)?

2. What is the probability that a hand of 13 cards contains no pairs?

3. What is the probability that a 13-card bridge hand contains

a) all 13 hearts?

b) 13 cards of the same suit?

c) seven spades and six clubs?

d) seven cards of one suit and six cards of a second suit?

e) four diamonds, six hearts, two spades, and one club?

f) four cards of one suit, six cards of a second suit, two cards of a third suit, and one card of the fourth suit?

4. What is the probability that a seven-card poker hand contains

    a) four cards of one kind and three cards of a second kind?

    b) three cards of one kind and pairs of each of two different kinds?

    c) pairs of each of three different kinds and a single card of a fourth kind?

    d) pairs of each of two different kinds and three cards of a third, fourth, and fifth kind?

    e) cards of seven different kinds?

    f) a seven-card flush?

    g) a seven-card straight?

    h) a seven-card straight flush?

An **octahedral die** has eight faces that are numbered 1 through 8.

5. a) What is the expected value of the number that comes up when a fair octahedral die is rolled?

    b) What is the variance of the number that comes up when a fair octahedral die is rolled?

A **dodecahedral die** has 12 faces that are numbered 1 through 12.

6. a) What is the expected value of the number that comes up when a fair dodecahedral die is rolled?

    b) What is the variance of the number that comes up when a fair dodecahedral die is rolled?

7. Suppose that a pair of fair octahedral dice is rolled.

    a) What is the expected value of the sum of the numbers that come up?

    b) What is the variance of the sum of the numbers that come up?

8. Suppose that a pair of fair dodecahedral dice is rolled.

    a) What is the expected value of the sum of the numbers that come up?

    b) What is the variance of the sum of the numbers that come up?

9. Suppose that a fair standard (cubic) die and a fair octahedral die are rolled together.

    a) What is the expected value of the sum of the numbers that come up?

    b) What is the variance of the sum of the numbers that come up?

10. Suppose that a fair octahedral die and a fair dodecahedral die are rolled together.

    a) What is the expected value of the sum of the numbers that come up?

    b) What is the variance of the sum of the numbers that come up?

11. Suppose n people, $n \geq 3$, play "odd person out" to decide who will buy the next round of refreshments. The n people each flip a fair coin simultaneously. If all the coins but one come up the same, the person whose coin comes up different buys the refreshments. Other-

wise, the people flip the coins again and continue until just one coin comes up different from all the others.

    a) What is the probability that the odd person out is decided in just one coin flip?

    b) What is the probability that the odd person out is decided with the kth flip?

    c) What is the expected number of flips needed to decide odd person out with n people?

12. Suppose that p and q are primes and $n = pq$. What is the probability that a randomly chosen positive integer less than n is not divisible by either p or q?

*13. Suppose that m and n are positive integers. What is the probability that a randomly chosen positive integer less than mn is not divisible by either m or n?

14. Suppose that $E_1, E_2, \ldots, E_n$ are n events with $p(E_i) > 0$ for $i = 1, 2, \ldots, n$. Show that

$$p(E_1 \cap E_2 \cap \cdots \cap E_n)$$
$$= p(E_1)p(E_2 \mid E_1)p(E_3 \mid E_1 \cap E_2)$$
$$\cdots p(E_n \mid E_1 \cap E_2 \cap \cdots \cap E_{n-1}).$$

15. There are three cards in a box. Both sides of one card are black, both sides of one card are red, and the third card has one black side and one red side. We pick a card at random and observe only one side.

    a) If the side is black, what is the probability that the other side is also black?

    b) What is the probability that the opposite side is the same color as the one we observed?

16. What is the probability that when a fair coin is flipped n times an equal number of heads and tails appear?

17. What is the probability that a randomly selected bit string of length 10 is a palindrome?

18. What is the probability that a randomly selected bit string of length 11 is a palindrome?

19. Consider the following game. A person flips a coin repeatedly until a head comes up. This person receives a payment of 2^n dollars if the first head comes up at the nth flip.

    a) Let X be a random variable equal to the amount of money the person wins. Show that the expected value of X does not exist (that is, it is infinite). Show that a person should be willing to wager any amount of money to play this game. (This is known as the **St. Petersburg Paradox.** Why do you suppose it is called a paradox?)

    b) Suppose that the person receives 2^n dollars if the first head comes up before the eighth flip and $2^8 = 256$ dollars if the first head comes up on or after the eighth flip. What is the expected value of the amount of money the person wins? How much money should a person be willing to pay to play this game?

20. Suppose that n balls are tossed into b bins so that each ball is equally likely to fall into any of the bins and that the tosses are independent.

a) Find the probability that a particular ball lands in a specified bin.

b) What is the expected number of balls that land in a particular bin?

c) What is the expected number of balls tossed until a particular bin contains a ball?

* **d)** What is the expected number of balls tossed until all bins contain a ball? [*Hint:* Let X_i denote the number of tosses required to have a ball land in an ith bin once $i - 1$ bins contain a ball. Find $E(X_i)$ and use the linearity of expectations.]

21. Suppose that A and B are events with probabilities $p(A) = 3/4$ and $p(B) = 1/3$.

a) What is the largest $p(A \cap B)$ can be? What is the smallest it can be? Give examples to show that both extremes for $p(A \cap B)$ are possible.

b) What is the largest $p(A \cup B)$ can be? What is the smallest it can be? Give examples to show that both extremes for $p(A \cup B)$ are possible.

22. Suppose that A and B are events with probabilities $p(A) = 2/3$ and $p(B) = 1/2$.

a) What is the largest $p(A \cap B)$ can be? What is the smallest it can be? Give examples to show that both extremes for $p(A \cap B)$ are possible.

b) What is the largest $p(A \cup B)$ can be? What is the smallest it can be? Give examples to show that both extremes for $p(A \cup B)$ are possible.

23. We say that the events $E_1, E_2, \ldots, E_n$ are **mutually independent** if $p(E_{i_1} \cap E_{i_2} \cap \cdots \cap E_{i_m}) = p(E_{i_1})p(E_{i_2}) \cdots p(E_{i_m})$ whenever i_j, $j = 1, 2, \ldots, m$, are integers with $1 \le i_1 < i_2 < \cdots < i_m \le n$ and $m \ge 2$.

a) Write out the conditions required for three events E_1, E_2, E_3 to be mutually independent.

b) Let E_1, E_2, and E_3 be the events that the first flip comes up heads, that the second flip comes up tails, and that the third flip comes up tails, respectively, when a fair coin is flipped three times. Are E_1, E_2, and E_3 mutually independent?

c) Let E_1, E_2, and E_3 be the events that the first flip comes up heads, that the third flip comes up heads, and that an even number of heads come up, respectively, when a fair coin is flipped three times. Are E_1, E_2, and E_3 mutually independent?

d) How many conditions must be checked to show that n events are mutually independent?

*24. Suppose that E and F are events with $p(F) \neq 0$. Show that the probability of E is the weighted average of the probability of E given F and the probability of E given the complement of F, $\overline{F}$, where the weights are the probabilities of F and $\overline{F}$, respectively. That is,

$$p(E) = p(E \mid F)p(F) + p(E \mid \overline{F})p(\overline{F}).$$

[*Hint:* Use the fact that $E = (E \cap F) \cup (E \cap \overline{F})$.]

*25. Suppose that E is an event from a sample space S and that $F_1, F_2, \ldots, F_n$ are mutually exclusive events such that $\bigcup_{i=1}^{n} F_i = S$. Assume that $p(E) \neq 0$ and $p(F_i) \neq 0$ for $i = 1, 2, \ldots, n$. Show that

$$p(F_j \mid E) = \frac{p(E \mid F_j)p(F_j)}{\sum_{i=1}^{n} p(E \mid F_i)p(F_i)}.$$

[*Hint:* Use the fact that $E = \bigcup_{i=1}^{n}(E \cap F_i)$.] This result is known as **Bayes' formula**, since it was developed by the English philosopher Thomas Bayes.

*26. A space probe near Neptune communicates with Earth using bit strings. Suppose that in its transmissions it sends a 1 one-third of the time and a 0 two-thirds of the time. When a 0 is sent, the probability it is received correctly is 0.9, and the probability it is received incorrectly (as a 1) is 0.1. When a 1 is sent, the probability it is received correctly is 0.8, and the probability it is received incorrectly (as a 0) is 0.2.

a) Use Exercise 24 to find the probability that a 0 is received.

b) Use Bayes' formula, given in Exercise 25, to find the probability that a 0 was transmitted, given that a 0 was received.

27. Let X be a random variable on a sample space S. Show that $V(aX + b) = a^2 V(X)$ whenever a and b are real numbers.

28. Use Chebyshev's Inequality to show that the probability that more than 10 people get the correct hat back when a hatcheck person returns hats at random does not exceed $1/100$ no matter how many people check their hats. (*Hint:* Use Example 6 and Exercise 35 in Section 5.3.)

29. Suppose that at least one of the events E_j, $j = 1, 2, \ldots, m$, is guaranteed to occur and no more than two can occur. Show that if $p(E_j) = q$ for $j = 1, 2, \ldots, m$ and $P(E_j \cap E_k) = r$ for $1 \le j < k \le m$, then $q \ge 1/m$ and $r \le 2/m$.

30. Show that if m is a positive integer, then the probability that the mth success occurs on the $(m + n)$th trial when independent Bernoulli trials, each with probability p of success, are run, is $\binom{n+m-1}{n} q^n p^m$.

31. There are n different types of collectible cards you can get as prizes when you buy a particular product. Suppose that every time you buy this product it is equally likely that you get any type of these cards. Let X be the random variable equal to the number of products that need to be purchased to obtain at least one of each type of card and let X_j be the random variable equal to the number of additional products that must be purchased after j different cards have been collected until a new card is obtained for $j = 0, 1, \ldots, n - 1$.

a) Show that $X = \sum_{j=0}^{n-1} X_j$.

b) Show that after j distinct types of cards have been obtained, the card obtained with the next purchase will be a card of a new type with probability $(n - j)/n$.

c) Show that X_j has a geometric distribution with parameter $(n - j)/n$.

d) Use parts (a) and (c) to show that $E(X) = n \sum_{j=1}^{n} 1/j$.

e) Use the approximation $\sum_{j=1}^{n} 1/j \approx \ln n + \gamma$, where $\gamma = 0.57721\ldots$ is Euler's constant, to find the expected number of products that you need to buy to get one card of each type if there are 50 different types of cards.

32. The **maximum satisfiability problem** asks for an assignment of truth values to the variables in a compound proposition in conjunctive normal form (which expresses a compound proposition as the conjunction of clauses where each clause is the disjunction of two or more variables or their negations) that makes as many of these clauses true as possible. For example, three but not four of the clauses in

$$(p \vee q) \wedge (p \vee \neg q) \wedge (\neg p \vee r) \wedge (\neg p \vee \neg r)$$

can be made true by an assignment of truth values to p, q, and r. We will show that probabilistic methods can provide a lower bound for the number of clauses that can be made true by an assignment of truth values to the variables.

a) Suppose that there are n variables in a compound proposition in conjunctive normal form. If we pick a truth value for each variable randomly by flipping a coin and assigning true to the variable if the coin comes up heads and false if it comes up tails, what is the probability of each possible assignment of truth values to the n variables?

b) Assuming that each clause is the disjunction of exactly two distinct variables or their negations, what is the probability that a given clause is true, given the random assignment of truth values from part (a)?

c) Suppose that there are D clauses in the compound proposition. What is the expected number of these clauses that are true, given the random assignment of truth values of the variables?

d) Use part (c) to show that for every compound proposition in conjunctive normal form there is an assignment of truth values to the variables that makes at least 3/4 of the clauses true.

Computer Projects

WRITE PROGRAMS WITH THESE INPUT AND OUTPUT.

1. Given a real number p with $0 \leq p \leq 1$, generate random numbers taken from a Bernoulli distribution with probability p.

2. Given a positive integer n, generate a random permutation of the set $\{1, 2, 3, \ldots, n\}$. (See Exercise 16 in Section 4.6.)

3. Given positive integers m and n, generate m random permutations of the first n positive integers. Find the number of inversions in each permutation and determine the average number of these inversions.

4. Given a positive integer n, simulate n repeated flips of a biased coin with probability p of heads and determine the number of heads that come up. Display the cumulative results.

5. Given positive integers n and m, generate m random permutations of the first n positive integers. Sort each permutation using the insertion sort, counting the number of comparisons used. Determine the average number of comparisons used over all m permutations.

6. Given positive integers n and m, generate m random permutations of the first n positive integers. Sort each permutation using the version of the bubble sort that terminates when a pass has been made with no interchanges, counting the number of comparisons used. Determine the average number of comparisons used over all m permutations.

7. Given a positive integer m, simulate the collection of cards that come with the purchase of products to find the number of products that must be purchased to obtain a full set of m different collector cards. (See Supplementary Exercise 31.)

8. Given positive integers m and n, simulate the placement of n keys, where a record with key k is placed at location $h(k) = k \bmod m$ and determine whether there is at least one collision.

9. Given a positive integer n, find the probability of selecting the six integers from the set $\{1, 2, \ldots, n\}$ that were mechanically selected in a lottery.

10. Simulate repeated trials of the Monty Hall Three Door problem (Example 11 in Section 5.1) to calculate the probability of winning with each strategy.

Computations and Explorations

USE A COMPUTATIONAL PROGRAM OR PROGRAMS YOU HAVE WRITTEN TO DO
THESE EXERCISES.

1. Find the probabilities of each type of hand in five-card poker and rank the types of hands by their probability.

2. Find some conditions so that the expected value of buying a $1 lottery ticket in the New Jersey pick-six lottery has an expected value of more than $1. To win you have to select the six numbers drawn, where order does not matter, from the positive integers 1 to 48, inclusive. The winnings are split evenly among holders of winning tickets. Be sure to consider the total size of the pot going into the drawing and the number of people buying tickets.

3. Estimate the probability that two integers selected at random are relatively prime by testing a large number of randomly selected pairs of integers. Look up the theorem that gives this probability and compare your results with the correct probability.

4. Determine the number of people needed to ensure that the probability at least two of them have the same day of the year as their birthday is at least 70 percent, at least 80 percent, at least 90 percent, at least 95 percent, at least 98 percent, and at least 99 percent.

5. Generate a list of 100 randomly selected permutations of the set of the first 100 positive integers. (See Exercise 16 in Section 4.6.)

6. Simulate the odd-person-out procedure (described in Exercise 11 of the Supplementary Exercises) for n people with $3 \leq n \leq 10$. Run a large number of trials for each value of n and use the results to estimate the expected number of flips needed to find the odd person out. Does your result agree with that found in Exercise 29 in Section 5.2? Vary the problem by supposing that exactly one person has a biased coin with probability of heads $p \neq 0.5$.

7. Given a positive integer n, simulate a hatcheck person randomly giving hats back to people. Determine the number who get the correct hat back.

Writing Projects

RESPOND TO THESE WITH ESSAYS USING OUTSIDE SOURCES.

1. Describe the origins of probability theory and the first uses of this theory.

2. Describe the different bets you can make when you play roulette. Find the probability of each of these bets in the American version where the wheel contains the numbers 0 and 00. Which is the best bet and which is the worst for you?

3. Discuss the probability of winning when you play the game of blackjack versus a casino. Is there a winning strategy for the person playing against the house?

4. Investigate the game of craps and discuss the probability that the shooter wins and how close to a fair game it is.

5. Discuss the history and solution of what is known as the Newton–Pepys problem, which asks which is most likely: rolling at least one six when six dice are rolled, rolling at least two sixes when 12 dice are rolled, or rolling at least three sixes when 18 dice are rolled.

6. Explain how Erdős and Rényi first used the probabilistic method and describe some other applications of this method.

7. Discuss the different types of probabilistic algorithms and describe some examples of each type.

6 Advanced Counting Techniques

Many counting problems cannot be solved easily using the methods discussed in Chapter 4. One such problem is: How many bit strings of length n do not contain two consecutive zeros? To solve this problem, let a_n be the number of such strings of length n. An argument can be given that shows $a_{n+1} = a_n + a_{n-1}$. This equation, called a recurrence relation, and the initial conditions $a_1 = 2$ and $a_2 = 3$ determine the sequence $\{a_n\}$. Moreover, an explicit formula can be found for a_n from the equation relating the terms of the sequence. As we will see, a similar technique can be used to solve many different types of counting problems.

We will also see that many counting problems can be solved using formal power series, called generating functions, where the coefficients of powers of x represent terms of sequence we are interested in. Besides solving counting problems, we will also be able to use generating functions to solve recurrence relations and to prove combinatorial identities.

Many other kinds of counting problems cannot be solved using the techniques discussed in Chapter 4, such as: How many ways are there to assign seven jobs to three employees so that each employee is assigned at least one job? How many primes are there less than 1000? Both of these problems can be solved by counting the number of elements in the union of sets. We will develop a technique, called the principle of inclusion–exclusion, that counts the number of elements in unions of sets, and we will show how this principle can be used to solve counting problems.

The techniques studied in this chapter, together with the basic techniques of Chapter 4, can be used to solve many counting problems.

6.1 Recurrence Relations

INTRODUCTION

The number of bacteria in a colony doubles every hour. If a colony begins with five bacteria, how many will be present in n hours? To solve this problem, let a_n be the number of bacteria at the end of n hours. Since the number of bacteria doubles every hour, the relationship $a_n = 2a_{n-1}$ holds whenever n is a positive integer. This relationship, together with the initial condition $a_0 = 5$, uniquely determines a_n for all nonnegative integers n. We can find a formula for a_n from this information.

Some of the counting problems that cannot be solved using the techniques discussed in Chapter 4 can be solved by finding relationships, called recurrence relations, between the terms of a sequence, as was done in the problem involving bacteria. We will study a variety of counting problems that can be modeled using recurrence relations. We will develop methods in this section and in the following section for finding explicit formulae for the terms of sequences that satisfy certain types of recurrence relations.

RECURRENCE RELATIONS

In Chapter 3 we discussed how sequences can be defined recursively. Recall that a recursive definition of a sequence specifies one or more initial terms and a rule for determining subsequent terms from those that precede them. Recursive definitions can be used to solve counting problems. When they are, the rule for finding terms from those that precede them is called a **recurrence relation.**

DEFINITION 1

> A *recurrence relation* for the sequence $\{a_n\}$ is an equation that expresses a_n in terms of one or more of the previous terms of the sequence, namely, $a_0, a_1, \ldots, a_{n-1}$, for all integers n with $n \geq n_0$, where n_0 is a nonnegative integer. A sequence is called a *solution* of a recurrence relation if its terms satisfy the recurrence relation.

There is an important connection between recursion and recurrence relations that we will exploit later in this chapter. Recursive algorithms provide the solution of a problem of size n in terms of the solutions of one or more instances of the same problem of smaller size. Consequently, when we analyze the complexity of a recursive algorithm, we obtain a recurrence relation that expresses the number of operations required to solve a problem of size n in terms of the number of operations required to solve the problem for one or more instances of smaller size.

EXAMPLE 1
Let $\{a_n\}$ be a sequence that satisfies the recurrence relation $a_n = a_{n-1} - a_{n-2}$ for $n = 2, 3, 4, \ldots$, and suppose that $a_0 = 3$ and $a_1 = 5$. What are a_2 and a_3?

Solution: We see from the recurrence relation that $a_2 = a_1 - a_0 = 5 - 3 = 2$ and $a_3 = a_2 - a_1 = 2 - 5 = -3$. We can find a_4, a_5, and each successive term in a similar way. ◄

EXAMPLE 2
Determine whether the sequence $\{a_n\}$ is a solution of the recurrence relation $a_n = 2a_{n-1} - a_{n-2}$ for $n = 2, 3, 4, \ldots$, where $a_n = 3n$ for every nonnegative integer n. Answer the same question where $a_n = 2^n$ and where $a_n = 5$.

Solution: Suppose that $a_n = 3n$ for every nonnegative integer n. Then, for $n \geq 2$, we see that $2a_{n-1} - a_{n-2} = 2[3(n-1)] - 3(n-2) = 3n = a_n$. Therefore, $\{a_n\}$, where $a_n = 3n$, is a solution of the recurrence relation.

Suppose that $a_n = 2^n$ for every nonnegative integer n. Note that $a_0 = 1, a_1 = 2$, and $a_2 = 4$. Since $2a_1 - a_0 = 2 \cdot 2 - 1 = 3 \neq a_2$, we see that $\{a_n\}$, where $a_n = 2^n$, is not a solution of the recurrence relation.

Suppose that $a_n = 5$ for every nonnegative integer n. Then for $n \geq 2$, we see that $a_n = 2a_{n-1} - a_{n-2} = 2 \cdot 5 - 5 = 5 = a_n$. Therefore, $\{a_n\}$, where $a_n = 5$, is a solution of the recurrence relation. ◄

The **initial conditions** for a sequence specify the terms that precede the first term where the recurrence relation takes effect. For instance, in Example 1, $a_0 = 3$ and $a_1 = 5$ are the initial conditions. The recurrence relation and initial conditions uniquely determine a sequence. This is the case since a recurrence relation, together with initial conditions, provide a recursive definition of the sequence. Any term of the sequence can be found from the initial conditions using the recurrence relation a sufficient number of times. However, there are better ways for computing the terms of certain classes of sequences defined by recurrence relations and initial conditions. We will discuss these methods in this section and in Section 6.2.

MODELING WITH RECURRENCE RELATIONS

We can use recurrence relations to model a wide variety of problems, such as finding compound interest, counting rabbits on an island, determining the number of moves in the Tower of Hanoi puzzle, and counting bit strings with certain properties.

EXAMPLE 3   **Compound Interest** Suppose that a person deposits $10,000 in a savings account at a bank yielding 11% per year with interest compounded annually. How much will be in the account after 30 years?

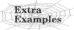

Solution: To solve this problem, let P_n denote the amount in the account after n years. Since the amount in the account after n years equals the amount in the account after $n - 1$ years plus interest for the nth year, we see that the sequence $\{P_n\}$ satisfies the recurrence relation

$$P_n = P_{n-1} + 0.11 P_{n-1} = (1.11) P_{n-1}.$$

The initial condition is $P_0 = 10{,}000$.

We can use an iterative approach to find a formula for P_n. Note that

$$P_1 = (1.11) P_0$$
$$P_2 = (1.11) P_1 = (1.11)^2 P_0$$
$$P_3 = (1.11) P_2 = (1.11)^3 P_0$$
$$\vdots$$
$$P_n = (1.11) P_{n-1} = (1.11)^n P_0.$$

When we insert the initial condition $P_0 = 10{,}000$, the formula $P_n = (1.11)^n 10{,}000$ is obtained. We can use mathematical induction to establish its validity. That the formula is valid for $n = 0$ is a consequence of the initial condition. Now assume that $P_n = (1.11)^n 10{,}000$. Then, from the recurrence relation and the induction hypothesis,

$$P_{n+1} = (1.11) P_n = (1.11)(1.11)^n 10{,}000 = (1.11)^{n+1} 10{,}000.$$

This shows that the explicit formula for P_n is valid.

Inserting $n = 30$ into the formula $P_n = (1.11)^n 10{,}000$ shows that after 30 years the account contains

$$P_{30} = (1.11)^{30} 10{,}000 = \$228{,}922.97. \qquad \blacktriangleleft$$

Example 4 shows how the population of rabbits on an island can be modeled using a recurrence relation.

EXAMPLE 4   **Rabbits and the Fibonacci Numbers** Consider this problem, which was originally posed by Leonardo di Pisa, also known as Fibonacci, in the thirteenth century in his book *Liber*

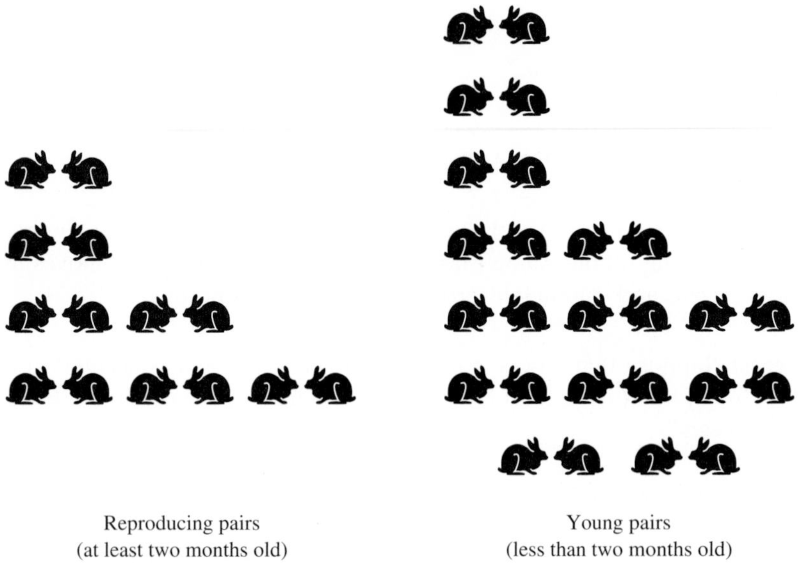

Month	Reproducing pairs	Young pairs	Total pairs
1	0	1	1
2	0	1	1
3	1	1	2
4	1	2	3
5	2	3	5
6	3	5	8

Reproducing pairs
(at least two months old)

Young pairs
(less than two months old)

FIGURE 1   Rabbits on an Island.

abaci. A young pair of rabbits (one of each sex) is placed on an island. A pair of rabbits does not breed until they are 2 months old. After they are 2 months old, each pair of rabbits produces another pair each month, as shown in Figure 1. Find a recurrence relation for the number of pairs of rabbits on the island after n months, assuming that no rabbits ever die.

Solution: Denote by f_n the number of pairs of rabbits after n months. We will show that $f_n, n = 1, 2, 3, \ldots$, are the terms of the Fibonacci sequence.

The rabbit population can be modeled using a recurrence relation. At the end of the first month, the number of pairs of rabbits on the island is $f_1 = 1$. Since this pair does not breed during the second month, $f_2 = 1$ also. To find the number of pairs after n months, add the number on the island the previous month, f_{n-1}, and the number of newborn pairs, which equals f_{n-2}, since each newborn pair comes from a pair at least 2 months old.

Consequently, the sequence $\{f_n\}$ satisfies the recurrence relation

$$f_n = f_{n-1} + f_{n-2}$$

for $n \geq 3$ together with the initial conditions $f_1 = 1$ and $f_2 = 1$. Since this recurrence relation and the initial conditions uniquely determine this sequence, the number of pairs of rabbits on the island after n months is given by the nth Fibonacci number.   ◀

Example 5 involves a famous puzzle.

EXAMPLE 5

The Tower of Hanoi A popular puzzle of the late nineteenth century invented by the French mathematician Édouard Lucas, called the Tower of Hanoi, consists of three pegs mounted on a board together with disks of different sizes. Initially these disks are placed on the first peg in order of size, with the largest on the bottom (as shown in Figure 2). The rules of the puzzle allow disks to be moved one at a time from one peg to another as long

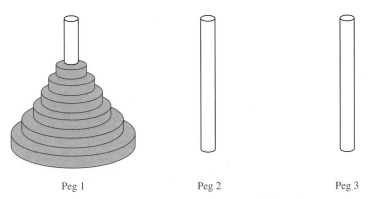

Peg 1 Peg 2 Peg 3

FIGURE 2 **The Initial Position in the Tower of Hanoi.**

as a disk is never placed on top of a smaller disk. The goal of the puzzle is to have all the disks on the second peg in order of size, with the largest on the bottom.

Let H_n denote the number of moves needed to solve the Tower of Hanoi problem with n disks. Set up a recurrence relation for the sequence $\{H_n\}$.

Solution: Begin with n disks on peg 1. We can transfer the top $n - 1$ disks, following the rules of the puzzle, to peg 3 using H_{n-1} moves (see Figure 3 for an illustration of the pegs and disks at this point). We keep the largest disk fixed during these moves. Then, we use one move to transfer the largest disk to the second peg. We can transfer the $n - 1$ disks on peg 3 to peg 2 using H_{n-1} additional moves, placing them on top of the largest disk, which always stays fixed on the bottom of peg 2. Moreover, it is easy to see that the puzzle cannot be solved using fewer steps. This shows that

$$H_n = 2H_{n-1} + 1.$$

The initial condition is $H_1 = 1$, since one disk can be transferred from peg 1 to peg 2, according to the rules of the puzzle, in one move.

We can use an iterative approach to solve this recurrence relation. Note that

$$
\begin{aligned}
H_n &= 2H_{n-1} + 1 \\
&= 2(2H_{n-2} + 1) + 1 = 2^2 H_{n-2} + 2 + 1 \\
&= 2^2(2H_{n-3} + 1) + 2 + 1 = 2^3 H_{n-3} + 2^2 + 2 + 1 \\
&\ \vdots \\
&= 2^{n-1} H_1 + 2^{n-2} + 2^{n-3} + \cdots + 2 + 1 \\
&= 2^{n-1} + 2^{n-2} + \cdots + 2 + 1 \\
&= 2^n - 1.
\end{aligned}
$$

We have used the recurrence relation repeatedly to express H_n in terms of previous terms of the sequence. In the next to last equality, the initial condition $H_1 = 1$ has been used. The last equality is based on the formula for the sum of the terms of a geometric series, which can be found in Theorem 1 in Section 3.2.

The iterative approach has produced the solution to the recurrence relation $H_n = 2H_{n-1} + 1$ with the initial condition $H_1 = 1$. This formula can be proved using mathematical induction. This is left as an exercise for the reader at the end of the section.

A myth created to accompany the puzzle tells of a tower in Hanoi where monks are transferring 64 gold disks from one peg to another, according to the rules of the puzzle.

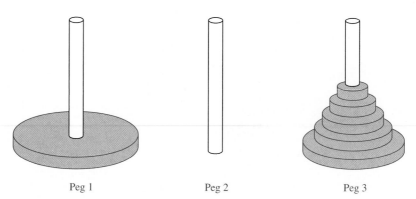

Peg 1   Peg 2   Peg 3

FIGURE 3   An Intermediate Position in the Tower of Hanoi.

The myth says that the world will end when they finish the puzzle. How long after the monks started will the world end if the monks take one second to move a disk?

From the explicit formula, the monks require

$$2^{64} - 1 = 18,446,744,073,709,551,615$$

moves to transfer the disks. Making one move per second, it will take them more than 500 billion years to solve the puzzle, so the world should survive a while longer than it already has.   ◄

Remark: Many people have studied variations of the original Tower of Hanoi puzzle discussed in Example 5. Some variations use more pegs, some allow disks to be of the same size, and some restrict the types of allowable disk moves. One of the oldest and most interesting variations is the **Reve's puzzle,**[*] proposed in 1907 by Henry Dudeney in his book *The Canterbury Puzzles*. The Reve's puzzle involves pilgrims challenged by the Reve to move a stack of cheeses of varying sizes from the first of four stools to another stool without ever placing a cheese on one of smaller diameter. The Reve's puzzle, expressed in terms of pegs and disks, follows the same rules as the Tower of Hanoi puzzle, except that four pegs are used. You may find it surprising that no one has been able to establish the minimum number of moves required to solve this puzzle for n disks. However, there is a conjecture, now more than 50 years old, that the minimum number of moves required equals the number of moves used by an algorithm invented by Frame and Stewart in 1939. (See Exercises 54–61 at the end of this section and [St94] for more information.)

Example 6 illustrates how recurrence relations can be used to count bit strings of a specified length that have a certain property.

EXAMPLE 6   Find a recurrence relation and give initial conditions for the number of bit strings of length n that do not have two consecutive 0s. How many such bit strings are there of length five?

Solution: Let a_n denote the number of bit strings of length n that do not have two consecutive 0s. To obtain a recurrence relation for $\{a_n\}$, note that by the sum rule, the number of bit strings of length n that do not have two consecutive 0s equals the number

---

[*]*Reve*, more commonly spelled *reeve*, is an archaic word for *governor*.

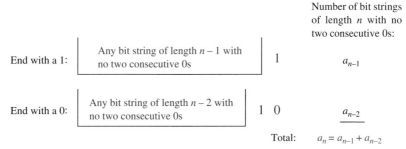

FIGURE 4 **Counting Bit Strings of Length n with No Two Consecutive 0s.**

of such bit strings ending with a 0 plus the number of such bit strings ending with a 1. We will assume that $n \geq 3$, so that the bit string has at least three bits.

The bit strings of length n ending with 1 that do not have two consecutive 0s are precisely the bit strings of length $n - 1$ with no two consecutive 0s with a 1 added at the end. Consequently, there are a_{n-1} such bit strings.

Bit strings of length n ending with a 0 that do not have two consecutive 0s must have 1 as their $(n - 1)$st bit; otherwise they would end with a pair of 0s. It follows that the bit strings of length n ending with a 0 that have no two consecutive 0s are precisely the bit strings of length $n - 2$ with no two consecutive 0s with 10 added at the end. Consequently, there are a_{n-2} such bit strings.

We conclude, as illustrated in Figure 4, that

$$a_n = a_{n-1} + a_{n-2}$$

for $n \geq 3$.

The initial conditions are $a_1 = 2$, since both bit strings of length one, 0 and 1 do not have consecutive 0s, and $a_2 = 3$, since the valid bit strings of length two are $01, 10$, and 11. To obtain a_5, we use the recurrence relation three times to find that

$$a_3 = a_2 + a_1 = 3 + 2 = 5,$$
$$a_4 = a_3 + a_2 = 5 + 3 = 8,$$
$$a_5 = a_4 + a_3 = 8 + 5 = 13.$$ ◄

Remark: Note that $\{a_n\}$ satisfies the same recurrence relation as the Fibonacci sequence. Since $a_1 = f_3$ and $a_2 = f_4$ it follows that $a_n = f_{n+2}$.

Example 7 shows how a recurrence relation can be used to model the number of codewords that are allowable using certain validity checks.

EXAMPLE 7 **Codeword Enumeration** A computer system considers a string of decimal digits a valid codeword if it contains an even number of 0 digits. For instance, 1230407869 is valid, whereas 120987045608 is not valid. Let a_n be the number of valid n-digit codewords. Find a recurrence relation for a_n.

Solution: Note that $a_1 = 9$ since there are 10 one-digit strings, and only one, namely, the string 0, is not valid. A recurrence relation can be derived for this sequence by considering how a valid n-digit string can be obtained from strings of $n - 1$ digits. There are two ways to form a valid string with n digits from a string with one fewer digit.

First, a valid string of n digits can be obtained by appending a valid string of $n - 1$ digits with a digit other than 0. This appending can be done in nine ways. Hence, a valid string with n digits can be formed in this manner in $9a_{n-1}$ ways.

Second, a valid string of n digits can be obtained by appending a 0 to a string of length $n - 1$ that is not valid. (This produces a string with an even number of 0 digits since the invalid string of length $n - 1$ has an odd number of 0 digits.) The number of ways that this can be done equals the number of invalid $(n - 1)$-digit strings. Since there are 10^{n-1} strings of length $n - 1$, and a_{n-1} are valid, there are $10^{n-1} - a_{n-1}$ valid n-digit strings obtained by appending an invalid string of length $n - 1$ with a 0.

Since all valid strings of length n are produced in one of these two ways, it follows that there are

$$
\begin{aligned}
a_n &= 9a_{n-1} + (10^{n-1} - a_{n-1}) \\
&= 8a_{n-1} + 10^{n-1}
\end{aligned}
$$

valid strings of length n. ◀

Example 8 establishes a recurrence relation that appears in many different contexts.

EXAMPLE 8 Find a recurrence relation for C_n, the number of ways to parenthesize the product of $n + 1$ numbers, $x_0 \cdot x_1 \cdot x_2 \cdots \cdot x_n$, to specify the order of multiplication. For example, $C_3 = 5$ since there are five ways to parenthesize $x_0 \cdot x_1 \cdot x_2 \cdot x_3$ to determine the order of multiplication:

$$((x_0 \cdot x_1) \cdot x_2) \cdot x_3 \qquad (x_0 \cdot (x_1 \cdot x_2)) \cdot x_3 \qquad (x_0 \cdot x_1) \cdot (x_2 \cdot x_3)$$
$$x_0 \cdot ((x_1 \cdot x_2) \cdot x_3) \qquad x_0 \cdot (x_1 \cdot (x_2 \cdot x_3)).$$

Solution: To develop a recurrence relation for C_n, we note that however we insert parentheses in the product $x_0 \cdot x_1 \cdot x_2 \cdots \cdot x_n$, one "$\cdot$" operator remains outside all parentheses, namely, the operator for the final multiplication to be performed. [For example, in $(x_0 \cdot (x_1 \cdot x_2)) \cdot x_3$, it is the final "$\cdot$", while in $(x_0 \cdot x_1) \cdot (x_2 \cdot x_3)$ it is the second "$\cdot$".] This final operator appears between two of the $n + 1$ numbers, say, x_k and x_{k+1}. There are $C_k C_{n-k-1}$ ways to insert parentheses to determine the order of the $n + 1$ numbers to be multiplied when the final operator appears between x_k and x_{k+1}, since there are C_k ways to insert parentheses in the product $x_0 \cdot x_1 \cdots \cdot x_k$ to determine the order in which these $k + 1$ numbers are to be multiplied and C_{n-k-1} ways to insert parentheses in the product $x_{k+1} \cdot x_{k+2} \cdots \cdot x_n$ to determine the order in which these $n - k$ numbers are to be multiplied. Since this final operator can appear between any two of the $n + 1$ numbers, it follows that

$$C_n = C_0 C_{n-1} + C_1 C_{n-2} + \cdots + C_{n-2} C_1 + C_{n-1} C_0$$

$$= \sum_{k=0}^{n-1} C_k C_{n-k-1}.$$

Note that the initial conditions are $C_0 = 1$ and $C_1 = 1$. This recurrence relation can be solved using the method of generating functions, which will be discussed in Section 6.4. It can be shown that $C_n = C(2n, n)/(n + 1)$. (See Exercise 41 at the end of that section.) ◀

Links The sequence $\{C_n\}$ is the sequence of **Catalan numbers.** This sequence appears as the solution of many different counting problems besides the one considered here (see the chapter on Catalan numbers in [MiRo91] or [Ro84a] for details).

Exercises

1. Find the first five terms of the sequence defined by each of these recurrence relations and initial conditions.

   a) $a_n = 6a_{n-1}, a_0 = 2$
   b) $a_n = a_{n-1}^2, a_1 = 2$
   c) $a_n = a_{n-1} + 3a_{n-2}, a_0 = 1, a_1 = 2$
   d) $a_n = na_{n-1} + n^2 a_{n-2}, a_0 = 1, a_1 = 1$
   e) $a_n = a_{n-1} + a_{n-3}, a_0 = 1, a_1 = 2, a_2 = 0$

2. Find the first six terms of the sequence defined by each of these recurrence relations and initial conditions.

   a) $a_n = -2a_{n-1}, a_0 = -1$
   b) $a_n = a_{n-1} - a_{n-2}, a_0 = 2, a_1 = -1$
   c) $a_n = 3a_{n-1}^2, a_0 = 1$
   d) $a_n = na_{n-1} + a_{n-2}^2, a_0 = -1, a_1 = 0$
   e) $a_n = a_{n-1} - a_{n-2} + a_{n-3}, a_0 = 1, a_1 = 1, a_2 = 2$

3. Let $a_n = 2^n + 5 \cdot 3^n$ for $n = 0, 1, 2, \ldots$.

   a) Find a_0, a_1, a_2, a_3, and a_4.
   b) Show that $a_2 = 5a_1 - 6a_0$, $a_3 = 5a_2 - 6a_1$, and $a_4 = 5a_3 - 6a_2$.
   c) Show that $a_n = 5a_{n-1} - 6a_{n-2}$ for all integers n with $n \geq 2$.

4. Show that the sequence $\{a_n\}$ is a solution of the recurrence relation $a_n = -3a_{n-1} + 4a_{n-2}$ if

   a) $a_n = 0$.
   b) $a_n = 1$.
   c) $a_n = (-4)^n$.
   d) $a_n = 2(-4)^n + 3$.

5. Is the sequence $\{a_n\}$ a solution of the recurrence relation $a_n = 8a_{n-1} - 16a_{n-2}$ if

   a) $a_n = 0$?   **b)** $a_n = 1$?
   c) $a_n = 2^n$?   **d)** $a_n = 4^n$?
   e) $a_n = n4^n$?
   f) $a_n = 2 \cdot 4^n + 3n4^n$?
   g) $a_n = (-4)^n$?
   h) $a_n = n^2 4^n$?

6. For each of these sequences find a recurrence relation satisfied by this sequence. (The answers are not unique since there are infinitely many different recurrence relations satisfied by any sequence.)

   a) $a_n = 3$ **b)** $a_n = 2n$
   c) $a_n = 2n + 3$ **d)** $a_n = 5^n$
   e) $a_n = n^2$ **f)** $a_n = n^2 + n$
   g) $a_n = n + (-1)^n$ **h)** $a_n = n!$

7. Show that the sequence $\{a_n\}$ is a solution of the recurrence relation $a_n = a_{n-1} + 2a_{n-2} + 2n - 9$ if

   a) $a_n = -n + 2$.
   b) $a_n = 5(-1)^n - n + 2$.
   c) $a_n = 3(-1)^n + 2^n - n + 2$.

   d) $a_n = 7 \cdot 2^n - n + 2$.

8. Find the solution to each of these recurrence relations with the given initial conditions. Use an iterative approach such as that used in Example 5.

   a) $a_n = -a_{n-1}, a_0 = 5$
   b) $a_n = a_{n-1} + 3, a_0 = 1$
   c) $a_n = a_{n-1} - n, a_0 = 4$
   d) $a_n = 2a_{n-1} - 3, a_0 = -1$
   e) $a_n = (n + 1)a_{n-1}, a_0 = 2$
   f) $a_n = 2na_{n-1}, a_0 = 3$
   g) $a_n = -a_{n-1} + n - 1, a_0 = 7$

9. Find the solution to each of these recurrence relations and initial conditions. Use an iterative approach such as that used in Example 5.

   a) $a_n = 3a_{n-1}, a_0 = 2$
   b) $a_n = a_{n-1} + 2, a_0 = 3$
   c) $a_n = a_{n-1} + n, a_0 = 1$
   d) $a_n = a_{n-1} + 2n + 3, a_0 = 4$
   e) $a_n = 2a_{n-1} - 1, a_0 = 1$
   f) $a_n = 3a_{n-1} + 1, a_0 = 1$
   g) $a_n = na_{n-1}, a_0 = 5$
   h) $a_n = 2na_{n-1}, a_0 = 1$

10. A person deposits \$1000 in an account that yields 9% interest compounded yearly.

   a) Set up a recurrence relation for the amount in the account at the end of n years.
   b) Find an explicit formula for the amount in the account at the end of n years.
   c) How much money will the account contain after 100 years?

11. Suppose that the number of bacteria in a colony triples every hour.

   a) Set up a recurrence relation for the number of bacteria after n hours have elapsed.
   b) If 100 bacteria are used to begin a new colony, how many bacteria will be in the colony in 10 hours?

12. Assume that the population of the world in 2002 is 6.2 billion and is growing at the rate of 1.3% a year.

   a) Set up a recurrence relation for the population of the world n years after 2002.
   b) Find an explicit formula for the population of the world n years after 2002.
   c) What will the population of the world be in 2022?

13. A factory makes custom sports cars at an increasing rate. In the first month only one car is made, in the second month two cars are made, and so on, with n cars made in the nth month.

   a) Set up a recurrence relation for the number of cars produced in the first n months by this factory.

b) How many cars are produced in the first year?

c) Find an explicit formula for the number of cars produced in the first n months by this factory.

14. An employee joined a company in 1999 with a starting salary of $50,000. Every year this employee receives a raise of $1000 plus 5% of the salary of the previous year.

   a) Set up a recurrence relation for the salary of this employee n years after 1999.

   b) What will the salary of this employee be in 2007?

   c) Find an explicit formula for the salary of this employee n years after 1999.

15. Find a recurrence relation for the balance $B(k)$ owed at the end of k months on a loan of $5000 at a rate of 7% if a payment of $100 is made each month. [*Hint:* Express $B(k)$ in terms of $B(k-1)$; the monthly interest is $(0.07/12)B(k-1)$.]

16. a) Find a recurrence relation for the balance $B(k)$ owed at the end of k months on a loan at a rate of r if a payment P is made on the loan each month. [*Hint:* Express $B(k)$ in terms of $B(k-1)$ and note that the monthly interest rate is $r/12$.]

   b) Determine what the monthly payment P should be so that the loan is paid off after T months.

17. Use mathematical induction to verify the formula derived in Example 5 for the number of moves required to complete the Tower of Hanoi puzzle.

18. a) Find a recurrence relation for the number of permutations of a set with n elements.

   b) Use this recurrence relation to find the number of permutations of a set with n elements using iteration.

19. A vending machine dispensing books of stamps accepts only dollar coins, $1 bills, and $5 bills.

   a) Find a recurrence relation for the number of ways to deposit n dollars in the vending machine, where the order in which the coins and bills are deposited matters.

   b) What are the initial conditions?

   c) How many ways are there to deposit $10 for a book of stamps?

20. A country uses as currency coins with values of 1 peso, 2 pesos, 5 pesos, and 10 pesos and bills with values of 5 pesos, 10 pesos, 20 pesos, 50 pesos, and 100 pesos. Find a recurrence relation for the number of ways to pay a bill of n pesos if the order in which the coins and bills are paid matters.

21. How many ways are there to pay a bill of 17 pesos using the currency described in Exercise 20, where the order in which coins and bills are paid matters?

*22. a) Find a recurrence relation for the number of strictly increasing sequences of positive integers that have 1 as their first term and n as their last term where n is a positive integer. That is, sequences $a_1, a_2, \ldots, a_k$, where $a_1 = 1$, $a_k = n$, and $a_j < a_{j+1}$ for $j = 1, 2, \ldots, k-1$.

b) What are the initial conditions?

c) How many sequences of the type described in (a) are there when n is a positive integer with $n \geq 2$?

23. a) Find a recurrence relation for the number of bit strings of length n that contain a pair of consecutive 0s.

   b) What are the initial conditions?

   c) How many bit strings of length seven contain two consecutive 0s?

24. a) Find a recurrence relation for the number of bit strings of length n that contain three consecutive 0s.

   b) What are the initial conditions?

   c) How many bit strings of length seven contain three consecutive 0s?

25. a) Find a recurrence relation for the number of bit strings of length n that do not contain three consecutive 0s.

   b) What are the initial conditions?

   c) How many bit strings of length seven do not contain three consecutive 0s?

*26. a) Find a recurrence relation for the number of bit strings that contain the string 01.

   b) What are the initial conditions?

   c) How many bit strings of length seven contain the string 01?

27. a) Find a recurrence relation for the number of ways to climb n stairs if the person climbing the stairs can take one stair or two stairs at a time.

   b) What are the initial conditions?

   c) How many ways can this person climb a flight of eight stairs?

28. a) Find a recurrence relation for the number of ways to climb n stairs if the person climbing the stairs can take one, two, or three stairs at a time.

   b) What are the initial conditions?

   c) How many ways can this person climb a flight of eight stairs?

A string that contains only 0s, 1s, and 2s is called a **ternary string.**

29. a) Find a recurrence relation for the number of ternary strings that do not contain two consecutive 0s.

   b) What are the initial conditions?

   c) How many ternary strings of length six do not contain two consecutive 0s?

30. a) Find a recurrence relation for the number of ternary strings that contain two consecutive 0s.

   b) What are the initial conditions?

   c) How many ternary strings of length six contain two consecutive 0s?

*31. a) Find a recurrence relation for the number of ternary strings that do not contain two consecutive 0s or two consecutive 1s.

   b) What are the initial conditions?

c) How many ternary strings of length six do not contain two consecutive 0s or two consecutive 1s?

*32. a) Find a recurrence relation for the number of ternary strings that contain either two consecutive 0s or two consecutive 1s.

b) What are the initial conditions?

c) How many ternary strings of length six contain two consecutive 0s or two consecutive 1s?

*33. a) Find a recurrence relation for the number of ternary strings that do not contain consecutive symbols that are the same.

b) What are the initial conditions?

c) How many ternary strings of length six do not contain consecutive symbols that are the same?

**34. a) Find a recurrence relation for the number of ternary strings that contain two consecutive symbols that are the same.

b) What are the initial conditions?

c) How many ternary strings of length six contain consecutive symbols that are the same?

35. Messages are transmitted over a communications channel using two signals. The transmittal of one signal requires 1 microsecond, and the transmittal of the other signal requires 2 microseconds.

a) Find a recurrence relation for the number of different messages consisting of sequences of these two signals, where each signal in the message is immediately followed by the next signal, that can be sent in n microseconds.

b) What are the initial conditions?

c) How many different messages can be sent in 10 microseconds using these two signals?

36. A bus driver pays all tolls, using only nickels and dimes, by throwing one coin at a time into the mechanical toll collector.

a) Find a recurrence relation for the number of different ways the bus driver can pay a toll of n cents (where the order in which the coins are used matters).

b) In how many different ways can the driver pay a toll of 45 cents?

37. a) Find the recurrence relation satisfied by R_n, where R_n is the number of regions that a plane is divided into by n lines, if no two of the lines are parallel and no three of the lines go through the same point.

b) Find R_n using iteration.

*38. a) Find the recurrence relation satisfied by R_n, where R_n is the number of regions into which the surface of a sphere is divided into by n great circles (which are the intersections of the sphere and planes passing through the center of the sphere), if no three of the great circles go through the same point.

b) Find R_n using iteration.

*39. a) Find the recurrence relation satisfied by S_n, where S_n is the number of regions into which three-dimensional space is divided by n planes if every three of the planes meet in one point, but no four of the planes go through the same point.

b) Find S_n using iteration.

40. Find a recurrence relation for the number of bit sequences of length n with an even number of 0s.

41. How many bit sequences of length seven contain an even number of 0s?

42. a) Find a recurrence relation for the number of ways to completely cover a $2 \times n$ chessboard with 1×2 dominos. (*Hint:* Consider separately the coverings where the position in the top right corner of the chessboard is covered by a domino positioned horizontally and where it is covered by a domino positioned vertically.)

b) What are the initial conditions for the recurrence relation in part (a)?

c) How many ways are there to completely cover a 2×17 chessboard with 1×2 dominos?

43. a) Find a recurrence relation for the number of ways to lay out a walkway with slate tiles if the tiles are red, green, or gray, so that no two red tiles are adjacent and tiles of the same color are considered indistinguishable.

b) What are the initial conditions for the recurrence relation in part (a)?

c) How many ways are there to lay out a path of seven tiles as described in part (a)?

44. Show that the Fibonacci numbers satisfy the recurrence relation $f_n = 5f_{n-4} + 3f_{n-5}$ for $n = 5, 6, 7, \ldots,$ together with the initial conditions $f_0 = 0$, $f_1 = 1$, $f_2 = 1$, $f_3 = 2$, and $f_4 = 3$. Use this recurrence relation to show that f_{5n} is divisible by 5, for $n = 1, 2, 3, \ldots.$

*45. Let $S(m, n)$ denote the number of onto functions from a set with m elements to a set with n elements. Show that $S(m, n)$ satisfies the recurrence relation

$$S(m, n) = n^m - \sum_{k=1}^{n-1} C(n, k) S(m, k)$$

whenever $m \geq n$ and $n > 1$, with the initial condition $S(m, 1) = 1$.

46. a) Write out all the ways the product $x_0 \cdot x_1 \cdot x_2 \cdot x_3 \cdot x_4$ can be parenthesized to determine the order of multiplication.

b) Use the recurrence relation developed in Example 8 to calculate C_4, the number of ways to parenthesize the product of five numbers so as to determine the order of multiplication. Verify that you listed the correct number of ways in part (a).

c) Check your result in part (b) by finding C_4, using the closed formula for C_n mentioned in the solution of Example 8.

47. a) Use the recurrence relation developed in Example 8 to determine C_5, the number of ways to parenthesize the product of six numbers so as to determine the order of multiplication.

b) Check your result with the closed formula for C_5 mentioned in the solution of Example 8.

*48. In the Tower of Hanoi puzzle, suppose our goal is to transfer all n disks from peg 1 to peg 3, but we cannot move a disk directly between pegs 1 and 3. Each move of a disk must be a move involving peg 2. As usual, we cannot place a disk on top of a smaller disk.

a) Find a recurrence relation for the number of moves required to solve the puzzle for n disks with this added restriction.

b) Solve this recurrence relation to find a formula for the number of moves required to solve the puzzle for n disks.

c) How many different arrangements are there of the n disks on three pegs so that no disk is on top of a smaller disk?

d) Show that every allowable arrangement of the n disks occurs in the solution of this variation of the puzzle.

Exercises 49–53 deal with a variation of the **Josephus problem** described by Graham, Knuth, and Patashnik in [GrKnPa94]. This problem is based on an account by the historian Flauvius Josephus, who was part of a band of 41 Jewish rebels trapped in a cave by the Romans during the Jewish-Roman war of the first century. The rebels preferred suicide to capture; they decided to form a circle and to repeatedly count off around the circle, killing every third rebel left alive. However, Josephus and another rebel did not want to be killed this way; they determined the positions where they should stand to be the last two rebels remaining alive. The variation we consider begins with n people, numbered 1 to n, standing around a circle. In each stage, every second person still left alive is eliminated until only one survives. We denote the number of the survivor by $J(n)$.

49. Determine the value of $J(n)$ for each integer n with $1 \le n \le 16$.

50. Use the values you found in Exercise 49 to conjecture a formula for $J(n)$. (*Hint:* Write $n = 2^m + k$, where m is a nonnegative integer and k is a nonnegative integer less than 2^m.)

51. Show that $J(n)$ satisfies the recurrence relation $J(2n) = 2J(n) - 1$ and $J(2n + 1) = 2J(n) + 1$, for $n \ge 1$, and $J(1) = 1$.

52. Use mathematical induction to prove the formula you conjectured in Exercise 50, making use of the recurrence relation from Exercise 51.

53. Determine $J(100)$, $J(1000)$, and $J(10000)$ from your formula for $J(n)$.

Exercises 54–61 involve the Reve's puzzle, the variation of the Tower of Hanoi puzzle with four pegs and n disks. Before presenting these exercises, we describe the Frame–Stewart algorithm for moving the disks from peg 1 to peg 4 so that no disk is ever on top of a smaller one. This algorithm, given the number of disks n as input, depends on a choice of an integer k with $1 \le k \le n$. When there is only one disk, move it from peg 1 to peg 4 and stop. For $n > 1$, the algorithm proceeds recursively, using these three steps. Recursively move the stack of the $n - k$ smallest disks from peg 1 to peg 2, using all four pegs. Next move the stack of the k largest disks from peg 1 to peg 4, using the three-peg algorithm from the Tower of Hanoi puzzle without using the peg holding the $n - k$ smallest disks. Finally, recursively move the smallest $n - k$ disks to peg 4, using all four pegs. Frame and Stewart showed that to produce the fewest moves using their algorithm, k should be chosen to be the smallest integer such that n does not exceed $t_k = k(k + 1)/2$, the kth triangular number, i.e., $t_{k-1} < n \le t_k$. The unsettled conjecture, known as **Frame's conjecture,** is that this algorithm uses the fewest number of moves required to solve the puzzle, no matter how the disks are moved.

54. Show that the Reve's puzzle with three disks can be solved using five, and no fewer, moves.

55. Show that the Reve's puzzle with four disks can be solved using nine, and no fewer, moves.

56. Describe the moves made by the Frame–Stewart algorithm, with k chosen so that the fewest moves are required, for

a) 5 disks. **b)** 6 disks. **c)** 7 disks. **d)** 8 disks.

*57. Show that if $R(n)$ is the number of moves used by the Frame–Stewart algorithm to solve the Reve's puzzle with n disks, where k is chosen to be the smallest integer with $n \le k(k + 1)/2$, then $R(n)$ satisfies the recurrence relation with $R(n) = 2R(n - k) + 2^k - 1$, with $R(0) = 0$ and $R(1) = 1$.

*58. Show that if k is as chosen in Exercise 57, then $R(n) - R(n - 1) = 2^{k-1}$.

*59. Show that if k is as chosen in Exercise 57, then $R(n) = \sum_{i=1}^{k} i 2^{i-1} - (t_k - n) 2^{k-1}$.

*60. Use Exercise 59 to give an upper bound on the number of moves required to solve the Reve's puzzle for all integers n with $1 \le n \le 25$.

*61. Show that $R(n)$ is $O(\sqrt{n} 2^{\sqrt{2n}})$.

Let $\{a_n\}$ be a sequence of real numbers. The **backward differences** of this sequence are defined recursively as shown next. The **first difference** ∇a_n is

$$\nabla a_n = a_n - a_{n-1}.$$

The **(k + 1)st difference** $\nabla^{k+1} a_n$ is obtained from $\nabla^k a_n$ by

$$\nabla^{k+1} a_n = \nabla^k a_n - \nabla^k a_{n-1}.$$

62. Find ∇a_n for the sequence $\{a_n\}$, where

    a) $a_n = 4$.
    b) $a_n = 2n$.
    c) $a_n = n^2$.
    d) $a_n = 2^n$.

63. Find $\nabla^2 a_n$ for the sequences in Exercise 62.

64. Show that $a_{n-1} = a_n - \nabla a_n$.

65. Show that $a_{n-2} = a_n - 2\nabla a_n + \nabla^2 a_n$.

*66. Prove that a_{n-k} can be expressed in terms of a_n, ∇a_n, $\nabla^2 a_n, \ldots, \nabla^k a_n$.

67. Express the recurrence relation $a_n = a_{n-1} + a_{n-2}$ in terms of a_n, ∇a_n, and $\nabla^2 a_n$.

68. Show that any recurrence relation for the sequence $\{a_n\}$ can be written in terms of a_n, ∇a_n, $\nabla^2 a_n, \ldots$. The resulting equation involving the sequences and its differences is called a **difference equation.**

6.2   Solving Recurrence Relations

INTRODUCTION

A wide variety of recurrence relations occur in models. Some of these recurrence relations can be solved using iteration or some other ad hoc technique. However, one important class of recurrence relations can be explicitly solved in a systematic way. These are recurrence relations that express the terms of a sequence as linear combinations of previous terms.

DEFINITION 1

A *linear homogeneous recurrence relation of degree k with constant coefficients* is a recurrence relation of the form

$$a_n = c_1 a_{n-1} + c_2 a_{n-2} + \cdots + c_k a_{n-k},$$

where $c_1, c_2, \ldots, c_k$ are real numbers, and $c_k \neq 0$.

The recurrence relation in the definition is **linear** since the right-hand side is a sum of multiples of the previous terms of the sequence. The recurrence relation is **homogeneous** since no terms occur that are not multiples of the a_js. The coefficients of the terms of the sequence are all **constants,** rather than functions that depend on n. The **degree** is k because a_n is expressed in terms of the previous k terms of the sequence.

    A consequence of the second principle of mathematical induction is that a sequence satisfying the recurrence relation in the definition is uniquely determined by this recurrence relation and the k initial conditions

$$a_0 = C_0, a_1 = C_1, \ldots, a_{k-1} = C_{k-1}.$$

EXAMPLE 1

The recurrence relation $P_n = (1.11)P_{n-1}$ is a linear homogeneous recurrence relation of degree one. The recurrence relation $f_n = f_{n-1} + f_{n-2}$ is a linear homogeneous recurrence relation of degree two. The recurrence relation $a_n = a_{n-5}$ is a linear homogeneous recurrence relation of degree five.   ◀

    Example 2 presents some examples of recurrence relations that are not linear homogeneous recurrence relations with constant coefficients.

EXAMPLE 2

The recurrence relation $a_n = a_{n-1} + a_{n-2}^2$ is not linear. The recurrence relation $H_n = 2H_{n-1} + 1$ is not homogeneous. The recurrence relation $B_n = nB_{n-1}$ does not have constant coefficients.   ◀

Linear homogeneous recurrence relations are studied for two reasons. First, they often occur in modeling of problems. Second, they can be systematically solved.

SOLVING LINEAR HOMOGENEOUS RECURRENCE RELATIONS WITH CONSTANT COEFFICIENTS

The basic approach for solving linear homogeneous recurrence relations is to look for solutions of the form $a_n = r^n$, where r is a constant. Note that $a_n = r^n$ is a solution of the recurrence relation $a_n = c_1 a_{n-1} + c_2 a_{n-2} + \cdots + c_k a_{n-k}$ if and only if

$$r^n = c_1 r^{n-1} + c_2 r^{n-2} + \cdots + c_k r^{n-k}.$$

When both sides of this equation are divided by r^{n-k} and the right-hand side is subtracted from the left, we obtain the equation

$$r^k - c_1 r^{k-1} - c_2 r^{k-2} - \cdots - c_{k-1} r - c_k = 0.$$

Consequently, the sequence $\{a_n\}$ with $a_n = r^n$ is a solution if and only if r is a solution of this last equation, which is called the **characteristic equation** of the recurrence relation. The solutions of this equation are called the **characteristic roots** of the recurrence relation. As we will see, these characteristic roots can be used to give an explicit formula for all the solutions of the recurrence relation.

We will first develop results that deal with linear homogeneous recurrence relations with constant coefficients of degree two. Then corresponding general results when the degree may be greater than two will be stated. Because the proofs needed to establish the results in the general case are more complicated, they will not be given in the text.

We now turn our attention to linear homogeneous recurrence relations of degree two. First, consider the case when there are two distinct characteristic roots.

THEOREM 1

Let c_1 and c_2 be real numbers. Suppose that $r^2 - c_1 r - c_2 = 0$ has two distinct roots r_1 and r_2. Then the sequence $\{a_n\}$ is a solution of the recurrence relation $a_n = c_1 a_{n-1} + c_2 a_{n-2}$ if and only if $a_n = \alpha_1 r_1^n + \alpha_2 r_2^n$ for $n = 0, 1, 2, \ldots$, where α_1 and α_2 are constants.

Proof: We must do two things to prove the theorem. First, it must be shown that if r_1 and r_2 are the roots of the characteristic equation, and α_1 and α_2 are constants, then the sequence $\{a_n\}$ with $a_n = \alpha_1 r_1^n + \alpha_2 r_2^n$ is a solution of the recurrence relation. Second, it must be shown that if the sequence $\{a_n\}$ is a solution, then $a_n = \alpha_1 r_1^n + \alpha_2 r_2^n$ for some constants α_1 and α_2.

Now we will show that if $a_n = \alpha_1 r_1^n + \alpha_2 r_2^n$, then the sequence $\{a_n\}$ is a solution of the recurrence relation. Since r_1 and r_2 are roots of $r^2 - c_1 r - c_2 = 0$, it follows that $r_1^2 = c_1 r_1 + c_2, r_2^2 = c_1 r_2 + c_2$.

From these equations, we see that

$$c_1 a_{n-1} + c_2 a_{n-2} = c_1 (\alpha_1 r_1^{n-1} + \alpha_2 r_2^{n-1}) + c_2 (\alpha_1 r_1^{n-2} + \alpha_2 r_2^{n-2})$$

$$= \alpha_1 r_1^{n-2} (c_1 r_1 + c_2) + \alpha_2 r_2^{n-2} (c_1 r_2 + c_2)$$

$$= \alpha_1 r_1^{n-2} r_1^2 + \alpha_2 r_2^{n-2} r_2^2$$

$$= \alpha_1 r_1^n + \alpha_2 r_2^n$$

$$= a_n.$$

This shows that the sequence $\{a_n\}$ with $a_n = \alpha_1 r_1^n + \alpha_2 r_2^n$ is a solution of the recurrence relation.

To show that every solution $\{a_n\}$ of the recurrence relation $a_n = c_1 a_{n-1} + c_2 a_{n-2}$ has $a_n = \alpha_1 r_1^n + \alpha_2 r_2^n$ for $n = 0, 1, 2, \ldots$, for some constants α_1 and α_2, suppose that $\{a_n\}$ is a solution of the recurrence relation, and the initial conditions $a_0 = C_0$ and $a_1 = C_1$ hold. It will be shown that there are constants α_1 and α_2 so that the sequence $\{a_n\}$ with $a_n = \alpha_1 r_1^n + \alpha_2 r_2^n$ satisfies these same initial conditions. This requires that

$$a_0 = C_0 = \alpha_1 + \alpha_2,$$
$$a_1 = C_1 = \alpha_1 r_1 + \alpha_2 r_2.$$

We can solve these two equations for α_1 and α_2. From the first equation it follows that $\alpha_2 = C_0 - \alpha_1$. Inserting this expression into the second equation gives

$$C_1 = \alpha_1 r_1 + (C_0 - \alpha_1) r_2.$$

Hence,

$$C_1 = \alpha_1 (r_1 - r_2) + C_0 r_2.$$

This shows that

$$\alpha_1 = \frac{(C_1 - C_0 r_2)}{r_1 - r_2}$$

and

$$\alpha_2 = C_0 - \alpha_1 = C_0 - \frac{(C_1 - C_0 r_2)}{r_1 - r_2} = \frac{C_0 r_1 - C_1}{r_1 - r_2},$$

where these expressions for α_1 and α_2 depend on the fact that $r_1 \neq r_2$. (When $r_1 = r_2$, this theorem is not true.) Hence, with these values for α_1 and α_2, the sequence $\{a_n\}$ with $\alpha_1 r_1^n + \alpha_2 r_2^n$ satisfies the two initial conditions. Since this recurrence relation and these initial conditions uniquely determine the sequence, it follows that $a_n = \alpha_1 r_1^n + \alpha_2 r_2^n$.
◁

The characteristic roots of a linear homogeneous recurrence relation with constant coefficients may be complex numbers. Theorem 1 (and also subsequent theorems in this section) still applies in this case. Recurrence relations with complex characteristic roots will not be discussed in the text. Readers familiar with complex numbers may wish to solve Exercises 38 and 39 at the end of this section.

Examples 3 and 4 show how to use Theorem 1 to solve recurrence relations.

EXAMPLE 3   What is the solution of the recurrence relation

$$a_n = a_{n-1} + 2a_{n-2}$$

with $a_0 = 2$ and $a_1 = 7$?

Solution: Theorem 1 can be used to solve this problem. The characteristic equation of the recurrence relation is $r^2 - r - 2 = 0$. Its roots are $r = 2$ and $r = -1$. Hence, the sequence $\{a_n\}$ is a solution to the recurrence relation if and only if

$$a_n = \alpha_1 2^n + \alpha_2 (-1)^n,$$

for some constants α_1 and α_2. From the initial conditions, it follows that

$$a_0 = 2 = \alpha_1 + \alpha_2,$$
$$a_1 = 7 = \alpha_1 \cdot 2 + \alpha_2 \cdot (-1).$$

Extra
Examples

Solving these two equations shows that $\alpha_1 = 3$ and $\alpha_2 = -1$. Hence, the solution to the recurrence relation and initial conditions is the sequence $\{a_n\}$ with

$$a_n = 3 \cdot 2^n - (-1)^n. \qquad \blacktriangleleft$$

EXAMPLE 4   Find an explicit formula for the Fibonacci numbers.

Solution: Recall that the sequence of Fibonacci numbers satisfies the recurrence relation $f_n = f_{n-1} + f_{n-2}$ and also satisfies the initial conditions $f_0 = 0$ and $f_1 = 1$. The roots of the characteristic equation $r^2 - r - 1 = 0$ are $r_1 = (1 + \sqrt{5})/2$ and $r_2 = (1 - \sqrt{5})/2$. Therefore, from Theorem 1 it follows that the Fibonacci numbers are given by

$$f_n = \alpha_1 \left(\frac{1 + \sqrt{5}}{2}\right)^n + \alpha_2 \left(\frac{1 - \sqrt{5}}{2}\right)^n,$$

for some constants α_1 and α_2. The initial conditions $f_0 = 0$ and $f_1 = 1$ can be used to find these constants. We have

$$f_0 = \alpha_1 + \alpha_2 = 0,$$

$$f_1 = \alpha_1 \left(\frac{1 + \sqrt{5}}{2}\right) + \alpha_2 \left(\frac{1 - \sqrt{5}}{2}\right) = 1.$$

The solution to these simultaneous equations for α_1 and α_2 is

$$\alpha_1 = 1/\sqrt{5}, \qquad \alpha_2 = -1/\sqrt{5}.$$

Consequently, the Fibonacci numbers are given by

$$f_n = \frac{1}{\sqrt{5}} \left(\frac{1 + \sqrt{5}}{2}\right)^n - \frac{1}{\sqrt{5}} \left(\frac{1 - \sqrt{5}}{2}\right)^n. \qquad \blacktriangleleft$$

Theorem 1 does not apply when there is one characteristic root of multiplicity two. If this happens, then $a_n = nr_0^n$ is another solution of the recurrence relation when r_0 is a root of multiplicity two of the characteristic equation. Theorem 2 shows how to handle this case.

THEOREM 2   Let c_1 and c_2 be real numbers with $c_2 \neq 0$. Suppose that $r^2 - c_1 r - c_2 = 0$ has only one root r_0. A sequence $\{a_n\}$ is a solution of the recurrence relation $a_n = c_1 a_{n-1} + c_2 a_{n-2}$ if and only if $a_n = \alpha_1 r_0^n + \alpha_2 n r_0^n$, for $n = 0, 1, 2, \ldots$, where α_1 and α_2 are constants.

The proof of Theorem 2 is left as an exercise at the end of the section. Example 5 illustrates the use of this theorem.

EXAMPLE 5   What is the solution of the recurrence relation

$$a_n = 6a_{n-1} - 9a_{n-2}$$

with initial conditions $a_0 = 1$ and $a_1 = 6$?

Solution: The only root of $r^2 - 6r + 9 = 0$ is $r = 3$. Hence, the solution to this recurrence relation is

$$a_n = \alpha_1 3^n + \alpha_2 n 3^n$$

for some constants α_1 and α_2. Using the initial conditions, it follows that

$$a_0 = 1 = \alpha_1,$$
$$a_1 = 6 = \alpha_1 \cdot 3 + \alpha_2 \cdot 3.$$

Solving these two equations shows that $\alpha_1 = 1$ and $\alpha_2 = 1$. Consequently, the solution to this recurrence relation and the initial conditions is

$$a_n = 3^n + n3^n. \qquad \blacktriangleleft$$

We will now state the general result about the solution of linear homogeneous recurrence relations with constant coefficients, where the degree may be greater than two, under the assumption that the characteristic equation has distinct roots. The proof of this result will be left as an exercise for the reader.

THEOREM 3

Let $c_1, c_2, \ldots, c_k$ be real numbers. Suppose that the characteristic equation

$$r^k - c_1 r^{k-1} - \cdots - c_k = 0$$

has k distinct roots $r_1, r_2, \ldots, r_k$. Then a sequence $\{a_n\}$ is a solution of the recurrence relation

$$a_n = c_1 a_{n-1} + c_2 a_{n-2} + \cdots + c_k a_{n-k}$$

if and only if

$$a_n = \alpha_1 r_1^n + \alpha_2 r_2^n + \cdots + \alpha_k r_k^n$$

for $n = 0, 1, 2, \ldots$, where $\alpha_1, \alpha_2, \ldots, \alpha_k$ are constants.

We illustrate the use of the theorem with Example 6.

EXAMPLE 6   Find the solution to the recurrence relation

$$a_n = 6a_{n-1} - 11a_{n-2} + 6a_{n-3}$$

with the initial conditions $a_0 = 2, a_1 = 5$, and $a_2 = 15$.

Solution: The characteristic polynomial of this recurrence relation is

$$r^3 - 6r^2 + 11r - 6.$$

The characteristic roots are $r = 1, r = 2$, and $r = 3$, since $r^3 - 6r^2 + 11r - 6 = (r-1)(r-2)(r-3)$. Hence, the solutions to this recurrence relation are of the form

$$a_n = \alpha_1 \cdot 1^n + \alpha_2 \cdot 2^n + \alpha_3 \cdot 3^n.$$

To find the constants α_1, α_2, and α_3, use the initial conditions. This gives

$$a_0 = 2 = \alpha_1 + \alpha_2 + \alpha_3,$$
$$a_1 = 5 = \alpha_1 + \alpha_2 \cdot 2 + \alpha_3 \cdot 3,$$
$$a_2 = 15 = \alpha_1 + \alpha_2 \cdot 4 + \alpha_3 \cdot 9.$$

When these three simultaneous equations are solved for α_1, α_2, and α_3, we find that $\alpha_1 = 1, \alpha_2 = -1$, and $\alpha_3 = 2$. Hence, the unique solution to this recurrence relation and the given initial conditions is the sequence $\{a_n\}$ with

$$a_n = 1 - 2^n + 2 \cdot 3^n. \qquad \blacktriangleleft$$

We now state the most general result about linear homogeneous recurrence relations with constant coefficients, allowing the characteristic equation to have multiple roots. The key point is that for each root r of the characteristic equation, the general solution has a summand of the form $P(n)r^n$, where $P(n)$ is a polynomial of degree $m - 1$, with m the multiplicity of this root. We leave the proof of this result as a challenging exercise for the reader.

THEOREM 4

Let $c_1, c_2, \ldots, c_k$ be real numbers. Suppose that the characteristic equation

$$r^k - c_1 r^{k-1} - \cdots - c_k = 0$$

has t distinct roots $r_1, r_2, \ldots, r_t$ with multiplicities $m_1, m_2, \ldots, m_t$, respectively, so that $m_i \geq 1$ for $i = 1, 2, \ldots, t$ and $m_1 + m_2 + \cdots + m_t = k$. Then a sequence $\{a_n\}$ is a solution of the recurrence relation

$$a_n = c_1 a_{n-1} + c_2 a_{n-2} + \cdots + c_k a_{n-k}$$

if and only if

$$
\begin{aligned}
a_n = \; & (\alpha_{1,0} + \alpha_{1,1} n + \cdots + \alpha_{1,m_1-1} n^{m_1-1}) r_1^n \\
& + (\alpha_{2,0} + \alpha_{2,1} n + \cdots + \alpha_{2,m_2-1} n^{m_2-1}) r_2^n \\
& + \cdots + (\alpha_{t,0} + \alpha_{t,1} n + \cdots + \alpha_{t,m_t-1} n^{m_t-1}) r_t^n
\end{aligned}
$$

for $n = 0, 1, 2, \ldots$, where $\alpha_{i,j}$ are constants for $1 \leq i \leq t$ and $0 \leq j \leq m_i - 1$.

Example 7 illustrates how Theorem 4 is used to find the general form of a solution of a linear homogeneous recurrence relation when the characteristic equation has several repeated roots.

EXAMPLE 7
Suppose that the roots of the characteristic equation of a linear homogeneous recurrence relation are 2, 2, 2, 5, 5, and 9 (that is, there are three roots, the root 2 with multiplicity three, the root 5 with multiplicity two, and the root 9 with multiplicity one). What is the form of the general solution?

Solution: By Theorem 4, the general form of the solution is

$$(\alpha_{1,0} + \alpha_{1,1} n + \alpha_{1,2} n^2) 2^n + (\alpha_{2,0} + \alpha_{2,1} n) 5^n + \alpha_{3,0} 9^n. \qquad \blacktriangleleft$$

We now illustrate the use of Theorem 4 to solve a linear homogeneous recurrence relation with constant coefficients when the characteristic equation has a root of multiplicity three.

EXAMPLE 8
Find the solution to the recurrence relation

$$a_n = -3a_{n-1} - 3a_{n-2} - a_{n-3}$$

with initial conditions $a_0 = 1, a_1 = -2$, and $a_2 = -1$.

Solution: The characteristic equation of this recurrence relation is

$$r^3 + 3r^2 + 3r + 1 = 0.$$

Since $r^3 + 3r^2 + 3r + 1 = (r+1)^3$, there is a single root $r = -1$ of multiplicity three of the characteristic equation. By Theorem 4 the solutions of this recurrence relation are of the form

$$a_n = \alpha_{1,0}(-1)^n + \alpha_{1,1}n(-1)^n + \alpha_{1,2}n^2(-1)^n.$$

To find the constants $\alpha_{1,0}$, $\alpha_{1,1}$, and $\alpha_{1,2}$, use the initial conditions. This gives

$$a_0 = 1 = \alpha_{1,0},$$
$$a_1 = -2 = -\alpha_{1,0} - \alpha_{1,1} - \alpha_{1,2},$$
$$a_2 = -1 = \alpha_{1,0} + 2\alpha_{1,1} + 4\alpha_{1,2}.$$

The simultaneous solution of these three equations is $\alpha_{1,0} = 1, \alpha_{1,1} = 3,$ and $\alpha_{1,2} = -2$. Hence, the unique solution to this recurrence relation and the given initial conditions is the sequence $\{a_n\}$ with

$$a_n = (1 + 3n - 2n^2)(-1)^n. \qquad \blacktriangleleft$$

LINEAR NONHOMOGENEOUS RECURRENCE RELATIONS WITH CONSTANT COEFFICIENTS

We have seen how to solve linear homogeneous recurrence relations with constant coefficients. Is there a relatively simple technique for solving a linear, but not homogeneous, recurrence relation with constant coefficients, such as $a_n = 3a_{n-1} + 2n$? We will see that the answer is yes for certain families of such recurrence relations.

The recurrence relation $a_n = 3a_{n-1} + 2n$ is an example of a **linear nonhomogeneous recurrence relation with constant coefficients,** that is, a recurrence relation of the form

$$a_n = c_1 a_{n-1} + c_2 a_{n-2} + \cdots + c_k a_{n-k} + F(n),$$

where $c_1, c_2, \ldots, c_k$ are real numbers and $F(n)$ is a function not identically zero depending only on n. The recurrence relation

$$a_n = c_1 a_{n-1} + c_2 a_{n-2} + \cdots + c_k a_{n-k}$$

is called the **associated homogeneous recurrence relation.** It plays an important role in the solution of the nonhomogeneous recurrence relation.

EXAMPLE 9   Each of the recurrence relations $a_n = a_{n-1} + 2^n$, $a_n = a_{n-1} + a_{n-2} + n^2 + n + 1$, $a_n = 3a_{n-1} + n3^n$, and $a_n = a_{n-1} + a_{n-2} + a_{n-3} + n!$ is a linear nonhomogeneous recurrence relation with constant coefficients. The associated linear homogeneous recurrence relations are $a_n = a_{n-1}, a_n = a_{n-1} + a_{n-2}, a_n = 3a_{n-1},$ and $a_n = a_{n-1} + a_{n-2} + a_{n-3}$, respectively. $\qquad \blacktriangleleft$

The key fact about linear nonhomogeneous recurrence relations with constant coefficients is that every solution is the sum of a particular solution and a solution of the associated linear homogeneous recurrence relation, as Theorem 5 shows.

THEOREM 5

If $\{a_n^{(p)}\}$ is a particular solution of the nonhomogeneous linear recurrence relation with constant coefficients

$$a_n = c_1 a_{n-1} + c_2 a_{n-2} + \cdots + c_k a_{n-k} + F(n),$$

then every solution is of the form $\{a_n^{(p)} + a_n^{(h)}\}$, where $\{a_n^{(h)}\}$ is a solution of the associated homogeneous recurrence relation

$$a_n = c_1 a_{n-1} + c_2 a_{n-2} + \cdots + c_k a_{n-k}.$$

Proof: Since $\{a_n^{(p)}\}$ is a particular solution of the nonhomogeneous recurrence relation, we know that

$$a_n^{(p)} = c_1 a_{n-1}^{(p)} + c_2 a_{n-2}^{(p)} + \cdots + c_k a_{n-k}^{(p)} + F(n).$$

Now suppose that $\{b_n\}$ is a second solution of the nonhomogeneous recurrence relation, so that

$$b_n = c_1 b_{n-1} + c_2 b_{n-2} + \cdots + c_k b_{n-k} + F(n).$$

Subtracting the first of these two equations from the second shows that

$$b_n - a_n^{(p)} = c_1(b_{n-1} - a_{n-1}^{(p)}) + c_2(b_{n-2} - a_{n-2}^{(p)}) + \cdots + c_k(b_{n-k} - a_{n-k}^{(p)}).$$

It follows that $\{b_n - a_n^{(p)}\}$ is a solution of the associated homogeneous linear recurrence, say, $\{a_n^{(h)}\}$. Consequently, $b_n = a_n^{(p)} + a_n^{(h)}$ for all n. ◁

By Theorem 5, we see that the key to solving nonhomogeneous recurrence relations with constant coefficients is finding a particular solution. Then every solution is a sum of this solution and a solution of the associated homogeneous recurrence relation. Although there is no general method for finding such a solution that works for every function $F(n)$, there are techniques that work for certain types of functions $F(n)$, such as polynomials and powers of constants. This is illustrated in Examples 10 and 11.

EXAMPLE 10

Find all solutions of the recurrence relation $a_n = 3a_{n-1} + 2n$. What is the solution with $a_1 = 3$?

Solution: To solve this linear nonhomogeneous recurrence relation with constant coefficients, we need to solve its associated linear homogeneous equation and to find a particular solution for the given nonhomogeneous equation. The associated linear homogeneous equation is $a_n = 3a_{n-1}$. Its solutions are $a_n^{(h)} = \alpha 3^n$, where α is a constant.

We now find a particular solution. Since $F(n) = 2n$ is a polynomial in n of degree one, a reasonable trial solution is a linear function in n, say, $p_n = cn + d$, where c and d are constants. To determine whether there are any solutions of this form, suppose that $p_n = cn + d$ is such a solution. Then the equation $a_n = 3a_{n-1} + 2n$ becomes $cn + d = 3(c(n-1) + d) + 2n$. Simplifying and combining like terms gives $(2 + 2c)n + (2d - 3c) = 0$. It follows that $cn + d$ is a solution if and only if $2 + 2c = 0$ and $2d - 3c = 0$. This shows that $cn + d$ is a solution if and only if $c = -1$ and $d = -3/2$. Consequently, $a_n^{(p)} = -n - 3/2$ is a particular solution.

By Theorem 5 all solutions are of the form

$$a_n = a_n^{(p)} + a_n^{(h)} = -n - \frac{3}{2} + \alpha \cdot 3^n,$$

where α is a constant.

To find the solution with $a_1 = 3$, let $n = 1$ in the formula we obtained for the general solution. We find that $3 = -1 - 3/2 + 3\alpha$, which implies that $\alpha = 11/6$. The solution we seek is $a_n = -n - 3/2 + (11/6)3^n$. ◄

EXAMPLE 11 Find all solutions of the recurrence relation

$$a_n = 5a_{n-1} - 6a_{n-2} + 7^n.$$

Solution: This is a linear nonhomogeneous recurrence relation. The solutions of its associated homogeneous recurrence relation

$$a_n = 5a_{n-1} - 6a_{n-2}$$

are $a_n^{(h)} = \alpha_1 \cdot 3^n + \alpha_2 \cdot 2^n$, where α_1 and α_2 are constants. Since $F(n) = 7^n$, a reasonable trial solution is $a_n^{(p)} = C \cdot 7^n$, where C is a constant. Substituting the terms of this sequence into the recurrence relation implies that $C \cdot 7^n = 5C \cdot 7^{n-1} - 6C \cdot 7^{n-2} + 7^n$. Factoring out 7^{n-2}, this equation becomes $49C = 35C - 6C + 49$, which implies that $20C = 49$, or that $C = 49/20$. Hence, $a_n^{(p)} = (49/20)7^n$ is a particular solution. By Theorem 5, all solutions are of the form

$$a_n = \alpha_1 \cdot 3^n + \alpha_2 \cdot 2^n + (49/20)7^n.$$ ◄

In Examples 10 and 11, we made an educated guess that there are solutions of a particular form. In both cases we were able to find particular solutions. This was not an accident. Whenever $F(n)$ is the product of a polynomial in n and the nth power of a constant, we know exactly what form a particular solution has, as stated in Theorem 6. We leave the proof of Theorem 6 as a challenging exercise for the reader.

THEOREM 6 Suppose that $\{a_n\}$ satisfies the linear nonhomogeneous recurrence relation

$$a_n = c_1 a_{n-1} + c_2 a_{n-2} + \cdots + c_k a_{n-k} + F(n),$$

where $c_1, c_2, \ldots, c_k$ are real numbers and

$$F(n) = (b_t n^t + b_{t-1} n^{t-1} + \cdots + b_1 n + b_0)s^n,$$

where $b_0, b_1, \ldots, b_t$ and s are real numbers. When s is not a root of the characteristic equation of the associated linear homogeneous recurrence relation, there is a particular solution of the form

$$(p_t n^t + p_{t-1} n^{t-1} + \cdots + p_1 n + p_0)s^n.$$

When s is a root of this characteristic equation and its multiplicity is m, there is a particular solution of the form

$$n^m(p_t n^t + p_{t-1} n^{t-1} + \cdots + p_1 n + p_0)s^n.$$

Note that in the case when s is a root of multiplicity m of the characteristic equation of the associated linear homogeneous recurrence relation, the factor n^m ensures that the proposed particular solution will not already be a solution of the associated linear homogeneous recurrence relation. We next provide an example to illustrate the form of a particular solution provided by Theorem 6.

EXAMPLE 12   What form does a particular solution of the linear nonhomogeneous recurrence relation $a_n = 6a_{n-1} - 9a_{n-2} + F(n)$ have when $F(n) = 3^n$, $F(n) = n3^n$, $F(n) = n^2 2^n$, and $F(n) = (n^2 + 1)3^n$?

Solution: The associated linear homogeneous recurrence relation is $a_n = 6a_{n-1} - 9a_{n-2}$. Its characteristic equation, $r^2 - 6r + 9 = (r - 3)^2 = 0$, has a single root, 3, of multiplicity two. To apply Theorem 6, with $F(n)$ of the form $P(n)s^n$, where $P(n)$ is a polynomial and s is a constant, we need to ask whether s is a root of this characteristic equation.

Since $s = 3$ is a root with multiplicity $m = 2$ but $s = 2$ is not a root, Theorem 6 tells us that a particular solution has the form $p_0 n^2 3^n$ if $F(n) = 3^n$, the form $n^2(p_1 n + p_0)3^n$ if $F(n) = n3^n$, the form $(p_2 n^2 + p_1 n + p_0)2^n$ if $F(n) = n^2 2^n$, and the form $n^2(p_2 n^2 + p_1 n + p_0)3^n$ if $F(n) = (n^2 + 1)3^n$.   ◀

Care must be taken when $s = 1$ when solving recurrence relations of the type covered by Theorem 6. In particular, to apply this theorem with $F(n) = b_t n^t + b_{t-1} n^{t-1} + \cdots + b_1 n + b_0$, the parameter s takes the value $s = 1$ (even though the term 1^n does not explicitly appear). By the theorem, the form of the solution then depends on whether 1 is a root of the characteristic equation of the associated linear homogeneous recurrence relation. This is illustrated in Example 13, which shows how Theorem 6 can be used to find a formula for the sum of the first n positive integers.

EXAMPLE 13   Let a_n be the sum of the first n positive integers, so that

$$a_n = \sum_{k=1}^{n} k.$$

Note that a_n satisfies the linear nonhomogeneous recurrence relation

$$a_n = a_{n-1} + n.$$

(To obtain a_n, the sum of the first n positive integers, from a_{n-1}, the sum of the first $n - 1$ positive integers, we add n.) Note that the initial condition is $a_1 = 1$.

The associated linear homogeneous recurrence relation for a_n is

$$a_n = a_{n-1}.$$

The solutions of this homogeneous recurrence relation are given by $a_n^{(h)} = c(1)^n = c$, where c is a constant. To find all solutions of $a_n = a_{n-1} + n$, we need find only a single particular solution. By Theorem 6, since $F(n) = n = n \cdot (1)^n$ and $s = 1$ is a root of degree one of the characteristic equation of the associated linear homogeneous recurrence relations, there is a particular solution of the form $n(p_1 n + p_0) = p_1 n^2 + p_0 n$.

Inserting this into the recurrence relation gives $p_1 n^2 + p_0 n = p_1(n - 1)^2 + p_0(n - 1) + n$. Simplifying, we see that $n(2p_1 - 1) + (p_0 - p_1) = 0$, which means that $2p_1 - 1 = 0$ and $p_0 - p_1 = 0$, so $p_0 = p_1 = 1/2$. Hence,

$$a_n^{(p)} = \frac{n^2}{2} + \frac{n}{2} = \frac{n(n + 1)}{2}$$

is a particular solution. Hence, all solutions of the original recurrence relation $a_n = a_{n-1} + n$ are given by $a_n = a_n^{(h)} + a_n^{(p)} = c + n(n + 1)/2$. Since $a_1 = 1$, we have $1 = a_1 = c + 1 \cdot 2/2 = c + 1$, so $c = 0$. It follows that $a_n = n(n + 1)/2$. (This is the same formula given in Table 2 in Section 3.2 and derived previously.)   ◀

Exercises

1. Determine which of these are linear homogeneous recurrence relations with constant coefficients. Also, find the degree of those that are.

   a) $a_n = 3a_{n-1} + 4a_{n-2} + 5a_{n-3}$
   b) $a_n = 2na_{n-1} + a_{n-2}$ **c)** $a_n = a_{n-1} + a_{n-4}$
   d) $a_n = a_{n-1} + 2$ **e)** $a_n = a_{n-1}^2 + a_{n-2}$
   f) $a_n = a_{n-2}$ **g)** $a_n = a_{n-1} + n$

2. Determine which of these are linear homogeneous recurrence relations with constant coefficients. Also, find the degree of those that are.

   a) $a_n = 3a_{n-2}$ **b)** $a_n = 3$
   c) $a_n = a_{n-1}^2$ **d)** $a_n = a_{n-1} + 2a_{n-3}$
   e) $a_n = a_{n-1}/n$
   f) $a_n = a_{n-1} + a_{n-2} + n + 3$
   g) $a_n = 4a_{n-2} + 5a_{n-4} + 9a_{n-7}$

3. Solve these recurrence relations together with the initial conditions given.

   a) $a_n = 2a_{n-1}$ for $n \geq 1, a_0 = 3$
   b) $a_n = a_{n-1}$ for $n \geq 1, a_0 = 2$
   c) $a_n = 5a_{n-1} - 6a_{n-2}$ for $n \geq 2, a_0 = 1, a_1 = 0$
   d) $a_n = 4a_{n-1} - 4a_{n-2}$ for $n \geq 2, a_0 = 6, a_1 = 8$
   e) $a_n = -4a_{n-1} - 4a_{n-2}$ for $n \geq 2, a_0 = 0, a_1 = 1$
   f) $a_n = 4a_{n-2}$ for $n \geq 2, a_0 = 0, a_1 = 4$
   g) $a_n = a_{n-2}/4$ for $n \geq 2, a_0 = 1, a_1 = 0$

4. Solve these recurrence relations together with the initial conditions given.

   a) $a_n = a_{n-1} + 6a_{n-2}$ for $n \geq 2, a_0 = 3, a_1 = 6$
   b) $a_n = 7a_{n-1} - 10a_{n-2}$ for $n \geq 2, a_0 = 2, a_1 = 1$
   c) $a_n = 6a_{n-1} - 8a_{n-2}$ for $n \geq 2, a_0 = 4, a_1 = 10$
   d) $a_n = 2a_{n-1} - a_{n-2}$ for $n \geq 2, a_0 = 4, a_1 = 1$
   e) $a_n = a_{n-2}$ for $n \geq 2, a_0 = 5, a_1 = -1$
   f) $a_n = -6a_{n-1} - 9a_{n-2}$ for $n \geq 2, a_0 = 3, a_1 = -3$
   g) $a_{n+2} = -4a_{n+1} + 5a_n$ for $n \geq 0, a_0 = 2, a_1 = 8$

5. How many different messages can be transmitted in n microseconds using the two signals described in Exercise 35 of Section 6.1?

6. How many different messages can be transmitted in n microseconds using three different signals if one signal requires 1 microsecond for transmittal, the other two signals require 2 microseconds each for transmittal, and a signal in a message is followed immediately by the next signal?

7. In how many ways can a $2 \times n$ rectangular board be tiled using 1×2 and 2×2 pieces?

8. A model for the number of lobsters caught per year is based on the assumption that the number of lobsters caught in a year is the average of the number caught in the two previous years.

   a) Find a recurrence relation for $\{L_n\}$, where L_n is the number of lobsters caught in year n, under the assumption for this model.

   b) Find L_n if 100,000 lobsters were caught in year 1 and 300,000 were caught in year 2.

9. A deposit of \$100,000 is made to an investment fund at the beginning of a year. On the last day of each year two dividends are awarded. The first dividend is 20% of the amount in the account during that year. The second dividend is 45% of the amount in the account in the previous year.

   a) Find a recurrence relation for $\{P_n\}$, where P_n is the amount in the account at the end of n years if no money is ever withdrawn.

   b) How much is in the account after n years if no money has been withdrawn?

*10. Prove Theorem 2.

11. The **Lucas numbers** satisfy the recurrence relation

    $$L_n = L_{n-1} + L_{n-2},$$

    and the initial conditions $L_0 = 2$ and $L_1 = 1$.

    a) Show that $L_n = f_{n-1} + f_{n+1}$ for $n = 2, 3, \ldots$, where f_n is the nth Fibonacci number.
    b) Find an explicit formula for the Lucas numbers.

12. Find the solution to $a_n = 2a_{n-1} + a_{n-2} - 2a_{n-3}$ for $n = 3, 4, 5, \ldots$, with $a_0 = 3, a_1 = 6$, and $a_2 = 0$.

13. Find the solution to $a_n = 7a_{n-2} + 6a_{n-3}$ with $a_0 = 9$, $a_1 = 10$, and $a_2 = 32$.

14. Find the solution to $a_n = 5a_{n-2} - 4a_{n-4}$ with $a_0 = 3$, $a_1 = 2, a_2 = 6$, and $a_3 = 8$.

15. Find the solution to $a_n = 2a_{n-1} + 5a_{n-2} - 6a_{n-3}$ with $a_0 = 7, a_1 = -4$, and $a_2 = 8$.

*16. Prove Theorem 3.

17. Prove this identity relating the Fibonacci numbers and the binomial coefficients:

    $$f_{n+1} = C(n, 0) + C(n-1, 1) + \cdots + C(n-k, k),$$

    where n is a positive integer and $k = \lfloor n/2 \rfloor$. [*Hint:* Let $a_n = C(n, 0) + C(n-1, 1) + \cdots + C(n-k, k)$. Show that the sequence $\{a_n\}$ satisfies the same recurrence relation and initial conditions satisfied by the sequence of Fibonacci numbers.]

18. Solve the recurrence relation $a_n = 6a_{n-1} - 12a_{n-2} + 8a_{n-3}$ with $a_0 = -5, a_1 = 4$, and $a_2 = 88$.

19. Solve the recurrence relation $a_n = -3a_{n-1} - 3a_{n-2} - a_{n-3}$ with $a_0 = 5, a_1 = -9$, and $a_2 = 15$.

20. Find the general form of the solutions of the recurrence relation $a_n = 8a_{n-2} - 16a_{n-4}$.

21. What is the general form of the solutions of a linear homogeneous recurrence relation if its characteristic equation has roots 1, 1, 1, 1, −2, −2, −2, 3, 3, −4?

22. What is the general form of the solutions of a linear homogeneous recurrence relation if its characteristic equation has the roots −1, −1, −1, 2, 2, 5, 5, 7?

23. Consider the nonhomogeneous linear recurrence relation $a_n = 3a_{n-1} + 2^n$.

a) Show that $a_n = -2^{n+1}$ is a solution of this recurrence relation.

b) Use Theorem 5 to find all solutions of this recurrence relation.

c) Find the solution with $a_0 = 1$.

24. Consider the nonhomogeneous linear recurrence relation $a_n = 2a_{n-1} + 2^n$.

a) Show that $a_n = n2^n$ is a solution of this recurrence relation.

b) Use Theorem 5 to find all solutions of this recurrence relation.

c) Find the solution with $a_0 = 2$.

25. a) Determine values of the constants A and B so that $a_n = An + B$ is a solution of the recurrence relation $a_n = 2a_{n-1} + n + 5$.

b) Use Theorem 5 to find all the solutions of this recurrence relation.

c) Find the solution of this recurrence relation with $a_0 = 4$.

26. What is the general form of the particular solution of the linear nonhomogeneous recurrence relation $a_n = 6a_{n-1} - 12a_{n-2} + 8a_{n-3} + F(n)$ if

a) $F(n) = n^2$? **b)** $F(n) = 2^n$?
c) $F(n) = n2^n$? **d)** $F(n) = (-2)^n$?
e) $F(n) = n^2 2^n$? **f)** $F(n) = n^3(-2)^n$?
g) $F(n) = 3$?

27. What is the general form of the particular solution of the linear nonhomogeneous recurrence relation $a_n = 8a_{n-2} - 16a_{n-4} + F(n)$ if

a) $F(n) = n^3$? **b)** $F(n) = (-2)^n$?
c) $F(n) = n2^n$? **d)** $F(n) = n^2 4^n$?
e) $F(n) = (n^2 - 2)(-2)^n$?
f) $F(n) = n^4 2^n$? **g)** $F(n) = 2$?

28. a) Find all solutions of the recurrence relation $a_n = 2a_{n-1} + 2n^2$.

b) Find the solution of the recurrence relation in part (a) with initial condition $a_1 = 4$.

29. a) Find all solutions of the recurrence relation $a_n = 2a_{n-1} + 3^n$.

b) Find the solution of the recurrence relation in part (a) with initial condition $a_1 = 5$.

30. a) Find all solutions of the recurrence relation $a_n = -5a_{n-1} - 6a_{n-2} + 42 \cdot 4^n$.

b) Find the solution of this recurrence relation with $a_1 = 56$ and $a_2 = 278$.

31. Find all solutions of the recurrence relation $a_n = 5a_{n-1} - 6a_{n-2} + 2^n + 3n$. (*Hint:* Look for a particular solution of the form $qn2^n + p_1n + p_2$, where q, p_1, and p_2 are constants.)

32. Find the solution of the recurrence relation $a_n = 2a_{n-1} + 3 \cdot 2^n$.

33. Find all solutions of the recurrence relation $a_n = 4a_{n-1} - 4a_{n-2} + (n + 1)2^n$.

34. Find all solutions of the recurrence relation $a_n = 7a_{n-1} - 16a_{n-2} + 12a_{n-3} + n4^n$ with $a_0 = -2$, $a_1 = 0$, and $a_2 = 5$.

35. Find the solution of the recurrence relation $a_n = 4a_{n-1} - 3a_{n-2} + 2^n + n + 3$ with $a_0 = 1$ and $a_1 = 4$.

36. Let a_n be the sum of the first n perfect squares, that is, $a_n = \sum_{k=1}^{n} k^2$. Show that the sequence $\{a_n\}$ satisfies the linear nonhomogeneous recurrence relation $a_n = a_{n-1} + n^2$ and the initial condition $a_1 = 1$. Use Theorem 6 to determine a formula for a_n by solving this recurrence relation.

37. Let a_n be the sum of the first n triangular numbers, that is, $a_n = \sum_{k=1}^{n} t_k$, where $t_k = k(k+1)/2$. Show that $\{a_n\}$ satisfies the linear nonhomogeneous recurrence relation $a_n = a_{n-1} + n(n + 1)/2$ and the initial condition $a_1 = 1$. Use Theorem 6 to determine a formula for a_n by solving this recurrence relation.

38. a) Find the characteristic roots of the linear homogeneous recurrence relation $a_n = 2a_{n-1} - 2a_{n-2}$. (*Note:* These are complex numbers.)

b) Find the solution of the recurrence relation in part (a) with $a_0 = 1$ and $a_1 = 2$.

*39. a) Find the characteristic roots of the linear homogeneous recurrence relation $a_n = a_{n-4}$. (*Note:* These include complex numbers.)

b) Find the solution of the recurrence relation in part (a) with $a_0 = 1, a_1 = 0, a_2 = -1$, and $a_3 = 1$.

*40. Solve the simultaneous recurrence relations

$$a_n = 3a_{n-1} + 2b_{n-1}$$
$$b_n = a_{n-1} + 2b_{n-1}$$

with $a_0 = 1$ and $b_0 = 2$.

*41. a) Use the formula found in Example 4 for f_n, the nth Fibonacci number, to show that f_n is the integer closest to

$$\frac{1}{\sqrt{5}} \left(\frac{1 + \sqrt{5}}{2} \right)^n.$$

b) Determine for which n f_n is greater than

$$\frac{1}{\sqrt{5}} \left(\frac{1 + \sqrt{5}}{2} \right)^n$$

and for which n f_n is less than

$$\frac{1}{\sqrt{5}} \left(\frac{1 + \sqrt{5}}{2} \right)^n.$$

42. Show that if $a_n = a_{n-1} + a_{n-2}, a_0 = s$ and $a_1 = t$, where s and t are constants, then $a_n = sf_{n-1} + tf_n$ for all positive integers n.

43. Express the solution of the linear nonhomogenous recurrence relation $a_n = a_{n-1} + a_{n-2} + 1$ for $n \geq 2$ where $a_0 = 0$ and $a_1 = 1$ in terms of the Fibonacci numbers. (*Hint:* Let $b_n = a_n + 1$ and apply Exercise 42 to the sequence b_n.)

*44. (Linear algebra required) Let $\mathbf{A}_n$ be the $n \times n$ matrix with 2s on its main diagonal, 1s in all positions next to a diagonal element, and 0s everywhere else. Find a recurrence relation for d_n, the determinant of $\mathbf{A}_n$. Solve this recurrence relation to find a formula for d_n.

45. Suppose that each pair of a genetically engineered species of rabbits left on an island produces two new pairs of rabbits at the age of 1 month and six new pairs of rabbits at the age of 2 months and every month afterward. None of the rabbits ever die or leave the island.

    a) Find a recurrence relation for the number of pairs of rabbits on the island n months after one newborn pair is left on the island.

    b) By solving the recurrence relation in (a) determine the number of pairs of rabbits on the island n months after one pair is left on the island.

46. Suppose that there are two goats on an island initially. The number of goats on the island doubles every year by natural reproduction, and some goats are either added or removed each year.

    a) Construct a recurrence relation for the number of goats on the island at the start of the nth year, assuming that during each year an extra 100 goats are put on the island.

    b) Solve the recurrence relation from part (a) to find the number of goats on the island at the start of the nth year.

    c) Construct a recurrence relation for the number of goats on the island at the start of the nth year, assuming that n goats are removed during the nth year for each $n \geq 3$.

    d) Solve the recurrence relation in part (c) for the number of goats on the island at the start of the nth year.

47. A new employee at an exciting new software company starts with a salary of \$50,000 and is promised that at the end of each year her salary will be double her salary the previous year, with an extra increment of \$10,000 for each year she has been with the company.

    a) Construct a recurrence relation for her salary for her nth year of employment.

    b) Solve this recurrence relation to find her salary for her nth year of employment.

Some linear recurrence relations that do not have constant coefficients can be systematically solved. This is the case for recurrence relations of the form $f(n)a_n = g(n)a_{n-1} + h(n)$. Exercises 48–50 illustrate this.

*48. a) Show that the recurrence relation

$$f(n)a_n = g(n)a_{n-1} + h(n),$$

for $n \geq 1$, and with $a_0 = C$, can be reduced to a recurrence relation of the form

$$b_n = b_{n-1} + Q(n)h(n),$$

where $b_n = g(n+1)Q(n+1)a_n$, with

$$Q(n) = (f(1)f(2) \cdots f(n-1))/(g(1)g(2) \cdots g(n)).$$

    b) Use part (a) to solve the original recurrence relation to obtain

$$a_n = \frac{C + \sum_{i=1}^{n} Q(i)h(i)}{g(n+1)Q(n+1)}.$$

*49. Use Exercise 48 to solve the recurrence relation $(n+1)a_n = (n+3)a_{n-1} + n$, for $n \geq 1$, with $a_0 = 1$.

50. It can be shown that the average number of comparisons made by the quick sort algorithm (described in the exercise set of Section 3.5), when sorting n elements in random order, satisfies the recurrence relation

$$C_n = n + 1 + \frac{2}{n} \sum_{k=0}^{n-1} C_k$$

for $n = 1, 2, \ldots$, with initial condition $C_0 = 0$.

    a) Show that $\{C_n\}$ also satisfies the recurrence relation $nC_n = (n+1)C_{n-1} + 2n$ for $n = 1, 2, \ldots$.

    b) Use Exercise 48 to solve the recurrence relation in part (a) to find an explicit formula for C_n.

**51. Prove Theorem 4.

**52. Prove Theorem 6.

53. Solve the recurrence relation $T(n) = nT^2(n/2)$ with initial condition $T(1) = 6$. [*Hint:* Let $n = 2^k$ and then make the substitution $a_k = \log T(2^k)$ to obtain a linear nonhomogeneous recurrence relation.]

6.3 Divide-and-Conquer Algorithms and Recurrence Relations

INTRODUCTION

Many recursive algorithms take a problem with a given input and divide it into one or more smaller problems. This reduction is successively applied until the solutions of the smaller problems can be found quickly. For instance, we perform a binary search by reducing the search for an element in a list to the search for this element in a list half as long. We successively apply this reduction until one element is left. When we sort a list of integers using the merge sort, we split the list into two halves of equal size and sort each half separately. We then merge the two sorted halves. Another example of this type of recursive algorithm is a procedure for multiplying integers that reduces the problem of the multiplication of two integers to three multiplications of pairs of integers with half as many bits. This reduction is successively applied until integers with one bit are obtained. These procedures are called **divide-and-conquer algorithms** because they *divide* a problem into one or more instances of the same problem of smaller size and they *conquer* the problem by using the solutions of the smaller problems to find a solution of the original problem, perhaps with some additional work.

In this section we will show how recurrence relations can be used to analyze the computational complexity of divide-and-conquer algorithms. We will use these recurrence relations to estimate the number of operations used by many different divide-and-conquer algorithms, including several that we introduce in this section.

DIVIDE-AND-CONQUER RECURRENCE RELATIONS

Suppose that a recursive algorithm divides a problem of size n into a subproblems, where each subproblem is of size n/b (for simplicity, assume that n is a multiple of b; in reality, the smaller problems are often of size equal to the nearest integers either less than or equal to, or greater than or equal to, n/b). Also, suppose that a total of $g(n)$ extra operations are required in the conquer step of the algorithm to combine the solutions of the subproblems into a solution of the original problem. Then, if $f(n)$ represents the number of operations required to solve the problem of size n, it follows that f satisfies the recurrence relation

$$f(n) = af(n/b) + g(n).$$

This is called a **divide-and-conquer recurrence relation.**

We will first set up the divide-and-conquer recurrence relations that can be used to study the complexity of some important algorithms. Then we will show how to use these divide-and-conquer recurrence relations to estimate the complexity of these algorithms.

EXAMPLE 1

Binary Search We introduced a binary search algorithm in Section 2.1. This binary search algorithm reduces the search for an element in a search sequence of size n to the binary search for this element in a search sequence of size $n/2$, when n is even. (Hence, the problem of size n has been reduced to *one* problem of size $n/2$.) Two comparisons are needed to implement this reduction (one to determine which half of the list to use and the other to determine whether any terms of the list remain). Hence, if $f(n)$ is the number of comparisons required to search for an element in a search sequence of size n, then

$$f(n) = f(n/2) + 2$$

when n is even. ◄

EXAMPLE 2

Finding the Maximum and Minimum of a Sequence Consider the following algorithm for locating the maximum and minimum elements of a sequence $a_1, a_2, \ldots, a_n$. If $n = 1$, then a_1 is the maximum and the minimum. If $n > 1$, split the sequence into two sequences, either where both have the same number of elements or where one of the sequences has one more element than the other. The problem is reduced to finding the maximum and minimum of each of the two smaller sequences. The solution to the original problem results from the comparison of the separate maxima and minima of the two smaller sequences to obtain the overall maximum and minimum.

Let $f(n)$ be the total number of comparisons needed to find the maximum and minimum elements of the sequence with n elements. We have shown that a problem of size n can be reduced into two problems of size $n/2$, when n is even, using two comparisons, one to compare the maxima of the two sequences and the other to compare the minima of the two sequences. This gives the recurrence relation

$$f(n) = 2f(n/2) + 2$$

when n is even. ◀

EXAMPLE 3 **Merge Sort** The merge sort algorithm (introduced in Section 3.5) splits a list to be sorted with n items, where n is even, into two lists with $n/2$ elements each, and uses fewer than n comparisons to merge the two sorted lists of $n/2$ items each into one sorted list. Consequently, the number of comparisons used by the merge sort to sort a list of n elements is less than $M(n)$, where the function $M(n)$ satisfies the divide-and-conquer recurrence relation

$$M(n) = 2M(n/2) + n.$$

◀

EXAMPLE 4 **Fast Multiplication of Integers** Surprisingly, there are more efficient algorithms than the conventional algorithm (described in Section 2.5) for multiplying integers. One of these algorithms, which uses a divide-and-conquer technique, will be described here. This fast multiplication algorithm proceeds by splitting each of two $2n$-bit integers into two blocks, each with n bits. Then, the original multiplication is reduced from the multiplication of two $2n$-bit integers to three multiplications of n-bit integers, plus shifts and additions.

Suppose that a and b are integers with binary expansions of length $2n$ (add initial bits of zero in these expansions if necessary to make them the same length). Let

$$a = (a_{2n-1}a_{2n-2}\cdots a_1a_0)_2 \quad \text{and} \quad b = (b_{2n-1}b_{2n-2}\cdots b_1b_0)_2.$$

Let

$$a = 2^n A_1 + A_0, \quad b = 2^n B_1 + B_0,$$

where

$$A_1 = (a_{2n-1}\cdots a_{n+1}a_n)_2, \quad A_0 = (a_{n-1}\cdots a_1a_0)_2,$$
$$B_1 = (b_{2n-1}\cdots b_{n+1}b_n)_2, \quad B_0 = (b_{n-1}\cdots b_1b_0)_2.$$

The algorithm for fast multiplication of integers is based on the fact that ab can be rewritten as

$$ab = (2^{2n} + 2^n)A_1B_1 + 2^n(A_1 - A_0)(B_0 - B_1) + (2^n + 1)A_0B_0.$$

The important fact about this identity is that it shows that the multiplication of two $2n$-bit integers can be carried out using three multiplications of n-bit integers, together with additions, subtractions, and shifts. This shows that if $f(n)$ is the total number of bit operations needed to multiply two n-bit integers, then

$$f(2n) = 3f(n) + Cn.$$

The reasoning behind this equation is as follows. The three multiplications of n-bit integers are carried out using $3f(n)$-bit operations. Each of the additions, subtractions, and shifts uses a constant multiple of n-bit operations, and Cn represents the total number of bit operations used by these operations. ◀

EXAMPLE 5 **Fast Matrix Multiplication** In Example 5 of Section 2.7 we showed that multiplying two $n \times n$ matrices using the definition of matrix multiplication required n^3 multiplications and $n^2(n-1)$ additions. Consequently, computing the product of two $n \times n$ matrices in this way requires $O(n^3)$ operations (multiplications and additions). Surprisingly, there are more efficient divide-and-conquer algorithms for multiplying two $n \times n$ matrices. Such an algorithm, invented by V. Strassen in 1969, reduces the multiplication of two $n \times n$ matrices, when n is even, to seven multiplications of two $(n/2) \times (n/2)$ matrices and 15 additions of $(n/2) \times (n/2)$ matrices. (See [CoLeRiSt01] for the details of this algorithm.) Hence, if $f(n)$ is the number of operations (multiplications and additions) used, it follows that

$$f(n) = 7f(n/2) + 15n^2/4$$

when n is even. ◀

As Examples 1–5 show, recurrence relations of the form $f(n) = af(n/b) + g(n)$ arise in many different situations. It is possible to derive estimates of the size of functions that satisfy such recurrence relations. Suppose that f satisfies this recurrence relation whenever n is divisible by b. Let $n = b^k$, where k is a positive integer. Then

$$
\begin{aligned}
f(n) &= af(n/b) + g(n) \\
&= a^2 f(n/b^2) + ag(n/b) + g(n) \\
&= a^3 f(n/b^3) + a^2 g(n/b^2) + ag(n/b) + g(n) \\
&\ \ \vdots \\
&= a^k f(n/b^k) + \sum_{j=0}^{k-1} a^j g(n/b^j).
\end{aligned}
$$

Since $n/b^k = 1$, it follows that

$$f(n) = a^k f(1) + \sum_{j=0}^{k-1} a^j g(n/b^j).$$

We can use this equation for $f(n)$ to estimate the size of functions that satisfy divide-and-conquer relations.

THEOREM 1 Let f be an increasing function that satisfies the recurrence relation

$$f(n) = af(n/b) + c$$

whenever n is divisible by b, where $a \geq 1$, b is an integer greater than 1, and c is a positive real number. Then

$$
f(n) \text{ is }
\begin{cases}
O(n^{\log_b a}) & \text{if } a > 1, \\
O(\log n) & \text{if } a = 1.
\end{cases}
$$

Furthermore, when $n = b^k$, where k is a positive integer,

$$f(n) = C_1 n^{\log_b a} + C_2,$$

where $C_1 = f(1) + c/(a-1)$ and $C_2 = -c/(a-1)$.

Proof: First let $n = b^k$. From the expression for $f(n)$ obtained in the discussion preceding the theorem, with $g(n) = c$, we have

$$f(n) = a^k f(1) + \sum_{j=0}^{k-1} a^j c = a^k f(1) + c \sum_{j=0}^{k-1} a^j.$$

First consider the case when $a = 1$. Then

$$f(n) = f(1) + ck.$$

Since $n = b^k$, we have $k = \log_b n$. Hence

$$f(n) = f(1) + c \log_b n.$$

When n is not a power of b, we have $b^k < n < b^{k+1}$, for a positive integer k. Since f is increasing, it follows that $f(n) \le f(b^{k+1}) = f(1) + c(k + 1) = (f(1) + c) + ck \le (f(1) + c) + c \log_b n$. Therefore, in both cases, $f(n)$ is $O(\log n)$ when $a = 1$.

Now suppose that $a > 1$. First assume that $n = b^k$ where k is a positive integer. From the formula for the sum of terms of a geometric progression (Theorem 1 in Section 3.2), it follows that

$$f(n) = a^k f(1) + c(a^k - 1)/(a - 1)$$

$$= a^k [f(1) + c/(a - 1)] - c/(a - 1)$$

$$= C_1 n^{\log_b a} + C_2,$$

since $a^k = a^{\log_b n} = n^{\log_b a}$ (see Exercise 4 in Appendix 1), where $C_1 = f(1) + c/(a-1)$ and $C_2 = -c/(a - 1)$.

Now suppose that n is not a power of b. Then $b^k < n < b^{k+1}$ where k is a nonnegative integer. Since f is increasing,

$$f(n) \le f(b^{k+1}) = C_1 a^{k+1} + C_2$$

$$\le (C_1 a) a^{\log_b n} + C_2$$

$$\le (C_1 a) n^{\log_b a} + C_2,$$

since $k \le \log_b n < k + 1$.

Hence, we have $f(n)$ is $O(n^{\log_b a})$.   ◁

Examples 6–9 illustrate how Theorem 1 is used.

EXAMPLE 6   Let $f(n) = 5 f(n/2) + 3$ and $f(1) = 7$. Find $f(2^k)$ where k is a positive integer. Also, estimate $f(n)$ if f is an increasing function.

Solution: From the proof of Theorem 1, with $a = 5, b = 2$, and $c = 3$, we see that if $n = 2^k$, then

$$f(n) = a^k [f(1) + c/(a - 1)] + [-c/(a - 1)]$$

$$= 5^k [7 + (3/4)] - 3/4$$

$$= 5^k (31/4) - 3/4.$$

Also, if $f(n)$ is increasing, Theorem 1 shows that $f(n)$ is $O(n^{\log_b a}) = O(n^{\log 5})$.   ◀

Extra Examples

We can use Theorem 1 to estimate the computational complexity of the binary search algorithm and the algorithm given in Example 2 for locating the minimum and maximum of a sequence.

EXAMPLE 7 Estimate the number of comparisons used by a binary search.

Solution: In Example 1 it was shown that $f(n) = f(n/2) + 2$ when n is even, where f is the number of comparisons required to perform a binary search on a sequence of size n. Hence, from Theorem 1, it follows that $f(n)$ is $O(\log n)$. ◀

EXAMPLE 8 Estimate the number of comparisons used to locate the maximum and minimum elements in a sequence using the algorithm given in Example 2.

Solution: In Example 2 we showed that $f(n) = 2f(n/2) + 2$, when n is even, where f is the number of comparisons needed by this algorithm. Hence, from Theorem 1, it follows that $f(n)$ is $O(n^{\log 2}) = O(n)$. ◀

We now state a more general, and more complicated, theorem, which has Theorem 1 as a special case. This theorem (or more powerful versions) is sometimes known as the Master Theorem because it is useful in analyzing the complexity of many important divide-and-conquer algorithms.

THEOREM 2 **MASTER THEOREM** Let f be an increasing function that satisfies the recurrence relation

$$f(n) = af(n/b) + cn^d$$

whenever $n = b^k$, where k is a positive integer, $a \geq 1$, b is an integer greater than 1, and c and d are real numbers with c positive and d nonnegative. Then

$$f(n) \text{ is } \begin{cases} O(n^d) & \text{if } a < b^d, \\ O(n^d \log n) & \text{if } a = b^d, \\ O(n^{\log_b a}) & \text{if } a > b^d. \end{cases}$$

The proof of Theorem 2 is left for the reader as Exercises 29–33 at the end of this section.

EXAMPLE 9 **Complexity of Merge Sort** In Example 3 we explained that the number of comparisons used by the merge sort to sort a list of n elements is less than $M(n)$, where $M(n) = 2M(n/2) + n$. By the Master Theorem (Theorem 2) we find that $M(n)$ is $O(n \log n)$, which agrees with the estimate found in Section 3.5. ◀

EXAMPLE 10 Estimate the number of bit operations needed to multiply two n-bit integers using the fast multiplication algorithm described in Example 4.

Solution: Example 4 shows that $f(n) = 3f(n/2) + Cn$, when n is even, where $f(n)$ is the number of bit operations required to multiply two n-bit integers using the fast multiplication algorithm. Hence, from the Master Theorem (Theorem 2), it follows that $f(n)$ is $O(n^{\log 3})$. Note that $\log 3 \sim 1.6$. Since the conventional algorithm for multiplication uses $O(n^2)$ bit operations, the fast multiplication algorithm is a substantial improvement over the conventional algorithm in terms of time complexity for sufficiently large integers, including large integers that occur in practical applications. ◀

EXAMPLE 11   Estimate the number of multiplications and additions required to multiply two $n \times n$ matrices using the matrix multiplication algorithm referred to in Example 5.

Solution: Let $f(n)$ denote the number of additions and multiplications used by the algorithm mentioned in Example 5 to multiply two $n \times n$ matrices. We have $f(n) = 7f(n/2) + 15n^2/4$, when n is even. Hence, from the Master Theorem (Theorem 2), it follows that $f(n)$ is $O(n^{\log 7})$. Note that $\log 7 \sim 2.8$. Since the conventional algorithm for multiplying two $n \times n$ matrices uses $O(n^3)$ additions and multiplications, it follows that for sufficiently large integers n, including those that occur in many practical applications, this algorithm is substantially more efficient in time complexity than the conventional algorithm.   ◀

THE CLOSEST-PAIR PROBLEM We conclude this section by introducing a divide-and-conquer algorithm from computational geometry, the part of discrete mathematics devoted to algorithms that solve geometric problems.

EXAMPLE 12   **The Closest-Pair Problem** Consider the problem of determining the closest pair of

Links

points in a set of n points $(x_1, y_1), \ldots, (x_n, y_n)$ in the plane, where the distance between two points (x_i, y_i) and (x_j, y_j) is the usual Euclidean distance $\sqrt{(x_i - x_j)^2 + (y_i - y_j)^2}$. This problem arises in many applications such as determining the closest pair of airplanes in the air space at a particular altitude being managed by an air traffic controller. How can this closest pair of points be found in an efficient way?

Solution: To solve this problem we can first determine the distance between every pair of points and then find the smallest of these distances. However, this approach requires $O(n^2)$ computations of distances and comparisons since there are $C(n, 2) = n(n - 1)/2$ pairs of points. Surprisingly, there is an elegant divide-and-conquer algorithm that can solve the closest-pair problem for n points using $O(n \log n)$ computations of distances and comparisons. The algorithm we describe here is due to Michael Samos (see [PrSa85]).

For simplicity, we assume that $n = 2^k$, where k is a positive integer. (We avoid some technical considerations that are needed when n is not a power of 2.) When $n = 2$, we have only one pair of points; the distance between these two points is the minimum distance.

The recursive part of the algorithm divides the problem into two subproblems, each involving half as many points. The first step is to use the merge sort algorithm to sort the points by their x coordinates. This requires $O(n \log n)$ operations. Once we have sorted the points by their x coordinates, we construct a vertical line ℓ dividing the n points into two parts, a left part and a right part of equal size, each containing $n/2$ points, as shown in Figure 1. (If any points fall on the dividing line ℓ, we divide them among the two parts if necessary.) At subsequent steps of the recursion we need not sort on x coordinates again, since we can select the corresponding sorted subset of all the points. This selection is a task that can be done with $O(n)$ comparisons.

There are three possibilities concerning the positions of the closest points: (*1*) they are both in the left region L, (*2*) they are both in the right region R, or (*3*) one point is in the left region and the other is in the right region. Apply the algorithm recursively to compute d_L and d_R, where d_L is the minimum distance between points in the left region and d_R is the minimum distance between points in the right region. Let $d = \min(d_L, d_R)$. To successfully divide the problem of finding the closest two points in the original set into the two problems of finding the shortest distances between points in the two regions separately, we have to handle the conquer part of the algorithm, which requires that we consider the case where the closest points lie in different regions, that is, one point is

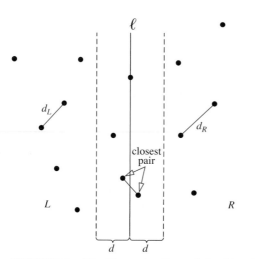

In this illustration the problem of finding the closest pair in a set of 16 points is reduced to two problems of finding the closest pair in a set of eight points *and* the problem of determining whether there are points closer than $d = \min(d_L, d_R)$ within the strip of width $2d$ centered at ℓ.

FIGURE 1   **The Recursive Step of the Algorithm for Solving the Closest-Pair Problem.**

in L and the other in R. Since there is a pair of points at distance d where both points lie in R or both points lie in L, for the closest points to lie in different regions requires that they must be a distance less than d apart.

For a point in the left region and a point in the right region to lie at a distance less than d apart, these points must lie in the vertical strip of width $2d$ that has the line ℓ as its center. (Otherwise, the distance between these points is greater than the difference in their x coordinates, which exceeds d.) To examine the points within this strip, we sort the points so that they are listed in order of increasing y coordinates. Again, this can be done in $O(n \log n)$ operations using merge sort and need only be done once at the start of the algorithm, and not at each recursive step. At each recursive step, we form a subset of the points in the region sorted by their y coordinates from the already sorted set of all points sorted by their y coordinates, which can be done with $O(n)$ comparisons.

Beginning with a point in the strip with the smallest y coordinate, we successively examine each point in the strip, computing the distance between this point and all other points in the strip that have larger y coordinates that could lie at a distance less than d from this point. Note that to examine a point p, we need only consider the distances between p and points in the set that lie within the rectangle of height d and width $2d$ with p on its base and with vertical sides at distance d from ℓ.

We can show that there are at most eight points from the set, including p, in or on this $2d \times d$ rectangle. To see this, note that there can be at most one point in each of the eight $d/2 \times d/2$ squares shown in Figure 2. This follows because the farthest apart points can be on or within one of these squares is the diagonal length $d/\sqrt{2}$ (which can be found using the Pythagorean theorem), which is less than d, and each of these $d/2 \times d/2$ squares lies entirely within the left region or the right region. This means that at this stage we need only compare at most seven distances, the distances between p and the seven or fewer other points in or on the rectangle, with d.

Because the total number of points in the strip of width $2d$ does not exceed n (the total number of points in the set), at most $7n$ distances need to be compared with d to find the minimum distance between points. That is, there are only $7n$ possible distances that could be less than d. Consequently, once the merge sort has been used to sort the pairs according to their x coordinates and according to their y coordinates, we find that

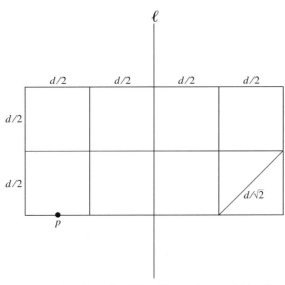

At most eight points, including p, can lie in or on the $2d \times d$ rectangle centered at ℓ since at most one point can lie in or on each of the eight $(d/2) \times (d/2)$ squares.

FIGURE 2   Showing That There Are at Most Seven Other Points to Consider for Each Point in the Strip.

the increasing function $f(n)$ satisfying the recurrence relation

$$f(n) = 2f(n/2) + 7n,$$

where $f(2) = 1$, exceeds the number of comparisons needed to solve the closest-pair problem for n points. By the Master Theorem (Theorem 2), it follows that $f(n)$ is $O(n \log n)$. The two sorts of points by their x coordinates and by their y coordinates each can be done using $O(n \log n)$ comparisons, by using the merge sort, and the sorted subsets of these coordinates at each of the $O(\log n)$ steps of the algorithm can be done using $O(n)$ comparisons each. Thus, we find that the closest-pair problem can be solved using $O(n \log n)$ comparisons. ◀

Exercises

1. How many comparisons are needed for a binary search in a set of 64 elements?

2. How many comparisons are needed to locate the maximum and minimum elements in a sequence with 128 elements using the algorithm in Example 2?

3. Multiply $(1110)_2$ and $(1010)_2$ using the fast multiplication algorithm.

4. Express the fast multiplication algorithm in pseudocode.

5. Determine a value for the constant C in Example 4 and use it to estimate the number of bit operations

needed to multiply two 64-bit integers using the fast multiplication algorithm.

6. How many operations are needed to multiply two 32×32 matrices using the algorithm referred to in Example 4?

7. Suppose that $f(n) = f(n/3) + 1$ when n is divisible by 3, and $f(1) = 1$. Find
   a) $f(3)$.   **b)** $f(27)$.   **c)** $f(729)$.

8. Suppose that $f(n) = 2f(n/2) + 3$ when n is even, and $f(1) = 5$. Find
   a) $f(2)$.   **b)** $f(8)$.   **c)** $f(64)$.   **d)** $f(1024)$.

9. Suppose that $f(n) = f(n/5) + 3n^2$ when n is divisible by 5, and $f(1) = 4$. Find

    a) $f(5)$.   **b)** $f(125)$.   **c)** $f(3125)$.

10. Find $f(n)$ when $n = 2^k$, where f satisfies the recurrence relation $f(n) = f(n/2) + 1$ with $f(1) = 1$.

11. Estimate the size of f in Exercise 10 if f is an increasing function.

12. Find $f(n)$ when $n = 3^k$, where f satisfies the recurrence relation $f(n) = 2f(n/3) + 4$ with $f(1) = 1$.

13. Estimate the size of f in Exercise 12 if f is an increasing function.

14. Suppose that there are $n = 2^k$ teams in an elimination tournament, where there are $n/2$ games in the first round, with the $n/2 = 2^{k-1}$ winners playing in the second round, and so on. Develop a recurrence relation for the number of rounds in the tournament.

15. How many rounds are in the elimination tournament described in Exercise 14 when there are 32 teams?

16. Solve the recurrence relation for the number of rounds in the tournament described in Exercise 14.

17. Suppose that the votes of n people for different candidates (where there can be more than two candidates) for a particular office are the elements of a sequence. A person wins the election if this person receives a majority of the votes.

    a) Devise a divide-and-conquer algorithm that determines whether a candidate received a majority and, if so, determine who this candidate is. (*Hint:* Assume that n is even and split the sequence of votes into two sequences, each with $n/2$ elements. Note that a candidate could not have received a majority of votes without receiving a majority of votes in at least one of the two halves.)

    b) Use the Master Theorem to estimate the number of comparisons needed by the algorithm you devised in part (a).

18. Suppose that each person in a group of n people votes for exactly two people from a slate of candidates to fill two positions on a committee. The top two finishers both win positions as long as each receives more than $n/2$ votes.

    a) Devise a divide-and-conquer algorithm that determines whether the two candidates who received the most votes each received at least $n/2$ votes and, if so, determine who these two candidates are.

    b) Use the Master Theorem to estimate the number of comparisons needed by the algorithm you devised in part (a).

19. a) Set up a divide-and-conquer recurrence relation for the number of multiplications required to compute x^n, where x is a real number and n is a positive integer, using the recursive algorithm from Exercise 18 in Section 3.5.

    b) Use the recurrence relation you found in part (a) to construct a big-O estimate for the number of

multiplications used to compute x^n using the recursive algorithm.

20. a) Set up a divide-and-conquer recurrence relation for the number of modular multiplications required to compute $a^n \bmod m$, where a, m, and n are positive integers, using the recursive algorithms from Example 2 in Section 3.5.

    b) Use the recurrence relation you found in part (a) to construct a big-O estimate for the number of modular multiplications used to compute $a^n \bmod m$ using the recursive algorithm.

21. Suppose that the function f satisfies the recurrence relation $f(n) = 2f(\sqrt{n}) + 1$ whenever n is a perfect square greater than 1 and $f(2) = 1$.

    a) Find $f(16)$.

    b) Find a big-O estimate for $f(n)$. (*Hint:* Make the substitution $m = \log n$.)

22. Suppose that the function f satisfies the recurrence relation $f(n) = 2f(\sqrt{n}) + \log n$ whenever n is a perfect square greater than 1 and $f(2) = 1$.

    a) Find $f(16)$.

    b) Find a big-O estimate for $f(n)$. (*Hint:* Make the substitution $m = \log n$.)

**23. This exercise deals with the problem of finding the largest sum of consecutive terms of a sequence of n real numbers. When all terms are positive, the sum of all terms provides the answer, but the situation is more complicated when some terms are negative. For example, the maximum sum of consecutive terms of the sequence $-2, 3, -1, 6, -7, 4$ is $3 + (-1) + 6 = 8$. (This exercise is based on [Be86].)

    a) Use pseudocode to describe an algorithm that solves this problem by finding the sums of consecutive terms starting with the first term, the sums of consecutive terms starting with the second term, and so on, keeping track of the maximum sum found so far as the algorithm proceeds.

    b) Determine the computational complexity of the algorithm in part (a) in terms of the number of sums computed and the number of comparisons made.

    c) Devise a divide-and-conquer algorithm to solve this problem. (*Hint:* Assume that there are an even number of terms in the sequence and split the sequence into two halves. Explain how to handle the case when the maximum sum of consecutive terms includes terms in both halves.)

    d) Use the algorithm from part (c) to find the maximum sum of consecutive terms of each of the following sequences: $-2, 4, -1, 3, 5, -6, 1, 2$; $4, 1, -3, 7, -1, -5, 3, -2$; and $-1, 6, 3, -4, -5, 8, -1, 7$.

    e) Find a recurrence relation for the number of sums and comparisons used by the divide-and-conquer algorithm from part (c).

f) Use the Master Theorem to estimate the computational complexity of the divide-and-conquer algorithm. How does it compare in terms of computational complexity with the algorithm from part (a)?

24. Apply the algorithm described in Example 12 for finding the closest pair of points, using the Euclidean distance between points, to find the closest pair of the points $(1, 3), (1, 7), (2, 4), (2, 9), (3, 1), (3, 5), (4, 3),$ and $(4, 7)$.

25. Apply the algorithm described in Example 12 for finding the closest pair of points, using the Euclidean distance between points, to find the closest pair of the points $(1, 2), (1, 6), (2, 4), (2, 8), (3, 1), (3, 6), (3, 10),$ $(4, 3), (5, 1), (5, 5), (5, 9), (6, 7), (7, 1), (7, 4), (7, 9),$ and $(8, 6)$.

*26. Use pseudocode to describe the recursive algorithm for solving the closest-pair problem as described in Example 12.

27. Construct a variation of the algorithm described in Example 12 along with justifications of the steps used by the algorithm to find the smallest distance between two points if the distance between two points is defined to be $d((x_i, y_i), (x_j, y_j)) = \max(|x_i - x_j|, |y_i - y_j|)$.

*28. Suppose someone picks a number x from a set of n numbers. A second person tries to guess the number by successively selecting subsets of the n numbers and asking the first person whether x is in each set. The first person answers either "yes" or "no." When the first person answers each query truthfully, we can find x using $\log n$ queries by successively splitting the sets used in each query in half. Ulam's problem, proposed by Stanislaw Ulam in 1976, asks for the number of queries required to find x, supposing that the first person is allowed to lie exactly once.

a) Show that by asking each question twice, given a number x and a set with n elements, and asking one more question when we find the lie, Ulam's problem can be solved using $2 \log n + 1$ queries.

b) Show that by dividing the initial set of n elements into four parts, each with $n/4$ elements, 1/4 of the elements can be eliminated using two queries. (*Hint:* Use two queries, where each of the queries asks whether the element is in the union of two of the subsets with $n/4$ elements and where one of the subsets of $n/4$ elements is used in both queries.)

c) Show from part (b) that if $f(n)$ equals the number of queries used to solve Ulam's problem using the method from part (b) and n is divisible by 4, then $f(n) = f(3n/4) + 2$.

d) Solve the recurrence relation in part (c) for $f(n)$.

e) Is the naive way to solve Ulam's problem by asking each question twice or the divide-and-conquer method based on part (b) more efficient? The most efficient way to solve Ulam's problem has been determined by A. Pelc [Pe87].

In Exercises 29–33, assume that f is an increasing function satisfying the recurrence relation $f(n) = af(n/b) + cn^d$, where $a \geq 1$, b is an integer greater than 1, and c and d are positive real numbers. These exercises supply a proof of Theorem 2.

*29. Show that if $a = b^d$ and n is a power of b, then $f(n) = f(1)n^d + cn^d \log_b n$.

30. Use Exercise 29 to show that if $a = b^d$, then $f(n)$ is $O(n^d \log n)$.

*31. Show that if $a \neq b^d$ and n is a power of b, then $f(n) = C_1 n^d + C_2 n^{\log_b a}$, where $C_1 = b^d c/(b^d - a)$ and $C_2 = f(1) + b^d c/(a - b^d)$.

32. Use Exercise 31 to show that if $a < b^d$, then $f(n)$ is $O(n^d)$.

33. Use Exercise 31 to show that if $a > b^d$, then $f(n)$ is $O(n^{\log_b a})$.

34. Find $f(n)$ when $n = 4^k$, where f satisfies the recurrence relation $f(n) = 5f(n/4) + 6n$, with $f(1) = 1$.

35. Estimate the size of f in Exercise 34 if f is an increasing function.

36. Find $f(n)$ when $n = 2^k$, where f satisfies the recurrence relation $f(n) = 8f(n/2) + n^2$ with $f(1) = 1$.

37. Estimate the size of f in Exercise 36 if f is an increasing function.

6.4 Generating Functions

INTRODUCTION

Generating functions are used to represent sequences efficiently by coding the terms of a sequence as coefficients of powers of a variable x in a formal power series. Generating functions can be used to solve many types of counting problems, such as the number of ways to select or distribute objects of different kinds, subject to a variety of constraints, and the number of ways to make change for a dollar using coins of different denominations. Generating functions can be used to solve recurrence relations by translating a recurrence relation for the terms of a sequence into an equation involving a generating function. This equation can then be solved to find a closed form for the generating function. From this closed form, the coefficients of the power series for the generating function can be found, solving the original recurrence relation. Generating functions can also be used to prove combinatorial identities by taking advantage of relatively simple relationships between functions that can be translated into identities involving the terms of sequences. Generating functions are a helpful tool for studying many properties of sequences besides those described in this section, such as their use for establishing asymptotic formulae for the terms of a sequence.

We begin with the definition of the generating function for a sequence.

DEFINITION 1

The *generating function for the sequence* $a_0, a_1, \ldots, a_k, \ldots$ of real numbers is the infinite series

$$G(x) = a_0 + a_1 x + \cdots + a_k x^k + \cdots = \sum_{k=0}^{\infty} a_k x^k.$$

Remark: The generating function for $\{a_k\}$ given in Definition 1 is sometimes called the **ordinary generating function** of $\{a_k\}$ to distinguish it from other types of generating functions for this sequence.

EXAMPLE 1   The generating functions for the sequences $\{a_k\}$ with $a_k = 3$, $a_k = k + 1$, and $a_k = 2^k$ are $\sum_{k=0}^{\infty} 3x^k$, $\sum_{k=0}^{\infty} (k + 1)x^k$, and $\sum_{k=0}^{\infty} 2^k x^k$, respectively.   ◄

We can define generating functions for finite sequences of real numbers by extending a finite sequence $a_0, a_1, \ldots, a_n$ into an infinite sequence by setting $a_{n+1} = 0$, $a_{n+2} = 0$, and so on. The generating function $G(x)$ of this infinite sequence $\{a_n\}$ is a polynomial of degree n since no terms of the form $a_j x^j$ with $j > n$ occur, that is,

$$G(x) = a_0 + a_1 x + \cdots + a_n x^n.$$

EXAMPLE 2   What is the generating function for the sequence $1, 1, 1, 1, 1, 1$?

Solution: The generating function of $1, 1, 1, 1, 1, 1$ is

$$1 + x + x^2 + x^3 + x^4 + x^5.$$

By Theorem 1 of Section 3.2 we have

$$(x^6 - 1)/(x - 1) = 1 + x + x^2 + x^3 + x^4 + x^5.$$

Consequently, $G(x) = (x^6 - 1)/(x - 1)$ is the generating function of the sequence $1, 1, 1, 1, 1, 1$.   ◄

EXAMPLE 3   Let m be a positive integer. Let $a_k = C(m, k)$, for $k = 0, 1, 2, \ldots, m$. What is the generating function for the sequence $a_0, a_1, \ldots, a_m$?

Solution: The generating function for this sequence is

$$G(x) = C(m, 0) + C(m, 1)x + C(m, 2)x^2 + \cdots + C(m, m)x^m.$$

The Binomial Theorem shows that $G(x) = (1 + x)^m$.   ◄

USEFUL FACTS ABOUT POWER SERIES

When generating functions are used to solve counting problems, they are usually considered to be **formal power series.** Questions about the convergence of these series are ignored. However, to apply some results from calculus, it is sometimes important to consider for which x the power series converges. We will not be concerned with questions of convergence in our discussions. Readers familiar with calculus can consult textbooks on this subject for details about the convergence of the series we consider here.

We will state now some important facts about infinite series used when working with generating functions. A discussion of these and related results can be found in calculus texts.

EXAMPLE 4 The function $f(x) = 1/(1 - x)$ is the generating function of the sequence $1, 1, 1, 1, \ldots,$ since

$$1/(1 - x) = 1 + x + x^2 + \cdots$$

for $|x| < 1$. ◀

EXAMPLE 5 The function $f(x) = 1/(1 - ax)$ is the generating function of the sequence $1, a, a^2, a^3, \ldots,$ since

$$1/(1 - ax) = 1 + ax + a^2 x^2 + \cdots$$

when $|ax| < 1$, or equivalently, for $|x| < 1/|a|$ for $a \neq 0$. ◀

We also will need some results on how to add and how to multiply two generating functions. Proofs of these results can be found in calculus texts.

THEOREM 1 Let $f(x) = \sum_{k=0}^{\infty} a_k x^k$ and $g(x) = \sum_{k=0}^{\infty} b_k x^k$. Then

$$f(x) + g(x) = \sum_{k=0}^{\infty} (a_k + b_k) x^k \quad \text{and} \quad f(x)g(x) = \sum_{k=0}^{\infty} \left(\sum_{j=0}^{k} a_j b_{k-j} \right) x^k.$$

We will illustrate how Theorem 1 can be used with Example 6.

EXAMPLE 6 Let $f(x) = 1/(1 - x)^2$. Use Example 4 to find the coefficients $a_0, a_1, a_2, \ldots$ in the expansion $f(x) = \sum_{k=0}^{\infty} a_k x^k$.

Solution: From Example 4 we see that

$$1/(1 - x) = 1 + x + x^2 + x^3 + \cdots.$$

Hence, from Theorem 1, we have

$$1/(1 - x)^2 = \sum_{k=0}^{\infty} \left(\sum_{j=0}^{k} 1 \right) x^k = \sum_{k=0}^{\infty} (k + 1) x^k.$$ ◀

Remark: This result also can be derived from Example 4 by differentiation. Taking derivatives is a useful technique for producing new identities from existing identities for generating functions.

To use generating functions to solve many important counting problems, we will need to apply the Binomial Theorem for exponents that are not positive integers. Before we state an extended version of the Binomial Theorem, we need to define extended binomial coefficients.

DEFINITION 2

Let u be a real number and k a nonnegative integer. Then the *extended binomial coefficient* $\binom{u}{k}$ is defined by

$$\binom{u}{k} = \begin{cases} u(u-1)\cdots(u-k+1)/k! & \text{if } k > 0, \\ 1 & \text{if } k = 0. \end{cases}$$

EXAMPLE 7

Find the values of the extended binomial coefficients $\binom{-2}{3}$ and $\binom{1/2}{3}$.

Solution: Taking $u = -2$ and $k = 3$ in Definition 2 gives us

$$\binom{-2}{3} = \frac{(-2)(-3)(-4)}{3!} = -4.$$

Similarly, taking $u = 1/2$ and $k = 3$ gives us

$$\binom{1/2}{3} = \frac{(1/2)(1/2-1)(1/2-2)}{3!}$$

$$= (1/2)(-1/2)(-3/2)/6$$

$$= 1/16. \qquad \blacktriangleleft$$

Example 8 provides a useful formula for extended binomial coefficients when the top parameter is a negative integer. It will be useful in our subsequent discussions.

EXAMPLE 8

When the top parameter is a negative integer, the extended binomial coefficient can be expressed in terms of an ordinary binomial coefficient. To see that this is the case, note that

$$\binom{-n}{r} = \frac{(-n)(-n-1)\cdots(-n-r+1)}{r!}$$

$$= \frac{(-1)^r n(n+1)\cdots(n+r-1)}{r!}$$

$$= \frac{(-1)^r (n+r-1)(n+r-2)\cdots n}{r!}$$

$$= \frac{(-1)^r (n+r-1)!}{r!(n-1)!}$$

$$= (-1)^r \binom{n+r-1}{r}$$

$$= (-1)^r C(n+r-1, r). \qquad \blacktriangleleft$$

We now state the extended Binomial Theorem.

THEOREM 2   **THE EXTENDED BINOMIAL THEOREM**   Let x be a real number with $|x| < 1$ and let u be a real number. Then

$$(1 + x)^u = \sum_{k=0}^{\infty} \binom{u}{k} x^k.$$

Theorem 2 can be proved using the theory of the Maclaurin series. We leave its proof to the reader with a familiarity with this part of calculus.

Remark: When u is a positive integer, the extended Binomial Theorem reduces to the Binomial Theorem presented in Section 4.4, since in that case $\binom{u}{k} = 0$ if $k > u$.

Example 9 illustrates the use of Theorem 2 when the exponent is a negative integer.

EXAMPLE 9   Find the generating functions for $(1 + x)^{-n}$ and $(1 - x)^{-n}$, where n is a positive integer, using the extended Binomial Theorem.

Solution: By the extended Binomial Theorem, it follows that

$$(1 + x)^{-n} = \sum_{k=0}^{\infty} \binom{-n}{k} x^k.$$

Using Example 8, which provides a simple formula for $\binom{-n}{k}$, we obtain

$$(1 + x)^{-n} = \sum_{k=0}^{\infty} (-1)^k C(n + k - 1, k) x^k.$$

Replacing x by $-x$, we find that

$$(1 - x)^{-n} = \sum_{k=0}^{\infty} C(n + k - 1, k) x^k. \qquad \blacktriangleleft$$

Table 1 presents a useful summary of some generating functions that arise frequently.

COUNTING PROBLEMS AND GENERATING FUNCTIONS

Generating functions can be used to solve a wide variety of counting problems. In particular, they can be used to count the number of combinations of various types. In Chapter 4 we developed techniques to count the r-combinations from a set with n elements when repetition is allowed and additional constraints may exist. Such problems are equivalent to counting the solutions to equations of the form

$$e_1 + e_2 + \cdots + e_n = C,$$

where C is a constant and each e_i is a nonnegative integer that may be subject to a specified constraint. Generating functions can also be used to solve counting problems of this type, as Examples 10–12 show.

TABLE 1 Useful Generating Functions.

$G(x)$	a_k
$(1+x)^n = \displaystyle\sum_{k=0}^{n} C(n,k)x^k$ $= 1 + C(n,1)x + C(n,2)x^n + \cdots + x^n$	$C(n,k)$
$(1+ax)^n = \displaystyle\sum_{k=0}^{n} C(n,k)a^k x^k$ $= 1 + C(n,1)ax + C(n,2)a^2 x^2 + \cdots + a^n x^n$	$C(n,k)a^k$
$(1+x^r)^n = \displaystyle\sum_{k=0}^{n} C(n,k)x^{rk}$ $= 1 + C(n,1)x^r + C(n,2)x^{2r} + \cdots + x^{rn}$	$C(n,k/r)$ if $r \mid k$; 0 otherwise
$\dfrac{1-x^{n+1}}{1-x} = \displaystyle\sum_{k=0}^{n} x^k = 1 + x + x^2 + \cdots + x^n$	1 if $k \le n$; 0 otherwise
$\dfrac{1}{1-x} = \displaystyle\sum_{k=0}^{\infty} x^k = 1 + x + x^2 + \cdots$	1
$\dfrac{1}{1-ax} = \displaystyle\sum_{k=0}^{\infty} a^k x^k = 1 + ax + a^2 x^2 + \cdots$	a^k
$\dfrac{1}{1-x^r} = \displaystyle\sum_{k=0}^{\infty} x^{rk} = 1 + x^r + x^{2r} + \cdots$	1 if $r \mid k$; 0 otherwise
$\dfrac{1}{(1-x)^2} = \displaystyle\sum_{k=0}^{\infty} (k+1)x^k = 1 + 2x + 3x^2 + \cdots$	$k+1$
$\dfrac{1}{(1-x)^n} = \displaystyle\sum_{k=0}^{\infty} C(n+k-1,k)x^k$ $= 1 + C(n,1)x + C(n+1,2)x^2 + \cdots$	$C(n+k-1,k) = C(n+k-1,n-1)$
$\dfrac{1}{(1+x)^n} = \displaystyle\sum_{k=0}^{\infty} C(n+k-1,k)(-1)^k x^k$ $= 1 - C(n,1)x + C(n+1,2)x^2 - \cdots$	$(-1)^k C(n+k-1,k) = (-1)^k C(n+k-1,n-1)$
$\dfrac{1}{(1-ax)^n} = \displaystyle\sum_{k=0}^{\infty} C(n+k-1,k)a^k x^k$ $= 1 + C(n,1)ax + C(n+1,2)a^2 x^2 + \cdots$	$C(n+k-1,k)a^k = C(n+k-1,n-1)a^k$
$e^x = \displaystyle\sum_{k=0}^{\infty} \dfrac{x^k}{k!} = 1 + x + \dfrac{x^2}{2!} + \dfrac{x^3}{3!} + \cdots$	$1/k!$
$\ln(1+x) = \displaystyle\sum_{k=0}^{\infty} \dfrac{(-1)^{k+1}}{k}x^k = x - \dfrac{x^2}{2} + \dfrac{x^3}{3} - \dfrac{x^4}{4} + \cdots$	$(-1)^{k+1}/k$

Note: The series for the last two generating functions can be found in most calculus books when power series are discussed.

EXAMPLE 10   Find the number of solutions of

$$e_1 + e_2 + e_3 = 17,$$

where e_1, e_2, and e_3 are nonnegative integers with $2 \le e_1 \le 5, 3 \le e_2 \le 6$, and $4 \le e_3 \le 7$.

Solution: The number of solutions with the indicated constraints is the coefficient of x^{17} in the expansion of

$$(x^2 + x^3 + x^4 + x^5)(x^3 + x^4 + x^5 + x^6)(x^4 + x^5 + x^6 + x^7).$$

This follows since we obtain a term equal to x^{17} in the product by picking a term in the first sum x^{e_1}, a term in the second sum x^{e_2}, and a term in the third sum x^{e_3}, where the exponents e_1, e_2, and e_3 satisfy the equation $e_1 + e_2 + e_3 = 17$ and the given constraints.

It is not hard to see that the coefficient of x^{17} in this product is 3. Hence, there are three solutions. (Note that the calculating of this coefficient involves about as much work as enumerating all the solutions of the equation with the given constraints. However, the method that this illustrates often can be used to solve wide classes of counting problems with special formulae, as we will see. Furthermore, a computer algebra system can be used to do such computations.)   ◀

EXAMPLE 11   In how many different ways can eight identical cookies be distributed among three distinct children if each child receives at least two cookies and no more than four cookies?

Solution: Since each child receives at least two but no more than four cookies, for each child there is a factor equal to

$$(x^2 + x^3 + x^4)$$

in the generating function for the sequence $\{c_n\}$, where c_n is the number of ways to distribute n cookies. Since there are three children, this generating function is

$$(x^2 + x^3 + x^4)^3.$$

We need the coefficient of x^8 in this product. The reason is that the x^8 terms in the expansion correspond to the ways that three terms can be selected, with one from each factor, that have exponents adding up to 8. Furthermore, the exponents of the term from the first, second, and third factors are the numbers of cookies the first, second, and third children receive, respectively. Computation shows that this coefficient equals 6. Hence, there are six ways to distribute the cookies so that each child receives at least two, but no more than four, cookies.   ◀

EXAMPLE 12   Use generating functions to determine the number of ways to insert tokens worth $1, $2, and $5 into a vending machine to pay for an item that costs r dollars in both the cases when the order in which the tokens are inserted does not matter and when the order does matter. (For example, there are two ways to pay for an item that costs $3 when the order in which the tokens are inserted does not matter: inserting three $1 tokens or one $1 token and a $2 token. When the order matters, there are three ways: inserting three $1 tokens, inserting a $1 token and then a $2 token, or inserting a $2 token and then a $1 token.)

Solution: Consider the case when the order in which the tokens are inserted does not matter. Here, all we care about is the number of each token used to produce a total of

r dollars. Since we can use any number of \$1 tokens, any number of \$2 tokens, and any number of \$5 tokens, the answer is the coefficient of x^r in the generating function

$$(1 + x + x^2 + x^3 + \cdots)(1 + x^2 + x^4 + x^6 + \cdots)(1 + x^5 + x^{10} + x^{15} + \cdots).$$

(The first factor in this product represents the \$1 tokens used, the second the \$2 tokens used, and the third the \$5 tokens used.) For example, the number of ways to pay for an item costing \$7 using \$1, \$2, and \$5 tokens is given by the coefficient of x^7 in this expansion, which equals 6.

When the order in which the tokens are inserted matters, the number of ways to insert exactly n tokens to produce a total of r dollars is the coefficient of x^r in

$$(x + x^2 + x^5)^n,$$

since each of the r tokens may be a \$1 token, a \$2 token, or a \$5 token. Since any number of tokens may be inserted, the number of ways to produce r dollars using \$1, \$2, or \$5 tokens, when the order in which the tokens are inserted matters, is the coefficient of x^r in

$$1 + (x + x^2 + x^5) + (x + x^2 + x^5)^2 + \cdots = \frac{1}{1 - (x + x^2 + x^5)}$$

$$= \frac{1}{1 - x - x^2 - x^5},$$

where we have added the number of ways to insert 0 tokens, 1 token, 2 tokens, 3 tokens, and so on, and where we have used the identity $1/(1 - x) = 1 + x + x^2 + \cdots$ with x replaced with $x + x^2 + x^5$. For example, the number of ways to pay for an item costing \$7 using \$1, \$2, and \$5 tokens, when the order in which the tokens are used matters, is the coefficient of x^7 in this expansion, which equals 26. [To see that this coefficient equals 26 requires the addition of the coefficients of x^7 in the expansions $(x + x^2 + x^5)^k$ for $2 \le k \le 7$. This can be done by hand with considerable computation, or a computer algebra system can be used.] ◀

Example 13 shows the versatility of generating functions when used to solve problems with differing assumptions.

EXAMPLE 13   Use generating functions to find the number of k-combinations of a set with n elements. Assume that the Binomial Theorem has already been established.

Solution: Each of the n elements in the set contributes the term $(1 + x)$ to the generating function $f(x) = \sum_{k=0}^{n} a_k x^k$. Here $f(x)$ is the generating function for $\{a_k\}$, where a_k represents the number of k-combinations of a set with n elements. Hence,

$$f(x) = (1 + x)^n.$$

But by the Binomial Theorem, we have

$$f(x) = \sum_{k=0}^{n} \binom{n}{k} x^k,$$

where

$$\binom{n}{k} = \frac{n!}{k!(n - k)!}.$$

Hence, $C(n, k)$, the number of k-combinations of a set with n elements, is

$$\frac{n!}{k!(n-k)!}.$$ ◀

Remark: We proved the Binomial Theorem in Section 4.4 using the formula for the number of r-combinations of a set with n elements. This example shows that the Binomial Theorem, which can be proved by mathematical induction, can be used to derive the formula for the number of r-combinations of a set with n elements.

EXAMPLE 14   Use generating functions to find the number of r-combinations from a set with n elements when repetition of elements is allowed.

Solution: Let $G(x)$ be the generating function for the sequence $\{a_r\}$, where a_r equals the number of r-combinations of a set with n elements with repetitions allowed. That is, $G(x) = \sum_{r=0}^{\infty} a_r x^r$. Since we can select any number of a particular member of the set with n elements when we form an r-combination with repetition allowed, each of the n elements contributes $(1 + x + x^2 + x^3 + \cdots)$ to a product expansion for $G(x)$. Each element contributes this factor since it may be selected zero times, one time, two times, three times, and so on, when an r-combination is formed (with a total of r elements selected). Since there are n elements in the set and each contributes this same factor to $G(x)$, we have

$$G(x) = (1 + x + x^2 + \cdots)^n.$$

As long as $|x| < 1$, we have $1 + x + x^2 + \cdots = 1/(1 - x)$, so

$$G(x) = 1/(1 - x)^n = (1 - x)^{-n}.$$

Applying the extended Binomial Theorem (Theorem 2), it follows that

$$(1 - x)^{-n} = (1 + (-x))^{-n} = \sum_{r=0}^{\infty} \binom{-n}{r} (-x)^r.$$

The number of r-combinations of a set with n elements with repetitions allowed, when r is a positive integer, is the coefficient a_r of x^r in this sum. Consequently, using Example 8 we find that a_r equals

$$\binom{-n}{r} (-1)^r = (-1)^r C(n + r - 1, r) \cdot (-1)^r$$

$$= C(n + r - 1, r).$$ ◀

Note that the result in Example 14 is the same result we stated as Theorem 2 in Section 4.5.

EXAMPLE 15   Use generating functions to find the number of ways to select r objects of n different kinds if we must select at least one object of each kind.

Solution: Since we need to select at least one object of each kind, each of the n kinds of objects contributes the factor $(x + x^2 + x^3 + \cdots)$ to the generating function $G(x)$ for the sequence $\{a_r\}$, where a_r is the number of ways to select r objects of n different kinds if we need at least one object of each kind. Hence,

$$G(x) = (x + x^2 + x^3 + \cdots)^n = x^n(1 + x + x^2 + \cdots)^n = x^n/(1 - x)^n.$$

Using the extended Binomial Theorem and Example 8, we have

$$
\begin{aligned}
G(x) &= x^n/(1-x)^n \\
&= x^n \cdot (1-x)^{-n} \\
&= x^n \sum_{r=0}^{\infty} \binom{-n}{r} (-x)^r \\
&= x^n \sum_{r=0}^{\infty} (-1)^r C(n+r-1,r)(-1)^r x^r \\
&= \sum_{r=0}^{\infty} C(n+r-1,r) x^{n+r} \\
&= \sum_{t=n}^{\infty} C(t-1,t-n) x^t \\
&= \sum_{r=n}^{\infty} C(r-1,r-n) x^r .
\end{aligned}
$$

We have shifted the summation in the next-to-last equality by setting $t = n + r$ so that $t = n$ when $r = 0$ and $n + r - 1 = t - 1$, and then we replaced t by r as the index of summation in the last equality to return to our original notation. Hence, there are $C(r-1, r-n)$ ways to select r objects of n different kinds if we must select at least one object of each kind. ◄

USING GENERATING FUNCTIONS TO SOLVE RECURRENCE RELATIONS

We can find the solution to a recurrence relation and its initial conditions by finding an explicit formula for the associated generating function. This is illustrated in Examples 16 and 17.

EXAMPLE 16   Solve the recurrence relation $a_k = 3a_{k-1}$ for $k = 1, 2, 3, \ldots$ and initial condition $a_0 = 2$.

Solution: Let $G(x)$ be the generating function for the sequence $\{a_k\}$, that is, $G(x) = \sum_{k=0}^{\infty} a_k x^k$. First note that

$$
xG(x) = \sum_{k=0}^{\infty} a_k x^{k+1} = \sum_{k=1}^{\infty} a_{k-1} x^k .
$$

Using the recurrence relation, we see that

$$
\begin{aligned}
G(x) - 3xG(x) &= \sum_{k=0}^{\infty} a_k x^k - 3 \sum_{k=1}^{\infty} a_{k-1} x^k \\
&= a_0 + \sum_{k=1}^{\infty} (a_k - 3a_{k-1}) x^k \\
&= 2,
\end{aligned}
$$

since $a_0 = 2$ and $a_k = 3a_{k-1}$. Thus,

$$G(x) - 3xG(x) = (1 - 3x)G(x) = 2.$$

Solving for $G(x)$ shows that $G(x) = 2/(1 - 3x)$. Using the identity $1/(1 - ax) = \sum_{k=0}^{\infty} a^k x^k$, from Table 1, we have

$$G(x) = 2 \sum_{k=0}^{\infty} 3^k x^k = \sum_{k=0}^{\infty} 2 \cdot 3^k x^k.$$

Consequently, $a_k = 2 \cdot 3^k$.   ◀

EXAMPLE 17   Suppose that a valid codeword is an n-digit number in decimal notation containing an even number of 0s. Let a_n denote the number of valid codewords of length n. In Example 7 of Section 6.1 we showed that the sequence $\{a_n\}$ satisfies the recurrence relation

$$a_n = 8a_{n-1} + 10^{n-1}$$

and the initial condition $a_1 = 9$. Use generating functions to find an explicit formula for a_n.

Solution: To make our work with generating functions simpler, we extend this sequence by setting $a_0 = 1$; when we assign this value to a_0 and use the recurrence relation, we have $a_1 = 8a_0 + 10^0 = 8 + 1 = 9$, which is consistent with our original initial condition. (It also makes sense because there is one code word of length 0—the empty string.)

We multiply both sides of the recurrence relation by x^n to obtain

$$a_n x^n = 8a_{n-1}x^n + 10^{n-1}x^n.$$

Let $G(x) = \sum_{n=0}^{\infty} a_n x^n$ be the generating function of the sequence $a_0, a_1, a_2, \ldots$. We sum both sides of the last equation starting with $n = 1$, to find that

$$G(x) - 1 = \sum_{n=1}^{\infty} a_n x^n = \sum_{n=1}^{\infty}(8a_{n-1}x^n + 10^{n-1}x^n)$$

$$= 8 \sum_{n=1}^{\infty} a_{n-1}x^n + \sum_{n=1}^{\infty} 10^{n-1}x^n$$

$$= 8x \sum_{n=1}^{\infty} a_{n-1}x^{n-1} + x \sum_{n=1}^{\infty} 10^{n-1}x^{n-1}$$

$$= 8x \sum_{n=0}^{\infty} a_n x^n + x \sum_{n=0}^{\infty} 10^n x^n$$

$$= 8xG(x) + x/(1 - 10x),$$

where we have used Example 5 to evaluate the second summation. Therefore, we have

$$G(x) - 1 = 8xG(x) + x/(1 - 10x).$$

Solving for $G(x)$ shows that

$$G(x) = \frac{1 - 9x}{(1 - 8x)(1 - 10x)}.$$

Expanding the right-hand side of this equation into partial fractions (as is done in the integration of rational functions studied in calculus) gives

$$G(x) = \frac{1}{2}\left(\frac{1}{1-8x} + \frac{1}{1-10x}\right).$$

Using Example 5 twice (once with $a = 8$ and once with $a = 10$) gives

$$G(x) = \frac{1}{2}\left(\sum_{n=0}^{\infty} 8^n x^n + \sum_{n=0}^{\infty} 10^n x^n\right)$$

$$= \sum_{n=0}^{\infty} \frac{1}{2}(8^n + 10^n)x^n.$$

Consequently, we have shown that

$$a_n = \frac{1}{2}(8^n + 10^n).$$

◀

PROVING IDENTITIES VIA GENERATING FUNCTIONS

In Chapter 4 we saw how combinatorial identities could be established using combinatorial proofs. Here we will show that such identities, as well as identities for extended binomial coefficients, can be proved using generating functions. Sometimes the generating function approach is simpler than other approaches, especially when it is simpler to work with the closed form of a generating function than with the terms of the sequence themselves. We illustrate how generating functions can be used to prove identities with Example 18.

EXAMPLE 18   Use generating functions to show that

$$\sum_{k=0}^{n} C(n, k)^2 = C(2n, n)$$

whenever n is a positive integer.

Solution: First note that by the Binomial Theorem $C(2n, n)$ is the coefficient of x^n in $(1 + x)^{2n}$. However, we also have

$$(1 + x)^{2n} = [(1 + x)^n]^2$$

$$= [C(n, 0) + C(n, 1)x + C(n, 2)x^2 + \cdots + C(n, n)x^n]^2.$$

The coefficient of x^n in this expression is $C(n, 0)C(n, n) + C(n, 1)C(n, n - 1) + C(n, 2)C(n, n - 2) + \cdots + C(n, n)C(n, 0)$. This equals $\sum_{k=0}^{n} C(n, k)^2$, since $C(n, n - k) = C(n, k)$. Since both $C(2n, n)$ and $\sum_{k=0}^{n} C(n, k)^2$ represent the coefficient of x^n in $(1 + x)^{2n}$, they must be equal. ◀

Exercises 42 and 43 at the end of this section ask that Pascal's identity and Vandermonde's identity be proved using generating functions.

Exercises

1. Find the generating function for the finite sequence $2, 2, 2, 2, 2, 2$.
2. Find the generating function for the finite sequence $1, 4, 16, 64, 256$.
3. Find a closed form for the generating function for each of these sequences. (Assume a general form for the terms of the sequence, using the most obvious choice of such a sequence.)
   a) $0, 2, 2, 2, 2, 2, 2, 0, 0, 0, 0, 0, \dots$
   b) $0, 0, 0, 1, 1, 1, 1, 1, 1, \dots$
   c) $0, 1, 0, 0, 1, 0, 0, 1, 0, 0, 1, \dots$
   d) $2, 4, 8, 16, 32, 64, 128, 256, \dots$
   e) $\binom{7}{0}, \binom{7}{1}, \binom{7}{2}, \dots, \binom{7}{7}, 0, 0, 0, 0, 0, \dots$
   f) $2, -2, 2, -2, 2, -2, 2, -2, \dots$
   g) $1, 1, 0, 1, 1, 1, 1, 1, 1, 1, 1, \dots$
   h) $0, 0, 0, 1, 2, 3, 4, \dots$
4. Find a closed form for the generating function for each of these sequences. (Assume a general form for the terms of the sequence, using the most obvious choice of such a sequence.)
   a) $-1, -1, -1, -1, -1, -1, -1, 0, 0, 0, 0, 0, 0, \dots$
   b) $1, 3, 9, 27, 81, 243, 729, \dots$
   c) $0, 0, 3, -3, 3, -3, 3, -3, \dots$
   d) $1, 2, 1, 1, 1, 1, 1, 1, 1, \dots$
   e) $\binom{7}{0}, 2\binom{7}{1}, 2^2\binom{7}{2}, \dots, 2^7\binom{7}{7}, 0, 0, 0, 0, \dots$
   f) $-3, 3, -3, 3, -3, 3, \dots$
   g) $0, 1, -2, 4, -8, 16, -32, 64, \dots$
   h) $1, 0, 1, 0, 1, 0, 1, 0, \dots$
5. Find a closed form for the generating function for the sequence $\{a_n\}$, where
   a) $a_n = 5$ for all $n = 0, 1, 2, \dots$.
   b) $a_n = 3^n$ for all $n = 0, 1, 2, \dots$.
   c) $a_n = 2$ for $n = 3, 4, 5, \dots$ and $a_0 = a_1 = a_2 = 0$.
   d) $a_n = 2n + 3$ for all $n = 0, 1, 2, \dots$.
   e) $a_n = \binom{8}{n}$ for all $n = 0, 1, 2, \dots$.
   f) $a_n = \binom{n+4}{n}$ for all $n = 0, 1, 2, \dots$.
6. Find a closed form for the generating function for the sequence $\{a_n\}$, where
   a) $a_n = -1$ for all $n = 0, 1, 2, \dots$.
   b) $a_n = 2^n$ for $n = 1, 2, 3, 4, \dots$ and $a_0 = 0$.
   c) $a_n = n - 1$ for $n = 0, 1, 2, \dots$.
   d) $a_n = 1/(n+1)!$ for $n = 0, 1, 2, \dots$.
   e) $a_n = \binom{n}{2}$ for $n = 0, 1, 2, \dots$.
   f) $a_n = \binom{10}{n+1}$ for $n = 0, 1, 2, \dots$.

7. For each of these generating functions, provide a closed formula for the sequence it determines.
   a) $(3x - 4)^3$
   b) $(x^3 + 1)^3$
   c) $1/(1 - 5x)$
   d) $x^3/(1 + 3x)$
   e) $x^2 + 3x + 7 + (1/(1 - x^2))$
   f) $(x^4/(1 - x^4)) - x^3 - x^2 - x - 1$
   g) $x^2/(1 - x)^2$
   h) $2e^{2x}$
8. For each of these generating functions, provide a closed formula for the sequence it determines.
   a) $(x^2 + 1)^3$
   b) $(3x - 1)^3$
   c) $1/(1 - 2x^2)$
   d) $x^2/(1 - x)^3$
   e) $x - 1 + (1/(1 - 3x))$
   f) $(1 + x^3)/(1 + x)^3$
   *g) $x/(1 + x + x^2)$
   h) $e^{3x^2} - 1$
9. Find the coefficient of x^{10} in the power series of each of these functions.
   a) $(1 + x^5 + x^{10} + x^{15} + \cdots)^3$
   b) $(x^3 + x^4 + x^5 + x^6 + x^7 + \cdots)^3$
   c) $(x^4 + x^5 + x^6)(x^3 + x^4 + x^5 + x^6 + x^7)(1 + x + x^2 + x^3 + x^4 + \cdots)$
   d) $(x^2 + x^4 + x^6 + x^8 + \cdots)(x^3 + x^6 + x^9 + \cdots)(x^4 + x^8 + x^{12} + \cdots)$
   e) $(1 + x^2 + x^4 + x^6 + x^8 + \cdots)(1 + x^4 + x^8 + x^{12} + \cdots)(1 + x^6 + x^{12} + x^{18} + \cdots)$
10. Find the coefficient of x^9 in the power series of each of these functions.
   a) $(1 + x^3 + x^6 + x^9 + \cdots)^3$
   b) $(x^2 + x^3 + x^4 + x^5 + x^6 + \cdots)^3$
   c) $(x^3 + x^5 + x^6)(x^3 + x^4)(x + x^2 + x^3 + x^4 + \cdots)$
   d) $(x + x^4 + x^7 + x^{10} + \cdots)(x^2 + x^4 + x^6 + x^8 + \cdots)$
   e) $(1 + x + x^2)^3$
11. Find the coefficient of x^{10} in the power series of each of these functions.
   a) $1/(1 - 2x)$
   b) $1/(1 + x)^2$
   c) $1/(1 - x)^3$
   d) $1/(1 + 2x)^4$
   e) $x^4/(1 - 3x)^3$
12. Find the coefficient of x^{12} in the power series of each of these functions.
   a) $1/(1 + 3x)$
   b) $1/(1 - 2x)^2$
   c) $1/(1 + x)^8$
   d) $1/(1 - 4x)^3$
   e) $x^3/(1 + 4x)^2$
13. Use generating functions to determine the number of different ways ten identical balloons can be given to four children if each child receives at least two balloons.
14. Use generating functions to determine the number of different ways 12 identical action figures can be given to five children so that each child receives at most three action figures.
15. Use generating functions to determine the number of different ways 15 identical stuffed animals can be

given to six children so that each child receives at least one but no more than three stuffed animals.

16. Use generating functions to find the number of ways to choose a dozen bagels from three varieties—egg, salty, and plain—if at least two bagels of each kind but no more than three salty bagels are chosen.

17. In how many ways can 25 identical donuts be distributed to four police officers so that each officer gets at least three but no more than seven donuts?

18. Use generating functions to find the number of ways to select 14 balls from a jar containing 100 red balls, 100 blue balls, and 100 green balls so that no fewer than 3 and no more than 10 blue balls are selected. Assume that the order in which the balls are drawn does not matter.

19. What is the generating function for the sequence $\{c_k\}$, where c_k is the number of ways to make change for k dollars using \$1 bills, \$2 bills, \$5 bills, and \$10 bills?

20. What is the generating function for the sequence $\{c_k\}$, where c_k represents the number of ways to make change for k pesos using bills worth 10 pesos, 20 pesos, 50 pesos, and 100 pesos?

21. Give a combinatorial interpretation of the coefficient of x^4 in the expansion $(1 + x + x^2 + x^3 + \cdots)^3$. Use this interpretation to find this number.

22. Give a combinatorial interpretation of the coefficient of x^6 in the expansion $(1 + x + x^2 + x^3 + \cdots)^n$. Use this interpretation to find this number.

23. a) What is the generating function for $\{a_k\}$, where a_k is the number of solutions of $x_1 + x_2 + x_3 = k$ when x_1, x_2, and x_3 are integers with $x_1 \geq 2$, $0 \leq x_2 \leq 3$, and $2 \leq x_3 \leq 5$?
    b) Use your answer to part (a) to find a_6.

24. a) What is the generating function for $\{a_k\}$, where a_k is the number of solutions of $x_1 + x_2 + x_3 + x_4 = k$ when x_1, x_2, x_3, and x_4 are integers with $x_1 \geq 3$, $1 \leq x_2 \leq 5$, $0 \leq x_3 \leq 4$, and $x_4 \geq 1$?
    b) Use your answer to part (a) to find a_7.

25. Explain how generating functions can be used to find the number of ways in which postage of r cents can be pasted on an envelope using 3-cent, 4-cent, and 20-cent stamps.

    a) Assume that the order the stamps are pasted on does not matter.
    b) Assume that the stamps are pasted in a row and the order in which they are pasted on matters.
    c) Use your answer to part (a) to determine the number of ways 46 cents of postage can be pasted on an envelope using 3-cent, 4-cent, and 20-cent stamps when the order the stamps are pasted on does not matter. (Use of a computer algebra program is advised.)
    d) Use your answer to part (b) to determine the number of ways 46 cents of postage can be pasted in a row on an envelope using 3-cent, 4-cent,

and 20-cent stamps when the order in which the stamps are pasted on matters. (Use of a computer algebra program is advised.)

26. a) Show that $1/(1 - x - x^2 - x^3 - x^4 - x^5 - x^6)$ is the generating function for the number of ways that the sum n can be obtained when a die is rolled repeatedly and the order of the rolls matters.
    b) Use part (a) to find the number of ways to roll a total of 8 when a die is rolled repeatedly, and the order of the rolls matters. (Use of a computer algebra package is advised.)

27. Use generating functions (and a computer algebra package, if available) to find the number of ways to make change for \$1 using

    a) dimes and quarters.
    b) nickels, dimes, and quarters.
    c) pennies, dimes, and quarters.
    d) pennies, nickels, dimes, and quarters.

28. Use generating functions (and a computer algebra package, if available) to find the number of ways to make change for \$1 using pennies, nickels, dimes, and quarters with

    a) no more than 10 pennies.
    b) no more than 10 pennies and no more than 10 nickels.
    * c) no more than 10 coins.

29. Use generating functions to find the number of ways to make change for \$100 using

    a) \$10, \$20, and \$50 bills.
    b) \$5, \$10, \$20, and \$50 bills.
    c) \$5, \$10, \$20, and \$50 bills if at least one bill of each denomination is used.
    d) \$5, \$10, and \$20 bills if at least one and no more than four of each denomination is used.

30. If $G(x)$ is the generating function for the sequence $\{a_k\}$, what is the generating function for each of these sequences?

    a) $2a_0, 2a_1, 2a_2, 2a_3, \ldots$ b) $0, a_0, a_1, a_2, a_3, \ldots$
    c) $0, 0, 0, 0, a_2, a_3, \ldots$ d) $a_2, a_3, a_4, \ldots$
    e) $a_1, 2a_2, 3a_3, 4a_4, \ldots$ (*Hint:* Calculus is required here.)
    f) $a_0^2, 2a_0a_1, a_1^2 + 2a_0a_2, 2a_0a_3 + 2a_1a_2, 2a_0a_4 + 2a_1a_3 + a_2^2, \ldots$

31. If $G(x)$ is the generating function for the sequence $\{a_k\}$, what is the generating function for each of these sequences?

    a) $0, 0, 0, a_3, a_4, a_5, \ldots$
    b) $a_0, 0, a_1, 0, a_2, 0, \ldots$
    c) $0, 0, 0, 0, a_0, a_1, a_2, \ldots$
    d) $a_0, 2a_1, 4a_2, 8a_3, 16a_4, \ldots$
    e) $a_0, a_1/2, a_2/3, a_3/4, \ldots$ (*Hint:* Calculus is required here.)
    f) $a_0, a_0 + a_1, a_0 + a_1 + a_2, a_0 + a_1 + a_2 + a_3, \ldots$

32. Use generating functions to solve the recurrence relation $a_k = 7a_{k-1}$ with the initial condition $a_0 = 5$.

33. Use generating functions to solve the recurrence relation $a_k = 3a_{k-1} + 2$ with the initial condition $a_0 = 1$.

34. Use generating functions to solve the recurrence relation $a_k = 3a_{k-1} + 4^{k-1}$ with the initial condition $a_0 = 1$.

35. Use generating functions to solve the recurrence relation $a_k = 5a_{k-1} - 6a_{k-2}$ with initial conditions $a_0 = 6$ and $a_1 = 30$.

36. Use generating functions to solve the recurrence relation $a_k = a_{k-1} + 2a_{k-2} + 2^k$ with initial conditions $a_0 = 4$ and $a_1 = 12$.

37. Use generating functions to solve the recurrence relation $a_k = 4a_{k-1} - 4a_{k-2} + k^2$ with initial conditions $a_0 = 2$ and $a_1 = 5$.

38. Use generating functions to solve the recurrence relation $a_k = 2a_{k-1} + 3a_{k-2} + 4^k + 6$ with initial conditions $a_0 = 20, a_1 = 60$.

39. Use generating functions to find an explicit formula for the Fibonacci numbers.

*40. a) Show that if n is a positive integer, then
$$\binom{-1/2}{n} = \frac{\binom{2n}{n}}{(-4)^n}.$$

b) Use the extended Binomial Theorem and part (a) to show that the coefficient of x^n in the expansion of $(1 - 4x)^{-1/2}$ is $\binom{2n}{n}$ for all nonnegative integers n.

*41. (Calculus required) Let $\{C_n\}$ be the sequence of Catalan numbers, that is, the solution to the recurrence relation $C_n = \sum_{k=0}^{n-1} C_k C_{n-1-k}$ with $C_0 = C_1 = 1$ (see Example 8 in Section 6.1).

a) Show that if $G(x)$ is the generating function for the sequence of Catalan numbers, then $xG(x)^2 - G(x) + 1 = 0$. Conclude (using the initial conditions) that $G(x) = (1 - \sqrt{1 - 4x})/(2x)$.

b) Use Exercise 40 to conclude that
$$G(x) = \sum_{n=0}^{\infty} \frac{1}{n+1} \binom{2n}{n} x^n,$$
so that
$$C_n = \frac{1}{n+1} \binom{2n}{n}.$$

42. Use generating functions to prove Pascal's identity: $C(n, r) = C(n-1, r) + C(n-1, r-1)$ when n and r are positive integers with $r < n$. [*Hint:* Use the identity $(1 + x)^n = (1 + x)^{n-1} + x(1 + x)^{n-1}$.]

43. Use generating functions to prove Vandermonde's identity: $C(m + n, r) = \sum_{k=0}^{r} C(m, r - k)C(n, k)$, whenever m, n, and r are nonnegative integers with r not exceeding either m or n. [*Hint:* Look at the coefficient of x^r in both sides of $(1 + x)^{m+n} = (1 + x)^m (1 + x)^n$.]

44. This exercise shows how to use generating functions to derive a formula for the sum of the first n squares.

a) Show that $(x^2 + x)/(1 - x)^4$ is the generating function for the sequence $\{a_n\}$, where $a_n = 1^2 + 2^2 + \cdots + n^2$.

b) Use part (a) to find an explicit formula for the sum $1^2 + 2^2 + \cdots + n^2$.

The **exponential generating function** for the sequence $\{a_n\}$ is the series
$$\sum_{n=0}^{\infty} \frac{a_n}{n!} x^n.$$

For example, the exponential generating function for the sequence $1, 1, 1, \ldots$ is the function $\sum_{n=0}^{\infty} x^n/n! = e^x$. (You will find this particular series useful in these exercises.) Note that e^x is the (ordinary) generating function for the sequence $1, 1, 1/2!, 1/3!, 1/4!, \ldots$.

45. Find a closed form for the exponential generating function for the sequence $\{a_n\}$, where

a) $a_n = 2$.   **b)** $a_n = (-1)^n$.
c) $a_n = 3^n$.   **d)** $a_n = n + 1$.
e) $a_n = 1/(n + 1)$.

46. Find a closed form for the exponential generating function for the sequence $\{a_n\}$, where

a) $a_n = (-2)^n$.   **b)** $a_n = -1$.   **c)** $a_n = n$.
d) $a_n = n(n - 1)$.   **e)** $a_n = 1/((n + 1)(n + 2))$.

47. Find the sequence with each of these functions as its exponential generating function.

a) $f(x) = e^{-x}$ **b)** $f(x) = 3x^{2x}$
c) $f(x) = e^{3x} - 3e^{2x}$
d) $f(x) = (1 - x) + e^{-2x}$
e) $f(x) = e^{-2x} - (1/(1 - x))$
f) $f(x) = e^{-3x} - (1 + x) + (1/(1 - 2x))$
g) $f(x) = e^{x^2}$

48. Find the sequence with each of these functions as its exponential generating function.

a) $f(x) = e^{3x}$ **b)** $f(x) = 2e^{-3x+1}$
c) $f(x) = e^{4x} + e^{-4x}$ **d)** $f(x) = (1 + 2x) + e^{3x}$
e) $f(x) = e^x - (1/(1 + x))$
f) $f(x) = xe^x$ **g)** $f(x) = e^{x^3}$

49. A coding system encodes messages using strings of octal (base 8) digits. A codeword is considered valid if and only if it contains an even number of 7s.

a) Find a linear nonhomogeneous recurrence relation for the number of valid codewords of length n. What are the initial conditions?

b) Solve this recurrence relation using Theorem 6 in Section 6.2.

c) Solve this recurrence relation using generating functions.

*50. A coding system encodes messages using strings of base 4 digits (that is, digits from the set $\{0, 1, 2, 3\}$). A codeword is valid if and only if it contains an even number of 0s and an even number of 1s. Let a_n equal the number of valid codewords of length n. Furthermore, let b_n, c_n, and d_n equal the number of strings of base 4 digits of length n with an even number of 0s and an odd number of 1s, with an odd number of 0s and an even number of 1s, and with an odd number of 0s and an odd number of 1s, respectively.

a) Show that $d_n = 4^n - a_n - b_n - c_n$. Use this to show that $a_{n+1} = 2a_n + b_n + c_n$, $b_{n+1} = b_n - c_n + 4^n$, and $c_{n+1} = c_n - b_n + 4^n$.

b) What are a_1, b_1, c_1, and d_1?

c) Use parts (a) and (b) to find a_3, b_3, c_3, and d_3.

d) Use the recurrence relations in part (a), together with the initial conditions in part (b), to set up three equations relating the generating functions $A(x)$, $B(x)$, and $C(x)$ for the sequences $\{a_n\}$, $\{b_n\}$, and $\{c_n\}$, respectively.

e) Solve the system of equations from part (d) to get explicit formulae for $A(x)$, $B(x)$, and $C(x)$ and use these to get explicit formulae for a_n, b_n, c_n, and d_n.

Generating functions are useful in studying the number of different types of partitions of an integer n. A **partition** of a positive integer is a way to write this integer as the sum of positive integers where repetition is allowed and the order of the integers in the sum does not matter. For example, the partitions of 5 (with no restrictions) are $1+1+1+1+1$, $1+1+1+2$, $1+1+3$, $1+2+2$, $1+4$, $2+3$, and 5. Exercises 51–56 illustrate some of these uses.

51. Show that the coefficient $p(n)$ of x^n in the formal power series expansion of $1/((1-x)(1-x^2)(1-x^3)\cdots)$ equals the number of partitions of n.

52. Show that the coefficient $p_o(n)$ of x^n in the formal power series expansion of $1/((1-x)(1-x^3)(1-x^5)\cdots)$ equals the number of partitions of n into odd integers, that is, the number of ways to write n as the sum of odd positive integers, where the order does not matter and repetitions are allowed.

53. Show that the coefficient $p_d(n)$ of x^n in the formal power series expansion of $(1+x)(1+x^2)(1+x^3)\cdots$ equals the number of partitions of n into distinct parts, that is, the number of ways to write n as the sum of positive integers, where the order does not matter but no repetitions are allowed.

54. Find $p_o(n)$, the number of partitions of n into odd parts with repetitions allowed, and $p_d(n)$, the number of partitions of n into distinct parts, for $1 \leq n \leq 8$, by writing each partition of each type for each integer.

55. Show that if n is a positive integer, then the number of partitions of n into distinct parts equals the number of partitions of n into odd parts with repetitions allowed; that is, $p_o(n) = p_d(n)$. [*Hint:* Show that the generating functions for $p_o(n)$ and $p_d(n)$ are equal.]

**56. (Calculus required) Use the generating function of $p(n)$ to show that $p(n) \leq e^{C\sqrt{n}}$ for some constant C. [Hardy and Ramanujan showed that $p(n) \sim e^{\pi\sqrt{2/3}\sqrt{n}}/(4\sqrt{3}n)$, which means that the ratio of $p(n)$ and the right-hand side approaches 1 as n approaches infinity.]

Suppose that X is a random variable on a sample space S such that $X(s)$ is a nonnegative integer for all $s \in S$. The **probability generating function** for X is

$$G_X(x) = \sum_{k=0}^{\infty} p(X(s) = k)x^k.$$

57. (Calculus required) Show that if G_X is the probability generating function for a random variable X such that $X(s)$ is a nonnegative integer for all $s \in S$, then

a) $G_X(1) = 1$.   b) $E(X) = G'_X(1)$.

c) $V(X) = G''_X(1) + G'_X(1) - G'_X(1)^2$.

58. Let X be the random variable whose value is n if the first success occurs on the nth trial when independent Bernoulli trials are performed, each with probability of success p.

a) Find a closed formula for the probability generating function G_X.

b) Find the expected value and the variance of X using Exercise 57 and the closed form for the probability generating function found in part (a).

59. Let m be a positive integer. Let X_m be the random variable whose value is n if the mth success occurs on the $(n+m)$th trial when independent Bernoulli trials are performed, each with probability of success p.

a) Using Exercise 30 in the Supplementary Exercises of Chapter 5, show that the probability generating function G_{X_m} is given by $G_{X_m}(x) = p^m/(1-qx)^m$, where $q = 1-p$.

b) Find the expected value and the variance of X_m using Exercise 57 and the closed form for the probability generating function in part (a).

60. Show that if X and Y are independent random variables on a sample space S such that $X(s)$ and $Y(s)$ are nonnegative integers for all $s \in S$, then $G_{X+Y}(x) = G_X(x)G_Y(x)$.

6.5   Inclusion–Exclusion

INTRODUCTION

A discrete mathematics class contains 30 women and 50 sophomores. How many students in the class are either women or sophomores? This question cannot be answered unless more information is provided. Adding the number of women in the class and the number of sophomores probably does not give the correct answer, because women sophomores are counted twice. This observation shows that the number of students in the class that are either sophomores or women is the sum of the number of women and the number of sophomores in the class minus the number of women sophomores. A technique for solving such counting problems was introduced in Section 4.1. In this section we will generalize the ideas introduced in that section to solve a wider range of counting problems.

THE PRINCIPLE OF INCLUSION–EXCLUSION

How many elements are in the union of two finite sets? In Section 1.7 we showed that the number of elements in the union of the two sets A and B is the sum of the numbers of elements in the sets minus the number of elements in their intersection. That is,

$$|A \cup B| = |A| + |B| - |A \cap B|.$$

As we showed in Section 4.1, the formula for the number of elements in the union of two sets is useful in counting problems. Examples 1–3 provide additional illustrations of the usefulness of this formula.

EXAMPLE 1   A discrete mathematics class contains 25 students majoring in computer science, 13 students majoring in mathematics, and eight joint mathematics and computer science majors. How many students are in this class, if every student is majoring in mathematics, computer science, or both mathematics and computer science?

Solution: Let A be the set of students in the class majoring in computer science and B be the set of students in the class majoring in mathematics. Then $A \cap B$ is the set of students in the class who are joint mathematics and computer science majors. Since every student in the class is majoring in either computer science or mathematics (or both), it follows that the number of students in the class is $|A \cup B|$. Therefore,

$$
\begin{aligned}
|A \cup B| &= |A| + |B| - |A \cap B| \\
&= 25 + 13 - 8 \\
&= 30.
\end{aligned}
$$

Therefore, there are 30 students in the class. This computation is illustrated in Figure 1. ◀

EXAMPLE 2   How many positive integers not exceeding 1000 are divisible by 7 or 11?

Solution: Let A be the set of positive integers not exceeding 1000 that are divisible by 7, and let B be the set of positive integers not exceeding 1000 that are divisible by 11. Then $A \cup B$ is the set of integers not exceeding 1000 that are divisible by either 7 or 11, and $A \cap B$ is the set of integers not exceeding 1000 that are divisible by both 7 and 11.

$$|A \cup B| = |A| + |B| - |A \cap B| = 25 + 13 - 8 = 30 \qquad |A \cup B| = |A| + |B| - |A \cap B| = 142 + 90 - 12 = 220$$

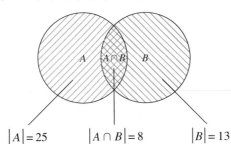

$|A| = 25 \qquad |A \cap B| = 8 \qquad |B| = 13$

**FIGURE 1   The Set of Students
in a Discrete Mathematics Class.**

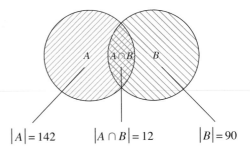

$|A| = 142 \qquad |A \cap B| = 12 \qquad |B| = 90$

**FIGURE 2   The Set of Positive Integers
Not Exceeding 1000 Divisible by Either 7 or 11.**

From Example 2 of Section 2.4, we know that among the positive integers not exceeding 1000 there are $\lfloor 1000/7 \rfloor$ integers divisible by 7 and $\lfloor 1000/11 \rfloor$ divisible by 11. Since 7 and 11 are relatively prime, the integers divisible by both 7 and 11 are those divisible by $7 \cdot 11$. Consequently, there are $\lfloor 1000/(11 \cdot 7) \rfloor$ positive integers not exceeding 1000 that are divisible by both 7 and 11. It follows that there are

$$\begin{aligned}
|A \cup B| &= |A| + |B| - |A \cap B| \\
&= \left\lfloor \frac{1000}{7} \right\rfloor + \left\lfloor \frac{1000}{11} \right\rfloor - \left\lfloor \frac{1000}{7 \cdot 11} \right\rfloor \\
&= 142 + 90 - 12 \\
&= 220
\end{aligned}$$

positive integers not exceeding 1000 that are divisible by either 7 or 11. This computation is illustrated in Figure 2. ◀

Example 3 shows how to find the number of elements in a finite universal set that are outside the union of two sets.

EXAMPLE 3   Suppose that there are 1807 freshmen at your school. Of these, 453 are taking a course in computer science, 567 are taking a course in mathematics, and 299 are taking courses in both computer science and mathematics. How many are not taking a course either in computer science or in mathematics?

Solution: To find the number of freshmen who are not taking a course in either mathematics or computer science, subtract the number that are taking a course in either of these subjects from the total number of freshmen. Let A be the set of all freshmen taking a course in computer science, and let B be the set of all freshmen taking a course in mathematics. It follows that $|A| = 453, |B| = 567$, and $|A \cap B| = 299$. The number of freshmen taking a course in either computer science or mathematics is

$$|A \cup B| = |A| + |B| - |A \cap B| = 453 + 567 - 299 = 721.$$

Consequently, there are $1807 - 721 = 1086$ freshmen who are not taking a course in computer science or mathematics. This is illustrated in Figure 3. ◀

Later in this section it will be shown how the number of elements in the union of a finite number of sets can be found. The result that will be developed is called the **principle of inclusion–exclusion.** Before considering unions of n sets, where n is any positive inte-

$$|A \cup B| = |A| + |B| - |A \cap B| = 453 + 567 - 299 = 721$$

$$|A \cap B| = 299 \qquad |U| = 1807$$

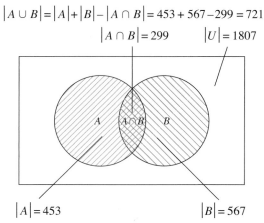

$$|A| = 453 \qquad |B| = 567$$

**FIGURE 3   The Set of Freshmen Not Taking
a Course in Either Computer Science or Mathematics.**

ger, a formula for the number of elements in the union of three sets A, B, and C will be derived. To construct this formula note that $|A| + |B| + |C|$ counts each element that is in exactly one of the three sets once, elements that are in exactly two of the sets twice, and elements in all three sets three times. This is illustrated in the first panel in Figure 4.

To remove the overcount of elements in more than one of the sets, subtract the number of elements in the intersections of all pairs of the three sets, obtaining

$$|A| + |B| + |C| - |A \cap B| - |A \cap C| - |B \cap C|.$$

This expression still counts elements that occur in exactly one of the sets once. An element that occurs in exactly two of the sets is also counted exactly once, since this element will occur in one of the three intersections of sets taken two at a time. However, those elements that occur in all three sets will be counted zero times by this expression, since they occur in all three intersections of sets taken two at a time. This is illustrated in the second panel in Figure 4.

To remedy this undercount, add the number of elements in the intersection of all three sets. This final expression counts each element once, whether it is in one, two, or

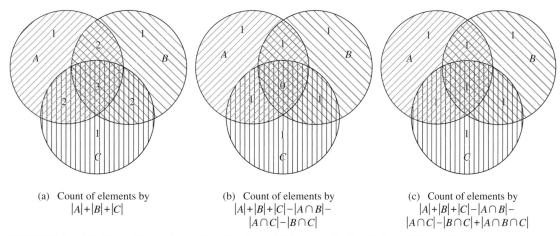

(a) Count of elements by
$|A| + |B| + |C|$

(b) Count of elements by
$|A| + |B| + |C| - |A \cap B| -$
$|A \cap C| - |B \cap C|$

(c) Count of elements by
$|A| + |B| + |C| - |A \cap B| -$
$|A \cap C| - |B \cap C| + |A \cap B \cap C|$

FIGURE 4   Finding a Formula for the Number of Elements in the Union of Three Sets.

three of the sets. Thus,

$$|A \cup B \cup C| = |A| + |B| + |C| - |A \cap B| - |A \cap C| - |B \cap C| + |A \cap B \cap C|.$$

This formula is illustrated in the third panel of Figure 4.

Example 4 illustrates how this formula can be used.

EXAMPLE 4   A total of 1232 students have taken a course in Spanish, 879 have taken a course in French, and 114 have taken a course in Russian. Further, 103 have taken courses in both Spanish and French, 23 have taken courses in both Spanish and Russian, and 14 have taken courses in both French and Russian. If 2092 students have taken at least one of Spanish, French, and Russian, how many students have taken a course in all three languages?

Solution: Let S be the set of students who have taken a course in Spanish, F the set of students who have taken a course in French, and R the set of students who have taken a course in Russian. Then

$$|S| = 1232, \qquad |F| = 879, \qquad |R| = 114,$$
$$|S \cap F| = 103, \quad |S \cap R| = 23, \quad |F \cap R| = 14,$$

and

$$|S \cup F \cup R| = 2092.$$

Inserting these quantities into the equation

$$|S \cup F \cup R| = |S| + |F| + |R| - |S \cap F| - |S \cap R| - |F \cap R| + |S \cap F \cap R|$$

gives

$$2092 = 1232 + 879 + 114 - 103 - 23 - 14 + |S \cap F \cap R|.$$

Solving for $|S \cap F \cap R|$ shows that $|S \cap F \cap R| = 7$. Therefore, there are seven students who have taken courses in Spanish, French, and Russian. This is illustrated in Figure 5.
◀

We will now state and prove the inclusion–exclusion principle, which tells us how many elements are in the union of a finite number of finite sets.

THEOREM 1   **THE PRINCIPLE OF INCLUSION–EXCLUSION** Let $A_1, A_2, \ldots, A_n$ be finite sets. Then

$$|A_1 \cup A_2 \cup \cdots \cup A_n| = \sum_{1 \le i \le n} |A_i| - \sum_{1 \le i < j \le n} |A_i \cap A_j|$$
$$+ \sum_{1 \le i < j < k \le n} |A_i \cap A_j \cap A_k| - \cdots + (-1)^{n+1} |A_1 \cap A_2 \cap \cdots \cap A_n|.$$

Proof: We will prove the formula by showing that an element in the union is counted exactly once by the right-hand side of the equation. Suppose that a is a member of exactly r of the sets $A_1, A_2, \ldots, A_n$ where $1 \le r \le n$. This element is counted $C(r, 1)$ times by $\Sigma |A_i|$. It is counted $C(r, 2)$ times by $\Sigma |A_i \cap A_j|$. In general, it is counted $C(r, m)$ times by the summation involving m of the sets A_i. Thus, this element is counted exactly

$$C(r, 1) - C(r, 2) + C(r, 3) - \cdots + (-1)^{r+1} C(r, r)$$

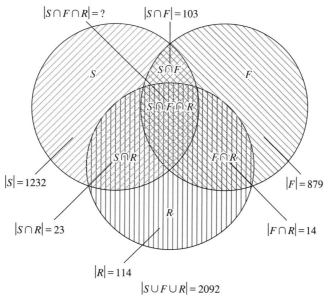

$|S \cap F \cap R| = ?$ $|S \cap F| = 103$

S $S \cap F$ F

$S \cap F \cap R$

$|S| = 1232$ $S \cap R$ $F \cap R$ $|F| = 879$

R

$|S \cap R| = 23$ $|F \cap R| = 14$

$|R| = 114$

$|S \cup F \cup R| = 2092$

FIGURE 5   **The Set of Students Who Have Taken Courses in Spanish, French, and Russian.**

times by the expression on the right-hand side of this equation. Our goal is to evaluate this quantity. By Corollary 2 of Section 4.4, we have

$$C(r, 0) - C(r, 1) + C(r, 2) - \cdots + (-1)^r C(r, r) = 0.$$

Hence,

$$1 = C(r, 0) = C(r, 1) - C(r, 2) + \cdots + (-1)^{r+1} C(r, r).$$

Therefore, each element in the union is counted exactly once by the expression on the right-hand side of the equation. This proves the principle of inclusion–exclusion.   ◁

The inclusion–exclusion principle gives a formula for the number of elements in the union of n sets for every positive integer n. There are terms in this formula for the number of elements in the intersection of every nonempty subset of the collection of the n sets. Hence, there are $2^n - 1$ terms in this formula.

EXAMPLE 5   Give a formula for the number of elements in the union of four sets.

Solution: The inclusion–exclusion principle shows that

$$
\begin{aligned}
|A_1 \cup A_2 \cup A_3 \cup A_4| = {} & |A_1| + |A_2| + |A_3| + |A_4| \\
& - |A_1 \cap A_2| - |A_1 \cap A_3| - |A_1 \cap A_4| - |A_2 \cap A_3| - |A_2 \cap A_4| \\
& - |A_3 \cap A_4| + |A_1 \cap A_2 \cap A_3| + |A_1 \cap A_2 \cap A_4| + |A_1 \cap A_3 \cap A_4| \\
& + |A_2 \cap A_3 \cap A_4| - |A_1 \cap A_2 \cap A_3 \cap A_4|.
\end{aligned}
$$

Note that this formula contains 15 different terms, one for each nonempty subset of $\{A_1, A_2, A_3, A_4\}$.   ◀

Exercises

1. How many elements are in $A_1 \cup A_2$ if there are 12 elements in A_1, 18 elements in A_2, and
   a) $A_1 \cap A_2 = \emptyset$?   **b)** $|A_1 \cap A_2| = 1$?
   c) $|A_1 \cap A_2| = 6$?   **d)** $A_1 \subseteq A_2$?

2. There are 345 students at a college who have taken a course in calculus, 212 who have taken a course in discrete mathematics, and 188 who have taken courses in both calculus and discrete mathematics. How many students have taken a course in either calculus or discrete mathematics?

3. A survey of households in the United States reveals that 96% have at least one television set, 98% have telephone service, and 95% have telephone service and at least one television set. What percentage of households in the United States have neither telephone service nor a television set?

4. A marketing report concerning personal computers states that 650,000 owners will buy a modem for their machines next year and 1,250,000 will buy at least one software package. If the report states that 1,450,000 owners will buy either a modem or at least one software package, how many will buy both a modem and at least one software package?

5. Find the number of elements in $A_1 \cup A_2 \cup A_3$ if there are 100 elements in each set if
   a) the sets are pairwise disjoint.
   b) there are 50 common elements in each pair of sets and no elements in all three sets.
   c) there are 50 common elements in each pair of sets and 25 elements in all three sets.
   d) the sets are equal.

6. Find the number of elements in $A_1 \cup A_2 \cup A_3$ if there are 100 elements in A_1, 1000 in A_2, and 10,000 in A_3 if
   a) $A_1 \subseteq A_2$ and $A_2 \subseteq A_3$.
   b) the sets are pairwise disjoint.
   c) there are two elements common to each pair of sets and one element in all three sets.

7. There are 2504 computer science students at a school. Of these, 1876 have taken a course in Pascal, 999 have taken a course in Fortran, and 345 have taken a course in C. Further, 876 have taken courses in both Pascal and Fortran, 231 have taken courses in both Fortran and C, and 290 have taken courses in both Pascal and C. If 189 of these students have taken courses in Fortran, Pascal, and C, how many of these 2504 students have not taken a course in any of these three programming languages?

8. In a survey of 270 college students, it is found that 64 like brussels sprouts, 94 like broccoli, 58 like cauliflower, 26 like both brussels sprouts and broccoli, 28 like both brussels sprouts and cauliflower, 22 like both broccoli and cauliflower, and 14 like all three vegetables. How many of the 270 students do not like any of these vegetables?

9. How many students are enrolled in a course either in calculus, discrete mathematics, data structures, or programming languages at a school if there are 507, 292, 312, and 344 students in these courses, respectively; 14 in both calculus and data structures; 213 in both calculus and programming languages; 211 in both discrete mathematics and data structures; 43 in both discrete mathematics and programming languages; and no student may take calculus and discrete mathematics, or data structures and programming languages, concurrently?

10. Find the number of positive integers not exceeding 100 that are not divisible by 5 or by 7.

11. Find the number of positive integers not exceeding 100 that are either odd or the square of an integer.

12. Find the number of positive integers not exceeding 1000 that are either the square or the cube of an integer.

13. How many bit strings of length eight do not contain six consecutive 0s?

*14. How many permutations of the 26 letters of the English alphabet do not contain any of the strings *fish*, *rat*, or *bird*?

15. How many permutations of the 10 digits either begin with the 3 digits 987, contain the digits 45 in the fifth and sixth positions, or end with the 3 digits 123?

16. How many elements are in the union of four sets if each of the sets has 100 elements, each pair of the sets shares 50 elements, each three of the sets share 25 elements, and there are 5 elements in all four sets?

17. How many elements are in the union of four sets if the sets have 50, 60, 70, and 80 elements, respectively, each pair of the sets has 5 elements in common, each triple of the sets has 1 common element, and no element is in all four sets?

18. How many terms are there in the formula for the number of elements in the union of 10 sets given by the principle of inclusion–exclusion?

19. Write out the explicit formula given by the principle of inclusion–exclusion for the number of elements in the union of five sets.

20. How many elements are in the union of five sets if the sets contain 10,000 elements each, each pair of sets has 1000 common elements, each triple of sets has 100 common elements, every four of the sets have 10 common elements, and there is 1 element in all five sets?

21. Write out the explicit formula given by the principle of inclusion–exclusion for the number of elements in the union of six sets when it is known that no three of these sets have a common intersection.

*22. Prove the principle of inclusion–exclusion using mathematical induction.

23. Let E_1, E_2, and E_3 be three events from a sample space S. Find a formula for the probability of $E_1 \cup E_2 \cup E_3$.

24. Find the probability that when a fair coin is flipped five times tails comes up exactly three times, the first and last flips come up tails, or the second and fourth flips come up heads.

25. Find the probability that when four numbers from 1 to 100, inclusive, are picked at random with no repetitions allowed, either all are odd, all are divisible by 3, or all are divisible by 5.

26. Find a formula for the probability of the union of four events in a sample space if no three of them can occur at the same time.

27. Find a formula for the probability of the union of five events in a sample space if no four of them can occur at the same time.

28. Find a formula for the probability of the union of n events in a sample space when no two of these events can occur at the same time.

29. Find a formula for the probability of the union of n events in a sample space.

6.6 Applications of Inclusion–Exclusion

INTRODUCTION

Many counting problems can be solved using the principle of inclusion–exclusion. For instance, we can use this principle to find the number of primes less than a positive integer. Many problems can be solved by counting the number of onto functions from one finite set to another. The inclusion–exclusion principle can be used to find the number of such functions. The famous hatcheck problem can be solved using the principle of inclusion–exclusion. This problem asks for the probability that no person is given the correct hat back by a hatcheck person who gives the hats back randomly.

AN ALTERNATIVE FORM OF INCLUSION–EXCLUSION

There is an alternative form of the principle of inclusion–exclusion that is useful in counting problems. In particular, this form can be used to solve problems that ask for the number of elements in a set that have none of n properties $P_1, P_2, \ldots, P_n$.

Let A_i be the subset containing the elements that have property P_i. The number of elements with all the properties $P_{i_1}, P_{i_2}, \ldots, P_{i_k}$ will be denoted by $N(P_{i_1} P_{i_2} \cdots P_{i_k})$. Writing these quantities in terms of sets, we have

$$|A_{i_1} \cap A_{i_2} \cap \cdots \cap A_{i_k}| = N(P_{i_1} P_{i_2} \cdots P_{i_k}).$$

If the number of elements with none of the properties $P_1, P_2, \ldots, P_n$ is denoted by $N(P'_1 P'_2 \cdots P'_n)$ and the number of elements in the set is denoted by N, it follows that

$$N(P'_1 P'_2 \cdots P'_n) = N - |A_1 \cup A_2 \cup \cdots \cup A_n|.$$

From the inclusion–exclusion principle, we see that

$$N(P'_1 P'_2 \cdots P'_n) = N - \sum_{1 \le i \le n} N(P_i) + \sum_{1 \le i < j \le n} N(P_i P_j)$$
$$- \sum_{1 \le i < j < k \le n} N(P_i P_j P_k) + \cdots + (-1)^n N(P_1 P_2 \cdots P_n).$$

Example 1 shows how the principle of inclusion–exclusion can be used to determine the number of solutions in integers of an equation with constraints.

EXAMPLE 1   How many solutions does

$$x_1 + x_2 + x_3 = 11$$

have, where $x_1, x_2,$ and x_3 are nonnegative integers with $x_1 \leq 3, x_2 \leq 4,$ and $x_3 \leq 6$?

Solution: To apply the principle of inclusion–exclusion, let a solution have property P_1 is $x_1 > 3$, property P_2 is $x_2 > 4$, and property P_3 is $x_3 > 6$. The number of solutions satisfying the inequalities $x_1 \leq 3, x_2 \leq 4,$ and $x_3 \leq 6$ is

$$N(P_1'P_2'P_3') = N - N(P_1) - N(P_2) - N(P_3) + N(P_1P_2)$$
$$+ N(P_1P_3) + N(P_2P_3) - N(P_1P_2P_3).$$

Using the same techniques as in Example 5 of Section 4.5, it follows that

- N = total number of solutions = $C(3 + 11 - 1, 11) = 78$,
- $N(P_1)$ = (number of solutions with $x_1 \geq 4$) = $C(3 + 7 - 1, 7) = C(9, 7) = 36$,
- $N(P_2)$ = (number of solutions with $x_2 \geq 5$) = $C(3 + 6 - 1, 6) = C(8, 6) = 28$,
- $N(P_3)$ = (number of solutions with $x_3 \geq 7$) = $C(3 + 4 - 1, 4) = C(6, 4) = 15$,
- $N(P_1P_2)$ = (number of solutions with $x_1 \geq 4$ and $x_2 \geq 5$) = $C(3 + 2 - 1, 2) = C(4, 2) = 6$,
- $N(P_1P_3)$ = (number of solutions with $x_1 \geq 4$ and $x_3 \geq 7$) = $C(3 + 0 - 1, 0) = 1$,
- $N(P_2P_3)$ = (number of solutions with $x_2 \geq 5$ and $x_3 \geq 7$) = 0,
- $N(P_1P_2P_3)$ = (number of solutions with $x_1 \geq 4, x_2 \geq 5,$ and $x_3 \geq 7$) = 0.

Inserting these quantities into the formula for $N(P_1'P_2'P_3')$ shows that the number of solutions with $x_1 \leq 3,$ $x_2 \leq 4,$ and $x_3 \leq 6$ equals

$$N(P_1'P_2'P_3') = 78 - 36 - 28 - 15 + 6 + 1 + 0 - 0 = 6. \qquad \blacktriangleleft$$

THE SIEVE OF ERATOSTHENES

The principle of inclusion–exclusion can be used to find the number of primes not exceeding a specified positive integer. Recall that a composite integer is divisible by a prime not exceeding its square root. So, to find the number of primes not exceeding 100, first note that composite integers not exceeding 100 must have a prime factor not exceeding 10. Because the only primes less than 10 are 2, 3, 5, and 7, the primes not exceeding 100 are these four primes and those positive integers greater than 1 and not exceeding 100 that are divisible by none of 2, 3, 5, or 7. To apply the principle of inclusion–exclusion, let P_1 be the property that an integer is divisible by 2, let P_2 be the property that an integer is divisible by 3, let P_3 be the property that an integer is divisible by 5, and let P_4 be the property that an integer is divisible by 7. Thus, the number of primes not exceeding 100 is

$$4 + N(P_1'P_2'P_3'P_4').$$

Since there are 99 positive integers greater than 1 and not exceeding 100, the principle of inclusion–exclusion shows that

$$N(P_1'P_2'P_3'P_4') = 99 - N(P_1) - N(P_2) - N(P_3) - N(P_4)$$
$$+ N(P_1P_2) + N(P_1P_3) + N(P_1P_4) + N(P_2P_3)$$
$$+ N(P_2P_4) + N(P_3P_4)$$

$$-N(P_1P_2P_3) - N(P_1P_2P_4) - N(P_1P_3P_4)$$
$$-N(P_2P_3P_4) + N(P_1P_2P_3P_4).$$

The number of integers not exceeding 100 (and greater than 1) that are divisible by all the primes in a subset of $\{2, 3, 5, 7\}$ is $\lfloor 100/N \rfloor$, where N is the product of the primes in this subset. (This follows since any two of these primes have no common factor.) Consequently,

$$N(P_1'P_2'P_3'P_4') = 99 - \left\lfloor \frac{100}{2} \right\rfloor - \left\lfloor \frac{100}{3} \right\rfloor - \left\lfloor \frac{100}{5} \right\rfloor - \left\lfloor \frac{100}{7} \right\rfloor$$

$$+ \left\lfloor \frac{100}{2 \cdot 3} \right\rfloor + \left\lfloor \frac{100}{2 \cdot 5} \right\rfloor + \left\lfloor \frac{100}{2 \cdot 7} \right\rfloor + \left\lfloor \frac{100}{3 \cdot 5} \right\rfloor$$

$$+ \left\lfloor \frac{100}{3 \cdot 7} \right\rfloor + \left\lfloor \frac{100}{5 \cdot 7} \right\rfloor$$

$$- \left\lfloor \frac{100}{2 \cdot 3 \cdot 5} \right\rfloor - \left\lfloor \frac{100}{2 \cdot 3 \cdot 7} \right\rfloor - \left\lfloor \frac{100}{2 \cdot 5 \cdot 7} \right\rfloor - \left\lfloor \frac{100}{3 \cdot 5 \cdot 7} \right\rfloor$$

$$+ \left\lfloor \frac{100}{2 \cdot 3 \cdot 5 \cdot 7} \right\rfloor$$

$$= 99 - 50 - 33 - 20 - 14 + 16 + 10 + 7 + 6 + 4 + 2$$

$$- 3 - 2 - 1 - 0 + 0$$

$$= 21.$$

Hence, there are $4 + 21 = 25$ primes not exceeding 100.

The **sieve of Eratosthenes** is used to find all primes not exceeding a specified positive integer. For instance, the following procedure is used to find the primes not exceeding 100. First the integers that are divisible by 2, other than 2, are deleted. Since 3 is the first integer greater than 2 that is left, all those integers divisible by 3, other than 3, are deleted. Since 5 is the next integer left after 3, those integers divisible by 5, other than 5, are deleted. The next integer left is 7, so those integers divisible by 7, other than 7, are deleted. Since all composite integers not exceeding 100 are divisible by 2, 3, 5, or 7, all remaining integers except 1 are prime. In Table 1, the panels display those integers deleted at each stage, where each integer divisible by 2, other than 2, is underlined in the first panel, each integer divisible by 3, other than 3, is underlined in the second panel, each integer divisible by 5, other than 5, is underlined in the third panel, and each integer divisible by 7, other than 7, is underlined in the fourth panel. The integers not underlined are the primes not exceeding 100.

ERATOSTHENES (276–194 B.C.E.)   It is known that Eratosthenes was born in Cyrene, a Greek colony west of Egypt, and spent time studying at Plato's Academy in Athens. We also know that King Ptolemy II invited Eratosthenes to Alexandria to tutor his son and that later Eratosthenes became chief librarian at the famous library at Alexandria, a central repository of ancient wisdom. Eratosthenes was an extremely versatile scholar, writing on mathematics, geography, astronomy, history, philosophy, and literary criticism. Besides his work in mathematics, he is most noted for his chronology of ancient history and for his famous measurement of the size of the earth.

TABLE 1 The Sieve of Eratosthenes.

Integers divisible by 2 other than 2 receive an underline.

1	2	3	4	5	6	7	8	9	10
11	12	13	14	15	16	17	18	19	20
21	22	23	24	25	26	27	28	29	30
31	32	33	34	35	36	37	38	39	40
41	42	43	44	45	46	47	48	49	50
51	52	53	54	55	56	57	58	59	60
61	62	63	64	65	66	67	68	69	70
71	72	73	74	75	76	77	78	79	80
81	82	83	84	85	86	87	88	89	90
91	92	93	94	95	96	97	98	99	100

Integers divisible by 3 other than 3 receive an underline.

1	2	3	4	5	6	7	8	9	10
11	12	13	14	15	16	17	18	19	20
21	22	23	24	25	26	27	28	29	30
31	32	33	34	35	36	37	38	39	40
41	42	43	44	45	46	47	48	49	50
51	52	53	54	55	56	57	58	59	60
61	62	63	64	65	66	67	68	69	70
71	72	73	74	75	76	77	78	79	80
81	82	83	84	85	86	87	88	89	90
91	92	93	94	95	96	97	98	99	100

Integers divisible by 5 other than 5 receive an underline.

1	2	3	4	5	6	7	8	9	10
11	12	13	14	15	16	17	18	19	20
21	22	23	24	25	26	27	28	29	30
31	32	33	34	35	36	37	38	39	40
41	42	43	44	45	46	47	48	49	50
51	52	53	54	55	56	57	58	59	60
61	62	63	64	65	66	67	68	69	70
71	72	73	74	75	76	77	78	79	80
81	82	83	84	85	86	87	88	89	90
91	92	93	94	95	96	97	98	99	100

Integers divisible by 7 other than 7 receive an underline; integers in color are prime.

1	2	3	4	5	6	7	8	9	10
11	12	13	14	15	16	17	18	19	20
21	22	23	24	25	26	27	28	29	30
31	32	33	34	35	36	37	38	39	40
41	42	43	44	45	46	47	48	49	50
51	52	53	54	55	56	57	58	59	60
61	62	63	64	65	66	67	68	69	70
71	72	73	74	75	76	77	78	79	80
81	82	83	84	85	86	87	88	89	90
91	92	93	94	95	96	97	98	99	100

THE NUMBER OF ONTO FUNCTIONS

The principle of inclusion–exclusion can also be used to determine the number of onto functions from a set with m elements to a set with n elements. First consider Example 2.

EXAMPLE 2 How many onto functions are there from a set with six elements to a set with three elements?

Solution: Suppose that the elements in the codomain are b_1, b_2, and b_3. Let P_1, P_2, and P_3 be the properties that b_1, b_2, and b_3 are not in the range of the function, respectively. Note that a function is onto if and only if it has none of the properties P_1, P_2, or P_3. By the inclusion–exclusion principle it follows that the number of onto functions from a set with six elements to a set with three elements is

$$N(P_1'P_2'P_3') = N - [N(P_1) + N(P_2) + N(P_3)]$$
$$+ [N(P_1P_2) + N(P_1P_3) + N(P_2P_3)] - N(P_1P_2P_3),$$

where N is the total number of functions from a set with six elements to one with three elements. We will evaluate each of the terms on the right-hand side of this equation.

From Example 5 of Section 4.1, it follows that $N = 3^6$. Note that $N(P_i)$ is the number of functions that do not have b_i in their range. Hence, there are two choices for the value of the function at each element of the domain. Therefore, $N(P_i) = 2^6$. Furthermore, there are $C(3, 1)$ terms of this kind. Note that $N(P_i P_j)$ is the number of functions that do not have b_i and b_j in their range. Hence, there is only one choice for the value of the function at each element of the domain. Therefore, $N(P_i P_j) = 1^6 = 1$. Furthermore, there are $C(3, 2)$ terms of this kind. Also, note that $N(P_1 P_2 P_3) = 0$, since this term is the number of functions that have none of b_1, b_2, and b_3 in their range. Clearly, there are no such functions. Therefore, the number of onto functions from a set with six elements to one with three elements is

$$3^6 - C(3, 1)2^6 + C(3, 2)1^6 = 729 - 192 + 3 = 540. \qquad \blacktriangleleft$$

The general result that tells us how many onto functions there are from a set with m elements to one with n elements will now be stated. The proof of this result is left as an exercise for the reader.

THEOREM 1   Let m and n be positive integers with $m \geq n$. Then, there are

$$n^m - C(n, 1)(n - 1)^m + C(n, 2)(n - 2)^m - \cdots + (-1)^{n-1}C(n, n - 1) \cdot 1^m$$

onto functions from a set with m elements to a set with n elements.

Remark: The number of onto functions from a set with m elements to a set with n elements equals $n! S_2(m, n)$, where $S_2(m, n)$ is a *Stirling number of the second kind.* See Chapter 6 of [MiRo91] for details.

One of the many different applications of Theorem 1 will now be described.

EXAMPLE 3   How many ways are there to assign five different jobs to four different employees if every employee is assigned at least one job?

Solution: Consider the assignment of jobs as a function from the set of five jobs to the set of four employees. An assignment where every employee gets at least one job is the same as an onto function from the set of jobs to the set of employees. Hence, by Theorem 1 it follows that there are

$$4^5 - C(4, 1)3^5 + C(4, 2)2^5 - C(4, 3)1^5 = 1024 - 972 + 192 - 4 = 240$$

ways to assign the jobs so that each employee is assigned at least one job. $\blacktriangleleft$

DERANGEMENTS

The principle of inclusion–exclusion will be used to count the permutations of n objects that leave no objects in their original positions. Consider Example 4.

EXAMPLE 4   **The Hatcheck Problem** A new employee checks the hats of n people at a restaurant, forgetting to put claim check numbers on the hats. When customers return for their hats, the checker gives them back hats chosen at random from the remaining hats. What is the probability that no one receives the correct hat?   ◄

Remark: The answer is the number of ways the hats can be arranged so that there is no hat in its original position divided by $n!$, the number of permutations of n hats. We will return to this example after we find the number of permutations of n objects that leave no objects in their original position.

A **derangement** is a permutation of objects that leaves no object in its original position. To solve the problem posed in Example 4 we will need to determine the number of derangements of a set of n objects.

EXAMPLE 5   The permutation 21453 is a derangement of 12345 because no number is left in its original position. However, 21543 is not a derangement of 12345, because this permutation leaves 4 fixed.   ◄

Let D_n denote the number of derangements of n objects. For instance, $D_3 = 2$, since the derangements of 123 are 231 and 312. We will evaluate D_n, for all positive integers n, using the principle of inclusion–exclusion.

THEOREM 2

The number of derangements of a set with n elements is

$$D_n = n! \left[1 - \frac{1}{1!} + \frac{1}{2!} - \frac{1}{3!} + \cdots + (-1)^n \frac{1}{n!} \right].$$

Proof: Let a permutation have property P_i if it fixes element i. The number of derangements is the number of permutations having none of the properties P_i for $i = 1, 2, \ldots, n$. This means that

$$D_n = N(P_1' P_2' \cdots P_n').$$

Using the principle of inclusion–exclusion, it follows that

$$D_n = N - \sum_i N(P_i) + \sum_{i<j} N(P_i P_j) - \sum_{i<j<k} N(P_i P_j P_k)$$
$$+ \cdots + (-1)^n N(P_1 P_2 \cdots P_n),$$

where N is the number of permutations of n elements. This equation states that the number of permutations that fix no elements equals the total number of permutations, less the number that fix at least one element, plus the number that fix at least two elements, less the number that fix at least three elements, and so on. All the quantities that occur on the right-hand side of this equation will now be found.

HISTORICAL NOTE   In *rencontres* (matches), an old French card game, the 52 cards in a deck are laid out in a row. The cards of a second deck are laid out with one card of the second deck on top of each card of the first deck. The score is determined by counting the number of matching cards in the two decks. In 1708 Pierre Raymond de Montmort (1678–1719) posed *le problème de rencontres:* What is the probability that no matches take place in the game of rencontres? The solution to Montmort's problem is the probability that a randomly selected permutation of 52 objects is a derangement, namely, $D_{52}/52!$, which, as we will see, is approximately $1/e$.

First, note that $N = n!$, since N is simply the total number of permutations of n elements. Also, $N(P_i) = (n - 1)!$. This follows from the product rule, since $N(P_i)$ is the number of permutations that fix element i, so that the ith position of the permutation is determined, but each of the remaining positions can be filled arbitrarily. Similarly,

$$N(P_i P_j) = (n - 2)!,$$

since this is the number of permutations that fix elements i and j, but where the other $n - 2$ elements can be arranged arbitrarily. In general, note that

$$N(P_{i_1} P_{i_2} \cdots P_{i_m}) = (n - m)!,$$

because this is the number of permutations that fix elements $i_1, i_2, \ldots, i_m$, but where the other $n - m$ elements can be arranged arbitrarily. Because there are $C(n, m)$ ways to choose m elements from n, it follows that

$$\sum_{1 \le i \le n} N(P_i) = C(n, 1)(n - 1)!,$$

$$\sum_{1 \le i < j \le n} N(P_i P_j) = C(n, 2)(n - 2)!,$$

and in general,

$$\sum_{1 \le i_1 < i_2 < \cdots < i_m \le n} N(P_{i_1} P_{i_2} \cdots P_{i_m}) = C(n, m)(n - m)!.$$

Consequently, inserting these quantities into our formula for D_n gives

$$D_n = n! - C(n, 1)(n - 1)! + C(n, 2)(n - 2)! - \cdots + (-1)^n C(n, n)(n - n)!$$

$$= n! - \frac{n!}{1!(n - 1)!}(n - 1)! + \frac{n!}{2!(n - 2)!}(n - 2)! - \cdots + (-1)^n \frac{n!}{n! \, 0!} 0!.$$

Simplifying this expression gives

$$D_n = n! \left[1 - \frac{1}{1!} + \frac{1}{2!} - \cdots + (-1)^n \frac{1}{n!} \right].$$

◁

It is now simple to find D_n for a given positive integer n. For instance, using Theorem 2, it follows that

$$D_3 = 3! \left[1 - \frac{1}{1!} + \frac{1}{2!} - \frac{1}{3!} \right] = 6 \left(1 - 1 + \frac{1}{2} - \frac{1}{6} \right) = 2,$$

as we have previously remarked.

The solution of the problem in Example 4 can now be given.

Solution: The probability that no one receives the correct hat is $D_n / n!$. By Theorem 2, this probability is

$$\frac{D_n}{n!} = 1 - \frac{1}{1!} + \frac{1}{2!} - \cdots + (-1)^n \frac{1}{n!}.$$

The values of this probability for $2 \le n \le 7$ are displayed in Table 2.

TABLE 2 The Probability of a Derangement.						
n	**2**	**3**	**4**	**5**	**6**	**7**
$D_n/n!$	0.50000	0.33333	0.37500	0.36667	0.36806	0.36786

Using methods from calculus it can be shown that

$$e^{-1} = 1 - \frac{1}{1!} + \frac{1}{2!} - \cdots + (-1)^n \frac{1}{n!} + \cdots \sim 0.368.$$

Since this is an alternating series with terms tending to zero, it follows that as n grows without bound, the probability that no one receives the correct hat converges to $e^{-1} \sim 0.368$. In fact, this probability can be shown to be within $1/(n+1)!$ of e^{-1}. ◀

Exercises

1. Suppose that in a bushel of 100 apples there are 20 that have worms in them and 15 that have bruises. Only those apples with neither worms nor bruises can be sold. If there are 10 bruised apples that have worms in them, how many of the 100 apples can be sold?

2. Of 1000 applicants for a mountain-climbing trip in the Himalayas, 450 get altitude sickness, 622 are not in good enough shape, and 30 have allergies. An applicant qualifies if and only if this applicant does not get altitude sickness, is in good shape, and does not have allergies. If there are 111 applicants who get altitude sickness and are not in good enough shape, 14 who get altitude sickness and have allergies, 18 who are not in good enough shape and have allergies, and 9 who get altitude sickness, are not in good enough shape, and have allergies, how many applicants qualify?

3. How many solutions does the equation $x_1 + x_2 + x_3 = 13$ have where x_1, x_2, and x_3 are nonnegative integers less than 6?

4. Find the number of solutions of the equation $x_1 + x_2 + x_3 + x_4 = 17$, where $x_i, i = 1, 2, 3, 4$, are nonnegative integers such that $x_1 \leq 3, x_2 \leq 4, x_3 \leq 5$, and $x_4 \leq 8$.

5. Find the number of primes less than 200 using the principle of inclusion–exclusion.

6. An integer is called **squarefree** if it is not divisible by the square of a positive integer greater than 1. Find the number of squarefree positive integers less than 100.

7. How many positive integers less than 10,000 are not the second or higher power of an integer?

8. How many onto functions are there from a set with seven elements to one with five elements?

9. How many ways are there to distribute six different toys to three different children such that each child gets at least one toy?

10. In how many ways can eight distinct balls be distributed into three distinct urns if each urn must contain at least one ball?

11. In how many ways can seven different jobs be assigned to four different employees so that each employee is assigned at least one job and the most difficult job is assigned to the best employee?

12. List all the derangements of $\{1, 2, 3, 4\}$.

13. How many derangements are there of a set with seven elements?

14. What is the probability that none of 10 people receives the correct hat if a hatcheck person hands their hats back randomly?

15. A machine that inserts letters into envelopes goes haywire and inserts letters randomly into envelopes. What is the probability that in a group of 100 letters

   a) no letter is put into the correct envelope?
   b) exactly 1 letter is put into the correct envelope?
   c) exactly 98 letters are put into the correct envelopes?
   d) exactly 99 letters are put into the correct envelopes?
   e) all letters are put into the correct envelope?

16. A group of n students is assigned seats for each of two classes in the same classroom. How many ways can these seats be assigned if no student is assigned the same seat for both classes?

*17. How many ways can the digits 0, 1, 2, 3, 4, 5, 6, 7, 8, 9 be arranged so that no even digit is in its original position?

*18. Use a combinatorial argument to show that the sequence $\{D_n\}$, where D_n denotes the number of derangements of n objects, satisfies the recurrence relation

$$D_n = (n-1)(D_{n-1} + D_{n-2})$$

for $n \geq 2$. (*Hint:* Note that there are $n-1$ choices for the first element k of a derangement. Consider separately the derangements that start with k that do and do not have 1 in the kth position.)

*19. Use Exercise 18 to show that

$$D_n = nD_{n-1} + (-1)^n$$

for $n \geq 1$.

20. Use Exercise 19 to find an explicit formula for D_n.

21. For which positive integers n is D_n, the number of derangements of n objects, even?

22. Suppose that p and q are distinct primes. Use the principle of inclusion–exclusion to find $\phi(pq)$, the number of positive integers not exceeding pq that are relatively prime to pq.

*23. Use the principle of inclusion–exclusion to derive a formula for $\phi(n)$ when the prime factorization of n is

$$n = p_1^{a_1} p_2^{a_2} \cdots p_m^{a_m}.$$

*24. Show that if n is a positive integer, then

$$n! = C(n, 0)D_n + C(n, 1)D_{n-1}$$
$$+ \cdots + C(n, n-1)D_1 + C(n, n)D_0,$$

where D_k is the number of derangements of k objects.

25. How many derangements of $\{1, 2, 3, 4, 5, 6\}$ begin with the integers 1, 2, and 3, in some order?

26. How many derangements of $\{1, 2, 3, 4, 5, 6\}$ end with the integers 1, 2, and 3, in some order?

27. Prove Theorem 1.

Key Terms and Results

TERMS

recurrence relation: a formula expressing terms of a sequence, except for some initial terms, as a function of one or more previous terms of the sequence

initial conditions for a recurrence relation: the values of the terms of a sequence satisfying the recurrence relation before this relation takes effect

linear homogeneous recurrence relation with constant coefficients: a recurrence relation that expresses the terms of a sequence, except initial terms, as a linear combination of previous terms

characteristic roots of a linear homogeneous recurrence relation with constant coefficients: the roots of the polynomial associated with a linear homogeneous recurrence relation with constant coefficients

linear nonhomogeneous recurrence relation with constant coefficients: a recurrence relation that expresses the terms of a sequence, except for initial terms, as a linear combination of previous terms plus a function that is not identically zero that depends only on the index

divide-and-conquer algorithm: an algorithm that solves a problem recursively by splitting it into a fixed number of smaller problems of the same type

generating function of a sequence: the formal series that has the nth term of the sequence as the coefficient of x^n

sieve of Eratosthenes: a procedure for finding the primes less than a specified positive integer

derangement: a permutation of objects such that no object is in its original place

RESULTS

the formula for the number of elements in the union of two finite sets:

$$|A \cup B| = |A| + |B| - |A \cap B|$$

the formula for the number of elements in the union of three finite sets:

$$|A \cup B \cup C| = |A| + |B| + |C| - |A \cap B| - |A \cap C|$$
$$- |B \cap C| + |A \cap B \cap C|$$

the principle of inclusion–exclusion:

$$|A_1 \cup A_2 \cup \cdots \cup A_n| = \sum_{1 \leq i \leq n} |A_i| - \sum_{1 \leq i < j \leq n} |A_i \cap A_j|$$
$$+ \sum_{1 \leq i < j < k \leq n} |A_i \cap A_j \cap A_k|$$
$$- \cdots + (-1)^{n+1} |A_1 \cap A_2 \cap \cdots \cap A_n|$$

the number of onto functions from a set with m elements to a set with n elements:

$$n^m - C(n, 1)(n-1)^m + C(n, 2)(n-2)^m$$
$$- \cdots + (-1)^{n-1} C(n, n-1) \cdot 1^m$$

the number of derangements of n objects:

$$D_n = n! \left[1 - \frac{1}{1!} + \frac{1}{2!} - \cdots + (-1)^n \frac{1}{n!} \right]$$

Review Questions

1. **a)** What is a recurrence relation?
   b) Find a recurrence relation for the amount of money that will be in an account after n years if $1,000,000 is deposited in an account yielding 9%.
2. Explain how the Fibonacci numbers are used to solve Fibonacci's problem about rabbits.
3. **a)** Find a recurrence relation for the number of steps needed to solve the Tower of Hanoi puzzle.
   b) Show how this recurrence relation can be solved using iteration.
4. **a)** Explain how to find a recurrence relation for the number of bit strings of length n not containing two consecutive 1s.
   b) Describe another counting problem that has a solution satisfying the same recurrence relation.
5. Define a linear homogeneous recurrence relation of degree k.
6. **a)** Explain how to solve linear homogeneous recurrence relations of degree 2.
   b) Solve the recurrence relation $a_n = 13a_{n-1} - 22a_{n-2}$ for $n \geq 2$ if $a_0 = 3$ and $a_1 = 15$.
   c) Solve the recurrence relation $a_n = 14a_{n-1} - 49a_{n-2}$ for $n \geq 2$ if $a_0 = 3$ and $a_1 = 35$.
7. **a)** Explain how to find $f(b^k)$ where k is a positive integer if $f(n)$ satisfies the divide-and-conquer recurrence relation $f(n) = af(n/b) + g(n)$ whenever b divides the positive integer n.
   b) Find $f(256)$ if $f(n) = 3f(n/4) + 5n/4$ and $f(1) = 7$.
8. **a)** Derive a divide-and-conquer recurrence relation for the number of comparisons used to find a number in a list using a binary search.
   b) Give a big-O estimate for the number of comparisons used by a binary search from the divide-and-conquer recurrence relation you gave in (a) using Theorem 1 in Section 6.3.
9. **a)** Give a formula for the number of elements in the union of three sets.

b) Explain why this formula is valid.
c) Explain how to use the formula from (a) to find the number of integers not exceeding 1000 that are divisible by 6, 10, or 15.
d) Explain how to use the formula from (a) to find the number of solutions in nonnegative integers to the equation $x_1 + x_2 + x_3 + x_4 = 22$ with $x_1 < 8, x_2 < 6$, and $x_3 < 5$.
10. **a)** Give a formula for the number of elements in the union of four sets and explain why it is valid.
   b) Suppose the sets A_1, A_2, A_3, and A_4 each contain 25 elements, the intersection of any two of these sets contains 5 elements, the intersection of any three of these sets contains 2 elements, and 1 element is in all four of the sets. How many elements are in the union of the four sets?
11. **a)** State the principle of inclusion–exclusion.
   b) Outline a proof of this principle.
12. Explain how the principle of inclusion–exclusion can be used to count the number of onto functions from a set with m elements to a set with n elements.
13. **a)** How can you count the number of ways to assign m jobs to n employees so that each employee is assigned at least one job?
   b) How many ways are there to assign seven jobs to three employees so that each employee is assigned at least one job?
14. Explain how the inclusion–exclusion principle can be used to count the number of primes not exceeding the positive integer n.
15. **a)** Define a derangement.
   b) Why is counting the number of ways a hatcheck person can return hats to n people, so that no one receives the correct hat, the same as counting the number of derangements of n objects?
   c) Explain how to count the number of derangements of n objects.

Supplementary Exercises

1. A group of ten people begin a chain letter, with each person sending the letter to four other people. Each of these people sends the letter to four additional people.

   a) Find a recurrence relation for the number of letters sent at the nth stage of this chain letter, if no person ever receives more than one letter.

b) What are the initial conditions for the recurrence relation in part (a)?
c) How many letters are sent at the nth stage of the chain letter?

2. A nuclear reactor has created 18 grams of a particular radioactive isotope. Every hour 1% of a radioactive isotope decays.

a) Set up a recurrence relation for the amount of this isotope left after n hours.

b) What are the initial conditions for the recurrence relation in part (a)?

c) Solve this recurrence relation.

3. Every hour the U.S. government prints 10,000 more $1 bills, 4000 more $5 bills, 3000 more $10 bills, 2500 more $20 bills, 1000 more $50 bills, and the same number of $100 bills as it did the previous hour. In the initial hour 1000 of each bill were produced.

a) Set up a recurrence relation for the amount of money produced in the nth hour.

b) What are the initial conditions for the recurrence relation in part (a)?

c) Solve the recurrence relation for the amount of money produced in the nth hour.

d) Set up a recurrence relation for the total amount of money produced in the first n hours.

e) Solve the recurrence relation for the total amount of money produced in the first n hours.

4. Suppose that every hour there are two new bacteria in a colony for each bacterium that was present the previous hour, and that all bacteria 2 hours old die. The colony starts with 100 new bacteria.

a) Set up a recurrence relation for the number of bacteria present after n hours.

b) What is the solution of this recurrence relation?

c) When will the colony contain more than 1 million bacteria?

5. Messages are sent over a communications channel using two different signals. One signal requires 2 microseconds for transmittal, and the other signal requires 3 microseconds for transmittal. Each signal of a message is followed immediately by the next signal.

a) Find a recurrence relation for the number of different signals that can be sent in n microseconds.

b) What are the initial conditions of the recurrence relation in part (a)?

c) How many different messages can be sent in 12 microseconds?

6. A small post office has only 4-cent stamps, 6-cent stamps, and 10-cent stamps. Find a recurrence relation for the number of ways to form postage of n cents with these stamps if the order that the stamps are used matters. What are the initial conditions for this recurrence relation?

7. How many ways are there to form these postages using the rules described in Exercise 6?

a) 12 cents
b) 14 cents
c) 18 cents
d) 22 cents

8. Find the solutions of the simultaneous system of congruences

$$a_n = a_{n-1} + b_{n-1}$$
$$b_n = a_{n-1} - b_{n-1}$$

with $a_0 = 1$ and $b_0 = 2$.

9. Solve the recurrence relation $a_n = a_{n-1}^2/a_{n-2}$ if $a_0 = 1$ and $a_1 = 2$. (*Hint:* Take logarithms of both sides to obtain a recurrence relation for the sequence $\log a_n$, $n = 0, 1, 2, \ldots$.)

*__10.__ Solve the recurrence relation $a_n = a_{n-1}^3 a_{n-2}^2$ if $a_0 = 2$ and $a_1 = 2$. (See the hint for Exercise 9.)

11. Find the solution of the recurrence relation $a_n = 3a_{n-1} - 3a_{n-2} + a_{n-3} + 1$ if $a_0 = 2, a_1 = 4$, and $a_2 = 8$.

12. Find the solution of the recurrence relation $a_n = 3a_{n-1} - 3a_{n-2} + a_{n-3}$ if $a_0 = 2, a_1 = 2$, and $a_2 = 4$.

*__13.__ Suppose that in Example 4 of Section 6.1 a pair of rabbits leaves the island after reproducing twice. Find a recurrence relation for the number of rabbits on the island in the middle of the nth month.

14. Find the solution to the recurrence relation $f(n) = 3f(n/5) + 2n^4$, when n is divisible by 5, for $n = 5^k$, where k is a positive integer and $f(1) = 1$.

15. Estimate the size of f in Exercise 14 if f is an increasing function.

16. Find a recurrence relation that describes the number of comparisons used by the following algorithm: Find the largest and second largest elements of a sequence of n numbers recursively by splitting the sequence into two subsequences with an equal number of terms, or where there is one more term in one subsequence than in the other, at each stage. Stop when subsequences with two terms are reached.

17. Estimate the number of comparisons used by the algorithm described in Exercise 16.

Let $\{a_n\}$ be a sequence of real numbers. The **forward differences** of this sequence are defined recursively as follows: The **first forward difference** is $\Delta a_n = a_{n+1} - a_n$; the **$(k+1)$th forward difference** $\Delta^{k+1} a_n$ is obtained from $\Delta^k a_n$ by $\Delta^{k+1} a_n = \Delta^k a_{n+1} - \Delta^k a_n$.

18. Find Δa_n where

a) $a_n = 3$.　　　　　　　**b)** $a_n = 4n + 7$.
c) $a_n = n^2 + n + 1$.

19. Let $a_n = 3n^3 + n + 2$. Find $\Delta^k a_n$ where k equals
a) 2.　　　　**b)** 3.　　　　**c)** 4.

*__20.__ Suppose that $a_n = P(n)$ where P is a polynomial of degree d. Prove that $\Delta^{d+1} a_n = 0$ for all nonnegative integers n.

21. Let $\{a_n\}$ and $\{b_n\}$ be sequences of real numbers. Show that

$$\Delta(a_n b_n) = a_{n+1}(\Delta b_n) + b_n(\Delta a_n).$$

22. Show that if $F(x)$ and $G(x)$ are the generating functions for the sequences $\{a_k\}$ and $\{b_k\}$, respectively, and c and d are real numbers, then $(cF + dG)(x)$ is the generating function for $\{ca_k + db_k\}$.

23. (Calculus required) This exercise shows how generating functions can be used to solve the recurrence relation $(n + 1)a_{n+1} = a_n + (1/n!)$ for $n \geq 0$ with initial condition $a_0 = 1$.

   a) Let $G(x)$ be the generating function for $\{a_n\}$. Show that $G'(x) = G(x) + e^x$ and $G(0) = 1$.

   b) Show from part (a) that $(e^{-x}G(x))' = 1$, and conclude that $G(x) = xe^x + e^x$.

   c) Use part (b) to find a closed form for a_n.

24. Suppose that 14 students get an A on the first exam in a discrete mathematics class, and 18 get an A on the second exam. If 22 students received an A on either the first exam or the second exam, how many students received an A on both exams?

25. There are 323 farms in Monmouth County that have at least one of horses, cows, and sheep. If 224 have horses, 85 have cows, 57 have sheep, and 18 farms have all three types of animals, how many farms have exactly two of these three types of animals?

26. Queries to a database of student records at a college produced the following data: There are 2175 students at the college, 1675 of these are not freshmen, 1074 students have taken a course in calculus, 444 students have taken a course in discrete mathematics, 607 students are not freshmen and have taken calculus, 350 students have taken calculus and discrete mathematics, 201 students are not freshmen and have taken discrete mathematics, and 143 students are not freshmen and have taken both calculus and discrete mathematics. Can all the responses to the queries be correct?

27. Students in the school of mathematics at a university major in one or more of the following four areas: applied mathematics (AM), pure mathematics (PM), operations research (OR), and computer science (CS). How many students are in this school if (including joint majors) there are 23 students majoring in AM; 17 students majoring in PM; 44 in OR; 63 in CS; 5 in AM and PM; 8 in AM and CS; 4 in AM and OR; 6 in PM and CS; 5 in PM and OR; 14 in OR and CS; 2 in PM, OR, and CS; 2 in AM, OR, and CS; 1 in PM, AM, and OR; 1 in PM, AM, and CS; and 1 in all four fields.

28. How many terms are needed when the inclusion–exclusion principle is used to express the number of elements in the union of seven sets if no more than five of these sets have a common element?

29. How many solutions in positive integers are there to the equation $x_1 + x_2 + x_3 = 20$ with $2 < x_1 < 6$, $6 < x_2 < 10$, and $0 < x_3 < 5$?

30. How many positive integers less than 1,000,000 are

   a) divisible by 2, 3, or 5?

   b) not divisible by 7, 11, or 13?

   c) divisible by 3 but not by 7?

31. How many positive integers less than 200 are

   a) second or higher powers of integers?

   b) either second or higher powers of integers or primes?

   c) not divisible by the square of an integer greater than 1?

   d) not divisible by the cube of an integer greater than 1?

   e) not divisible by three or more primes?

*32. How many ways are there to assign six different jobs to three different employees if the hardest job is assigned to the most experienced employee and the easiest job is assigned to the least experienced employee?

33. What is the probability that exactly one person is given back the correct hat by a hatcheck person who gives n people their hats back at random?

34. How many bit strings of length six do not contain four consecutive 1s?

35. What is the probability that a bit string of length six contains at least four 1s?

Computer Projects

WRITE PROGRAMS WITH THESE INPUT AND OUTPUT.

1. Given a positive integer n, list all the moves required in the Tower of Hanoi puzzle to move n disks from one peg to another according to the rules of the puzzle.

2. Given a positive integer n and an integer k with $1 \leq k \leq n$, list all the moves used by the Frame–Stewart algorithm (described in the preamble to Exercise 54 of Section 6.1) to move n disks from one peg to another using four pegs according to the rules of the puzzle.

3. Given a positive integer n, list all the bit sequences of length n that do not have a pair of consecutive 0s.

4. Given a positive integer n, write out all ways to parenthesize the product of $n + 1$ variables.

5. Given a recurrence relation $a_n = c_1 a_{n-1} + c_2 a_{n-2}$ where c_1 and c_2 are real numbers, initial conditions $a_0 = C_0$ and $a_1 = C_1$, and a positive integer k, find a_k using iteration.

6. Given a recurrence relation $a_n = c_1 a_{n-1} + c_2 a_{n-2}$ and initial conditions $a_0 = C_0$ and $a_1 = C_1$, determine the unique solution.

7. Given a recurrence relation of the form $f(n) = af(n/b) + c$, where a is a real number, b is a positive integer, and c is a real number, and a positive integer k, find $f(b^k)$ using iteration.

8. Given the number of elements in the intersection of three sets, the number of elements in each pairwise intersection of these sets, and the number of elements

in each set, find the number of elements in their union.

9. Given a positive integer n, produce the formula for the number of elements in the union of n sets.

10. Given positive integers m and n, find the number of onto functions from a set with m elements to a set with n elements.

11. Given a positive integer n, list all the derangements of the set $\{1, 2, 3, \ldots, n\}$.

Computations and Explorations

USE A COMPUTATIONAL PROGRAM OR PROGRAMS YOU HAVE WRITTEN TO DO THESE EXERCISES.

1. Find the exact value of f_{100}, f_{500}, and f_{1000} where f_n is the nth Fibonacci number.

2. Find the smallest Fibonacci number greater than 1,000,000, greater than 1,000,000,000, and greater than 1,000,000,000,000.

3. Find as many prime Fibonacci numbers as you can. It is unknown whether there are infinitely many of these.

4. Write out all the moves required to solve the Tower of Hanoi puzzle with 10 disks.

5. Write out all the moves required to use the Frame–Stewart algorithm to move 20 disks from one peg to another peg using four pegs according to the rules of the Reve's puzzle.

6. Verify the Frame conjecture for solving the Reve's puzzle for n disks for as many integers n as possible by showing that the puzzle cannot be solved using fewer moves than are made by the Frame–Stewart algorithm with the optimal choice of k.

7. Compute the number of operations required to mul-

tiply two integers with n bits for various integers n including 16, 64, 256, and 1024 using the fast multiplication described in Section 6.3 and the standard algorithm for multiplying integers (Algorithm 3 in Section 2.5).

8. Compute the number of operations required to multiply two $n \times n$ matrices for various integers n including 4, 16, 64, and 128 using the fast matrix multiplication described in Section 6.3 and the standard algorithm for multiplying matrices (Algorithm 1 in Section 2.7).

9. Use the sieve of Eratosthenes to find all the primes not exceeding 1000.

10. Find the number of primes not exceeding 10,000 using the method described in Section 6.6 to find the number of primes not exceeding 100.

11. List all the derangements of $\{1, 2, 3, 4, 5, 6, 7, 8\}$.

12. Compute the probability that a permutation of n objects is a derangement for all positive integers not exceeding 20 and determine how quickly these probabilities approach the number e.

Writing Projects

RESPOND WITH ESSAYS USING OUTSIDE SOURCES.

1. Find the original source where Fibonacci presented his puzzle about modeling rabbit populations. Discuss this problem and other problems posed by Fibonacci and give some information about Fibonacci himself.

2. Explain how the Fibonacci numbers arise in a variety of applications, such as in phyllotaxis, the study of arrangement of leaves in plants, in the study of reflections by mirrors, and so on.

3. Describe different variations of the Tower of Hanoi puzzle, including those with more than three pegs (including the Reve's puzzle discussed in the text and exercises), those where disk moves are restricted, and those where disks may have the same size. Include what is known about the number of moves required to solve each variation.

4. Discuss as many different problems as possible where the Catalan numbers arise.

5. Describe the solution of Ulam's problem (see Exercise 28 in Section 6.3) involving searching with one lie found by Pelc.

6. Discuss variations of Ulam's problem (see Exercise 28 in Section 6.3) involving searching with more than one lie and what is known about this problem.

7. Discuss divide-and-conquer algorithms for finding the convex hull of a set of points in the plane.

8. Look up the definition of the *lucky numbers*. Explain how they are found using a sieve technique similar to the sieve of Eratosthenes. Find all the lucky numbers less than 1000.

9. Describe how sieve methods are used in number theory. What kind of results have been established using such methods?

10. Look up the rules of the old French card game of *rencontres*. Describe these rules and describe the work of Pierre Raymond de Montmort on *le problème de rencontres*.

11. Describe how exponential generating functions can be used to solve a variety of counting problems.

12. Describe the Polyá theory of counting and the kind of counting problems that can be solved using this theory.

13. The *problème des ménages* (the problem of the households) asks for the number of ways to arrange *n* couples around a table so that the sexes alternate and no husband and wife are seated together. Explain the method used by E. Lucas to solve this problem.

14. Explain how *rook polynomials* can be used to solve counting problems.

Relations

Relationships between elements of sets occur in many contexts. Every day we deal with relationships such as those between a business and its telephone number, an employee and his or her salary, a person and a relative, and so on. In mathematics we study relationships such as those between a positive integer and one that it divides, an integer and one that it is congruent to modulo 5, a real number and one that is larger than it, and so on. Relationships such as that between a program and a variable it uses and that between a computer language and a valid statement in this language often arise in computer science.

Relationships between elements of sets are represented using the structure called a relation. Relations can be used to solve problems such as determining which pairs of cities are linked by airline flights in a network, finding a viable order for the different phases of a complicated project, or producing a useful way to store information in computer databases.

7.1 Relations and Their Properties

INTRODUCTION

The most direct way to express a relationship between elements of two sets is to use ordered pairs made up of two related elements. For this reason, sets of ordered pairs are called binary relations. In this section we introduce the basic terminology used to describe binary relations. Later in this chapter we will use relations to solve problems involving communications networks, project scheduling, and identifying elements in sets with common properties.

DEFINITION 1   Let A and B be sets. A *binary relation from A to B* is a subset of $A \times B$.

In other words, a binary relation from A to B is a set R of ordered pairs where the first element of each ordered pair comes from A and the second element comes from B. We use the notation $a\,R\,b$ to denote that $(a, b) \in R$ and $a\,\not{R}\,b$ to denote that $(a, b) \notin R$. Moreover, when (a, b) belongs to R, a is said to be **related to** b by R.

Binary relations represent relationships between the elements of two sets. We will introduce *n*-ary relations, which express relationships among elements of more than two sets, later in this chapter. We will omit the word *binary* when there is no danger of confusion.

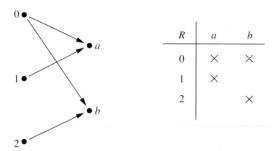

R	a	b
0	×	×
1	×	
2		×

FIGURE 1   **Displaying the Ordered Pairs in the Relation R from Example 3.**

Examples 1–3 illustrate the notion of a relation.

EXAMPLE 1   Let A be the set of students in your school, and let B be the set of courses. Let R be the relation that consists of those pairs (a, b) where a is a student enrolled in course b. For instance, if Jason Goodfriend and Deborah Sherman are enrolled in CS518, which is Discrete Mathematics, the pairs (Jason Goodfriend, CS518) and (Deborah Sherman, CS518) belong to R. If Jason Goodfriend is also enrolled in CS510, which is Data Structures, then the pair (Jason Goodfriend, CS510) is also in R. However, if Deborah Sherman is not enrolled in CS510, then the pair (Deborah Sherman, CS510) is not in R.

Note that if a student is not currently enrolled in any courses there will be no pairs in R that have this student as the first element. Similarly, if a course is not currently being offered there will be no pairs in R that have this course as their second element.   ◀

EXAMPLE 2   Let A be the set of all cities, and let B be the set of the 50 states in the United States of America. Define the relation R by specifying that (a, b) belongs to R if city a is in state b. For instance, (Boulder, Colorado), (Bangor, Maine), (Ann Arbor, Michigan), (Cupertino, California), and (Red Bank, New Jersey) are in R.   ◀

EXAMPLE 3   Let $A = \{0, 1, 2\}$ and $B = \{a, b\}$. Then $\{(0, a), (0, b), (1, a), (2, b)\}$ is a relation from A to B. This means, for instance, that $0\,R\,a$, but that $1\,\not{R}\,b$. Relations can be represented graphically, as shown in Figure 1, using arrows to represent ordered pairs. Another way to represent this relation is to use a table, which is also done in Figure 1. We will discuss representations of relations in more detail in Section 7.3.   ◀

FUNCTIONS AS RELATIONS

Recall that a function f from a set A to a set B (as defined in Section 1.8) assigns exactly one element of B to each element of A. The graph of f is the set of ordered pairs (a, b) such that $b = f(a)$. Since the graph of f is a subset of $A \times B$, it is a relation from A to B. Moreover, the graph of a function has the property that every element of A is the first element of exactly one ordered pair of the graph.

Conversely, if R is a relation from A to B such that every element in A is the first element of exactly one ordered pair of R, then a function can be defined with R as its graph. This can be done by assigning to an element a of A the unique element $b \in B$ such that $(a, b) \in R$.

A relation can be used to express a one-to-many relationship between the elements of the sets A and B, where an element of A may be related to more than one element of B. A function represents a relation where exactly one element of B is related to each element of A.

Relations are a generalization of functions; they can be used to express a much wider class of relationships between sets.

RELATIONS ON A SET

Relations from a set A to itself are of special interest.

DEFINITION 2   A *relation on* the set A is a relation from A to A.

In other words, a relation on a set A is a subset of $A \times A$.

EXAMPLE 4   Let A be the set $\{1, 2, 3, 4\}$. Which ordered pairs are in the relation $R = \{(a, b) \mid a$ divides $b\}$?

Solution: Since (a, b) is in R if and only if a and b are positive integers not exceeding 4 such that a divides b, we see that

$$R = \{(1, 1), (1, 2), (1, 3), (1, 4), (2, 2), (2, 4), (3, 3), (4, 4)\}.$$

The pairs in this relation are displayed both graphically and in tabular form in Figure 2. ◀

Next, some examples of relations on the set of integers will be given in Example 5.

EXAMPLE 5   Consider these relations on the set of integers:

$R_1 = \{(a, b) \mid a \leq b\}$,
$R_2 = \{(a, b) \mid a > b\}$,
$R_3 = \{(a, b) \mid a = b$ or $a = -b\}$,
$R_4 = \{(a, b) \mid a = b\}$,
$R_5 = \{(a, b) \mid a = b + 1\}$,
$R_6 = \{(a, b) \mid a + b \leq 3\}$.

Which of these relations contain each of the pairs $(1, 1), (1, 2), (2, 1), (1, -1)$, and $(2, 2)$?

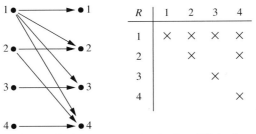

R	1	2	3	4
1	×	×	×	×
2		×		×
3			×	
4				×

FIGURE 2   Displaying the Ordered Pairs in the Relation R from Example 4.

Remark: Unlike the relations in Examples 1–4, these are relations on an infinite set.

Solution: The pair $(1, 1)$ is in R_1, R_3, R_4, and R_6; $(1, 2)$ is in R_1 and R_6; $(2, 1)$ is in R_2, R_5, and R_6; $(1, -1)$ is in R_2, R_3, and R_6; and finally, $(2, 2)$ is in R_1, R_3, and R_4. ◀

It is not hard to determine the number of relations on a finite set, since a relation on a set A is simply a subset of $A \times A$.

EXAMPLE 6 How many relations are there on a set with n elements?

Solution: A relation on a set A is a subset of $A \times A$. Since $A \times A$ has n^2 elements when A has n elements, and a set with m elements has 2^m subsets, there are 2^{n^2} subsets of $A \times A$. Thus, there are 2^{n^2} relations on a set with n elements. For example, there are $2^{3^2} = 2^9 = 512$ relations on the set $\{a, b, c\}$. ◀

PROPERTIES OF RELATIONS

There are several properties that are used to classify relations on a set. We will introduce the most important of these here.

In some relations an element is always related to itself. For instance, let R be the relation on the set of all people consisting of pairs (x, y) where x and y have the same mother and the same father. Then $x \, R \, x$ for every person x.

DEFINITION 3

> A relation R on a set A is called *reflexive* if $(a, a) \in R$ for every element $a \in A$.

Remark: Using quantifiers we see that the relation R on the set A is reflexive if $\forall a((a, a) \in R)$, where the universe of discourse is the set of all elements in A.

We see that a relation on A is reflexive if every element of A is related to itself. Examples 7–9 illustrate the concept of a reflexive relation.

EXAMPLE 7 Consider the following relations on $\{1, 2, 3, 4\}$:

$R_1 = \{(1, 1), (1, 2), (2, 1), (2, 2), (3, 4), (4, 1), (4, 4)\}$,

$R_2 = \{(1, 1), (1, 2), (2, 1)\}$,

$R_3 = \{(1, 1), (1, 2), (1, 4), (2, 1), (2, 2), (3, 3), (4, 1), (4, 4)\}$,

$R_4 = \{(2, 1), (3, 1), (3, 2), (4, 1), (4, 2), (4, 3)\}$,

$R_5 = \{(1, 1), (1, 2), (1, 3), (1, 4), (2, 2), (2, 3), (2, 4), (3, 3), (3, 4), (4, 4)\}$,

$R_6 = \{(3, 4)\}$.

Which of these relations are reflexive?

Solution: The relations R_3 and R_5 are reflexive since they both contain all pairs of the form (a, a), namely, $(1, 1), (2, 2), (3, 3)$, and $(4, 4)$. The other relations are not reflexive since they do not contain all of these ordered pairs. In particular, R_1, R_2, R_4, and R_6 are not reflexive since $(3, 3)$ is not in any of these relations. ◀

EXAMPLE 8   Which of the relations from Example 5 are reflexive?

Solution: The reflexive relations from Example 5 are R_1 (since $a \leq a$ for every integer a), R_3, and R_4. For each of the other relations in this example it is easy to find a pair of the form (a, a) that is not in the relation. (This is left as an exercise for the reader.) ◄

EXAMPLE 9   Is the "divides" relation on the set of positive integers reflexive?

Solution: Since $a \mid a$ whenever a is a positive integer, the "divides" relation is reflexive. ◄

In some relations an element is related to a second element if and only if the second element is also related to the first element. The relation consisting of pairs (x, y) where x and y are students at your school with at least one common class has this property. Other relations have the property that if an element is related to a second element, then this second element is not related to the first. The relation consisting of the pairs (x, y) where x and y are students at your school where x has a higher grade point average than y has this property.

DEFINITION 4

A relation R on a set A is called *symmetric* if $(b, a) \in R$ whenever $(a, b) \in R$, for all $a, b \in A$. A relation R on a set A such that $(a, b) \in R$ and $(b, a) \in R$ only if $a = b$, for all $a, b \in A$, is called *antisymmetric*.

Remark: Using quantifiers, we see that the relation R on the set A is symmetric if $\forall a \forall b ((a, b) \in R \rightarrow (b, a) \in R)$. Similarly, the relation R on the set A is antisymmetric if $\forall a \forall b (((a, b) \in R \land (b, a) \in R) \rightarrow (a = b))$.

That is, a relation is symmetric if and only if a is related to b implies that b is related to a. A relation is antisymmetric if and only if there are no pairs of distinct elements a and b with a related to b and b related to a. The terms *symmetric* and *antisymmetric* are not opposites, since a relation can have both of these properties or may lack both of them (see Exercise 8 at the end of this section). A relation cannot be both symmetric and antisymmetric if it contains some pair of the form (a, b) where $a \neq b$.

Remark: Although relatively few relations are symmetric or antisymmetric, as counting arguments can show, many important relations have one of these properties. (See Exercise 45.)

EXAMPLE 10   Which of the relations from Example 7 are symmetric and which are antisymmetric?

Solution: The relations R_2 and R_3 are symmetric, because in each case (b, a) belongs to the relation whenever (a, b) does. For R_2, the only thing to check is that both $(2, 1)$ and $(1, 2)$ are in the relation. For R_3, it is necessary to check that both $(1, 2)$ and $(2, 1)$ belong to the relation, and $(1, 4)$ and $(4, 1)$ belong to the relation. The reader should verify that none of the other relations is symmetric. This is done by finding a pair (a, b) so that it is in the relation but (b, a) is not.

R_4, R_5, and R_6 are all antisymmetric. For each of these relations there is no pair of elements a and b with $a \neq b$ such that both (a, b) and (b, a) belong to the relation. The reader should verify that none of the other relations is antisymmetric. This is done by finding a pair (a, b) with $a \neq b$ so that (a, b) and (b, a) are both in the relation. ◄

EXAMPLE 11   Which of the relations from Example 5 are symmetric and which are antisymmetric?

Solution: The relations R_3, R_4, and R_6 are symmetric. R_3 is symmetric, for if $a = b$ or $a = -b$, then $b = a$ or $b = -a$. R_4 is symmetric since $a = b$ implies that $b = a$. R_6 is symmetric since $a + b \leq 3$ implies that $b + a \leq 3$. The reader should verify that none of the other relations is symmetric.

The relations R_1, R_2, R_4, and R_5 are antisymmetric. R_1 is antisymmetric because the inequalities $a \leq b$ and $b \leq a$ imply that $a = b$. R_2 is antisymmetric since it is impossible for $a > b$ and $b > a$. R_4 is antisymmetric, since two elements are related with respect to R_4 if and only if they are equal. R_5 is antisymmetric since it is impossible that $a = b + 1$ and $b = a + 1$. The reader should verify that none of the other relations is antisymmetric. ◄

EXAMPLE 12   Is the "divides" relation on the set of positive integers symmetric? Is it antisymmetric?

Solution: This relation is not symmetric since $1 \mid 2$, but $2 \nmid 1$. It is antisymmetric, for if a and b are positive integers with $a \mid b$ and $b \mid a$, then $a = b$ (the verification of this is left as an exercise for the reader). ◄

Let R be the relation consisting of all pairs (x, y) of students at your school, where x has taken more credits than y. Suppose that x is related to y and y is related to z. This means that x has taken more credits than y and y has taken more credits than z. We can conclude that x has taken more credits than z, so that x is related to z. What we have shown is that R has the transitive property, which is defined as follows.

DEFINITION 5

> A relation R on a set A is called *transitive* if whenever $(a, b) \in R$ and $(b, c) \in R$, then $(a, c) \in R$, for all $a, b, c \in A$.

Remark: Using quantifiers we see that the relation R on a set A is transitive if we have $\forall a \forall b \forall c (((a, b) \in R \wedge (b, c) \in R) \rightarrow (a, c) \in R)$.

EXAMPLE 13   Which of the relations in Example 7 are transitive?

Solution: R_4, R_5, and R_6 are transitive. For each of these relations, we can show that it is transitive by verifying that if (a, b) and (b, c) belong to this relation, then (a, c) also does. For instance, R_4 is transitive, since $(3, 2)$ and $(2, 1)$, $(4, 2)$ and $(2, 1)$, $(4, 3)$ and $(3, 1)$, and $(4, 3)$ and $(3, 2)$ are the only such sets of pairs, and $(3, 1)$, $(4, 1)$, and $(4, 2)$ belong to R_4. The reader should verify that R_5 and R_6 are transitive.

R_1 is not transitive since $(3, 4)$ and $(4, 1)$ belong to R_1, but $(3, 1)$ does not. R_2 is not transitive since $(2, 1)$ and $(1, 2)$ belong to R_2, but $(2, 2)$ does not. R_3 is not transitive since $(4, 1)$ and $(1, 2)$ belong to R_3, but $(4, 2)$ does not. ◄

EXAMPLE 14   Which of the relations in Example 5 are transitive?

Solution: The relations R_1, R_2, R_3, and R_4 are transitive. R_1 is transitive since $a \leq b$ and $b \leq c$ imply that $a \leq c$. R_2 is transitive since $a > b$ and $b > c$ imply that $a > c$.

R_3 is transitive since $a = \pm b$ and $b = \pm c$ imply that $a = \pm c$. R_4 is clearly transitive, as the reader should verify. R_5 is not transitive since $(2, 1)$ and $(1, 0)$ belong to R_5, but $(2, 0)$ does not. R_6 is not transitive since $(2, 1)$ and $(1, 2)$ belong to R_6, but $(2, 2)$ does not. ◀

EXAMPLE 15 Is the "divides" relation on the set of positive integers transitive?

Solution: Suppose that a divides b and b divides c. Then there are positive integers k and l such that $b = ak$ and $c = bl$. Hence, $c = a(kl)$, so that a divides c. It follows that this relation is transitive. ◀

We can use counting techniques to determine the number of relations with specific properties. Finding the number of relations with a particular property provides information about how common this property is in the set of all relations on a set with n elements.

EXAMPLE 16 How many reflexive relations are there on a set with n elements?

Solution: A relation R on a set A is a subset of $A \times A$. Consequently, a relation is determined by specifying whether each of the n^2 ordered pairs in $A \times A$ is in R. However, if R is reflexive, each of the n ordered pairs (a, a) for $a \in A$ must be in R. Each of the other $n(n - 1)$ ordered pairs of the form (a, b) where $a \neq b$ may or may not be in R. Hence, by the product rule for counting, there are $2^{n(n-1)}$ reflexive relations [this is the number of ways to choose whether each element (a, b) with $a \neq b$ belongs to R]. ◀

The number of symmetric relations and the number of antisymmetric relations on a set with n elements can be found using reasoning similar to that in Example 16 (see Exercise 45 at the end of this section). Counting the transitive relations on a set with n elements is a problem beyond the scope of this book.

COMBINING RELATIONS

Since relations from A to B are subsets of $A \times B$, two relations from A to B can be combined in any way two sets can be combined. Consider Examples 17–19.

EXAMPLE 17 Let $A = \{1, 2, 3\}$ and $B = \{1, 2, 3, 4\}$. The relations $R_1 = \{(1, 1), (2, 2), (3, 3)\}$ and $R_2 = \{(1, 1), (1, 2), (1, 3), (1, 4)\}$ can be combined to obtain

$$R_1 \cup R_2 = \{(1, 1), (1, 2), (1, 3), (1, 4), (2, 2), (3, 3)\},$$
$$R_1 \cap R_2 = \{(1, 1)\},$$
$$R_1 - R_2 = \{(2, 2), (3, 3)\},$$
$$R_2 - R_1 = \{(1, 2), (1, 3), (1, 4)\}.$$ ◀

EXAMPLE 18 Let A and B be the set of all students and the set of all courses at a school, respectively. Suppose that R_1 consists of all ordered pairs (a, b), where a is a student who has taken course b, and R_2 consists of all ordered pairs (a, b), where a is a student who requires course b to graduate. What are the relations $R_1 \cup R_2$, $R_1 \cap R_2$, $R_1 \oplus R_2$, $R_1 - R_2$, and $R_2 - R_1$?

Solution: The relation $R_1 \cup R_2$ consists of all ordered pairs (a, b), where a is a student who either has taken course b or needs course b to graduate, and $R_1 \cap R_2$ is the set of all ordered pairs (a, b), where a is a student who has taken course b and needs this course to graduate. Also, $R_1 \oplus R_2$ consists of all ordered pairs (a, b), where student a has taken course b but does not need it to graduate or needs course b to graduate but has not taken it. $R_1 - R_2$ is the set of ordered pairs (a, b), where a has taken course b but does not need it to graduate; that is, b is an elective course that a has taken. $R_2 - R_1$ is the set of all ordered pairs (a, b), where b is a course that a needs to graduate but has not taken. ◀

EXAMPLE 19   Let R_1 be the "less than" relation on the set of real numbers and let R_2 be the "greater than" relation on the set of real numbers, that is, $R_1 = \{(x, y) \mid x < y\}$ and $R_2 = \{(x, y) \mid x > y\}$. What are $R_1 \cup R_2$, $R_1 \cap R_2$, $R_1 - R_2$, $R_2 - R_1$, and $R_1 \oplus R_2$?

Solution: We note that $(x, y) \in R_1 \cup R_2$ if and only if $(x, y) \in R_1$ or $(x, y) \in R_2$. Hence $(x, y) \in R_1 \cup R_2$ if and only if $x < y$ or $x > y$. Because the condition $x < y$ or $x > y$ is the same as the condition $x \neq y$, it follows that $R_1 \cup R_2 = \{(x, y) \mid x \neq y\}$. In other words, the union of the "less than" relation and the "greater than" relation is the "not equals" relation.

Next, note that it is impossible for a pair (x, y) to belong to both R_1 and R_2 since it is impossible for $x < y$ and $x > y$. It follows that $R_1 \cap R_2 = \emptyset$. We also see that $R_1 - R_2 = R_1$, $R_2 - R_1 = R_2$, and $R_1 \oplus R_2 = R_1 \cup R_2 - R_1 \cap R_2 = \{(x, y) \mid x \neq y\}$. ◀

There is another way that relations are combined that is analogous to the composition of functions.

DEFINITION 6   Let R be a relation from a set A to a set B and S a relation from B to a set C. The *composite* of R and S is the relation consisting of ordered pairs (a, c), where $a \in A, c \in C$, and for which there exists an element $b \in B$ such that $(a, b) \in R$ and $(b, c) \in S$. We denote the composite of R and S by $S \circ R$.

Computing the composite of two relations requires that we find elements that are the second element of ordered pairs in the first relation and the first element of ordered pairs in the second relation, as Examples 20 and 21 illustrate.

EXAMPLE 20   What is the composite of the relations R and S where R is the relation from $\{1, 2, 3\}$ to $\{1, 2, 3, 4\}$ with $R = \{(1, 1), (1, 4), (2, 3), (3, 1), (3, 4)\}$ and S is the relation from $\{1, 2, 3, 4\}$ to $\{0, 1, 2\}$ with $S = \{(1, 0), (2, 0), (3, 1), (3, 2), (4, 1)\}$?

Solution: $S \circ R$ is constructed using all ordered pairs in R and ordered pairs in S, where the second element of the ordered pair in R agrees with the first element of the ordered pair in S. For example, the ordered pairs $(2, 3)$ in R and $(3, 1)$ in S produce the ordered pair $(2, 1)$ in $S \circ R$. Computing all the ordered pairs in the composite, we find

$$S \circ R = \{(1, 0), (1, 1), (2, 1), (2, 2), (3, 0), (3, 1)\}.$$ ◀

EXAMPLE 21   **Composing the Parent Relation with Itself** Let R be the relation on the set of all people such that $(a, b) \in R$ if person a is a parent of person b. Then $(a, c) \in R \circ R$ if and only if there is a person b such that $(a, b) \in R$ and $(b, c) \in R$, that is, if and only if there is a person b such that a is a parent of b and b is a parent of c. In other words, $(a, c) \in R \circ R$ if and only if a is a grandparent of c. ◀

The powers of a relation R can be recursively defined from the definition of a composite of two relations.

DEFINITION 7

Let R be a relation on the set A. The powers $R^n, n = 1, 2, 3, \ldots,$ are defined recursively by

$$R^1 = R \quad \text{and} \quad R^{n+1} = R^n \circ R.$$

The definition shows that $R^2 = R \circ R, R^3 = R^2 \circ R = (R \circ R) \circ R,$ and so on.

EXAMPLE 22

Let $R = \{(1, 1), (2, 1), (3, 2), (4, 3)\}$. Find the powers $R^n, n = 2, 3, 4, \ldots.$

Solution: Since $R^2 = R \circ R,$ we find that $R^2 = \{(1, 1), (2, 1), (3, 1), (4, 2)\}$. Furthermore, since $R^3 = R^2 \circ R, R^3 = \{(1, 1), (2, 1), (3, 1), (4, 1)\}$. Additional computation shows that R^4 is the same as $R^3,$ so $R^4 = \{(1, 1), (2, 1), (3, 1), (4, 1)\}$. It also follows that $R^n = R^3$ for $n = 5, 6, 7, \ldots.$ The reader should verify this. ◀

The following theorem shows that the powers of a transitive relation are subsets of this relation. It will be used in Section 7.4.

THEOREM 1

The relation R on a set A is transitive if and only if $R^n \subseteq R$ for $n = 1, 2, 3, \ldots.$

Proof: We first prove the "if" part of the theorem. We suppose that $R^n \subseteq R$ for $n = 1, 2, 3, \ldots.$ In particular, $R^2 \subseteq R$. To see that this implies R is transitive, note that if $(a, b) \in R$ and $(b, c) \in R,$ then by the definition of composition, $(a, c) \in R^2$. Since $R^2 \subseteq R,$ this means that $(a, c) \in R$. Hence R is transitive.

We will use mathematical induction to prove the only if part of the theorem. Note that this part of the theorem is trivially true for $n = 1.$

Assume that $R^n \subseteq R$ where n is a positive integer. This is the inductive hypothesis. To complete the inductive step we must show that this implies that R^{n+1} is also a subset of R. To show this, assume that $(a, b) \in R^{n+1}$. Then, since $R^{n+1} = R^n \circ R,$ there is an element x with $x \in A$ such that $(a, x) \in R$ and $(x, b) \in R^n$. The inductive hypothesis, namely, that $R^n \subseteq R,$ implies that $(x, b) \in R$. Furthermore, since R is transitive, and $(a, x) \in R$ and $(x, b) \in R,$ it follows that $(a, b) \in R$. This shows that $R^{n+1} \subseteq R,$ completing the proof. ◁

Exercises

1. List the ordered pairs in the relation R from $A = \{0, 1, 2, 3, 4\}$ to $B = \{0, 1, 2, 3\}$ where $(a, b) \in R$ if and only if

a) $a = b.$
b) $a + b = 4.$
c) $a > b.$
d) $a \mid b.$
e) $\gcd(a, b) = 1.$
f) $\text{lcm}(a, b) = 2.$

2. a) List all the ordered pairs in the relation $R = \{(a, b) \mid a \text{ divides } b\}$ on the set $\{1, 2, 3, 4, 5, 6\}$.

b) Display this relation graphically, as was done in Example 4.

c) Display this relation in tabular form, as was done in Example 4.

3. For each of these relations on the set {1, 2, 3, 4}, decide whether it is reflexive, whether it is symmetric, whether it is antisymmetric, and whether it is transitive.

   a) {(2, 2), (2, 3), (2, 4), (3, 2), (3, 3), (3, 4)}
   b) {(1, 1), (1, 2), (2, 1), (2, 2), (3, 3), (4, 4)}
   c) {(2, 4), (4, 2)}
   d) {(1, 2), (2, 3), (3, 4)}
   e) {(1, 1), (2, 2), (3, 3), (4, 4)}
   f) {(1, 3), (1, 4), (2, 3), (2, 4), (3, 1), (3, 4)}

4. Determine whether the relation R on the set of all people is reflexive, symmetric, antisymmetric, and/or transitive, where $(a, b) \in R$ if and only if

   a) a is taller than b.
   b) a and b were born on the same day.
   c) a has the same first name as b.
   d) a and b have a common grandparent.

5. Determine whether the relation R on the set of all Web pages is reflexive, symmetric, antisymmetric, and/or transitive, where $(a, b) \in R$ if and only if

   a) everyone who has visited Web page a has also visited Web page b.
   b) there are no common links found on both Web page a and Web page b.
   c) there is at least one common link on Web page a and Web page b.
   d) there is a Web page that includes links to both Web page a and Web page b.

6. Determine whether the relation R on the set of all real numbers is reflexive, symmetric, antisymmetric, and/or transitive, where $(x, y) \in R$ if and only if

   a) $x + y = 0$.   b) $x = \pm y$.
   c) $x - y$ is a rational number.
   d) $x = 2y$.   e) $xy \geq 0$.
   f) $xy = 0$.   g) $x = 1$.
   h) $x = 1$ or $y = 1$.

7. Determine whether the relation R on the set of all integers is reflexive, symmetric, antisymmetric, and/or transitive, where $(x, y) \in R$ if and only if

   a) $x \neq y$.   b) $xy \geq 1$.
   c) $x = y + 1$ or $x = y - 1$.
   d) $x \equiv y \pmod 7$.   e) x is a multiple of y.
   f) x and y are both negative or both nonnegative.
   g) $x = y^2$.   h) $x \geq y^2$.

8. Give an example of a relation on a set that is

   a) symmetric and antisymmetric.
   b) neither symmetric nor antisymmetric.

☞ A relation R on the set A is **irreflexive** if for every $a \in A$, $(a, a) \notin R$. That is, R is irreflexive if no element in A is related to itself.

9. Which relations in Exercise 3 are irreflexive?
10. Which relations in Exercise 4 are irreflexive?

11. Which relations in Exercise 5 are irreflexive?
12. Which relations in Exercise 6 are irreflexive?
13. Can a relation on a set be neither reflexive nor irreflexive?
14. Use quantifiers to express what it means for a relation to be irreflexive.
15. Give an example of an irreflexive relation on the set of all people.

A relation R is called **asymmetric** if $(a, b) \in R$ implies that $(b, a) \notin R$.

16. Which relations in Exercise 3 are asymmetric?
17. Which relations in Exercise 4 are asymmetric?
18. Which relations in Exercise 5 are asymmetric?
19. Which relations in Exercise 6 are asymmetric?
20. Must an asymmetric relation also be antisymmetric? Must an antisymmetric relation be asymmetric? Give reasons for your answers.
21. Use quantifiers to express what it means for a relation to be asymmetric.
22. Give an example of an asymmetric relation on the set of all people.
23. How many different relations are there from a set with m elements to a set with n elements?

☞ Let R be a relation from a set A to a set B. The **inverse relation** from B to A, denoted by R^{-1}, is the set of ordered pairs $\{(b, a) \mid (a, b) \in R\}$. The **complementary relation** $\overline{R}$ is the set of ordered pairs $\{(a, b) \mid (a, b) \notin R\}$.

24. Let R be the relation $R = \{(a, b) \mid a < b\}$ on the set of integers. Find

    a) R^{-1}.   b) $\overline{R}$.

25. Let R be the relation $R = \{(a, b) \mid a$ divides $b\}$ on the set of positive integers. Find

    a) R^{-1}.   b) $\overline{R}$.

26. Let R be the relation on the set of all states in the United States consisting of pairs (a, b) where state a borders state b. Find

    a) R^{-1}.   b) $\overline{R}$.

27. Suppose that the function f from A to B is a one-to-one correspondence. Let R be the relation that equals the graph of f. That is, $R = \{(a, f(a)) \mid a \in A\}$. What is the inverse relation R^{-1}?

28. Let $R_1 = \{(1, 2), (2, 3), (3, 4)\}$ and $R_2 = \{(1, 1), (1, 2), (2, 1), (2, 2), (2, 3), (3, 1), (3, 2), (3, 3), (3, 4)\}$ be relations from {1, 2, 3} to {1, 2, 3, 4}. Find

    a) $R_1 \cup R_2$.   b) $R_1 \cap R_2$.
    c) $R_1 - R_2$.   d) $R_2 - R_1$.

29. Let A be the set of students at your school and B the set of books in the school library. Let R_1 and R_2 be the relations consisting of all ordered pairs (a, b) where student a is required to read book b in a course, and where student a has read book b, respectively. Describe the ordered pairs in each of these relations.

a) $R_1 \cup R_2$ **b)** $R_1 \cap R_2$ **c)** $R_1 \oplus R_2$
d) $R_1 - R_2$ **e)** $R_2 - R_1$

30. Let R be the relation $\{(1, 2), (1, 3), (2, 3), (2, 4), (3, 1)\}$, and let S be the relation $\{(2, 1), (3, 1), (3, 2), (4, 2)\}$. Find $S \circ R$.

31. Let R be the relation on the set of people consisting of pairs (a, b) where a is a parent of b. Let S be the relation on the set of people consisting of pairs (a, b) where a and b are siblings (brothers or sisters). What are $S \circ R$ and $R \circ S$?

Exercises 32–35 deal with these relations on the set of real numbers:

$R_1 = \{(a, b) \in \mathbf{R}^2 \mid a > b\}$, the "greater than" relation,
$R_2 = \{(a, b) \in \mathbf{R}^2 \mid a \geq b\}$, the "greater than or equal to" relation,
$R_3 = \{(a, b) \in \mathbf{R}^2 \mid a < b\}$, the "less than" relation,
$R_4 = \{(a, b) \in \mathbf{R}^2 \mid a \leq b\}$, the "less than or equal to" relation,
$R_5 = \{(a, b) \in \mathbf{R}^2 \mid a = b\}$, the "equal to" relation,
$R_6 = \{(a, b) \in \mathbf{R}^2 \mid a \neq b\}$, the "unequal to" relation.

32. Find

a) $R_1 \cup R_3$.   **b)** $R_1 \cup R_5$.   **c)** $R_2 \cap R_4$.
d) $R_3 \cap R_5$.   **e)** $R_1 - R_2$.   **f)** $R_2 - R_1$.
g) $R_1 \oplus R_3$.   **h)** $R_2 \oplus R_4$.

33. Find

a) $R_2 \cup R_4$.   **b)** $R_3 \cup R_6$.   **c)** $R_3 \cap R_6$.
d) $R_4 \cap R_6$.   **e)** $R_3 - R_6$.   **f)** $R_6 - R_3$.
g) $R_2 \oplus R_6$.   **h)** $R_3 \oplus R_5$.

34. Find

a) $R_1 \circ R_1$.   **b)** $R_1 \circ R_2$.   **c)** $R_1 \circ R_3$.
d) $R_1 \circ R_4$.   **e)** $R_1 \circ R_5$.   **f)** $R_1 \circ R_6$.
g) $R_2 \circ R_3$.   **h)** $R_3 \circ R_3$.

35. Find

a) $R_2 \circ R_1$.   **b)** $R_2 \circ R_2$.   **c)** $R_3 \circ R_5$.
d) $R_4 \circ R_1$.   **e)** $R_5 \circ R_3$.   **f)** $R_3 \circ R_6$.
g) $R_4 \circ R_6$.   **h)** $R_6 \circ R_6$.

36. Let R be the parent relation on the set of all people (see Example 21). When is an ordered pair in the relation R^3?

37. Let R be the relation on the set of people with doctorates such that $(a, b) \in R$ if and only if a was the thesis advisor of b. When is an ordered pair (a, b) in R^2? When is an ordered pair (a, b) in R^n when n is a positive integer? (Note that every person with a doctorate has a thesis advisor.)

38. Let R_1 and R_2 be the "divides" and "is a multiple of" relations on the set of all positive integers, respectively. That is, $R_1 = \{(a, b) \mid a \text{ divides } b\}$ and $R_2 = \{(a, b) \mid a \text{ is a multiple of } b\}$. Find

a) $R_1 \cup R_2$.   **b)** $R_1 \cap R_2$.   **c)** $R_1 - R_2$.
d) $R_2 - R_1$.   **e)** $R_1 \oplus R_2$.

39. Let R_1 and R_2 be the "congruent modulo 3" and the "congruent modulo 4" relations, respectively, on the set of integers. That is, $R_1 = \{(a, b) \mid a \equiv b \,(\mathrm{mod}\ 3)\}$ and $R_2 = \{(a, b) \mid a \equiv b \,(\mathrm{mod}\ 4)\}$. Find

a) $R_1 \cup R_2$.   **b)** $R_1 \cap R_2$.   **c)** $R_1 - R_2$.
d) $R_2 - R_1$.   **e)** $R_1 \oplus R_2$.

40. List the 16 different relations on the set $\{0, 1\}$.

41. How many of the 16 different relations on $\{0, 1\}$ contain the pair $(0, 1)$?

42. Which of the 16 relations on $\{0, 1\}$, which you listed in Exercise 40, are

a) reflexive?   **b)** irreflexive?
c) symmetric?   **d)** antisymmetric?
e) asymmetric?   **f)** transitive?

43. a) How many relations are there on the set $\{a, b, c, d\}$?

b) How many relations are there on the set $\{a, b, c, d\}$ that contain the pair (a, a)?

44. Let S be a set with n elements and let a and b be distinct elements of S. How many relations are there on S such that

a) $(a, b) \in S$?   **b)** $(a, b) \notin S$?
c) there are no ordered pairs in the relation that have a as their first element?
d) there is at least one ordered pair in the relation that has a as its first element?
e) there are no ordered pairs in the relation that have a as their first element and there are no ordered pairs in the relation that have b as their second element?
f) there is at least one ordered pair in the relation that either has a as its first element or has b as its second element?

*45. How many relations are there on a set with n elements that are

a) symmetric?   **b)** antisymmetric?
c) asymmetric?   **d)** irreflexive?
e) reflexive and symmetric?
f) neither reflexive nor irreflexive?

*46. How many transitive relations are there on a set with n elements if

a) $n = 1$?   **b)** $n = 2$?   **c)** $n = 3$?

47. Find the error in the "proof" of the following "theorem."

"Theorem": Let R be a relation on a set A that is symmetric and transitive. Then R is reflexive.

"Proof": Let $a \in A$. Take an element $b \in A$ such that $(a, b) \in R$. Since R is symmetric, we also have $(b, a) \in R$. Now using the transitive property, we can conclude that $(a, a) \in R$ since $(a, b) \in R$ and $(b, a) \in R$.

48. Suppose that R and S are reflexive relations on a set A. Prove or disprove each of these statements.

a) $R \cup S$ is reflexive.
b) $R \cap S$ is reflexive.
c) $R \oplus S$ is irreflexive.
d) $R - S$ is irreflexive.
e) $S \circ R$ is reflexive.

49. Show that the relation R on a set A is symmetric if and only if $R = R^{-1}$, where R^{-1} is the inverse relation.
50. Show that the relation R on a set A is antisymmetric if and only if $R \cap R^{-1}$ is a subset of the diagonal relation $\Delta = \{(a, a) \mid a \in A\}$.
51. Show that the relation R on a set A is reflexive if and only if the inverse relation R^{-1} is reflexive.
52. Show that the relation R on a set A is reflexive if and only if the complementary relation $\overline{R}$ is irreflexive.

53. Let R be a relation that is reflexive and transitive. Prove that $R^n = R$ for all positive integers n.
54. Let R be the relation on the set $\{1, 2, 3, 4, 5\}$ containing the ordered pairs $(1, 1), (1, 2), (1, 3), (2, 3), (2, 4), (3, 1), (3, 4), (3, 5), (4, 2), (4, 5), (5, 1), (5, 2)$, and $(5, 4)$. Find
a) R^2.   b) R^3.
c) R^4.   d) R^5.
55. Let R be a reflexive relation on a set A. Show that R^n is reflexive for all positive integers n.
*56. Let R be a symmetric relation. Show that R^n is symmetric for all positive integers n.
57. Suppose that the relation R is irreflexive. Is R^2 necessarily irreflexive? Give a reason for your answer.

7.2   *n*-ary Relations and Their Applications

INTRODUCTION

Relationships among elements of more than two sets often arise. For instance, there is a relationship involving the name of a student, the student's major, and the student's grade point average. Similarly, there is a relationship involving the airline, flight number, starting point, destination, departure time, and arrival time of a flight. An example of such a relationship in mathematics involves three integers where the first integer is larger than the second integer, which is larger than the third. Another example is the betweenness relationship involving points on a line, such that three points are related when the second point is between the first and the third.

We will study relationships among elements from more than two sets in this section. These relationships are called *n*-ary relations. These relations are used to represent computer databases. These representations help us answer queries about the information stored in databases, such as: Which flights land at O'Hare Airport between 3 A.M. and 4 A.M.? Which students at your school are sophomores majoring in mathematics or computer science and have greater than a 3.0 average? Which employees of a company have worked for the company less than 5 years and make more than $50,000?

n-ARY RELATIONS

We begin with the basic definition on which the theory of relational databases rests.

DEFINITION 1   Let $A_1, A_2, \ldots, A_n$ be sets. An *n-ary relation* on these sets is a subset of $A_1 \times A_2 \times \cdots \times A_n$. The sets $A_1, A_2, \ldots, A_n$ are called the *domains* of the relation, and n is called its *degree*.

EXAMPLE 1   Let R be the relation on $\mathbf{N} \times \mathbf{N} \times \mathbf{N}$ consisting of triples (a, b, c) where a, b, and c are integers with $a < b < c$. Then $(1, 2, 3) \in R$, but $(2, 4, 3) \notin R$. The degree of this relation is 3. Its domains are all equal to the set of integers.   ◄

EXAMPLE 2   Let R be the relation consisting of 5-tuples (A, N, S, D, T) representing airplane flights, where A is the airline, N is the flight number, S is the starting point, D is the destination, and T is the departure time. For instance, if Nadir Express Airlines has flight 963 from Newark to Bangor at 15:00, then (Nadir, 963, Newark, Bangor, 15:00) belongs to R. The degree of this relation is 5, and its domains are the set of all airlines, the set of flight numbers, the set of cities, the set of cities (again), and the set of times. ◀

DATABASES AND RELATIONS

The time required to manipulate information in a database depends on how this information is stored. The operations of adding and deleting records, updating records, searching for records, and combining records from overlapping databases are performed millions of times each day in a large database. Because of the importance of these operations, various methods for representing databases have been developed. We will discuss one of these methods, called the **relational data model,** based on the concept of a relation.

A database consists of **records,** which are *n*-tuples, made up of **fields.** The fields are the entries of the *n*-tuples. For instance, a database of student records may be made up of fields containing the name, student number, major, and grade point average of the student. The relational data model represents a database of records as an *n*-ary relation. Thus, student records are represented as 4-tuples of the form (*STUDENT NAME, ID NUMBER, MAJOR, GPA*). A sample database of six such records is:

(Ackermann, 231455, Computer Science, 3.88)

(Adams, 888323, Physics, 3.45)

(Chou, 102147, Computer Science, 3.49)

(Goodfriend, 453876, Mathematics, 3.45)

(Rao, 678543, Mathematics, 3.90)

(Stevens, 786576, Psychology, 2.99).

Relations used to represent databases are also called **tables,** since these relations are often displayed as tables. Each column of the table corresponds to an *attribute* of the database. For instance, the same database of students is displayed in Table 1. The attributes of this database are Student Name, ID Number, Major, and GPA.

A domain of an *n*-ary relation is called a **primary key** when the value of the *n*-tuple from this domain determines the *n*-tuple. That is, a domain is a primary key when no two *n*-tuples in the relation have the same value from this domain.

Records are often added to or deleted from databases. Because of this, the property that a domain is a primary key is time-dependent. Consequently, a primary key should be chosen that remains one whenever the database is changed. This can be done by using

TABLE 1 Students.			
Student_name	*ID_number*	*Major*	*GPA*
Ackermann	231455	Computer Science	3.88
Adams	888323	Physics	3.45
Chou	102147	Computer Science	3.49
Goodfriend	453876	Mathematics	3.45
Rao	678543	Mathematics	3.90
Stevens	786576	Psychology	2.99

a primary key of the **intension** of the database, which contains all the *n*-tuples that can ever be included in an *n*-ary relation representing this database.

EXAMPLE 3 Which domains are primary keys for the *n*-ary relation displayed in Table 1, assuming that no *n*-tuples will be added in the future?

Solution: Since there is only one 4-tuple in this table for each student name, the domain of student names is a primary key. Similarly, the ID numbers in this table are unique, so that the domain of ID numbers is also a primary key. However, the domain of major fields of study is not a primary key, since more than one 4-tuple contains the same major field of study. The domain of grade point averages is also not a primary key, since there are two 4-tuples containing the same GPA. ◀

Combinations of domains can also uniquely identify *n*-tuples in an *n*-ary relation. When the values of a set of domains determine an *n*-tuple in a relation, the Cartesian product of these domains is called a **composite key.**

EXAMPLE 4 Is the Cartesian product of the domain of major fields of study and the domain of GPAs a composite key for the *n*-ary relation from Table 1, assuming that no *n*-tuples are ever added?

Solution: Since no two 4-tuples from this table have both the same major and the same GPA, this Cartesian product is a composite key. ◀

Since primary and composite keys are used to identify records uniquely in a database, it is important that keys remain valid when new records are added to the database. Hence, checks should be made to ensure that every new record has values that are different in the appropriate field, or fields, from all other records in this table. For instance, it makes sense to use the student identification number as a key for student records if no two students ever have the same student identification number. A university should not use the name field as a key, since two students may have the same name (such as John Smith).

OPERATIONS ON *n*-ARY RELATIONS

There are a variety of operations on *n*-ary relations that can be used to form new *n*-ary relations. Applied together, these operations can answer queries on databases that ask for all tuples that satisfy certain conditions.

The most basic operation on an *n*-ary relation is determining all *n*-tuples in the *n*-ary relation that satisfy certain conditions. For example, we may want to find all the records of all computer science majors in a database of student records. We may want to find all students who have a grade point average above 3.5 in this same database. We may want to find the records of all computer science majors who have a grade point average above 3.5 in this database. To perform such tasks we use the selection operator.

DEFINITION 2 Let *R* be an *n*-ary relation and *C* a condition that elements in *R* may satisfy. Then the *selection operator* s_C maps the *n*-ary relation *R* to the *n*-ary relation of all *n*-tuples from *R* that satisfy the condition *C*.

EXAMPLE 5 To find the records of computer science majors in the *n*-ary relation R shown in Table 1, we use the operator s_{C_1}, where C_1 is the condition Major = "Computer Science." The result is the two 4-tuples (Ackermann, 231455, Computer Science, 3.88) and (Chou, 102147, Computer Science, 3.49). Similarly, to find the records of students who have a grade point average above 3.5 in this database, we use the operator s_{C_2}, where C_2 is the condition GPA > 3.5. The result is the two 4-tuples (Ackermann, 231455, Computer Science, 3.88) and (Rao, 678543, Mathematics, 3.90). Finally, to find the records of computer science majors who have a GPA above 3.5, we use the operator s_{C_3}, where C_3 is the condition (Major = "Computer Science" $\land$ GPA > 3.5). The result consists of the single 4-tuple (Ackermann, 231455, Computer Science, 3.88). ◀

Projections are used to form new *n*-ary relations by deleting the same fields in every record of the relation.

DEFINITION 3 The *projection* $P_{i_1 i_2,\ldots,i_m}$ maps the *n*-tuple $(a_1, a_2, \ldots, a_n)$ to the *m*-tuple $(a_{i_1}, a_{i_2}, \ldots, a_{i_m})$, where $m \leq n$.

In other words, the projection $P_{i_1,i_2,\ldots,i_m}$ deletes $n - m$ of the components of an *n*-tuple, leaving the i_1th, i_2th, ..., and i_mth components.

EXAMPLE 6 What results when the projection $P_{1,3}$ is applied to the 4-tuples $(2, 3, 0, 4)$, (Jane Doe, 234111001, Geography, 3.14), and (a_1, a_2, a_3, a_4)?

Solution: The projection $P_{1,3}$ sends these 4-tuples to $(2, 0)$, (Jane Doe, Geography), and (a_1, a_3), respectively. ◀

Example 7 illustrates how new relations are produced using projections.

EXAMPLE 7 What relation results when the projection $P_{1,4}$ is applied to the relation in Table 1?

Solution: When the projection $P_{1,4}$ is used, the second and third columns of the table are deleted, and pairs representing student names and grade point averages are obtained. Table 2 displays the results of this projection. ◀

Fewer rows may result when a projection is applied to the table for a relation. This happens when some of the *n*-tuples in the relation have identical values in each of the *m* components of the projection, and only disagree in components deleted by the projection. For instance, consider the following example.

EXAMPLE 8 What is the table obtained when the projection $P_{1,2}$ is applied to the relation in Table 3?

Solution: Table 4 displays the relation obtained when $P_{1,2}$ is applied to Table 3. Note that there are fewer rows after this projection is applied. ◀

The **join** operation is used to combine two tables into one when these tables share some identical fields. For instance, a table containing fields for airline, flight number, and gate, and another table containing fields for flight number, gate, and departure time can be combined into a table containing fields for airline, flight number, gate, and departure time.

TABLE 2 GPAs.	
Student_ name	*GPA*
Ackermann	3.88
Adams	3.45
Chou	3.49
Goodfriend	3.45
Rao	3.90
Stevens	2.99

TABLE 3 Enrollments.

Student	*Major*	*Course*
Glauser	Biology	BI 290
Glauser	Biology	MS 475
Glauser	Biology	PY 410
Marcus	Mathematics	MS 511
Marcus	Mathematics	MS 603
Marcus	Mathematics	CS 322
Miller	Computer Science	MS 575
Miller	Computer Science	CS 455

TABLE 4 Majors.

Student	*Major*
Glauser	Biology
Marcus	Mathematics
Miller	Computer Science

DEFINITION 4

Let R be a relation of degree m and S a relation of degree n. The *join* $J_p(R, S)$, where $p \leq m$ and $p \leq n$, is a relation of degree $m + n - p$ that consists of all $(m + n - p)$-tuples $(a_1, a_2, \ldots, a_{m-p}, c_1, c_2, \ldots, c_p, b_1, b_2, \ldots, b_{n-p})$, where the m-tuple $(a_1, a_2, \ldots, a_{m-p}, c_1, c_2, \ldots, c_p)$ belongs to R and the n-tuple $(c_1, c_2, \ldots, c_p, b_1, b_2, \ldots, b_{n-p})$ belongs to S.

In other words, the join operator J_p produces a new relation from two relations by combining all m-tuples of the first relation with all n-tuples of the second relation, where the last p components of the m-tuples agree with the first p components of the n-tuples.

EXAMPLE 9 What relation results when the join operator J_2 is used to combine the relation displayed in Tables 5 and 6?

Solution: The join J_2 produces the relation shown in Table 7. ◀

There are other operators besides projections and joins that produce new relations from existing relations. A description of these operations can be found in books on database theory.

TABLE 5 Teaching_assignments.

Professor	*Department*	*Course_ number*
Cruz	Zoology	335
Cruz	Zoology	412
Farber	Psychology	501
Farber	Psychology	617
Grammer	Physics	544
Grammer	Physics	551
Rosen	Computer Science	518
Rosen	Mathematics	575

TABLE 6 Class_schedule.

Department	*Course_ number*	*Room*	*Time*
Computer Science	518	N521	2:00 P.M.
Mathematics	575	N502	3:00 P.M.
Mathematics	611	N521	4:00 P.M.
Physics	544	B505	4:00 P.M.
Psychology	501	A100	3:00 P.M.
Psychology	617	A110	11:00 A.M.
Zoology	335	A100	9:00 A.M.
Zoology	412	A100	8:00 A.M.

TABLE 7 **Teaching_schedule.**				
Professor	*Department*	*Course_number*	*Room*	*Time*
Cruz	Zoology	335	A100	9:00 A.M.
Cruz	Zoology	412	A100	8:00 A.M.
Farber	Psychology	501	A100	3:00 P.M.
Farber	Psychology	617	A110	11:00 A.M.
Grammer	Physics	544	B505	4:00 P.M.
Rosen	Computer Science	518	N521	2:00 P.M.
Rosen	Mathematics	575	N502	3:00 P.M.

SQL

The database query language SQL (short for Structured Query Language) can be used to carry out the operations we have described in this section. Example 10 illustrates how SQL commands are related to operations on *n*-ary relations.

EXAMPLE 10   We will illustrate how SQL is used to express queries by showing how SQL can be employed to make a query about airline flights using Table 8. The SQL statements

```
SELECT Departure_time
FROM Flights
WHERE destination='Detroit'
```

are used to find the projection P_5 (on the departure_time attribute) of the selection of 5-tuples in the Flights database that satisfy the condition: destination = 'Detroit'. The output would be a list containing the times of flights that have Detroit as their destination, namely, 08:10, 08:47, and 09:44. SQL uses the FROM clause to identify the *n*-ary relation the query is applied to, the WHERE clause to specify the condition of the selection operation, and the SELECT clause to specify the projection operation that is to be applied. (*Beware:* SQL uses SELECT to represent a projection, rather than a selection operation. This is an unfortunate example of conflicting terminology.)   ◀

Example 11 shows how SQL queries can be made involving more than one table.

TABLE 8 **Flights.**				
Airline	*Flight_number*	*Gate*	*Destination*	*Departure_time*
Nadir	122	34	Detroit	08:10
Acme	221	22	Denver	08:17
Acme	122	33	Anchorage	08:22
Acme	323	34	Honolulu	08:30
Nadir	199	13	Detroit	08:47
Acme	222	22	Denver	09:10
Nadir	322	34	Detroit	09:44

EXAMPLE 11   The SQL statements

```
SELECT professor, time
FROM teaching_assignments, class_schedule
WHERE department='mathematics'
```

are used to find the projection $P_{1,5}$ of the 5-tuples in the database (shown in Table 7), which is the join J_2 of the teaching_assignments and class_schedule databases in Tables 5 and 6, respectively, which satisfy the condition: department = mathematics. The output would consist of the single 2-tuple (Rosen, 3:00 P.M.). The SQL FROM clause is used here to find the join of two different databases.   ◀

We have only touched on the basic concepts of relational databases in this section. More information can be found in [AhUl95].

Exercises

1. List the triples in the relation $\{(a, b, c) \mid a, b,$ and c are integers with $0 < a < b < c < 5\}$.

2. Which 4-tuples are in the relation $\{(a, b, c, d) \mid a, b, c,$ and d are positive integers with $abcd = 6\}$?

3. List the 5-tuples in the relation in Table 8.

4. Assuming that no new n-tuples are added, find all the primary keys for the relations displayed in

    a) Table 3.   **b)** Table 5.
    c) Table 6.   **d)** Table 8.

5. Assuming that no new n-tuples are added, find a composite key with two fields containing the *Airline* field for the database in Table 8.

6. Assuming that no new n-tuples are added, find a composite key with two fields containing the Professor field for the database in Table 7.

7. The 3-tuples in a 3-ary relation represent the following attributes of a student database: student ID number, name, phone number.

    a) Is student ID number likely to be a primary key?
    b) Is name likely to be a primary key?
    c) Is phone number likely to be a primary key?

8. The 4-tuples in a 4-ary relation represent these attributes of published books: title, ISBN, publication date, number of pages.

    a) What is a likely primary key for this relation?
    b) Under what conditions would (title, publication date) be a composite key?
    c) Under what conditions would (title, number of pages) be a composite key?

9. The 5-tuples in a 5-ary relation represent these attributes of all people in the United States: name, Social Security number, street address, city, state.

    a) Determine a primary key for this relation.
    b) Under what conditions would (name, street address) be a composite key?
    c) Under what conditions would (name, street address, city) be a composite key?

10. What do you obtain when you apply the selection operator s_C, where C is the condition Room = A100, to the database in Table 7?

11. What do you obtain when you apply the selection operator s_C, where C is the condition Destination = Detroit, to the database in Table 8?

12. What do you obtain when you apply the selection operator s_C, where C is the condition (Project = 2) ∧ (Quantity ≥ 50), to the database in Table 10?

13. What do you obtain when you apply the selection operator s_C, where C is the condition (Airline = Nadir) ∨ (Destination = Denver), to the database in Table 8?

14. What do you obtain when you apply the projection $P_{2,3,5}$ to the 5-tuple (a, b, c, d, e)?

15. Which projection mapping is used to delete the first, second, and fourth components of a 6-tuple?

16. Display the table produced by applying the projection $P_{1,2,4}$ to Table 8.

17. Display the table produced by applying the projection $P_{1,4}$ to Table 8.

18. How many components are there in the n-tuples in the table obtained by applying the join operator J_3 to two tables with 5-tuples and 8-tuples, respectively?

19. Construct the table obtained by applying the join operator J_2 to the relations in Tables 9 and 10.

20. Show that if C_1 and C_2 are conditions that elements of the n-ary relation R may satisfy, then $s_{C_1 \wedge C_2}(R) = s_{C_1}(s_{C_2}(R))$.

TABLE 9 Part_needs.		
Supplier	*Part number*	*Project*
23	1092	1
23	1101	3
23	9048	4
31	4975	3
31	3477	2
32	6984	4
32	9191	2
33	1001	1

TABLE 10 Parts_inventory.			
Part_ number	*Project*	*Quantity*	*Color_ code*
1001	1	14	8
1092	1	2	2
1101	3	1	1
3477	2	25	2
4975	3	6	2
6984	4	10	1
9048	4	12	2
9191	2	80	4

21. Show that if C_1 and C_2 are conditions that elements of the n-ary relation R may satisfy, then $s_{C_1}(s_{C_2}(R)) = s_{C_2}(s_{C_1}(R))$.

22. Show that if C is a condition that elements of the n-ary relations R and S may satisfy, then $s_C(R \cup S) = s_C(R) \cup s_C(S)$.

23. Show that if C is a condition that elements of the n-ary relations R and S may satisfy, then $s_C(R \cap S) = s_C(R) \cap s_C(S)$.

24. Show that if C is a condition that elements of the n-ary relations R and S may satisfy, then $s_C(R - S) = s_C(R) - s_C(S)$.

25. Show that if R and S are both n-ary relations, then $P_{i_1, i_2, \ldots, i_m}(R \cup S) = P_{i_1, i_2, \ldots, i_m}(R) \cup P_{i_1, i_2, \ldots, i_m}(S)$.

26. Give an example to show that if R and S are both n-ary relations, then $P_{i_1, i_2, \ldots, i_m}(R \cap S)$ may be different from $P_{i_1, i_2, \ldots, i_m}(R) \cap P_{i_1, i_2, \ldots, i_m}(S)$.

27. Give an example to show that if R and S are both

n-ary relations, then $P_{i_1, i_2, \ldots, i_m}(R - S)$ may be different from $P_{i_1, i_2, \ldots, i_m}(R) - P_{i_1, i_2, \ldots, i_m}(S)$.

28. a) What are the operations that correspond to the query expressed using these SQL statements?

```
SELECT Supplier
FROM Part_needs
WHERE 1000 ≤ Part_number ≤ 5000
```

b) What is the output of this query given the database in Table 9 as input?

29. a) What are the operations that correspond to the query expressed using these SQL statements?

```
SELECT Supplier, Project
FROM Part_needs, Parts_inventory
WHERE Quantity ≤ 10
```

b) What is the output of this query given the databases in Tables 9 and 10 as input?

7.3 **Representing Relations**

INTRODUCTION

There are many ways to represent a relation between finite sets. As we have seen, one way is to list its ordered pairs. In this section we will discuss two alternative methods for representing relations. One method uses zero–one matrices. The other method uses directed graphs.

Generally, matrices are appropriate for the representation of relations in computer programs. On the other hand, people often find the representation of relations using directed graphs useful for understanding the properties of these relations.

REPRESENTING RELATIONS USING MATRICES

A relation between finite sets can be represented using a zero–one matrix. Suppose that R is a relation from $A = \{a_1, a_2, \ldots, a_m\}$ to $B = \{b_1, b_2, \ldots, b_n\}$. (Here the elements of the sets A and B have been listed in a particular, but arbitrary, order. Furthermore, when $A = B$ we use the same ordering for A and B.) The relation R can be represented by the matrix $\mathbf{M}_R = [m_{ij}]$, where

$$m_{ij} = \begin{cases} 1 & \text{if } (a_i, b_j) \in R, \\ 0 & \text{if } (a_i, b_j) \notin R. \end{cases}$$

In other words, the zero–one matrix representing R has a 1 as its (i, j) entry when a_i is related to b_j, and a 0 in this position if a_i is not related to b_j. (Such a representation depends on the orderings used for A and B.)

The use of matrices to represent relations is illustrated in Examples 1–6.

EXAMPLE 1 Suppose that $A = \{1, 2, 3\}$ and $B = \{1, 2\}$. Let R be the relation from A to B containing (a, b) if $a \in A, b \in B$, and $a > b$. What is the matrix representing R if $a_1 = 1, a_2 = 2$, and $a_3 = 3$, and $b_1 = 1$ and $b_2 = 2$?

Solution: Since $R = \{(2, 1), (3, 1), (3, 2)\}$, the matrix for R is

$$\mathbf{M}_R = \begin{bmatrix} 0 & 0 \\ 1 & 0 \\ 1 & 1 \end{bmatrix}.$$

The 1s in $\mathbf{M}_R$ show that the pairs $(2, 1), (3, 1)$, and $(3, 2)$ belong to R. The 0s show that no other pairs belong to R. ◄

EXAMPLE 2 Let $A = \{a_1, a_2, a_3\}$ and $B = \{b_1, b_2, b_3, b_4, b_5\}$. Which ordered pairs are in the relation R represented by the matrix

$$\mathbf{M}_R = \begin{bmatrix} 0 & 1 & 0 & 0 & 0 \\ 1 & 0 & 1 & 1 & 0 \\ 1 & 0 & 1 & 0 & 1 \end{bmatrix} ?$$

Solution: Since R consists of those ordered pairs (a_i, b_j) with $m_{ij} = 1$, it follows that

$$R = \{(a_1, b_2), (a_2, b_1), (a_2, b_3), (a_2, b_4), (a_3, b_1), (a_3, b_3), (a_3, b_5)\}. \qquad ◄$$

FIGURE 1
The Zero–One Matrix for a Reflexive Relation.

The matrix of a relation on a set, which is a square matrix, can be used to determine whether the relation has certain properties. Recall that a relation R on A is reflexive if $(a, a) \in R$ whenever $a \in A$. Thus, R is reflexive if and only if $(a_i, a_i) \in R$ for $i = 1, 2, \ldots, n$. Hence, R is reflexive if and only if $m_{ii} = 1$, for $i = 1, 2, \ldots, n$. In other words, R is reflexive if all the elements on the main diagonal of $\mathbf{M}_R$ are equal to 1, as shown in Figure 1.

The relation R is symmetric if $(a, b) \in R$ implies that $(b, a) \in R$. Consequently, the relation R on the set $A = \{a_1, a_2, \ldots, a_n\}$ is symmetric if and only if $(a_j, a_i) \in R$ whenever $(a_i, a_j) \in R$. In terms of the entries of $\mathbf{M}_R$, R is symmetric if and only if $m_{ji} = 1$ whenever $m_{ij} = 1$. This also means $m_{ji} = 0$ whenever $m_{ij} = 0$. Consequently, R is symmetric if and only if $m_{ij} = m_{ji}$, for all pairs of integers i and j

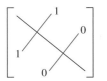 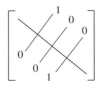

(a) Symmetric   (b) Antisymmetric

**FIGURE 2   The Zero–One Matrices for
Symmetric and Antisymmetric Relations.**

with $i = 1, 2, \ldots, n$ and $j = 1, 2, \ldots, n$. Recalling the definition of the transpose of a matrix from Section 2.7, we see that R is symmetric if and only if

$$\mathbf{M}_R = (\mathbf{M}_R)^t,$$

that is, if $\mathbf{M}_R$ is a symmetric matrix. The form of the matrix for a symmetric relation is illustrated in Figure 2(a).

The relation R is antisymmetric if and only if $(a, b) \in R$ and $(b, a) \in R$ imply that $a = b$. Consequently, the matrix of an antisymmetric relation has the property that if $m_{ij} = 1$ with $i \neq j$, then $m_{ji} = 0$. Or, in other words, either $m_{ij} = 0$ or $m_{ji} = 0$ when $i \neq j$. The form of the matrix for an antisymmetric relation is illustrated in Figure 2(b).

EXAMPLE 3   Suppose that the relation R on a set is represented by the matrix

$$\mathbf{M}_R = \begin{bmatrix} 1 & 1 & 0 \\ 1 & 1 & 1 \\ 0 & 1 & 1 \end{bmatrix}.$$

Is R reflexive, symmetric, and/or antisymmetric?

Solution: Since all the diagonal elements of this matrix are equal to 1, R is reflexive. Moreover, since $\mathbf{M}_R$ is symmetric, it follows that R is symmetric. It is also easy to see that R is not antisymmetric.   ◀

The Boolean operations join and meet (discussed in Section 2.7) can be used to find the matrices representing the union and the intersection of two relations. Suppose that R_1 and R_2 are relations on a set A represented by the matrices $\mathbf{M}_{R_1}$ and $\mathbf{M}_{R_2}$, respectively. The matrix representing the union of these relations has a 1 in the positions where either $\mathbf{M}_{R_1}$ or $\mathbf{M}_{R_2}$ has a 1. The matrix representing the intersection of these relations has a 1 in the positions where both $\mathbf{M}_{R_1}$ and $\mathbf{M}_{R_2}$ have a 1. Thus, the matrices representing the union and intersection of these relations are

$$\mathbf{M}_{R_1 \cup R_2} = \mathbf{M}_{R_1} \vee \mathbf{M}_{R_2} \quad \text{and} \quad \mathbf{M}_{R_1 \cap R_2} = \mathbf{M}_{R_1} \wedge \mathbf{M}_{R_2}.$$

EXAMPLE 4   Suppose that the relations R_1 and R_2 on a set A are represented by the matrices

$$\mathbf{M}_{R_1} = \begin{bmatrix} 1 & 0 & 1 \\ 1 & 0 & 0 \\ 0 & 1 & 0 \end{bmatrix} \quad \text{and} \quad \mathbf{M}_{R_2} = \begin{bmatrix} 1 & 0 & 1 \\ 0 & 1 & 1 \\ 1 & 0 & 0 \end{bmatrix}.$$

What are the matrices representing $R_1 \cup R_2$ and $R_1 \cap R_2$?

Solution: The matrices of these relations are

$$\mathbf{M}_{R_1 \cup R_2} = \mathbf{M}_{R_1} \vee \mathbf{M}_{R_2} = \begin{bmatrix} 1 & 0 & 1 \\ 1 & 1 & 1 \\ 1 & 1 & 0 \end{bmatrix},$$

$$\mathbf{M}_{R_1 \cap R_2} = \mathbf{M}_{R_1} \wedge \mathbf{M}_{R_2} = \begin{bmatrix} 1 & 0 & 1 \\ 0 & 0 & 0 \\ 0 & 0 & 0 \end{bmatrix}. \qquad \blacktriangleleft$$

We now turn our attention to determining the matrix for the composite of relations. This matrix can be found using the Boolean product of the matrices (discussed in Section 2.7) for these relations. In particular, suppose that R is a relation from A to B and S is a relation from B to C. Suppose that A, B, and C have $m, n,$ and p elements, respectively. Let the zero–one matrices for $S \circ R$, R, and S be $\mathbf{M}_{S \circ R} = [t_{ij}]$, $\mathbf{M}_R = [r_{ij}]$, and $\mathbf{M}_S = [s_{ij}]$, respectively (these matrices have sizes $m \times p, m \times n,$ and $n \times p$, respectively). The ordered pair (a_i, c_j) belongs to $S \circ R$ if and only if there is an element b_k such that (a_i, b_k) belongs to R and (b_k, c_j) belongs to S. It follows that $t_{ij} = 1$ if and only if $r_{ik} = s_{kj} = 1$ for some k. From the definition of the Boolean product, this means that

$$\mathbf{M}_{S \circ R} = \mathbf{M}_R \odot \mathbf{M}_S.$$

EXAMPLE 5 Find the matrix representing the relations $S \circ R$ where the matrices representing R and S are

$$\mathbf{M}_R = \begin{bmatrix} 1 & 0 & 1 \\ 1 & 1 & 0 \\ 0 & 0 & 0 \end{bmatrix} \quad \text{and} \quad \mathbf{M}_S = \begin{bmatrix} 0 & 1 & 0 \\ 0 & 0 & 1 \\ 1 & 0 & 1 \end{bmatrix}.$$

Solution: The matrix for $S \circ R$ is

$$\mathbf{M}_{S \circ R} = \mathbf{M}_R \odot \mathbf{M}_S = \begin{bmatrix} 1 & 1 & 1 \\ 0 & 1 & 1 \\ 0 & 0 & 0 \end{bmatrix}. \qquad \blacktriangleleft$$

The matrix representing the composite of two relations can be used to find the matrix for $\mathbf{M}_{R^n}$. In particular,

$$\mathbf{M}_{R^n} = \mathbf{M}_R^{[n]},$$

from the definition of Boolean powers. Exercise 35 at the end of this section asks for a proof of this formula.

EXAMPLE 6 Find the matrix representing the relation R^2, where the matrix representing R is

$$\mathbf{M}_R = \begin{bmatrix} 0 & 1 & 0 \\ 0 & 1 & 1 \\ 1 & 0 & 0 \end{bmatrix}.$$

Solution: The matrix for R^2 is

$$\mathbf{M}_{R^2} = \mathbf{M}_R^{[2]} = \begin{bmatrix} 0 & 1 & 1 \\ 1 & 1 & 1 \\ 0 & 1 & 0 \end{bmatrix}. \qquad \blacktriangleleft$$

REPRESENTING RELATIONS USING DIGRAPHS

We have shown that a relation can be represented by listing all of its ordered pairs or by using a zero–one matrix. There is another important way of representing a relation using a pictorial representation. Each element of the set is represented by a point, and each ordered pair is represented using an arc with its direction indicated by an arrow. We use such pictorial representations when we think of relations on a finite set as **directed graphs,** or **digraphs.**

DEFINITION 1

A *directed graph,* or *digraph,* consists of a set V of *vertices* (or *nodes*) together with a set E of ordered pairs of elements of V called *edges* (or *arcs*). The vertex a is called the *initial vertex* of the edge (a, b), and the vertex b is called the *terminal vertex* of this edge.

An edge of the form (a, a) is represented using an arc from the vertex a back to itself. Such an edge is called a **loop.**

EXAMPLE 7   The directed graph with vertices a, b, c, and d, and edges (a, b), (a, d), (b, b), (b, d), (c, a), (c, b), and (d, b) is displayed in Figure 3.   ◄

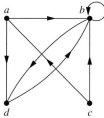

FIGURE 3
A Directed Graph.

The relation R on a set A is represented by the directed graph that has the elements of A as its vertices and the ordered pairs (a, b), where $(a, b) \in R$, as edges. This assignment sets up a one-to-one correspondence between the relations on a set A and the directed graphs with A as their set of vertices. Thus, every statement about relations corresponds to a statement about directed graphs, and vice versa. Directed graphs give a visual display of information about relations. As such, they are often used to study relations and their properties. (Note that relations from a set A to a set B can be represented by a directed graph where there is a vertex for each element of A and a vertex for each element of B, as shown in Section 7.1. However, when $A = B$, such representation provides much less insight than the digraph representations described here.) The use of directed graphs to represent relations on a set is illustrated in Examples 8–10.

EXAMPLE 8   The directed graph of the relation

$$R = \{(1, 1), (1, 3), (2, 1), (2, 3), (2, 4), (3, 1), (3, 2), (4, 1)\}$$

on the set $\{1, 2, 3, 4\}$ is shown in Figure 4.   ◄

EXAMPLE 9   What are the ordered pairs in the relation R represented by the directed graph shown in Figure 5?

Solution: The ordered pairs (x, y) in the relation are

$$R = \{(1, 3), (1, 4), (2, 1), (2, 2), (2, 3), (3, 1), (3, 3), (4, 1), (4, 3)\}.$$

Each of these pairs corresponds to an edge of the directed graph, with $(2, 2)$ and $(3, 3)$ corresponding to loops.   ◄

The directed graph representing a relation can be used to determine whether the relation has various properties. For instance, a relation is reflexive if and only if there is a loop at every vertex of the directed graph, so that every ordered pair of the

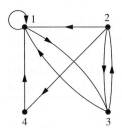

FIGURE 4 The Directed Graph of the Relation *R*.

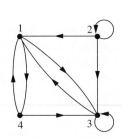

FIGURE 5 The Directed Graph of the Relation *R*.

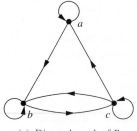

(a) Directed graph of *R*

(b) Directed graph of *S*

FIGURE 6 The Directed Graphs of the Relations *R* and *S*.

form (x, x) occurs in the relation. A relation is symmetric if and only if for every edge between distinct vertices in its digraph there is an edge in the opposite direction, so that (y, x) is in the relation whenever (x, y) is in the relation. Similarly, a relation is antisymmetric if and only if there are never two edges in opposite directions between distinct vertices. Finally, a relation is transitive if and only if whenever there is an edge from a vertex x to a vertex y and an edge from a vertex y to a vertex z, there is an edge from x to z (completing a triangle where each side is a directed edge with the correct direction).

Remark: Note that a symmetric relation can be represented by an undirected graph, which is a graph where edges do not have directions. We will study undirected graphs in Chapter 8.

EXAMPLE 10 Determine whether the relations for the directed graphs shown in Figure 6 are reflexive, symmetric, antisymmetric, and/or transitive.

Solution: Since there are loops at every vertex of the directed graph of R, it is reflexive. R is neither symmetric nor antisymmetric since there is an edge from a to b but not one from b to a, but there are edges in both directions connecting b and c. Finally, R is not transitive since there is an edge from a to b and an edge from b to c, but no edge from a to c.

Since loops are not present at all the vertices of the directed graph of S, this relation is not reflexive. It is symmetric and not antisymmetric, since every edge between distinct vertices is accompanied by an edge in the opposite direction. It is also not hard to see from the directed graph that S is not transitive, since (c, a) and (a, b) belong to S, but (c, b) does not belong to S. ◀

Exercises

1. Represent each of these relations on $\{1, 2, 3\}$ with a matrix (with the elements of this set listed in increasing order).

   a) $\{(1, 1), (1, 2), (1, 3)\}$
   b) $\{(1, 2), (2, 1), (2, 2), (3, 3)\}$
   c) $\{(1, 1), (1, 2), (1, 3), (2, 2), (2, 3), (3, 3)\}$
   d) $\{(1, 3), (3, 1)\}$

2. Represent each of these relations on $\{1, 2, 3, 4\}$ with a matrix (with the elements of this set listed in increasing order).

a) $\{(1,2),(1,3),(1,4),(2,3),(2,4),(3,4)\}$
b) $\{(1,1),(1,4),(2,2),(3,3),(4,1)\}$
c) $\{(1,2),(1,3),(1,4),(2,1),(2,3),(2,4),(3,1),$
$(3,2),(3,4),(4,1),(4,2),(4,3)\}$
d) $\{(2,4),(3,1),(3,2),(3,4)\}$

3. List the ordered pairs in the relations on $\{1,2,3\}$ corresponding to these matrices (where the rows and columns correspond to the integers listed in increasing order).

a) $\begin{bmatrix} 1 & 0 & 1 \\ 0 & 1 & 0 \\ 1 & 0 & 1 \end{bmatrix}$ **b)** $\begin{bmatrix} 0 & 1 & 0 \\ 0 & 1 & 0 \\ 0 & 1 & 0 \end{bmatrix}$

c) $\begin{bmatrix} 1 & 1 & 1 \\ 1 & 0 & 1 \\ 1 & 1 & 1 \end{bmatrix}$

4. List the ordered pairs in the relations on $\{1,2,3,4\}$ corresponding to these matrices (where the rows and columns correspond to the integers listed in increasing order).

a) $\begin{bmatrix} 1 & 1 & 0 & 1 \\ 1 & 0 & 1 & 0 \\ 0 & 1 & 1 & 1 \\ 1 & 0 & 1 & 1 \end{bmatrix}$ **b)** $\begin{bmatrix} 1 & 1 & 1 & 0 \\ 0 & 1 & 0 & 0 \\ 0 & 0 & 1 & 1 \\ 1 & 0 & 0 & 1 \end{bmatrix}$

c) $\begin{bmatrix} 0 & 1 & 0 & 1 \\ 1 & 0 & 1 & 0 \\ 0 & 1 & 0 & 1 \\ 1 & 0 & 1 & 0 \end{bmatrix}$

5. How can the directed graph representing a relation R on a set A be used to determine whether the relation is irreflexive?

6. How can the directed graph representing a relation R on a set A be used to determine whether the relation is asymmetric?

7. Determine whether the relations represented by the matrices in Exercise 3 are reflexive, irreflexive, symmetric, antisymmetric, and/or transitive.

8. Determine whether the relations represented by the matrices in Exercise 4 are reflexive, irreflexive, symmetric, antisymmetric, and/or transitive.

9. How many nonzero entries does the matrix representing the relation R on $A = \{1,2,3,\dots,100\}$ consisting of the first 100 positive integers have if R is

a) $\{(a,b) \mid a > b\}$?
b) $\{(a,b) \mid a \neq b\}$?
c) $\{(a,b) \mid a = b+1\}$?
d) $\{(a,b) \mid a = 0\}$?
e) $\{(a,b) \mid ab = 0\}$?

10. How many nonzero entries does the matrix representing the relation R on $A = \{1,2,3,\dots,1000\}$ consisting of the first 1000 positive integers have if R is

a) $\{(a,b) \mid a \leq b\}$?
b) $\{(a,b) \mid a = b \pm 1\}$?
c) $\{(a,b) \mid a+b = 1000\}$?
d) $\{(a,b) \mid a+b \leq 1001\}$?
e) $\{(a,b) \mid a \neq 0\}$?

11. How can the matrix for $\overline{R}$, the complement of the relation R, be found from the matrix representing R, when R is a relation on a finite set A?

12. How can the matrix for R^{-1}, the inverse of the relation R, be found from the matrix representing R, when R is a relation on a finite set A?

13. Let R be the relation represented by the matrix

$$\mathbf{M}_R = \begin{bmatrix} 0 & 1 & 1 \\ 1 & 1 & 0 \\ 1 & 0 & 1 \end{bmatrix}.$$

Find the matrix representing

a) R^{-1}.   **b)** $\overline{R}$.   **c)** R^2.

14. Let R_1 and R_2 be relations on a set A represented by the matrices

$$\mathbf{M}_{R_1} = \begin{bmatrix} 0 & 1 & 0 \\ 1 & 1 & 1 \\ 1 & 0 & 0 \end{bmatrix} \quad \text{and} \quad \mathbf{M}_{R_2} = \begin{bmatrix} 0 & 1 & 0 \\ 0 & 1 & 1 \\ 1 & 1 & 1 \end{bmatrix}.$$

Find the matrices that represent

a) $R_1 \cup R_2$.   **b)** $R_1 \cap R_2$.   **c)** $R_2 \circ R_1$.
d) $R_1 \circ R_1$.   **e)** $R_1 \oplus R_2$.

15. Let R be the relation represented by the matrix

$$\mathbf{M}_R = \begin{bmatrix} 0 & 1 & 0 \\ 0 & 0 & 1 \\ 1 & 1 & 0 \end{bmatrix}.$$

Find the matrices that represent

a) R^2.   **b)** R^3.   **c)** R^4.

16. Let R be a relation on a set A with n elements. If there are k nonzero entries in $\mathbf{M}_R$, the matrix representing R, how many nonzero entries are there in $\mathbf{M}_{R^{-1}}$, the matrix representing R^{-1}, the inverse of R?

17. Let R be a relation on a set A with n elements. If there are k nonzero entries in $\mathbf{M}_R$, the matrix representing R, how many nonzero entries are there in $\mathbf{M}_{\overline{R}}$, the matrix representing $\overline{R}$, the complement of R?

18. Draw the directed graphs representing each of the relations from Exercise 1.

19. Draw the directed graphs representing each of the relations from Exercise 2.

20. Draw the directed graph representing each of the relations from Exercise 3.

21. Draw the directed graph representing each of the relations from Exercise 4.

22. Draw the directed graph that represents the relation $\{(a,a),(a,b),(b,c),(c,b),(c,d),(d,a),(d,b)\}$.

In Exercises 23–28 list the ordered pairs in the relations represented by the directed graphs.

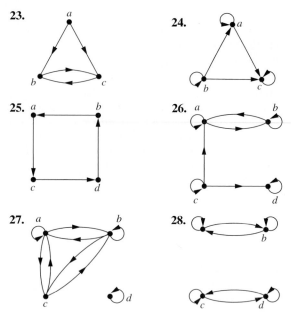

23.

24.

25.

26.

27.

28.

29. How can the directed graph of a relation R on a finite set A be used to determine whether a relation is asymmetric?

30. How can the directed graph of a relation R on a finite set A be used to determine whether a relation is irreflexive?

31. Determine whether the relations represented by the directed graphs shown in Exercises 23–25 are reflexive, irreflexive, symmetric, antisymmetric, and/or transitive.

32. Determine whether the relations represented by the directed graphs shown in Exercises 26–28 are reflexive, irreflexive, symmetric, antisymmetric, asymmetric, and/or transitive.

33. Let R be a relation on a set A. Explain how to use the directed graph representing R to obtain the directed graph representing the inverse relation R^{-1}.

34. Let R be a relation on a set A. Explain how to use the directed graph representing R to obtain the directed graph representing the complementary relation $\overline{R}$.

35. Show that if $\mathbf{M}_R$ is the matrix representing the relation R, then $\mathbf{M}_R^{[n]}$ is the matrix representing the relation R^n.

36. Given the directed graphs representing two relations, how can the directed graph of the union, intersection, symmetric difference, difference, and composition of these relations be found?

7.4 Closures of Relations

INTRODUCTION

A computer network has data centers in Boston, Chicago, Denver, Detroit, New York, and San Diego. There are direct, one-way telephone lines from Boston to Chicago, from Boston to Detroit, from Chicago to Detroit, from Detroit to Denver, and from New York to San Diego. Let R be the relation containing (a, b) if there is a telephone line from the data center in a to that in b. How can we determine if there is some (possibly indirect) link composed of one or more telephone lines from one center to another? Since not all links are direct, such as the link from Boston to Denver that goes through Detroit, R cannot be used directly to answer this. In the language of relations, R is not transitive, so it does not contain all the pairs that can be linked. As we will show in this section, we can find all pairs of data centers that have a link by constructing the smallest transitive relation that contains R. This relation is called the **transitive closure** of R.

In general, let R be a relation on a set A. R may or may not have some property **P**, such as reflexivity, symmetry, or transitivity. If there is a relation S with property **P** containing R such that S is a subset of every relation with property **P** containing R, then S is called the **closure** of R with respect to **P**. (Note that the closure of a relation with respect to a property may not exist; see Exercises 15 and 35 at the end of this section.) We will show how reflexive, symmetric, and transitive closures of relations can be found.

CLOSURES

The relation $R = \{(1, 1), (1, 2), (2, 1), (3, 2)\}$ on the set $A = \{1, 2, 3\}$ is not reflexive. How can we produce a reflexive relation containing R that is as small as possible? This can be done by adding $(2, 2)$ and $(3, 3)$ to R, since these are the only pairs of the form (a, a) that are not in R. Clearly, this new relation contains R. Furthermore, *any* reflexive relation that contains R must also contain $(2, 2)$ and $(3, 3)$. Because this relation contains R, is reflexive, and is contained within every reflexive relation that contains R, it is called the **reflexive closure** of R.

As this example illustrates, given a relation R on a set A, the reflexive closure of R can be formed by adding to R all pairs of the form (a, a) with $a \in A$, not already in R. The addition of these pairs produces a new relation that is reflexive, contains R, and is contained within any reflexive relation containing R. We see that the reflexive closure of R equals $R \cup \Delta$, where $\Delta = \{(a, a) \mid a \in A\}$ is the **diagonal relation** on A. (The reader should verify this.)

EXAMPLE 1   What is the reflexive closure of the relation $R = \{(a, b) \mid a < b\}$ on the set of integers?

Solution: The reflexive closure of R is

$$R \cup \Delta = \{(a, b) \mid a < b\} \cup \{(a, a) \mid a \in \mathbf{Z}\} = \{(a, b) \mid a \leq b\}. \qquad \blacktriangleleft$$

The relation $\{(1, 1), (1, 2), (2, 2), (2, 3), (3, 1), (3, 2)\}$ on $\{1, 2, 3\}$ is not symmetric. How can we produce a symmetric relation that is as small as possible and contains R? To do this, we need only add $(2, 1)$ and $(1, 3)$, since these are the only pairs of the form (b, a) with $(a, b) \in R$ that are not in R. This new relation is symmetric and contains R. Furthermore, *any* symmetric relation that contains R must contain this new relation, since a symmetric relation that contains R must contain $(2, 1)$ and $(1, 3)$. Consequently, this new relation is called the **symmetric closure** of R.

As this example illustrates, the symmetric closure of a relation R can be constructed by adding all ordered pairs of the form (b, a), where (a, b) is in the relation, that are not already present in R. Adding these pairs produces a relation that is symmetric, that contains R, and that is contained in any symmetric relation that contains R. The symmetric closure of a relation can be constructed by taking the union of a relation with its inverse; that is, $R \cup R^{-1}$ is the symmetric closure of R, where $R^{-1} = \{(b, a) \mid (a, b) \in R\}$. The reader should verify this statement.

EXAMPLE 2   What is the symmetric closure of the relation $R = \{(a, b) \mid a > b\}$ on the set of positive integers?

Solution: The symmetric closure of R is the relation

$$R \cup R^{-1} = \{(a, b) \mid a > b\} \cup \{(b, a) \mid a > b\} = \{(a, b) \mid a \neq b\}.$$

This last equality follows because R contains all ordered pairs of positive integers where the first element is greater than the second element and R^{-1} contains all ordered pairs of positive integers where the first element is less than the second. $\qquad \blacktriangleleft$

Suppose that a relation R is not transitive. How can we produce a transitive relation that contains R such that this new relation is contained within any transitive relation that contains R? Can the transitive closure of a relation R be produced by adding all the pairs of the form (a, c) where (a, b) and (b, c) are already in the relation? Consider the relation $R = \{(1, 3), (1, 4), (2, 1), (3, 2)\}$ on the set $\{1, 2, 3, 4\}$. This relation is not transitive since it does not contain all pairs of the form (a, c) where (a, b) and (b, c) are in R. The pairs of this form not in R are $(1, 2), (2, 3), (2, 4)$, and $(3, 1)$. Adding these pairs does *not* produce a transitive relation, since the resulting relation contains $(3, 1)$ and

(1, 4) but does not contain (3, 4). This shows that constructing the transitive closure of a relation is more complicated than constructing either the reflexive or symmetric closure. The rest of this section develops algorithms for constructing transitive closures. As will be shown later in this section, the transitive closure of a relation can be found by adding new ordered pairs that must be present and then repeating this process until no new ordered pairs are needed.

PATHS IN DIRECTED GRAPHS

We will see that representing relations by directed graphs helps in the construction of transitive closures. We now introduce some terminology that we will use for this purpose.

A path in a directed graph is obtained by traversing along edges (in the same direction as indicated by the arrow on the edge).

DEFINITION 1

> A *path* from a to b in the directed graph G is a sequence of edges $(x_0, x_1), (x_1, x_2),$ $(x_2, x_3), \ldots, (x_{n-1}, x_n)$ in G, where n is a nonnegative integer, and $x_0 = a$ and $x_n = b$, that is, a sequence of edges where the terminal vertex of an edge is the same as the initial vertex in the next edge in the path. This path is denoted by $x_0, x_1, x_2, \ldots, x_{n-1}, x_n$ and has *length n*. We view the empty set of edges as a path from a to a. A path of length $n \geq 1$ that begins and ends at the same vertex is called a *circuit* or *cycle*.

A path in a directed graph can pass through a vertex more than once. Moreover, an edge in a directed graph can occur more than once in a path.

EXAMPLE 3   Which of the following are paths in the directed graph shown in Figure 1: a, b, e, d; a, e, c, d, b; b, a, c, b, a, a, b; d, c; c, b, a; e, b, a, b, a, b, e? What are the lengths of those that are paths? Which of the paths in this list are circuits?

Solution: Since each of $(a, b), (b, e),$ and (e, d) is an edge, a, b, e, d is a path of length three. Since (c, d) is not an edge, a, e, c, d, b is not a path. Also, b, a, c, b, a, a, b is a path of length six since $(b, a), (a, c), (c, b), (b, a), (a, a),$ and (a, b) are all edges. We see that d, c is a path of length one, since (d, c) is an edge. Also c, b, a is a path of length two, since (c, b) and (b, a) are edges. All of $(e, b), (b, a), (a, b), (b, a), (a, b),$ and (b, e) are edges, so that e, b, a, b, a, b, e is a path of length six.

The two paths b, a, c, b, a, a, b and e, b, a, b, a, b, e are circuits since they begin and end at the same vertex. The paths a, b, e, d; c, b, a; and d, c are not circuits. ◀

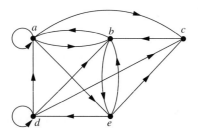

FIGURE 1   A Directed Graph.

The term *path* also applies to relations. Carrying over the definition from directed graphs to relations, there is a **path** from a to b in R if there is a sequence of elements a, $x_1, x_2, \ldots, x_{n-1}, b$ with $(a, x_1) \in R, (x_1, x_2) \in R, \ldots,$ and $(x_{n-1}, b) \in R$. Theorem 1 can be obtained from the definition of a path in a relation.

THEOREM 1   Let R be a relation on a set A. There is a path of length n, where n is a positive integer, from a to b if and only if $(a, b) \in R^n$.

Proof: We will use mathematical induction. By definition, there is a path from a to b of length one if and only if $(a, b) \in R$, so the theorem is true when $n = 1$.

Assume that the theorem is true for the positive integer n. This is the inductive hypothesis. There is a path of length $n + 1$ from a to b if and only if there is an element $c \in A$ such that there is a path of length one from a to c, so $(a, c) \in R$, and a path of length n from c to b, that is, $(c, b) \in R^n$. Consequently, by the induction hypothesis, there is a path of length $n+1$ from a to b if and only if there is an element c with $(a, c) \in R$ and $(c, b) \in R^n$. But there is such an element if and only if $(a, b) \in R^{n+1}$. Therefore, there is a path of length $n + 1$ from a to b if and only if $(a, b) \in R^{n+1}$. This completes the proof.   ◁

TRANSITIVE CLOSURES

We now show that finding the transitive closure of a relation is equivalent to determining which pairs of vertices in the associated directed graph are connected by a path. With this in mind, we define a new relation.

DEFINITION 2   Let R be a relation on a set A. The *connectivity relation* R^* consists of the pairs (a, b) such that there is a path of length at least one from a to b in R.

Since R^n consists of the pairs (a, b) such that there is a path of length n from a to b, it follows that R^* is the union of all the sets R^n. In other words,

$$R^* = \bigcup_{n=1}^{\infty} R^n.$$

The connectivity relation is useful in many models.

EXAMPLE 4   Let R be the relation on the set of all people in the world that contains (a, b) if a has met b. What is R^n, where n is a positive integer greater than one? What is R^*?

Solution: The relation R^2 contains (a, b) if there is a person c such that $(a, c) \in R$ and $(c, b) \in R$, that is, if there is a person c such that a has met c and c has met b. Similarly, R^n consists of those pairs (a, b) such that there are people $x_1, x_2, \ldots, x_{n-1}$ such that a has met x_1, x_1 has met x_2, $\ldots$, and x_{n-1} has met b.

The relation R^* contains (a, b) if there is a sequence of people, starting with a and ending with b, such that each person in the sequence has met the next person in the sequence. (There are many interesting conjectures about R^*. Do you think that this connectivity relation includes the pair with you as the first element and the president of Mongolia as the second element? We will use graphs to model this application in Chapter 8.) ◀

EXAMPLE 5 Let R be the relation on the set of all subway stops in New York City that contains (a, b) if it is possible to travel from stop a to stop b without changing trains. What is R^n when n is a positive integer? What is R^*?

Solution: The relation R^n contains (a, b) if it is possible to travel from stop a to stop b by making $n - 1$ changes of trains. The relation R^* consists of the ordered pairs (a, b) where it is possible to travel from stop a to stop b making as many changes of trains as necessary. (The reader should verify these statements.) ◀

EXAMPLE 6 Let R be the relation on the set of all states in the United States that contains (a, b) if state a and state b have a common border. What is R^n where n is a positive integer? What is R^*?

Solution: The relation R^n consists of the pairs (a, b) where it is possible to go from state a to state b by crossing exactly n state borders. R^* consists of the ordered pairs (a, b) where it is possible to go from state a to state b crossing as many borders as necessary. (The reader should verify these statements.) The only ordered pairs not in R^* are those containing states that are not connected to the continental United States (i.e., those pairs containing Alaska or Hawaii). ◀

Theorem 2 shows that the transitive closure of a relation and the associated connectivity relation are the same.

THEOREM 2 The transitive closure of a relation R equals the connectivity relation R^*.

Proof: Note that R^* contains R by definition. To show that R^* is the transitive closure of R we must also show that R^* is transitive and that $R^* \subseteq S$ whenever S is a transitive relation that contains R.

First, we show that R^* is transitive. If $(a, b) \in R^*$ and $(b, c) \in R^*$, then there are paths from a to b and from b to c in R. We obtain a path from a to c by starting with the path from a to b and following it with the path from b to c. Hence, $(a, c) \in R^*$. It follows that R^* is transitive.

Now suppose that S is a transitive relation containing R. Since S is transitive, S^n also is transitive (the reader should verify this) and $S^n \subseteq S$ (by Theorem 1 of Section 7.1). Furthermore, since

$$S^* = \bigcup_{k=1}^{\infty} S^k,$$

and $S^k \subseteq S$, it follows that $S^* \subseteq S$. Now note that if $R \subseteq S$, then $R^* \subseteq S^*$, because any path in R is also a path in S. Consequently, $R^* \subseteq S^* \subseteq S$. Thus, any transitive relation that contains R must also contain R^*. Therefore, R^* is the transitive closure of R. ◁

Now that we know that the transitive closure equals the connectivity relation, we turn our attention to the problem of computing this relation. We do not need to examine arbitrarily long paths to determine whether there is a path between two vertices in a finite directed graph. As Lemma 1 shows, it is sufficient to examine paths containing no more than n edges, where n is the number of elements in the set.

LEMMA 1

Let A be a set with n elements, and let R be a relation on A. If there is a path of length at least one in R from a to b, then there is such a path with length not exceeding n. Moreover, when $a \neq b$, if there is a path of length at least one in R from a to b, then there is such a path with length not exceeding $n - 1$.

Proof: Suppose there is a path from a to b in R. Let m be the length of the shortest such path. Suppose that $x_0, x_1, x_2, \ldots, x_{m-1}, x_m$, where $x_0 = a$ and $x_m = b$, is such a path.

Suppose that $a = b$ and that $m > n$, so that $m \geq n + 1$. By the pigeonhole principle, since there are n vertices in A, among the m vertices $x_0, x_1, \ldots, x_{m-1}$, at least two are equal (see Figure 2).

Suppose that $x_i = x_j$ with $0 \leq i < j \leq m - 1$. Then the path contains a circuit from x_i to itself. This circuit can be deleted from the path from a to b, leaving a path, namely, $x_0, x_1, \ldots, x_i, x_{j+1}, \ldots, x_{m-1}, x_m$, from a to b of shorter length. Hence, the path of shortest length must have length less than or equal to n.

The case where $a \neq b$ is left as an exercise for the reader. ◁

From Lemma 1, we see that the transitive closure of R is the union of $R, R^2, R^3, \ldots,$ and R^n. This follows since there is a path in R^* between two vertices if and only if there is a path between these vertices in R^i, for some positive integer i with $i \leq n$. Since

$$R^* = R \cup R^2 \cup R^3 \cup \cdots \cup R^n,$$

and the zero–one matrix representing a union of relations is the join of the zero–one matrices of these relations, the zero–one matrix for the transitive closure is the join of the zero–one matrices of the first n powers of the zero–one matrix of R.

THEOREM 3

Let $\mathbf{M}_R$ be the zero–one matrix of the relation R on a set with n elements. Then the zero–one matrix of the transitive closure R^* is

$$\mathbf{M}_{R^*} = \mathbf{M}_R \vee \mathbf{M}_R^{[2]} \vee \mathbf{M}_R^{[3]} \vee \cdots \vee \mathbf{M}_R^{[n]}.$$

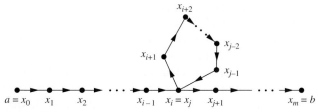

FIGURE 2 **Producing a Path with Length Not Exceeding n.**

EXAMPLE 7 Find the zero–one matrix of the transitive closure of the relation R where

$$\mathbf{M}_R = \begin{bmatrix} 1 & 0 & 1 \\ 0 & 1 & 0 \\ 1 & 1 & 0 \end{bmatrix}.$$

Solution: From Theorem 3, it follows that the zero–one matrix of R^* is

$$\mathbf{M}_{R^*} = \mathbf{M}_R \vee \mathbf{M}_R^{[2]} \vee \mathbf{M}_R^{[3]}.$$

Since

$$\mathbf{M}_R^{[2]} = \begin{bmatrix} 1 & 1 & 1 \\ 0 & 1 & 0 \\ 1 & 1 & 1 \end{bmatrix} \quad \text{and} \quad \mathbf{M}_R^{[3]} = \begin{bmatrix} 1 & 1 & 1 \\ 0 & 1 & 0 \\ 1 & 1 & 1 \end{bmatrix},$$

it follows that

$$\mathbf{M}_{R^*} = \begin{bmatrix} 1 & 0 & 1 \\ 0 & 1 & 0 \\ 1 & 1 & 0 \end{bmatrix} \vee \begin{bmatrix} 1 & 1 & 1 \\ 0 & 1 & 0 \\ 1 & 1 & 1 \end{bmatrix} \vee \begin{bmatrix} 1 & 1 & 1 \\ 0 & 1 & 0 \\ 1 & 1 & 1 \end{bmatrix} = \begin{bmatrix} 1 & 1 & 1 \\ 0 & 1 & 0 \\ 1 & 1 & 1 \end{bmatrix}. \quad \blacktriangleleft$$

Theorem 3 can be used as a basis for an algorithm for computing the matrix of the relation R^*. To find this matrix, the successive Boolean powers of $\mathbf{M}_R$, up to the nth power, are computed. As each power is calculated, its join with the join of all smaller powers is formed. When this is done with the nth power, the matrix for R^* has been found. This procedure is displayed as Algorithm 1.

ALGORITHM 1 A Procedure for Computing the Transitive Closure.

procedure *transitive closure* ($\mathbf{M}_R$: zero–one $n \times n$ matrix)
$\mathbf{A} := \mathbf{M}_R$
$\mathbf{B} := \mathbf{A}$
for $i := 2$ **to** n
begin
  $\mathbf{A} := \mathbf{A} \odot \mathbf{M}_R$
  $\mathbf{B} := \mathbf{B} \vee \mathbf{A}$
end {$\mathbf{B}$ is the zero–one matrix for R^*}

We can easily find the number of bit operations used by Algorithm 1 to determine the transitive closure of a relation. Computing the Boolean powers $\mathbf{M}_R, \mathbf{M}_R^{[2]}, \ldots, \mathbf{M}_R^{[n]}$ requires that $n-1$ Boolean products of $n \times n$ zero–one matrices be found. Each of these Boolean products can be found using $n^2(2n-1)$ bit operations. Hence, these products can be computed using $n^2(2n-1)(n-1)$ bit operations.

To find $\mathbf{M}_{R^*}$ from the n Boolean powers of $\mathbf{M}_R$, $n-1$ joins of zero–one matrices need to be found. Computing each of these joins uses n^2 bit operations. Hence, $(n-1)n^2$ bit operations are used in this part of the computation. Therefore, when Algorithm 1 is used, the matrix of the transitive closure of a relation on a set with n elements can be found using $n^2(2n-1)(n-1) + (n-1)n^2 = 2n^3(n-1)$, which is $O(n^4)$ bit operations. The remainder of this section describes a more efficient algorithm for finding transitive closures.

WARSHALL'S ALGORITHM

Warshall's algorithm, named after Stephen Warshall, who described this algorithm in 1960, is an efficient method for computing the transitive closure of a relation. Algorithm 1 can find the transitive closure of a relation on a set with n elements using $2n^3(n-1)$ bit operations. However, the transitive closure can be found by Warshall's algorithm using only $2n^3$ bit operations.

Remark: Warshall's algorithm is sometimes called the Roy–Warshall algorithm, since Bernard Roy described this algorithm in 1959.

Suppose that R is a relation on a set with n elements. Let $v_1, v_2, \ldots, v_n$ be an arbitrary listing of these n elements. The concept of the **interior vertices** of a path is used in Warshall's algorithm. If $a, x_1, x_2, \ldots, x_{m-1}, b$ is a path, its interior vertices are $x_1, x_2, \ldots, x_{m-1}$, that is, all the vertices of the path that occur somewhere other than as the first and last vertices in the path. For instance, the interior vertices of a path a, c, d, f, g, h, b, j in a directed graph are $c, d, f, g, h,$ and b. The interior vertices of a, c, d, a, f, b are $c, d, a,$ and f. (Note that the first vertex in the path is not an interior vertex unless it is visited again by the path, except as the last vertex. Similarly, the last vertex in the path is not an interior vertex unless it was visited previously by the path, except as the first vertex.)

Warshall's algorithm is based on the construction of a sequence of zero–one matrices. These matrices are $\mathbf{W}_0, \mathbf{W}_1, \ldots, \mathbf{W}_n$, where $\mathbf{W}_0 = \mathbf{M}_R$ is the zero–one matrix of this relation, and $\mathbf{W}_k = [w_{ij}^{(k)}]$, where $w_{ij}^{(k)} = 1$ if there is a path from v_i to v_j such that all the interior vertices of this path are in the set $\{v_1, v_2, \ldots, v_k\}$ (the first k vertices in the list) and is 0 otherwise. (The first and last vertices in the path may be outside the set of the first k vertices in the list.) Note that $\mathbf{W}_n = \mathbf{M}_{R^*}$, since the (i, j)th entry of $\mathbf{M}_{R^*}$ is 1 if and only if there is a path from v_i to v_j, with all interior vertices in the set $\{v_1, v_2, \ldots, v_n\}$ (but these are the only vertices in the directed graph). Example 8 illustrates what the matrix $\mathbf{W}_k$ represents.

STEPHEN WARSHALL (BORN 1935) Stephen Warshall, born in New York City, went to public school in Brooklyn. He attended Harvard University, receiving his degree in mathematics in 1956. He never received an advanced degree, since at that time no programs were available in his areas of interest. However, he took graduate courses at several different universities and contributed to the development of computer science and software engineering.

After graduating from Harvard, Warshall worked at ORO (Operation Research Office), which was set up by Johns Hopkins to do research and development for the U.S. Army. In 1958 he left ORO to take a position at a company called Technical Operations, where he helped build a research and development laboratory for military software projects. In 1961 he left Technical Operations to found Massachusetts Computer Associates. Later, this company became part of Applied Data Research (ADR). After the merger, Warshall sat on the board of directors of ADR and managed a variety of projects and organizations. He retired from ADR in 1982.

During his career Warshall carried out research and development in operating systems, compiler design, language design, and operations research. In the 1971–1972 academic year he presented lectures on software engineering at French universities. There is an interesting anecdote about his proof that the transitive closure algorithm, now known as Warshall's algorithm, is correct. He and a colleague at Technical Operations bet a bottle of rum on who first could determine whether this algorithm always works. Warshall came up with his proof overnight, winning the bet and the rum, which he shared with the loser of the bet. Because Warshall did not like sitting at a desk, he did much of his creative work in unconventional places, such as on a sailboat in the Indian Ocean or in a Greek lemon orchard.

EXAMPLE 8 Let R be the relation with directed graph shown in Figure 3. Let a, b, c, d be a listing of the elements of the set. Find the matrices $\mathbf{W}_0, \mathbf{W}_1, \mathbf{W}_2, \mathbf{W}_3,$ and $\mathbf{W}_4$. The matrix $\mathbf{W}_4$ is the transitive closure of R.

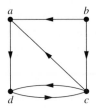

FIGURE 3
The Directed
Graph of the
Relation R.

Solution: Let $v_1 = a, v_2 = b, v_3 = c,$ and $v_4 = d$. $\mathbf{W}_0$ is the matrix of the relation. Hence,

$$\mathbf{W}_0 = \begin{bmatrix} 0 & 0 & 0 & 1 \\ 1 & 0 & 1 & 0 \\ 1 & 0 & 0 & 1 \\ 0 & 0 & 1 & 0 \end{bmatrix}.$$

$\mathbf{W}_1$ has 1 as its (i, j)th entry if there is a path from v_i to v_j that has only $v_1 = a$ as an interior vertex. Note that all paths of length one can still be used since they have no interior vertices. Also, there is now an allowable path from b to d, namely, b, a, d. Hence,

$$\mathbf{W}_1 = \begin{bmatrix} 0 & 0 & 0 & 1 \\ 1 & 0 & 1 & 1 \\ 1 & 0 & 0 & 1 \\ 0 & 0 & 1 & 0 \end{bmatrix}.$$

$\mathbf{W}_2$ has 1 as its (i, j)th entry if there is a path from v_i to v_j that has only $v_1 = a$ and/or $v_2 = b$ as its interior vertices, if any. Since there are no edges that have b as a terminal vertex, no new paths are obtained when we permit b to be an interior vertex. Hence, $\mathbf{W}_2 = \mathbf{W}_1$.

$\mathbf{W}_3$ has 1 as its (i, j)th entry if there is a path from v_i to v_j that has only $v_1 = a$, $v_2 = b$, and/or $v_3 = c$ as its interior vertices, if any. We now have paths from d to a, namely, d, c, a, and from d to d, namely, d, c, d. Hence,

$$\mathbf{W}_3 = \begin{bmatrix} 0 & 0 & 0 & 1 \\ 1 & 0 & 1 & 1 \\ 1 & 0 & 0 & 1 \\ 1 & 0 & 1 & 1 \end{bmatrix}.$$

Finally, $\mathbf{W}_4$ has 1 as its (i, j)th entry if there is a path from v_i to v_j that has $v_1 = a$, $v_2 = b, v_3 = c,$ and/or $v_4 = d$ as interior vertices, if any. Since these are all the vertices of the graph, this entry is 1 if and only if there is a path from v_i to v_j. Hence,

$$\mathbf{W}_4 = \begin{bmatrix} 1 & 0 & 1 & 1 \\ 1 & 0 & 1 & 1 \\ 1 & 0 & 1 & 1 \\ 1 & 0 & 1 & 1 \end{bmatrix}.$$

This last matrix, $\mathbf{W}_4$, is the matrix of the transitive closure. ◄

Warshall's algorithm computes $\mathbf{M}_{R^*}$ by efficiently computing $\mathbf{W}_0 = \mathbf{M}_R, \mathbf{W}_1,$ $\mathbf{W}_2, \ldots, \mathbf{W}_n = \mathbf{M}_{R^*}$. This observation shows that we can compute $\mathbf{W}_k$ directly from $\mathbf{W}_{k-1}$: There is a path from v_i to v_j with no vertices other than $v_1, v_2, \ldots, v_k$ as interior vertices if and only if either there is a path from v_i to v_j with its interior vertices among the first $k - 1$ vertices in the list, or there are paths from v_i to v_k and from v_k to v_j that have interior vertices only among the first $k - 1$ vertices in the list. That is, either a path from v_i to v_j already existed before v_k was permitted as an interior vertex, or allowing v_k as an interior vertex produces a path that goes from v_i to v_k and then from v_k to v_j. These two cases are shown in Figure 4.

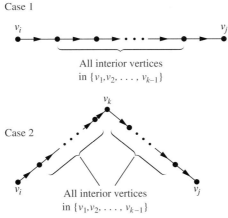

Case 1

All interior vertices
in $\{v_1, v_2, \ldots, v_{k-1}\}$

Case 2

All interior vertices
in $\{v_1, v_2, \ldots, v_{k-1}\}$

FIGURE 4 **Adding v_k to the Set of
Allowable Interior Vertices.**

The first type of path exists if and only if $w_{ij}^{[k-1]} = 1$, and the second type of path exists if and only if both $w_{ik}^{[k-1]}$ and $w_{kj}^{[k-1]}$ are 1. Hence, $w_{ij}^{[k]}$ is 1 if and only if either $w_{ij}^{[k-1]}$ is 1 or both $w_{ik}^{[k-1]}$ and $w_{kj}^{[k-1]}$ are 1. This gives us Lemma 2.

LEMMA 2

Let $\mathbf{W}_k = [w_{ij}^{[k]}]$ be the zero–one matrix that has a 1 in its (i, j)th position if and only if there is a path from v_i to v_j with interior vertices from the set $\{v_1, v_2, \ldots, v_k\}$. Then

$$w_{ij}^{[k]} = w_{ij}^{[k-1]} \vee (w_{ik}^{[k-1]} \wedge w_{kj}^{[k-1]}),$$

whenever i, j, and k are positive integers not exceeding n.

Lemma 2 gives us the means to compute efficiently the matrices $\mathbf{W}_k$, $k = 1, 2, \ldots, n$. We display the pseudocode for Warshall's algorithm, using Lemma 2, as Algorithm 2.

ALGORITHM 2 Warshall Algorithm.

```
procedure Warshall (M_R : n × n zero–one matrix)
W := M_R
for k := 1 to n
begin
    for i := 1 to n
    begin
        for j := 1 to n
        w_ij := w_ij ∨ (w_ik ∧ w_kj)
    end
end {W = [w_ij] is M_R*}
```

The computational complexity of Warshall's algorithm can easily be computed in terms of bit operations. To find the entry $w_{ij}^{[k]}$ from the entries $w_{ij}^{[k-1]}$, $w_{ik}^{[k-1]}$, and $w_{kj}^{[k-1]}$ using Lemma 2 requires two bit operations. To find all n^2 entries of $\mathbf{W}_k$ from those of $\mathbf{W}_{k-1}$ requires $2n^2$ bit operations. Since Warshall's algorithm begins with $\mathbf{W}_0 = \mathbf{M}_R$ and computes the sequence of n zero–one matrices $\mathbf{W}_1, \mathbf{W}_2, \ldots, \mathbf{W}_n = \mathbf{M}_{R^*}$, the total number of bit operations used is $n \cdot 2n^2 = 2n^3$.

Exercises

1. Let R be the relation on the set $\{0, 1, 2, 3\}$ containing the ordered pairs $(0, 1), (1, 1), (1, 2), (2, 0), (2, 2),$ and $(3, 0)$. Find the
  a) reflexive closure of R.   **b)** symmetric closure of R.

2. Let R be the relation $\{(a, b) \mid a \neq b\}$ on the set of integers. What is the reflexive closure of R?

3. Let R be the relation $\{(a, b) \mid a \text{ divides } b\}$ on the set of integers. What is the symmetric closure of R?

4. How can the directed graph representing the reflexive closure of a relation on a finite set be constructed from the directed graph of the relation?

In Exercises 5–7 draw the directed graph of the reflexive closure of the relations with the directed graph shown.

5.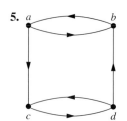

6.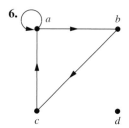

7.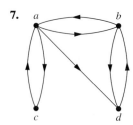

8. How can the directed graph representing the symmetric closure of a relation on a finite set be constructed from the directed graph for this relation?

9. Find the directed graphs of the symmetric closures of the relations with directed graphs shown in Exercises 5–7.

10. Find the smallest relation containing the relation in Example 2 that is both reflexive and symmetric.

11. Find the directed graph of the smallest relation that is both reflexive and symmetric for each of the relations with directed graphs shown in Exercises 5–7.

12. Suppose that the relation R on the finite set A is represented by the matrix $\mathbf{M}_R$. Show that the matrix that represents the reflexive closure of R is $\mathbf{M}_R \vee \mathbf{I}_n$.

13. Suppose that the relation R on the finite set A is represented by the matrix $\mathbf{M}_R$. Show that the matrix that represents the symmetric closure of R is $\mathbf{M}_R \vee \mathbf{M}_R^t$.

14. Show that the closure of a relation R with respect to a property $\mathbf{P}$, if it exists, is the intersection of all the relations with property $\mathbf{P}$ that contain R.

15. When is it possible to define the "irreflexive closure" of a relation R, that is, a relation that contains R, is irreflexive, and is contained in every irreflexive relation that contains R?

16. Determine whether these sequences of vertices are paths in these directed graph.

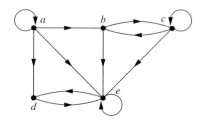

  a) a, b, c, e
  b) b, e, c, b, e
  c) a, a, b, e, d, e
  d) b, c, e, d, a, a, b
  e) b, c, c, b, e, d, e, d
  f) $a, a, b, b, c, c, b, e, d$

17. Find all circuits of length three in the directed graph in Exercise 16.

18. Determine whether there is a path in the directed graph in Exercise 16 beginning at the first vertex given and ending at the second vertex given.

  a) a, b **b)** b, a **c)** b, b
  d) a, e **e)** b, d **f)** c, d
  g) d, d **h)** e, a **i)** e, c

19. Let R be the relation on the set $\{1, 2, 3, 4, 5\}$ containing the ordered pairs $(1, 3), (2, 4), (3, 1), (3, 5), (4, 3)$, $(5, 1), (5, 2)$, and $(5, 4)$. Find

a) R^2.　　　　　　　**b)** R^3.
c) R^4.　　　　　　　**d)** R^5.
e) R^6.　　　　　　　**f)** R^*.

20. Let R be the relation that contains the pair (a, b) if a and b are cities such that there is a direct non-stop airline flight from a to b. When is (a, b) in

a) R^2?　　　**b)** R^3?　　　**c)** R^*?

21. Let R be the relation on the set of all students containing the ordered pair (a, b) if a and b are in at least one common class and $a \neq b$. When is (a, b) in

a) R^2?　　　**b)** R^3?　　　**c)** R^*?

22. Suppose that the relation R is reflexive. Show that R^* is reflexive.

23. Suppose that the relation R is symmetric. Show that R^* is symmetric.

24. Suppose that the relation R is irreflexive. Is the relation R^2 necessarily irreflexive?

25. Use Algorithm 1 to find the transitive closures of these relations on $\{1, 2, 3, 4\}$.

a) $\{(1, 2), (2, 1), (2, 3), (3, 4), (4, 1)\}$
b) $\{(2, 1), (2, 3), (3, 1), (3, 4), (4, 1), (4, 3)\}$
c) $\{(1, 2), (1, 3), (1, 4), (2, 3), (2, 4), (3, 4)\}$
d) $\{(1, 1), (1, 4), (2, 1), (2, 3), (3, 1), (3, 2), (3, 4),$
$(4, 2)\}$

26. Use Algorithm 1 to find the transitive closures of these relations on $\{a, b, c, d, e\}$.

a) $\{(a, c), (b, d), (c, a), (d, b), (e, d)\}$
b) $\{(b, c), (b, e), (c, e), (d, a), (e, b), (e, c)\}$
c) $\{(a, b), (a, c), (a, e), (b, a), (b, c), (c, a), (c, b),$
$(d, a), (e, d)\}$

d) $\{(a, e), (b, a), (b, d), (c, d), (d, a), (d, c), (e, a),$
$(e, b), (e, c), (e, e)\}$

27. Use Warshall's algorithm to find the transitive closures of the relations in Exercise 25.

28. Use Warshall's algorithm to find the transitive closures of the relations in Exercise 26.

29. Find the smallest relation containing the relation $\{(1, 2), (1, 4), (3, 3), (4, 1)\}$ that is

a) reflexive and transitive.
b) symmetric and transitive.
c) reflexive, symmetric, and transitive.

30. Finish the proof of the case when $a \neq b$ in Lemma 1.

31. Algorithms have been devised that use $O(n^{2.8})$ bit operations to compute the Boolean product of two $n \times n$ zero–one matrices. Assuming that these algorithms can be used, give big-O estimates for the number of bit operations using Algorithm 1 and using Warshall's algorithm to find the transitive closure of a relation on a set with n elements.

*32. Devise an algorithm using the concept of interior vertices in a path to find the length of the shortest path between two vertices in a directed graph, if such a path exists.

33. Adapt Algorithm 1 to find the reflexive closure of the transitive closure of a relation on a set with n elements.

34. Adapt Warshall's algorithm to find the reflexive closure of the transitive closure of a relation on a set with n elements.

35. Show that the closure with respect to the property **P** of the relation $R = \{(0, 0), (0, 1), (1, 1), (2, 2)\}$ on the set $\{0, 1, 2\}$ does not exist if **P** is the property

a) "is not reflexive."
b) "has an odd number of elements."

7.5　Equivalence Relations

INTRODUCTION

Students at a college register for classes the day before the start of a semester. Those with last names beginning with a letter from A to G, from H to N, and from O to Z may register at any time in the periods 8 A.M. to 11 A.M., 11 A.M. to 2 P.M., and 2 P.M. to 5 P.M., respectively. Let R be the relation containing (x, y) if and only if x and y are students with last names beginning with letters in the same block. Consequently, x and y can register at the same time if and only if (x, y) belongs to R. It is easy to see that R is reflexive, symmetric, and transitive. Furthermore, R divides the set of students into three classes, depending on the first letters of their last names. To know when a student can register we are concerned only with which of the three classes the student is in, and we do not care about the identity of the student.

The integers a and b are related by the "congruence modulo 4" relation when 4 divides $a - b$. We will show later that this relation is reflexive, symmetric, and transitive. It is not hard to see that a is related to b if and only if a and b have the same remainder when divided by 4. It follows that this relation splits the set of integers into four different classes. When we care only what remainder an integer leaves when it is divided by 4, we need only know which class it is in, not its particular value.

These two relations, R and congruence modulo 4, are examples of equivalence relations, namely, relations that are reflexive, symmetric, and transitive. In this section we will show that such relations split sets into disjoint classes of equivalent elements. Equivalence relations arise whenever we care only whether an element of a set is in a certain class of elements, instead of caring about its particular identity.

EQUIVALENCE RELATIONS

In this section we will study relations with a particular combination of properties that allows them to be used to relate objects that are similar in some way.

DEFINITION 1

A relation on a set A is called an *equivalence relation* if it is reflexive, symmetric, and transitive.

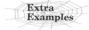

Two elements that are related by an equivalence relation are called **equivalent.** (This definition makes sense since an equivalence relation is symmetric.) Since an equivalence relation is reflexive, in an equivalence relation every element is equivalent to itself. Furthermore, since an equivalence relation is transitive, if a and b are equivalent and b and c are equivalent, it follows that a and c are equivalent.

Examples 1–4 illustrate the notion of an equivalence relation.

EXAMPLE 1 Suppose that R is the relation on the set of strings of English letters such that $a R b$ if and only if $l(a) = l(b)$, where $l(x)$ is the length of the string x. Is R an equivalence relation?

Solution: Since $l(a) = l(a)$, it follows that $a R a$ whenever a is a string, so that R is reflexive. Next, suppose that $a R b$, so that $l(a) = l(b)$. Then $b R a$, since $l(b) = l(a)$. Hence, R is symmetric. Finally, suppose that $a R b$ and $b R c$. Then $l(a) = l(b)$ and $l(b) = l(c)$. Hence, $l(a) = l(c)$, so that $a R c$. Consequently, R is transitive. Since R is reflexive, symmetric, and transitive, it is an equivalence relation. ◄

EXAMPLE 2 Let R be the relation on the set of integers such that $a R b$ if and only if $a = b$ or $a = -b$. In Section 7.1 we showed that R is reflexive, symmetric, and transitive. It follows that R is an equivalence relation. ◄

EXAMPLE 3 Let R be the relation on the set of real numbers such that $a R b$ if and only if $a - b$ is an integer. Is R an equivalence relation?

Solution: Since $a - a = 0$ is an integer for all real numbers a, $a R a$ for all real numbers a. Hence, R is reflexive. Now suppose that $a R b$. Then $a - b$ is an integer, so that $b - a$ is also an integer. Hence, $b R a$. It follows that R is symmetric. If $a R b$ and $b R c$, then

$a - b$ and $b - c$ are integers. Therefore, $a - c = (a - b) + (b - c)$ is also an integer. Hence, $a\,R\,c$. Thus, R is transitive. Consequently, R is an equivalence relation. ◀

One of the most widely used equivalence relations is congruence modulo m, where m is a positive integer greater than 1.

EXAMPLE 4 **Congruence Modulo m** Let m be a positive integer with $m > 1$. Show that the relation

$$R = \{(a, b) \mid a \equiv b \pmod{m}\}$$

is an equivalence relation on the set of integers.

Solution: Recall from Section 2.4 that $a \equiv b \pmod{m}$ if and only if m divides $a - b$. Note that $a - a = 0$ is divisible by m, since $0 = 0 \cdot m$. Hence, $a \equiv a \pmod{m}$, so that congruence modulo m is reflexive. Now suppose that $a \equiv b \pmod{m}$. Then $a - b$ is divisible by m, so that $a - b = km$, where k is an integer. It follows that $b - a = (-k)m$, so that $b \equiv a \pmod{m}$. Hence, congruence modulo m is symmetric. Next, suppose that $a \equiv b \pmod{m}$ and $b \equiv c \pmod{m}$. Then m divides both $a - b$ and $b - c$. Therefore, there are integers k and l with $a - b = km$ and $b - c = lm$. Adding these two equations shows that $a - c = (a - b) + (b - c) = km + lm = (k + l)m$. Thus, $a \equiv c \pmod{m}$. Therefore, congruence modulo m is transitive. It follows that congruence modulo m is an equivalence relation. ◀

EQUIVALENCE CLASSES

Let A be the set of all students in your school who graduated from high school. Consider the relation R on A that consists of all pairs (x, y) where x and y graduated from the same high school. Given a student x, we can form the set of all students equivalent to x with respect to R. This set consists of all students who graduated from the same high school as x did. This subset of A is called an equivalence class of the relation.

DEFINITION 2 Let R be an equivalence relation on a set A. The set of all elements that are related to an element a of A is called the *equivalence class* of a. The equivalence class of a with respect to R is denoted by $[a]_R$. When only one relation is under consideration, we will delete the subscript R and write $[a]$ for this equivalence class.

In other words, if R is an equivalence relation on a set A, the equivalence class of the element a is

$$[a]_R = \{s \mid (a, s) \in R\}.$$

If $b \in [a]_R$, then b is called a **representative** of this equivalence class. Any element of a class can be used as a representative of this class. That is, there is nothing special about the particular element chosen as the representative of the class.

EXAMPLE 5 What is the equivalence class of an integer for the equivalence relation of Example 2?

Solution: Since an integer is equivalent to itself and its negative in this equivalence relation, it follows that $[a] = \{-a, a\}$. This set contains two distinct integers unless $a = 0$. For instance, $[7] = \{-7, \ 7\}, [-5] = \{-5, \ 5\}$, and $[0] = \{0\}$. ◀

EXAMPLE 6 What are the equivalence classes of 0 and 1 for congruence modulo 4?

Solution: The equivalence class of 0 contains all integers a such that $a \equiv 0 \pmod 4$. The integers in this class are those divisible by 4. Hence, the equivalence class of 0 for this relation is

$$[0] = \{\ldots, -8, -4, 0, 4, 8, \ldots\}.$$

The equivalence class of 1 contains all the integers a such that $a \equiv 1 \pmod 4$. The integers in this class are those that have a remainder of 1 when divided by 4. Hence, the equivalence class of 1 for this relation is

$$[1] = \{\ldots, -7, -3, 1, 5, 9, \ldots\}. \qquad \blacktriangleleft$$

In Example 6 the equivalence classes of 0 and 1 with respect to congruence modulo 4 were found. Example 6 can easily be generalized, replacing 4 with any positive integer m. The equivalence classes of the relation congruence modulo m are called the **congruence classes modulo** m. The congruence class of an integer a modulo m is denoted by $[a]_m$ so that $[a]_m = \{\ldots, a - 2m, a - m, a, a + m, a + 2m, \ldots\}$. For instance, from Example 6 it follows that $[0]_4 = \{\ldots, -8, -4, 0, 4, 8, \ldots\}$ and $[1]_4 = \{\ldots, -7, -3, 1, 5, 9, \ldots\}$.

EQUIVALENCE CLASSES AND PARTITIONS

Let A be the set of students at your school who are majoring in exactly one subject, and let R be the relation on A consisting of pairs (x, y) where x and y are students with the same major. Then R is an equivalence relation, as the reader should verify. We can see that R splits all students in A into a collection of disjoint subsets, where each subset contains students with a specified major. For instance, one subset contains all students majoring (just) in computer science, and a second subset contains all students majoring in history. Furthermore, these subsets are equivalence classes of R. This example illustrates how the equivalence classes of an equivalence relation partition a set into disjoint, nonempty subsets. We will make these notions more precise in the following discussion.

Let R be a relation on the set A. Theorem 1 shows that the equivalence classes of two elements of A are either identical or disjoint.

THEOREM 1

Let R be an equivalence relation on a set A. These statements are equivalent:

(*i*) $a \, R \, b$ (*ii*) $[a] = [b]$ (*iii*) $[a] \cap [b] \neq \emptyset$

Proof: We first show that (*i*) implies (*ii*). Assume that $a \, R \, b$. We will prove that $[a] = [b]$ by showing $[a] \subseteq [b]$ and $[b] \subseteq [a]$. Suppose $c \in [a]$. Then $a \, R \, c$. Since $a \, R \, b$ and R is symmetric, we know that $b \, R \, a$. Furthermore, since R is transitive and $b \, R \, a$ and $a \, R \, c$, it follows that $b \, R \, c$. Hence, $c \in [b]$. This shows that $[a] \subseteq [b]$. The proof that $[b] \subseteq [a]$ is similar; it is left as an exercise for the reader.

Second, we will show that (*ii*) implies (*iii*). Assume that $[a] = [b]$. It follows that $[a] \cap [b] \neq \emptyset$ since $[a]$ is nonempty (since $a \in [a]$ because R is reflexive).

Next, we will show that (*iii*) implies (*i*). Suppose that $[a] \cap [b] \neq \emptyset$. Then there is an element c with $c \in [a]$ and $c \in [b]$. In other words, $a \, R \, c$ and $b \, R \, c$. By the symmetric property, $c \, R \, b$. Then by transitivity, since $a \, R \, c$ and $c \, R \, b$, we have $a \, R \, b$.

Since (*i*) implies (*ii*), (*ii*) implies (*iii*), and (*iii*) implies (*i*), the three statements, (*i*), (*ii*), and (*iii*), are equivalent. ◁

We are now in a position to show how an equivalence relation *partitions* a set. Let R be an equivalence relation on a set A. The union of the equivalence classes of R is all of A, since an element a of A is in its own equivalence class, namely, $[a]_R$. In other words,

$$\bigcup_{a \in A} [a]_R = A.$$

In addition, from Theorem 1, it follows that these equivalence classes are either equal or disjoint, so

$$[a]_R \cap [b]_R = \emptyset$$

when $[a]_R \neq [b]_R$.

These two observations show that the equivalence classes form a partition of A, since they split A into disjoint subsets. More precisely, a **partition** of a set S is a collection of disjoint nonempty subsets of S that have S as their union. In other words, the collection of subsets $A_i, i \in I$ (where I is an index set) forms a partition of S if and only if

$$A_i \neq \emptyset \text{ for } i \in I,$$

$$A_i \cap A_j = \emptyset \text{ when } i \neq j,$$

and

$$\bigcup_{i \in I} A_i = S.$$

(Here the notation $\bigcup_{i \in I} A_i$ represents the union of the sets A_i for all $i \in I$.) Figure 1 illustrates the concept of a partition of a set.

EXAMPLE 7   Suppose that $S = \{1, 2, 3, 4, 5, 6\}$. The collection of sets $A_1 = \{1, 2, 3\}$, $A_2 = \{4, 5\}$, and $A_3 = \{6\}$ forms a partition of S, since these sets are disjoint and their union is S. ◀

We have seen that the equivalence classes of an equivalence relation on a set form a partition of the set. The subsets in this partition are the equivalence classes. Conversely, every partition of a set can be used to form an equivalence relation. Two elements are equivalent with respect to this relation if and only if they are in the same subset of the partition.

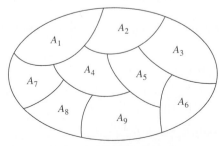

FIGURE 1   A Partition of a Set.

To see this, assume that $\{A_i \mid i \in I\}$ is a partition on S. Let R be the relation on S consisting of the pair (x, y) where x and y belong to the same subset A_i in the partition. To show that R is an equivalence relation we must show that R is reflexive, symmetric, and transitive.

We see that $(a, a) \in R$ for every $a \in S$, since a is in the same subset as itself. Hence, R is reflexive. If $(a, b) \in R$, then b and a are in the same subset of the partition, so that $(b, a) \in R$ as well. Hence, R is symmetric. If $(a, b) \in R$ and $(b, c) \in R$, then a and b are in the same subset in the partition, X, and b and c are in the same subset of the partition, Y. Since the subsets of the partition are disjoint, and b belongs to X and Y, it follows that $X = Y$. Consequently, a and c belong to the same subset of the partition, so that $(a, c) \in R$. Thus, R is transitive.

It follows that R is an equivalence relation. The equivalence classes of R consist of subsets of S containing related elements, and by the definition of R, these are the subsets of the partition. Theorem 2 summarizes the connections we have established between equivalence relations and partitions.

THEOREM 2

Let R be an equivalence relation on a set S. Then the equivalence classes of R form a partition of S. Conversely, given a partition $\{A_i \mid i \in I\}$ of the set S, there is an equivalence relation R that has the sets $A_i, i \in I$, as its equivalence classes.

Example 8 shows how to construct an equivalence relation from a partition.

EXAMPLE 8   List the ordered pairs in the equivalence relation R produced by the partition $A_1 = \{1, 2, 3\}$, $A_2 = \{4, 5\}$, and $A_3 = \{6\}$ of $S = \{1, 2, 3, 4, 5, 6\}$, given in Example 7.

Solution: The subsets in the partition are the equivalence classes of R. The pair $(a, b) \in R$ if and only if a and b are in the same subset of the partition. The pairs $(1, 1), (1, 2), (1, 3)$, $(2, 1), (2, 2), (2, 3), (3, 1), (3, 2)$, and $(3, 3)$ belong to R since $A_1 = \{1, 2, 3\}$ is an equivalence class; the pairs $(4, 4), (4, 5), (5, 4)$, and $(5, 5)$ belong to R since $A_2 = \{4, 5\}$ is an equivalence class; and finally the pair $(6, 6)$ belongs to R since $\{6\}$ is an equivalence class. No pair other than those listed belongs to R.   ◄

The congruence classes modulo m provide a useful illustration of Theorem 2. There are m different congruence classes modulo m, corresponding to the m different remainders possible when an integer is divided by m. These m congruence classes are denoted by $[0]_m, [1]_m, \ldots, [m-1]_m$. They form a partition of the set of integers.

EXAMPLE 9   What are the sets in the partition of the integers arising from congruence modulo 4?

Solution: There are four congruence classes, corresponding to $[0]_4, [1]_4, [2]_4$, and $[3]_4$. They are the sets

$$[0]_4 = \{\ldots, -8, -4, 0, 4, 8, \ldots\}$$
$$[1]_4 = \{\ldots, -7, -3, 1, 5, 9, \ldots\}$$
$$[2]_4 = \{\ldots, -6, -2, 2, 6, 10, \ldots\}$$
$$[3]_4 = \{\ldots, -5, -1, 3, 7, 11, \ldots\}.$$

These congruence classes are disjoint, and every integer is in exactly one of them. In other words, as Theorem 2 says, these congruence classes form a partition.   ◄

Exercises

1. Which of these relations on $\{0, 1, 2, 3\}$ are equivalence relations? Determine the properties of an equivalence relation that the others lack.
   a) $\{(0, 0), (1, 1), (2, 2), (3, 3)\}$
   b) $\{(0, 0), (0, 2), (2, 0), (2, 2), (2, 3), (3, 2), (3, 3)\}$
   c) $\{(0, 0), (1, 1), (1, 2), (2, 1), (2, 2), (3, 3)\}$
   d) $\{(0, 0), (1, 1), (1, 3), (2, 2), (2, 3), (3, 1), (3, 2),$
      $(3, 3)\}$
   e) $\{(0, 0), (0, 1), (0, 2), (1, 0), (1, 1), (1, 2), (2, 0),$
      $(2, 2), (3, 3)\}$

2. Which of these relations on the set of all people are equivalence relations? Determine the properties of an equivalence relation that the others lack.
   a) $\{(a, b) \mid a$ and b are the same age$\}$
   b) $\{(a, b) \mid a$ and b have the same parents$\}$
   c) $\{(a, b) \mid a$ and b share a common parent$\}$
   d) $\{(a, b) \mid a$ and b have met$\}$
   e) $\{(a, b) \mid a$ and b speak a common language$\}$

3. Which of these relations on the set of all functions from $\mathbf{Z}$ to $\mathbf{Z}$ are equivalence relations? Determine the properties of an equivalence relation that the others lack.
   a) $\{(f, g) \mid f(1) = g(1)\}$
   b) $\{(f, g) \mid f(0) = g(0)$ or $f(1) = g(1)\}$
   c) $\{(f, g) \mid f(x) - g(x) = 1$ for all $x \in \mathbf{Z}\}$
   d) $\{(f, g) \mid f(x) - g(x) = C$ for some $C \in \mathbf{Z}$ for all $x \in \mathbf{Z}\}$
   e) $\{(f, g) \mid f(0) = g(1)$ and $f(1) = g(0)\}$

4. Define three equivalence relations on the set of students in your discrete mathematics class different from the relations discussed in the text. Determine the equivalence classes for these equivalence relations.

5. Suppose that A is a nonempty set, and f is a function that has A as its domain. Let R be the relation on A consisting of all ordered pairs (x, y) where $f(x) = f(y)$.
   a) Show that R is an equivalence relation on A.
   b) What are the equivalence classes of R?

6. Suppose that A is a nonempty set and R is an equivalence relation on A. Show that there is a function f with A as its domain such that $(x, y) \in R$ if and only if $f(x) = f(y)$.

7. Show that the relation R, consisting of all pairs (x, y) where x and y are bit strings of length three or more that agree in their first three bits, is an equivalence relation on the set of all bit strings of length three or more.

8. Show that the relation R, consisting of all pairs (x, y) where x and y are bit strings of length three or more that agree except perhaps in their first three bits, is an equivalence relation on the set of all bit strings.

9. Show that propositional equivalence is an equivalence relation on the set of all compound propositions.

10. Let R be the relation on the set of ordered pairs of positive integers such that $((a, b), (c, d)) \in R$ if and only if $ad = bc$. Show that R is an equivalence relation.

11. (Requires calculus)
    a) Show that the relation R on the set of all differentiable functions from $\mathbf{R}$ to $\mathbf{R}$ consisting of all pairs (f, g), where $f'(x) = g'(x)$ for all real numbers x, is an equivalence relation.
    b) Which functions are in the same equivalence class as the function $f(x) = x^2$?

12. (Requires calculus)
    a) Let n be a positive integer. Show that the relation R on the set of all polynomials with real-valued coefficients consisting of all pairs (f, g), where $f^{(n)}(x) = g^{(n)}(x)$, is an equivalence relation. [Here $f^{(n)}(x)$ is the nth derivative of $f(x)$.]
    b) Which functions are in the same equivalence class as the function $f(x) = x^4$, where $n = 3$?

13. Let R be the relation on the set of all URLs (or Web addresses) such that $x R y$ if and only if the Web page at x is the same as the Web page at y. Show that R is an equivalence relation.

14. Let R be the relation on the set of all people who have visited a particular Web page such that $x R y$ if and only if person x and person y have followed the same set of links starting at this Web page (going from Web page to Web page until they stop using the Web). Show that R is an equivalence relation.

In Exercises 15–17 determine whether the relation with the directed graphs shown is an equivalence relation.

15. 16.

17.

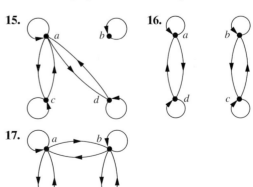

18. Determine whether the relations represented by these zero–one matrices are equivalence relations.

a) $\begin{bmatrix} 1 & 1 & 1 \\ 0 & 1 & 1 \\ 1 & 1 & 1 \end{bmatrix}$ **b)** $\begin{bmatrix} 1 & 0 & 1 & 0 \\ 0 & 1 & 0 & 1 \\ 1 & 0 & 1 & 0 \\ 0 & 1 & 0 & 1 \end{bmatrix}$ **c)** $\begin{bmatrix} 1 & 1 & 1 & 0 \\ 1 & 1 & 1 & 0 \\ 1 & 1 & 1 & 0 \\ 0 & 0 & 0 & 1 \end{bmatrix}$

19. Show that the relation R on the set of all bit strings such that $s\, R\, t$ if and only if s and t contain the same number of 1s is an equivalence relation.

20. What are the equivalence classes of the equivalence relations in Exercise 1?

21. What are the equivalence classes of the equivalence relations in Exercise 2?

22. What are the equivalence classes of the equivalence relations in Exercise 3?

23. What is the equivalence class of the bit string 011 for the equivalence relation in Exercise 19?

24. What are the equivalence classes of these bit strings for the equivalence relation in Exercise 7?
a) 010   **b)** 1011   **c)** 11111   **d)** 01010101

25. Describe the equivalence classes of the bit strings in Exercise 24 for the equivalence relation from Exercise 8.

26. What is the congruence class $[4]_m$ when m is
a) 2?   **b)** 3?   **c)** 6?   **d)** 8?

27. Give a description of each of the congruence classes modulo 6.

28. a) What is the equivalence class of $(1, 2)$ with respect to the equivalence relation in Exercise 10?
b) Give an interpretation of the equivalence classes for the equivalence relation R in Exercise 10.

29. Which of these collections of subsets are partitions of $\{1, 2, 3, 4, 5, 6\}$?

a) $\{1, 2\}, \{2, 3, 4\}, \{4, 5, 6\}$
b) $\{1\}, \{2, 3, 6\}, \{4\}, \{5\}$
c) $\{2, 4, 6\}, \{1, 3, 5\}$ **d)** $\{1, 4, 5\}, \{2, 6\}$

30. Which of these collections of subsets are partitions of $\{-3, -2, -1, 0, 1, 2, 3\}$?

a) $\{-3, -1, 1, 3\}, \{-2, 0, 2\}$
b) $\{-3, -2, -1, 0\}, \{0, 1, 2, 3\}$
c) $\{-3, 3\}, \{-2, 2\}, \{-1, 1\}, \{0\}$
d) $\{-3, -2, 2, 3\}, \{-1, 1\}$

31. Which of these collections of subsets are partitions on the set of bit strings of length 8?

a) the set of bit strings that begin with 1, the set of bit strings that begin with 00, and the set of bit strings that begin with 01
b) the set of bit strings that contain the string 00, the set of bit strings that contain the string 01, the set of bit strings that contain the string 10, and the set of bit strings that contain the string 11
c) the set of bit strings that end with 00, the set of bit strings that end with 01, the set of bit strings

that end with 10, and the set of bit strings that end with 11
d) the set of bit strings that end with 111, the set of bit strings that end with 011, and the set of bit strings that end with 00
e) the set of bit strings that have $3k$ ones, where k is a nonnegative integer; the set of bit strings that contain $3k + 1$ ones, where k is a nonnegative integer; and the set of bit strings that contain $3k + 2$ ones, where k is a positive integer

32. Which of these collections of subsets are partitions of the set of integers?

a) the set of even integers and the set of odd integers
b) the set of positive integers and the set of negative integers
c) the set of integers divisible by 3, the set of integers leaving a remainder of 1 when divided by 3, and the set of integers leaving a remainder of 2 when divided by 3
d) the set of integers less than -100, the set of integers with absolute value not exceeding 100, and the set of integers greater than 100
e) the set of integers not divisible by 3, the set of even integers, and the set of integers that leave a remainder of 3 when divided by 6

33. Which of these are partitions of the set $\mathbf{Z} \times \mathbf{Z}$ of ordered pairs of integers?

a) the set of pairs (x, y) where x or y is odd; the set of pairs (x, y) where x is even; and the set of pairs (x, y) where y is even
b) the set of pairs (x, y) where both x and y are odd; the set of pairs (x, y) where exactly one of x and y is odd; and the set of pairs (x, y) where both x and y are even
c) the set of pairs (x, y) where x is positive; the set of pairs (x, y) where y is positive; and the set of pairs (x, y) where both x and y are negative
d) the set of pairs (x, y) where $3 \mid x$ and $3 \mid y$; the set of pairs (x, y) where $3 \mid x$ and $3 \nmid y$; the set of pairs (x, y) where $3 \nmid x$ and $3 \mid y$; and the set of pairs (x, y), where $3 \nmid x$ and $3 \nmid y$
e) the set of pairs (x, y) where $x > 0$ and $y > 0$; the set of pairs (x, y) where $x > 0$ and $y \le 0$; the set of pairs (x, y) where $x \le 0$ and $y > 0$; and the set of pairs (x, y) where $x \le 0$ and $y \le 0$
f) the set of pairs (x, y) where $x \ne 0$ and $y \ne 0$; the set of pairs (x, y) where $x = 0$ and $y \ne 0$; and the set of pairs (x, y) where $x \ne 0$ and $y = 0$

34. Which of these are partitions of the set of real numbers?

a) the negative real numbers, $\{0\}$, the positive real numbers
b) the set of irrational numbers, the set of rational numbers

c) the set of intervals $[k, k+1], k = \ldots, -2, -1, 0,$ $1, 2, \ldots$

d) the set of intervals $(k, k+1), k = \ldots, -2, -1, 0,$ $1, 2, \ldots$

e) the set of intervals $(k, k+1], k = \ldots, -2, -1, 0,$ $1, 2, \ldots$

f) the sets $\{x + n \mid n \in \mathbf{Z}\}$ for all $x \in [0, 1)$

35. List the ordered pairs in the equivalence relations produced by these partitions of $\{0, 1, 2, 3, 4, 5\}$.

a) $\{0\}, \{1, 2\}, \{3, 4, 5\}$

b) $\{0, 1\}, \{2, 3\}, \{4, 5\}$

c) $\{0, 1, 2\}, \{3, 4, 5\}$

d) $\{0\}, \{1\}, \{2\}, \{3\}, \{4\}, \{5\}$

36. List the ordered pairs in the equivalence relations produced by these partitions of $\{a, b, c, d, e, f, g\}$.

a) $\{a, b\}, \{c, d\}, \{e, f, g\}$

b) $\{a\}, \{b\}, \{c, d\}, \{e, f\}, \{g\}$

c) $\{a, b, c, d\}, \{e, f, g\}$

d) $\{a, c, e, g\}, \{b, d\}, \{f\}$

A partition P_1 is called a **refinement** of the partition P_2 if every set in P_1 is a subset of one of the sets in P_2.

37. Show that the partition formed from the congruence classes modulo 6 is a refinement of the partition formed from the congruence classes modulo 3.

38. Show that the partition of the set of people living in the United States consisting of subsets of people living in the same county (or parish) and same state is a refinement of the partition consisting of subsets of people living in the same state.

39. Show that the partition of the set of bit strings of length 16 formed by equivalence classes of bit strings that agree on the last eight bits is a refinement of the partition formed from the equivalence classes of bit strings that agree on the last four bits.

40. Suppose that R_1 and R_2 are equivalence relations on a set A. Let P_1 and P_2 be the partitions that correspond to R_1 and R_2, respectively. Show that $R_1 \subseteq R_2$ if and only if P_1 is a refinement of P_2.

41. Find the smallest equivalence relation on the set $\{a, b, c, d, e\}$ containing the relation $\{(a, b), (a, c), (d, e)\}$.

42. Suppose that R_1 and R_2 are equivalence relations on the set S. Determine whether each of these combinations of R_1 and R_2 must be an equivalence relation.
a) $R_1 \cup R_2$ **b)** $R_1 \cap R_2$ **c)** $R_1 \oplus R_2$

43. Consider the equivalence relation from Example 3, namely, $R = \{(x, y) \mid x - y \text{ is an integer}\}$.

a) What is the equivalence class of 1 for this equivalence relation?

b) What is the equivalence class of 1/2 for this equivalence relation?

*44. Each bead on a bracelet with three beads is either red, white, or blue, as illustrated in the figure shown.

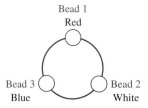

Bead 1
Red

Bead 3   Bead 2
Blue   White

Define the relation R between bracelets as: (B_1, B_2), where B_1 and B_2 are bracelets, belongs to R if and only if B_2 can be obtained from B_1 by rotating it or rotating it and then reflecting it.

a) Show that R is an equivalence relation.

b) What are the equivalence classes of R?

*45. Let R be the relation on the set of all colorings of the 2×2 chessboard where each of the four squares is colored either red or blue so that (C_1, C_2), where C_1 and C_2 are 2×2 chessboards with each of their four squares colored blue or red, belongs to R if and only if C_2 can be obtained from C_1 either by rotating the chessboard or by rotating it and then reflecting it.

a) Show that R is an equivalence relation.

b) What are the equivalence classes of R?

46. a) Let R be the relation on the set of functions from $\mathbf{Z}^+$ to $\mathbf{Z}^+$ such that (f, g) belongs to R if and only if f is $\Theta(g)$ (see Section 2.2). Show that R is an equivalence relation.

b) Describe the equivalence class containing $f(n) = n^2$ for the equivalence relation of part (a).

47. Determine the number of different equivalence relations on a set with three elements by listing them.

48. Determine the number of different equivalence relations on a set with four elements by listing them.

*49. Do we necessarily get an equivalence relation when we form the transitive closure of the symmetric closure of the reflexive closure of a relation?

*50. Do we necessarily get an equivalence relation when we form the symmetric closure of the reflexive closure of the transitive closure of a relation?

51. Suppose we use Theorem 2 to form a partition P from an equivalence relation R. What is the equivalence relation R' that results if we use Theorem 2 again to form an equivalence relation from P?

52. Suppose we use Theorem 2 to form an equivalence relation R from a partition P. What is the partition P' that results if we use Theorem 2 again to form a partition from R?

53. Devise an algorithm to find the smallest equivalence relation containing a given relation.

*54. Let $p(n)$ denote the number of different equivalence relations on a set with n elements (and by Theorem 2 the number of partitions of a set with n elements). Show that $p(n)$ satisfies the recurrence relation $p(n) = \sum_{j=0}^{n-1} C(n-1, j) p(n - j - 1)$ and the initial condition $p(0) = 1$. [*Note:* The numbers $p(n)$ are called **Bell numbers** after the American mathematician E. T. Bell.]

55. Use Exercise 54 to find the number of different equivalence relations on a set with n elements where n is a positive integer not exceeding 10.

7.6 Partial Orderings

INTRODUCTION

We often use relations to order some or all of the elements of sets. For instance, we order words using the relation containing pairs of words (x, y) where x comes before y in the dictionary. We schedule projects using the relation consisting of pairs (x, y) where x and y are tasks in a project such that x must be completed before y begins. We order the set of integers using the relation containing the pairs (x, y) where x is less than y. When we add all of the pairs of the form (x, x) to these relations, we obtain a relation that is reflexive, antisymmetric, and transitive. These are properties that characterize relations used to order the elements of sets using their relative size.

DEFINITION 1

> A relation R on a set S is called a *partial ordering* or *partial order* if it is reflexive, antisymmetric, and transitive. A set S together with a partial ordering R is called a *partially ordered set,* or *poset,* and is denoted by (S, R).

EXAMPLE 1 Show that the "greater than or equal" relation ($\geq$) is a partial ordering on the set of integers.

Solution: Since $a \geq a$ for every integer a, $\geq$ is reflexive. If $a \geq b$ and $b \geq a$, then $a = b$. Hence, $\geq$ is antisymmetric. Finally, $\geq$ is transitive since $a \geq b$ and $b \geq c$ imply that $a \geq c$. It follows that $\geq$ is a partial ordering on the set of integers and $(\mathbf{Z}, \geq)$ is a poset. ◀

EXAMPLE 2 The divisibility relation | is a partial ordering on the set of positive integers, since it is reflexive, antisymmetric, and transitive, as was shown in Section 7.1. We see that $(\mathbf{Z}^+, |)$ is a poset. ($\mathbf{Z}^+$ denotes the set of positive integers.) ◀

EXAMPLE 3 Show that the inclusion relation $\subseteq$ is a partial ordering on the power set of a set S.

Solution: Since $A \subseteq A$ whenever A is a subset of S, $\subseteq$ is reflexive. It is antisymmetric since $A \subseteq B$ and $B \subseteq A$ imply that $A = B$. Finally, $\subseteq$ is transitive, since $A \subseteq B$ and $B \subseteq C$ imply that $A \subseteq C$. Hence, $\subseteq$ is a partial ordering on $P(S)$, and $(P(S), \subseteq)$ is a poset. ◀

In a poset the notation $a \preccurlyeq b$ denotes that $(a, b) \in R$. This notation is used because the "less than or equal to" relation is a paradigm for a partial ordering. (Note that the symbol $\preccurlyeq$ is used to denote the relation in *any* poset, not just the "less than or equals" relation.) The notation $a \prec b$ denotes that $a \preccurlyeq b$, but $a \neq b$. Also, we say "a is less than b" or "b is greater than a" if $a \prec b$.

When a and b are elements of the poset $(S, \preccurlyeq)$, it is not necessary that either $a \preccurlyeq b$ or $b \preccurlyeq a$. For instance, in $(P(\mathbf{Z}), \subseteq)$, $\{1, 2\}$ is not related to $\{1, 3\}$, and vice versa, since neither set is contained within the other. Similarly, in $(\mathbf{Z}, |)$, 2 is not related to 3 and 3 is not related to 2, since $2 \nmid 3$ and $3 \nmid 2$. This leads to Definition 2.

DEFINITION 2   The elements a and b of a poset $(S, \preccurlyeq)$ are called *comparable* if either $a \preccurlyeq b$ or $b \preccurlyeq a$. When a and b are elements of S such that neither $a \preccurlyeq b$ nor $b \preccurlyeq a$, a and b are called *incomparable*.

EXAMPLE 4   In the poset $(\mathbf{Z}^+, |)$, are the integers 3 and 9 comparable? Are 5 and 7 comparable?

Solution: The integers 3 and 9 are comparable, since $3 \mid 9$. The integers 5 and 7 are incomparable, because $5 \nmid 7$ and $7 \nmid 5$.   ◀

The adjective "partial" is used to describe partial orderings since pairs of elements may be incomparable. When every two elements in the set are comparable, the relation is called a **total ordering.**

DEFINITION 3   If $(S, \preccurlyeq)$ is a poset and every two elements of S are comparable, S is called a *totally ordered* or *linearly ordered set,* and $\preccurlyeq$ is called a *total order* or a *linear order.* A totally ordered set is also called a *chain.*

EXAMPLE 5   The poset $(\mathbf{Z}, \leq)$ is totally ordered, since $a \leq b$ or $b \leq a$ whenever a and b are integers.   ◀

EXAMPLE 6   The poset $(\mathbf{Z}^+, |)$ is not totally ordered since it contains elements that are incomparable, such as 5 and 7.   ◀

In Chapter 3 we noted that $(\mathbf{Z}^+, \leq)$ is well-ordered, where $\leq$ is the usual "less than or equals" relation. We now define well-ordered sets.

DEFINITION 4 $(S, \preccurlyeq)$ is a *well-ordered set* if it is a poset such that $\preccurlyeq$ is a total ordering and such that every nonempty subset of S has a least element.

EXAMPLE 7   The set of ordered pairs of positive integers, $\mathbf{Z}^+ \times \mathbf{Z}^+$, with $(a_1, a_2) \preccurlyeq (b_1, b_2)$ if $a_1 < b_1$, or if $a_1 = b_1$ and $a_2 \leq b_2$ (the lexicographic ordering), is a well-ordered set. The verification of this is left as an exercise at the end of this section. The set $\mathbf{Z}$, with the usual $\leq$ ordering, is not well-ordered since the set of negative integers, which is a subset of $\mathbf{Z}$, has no least element.   ◀

In Section 3.4 we showed how to use the principle of well-ordered induction to prove results about a well-ordered set. We now state and prove that this proof technique is valid.

THEOREM 1   **THE PRINCIPLE OF WELL-ORDERED INDUCTION** Suppose that S is a well-ordered set. Then $P(x)$ is true for all $x \in S$, if:

BASIS STEP: $P(x_0)$ is true for the least element of S, and

INDUCTION STEP: For every $y \in S$ if $P(x)$ is true for all $x \prec y$, then $P(y)$ is true.

Proof: Suppose it is not the case that $P(x)$ is true for all $x \in S$. Then there is an element $y \in S$ such that $P(y)$ is false. Consequently, the set $A = \{x \in S \mid P(x) \text{ is false}\}$ is nonempty. Since S is well-ordered, the set A has a least element a. We know that $a \neq x_0$, since by the basis step we know that $P(x_0)$ is true. By the choice of a as a least element of A, we know that $P(x)$ is true for all $x \in S$ with $x \prec a$. This implies by the induction step that $P(a)$ is true. This contradiction shows that $P(x)$ must be true for all $x \in S$. ◁

The principle of well-ordered induction is a versatile technique for proving results about well-ordered sets. Even when it is possible to use mathematical induction for the set of positive integers to prove a theorem, it may be simpler to use the principle of well-ordered induction, as we saw in Example 15 in Section 3.4, where we proved a result about the well-ordered set $(\mathbf{N} \times \mathbf{N}, \preccurlyeq \}$ where $\preccurlyeq$ is lexicographic ordering on $\mathbf{N} \times \mathbf{N}$.

LEXICOGRAPHIC ORDER

The words in a dictionary are listed in alphabetic, or lexicographic, order, which is based on the ordering of the letters in the alphabet. This is a special case of an ordering of strings on a set constructed from a partial ordering on the set. We will show how this construction works in any poset.

First, we will show how to construct a partial ordering on the Cartesian product of two posets, $(A_1, \preccurlyeq_1)$ and $(A_2, \preccurlyeq_2)$. The **lexicographic ordering** $\preccurlyeq$ on $A_1 \times A_2$ is defined by specifying that one pair is less than a second pair if the first entry of the first pair is less than (in A_1) the first entry of the second pair, or if the first entries are equal, but the second entry of this pair is less than (in A_2) the second entry of the second pair. In other words, (a_1, a_2) is less than (b_1, b_2), that is,

$$(a_1, a_2) \prec (b_1, b_2),$$

either if $a_1 \prec_1 b_1$ or if both $a_1 = b_1$ and $a_2 \prec_2 b_2$.

We obtain a partial ordering $\preccurlyeq$ by adding equality to the ordering $\prec$ on $A_1 \times A_2$. The verification of this is left as an exercise.

EXAMPLE 8   Determine whether $(3, 5) \prec (4, 8)$, whether $(3, 8) \prec (4, 5)$, and whether $(4, 9) \prec (4, 11)$ in the poset $(\mathbf{Z} \times \mathbf{Z}, \preccurlyeq)$, where $\preccurlyeq$ is the lexicographic ordering constructed from the usual $\leq$ relation on $\mathbf{Z}$.

Solution: Since $3 < 4$, it follows that $(3, 5) \prec (4, 8)$ and that $(3, 8) \prec (4, 5)$. We have $(4, 9) \prec (4, 11)$, since the first entries of $(4, 9)$ and $(4, 11)$ are the same, but $9 < 11$. ◀

In Figure 1 the ordered pairs in $\mathbf{Z}^+ \times \mathbf{Z}^+$ that are less than $(3, 4)$ are highlighted.

A lexicographic ordering can be defined on the Cartesian product of n posets $(A_1, \preccurlyeq_1), (A_2, \preccurlyeq_2), \ldots, (A_n, \preccurlyeq_n)$. Define the partial ordering $\preccurlyeq$ on $A_1 \times A_2 \times \cdots \times A_n$ by

$$(a_1, a_2, \ldots, a_n) \prec (b_1, b_2, \ldots, b_n)$$

if $a_1 \prec_1 b_1$, or if there is an integer $i > 0$ such that $a_1 = b_1, \ldots, a_i = b_i$, and $a_{i+1} \prec_{i+1} b_{i+1}$. In other words, one n-tuple is less than a second n-tuple if the entry of the first n-tuple in the first position where the two n-tuples disagree is less than the entry in that position in the second n-tuple.

EXAMPLE 9   Note that $(1, 2, 3, 5) \prec (1, 2, 4, 3)$, since the entries in the first two positions of these 4-tuples agree, but in the third position the entry in the first 4-tuple, 3, is less than that in

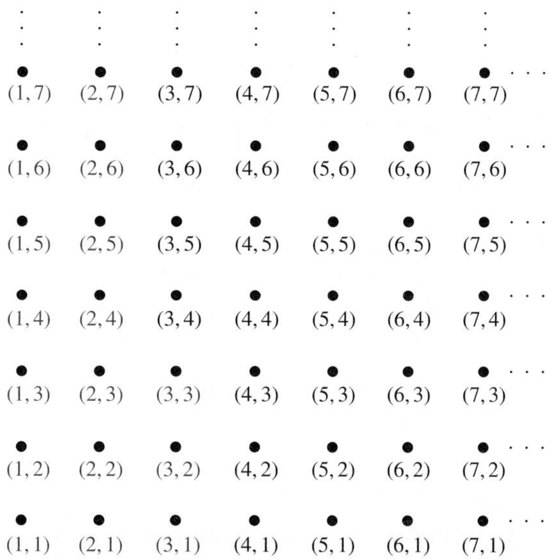

$$\vdots \quad \vdots \quad \vdots \quad \vdots \quad \vdots \quad \vdots \quad \vdots$$

$\bullet$	$\bullet$	$\bullet$	$\bullet$	$\bullet$	$\bullet$	$\bullet \cdots$
$(1,7)$	$(2,7)$	$(3,7)$	$(4,7)$	$(5,7)$	$(6,7)$	$(7,7)$
$\bullet$	$\bullet$	$\bullet$	$\bullet$	$\bullet$	$\bullet$	$\bullet \cdots$
$(1,6)$	$(2,6)$	$(3,6)$	$(4,6)$	$(5,6)$	$(6,6)$	$(7,6)$
$\bullet$	$\bullet$	$\bullet$	$\bullet$	$\bullet$	$\bullet$	$\bullet \cdots$
$(1,5)$	$(2,5)$	$(3,5)$	$(4,5)$	$(5,5)$	$(6,5)$	$(7,5)$
$\bullet$	$\bullet$	$\bullet$	$\bullet$	$\bullet$	$\bullet$	$\bullet \cdots$
$(1,4)$	$(2,4)$	$(3,4)$	$(4,4)$	$(5,4)$	$(6,4)$	$(7,4)$
$\bullet$	$\bullet$	$\bullet$	$\bullet$	$\bullet$	$\bullet$	$\bullet \cdots$
$(1,3)$	$(2,3)$	$(3,3)$	$(4,3)$	$(5,3)$	$(6,3)$	$(7,3)$
$\bullet$	$\bullet$	$\bullet$	$\bullet$	$\bullet$	$\bullet$	$\bullet \cdots$
$(1,2)$	$(2,2)$	$(3,2)$	$(4,2)$	$(5,2)$	$(6,2)$	$(7,2)$
$\bullet$	$\bullet$	$\bullet$	$\bullet$	$\bullet$	$\bullet$	$\bullet \cdots$
$(1,1)$	$(2,1)$	$(3,1)$	$(4,1)$	$(5,1)$	$(6,1)$	$(7,1)$

FIGURE 1 **The Ordered Pairs Less Than $(3, 4)$ in Lexicographic Order.**

the second 4-tuple, 4. (Here the ordering on 4-tuples is the lexicographic ordering that comes from the usual "less than or equals" relation on the set of integers.) ◀

We can now define lexicographic ordering of strings. Consider the strings $a_1 a_2 \cdots a_m$ and $b_1 b_2 \cdots b_n$ on a partially ordered set S. Suppose these strings are not equal. Let t be the minimum of m and n. The definition of lexicographic ordering is that the string $a_1 a_2 \cdots a_m$ is less than $b_1 b_2 \cdots b_n$ if and only if

$$(a_1, a_2, \ldots, a_t) \prec (b_1, b_2, \ldots, b_t), \text{ or}$$
$$(a_1, a_2, \ldots, a_t) = (b_1, b_2, \ldots, b_t) \text{ and } m < n,$$

where $\prec$ in this inequality represents the lexicographic ordering of S^t. In other words, to determine the ordering of two different strings, the longer string is truncated to the length of the shorter string, namely, to $t = \min(m, n)$ terms. Then the t-tuples made up of the first t terms of each string are compared using the lexicographic ordering on S^t. One string is less than another string if the t-tuple corresponding to the first string is less than the t-tuple of the second string, or if these two t-tuples are the same, but the second string is longer. The verification that this is a partial ordering is left as an exercise for the reader.

EXAMPLE 10 Consider the set of strings of lowercase English letters. Using the ordering of letters in the alphabet, a lexicographic ordering on the set of strings can be constructed. A string is less than a second string if the letter in the first string in the first position where the strings differ comes before the letter in the second string in this position, or if the first string and the second string agree in all positions, but the second string has more letters. This ordering is the same as that used in dictionaries. For example,

    discreet $\prec$ *discrete*,

since these strings differ first in the seventh position, and $e \prec t$. Also,

    discreet $\prec$ *discreetness*,

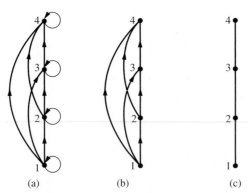

FIGURE 2 **Constructing the Hasse Diagram for** $(\{1, 2, 3, 4\}, \leq)$.

since the first eight letters agree, but the second string is longer. Furthermore,

$discrete \prec discretion,$

since

$discrete \prec discreti.$ ◀

HASSE DIAGRAMS

Many edges in the directed graph for a finite poset do not have to be shown since they must be present. For instance, consider the directed graph for the partial ordering $\{(a, b) \mid a \leq b\}$ on the set $\{1, 2, 3, 4\}$, shown in Figure 2(a). Since this relation is a partial ordering, it is reflexive, and its directed graph has loops at all vertices. Consequently, we do not have to show these loops since they must be present; in Figure 2(b) loops are not shown. Because a partial ordering is transitive, we do not have to show those edges that must be present because of transitivity. For example, in Figure 2(c) the edges (1, 3), (1, 4), and (2, 4) are not shown since they must be present. If we assume that all edges are pointed "upward" (as they are drawn in the figure), we do not have to show the directions of the edges; Figure 2(c) does not show directions.

In general, we can represent a partial ordering on a finite set using this procedure: Start with the directed graph for this relation. Because a partial ordering is reflexive, a loop is present at every vertex. Remove these loops. Remove all edges that must be

HELMUT HASSE (1898–1979) Helmut Hasse was born in Kassel, Germany. He served in the German navy after high school. He began his university studies at Göttingen University in 1918, moving in 1920 to Marburg University to study under the number theorist Kurt Hensel. During this time, Hasse made fundamental contributions to algebraic number theory. He became Hensel's successor at Marburg, later becoming director of the famous mathematical institute at Göttingen in 1934, and took a position at Hamburg University in 1950. Hasse served for 50 years as an editor of *Crelle's Journal,* a famous German mathematics periodical, taking over the job of chief editor in 1936 when the Nazis forced Hensel to resign. During World War II Hasse worked on applied mathematics research for the German navy. He was noted for the clarity and personal style of his lectures and was devoted both to number theory and to his students. (Hasse has been controversial for connections with the Nazi party. Investigations have shown he was a strong German nationalist but not an ardent Nazi.)

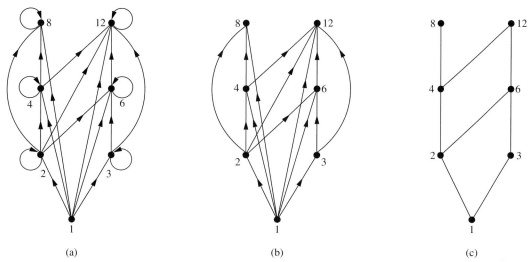

(a)   (b)   (c)

FIGURE 3   **Constructing the Hasse Diagram of ({1, 2, 3, 4, 6, 8, 12}, |).**

present because of the transitivity, since they must be present because a partial ordering is transitive. For instance, if (a, b) and (b, c) are in the partial ordering, remove the edge (a, c), since it must be present also. Furthermore, if (c, d) is also in the partial ordering, remove the edge (a, d), since it must be present also. Finally, arrange each edge so that its initial vertex is below its terminal vertex (as it is drawn on paper). Remove all the arrows on the directed edges, since all edges point "upward" toward their terminal vertex. (The edges left correspond to pairs in the covering relation of the poset. See the preamble to Exercise 22.)

These steps are well defined, and only a finite number of steps need to be carried out for a finite poset. When all the steps have been taken, the resulting diagram contains sufficient information to find the partial ordering. This diagram is called a **Hasse diagram,** named after the twentieth-century German mathematician Helmut Hasse.

EXAMPLE 11   Draw the Hasse diagram representing the partial ordering $\{(a, b) \mid a$ divides $b\}$ on $\{1, 2, 3, 4, 6, 8, 12\}$.

Solution: Begin with the digraph for this partial order, as shown in Figure 3(a). Remove all loops, as shown in Figure 3(b). Then delete all the edges implied by the transitive property. These are $(1, 4), (1, 6), (1, 8), (1, 12), (2, 8), (2, 12)$, and $(3, 12)$. Arrange all edges to point upward, and delete all arrows to obtain the Hasse diagram. The resulting Hasse diagram is shown in Figure 3(c). ◄

EXAMPLE 12   Draw the Hasse diagram for the partial ordering $\{(A, B) \mid A \subseteq B\}$ on the power set $P(S)$ where $S = \{a, b, c\}$.

Solution: The Hasse diagram for this partial ordering is obtained from the associated digraph by deleting all the loops and all the edges that occur from transitivity, namely, $(\emptyset, \{a, b\})$, $(\emptyset, \{a, c\})$, $(\emptyset, \{b, c\})$, $(\emptyset, \{a, b, c\})$, $(\{a\}, \{a, b, c\})$, $(\{b\}, \{a, b, c\})$, and $(\{c\}, \{a, b, c\})$. Finally all edges point upward, and arrows are deleted. The resulting Hasse diagram is illustrated in Figure 4. ◄

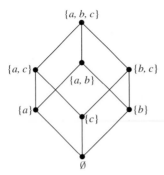

FIGURE 4   The Hasse Diagram of $(P(\{a, b, c\}), \subseteq)$.

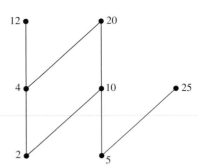

FIGURE 5   The Hasse Diagram of a Poset.

MAXIMAL AND MINIMAL ELEMENTS

Elements of posets that have certain extremal properties are important for many applications. An element of a poset is called maximal if it is not less than any element of the poset. That is, a is **maximal** in the poset $(S, \preccurlyeq)$ if there is no $b \in S$ such that $a \prec b$. Similarly, an element of a poset is called minimal if it is not greater than any element of the poset. That is, a is **minimal** if there is no element $b \in S$ such that $b \prec a$. Maximal and minimal elements are easy to spot using a Hasse diagram. They are the "top" and "bottom" elements in the diagram.

EXAMPLE 13   Which elements of the poset $(\{2, 4, 5, 10, 12, 20, 25\}, |)$ are maximal, and which are minimal?

Solution: The Hasse diagram in Figure 5 for this poset shows that the maximal elements are 12, 20, and 25, and the minimal elements are 2 and 5. As this example shows, a poset can have more than one maximal element and more than one minimal element.   ◄

Sometimes there is an element in a poset that is greater than every other element. Such an element is called the greatest element. That is, a is the **greatest element** of the poset $(S, \preccurlyeq)$ if $b \preccurlyeq a$ for all $b \in S$. The greatest element is unique when it exists [see Exercise 34(a) at the end of this section]. Likewise, an element is called the least element if it is less than all the other elements in the poset. That is, a is the **least element** of $(S, \preccurlyeq)$ if $a \preccurlyeq b$ for all $b \in S$. The least element is unique when it exists [see Exercise 34(b) at the end of the section].

EXAMPLE 14   Determine whether the posets represented by each of the Hasse diagrams in Figure 6 have a greatest element and a least element.

Solution: The least element of the poset with Hasse diagram (a) is a. This poset has no greatest element. The poset with Hasse diagram (b) has neither a least nor a greatest element. The poset with Hasse diagram (c) has no least element. Its greatest element is d. The poset with Hasse diagram (d) has least element a and greatest element d.   ◄

EXAMPLE 15   Let S be a set. Determine whether there is a greatest element and a least element in the poset $(P(S), \subseteq)$.

Solution: The least element is the empty set, since $\emptyset \subseteq T$ for any subset T of S. The set S is the greatest element in this poset, since $T \subseteq S$ whenever T is a subset of S.   ◄

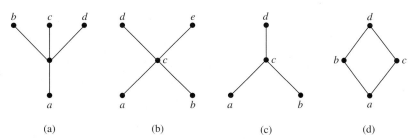

FIGURE 6   **Hasse Diagrams of Four Posets.**

EXAMPLE 16   Is there a greatest element and a least element in the poset $(\mathbf{Z}^+, |)$?

Solution: The integer 1 is the least element since $1 | n$ whenever n is a positive integer. Since there is no integer that is divisible by all positive integers, there is no greatest element.   ◀

Sometimes it is possible to find an element that is greater than all the elements in a subset A of a poset $(S, \preccurlyeq)$. If u is an element of S such that $a \preccurlyeq u$ for all elements $a \in A$, then u is called an **upper bound** of A. Likewise, there may be an element less than all the elements in A. If l is an element of S such that $l \preccurlyeq a$ for all elements $a \in A$, then l is called a **lower bound** of A.

EXAMPLE 17   Find the lower and upper bounds of the subsets $\{a, b, c\}$, $\{j, h\}$, and $\{a, c, d, f\}$ in the poset with the Hasse diagram shown in Figure 7.

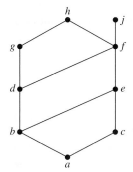

FIGURE 7
The Hasse Diagram of a Poset.

Solution: The upper bounds of $\{a, b, c\}$ are e, f, j, and h, and its only lower bound is a. There are no upper bounds of $\{j, h\}$, and its lower bounds are a, b, c, d, e, and f. The upper bounds of $\{a, c, d, f\}$ are f, h, and j, and its lower bound is a.   ◀

The element x is called the **least upper bound** of the subset A if x is an upper bound that is less than every other upper bound of A. Since there is only one such element, if it exists, it makes sense to call this element *the* least upper bound [see Exercise 36(a) at the end of this section]. That is, x is the least upper bound of A if $a \preccurlyeq x$ whenever $a \in A$, and $x \preccurlyeq z$ whenever z is an upper bound of A. Similarly, the element y is called the **greatest lower bound** of A if y is a lower bound of A and $z \preccurlyeq y$ whenever z is a lower bound of A. The greatest lower bound of A is unique if it exists [see Exercise 36(b) at the end of this section]. The greatest lower bound and least upper bound of a subset A are denoted by $\mathrm{glb}(A)$ and $\mathrm{lub}(A)$, respectively.

EXAMPLE 18   Find the greatest lower bound and the least upper bound of $\{b, d, g\}$, if they exist, in the poset shown in Figure 7.

Solution: The upper bounds of $\{b, d, g\}$ are g and h. Since $g \prec h$, g is the least upper bound. The lower bounds of $\{b, d, g\}$ are a and b. Since $a \prec b$, b is the greatest lower bound.   ◀

EXAMPLE 19   Find the greatest lower bound and the least upper bound of the sets $\{3, 9, 12\}$ and $\{1, 2, 4, 5, 10\}$, if they exist, in the poset $(\mathbf{Z}^+, |)$.

Solution: An integer is a lower bound of {3, 9, 12} if 3, 9, and 12 are divisible by this integer. The only such integers are 1 and 3. Since 1 | 3, 3 is the greatest lower bound of {3, 9, 12}. The only lower bound for the set {1, 2, 4, 5, 10} with respect to | is the element 1. Hence, 1 is the greatest lower bound for {1, 2, 4, 5, 10}.

An integer is an upper bound for {3, 9, 12} if and only if it is divisible by 3, 9, and 12. The integers with this property are those divisible by the least common multiple of 3, 9, and 12, which is 36. Hence, 36 is the least upper bound of {3, 9, 12}. A positive integer is an upper bound for the set {1, 2, 4, 5, 10} if and only if it is divisible by 1, 2, 4, 5, and 10. The integers with this property are those integers divisible by the least common multiple of these integers, which is 20. Hence, 20 is the least upper bound of {1, 2, 4, 5, 10}. ◀

LATTICES

A partially ordered set in which every pair of elements has both a least upper bound and a greatest lower bound is called a **lattice.** Lattices have many special properties. Furthermore, lattices are used in many different applications such as models of information flow and play an important role in Boolean algebra.

EXAMPLE 20　Determine whether the posets represented by each of the Hasse diagrams in Figure 8 are lattices.

Solution: The posets represented by the Hasse diagrams in (a) and (c) are both lattices because in each poset every pair of elements has both a least upper bound and a greatest lower bound, as the reader should verify. On the other hand, the poset with the Hasse diagram shown in (b) is not a lattice, since the elements b and c have no least upper bound. To see this note that each of the elements d, e, and f is an upper bound, but none of these three elements precedes the other two with respect to the ordering of this poset. ◀

EXAMPLE 21　Is the poset $(\mathbf{Z}^+, |)$ a lattice?

Solution: Let a and b be two positive integers. The least upper bound and greatest lower bound of these two integers are the least common multiple and the greatest common divisor of these integers, respectively, as the reader should verify. It follows that this poset is a lattice. ◀

EXAMPLE 22　Determine whether the posets ({1, 2, 3, 4, 5}, |) and ({1, 2, 4, 8, 16}, |) are lattices.

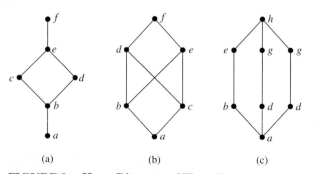

(a)　　　　(b)　　　　(c)

FIGURE 8　**Hasse Diagrams of Three Posets.**

Solution: Since 2 and 3 have no upper bounds in ({1, 2, 3, 4, 5}, |), they certainly do not have a least upper bound. Hence, the first poset is not a lattice.

Every two elements of the second poset have both a least upper bound and a greatest lower bound. The least upper bound of two elements in this poset is the larger of the elements and the greatest lower bound of two elements is the smaller of the elements, as the reader should verify. Hence this second poset is a lattice. ◄

EXAMPLE 23 Determine whether $(P(S), \subseteq)$ is a lattice where S is a set.

Solution: Let A and B be two subsets of S. The least upper bound and the greatest lower bound of A and B are $A \cup B$ and $A \cap B$, respectively, as the reader can show. Hence $(P(S), \subseteq)$ is a lattice. ◄

EXAMPLE 24

The Lattice Model of Information Flow In many settings the flow of information from one person or computer program to another is restricted via security clearances. We can use a lattice model to represent different information flow policies. For example, one common information flow policy is the *multilevel security policy* used in government and military systems. Each piece of information is assigned to a security class, and each security class is represented by a pair (A, C) where A is an *authority level* and C is a *category*. People and computer programs are then allowed access to information from a specific restricted set of security classes.

The typical authority levels used in the U.S. government are unclassified (0), confidential (1), secret (2), and top secret (3). Categories used in security classes are the subsets of a set of all *compartments* relevant to a particular area of interest. Each compartment represents a particular subject area. For example, if the set of compartments is {*spies, moles, double agents*}, then there are eight different categories, one for each of the eight subsets of the set of compartments, such as {*spies, moles*}.

We can order security classes by specifying that $(A_1, C_1) \preccurlyeq (A_2, C_2)$ if and only if $A_1 \leq A_2$ and $C_1 \subseteq C_2$. Information is permitted to flow from security class (A_1, C_1) into security class (A_2, C_2) if and only if $(A_1, C_1) \preccurlyeq (A_2, C_2)$. For example, information is permitted to flow from the security class (*secret*, {*spies, moles*}) into the security class (*top secret*, {*spies, moles, double agents*}), whereas information is not allowed to flow from the security class (*top secret*, {*spies, moles*}) into either of the security classes (*secret*, {*spies, moles, double agents*}) or (*top secret*, {*spies*}).

We leave it to the reader (see Exercise 42 at the end of this section) to show that the set of all security classes with the ordering defined in this example forms a lattice. ◄

TOPOLOGICAL SORTING

Suppose that a project is made up of 20 different tasks. Some tasks can be completed only after others have been finished. How can an order be found for these tasks? To model this problem we set up a partial order on the set of tasks, so that $a \prec b$ if and only if a and b are tasks where b cannot be started until a has been completed. To produce a schedule for the project, we need to produce an order for all 20 tasks that is compatible with this partial order. We will show how this can be done.

We begin with a definition. A total ordering $\preccurlyeq$ is said to be **compatible** with the partial ordering R if $a \preccurlyeq b$ whenever $a \, R \, b$. Constructing a compatible total ordering from a partial ordering is called **topological sorting.** We will need to use Lemma 1.

LEMMA 1 Every finite nonempty poset $(S, \preccurlyeq)$ has a minimal element.

Proof: Choose an element a_0 of S. If a_0 is not minimal, then there is an element a_1 with $a_1 \prec a_0$. If a_1 is not minimal, there is an element a_2 with $a_2 \prec a_1$. Continue this process, so that if a_n is not minimal, there is an element a_{n+1} with $a_{n+1} \prec a_n$. Since there are only a finite number of elements in the poset, this process must end with a minimal element a_n. ◁

The topological sorting algorithm we will describe works for any finite nonempty poset. To define a total ordering on the poset $(A, \preccurlyeq)$, first choose a minimal element a_1; such an element exists by Lemma 1. Next, note that $(A - \{a_1\}, \preccurlyeq)$ is also a poset, as the reader should verify. If it is nonempty, choose a minimal element a_2 of this poset. Then remove a_2 as well, and if there are additional elements left, choose a minimal element a_3 in $A - \{a_1, a_2\}$. Continue this process by choosing a_{k+1} to be a minimal element in $A - \{a_1, a_2, \ldots, a_k\}$, as long as elements remain.

Since A is a finite set, this process must terminate. The end product is a sequence of elements $a_1, a_2, \ldots, a_n$. The desired total ordering is defined by

$$a_1 \preccurlyeq a_2 \preccurlyeq \cdots \preccurlyeq a_n.$$

This total ordering is compatible with the original partial ordering. To see this, note that if $b \prec c$ in the original partial ordering, c is chosen as the minimal element at a phase of the algorithm where b has already been removed, for otherwise c would not be a minimal element. Pseudocode for this topological sorting algorithm is shown in Algorithm 1.

ALGORITHM 1 Topological Sorting

---

procedure *topological sort* (S: finite poset)
$k := 1$
while $S \neq \emptyset$
begin
    $a_k := $ a minimal element of S {such an element exists by Lemma 1}
    $S := S - \{a_k\}$
    $k := k + 1$
end $\{a_1, a_2, \ldots, a_n$ is a compatible total ordering of $S\}$

EXAMPLE 25 Find a compatible total ordering for the poset $(\{1, 2, 4, 5, 12, 20\}, |)$.

Solution: The first step is to choose a minimal element. This must be 1, since it is the only minimal element. Next, select a minimal element of $(\{2, 4, 5, 12, 20\}, |)$. There are two minimal elements in this poset, namely, 2 and 5. We select 5. The remaining elements are $\{2, 4, 12, 20\}$. The only minimal element at this stage is 2. Next, 4 is chosen since it is the only minimal element of $(\{4, 12, 20\}, |)$. Since both 12 and 20 are minimal elements of $(\{12, 20\}, |)$, either can be chosen next. We select 20, which leaves 12 as the last element left. This produces the total ordering

$$1 \prec 5 \prec 2 \prec 4 \prec 20 \prec 12.$$

The steps used by this sorting algorithm are displayed in Figure 9. ◀

Topological sorting has an application to the scheduling of projects.

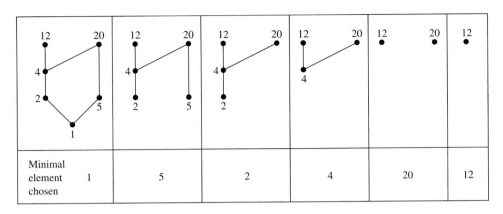

FIGURE 9   A Topological Sort of ({1, 2, 4, 5, 12, 20}, |).

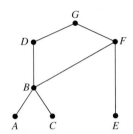

FIGURE 10   The Hasse Diagram for Seven Tasks.

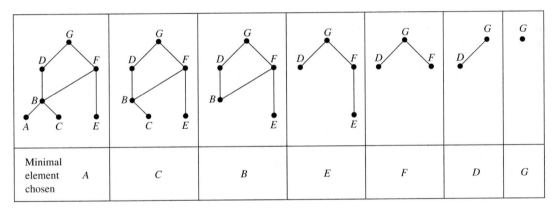

FIGURE 11   A Topological Sort of the Tasks.

EXAMPLE 26   A development project at a computer company requires the completion of seven tasks. Some of these tasks can be started only after other tasks are finished. A partial ordering on tasks is set up by considering task $X \prec$ task Y if task Y cannot be started until task X has been completed. The Hasse diagram for the seven tasks, with respect to this partial ordering, is shown in Figure 10. Find an order in which these tasks can be carried out to complete the project.

Solution: An ordering of the seven tasks can be obtained by performing a topological sort. The steps of a sort are illustrated in Figure 11. The result of this sort, $A \prec C \prec B \prec E \prec F \prec D \prec G$, gives one possible order for the tasks.   ◀

Exercises

1. Which of these are posets?

    a) $(\mathbf{Z}, =)$ **b)** $(\mathbf{Z}, \neq)$

    c) $(\mathbf{Z}, \geq)$ **d)** $(\mathbf{Z}, \not|)$

2. Determine whether the relations represented by these zero–one matrices are partial orders.

a) $\begin{bmatrix} 1 & 0 & 1 \\ 1 & 1 & 0 \\ 0 & 0 & 1 \end{bmatrix}$ **b)** $\begin{bmatrix} 1 & 0 & 0 \\ 0 & 1 & 0 \\ 1 & 0 & 1 \end{bmatrix}$

c)
$$\begin{bmatrix} 1 & 0 & 1 & 0 \\ 0 & 1 & 1 & 0 \\ 0 & 0 & 1 & 1 \\ 1 & 1 & 0 & 1 \end{bmatrix}$$

In Exercises 3–5 determine whether the relation with the directed graph shown is a partial order.

3.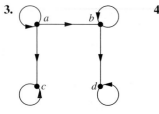

4.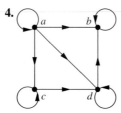

5.

6. Let (S, R) be a poset. Show that (S, R^{-1}) is also a poset, where R^{-1} is the inverse of R. The poset (S, R^{-1}) is called the **dual** of (S, R).

7. Find the duals of these posets.

a) $(\{0, 1, 2\}, \leq)$ **b)** $(\mathbf{Z}, \geq)$
c) $(P(\mathbf{Z}), \supseteq)$ **d)** $(\mathbf{Z}^+, |)$

8. Which of these pairs of elements are comparable in the poset $(\mathbf{Z}^+, |)$?

a) 5, 15   **b)** 6, 9   **c)** 8, 16   **d)** 7, 7

9. Find two incomparable elements in these posets.

a) $(P(\{0, 1, 2\}), \subseteq)$ **b)** $(\{1, 2, 4, 6, 8\}, |)$

10. Let $S = \{1, 2, 3, 4\}$. With respect to the lexicographic order based on the usual "less than" relation,

a) find all pairs in $S \times S$ less than $(2, 3)$.
b) find all pairs in $S \times S$ greater than $(3, 1)$.
c) draw the Hasse diagram of the poset $(S \times S, \preccurlyeq)$.

11. Find the lexicographic ordering of these n-tuples:

a) $(1, 1, 2), (1, 2, 1)$
b) $(0, 1, 2, 3), (0, 1, 3, 2)$
c) $(1, 0, 1, 0, 1), (0, 1, 1, 1, 0)$

12. Find the lexicographic ordering of these strings of lowercase English letters:

a) *quack, quick, quicksilver, quicksand, quacking*
b) *open, opener, opera, operand, opened*
c) *zoo, zero, zoom, zoology, zoological*

13. Find the lexicographic ordering of the bit strings $0, 01, 11, 001, 010, 011, 0001,$ and 0101 based on the ordering $0 < 1$.

14. Draw the Hasse diagram for the "greater than or equal to" relation on $\{0, 1, 2, 3, 4, 5\}$.

15. Draw the Hasse diagram for the "less than or equal to" relation on $\{0, 2, 5, 10, 11, 15\}$.

16. Draw the Hasse diagram for divisibility on the set

a) $\{1, 2, 3, 4, 5, 6\}$.
b) $\{3, 5, 7, 11, 13, 16, 17\}$.

c) $\{2, 3, 5, 10, 11, 15, 25\}$.
d) $\{1, 3, 9, 27, 81, 243\}$.

17. Draw the Hasse diagram for divisibility on the set

a) $\{1, 2, 3, 4, 5, 6, 7, 8\}$.
b) $\{1, 2, 3, 5, 7, 11, 13\}$.
c) $\{1, 2, 3, 6, 12, 24, 36, 48\}$.
d) $\{1, 2, 4, 8, 16, 32, 64\}$.

18. Draw the Hasse diagram for inclusion on the set $P(S)$ where $S = \{a, b, c, d\}$.

In Exercises 19–21 list all ordered pairs in the partial ordering with the accompanying Hasse diagram.

19. 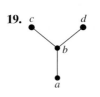   **20.**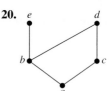

21.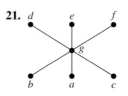

Let $(S, \preccurlyeq)$ be a poset. We say that an element $y \in S$ **covers** an element $x \in S$ if $x \prec y$ and there is no element $z \in S$ such that $x \prec z \prec y$. The set of pairs (x, y) such that y covers x is called the **covering relation** of $(S, \preccurlyeq)$.

22. What is the covering relation of the partial ordering $\{(a, b) \mid a$ divides $b\}$ on $\{1, 2, 3, 4, 6, 12\}$?

23. What is the covering relation of the partial ordering $\{(A, B) \mid A \subseteq B\}$ on the power set of S where $S = \{a, b, c\}$?

24. Show that the pair (x, y) belongs to the covering relation of the finite poset $(S, \preccurlyeq)$ if and only if x is lower than y and there is an edge joining x and y in the Hasse diagram of this poset.

25. Show that a finite poset can be reconstructed from its covering relation.

26. Answer these questions for the partial order represented by this Hasse diagram.

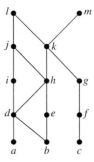

a) Find the maximal elements.

b) Find the minimal elements.

c) Is there a greatest element?

d) Is there a least element?

e) Find all upper bounds of $\{a, b, c\}$.

f) Find the least upper bound of $\{a, b, c\}$, if it exists.

g) Find all lower bounds of $\{f, g, h\}$.

h) Find the greatest lower bound of $\{f, g, h\}$, if it exists.

27. Answer these questions for the poset $(\{3, 5, 9, 15, 24, 45\}, |)$.

a) Find the maximal elements.

b) Find the minimal elements.

c) Is there a greatest element?

d) Is there a least element?

e) Find all upper bounds of $\{3, 5\}$.

f) Find the least upper bound of $\{3, 5\}$, if it exists.

g) Find all lower bounds of $\{15, 45\}$.

h) Find the greatest lower bound of $\{15, 45\}$, if it exists.

28. Answer these questions for the poset $(\{2, 4, 6, 9, 12, 18, 27, 36, 48, 60, 72\}, |)$.

a) Find the maximal elements.

b) Find the minimal elements.

c) Is there a greatest element?

d) Is there a least element?

e) Find all upper bounds of $\{2, 9\}$.

f) Find the least upper bound of $\{2, 9\}$, if it exists.

g) Find all lower bounds of $\{60, 72\}$.

h) Find the greatest lower bound of $\{60, 72\}$, if it exists.

29. Answer these questions for the poset $(\{\{1\}, \{2\}, \{4\}, \{1, 2\}, \{1, 4\}, \{2, 4\}, \{3, 4\}, \{1, 3, 4\}, \{2, 3, 4\}\}, \subseteq)$.

a) Find the maximal elements.

b) Find the minimal elements.

c) Is there a greatest element?

d) Is there a least element?

e) Find all upper bounds of $\{\{2\}, \{4\}\}$.

f) Find the least upper bound of $\{\{2\}, \{4\}\}$, if it exists.

g) Find all lower bounds of $\{\{1, 3, 4\}, \{2, 3, 4\}\}$.

h) Find the greatest lower bound of $\{\{1, 3, 4\}, \{2, 3, 4\}\}$, if it exists.

30. Give a poset that has

a) a minimal element but no maximal element.

b) a maximal element but no minimal element.

c) neither a maximal nor a minimal element.

31. Show that lexicographic order is a partial ordering on the Cartesian product of two posets.

32. Show that lexicographic order is a partial ordering on the set of strings from a poset.

33. Suppose that $(S, \preccurlyeq_1)$ and $(T, \preccurlyeq_2)$ are posets. Show that $(S \times T, \preccurlyeq)$ is a poset where $(s, t) \preccurlyeq (u, v)$ if and only if $s \preccurlyeq_1 u$ and $t \preccurlyeq_2 v$.

34. a) Show that there is exactly one greatest element of a poset, if such an element exists.

b) Show that there is exactly one least element of a poset, if such an element exists.

35. a) Show that there is exactly one maximal element in a poset with a greatest element.

b) Show that there is exactly one minimal element in a poset with a least element.

36. a) Show that the least upper bound of a set in a poset is unique if it exists.

b) Show that the greatest lower bound of a set in a poset is unique if it exists.

37. Determine whether the posets with these Hasse diagrams are lattices.

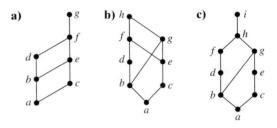

38. Determine whether these posets are lattices.

a) $(\{1, 3, 6, 9, 12\}, |)$

b) $(\{1, 5, 25, 125\}, |)$

c) $(\mathbf{Z}, \geq)$

d) $(P(S), \supseteq)$, where $P(S)$ is the power set of a set S

39. Show that every nonempty subset of a lattice has a least upper bound and a greatest lower bound.

40. Show that if the poset (S, R) is a lattice then the dual poset (S, R^{-1}) is also a lattice.

41. In a company, the lattice model of information flow is used to control sensitive information with security classes represented by ordered pairs (A, C). Here A is an authority level which may be nonproprietary (0), proprietary (1), restricted (2), or registered (3). A category C is a subset of the set of all projects $\{Cheetah, Impala, Puma\}$. (Names of animals are often used as code names for projects in companies.)

a) Is information permitted to flow from $(Proprietary, \{Cheetah, Puma\})$ into $(Restricted, \{Puma\})$?

b) Is information permitted to flow from $(Restricted, \{Cheetah\})$ into $(Registered, \{Cheetah, Impala\})$?

c) Into which classes is information from $(Proprietary, \{Cheetah, Puma\})$ permitted to flow?

d) From which classes is information permitted to flow into the security class $(Restricted, \{Impala, Puma\})$?

42. Show that the set S of security classes (A, C) is a lattice, where A is a positive integer representing an authority class and C is a subset of a finite set of compartments, with $(A_1, C_1) \preccurlyeq (A_2, C_2)$ if and only if $A_1 \leq A_2$ and $C_1 \subseteq C_2$. [*Hint:* First show that $(S, \preccurlyeq)$

is a poset and then show that the least upper bound and greatest lower bound of (A_1, C_1) and (A_2, C_2) are $(\max(A_1, A_2), C_1 \cup C_2)$ and $(\min(A_1, A_2), C_1 \cap C_2)$, respectively.]

*43. Show that the set of all partitions of a set S with the relation $P_1 \preccurlyeq P_2$ if the partition P_1 is a refinement of the partition P_2 is a lattice. (See the preamble to Exercise 37 of Section 7.5.)

44. Show that every totally ordered set is a lattice.

45. Show that every finite lattice has a least element and a greatest element.

46. Give an example of an infinite lattice with

   a) neither a least nor a greatest element.
   b) a least but not a greatest element.
   c) a greatest but not a least element.
   d) both a least and a greatest element.

47. Verify that $(\mathbf{Z}^+ \times \mathbf{Z}^+, \preccurlyeq)$ is a well-ordered set, where $\preccurlyeq$ is lexicographic order, as claimed in Example 7.

48. Determine whether each of these posets is well-ordered.

   a) $(S, \leq)$, where $\{S = 10, 11, 12, \ldots\}$
   b) $(\mathbf{Q} \cap [0, 1], \leq)$ (the set of rational numbers between 0 and 1 inclusive)
   c) $(S, \leq)$, where S is the set of positive rational numbers with denominators not exceeding 3
   d) $(\mathbf{Z}^-, \geq)$, where $\mathbf{Z}^-$ is the set of negative integers

A poset $(R, \preccurlyeq)$ is **well-founded** if there is no infinite decreasing sequence of elements in the poset, that is, elements $x_1, x_2, \ldots, x_n$ such that $\cdots \prec x_n \prec \cdots \prec x_2 \prec x_1$. A poset $(R, \preccurlyeq)$ is **dense** if for all $x \in S$ and $y \in S$ with $x \prec y$, there is an element $z \in R$ such that $x \prec z \prec y$.

49. Show that the poset $(\mathbf{Z}, \preccurlyeq)$ where $x \prec y$ if and only if $|x| < |y|$ is well-founded but is not a totally ordered set.

50. Show that a dense poset with at least two elements that are related is not well-founded.

51. Show that the poset of rational numbers with the usual "less than or equal to" relation, $(\mathbf{Q}, \leq)$, is a dense poset.

*52. Show that the set of strings of lowercase English letters with lexicographic order is neither well-founded nor dense.

53. Show that a poset is well-ordered if and only if it is totally ordered and well-founded.

54. Show that a finite nonempty poset has a maximal element.

55. Find a compatible total order for the poset with the Hasse diagram shown in Exercise 26.

56. Find a compatible total order for the divisibility relation on the set $\{1, 2, 3, 6, 8, 12, 24, 36\}$.

57. Find an order different from that constructed in Example 26 for completing the tasks in the development project.

58. Schedule the tasks needed to build a house, by specifying their order, if the Hasse diagram representing these tasks is as shown in the figure.

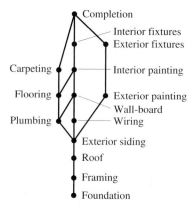

59. Find an ordering of the tasks of a software project if the Hasse diagram for the tasks of the project is as shown.

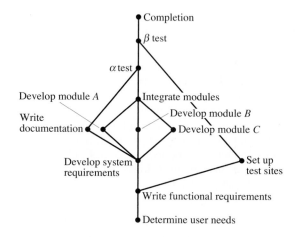

Key Terms and Results

TERMS

binary relation from A to B: a subset of $A \times B$

relation on A: a binary relation from A to itself (i.e., a subset of $A \times A$)

$S \circ R$: composite of R and S
R^{-1}: inverse relation of R
R^n: nth power of R
reflexive: a relation R on A is reflexive if $(a, a) \in R$ for all $a \in A$

symmetric: a relation R on A is symmetric if $(b, a) \in R$ whenever $(a, b) \in R$

antisymmetric: a relation R on A is antisymmetric if $a = b$ whenever $(a, b) \in R$ and $(b, a) \in R$

transitive: a relation R on A is transitive if $(a, b) \in R$ and $(b, c) \in R$ implies that $(a, c) \in R$

n-ary relation on $A_1, A_2, \ldots, A_n$: a subset of $A_1 \times A_2 \times \cdots \times A_n$

relational data model: a model for representing databases using n-ary relations

primary key: a domain of an n-ary relation such that an n-tuple is uniquely determined by its value for this domain

composite key: the Cartesian product of domains of an n-ary relation such that an n-tuple is uniquely determined by its values in these domains

selection operator: a function that selects the n-tuples in an n-ary relation that satisfy a specified condition

projection: a function that produces relations of smaller degree from an n-ary relation by deleting fields

join: a function that combines n-ary relations that agree on certain fields

directed graph or digraph: a set of elements called vertices and ordered pairs of these elements, called edges

loop: an edge of the form (a, a)

closure of a relation R with respect to a property P: the relation S (if it exists) that contains R, has property **P**, and is contained within any relation that contains R and has property **P**

path in a digraph: a sequence of edges $(a, x_1), (x_1, x_2), \ldots, (x_{n-2}, x_{n-1}), (x_{n-1}, b)$ such that the terminal vertex of each edge is the initial vertex of the succeeding edge in the sequence

circuit (or cycle) in a digraph: a path that begins and ends at the same vertex

R^* (connectivity relation): the relation consisting of those ordered pairs (a, b) such that there is a path from a to b

equivalence relation: a reflexive, symmetric, and transitive relation

equivalent: if R is an equivalence relation, a is equivalent to b if $a R b$

$[a]_R$ (equivalence class of a with respect to R): the set of all elements of A that are equivalent to a

$[a]_m$ (congruence class modulo m): the set of integers congruent to a modulo m

partition of a set S: a collection of pairwise disjoint nonempty subsets that have S as their union

partial ordering: a relation that is reflexive, antisymmetric, and transitive

poset (S, R): a set S and a partial ordering R on this set

comparable: the elements a and b in the poset $(A, \preccurlyeq)$ are comparable if $a \preccurlyeq b$ or $b \preccurlyeq a$

incomparable: elements in a poset that are not comparable

total (or linear) ordering: a partial ordering for which every pair of elements are comparable

totally (or linearly) ordered set: a poset with a total (or linear) ordering

well-ordered set: a poset $(S, \preccurlyeq)$ where $\preccurlyeq$ is a total order and every nonempty subset of S has a least element

lexicographic order: a partial ordering of Cartesian products or strings (see pages 518–519)

Hasse diagram: a graphical representation of a poset where loops and all edges resulting from the transitive property are not shown, and the direction of the edges is indicated by the position of the vertices

maximal element: an element of a poset that is not less than any other element of the poset

minimal element: an element of a poset that is not greater than any other element of the poset

greatest element: an element of a poset greater than or equal to all other elements in this set

least element: an element of a poset less than or equal to all other elements in this set

upper bound of a set: an element in a poset greater than all other elements in the set

lower bound of a set: an element in a poset less than all other elements in the set

least upper bound of a set: an upper bound of the set that is less than all other upper bounds

greatest lower bound of a set: a lower bound of the set that is greater than all other lower bounds

lattice: a partially ordered set in which every two elements have a greatest lower bound and a least upper bound

compatible total ordering for a partial ordering: a total ordering that contains the given partial ordering

topological sort: the construction of a total ordering compatible with a given partial ordering

RESULTS

The reflexive closure of a relation R on the set A equals $R \cup \Delta$, where $\Delta = \{(a, a) \mid a \in A\}$.

The symmetric closure of a relation R on the set A equals $R \cup R^{-1}$, where $R^{-1} = \{(b, a) \mid (a, b) \in R\}$.

The transitive closure of a relation equals the connectivity relation formed from this relation.

Warshall's algorithm for finding the transitive closure of a relation (see pages 503–505).

Let R be an equivalence relation. Then the following three statements are equivalent: (1) $a R b$; (2) $[a]_R \cap [b]_R \neq \emptyset$; (3) $[a]_R = [b]_R$.

The equivalence classes of an equivalence relation on a set A form a partition of A. Conversely, an equivalence relation can be constructed from any partition so that the equivalence classes are the subsets in the partition.

The principle of well-ordered induction.

The topological sorting algorithm (see pages 525–526).

Review Questions

1. a) What is a relation on a set?

b) How many relations are there on a set with n elements?

2. a) What is a reflexive relation?

b) What is a symmetric relation?

c) What is an antisymmetric relation?

d) What is a transitive relation?

3. Give an example of a relation on the set $\{1, 2, 3, 4\}$ that is

a) reflexive, symmetric, and not transitive.

b) not reflexive, symmetric, and transitive.

c) reflexive, antisymmetric, and not transitive.

d) reflexive, symmetric, and transitive.

e) reflexive, antisymmetric, and transitive.

4. a) How many reflexive relations are there on a set with n elements?

b) How many symmetric relations are there on a set with n elements?

c) How many antisymmetric relations are there on a set with n elements?

5. a) Explain how an n-ary relation can be used to represent information about students at a university.

b) How can the 5-ary relation containing names of students, their addresses, telephone numbers, majors, and grade point averages be used to form a 3-ary relation containing the names of students, their majors, and their grade point averages?

c) How can the 4-ary relation containing names of students, their addresses, telephone numbers, and majors and the 4-ary relation containing names of students, their student numbers, majors, and numbers of credit hours be combined into a single n-ary relation?

6. a) Explain how to use a zero–one matrix to represent a relation on a finite set.

b) Explain how to use the zero–one matrix representing a relation to determine whether the relation is reflexive, symmetric, and/or antisymmetric.

7. a) Explain how to use a directed graph to represent a relation on a finite set.

b) Explain how to use the directed graph representing a relation to determine whether a relation is reflexive, symmetric, and/or antisymmetric.

8. a) Define the reflexive closure and the symmetric closure of a relation.

b) How can you construct the reflexive closure of a relation?

c) How can you construct the symmetric closure of a relation?

d) Find the reflexive closure and the symmetric closure of the relation $\{(1, 2), (2, 3), (2, 4), (3, 1)\}$ on the set $\{1, 2, 3, 4\}$.

9. a) Define the transitive closure of a relation.

b) Can the transitive closure of a relation be obtained by including all pairs (a, c) such that (a, b) and (b, c) belong to the relation?

c) Describe two algorithms for finding the transitive closure of a relation.

d) Find the transitive closure of the relation $\{(1,1), (1,3), (2,1), (2,3), (2,4), (3,2), (3,4), (4,1)\}$.

10. a) Define an equivalence relation.

b) Which relations on the set $\{a, b, c, d\}$ are equivalence relations and contain (a, b) and (b, d)?

11. a) Show that congruence modulo m is an equivalence relation whenever m is a positive integer.

b) Show that the relation $\{(a, b) \mid a \equiv \pm b \pmod 7\}$ is an equivalence relation on the set of integers.

12. a) What are the equivalence classes of an equivalence relation?

b) What are the equivalence classes of the "congruent modulo 5" relation?

c) What are the equivalence classes of the equivalence relation in Question 11(b)?

13. Explain the relationship between equivalence relations on a set and partitions of this set.

14. a) Define a partial ordering.

b) Show that the divisibility relation on the set of positive integers is a partial order.

15. Explain how partial orderings on the sets A_1 and A_2 can be used to define a partial ordering on the set $A_1 \times A_2$.

16. a) Explain how to construct the Hasse diagram of a partial order on a finite set.

b) Draw the Hasse diagram of the divisibility relation on the set $\{2, 3, 5, 9, 12, 15, 18\}$.

17. a) Define a maximal element of a poset and the greatest element of a poset.

b) Give an example of a poset that has three maximal elements.

c) Give an example of a poset with a greatest element.

18. a) Define a lattice.

b) Give an example of a poset with five elements that is a lattice and an example of a poset with five elements that is not a lattice.

19. a) Show that every finite subset of a lattice has a greatest lower bound and a least upper bound.

b) Show that every lattice with a finite number of elements has a least element and a greatest element.

20. a) Define a well-ordered set.

b) Describe an algorithm for producing a well-ordered set from a partially ordered set.

c) Explain how the algorithm from (b) can be used to order the tasks in a project if each task can be done only after one or more of the other tasks have been completed.

Supplementary Exercises

1. Let S be the set of all strings of English letters. Determine whether these relations are reflexive, irreflexive, symmetric, antisymmetric, and/or transitive.
   a) $R_1 = \{(a, b) \mid a$ and b have no letters in common$\}$
   b) $R_2 = \{(a, b) \mid a$ and b are not the same length$\}$
   c) $R_3 = \{(a, b) \mid a$ is longer than $b\}$

2. Construct a relation on the set $\{a, b, c, d\}$ that is
   a) reflexive, symmetric, but not transitive.
   b) irreflexive, symmetric, and transitive.
   c) irreflexive, antisymmetric, and not transitive.
   d) reflexive, neither symmetric nor antisymmetric, and transitive.
   e) neither reflexive, irreflexive, symmetric, antisymmetric, nor transitive.

3. Show that the relation R on $\mathbf{Z} \times \mathbf{Z}$ defined by $(a, b) \, R \, (c, d)$ if and only if $a + d = b + c$ is an equivalence relation.

4. Show that a subset of an antisymmetric relation is also antisymmetric.

5. Let R be a reflexive relation on a set A. Show that $R \subseteq R^2$.

6. Suppose that R_1 and R_2 are reflexive relations on a set A. Show that $R_1 \oplus R_2$ is irreflexive.

7. Suppose that R_1 and R_2 are reflexive relations on a set A. Is $R_1 \cap R_2$ also reflexive? Is $R_1 \cup R_2$ also reflexive?

8. Suppose that R is a symmetric relation on a set A. Is $\overline{R}$ also symmetric?

9. Let R_1 and R_2 be symmetric relations. Is $R_1 \cap R_2$ also symmetric? Is $R_1 \cup R_2$ also symmetric?

10. A relation R is called **circular** if $a \, R \, b$ and $b \, R \, c$ imply that $c \, R \, a$. Show that R is reflexive and circular if and only if it is an equivalence relation.

11. Show that a primary key in an n-ary relation is a primary key in any projection of this relation that contains this key as one of its fields.

12. Is the primary key in an n-ary relation also a primary key in a larger relation obtained by taking the join of this relation with a second relation?

13. Show that the reflexive closure of the symmetric closure of a relation is the same as the symmetric closure of its reflexive closure.

14. Let R be the relation on the set of all mathematicians that contains the ordered pair (a, b) if and only if a and b have written a paper together.
    a) Describe the relation R^2.
    b) Describe the relation R^*.
    c) The **Erdős number** of a mathematician is 1 if this mathematician wrote a paper with the prolific Hungarian mathematician Paul Erdős, it is 2 if this mathematician did not write a joint paper with Erdős but wrote a joint paper with someone who wrote a joint paper with Erdős, and so on (except that the Erdős number of Erdős himself is 0). Give a definition of the Erdős number in terms of paths in R.

PAUL ERDŐS (1913–1996) Paul Erdős, born in Budapest, Hungary, was the son of two high school mathematics teachers. He was a child prodigy; at age 3 he could multiply three-digit numbers in his head, and at 4 he discovered negative numbers on his own. Because his mother did not want to expose him to contagious diseases, he was mostly home-schooled. At 17 Erdős entered Eőtvős University, graduating four years later with a Ph.D. in mathematics. After graduating he spent four years at Manchester, England, on a postdoctoral fellowship. In 1938 he went to the United States because of the difficult political situation in Hungary, especially for Jews. He spent much of his time in the United States, except for 1954 to 1962, when he was banned as part of the paranoia of the McCarthy era. He also spent considerable time in Israel.

Erdős made many significant contributions to combinatorics and to number theory. One of the discoveries of which he was most proud is his elementary proof (in the sense that it does not use any complex analysis) of the Prime Number Theorem, which provides an estimate for the number of primes not exceeding a fixed positive integer. He also participated in the modern development of Ramsey theory.

Erdős traveled extensively throughout the world to work with other mathematicians, visiting conferences, universities, and research laboratories. He almost entirely devoted himself to mathematics, traveling from one mathematician to the next, proclaiming "My brain is open." Erdős was the author or coauthor of more than 1500 papers and had more than 500 coauthors. He had no permanent home. Copies of his articles are kept by Ron Graham, a famous discrete mathematician with whom he collaborated extensively and who took care of many of his worldly needs.

Erdős offered rewards, ranging from $10 to $10,000, for the solution of problems that he found particularly interesting, with the size of the reward depending on the difficulty of the problem. He paid out close to $4000. Erdős had his own special language, using such terms as "epsilon" (child), "boss" (woman), "slave" (man), "captured" (married), "liberated" (divorced), "Supreme Fascist" (God), "Sam" (United States), and "Joe" (Soviet Union). Although he was curious about many things, he concentrated almost all his energy on mathematical research. He had no hobbies and no full-time job. He never married and apparently remained celibate. Erdős was extremely generous, donating much of the money he collected from prizes, awards, and stipends for scholarships and to worthwhile causes. He traveled extremely lightly and did not like having many material possessions.

15. a) Give an example to show that the transitive closure of the symmetric closure of a relation is not necessarily the same as the symmetric closure of the transitive closure of this relation.

b) Show, however, that the transitive closure of the symmetric closure of a relation must contain the symmetric closure of the transitive closure of this relation.

16. a) Let S be the set of subroutines of a computer program. Define the relation R by **P** R **Q** if subroutine **P** calls subroutine **Q** during its execution. Describe the transitive closure of R.

b) For which subroutines **P** does (**P**, **P**) belong to the transitive closure of R?

c) Describe the reflexive closure of the transitive closure of R.

17. Suppose that R and S are relations on a set A with $R \subseteq S$ such that the closures of R and S with respect to a property **P** both exist. Show that the closure of R with respect to **P** is a subset of the closure of S with respect to **P**.

18. Show that the symmetric closure of the union of two relations is the union of their symmetric closures.

*19. Devise an algorithm, based on the concept of interior vertices, that finds the length of the longest path between two vertices in a directed graph, or determines that there are arbitrarily long paths between these vertices.

20. Which of these are equivalence relations on the set of all people?

a) $\{(x, y) \mid x$ and y have the same sign of the zodiac$\}$

b) $\{(x, y) \mid x$ and y were born in the same year$\}$

c) $\{(x, y) \mid x$ and y have been in the same city$\}$

*21. How many different equivalence relations with exactly three different equivalence classes are there on a set with five elements?

22. Show that $\{(x, y) \mid x - y \in \mathbf{Q}\}$ is an equivalence relation on the set of real numbers where $\mathbf{Q}$ denotes the set of rational numbers. What are $[1]$, $[\frac{1}{2}]$, and $[\pi]$?

23. Suppose that $P_1 = \{A_1, A_2, \ldots, A_m\}$ and $P_2 = \{B_1, B_2, \ldots, B_n\}$ are both partitions of the set S. Show that the collection of nonempty subsets of the form $A_i \cap B_j$ is a partition of S that is a refinement of both P_1 and P_2 (see the preamble to Exercise 37 of Section 7.5).

*24. Show that the transitive closure of the symmetric closure of the reflexive closure of a relation R is the smallest equivalence relation that contains R.

25. Let $\mathbf{R}(S)$ be the set of all relations on a set S. Define the relation $\preccurlyeq$ on $\mathbf{R}(S)$ by $R_1 \preccurlyeq R_2$ if $R_1 \subseteq R_2$, where R_1 and R_2 are relations on S. Show that $(\mathbf{R}(S), \preccurlyeq)$ is a poset.

26. Let $\mathbf{P}(S)$ be the set of all partitions of the set S. Define the relation $\preccurlyeq$ on $\mathbf{P}(S)$ by $P_1 \preccurlyeq P_2$ if P_1 is a refinement of P_2 (see Exercise 37 of Section 7.5). Show that $(\mathbf{P}(S), \preccurlyeq)$ is a poset.

27. Schedule the tasks needed to cook a Chinese meal by specifying their order, if the Hasse diagram representing these tasks is as shown here.

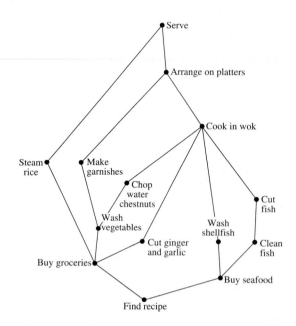

A subset of a poset such that every two elements of this subset are comparable is called a **chain.** A subset of a poset is called an **antichain** if every two elements of this subset are incomparable.

28. Find all chains in the posets with the Hasse diagrams shown in Exercises 19–21 in Section 7.6.

29. Find all antichains in the posets with the Hasse diagrams shown in Exercises 19–21 in Section 7.6.

30. Find an antichain with the greatest number of elements in the poset with the Hasse diagram of Exercise 26 in Section 7.6.

31. Show that every maximal chain in a finite poset $(S, \preccurlyeq)$ contains a minimal element of S. (A maximal chain is a chain that is not a subset of a larger chain.)

**32. Show that a poset can be partitioned into k chains, where k is the largest number of elements in an antichain in this poset.

*33. Show that in any group of $mn + 1$ people there is either a list of $m + 1$ people where a person in the list (except for the first person listed) is a descendant of the previous person on the list, or there are $n + 1$ people such that none of these people is a descendant of any of the other n people. (*Hint:* Use Exercise 32.)

Suppose that $(S, \preccurlyeq)$ is a well-founded partially ordered set. The *principle of well-founded induction* states that $P(x)$ is true for all $x \in S$ if $\forall x(\forall y(y \prec x \rightarrow P(y)) \rightarrow P(x))$.

34. Show that no separate basis case is needed for the principle of well-founded induction. That is, $P(u)$ is true for all minimal elements u in S if $\forall x (\forall y (y \prec x \rightarrow P(y)) \rightarrow P(x))$.

*35. Show that the principle of well-founded induction is valid.

A relation R on a set A is a **quasi-ordering** on A if R is reflexive and transitive.

36. Let R be the relation on the set of all functions from $\mathbf{Z}^+$ to $\mathbf{Z}^+$ such that (f, g) belongs to R if and only if f is $O(g)$. Show that R is a quasi-ordering.

37. Let R be a quasi-ordering on a set A. Show that $R \cap R^{-1}$ is an equivalence relation.

*38. Let R be a quasi-ordering and let S be the relation on the set of equivalence classes of $R \cap R^{-1}$ such that (C, D) belongs to S, where C and D are equivalence classes of R, if and only if there are elements c of C and d of D such that (c, d) belongs to R. Show that S is a partial ordering.

Let L be a lattice. Define the **meet** ($\wedge$) and **join** ($\vee$) operations by $x \wedge y = \text{glb}(x, y)$ and $x \vee y = \text{lub}(x, y)$.

39. Show that the following properties hold for all elements x, y, and z of a lattice L.

   a) $x \wedge y = y \wedge x$ and $x \vee y = y \vee x$ (**commutative laws**)

   b) $(x \wedge y) \wedge z = x \wedge (y \wedge z)$ and $(x \vee y) \vee z = x \vee (y \vee z)$ (**associative laws**)

   c) $x \wedge (x \vee y) = x$ and $x \vee (x \wedge y) = x$ (**absorption laws**)

   d) $x \wedge x = x$ and $x \vee x = x$ (**idempotent laws**)

40. Show that if x and y are elements of a lattice L, then $x \vee y = y$ if and only if $x \wedge y = x$.

A lattice L is **bounded** if it has both an **upper bound**, denoted by 1, such that $x \preccurlyeq 1$ for all $x \in L$ and a **lower bound**, denoted by 0, such that $0 \preccurlyeq x$ for all $x \in L$.

41. Show that if L is a bounded lattice with upper bound 1 and lower bound 0 then these properties hold for all elements $x \in L$.

   a) $x \vee 1 = 1$ **b)** $x \wedge 1 = x$
   c) $x \vee 0 = x$ **d)** $x \wedge 0 = 0$

42. Show that every finite lattice is bounded.

A lattice is called **distributive** if $x \vee (y \wedge z) = (x \vee y) \wedge (x \vee z)$ and $x \wedge (y \vee z) = (x \wedge y) \vee (x \wedge z)$ for all x, y, and z in L.

*43. Give an example of a lattice that is not distributive.

44. Show that the lattice $(P(S), \subseteq)$ where $P(S)$ is the power set of a finite set S is distributive.

45. Is the lattice $(\mathbf{Z}^+, |)$ distributive?

The **complement** of an element a of a bounded lattice L with upper bound 1 and lower bound 0 is an element b such that $a \vee b = 1$ and $a \wedge b = 0$. Such a lattice is **complemented** if every element of the lattice has a complement.

46. Give an example of a finite lattice where at least one element has more than one complement and at least one element has no complement.

47. Show that the lattice $(P(S), \subseteq)$ where $P(S)$ is the power set of a finite set S is complemented.

*48. Show that if L is a finite distributive lattice, then an element of L has at most one complement.

Computer Projects

WRITE PROGRAMS WITH THESE INPUT AND OUTPUT.

1. Given the matrix representing a relation on a finite set, determine whether the relation is reflexive and/or irreflexive.

2. Given the matrix representing a relation on a finite set, determine whether the relation is symmetric and/or antisymmetric.

3. Given the matrix representing a relation on a finite set, determine whether the relation is transitive.

4. Given a positive integer n, display all the relations on a set with n elements.

* 5. Given a positive integer n, determine the number of transitive relations on a set with n elements.

* 6. Given a positive integer n, determine the number of equivalence relations on a set with n elements.

* 7. Given a positive integer n, display all the equivalence relations on the set of the n smallest positive integers.

8. Given an n-ary relation, find the projection of this relation when specified fields are deleted.

9. Given an m-ary relation and an n-ary relation, and a set of common fields, find the join of these relations with respect to these common fields.

10. Given the matrix representing a relation on a finite set, find the matrix representing the reflexive closure of this relation.

11. Given the matrix representing a relation on a finite set, find the matrix representing the symmetric closure of this relation.

12. Given the matrix representing a relation on a finite set, find the matrix representing the transitive closure of this relation by computing the join of the powers of the matrix representing the relation.

13. Given the matrix representing a relation on a finite set, find the matrix representing the transitive closure of this relation using Warshall's algorithm.

14. Given the matrix representing a relation on a finite set, find the matrix representing the smallest equivalence relation containing this relation.

15. Given a partial ordering on a finite set, find a total ordering compatible with it using topological sorting.

Computations and Explorations

USE A COMPUTATIONAL PROGRAM OR PROGRAMS YOU HAVE WRITTEN TO DO THESE EXERCISES.

1. Display all the different relations on a set with four elements.

2. Display all the different reflexive and symmetric relations on a set with six elements.

3. Display all the reflexive and transitive relations on a set with five elements.

*4. Determine how many transitive relations there are on a set with n elements for all positive integers n with $n \leq 7$.

5. Find the transitive closure of a relation of your choice on a set with at least 20 elements. Either use a relation that corresponds to direct links in a particular transportation or communications network or use a randomly generated relation.

6. Compute the number of different equivalence relations on a set with n elements for all positive integers n not exceeding 20.

7. Display all the equivalence relations on a set with seven elements.

*8. Display all the partial orders on a set with five elements.

*9. Display all the lattices on a set with five elements.

Writing Projects

RESPOND TO THESE QUESTIONS WITH ESSAYS USING OUTSIDE SOURCES.

1. Discuss the concept of a fuzzy relation. How are fuzzy relations used?

2. Describe the basic principles of relational databases, going beyond what was covered in Section 7.2. How widely used are relational databases as compared with other types of databases?

3. Look up the original papers by Warshall and by Roy (in French) in which they develop algorithms for finding transitive closures. Discuss their approaches. Why do you suppose that what we call Warshall's algorithm was discovered independently by more than one person?

4. Describe how equivalence classes can be used to define the rational numbers as classes of pairs of integers and how the basic arithmetic operations on rational numbers can be defined following this approach. (See Exercise 10 in Section 7.5.)

5. Explain how Helmut Hasse used what we now call Hasse diagrams.

6. Describe some of the mechanisms used to enforce information flow policies in computer operating systems.

7. Discuss the use of the Program Evaluation and Review Technique (PERT) to schedule the tasks of a large complicated project. How widely is PERT used?

8. Discuss the use of the Critical Path Method (CPM) to find the shortest time for the completion of a project. How widely is CPM used?

9. Discuss the concept of *duality* in a lattice. Explain how duality can be used to establish new results.

10. Explain what is meant by a *modular lattice*. Describe some of the properties of modular lattices and describe how modular lattices arise in the study of projective geometry.

8 Graphs

Graph theory is an old subject with many modern applications. Its basic ideas were introduced in the eighteenth century by the great Swiss mathematician Leonhard Euler. He used graphs to solve the famous Königsberg bridge problem, which we will discuss in this chapter.

Graphs are used to solve problems in many fields. For instance, graphs can be used to determine whether a circuit can be implemented on a planar circuit board. We can distinguish between two chemical compounds with the same molecular formula but different structures using graphs. Graphs can be used to study the structure of the World Wide Web. We can determine whether two computers are connected by a communications link using graph models of computer networks. Graphs with weights assigned to their edges can be used to solve problems such as finding the shortest path between two cities in a transportation network. We can also use graphs to schedule exams and assign channels to television stations.

8.1 Introduction to Graphs

Links

Graphs are discrete structures consisting of vertices and edges that connect these vertices. There are several different types of graphs that differ with respect to the kind and number of edges that can connect a pair of vertices. Problems in almost every conceivable discipline can be solved using graph models. We will give examples to show how graphs are used as models in a variety of areas. For instance, we will show how graphs are used to represent the competition of different species in an ecological niche, how graphs are used to represent who influences whom in an organization, and how graphs are used to represent the outcome of tournaments. Later we will show how graphs can be used to solve many types of problems, such as computing the number of different combinations of flights between two cities in an airline network, determining whether it is possible to walk down all the streets in a city without going down a street twice, and finding the number of colors needed to color the regions of a map.

TYPES OF GRAPHS

We will introduce the different types of graphs by showing how each can be used to model a computer network. Suppose that a network is made up of computers and telephone lines between computers. We can represent the location of each computer by a point and each telephone line by an arc, as shown in Figure 1.

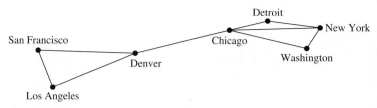

FIGURE 1   A Computer Network.

In the network in Figure 1 there is at most one telephone line between two computers, each line operates in both directions, and no computer has a telephone line to itself. Consequently this network can be modeled using a **simple graph,** consisting of **vertices** that represent the computers and **undirected edges** that represent telephone lines. Each edge connects two distinct vertices and no two edges connect the same pair of vertices.

DEFINITION 1

A *simple graph* $G = (V, E)$ consists of V, a nonempty set of *vertices,* and E, a set of unordered pairs of distinct elements of V called *edges.*

When there is heavy traffic, there may be multiple telephone lines between computers in a network, as illustrated in Figure 2. Simple graphs are not sufficient to model such networks. Instead, we use **multigraphs,** which consist of vertices and undirected edges between these vertices, with multiple edges between pairs of vertices allowed. Every simple graph is also a multigraph. However, not all multigraphs are simple graphs, since in a multigraph two or more edges may connect the same pair of vertices.

We cannot use a pair of vertices to specify an edge of a graph when multiple edges are present. This makes the formal definition of multigraphs somewhat complicated.

DEFINITION 2

A *multigraph* $G = (V, E)$ consists of a set V of vertices, a set E of edges, and a function f from E to $\{\{u, v\} \mid u, v \in V, u \neq v\}$. The edges e_1 and e_2 are called *multiple* or *parallel edges* if $f(e_1) = f(e_2)$.

The reader should note that multiple edges in a multigraph are associated to the same pair of vertices. However, we will say that $\{u, v\}$ is an edge of a graph $G = (V, E)$ if there is at least one edge e with $f(e) = \{u, v\}$. We will not distinguish between the edge e and the set $\{u, v\}$ associated to it unless the identity of individual multiple edges is important.

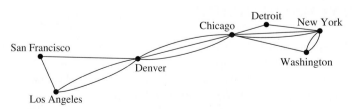

FIGURE 2   A Computer Network with Multiple Lines.

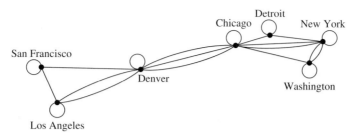

FIGURE 3   **A Computer Network with Diagnostic Lines.**

A computer network may contain a telephone line from a computer to itself (perhaps for diagnostic purposes), as illustrated in Figure 3. We cannot use multigraphs to model such networks, since **loops,** which are edges from a vertex to itself, are not allowed in multigraphs. Instead, we use **pseudographs,** which are more general than multigraphs, since an edge in a pseudograph may connect a vertex with itself.

To formally define pseudograph we must be able to associate edges to sets containing just one vertex.

DEFINITION 3   A *pseudograph* $G = (V, E)$ consists of a set V of vertices, a set E of edges, and a function f from E to $\{\{u, v\} \mid u, v \in V\}$. An edge is a *loop* if $f(e) = \{u, u\} = \{u\}$ for some $u \in V$.

Telephone lines in a computer network may not operate in both directions. For instance, in Figure 4 the host computer in New York can only receive data from other computers and cannot send out data. Other telephone lines operating in both directions are represented by pairs of edges in opposite directions.

We use directed graphs (which were studied in Chapter 7) to model such networks. The edges of a directed graph are ordered pairs. Loops, ordered pairs of the same element, are allowed, but multiple edges in the same direction between two vertices are not. Recall this definition.

DEFINITION 4   A *directed graph* (V, E) consists of a set of vertices V and a set of edges E that are ordered pairs of elements of V.

We use an arrow pointing from u to v to indicate the direction of the edge (u, v).

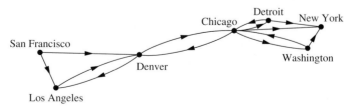

FIGURE 4   **A Communications Network with One-Way Telephone Lines.**

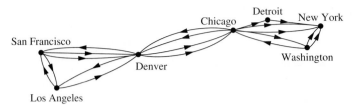

FIGURE 5   A Computer Network with Multiple One-Way Lines.

Finally, multiple lines may be present in the computer network, so that there may be several one-way lines to the host in New York from each location, and perhaps more than one line back to each remote computer from the host, as illustrated in Figure 5. We use **directed multigraphs,** which may have multiple **directed edges** from a vertex to a second (possibly the same) vertex, to model such networks. We now present the formal definition of a directed multigraph.

DEFINITION 5   A *directed multigraph* $G = (V, E)$ consists of a set V of vertices, a set E of edges, and a function f from E to $\{(u, v) \mid u, v \in V\}$. The edges e_1 and e_2 are *multiple edges* if $f(e_1) = f(e_2)$.

The reader should note that multiple directed edges are associated to the same pair of vertices. However, we will say that (u, v) is an edge of $G = (V, E)$ as long as there is at least one edge e with $f(e) = (u, v)$. We will not make the distinction between the edge e and the ordered pair (u, v) associated to it unless the identity of individual multiple edges is important.

This terminology for the various types of graphs is summarized in Table 1. We will use **graph** to describe graphs with directed or undirected edges, with or without loops and multiple edges. We will use the terms **undirected graph** or **pseudograph** for an undirected graph that may have multiple edges and loops. We will always use the adjective **directed** when referring to graphs that have ordered pairs associated to their edges.

Because of the relatively modern interest in graph theory, and because it has applications to a wide variety of disciplines, many different terminologies of graph theory are commonly used. The reader should determine how such terms are being used whenever they are encountered.

Type	Edges	Multiple Edges Allowed?	Loops Allowed?
Simple graph	Undirected	No	No
Multigraph	Undirected	Yes	No
Pseudograph	Undirected	Yes	Yes
Directed graph	Directed	No	Yes
Directed multigraph	Directed	Yes	Yes

TABLE 1 Graph Terminology.

GRAPH MODELS

Links

Graphs are used in a wide variety of models. We will present a few graph models from diverse fields here. Others will be introduced in subsequent sections of this and later chapters.

EXAMPLE 1

Links

Niche Overlap Graphs in Ecology Graphs are used in many models involving the interaction of different species of animals. For instance, the competition between species in an ecosystem can be modeled using a **niche overlap graph.** Each species is represented by a vertex. An undirected edge connects two vertices if the two species represented by these vertices compete (that is, some of the food resources they use are the same). The graph in Figure 6 models the ecosystem of a forest. We see from this graph that squirrels and raccoons compete but that crows and shrews do not. ◄

EXAMPLE 2

Links

Acquaintanceship Graphs We can use graph models to represent various relationships between people. For example, we can use a graph to represent whether two people know each other, that is, whether they are acquainted. Each person in a particular group of people is represented by a vertex. An undirected edge is used to connect two people when these people know each other. A small acquaintanceship graph is shown in Figure 7. The acquaintanceship graph of all people in the world has more than 6 billion vertices and probably more than 1 trillion edges! We will discuss this graph further in Section 8.4. ◄

EXAMPLE 3

Influence Graphs In studies of group behavior it is observed that certain people can influence the thinking of others. A directed graph called an **influence graph** can be used to model this behavior. Each person of the group is represented by a vertex. There is a directed edge from vertex a to vertex b when the person represented by vertex a influences the person represented by vertex b. An example of an influence graph for members of a group is shown in Figure 8. In the group modeled by this influence graph, Deborah can influence Brian, Fred, and Linda, but no one can influence her. Also, Yvonne and Brian can influence each other. ◄

EXAMPLE 4

Links

The Hollywood Graph The **Hollywood graph** represents actors by vertices and connects two vertices when the actors represented by these vertices have worked together on a movie. According to the Internet Movie Database, in November 2001 the Hollywood graph had 574,724 vertices representing actors who have appeared in 292,609 films, and had more than 16 million edges. We will discuss some aspects of the Hollywood graph later in Section 8.4. ◄

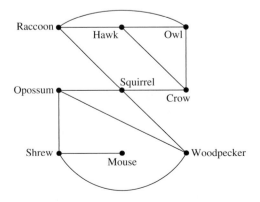

FIGURE 6   A Niche Overlap Graph.

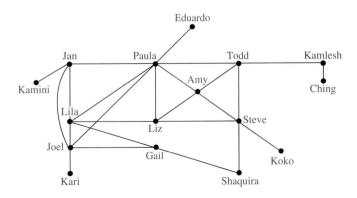

FIGURE 7   An Acquaintanceship Graph.

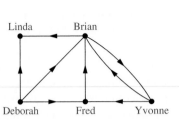

FIGURE 8   An Influence Graph.

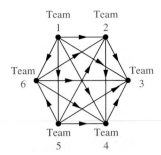

**FIGURE 9   A Graph Model of
a Round-Robin Tournament.**

EXAMPLE 5   **Round-Robin Tournaments** A tournament where each team plays each other team exactly once is called a **round-robin tournament.** Such tournaments can be modeled using directed graphs where each team is represented by a vertex. Note that (a, b) is an edge if team a beats team b. Such a directed graph model is presented in Figure 9. We see that Team 1 is undefeated in this tournament, and Team 3 is winless.   ◄

EXAMPLE 6   **Collaboration Graphs** A graph called a **collaboration graph** can be used to model joint authorship of academic papers. In a collaboration graph vertices represent people (perhaps restricted to members of a certain academic community) and edges link two people if they have jointly written a paper. The collaboration graph for people working together on research papers in mathematics has been found to have more than 337,000 vertices and 496,000 edges. We will have more to say about this graph in Section 8.4.   ◄

EXAMPLE 7   **Call Graphs** Graphs can be used to model telephone calls made in a network, such as a long-distance telephone network. In particular, a directed multigraph can be used to model calls where each telephone number is represented by a vertex and each telephone call is represented by a directed edge. The edge representing a call starts at a telephone number from which the call was made and ends at the telephone number to which the call was made.

A small telephone call graph is displayed in Figure 10(a), representing seven telephone numbers. This graph shows, for instance, that three calls have been made from 732-555-1234 to 732-555-9876 and two in the other direction, but no calls have been made from 732-555-4444 to any of the other six numbers except 732-555-0011. When we care only whether there has been a call connecting two telephone numbers, we use an undirected graph with an edge connecting telephone numbers when there has been a call between these numbers. This version of the call graph is displayed in Figure 10(b).   ◄

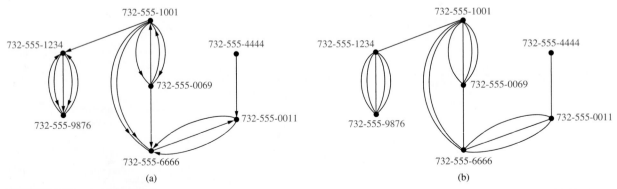

FIGURE 10   A Call Graph.

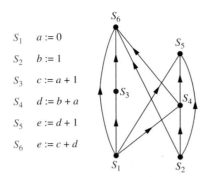

S_1 $a := 0$

S_2 $b := 1$

S_3 $c := a + 1$

S_4 $d := b + a$

S_5 $e := d + 1$

S_6 $e := c + d$

FIGURE 11 **A Precedence Graph.**

Call graphs that model actual calling activities can be huge. For example, one call graph studied at AT&T, which models calls made in a 20-day period, has about 290 million vertices and 4 billion edges. We will discuss call graphs further in Section 8.4.

EXAMPLE 8 **The Web Graph** The World Wide Web can be modeled as a directed graph where each Web page is represented by a vertex and where an edge starts at the Web page a and ends at the Web page b if there is a link on a pointing to b. Since new Web pages are created and others removed somewhere on the Web almost every second, the Web graph changes on an almost continual basis. Currently the Web graph has more than a billion vertices and tens of billions of edges. Many people are studying the properties of the Web graph to better understand the nature of the Web. We will return to Web graphs in Section 8.4, and in Chapter 9 we will explain how the Web graph is used by the Web crawlers that search engines use to create indices of Web pages. ◀

EXAMPLE 9 **Precedence Graphs and Concurrent Processing** Computer programs can be executed more rapidly by executing certain statements concurrently. It is important not to execute a statement that requires results of statements not yet executed. The dependence of statements on previous statements can be represented by a directed graph. Each statement is represented by a vertex, and there is an edge from one vertex to a second vertex if the statement represented by the second vertex cannot be executed before the statement represented by the first vertex has been executed. This graph is called a **precedence graph.** A computer program and its graph are displayed in Figure 11. For instance, the graph shows that statement S_5 cannot be executed before statements S_1, S_2, and S_4 are executed. ◀

Exercises

1. Draw graph models, stating the type of graph used, to represent airline routes where every day there are four flights from Boston to Newark, two flights from Newark to Boston, three flights from Newark to Miami, two flights from Miami to Newark, one flight from Newark to Detroit, two flights from Detroit to Newark, three flights from Newark to Washington, two flights from Washington to Newark, and one flight from Washington to Miami, with

a) an edge between vertices representing cities that have a flight between them (in either direction).

b) an edge between vertices representing cities for each flight that operates between them (in either direction).

c) an edge between vertices representing cities for each flight that operates between them (in either direction), plus a loop for a special sightseeing trip that takes off and lands in Miami.

d) an edge from a vertex representing a city where a flight starts to the vertex representing the city where it ends.

e) an edge for each flight from a vertex representing a city where the flight begins to the vertex representing the city where the flight ends.

2. What kind of graph can be used to model a highway system between major cities where

a) there is an edge between the vertices representing cities if there is an interstate highway between them?

b) there is an edge between the vertices representing cities for each interstate highway between them?

c) there is an edge between the vertices representing cities for each interstate highway between them, and there is a loop at the vertex representing a city if there is an interstate highway that circles this city?

For Exercises 3–9, determine whether the graph shown is a simple graph, a multigraph (and not a simple graph), a pseudograph (and not a multigraph), a directed graph, or a directed multigraph (and not a directed graph).

3.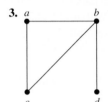

4.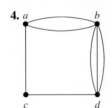

5.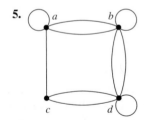

6.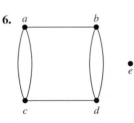

7.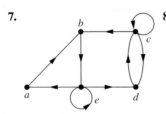

8.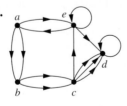

9.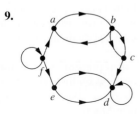

10. For each undirected graph in Exercises 3–9 that is not simple, find a set of edges to remove to make it simple.

11. The **intersection graph** of a collection of sets A_1, $A_2, \ldots, A_n$ is the graph that has a vertex for each of these sets and has an edge connecting the vertices representing two sets if these sets have a nonempty intersection. Construct the intersection graph of these collections of sets.

a) $A_1 = \{0, 2, 4, 6, 8\}$, $A_2 = \{0, 1, 2, 3, 4\}$,
$A_3 = \{1, 3, 5, 7, 9\}$, $A_4 = \{5, 6, 7, 8, 9\}$,
$A_5 = \{0, 1, 8, 9\}$

b) $A_1 = \{\ldots, -4, -3, -2, -1, 0\}$,
$A_2 = \{\ldots, -2, -1, 0, 1, 2, \ldots\}$,
$A_3 = \{\ldots, -6, -4, -2, 0, 2, 4, 6, \ldots\}$,
$A_4 = \{\ldots, -5, -3, -1, 1, 3, 5, \ldots\}$,
$A_5 = \{\ldots, -6, -3, 0, 3, 6, \ldots\}$

c) $A_1 = \{x \mid x < 0\}$,
$A_2 = \{x \mid -1 < x < 0\}$,
$A_3 = \{x \mid 0 < x < 1\}$,
$A_4 = \{x \mid -1 < x < 1\}$,
$A_5 = \{x \mid x > -1\}$,
$A_6 = \mathbf{R}$

12. Use the niche overlap graph in Figure 6 to determine the species that compete with hawks.

13. Construct a niche overlap graph for six species of birds, where the hermit thrush competes with the robin and with the blue jay, the robin also competes with the mockingbird, the mockingbird also competes with the blue jay, and the nuthatch competes with the hairy woodpecker.

14. Draw the acquaintanceship graph that represents that Tom and Patricia, Tom and Hope, Tom and Sandy, Tom and Amy, Tom and Marika, Jeff and Patricia, Jeff and Mary, Patricia and Hope, Amy and Hope, and Amy and Marika know each other, but none of the other pairs of people listed know each other.

15. We can use a graph to represent whether two people were alive at the same time. Draw such a graph to represent whether each pair of the mathematicians and computer scientists with biographies in the first four chapters of this book who died before 1900 were contemporaneous. (Assume two people lived at the same time if they were alive during the same year.)

16. Who can influence Fred and whom can Fred influence in the influence graph in Example 3?

17. Construct an influence graph for the board members of a company if the President can influence the Director of Research and Development, the Director of Marketing, and the Director of Operations; the Director of Research and Development can influence the Director of Operations; the Director of Marketing can influence the Director of Operations; and no one can influence, or be influenced by, the Chief Financial Officer.

18. Which other teams did Team 4 beat and which teams beat Team 4 in the round-robin tournament represented by the graph in Figure 9?

19. In a round-robin tournament the Tigers beat the Blue Jays, the Tigers beat the Cardinals, the Tigers beat the Orioles, the Blue Jays beat the Cardinals, the Blue Jays beat the Orioles, and the Cardinals beat the Orioles. Model this outcome with a directed graph.

20. Draw the call graph for the telephone numbers 555-0011, 555-1221, 555-1333, 555-8888, 555-2222, 555-0091, and 555-1200 if there were three calls from 555-0011 to 555-8888 and two calls from 555-8888 to 555-0011, two calls from 555-2222 to 555-0091, two calls from 555-1221 to each of the other numbers, and one call from 555-1333 to each of 555-0011, 555-1221, and 555-1200.

21. Explain how the two telephone call graphs for calls made during the month of January and calls made during the month of February can be used to determine the new telephone numbers of people who have changed their telephone numbers.

22. a) Explain how graphs can be used to model electronic mail messages in a network.
  b) Describe a graph that models the electronic mail sent in a network in a particular week.

23. How can a graph that models e-mail messages sent in a network be used to find people who have recently changed their primary e-mail address?

24. How can a graph that models e-mail messages sent in a network be used to find electronic mail mailing lists used to send the same message to many different e-mail addresses?

25. Describe a graph model that represents marriages between people. Does this graph have any special properties?

26. Which statements must be executed before S_6 is executed in the program in Example 9? (Use the precedence graph in Figure 11.)

27. Construct a precedence graph for the following program:

$$S_1: x := 0$$
$$S_2: x := x + 1$$
$$S_3: y := 2$$
$$S_4: z := y$$
$$S_5: x := x + 2$$
$$S_6: y := x + z$$
$$S_7: z := 4$$

28. Describe a discrete structure based on a graph that can be used to model airline routes and their flight times. (*Hint:* Add structure to a directed graph.)

29. Describe a discrete structure based on a graph that can be used to model relationships between pairs of individuals in a group, where each individual may either like, dislike, or be neutral about another individual, and the reverse relationship may be different. (*Hint:* Add structure to a directed graph. Treat separately the edges in opposite directions between vertices representing two individuals.)

8.2 Graph Terminology

INTRODUCTION

Links

We introduce some of the basic vocabulary of graph theory in this section. We will use this vocabulary when we solve many different types of problems. One such problem involves determining whether a graph can be drawn in the plane so that no two of its edges cross. Another example is deciding whether there is a one-to-one correspondence between the vertices of two graphs that produces a one-to-one correspondence between the edges of the graphs. We will also introduce several important families of graphs often used as examples and in models.

BASIC TERMINOLOGY

First, we give some terminology that describes the vertices and edges of undirected graphs.

DEFINITION 1

Two vertices u and v in an undirected graph G are called *adjacent* (or *neighbors*) in G if $\{u, v\}$ is an edge of G. If $e = \{u, v\}$, the edge e is called *incident with* the vertices u and v. The edge e is also said to *connect* u and v. The vertices u and v are called *endpoints* of the edge $\{u, v\}$.

To keep track of how many edges are incident to a vertex, we make the following definition.

DEFINITION 2

The *degree of a vertex in an undirected graph* is the number of edges incident with it, except that a loop at a vertex contributes twice to the degree of that vertex. The degree of the vertex v is denoted by $\deg(v)$.

EXAMPLE 1

What are the degrees of the vertices in the graphs G and H displayed in Figure 1?

Solution: In G, $\deg(a) = 2$, $\deg(b) = \deg(c) = \deg(f) = 4$, $\deg(d) = 1$, $\deg(e) = 3$, and $\deg(g) = 0$. In H, $\deg(a) = 4$, $\deg(b) = \deg(e) = 6$, $\deg(c) = 1$, and $\deg(d) = 5$.
◀

A vertex of degree zero is called **isolated.** It follows that an isolated vertex is not adjacent to any vertex. Vertex g in graph G in Example 1 is isolated. A vertex is **pendant** if and only if it has degree one. Consequently, a pendant vertex is adjacent to exactly one other vertex. Vertex d in graph G in Example 1 is pendant.

What do we get when we add the degrees of all the vertices of a graph $G = (V, E)$? Each edge contributes two to the sum of the degrees of the vertices since an edge is incident with exactly two (possibly equal) vertices. This means that the sum of the degrees of the vertices is twice the number of edges. We have the result in Theorem 1, which is sometimes called the Handshaking Theorem, because of the analogy between an edge having two endpoints and a handshake involving two hands.

THEOREM 1

THE HANDSHAKING THEOREM Let $G = (V, E)$ be an undirected graph with e edges. Then

$$2e = \sum_{v \in V} \deg(v).$$

(Note that this applies even if multiple edges and loops are present.)

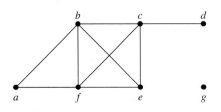

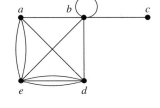

FIGURE 1 **The Undirected Graphs G and H.**

EXAMPLE 2 How many edges are there in a graph with ten vertices each of degree six?

Solution: Since the sum of the degrees of the vertices is $6 \cdot 10 = 60$, it follows that $2e = 60$. Therefore, $e = 30$. ◄

Theorem 1 shows that the sum of the degrees of the vertices of an undirected graph is even. This simple fact has many consequences, one of which is given as Theorem 2.

THEOREM 2   An undirected graph has an even number of vertices of odd degree.

Proof: Let V_1 and V_2 be the set of vertices of even degree and the set of vertices of odd degree, respectively, in an undirected graph $G = (V, E)$. Then

$$2e = \sum_{v \in V} \deg(v) = \sum_{v \in V_1} \deg(v) + \sum_{v \in V_2} \deg(v).$$

Since $\deg(v)$ is even for $v \in V_1$, the first term in the right-hand side of the last equality is even. Furthermore, the sum of the two terms on the right-hand side of the last equality is even, since this sum is $2e$. Hence, the second term in the sum is also even. Since all the terms in this sum are odd, there must be an even number of such terms. Thus, there are an even number of vertices of odd degree. ◁

Terminology for graphs with directed edges reflects the fact that edges in directed graphs have directions.

DEFINITION 3   When (u, v) is an edge of the graph G with directed edges, u is said to be *adjacent to* v and v is said to be *adjacent from u.* The vertex u is called the *initial vertex* of (u, v), and v is called the *terminal* or *end vertex* of (u, v). The initial vertex and terminal vertex of a loop are the same.

Since the edges in graphs with directed edges are ordered pairs, the definition of the degree of a vertex can be refined to reflect the number of edges with this vertex as the initial vertex and as the terminal vertex.

DEFINITION 4   In a graph with directed edges the *in-degree of a vertex v*, denoted by $\deg^-(v)$, is the number of edges with v as their terminal vertex. The *out-degree of v*, denoted by $\deg^+(v)$, is the number of edges with v as their initial vertex. (Note that a loop at a vertex contributes 1 to both the in-degree and the out-degree of this vertex.)

EXAMPLE 3 Find the in-degree and out-degree of each vertex in the graph G with directed edges shown in Figure 2.

Solution: The in-degrees in G are: $\deg^-(a) = 2$, $\deg^-(b) = 2$, $\deg^-(c) = 3$, $\deg^-(d) = 2$, $\deg^-(e) = 3$, and $\deg^-(f) = 0$. The out-degrees are: $\deg^+(a) = 4$, $\deg^+(b) = 1$, $\deg^+(c) = 2$, $\deg^+(d) = 2$, $\deg^+(e) = 3$, and $\deg^+(f) = 0$. ◄

Since each edge has an initial vertex and a terminal vertex, the sum of the in-degrees and the sum of the out-degrees of all vertices in a graph with directed edges are the same. Both of these sums are the number of edges in the graph. This result is stated as Theorem 3.

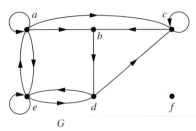

G

FIGURE 2   The Directed Graph G.

THEOREM 3   Let $G = (V, E)$ be a graph with directed edges. Then

$$\sum_{v \in V} \deg^-(v) = \sum_{v \in V} \deg^+(v) = |E|.$$

There are many properties of a graph with directed edges that do not depend on the direction of its edges. Consequently, it is often useful to ignore these directions. The undirected graph that results from ignoring directions of edges is called the **underlying undirected graph.** A graph with directed edges and its underlying undirected graph have the same number of edges.

SOME SPECIAL SIMPLE GRAPHS

We will now introduce several classes of simple graphs. These graphs are often used as examples and arise in many applications.

EXAMPLE 4   **Complete Graphs** The **complete graph on *n* vertices,** denoted by K_n, is the simple graph that contains exactly one edge between each pair of distinct vertices. The graphs K_n, for $n = 1, 2, 3, 4, 5, 6$, are displayed in Figure 3.   ◀

EXAMPLE 5   **Cycles** The **cycle C_n,** $n \geq 3$, consists of n vertices $v_1, v_2, \ldots, v_n$ and edges $\{v_1, v_2\}$, $\{v_2, v_3\}, \ldots, \{v_{n-1}, v_n\}$, and $\{v_n, v_1\}$. The cycles C_3, C_4, C_5, and C_6 are displayed in Figure 4.   ◀

EXAMPLE 6   **Wheels** We obtain the **wheel W_n** when we add an additional vertex to the cycle C_n, for $n \geq 3$, and connect this new vertex to each of the n vertices in C_n, by new edges. The wheels W_3, W_4, W_5, and W_6 are displayed in Figure 5.   ◀

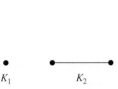

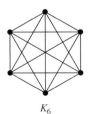

K_1 K_2 K_3 K_4 K_5 K_6

FIGURE 3   The Graphs K_n for $1 \leq n \leq 6$.

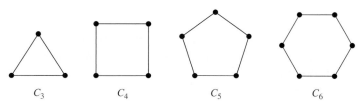

FIGURE 4 **The Cycles C_3, C_4, C_5, and C_6.**

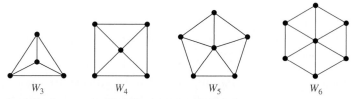

FIGURE 5 **The Wheels W_3, W_4, W_5, and W_6.**

EXAMPLE 7 n**-Cubes** The n-dimensional cube, or **n-cube,** denoted by Q_n, is the graph that has vertices representing the 2^n bit strings of length n. Two vertices are adjacent if and only if the bit strings that they represent differ in exactly one bit position. The graphs Q_1, Q_2, and Q_3 are displayed in Figure 6. Note that you can construct the $(n + 1)$-cube Q_{n+1} from the n-cube Q_n by making two copies of Q_n, prefacing the labels on the vertices with a 0 in one copy of Q_n and with a 1 in the other copy of Q_n, and adding edges connecting two vertices that have labels differing only in the first bit. In Figure 6, Q_3 is constructed from Q_2 by drawing two copies of Q_2 as the top and bottom faces of Q_3, adding 0 at the beginning of the label of each vertex in the bottom face and 1 at the beginning of the label of each vertex in the top face. ◀

BIPARTITE GRAPHS

Sometimes a graph has the property that its vertex set can be divided into two disjoint subsets such that each edge connects a vertex in one of these subsets to a vertex in the other subset. For example, consider the graph representing marriages between people in a village, where each person is represented by a vertex and a marriage is represented by an edge. In this graph, each edge connects a vertex in the subset of vertices representing males and a vertex in the subset of vertices representing females. This leads us to Definition 5.

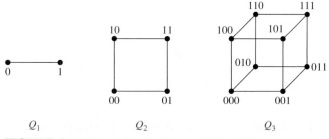

FIGURE 6 **The n-cube Q_n for $n = 1, 2,$ and 3.**

DEFINITION 5   A simple graph G is called *bipartite* if its vertex set V can be partitioned into two disjoint sets V_1 and V_2 such that every edge in the graph connects a vertex in V_1 and a vertex in V_2 (so that no edge in G connects either two vertices in V_1 or two vertices in V_2).

In Example 8 we will show that C_6 is bipartite, and in Example 9 we will show that K_3 is not bipartite.

EXAMPLE 8 C_6 is bipartite, as shown in Figure 7, since its vertex set can be partitioned into the two sets $V_1 = \{v_1, v_3, v_5\}$ and $V_2 = \{v_2, v_4, v_6\}$, and every edge of C_6 connects a vertex in V_1 and a vertex in V_2.   ◀

EXAMPLE 9 K_3 is not bipartite. To see this, note that if we divide the vertex set of K_3 into two disjoint sets, one of the two sets must contain two vertices. If the graph were bipartite, these two vertices could not be connected by an edge, but in K_3 each vertex is connected to every other vertex by an edge.   ◀

EXAMPLE 10   Are the graphs G and H displayed in Figure 8 bipartite?

Solution: Graph G is bipartite, since its vertex set is the union of two disjoint sets, $\{a, b, d\}$ and $\{c, e, f, g\}$, and each edge connects a vertex in one of these subsets to a vertex in the other subset. (Note that for G to be bipartite it is not necessary that every vertex in $\{a, b, d\}$ be adjacent to every vertex in $\{c, e, f, g\}$. For instance, b and g are not adjacent.)

Graph H is not bipartite since its vertex set cannot be partitioned into two subsets so that edges do not connect two vertices from the same subset. (The reader should verify this by considering the vertices a, b, and f.)   ◀

A graph is bipartite if and only if it is possible to color the vertices of the graph with at most two colors so that no two adjacent vertices have the same color. We will study graph colorings in Section 8.8. Also, a graph is bipartite if and only if it is not possible to start at a vertex and return to this vertex by traversing an odd number of distinct edges. We will make this notion more precise when we discuss paths and circuits in graphs in Section 8.4.

EXAMPLE 11   **Complete Bipartite Graphs** The **complete bipartite graph** $K_{m,n}$ is the graph that has its vertex set partitioned into two subsets of m and n vertices, respectively. There is an

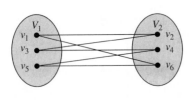

FIGURE 7 **Showing That C_6 Is Bipartite.**

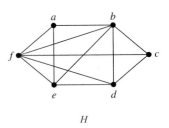

G H

FIGURE 8 **The Undirected Graphs G and H.**

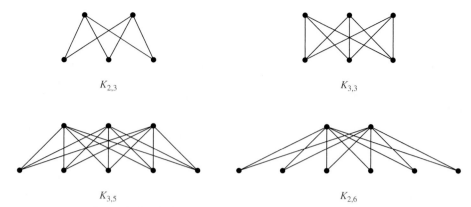

FIGURE 9   **Some Complete Bipartite Graphs.**

edge between two vertices if and only if one vertex is in the first subset and the other vertex is in the second subset. The complete bipartite graphs $K_{2,3}$, $K_{3,3}$, $K_{3,5}$, and $K_{2,6}$ are displayed in Figure 9.   ◀

SOME APPLICATIONS OF SPECIAL TYPES OF GRAPHS

We will show how special types of graphs are used in models for data communications and parallel processing.

EXAMPLE 12   **Local Area Networks**   The various computers in a building, such as minicomputers and personal computers, as well as peripheral devices such as printers and plotters, can be connected using a *local area network*. Some of these networks are based on a *star topology*, where all devices are connected to a central control device. A local area network can be represented using a complete bipartite graph $K_{1,n}$, as shown in Figure 10(a). Messages are sent from device to device through the central control device.

Other local area networks are based on a *ring topology*, where each device is connected to exactly two others. Local area networks with a ring topology are modeled using n-cycles, C_n, as shown in Figure 10(b). Messages are sent from device to device around the cycle until the intended recipient of a message is reached.

Finally, some local area networks use a hybrid of these two topologies. Messages may be sent around the ring, or through a central device. This redundancy makes the network more reliable. Local area networks with this redundancy can be modeled using wheels W_n, as shown in Figure 10(c).   ◀

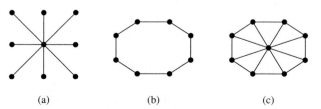

(a)   (b)   (c)

FIGURE 10   **Star, Ring, and Hybrid Topologies for Local Area Networks.**

EXAMPLE 13   **Interconnection Networks for Parallel Computation** Until recently, computers executed programs one operation at a time. Consequently, the algorithms written to solve problems were designed to perform one step at a time; such algorithms are called **serial.** (Almost all algorithms described in this book are serial.) However, many computationally intense problems, such as weather simulations, medical imaging, and cryptanalysis, cannot be solved in a reasonable amount of time using serial operations, even on a supercomputer. Furthermore, there is a physical limit to how fast a computer can carry out basic operations, so that there will always be problems that cannot be solved in a reasonable length of time using serial operations.

Parallel processing, which uses computers made up of many separate processors, each with its own memory, helps overcome the limitations of serial computers. **Parallel algorithms,** which break a problem into a number of subproblems that can be solved concurrently, can then be devised to rapidly solve problems using a computer with multiple processors. In a parallel algorithm, a single instruction stream controls the execution of the algorithm, sending subproblems to different processors, and directs the input and output of these subproblems to the appropriate processors.

When parallel processing is used, one processor may need output generated by another processor. Consequently, these processors need to be interconnected. We can use the appropriate type of graph to represent the interconnection network of the processors in a computer with multiple processors. In the following discussion, we will describe the most commonly used types of interconnection networks for parallel processors. The type of interconnection network used to implement a particular parallel algorithm depends on the requirements for exchange of data between processors, the desired speed, and, of course, the available hardware.

The simplest, but most expensive, network-interconnecting processors include a two-way link between each pair of processors. This network can be represented by K_n, the complete graph on n vertices, when there are n processors. However, there are serious problems with this type of interconnection network because the required number of connections is so large. In reality, the number of direct connections to a processor is limited, so when there are a large number of processors, a processor cannot be linked directly to all others. For example, when there are 64 processors, $C(64, 2) = 2016$ connections would be required, and each processor would have to be directly connected to 63 others.

On the other hand, perhaps the simplest way to interconnect n processors is to use an arrangement known as a **linear array.** Each processor P_i, other than P_1 and P_n, is connected to its neighbors P_{i-1} and P_{i+1} via a two-way link. P_1 is connected only to P_2, and P_n is connected only to P_{n-1}. The linear array for six processors is shown in Figure 11. The advantage of a linear array is that each processor has at most two direct connections to other processors. The disadvantage is that it is sometimes necessary to use a large number of intermediate links, called **hops,** for processors to share information.

The **mesh network** (or **two-dimensional array**) is a commonly used interconnection network. In such a network, the number of processors is a perfect square, say $n = m^2$. The n processors are labeled $P(i, j), 0 \le i \le m - 1, 0 \le j \le m - 1$. Two-way links connect processor $P(i, j)$ with its four neighbors, processors $P(i \pm 1, j)$ and $P(i, j \pm 1)$, as long as these are processors in the mesh. (Note that four processors, on the corners of the mesh, have only two adjacent processors, and other processors on the boundaries have only three neighbors. Sometimes a variant of a mesh network in which every processor has exactly four connections is used; see Exercise 52 at the end of this section.) The mesh network limits the number of links for each processor. Communication between some pairs of processors requires $O(\sqrt{n}) = O(m)$ intermediate links. (See Exercise 53 at the end of this section.) The graph representing the mesh network for 16 processors is shown in Figure 12.

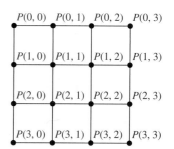

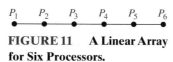

FIGURE 11   **A Linear Array**
for Six Processors.

FIGURE 12   **A Mesh**
Network for 16 Processors.

One important type of interconnection network is the hypercube. For such a network, the number of processors is a power of 2, $n = 2^m$. The n processors are labeled $P_0, P_1, \ldots, P_{n-1}$. Each processor has two-way connections to m other processors. Processor P_i is linked to the processors with indices whose binary representations differ from the binary representation of i in exactly one bit. The hypercube network balances the number of direct connections for each processor and the number of intermediate connections required so that processors can communicate. Many computers have been built using a hypercube network, and many parallel algorithms have been devised that use a hypercube network. The graph Q_n, the n-cube, represents the hypercube network with n processors. Figure 13 displays the hypercube network for eight processors. (Figure 13 displays a different way to draw Q_3 than was shown in Figure 6.) ◀

NEW GRAPHS FROM OLD

Sometimes we need only part of a graph to solve a problem. For instance, we may care only about the part of a large computer network that involves the computer centers in New York, Denver, Detroit, and Atlanta. Then we can ignore the other computer centers and all telephone lines not linking two of these specific four computer centers. In the graph model for the large network, we can remove the vertices corresponding to the computer centers other than the four of interest, and we can remove all edges incident with a vertex that was removed. When edges and vertices are removed from a graph, without removing endpoints of any remaining edges, a smaller graph is obtained. Such a graph is called a **subgraph** of the original graph.

DEFINITION 6   A *subgraph of a graph* $G = (V, E)$ is a graph $H = (W, F)$ where $W \subseteq V$ and $F \subseteq E$.

EXAMPLE 14   The graph G shown in Figure 14 is a subgraph of K_5. ◀

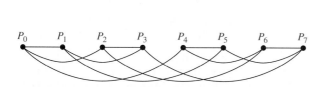

FIGURE 13   **A Hypercube Network for Eight**
Processors.

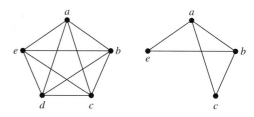

FIGURE 14   **A Subgraph of K_5.**

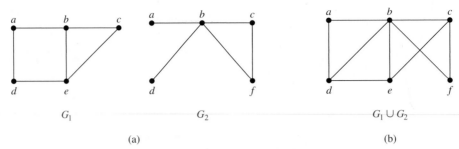

FIGURE 15   (a) The Simple Graphs G_1 and G_2; (b) Their Union $G_1 \cup G_2$.

Two or more graphs can be combined in various ways. The new graph that contains all the vertices and edges of these graphs is called the **union** of the graphs. We will give a more formal definition for the union of two simple graphs.

DEFINITION 7   The *union* of two simple graphs $G_1 = (V_1, E_1)$ and $G_2 = (V_2, E_2)$ is the simple graph with vertex set $V_1 \cup V_2$ and edge set $E_1 \cup E_2$. The union of G_1 and G_2 is denoted by $G_1 \cup G_2$.

EXAMPLE 15   Find the union of the graphs G_1 and G_2 shown in Figure 15(a).   ◄

Solution: The vertex set of the union $G_1 \cup G_2$ is the union of the two vertex sets, namely, $\{a, b, c, d, e, f\}$. The edge set of the union is the union of the two edge sets. The union is displayed in Figure 15(b).

Exercises

In Exercises 1–3 find the number of vertices, the number of edges, and the degree of each vertex in the given undirected graph. Identify all isolated and pendant vertices.

1.
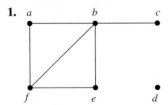

2.
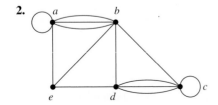

3.
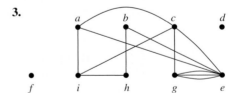

4. Find the sum of the degrees of the vertices of each graph in Exercises 1–3 and verify that it equals twice the number of edges in the graph.

5. Can a simple graph exist with 15 vertices each of degree five?

6. Show that the sum, over the set of people at a party, of the number of people a person has shaken hands with, is even. Assume that no one shakes his or her own hand.

In Exercises 7–9 determine the number of vertices and edges and find the in-degree and out-degree of each vertex for the given directed multigraph.

7. 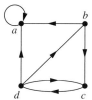   **8.**

9.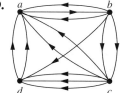

10. For each of the graphs in Exercises 7–9 determine the sum of the in-degrees of the vertices and the sum of the out-degrees of the vertices directly. Show that they are both equal to the number of edges in the graph.

11. Construct the underlying undirected graph for the graph with directed edges in Figure 2.

12. What does the degree of a vertex represent in the acquaintanceship graph, where vertices represent all the people in the world? What do isolated and pendant vertices in this graph represent? In one study it was estimated that the average degree of a vertex in this graph is 1000. What does this mean in terms of the model?

13. What does the degree of a vertex represent in a collaboration graph? What do isolated and pendant vertices represent?

14. What does the degree of a vertex in the Hollywood graph represent? What do the isolated and pendant vertices represent?

15. What do the in-degree and the out-degree of a vertex in a telephone call graph, as described in Example 7 of Section 8.1, represent? What does the degree of a vertex in the undirected version of this graph represent?

16. What do the in-degree and the out-degree of a vertex in the Web graph, as described in Example 8 of Section 8.1, represent?

17. What do the in-degree and the out-degree of a vertex in a directed graph modeling a round-robin tournament represent?

18. Draw these graphs.

    a) K_7 **b)** $K_{1,8}$ **c)** $K_{4,4}$
    d) C_7 **e)** W_7 **f)** Q_4

In Exercises 19–23 determine whether the graph is bipartite.

19. 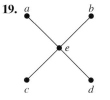   **20.**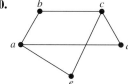

21.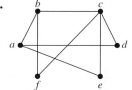

22.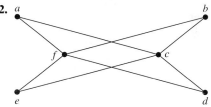

23.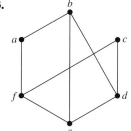

24. For which values of n are these graphs bipartite?

    a) K_n **b)** C_n **c)** W_n **d)** Q_n

25. How many vertices and how many edges do these graphs have?

    a) K_n **b)** C_n **c)** W_n
    d) $K_{m,n}$ **e)** Q_n

26. How many edges does a graph have if it has vertices of degree 4, 3, 3, 2, 2? Draw such a graph.

27. How many edges does a graph have if it has vertices of degree 5, 2, 2, 2, 2, 1? Draw such a graph.

28. Does there exist a simple graph with six vertices of these degrees? If so, draw such a graph.

    a) 0, 1, 2, 3, 4, 5   **b)** 1, 2, 3, 4, 5, 6
    c) 2, 2, 2, 2, 2, 2   **d)** 3, 2, 3, 2, 3, 2
    e) 3, 2, 2, 2, 2, 3   **f)** 1, 1, 1, 1, 1, 1
    g) 3, 3, 3, 3, 3, 5   **h)** 1, 2, 3, 4, 5, 5

29. Does there exist a simple graph with five vertices of these degrees? If so, draw such a graph.

a) $3, 3, 3, 3, 2$
b) $1, 2, 3, 4, 5$
c) $1, 2, 3, 4, 4$
d) $3, 4, 3, 4, 3$
e) $0, 1, 2, 2, 3$
f) $1, 1, 1, 1, 1$

30. How many subgraphs with at least one vertex does K_2 have?

31. How many subgraphs with at least one vertex does K_3 have?

32. How many subgraphs with at least one vertex does W_3 have?

33. Draw all subgraphs of this graph.

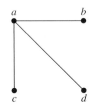

34. Let G be a graph with v vertices and e edges. Let M be the maximum degree of the vertices of G, and let m be the minimum degree of the vertices of G. Show that

a) $2e/v \geq m$.
b) $2e/v \leq M$.

A simple graph is called **regular** if every vertex of this graph has the same degree. A regular graph is called n-**regular** if every vertex in this graph has degree n.

35. For which values of n are these graphs regular?

a) K_n **b)** C_n **c)** W_n **d)** Q_n

36. For which values of m and n is $K_{m,n}$ regular?

37. How many vertices does a regular graph of degree 4 with 10 edges have?

In Exercises 38–40 find the union of the given pair of simple graphs. (Assume edges with the same endpoints are the same.)

38.

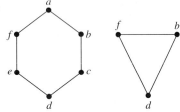

39.

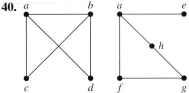

40.

41. The **complementary graph** $\overline{G}$ of a simple graph G has the same vertices as G. Two vertices are adjacent in $\overline{G}$ if and only if they are not adjacent in G. Find these graphs.

a) $\overline{K_n}$ **b)** $\overline{K_{m,n}}$ **c)** $\overline{C_n}$ **d)** $\overline{Q_n}$

42. If G is a simple graph with 15 edges and $\overline{G}$ has 13 edges, how many vertices does G have?

43. If the simple graph G has v vertices and e edges, how many edges does $\overline{G}$ have?

*44. Show that if G is a bipartite simple graph with v vertices and e edges, then $e \leq v^2/4$.

45. Show that if G is a simple graph with n vertices, then the union of G and $\overline{G}$ is K_n.

*46. Describe an algorithm to decide whether a graph is bipartite.

The **converse** of a directed graph $G = (V, E)$, denoted by G^c, is the directed graph (V, F) where $(u, v) \in F$ if and only if $(v, u) \in E$.

47. Draw the converse of each of the graphs in Exercises 7–9 in Section 8.1.

48. Show that $(G^c)^c = G$ whenever G is a directed graph.

49. Show that the graph G is its own converse if and only if the relation associated with G (see Section 7.3) is symmetric.

50. Extend the definition of the converse of a directed graph to the notion of the converse of a directed multigraph.

51. Draw the mesh network for interconnecting nine parallel processors.

52. In a variant of a mesh network for interconnecting $n = m^2$ processors, processor $P(i, j)$ is connected to the four processors $P((i \pm 1) \bmod m, j)$, $P(i, (j \pm 1) \bmod m)$, so that connections wrap around the edges of the mesh. Draw this variant of the mesh network for 16 processors.

53. Show that every pair of processors in a mesh network of $n = m^2$ processors can communicate using $O(\sqrt{n}) = O(m)$ hops between directly connected processors.

8.3 Representing Graphs and Graph Isomorphism

INTRODUCTION

There are many useful ways to represent graphs. As we will see throughout this chapter, in working with a graph it is helpful to be able to choose its most convenient representation. In this section we will show how to represent graphs in several different ways.

Sometimes, two graphs have exactly the same form, in the sense that there is a one-to-one correspondence between their vertex sets that preserves edges. In such a case, we say that the two graphs are **isomorphic.** Determining whether two graphs are isomorphic is an important problem of graph theory that we will study in this section.

REPRESENTING GRAPHS

One way to represent a graph without multiple edges is to list all the edges of this graph. Another way to represent a graph with no multiple edges is to use **adjacency lists,** which specify the vertices that are adjacent to each vertex of the graph.

EXAMPLE 1   Use adjacency lists to describe the simple graph given in Figure 1.

Solution: Table 1 lists those vertices adjacent to each of the vertices of the graph.   ◀

EXAMPLE 2   Represent the directed graph shown in Figure 2 by listing all the vertices that are the terminal vertices of edges starting at each vertex of the graph.

Solution: Table 2 represents the directed graph shown in Figure 2.   ◀

ADJACENCY MATRICES

Carrying out graph algorithms using the representation of graphs by lists of edges, or by adjacency lists, can be cumbersome if there are many edges in the graph. To simplify computation, graphs can be represented using matrices. Two types of matrices commonly used to represent graphs will be presented here. One is based on the adjacency of vertices, and the other is based on incidence of vertices and edges.

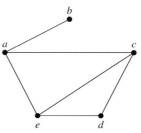

FIGURE 1   A Simple Graph.

TABLE 1 An Edge List for a Simple Graph.	
Vertex	*Adjacent Vertices*
a	b, c, e
b	a
c	a, d, e
d	c, e
e	a, c, d

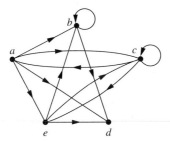

FIGURE 2   A Directed Graph.

TABLE 2 An Edge List for a Directed Graph.

Initial Vertex	Terminal Vertices
a	b, c, d, e
b	b, d
c	a, c, e
d	
e	b, c, d

Suppose that $G = (V, E)$ is a simple graph where $|V| = n$. Suppose that the vertices of G are listed arbitrarily as $v_1, v_2, \ldots, v_n$. The **adjacency matrix A** (or $\mathbf{A}_G$) of G, with respect to this listing of the vertices, is the $n \times n$ zero–one matrix with 1 as its (i, j)th entry when v_i and v_j are adjacent, and 0 as its (i, j)th entry when they are not adjacent. In other words, if its adjacency matrix is $\mathbf{A} = [a_{ij}]$, then

$$a_{ij} = \begin{cases} 1 & \text{if } \{v_i, v_j\} \text{ is an edge of } G, \\ 0 & \text{otherwise.} \end{cases}$$

Note that an adjacency matrix of a graph is based on the ordering chosen for the vertices. Hence, there are as many as $n!$ different adjacency matrices for a graph with n vertices, since there are $n!$ different orderings of n vertices.

The adjacency matrix of a simple graph is symmetric, that is, $a_{ij} = a_{ji}$, since both of these entries are 1 when v_i and v_j are adjacent, and both are 0 otherwise. Furthermore, since a simple graph has no loops, each entry a_{ii}, $i = 1, 2, 3, \ldots, n$, is 0.

Note that when there are relatively few edges in a graph, the adjacency matrix is a **sparse matrix,** that is, a matrix with few nonzero entries. There are special techniques for representing, and computing with, such matrices. Also, it sometimes may be more efficient to work with lists of edges when representing and working with such graphs.

EXAMPLE 3   Use an adjacency matrix to represent the graph shown in Figure 3.

Solution: We order the vertices as a, b, c, d. The matrix representing this graph is

$$\begin{bmatrix} 0 & 1 & 1 & 1 \\ 1 & 0 & 1 & 0 \\ 1 & 1 & 0 & 0 \\ 1 & 0 & 0 & 0 \end{bmatrix}.$$

◀

FIGURE 3 A Simple Graph.

EXAMPLE 4   Draw a graph with the adjacency matrix

$$\begin{bmatrix} 0 & 1 & 1 & 0 \\ 1 & 0 & 0 & 1 \\ 1 & 0 & 0 & 1 \\ 0 & 1 & 1 & 0 \end{bmatrix}$$

with respect to the ordering of vertices a, b, c, d.

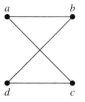

FIGURE 4   A Graph with the Given Adjacency Matrix.

Solution: A graph with this adjacency matrix is shown in Figure 4.   ◀

Adjacency matrices can also be used to represent undirected graphs with loops and with multiple edges. A loop at the vertex a_i is represented by a 1 at the (i, i)th position of the adjacency matrix. When multiple edges are present, the adjacency matrix is no longer a zero–one matrix, since the (i, j)th entry of this matrix equals the number of edges that are associated to $\{a_i, a_j\}$. All undirected graphs, including multigraphs and pseudographs, have symmetric adjacency matrices.

EXAMPLE 5   Use an adjacency matrix to represent the pseudograph shown in Figure 5.

FIGURE 5   A Pseudograph.

Solution: The adjacency matrix using the ordering of vertices a, b, c, d is

$$\begin{bmatrix} 0 & 3 & 0 & 2 \\ 3 & 0 & 1 & 1 \\ 0 & 1 & 1 & 2 \\ 2 & 1 & 2 & 0 \end{bmatrix}.$$

◀

We used zero–one matrices in Chapter 7 to represent directed graphs. The matrix for a directed graph $G = (V, E)$ has a 1 in its (i, j)th position if there is an edge from v_i to v_j, where $v_1, v_2, \ldots, v_n$ is an arbitrary listing of the vertices of the directed graph. In other words, if $\mathbf{A} = [a_{ij}]$ is the adjacency matrix for the directed graph with respect to this listing of the vertices, then

$$a_{ij} = \begin{cases} 1 & \text{if } (v_i, v_j) \text{ is an edge of } G, \\ 0 & \text{otherwise.} \end{cases}$$

The adjacency matrix for a directed graph does not have to be symmetric, since there may not be an edge from a_j to a_i when there is an edge from a_i to a_j.

Adjacency matrices can also be used to represent directed multigraphs. Again, such matrices are not zero–one matrices when there are multiple edges in the same direction connecting two vertices. In the adjacency matrix for a directed multigraph, a_{ij} equals the number of edges that are associated to (v_i, v_j).

INCIDENCE MATRICES

Another common way to represent graphs is to use **incidence matrices.** Let $G = (V, E)$ be an undirected graph. Suppose that $v_1, v_2, \ldots, v_n$ are the vertices and $e_1, e_2, \ldots, e_m$ are the edges of G. Then the incidence matrix with respect to this ordering of V and E is the $n \times m$ matrix $\mathbf{M} = [m_{ij}]$, where

$$m_{ij} = \begin{cases} 1 & \text{when edge } e_j \text{ is incident with } v_i, \\ 0 & \text{otherwise.} \end{cases}$$

EXAMPLE 6   Represent the graph shown in Figure 6 with an incidence matrix.

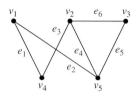

FIGURE 6 An Undirected Graph.

Solution: The incidence matrix is

$$
\begin{array}{c}
\\
v_1 \\
v_2 \\
v_3 \\
v_4 \\
v_5
\end{array}
\begin{array}{cccccc}
e_1 & e_2 & e_3 & e_4 & e_5 & e_6 \\
\left[\begin{array}{cccccc}
1 & 1 & 0 & 0 & 0 & 0 \\
0 & 0 & 1 & 1 & 0 & 1 \\
0 & 0 & 0 & 0 & 1 & 1 \\
1 & 0 & 1 & 0 & 0 & 0 \\
0 & 1 & 0 & 1 & 1 & 0
\end{array}\right].
\end{array}
$$

◄

Incidence matrices can also be used to represent multiple edges and loops. Multiple edges are represented in the incidence matrix using columns with identical entries, since these edges are incident with the same pair of vertices. Loops are represented using a column with exactly one entry equal to 1, corresponding to the vertex that is incident with this loop.

EXAMPLE 7   Represent the pseudograph shown in Figure 7 using an incidence matrix.

Solution: The incidence matrix for this graph is

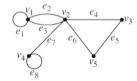

FIGURE 7 A Pseudograph.

$$
\begin{array}{c}
\\
v_1 \\
v_2 \\
v_3 \\
v_4 \\
v_5
\end{array}
\begin{array}{cccccccc}
e_1 & e_2 & e_3 & e_4 & e_5 & e_6 & e_7 & e_8 \\
\left[\begin{array}{cccccccc}
1 & 1 & 1 & 0 & 0 & 0 & 0 & 0 \\
0 & 1 & 1 & 1 & 0 & 1 & 1 & 0 \\
0 & 0 & 0 & 1 & 1 & 0 & 0 & 0 \\
0 & 0 & 0 & 0 & 0 & 0 & 1 & 1 \\
0 & 0 & 0 & 0 & 1 & 1 & 0 & 0
\end{array}\right].
\end{array}
$$

◄

ISOMORPHISM OF GRAPHS

We often need to know whether it is possible to draw two graphs in the same way. For instance, in chemistry, graphs are used to model compounds. Different compounds can have the same molecular formula but can differ in structure. Such compounds will be represented by graphs that cannot be drawn in the same way. The graphs representing previously known compounds can be used to determine whether a supposedly new compound has been studied before.

There is a useful terminology for graphs with the same structure.

DEFINITION 1

The simple graphs $G_1 = (V_1, E_1)$ and $G_2 = (V_2, E_2)$ are *isomorphic* if there is a one-to-one and onto function f from V_1 to V_2 with the property that a and b are adjacent in G_1 if and only if $f(a)$ and $f(b)$ are adjacent in G_2, for all a and b in V_1. Such a function f is called an *isomorphism*.*

In other words, when two simple graphs are isomorphic, there is a one-to-one correspondence between vertices of the two graphs that preserves the adjacency relationship. Isomorphism of simple graphs is an equivalence relation. (We leave the verification of this as Exercise 45 at the end of this section.)

EXAMPLE 8   Show that the graphs $G = (V, E)$ and $H = (W, F)$, displayed in Figure 8, are isomorphic.

*The word *isomorphism* comes from the Greek roots *isos* for "equal" and *morphe* for "form."

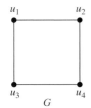

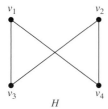

FIGURE 8 **The Graphs *G* and *H*.**

Solution: The function f with $f(u_1) = v_1, f(u_2) = v_4, f(u_3) = v_3,$ and $f(u_4) = v_2$ is a one-to-one correspondence between V and W. To see that this correspondence preserves adjacency, note that adjacent vertices in G are u_1 and u_2, u_1 and u_3, u_2 and u_4, and u_3 and u_4, and each of the pairs $f(u_1) = v_1$ and $f(u_2) = v_4$, $f(u_1) = v_1$ and $f(u_3) = v_3$, $f(u_2) = v_4$ and $f(u_4) = v_2$, and $f(u_3) = v_3$ and $f(u_4) = v_2$ are adjacent in H. ◀

It is often difficult to determine whether two simple graphs are isomorphic. There are $n!$ possible one-to-one correspondences between the vertex sets of two simple graphs with n vertices. Testing each such correspondence to see whether it preserves adjacency and nonadjacency is impractical if n is at all large.

However, we can often show that two simple graphs are not isomorphic by showing that they do not share a property that isomorphic simple graphs must both have. Such a property is called an **invariant** with respect to isomorphism of simple graphs. For instance, isomorphic simple graphs must have the same number of vertices, since there is a one-to-one correspondence between the sets of vertices of the graphs. Isomorphic simple graphs also must have the same number of edges, because the one-to-one correspondence between vertices establishes a one-to-one correspondence between edges. In addition, the degrees of the vertices in isomorphic simple graphs must be the same. That is, a vertex v of degree d in G must correspond to a vertex $f(v)$ of degree d in H, since a vertex w in G is adjacent to v if and only if $f(v)$ and $f(w)$ are adjacent in H.

EXAMPLE 9 Show that the graphs displayed in Figure 9 are not isomorphic.

Solution: Both G and H have five vertices and six edges. However, H has a vertex of degree one, namely, e, whereas G has no vertices of degree one. It follows that G and H are not isomorphic. ◀

The number of vertices, the number of edges, and the degrees of the vertices are all invariants under isomorphism. If any of these quantities differ in two simple graphs, these graphs cannot be isomorphic. However, when these invariants are the same, it does not necessarily mean that the two graphs are isomorphic. There are no useful sets of invariants currently known that can be used to determine whether simple graphs are isomorphic.

EXAMPLE 10 Determine whether the graphs shown in Figure 10 are isomorphic.

Solution: The graphs G and H both have eight vertices and ten edges. They also both have four vertices of degree two and four of degree three. Since these invariants all agree, it is still conceivable that these graphs are isomorphic.

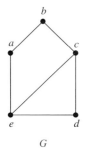

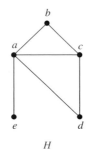

FIGURE 9 **The Graphs *G* and *H*.**

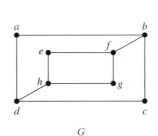

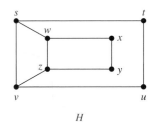

FIGURE 10 **The Graphs *G* and *H*.**

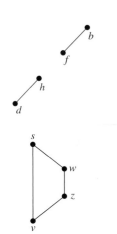

FIGURE 11   The
Subgraphs of *G*
and *H* Made Up
of Vertices of Degree
Three and the Edges
Connecting Them.

However, *G* and *H* are not isomorphic. To see this, note that since $\deg(a) = 2$ in *G*, *a* must correspond to either *t*, *u*, *x*, or *y* in *H*, since these are the vertices of degree two in *H*. However, each of these four vertices in *H* is adjacent to another vertex of degree two in *H*, which is not true for *a* in *G*.

Another way to see that *G* and *H* are not isomorphic is to note that the subgraphs of *G* and *H* made up of vertices of degree three and the edges connecting them must be isomorphic if these two graphs are isomorphic (the reader should verify this). However, these subgraphs, shown in Figure 11, are not isomorphic.   ◄

To show that a function *f* from the vertex set of a graph *G* to the vertex set of a graph *H* is an isomorphism, we need to show that *f* preserves edges. One helpful way to do this is to use adjacency matrices. In particular, to show that *f* is an isomorphism, we can show that the adjacency matrix of *G* is the same as the adjacency matrix of *H*, when rows and columns are labeled to correspond to the images under *f* of the vertices in *G* that are the labels of these rows and columns in the adjacency matrix of *G*. We illustrate how this is done in Example 11.

EXAMPLE 11   Determine whether the graphs *G* and *H* displayed in Figure 12 are isomorphic.

Solution: Both *G* and *H* have six vertices and seven edges. Both have four vertices of degree two and two vertices of degree three. It is also easy to see that the subgraphs of *G* and *H* consisting of all vertices of degree two and the edges connecting them are isomorphic (as the reader should verify). Since *G* and *H* agree with respect to these invariants, it is reasonable to try to find an isomorphism *f*.

We now will define a function *f* and then determine whether it is an isomorphism. Since $\deg(u_1) = 2$ and since u_1 is not adjacent to any other vertex of degree two, the image of u_1 must be either v_4 or v_6, the only vertices of degree two in *H* not adjacent to a vertex of degree two. We arbitrarily set $f(u_1) = v_6$. [If we found that this choice did not lead to isomorphism, we would then try $f(u_1) = v_4$.] Since u_2 is adjacent to u_1, the possible images of u_2 are v_3 and v_5. We arbitrarily set $f(u_2) = v_3$. Continuing in this way, using adjacency of vertices and degrees as a guide, we set $f(u_3) = v_4$, $f(u_4) = v_5$, $f(u_5) = v_1$, and $f(u_6) = v_2$. We now have a one-to-one correspondence between the vertex set of *G* and the vertex set of *H*, namely: $f(u_1) = v_6$, $f(u_2) = v_3$, $f(u_3) = v_4$, $f(u_4) = v_5$, $f(u_5) = v_1$, $f(u_6) = v_2$. To see whether *f* preserves edges, we examine the adjacency matrix of *G*,

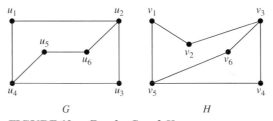

FIGURE 12   Graphs *G* and *H*.

$$\mathbf{A}_G = \begin{array}{c} \\ u_1 \\ u_2 \\ u_3 \\ u_4 \\ u_5 \\ u_6 \end{array} \overset{\begin{array}{cccccc} u_1 & u_2 & u_3 & u_4 & u_5 & u_6 \end{array}}{\begin{bmatrix} 0 & 1 & 0 & 1 & 0 & 0 \\ 1 & 0 & 1 & 0 & 0 & 1 \\ 0 & 1 & 0 & 1 & 0 & 0 \\ 1 & 0 & 1 & 0 & 1 & 0 \\ 0 & 0 & 0 & 1 & 0 & 1 \\ 0 & 1 & 0 & 0 & 1 & 0 \end{bmatrix}},$$

and the adjacency matrix of H with the rows and columns labeled by the images of the corresponding vertices in G,

$$\mathbf{A}_H = \begin{array}{c} \\ v_6 \\ v_3 \\ v_4 \\ v_5 \\ v_1 \\ v_2 \end{array} \overset{\begin{array}{cccccc} v_6 & v_3 & v_4 & v_5 & v_1 & v_2 \end{array}}{\begin{bmatrix} 0 & 1 & 0 & 1 & 0 & 0 \\ 1 & 0 & 1 & 0 & 0 & 1 \\ 0 & 1 & 0 & 1 & 0 & 0 \\ 1 & 0 & 1 & 0 & 1 & 0 \\ 0 & 0 & 0 & 1 & 0 & 1 \\ 0 & 1 & 0 & 0 & 1 & 0 \end{bmatrix}}.$$

Since $\mathbf{A}_G = \mathbf{A}_H$, it follows that f preserves edges. We conclude that f is an isomorphism, so that G and H are isomorphic. Note that if f turned out not to be an isomorphism, we would *not* have established that G and H are not isomorphic, since another correspondence of the vertices in G and H may be an isomorphism. ◄

The best algorithms known for determining whether two graphs are isomorphic have exponential worst-case time complexity (in the number of vertices of the graphs). However, linear average-case time complexity algorithms are known that solve this problem, and there is some hope that an algorithm with polynomial worst-case time complexity for determining whether two graphs are isomorphic can be found. The best practical algorithm, called NAUTY, can be used to determine whether two graphs with as many as 100 vertices are isomorphic in less than one second on a modern PC. The software for NAUTY can be downloaded over the Internet and experimented with.

Exercises

In Exercises 1–4 use an adjacency list to represent the given graph.

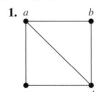

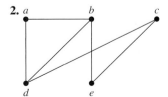

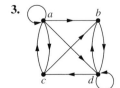

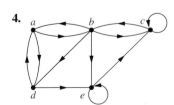

5. Represent the graph in Exercise 1 with an adjacency matrix.

6. Represent the graph in Exercise 2 with an adjacency matrix.

7. Represent the graph in Exercise 3 with an adjacency matrix.

8. Represent the graph in Exercise 4 with an adjacency matrix.

9. Represent each of these graphs with an adjacency matrix.

    a) K_4 **b)** $K_{1,4}$ **c)** $K_{2,3}$

    d) C_4 **e)** W_4 **f)** Q_3

In Exercises 10–12 draw a graph with the given adjacency matrix.

10. $\begin{bmatrix} 0 & 1 & 0 \\ 1 & 0 & 1 \\ 0 & 1 & 0 \end{bmatrix}$ **11.** $\begin{bmatrix} 0 & 0 & 1 & 1 \\ 0 & 0 & 1 & 0 \\ 1 & 1 & 0 & 1 \\ 1 & 1 & 1 & 0 \end{bmatrix}$

12. $\begin{bmatrix} 1 & 1 & 1 & 0 \\ 0 & 0 & 1 & 0 \\ 1 & 0 & 1 & 0 \\ 1 & 1 & 1 & 0 \end{bmatrix}$

In Exercises 13–15 represent the given graph using an adjacency matrix.

13.

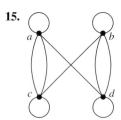

14.

15.

In Exercises 16–18 draw an undirected graph represented by the given adjacency matrix.

16. $\begin{bmatrix} 1 & 3 & 2 \\ 3 & 0 & 4 \\ 2 & 4 & 0 \end{bmatrix}$

17. $\begin{bmatrix} 1 & 2 & 0 & 1 \\ 2 & 0 & 3 & 0 \\ 0 & 3 & 1 & 1 \\ 1 & 0 & 1 & 0 \end{bmatrix}$ **18.** $\begin{bmatrix} 0 & 1 & 3 & 0 & 4 \\ 1 & 2 & 1 & 3 & 0 \\ 3 & 1 & 1 & 0 & 1 \\ 0 & 3 & 0 & 0 & 2 \\ 4 & 0 & 1 & 2 & 3 \end{bmatrix}$

In Exercises 19–21 find the adjacency matrix of the given directed multigraph.

19.

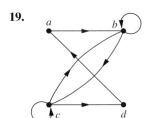

20.

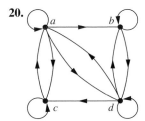

21.

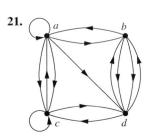

In Exercises 22–24 draw the graph represented by the given adjacency matrix.

22. $\begin{bmatrix} 1 & 0 & 1 \\ 0 & 0 & 1 \\ 1 & 1 & 1 \end{bmatrix}$ **23.** $\begin{bmatrix} 1 & 2 & 1 \\ 2 & 0 & 0 \\ 0 & 2 & 2 \end{bmatrix}$ **24.** $\begin{bmatrix} 0 & 2 & 3 & 0 \\ 1 & 2 & 2 & 1 \\ 2 & 1 & 1 & 0 \\ 1 & 0 & 0 & 2 \end{bmatrix}$

25. Is every zero–one square matrix that is symmetric and has zeros on the diagonal the adjacency matrix of a simple graph?

26. Use an incidence matrix to represent the graphs in Exercises 1 and 2.

27. Use an incidence matrix to represent the graphs in Exercises 13–15.

*28. What is the sum of the entries in a row of the adjacency matrix for an undirected graph? For a directed graph?

*29. What is the sum of the entries in a column of the adjacency matrix for an undirected graph? For a directed graph?

30. What is the sum of the entries in a row of the incidence matrix for an undirected graph?

31. What is the sum of the entries in a column of the incidence matrix for an undirected graph?

*32. Find an adjacency matrix for each of these graphs.
    a) K_n **b)** C_n **c)** W_n **d)** $K_{m,n}$ **e)** Q_n

*33. Find incidence matrices for the graphs in parts (a)–(d) of Exercise 32.

In Exercises 34–44 determine whether the given pair of graphs is isomorphic. Exhibit an isomorphism or provide a rigorous argument that none exists.

34.

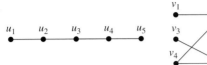

35.

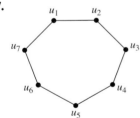

36.

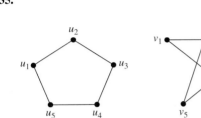

37.

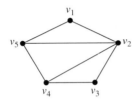

38.

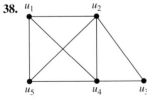

39.

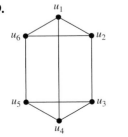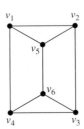

40.

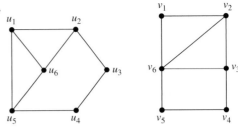

41.

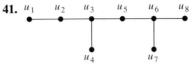

42.

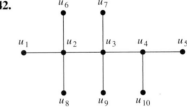

43.

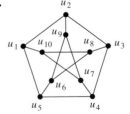

44.

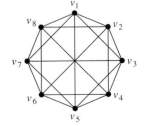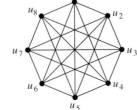

45. Show that isomorphism of simple graphs is an equivalence relation.

46. Suppose that G and H are isomorphic simple graphs. Show that their complementary graphs $\overline{G}$ and $\overline{H}$ are also isomorphic.

47. Describe the row and column of an adjacency matrix of a graph corresponding to an isolated vertex.

48. Describe the row of an incidence matrix of a graph corresponding to an isolated vertex.

49. Show that the vertices of a bipartite graph with two or more vertices can be ordered so that its adjacency matrix has the form

$$\begin{bmatrix} \mathbf{0} & \mathbf{A} \\ \mathbf{B} & \mathbf{0} \end{bmatrix}$$

where the four entries shown are rectangular blocks.

A simple graph G is called **self-complementary** if G and $\overline{G}$ are isomorphic.

50. Show that this graph is self-complementary.

51. Find a self-complementary simple graph with five vertices.

*52. Show that if G is a self-complementary simple graph with v vertices, then $v \equiv 0$ or 1 (mod 4).

53. For which integers n is C_n self-complementary?

54. How many nonisomorphic simple graphs are there with n vertices, when n is

a) 2?   **b)** 3?   **c)** 4?

55. How many nonisomorphic simple graphs are there with five vertices and three edges?

56. How many nonisomorphic simple graphs are there with six vertices and four edges?

57. Are the simple graphs with the following adjacency matrices isomorphic?

a) $\begin{bmatrix} 0 & 0 & 1 \\ 0 & 0 & 1 \\ 1 & 1 & 0 \end{bmatrix}$, $\begin{bmatrix} 0 & 1 & 1 \\ 1 & 0 & 0 \\ 1 & 0 & 0 \end{bmatrix}$

b) $\begin{bmatrix} 0 & 1 & 0 & 1 \\ 1 & 0 & 0 & 1 \\ 0 & 0 & 0 & 1 \\ 1 & 1 & 1 & 0 \end{bmatrix}$, $\begin{bmatrix} 0 & 1 & 1 & 1 \\ 1 & 0 & 0 & 1 \\ 1 & 0 & 0 & 1 \\ 1 & 1 & 1 & 0 \end{bmatrix}$

c) $\begin{bmatrix} 0 & 1 & 1 & 0 \\ 1 & 0 & 0 & 1 \\ 1 & 0 & 0 & 1 \\ 0 & 1 & 1 & 0 \end{bmatrix}$, $\begin{bmatrix} 0 & 1 & 0 & 1 \\ 1 & 0 & 0 & 0 \\ 0 & 0 & 0 & 1 \\ 1 & 0 & 1 & 0 \end{bmatrix}$

58. Determine whether the graphs without loops with these incidence matrices are isomorphic.

a) $\begin{bmatrix} 1 & 0 & 1 \\ 0 & 1 & 1 \\ 1 & 1 & 0 \end{bmatrix}$, $\begin{bmatrix} 1 & 1 & 0 \\ 1 & 0 & 1 \\ 0 & 1 & 1 \end{bmatrix}$

b) $\begin{bmatrix} 1 & 1 & 0 & 0 & 0 \\ 1 & 0 & 1 & 0 & 1 \\ 0 & 0 & 0 & 1 & 1 \\ 0 & 1 & 1 & 1 & 0 \end{bmatrix}$, $\begin{bmatrix} 0 & 1 & 0 & 0 & 1 \\ 0 & 1 & 1 & 1 & 0 \\ 1 & 0 & 0 & 1 & 0 \\ 1 & 0 & 1 & 0 & 1 \end{bmatrix}$

59. Extend the definition of isomorphism of simple graphs to undirected graphs containing loops and multiple edges.

60. Define isomorphism of directed graphs.

In Exercises 61–64 determine whether the given pair of directed graphs is isomorphic.

61.

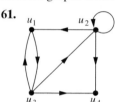

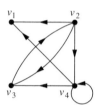

62.

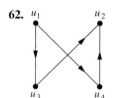

63.

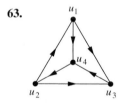

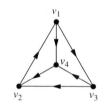

64.

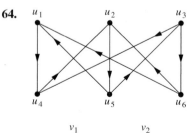

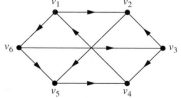

65. Show that if G and H are isomorphic directed graphs, then the converses of G and H (defined in the preamble of Exercise 47 of Section 8.2) are also isomorphic.

*66. How many nonisomorphic directed simple graphs are there with n vertices, when n is

    a) 2?
    b) 3?
    c) 4?

*67. What is the product of the incidence matrix and its transpose for an undirected graph?

*68. How much storage is needed to represent a simple graph with v vertices and e edges using

    a) adjacency lists?
    b) an adjacency matrix?
    c) an incidence matrix?

A **devil's pair** for a purported isomorphism test is a pair of nonisomorphic graphs that the test fails to show are not isomorphic.

69. Find a devil's pair for the test that checks the sequence of degrees of the vertices in the two graphs to make sure they agree.

8.4 **Connectivity**

INTRODUCTION

Many problems can be modeled with paths formed by traveling along the edges of graphs. For instance, the problem of determining whether a message can be sent between two computers using intermediate links can be studied with a graph model. Problems of efficiently planning routes for mail delivery, garbage pickup, diagnostics in computer networks, and so on, can be solved using models that involve paths in graphs.

PATHS

Informally, a **path** is a sequence of edges that begins at a vertex of a graph and travels along edges of the graph, always connecting pairs of adjacent vertices. A formal definition of paths and related terminology is given in Definition 1.

DEFINITION 1

Let n be a nonnegative integer and G an undirected graph. A *path* of *length* n from u to v in G is a sequence of n edges $e_1, \ldots, e_n$ of G such that $f(e_1) = \{x_0, x_1\}$, $f(e_2) = \{x_1, x_2\}, \ldots, f(e_n) = \{x_{n-1}, x_n\}$, where $x_0 = u$ and $x_n = v$. When the graph is simple, we denote this path by its vertex sequence $x_0, x_1, \ldots, x_n$ (since listing these vertices uniquely determines the path). The path is a *circuit* if it begins and ends at the same vertex, that is, if $u = v$, and has length greater than zero. The path or circuit is said to *pass through* the vertices $x_1, x_2, \ldots, x_{n-1}$ or *traverse* the edges $e_1, e_2, \ldots, e_n$. A path or circuit is *simple* if it does not contain the same edge more than once.

When it is not necessary to distinguish between multiple edges, we will denote a path $e_1, e_2, \ldots, e_n$, where $f(e_i) = \{x_{i-1}, x_i\}$ for $i = 1, 2, \ldots, n$ by its vertex sequence $x_0, x_1, \ldots, x_n$. This notation identifies a path only up to the vertices it passes through. There may be more than one path that passes through this sequence of vertices. Note that a path of length zero consists of a single vertex.

EXAMPLE 1   In the simple graph shown in Figure 1, a, d, c, f, e is a simple path of length 4, since $\{a, d\}$, $\{d, c\}, \{c, f\}$, and $\{f, e\}$ are all edges. However, d, e, c, a is not a path, since $\{e, c\}$ is not

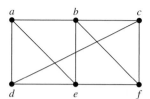

FIGURE 1   A Simple Graph.

an edge. Note that b, c, f, e, b is a circuit of length 4 since $\{b, c\}, \{c, f\}, \{f, e\}$, and $\{e, b\}$ are edges, and this path begins and ends at b. The path a, b, e, d, a, b, which is of length 5, is not simple since it contains the edge $\{a, b\}$ twice.   ◀

Paths and circuits in directed graphs were introduced in Chapter 7. We now define such paths for directed multigraphs.

DEFINITION 2

Let n be a nonnegative integer and G a directed multigraph. A *path* of length n from u to v in G is a sequence of edges $e_1, e_2, \ldots, e_n$ of G such that $f(e_1) = (x_0, x_1)$, $f(e_2) = (x_1, x_2), \ldots, f(e_n) = (x_{n-1}, x_n)$, where $x_0 = u$ and $x_n = v$. When there are no multiple edges in the directed graph, this path is denoted by its vertex sequence $x_0, x_1, x_2, \ldots, x_n$. A path of length greater than zero that begins and ends at the same vertex is called a *circuit* or *cycle*. A path or circuit is called *simple* if it does not contain the same edge more than once.

Note that the terminal vertex of an edge in a path is the initial vertex of the next edge in the path. When it is not necessary to distinguish between multiple edges, we will denote a path $e_1, e_2, \ldots, e_n$ where $f(e_i) = (x_{i-1}, x_i)$ for $i = 1, 2, \ldots, n$ by its vertex sequence $x_0, x_1, \ldots, x_n$. The notation identifies a path only up to the vertices it passes through. There may be more than one path that passes through this sequence of vertices.

Paths represent useful information in many graph models, as Examples 2–4 demonstrate.

EXAMPLE 2

Paths in Acquaintanceship Graphs In an acquaintanceship graph there is a path between two people if there is a chain of people linking these people, where two people adjacent in the chain know one another. For example, in Figure 7 in Section 8.1, there is a chain of six people linking Kamini and Ching. Many social scientists have conjectured that almost every pair of people in the world are linked by a small chain of people, perhaps containing just five or fewer people. This would mean that almost every pair of vertices in the acquaintanceship graph containing all people in the world is linked by a path of length not exceeding four. The play *Six Degrees of Separation* by John Guare is based on this notion.   ◀

EXAMPLE 3

Paths in Collaboration Graphs In a collaboration graph two vertices a and b, which represent authors are connected by a path when there is a sequence of authors beginning at a and ending at b such that the two authors represented by the endpoints of each edge have written a joint paper. In the collaboration graph of all mathematicians, the **Erdős number** of a mathematician m (defined in terms of relations in Supplementary Exercise 14 in Chapter 7) is the length of the shortest path between m and the vertex representing

TABLE 1 The Number of Mathematicians with a Given Erdős Number.	
Erdős Number	**Number of People**
0	1
1	502
2	713
3	26,422
4	62,136
5	66,157
6	32,280
7	10,431
8	3,214
9	953
10	262
11	94
12	23
13	4
14	7
15	1

TABLE 2 The Number of Actors with a Given Bacon Number.	
Bacon Number	**Number of People**
0	1
1	1,479
2	115,204
3	285,929
4	65,021
5	4,535
6	534
7	81
8	28
9	1
10	1

the extremely prolific mathematician Paul Erdős (who died in 1996). That is, the Erdős number of a mathematician is the length of the shortest chain of mathematicians that begins with Paul Erdős and ends with this mathematician, where each adjacent pair of mathematicians have written a joint paper. The number of mathematicians with each Erdős number as of November 2001, according to the Erdős Number Project, is shown in Table 1. ◀

EXAMPLE 4

Links

Paths in the Hollywood Graph In the Hollywood graph (see Example 4 in Section 8.1) two vertices a and b are linked when there is a chain of actors linking a and b where every two actors adjacent in the chain have acted in the same movie. In the Hollywood graph, the **Bacon number** of an actor c is defined to be the length of the shortest path connecting c and the well-known actor Kevin Bacon. As new movies are made, including new ones with Kevin Bacon, the Bacon number of actors can change. In Table 2 we show the number of actors with each Bacon number as of November 2001 using data from the Oracle of Bacon website. ◀

CONNECTEDNESS IN UNDIRECTED GRAPHS

When does a computer network have the property that every pair of computers can share information, if messages can be sent through one or more intermediate computers? When a graph is used to represent this computer network, where vertices represent the computers and edges represent the communication links, this question becomes: When is there always a path between two vertices in the graph?

DEFINITION 3

An undirected graph is called *connected* if there is a path between every pair of distinct vertices of the graph.

Thus, any two computers in the network can communicate if and only if the graph of this network is connected.

EXAMPLE 5 The graph G_1 in Figure 2 is connected, since for every pair of distinct vertices there is a path between them (the reader should verify this). However, the graph G_2 in Figure 2 is not connected. For instance, there is no path in G_2 between vertices a and d. ◀

We will need the following theorem in Chapter 9.

THEOREM 1

There is a simple path between every pair of distinct vertices of a connected undirected graph.

Proof: Let u and v be two distinct vertices of the connected undirected graph $G = (V, E)$. Since G is connected, there is at least one path between u and v. Let $x_0, x_1, \ldots, x_n$, where $x_0 = u$ and $x_n = v$, be the vertex sequence of a path of least length. This path of least length is simple. To see this, suppose it is not simple. Then $x_i = x_j$ for some i and j with $0 \leq i < j$. This means that there is a path from u to v of shorter length with vertex sequence $x_0, x_1, \ldots, x_{i-1}, x_j, \ldots, x_n$ obtained by deleting the edges corresponding to the vertex sequence $x_i, \ldots, x_{j-1}$. ◁

A graph that is not connected is the union of two or more connected subgraphs, each pair of which has no vertex in common. These disjoint connected subgraphs are called the **connected components** of the graph.

EXAMPLE 6 What are the connected components of the graph H shown in Figure 3?

Solution: The graph H is the union of three disjoint connected subgraphs H_1, H_2, and H_3, shown in Figure 3. These three subgraphs are the connected components of H. ◀

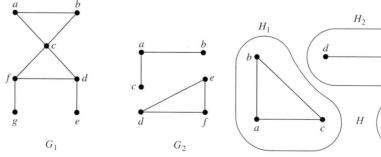

FIGURE 2   **The Graphs G_1 and G_2.**

FIGURE 3   **The Graph H and Its Connected Components H_1, H_2, and H_3.**

EXAMPLE 7   **Connected Components of Call Graphs** Two vertices x and y are in the same component of a telephone call graph (see Example 7 in Section 8.1) when there is a sequence of telephone calls beginning at x and ending at y. When a call graph for telephone calls made during a particular day in the AT&T network was analyzed, this graph was found to have 53,767,087 vertices, more than 170 million edges, and more than 3.7 million connected components. Most of these components were small; approximately three-fourths consisted of two vertices representing pairs of telephone numbers that called only each other. This graph has one huge connected component with 44,989,297 vertices comprising more than 80% of the total. Furthermore, every vertex in this component can be linked to any other vertex by a chain of no more than 20 calls.   ◄

Sometimes the removal of a vertex and all edges incident with it produces a subgraph with more connected components than in the original graph. Such vertices are called **cut vertices** (or **articulation points**). The removal of a cut vertex from a connected graph produces a subgraph that is not connected. Analogously, an edge whose removal produces a graph with more connected components than in the original graph is called a **cut edge** or **bridge.**

EXAMPLE 8 Find the cut vertices and cut edges in the graph G shown in Figure 4.

Solution: The cut vertices of G are b, c, and e. The removal of one of these vertices (and its adjacent edges) disconnects the graph. The cut edges are $\{a, b\}$ and $\{c, e\}$. Removing either one of these edges disconnects G.   ◄

CONNECTEDNESS IN DIRECTED GRAPHS

There are two notions of connectedness in directed graphs, depending on whether the directions of the edges are considered.

DEFINITION 4 A directed graph is *strongly connected* if there is a path from a to b and from b to a whenever a and b are vertices in the graph.

For a directed graph to be strongly connected there must be a sequence of directed edges from any vertex in the graph to any other vertex. A directed graph can fail to be strongly connected but still be in "one piece." Definition 5 makes this notion precise.

DEFINITION 5 A directed graph is *weakly connected* if there is a path between every two vertices in the underlying undirected graph.

That is, a directed graph is weakly connected if and only if there is always a path between two vertices when the directions of the edges are disregarded. Clearly, any strongly connected directed graph is also weakly connected.

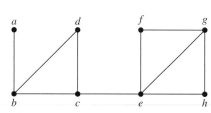

FIGURE 4   The Graph G.

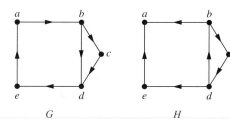

FIGURE 5   The Directed Graphs G and H.

EXAMPLE 9   Are the directed graphs G and H shown in Figure 5 strongly connected? Are they weakly connected?

Solution: G is strongly connected because there is a path between any two vertices in this directed graph (the reader should verify this). Hence, G is also weakly connected. The graph H is not strongly connected. There is no directed path from a to b in this graph. However, H is weakly connected, since there is a path between any two vertices in the underlying undirected graph of H (the reader should verify this). ◄

The subgraphs of a directed graph G that are strongly connected but not contained in larger strongly connected subgraphs, that is, the maximal strongly connected subgraphs, are called the **strongly connected components** or **strong components** of G.

EXAMPLE 10   The graph H in Figure 5 has three strongly connected components, consisting of the vertex a; the vertex e; and the graph consisting of the vertices $b, c,$ and d and edges $(b, c), (c, d),$ and (d, b). ◄

EXAMPLE 11   **The Strongly Connected Components of the Web Graph** The Web graph introduced in Example 8 of Section 8.1 represents Web pages with vertices and links with directed edges. A snapshot of the Web in 1999 produced a Web graph with over 200 million vertices and over 1.5 billion edges. (See [Br00] for details.) The underlying undirected graph of this Web graph is not connected and has a connected component that includes approximately 90% of the vertices in the graph. The subgraph of the original directed graph corresponding to this connected component of the underlying undirected graph (that is, with the same vertices and all directed edges connecting vertices in this graph) has one very large strongly connected component and many small ones. The former is called the **giant strongly connected component (GSCC)** of the directed graph. A Web page in this component can be reached following links starting at any other page in this component. The GSCC in the Web graph produced by this study was found to have over 53 million vertices. The remaining vertices in the large connected component of the undirected graph represent three different types of Web pages: pages that can be reached from a page in the GSCC, but do not link back to these pages following a series of links; pages that link back to pages in the GSCC following a series of links, but cannot be reached by following links on pages in the GSCC; and pages that cannot reach pages in the GSCC and cannot be reached from pages in the GSCC following a series of links. In this study, each of these three other sets was found to have approximately 44 million vertices. (It is rather surprising that these three sets are close to the same size.) ◄

PATHS AND ISOMORPHISM

There are several ways that paths and circuits can help determine whether two graphs are isomorphic. For example, the existence of a simple circuit of a particular length is a useful invariant that can be used to show that two graphs are not isomorphic. In addition, paths can be used to construct mappings that may be isomorphisms.

As we mentioned, a useful isomorphic invariant for simple graphs is the existence of a simple circuit of length k, where k is a positive integer greater than 2. (The proof that this is an invariant is left as Exercise 44 at the end of this section.) Example 12 illustrates how this invariant can be used to show that two graphs are not isomorphic.

EXAMPLE 12   Determine whether the graphs G and H shown in Figure 6 are isomorphic.

Solution: Both G and H have six vertices and eight edges. Each has four vertices of degree three, and two vertices of degree two. So, the three invariants—number of vertices, number of edges, and degrees of vertices—all agree for the two graphs. However, H has a simple circuit of length three, namely, v_1, v_2, v_6, v_1 whereas G has no simple circuit of length three, as can be determined by inspection (all simple circuits in G have length at least four). Since the existence of a simple circuit of length three is an isomorphic invariant, G and H are not isomorphic.   ◄

We have shown how the existence of a type of path, namely, a simple circuit of a particular length, can be used to show that two graphs are not isomorphic. We can also use paths to find mappings that are potential isomorphisms.

EXAMPLE 13   Determine whether the graphs G and H shown in Figure 7 are isomorphic.

Solution: Both G and H have five vertices and six edges, both have two vertices of degree three and three vertices of degree two, and both have a simple circuit of length three, a simple circuit of length four, and a simple circuit of length five. Since all these isomorphic invariants agree, G and H may be isomorphic. To find a possible isomorphism, we can follow paths that go through all vertices so that the corresponding vertices in the two graphs have the same degree. For example, the paths u_1, u_4, u_3, u_2, u_5 in G and v_3, v_2, v_1, v_5, v_4 in H both go through every vertex in the graph, start at a vertex of degree three, go through vertices of degrees two, three, and two, respectively, and end at a vertex of degree two. By following these paths through the graphs, we define the mapping f with $f(u_1) = v_3$, $f(u_4) = v_2$, $f(u_3) = v_1$, $f(u_2) = v_5$, and $f(u_5) = v_4$. The reader can show that f is an isomorphism, so that G and H are isomorphic, either by showing that f preserves edges, or by showing that with the appropriate orderings of vertices the adjacency matrices of G and H are the same.   ◄

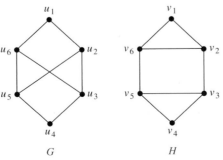

FIGURE 6   **The Graphs G and H.**

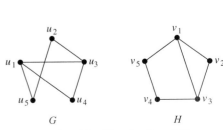

FIGURE 7   **The Graphs G and H.**

COUNTING PATHS BETWEEN VERTICES

The number of paths between two vertices in a graph can be determined using its adjacency matrix.

THEOREM 2

Let G be a graph with adjacency matrix $\mathbf{A}$ with respect to the ordering $v_1, v_2, \ldots, v_n$ (with directed or undirected edges, with multiple edges and loops allowed). The number of different paths of length r from v_i to v_j, where r is a positive integer, equals the (i, j)th entry of $\mathbf{A}^r$.

Proof: The theorem will be proved using mathematical induction. Let G be a graph with adjacency matrix $\mathbf{A}$ (assuming an ordering $v_1, v_2, \ldots, v_n$ of the vertices of G). The number of paths from v_i to v_j of length 1 is the (i, j)th entry of $\mathbf{A}$, since this entry is the number of edges from v_i to v_j.

Assume that the (i, j)th entry of $\mathbf{A}^r$ is the number of different paths of length r from v_i to v_j. This is the induction hypothesis. Since $\mathbf{A}^{r+1} = \mathbf{A}^r \mathbf{A}$, the (i, j)th entry of $\mathbf{A}^{r+1}$ equals

$$b_{i1}a_{1j} + b_{i2}a_{2j} + \cdots + b_{in}a_{nj},$$

where b_{ik} is the (i, k)th entry of $\mathbf{A}^r$. By the induction hypothesis, b_{ik} is the number of paths of length r from v_i to v_k.

A path of length $r + 1$ from v_i to v_j is made up of a path of length r from v_i to some intermediate vertex v_k, and an edge from v_k to v_j. By the product rule for counting, the number of such paths is the product of the number of paths of length r from v_i to v_k, namely, b_{ik}, and the number of edges from v_k to v_j, namely, a_{kj}. When these products are added for all possible intermediate vertices v_k, the desired result follows by the sum rule for counting. ◁

EXAMPLE 14

How many paths of length 4 are there from a to d in the simple graph G in Figure 8?

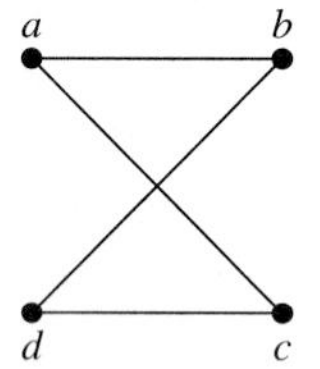

Solution: The adjacency matrix of G (ordering the vertices as a, b, c, d) is

$$\mathbf{A} = \begin{bmatrix} 0 & 1 & 1 & 0 \\ 1 & 0 & 0 & 1 \\ 1 & 0 & 0 & 1 \\ 0 & 1 & 1 & 0 \end{bmatrix}.$$

FIGURE 8   The Graph *G*.

Hence, the number of paths of length 4 from a to d is the $(1, 4)$th entry of $\mathbf{A}^4$. Since

$$\mathbf{A}^4 = \begin{bmatrix} 8 & 0 & 0 & 8 \\ 0 & 8 & 8 & 0 \\ 0 & 8 & 8 & 0 \\ 8 & 0 & 0 & 8 \end{bmatrix},$$

there are exactly eight paths of length 4 from a to d. By inspection of the graph, we see that a, b, a, b, d; a, b, a, c, d; a, b, d, b, d; a, b, d, c, d; a, c, a, b, d; a, c, a, c, d; a, c, d, b, d; and a, c, d, c, d are the eight paths from a to d. ◄

Theorem 2 can be used to find the length of the shortest path between two vertices of a graph (see Exercise 40), and it can also be used to determine whether a graph is connected (see Exercises 45 and 46).

Exercises

1. Does each of these lists of vertices form a path in the following graph? Which paths are simple? Which are circuits? What are the lengths of those that are paths?

a) a, e, b, c, b **b)** a, e, a, d, b, c, a
c) e, b, a, d, b, e **d)** c, b, d, a, e, c

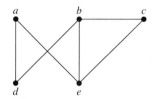

2. Does each of these lists of vertices form a path in the following graph? Which paths are simple? Which are circuits? What are the lengths of those that are paths?

a) a, b, e, c, b **b)** a, d, a, d, a
c) a, d, b, e, a **d)** a, b, e, c, b, d, a

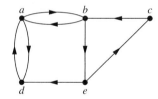

In Exercises 3–5 determine whether the given graph is connected.

3.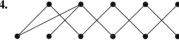

4.

5.

6. How many connected components does each of the graphs in Exercises 3–5 have? For each graph find each of its connected components.

7. What do the connected components of acquaintance-ship graphs represent?

8. What do the connected components of a collaboration graph represent?

9. Explain why in the collaboration graph of mathematicians a vertex representing a mathematician is in the same connected component as the vertex representing Paul Erdős if and only if that mathematician has a finite Erdős number.

10. In the Hollywood graph (see Example 4 in Section 8.1), when is the vertex representing an actor in the same connected component as the vertex representing Kevin Bacon?

11. What do the strongly connected components of a telephone call graph represent?

12. Find the strongly connected components of each of these graphs.

a)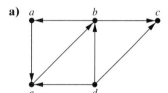

b)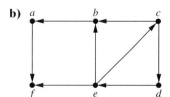

c)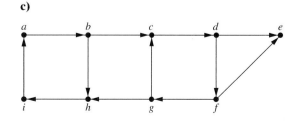

13. Find the strongly connected components of each of these graphs.

a)

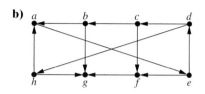

b)

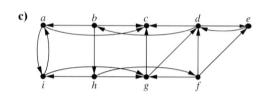

c)

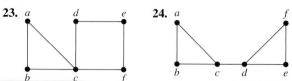

14. Show that all vertices visited in a directed path connecting two vertices in the same strongly connected component of a directed graph are also in this strongly connected component.

15. Find the number of paths of length n between two different vertices in K_4 if n is

a) 2. **b)** 3. **c)** 4. **d)** 5.

16. Find the number of paths of length n between any two adjacent vertices in $K_{3,3}$ for the values of n in Exercise 15.

17. Find the number of paths of length n between any two nonadjacent vertices in $K_{3,3}$ for the values of n in Exercise 15.

18. Find the number of paths between c and d in the graph in Figure 1 of length

a) 2. **b)** 3. **c)** 4. **d)** 5. **e)** 6. **f)** 7.

19. Find the number of paths from a to e in the directed graph in Exercise 2 of length

a) 2. **b)** 3. **c)** 4. **d)** 5. **e)** 6. **f)** 7.

*20. Show that a connected graph with n vertices has at least $n - 1$ edges.

21. Let $G = (V, E)$ be a simple graph. Let R be the relation on V consisting of pairs of vertices (u, v) such that there is a path from u to v or such that $u = v$. Show that R is an equivalence relation.

*22. Show that in any simple graph there is a path from any vertex of odd degree to some other vertex of odd degree.

In Exercises 23–25 find all the cut vertices of the given graph.

23.

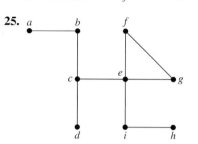

24.

25.

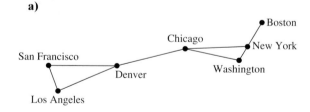

26. Find all the cut edges in the graphs in Exercises 23–25.

*27. Suppose that v is an endpoint of a cut edge. Prove that v is a cut vertex if and only if this vertex is not pendant.

*28. Show that a vertex c in the connected simple graph G is a cut vertex if and only if there are vertices u and v, both different from c, such that every path between u and v passes through c.

*29. Show that a simple graph with at least two vertices has at least two vertices that are not cut vertices.

*30. Show that an edge in a simple graph is a cut edge if and only if this edge is not part of any simple circuit in the graph.

31. A communications link in a network should be provided with a backup link if its failure makes it impossible for some message to be sent. For each of the communications networks shown here in (a) and (b), determine those links that should be backed up.

a)

b)

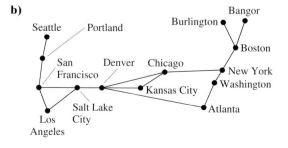

A **vertex basis** in a directed graph is a set of vertices such that there is a path to every vertex in the directed graph not in the set from some vertex in this set and there is no path from any vertex in the set to another vertex in the set.

32. Find a vertex basis for each of the directed graphs in Exercises 7–9 of Section 8.2.

33. What is the significance of a vertex basis in an influence graph (described in Example 3 of Section 8.1)? Find a vertex basis in the influence graph in this example.

☞ **34.** Show that if a connected simple graph G is the union of the graphs G_1 and G_2, then G_1 and G_2 have at least one common vertex.

***35.** Show that if a simple graph G has k connected components and these components have $n_1, n_2, \ldots, n_k$ vertices, respectively, then the number of edges of G does not exceed

$$\sum_{i=1}^{k} C(n_i, 2).$$

***36.** Use Exercise 35 to show that a simple graph with n vertices and k connected components has at most $(n - k)(n - k + 1)/2$ edges. [*Hint:* First show that

$$\sum_{i=1}^{k} n_i^2 \leq n^2 - (k - 1)(2n - k)$$

where n_i is the number of vertices in the ith connected component.]

***37.** Show that a simple graph G with n vertices is connected if it has more than $(n - 1)(n - 2)/2$ edges.

38. Describe the adjacency matrix of a graph with n connected components when the vertices of the graph are listed so that vertices in each connected component are listed successively.

39. How many nonisomorphic connected simple graphs are there with n vertices when n is

a) 2? **b)** 3? **c)** 4? **d)** 5?

40. Explain how Theorem 2 can be used to find the length of the shortest path from a vertex v to a vertex w in a graph.

41. Use Theorem 2 to find the length of the shortest path between a and f in the multigraph in Figure 1.

42. Use Theorem 2 to find the length of the shortest path from a to c in the directed graph in Exercise 2.

☞ **43.** Let P_1 and P_2 be two simple paths between the vertices u and v in the simple graph G that do not contain the same set of edges. Show that there is a simple circuit in G.

44. Show that the existence of a simple circuit of length k, where k is a positive integer greater than 2, is an isomorphic invariant.

45. Explain how Theorem 2 can be used to determine whether a graph is connected.

46. Use Exercise 45 to show that the graph G in Figure 2 is connected whereas the graph H in that figure is not connected.

47. Show that a simple graph G is bipartite if and only if it has no circuits with an odd number of edges.

8.5   Euler and Hamilton Paths

INTRODUCTION

Can we travel along the edges of a graph starting at a vertex and returning to it by traversing each edge of the graph exactly once? Similarly, can we travel along the edges of a graph starting at a vertex and returning to it while visiting each vertex of the graph exactly once? Although these questions seem to be similar, the first question, which asks whether a graph has an *Euler circuit*, can be easily answered for all graphs, while the second question, which asks whether a graph has a *Hamilton circuit*, is quite difficult to solve. In this section we will study these questions and discuss the difficulty of solving them. Although both questions have many practical applications in many different areas, both arose in old puzzles. We will learn about these old puzzles as well as modern practical applications.

EULER PATHS AND CIRCUITS

Links

The town of Königsberg, Prussia (now called Kaliningrad and part of the Russian republic), was divided into four sections by the branches of the Pregel River. These four

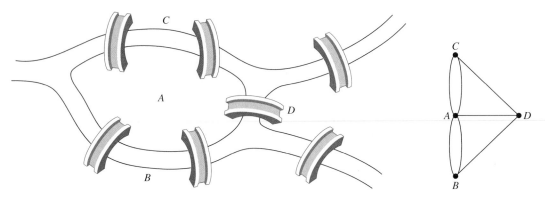

FIGURE 1 **The Seven Bridges of Königsberg.**

FIGURE 2 **Multigraph Model of the Town of Königsberg.**

sections included the two regions on the banks of the Pregel, Kneiphof Island, and the region between the two branches of the Pregel. In the eighteenth century seven bridges connected these regions. Figure 1 depicts these regions and bridges.

The townspeople took long walks through town on Sundays. They wondered whether it was possible to start at some location in the town, travel across all the bridges without crossing any bridge twice, and return to the starting point.

The Swiss mathematician Leonhard Euler solved this problem. His solution, published in 1736, may be the first use of graph theory. Euler studied this problem using the multigraph obtained when the four regions are represented by vertices and the bridges by edges. This multigraph is shown in Figure 2.

The problem of traveling across every bridge without crossing any bridge more than once can be rephrased in terms of this model. The question becomes: Is there a simple circuit in this multigraph that contains every edge?

DEFINITION 1

> An *Euler circuit* in a graph G is a simple circuit containing every edge of G. An *Euler path* in G is a simple path containing every edge of G.

Examples 1 and 2 illustrate the concept of Euler circuits and paths.

EXAMPLE 1 Which of the undirected graphs in Figure 3 have an Euler circuit? Of those that do not, which have an Euler path?

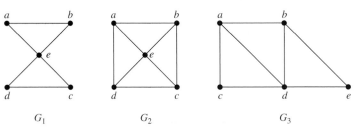

FIGURE 3 **The Undirected Graphs G_1, G_2, and G_3.**

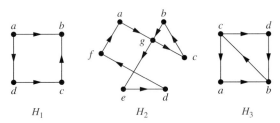

FIGURE 4   **The Directed Graphs $H_1, H_2,$ and H_3.**

Solution: The graph G_1 has an Euler circuit, for example a, e, c, d, e, b, a. Neither of the graphs G_2 or G_3 has an Euler circuit (the reader should verify this). However, G_3 has an Euler path, namely, a, c, d, e, b, d, a, b. G_2 does not have an Euler path (as the reader should verify). ◀

EXAMPLE 2   Which of the directed graphs in Figure 4 have an Euler circuit? Of those that do not, which have an Euler path?

Solution: The graph H_2 has an Euler circuit, for example, $a, g, c, b, g, e, d, f, a$. Neither H_1 nor H_3 has an Euler circuit (as the reader should verify). H_3 has an Euler path, namely c, a, b, c, d, b, but H_1 does not (as the reader should verify). ◀

NECESSARY AND SUFFICIENT CONDITIONS FOR EULER CIRCUITS AND PATHS There are simple criteria for determining whether a multigraph has an Euler circuit or an Euler path. Euler discovered them when he solved the famous Königsberg bridge problem. We will assume that all graphs discussed in this section have a finite number of vertices and edges.

What can we say if a connected multigraph has an Euler circuit? What we can show is that every vertex must have even degree. To do this, first note that an Euler circuit begins with a vertex a and continues with an edge incident to a, say $\{a, b\}$. The edge $\{a, b\}$ contributes one to $\deg(a)$. Each time the circuit passes through a vertex it contributes two to the vertex's degree, since the circuit enters via an edge incident with this vertex and leaves via another such edge. Finally, the circuit terminates where it started, contributing one

LEONHARD EULER (1707–1783) Leonhard Euler was the son of a Calvinist minister from the vicinity of Basel, Switzerland. At 13 he entered the University of Basel, pursuing a career in theology, as his father wished. At the university Euler was tutored by Johann Bernoulli of the famous Bernoulli family of mathematicians. His interest and skills led him to abandon his theological studies and take up mathematics. Euler obtained his master's degree in philosophy at the age of 16. In 1727 Peter the Great invited him to join the Academy at St. Petersburg. In 1741 he moved to the Berlin Academy, where he stayed until 1766. He then returned to St. Petersburg, where he remained for the rest of his life.

Euler was incredibly prolific, contributing to many areas of mathematics, including number theory, combinatorics, and analysis, as well as its applications to such areas as music and naval architecture. He wrote over 1100 books and papers and left so much unpublished work that it took 47 years after he died for all his work to be published. During his life his papers accumulated so quickly that he kept a large pile of articles awaiting publication. The Berlin Academy published the papers on top of this pile so later results were often published before results they depended on or superseded. Euler had 13 children and was able to continue his work while a child or two bounced on his knees. He was blind for the last 17 years of his life, but because of his fantastic memory this did not diminish his mathematical output. The project of publishing his collected works, undertaken by the Swiss Society of Natural Science, is still going on and will require more than 75 volumes.

to deg(a). Therefore, deg(a) must be even, because the circuit contributes one when it begins, one when it ends, and two every time it passes through a (if it ever does). A vertex other than a has even degree because the circuit contributes two to its degree each time it passes through the vertex. We conclude that if a connected graph has an Euler circuit, then every vertex must have even degree.

Is this necessary condition for the existence of an Euler circuit also sufficient? That is, must an Euler circuit exist in a connected multigraph if all vertices have even degree? This question can be settled affirmatively with a construction.

Suppose that G is a connected multigraph and the degree of every vertex of G is even. We will form a simple circuit that begins at an arbitrary vertex a of G. Let $x_0 = a$. First, we arbitrarily choose an edge $\{x_0, x_1\}$ incident with a. We continue by building a simple path $\{x_0, x_1\}, \{x_1, x_2\}, \ldots, \{x_{n-1}, x_n\}$ that is as long as possible. For instance, in the graph G in Figure 5 we begin at a and choose in succession the edges $\{a, f\}, \{f, c\},$ $\{c, b\},$ and $\{b, a\}$.

The path terminates since the graph has a finite number of edges. It begins at a with an edge of the form $\{a, x\}$, and it terminates at a with an edge of the form $\{y, a\}$. This follows because each time the path goes through a vertex with even degree, it uses only one edge to enter this vertex, so that at least one edge remains for the path to leave the vertex. This path may use all the edges, or it may not.

An Euler circuit has been constructed if all the edges have been used. Otherwise, consider the subgraph H obtained from G by deleting the edges already used and vertices that are not incident with any remaining edges. When we delete the circuit a, f, c, b, a from the graph in Figure 5, we obtain the subgraph labeled as H.

Since G is connected, H has at least one vertex in common with the circuit that has been deleted. Let w be such a vertex. (In our example, c is the vertex.)

Every vertex in H has even degree (because in G all vertices had even degree, and for each vertex, pairs of edges incident with this vertex have been deleted to form H). Note that H may not be connected. Beginning at w, construct a simple path in H by choosing edges as long as possible, as was done in G. This path must terminate at w. For instance, in Figure 5, c, d, e, c is a path in H. Next, form a circuit in G by splicing the circuit in H with the original circuit in G (this can be done since w is one of the vertices in this circuit). When this is done in the graph in Figure 5, we obtain the circuit a, f, c, d, e, c, b, a.

Continue this process until all edges have been used. (The process must terminate since there are only a finite number of edges in the graph.) This produces an Euler circuit. The construction shows that if the vertices of a connected multigraph all have even degree, then the graph has an Euler circuit.

We summarize these results in Theorem 1.

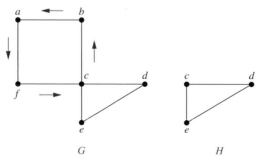

FIGURE 5   Constructing an Euler Circuit in G.

THEOREM 1   A connected multigraph has an Euler circuit if and only if each of its vertices has even degree.

We can now solve the Königsberg bridge problem. Since the multigraph representing these bridges, shown in Figure 2, has four vertices of odd degree, it does not have an Euler circuit. There is no way to start at a given point, cross each bridge exactly once, and return to the starting point.

Algorithm 1 gives the constructive procedure for finding Euler circuits given in the discussion preceding Theorem 1. (Since the circuits in the procedure are chosen arbitrarily, there is some ambiguity. We will not bother to remove this ambiguity by specifying the steps of the procedure more precisely.)

ALGORITHM 1 Constructing Euler Circuits.

---

procedure *Euler*(G: connected multigraph with all vertices of
        even degree)
circuit := a circuit in G beginning at an arbitrarily chosen
        vertex with edges successively added to form a path that
        returns to this vertex
H := G with the edges of this circuit removed
while H has edges
begin
        subcircuit := a circuit in H beginning at a vertex in H that
                also is an endpoint of an edge of *circuit*
        H := H with edges of *subcircuit* and all isolated vertices
                removed
        circuit := *circuit* with *subcircuit* inserted at the appropriate
                vertex
end {*circuit* is an Euler circuit}

The next example shows how Euler paths and circuits can be used to solve a type of puzzle.

EXAMPLE 3   Many puzzles ask you to draw a picture in a continuous motion without lifting a pencil so that no part of the picture is retraced. We can solve such puzzles using Euler circuits and paths. For example, can *Mohammed's scimitars,* shown in Figure 6, be drawn in this way, where the drawing begins and ends at the same point?

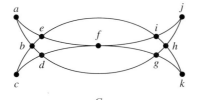

G

FIGURE 6   Mohammed's Scimitars.

Solution: We can solve this problem since the graph G shown in Figure 6 has an Euler circuit. It has such a circuit since all its vertices have even degree. We will use Algorithm 1 to construct an Euler circuit. First, we form the circuit $a, b, d, c, b, e, i, f, e, a$. We obtain the subgraph H by deleting the edges in this circuit and all vertices that become isolated when these edges are removed. Then we form the circuit $d, g, h, j, i, h, k, g, f, d$ in H. After forming this circuit we have used all edges in G. Splicing this new circuit into the first circuit at the appropriate place produces the Euler circuit $a, b, d, g, h, j, i, h, k, g, f, d, c, b, e, i, f, e, a$. This circuit gives a way to draw the scimitars without lifting the pencil or retracing part of the picture. ◀

Another algorithm for constructing Euler circuits, called Fleury's algorithm, is described in the exercises at the end of this section.

We will now show that a connected multigraph has an Euler path (and not an Euler circuit) if and only if it has exactly two vertices of odd degree. First, suppose that a connected multigraph does have an Euler path from a to b, but not an Euler circuit. The first edge of the path contributes one to the degree of a. A contribution of two to the degree of a is made every time the path passes through a. The last edge in the path contributes one to the degree of b. Every time the path goes through b there is a contribution of two to its degree. Consequently, both a and b have odd degree. Every other vertex has even degree, since the path contributes two to the degree of a vertex whenever it passes through it.

Now consider the converse. Suppose that a graph has exactly two vertices of odd degree, say a and b. Consider the larger graph made up of the original graph with the addition of an edge $\{a, b\}$. Every vertex of this larger graph has even degree, so that there is an Euler circuit. The removal of the new edge produces an Euler path in the original graph. Theorem 2 summarizes these results.

THEOREM 2   A connected multigraph has an Euler path but not an Euler circuit if and only if it has exactly two vertices of odd degree.

EXAMPLE 4   Which graphs shown in Figure 7 have an Euler path?

Solution: G_1 contains exactly two vertices of odd degree, namely, b and d. Hence, it has an Euler path that must have b and d as its endpoints. One such Euler path is d, a, b, c, d, b. Similarly, G_2 has exactly two vertices of odd degree, namely, b and d. So it has an Euler path that must have b and d as endpoints. One such Euler path is $b, a, g, f, e, d, c, g, b, c, f, d$. G_3 has no Euler path since it has six vertices of odd degree. ◀

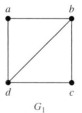
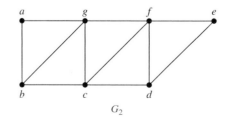
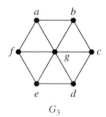

FIGURE 7   Three Undirected Graphs.

Returning to eighteenth-century Königsberg, is it possible to start at some point in the town, travel across all the bridges, and end up at some other point in town? This question can be answered by determining whether there is an Euler path in the multigraph representing the bridges in Königsberg. Since there are four vertices of odd degree in this multigraph, there is no Euler path, so such a trip is impossible.

Necessary and sufficient conditions for Euler paths and circuits in directed graphs are discussed in the exercises at the end of this section.

Euler paths and circuits can be used to solve many practical problems. For example, many applications ask for a path or circuit that traverses each street in a neighborhood, each road in a transportation network, each connection in a utility grid, or each link in a communications network exactly once. Finding an Euler path or circuit in the appropriate graph model can solve such problems. For example, if a postman can find an Euler path in the graph that represents the streets the postman needs to cover, this path produces a route that traverses each street of the route exactly once. If no Euler path exists, some streets will have to be traversed more than once. This problem is known as the *Chinese postman problem* in honor of Guan Meigu, who posed it in 1962. See [MiRo91] for more information on the solution of the Chinese postman problem when no Euler path exists.

Among the other areas where Euler circuits and paths are applied is in layout of circuits, in network multicasting, and in molecular biology, where Euler paths are used in the sequencing of DNA.

HAMILTON PATHS AND CIRCUITS

We have developed necessary and sufficient conditions for the existence of paths and circuits that contain every edge of a multigraph exactly once. Can we do the same for simple paths and circuits that contain every vertex of the graph exactly once?

DEFINITION 2

A path $x_0, x_1, \ldots, x_{n-1}, x_n$ in the graph $G = (V, E)$ is called a *Hamilton path* if $V = \{x_0, x_1, \ldots, x_{n-1}, x_n\}$ and $x_i \neq x_j$ for $0 \leq i < j \leq n$. A circuit $x_0, x_1, \ldots, x_{n-1}, x_n, x_0$ (with $n > 1$) in a graph $G = (V, E)$ is called a *Hamilton circuit* if $x_0, x_1, \ldots, x_{n-1}, x_n$ is a Hamilton path.

This terminology comes from a game, called the *Icosian puzzle*, invented in 1857 by the Irish mathematician Sir William Rowan Hamilton. It consisted of a wooden dodecahedron [a polyhedron with 12 regular pentagons as faces, as shown in Figure 8(a)], with a peg at each vertex of the dodecahedron, and string. The 20 vertices of the dodecahedron were labeled with different cities in the world. The object of the puzzle was to start at a city and travel along the edges of the dodecahedron, visiting each of the other 19 cities exactly once, and end back at the first city. The circuit traveled was marked off using the strings and pegs.

Since the author cannot supply each reader with a wooden solid with pegs and string, we will consider the equivalent question: Is there a circuit in the graph shown in Figure 8(b) that passes through each vertex exactly once? This solves the puzzle since this graph is isomorphic to the graph consisting of the vertices and edges of the dodecahedron. A solution of Hamilton's puzzle is shown in Figure 9.

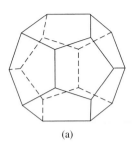

(a)

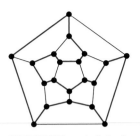

(b)

FIGURE 8 Hamilton's "A Voyage Round the World" Puzzle.

FIGURE 9 A Solution to the "A Voyage Round the World" Puzzle.

EXAMPLE 5 Which of the simple graphs in Figure 10 have a Hamilton circuit or, if not, a Hamilton path?

Solution: G_1 has a Hamilton circuit: a, b, c, d, e, a. There is no Hamilton circuit in G_2 (this can be seen by noting that any circuit containing every vertex must contain the edge $\{a, b\}$ twice), but G_2 does have a Hamilton path, namely a, b, c, d. G_3 has neither a Hamilton circuit nor a Hamilton path, since any path containing all vertices must contain one of the edges $\{a, b\}, \{e, f\}$, and $\{c, d\}$ more than once. ◀

Is there a simple way to determine whether a graph has a Hamilton circuit or path? At first, it might seem that there should be an easy way to determine this, since there

WILLIAM ROWAN HAMILTON (1805–1865) William Rowan Hamilton, the most famous Irish scientist ever to have lived, was born in 1805 in Dublin. His father was a successful lawyer, his mother came from a family noted for their intelligence, and he was a child prodigy. By the age of 3 he was an excellent reader and had mastered advanced arithmetic. Because of his brilliance, he was sent off to live with his uncle James, a noted linguist. By age 8 Hamilton had learned Latin, Greek, and Hebrew; by 10 he had also learned Italian and French and he began his study of oriental languages, including Arabic, Sanskrit, and Persian. During this period he took pride in knowing as many languages as his age. At 17, no longer devoted to learning new languages and having mastered calculus and much mathematical astronomy, he began original work in optics, and he also found an important mistake in Laplace's work on celestial mechanics. Before entering Trinity College, Dublin, at 18, Hamilton had not attended school; rather, he received private tutoring. At Trinity, he was a superior student in both the sciences and the classics. Prior to receiving his degree, because of his brilliance he was appointed the Astronomer Royal of Ireland, beating out several famous astronomers for the post. He held this position until his death, living and working at Dunsink Observatory outside of Dublin. Hamilton made important contributions to optics, abstract algebra, and dynamics. Hamilton invented algebraic objects called quaternions as an example of a noncommutative system. He discovered the appropriate way to multiply quaternions while walking along a canal in Dublin. In his excitement, he carved the formula in the stone of a bridge crossing the canal, a spot marked today by a plaque. Later, Hamilton remained obsessed with quaternions, working to apply them to other areas of mathematics, instead of moving to new areas of research.

In 1857 Hamilton invented "The Icosian Game" based on his work in noncommutative algebra. He sold the idea for 25 pounds to a dealer in games and puzzles. (Since the game never sold well, this turned out to be a bad investment for the dealer.) The "Traveler's Dodecahedron," also called "A Voyage Round the World," the puzzle described in this section, is a variant of that game.

Hamilton married his third love in 1833, but his marriage worked out poorly, since his wife, a semi-invalid, was unable to cope with his household affairs. He suffered from alcoholism and lived reclusively for the last two decades of his life. He died from gout in 1865, leaving masses of papers containing unpublished research. Mixed in with these papers were a large number of dinner plates, many containing the remains of desiccated, uneaten chops.

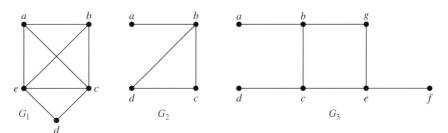

FIGURE 10   Three Simple Graphs.

is a simple way to answer the similar question of whether a graph has an Euler circuit. Surprisingly, there are no known simple necessary and sufficient criteria for the existence of Hamilton circuits. However, many theorems are known that give sufficient conditions for the existence of Hamilton circuits. Also, certain properties can be used to show that a graph has no Hamilton circuit. For instance, a graph with a vertex of degree one cannot have a Hamilton circuit, since in a Hamilton circuit each vertex is incident with two edges in the circuit. Moreover, if a vertex in the graph has degree two, then both edges that are incident with this vertex must be part of any Hamilton circuit. Also, note that when a Hamilton circuit is being constructed and this circuit has passed through a vertex, then all remaining edges incident with this vertex, other than the two used in the circuit, can be removed from consideration. Furthermore, a Hamilton circuit cannot contain a smaller circuit within it.

EXAMPLE 6   Show that neither graph displayed in Figure 11 has a Hamilton circuit.

Solution: There is no Hamilton circuit in G since G has a vertex of degree one, namely, e. Now consider H. Since the degrees of the vertices a, b, d, and e are all two, every edge incident with these vertices must be part of any Hamilton circuit. It is now easy to see that no Hamilton circuit can exist in H, for any Hamilton circuit would have to contain four edges incident with c, which is impossible.   ◀

EXAMPLE 7   Show that K_n has a Hamilton circuit whenever $n \geq 3$.

Solution: We can form a Hamilton circuit in K_n beginning at any vertex. Such a circuit can be built by visiting vertices in any order we choose, as long as the path begins and

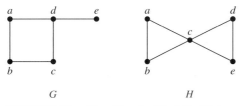

FIGURE 11   Two Graphs That Do Not Have a Hamilton Circuit.

ends at the same vertex and visits each other vertex exactly once. This is possible since there are edges in K_n between any two vertices. ◀

Although no useful necessary and sufficient conditions for the existence of Hamilton circuits are known, quite a few sufficient conditions have been found. Note that the more edges a graph has, the more likely it is to have a Hamilton circuit. Furthermore, adding edges (but not vertices) to a graph with a Hamilton circuit produces a graph with the same Hamilton circuit. Consequently, as we add edges to a graph, especially when we make sure to add edges to each vertex, we make it increasingly likely that a Hamilton circuit exists in this graph. Consequently, we would expect there to be sufficient conditions for the existence of Hamilton circuits that depend on the degrees of vertices being sufficiently large. We state two of the most important sufficient conditions here. These conditions were found by Gabriel A. Dirac in 1952 and Oystein Ore in 1960.

THEOREM 3 **DIRAC'S THEOREM** If G is a simple graph with n vertices with $n \geq 3$ such that the degree of every vertex in G is at least $n/2$, then G has a Hamilton circuit.

THEOREM 4 **ORE'S THEOREM** If G is a simple graph with n vertices with $n \geq 3$ such that $\deg(u) + \deg(v) \geq n$ for every pair of nonadjacent vertices u and v in G, then G has a Hamilton circuit.

The proof of Ore's Theorem is outlined in Exercise 65 at the end of this section. Dirac's Theorem can be proved as a corollary to Ore's Theorem since the conditions of Dirac's Theorem imply those of Ore's Theorem.

Both Ore's Theorem and Dirac's Theorem provide sufficient conditions for a connected simple graph to have a Hamilton circuit. However, these theorems do not provide necessary conditions for the existence of a Hamilton circuit. For example, the graph C_5 has a Hamilton circuit but does not satisfy the hypotheses of either Ore's Theorem or Dirac's Theorem, as the reader can verify.

The best algorithms known for finding a Hamilton circuit in a graph or determining that no such circuit exists have exponential worst-case time complexity (in the number of vertices of the graph). Finding an algorithm that solves this problem with polynomial worst-case time complexity would be a major accomplishment because it has been shown that this problem is NP-complete (see Section 2.3). Consequently, the existence of such an algorithm would imply that many other seemingly intractable problems could be solved using algorithms with polynomial worst-case time complexity.

Hamilton paths and circuits can be used to solve practical problems. For example, many applications ask for a path or circuit that visits each road intersection in a city, each place pipelines intersect in a utility grid, or each node in a communications network exactly once. Finding a Hamilton path or circuit in the appropriate graph model can solve such problems. The famous **traveling salesman problem** asks for the shortest route a traveling salesman should take to visit a set of cities. This problem reduces to finding a Hamilton circuit in a complete graph such that the total weight of its edges is as small as possible. We will return to this question in Section 8.6.

We now describe a less obvious application of Hamilton circuits to coding.

EXAMPLE 8 **Gray Codes** The position of a rotating pointer can be represented in digital form. One way to do this is to split the circle into 2^n arcs of equal length and to assign a bit string of length n to each arc. Two ways to do this using bit strings of length three are shown in Figure 12.

The digital representation of the position of the pointer can be determined using a set of n contacts. Each contact is used to read one bit in the digital representation of the position. This is illustrated in Figure 13 for the two assignments from Figure 12.

When the pointer is near the boundary of two arcs, a mistake may be made in reading its position. This may result in a major error in the bit string read. For instance, in the coding scheme in Figure 12(a), if a small error is made in determining the position of the pointer, the bit string 100 is read instead of 011. All three bits are incorrect! To minimize the effect of an error in determining the position of the pointer, the assignment

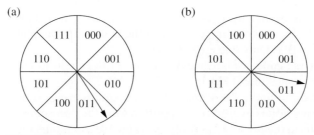

(a)   (b)

FIGURE 12 **Converting the Position of a Pointer into Digital Form.**

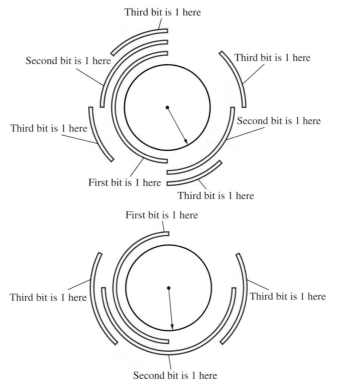

FIGURE 13 **The Digital Representation of the Position of the Pointer.**

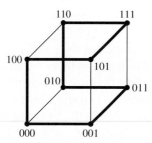

FIGURE 14   A Hamilton Circuit for Q_3.

of the bit strings to the 2^n arcs should be made so that only one bit is different in the bit strings represented by adjacent arcs. This is exactly the situation in the coding scheme in Figure 12(b). An error in determining the position of the pointer gives the bit string 010 instead of 011. Only one bit is wrong.

Links

A **Gray code** is a labeling of the arcs of the circle so that adjacent arcs are labeled with bit strings that differ in exactly one bit. The assignment in Figure 12(b) is a Gray code. We can find a Gray code by listing all bit strings of length n in such a way that each string differs in exactly one position from the preceding bit string, and the last string differs from the first in exactly one position. We can model this problem using the n-cube Q_n. What is needed to solve this problem is a Hamilton circuit in Q_n. Such Hamilton circuits are easily found. For instance, a Hamilton circuit for Q_3 is displayed in Figure 14. The sequence of bit strings differing in exactly one bit produced by this Hamilton circuit is 000, 001, 011, 010, 110, 111, 101, 100.

Gray codes are named after Frank Gray, who invented them in the 1940s at AT&T Bell Laboratories to minimize the effect of errors in transmitting digital signals.   ◄

Exercises

In Exercises 1–8 determine whether the given graph has an Euler circuit. Construct such a circuit when one exists. If no Euler circuit exists, determine whether the graph has an Euler path and construct such a path if one exists.

1.

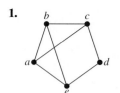

2.

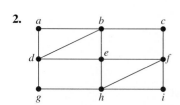

3.

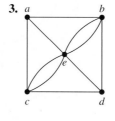

4.

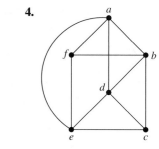

5.

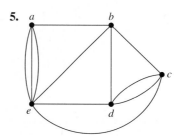

6.

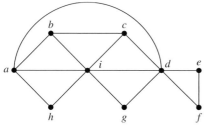

7.

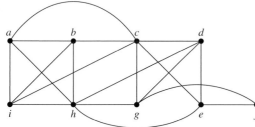

8.

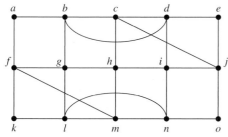

9. In Kaliningrad (the Russian name for Königsberg) there are two additional bridges, besides the seven that were present in the eighteenth century. These new bridges connect regions B and C and regions B and D, respectively. Can someone cross all nine bridges in Kaliningrad exactly once and return to the starting point?

10. Can someone cross all the bridges shown in this map exactly once and return to the starting point?

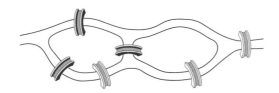

11. When can the centerlines of the streets in a city be painted without traveling a street more than once? (Assume that all the streets are two-way streets.)

12. Devise a procedure, similar to Algorithm 1, for constructing Euler paths in multigraphs.

In Exercises 13–15 determine whether the picture shown can be drawn with a pencil in a continuous motion without lifting the pencil or retracing part of the picture.

13.

14.

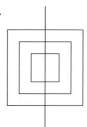

15.

*16. Show that a directed multigraph having no isolated vertices has an Euler circuit if and only if the graph is weakly connected and the in-degree and out-degree of each vertex are equal.

*17. Show that a directed multigraph having no isolated vertices has an Euler path but not an Euler circuit if and only if the graph is weakly connected and the in-degree and out-degree of each vertex are equal for all but two vertices, one that has in-degree one larger than its out-degree and the other that has out-degree one larger than its in-degree.

In Exercises 18–23 determine whether the directed graph shown has an Euler circuit. Construct an Euler circuit if one exists. If no Euler circuit exists, determine when the directed graph has an Euler path. Construct an Euler path if one exists.

18.

19.

20.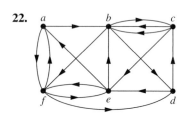

21.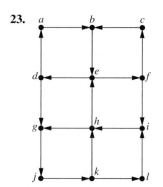

22.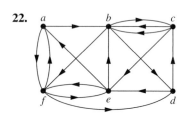

23.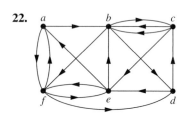

*24. Devise an algorithm for constructing Euler circuits in directed graphs.

25. Devise an algorithm for constructing Euler paths in directed graphs.

26. For which values of n do these graphs have an Euler circuit?

    a) K_n **b)** C_n **c)** W_n **d)** Q_n

27. For which values of n do the graphs in Exercise 26 have an Euler path but no Euler circuit?

28. For which values of m and n does the complete bipartite graph $K_{m,n}$ have an

    a) Euler circuit?
    b) Euler path?

29. Find the least number of times it is necessary to lift a pencil from the paper when drawing each of the graphs in Exercises 1–7 without retracing any part of the graph.

In Exercises 30–36 determine whether the given graph has a Hamilton circuit. If it does, find such a circuit. If it does not, give an argument to show why no such circuit exists.

30.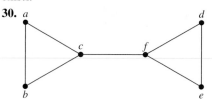

31.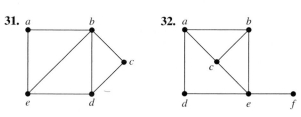

32.

33.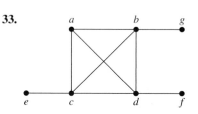

34.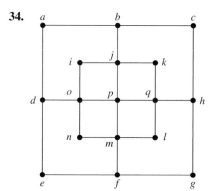

35.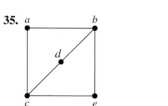

36.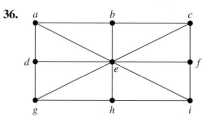

37. Does the graph in Exercise 30 have a Hamilton path? If so, find such a path. If it does not, give an argument to show why no such path exists.

38. Does the graph in Exercise 31 have a Hamilton path? If so, find such a path. If it does not, give an argument to show why no such path exists.

39. Does the graph in Exercise 32 have a Hamilton path? If so, find such a path. If it does not, give an argument to show why no such path exists.

40. Does the graph in Exercise 33 have a Hamilton path? If so, find such a path. If it does not, give an argument to show why no such path exists.

*41. Does the graph in Exercise 34 have a Hamilton path? If so, find such a path. If it does not, give an argument to show why no such path exists.

42. Does the graph in Exercise 35 have a Hamilton path? If so, find such a path. If it does not, give an argument to show why no such path exists.

43. Does the graph in Exercise 36 have a Hamilton path? If so, find such a path. If it does not, give an argument to show why no such path exists.

44. For which values of n do the graphs in Exercise 26 have a Hamilton circuit?

45. For which values of m and n does the complete bipartite graph $K_{m,n}$ have a Hamilton circuit?

*46. Show that the **Petersen graph,** shown here, does not have a Hamilton circuit, but that the subgraph obtained by deleting a vertex v, and all edges incident with v, does have a Hamilton circuit.

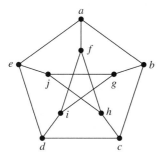

47. For each of these graphs, determine (i) whether Dirac's Theorem can be used to show that the graph has a Hamilton circuit, (ii) whether Ore's Theorem can be used to show that the graph has a Hamilton circuit, and (iii) whether the graph has a Hamilton circuit.

a) **b)**

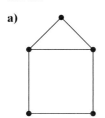

c) **d)**

48. Can you find a simple graph with n vertices with $n \geq 3$ that does not have a Hamilton circuit, yet the degree of every vertex in the graph is at least $(n - 1)/2$?

*49. Show that there is a Gray code of order n whenever n is a positive integer, or equivalently, show that the n-cube Q_n, $n > 1$, always has a Hamilton circuit. (*Hint:* Use mathematical induction. Show how to produce a Gray code of order n from one of order $n - 1$.)

Fleury's algorithm for constructing Euler circuits begins with an arbitrary vertex of a connected multigraph and forms a circuit by choosing edges successively. Once an edge is chosen, it is removed. Edges are chosen successively so that each edge begins where the last edge ends, and so that this edge is not a cut edge unless there is no alternative.

50. Use Fleury's algorithm to find an Euler circuit in the graph G in Figure 5.

*51. Express Fleury's algorithm in pseudocode.

**52. Prove that Fleury's algorithm always produces an Euler circuit.

*53. Give a variant of Fleury's algorithm to produce Euler paths.

54. A diagnostic message can be sent out over a computer network to perform tests over all links and in all devices. What sort of paths should be used to test all links? To test all devices?

55. Show that a bipartite graph with an odd number of vertices does not have a Hamilton circuit.

A **knight** is a chess piece that can move either two spaces horizontally and one space vertically or one space horizontally and two spaces vertically. That is, a knight on square (x, y) can move to any of the eight squares $(x \pm 2, y \pm 1)$, $(x \pm 1, y \pm 2)$, if these squares are on the chessboard, as illustrated here.

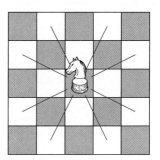

A **knight's tour** is a sequence of legal moves by a knight starting at some square and visiting each square exactly once. A knight's tour is called **reentrant** if there is a legal move that takes the knight from the last square of the tour back to where the tour began. We can model knight's tours using the graph that has a vertex for each square on the board, with an edge connecting two vertices if a knight can legally move between the squares represented by these vertices.

56. Draw the graph that represents the legal moves of a knight on a 3 × 3 chessboard.

57. Draw the graph that represents the legal moves of a knight on a 3 × 4 chessboard.

58. a) Show that finding a knight's tour on an $m \times n$ chessboard is equivalent to finding a Hamilton path on the graph representing the legal moves of a knight on that board.

b) Show that finding a reentrant knight's tour on an $m \times n$ chessboard is equivalent to finding a Hamilton circuit on the corresponding graph.

*59. Show that there is a knight's tour on a 3 × 4 chessboard.

*60. Show that there is no knight's tour on a 3 × 3 chessboard.

*61. Show that there is no knight's tour on a 4 × 4 chessboard.

62. Show that the graph representing the legal moves of a knight on an $m \times n$ chessboard, whenever m and n are positive integers, is bipartite.

63. Show that there is no reentrant knight's tour on an $m \times n$ chessboard when m and n are both odd. (*Hint:* Use Exercises 55, 58b, and 62.)

*64. Show that there is a knight's tour on an 8 × 8 chessboard. (*Hint:* You can construct a knight's tour using a method invented by H. C. Warnsdorff in 1823:

Start in any square, and then always move to a square connected to the fewest number of unused squares. Although this method may not always produce a knight's tour, it often does.)

65. The parts of this exercise outline a proof of Ore's Theorem. Suppose that G is a simple graph with n vertices, $n \geq 3$, and $\deg(x) + \deg(y) \geq n$ whenever x and y are nonadjacent vertices in G. Ore's Theorem states that under these conditions, G has a Hamilton circuit.

a) Show that if G does not have a Hamilton circuit, then there exists another graph H with the same vertices as G, which can be constructed by adding edges to G such that the addition of a single edge would produce a Hamilton circuit in H. (*Hint:* Add as many edges as possible at each successive vertex of G without producing a Hamilton circuit.)

b) Show that there is a Hamilton path in H.

c) Let $v_1, v_2, \ldots, v_n$ be a Hamilton path in H. Show that $\deg(v_1) + \deg(v_n) \geq n$ and that there are at most $\deg(v_1)$ vertices not adjacent to v_n (including v_n itself).

d) Let S be the set of vertices preceding each vertex adjacent to v_1 in the Hamilton path. Show that S contains $\deg(v_1)$ vertices and $v_n \notin S$.

e) Show that S contains a vertex v_k, which is adjacent to v_n, implying that there are edges connecting v_1 and v_{k+1} and v_k and v_n.

f) Show that part (e) implies that $v_1, v_2, \ldots, v_{k-1}, v_k, v_n, v_{n-1}, \ldots, v_{k+1}, v_1$ is a Hamilton circuit in G. Conclude from this contradiction that Ore's Theorem holds.

Links

JULIUS PETER CHRISTIAN PETERSEN (1839–1910) Julius Petersen was born in the Danish town of Sorø. His father was a dyer. In 1854 his parents were no longer able to pay for his schooling, so he became an apprentice in an uncle's grocery store. When this uncle died, he left Petersen enough money to return to school. After graduating, he began studying engineering at the Polytechnical School in Copenhagen, later deciding to concentrate on mathematics. He published his first textbook, a book on logarithms, in 1858. When his inheritance ran out, he had to teach to make a living. From 1859 until 1871 Petersen taught at a prestigious private high school in Copenhagen. While teaching high school he continued his studies, entering Copenhagen University in 1862. He married Laura Bertelsen in 1862; they had three children, two sons and a daughter.

Petersen obtained a mathematics degree from Copenhagen University in 1866 and finally obtained his doctorate in 1871 from that school. After receiving his doctorate, he taught at a polytechnic and military academy. In 1887 he was appointed to a professorship at the University of Copenhagen. Petersen was well known in Denmark as the author of a large series of textbooks for high schools and universities. One of his books, *Methods and Theories for the Solution of Problems of Geometrical Construction,* was translated into eight languages, with the English language version last reprinted in 1960 and the French version reprinted as recently as 1990, more than a century after the original publication date.

Petersen worked in a wide range of areas, including algebra, analysis, cryptography, geometry, mechanics, mathematical economics, and number theory. His contributions to graph theory, including results on regular graphs, are his best-known work. He was noted for his clarity of exposition, originality, sense of humor, vigor, and teaching. One interesting fact about Petersen was that he preferred not to read the writings of other mathematicians. This led him often to rediscover results already proved by others, often with embarrassing consequences. However, he was often angry when other mathematicians did not read his writings!

Petersen's death was front-page news in Copenhagen. A newspaper of the time described him as the Hans Christian Andersen of science—a child of the people who made good in the academic world.

8.6 Shortest-Path Problems

INTRODUCTION

Many problems can be modeled using graphs with weights assigned to their edges. As an illustration, consider how an airline system can be modeled. We set up the basic graph model by representing cities by vertices and flights by edges. Problems involving distances can be modeled by assigning distances between cities to the edges. Problems involving flight time can be modeled by assigning flight times to edges. Problems involving fares can be modeled by assigning fares to the edges. Figure 1 displays three different assignments of weights to the edges of a graph representing distances, flight times, and fares, respectively.

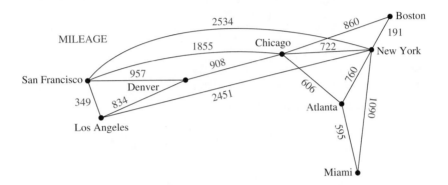

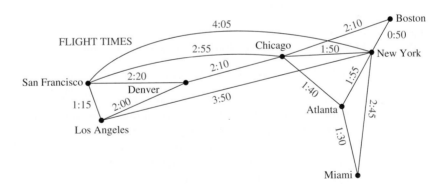

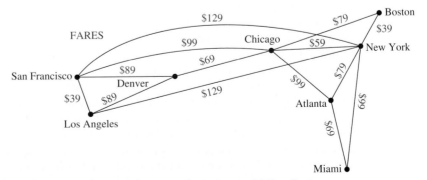

FIGURE 1   Weighted Graphs Modeling an Airline System.

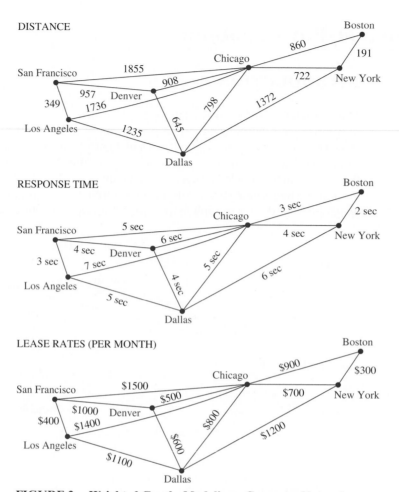

FIGURE 2 **Weighted Graphs Modeling a Computer Network.**

Graphs that have a number assigned to each edge are called **weighted graphs.** Weighted graphs are used to model computer networks. Communications costs (such as the monthly cost of leasing a telephone line), the response times of the computers over these lines, or the distance between computers, can all be studied using weighted graphs. Figure 2 displays weighted graphs that represent three ways to assign weights to the edges of a graph of a computer network, corresponding to costs, response times over the lines, and distance.

Several types of problems involving weighted graphs arise frequently. Determining a path of least length between two vertices in a network is one such problem. To be more specific, let the **length** of a path in a weighted graph be the sum of the weights of the edges of this path. (The reader should note that this use of the term *length* is different from the use of *length* to denote the number of edges in a path in a graph without weights.) The question is: What is a shortest path, that is, a path of least length, between two given vertices? For instance, in the airline system represented by the weighted graph shown in Figure 1, what is a shortest path in air distance between Boston and Los Angeles? What combinations of flights has the smallest total flight time (that is, total time in the air, not including time between flights) between Boston and Los Angeles? What is the cheapest fare between these two cities? In the computer network shown in Figure 2, what is a least expensive set of telephone lines needed to connect the computers in San Francisco with those in New York? Which set of telephone lines gives a fastest response time for communications between San Francisco and New York? Which set of lines has a shortest overall distance?

Another important problem involving weighted graphs asks for a circuit of shortest total length that visits every vertex of a complete graph exactly once. This is the famous *traveling salesman problem,* which asks for an order in which a salesman should visit each of the cities on his route exactly once so that he travels the minimum total distance. We will discuss the traveling salesman problem later in this section.

A SHORTEST-PATH ALGORITHM

There are several different algorithms that find a shortest path between two vertices in a weighted graph. We will present an algorithm discovered by the Dutch mathematician Edsger Dijkstra in 1959. The version we will describe solves this problem in undirected weighted graphs where all the weights are positive. It is easy to adapt it to solve shortest-path problems in directed graphs.

Before giving a formal presentation of the algorithm, we will give a motivating example.

EXAMPLE 1   What is the length of a shortest path between a and z in the weighted graph shown in Figure 3?

Solution: Although a shortest path is easily found by inspection, we will develop some ideas useful in understanding Dijkstra's algorithm. We will solve this problem by finding the length of a shortest path from a to successive vertices, until z is reached.

The only paths starting at a that contain no vertex other than a (until the terminal vertex is reached) are a, b and a, d. Since the lengths of a, b and a, d are 4 and 2, respectively, it follows that d is the closest vertex to a.

We can find the next closest vertex by looking at all paths that go through only a and d (until the terminal vertex is reached). The shortest such path to b is still a, b, with length 4, and the shortest such path to e is a, d, e, with length 5. Consequently, the next closest vertex to a is b.

EDSGER WYBE DIJKSTRA (1930–2002)   Edsger Dijkstra, born in the Netherlands, began programming computers in the early 1950s while studying theoretical physics at the University of Leiden. In 1952, realizing that he was more interested in programming than in physics, he quickly completed the requirements for his physics degree and began his career as a programmer, even though programming was not a recognized profession. (In 1957, the authorities in Amsterdam refused to accept "programming" as his profession on his marriage license. However, they did accept "theoretical physicist" when he changed his entry to this.)

Dijkstra has been one of the most forceful proponents of programming as a scientific discipline. He has made fundamental contributions to the areas of operating systems, including deadlock avoidance; programming languages, including the notion of structured programming; and algorithms. In 1972 Dijkstra received the Turing Award from the Association for Computing Machinery, one of the most prestigious awards in computer science. Dijkstra became a Burroughs Research Fellow in 1973, and in 1984 he was appointed to a chair in Computer Science at the University of Texas, Austin.

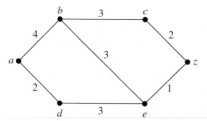

FIGURE 3   A Weighted Simple Graph.

To find the third closest vertex to a, we need to examine only paths that go through only a, d, and b (until the terminal vertex is reached). There is a path of length 7 to c, namely, a, b, c, and a path of length 6 to z, namely, a, d, e, z. Consequently, z is the next closest vertex to a, and the length of a shortest path to z is 6.   ◀

Example 1 illustrates the general principles used in Dijkstra's algorithm. Note that a shortest path from a to z could have been found by inspection. However, inspection is impractical for both humans and computers for graphs with large numbers of edges.

We will now consider the general problem of finding the length of a shortest path between a and z in an undirected connected simple weighted graph. Dijkstra's algorithm proceeds by finding the length of a shortest path from a to a first vertex, the length of a shortest path from a to a second vertex, and so on, until the length of a shortest path from a to z is found.

The algorithm relies on a series of iterations. A distinguished set of vertices is constructed by adding one vertex at each iteration. A labeling procedure is carried out at each iteration. In this labeling procedure, a vertex w is labeled with the length of a shortest path from a to w that contains only vertices already in the distinguished set. The vertex added to the distinguished set is one with a minimal label among those vertices not already in the set.

We now give the details of Dijkstra's algorithm. It begins by labeling a with 0 and the other vertices with ∞. We use the notation $L_0(a) = 0$ and $L_0(v) = \infty$ for these labels before any iterations have taken place (the subscript 0 stands for the "0th" iteration). These labels are the lengths of shortest paths from a to the vertices, where the paths contain only the vertex a. (Since no path from a to a vertex different from a exists, ∞ is the length of a shortest path between a and this vertex.)

Dijkstra's algorithm proceeds by forming a distinguished set of vertices. Let S_k denote this set after k iterations of the labeling procedure. We begin with $S_0 = \emptyset$. The set S_k is formed from S_{k-1} by adding a vertex u not in S_{k-1} with the smallest label. Once u is added to S_k, we update the labels of all vertices not in S_k, so that $L_k(v)$, the label of the vertex v at the kth stage, is the length of a shortest path from a to v that contains vertices only in S_k (that is, vertices that were already in the distinguished set together with u).

Let v be a vertex not in S_k. To update the label of v, note that $L_k(v)$ is the length of a shortest path from a to v containing only vertices in S_k. The updating can be carried out efficiently when this observation is used: A shortest path from a to v containing only elements of S_k is either a shortest path from a to v that contains only elements of S_{k-1} (that is, the distinguished vertices not including u), or it is a shortest path from a to u at the $(k-1)$st stage with the edge (u, v) added. In other words,

$$L_k(a, v) = \min\{L_{k-1}(a, v), L_{k-1}(a, u) + w(u, v)\}.$$

This procedure is iterated by successively adding vertices to the distinguished set until z is added. When z is added to the distinguished set, its label is the length of a shortest path from a to z. Dijkstra's algorithm is given in Algorithm 1. Later we will give a proof that this algorithm is correct.

ALGORITHM 1 Dijkstra's Algorithm.

---

procedure $Dijkstra(G$: weighted connected simple graph, with
     all weights positive)
$\{G$ has vertices $a = v_0, v_1, \ldots, v_n = z$ and weights $w(v_i, v_j)$
     where $w(v_i, v_j) = \infty$ if $\{v_i, v_j\}$ is not an edge in $G\}$
for $i := 1$ **to** n
    $L(v_i) := \infty$
$L(a) := 0$
$S := \emptyset$
$\{$the labels are now initialized so that the label of a is 0 and all
    other labels are ∞, and S is the empty set$\}$
while $z \notin S$
begin
    $u :=$ a vertex not in S with $L(u)$ minimal
    $S := S \cup \{u\}$
    for all vertices v not in S
        if $L(u) + w(u, v) < L(v)$ **then** $L(v) := L(u) + w(u, v)$
    $\{$this adds a vertex to S with minimal label and updates the
        labels of vertices not in $S\}$
end $\{L(z) =$ length of a shortest path from a to $z\}$

Example 2 illustrates how Dijkstra's algorithm works. Afterward, we will show that this algorithm always produces the length of a shortest path between two vertices in a weighted graph.

EXAMPLE 2   Use Dijkstra's algorithm to find the length of a shortest path between the vertices a and z in the weighted graph displayed in Figure 4(a).

Solution: The steps used by Dijkstra's algorithm to find a shortest path between a and z are shown in Figure 4. At each iteration of the algorithm the vertices of the set S_k are circled. A shortest path from a to each vertex containing only vertices in S_k is indicated for each iteration. The algorithm terminates when z is circled. We find that a shortest path from a to z is $a, c, b, d, e, z,$ with length 13.   ◀

Remark: In performing Dijkstra's algorithm it is sometimes more convenient to keep track of labels of vertices in each step using a table instead of redrawing the graph for each step.

Next, we use an inductive argument to show that Dijkstra's algorithm produces the length of a shortest path between two vertices a and z in an undirected connected weighted graph. Take as the induction hypothesis the following assertion: At the kth iteration

    (i) the label of every vertex v in S is the length of a shortest path from a to this vertex, and

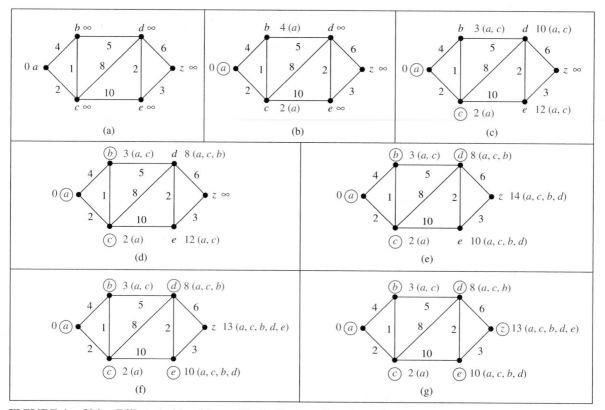

FIGURE 4 Using Dijkstra's Algorithm to Find a Shortest Path from a to z.

(ii) the label of every vertex not in S is the length of a shortest path from a to this vertex that contains only (besides the vertex itself) vertices in S.

When $k = 0$, before any iterations are carried out, $S = \emptyset$, so the length of a shortest path from a to a vertex other than a is ∞. Hence, the basis case is true.

Assume that the inductive hypothesis holds for the kth iteration. Let v be the vertex added to S at the $(k + 1)$st iteration so that v is a vertex not in S at the end of the kth iteration with the smallest label (in the case of ties, any vertex with smallest label may be used).

From the inductive hypothesis we see that the vertices in S before the $(k + 1)$st iteration are labeled with the length of a shortest path from a. Also, v must be labeled with the length of a shortest path to it from a. If this were not the case, at the end of the kth iteration there would be a path of length less than $L_k(v)$ containing a vertex not in S (because $L_k(v)$ is the length of a shortest path from a to v containing only vertices in S after the kth iteration). Let u be the first vertex not in S in such a path. There is a path with length less than $L_k(v)$ from a to u containing only vertices of S. This contradicts the choice of v. Hence, (i) holds at the end of the $(k + 1)$st iteration.

Let u be a vertex not in S after $k + 1$ iterations. A shortest path from a to u containing only elements of S either contains v or it does not. If it does not contain v, then by the inductive hypothesis its length is $L_k(u)$. If it does contain v, then it must be made up of a path from a to v of shortest possible length containing elements of S other than v, followed by the edge from v to u. In this case its length would be $L_k(v) + w(v, u)$. This shows that (ii) is true, since $L_{k+1}(u) = \min\{L_k(u), L_k(v) + w(v, u)\}$.

Theorem 1 has been proved.

THEOREM 1 Dijkstra's algorithm finds the length of a shortest path between two vertices in a connected simple undirected weighted graph.

We can now estimate the computational complexity of Dijkstra's algorithm (in terms of additions and comparisons). The algorithm uses no more than $n - 1$ iterations, since one vertex is added to the distinguished set at each iteration. We are done if we can estimate the number of operations used for each iteration. We can identify the vertex not in S_k with the smallest label using no more than $n - 1$ comparisons. Then we use an addition and a comparison to update the label of each vertex not in S_k. It follows that no more than $2(n - 1)$ operations are used at each iteration, since there are no more than $n - 1$ labels to update at each iteration. Since we use no more than $n - 1$ iterations, each using no more than $2(n - 1)$ operations, we have the following theorem.

THEOREM 2 Dijkstra's algorithm uses $O(n^2)$ operations (additions and comparisons) to find the length of a shortest path between two vertices in a connected simple undirected weighted graph with n vertices.

THE TRAVELING SALESMAN PROBLEM

We now discuss an important problem involving weighted graphs. Consider the following problem: A traveling salesman wants to visit each of n cities exactly once and return to his starting point. For example, suppose that the salesman wants to visit Detroit, Toledo, Saginaw, Grand Rapids, and Kalamazoo (see Figure 5). In which order should he visit these cities to travel the minimum total distance? To solve this problem we can assume the salesman starts in Detroit (since this must be part of the circuit) and examine all possible ways for him to visit the other four cities and then return to Detroit (starting elsewhere will produce the same circuits). There are a total of 24 such circuits, but since we travel the same distance when we travel a circuit in reverse order, we need only consider 12 different circuits to find the minimum total distance he must travel. We list these 12 different circuits and the total distance traveled for each circuit. As can be seen from the list, the minimum total distance of 458 miles is traveled using the circuit Detroit–Toledo–Kalamazoo–Grand Rapids–Saginaw–Detroit (or its reverse).

Route	*Total Distance (miles)*
Detroit–Toledo–Grand Rapids–Saginaw–Kalamazoo–Detroit	610
Detroit–Toledo–Grand Rapids–Kalamazoo–Saginaw–Detroit	516
Detroit–Toledo–Kalamazoo–Saginaw–Grand Rapids–Detroit	588
Detroit–Toledo–Kalamazoo–Grand Rapids–Saginaw–Detroit	458
Detroit–Toledo–Saginaw–Kalamazoo–Grand Rapids–Detroit	540
Detroit–Toledo–Saginaw–Grand Rapids–Kalamazoo–Detroit	504
Detroit–Saginaw–Toledo–Grand Rapids–Kalamazoo–Detroit	598
Detroit–Saginaw–Toledo–Kalamazoo–Grand Rapids–Detroit	576
Detroit–Saginaw–Kalamazoo–Toledo–Grand Rapids–Detroit	682
Detroit–Saginaw–Grand Rapids–Toledo–Kalamazoo–Detroit	646
Detroit–Grand Rapids–Saginaw–Toledo–Kalamazoo–Detroit	670
Detroit–Grand Rapids–Toledo–Saginaw–Kalamazoo–Detroit	728

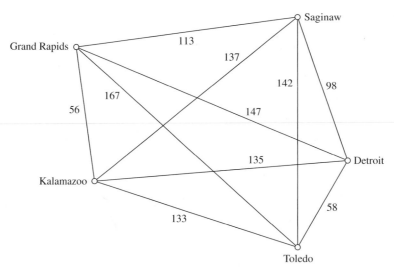

FIGURE 5   The Graph Showing the Distances Between Five Cities.

We just described an instance of the **traveling salesman problem.** The traveling sales-man problem asks for the circuit of minimum total weight in a weighted, complete, un-directed graph that visits each vertex exactly once and returns to its starting point. This is equivalent to asking for a Hamilton circuit with minimum total weight in the complete graph, since each vertex is visited exactly once in the circuit.

The most straightforward way to solve an instance of the traveling salesman problem is to examine all possible Hamilton circuits and select one of minimum total length. How many circuits do we have to examine to solve the problem if there are n vertices in the graph? Once a starting point is chosen, there are $(n - 1!)$ different Hamilton circuits to examine, since there are $n - 1$ choices for the second vertex, $n - 2$ choices for the third vertex, and so on. Since a Hamilton circuit can be traveled in reverse order, we need only examine $(n - 1)!/2$ circuits to find our answer. Note that $(n - 1)!/2$ grows extremely rapidly. Trying to solve a traveling salesman problem in this way when there are only a few dozen vertices is impractical. For example, with 25 vertices, a total of $24!/2$ (approximately 3.1×10^{23}) different Hamilton circuits would have to be considered. If it took just one nanosecond (10^{-9} second) to examine each Hamilton circuit, a total of approximately ten million years would be required to find a minimum-length Hamilton circuit in this graph by exhaustive search techniques.

Since the traveling salesman problem has both practical and theoretical importance, a great deal of effort has been devoted to devising efficient algorithms that solve it. However, no algorithm with polynomial worst-case time complexity is known for solving this problem. Furthermore, if a polynomial worst-case time complexity algorithm were discovered for the traveling salesman problem, many other difficult problems would also be solvable using polynomial worst-case time complexity algorithms (such as de-termining whether a proposition in n variables is a tautology, discussed in Chapter 1). This follows from the theory of NP-completeness. (For more information about this, consult [GaJo79].)

A practical approach to the traveling salesman problem when there are many vertices to visit is to use an **approximation algorithm.** These are algorithms that do not necessarily produce the exact solution to the problem but instead are guaranteed to produce a solution that is close to an exact solution. That is, they may produce a Hamilton circuit with total weight W' so that $W \leq W' \leq cW$, where W is the total length of an exact solution and c is a constant. For example, there is an algorithm with polynomial worst-case time complexity that works if the weighted graph satisfies the triangle inequality such that $c = 3/2$. For general weighted graphs for every positive real number k no algorithm is known that will always produce a solution at most k times a best solution. If such an algorithm existed, this would show that the class P would be the same as the class NP, perhaps the most famous open question about the complexity of algorithms (see Section 2.3).

In practice, algorithms have been developed that can solve traveling salesman problems with as many as 1000 vertices within 2% of an exact solution using only a few minutes of computer time. For more information about the traveling salesman problem, including history, applications, and algorithms, see the chapter on this topic in *Applications of Discrete Mathematics* [MiRo91].

Exercises

1. For each of these problems about a subway system, describe a weighted graph model that can be used to solve the problem.

   a) What is the least amount of time required to travel between two stops?
   b) What is the minimum distance that can be traveled to reach a stop from another stop?
   c) What is the least fare required to travel between two stops if fares between stops are added to give the total fare?

In Exercises 2–4 find the length of a shortest path between a and z in the given weighted graph.

2.

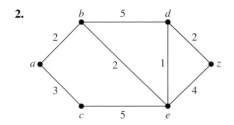

3.

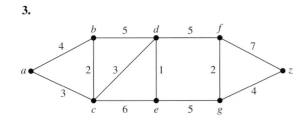

4.

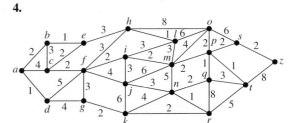

5. Find a shortest path between a and z in each of the weighted graphs in Exercises 2–4.

6. Find the length of a shortest path between these pairs of vertices in the weighted graph in Exercise 3.

   a) a and d
   b) a and f
   c) c and f
   d) b and z

7. Find shortest paths in the weighted graph in Exercise 3 between the pairs of vertices in Exercise 6.

8. Find a shortest path (in mileage) between each of the following pairs of cities in the airline system shown in Figure 1.

   a) New York and Los Angeles
   b) Boston and San Francisco
   c) Miami and Denver
   d) Miami and Los Angeles

9. Find a combination of flights with the least total air time between the pairs of cities in Exercise 8, using the flight times shown in Figure 1.

10. Find a least expensive combination of flights connecting the pairs of cities in Exercise 8, using the fares shown in Figure 1.

11. Find a shortest route (in distance) between computer centers in each of these pairs of cities in the communications network shown in Figure 2.

    a) Boston and Los Angeles
    b) New York and San Francisco
    c) Dallas and San Francisco
    d) Denver and New York

12. Find a route with the shortest response time between the pairs of computer centers in Exercise 11 using the response times given in Figure 2.

13. Find a least expensive route, in monthly lease charges, between the pairs of computer centers in Exercise 11 using the lease charges given in Figure 2.

14. Explain how to find a path with the least number of edges between two vertices in an undirected graph by considering it as a shortest path problem in a weighted graph.

15. Extend Dijkstra's algorithm for finding the length of a shortest path between two vertices in a weighted simple connected graph so that the length of a shortest path between the vertex a and every other vertex of the graph is found.

16. Extend Dijkstra's algorithm for finding the length of a shortest path between two vertices in a weighted simple connected graph so that a shortest path between these vertices is constructed.

17. The weighted graphs in the figures here show some major roads in New Jersey. Part (a) shows the distances between cities on these roads; part (b) shows the tolls.

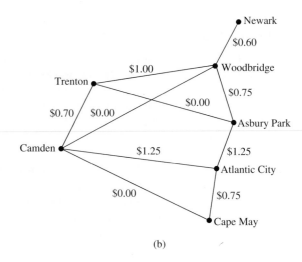

(b)

    a) Find a shortest route in distance between Newark and Camden, and between Newark and Cape May, using these roads.

    b) Find a least expensive route in terms of total tolls using the roads in the graph between the pairs of cities in part (a) of this exercise.

18. Is a shortest path between two vertices in a weighted graph unique if the weights of edges are distinct?

19. What are some applications where it is necessary to find the length of a longest simple path between two vertices in a weighted graph?

20. What is the length of a longest simple path in the weighted graph in Figure 4 between a and z? Between c and z?

Floyd's algorithm, displayed as Algorithm 2, can be used to find the length of a shortest path between all pairs of vertices in a weighted connected simple graph. However, this algorithm cannot be used to construct shortest paths. (We assign an infinite weight to any pair of vertices not connected by an edge in the graph.)

21. Use Floyd's algorithm to find the distance between all pairs of vertices in the weighted graph in Figure 4(a).

*22. Prove that Floyd's algorithm determines the shortest distance between all pairs of vertices in a weighted simple graph.

*23. Give a big-O estimate of the number of operations (comparisons and additions) used by Floyd's algorithm to determine the shortest distance between every pair of vertices in a weighted simple graph with n vertices.

*24. Show that Dijkstra's algorithm may not work if edges can have negative weights.

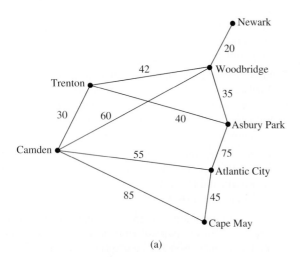

(a)

ALGORITHM 2 Floyd's Algorithm.

procedure *Floyd*(*G*: weighted simple graph)
{*G* has vertices $v_1, v_2, \ldots, v_n$ and weights
$w(v_i, v_j)$ with $w(v_i, v_j) = \infty$ if (v_i, v_j) is
not an edge}
for $i := 1$ **to** n
  for $j := 1$ **to** n
    $d(v_i, v_j) := w(v_i, v_j)$
for $i := 1$ **to** n
  for $j := 1$ **to** n
    for $k := 1$ **to** n
      if $d(v_j, v_i) + d(v_i, v_k) < d(v_j, v_k)$
      then $d(v_j, v_k) :=$
        $d(v_j, v_i) + d(v_i, v_k)$
{$d(v_i, v_j)$ is the length of a shortest path
between v_i and v_j}

25. Solve the traveling salesman problem for this graph
by finding the total weight of all Hamilton circuits and
determining a circuit with minimum total weight.

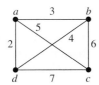

26. Solve the traveling salesman problem for this graph
by finding the total weight of all Hamilton circuits and
determining a circuit with minimum total weight.

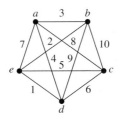

27. Find a route with the least total airfare that visits each
of the cities in this graph, where the weight on an edge
is the least price available for a flight between the
two cities.

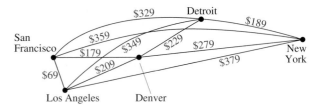

28. Find a route with the least total airfare that visits each
of the cities in this graph, where the weight on an edge
is the least price available for a flight between the
two cities.

29. Construct a weighted undirected graph such that the
total weight of a circuit that visits every vertex at least
once is minimized for a circuit that visits some vertices more than once. (*Hint:* There are examples with
three vertices.)

30. Show that the problem of finding a circuit of minimum total weight that visits every vertex of a
weighted graph at least once can be reduced to the
problem of finding a circuit of minimum total weight
that visits each vertex of a weighted graph exactly
once. Do so by constructing a new weighted graph
with the same vertices and edges as the original graph
but whose weight of the edge connecting the vertices
u and *v* is equal to the minimum total weight of a path
from *u* to *v* in the original graph.

8.7 **Planar Graphs**

INTRODUCTION

Consider the problem of joining three houses to each of three separate utilities, as shown
in Figure 1. Is it possible to join these houses and utilities so that none of the connections cross? This problem can be modeled using the complete bipartite graph $K_{3,3}$. The

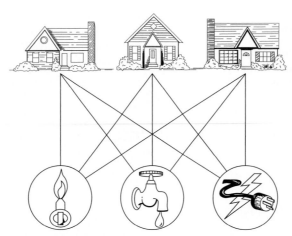

FIGURE 1   Three Houses and Three Utilities.

original question can be rephrased as: Can $K_{3,3}$ be drawn in the plane so that no two of its edges cross?

In this section we will study the question of whether a graph can be drawn in the plane without edges crossing. In particular, we will answer the houses-and-utilities problem.

There are always many ways to represent a graph. When is it possible to find at least one way to represent this graph in a plane without any edges crossing?

DEFINITION 1

A graph is called *planar* if it can be drawn in the plane without any edges crossing (where a crossing of edges is the intersection of the lines or arcs representing them at a point other than their common endpoint). Such a drawing is called a *planar representation* of the graph.

A graph may be planar even if it is usually drawn with crossings, since it may be possible to draw it in a different way without crossings.

EXAMPLE 1   Is K_4 (shown in Figure 2 with two edges crossing) planar?

Solution: K_4 is planar because it can be drawn without crossings, as shown in Figure 3. ◄

EXAMPLE 2   Is Q_3, shown in Figure 4, planar?

Solution: Q_3 is planar, because it can be drawn without any edges crossing, as shown in Figure 5. ◄

We can show that a graph is planar by displaying a planar representation. It is harder to show that a graph is nonplanar. We will give an example to show how this can be done in an ad hoc fashion. Later we will develop some general results that can be used to do this.

EXAMPLE 3   Is $K_{3,3}$, shown in Figure 6, planar?

Solution: Any attempt to draw $K_{3,3}$ in the plane with no edges crossing is doomed. We now show why. In any planar representation of $K_{3,3}$, the vertices v_1 and v_2 must be

FIGURE 2 The Graph K_4.

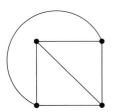

FIGURE 3 K_4 Drawn with No Crossings.

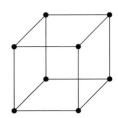

FIGURE 4 The Graph Q_3.

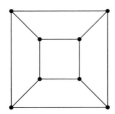

FIGURE 5 A Planar Representation of Q_3.

connected to both v_4 and v_5. These four edges form a closed curve that splits the plane into two regions, R_1 and R_2, as shown in Figure 7(a). The vertex v_3 is in either R_1 or R_2. When v_3 is in R_2, the inside of the closed curve, the edges between v_3 and v_4 and between v_3 and v_5 separate R_2 into two subregions, R_{21} and R_{22}, as shown in Figure 7(b).

Next, note that there is no way to place the final vertex v_6 without forcing a crossing. For if v_6 is in R_1, then the edge between v_6 and v_3 cannot be drawn without a crossing. If v_6 is in R_{21}, then the edge between v_2 and v_6 cannot be drawn without a crossing. If v_6 is in R_{22}, then the edge between v_1 and v_6 cannot be drawn without a crossing.

A similar argument can be used when v_3 is in R_1. The completion of this argument is left for the reader (see Exercise 10 at the end of this section). It follows that $K_{3,3}$ is not planar. ◄

Example 3 solves the utilities-and-houses problem that was described at the beginning of this section. The three houses and three utilities cannot be connected in the plane without a crossing. A similar argument can be used to show that K_5 is nonplanar. (See Exercise 11 at the end of this section.)

Planarity of graphs plays an important role in the design of electronic circuits. We can model a circuit with a graph by representing components of the circuit by vertices and connections between them by edges. We can print a circuit on a single board with no connections crossing if the graph representing the circuit is planar. When this graph is not planar, we must turn to more expensive options. For example, we can partition the vertices in the graph representing the circuit into planar subgraphs. We then construct the circuit using multiple layers. (See the preamble to Exercise 30 to learn about the thickness of a graph.) We can construct the circuit using insulated wires whenever connections cross. In this case, drawing the graph with the fewest possible crossings is important. (See the preamble to Exercise 26 to learn about the crossing number of a graph.)

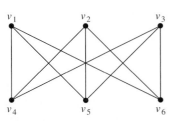

FIGURE 6 The Graph $K_{3,3}$.

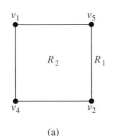

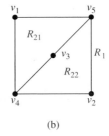

(a)   (b)

FIGURE 7 Showing $K_{3,3}$ Is Nonplanar.

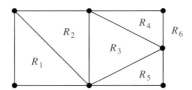

FIGURE 8 **The Regions of the Planar Representation of a Graph.**

EULER'S FORMULA

A planar representation of a graph splits the plane into **regions,** including an unbounded region. For instance, the planar representation of the graph shown in Figure 8 splits the plane into six regions. These are labeled in the figure. Euler showed that all planar representations of a graph split the plane into the same number of regions. He accomplished this by finding a relationship among the number of regions, the number of vertices, and the number of edges of a planar graph.

THEOREM 1

EULER'S FORMULA Let G be a connected planar simple graph with e edges and v vertices. Let r be the number of regions in a planar representation of G. Then $r = e - v + 2$.

Proof: First, we specify a planar representation of G. We will prove the theorem by constructing a sequence of subgraphs $G_1, G_2, \ldots, G_e = G$, successively adding an edge at each stage. This is done using the following inductive definition. Arbitrarily pick one edge of G to obtain G_1. Obtain G_n from G_{n-1} by arbitrarily adding an edge that is incident with a vertex already in G_{n-1}, adding the other vertex incident with this edge if it is not already in G_{n-1}. This construction is possible since G is connected. G is obtained after e edges are added. Let r_n, e_n, and v_n represent the number of regions, edges, and vertices of the planar representation of G_n induced by the planar representation of G, respectively.

The proof will now proceed by induction. The relationship $r_1 = e_1 - v_1 + 2$ is true for G_1, since $e_1 = 1, v_1 = 2$, and $r_1 = 1$. This is shown in Figure 9.

Now assume that $r_n = e_n - v_n + 2$. Let $\{a_{n+1}, b_{n+1}\}$ be the edge that is added to G_n to obtain G_{n+1}. There are two possibilities to consider. In the first case, both a_{n+1} and b_{n+1} are already in G_n. These two vertices must be on the boundary of a common region R, or else it would be impossible to add the edge $\{a_{n+1}, b_{n+1}\}$ to G_n without two edges crossing (and G_{n+1} is planar). The addition of this new edge splits R into two regions. Consequently, in this case, $r_{n+1} = r_n + 1, e_{n+1} = e_n + 1$, and $v_{n+1} = v_n$. Thus, each side of the formula relating the number of regions, edges, and vertices increases by exactly one, so this formula is still true. In other words, $r_{n+1} = e_{n+1} - v_{n+1} + 2$. This case is illustrated in Figure 10(a).

In the second case, one of the two vertices of the new edge is not already in G_n. Suppose that a_{n+1} is in G_n but that b_{n+1} is not. Adding this new edge does not produce any new regions, since b_{n+1} must be in a region that has a_{n+1} on its boundary. Consequently, $r_{n+1} = r_n$. Moreover, $e_{n+1} = e_n + 1$ and $v_{n+1} = v_n + 1$. Each side of the formula relating the number of regions, edges, and vertices remains the same, so the formula is still true. In other words, $r_{n+1} = e_{n+1} - v_{n+1} + 2$. This case is illustrated in Figure 10(b).

R_1

$u_1 \qquad v_1$

FIGURE 9 The Basis Case of the Proof of Euler's Formula.

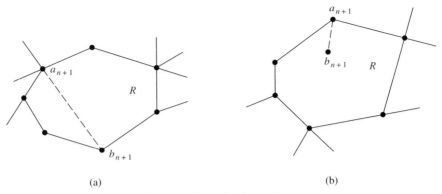

(a)   (b)

FIGURE 10   **Adding an Edge to G_n to Produce G_{n+1}.**

We have completed the induction argument. Hence $r_n = e_n - v_n + 2$ for all n. Since the original graph is the graph G_e, obtained after e edges have been added, the theorem is true.   ◁

Euler's formula is illustrated in Example 4.

EXAMPLE 4   Suppose that a connected planar simple graph has 20 vertices, each of degree 3. Into how many regions does a representation of this planar graph split the plane?

Solution: This graph has 20 vertices, each of degree 3, so that $v = 20$. Since the sum of the degrees of the vertices, $3v = 3 \cdot 20 = 60$, is equal to twice the number of edges, $2e$, we have $2e = 60$, or $e = 30$. Consequently, from Euler's formula, the number of regions is

$$r = e - v + 2 = 30 - 20 + 2 = 12.$$ ◀

Euler's formula can be used to establish some inequalities that must be satisfied by planar graphs. One such inequality is given in the following corollary.

COROLLARY 1   If G is a connected planar simple graph with e edges and v vertices where $v \geq 3$, then $e \leq 3v - 6$.

Before we prove Corollary 1 we will use it to prove the following useful result.

COROLLARY 2   If G is a connected planar simple graph, then G has a vertex of degree not exceeding five.

Proof: If G has one or two vertices, the result is true. If G has at least three vertices, by Corollary 1 we know that $e \leq 3v - 6$, so $2e \leq 6v - 12$. If the degree of every vertex were at least six, then since $2e = \sum_{v \in V} \deg(v)$ (by the Handshaking Theorem), we would have $2e \geq 6v$. But this contradicts the inequality $2e \leq 6v - 12$. It follows that there must be a vertex with degree no greater than five.   ◁

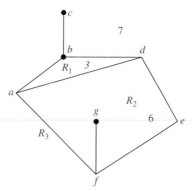

FIGURE 11   The Degrees of Regions.

The proof of Corollary 1 is based on the concept of the **degree** of a region, which is defined to be the number of edges on the boundary of this region. When an edge occurs twice on the boundary (so that it is traced out twice when the boundary is traced out), it contributes two to the degree. The degrees of the regions of the graph shown in Figure 11 are displayed in the figure.

The proof of Corollary 1 can now be given.

Proof: A connected planar simple graph drawn in the plane divides the plane into regions, say r of them. The degree of each region is at least three. (Since the graphs discussed here are simple graphs, no multiple edges that could produce regions of degree two, or loops that could produce regions of degree one, are permitted.) In particular, note that the degree of the unbounded region is at least three since there are at least three vertices in the graph.

Note that the sum of the degrees of the regions is exactly twice the number of edges in the graph, because each edge occurs on the boundary of a region exactly twice (either in two different regions, or twice in the same region). Since each region has degree greater than or equal to three, it follows that

$$2e = \sum_{\text{all regions } R} \deg(R) \geq 3r.$$

Hence,

$(2/3)e \geq r.$

Using $r = e - v + 2$ (Euler's formula), we obtain

$e - v + 2 \leq (2/3)e.$

It follows that $e/3 \leq v - 2$. This shows that $e \leq 3v - 6$.   ◁

This corollary can be used to demonstrate that K_5 is nonplanar.

EXAMPLE 5   Show that K_5 is nonplanar using Corollary 1.

Solution: The graph K_5 has five vertices and ten edges. However, the inequality $e \leq 3v - 6$ is not satisfied for this graph since $e = 10$ and $3v - 6 = 9$. Therefore, K_5 is not planar.
◀

It was previously shown that $K_{3,3}$ is not planar. Note, however, that this graph has six vertices and nine edges. This means that the inequality $e = 9 \leq 12 = 3 \cdot 6 - 6$ is satisfied. Consequently, the fact that the inequality $e \leq 3v - 6$ is satisfied does *not* imply that a graph is planar. However, the following corollary of Theorem 1 can be used to show that $K_{3,3}$ is nonplanar.

COROLLARY 3   If a connected planar simple graph has e edges and v vertices with $v \geq 3$ and no circuits of length three, then $e \leq 2v - 4$.

The proof of Corollary 3 is similar to that of Corollary 1, except that in this case the fact that there are no circuits of length three implies that the degree of a region must be at least four. The details of this proof are left for the reader (see Exercise 15 at the end of this section).

EXAMPLE 6   Use Corollary 3 to show that $K_{3,3}$ is nonplanar.

Solution: Since $K_{3,3}$ has no circuits of length three (this is easy to see since it is bipartite), Corollary 3 can be used. $K_{3,3}$ has six vertices and nine edges. Since $e = 9$ and $2v - 4 = 8$, Corollary 3 shows that $K_{3,3}$ is nonplanar.   ◄

KURATOWSKI'S THEOREM

We have seen that $K_{3,3}$ and K_5 are not planar. Clearly, a graph is not planar if it contains either of these two graphs as a subgraph. Surprisingly, all nonplanar graphs must contain a subgraph that can be obtained from $K_{3,3}$ or K_5 using certain permitted operations.

If a graph is planar, so will be any graph obtained by removing an edge $\{u, v\}$ and adding a new vertex w together with edges $\{u, w\}$ and $\{w, v\}$. Such an operation is called an **elementary subdivision.** The graphs $G_1 = (V_1, E_1)$ and $G_2 = (V_2, E_2)$ are called **homeomorphic** if they can be obtained from the same graph by a sequence of elementary subdivisions. The three graphs displayed in Figure 12 are homeomorphic, since all can be obtained from the first graph by elementary subdivisions. (The reader should determine the sequences of elementary subdivisions needed to obtain G_2 and G_3 from G_1.)

KAZIMIERZ KURATOWSKI (1896–1980)   Kazimierz Kuratowski, the son of a famous Warsaw lawyer, attended secondary school in Warsaw. He studied in Glasgow, Scotland, from 1913 to 1914 but could not return there after the outbreak of World War I. In 1915 he entered Warsaw University, where he was active in the Polish patriotic student movement. He published his first paper in 1919 and received his Ph.D. in 1921. He was an active member of the group known as the Warsaw School of Mathematics, working in the areas of the foundations of set theory and topology. He was appointed associate professor at the Lwów Polytechnical University, where he stayed for seven years, collaborating with the important Polish mathematicians Banach and Ulam. In 1930, while at Lwów, Kuratowski completed his work characterizing planar graphs.

In 1934 he returned to Warsaw University as a full professor. Until the start of World War II, he was active in research and teaching. During the war, because of the persecution of educated Poles, Kuratowski went into hiding under an assumed name and taught at the clandestine Warsaw University. After the war he helped revive Polish mathematics, serving as director of the Polish National Mathematics Institute. He wrote over 180 papers and three widely used textbooks.

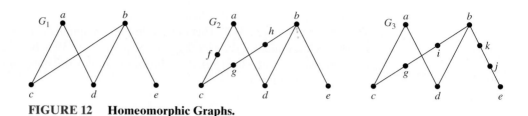

FIGURE 12   Homeomorphic Graphs.

The Polish mathematician Kazimierz Kuratowski established the following theorem in 1930, which characterizes planar graphs using the concept of graph homeomorphism.

THEOREM 2

A graph is nonplanar if and only if it contains a subgraph homeomorphic to $K_{3,3}$ or K_5.

It is clear that a graph containing a subgraph homeomorphic to $K_{3,3}$ or K_5 is nonplanar. However, the proof of the converse, namely that every nonplanar graph contains a subgraph homeomorphic to $K_{3,3}$ or K_5, is complicated and will not be given here. Examples 7 and 8 illustrate how Kuratowski's Theorem is used.

EXAMPLE 7   Determine whether the graph G shown in Figure 13 is planar.

Solution: G has a subgraph H homeomorphic to K_5. H is obtained by deleting h, j, and k and all edges incident with these vertices. H is homeomorphic to K_5 since it can be obtained from K_5 (with vertices a, b, c, g, and i) by a sequence of elementary subdivisions, adding the vertices d, e, and f. (The reader should construct such a sequence of elementary subdivisions.) Hence, G is nonplanar.   ◀

EXAMPLE 8   Is the Petersen graph, shown in Figure 14(a), planar? (The Danish mathematician Julius Petersen studied this graph in 1891; it is often used to illustrate various theoretical properties of graphs.)

Solution: The subgraph H of the Petersen graph obtained by deleting b and the three edges that have b as an endpoint, shown in Figure 14(b), is homeomorphic to $K_{3,3}$, with

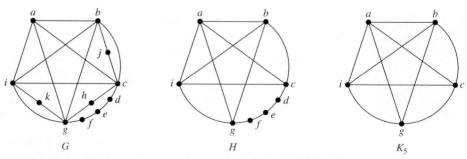

FIGURE 13   The Undirected Graph G, a Subgraph H Homeomorphic to K_5, and K_5.

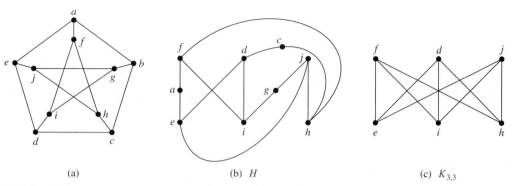

(a)   (b) *H*   (c) $K_{3,3}$

FIGURE 14   **(a) The Petersen Graph, (b) a Subgraph *H* Homeomorphic to $K_{3,3}$, and (c) $K_{3,3}$.**

vertex sets $\{f, d, j\}$ and $\{e, i, h\}$, since it can be obtained by a sequence of elementary subdivisions, deleting $\{d, h\}$ and adding $\{c, h\}$ and $\{c, d\}$, deleting $\{e, f\}$ and adding $\{a, e\}$ and $\{a, f\}$, and deleting $\{i, j\}$ and adding $\{g, i\}$ and $\{g, j\}$. Hence, the Petersen graph is not planar. ◀

Exercises

1. Can five houses be connected to two utilities without connections crossing?

In Exercises 2–4 draw the given planar graph without any crossings.

2.

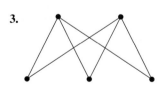

3.

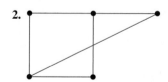

4.

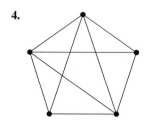

In Exercises 5–9 determine whether the given graph is planar. If so, draw it so that no edges cross.

5.

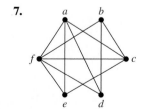

6.

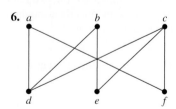

7.

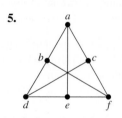

8.

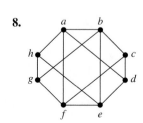

9.

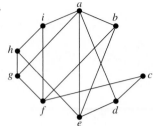

10. Complete the argument in Example 3.

11. Show that K_5 is nonplanar using an argument similar to that given in Example 3.

12. Suppose that a connected planar graph has eight vertices, each of degree three. Into how many regions is the plane divided by a planar representation of this graph?

13. Suppose that a connected planar graph has six vertices, each of degree four. Into how many regions is the plane divided by a planar representation of this graph?

14. Suppose that a connected planar graph has 30 edges. If a planar representation of this graph divides the plane into 20 regions, how many vertices does this graph have?

15. Prove Corollary 3.

16. Suppose that a connected bipartite planar simple graph has e edges and v vertices. Show that $e \le 2v - 4$ if $v \ge 3$.

*17. Suppose that a connected planar simple graph with e edges and v vertices contains no simple circuits of length 4 or less. Show that $e \le (5/3)v - (10/3)$ if $v \ge 4$.

18. Suppose that a planar graph has k connected components, e edges, and v vertices. Also suppose that the plane is divided into r regions by a planar representation of the graph. Find a formula for r in terms of e, v, and k.

19. Which of these nonplanar graphs have the property that the removal of any vertex and all edges incident with that vertex produces a planar graph?

    a) K_5 **b)** K_6 **c)** $K_{3,3}$ **d)** $K_{3,4}$

In Exercises 20–22 determine whether the given graph is homeomorphic to $K_{3,3}$.

20.

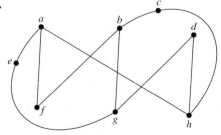

21.

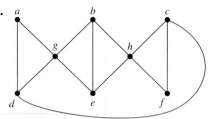

22.

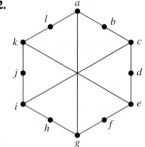

In Exercises 23–25 use Kuratowski's Theorem to determine whether the given graph is planar.

23.

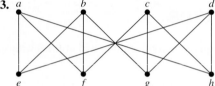

24.

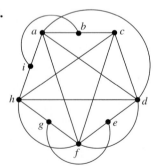

25.

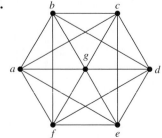

The **crossing number** of a simple graph is the minimum number of crossings that can occur when this graph is drawn in the plane where no three arcs representing edges are permitted to cross at the same point.

26. Show that $K_{3,3}$ has 1 as its crossing number.

**27. Find the crossing numbers of each of these nonplanar graphs.

a) K_5 **b)** K_6 **c)** K_7

d) $K_{3,4}$ **e)** $K_{4,4}$ **f)** $K_{5,5}$

*28. Find the crossing number of the Petersen graph.

**29. Show that if m and n are even positive integers, the crossing number of $K_{m,n}$ is less than or equal to $mn(m-2)(n-2)/16$. (*Hint:* Place m vertices along the x axis so that they are equally spaced and symmetric about the origin and place n vertices along the y axis so that they are equally spaced and symmetric about the origin. Now connect each of the m vertices on the x axis to each of the vertices on the y axis and count the crossings.)

The **thickness** of a simple graph G is the smallest number of planar subgraphs of G that have G as their union.

30. Show that $K_{3,3}$ has 2 as its thickness.

*31. Find the thickness of the graphs in Exercise 27.

32. Show that if G is a connected simple graph with v vertices and e edges, then the thickness of G is at least $\lceil e/(3v-6) \rceil$.

*33. Use Exercise 32 to show that the thickness of K_n is at least $\lfloor (n+7)/6 \rfloor$ whenever n is a positive integer.

34. Show that if G is a connected simple graph with v vertices and e edges and no circuits of length three, then the thickness of G is at least $\lceil e/(2v-4) \rceil$.

35. Use Exercise 34 to show that the thickness of $K_{m,n}$ is at least $\lceil mn/(2m+2n-4) \rceil$ whenever m and n are positive integers.

*36. Draw K_5 on the surface of a torus (a doughnut-shaped solid) so that no edges cross.

*37. Draw $K_{3,3}$ on the surface of a torus so that no edges cross.

8.8   Graph Coloring

INTRODUCTION

Problems related to the coloring of maps of regions, such as maps of parts of the world, have generated many results in graph theory. When a map* is colored, two regions with a common border are customarily assigned different colors. One way to ensure that two adjacent regions never have the same color is to use a different color for each region. However, this is inefficient, and on maps with many regions it would be hard to distinguish similar colors. Instead, a small number of colors should be used whenever possible. Consider the problem of determining the least number of colors that can be used to color a map so that adjacent regions never have the same color. For instance, for the map shown on the left in Figure 1, four colors suffice, but three colors are not enough. (The reader should check this.) In the map on the right in Figure 1, three colors are sufficient (but two are not).

    Each map in the plane can be represented by a graph. To set up this correspondence, each region of the map is represented by a vertex. Edges connect two vertices if the regions represented by these vertices have a common border. Two regions that touch at only one point are not considered adjacent. The resulting graph is called the **dual graph** of the map. By the way in which dual graphs of maps are constructed, it is clear that any map in the plane has a planar dual graph. Figure 2 displays the dual graphs that correspond to the maps shown in Figure 1.

    The problem of coloring the regions of a map is equivalent to the problem of coloring the vertices of the dual graph so that no two adjacent vertices in this graph have the same color. We now define a graph coloring.

---

*We will assume that all regions in a map are connected. This eliminates any problems presented by such geographical entities as Michigan.

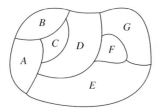
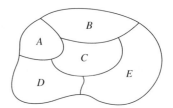

FIGURE 1   Two Maps.

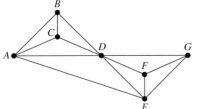
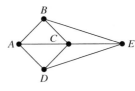

FIGURE 2   Dual Graphs of the Maps in Figure 1.

DEFINITION 1

A *coloring* of a simple graph is the assignment of a color to each vertex of the graph so that no two adjacent vertices are assigned the same color.

A graph can be colored by assigning a different color to each of its vertices. However, for most graphs a coloring can be found that uses fewer colors than the number of vertices in the graph. What is the least number of colors necessary?

DEFINITION 2

The *chromatic number* of a graph is the least number of colors needed for a coloring of this graph.

Note that asking for the chromatic number of a planar graph is the same as asking for the minimum number of colors required to color a planar map so that no two adjacent regions are assigned the same color. This question has been studied for more than 100 years. The answer is provided by one of the most famous theorems in mathematics.

THEOREM 1

THE FOUR COLOR THEOREM   The chromatic number of a planar graph is no greater than four.

ALFRED BRAY KEMPE (1849–1922)   Kempe was a barrister and a leading authority on ecclesiastical law. However, having studied mathematics at Cambridge University, he retained his interest in it, and later in life he devoted considerable time to mathematical research. Kempe made contributions to kinematics, the branch of mathematics dealing with motion, and to mathematical logic. However, Kempe is best remembered for his fallacious proof of the Four Color Theorem.

The Four Color Theorem was originally posed as a conjecture in the 1850s. It was finally proved by the American mathematicians Kenneth Appel and Wolfgang Haken in 1976. Prior to 1976, many incorrect proofs were published, often with hard-to-find errors. In addition, many futile attempts were made to construct counterexamples by drawing maps that require more than four colors. (Proving the Five Color Theorem is not that difficult; see Exercise 35.)

Perhaps the most notorious fallacious proof in all of mathematics is the incorrect proof of the Four Color Theorem published in 1879 by a London barrister and amateur mathematician, Alfred Kempe. Mathematicians accepted his proof as correct until 1890, when Percy Heawood found an error that made Kempe's argument incomplete. However, Kempe's line of reasoning turned out to be the basis of the successful proof given by Appel and Haken. Their proof relies on a careful case-by-case analysis carried out by computer. They showed that if the Four Color Theorem were false, there would have to be a counterexample of one of approximately 2000 different types, and they then showed that none of these types exists. They used over 1000 hours of computer time in their proof. This proof generated a large amount of controversy, since computers played such an important role in it. For example, could there be an error in a computer program that led to incorrect results? Was their argument really a proof if it depended on what could be unreliable computer output?

Note that the Four Color Theorem applies only to planar graphs. Nonplanar graphs can have arbitrarily large chromatic numbers, as will be shown in Example 2.

Two things are required to show that the chromatic number of a graph is k. First, we must show that the graph can be colored with k colors. This can be done by constructing such a coloring. Second, we must show that the graph cannot be colored using fewer than k colors. Examples 1–4 illustrate how chromatic numbers can be found.

EXAMPLE 1   What are the chromatic numbers of the graphs G and H shown in Figure 3?

Solution: The chromatic number of G is at least three, since the vertices a, b, and c must be assigned different colors. To see if G can be colored with three colors, assign red to a, blue to b, and green to c. Then, d can (and must) be colored red since it is adjacent to b and c. Furthermore, e can (and must) be colored green since it is adjacent only to vertices colored red and blue, and f can (and must) be colored blue since it is adjacent only to vertices colored red and green. Finally, g can (and must) be colored red since it is adjacent only to vertices colored blue and green. This produces a coloring of G using exactly three colors. Figure 4 displays such a coloring.

HISTORICAL NOTE   In 1852, an ex-student of Augustus De Morgan, Francis Guthrie, noticed that the counties in England could be colored using four colors so that no adjacent counties were assigned the same color. On this evidence, he conjectured that the Four Color Theorem was true. Francis told his brother Frederick, at that time a student of De Morgan, about this problem. Frederick in turn asked his teacher De Morgan about his brother's conjecture. De Morgan was extremely interested in this problem and publicized it throughout the mathematical community. In fact, the first written reference to the conjecture can be found in a letter from De Morgan to Sir William Rowan Hamilton. Although De Morgan thought Hamilton would be interested in this problem, Hamilton apparently was not interested in it, since it had nothing to do with quaternions.

HISTORICAL NOTE   Although a simpler proof of the Four Color Theorem was found by Robertson, Sanders, Seymour, and Thomas in 1996, reducing the computational part of the proof to examining 633 configurations, no proof that does not rely on extensive computation has yet been found.

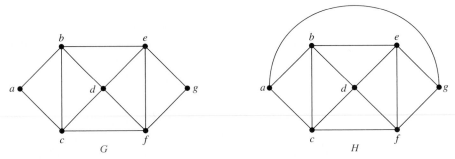

FIGURE 3   The Simple Graphs G and H.

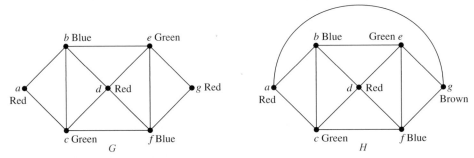

FIGURE 4   Colorings of the Graphs G and H.

The graph H is made up of the graph G with an edge connecting a and g. Any attempt to color H using three colors must follow the same reasoning as that used to color G, except at the last stage, when all vertices other than g have been colored. Then, since g is adjacent (in H) to vertices colored red, blue, and green, a fourth color, say brown, needs to be used. Hence, H has a chromatic number equal to 4. A coloring of H is shown in Figure 4.   ◀

EXAMPLE 2   What is the chromatic number of K_n?

Solution: A coloring of K_n can be constructed using n colors by assigning a different color to each vertex. Is there a coloring using fewer colors? The answer is no. No two vertices can be assigned the same color, since every two vertices of this graph are adjacent. Hence, the chromatic number of $K_n = n$. (Recall that K_n is not planar when $n \geq 5$, so this result does not contradict the Four Color Theorem.) A coloring of K_5 using five colors is shown in Figure 5.   ◀

EXAMPLE 3   What is the chromatic number of the complete bipartite graph $K_{m,n}$, where m and n are positive integers?

Solution: The number of colors needed may seem to depend on m and n. However, only two colors are needed. Color the set of m vertices with one color and the set of n vertices with a second color. Since edges connect only a vertex from the set of m vertices and a vertex from the set of n vertices, no two adjacent vertices have the same color. A coloring of $K_{3,4}$ with two colors is displayed in Figure 6.   ◀

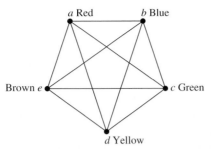

FIGURE 5 **A Coloring of K_5.**

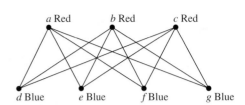

FIGURE 6 **A Coloring of $K_{3,4}$.**

Every connected bipartite simple graph has a chromatic number of 2, or 1, since the reasoning used in Example 3 applies to any such graph. Conversely, every graph with a chromatic number of 2 is bipartite. (See Exercises 25 and 26 at the end of this section.)

EXAMPLE 4   What is the chromatic number of the graph C_n? (Recall that C_n is the cycle with n vertices.)

Solution: We will first consider some individual cases. To begin, let $n = 6$. Pick a vertex and color it red. Proceed clockwise in the planar depiction of C_6 shown in Figure 7. It is necessary to assign a second color, say blue, to the next vertex reached. Continue in the clockwise direction; the third vertex can be colored red, the fourth vertex blue, and the fifth vertex red. Finally, the sixth vertex, which is adjacent to the first, can be colored blue. Hence, the chromatic number of C_6 is 2. Figure 7 displays the coloring constructed here.

Next, let $n = 5$ and consider C_5. Pick a vertex and color it red. Proceeding clockwise, it is necessary to assign a second color, say blue, to the next vertex reached. Continuing in the clockwise direction, the third vertex can be colored red, and the fourth vertex can be colored blue. The fifth vertex cannot be colored either red or blue, since it is adjacent to the fourth vertex and the first vertex. Consequently, a third color is required for this vertex. Note that we would have also needed three colors if we had colored vertices in the counterclockwise direction. Thus, the chromatic number of C_5 is 3. A coloring of C_5 using three colors is displayed in Figure 7.

In general, two colors are needed to color C_n when n is even. To construct such a coloring, simply pick a vertex and color it red. Proceed around the graph in a clockwise direction (using a planar representation of the graph) coloring the second vertex blue, the third vertex red, and so on. The nth vertex can be colored blue, since the two vertices adjacent to it, namely the $(n - 1)$st and the first vertices, are both colored red.

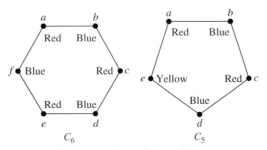

FIGURE 7   **Colorings of C_5 and C_6.**

When n is odd and $n > 1$, the chromatic number of C_n is 3. To see this, pick an initial vertex. To use only two colors, it is necessary to alternate colors as the graph is traversed in a clockwise direction. However, the nth vertex reached is adjacent to two vertices of different colors, namely, the first and $(n - 1)$st. Hence, a third color must be used. ◀

Links

The best algorithms known for finding the chromatic number of a graph have exponential worst-case time complexity (in the number of vertices of the graph). Even the problem of finding an approximation to the chromatic number of a graph is difficult. It has been shown that if there were an algorithm with polynomial worst-case time complexity that could approximate the chromatic number of a graph up to a factor of 2 (that is, construct a bound which was no more than double the chromatic number of the graph), then an algorithm with polynomial worst-case time complexity for finding the chromatic number of the graph would also exist.

APPLICATIONS OF GRAPH COLORINGS

Graph coloring has a variety of applications to problems involving scheduling and assignments. (Note that since no efficient algorithm is known for graph coloring, this does not lead to efficient algorithms for scheduling and assignments.) Examples of such applications will be given here. The first application deals with the scheduling of final exams.

EXAMPLE 5 **Scheduling Final Exams** How can the final exams at a university be scheduled so that no student has two exams at the same time?

Solution: This scheduling problem can be solved using a graph model, with vertices representing courses and with an edge between two vertices if there is a common student in the courses they represent. Each time slot for a final exam is represented by a different color. A scheduling of the exams corresponds to a coloring of the associated graph.

For instance, suppose there are seven finals to be scheduled. Suppose the courses are numbered 1 through 7. Suppose that the following pairs of courses have common students: 1 and 2, 1 and 3, 1 and 4, 1 and 7, 2 and 3, 2 and 4, 2 and 5, 2 and 7, 3 and 4, 3 and 6, 3 and 7, 4 and 5, 4 and 6, 5 and 6, 5 and 7, and 6 and 7. In Figure 8 the graph associated with this set of classes is shown. A scheduling consists of a coloring of this graph.

Since the chromatic number of this graph is 4 (the reader should verify this), four time slots are needed. A coloring of the graph using four colors and the associated schedule are shown in Figure 9. ◀

Now consider an application to the assignment of television channels.

EXAMPLE 6 **Frequency Assignments** Television channels 2 through 13 are assigned to stations in North America that no two stations within 150 miles can operate on the same channel. How can the assignment of channels be modeled by graph coloring?

Solution: Construct a graph by assigning a vertex to each station. Two vertices are connected by an edge if they are located within 150 miles of each other. An assignment of channels corresponds to a coloring of the graph, where each color represents a different channel. ◀

An application of graph coloring to compilers is considered in Example 7.

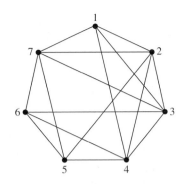

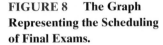

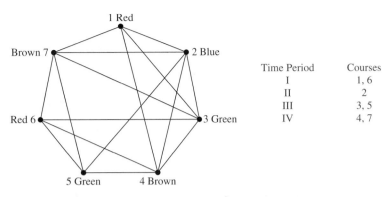

Time Period	Courses
I	1, 6
II	2
III	3, 5
IV	4, 7

FIGURE 8   The Graph Representing the Scheduling of Final Exams.

FIGURE 9   Using a Coloring to Schedule Final Exams.

EXAMPLE 7   Index Registers In efficient compilers the execution of loops is speeded up when frequently used variables are stored temporarily in index registers in the central processing unit, instead of in regular memory. For a given loop, how many index registers are needed? This problem can be addressed using a graph coloring model. To set up the model, let each vertex of a graph represent a variable in the loop. There is an edge between two vertices if the variables they represent must be stored in index registers at the same time during the execution of the loop. Thus, the chromatic number of the graph gives the number of index registers needed, since different registers must be assigned to variables when the vertices representing these variables are adjacent in the graph.   ◄

Exercises

In Exercises 1–4 construct the dual graph for the map shown. Then find the number of colors needed to color the map so that no two adjacent regions have the same color.

1.

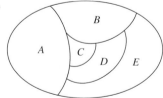

2.

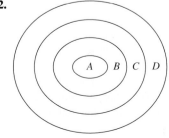

3.

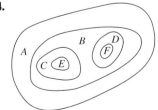

4.

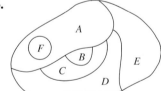

In Exercises 5–11 find the chromatic number of the given graph.

5.

6.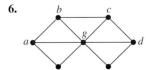

7.

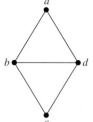

8.

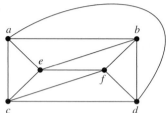

9.

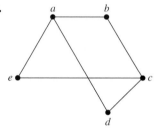

10.

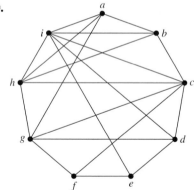

11.

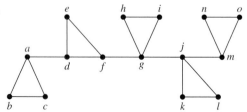

12. For the graphs in Exercises 5–11, decide whether it is possible to decrease the chromatic number by removing a single vertex and all edges incident with it.

13. Which graphs have a chromatic number of 1?

14. What is the least number of colors needed to color a map of the United States? Do not consider adjacent states that meet only at a corner. Suppose that Michigan is one region. Consider the vertices representing Alaska and Hawaii as isolated vertices.

15. What is the chromatic number of W_n?

16. Show that a simple graph that has a circuit with an odd number of vertices in it cannot be colored using two colors.

17. Schedule the final exams for Math 115, Math 116, Math 185, Math 195, CS 101, CS 102, CS 273, and CS 473, using the fewest number of different time slots, if there are no students taking both Math 115 and CS 473, both Math 116 and CS 473, both Math 195 and CS 101, both Math 195 and CS 102, both Math 115 and Math 116, both Math 115 and Math 185, and both Math 185 and Math 195, but there are students in every other combination of courses.

18. How many different channels are needed for six stations located at the distances shown in the table, if two stations cannot use the same channel when they are within 150 miles of each other?

	1	**2**	**3**	**4**	**5**	**6**
1	—	85	175	200	50	100
2	85	—	125	175	100	160
3	175	125	—	100	200	250
4	200	175	100	—	210	220
5	50	100	200	210	—	100
6	100	160	250	220	100	—

19. The mathematics department has six committees each meeting once a month. How many different meeting times must be used to ensure that no member is scheduled to attend two meetings at the same time if the committees are $C_1 = \{$Arlinghaus, Brand, Zaslavsky$\}$, $C_2 = \{$Brand, Lee, Rosen$\}$, $C_3 = \{$Arlinghaus, Rosen, Zaslavsky$\}$, $C_4 = \{$Lee, Rosen, Zaslavsky$\}$, $C_5 = \{$Arlinghaus, Brand$\}$, and $C_6 = \{$Brand, Rosen, Zaslavsky$\}$?

20. A zoo wants to set up natural habitats in which to exhibit its animals. Unfortunately, some animals will eat some of the others when given the opportunity. How can a graph model and a coloring be used to determine the number of different habitats needed and the placement of the animals in these habitats?

An **edge coloring** of a graph is an assignment of colors to edges so that edges incident with a common vertex are assigned different colors. The **edge chromatic number** of a graph is the smallest number of colors that can be used in an edge coloring of the graph.

21. Find the edge chromatic number of each of the graphs in Exercises 5–11.

*****22.** Find the edge chromatic numbers of

    a) K_n.

    b) $K_{m,n}$.

    c) C_n.

    d) W_n.

23. Seven variables occur in a loop of a computer program. The variables and the steps during which they must be stored are: t: steps 1 through 6; u: step 2; v: steps 2 through 4; w: steps 1, 3, and 5; x: steps 1 and 6; y: steps 3 through 6; and z: steps 4 and 5. How many different index registers are needed to store these variables during execution?

24. What can be said about the chromatic number of a graph that has K_n as a subgraph?

25. Show that a simple graph with a chromatic number of 2 is bipartite.

26. Show that a connected bipartite graph has a chromatic number of 2.

This algorithm can be used to color a simple graph: First, list the vertices $v_1, v_2, v_3, \ldots, v_n$ in order of decreasing degree so that $\deg(v_1) \geq \deg(v_2) \geq \cdots \geq \deg(v_n)$. Assign color 1 to v_1 and to the next vertex in the list not adjacent to v_1 (if one exists), and successively to each vertex in the list not adjacent to a vertex already assigned color 1. Then assign color 2 to the first vertex in the list not already colored. Successively assign color 2 to vertices in the list that have not been already colored and are not adjacent to vertices assigned color 2. If uncolored vertices remain, assign color 3 to the first vertex in the list not yet colored, and use color 3 to successively color those vertices not already colored and not adjacent to vertices assigned color 3. Continue this process until all vertices are colored.

27. Construct a coloring of the graph shown using this algorithm.

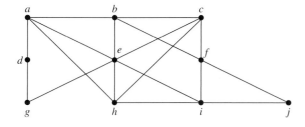

*****28.** Use pseudocode to describe this coloring algorithm.

*****29.** Show that the coloring produced in this algorithm may use more colors than are necessary to color a graph.

A **k-tuple coloring** of a graph G is an assignment of a set of k different colors to each of the vertices of G so that no two adjacent vertices are assigned a common color. We denote by $\chi_k(G)$ the smallest positive integer n such that G has a k-tuple coloring using n colors. For example, $\chi_2(C_4) = 4$. To see this, note that using only four colors we can assign two colors to each vertex of C_4, as illustrated, so that no two adjacent vertices are assigned the same color. Furthermore, no fewer than four colors suffice because the vertices v_1 and v_2 each must be assigned two colors, and a common color cannot be assigned to both v_1 and v_2. (For more information about k-tuple coloring, see [MiRo91].)

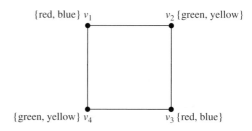

30. Find these values:

    a) $\chi_2(K_3)$

    b) $\chi_2(K_4)$

    c) $\chi_2(W_4)$

    d) $\chi_2(C_5)$

    e) $\chi_2(K_{3,4})$

    f) $\chi_3(K_5)$

  *****g)** $\chi_3(C_5)$

    h) $\chi_3(K_{4,5})$

*****31.** Let G and H be the graphs displayed in Figure 3. Find

    a) $\chi_2(G)$.

    b) $\chi_2(H)$.

    c) $\chi_3(G)$.

    d) $\chi_3(H)$.

32. What is $\chi_k(G)$ if G is a bipartite graph and k is a positive integer?

33. Frequencies for mobile radio (or cellular) telephones are assigned by zones. Each zone is assigned a set of frequencies to be used by vehicles in that zone. The same frequency cannot be used in zones where interference is a problem. Explain how a k-tuple coloring can be used to assign k frequencies to each mobile radio zone in a region.

*****34.** Show that every planar graph G can be colored using six or fewer colors. (*Hint:* Use mathematical induction on the number of vertices of the graph. Apply Corollary 2 of Section 8.7 to find a vertex v with $\deg(v) \leq 5$. Consider the subgraph of G obtained by deleting v and all edges incident to it.)

******35.** Show that every planar graph G can be colored using five or fewer colors. (*Hint:* Use the hint provided for Exercise 34.)

Key Terms and Results

TERMS

undirected edge: an edge associated to a set $\{u, v\}$, where u and v are vertices

directed edge: an edge associated to an ordered pair (u, v), where u and v are vertices

multiple edges: distinct edges connecting the same vertices

loop: an edge connecting a vertex with itself

undirected graph: a set of vertices together with a set of undirected edges that connect these vertices

simple graph: an undirected graph with no multiple edges or loops

multigraph: an undirected graph that may contain multiple edges but no loops

pseudograph: an undirected graph that may contain multiple edges and loops

directed graph: a set of vertices together with a set of directed edges that connect these vertices

directed multigraph: a graph with directed edges that may contain multiple directed edges

adjacent: two vertices are adjacent if there is an edge between them

incident: an edge is incident with a vertex if the vertex is an endpoint of that edge

deg (v) (degree of the vertex v in an undirected graph): the number of edges incident with v

deg$^-$ (v) (the in-degree of the vertex v in a graph with directed edges): the number of edges with v as their terminal vertex

deg$^+$ (v) (the out-degree of the vertex v in a graph with directed edges): the number of edges with v as their initial vertex

underlying undirected graph of a graph with directed edges: the undirected graph obtained by ignoring the directions of the edges

K_n (complete graph on n vertices): the undirected graph with n vertices where each pair of vertices is connected by an edge

bipartite graph: a graph with vertex set partitioned into subsets V_1 and V_2 such that each edge connects a vertex in V_1 and a vertex in V_2

$K_{m,n}$ (complete bipartite graph): the graph with vertex set partitioned into a subset of m elements and a subset of n elements such that two vertices are connected by an edge if and only if one is in the first subset and the other is in the second subset

C_n (cycle of size n), $n \geq 3$: the graph with n vertices $v_1, v_2, \ldots, v_n$ and edges $\{v_1, v_2\}$, $\{v_2, v_3\}, \ldots$, $\{v_{n-1}, v_n\}$, $\{v_n, v_1\}$

W_n (wheel of size n), $n \geq 3$: the graph obtained from C_n by adding a vertex and edges from this vertex to the original vertices in C_n

Q_n (n-cube), $n \geq 1$: the graph that has the 2^n bit strings of length n as its vertices and edges connecting every pair of bit strings that differ by exactly one bit

isolated vertex: a vertex of degree zero

pendant vertex: a vertex of degree one

regular graph: a graph where all vertices have the same degree

subgraph of a graph $G = (V, E)$: a graph (W, F) where W is a subset of V and F is a subset of E

$G_1 \cup G_2$ (union of G_1 and G_2): the graph $(V_1 \cup V_2, E_1 \cup E_2)$ where $G_1 = (V_1, E_1)$ and $G_2 = (V_2, E_2)$

adjacency matrix: a matrix representing a graph using the adjacency of vertices

incidence matrix: a matrix representing a graph using the incidence of edges and vertices

isomorphic simple graphs: the simple graphs $G_1 = (V_1, E_1)$ and $G_2 = (V_2, E_2)$ are isomorphic if there is a one-to-one correspondence f from V_1 to V_2 such that $\{f(v_1), f(v_2)\} \in E_2$ if and only if $\{v_1, v_2\} \in E_1$ for all v_1 and v_2 in V_1

invariant: a property that isomorphic graphs either both have or both do not have

path from u to v in an undirected graph: a sequence of edges $e_1, e_2, \ldots, e_n$ where e_i is associated to $\{x_i, x_{i+1}\}$ for $i = 0, 1, \ldots, n$ where $x_0 = u$ and $x_{n+1} = v$

path from u to v in a graph with directed edges: a sequence of edges $e_1, e_2, \ldots, e_n$ where e_i is associated to (x_i, x_{i+1}) for $i = 0, 1, \ldots, n$ where $x_0 = u$ and $x_{n+1} = v$

simple path: a path that does not contain an edge more than once

circuit: a path of length $n \geq 1$ that begins and ends at the same vertex

connected graph: an undirected graph with the property that there is a path between every pair of vertices

connected component of a graph G: a maximal connected subgraph of G

strongly connected directed graph: a directed graph with the property that there is a directed path from every vertex to every vertex

strongly connected component of a directed graph G: a maximal strongly connected subgraph of G

Euler circuit: a circuit that contains every edge of a graph exactly once

Euler path: a path that contains every edge of a graph exactly once

Hamilton path: a path $x_0, x_1, \ldots, x_n$ in a simple graph $G = (V, E)$ such that $\{x_0, x_1, \ldots, x_n\} = V$ and $x_i \neq x_j$ for $0 \leq i < j \leq n$

Hamilton circuit: a circuit $x_0, x_1, \ldots, x_n, x_0$ in a simple graph such that $x_0, x_1, \ldots, x_n$ is a Hamilton path

weighted graph: a graph with numbers assigned to its edges

shortest-path problem: the problem of determining the path in a weighted graph such that the sum of the weights of the edges in this path is a minimum over all paths between specified vertices

traveling salesman problem: the problem that asks for the circuit of shortest total length which visits every vertex of the graph exactly once

planar graph: a graph that can be drawn in the plane with no crossings

regions of a representation of a planar graph: the regions the plane is divided into by the planar representation of the graph

elementary subdivision: the removal of an edge $\{u, v\}$ of an undirected graph and the addition of a new vertex w together with edges $\{u, w\}$ and $\{w, v\}$

homeomorphic: two undirected graphs are homeomorphic if they can be obtained from the same graph by a sequence of elementary subdivisions

graph coloring: an assignment of colors to the vertices of a graph so that no two adjacent vertices have the same color

chromatic number: the minimum number of colors needed in a coloring of a graph

RESULTS

There is an Euler circuit in a connected multigraph if and only if every vertex has even degree.

There is an Euler path in a connected multigraph if and only if at most two vertices have odd degree.

Dijkstra's algorithm: a procedure for finding a shortest path between two vertices in a weighted graph (see Section 8.6).

Euler's formula: $r = e - v + 2$ where $r, e,$ and v are the number of regions of a planar representation, the number of edges, and the number of vertices, respectively, of a planar graph.

Kuratowski's Theorem: A graph is nonplanar if and only if it contains a subgraph homeomorphic to $K_{3,3}$ or K_5. (Proof beyond scope of this book.)

the Four Color Theorem: Every planar graph can be colored using no more than four colors. (Proof far beyond the scope of this book!)

Review Questions

1. a) Define a simple graph, a multigraph, a pseudo-graph, a directed graph, and a directed multi-graph.

 b) Use an example to show how each of the types of graph in part (a) can be used in modeling. For example, explain how to model different aspects of a computer network or airline routes.

2. Give at least four examples of how graphs are used in modeling.

3. What is the relationship between the sum of the degrees of the vertices in an undirected graph and the number of edges in this graph? Explain why this relationship holds.

4. Why must there be an even number of vertices of odd degree in an undirected graph?

5. What is the relationship between the sum of the in-degrees and the sum of the out-degrees of the vertices in a directed graph? Explain why this relationship holds.

6. Describe the following families of graphs.

 a) K_n, the complete graph on n vertices

 b) $K_{m,n}$, the complete bipartite graph on m and n vertices

 c) C_n, the cycle with n vertices

 d) W_n, the wheel of size n

 e) Q_n, the n-cube

7. How many vertices and how many edges are there in each of the graphs in the families in Question 6?

8. a) What is a bipartite graph?

 b) Which of the graphs $K_n, C_n,$ and W_n are bipartite?

 c) How can you determine whether an undirected graph is bipartite?

9. a) Describe three different methods that can be used to represent a graph.

 b) Draw a simple graph with at least five vertices and eight edges. Illustrate how it can be represented using the methods you described in part (a).

10. a) What does it mean for two simple graphs to be isomorphic?

 b) What is meant by an invariant with respect to isomorphism for simple graphs? Give at least five examples of such invariants.

 c) Give an example of two graphs that have the same numbers of vertices, edges, and degrees of vertices, but that are not isomorphic.

 d) Is a set of invariants known that can be used to efficiently determine whether two simple graphs are isomorphic?

11. a) What does it mean for a graph to be connected?

 b) What are the connected components of a graph?

12. a) Explain how an adjacency matrix can be used to represent a graph.

b) How can adjacency matrices be used to determine whether a function from the vertex set of a graph G to the vertex set of a graph H is an isomorphism?

c) How can the adjacency matrix of a graph be used to determine the number of paths of length r, where r is a positive integer, between two vertices of a graph?

13. a) Define an Euler circuit and an Euler path in an undirected graph.

b) Describe the famous Königsberg bridge problem and explain how to rephrase it in terms of an Euler circuit.

c) How can it be determined whether an undirected graph has an Euler path?

d) How can it be determined whether an undirected graph has an Euler circuit?

14. a) Define a Hamilton circuit in a simple graph.

b) Give some properties of a simple graph that imply that it does not have a Hamilton circuit.

15. Give examples of at least two problems that can be solved by finding the shortest path in a weighted graph.

16. a) Describe Dijkstra's algorithm for finding the shortest path in a weighted graph between two vertices.

b) Draw a weighted graph with at least ten vertices and 20 edges. Use Dijkstra's algorithm to find the shortest path between two vertices of your choice in the graph.

17. a) What does it mean for a graph to be planar?

b) Give an example of a nonplanar graph.

18. a) What is Euler's formula for planar graphs?

b) How can Euler's formula for planar graphs be used to show that a simple graph is nonplanar?

19. State Kuratowski's Theorem on the planarity of graphs and explain how it characterizes which graphs are planar.

20. a) Define the chromatic number of a graph.

b) What is the chromatic number of the graph K_n when n is a positive integer?

c) What is the chromatic number of the graph C_n when n is a positive integer greater than 2?

d) What is the chromatic number of the graph $K_{m,n}$ when m and n are positive integers?

21. State the Four Color Theorem. Are there graphs that cannot be colored with four colors?

22. Explain how graph coloring can be used in modeling. Use at least two different examples.

Supplementary Exercises

1. How many edges does a 50-regular graph with 100 vertices have?

2. How many nonisomorphic subgraphs does K_3 have?

In Exercises 3–5 determine whether the given pair of graphs is isomorphic.

3.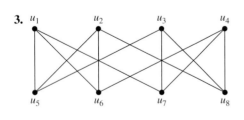

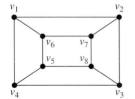

4.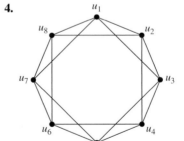

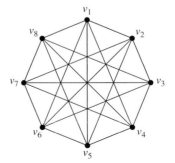

*5.

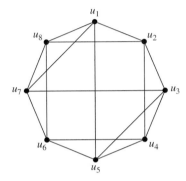

The **complete *m*-partite graph** $K_{n_1,n_2,\ldots,n_m}$ has vertices partitioned into m subsets of $n_1, n_2, \ldots, n_m$ elements each, and vertices are adjacent if and only if they are in different subsets in the partition.

6. Draw these graphs.

a) $K_{1,2,3}$
b) $K_{2,2,2}$
c) $K_{1,2,2,3}$

*7. How many vertices and how many edges does the complete *m*-partite graph $K_{n_1,n_2,\ldots,n_m}$ have?

*8. a) Prove or disprove that there are always two vertices with the same degree in a finite simple graph having at least two vertices.

b) Do the same as in part (a) for finite multigraphs.

Let $G = (V, E)$ be a simple graph. The **subgraph induced** by a subset W of the vertex set V is the graph (W, F), where the edge set F contains an edge in E if and only if both endpoints of this edge are in W.

9. Consider the graph shown in Figure 3 of Section 8.4. Find the subgraphs induced by

a) $\{a, b, c\}$.
b) $\{a, e, g\}$.
c) $\{b, c, f, g, h\}$.

10. Let n be a positive integer. Show that a subgraph induced by a nonempty subset of the vertex set of K_n is a complete graph.

A **clique** in a simple undirected graph is a complete subgraph that is not contained in any larger complete subgraph. In Exercises 11–13 find all cliques in the graph shown.

11.

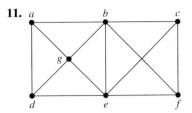

12.

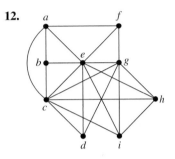

13.

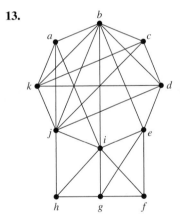

A **dominating set** of vertices in a simple graph is a set of vertices such that every other vertex is adjacent to at least one vertex of this set. A dominating set with the least number of vertices is called a **minimum dominating set**. In Exercises 14–16 find a minimum dominating set for the given graph.

14.   **15.**

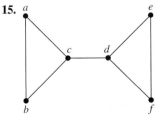

16.

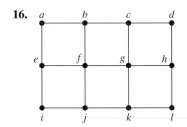

A simple graph can be used to determine the minimum number of queens on a chessboard that control the entire chessboard. An $n \times n$ chessboard has n^2 squares in an $n \times n$ configuration. A queen in a given position controls all squares in the same row, the same column, and on the two diagonals containing this square, as illustrated. The appropriate simple graph has n^2 vertices, one for each square, and two vertices are adjacent if a queen in the square represented by one of the vertices controls the square represented by the other vertex.

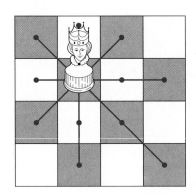

The Squares
Controlled
by a Queen

17. Construct the simple graph representing the $n \times n$ chessboard with edges representing the control of squares by queens for

    a) $n = 3$.

    b) $n = 4$.

18. Explain how the concept of a minimum dominating set applies to the problem of determining the minimum number of queens controlling an $n \times n$ chessboard.

**19. Find the minimum number of queens controlling an $n \times n$ chessboard for

    a) $n = 3$.

    b) $n = 4$.

    c) $n = 5$.

20. Suppose that G_1 and H_1 are isomorphic and that G_2 and H_2 are isomorphic. Prove or disprove that $G_1 \cup G_2$ and $H_1 \cup H_2$ are isomorphic.

21. Show that each of these properties is an invariant that isomorphic simple graphs either both have or both do not have.

    a) connectedness

    b) the existence of a Hamilton circuit

    c) the existence of an Euler circuit

    d) having crossing number C

    e) having n isolated vertices

    f) being bipartite

22. How can the adjacency matrix of $\overline{G}$ be found from the adjacency matrix of G, where G is a simple graph?

23. How many nonisomorphic connected bipartite simple graphs are there with four vertices?

*24. How many nonisomorphic simple connected graphs with five vertices are there

    a) with no vertex of degree more than two?

    b) with chromatic number equal to four?

    c) that are nonplanar?

A directed graph is **self-converse** if it is isomorphic to its converse.

25. Determine whether the following graphs are self-converse.

    a)

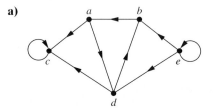

    b)

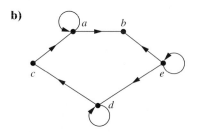

26. Show that if the directed graph G is self-converse and H is a directed graph isomorphic to G, then H is also self-converse.

An **orientation** of an undirected simple graph is an assignment of directions to its edges so that the resulting directed graph is strongly connected. When an orientation of an undirected graph exists, this graph is called **orientable.** In Exercises 27–29 determine whether the given simple graph is orientable.

27.

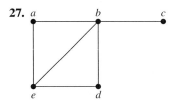

28.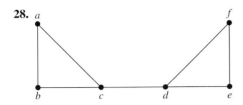

29.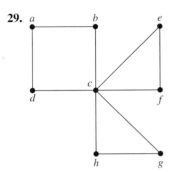

30. Because traffic is growing heavy in the central part of a city, traffic engineers are planning to change all the streets, which are currently two-way, into one-way streets. Explain how to model this problem.

*31. Show that a graph is not orientable if it has a cut edge.

A **tournament** is a simple directed graph such that if u and v are distinct vertices in the graph, exactly one of (u, v) and (v, u) is an edge of the graph.

32. How many different tournaments are there with n vertices?

33. What is the sum of the in-degree and out-degree of a vertex in a tournament?

*34. Show that every tournament has a Hamilton path.

35. Given two chickens in a flock, one of them is dominant. This defines the **pecking order** of the flock. How can a tournament be used to model pecking order?

36. Suppose that G is a connected multigraph with $2k$ vertices of odd degree. Show that there exist k subgraphs that have G as their union, where each of these subgraphs has an Euler path and where no two of these subgraphs have an edge in common. (*Hint:* Add k edges to the graph connecting pairs of vertices of odd degree and use an Euler circuit in this larger graph.)

*37. Let G be a simple graph with n vertices. The **bandwidth** of G, denoted by $B(G)$, is the minimum, over all permutations, $a_1, a_2, \ldots, a_n$ of the vertices of G, of $\max\{|i - j| \mid a_i \text{ and } a_j \text{ are adjacent}\}$. That is, the bandwidth is the minimum over all listings of the vertices of the maximum difference in the indices assigned to adjacent vertices. Find the bandwidths of these graphs.

   a) K_5 **b)** $K_{1,3}$ **c)** $K_{2,3}$
   d) $K_{3,3}$ **e)** Q_3 **f)** C_5

*38. The **distance** between two distinct vertices v_1 and v_2 of a connected simple graph is the length (number of edges) of the shortest path between v_1 and v_2. The

radius of a graph is the minimum overall vertices v of the maximum distance from v to another vertex. The **diameter** of a graph is the maximum distance between two distinct vertices. Find the radius and diameter of

   a) K_6.   **b)** $K_{4,5}$.   **c)** Q_3.   **d)** C_6.

*39. a) Show that if the diameter of the simple graph G is at least four, then the diameter of its complement $\overline{G}$ is no more than two.

   b) Show that if the diameter of the simple graph G is at least three, then the diameter of its complement $\overline{G}$ is no more than three.

*40. Suppose that a multigraph has $2m$ vertices of odd degree. Show that any circuit that contains every edge of the graph must contain at least m edges more than once.

41. Find the second shortest path between the vertices a and z in Figure 3 of Section 8.6.

42. Devise an algorithm for finding the second shortest path between two vertices in a simple connected weighted graph.

43. Find the shortest path between the vertices a and z that passes through the vertex e in the weighted graph in Figure 4 in Section 8.6.

44. Devise an algorithm for finding the shortest path between two vertices in a simple connected weighted graph that passes through a specified third vertex.

*45. Show that if G is a simple graph with at least 11 vertices, then either G or $\overline{G}$, the complement of G, is nonplanar.

A set of vertices in a graph is called **independent** if no two vertices in the set are adjacent. The **independence number** of a graph is the maximum number of vertices in an independent set of vertices for the graph.

*46. What is the independence number of

   a) K_n?   **b)** C_n?   **c)** Q_n?   **d)** $K_{m,n}$?

47. Show that the number of vertices in a simple graph is less than or equal to the product of the independence number and the chromatic number of the graph.

48. Show that the chromatic number of a graph is less than or equal to $v - i + 1$, where v is the number of vertices in the graph and i is the independence number of this graph.

49. Suppose that to generate a random simple graph with n vertices we first choose a real number p with $0 \le p \le 1$. For each of the $C(n, 2)$ pairs of distinct vertices we generate a random number x between 0 and 1. If $0 \le x \le p$, we connect these two vertices with an edge; otherwise these vertices are not connected.

   a) What is the probability that a graph with m edges where $0 \le m \le C(n, 2)$ is generated?

   b) What is the expected number of edges in a randomly generated graph with n vertices if each edge is included with probability p?

c) Show that if $p = 1/2$ then every simple graph with n vertices is equally likely to be generated.

A property retained whenever additional edges are added to a simple graph (without adding vertices) is called **monotone increasing,** and a property that is retained whenever edges are removed from a simple graph (without removing vertices) is called **monotone decreasing.**

50. For each of these properties, determine whether it is monotone increasing and determine whether it is monotone decreasing.

a) The graph G is connected.
b) The graph G is not connected.
c) The graph G has an Euler circuit.

d) The graph G has a Hamilton circuit.
e) The graph G is planar.
f) The graph G has chromatic number four.
g) The graph G has radius three.
h) The graph G has diameter three.

51. Show that the graph property P is monotone increasing if and only if the graph property Q is monotone decreasing where Q is the property of not having property P.

∗∗52. Suppose that P is a monotone increasing property of simple graphs. Show that the probability a random graph with n vertices has property P is a monotonic nondecreasing function of p, the probability an edge is chosen to be in the graph.

Computer Projects

WRITE PROGRAMS WITH THESE INPUT AND OUTPUT.

1. Given the vertex pairs associated to the edges of an undirected graph, find the degree of each vertex.

2. Given the ordered pairs of vertices associated to the edges of a directed graph, determine the in-degree and out-degree of each vertex.

3. Given the list of edges of a simple graph, determine whether the graph is bipartite.

4. Given the vertex pairs associated to the edges of a graph, construct an adjacency matrix for the graph. (Produce a version that works when loops, multiple edges, or directed edges are present.)

5. Given an adjacency matrix of a graph, list the edges of this graph and give the number of times each edge appears.

6. Given the vertex pairs associated to the edges of an undirected graph and the number of times each edge appears, construct an incidence matrix for the graph.

7. Given an incidence matrix of an undirected graph, list its edges and give the number of times each edge appears.

8. Given a positive integer n, generate an undirected graph by producing an adjacency matrix for the graph so that all simple graphs are equally likely to be generated.

9. Given a positive integer n, generate a directed graph by producing an adjacency matrix for the graph so that all directed graphs are equally likely to be generated.

10. Given the lists of edges of two simple graphs with no more than six vertices, determine whether the graphs are isomorphic.

11. Given an adjacency matrix of a graph and a positive integer n, find the number of paths of length n between two vertices. (Produce a version that works for directed and undirected graphs.)

∗12. Given the list of edges of a simple graph, determine whether it is connected and find the number of connected components if it is not connected.

13. Given the vertex pairs associated to the edges of a multigraph, determine whether it has an Euler circuit and, if not, whether it has an Euler path. Construct an Euler path or circuit if it exists.

∗14. Given the ordered pairs of vertices associated to the edges of a directed multigraph, construct an Euler path or Euler circuit, if such a path or circuit exists.

∗∗15. Given the list of edges of a simple graph, produce a Hamilton circuit, or determine that the graph does not have such a circuit.

∗∗16. Given the list of edges of a simple graph, produce a Hamilton path, or determine that the graph does not have such a path.

17. Given the list of edges and weights of these edges of a weighted connected simple graph and two vertices in this graph, find the length of a shortest path between them using Dijkstra's algorithm. Also, find a shortest path.

18. Given the list of edges of an undirected graph, find a coloring of this graph using the algorithm given in the exercise set of Section 8.8.

19. Given a list of students and the courses that they are enrolled in, construct a schedule of final exams.

20. Given the distances between pairs of television stations, assign frequencies to these stations.

Computations and Explorations

USE A COMPUTATIONAL PROGRAM OR PROGRAMS YOU HAVE WRITTEN TO DO THESE EXERCISES.

1. Display all the simple graphs with four vertices.
2. Display a full set of nonisomorphic simple graphs with six vertices.
3. Display a full set of nonisomorphic directed graphs with four vertices.
4. Generate at random 10 different simple graphs each with 20 vertices so that each such graph is equally likely to be generated.
5. Construct a Gray code where the code words are bit strings of length six.
6. Construct knight's tours on chessboards of various sizes.
7. Determine whether each of the graphs you generated in Exercise 4 of this set is planar. If you can, determine the thickness of each of the graphs that are not planar.
8. Determine whether each of the graphs you generated in Exercise 4 of this set is connected. If a graph is not connected, determine the number of connected components of the graph.
9. Generate at random simple graphs with ten vertices. Stop when you have constructed one with an Euler circuit. Display an Euler circuit in this graph.
10. Generate at random simple graphs with ten vertices. Stop when you have constructed one with a Hamilton circuit. Display a Hamilton circuit in this graph.
11. Find the chromatic number of each of the graphs you generated in Exercise 4 of this set.
**12. Find the shortest path a traveling salesperson can take to visit each of the capitals of the 50 states in the United States, traveling by air between cities in a straight line.
*13. Estimate the probability that a randomly generated simple graph with n vertices is connected for each positive integer n not exceeding ten by generating a set of random simple graphs and determining whether each is connected.
**14. Work on the problem of determining whether the crossing number of $K(7, 7)$ is 77, 79, or 81. It is known that it equals one of these three values.

Writing Projects

RESPOND TO THESE QUESTIONS WITH ESSAYS USING OUTSIDE SOURCES.

1. Describe the origins and development of graph theory prior to the year 1900.
2. Discuss the applications of graph theory to the study of ecosystems.
3. Discuss the applications of graph theory to sociology and psychology.
4. Discuss what can be learned by investigating the properties of the Web graph.
5. Describe algorithms for drawing a graph on paper or on a display given the vertices and edges of the graph. What considerations arise in drawing a graph so that it has the best appearance for understanding its properties?
6. What are some of the capabilities that a software tool for inputting, displaying, and manipulating graphs should have? Which of these capabilities do available tools have?
7. Describe some of the algorithms available for determining whether two graphs are isomorphic and the computational complexity of these algorithms. What is the most efficient such algorithm currently known?
8. Describe how Euler paths can be used to help determine DNA sequences.
9. Define *de Bruijn sequences* and discuss how they arise in applications. Explain how de Bruijn sequences can be constructed using Euler circuits.
10. Describe the *Chinese postman problem* and explain how to solve this problem.
11. Describe some of the different conditions that imply that a graph has a Hamilton circuit.
12. Describe some of the strategies and algorithms used to solve the traveling salesman problem.
13. Describe several different algorithms for determining whether a graph is planar. What is the computational complexity of each of these algorithms?
14. In modeling, very large scale integration (VLSI) graphs are sometimes embedded in a book, with the vertices on the spine and the edges on pages. Define the *book number* of a graph and find the book number of various graphs including K_n for $n = 3, 4, 5,$ and 6.
15. Discuss the history of the Four Color Theorem.

16. Describe the role computers played in the proof of the Four Color Theorem. How can we be sure that a proof that relies on a computer is correct?

17. Describe and compare several different algorithms for coloring a graph, in terms of whether they produce a coloring with the least number of colors pos-

sible and in terms of their complexity.

18. Explain how graph multicolorings can be used in a variety of different models.

19. Explain how the theory of random graphs can be used in nonconstructive existence proofs of graphs with certain properties.

9

Trees

A connected graph that contains no simple circuits is called a tree. Trees were used as long ago as 1857, when the English mathematician Arthur Cayley used them to count certain types of chemical compounds. Since that time, trees have been employed to solve problems in a wide variety of disciplines, as the examples in this chapter will show.

Trees are particularly useful in computer science, where they are employed in a wide range of algorithms. For instance, trees are used to construct efficient algorithms for locating items in a list. They can be used in algorithms, such as Huffman coding, that construct efficient codes saving costs in data transmission and storage. Trees can be used to study games such as checkers and chess and can help determine winning strategies for playing these games. Trees can be used to model procedures carried out using a sequence of decisions. Constructing these models can help determine the computational complexity of algorithms based on a sequence of decisions, such as sorting algorithms.

Procedures for building trees containing every vertex of a graph, including depth-first search and breadth-first search, can be used to systematically explore the vertices of a graph. Exploring the vertices of a graph via depth-first search, also known as backtracking, allows for the systematic search for solutions to a wide variety of problems, such as determining how eight queens can be placed on a chessboard so that no queen can attack another.

We can assign weights to the edges of a tree to model many problems. For example, using weighted trees we can develop algorithms to construct networks containing the least expensive set of telephone lines linking different network nodes.

9.1 Introduction to Trees

Links

In Chapter 8 we showed how graphs can be used to model and solve many problems. In this chapter we will focus on a particular type of graph called a **tree,** so named because such graphs resemble trees. For example, *family trees* are graphs that represent genealogical charts. Family trees use vertices to represent the members of a family and edges to represent parent–child relationships. The family tree of the Bernoulli family of Swiss mathematicians is shown in Figure 1. The undirected graph representing a family tree is an example of a tree.

DEFINITION 1	A *tree* is a connected undirected graph with no simple circuits.

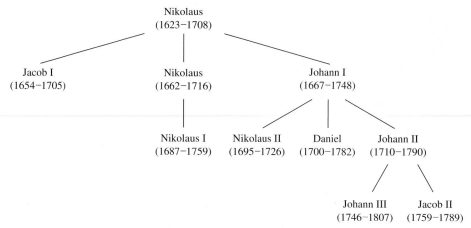

FIGURE 1 **The Bernoulli Family of Mathematicians.**

Since a tree cannot have a simple circuit, a tree cannot contain multiple edges or loops. Therefore any tree must be a simple graph.

EXAMPLE 1 Which of the graphs shown in Figure 2 are trees?

Solution: G_1 and G_2 are trees, since both are connected graphs with no simple circuits. G_3 is not a tree because e, b, a, d, e is a simple circuit in this graph. Finally, G_4 is not a tree since it is not connected. ◀

Any connected graph that contains no simple circuits is a tree. What about graphs containing no simple circuits that are not necessarily connected? These graphs are called **forests** and have the property that each of their connected components is a tree. Figure 3 displays a forest.

Trees are often defined as undirected graphs with the property that there is a unique simple path between every pair of vertices. The following theorem shows that this alternative definition is equivalent to our definition.

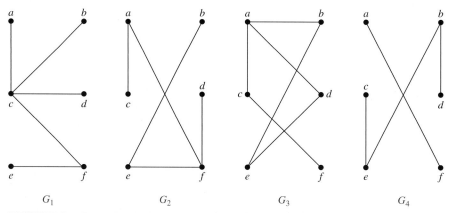

FIGURE 2 **Examples of Trees and Graphs That Are Not Trees.**

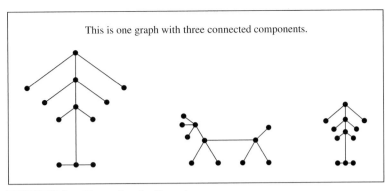

This is one graph with three connected components.

FIGURE 3   Example of a Forest.

THEOREM 1   An undirected graph is a tree if and only if there is a unique simple path between any two of its vertices.

Proof: First assume that T is a tree. Then T is a connected graph with no simple circuits. Let x and y be two vertices of T. Since T is connected, by Theorem 1 of Section 8.4 there is a simple path between x and y. Moreover, this path must be unique, for if there were a second such path, the path formed by combining the first path from x to y followed by the path from y to x obtained by reversing the order of the second path from x to y would form a circuit. This implies, using Exercise 43 of Section 8.4, that there is a simple circuit in T. Hence, there is a unique simple path between any two vertices of a tree.

Now assume that there is a unique simple path between any two vertices of a graph T. Then T is connected, since there is a path between any two of its vertices. Furthermore, T can have no simple circuits. To see that this is true, suppose T had a simple circuit that contained the vertices x and y. Then there would be two simple paths between x and y, since the simple circuit is made up of a simple path from x to y and a second simple path from y to x. Hence, a graph with a unique simple path between any two vertices is a tree. ◁

In many applications of trees a particular vertex of a tree is designated as the **root.** Once we specify a root, we can assign a direction to each edge as follows. Since there is a unique path from the root to each vertex of the graph (by Theorem 1), we direct each edge away from the root. Thus, a tree together with its root produces a directed graph called a **rooted tree.**

DEFINITION 2   A *rooted tree* is a tree in which one vertex has been designated as the root and every edge is directed away from the root.

Rooted trees can also be defined recursively. Refer to Section 3.4 to see how this can be done. We can change an unrooted tree into a rooted tree by choosing any vertex as the root. Note that different choices of the root produce different rooted trees. For instance, Figure 4 displays the rooted trees formed by designating a to be the root and c to be the

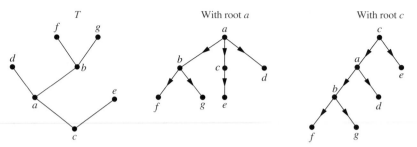

FIGURE 4   **A Tree and Rooted Trees Formed by Designating Two Roots.**

root, respectively, in the tree T. We usually draw a rooted tree with its root at the top of the graph. The arrows indicating the directions of the edges in a rooted tree can be omitted, since the choice of root determines the directions of the edges.

The terminology for trees has botanical and genealogical origins. Suppose that T is a rooted tree. If v is a vertex in T other than the root, the **parent** of v is the unique vertex u such that there is a directed edge from u to v (the reader should show that such a vertex is unique). When u is the parent of v, v is called a **child** of u. Vertices with the same parent are called **siblings.** The **ancestors** of a vertex other than the root are the vertices in the path from the root to this vertex, excluding the vertex itself and including the root (that is, its parent, its parent's parent, and so on, until the root is reached). The **descendants** of a vertex v are those vertices that have v as an ancestor. A vertex of a tree is called a **leaf** if it has no children. Vertices that have children are called **internal vertices.** The root is an internal vertex unless it is the only vertex in the graph, in which case it is a leaf.

If a is a vertex in a tree, the **subtree** with a as its root is the subgraph of the tree consisting of a and its descendants and all edges incident to these descendants.

EXAMPLE 2   In the rooted tree T (with root a) shown in Figure 5, find the parent of c, the children of g, the siblings of h, all ancestors of e, all descendants of b, all internal vertices, and all leaves. What is the subtree rooted at g?

Solution: The parent of c is b. The children of g are h, i, and j. The siblings of h are i and j. The ancestors of e are c, b, and a. The descendants of b are c, d, and e. The internal vertices are a, b, c, g, h, and j. The leaves are d, e, f, i, k, l, and m. The subtree rooted at g is shown in Figure 6.   ◀

Rooted trees with the property that all of their internal vertices have the same number of children are used in many different applications. Later in this chapter we will use such trees to study problems involving searching, sorting, and coding.

DEFINITION 3   A rooted tree is called an *m-ary tree* if every internal vertex has no more than m children. The tree is called a *full m-ary tree* if every internal vertex has exactly m children. An m-ary tree with $m = 2$ is called a *binary tree.*

EXAMPLE 3   Are the rooted trees in Figure 7 full m-ary trees for some positive integer m?

Solution: T_1 is a full binary tree since each of its internal vertices has two children. T_2 is a full 3-ary tree since each of its internal vertices has three children. In T_3 each internal

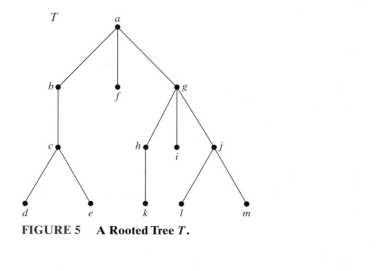

FIGURE 5   A Rooted Tree T.

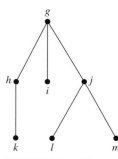

FIGURE 6   The Subtree Rooted at g.

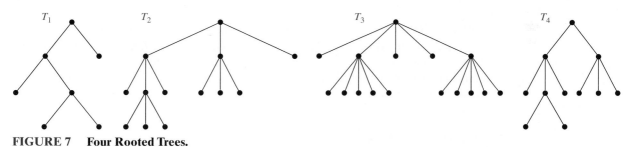

FIGURE 7   Four Rooted Trees.

vertex has five children, so T_3 is a full 5-ary tree. T_4 is not a full m-ary tree for any m since some of its internal vertices have two children and others have three children.   ◀

An **ordered rooted tree** is a rooted tree where the children of each internal vertex are ordered. Ordered rooted trees are drawn so that the children of each internal vertex are shown in order from left to right. Note that a representation of a rooted tree in the conventional way determines an ordering for its edges. We will use such orderings of edges in drawings without explicitly mentioning that we are considering a rooted tree to be ordered.

In an ordered binary tree (usually called just a **binary tree**), if an internal vertex has two children, the first child is called the **left child** and the second child is called the **right child.** The tree rooted at the left child of a vertex is called the **left subtree** of this vertex, and the tree rooted at the right child of a vertex is called the **right subtree** of the vertex. The reader should note that for some applications every vertex of a binary tree, other than the root, is designated as a right or a left child of its parent. This is done even when some vertices have only one child. We will make such designations whenever it is necessary, but not otherwise.

Ordered rooted trees can be defined recursively. Binary trees, a type of ordered rooted trees, were defined this way in Section 3.4.

EXAMPLE 4   What are the left and right children of d in the binary tree T shown in Figure 8(a) (where the order is that implied by the drawing)? What are the left and right subtrees of c?

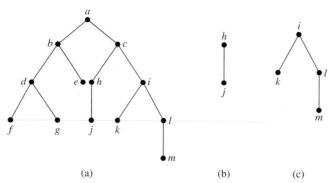

FIGURE 8   A Binary Tree T and Left and Right Subtrees of
the Vertex c.

Solution: The left child of d is f and the right child is g. We show the left and right
subtrees of c in Figures 8(b) and 8(c), respectively.   ◀

Just as in the case of graphs, there is no standard terminology used to describe trees,
rooted trees, ordered rooted trees, and binary trees. This nonstandard terminology occurs since trees are used extensively throughout computer science, which is a relatively
young field. The reader should carefully check meanings given to terms dealing with trees
whenever they occur.

TREES AS MODELS

Trees are used as models in such diverse areas as computer science, chemistry, geology,
botany, and psychology. We will describe a variety of such models based on trees.

EXAMPLE 5   **Saturated Hydrocarbons and Trees** Graphs can be used to represent molecules, where
atoms are represented by vertices and bonds between them by edges. The English mathematician Arthur Cayley discovered trees in 1857 when he was trying to enumerate the
isomers of compounds of the form C_nH_{2n+2}, which are called *saturated hydrocarbons*.

In graph models of saturated hydrocarbons, each carbon atom is represented by a
vertex of degree 4, and each hydrogen atom is represented by a vertex of degree 1. There
are $3n + 2$ vertices in a graph representing a compound of the form C_nH_{2n+2}. The number of edges in such a graph is half the sum of the degrees of the vertices. Hence, there

ARTHUR CAYLEY (1821–1895) Arthur Cayley, the son of a merchant, displayed his mathematical
talents at an early age with amazing skill in numerical calculations. Cayley entered Trinity College, Cambridge, when he was 17. While in college he developed a passion for reading novels. Cayley excelled at
Cambridge and was elected to a 3-year appointment as Fellow of Trinity and assistant tutor. During this
time Cayley began his study of n-dimensional geometry and made a variety of contributions to geometry
and to analysis. He also developed an interest in mountaineering, which he enjoyed during vacations in
Switzerland. Since no position as a mathematician was available to him, Cayley left Cambridge, entering
the legal profession and gaining admittance to the bar in 1849. Although Cayley limited his legal work
to be able to continue his mathematics research, he developed a reputation as a legal specialist. During
his legal career he was able to write more than 300 mathematical papers. In 1863 Cambridge University
established a new post in mathematics and offered it to Cayley. He took this job, even though it paid less
money than he made as a lawyer.

$$
\begin{array}{ccc}
& H & \\
& | & \\
H-&C&-H \\
& | & \\
H-&C&-H \\
& | & \\
H-&C&-H \\
& | & \\
H-&C&-H \\
& | & \\
& H &
\end{array}
$$

Butane   Isobutane

FIGURE 9   The Two Isomers of Butane.

are $(4n + 2n + 2)/2 = 3n + 1$ edges in this graph. Since the graph is connected and the number of edges is one less than the number of vertices, it must be a tree (see Exercise 15 at the end of this section).

The nonisomorphic trees with n vertices of degree 4 and $2n + 2$ of degree 1 represent the different isomers of $C_n H_{2n+2}$. For instance, when $n = 4$, there are exactly two nonisomorphic trees of this type (the reader should verify this). Hence, there are exactly two different isomers of $C_4 H_{10}$. Their structures are displayed in Figure 9. These two isomers are called butane and isobutane.   ◀

EXAMPLE 6   Representing Organizations The structure of a large organization can be modeled using a rooted tree. Each vertex in this tree represents a position in the organization. An edge from one vertex to another indicates that the person represented by the initial vertex is the (direct) boss of the person represented by the terminal vertex. The graph shown in Figure 10 displays such a tree. In the organization represented by this tree, the Director of Hardware Development works directly for the Vice President of R&D.   ◀

EXAMPLE 7   Computer File Systems Files in computer memory can be organized into directories. A directory can contain both files and subdirectories. The root directory contains the entire file system. Thus, a file system may be represented by a rooted tree, where the root represents the root directory, internal vertices represent subdirectories, and leaves represent ordinary files or empty directories. One such file system is shown in Figure 11.

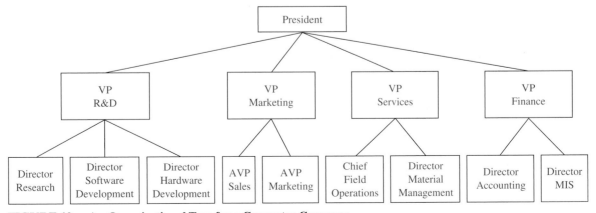

FIGURE 10   An Organizational Tree for a Computer Company.

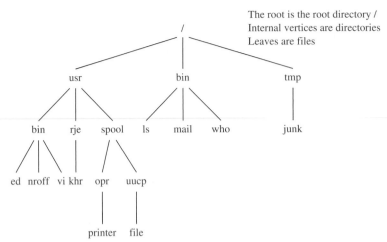

The root is the root directory /
Internal vertices are directories
Leaves are files

FIGURE 11   A Computer File System.

In this system, the file khr is in the directory rje. (Note that links to files where the same file may have more than one pathname can lead to circuits in computer file systems.) ◀

EXAMPLE 8

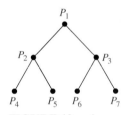

FIGURE 12   A Tree-Connected Network of Seven Processors.

Tree-Connected Parallel Processors In Example 13 of Section 8.2 we described several interconnection networks for parallel processing. A **tree-connected network** is another important way to interconnect processors. The graph representing such a network is a complete binary tree. Such a network interconnects $n = 2^k - 1$ processors, where k is a positive integer. A processor represented by the vertex v that is not a root or a leaf has three two-way connections—one to the processor represented by the parent of v and two to the processors represented by the two children of v. The processor represented by the root has two two-way connections to the processors represented by its two children. A processor represented by a leaf v has a single two-way connection to the parent of v. We display a tree-connected network with seven processors in Figure 12.

We will illustrate how a tree-connected network can be used for parallel computation. In particular, we will show how the processors in Figure 12 can be used to add eight numbers, using three steps. In the first step, we add x_1 and x_2 using P_4, x_3 and x_4 using P_5, x_5 and x_6 using P_6, and x_7 and x_8 using P_7. In the second step, we add $x_1 + x_2$ and $x_3 + x_4$ using P_2 and $x_5 + x_6$ and $x_7 + x_8$ using P_3. Finally, in the third step, we add $x_1 + x_2 + x_3 + x_4$ and $x_5 + x_6 + x_7 + x_8$ using P_1. The three steps used to add eight numbers compares favorably to the seven steps required to add eight numbers serially, where the steps are the addition of one number to the sum of the previous numbers in the list. ◀

PROPERTIES OF TREES

We will often need results relating the numbers of edges and vertices of various types in trees.

THEOREM 2

A tree with n vertices has $n - 1$ edges.

Proof: We will use mathematical induction to prove this theorem. Note that for all the trees here we can choose a root and consider the tree rooted.

BASIS STEP: When $n = 1$, a tree with $n = 1$ vertex has no edges. It follows that the theorem is true for $n = 1$.

INDUCTIVE STEP: The induction hypothesis states that every tree with k vertices has $k - 1$ edges, where k is a positive integer. Suppose that a tree T has $k + 1$ vertices and that v is a leaf of T (which must exist since the tree is finite), and let w be the parent of v. Removing from T the vertex v and the edge connecting w to v produces a tree T' with k vertices, since the resulting graph is still connected and has no simple circuits. By the induction hypothesis, T' has $k - 1$ edges. It follows that T has k edges since it has one more edge than T', the edge connecting v and w. This completes the induction step. ◁

The number of vertices in a full m-ary tree with a specified number of internal vertices is determined, as the following theorem shows. As in Theorem 2, we will use n to denote the number of vertices in a tree.

THEOREM 3   A full m-ary tree with i internal vertices contains $n = mi + 1$ vertices.

Proof: Every vertex, except the root, is the child of an internal vertex. Since each of the i internal vertices has m children, there are mi vertices in the tree other than the root. Therefore, the tree contains $n = mi + 1$ vertices. ◁

Suppose that T is a full m-ary tree. Let i be the number of internal vertices and l the number of leaves in this tree. Once one of n, i, and l is known, the other two quantities are determined. How to find the other two quantities from the one that is known is given in the following theorem.

THEOREM 4   A full m-ary tree with

- (*i*) n vertices has $i = (n - 1)/m$ internal vertices and $l = [(m - 1)n + 1]/m$ leaves,
- (*ii*) i internal vertices has $n = mi + 1$ vertices and $l = (m - 1)i + 1$ leaves,
- (*iii*) l leaves has $n = (ml - 1)/(m - 1)$ vertices and $i = (l - 1)/(m - 1)$ internal vertices.

Proof: Let n represent the number of vertices, i the number of internal vertices, and l the number of leaves. The three parts of the theorem can all be proved using the equality given in Theorem 3, that is, $n = mi + 1$, together with the equality $n = l + i$, which is true because each vertex is either a leaf or an internal vertex. We will prove part (*i*) here. The proofs of parts (*ii*) and (*iii*) are left as exercises for the reader.

Solving for i in $n = mi + 1$ gives $i = (n - 1)/m$. Then inserting this expression for i into the equation $n = l + i$ shows that $l = n - i = n - (n - 1)/m = [(m - 1)n + 1]/m$. ◁

Example 9 illustrates how Theorem 4 can be used.

EXAMPLE 9   Suppose that someone starts a chain letter. Each person who receives the letter is asked to send it on to four other people. Some people do this, but others do not send any letters. How many people have seen the letter, including the first person, if no one receives more than one letter and if the chain letter ends after there have been 100 people who read it but did not send it out? How many people sent out the letter?

Solution: The chain letter can be represented using a 4-ary tree. The internal vertices correspond to people who sent out the letter, and the leaves correspond to people who did not send it out. Since 100 people did not send out the letter, the number of leaves in this rooted tree is $l = 100$. Hence, part (*iii*) of Theorem 4 shows that the number of people who have seen the letter is $n = (4 \cdot 100 - 1)/(4 - 1) = 133$. Also, the number of internal vertices is $133 - 100 = 33$, so that 33 people sent out the letter.   ◄

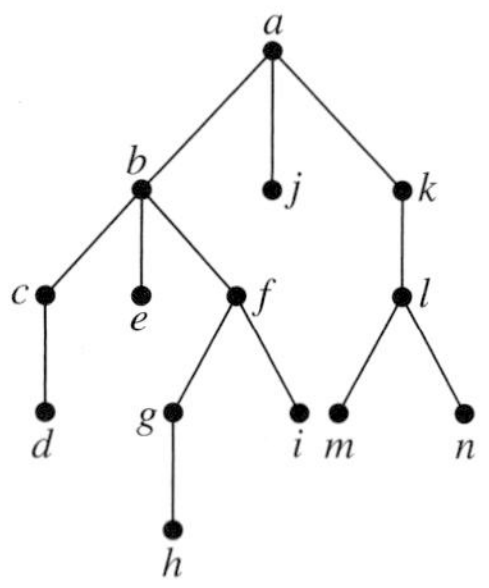

It is often desirable to use rooted trees that are "balanced" so that the subtrees at each vertex contain paths of approximately the same length. Some definitions will make this concept clear. The **level** of a vertex v in a rooted tree is the length of the unique path from the root to this vertex. The level of the root is defined to be zero. The **height** of a rooted tree is the maximum of the levels of vertices. In other words, the height of a rooted tree is the length of the longest path from the root to any vertex.

FIGURE 13   A Rooted Tree.

EXAMPLE 10   Find the level of each vertex in the rooted tree shown in Figure 13. What is the height of this tree?

Solution: The root a is at level 0. Vertices b, j, and k are at level 1. Vertices c, e, f, and l are at level 2. Vertices d, g, i, m, and n are at level 3. Finally, vertex h is at level 4. Since the largest level of any vertex is 4, this tree has height 4.   ◄

A rooted m-ary tree of height h is **balanced** if all leaves are at levels h or $h - 1$.

EXAMPLE 11   Which of the rooted trees shown in Figure 14 are balanced?

Solution: T_1 is balanced, since all its leaves are at levels 3 and 4. However, T_2 is not balanced, since it has leaves at levels 2, 3, and 4. Finally, T_3 is balanced, since all its leaves are at level 3.   ◄

The results in Theorem 5 relate the height and the number of leaves in m-ary trees.

THEOREM 5   There are at most m^h leaves in an m-ary tree of height h.

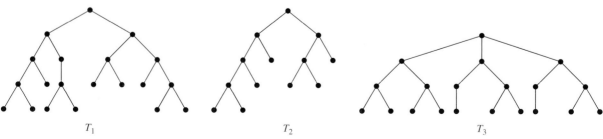

FIGURE 14   Some Rooted Trees.

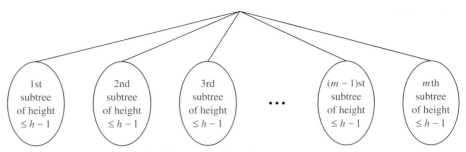

FIGURE 15   The Inductive Step of the Proof.

Proof: The proof uses mathematical induction on the height. First, consider m-ary trees of height 1. These trees consist of a root with no more than m children, each of which is a leaf. Hence there are no more than $m^1 = m$ leaves in an m-ary tree of height 1. This is the basis step of the inductive argument.

Now assume that the result is true for all m-ary trees of height less than h; this is the inductive hypothesis. Let T be an m-ary tree of height h. The leaves of T are the leaves of the subtrees of T obtained by deleting the edges from the root to each of the vertices at level 1, as shown in Figure 15.

Each of these subtrees has height less than or equal to $h - 1$. So by the inductive hypothesis, each of these rooted trees has at most m^{h-1} leaves. Since there are at most m such subtrees, each with a maximum of m^{h-1} leaves, there are at most $m \cdot m^{h-1} = m^h$ leaves in the rooted tree. This finishes the inductive argument.   ◁

COROLLARY 1

> If an m-ary tree of height h has l leaves, then $h \geq \lceil \log_m l \rceil$. If the m-ary tree is full and balanced, then $h = \lceil \log_m l \rceil$. (We are using the ceiling function here. Recall that $\lceil x \rceil$ is the smallest integer greater than or equal to x.)

Proof: We know that $l \leq m^h$ from Theorem 5. Taking logarithms to the base m shows that $\log_m l \leq h$. Since h is an integer, we have $h \geq \lceil \log_m l \rceil$. Now suppose that the tree is balanced. Then each leaf is at level h or $h - 1$, and since the height is h, there is at least one leaf at level h. It follows that there must be more than m^{h-1} leaves (see Exercise 30 at the end of this section). Since $l \leq m^h$, we have $m^{h-1} < l \leq m^h$. Taking logarithms to the base m in this inequality gives $h - 1 < \log_m l \leq h$. Hence, $h = \lceil \log_m l \rceil$.   ◁

Exercises

1. Which of these graphs are trees?

a) **b)**

c) **d)**

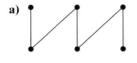

e) **f)**

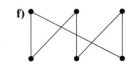

2. Which of these graphs are trees?

a) **b)**

c)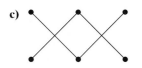

d)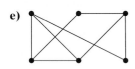

e)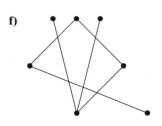

f)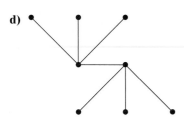

3. Answer these questions about the rooted tree illustrated.

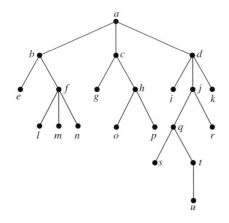

a) Which vertex is the root?
b) Which vertices are internal?
c) Which vertices are leaves?
d) Which vertices are children of j?
e) Which vertex is the parent of h?
f) Which vertices are siblings of o?
g) Which vertices are ancestors of m?
h) Which vertices are descendants of b?

4. Answer the same questions as listed in Exercise 3 for the rooted tree illustrated.

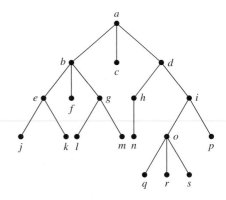

5. Is the rooted tree in Exercise 3 a full m-ary tree for some positive integer m?

6. Is the rooted tree in Exercise 4 a full m-ary tree for some positive integer m?

7. What is the level of each vertex of the rooted tree in Exercise 3?

8. What is the level of each vertex of the rooted tree in Exercise 4?

9. Draw the subtree of the tree in Exercise 3 that is rooted at

  a) a.   **b)** c.   **c)** e.

10. Draw the subtree of the tree in Exercise 4 that is rooted at

  a) a.   **b)** c.   **c)** e.

11. a) How many nonisomorphic unrooted trees are there with three vertices?

  b) How many nonisomorphic rooted trees are there with three vertices (using isomorphism for directed graphs)?

*12. a) How many nonisomorphic unrooted trees are there with four vertices?

  b) How many nonisomorphic rooted trees are there with four vertices (using isomorphism for directed graphs)?

*13. a) How many nonisomorphic unrooted trees are there with five vertices?

  b) How many nonisomorphic rooted trees are there with five vertices (using isomorphism for directed graphs)?

*14. Show that a simple graph is a tree if and only if it is connected, but the deletion of any of its edges produces a graph that is not connected.

*15. Let G be a simple graph with n vertices. Show that G is a tree if and only if G is connected and has $n - 1$ edges.

16. Which complete bipartite graphs $K_{m,n}$, where m and n are positive integers, are trees?

17. How many edges does a tree with 10,000 vertices have?

18. How many vertices does a full 5-ary tree with 100 internal vertices have?

19. How many edges does a full binary tree with 1000 internal vertices have?

20. How many leaves does a full 3-ary tree with 100 vertices have?

21. Suppose 1000 people enter a chess tournament. Use a rooted tree model of the tournament to determine how many games must be played to determine a champion, if a player is eliminated after one loss and games are played until only one entrant has not lost. (Assume there are no ties.)

22. A chain letter starts when a person sends a letter to five others. Each person who receives the letter either sends it to five other people who have never received it or does not send it to anyone. Suppose that 10,000 people send out the letter before the chain ends and that no one receives more than one letter. How many people receive the letter, and how many do not send it out?

23. A chain letter starts with a person sending a letter out to ten others. Each person is asked to send the letter out to ten others, and each letter contains a list of the previous six people in the chain. Unless there are fewer than six names in the list, each person sends one dollar to the first person in this list, removes the name of this person from the list, moves up each of the other five names one position, and inserts his or her name at the end of this list. If no person breaks the chain and no one receives more than one letter, how much money will a person in the chain ultimately receive?

*24. Either draw a full m-ary tree with 76 leaves and height 3 where m is a positive integer or show that no such tree exists.

*25. Either draw a full m-ary tree with 84 leaves and height 3 where m is a positive integer or show that no such tree exists.

*26. A full m-ary tree T has 81 leaves and height 4.

   a) Give the upper and lower bounds for m.
   b) What is m if T is also balanced?

A **complete m-ary tree** is a full m-ary tree where every leaf is at the same level.

27. Construct a complete binary tree of height 4 and a complete 3-ary tree of height 3.

28. How many vertices and how many leaves does a complete m-ary tree of height h have?

29. Prove

   a) part (*ii*) of Theorem 4.
   b) part (*iii*) of Theorem 4.

☞ 30. Show that a full m-ary balanced tree of height h has more than m^{h-1} leaves.

31. How many edges are there in a forest of t trees containing a total of n vertices?

32. Explain how a tree can be used to represent the table of contents of a book organized into chapters, where each chapter is organized into sections, and each section is organized into subsections.

33. How many different isomers do these saturated hydrocarbons have?

   a) C_3H_8 **b)** C_5H_{12} **c)** C_6H_{14}

34. What does each of these represent in an organizational tree?

   a) the parent of a vertex
   b) a child of a vertex
   c) a sibling of a vertex
   d) the ancestors of a vertex
   e) the descendants of a vertex
   f) the level of a vertex
   g) the height of the tree

35. Answer the same questions as those given in Exercise 34 for a rooted tree representing a computer file system.

36. a) Draw the complete binary tree with 15 vertices that represents a tree-connected network of 15 processors.
   b) Show how 16 numbers can be added using the 15 processors in part (a) using four steps.

37. Let n be a power of 2. Show that n numbers can be added in $\log n$ steps using a tree-connected network of $n - 1$ processors.

*38. A **labeled tree** is a tree where each vertex is assigned a label. Two labeled trees are considered isomorphic when there is an isomorphism between them that preserves the labels of vertices. How many nonisomorphic trees are there with three vertices labeled with different integers from the set $\{1, 2, 3\}$? How many nonisomorphic trees are there with four vertices labeled with different integers from the set $\{1, 2, 3, 4\}$?

The **eccentricity** of a vertex in an unrooted tree is the length of the longest simple path beginning at this vertex. A vertex is called a **center** if no vertex in the tree has smaller eccentricity than this vertex. In Exercises 39–41 find every vertex that is a center in the given tree.

39. 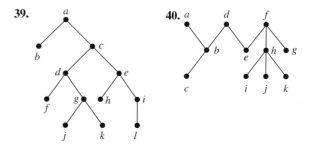   **40.**

41.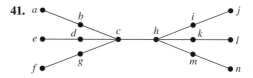

42. Show that a center should be chosen as the root to produce a rooted tree of minimal height from an unrooted tree.

*43. Show that a tree has either one center or two centers that are adjacent.

44. Show that every tree can be colored using two colors.

The **rooted Fibonacci trees** T_n are defined recursively in the following way. T_1 and T_2 are both the rooted tree consisting of a single vertex, and for $n = 3, 4, \ldots$, the rooted tree T_n is constructed from a root with T_{n-1} as its left subtree and T_{n-2} as its right subtree.

45. Draw the first seven rooted Fibonacci trees.

*46. How many vertices, leaves, and internal vertices does the rooted Fibonacci tree T_n have, where n is a positive integer? What is its height?

47. What is wrong with the following "proof" using mathematical induction of the statement that every tree with n vertices has a path of length $n - 1$. *Basis step:* Every tree with one vertex clearly has a path of length 0. *Inductive step:* Assume that a tree with n vertices has a path of length $n - 1$, which has u as its terminal vertex. Add a vertex v and the edge from u to v. The resulting tree has $n + 1$ vertices and has a path of length n. This completes the induction step.

*48. Show that the average depth of a leaf in a binary tree with n vertices is $\Omega(\log n)$.

9.2 Applications of Trees

INTRODUCTION

We will discuss three problems that can be studied using trees. The first problem is: How should items in a list be stored so that an item can be easily located? The second problem is: What series of decisions should be made to find an object with a certain property in a collection of objects of a certain type? The third problem is: How should a set of characters be efficiently coded by bit strings?

BINARY SEARCH TREES

Searching for items in a list is one of the most important tasks that arises in computer science. Our primary goal is to implement a searching algorithm that finds items efficiently when the items are totally ordered. This can be accomplished through the use of a **binary search tree,** which is a binary tree in which each child of a vertex is designated as a right or left child, no vertex has more than one right child or left child, and each vertex is labeled with a key, which is one of the items. Furthermore, vertices are assigned keys so that the key of a vertex is both larger than the keys of all vertices in its left subtree and smaller than the keys of all vertices in its right subtree.

This recursive procedure is used to form the binary search tree for a list of items. Start with a tree containing just one vertex, namely, the root. The first item in the list is assigned as the key of the root. To add a new item, first compare it with the keys of vertices already in the tree, starting at the root and moving to the left if the item is less than the key of the respective vertex if this vertex has a left child, or moving to the right if the item is greater than the key of the respective vertex if this vertex has a right child. When the item is less than the respective vertex and this vertex has no left child, then a new vertex with this item as its key is inserted as a new left child. Similarly, when the item is greater than the respective vertex and this vertex has no right child, then a new vertex with this item as its key is inserted as a new right child. We illustrate this procedure with Example 1.

EXAMPLE 1 Form a binary search tree for the words *mathematics, physics, geography, zoology, meteorology, geology, psychology,* and *chemistry* (using alphabetical order).

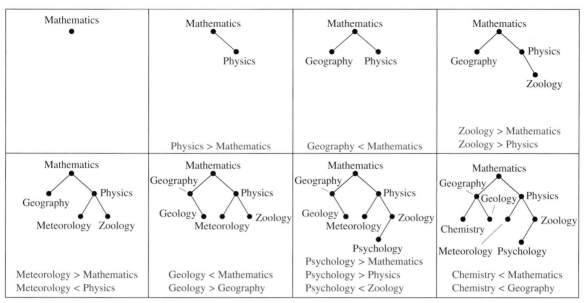

FIGURE 1   **Constructing a Binary Search Tree.**

Solution: Figure 1 displays the steps used to construct this binary search tree. The word *mathematics* is the key of the root. Since *physics* comes after *mathematics* (in alphabetical order), add a right child of the root with key *physics.* Since *geography* comes before *mathematics,* add a left child of the root with key *geography.* Next, add a right child of the vertex with key *physics,* and assign it the key *zoology,* since *zoology* comes after *mathematics* and after *physics.* Similarly, add a left child of the vertex with key *physics* and assign this new vertex the key *meteorology.* Add a right child of the vertex with key *geography* and assign this new vertex the key *geology.* Add a left child of the vertex with key *zoology* and assign it the key *psychology.* Add a left child of the vertex with key *geography* and assign it the key *chemistry.* (The reader should work through all the comparisons needed at each step.)   ◄

To locate an item we try to add it to a binary search tree. We locate it if it is present. Algorithm 1 gives pseudocode for locating an item in a binary search tree and adding a new vertex with this item as its key if the item is not found. Algorithm 1 locates x if it is already the key of a vertex. When x is not a key, a new vertex with key x is added to the tree. In the pseudocode, v is the vertex that has x as its key, and *label*(v) represents the key of vertex v.

We will now determine the computational complexity of this procedure. Suppose we have a binary search tree T for a list of n items. We can form a full binary tree U from T by adding unlabeled vertices whenever necessary so that every vertex with a key has two children. This is illustrated in Figure 2. Once we have done this, we can easily locate or add a new item as a key without adding a vertex.

The most comparisons needed to add a new item is the length of the longest path in U from the root to a leaf. The internal vertices of U are the vertices of T. It follows that U has n internal vertices. We can now use part (*ii*) of Theorem 4 in Section 9.1 to conclude that U has $n + 1$ leaves. Using Corollary 1 of Section 9.1, we see that the height of U is greater than or equal to $h = \lceil \log(n + 1) \rceil$. Consequently, it is necessary to perform at least $\lceil \log(n + 1) \rceil$ comparisons to add some item. Note that if U is balanced,

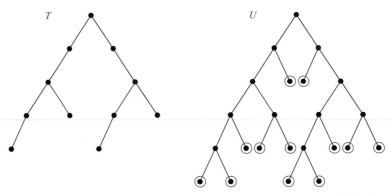

Unlabeled vertices circled

FIGURE 2   Adding Unlabeled Vertices to Make a Binary Search Tree Full.

its height is $\lceil \log(n + 1) \rceil$ (by Corollary 1 of Section 9.1). Thus, if a binary search tree is balanced, locating or adding an item requires no more than $\lceil \log(n + 1) \rceil$ comparisons. A binary search tree can become unbalanced as items are added to it. Since balanced binary search trees give optimal worst-case complexity for binary searching, algorithms have been devised that rebalance binary search trees as items are added. The interested reader can consult references on data structures for the description of such algorithms.

ALGORITHM 1 Binary Search Tree Algorithm.

---

procedure *insertion*(T: binary search tree, x: item)
$v :=$ root of T
{a vertex not present in T has the value *null*}
while $v \neq null$ and *label*(v) $\neq x$
begin
    if $x < label(v)$ **then**
    if left child of $v \neq null$ **then** $v :=$ left child of v
    else add *new vertex* as a left child of v and set $v := null$
    else
    if right child of $v \neq null$ **then** $v :=$ right child of v
    else add *new vertex* as a right child of v to T and set $v := null$
end
if root of $T = null$ **then** add a vertex v to the tree and label it with x
else if v is null or *label*(v) $\neq x$ **then** label *new vertex* with x and let v be this new vertex
{$v =$ location of x}

DECISION TREES

Rooted trees can be used to model problems in which a series of decisions leads to a solution. For instance, a binary search tree can be used to locate items based on a series of comparisons, where each comparison tells us whether we have located the item, or whether we should go right or left in a subtree. A rooted tree in which each internal vertex corresponds to a decision, with a subtree at these vertices for each possible outcome of the decision, is called a **decision tree.** The possible solutions of the problem correspond to the paths to the leaves of this rooted tree. Example 2 illustrates an application of decision trees.

EXAMPLE 2   Suppose there are seven coins, all with the same weight, and a counterfeit coin that weighs less than the others. How many weighings are necessary using a balance scale to determine which of the eight coins is the counterfeit one? Give an algorithm for finding this counterfeit coin.

Solution: There are three possibilities for each weighing on a balance scale. The two pans can have equal weight, the first pan can be heavier, or the second pan can be heavier. Consequently, the decision tree for the sequence of weighings is a 3-ary tree. There are at least eight leaves in the decision tree since there are eight possible outcomes (since each of the eight coins can be the counterfeit lighter coin), and each possible outcome must be represented by at least one leaf. The largest number of weighings needed to determine the counterfeit coin is the height of the decision tree. From Corollary 1 of Section 9.1 it follows that the height of the decision tree is at least $\lceil \log_3 8 \rceil = 2$. Hence, at least two weighings are needed.

It is possible to determine the counterfeit coin using two weighings. The decision tree that illustrates how this is done is shown in Figure 3.   ◀

THE COMPLEXITY OF SORTING ALGORITHMS Many different sorting algorithms have been developed. To decide whether a particular sorting algorithm is efficient, its complexity is determined. Using decision trees as models, a lower bound for the worst-case complexity of sorting algorithms can be found.

We can use decision trees to model sorting algorithms and to determine an estimate for the worst-case complexity of these algorithms. Note that given n elements, there are $n!$ possible orderings of these elements, since each of the $n!$ permutations of these elements can be the correct order. The sorting algorithms studied in this book, and most commonly used sorting algorithms, are based on binary comparisons, that is, the comparison of two elements at a time. The result of each such comparison narrows down the set of possible orderings. Thus, a sorting algorithm based on binary comparisons can be represented by a binary decision tree in which each internal vertex represents a comparison of two elements. Each leaf represents one of the $n!$ permutations of n elements.

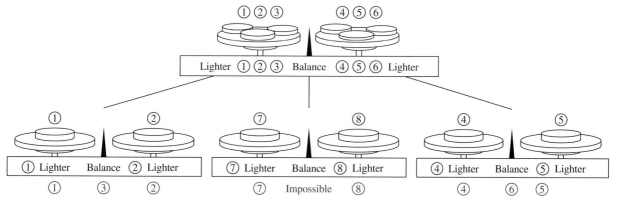

FIGURE 3   **A Decision Tree for Locating a Counterfeit Coin. The counterfeit coin is shown in color below each final weighing.**

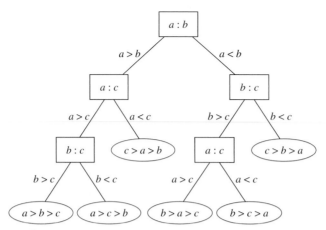

FIGURE 4   A Decision Tree for Sorting Three Distinct Elements.

EXAMPLE 3   We display in Figure 4 a decision tree that orders the elements of the list a, b, c.   ◀

The complexity of a sort based on binary comparisons is measured in terms of the number of such comparisons used. The largest number of binary comparisons ever needed to sort a list with n elements gives the worst-case performance of the algorithm. The most comparisons used equals the longest path length in the decision tree representing the sorting procedure. In other words, the largest number of comparisons ever needed is equal to the height of the decision tree. Since the height of a binary tree with $n!$ leaves is at least $\lceil \log n! \rceil$ (using Corollary 1 in Section 9.1), at least $\lceil \log n! \rceil$ comparisons are needed, as stated in Theorem 1.

THEOREM 1   A sorting algorithm based on binary comparisons requires at least $\lceil \log n! \rceil$ comparisons.

We can use Theorem 1 to provide a big-Omega estimate for the number of comparisons used by a sorting algorithm based on binary comparison. We need only note that $\lceil \log n! \rceil$ is $\Theta(n \log n)$, one of the commonly used reference functions for the computational complexity of algorithms. This estimate is correct because by Example 5 of Section 2.2 $\lceil \log n! \rceil$ is $O(n \log n)$. By Exercise 62 of Section 2.2 $\lceil \log n! \rceil$ is $\Omega(n \log n)$, since $\log n! > (n \log n)/4$ for $n > 4$. Consequently, $\lceil \log n! \rceil$ is $\Theta(n \log n)$. Corollary 1 is a consequence of this estimate.

COROLLARY 1   The number of comparisons used by a sorting algorithm to sort n elements based on binary comparisons is $\Omega(n \log n)$.

A consequence of Corollary 1 is that a sorting algorithm based on binary comparisons that uses $\Theta(n \log n)$ comparisons, in the worst case, to sort n elements is optimal, in the sense that no other such algorithm has better worst-case complexity. Note that by Theorem 1 in Section 3.5 we see that the merge sort algorithm is optimal in this sense.

We can also establish a similar result for the average-case complexity of sorting algorithms. The average number of comparisons used by a sorting algorithm based on binary comparisons is the average depth of a leaf in the decision tree representing the sorting algorithm. By Exercise 48 in Section 9.1 we know that the average depth of a leaf in a binary tree with N vertices is $\Omega(\log N)$. We obtain the following estimate when we let $N = n!$ and note that a function that is $\Omega(\log n!)$ is also $\Omega(n \log n)$ because $\log n! = \Theta(n \log n)$.

THEOREM 2

> The average number of comparisons used by a sorting algorithm to sort n elements based on binary comparisons is $\Omega(n \log n)$.

PREFIX CODES

Consider the problem of using bit strings to encode the letters of the English alphabet (where no distinction is made between lowercase and uppercase letters). We can represent each letter with a bit string of length five, since there are only 26 letters and there are 32 bit strings of length five. The total number of bits used to encode data is five times the number of characters in the text when each character is encoded with five bits. Is it possible to find a coding scheme of these letters so that, when data are coded, fewer bits are used? We can save memory and reduce transmittal time if this can be done.

Consider using bit strings of different lengths to encode letters. Letters that occur more frequently should be encoded using short bit strings, and longer bit strings should be used to encode rarely occurring letters. When letters are encoded using varying numbers of bits, some method must be used to determine where the bits for each character start and end. For instance, if e were encoded with 0, a with 1, and t with 01, then the bit string 0101 could correspond to *eat, tea, eaea,* or *tt.*

One way to ensure that no bit string corresponds to more than one sequence of letters is to encode letters so that the bit string for a letter never occurs as the first part of the bit string for another letter. Codes with this property are called **prefix codes.** For instance, the encoding of e as 0, a as 10, and t as 11 is a prefix code. A word can be recovered from the unique bit string that encodes its letters. For example, the string 10110 is the encoding of *ate.* To see this, note that the initial 1 does not represent a character, but 10 does represent a (and could not be the first part of the bit string of another letter). Then, the next 1 does not represent a character, but 11 does represent t. The final bit, 0, represents e.

A prefix code can be represented using a binary tree, where the characters are the labels of the leaves in the tree. The edges of the tree are labeled so that an edge leading to a left child is assigned a 0 and an edge leading to a right child is assigned a 1. The bit string used to encode a character is the sequence of labels of the edges in the unique path from the root to the leaf that has this character as its label. For instance, the tree in Figure 5 represents the encoding of e by 0, a by 10, t by 110, n by 1110, and s by 1111.

The tree representing a code can be used to decode a bit string. For instance, consider the word encoded by 11111011100 using the code in Figure 5. This bit string can be decoded by starting at the root, using the sequence of bits to form a path that stops when a leaf is reached. Each 0 bit takes the path down the edge leading to the left child of the last vertex in the path, and each 1 bit corresponds to the right child of this vertex. Consequently, the initial 1111 corresponds to the path starting at the root, going right four

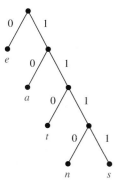

FIGURE 5   The Binary Tree with a Prefix Code.

times, leading to a leaf in the graph that has *s* as its label, since the string 1111 is the code for *s*. Continuing with the fifth bit, we reach a leaf next after going right then left, when the vertex labeled with *a*, which is encoded by 10, is visited. Starting with the seventh bit, we reach a leaf next after going right three times and then left, when the vertex labeled with *n*, which is encoded by 1110, is visited. Finally, the last bit, 0, leads to the leaf that is labeled with *e*. Therefore, the original word is *sane*.

We can construct a prefix code from any binary tree where the left edge at each internal vertex is labeled by 0 and the right edge by a 1 and where the leaves are labeled by characters. Characters are encoded with the bit string constructed using the labels of the edges in the unique path from the root to the leaves.

HUFFMAN CODING We now introduce an algorithm that takes as input the frequencies (which are the probabilities of occurrences) of symbols in a string and produces as output a prefix code that encodes the string using the fewest possible bits, among all possible binary prefix codes for these symbols. This algorithm, known as **Huffman coding,** was developed by David Huffman in a term paper he wrote in 1951 while a graduate student at MIT. (Note that this algorithm assumes that we already know how many times each symbol occurs in the string so that we can compute the frequency of each symbol by dividing the number of times this symbol occurs by the length of the string.) Huffman coding is a fundamental algorithm in *data compression,* the subject devoted to reducing the number of bits required to represent information. Huffman coding is extensively used to compress bit strings representing text and it also plays an important role in compressing audio and image files.

Algorithm 2 presents the Huffman coding algorithm. Given symbols and their frequencies, our goal is to construct a rooted binary tree where the symbols are the labels of the leaves. The algorithm begins with a forest of trees each consisting of one vertex, where each vertex has a symbol as its label and where the weight of this vertex equals the frequency of the symbol that is its label. At each step we combine two trees having the least total weight into a single tree by introducing a new root and placing the tree with larger weight as its left subtree and the tree with smaller weight as its right subtree. Furthermore, we assign the sum of the weights of the two subtrees of this tree as the total weight of the tree. (Although procedures for breaking ties by choosing between trees with equal weights can be specified, we will not specify such procedures here.) The algorithm is finished when it has constructed a tree, that is, when the forest is reduced to a single tree.

DAVID A. HUFFMAN (1925–1999) David Huffman grew up in Ohio. At the age of 18 he received his B.S. in electrical engineering from The Ohio State University. Afterward he served in the U.S. Navy as a radar maintenance officer on a destroyer that had the mission of clearing mines in Asian waters after World War II. Later, he earned his M.S. from Ohio State and his Ph.D. in electrical engineering from MIT. Huffman joined the MIT faculty in 1953, where he remained until 1967 when he became the founding member of the computer science department at the University of California at Santa Cruz. He played an important role in developing this department and spent the remainder of his career there, retiring in 1994.

Huffman is noted for his contributions to information theory and coding, signal designs for radar and for communications, and design procedures for asynchronous logical circuits. His work on surfaces with zero curvature led him to develop original techniques for folding paper and vinyl into unusual shapes considered works of art by many and publicly displayed in several exhibits. However, Huffman is best known for his development of what is now called Huffman coding, a result of a term paper he wrote during his graduate work at MIT.

Huffman enjoyed exploring the outdoors, hiking, and traveling extensively. He became certified as a scuba diver when he was in his late 60s. He kept poisonous snakes as pets.

ALGORITHM 2 Huffman Coding.

---

procedure *Huffman*(*C*: symbols a_i with frequencies $w_i, i = 1, \ldots, n$)
F := forest of n rooted trees, each consisting of the single vertex a_i and assigned weight w_i
while F is not a tree
begin
    Replace the rooted trees T and T' of least weights from F with
    $w(T) \geq w(T')$ with a tree having a new root that has T as its
    left subtree and T' as its right subtree. Label the new edge to T with 0
    and the new edge to T' with 1.
    Assign $w(T) + w(T')$ as the weight of the new tree.
end
{the Huffman coding for the symbol a_i is the concatenation of the labels
of the edges in the unique path from the root to the vertex a_i}

Example 4 illustrates how Algorithm 2 is used to encode a set of five symbols.

EXAMPLE 4

Extra Examples

Use Huffman coding to encode the following symbols with the frequencies listed: A: 0.08, B: 0.10, C: 0.12, D: 0.15, E: 0.20, F: 0.35. What is the average number of bits used to encode a character?

Solution: Figure 6 displays the steps used to encode these symbols. The encoding produced encodes A by 111, B by 110, C by 011, D by 010, E by 10, and F by 00. The average number of bits used to encode a symbol using this encoding is $3 \cdot 0.08 + 3 \cdot 0.10 + 3 \cdot 0.12 + 3 \cdot 0.15 + 2 \cdot 0.20 + 2 \cdot 0.35 = 2.45$. ◀

Note that Huffman coding is a greedy algorithm. Replacing the two subtrees with the smallest weight at each step leads to an optimal code in the sense that no binary prefix code for these symbols can encode these symbols using fewer bits. We leave the proof that Huffman codes are optimal as Exercise 32 at the end of this section.

There are many variations of Huffman coding. For example, instead of encoding single symbols, we can encode blocks of symbols of a specified length, such as blocks of two symbols. Doing so may reduce the number of bits required to encode the string (see Exercise 30 at the end of this section). We can also use more than two symbols to encode the original symbols in the string (see the preamble to Exercise 28 at the end of this section). Furthermore, a variation known as adaptive Huffman coding (see [Sa00]) can be used when the frequency of each symbol in a string is not known in advance, so that encoding is done at the same time the string is being read.

GAME TREES

Links

Trees can be used to analyze certain types of games such as tic-tac-toe, nim, checkers, and chess. In each of these games two players take turns making moves. Each player knows the moves made by the other player and no element of chance enters into the game. We model such games using **game trees;** the vertices of these trees represent the positions that a game can be in as it progresses; the edges represent legal moves between these

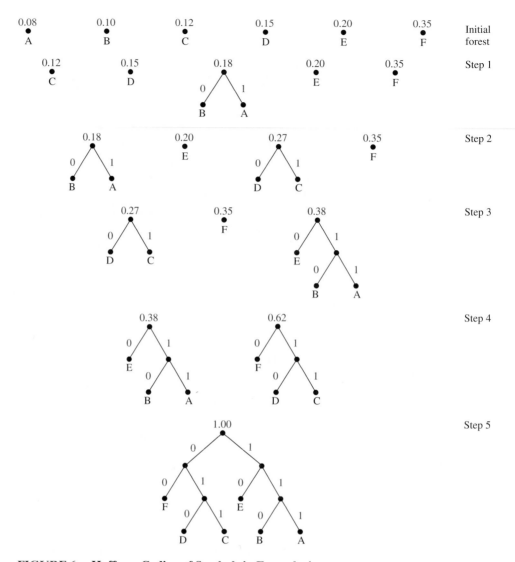

FIGURE 6   Huffman Coding of Symbols in Example 4.

positions. Because game trees are usually large, we simplify game trees by representing all symmetric positions of a game by the same vertex. However, the same position of a game may be represented by different vertices if different sequences of moves lead to this position. The root represents the starting position. The usual convention is to represent vertices at even levels by boxes and vertices at odd levels by circles. When the game is in a position represented by a vertex at an even level, it is the first player's move; when the game is in a position represented by a vertex at an odd level, it is the second player's move. Game trees may be infinite when the games they represent never end, such as games that can enter infinite loops, but for most games there are rules that lead to finite game trees.

The leaves of a game tree represent the final positions of a game. We assign a value to each leaf indicating the payoff to the first player if the game terminates in the position represented by this leaf. For games that are win–lose, we label a terminal vertex represented by a circle with a 1 to indicate a win by the first player and we label a terminal vertex represented by a box with a −1 to indicate a win by the second player. For games

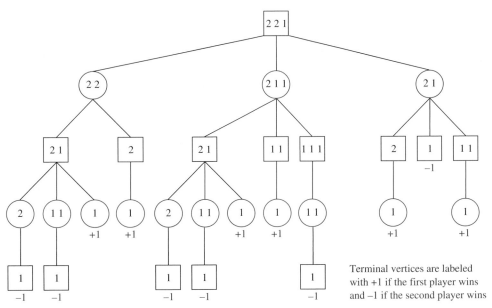

FIGURE 7   **The Game Tree for a Game of Nim.**

where draws are allowed, we label a terminal vertex corresponding to a draw position with a 0. Note that for win–lose games we have assigned values to terminal vertices so that the larger the value, the better the outcome for the first player.

In Example 5 we display a game tree for a well-known and well-studied game.

EXAMPLE 5   **Nim** In a version of the game of **nim,** at the start of a game there are a number of piles of stones. Two players take turns making moves; a legal move consists of removing one or more stones from one of the piles, without removing all the stones left. A player without a legal move loses. (Another way to look at this is that the player removing the last stone loses since the position with no piles of stones is not allowed.) The game tree shown in Figure 7 represents this version of nim given the starting position where there are three piles of stones containing two, two, and one stone each, respectively. We represent each position with an unordered list of the number of stones in the different piles (the order of the piles does not matter). The initial move by the first player can lead to three possible positions because this player can remove one stone from a pile with two stones (leaving three piles containing one, one, and two stones); two stones from a pile containing two stones (leaving two piles containing two stones and one stone); or one stone from the pile containing one stone (leaving two piles of two stones). When only one pile with one stone is left, no legal moves are possible, so such positions are terminal positions. Since nim is a win–lose game, we label the terminal vertices with $+1$ when they represent wins for the first player and -1 when they represent wins for the second player.   ◀

EXAMPLE 6   **Tic-tac-toe** The game tree for tic-tac-toe is extremely large and cannot be drawn here, although a computer could easily build such a tree. We show a portion of the game tic-tac-toe in Figure 8(a). Note that by considering symmetric positions equivalent, we need only consider three possible initial moves as shown in Figure 8(a). We also show a subtree of this game tree leading to terminal positions in Figure 8(b), where a player who can win makes a winning move.   ◀

(a)   (b)

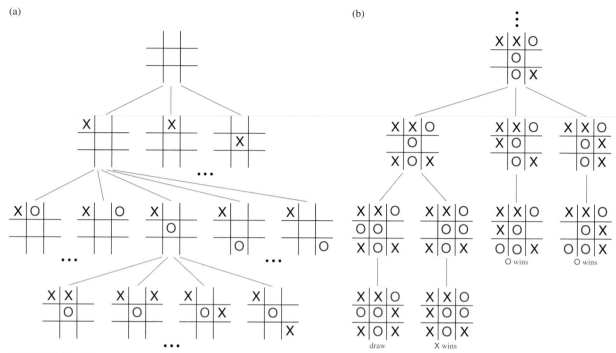

FIGURE 8   Some of the Game Tree for Tic-Tac-Toe.

We can recursively define the values of all vertices in a game tree in a way that enables us to determine the outcome of this game when both players follow optimal strategies. By a **strategy** we mean a set of rules that tells a player how to select moves to win the game. An optimal strategy for the first player is a strategy that maximizes the payoff to this player and for the second player is a strategy that minimizes this payoff. We now recursively define the value of a vertex.

DEFINITION 1   The *value of a vertex in a game tree* is defined recursively as:

(*i*)   the value of a leaf is the payoff to the first player when the game terminates in the position represented by this leaf.

(*ii*)   the value of an internal vertex at an even level is the maximum of the values of its children, and the value of an internal vertex at an odd level is the minimum of the values of its children.

The strategy where the first player moves to a position represented by a child with maximum value and the second player moves to a position of a child with minimum value is called the **minmax strategy.** We can determine who will win the game when both players follow the minmax strategy by calculating the value of the root of the tree; this value is called the **value** of the tree. This is a consequence of Theorem 3.

THEOREM 3   The value of a vertex of a game tree tells us the payoff to the first player if both players follow the minmax strategy and play starts from the position represented by this vertex.

Proof: We will use induction to prove this theorem.

BASIS STEP: If the vertex is a leaf, by definition the value assigned to this vertex is the payoff to the first player.

INDUCTIVE STEP: The inductive hypothesis is the assumption that the values of the children of a vertex are the payoffs to the first player, assuming that play starts at each of the positions represented by these vertices. We need to consider two cases, when it is the first player's turn and when it is the second player's turn.

When it is the first player's turn, this player follows the minmax strategy and moves to the position represented by the child with the largest value. By the inductive hypothesis, this value is the payoff to the first player when play starts at the position represented by this child and follows the minmax strategy. By the recursive step in the definition of the value of an internal vertex at an even level (as the maximum value of its children), the value of this vertex is the payoff when play begins at the position represented by this vertex.

When it is the second player's turn, this player follows the minmax strategy and moves to the position represented by the child with the least value. By the inductive hypothesis, this value is the payoff to the first player when play starts at the position represented by this child and both players follow the minmax strategy. By the recursive definition of the value of an internal vertex at an odd level as the minimum value of its children, the value of this vertex is the payoff when play begins at the position represented by this vertex. ◁

Remark: By extending the proof of Theorem 3, it can be shown that the minmax strategy is the optimal strategy for both players.

Example 7 illustrates how the minmax procedure works. It displays the values assigned to the internal vertices in the game tree from Example 5. Note that we can shorten the computation required by noting that for win–lose games, once a child of a square vertex with value $+1$ is found, the value of the square vertex is also $+1$ since $+1$ is the largest possible payoff. Similarly, once a child of a circle vertex with value -1 is found, this is the value of the circle vertex also.

EXAMPLE 7   In Example 5 we constructed the game tree for nim with a starting position where there are three piles containing two, two, and one stones. In Figure 9 we show the values of the vertices of this game tree. The values of the vertices are computed using the values of the leaves and working one level up at a time. In the right margin of this figure we indicate whether we use the maximum or minimum of the values of the children to find the value of an internal vertex at each level. For example, once we have found the values of the three children of the root, which are $1, -1$, and -1, we find the value of the root by computing $\max(1, -1, -1) = 1$. Since the value of the root is 1, it follows that the first player wins when both players follow a minmax strategy.   ◀

Game trees for some well-known games can be extraordinarily large, since these games have many different possible moves. For example, the game tree for chess has been estimated to have as many as 10^{100} vertices! Since it may be impossible to use Theorem 3 directly to study a game because of the size of the game tree, various approaches have been devised to help determine good strategies and to determine the outcome of such games. One useful technique, called *alpha-beta pruning*, eliminates much computation by pruning portions of the game tree that cannot affect the values of ancestor

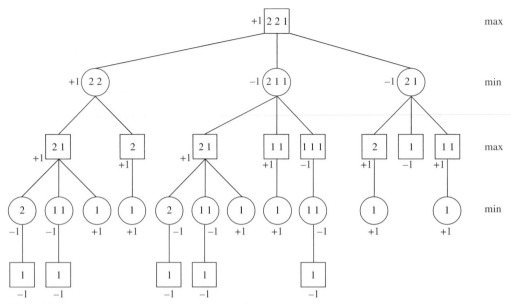

FIGURE 9 **Showing the Values of Vertices in the Game of Nim.**

vertices. (For information about alpha-beta pruning, consult [Gr90].) Another useful approach is to use *evaluation functions*, which estimate the value of internal vertices in the game tree when it is not feasible to compute these values exactly. For example, in the game of tic-tac-toe, as an evaluation function for a position, we may use the number of files (rows, columns, and diagonals) containing no Os (used to indicate moves of the second player) minus the number of files containing no Xs (used to indicate moves of the first player). This evaluation function provides some indication of which player has the advantage in the game. Once the values of an evaluation function are inserted, the value of the game can be computed following the rules used for the minmax strategy. Computer programs created to play chess, such as the famous Deep Blue program, are based on sophisticated evaluation functions. For more information about how computers play chess see [Le91].

Exercises

1. Build a binary search tree for the words *banana, peach, apple, pear, coconut, mango,* and *papaya* using alphabetical order.

2. Build a binary search tree for the words *oenology, phrenology, campanology, ornithology, ichthyology, limnology, alchemy,* and *astrology* using alphabetical order.

3. How many comparisons are needed to locate or to add each of these words in the search tree for Exercise 1, starting fresh each time?

    a) *pear*   **b)** *banana*
    c) *kumquat*   **d)** *orange*

4. How many comparisons are needed to locate or to add each of the words in the search tree for Exercise 2, starting fresh each time?

    a) *palmistry*
    b) *etymology*
    c) *paleontology*
    d) *glaciology*

5. Using alphabetical order, construct a binary search tree for the words in the sentence "*The quick brown fox jumps over the lazy dog.*"

6. How many weighings of a balance scale are needed to find a lighter counterfeit coin among four coins? Describe an algorithm to find the lighter coin using this number of weighings.

7. How many weighings of a balance scale are needed to find a counterfeit coin among four coins if the counterfeit coin may be either heavier or lighter than the others? Describe an algorithm to find the counterfeit coin using this number of weighings.

****8.** How many weighings of a balance scale are needed to find a counterfeit coin among eight coins if the counterfeit coin is either heavier or lighter than the others? Describe an algorithm to find the counterfeit coin using this number of weighings.

****9.** How many weighings of a balance scale are needed to find a counterfeit coin among 12 coins if the counterfeit coin is lighter than the others? Describe an algorithm to find the lighter coin using this number of weighings.

****10.** One of four coins may be counterfeit. If it is counterfeit, it may be lighter or heavier than the others. How many weighings are needed, using a balance scale, to determine whether there is a counterfeit coin, and if there is, whether it is lighter or heavier than the others? Describe an algorithm to find the counterfeit coin and determine whether it is lighter or heavier using this number of weighings.

11. Find the least number of comparisons needed to sort four elements and devise an algorithm that sorts these elements using this number of comparisons.

****12.** Find the least number of comparisons needed to sort five elements and devise an algorithm that sorts these elements using this number of comparisons.

The **tournament sort** is a sorting algorithm that works by building an ordered binary tree. We represent the elements to be sorted by vertices that will become the leaves. We build up the tree one level at a time as we would construct the tree representing the winners of matches in a tournament. Working left to right, we compare pairs of consecutive elements, adding a parent vertex labeled with the larger of the two elements under comparison. We make similar comparisons between labels of vertices at each level until we reach the root of the tree that is labeled with the largest element. The tree constructed by the tournament sort of $22, 8, 14, 17, 3, 9, 27, 11$ is illustrated in part (a) of the figure. Once the largest element has been determined, the leaf with this label is relabeled by $-\infty$, which is defined to be less than every element. The labels of all vertices on the path from this vertex up to the root of the tree are recalculated, as shown in part (b) of the figure. This produces the second largest element. This process continues until the entire list has been sorted.

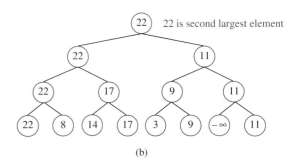

(b)

13. Complete the tournament sort of the list $22, 8, 14, 17, 3, 9, 27, 11$. Show the labels of the vertices at each step.

14. Use the tournament sort to sort the list $17, 4, 1, 5, 13, 10, 14, 6$.

15. Describe the tournament sort using pseudocode.

16. Assuming that n, the number of elements to be sorted, equals 2^k for some positive integer k, determine the number of comparisons used by the tournament sort to find the largest element of the list using the tournament sort.

17. How many comparisons does the tournament sort use to find the second largest, the third largest, and so on, up to the $(n-1)$st largest (or second smallest) element?

18. Show that the tournament sort requires $\Theta(n \log n)$ comparisons to sort a list of n elements. (*Hint:* By inserting the appropriate number of dummy elements defined to be smaller than all integers, such as $-\infty$, assume that $n = 2^k$ for some positive integer k.)

19. Which of these codes are prefix codes?

a) $a: 11, e: 00, t: 10, s: 01$
b) $a: 0, e: 1, t: 01, s: 001$
c) $a: 101, e: 11, t: 001, s: 011, n: 010$
d) $a: 010, e: 11, t: 011, s: 1011, n: 1001, i: 10101$

20. Construct the binary tree with prefix codes representing these coding schemes.

a) $a: 11, e: 0, t: 101, s: 100$
b) $a: 1, e: 01, t: 001, s: 0001, n: 00001$
c) $a: 1010, e: 0, t: 11, s: 1011, n: 1001, i: 10001$

21. What are the codes for a, e, i, k, o, p, and u if the coding scheme is represented by this tree?

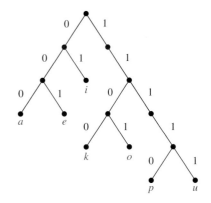

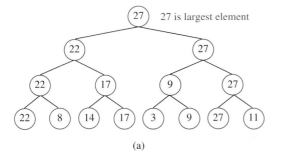

(a)

22. Given the coding scheme $a: 001, b: 0001, e: 1, r: 0000,$ $s: 0100, t: 011, x: 01010,$ find the word represented by

   a) 01110100011.

   b) 0001110000.

   c) 0100101010.

   d) 01100101010.

23. Use Huffman coding to encode these symbols with given frequencies: a: 0.20, b: 0.10, c: 0.15, d: 0.25, e: 0.30. What is the average number of bits required to encode a character?

24. Use Huffman coding to encode these symbols with given frequencies: A: 0.10, B: 0.25, C: 0.05, D: 0.15, E: 0.30, F: 0.07, G: 0.08. What is the average number of bits required to encode a symbol?

25. Construct two different Huffman codes for these symbols and frequencies: t: 0.2, u: 0.3, v: 0.2, w: 0.3.

26. a) Use Huffman coding to encode these symbols with frequencies a: 0.4, b: 0.2, c: 0.2, d: 0.1, e: 0.1 in two different ways by breaking ties in the algorithm differently. First, among the trees of minimum weight select two trees with the largest number of vertices to combine at each stage of the algorithm. Second, among the trees of minimum weight select two trees with the smallest number of vertices at each stage.

   b) Compute the average number of bits required to encode a symbol with each code and compute the variances of this number of bits for each code. Which tie-breaking procedure produced the smaller variance in the number of bits required to encode a symbol?

27. Construct a Huffman code for the letters of the English alphabet where the frequencies of letters in typical English text are as shown in this table.

Letter	Frequency	Letter	Frequency
A	0.0817	N	0.0662
B	0.0145	O	0.0781
C	0.0248	P	0.0156
D	0.0431	Q	0.0009
E	0.1232	R	0.0572
F	0.0209	S	0.0628
G	0.0182	T	0.0905
H	0.0668	U	0.0304
I	0.0689	V	0.0102
J	0.0010	W	0.0264
K	0.0080	X	0.0015
L	0.0397	Y	0.0211
M	0.0277	Z	0.0005

Suppose that m is a positive integer with $m \geq 2$. An m-ary Huffman code for a set of N symbols can be constructed analogously to the construction of a binary Huffman code. At the initial step, $((N - 1) \bmod (m - 1)) + 1$ trees consisting of a single vertex with least weights are combined into a rooted tree with these vertices as leaves. At each subsequent step, the m trees of least weight are combined into an m-ary tree.

28. Describe the m-ary Huffman coding algorithm in pseudocode.

29. Using the symbols 0, 1, and 2 use ternary ($m = 3$) Huffman coding to encode these letters with the given frequencies: A: 0.25, E: 0.30, N: 0.10, R: 0.05, T: 0.12, Z: 0.18.

30. Consider the three symbols A, B, and C with frequencies A: 0.80, B: 0.19, C: 0.01.

   a) Construct a Huffman code for these three symbols.

   b) Form a new set of nine symbols by grouping together blocks of two symbols, AA, AB, AC, BA, BB, BC, CA, CB, and CC. Construct a Huffman code for these nine symbols, assuming that the occurrences of symbols in the original text are independent.

   c) Compare the average number of bits required to encode text using the Huffman code for the three symbols in part (a) and the Huffman code for the nine blocks of two symbols constructed in part (b). Which is more efficient?

31. Given $n + 1$ symbols $x_1, x_2, \ldots, x_n, x_{n+1}$ appearing 1, $f_1, f_2, \ldots, f_n$ times in a symbol string, respectively, where f_j is the jth Fibonacci number, what is the maximum number of bits used to encode a symbol when all possible tie-breaking selections are considered at each stage of the Huffman coding algorithm?

*32. Show that Huffman codes are optimal in the sense that they represent a string of symbols using the fewest bits among all binary prefix codes.

33. Draw a game tree for nim if the starting position consists of two piles with two and three stones, respectively. When drawing the tree represent by the same vertex symmetric positions that result from the same move. Find the value of each vertex of the game tree. Who wins the game if both players follow an optimal strategy?

34. Draw a game tree for nim if the starting position consists of three piles with one, two, and three stones, respectively. When drawing the tree represent by the same vertex symmetric positions that result from the same move. Find the value of each vertex of the game tree. Who wins the game if both players follow an optimal strategy?

35. Suppose that we vary the payoff to the winning player in the game of nim so that the payoff is n dollars when n is the number of legal moves made before a terminal position is reached. Find the payoff to the first player if the initial position consists of

   a) two piles with one and three stones, respectively.

   b) two piles with two and four stones, respectively.

   c) three piles with one, two, and three stones, respectively.

36. Suppose that in a variation of the game of nim we allow a player to either remove one or more stones from a pile or merge the stones from two piles into one pile as long as at least one stone remains. Draw the game tree for this variation of nim if the starting position consists of three piles containing two, two, and one stone, respectively. Find the values of each vertex in the game tree and determine the winner if both players follow an optimal strategy.

37. Draw the subtree of the game tree for tic-tac-toe beginning at each of these positions. Determine the value of each of these subtrees.

   a)

O	X	X
X	O	O
		X

   b)

X	O	X
O	X	X
		O

   c)

X		O
O	O	
X		X

   d)

	O	X
	X	O
X	O	

38. Suppose that the first four moves of a tic-tac-toe game are as shown. Does the first player (whose moves are marked by Xs) have a strategy that will always win?

   a)

X	O	X
	O	

   b)

X	O	X
	O	

   c)

	O	X
	X	
O		

   d)

X		
X	O	
O		

39. Show that if a game of nim begins with two piles containing the same number of stones, as long as this number is at least two, then the second player wins when both players follow optimal strategies.

40. Show that if a game of nim begins with two piles containing different numbers of stones, the first player wins when both players follow optimal strategies.

41. How many children does the root of the game tree for checkers have? How many grandchildren does it have?

42. How many children does the root of the game tree for nim have and how many grandchildren does it have if the starting position is

   a) piles with four and five stones, respectively.

   b) piles with two, three, and four stones, respectively.

   c) piles with one, two, three, and four stones, respectively.

   d) piles with two, two, three, three, and five stones, respectively.

43. Draw the game tree for the game of tic-tac-toe for the levels corresponding to the first two moves. Assign the value of the evaluation function mentioned in the text that assigns to a position the number of files containing no Os minus the number of files containing no Xs as the value of each vertex at this level and compute the value of the tree for vertices as if the evaluation function gave the correct values for these vertices.

44. Use pseudocode to describe an algorithm for determining the value of a game tree when both players follow a minmax strategy.

9.3 Tree Traversal

INTRODUCTION

Ordered rooted trees are often used to store information. We need procedures for visiting each vertex of an ordered rooted tree to access data. We will describe several important algorithms for visiting all the vertices of an ordered rooted tree. Ordered rooted trees can also be used to represent various types of expressions, such as arithmetic expressions involving numbers, variables, and operations. The different listings of the vertices of ordered rooted trees used to represent expressions are useful in the evaluation of these expressions.

UNIVERSAL ADDRESS SYSTEMS

Procedures for traversing all vertices of an ordered rooted tree rely on the orderings of children. In ordered rooted trees, the children of an internal vertex are shown from left to right in the drawings representing these directed graphs.

We will describe one way to order totally the vertices of an ordered rooted tree. To produce this ordering, we must first label all the vertices. We do this recursively:

1. Label the root with the integer 0. Then label its k children (at level 1) from left to right with $1, 2, 3, \ldots, k$.

2. For each vertex v at level n with label A, label its k_v children, as they are drawn from left to right, with $A.1, A.2, \ldots, A.k_v$.

Following this procedure, a vertex v at level n, for $n \geq 1$, is labeled $x_1.x_2.\cdots.x_n$, where the unique path from the root to v goes through the x_1th vertex at level 1, the x_2th vertex at level 2, and so on. This labeling is called the **universal address system** of the ordered rooted tree.

We can totally order the vertices using the lexicographic ordering of their labels in the universal address system. The vertex labeled $x_1.x_2.\cdots.x_n$ is less than the vertex labeled $y_1.y_2.\cdots.y_m$ if there is an $i, 0 \leq i \leq n$, with $x_1 = y_1, x_2 = y_2, \ldots, x_{i-1} = y_{i-1}$, and $x_i < y_i$; or if $n < m$ and $x_i = y_i$ for $i = 1, 2, \ldots, n$.

EXAMPLE 1 We display the labelings of the universal address system next to the vertices in the ordered rooted tree shown in Figure 1. The lexicographic ordering of the labelings is

$$0 < 1 < 1.1 < 1.2 < 1.3 < 2 < 3 < 3.1 < 3.1.1 < 3.1.2 < 3.1.2.1 < 3.1.2.2$$
$$< 3.1.2.3 < 3.1.2.4 < 3.1.3 < 3.2 < 4 < 4.1 < 5 < 5.1 < 5.1.1 < 5.2 < 5.3 \quad \blacktriangleleft$$

TRAVERSAL ALGORITHMS

Procedures for systematically visiting every vertex of an ordered rooted tree are called **traversal algorithms.** We will describe three of the most commonly used such algorithms, **preorder traversal, inorder traversal,** and **postorder traversal.** Each of these algorithms can be defined recursively. We first define preorder traversal.

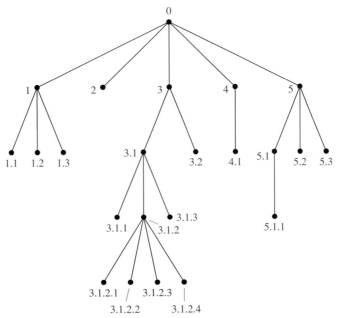

FIGURE 1   **The Universal Address System of an Ordered Rooted Tree.**

DEFINITION 1

Let T be an ordered rooted tree with root r. If T consists only of r, then r is the *preorder traversal* of T. Otherwise, suppose that $T_1, T_2, \ldots, T_n$ are the subtrees at r from left to right in T. The *preorder traversal* begins by visiting r. It continues by traversing T_1 in preorder, then T_2 in preorder, and so on, until T_n is traversed in preorder.

The reader should verify that the preorder traversal of an ordered rooted tree gives the same ordering of the vertices as the ordering obtained using a universal address system. Figure 2 indicates how a preorder traversal is carried out.

    Example 2 illustrates preorder traversal.

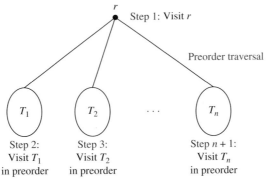

FIGURE 2   **Preorder Traversal.**

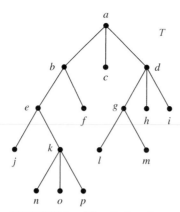

**FIGURE 3   The Ordered
Rooted Tree *T*.**

EXAMPLE 2   In which order does a preorder traversal visit the vertices in the ordered rooted tree *T* shown in Figure 3?

Solution: The steps of the preorder traversal of *T* are shown in Figure 4. We traverse *T* in preorder by first listing the root *a*, followed by the preorder list of the subtree with root *b*, the preorder list of the subtree with root *c* (which is just *c*) and the preorder list of the subtree with root *d*.

The preorder list of the subtree with root *b* begins by listing *b*, then the vertices of the subtree with root *e* in preorder, and then the subtree with root *f* in preorder (which is just *f*). The preorder list of the subtree with root *d* begins by listing *d*, followed by the preorder list of the subtree with root *g*, followed by the subtree with root *h* (which is just *h*), followed by the subtree with root *i* (which is just *i*).

The preorder list of the subtree with root *e* begins by listing *e*, followed by the preorder listing of the subtree with root *j* (which is just *j*), followed by the preorder listing of the subtree with root *k*. The preorder listing of the subtree with root *g* is *g* followed by *l*, followed by *m*. The preorder listing of the subtree with root *k* is *k, n, o, p*. Consequently, the preorder traversal of *T* is *a, b, e, j, k, n, o, p, f, c, d, g, l, m, h, i*.   ◀

We will now define inorder traversal.

DEFINITION 2   Let *T* be an ordered rooted tree with root *r*. If *T* consists only of *r*, then *r* is the *inorder traversal* of *T*. Otherwise, suppose that $T_1, T_2, \ldots, T_n$ are the subtrees at *r* from left to right. The *inorder traversal* begins by traversing T_1 in inorder, then visiting *r*. It continues by traversing T_2 in inorder, then T_3 in inorder, ..., and finally T_n in inorder.

Figure 5 indicates how inorder traversal is carried out. Example 3 illustrates how inorder traversal is carried out for a particular tree.

EXAMPLE 3   In which order does an inorder traversal visit the vertices of the ordered rooted tree *T* in Figure 3?

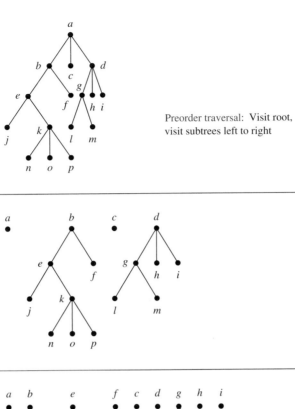

Preorder traversal: Visit root,
visit subtrees left to right

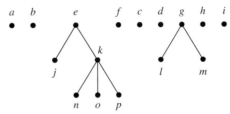

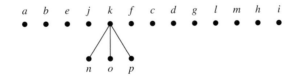

FIGURE 4   **The Preorder Traversal of T.**

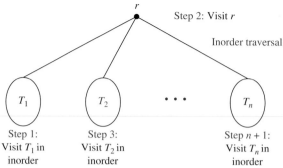

Step 2: Visit r

Inorder traversal

Step 1:
Visit T_1 in
inorder

Step 3:
Visit T_2 in
inorder

Step $n + 1$:
Visit T_n in
inorder

FIGURE 5   Inorder Traversal.

Solution: The steps of the inorder traversal of the ordered rooted tree T are shown in Figure 6. The inorder traversal begins with an inorder traversal of the subtree with root b, the root a, the inorder listing of the subtree with root c, which is just c, and the inorder listing of the subtree with root d.

The inorder listing of the subtree with root b begins with the inorder listing of the subtree with root e, the root b, and f. The inorder listing of the subtree with root d begins with the inorder listing of the subtree with root g, followed by the root d, followed by h, followed by i.

The inorder listing of the subtree with root e is j, followed by the root e, followed by the inorder listing of the subtree with root k. The inorder listing of the subtree with root g is l, g, m. The inorder listing of the subtree with root k is n, k, o, p. Consequently, the inorder listing of the ordered rooted tree is $j, e, n, k, o, p, b, f, a, c, l, g, m, d, h, i$.   ◀

We now define postorder traversal.

DEFINITION 3

Let T be an ordered rooted tree with root r. If T consists only of r, then r is the *postorder traversal* of T. Otherwise, suppose that $T_1, T_2, \ldots, T_n$ are the subtrees at r from left to right. The *postorder traversal* begins by traversing T_1 in postorder, then T_2 in postorder, $\ldots$, then T_n in postorder, and ends by visiting r.

Figure 7 illustrates how postorder traversal is done. Example 4 illustrates how postorder traversal works.

EXAMPLE 4   In which order does a postorder traversal visit the vertices of the ordered rooted tree T shown in Figure 3?

Solution: The steps of the postorder traversal of the ordered rooted tree T are shown in Figure 8. The postorder traversal begins with the postorder traversal of the subtree with root b, the postorder traversal of the subtree with root c, which is just c, the postorder traversal of the subtree with root d, followed by the root a.

Extra
Examples

Extra
Examples

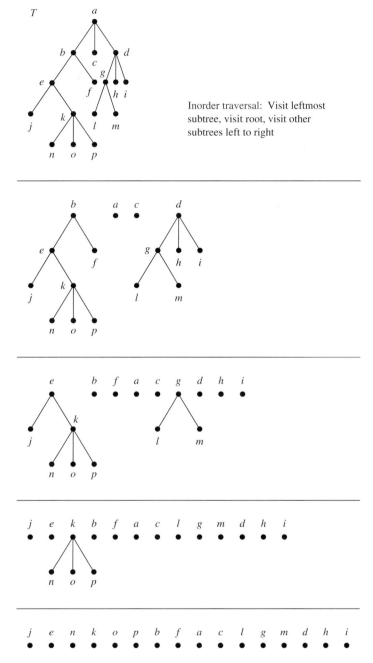

FIGURE 6 **The Inorder Traversal of *T*.**

The postorder traversal of the subtree with root *b* begins with the postorder traversal of the subtree with root *e*, followed by *f*, followed by the root *b*. The postorder traversal of the rooted tree with root *d* begins with the postorder traversal of the subtree with root *g*, followed by *h*, followed by *i*, followed by the root *d*.

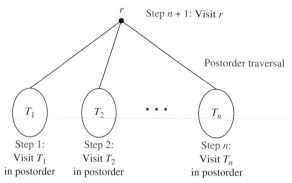

Step $n + 1$: Visit r

Postorder traversal

Step 1:
Visit T_1
in postorder

Step 2:
Visit T_2
in postorder

Step n:
Visit T_n
in postorder

FIGURE 7   **Postorder Traversal.**

The postorder traversal of the subtree with root e begins with j, followed by the postorder traversal of the subtree with root k, followed by the root e. The postorder traversal of the subtree with root g is l, m, g. The postorder traversal of the subtree with root k is n, o, p, k. Therefore, the postorder traversal of T is $j, n, o, p, k, e, f, b, c, l, m,$ g, h, i, d, a. ◀

There are easy ways to list the vertices of an ordered rooted tree in preorder, inorder, and postorder. To do this, first draw a curve around the ordered rooted tree starting at the root, moving along the edges as shown in the example in Figure 9. We can list the vertices in preorder by listing each vertex the first time this curve passes it. We can list the vertices in inorder by listing a leaf the first time the curve passes it and listing each internal vertex the second time the curve passes it. We can list the vertices in postorder by listing a vertex the last time it is passed on the way back up to its parent. When this is done in the rooted tree in Figure 9, it follows that the preorder traversal gives $a, b, d, h,$ e, i, j, c, f, g, k, the inorder traversal gives $h, d, b, i, e, j, a, f, c, k, g$, and the postorder traversal gives $h, d, i, j, e, b, f, k, g, c, a$.

Algorithms for traversing ordered rooted trees in preorder, inorder, or postorder are most easily expressed recursively.

ALGORITHM 1 Preorder Traversal.

---

procedure *preorder*(T: ordered rooted tree)
$r :=$ root of T
list r
for each child c of r from left to right
begin
    $T(c) :=$ subtree with c as its root
    preorder($T(c)$)
end

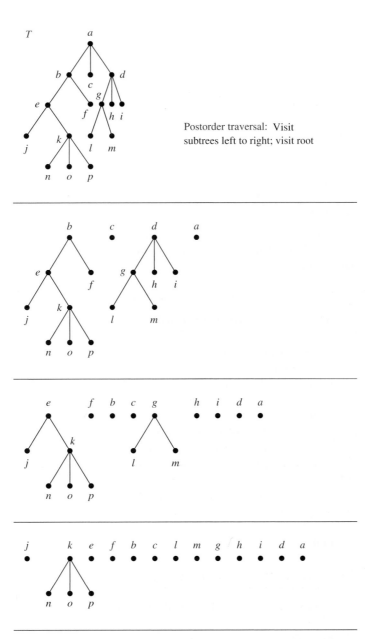

Postorder traversal: Visit subtrees left to right; visit root

FIGURE 8   **The Postorder Traversal of T.**

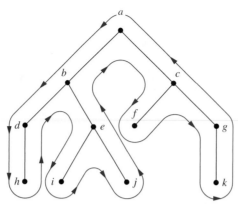

FIGURE 9   A Shortcut for Traversing an Ordered Rooted Tree in Preorder, Inorder, and Postorder.

ALGORITHM 2 Inorder Traversal.

procedure *inorder*(T: ordered rooted tree)
$r := $ root of T
if r is a leaf **then** list r
else
begin
  $l := $ first child of r from left to right
  $T(l) := $ subtree with l as its root
  inorder$(T(l))$
  list r
  for each child c of r except for l from left to right
    $T(c) := $ subtree with c as its root
    inorder$(T(c))$
end

ALGORITHM 3 Postorder Traversal.

procedure *postorder*(T: ordered rooted tree)
$r := $ root of T
for each child c of r from left to right
begin
  $T(c) := $ subtree with c as its root
  postorder$(T(c))$
end
list r

INFIX, PREFIX, AND POSTFIX NOTATION

We can represent complicated expressions, such as compound propositions, combinations of sets, and arithmetic expressions using ordered rooted trees. For instance, consider

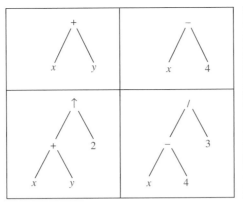
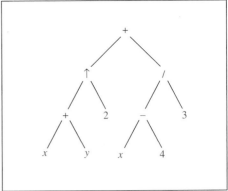

FIGURE 10   A Binary Tree Representing $((x + y) \uparrow 2) + ((x - 4)/3)$.

the representation of an arithmetic expression involving the operators + (addition), − (subtraction), ∗ (multiplication), / (division), and ↑ (exponentiation). We will use parentheses to indicate the order of the operations. An ordered rooted tree can be used to represent such expressions, where the internal vertices represent operations, and the leaves represent the variables or numbers. Each operation operates on its left and right subtrees (in that order).

EXAMPLE 5   What is the ordered rooted tree that represents the expression $((x+y) \uparrow 2)+((x-4)/3)$?

Solution: The binary tree for this expression can be built from the bottom up. First, a subtree for the expression $x + y$ is constructed. Then this is incorporated as part of the larger subtree representing $(x + y) \uparrow 2$. Also, a subtree for $x - 4$ is constructed, and then this is incorporated into a subtree representing $(x - 4)/3$. Finally the subtrees representing $(x + y) \uparrow 2$ and $(x - 4)/3$ are combined to form the ordered rooted tree representing $((x + y) \uparrow 2) + ((x - 4)/3)$. These steps are shown in Figure 10.   ◄

An inorder traversal of the binary tree representing an expression produces the original expression with the elements and operations in the same order as they originally occurred, except for unary operations, which instead immediately follow their operands. For instance, inorder traversals of the binary trees in Figure 11, which represent the expressions $(x + y)/(x + 3)$, $(x + (y/x)) + 3$, and $x + (y/(x + 3))$, all lead to the infix expression $x + y/x + 3$. To make such expressions unambiguous it is necessary to include parentheses in the inorder traversal whenever we encounter an operation. The fully parenthesized expression obtained in this way is said to be in **infix form.**

We obtain the **prefix form** of an expression when we traverse its rooted tree in preorder. Expressions written in prefix form are said to be in **Polish notation,** which is named after the Polish logician Jan Łukasiewicz. An expression in prefix notation (where each operation has a specified number of operands), is unambiguous, so no parentheses are needed in such an expression. The verification of this is left as an exercise for the reader.

EXAMPLE 6   What is the prefix form for $((x + y) \uparrow 2) + ((x - 4)/3)$?

Solution: We obtain the prefix form for this expression by traversing the binary tree that represents it, shown in Figure 10. This produces $+ \uparrow + x\, y\, 2 / - x\, 4\, 3$.   ◄

In the prefix form of an expression, a binary operator, such as +, precedes its two operands. Hence, we can evaluate an expression in prefix form by working from right

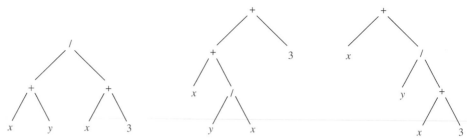

FIGURE 11 **Rooted Trees Representing** $(x + y)/(x + 3)$, $(x + (y/x)) + 3$, **and** $x +$ $(y/(x + 3))$**.**

to left. When we encounter an operator, we perform the corresponding operation with the two operands immediately to the right of this operand. Also, whenever an operation is performed, we consider the result a new operand.

EXAMPLE 7   What is the value of the prefix expression $+ - * 2\ 3\ 5\ /\ \uparrow 2\ 3\ 4$?

Solution: The steps used to evaluate this expression by working right to left, and performing operations using the operands on the right, are shown in Figure 12. The value of this expression is 3.   ◀

We obtain the **postfix form** of an expression by traversing its binary tree in postorder. Expressions written in postfix form are said to be in **reverse Polish notation.** Expressions in reverse Polish notation are unambiguous, so parentheses are not needed. The verification of this is left to the reader.

EXAMPLE 8   What is the postfix form of the expression $((x + y) \uparrow 2) + ((x - 4)/3)$?

JAN ŁUKASIEWICZ (1878–1956) Jan Łukasiewicz was born into a Polish-speaking family in Lvov. At that time Lvov was part of Austria, but it is now in the Ukraine. His father was a captain in the Austrian army. Łukasiewicz became interested in mathematics while in high school. He studied mathematics and philosophy at the University of Lvov at both the undergraduate and graduate levels. After completing his doctoral work he became a lecturer there, and in 1911 he was appointed to a professorship. When the University of Warsaw was reopened as a Polish university in 1915, Łukasiewicz accepted an invitation to join the faculty. In 1919 he served as the Polish Minister of Education. He returned to the position of professor at Warsaw University where he remained from 1920 to 1939, serving as rector of the university twice.

Łukasiewicz was one of the cofounders of the famous Warsaw School of Logic. He published his famous text, *Elements of Mathematical Logic,* in 1928. With his influence mathematical logic was made a required course for mathematics and science undergraduates in Poland. His lectures were considered excellent, even attracting students of the humanities.

Łukasiewicz and his wife experienced great suffering during World War II, which he documented in a posthumously published autobiography. After the war they lived in exile in Belgium. Fortunately, in 1949 he was offered a position at the Royal Irish Academy in Dublin.

Łukasiewicz worked on mathematical logic throughout his career. His work on a three-valued logic was an important contribution to the subject. Nevertheless, he is best known in the mathematical and computer science communities for his introduction of parenthesis-free notation, now called Polish notation.

$$+ \quad - \quad * \quad 2 \quad 3 \quad 5 \quad / \quad \underbrace{\uparrow \quad 2 \quad 3}_{} \quad 4$$
$$2 \uparrow 3 = 8$$

$$+ \quad - \quad * \quad 2 \quad 3 \quad 5 \quad \underbrace{/ \quad 8 \quad 4}_{}$$
$$8 / 4 = 2$$

$$+ \quad - \quad \underbrace{* \quad 2 \quad 3}_{} \quad 5 \quad 2$$
$$2 * 3 = 6$$

$$+ \quad \underbrace{- \quad 6 \quad 5}_{} \quad 2$$
$$6 - 5 = 1$$

$$\underbrace{+ \quad 1 \quad 2}_{}$$
$$1 + 2 = 3$$

Value of expression: 3

$$7 \quad \underbrace{2 \quad 3 \quad *}_{} \quad - \quad 4 \quad \uparrow \quad 9 \quad 3 \quad / \quad +$$
$$2 * 3 = 6$$

$$\underbrace{7 \quad 6 \quad -}_{} \quad 4 \quad \uparrow \quad 9 \quad 3 \quad / \quad +$$
$$7 - 6 = 1$$

$$\underbrace{1 \quad 4 \quad \uparrow}_{} \quad 9 \quad 3 \quad / \quad +$$
$$1^4 = 1$$

$$1 \quad \underbrace{9 \quad 3 \quad /}_{} \quad +$$
$$9 / 3 = 3$$

$$\underbrace{1 \quad 3 \quad +}_{}$$
$$1 + 3 = 4$$

Value of expression: 4

FIGURE 12   Evaluating a Prefix Expression.   **FIGURE 13   Evaluating a Postfix Expression.**

Solution: The postfix form of the expression is obtained by carrying out a postorder traversal of the binary tree for this expression, shown in Figure 10. This produces the postfix expression: $x \; y + 2 \uparrow x \; 4 - 3 / +$. ◀

In the postfix form of an expression, a binary operator follows its two operands. So, to evaluate an expression from its postfix form, work from left to right, carrying out operations whenever an operator follows two operands. After an operation is carried out, the result of this operation becomes a new operand.

EXAMPLE 9   What is the value of the postfix expression $7 \; 2 \; 3 * - 4 \uparrow 9 \; 3 / +$?

Solution: The steps used to evaluate this expression by starting at the left and carrying out operations when two operands are followed by an operator are shown in Figure 13. The value of this expression is 4. ◀

Rooted trees can be used to represent other types of expressions, such as those representing compound propositions and combinations of sets. In these examples unary operators, such as the negation of a proposition, occur. To represent such operators and their operands, a vertex representing the operator and a child of this vertex representing the operand are used.

EXAMPLE 10   Find the ordered rooted tree representing the compound proposition $(\neg(p \wedge q)) \leftrightarrow (\neg p \vee \neg q)$. Then use this rooted tree to find the prefix, postfix, and infix forms of this expression.

Extra
Examples

Solution: The rooted tree for this compound proposition is constructed from the bottom up. First, subtrees for $\neg p$ and $\neg q$ are formed (where $\neg$ is considered a unary operator). Also, a subtree for $p \wedge q$ is formed. Then subtrees for $\neg(p \wedge q)$ and $(\neg p) \vee (\neg q)$ are constructed. Finally, these two subtrees are used to form the final rooted tree. The steps of this procedure are shown in Figure 14.

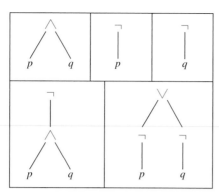
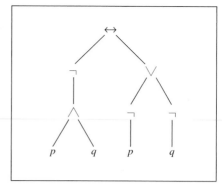

FIGURE 14   Constructing the Rooted Tree for a Compound Proposition.

The prefix, postfix, and infix forms of this expression are found by traversing this rooted tree in preorder, postorder, and inorder (including parentheses), respectively. These traversals give $\leftrightarrow \neg \wedge pq \vee \neg p \neg q$, $pq \wedge \neg p \neg q \neg \vee \leftrightarrow$, and $(\neg(p \wedge q)) \leftrightarrow ((\neg p) \vee (\neg q))$, respectively. ◄

Because prefix and postfix expressions are unambiguous and because they can easily be evaluated without scanning back and forth, they are used extensively in computer science. Such expressions are especially useful in the construction of compilers.

Exercises

In Exercises 1–3 construct the universal address system for the given ordered rooted tree. Then use this to order its vertices using the lexicographic order of their labels.

1.

2.

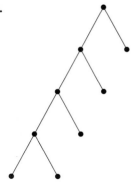

3.

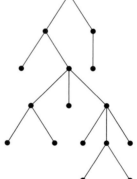

4. Suppose that the address of the vertex v in the ordered rooted tree T is 3.4.5.2.4.

a) At what level is v?

b) What is the address of the parent of v?

c) What is the least number of siblings v can have?

d) What is the smallest possible number of vertices in T if v has this address?

e) Find the other addresses that must occur.

5. Suppose that the vertex with the largest address in an ordered rooted tree T has address 2.3.4.3.1. Is it possible to determine the number of vertices in T?

6. Can the leaves of an ordered rooted tree have the following list of universal addresses? If so, construct such an ordered rooted tree.

a) 1.1.1, 1.1.2, 1.2, 2.1.1.1, 2.1.2, 2.1.3, 2.2, 3.1.1, 3.1.2.1, 3.1.2.2, 3.2

b) 1.1, 1.2.1, 1.2.2, 1.2.3, 2.1, 2.2.1, 2.3.1, 2.3.2, 2.4.2.1, 2.4.2.2, 3.1, 3.2.1, 3.2.2

c) 1.1, 1.2.1, 1.2.2, 1.2.2.1, 1.3, 1.4, 2, 3.1, 3.2, 4.1.1.1

In Exercises 7–9 determine the order in which a preorder traversal visits the vertices of the given ordered rooted tree.

7.

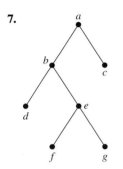

8.

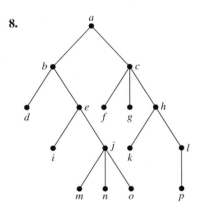

9.

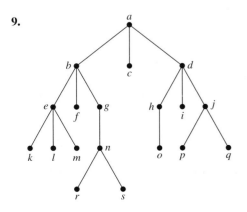

10. In which order are the vertices of the ordered rooted tree in Exercise 7 visited using an inorder traversal?

11. In which order are the vertices of the ordered rooted tree in Exercise 8 visited using an inorder traversal?

12. In which order are the vertices of the ordered rooted tree in Exercise 9 visited using an inorder traversal?

13. In which order are the vertices of the ordered rooted tree in Exercise 7 visited using a postorder traversal?

14. In which order are the vertices of the ordered rooted tree in Exercise 8 visited using a postorder traversal?

15. In which order are the vertices of the ordered rooted tree in Exercise 9 visited using a postorder traversal?

16. a) Represent the expression $((x + 2) \uparrow 3) * (y - (3 + x)) - 5$ using a binary tree.
Write this expression in
b) prefix notation.
c) postfix notation.
d) infix notation.

17. a) Represent the expressions $(x + xy) + (x/y)$ and $x + ((xy + x)/y)$ using binary trees.
Write these expressions in
b) prefix notation.
c) postfix notation.
d) infix notation.

18. a) Represent the compound propositions $\neg(p \wedge q) \leftrightarrow (\neg p \vee \neg q)$ and $(\neg p \wedge (q \leftrightarrow \neg p)) \vee \neg q$ using ordered rooted trees.
Write these expressions in
b) prefix notation.
c) postfix notation.
d) infix notation.

19. a) Represent $(A \cap B) - (A \cup (B - A))$ using an ordered rooted tree.
Write this expression in
b) prefix notation.
c) postfix notation.
d) infix notation.

*20. In how many ways can the string $\neg p \wedge q \leftrightarrow \neg p \vee \neg q$ be fully parenthesized to yield an infix expression?

*21. In how many ways can the string $A \cap B - A \cap B - A$ be fully parenthesized to yield an infix expression?

22. Draw the ordered rooted tree corresponding to each of these arithmetic expressions written in prefix notation. Then write each expression using infix notation.

a) $+ * + - 5\ 3\ 2\ 1\ 4$
b) $\uparrow + 2\ 3 - 5\ 1$
c) $* / 9\ 3 + * 2\ 4 - 7\ 6$

23. What is the value of each of these prefix expressions?

a) $- * 2 / 8\ 4\ 3$
b) $\uparrow - * 3\ 3 * 4\ 2\ 5$
c) $+ - \uparrow 3\ 2 \uparrow 2\ 3 / 6 - 4\ 2$
d) $* + 3 + 3 \uparrow 3 + 3\ 3\ 3$

24. What is the value of each of these postfix expressions?

  a) $5\,2\,1\,-\,-\,3\,1\,4\,+\,+\,*$
  b) $9\,3\,/\,5\,+\,7\,2\,-\,*$
  c) $3\,2\,*\,2\,\uparrow\,5\,3\,-\,8\,4\,/\,*\,-$

25. Construct the ordered rooted tree whose preorder traversal is $a, b, f, c, g, h, i, d, e, j, k, l$, where a has four children, c has three children, j has two children, b and e have one child each, and all other vertices are leaves.

*26. Show that an ordered rooted tree is uniquely determined when a list of vertices generated by a preorder traversal of the tree and the number of children of each vertex are specified.

*27. Show that an ordered rooted tree is uniquely determined when a list of vertices generated by a postorder traversal of the tree and the number of children of each vertex are specified.

28. Show that preorder traversals of the two ordered rooted trees displayed below produce the same list of vertices. Note that this does not contradict the statement in Exercise 26, since the numbers of children of internal vertices in the two ordered rooted trees differ.

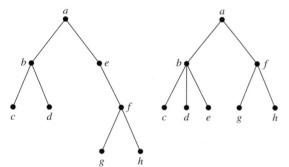

29. Show that postorder traversals of these two ordered rooted trees produce the same list of vertices. Note that this does not contradict the statement in Exercise 27, since the numbers of children of internal vertices in the two ordered rooted trees differ.

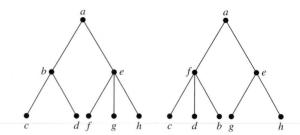

Well-formed formulae in prefix notation over a set of symbols and a set of binary operators are defined recursively by these rules:

  i) if x is a symbol, then x is a well-formed formula in prefix notation;
  ii) if X and Y are well-formed formulae and $*$ is an operator, then $*\,XY$ is a well-formed formula.

30. Which of these are well-formed formulae over the symbols $\{x, y, z\}$ and the set of binary operators $\{\times, +, \circ\}$?

  a) $\times\,+\,+\,x\,y\,x$
  b) $\circ\,x\,y\,\times\,x\,z$
  c) $\times\,\circ\,x\,z\,\times\,\times\,x\,y$
  d) $\times\,+\,\circ\,x\,x\,\circ\,x\,x\,x$

*31. Show that any well-formed formula in prefix notation over a set of symbols and a set of binary operators contains exactly one more symbol than the number of operators.

32. Give a definition of well-formed formulae in postfix notation over a set of symbols and a set of binary operators.

33. Give six examples of well-formed formulae with three or more operators in postfix notation over the set of symbols $\{x, y, z\}$ and the set of operators $\{+, \times, \circ\}$.

34. Extend the definition of well-formed formulae in prefix notation to sets of symbols and operators where the operators may not be binary.

9.4 Spanning Trees

INTRODUCTION

Consider the system of roads in Maine represented by the simple graph shown in Figure 1(a). The only way the roads can be kept open in the winter is by frequently plowing them. The highway department wants to plow the fewest roads so that there will always be cleared roads connecting any two towns. How can this be done?

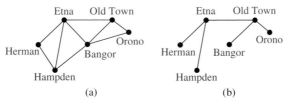

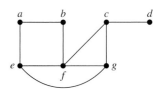

FIGURE 1 **(a) A Road System and (b) a Set of Roads to Plow.**

FIGURE 2 **The Simple Graph G.**

At least five roads must be plowed to ensure that there is a path between any two towns. Figure 1(b) shows one such set of roads. Note that the subgraph representing these roads is a tree, since it is connected and contains six vertices and five edges.

This problem was solved with a connected subgraph with the minimum number of edges containing all vertices of the original simple graph. Such a graph must be a tree.

DEFINITION 1
> Let G be a simple graph. A *spanning tree* of G is a subgraph of G that is a tree containing every vertex of G.

A simple graph with a spanning tree must be connected, since there is a path in the spanning tree between any two vertices. The converse is also true; that is, every connected simple graph has a spanning tree. We will give an example before proving this result.

EXAMPLE 1 Find a spanning tree of the simple graph G shown in Figure 2.

Solution: The graph G is connected, but it is not a tree because it contains simple circuits. Remove the edge $\{a, e\}$. This eliminates one simple circuit, and the resulting subgraph is still connected and still contains every vertex of G. Next remove the edge $\{e, f\}$ to eliminate a second simple circuit. Finally, remove edge $\{c, g\}$ to produce a simple graph with no simple circuits. This subgraph is a spanning tree, since it is a tree that contains every vertex of G. The sequence of edge removals used to produce the spanning tree is illustrated in Figure 3.

The tree shown in Figure 3 is not the only spanning tree of G. For instance, each of the trees shown in Figure 4 is a spanning tree of G. ◀

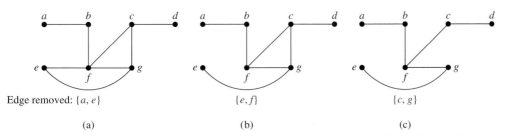

FIGURE 3 **Producing a Spanning Tree for G by Removing Edges That Form Simple Circuits.**

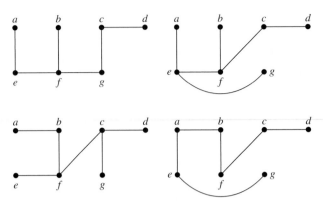

FIGURE 4   **Spanning Trees of G.**

THEOREM 1   A simple graph is connected if and only if it has a spanning tree.

Proof: First, suppose that a simple graph G has a spanning tree T. T contains every vertex of G. Furthermore, there is a path in T between any two of its vertices. Since T is a subgraph of G, there is a path in G between any two of its vertices. Hence, G is connected.

Now suppose that G is connected. If G is not a tree, it must contain a simple circuit. Remove an edge from one of these simple circuits. The resulting subgraph has one fewer edge but still contains all the vertices of G and is connected. This subgraph is still connected because when two vertices are connected by a path containing the removed edge, they are connected by a path not containing this edge. We can construct such a path by inserting into the original path, at the point where the removed edge once was, the simple circuit with this edge removed. If this subgraph is not a tree, it has a simple circuit; so as before, remove an edge that is in a simple circuit. Repeat this process until no simple circuits remain. This is possible because there are only a finite number of edges in the graph. The process terminates when no simple circuits remain. A tree is produced since the graph stays connected as edges are removed. This tree is a spanning tree since it contains every vertex of G. ◁

Spanning trees are important in data networking, as Example 2 shows.

EXAMPLE 2   **IP Multicasting** Spanning trees play an important role in multicasting over Internet Protocol (IP) networks. To send data from a source computer to multiple receiving computers, each of which is a subnetwork, data could be sent separately to each computer. This type of networking, called unicasting, is inefficient, since many copies of the same data are transmitted over the network. To make the transmission of data to multiple receiving computers more efficient, IP multicasting is used. With IP multicasting, a computer sends a single copy of data over the network, and as data reaches intermediate routers the data are forwarded to one or more other routers so that ultimately all receiving computers in their various subnetworks receive these data. (Routers are computers that are dedicated to forwarding IP datagrams between subnetworks in a network. In multicasting, routers use Class D addresses, each representing a session that receiving computers may join; see Example 15 in Section 4.1.)

For data to reach receiving computers as quickly as possible, there should be no loops (which in graph theory terminology are circuits or cycles) in the path that data take

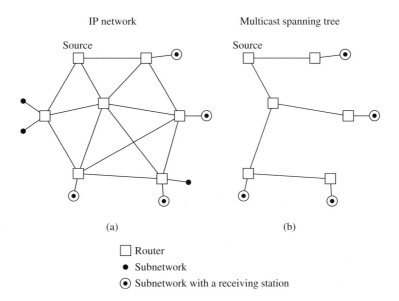

IP network   Multicast spanning tree

Source   Source

(a)   (b)

☐ Router
● Subnetwork
◉ Subnetwork with a receiving station

FIGURE 5   **A Multicast Spanning Tree.**

through the network. That is, once data have reached a particular router, data should never return to this router. To avoid loops, the multicast routers use network algorithms to construct a spanning tree in the graph that has the multicast source, the routers, and the subnetworks containing receiving computers as vertices, with edges representing the links between computers and/or routers. The root of this spanning tree is the multicast source. The subnetworks containing receiving computers are leaves of the tree. (Note that subnetworks not containing receiving stations are not included in the graph.) This is illustrated in Figure 5. ◀

DEPTH-FIRST SEARCH

Links

Demo

The proof of Theorem 1 gives an algorithm for finding spanning trees by removing edges from simple circuits. This algorithm is inefficient, since it requires that simple circuits be identified. Instead of constructing spanning trees by removing edges, spanning trees can be built up by successively adding edges. Two algorithms based on this principle will be presented here.

    We can build a spanning tree for a connected simple graph using a **depth-first search.** We will form a rooted tree, and the spanning tree will be the underlying undirected graph of this rooted tree. Arbitrarily choose a vertex of the graph as the root. Form a path starting at this vertex by successively adding vertices and edges, where each new edge is incident with the last vertex in the path and a vertex not already in the path. Continue adding vertices and edges to this path as long as possible. If the path goes through all vertices of the graph, the tree consisting of this path is a spanning tree. However, if the path does not go through all vertices, more vertices and edges must be added. Move back to the next to last vertex in the path, and, if possible, form a new path starting at this vertex passing through vertices that were not already visited. If this cannot be done, move back another vertex in the path, that is, two vertices back in the path, and try again.

    Repeat this procedure, beginning at the last vertex visited, moving back up the path one vertex at a time, forming new paths that are as long as possible until no more edges can be added. Since the graph has a finite number of edges and is connected, this process ends with the production of a spanning tree. Each vertex that ends a path at a stage of the algorithm will be a leaf in the rooted tree, and each vertex where a path is constructed starting at this vertex will be an internal vertex.

The reader should note the recursive nature of this procedure. Also, note that if the vertices in the graph are ordered, the choices of edges at each stage of the procedure are all determined when we always choose the first vertex in the ordering that is available. However, we will not always explicitly order the vertices of a graph.

Depth-first search is also called **backtracking,** since the algorithm returns to vertices previously visited to add paths. Example 3 illustrates backtracking.

EXAMPLE 3   Use a depth-first search to find a spanning tree for the graph G shown in Figure 6.

Solution: The steps used by a depth-first search to produce a spanning tree of G are shown in Figure 7. We arbitrarily start with the vertex f. A path is built by successively adding edges incident with vertices not already in the path, as long as this is possible. This produces a path f, g, h, k, j (note that other paths could have been built). Next, backtrack to k. There is no path beginning at k containing vertices not already visited. So we backtrack to h. Form the path h, i. Then backtrack to h, and then to f. From f build the path f, d, e, c, a. Then backtrack to c and form the path c, b. This produces the spanning tree.   ◀

The edges selected by a depth-first search of a graph are called **tree edges.** All other edges of the graph must connect a vertex to an ancestor or descendant of this vertex in the tree. These edges are called **back edges.** (Exercise 39 asks for a proof of this fact.)

EXAMPLE 4   In Figure 8 we highlight the tree edges found by a depth-first search starting at vertex f by showing them with heavy colored lines. The back edges (e, f) and (f, h) are shown with thinner black lines.   ◀

We have explained how to find a spanning tree of a graph using depth-first search. However, our discussion so far has not brought out the recursive nature of depth-first search. To help make the recursive nature of the algorithm clear, we need a little terminology. We say that we *explore* from a vertex v when we carry out the steps of depth-first search beginning when v is added to the tree and ending when we have backtracked back to v for the last time. The key observation needed to understand the recursive nature of the algorithm is that when we add an edge connecting a vertex v to a vertex w, we finish exploring from w before we return to v to complete exploring from v.

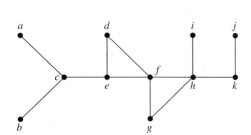

FIGURE 6   The Graph G.

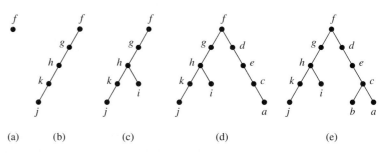

FIGURE 7   A Depth-First Search of G.

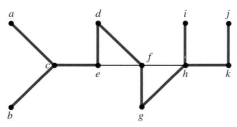

**FIGURE 8   The Tree Edges and Back
Edges of the Depth-First Search
in Example 4.**

In Algorithm 1 we construct the spanning tree of a graph G with vertices $v_1, \ldots, v_n$ by first selecting the vertex v_1 to be the root. We initially set T to be the tree with just this one vertex. At each step we add a new vertex to the tree T together with an edge from a vertex already in T to this new vertex and we explore from this new vertex. Note that at the completion of the algorithm, T contains no simple circuits since no edge is added that connects a vertex already in the tree. Moreover, T remains connected as it is built. (These last two observations can be easily proved via mathematical induction.) Since G is connected, every vertex in G is visited by the algorithm and is added to the tree (as the reader should verify). It follows that T is a spanning tree of G.

ALGORITHM 1 Depth-First Search.

---

procedure $DFS(G$: connected graph with vertices $v_1, v_2, \ldots, v_n)$
$T :=$ tree consisting only of the vertex v_1
$visit(v_1)$
procedure $visit(v$: vertex of $G)$
for each vertex w adjacent to v and not yet in T
begin
    add vertex w and edge $\{v, w\}$ to T
    $visit(w)$
end

We now analyze the computational complexity of the depth-first search algorithm. The key observation is that for each vertex v, the procedure $visit(v)$ is called when the vertex v is first encountered in the search and it is not called again. Assuming that the adjacency lists for G are available (see Section 8.3), no computations are required to find the vertices adjacent to v. As we follow the steps of the algorithm, we examine each edge at most twice to determine whether to add this edge and one of its endpoints to the tree. Consequently, the procedure DFS constructs a spanning tree using $O(e)$, or $O(n^2)$, steps where e and n are the number of edges and vertices in G, respectively. [Note that a step involves examining a vertex to see whether it is already in the spanning tree as it is being built and adding this vertex and the corresponding edge if the vertex is not already in the tree. We have also made use of the inequality $e \leq n(n-1)/2$, which holds for any simple graph.]

Depth-first search can be used as the basis for algorithms that solve many different problems. For example, it can be used to find the components of a graph and it can be used to find the cut vertices of a connected graph. As we will see, depth-first search is the basis of backtracking techniques used to search for solutions of computationally difficult problems. (See [GrYe99], [Ma89], and [CoLeRiSt01] for a discussion of algorithms based on depth-first search.)

BREADTH-FIRST SEARCH

We can also produce a spanning tree of a simple graph by the use of a **breadth-first search.** Again, a rooted tree will be constructed, and the underlying undirected graph of this rooted tree forms the spanning tree. Arbitrarily choose a root from the vertices of the graph. Then add all edges incident to this vertex. The new vertices added at this stage become the vertices at level 1 in the spanning tree. Arbitrarily order them. Next, for each vertex at level 1, visited in order, add each edge incident to this vertex to the tree as long as it does not produce a simple circuit. Arbitrarily order the children of each vertex at level 1. This produces the vertices at level 2 in the tree. Follow the same procedure until all the vertices in the tree have been added. The procedure ends since there are only a finite number of edges in the graph. A spanning tree is produced since we have produced a tree containing every vertex of the graph. An example of a breadth-first search is given in Example 5.

EXAMPLE 5 Use a breadth-first search to find a spanning tree for the graph shown in Figure 9.

Solution: The steps of the breadth-first search procedure are shown in Figure 10. We choose the vertex e to be the root. Then we add edges incident with all vertices adjacent to e, so that edges from e to b, d, f, and i are added. These four vertices are at level 1 in the tree. Next, add the edges from these vertices at level 1 to adjacent vertices not already in the tree. Hence, the edges from b to a and c are added, as are edges from d to h, from f to j and g, and from i to k. The new vertices a, c, h, j, g, and k are at level 2. Next, add edges from these vertices to adjacent vertices not already in the graph. This adds edges from g to l and from k to m. ◀

We describe breadth-first search in pseudocode as Algorithm 2. In this algorithm we assume the vertices of the connected graph G are ordered as $v_1, v_2, \ldots, v_n$. In the algorithm we use the term "process" to describe the procedure of adding new vertices,

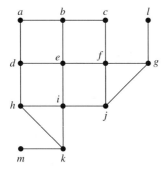

FIGURE 9 **A Graph G.**

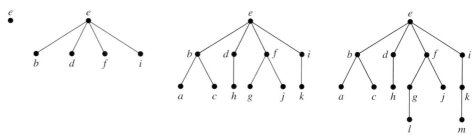

FIGURE 10   A Breadth-First Search of G.

and corresponding edges, to the tree adjacent to the current vertex being processed as long as a simple circuit is not produced.

ALGORITHM 2 Breadth-First Search.

procedure *BFS* (*G*: connected graph with vertices $v_1, v_2, \ldots, v_n$)
$T :=$ tree consisting only of vertex v_1
$L :=$ empty list
put v_1 in the list L of unprocessed vertices
while L is not empty
begin
   remove the first vertex, v, from L
   for each neighbor w of v
     if w is not in L and not in T **then**
     begin
       add w to the end of the list L
       add w and edge $\{v, w\}$ to T
     end
end

We now analyze the computational complexity of breadth-first search. For each vertex v in the graph we examine all vertices adjacent to v and we add each vertex not yet visited to the tree T. Assuming we have the adjacency lists for the graph available, no computation is required to determine which vertices are adjacent to a given vertex. As in the analysis of the depth-first search algorithm, we see that we examine each edge at most twice to determine whether we should add this edge and its endpoint not already in the tree. It follows that the breadth-first search algorithm uses $O(e)$ or $O(n^2)$ steps.

BACKTRACKING APPLICATIONS

There are problems that can be solved only by performing an exhaustive search of all possible solutions. One way to search systematically for a solution is to use a decision tree, where each internal vertex represents a decision and each leaf a possible solution. To find a solution via backtracking, first make a sequence of decisions in an attempt to reach a solution as long as this is possible. The sequence of decisions can be represented by a path in the decision tree. Once it is known that no solution can result from any further sequence of decisions, backtrack to the parent of the current vertex and work toward a

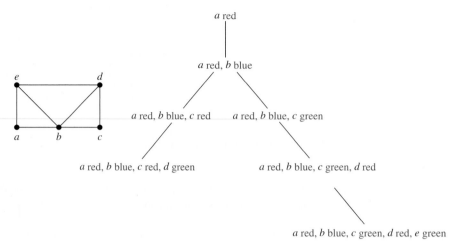

FIGURE 11   Coloring a Graph Using Backtracking.

solution with another series of decisions, if this is possible. The procedure continues until a solution is found, or it is established that no solution exists. The following examples illustrate the usefulness of backtracking.

EXAMPLE 6   **Graph Colorings** How can backtracking be used to decide whether a graph can be colored using n colors?

Solution: We can solve this problem using backtracking in the following way. First pick some vertex a and assign it color 1. Then pick a second vertex b, and if b is not adjacent to a, assign it color 1. Otherwise, assign color 2 to b. Then go on to a third vertex c. Use color 1, if possible, for c. Otherwise use color 2, if this is possible. Only if neither color 1 nor color 2 can be used should color 3 be used. Continue this process as long as it is possible to assign one of the n colors to each additional vertex, always using the first allowable color in the list. If a vertex is reached that cannot be colored by any of the n colors, backtrack to the last assignment made and change the coloring of the last vertex colored, if possible, using the next allowable color in the list. If it is not possible to change this coloring, backtrack further to previous assignments, one step back at a time, until it is possible to change a coloring of a vertex. Then continue assigning colors of additional vertices as long as possible. If a coloring using n colors exists, backtracking will produce it. (Unfortunately this procedure can be extremely inefficient.)

In particular, consider the problem of coloring the graph shown in Figure 11 with three colors. The tree shown in Figure 11 illustrates how backtracking can be used to construct a 3-coloring. In this procedure, red is used first, then blue, and finally green. This simple example can obviously be done without backtracking, but it is a good illustration of the technique.

In this tree, the initial path from the root, which represents the assignment of red to a, leads to a coloring with a red, b blue, c red, and d green. It is impossible to color e using any of the three colors when $a, b, c,$ and d are colored in this way. So, backtrack to the parent of the vertex representing this coloring. Since no other color can be used for d, backtrack one more level. Then change the color of c to green. We obtain a coloring of the graph by then assigning red to d and green to e.   ◀

EXAMPLE 7

The n-Queens Problem The n-queens problem asks how n queens can be placed on an $n \times n$ chessboard so that no two queens can attack one another. How can backtracking be used to solve the n-queens problem?

Solution: To solve this problem we must find n positions on an $n \times n$ chessboard so that no two of these positions are in the same row, same column, or in the same diagonal [a diagonal consists of all positions (i, j) with $i + j = m$ for some m, or $i - j = m$ for some m]. We will use backtracking to solve the n-queens problem. We start with an empty chessboard. At stage $k + 1$ we attempt putting an additional queen on the board in the $(k + 1)$st column, where there are already queens in the first k columns. We examine squares in the $(k + 1)$st column starting with the square in the first row, looking for a position to place this queen so that it is not in the same row or on the same diagonal as a queen already on the board. (We already know it is not in the same column.) If it is impossible to find a position to place the queen in the $(k + 1)$st column, backtrack to the placement of the queen in the kth column, and place this queen in the next allowable row in this column, if such a row exists. If no such row exists, backtrack further.

In particular, Figure 12 displays a backtracking solution to the four-queens problem. In this solution, we place a queen in the first row and column. Then we put a queen in the third row of the second column. However, this makes it impossible to place a queen in the third column. So we backtrack and put a queen in the fourth row of the second column. When we do this, we can place a queen in the second row of the third column. But there is no way to add a queen to the fourth column. This shows that no solution results when a queen is placed in the first row and column. We backtrack to the empty chessboard, and place a queen in the second row of the first column. This leads to a solution as shown in Figure 12. ◄

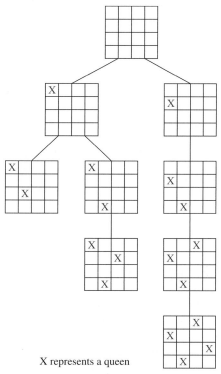

X represents a queen

FIGURE 12   A Backtracking Solution of the Four-Queens Problem.

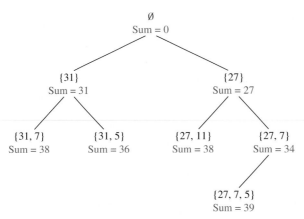

FIGURE 13   Find a Sum Equal to 39 Using Backtracking.

EXAMPLE 8   **Sums of Subsets** Consider this problem. Given a set of positive integers $x_1, x_2, \ldots, x_n$, find a subset of this set of integers that has M as its sum. How can backtracking be used to solve this problem?

Solution: We start with a sum with no terms. We build up the sum by successively adding terms. An integer in the sequence is included if the sum remains less than M when this integer is added to the sum. If a sum is reached so that the addition of any term is greater than M, backtrack by dropping the last term of the sum.

Figure 13 displays a backtracking solution to the problem of finding a subset of $\{31, 27, 15, 11, 7, 5\}$ with sum equal to 39.   ◀

DEPTH-FIRST SEARCH IN DIRECTED GRAPHS

We can easily modify both depth-first search and breadth-first search so that they can run given a directed graph as input. However, the output will not necessarily be a spanning tree, but rather a spanning forest. In both algorithms we can add an edge only when it is directed away from the vertex that is being visited and to a vertex not yet added. If at a stage of either algorithm we find that no edge exists starting at a vertex already added to one not yet added, the next vertex added by the algorithm becomes the root of a new tree in the spanning forest. This is illustrated in Example 9.

EXAMPLE 9   What is the output of depth-first search given the graph G shown in Figure 14(a) as input?

Solution: We begin the depth-first search at vertex a and add vertices b, c, and g and the corresponding edges where we are blocked. We backtrack to c but we are still blocked, and then backtrack to b, where we add vertices f and e and the corresponding edges. Backtracking takes us all the way back to a. We then start a new tree at d and add vertices h, l, k, and j and the corresponding edges. We backtrack to k, then l, then h, and back to d. Finally, we start a new tree at i, completing the depth-first search. The output is shown in Figure 14(b).   ◀

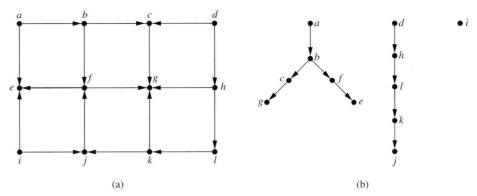

(a)   (b)

FIGURE 14   **A Depth-First Search of a Directed Graph.**

Depth-first search in directed graphs is the basis of many algorithms (see [GrYe99], [Ma89], and [CoLeRiSt01]). It can be used to determine whether a directed graph has a circuit, it can be used to carry out a topological sort of a graph, and it can also be used to find the strongly connected components of a directed graph.

We conclude this section with an application of depth-first search and breadth-first search to search engines on the Web.

EXAMPLE 10   **Web Spiders** To index websites, search engines such as Google™, Hotbot®, and Lycos® systematically explore the Web starting at known sites. These search engines use programs called Web spiders (or crawlers or bots) to visit websites and analyze their contents. Web spiders use both depth-first searching and breadth-first searching to create indices. As described in Example 8 in Section 8.1, Web pages and links between them can be modeled by a directed graph called the Web graph. Web pages are represented by vertices and links are represented by directed edges. Using a depth-first search, an initial Web page is selected, a link is followed to a second Web page (if there is such a link), a link on the second Web page is followed to a third Web page, if there is such a link, and so on, until a page with no new links is found. Backtracking is then used to examine links at the previous level to look for new links, and so on. (Because of practical limitations, Web spiders have limits to the depth they search in a depth-first search.) Using a breadth-first search, an initial Web page is selected and a link on this page is followed to a second Web page, then a second link on the initial page is followed (if it exists), and so on, until all links of the initial page have been followed. Then links on the pages one level down are followed, page by page, and so on. ◀

Exercises

1. How many edges must be removed from a connected graph with n vertices and m edges to produce a spanning tree?

In Exercises 2–6 find a spanning tree for the graph shown by removing edges in simple circuits.

2.
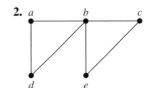

3.

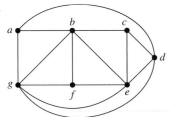

4.

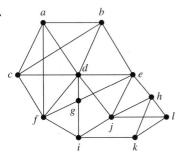

5.

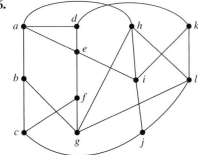

6.

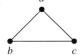

7. Find a spanning tree for each of these graphs.

a) K_5 **b)** $K_{4,4}$ **c)** $K_{1,6}$
d) Q_3 **e)** C_5 **f)** W_5

In Exercises 8–10 draw all the spanning trees of the given simple graphs.

8.

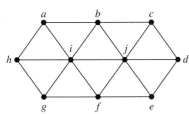

9.

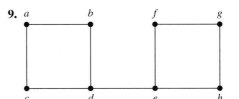

10.

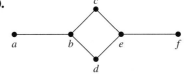

*11. How many different spanning trees does each of these simple graphs have?

a) K_3 **b)** K_4 **c)** $K_{2,2}$ **d)** C_5

*12. How many nonisomorphic spanning trees does each of these simple graphs have?

a) K_3 **b)** K_4 **c)** K_5

In Exercises 13–15 use a depth-first search to produce a spanning tree for the given simple graph. Choose a as the root of this spanning tree and assume that the vertices are ordered alphabetically.

13.

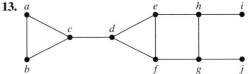

14.

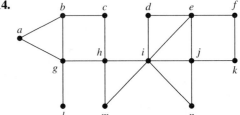

15.

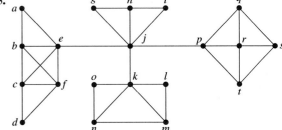

16. Use a breadth-first search to produce a spanning tree for each of the simple graphs in Exercises 13–15. Choose a as the root of each spanning tree.

17. Use depth-first search to find a spanning tree of each of these graphs.

   a) W_6 (see Example 6 of Section 8.2), starting at the vertex of degree 6
   b) K_5
   c) $K_{3,4}$, starting at a vertex of degree 3
   d) Q_3

18. Use breadth-first search to find a spanning tree of each of the graphs in Exercise 17.

19. Describe the trees produced by a breadth-first search and a depth-first search of the wheel graph W_n, starting at the vertex of degree n, where n is an integer with $n \geq 3$. (See Example 6 of Section 8.2.) Justify your answers.

20. Describe the trees produced by a breadth-first search and a depth-first search of the complete graph K_n, where n is a positive integer. Justify your answers.

21. Describe the trees produced by a breadth-first search and a depth-first search of the complete bipartite graph $K_{m,n}$, starting at a vertex of degree m, where m and n are positive integers. Justify your answers.

22. Explain how breadth-first search and how depth-first search can be used to determine whether a graph is bipartite.

23. Suppose that an airline must reduce its flight schedule to save money. If its original routes are as illustrated here, which flights can be discontinued to retain service between all pairs of cities (where it may be necessary to combine flights to fly from one city to another)?

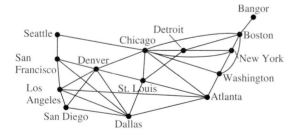

24. When must an edge of a connected simple graph be in every spanning tree for this graph?

25. Which connected simple graphs have exactly one spanning tree?

26. Explain how a breadth-first search or a depth-first search can be used to order the vertices of a connected graph.

*__27.__ Show that the length of the shortest path between vertices v and u in a connected simple graph equals the level number of u in the breadth-first spanning tree of G with root v.

28. Use backtracking to try to find a coloring of each of the graphs in Exercises 7–9 of Section 8.8 using three colors.

29. Use backtracking to solve the n-queens problem for these values of n.

   a) $n = 3$ **b)** $n = 5$ **c)** $n = 6$

30. Use backtracking to find a subset, if it exists, of the set $\{27, 24, 19, 14, 11, 8\}$ with sum

   a) 20.   **b)** 41.   **c)** 60.

31. Explain how backtracking can be used to find a Hamilton path or circuit in a graph.

32. a) Explain how backtracking can be used to find the way out of a maze, given a starting position and the exit position. Consider the maze divided into positions, where at each position the set of available moves includes one to four possibilities (up, down, right, left).

   b) Find a path from the starting position marked by X to the exit in this maze.

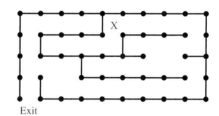

A **spanning forest** of graph G is a forest that contains every vertex of G such that two vertices are in the same tree of the forest when there is a path in G between these two vertices.

33. Show that every finite simple graph has a spanning forest.

34. How many trees are in the spanning forest of a graph?

35. How many edges must be removed to produce the spanning forest of a graph with n vertices, m edges, and c connected components?

36. Devise an algorithm for constructing the spanning forest of a graph based on deleting edges that form simple circuits.

37. Devise an algorithm for constructing the spanning forest of a graph based on depth-first searching.

38. Devise an algorithm for constructing the spanning forest of a graph based on breadth-first searching.

39. Let G be a connected graph. Show that if T is a spanning tree of G constructed using depth-first search, then an edge of G not in T must be a back edge, that is, it must connect a vertex to one of its ancestors or one of its descendants in T.

40. Let G be a connected graph. Show that if T is a spanning tree of G constructed using breadth-first search, then an edge of G not in T must connect vertices at the same level or at levels that differ by 1 in this spanning tree.

41. For which graphs do depth-first search and breadth-first search produce identical spanning trees no matter which vertex is selected as the root of the tree? Justify your answer.

42. Use Exercise 39 to prove that if G is a connected, simple graph with n vertices and G does not contain a simple path of length k then it contains at most $(k-1)n$ edges.

43. Use mathematical induction to prove that breadth-first search visits vertices in order of their level in the resulting spanning tree.

44. Use pseudocode to describe a variation of depth-first search that assigns the integer n to the nth vertex visited in the search. Show that this numbering corresponds to the numbering of the vertices created by a preorder traversal of the spanning tree.

45. Use pseudocode to describe a variation of breadth-first search that assigns the integer m to the mth vertex visited in the search.

*46. Suppose that G is a directed graph and T is a spanning tree constructed using breadth-first search. Show that every edge of G connects two vertices at the same level, a vertex to a vertex at one level lower, or a vertex to a vertex at some higher level.

47. Show that if G is a directed graph and T is a spanning tree constructed using depth-first search, then every edge not in the spanning tree is a **forward edge** connecting an ancestor to a descendant, a **back edge** connecting a descendant to an ancestor, or a **cross edge** connecting a vertex to a vertex in a previously visited subtree.

*48. Describe a variation of depth-first search that assigns the smallest available positive integer to a vertex when the algorithm is totally finished with this vertex. Show that in this numbering, each vertex has a larger number than its children and that the children have increasing numbers from left to right.

Let T_1 and T_2 be spanning trees of a graph. The **distance** between T_1 and T_2 is the number of edges in T_1 and T_2 that are not common to T_1 and T_2.

49. Find the distance between each pair of spanning trees shown in Figures 3(c) and 4 of the graph G shown in Figure 2.

*50. Suppose that T_1, T_2, and T_3 are spanning trees of the simple graph G. Show that the distance between T_1 and T_3 does not exceed the sum of the distance between T_1 and T_2 and the distance between T_2 and T_3.

**51. Suppose that T_1 and T_2 are spanning trees of a simple graph G. Moreover, suppose that e_1 is an edge in T_1 that is not in T_2. Show that there is an edge e_2 in T_2 that is not in T_1 such that T_1 remains a spanning tree if e_1 is removed from it and e_2 is added to it, and T_2 remains a spanning tree if e_2 is removed from it and e_1 is added to it.

*52. Show that it is possible to find a sequence of spanning trees leading from any spanning tree to any other by successively removing one edge and adding another.

A **rooted spanning tree** of a directed graph is a rooted tree containing edges of the graph such that every vertex of the graph is an endpoint of one of the edges in the tree.

53. For each of the directed graphs in Exercises 18–23 of Section 8.5 either find a rooted spanning tree of the graph or determine that no such tree exists.

*54. Show that a connected directed graph in which each vertex has the same in-degree and out-degree has a rooted spanning tree. (*Hint:* Use an Euler circuit.)

*55. Give an algorithm to build a rooted spanning tree for connected directed graphs in which each vertex has the same in-degree and out-degree.

*56. Show that if G is a directed graph and T is a spanning tree constructed using depth-first search, then G contains a circuit if and only if G contains a back edge (see Exercise 47) relative to the spanning tree T.

*57. Use Exercise 56 to construct an algorithm for determining whether a directed graph contains a circuit.

9.5   Minimum Spanning Trees

INTRODUCTION

A company plans to build a communications network connecting its five computer centers. Any pair of these centers can be linked with a leased telephone line. Which links should be made to ensure that there is a path between any two computer centers so that the total cost of the network is minimized? We can model this problem using the weighted graph shown in Figure 1, where vertices represent computer centers, edges represent possible leased lines, and the weights on edges are the monthly lease rates of the lines represented by the edges. We can solve this problem by finding a spanning tree so that the sum of the

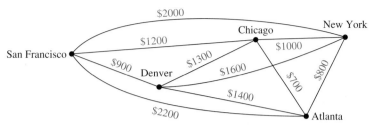

FIGURE 1 **A Weighted Graph Showing Monthly Lease Costs for Lines in a Computer Network.**

weights of the edges of the tree is minimized. Such a spanning tree is called a **minimum spanning tree.**

ALGORITHMS FOR MINIMUM SPANNING TREES

A wide variety of problems are solved by finding a spanning tree in a weighted graph such that the sum of the weights of the edges in the tree is a minimum.

DEFINITION 1

> A *minimum spanning tree* in a connected weighted graph is a spanning tree that has the smallest possible sum of weights of its edges.

We will present two algorithms for constructing minimum spanning trees. Both proceed by successively adding edges of smallest weight from those edges with a specified property that have not already been used. Both are greedy algorithms. Recall from Section 2.10 that a greedy algorithm is a procedure that makes an optimal choice at each of its steps. Optimizing at each step does not guarantee that the optimal overall solution is produced. However, the two algorithms presented in this section for constructing minimum spanning trees are greedy algorithms that do produce optimal solutions.

The first algorithm that we will discuss was given by Robert Prim in 1957, although the basic ideas of this algorithm have an earlier origin. To carry out **Prim's algorithm,** begin by choosing any edge with smallest weight, putting it into the spanning tree. Successively add to the tree edges of minimum weight that are incident to a vertex already in the tree and not forming a simple circuit with those edges already in the tree. Stop when $n - 1$ edges have been added.

Later in this section, we will prove that this algorithm produces a minimum spanning tree for any connected weighted graph. Algorithm 1 gives a pseudocode description of Prim's algorithm.

ROBERT CLAY PRIM (BORN 1921) Robert Prim, born in Sweetwater, Texas, received his B.S. in electrical engineering in 1941 and his Ph.D. in mathematics from Princeton University in 1949. He was an engineer at the General Electric Company from 1941 until 1944, an engineer and mathematician at the United States Naval Ordnance Lab from 1944 until 1949, and a research associate at Princeton University from 1948 until 1949. Among the other positions he has held are director of mathematics and mechanics research at Bell Telephone Laboratories from 1958 until 1961 and vice president of research at Sandia Corporation. He is currently retired.

ALGORITHM 1 Prim's Algorithm.

---

procedure *Prim*(*G*: weighted connected undirected graph with *n* vertices)
T := a minimum-weight edge
for *i* := 1 **to** *n* − 2
begin
    e := an edge of minimum weight incident to a vertex in *T* and not
        forming a simple circuit in *T* if added to *T*
    T := *T* with *e* added
end {*T* is a minimum spanning tree of *G*}

Note that the choice of an edge to add at a stage of the algorithm is not determined when there is more than one edge with the same weight that satisfies the appropriate criteria. We need to order the edges to make the choices deterministic. We will not worry about this in the remainder of the section. Also note that there may be more than one minimum spanning tree for a given connected weighted simple graph. (See Exercise 9.) Examples 1 and 2 illustrate how Prim's algorithm is used.

EXAMPLE 1 Use Prim's algorithm to design a minimum-cost communications network connecting all the computers represented by the graph in Figure 1.

Solution: We solve this problem by finding a minimum spanning tree in the graph in Figure 1. Prim's algorithm is carried out by choosing an initial edge of minimum weight and successively adding edges of minimum weight that are incident to a vertex in the tree and that do not form simple circuits. The edges in color in Figure 2 show a minimum spanning tree produced by Prim's algorithm, with the choice made at each step displayed. ◄

EXAMPLE 2 Use Prim's algorithm to find a minimum spanning tree in the graph shown in Figure 3.

Solution: A minimum spanning tree constructed using Prim's algorithm is shown in Figure 4. The successive edges chosen are displayed. ◄

The second algorithm we will discuss was discovered by Joseph Kruskal in 1956, although the basic ideas it uses were described much earlier. To carry out **Kruskal's algorithm,** choose an edge in the graph with minimum weight.

Successively add edges with minimum weight that do not form a simple circuit with those edges already chosen. Stop after *n* − 1 edges have been selected.

The proof that Kruskal's algorithm produces a minimum spanning tree for every connected weighted graph is left as an exercise at the end of this section. Pseudocode for Kruskal's algorithm is given in Algorithm 2.

JOSEPH BERNARD KRUSKAL (BORN 1928) Joseph Kruskal, born in New York City, attended the University of Chicago and received his Ph.D. from Princeton University in 1954. He was an instructor in mathematics at Princeton and at the University of Wisconsin, and later he was an assistant professor at the University of Michigan. In 1959 he became a member of the technical staff at Bell Laboratories, a position he continues to hold. His current research interests include statistical linguistics and psychometrics. Besides his work on minimum spanning trees, Kruskal is also known for contributions to multidimensional scaling. Kruskal discovered his algorithm for producing minimum spanning trees when he was a second-year graduate student. He was not sure his $2\frac{1}{2}$-page paper on this subject was worthy of publication, but was convinced by others to submit it.

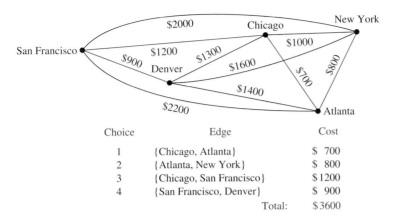

Choice	Edge	Cost
1	{Chicago, Atlanta}	$ 700
2	{Atlanta, New York}	$ 800
3	{Chicago, San Francisco}	$1200
4	{San Francisco, Denver}	$ 900
	Total:	$3600

FIGURE 2 **A Minimum Spanning Tree for the Weighted Graph in Figure 1.**

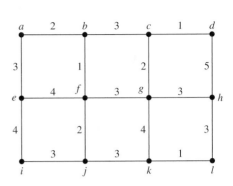

FIGURE 3 **A Weighted Graph.**

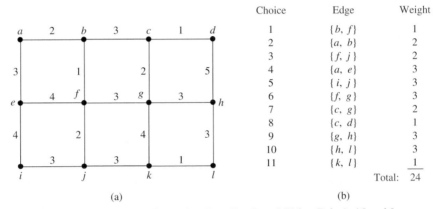

(a)

Choice	Edge	Weight
1	{b, f}	1
2	{a, b}	2
3	{f, j}	2
4	{a, e}	3
5	{i, j}	3
6	{f, g}	3
7	{c, g}	2
8	{c, d}	1
9	{g, h}	3
10	{h, l}	3
11	{k, l}	1
	Total:	24

(b)

FIGURE 4 **A Minimum Spanning Tree Produced Using Prim's Algorithm.**

ALGORITHM 2 Kruskal's Algorithm.

procedure *Kruskal*(*G*: weighted connected undirected graph with *n* vertices)
T := empty graph
for *i* := 1 **to** *n* − 1
begin
    e := any edge in *G* with smallest weight that does not form a simple circuit when added to *T*
    T := *T* with *e* added
end {*T* is a minimum spanning tree of *G*}

HISTORICAL NOTE Joseph Kruskal and Robert Prim developed their algorithms for constructing minimum spanning trees in the mid-1950s. However, they were not the first people to discover such algorithms. For example, the work of the anthropologist Jan Czekanowski, in 1909, contains many of the ideas required to find minimum spanning trees. In 1926, Otakar Boruvka described methods for constructing minimum spanning trees in work relating to the construction of electric power networks.

The reader should note the difference between Prim's and Kruskal's algorithms. In Prim's algorithm edges of minimum weight that are incident to a vertex already in the tree, and not forming a circuit, are chosen; whereas in Kruskal's algorithm edges of minimum weight that are not necessarily incident to a vertex already in the tree, and that do not form a circuit, are chosen. Note that as in Prim's algorithm, if the edges are not ordered, there may be more than one choice for the edge to add at a stage of this procedure. Consequently, the edges need to be ordered for the procedure to be deterministic. Example 3 illustrates how Kruskal's algorithm is used.

EXAMPLE 3   Use Kruskal's algorithm to find a minimum spanning tree in the weighted graph shown in Figure 3.

Solution: A minimum spanning tree and the choices of edges at each stage of Kruskal's algorithm are shown in Figure 5.   ◀

We will now prove that Prim's algorithm produces a minimum spanning tree of a connected weighted graph.

Proof: Let G be a connected weighted graph. Suppose that the successive edges chosen by Prim's algorithm are $e_1, e_2, \ldots, e_{n-1}$. Let S be the tree with $e_1, e_2, \ldots, e_{n-1}$ as its edges, and let S_k be the tree with $e_1, e_2, \ldots, e_k$ as its edges. Let T be a minimum spanning tree of G containing the edges $e_1, e_2, \ldots, e_k$, where k is the maximum integer with the property that a minimum spanning tree exists containing the first k edges chosen by Prim's algorithm. The theorem follows if we can show that $S = T$.

Suppose that $S \neq T$, so that $k < n - 1$. Consequently, T contains $e_1, e_2, \ldots, e_k$, but not e_{k+1}. Consider the graph made up of T together with e_{k+1}. Since this graph is connected and has n edges, too many edges to be a tree, it must contain a simple circuit. This simple circuit must contain e_{k+1} since there was no simple circuit in T. Furthermore, there must be an edge in the simple circuit that does not belong to S_{k+1} since S_{k+1} is a tree. By starting at an endpoint of e_{k+1} that is also an endpoint of one of the edges $e_1, \ldots, e_k$, and following the circuit until it reaches an edge not in S_{k+1}, we can find an edge e not in S_{k+1} that has an endpoint that is also an endpoint of one of the edges $e_1, e_2, \ldots, e_k$. By deleting e from T and adding e_{k+1}, we obtain a tree T' with $n - 1$ edges (it is a tree since it has no simple circuits). Note that the tree T' contains $e_1, e_2, \ldots, e_k, e_{k+1}$. Fur-

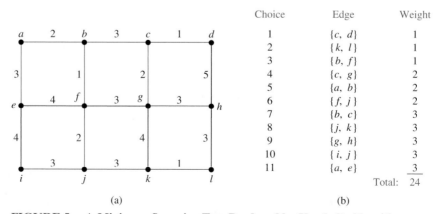

Choice	Edge	Weight
1	$\{c, d\}$	1
2	$\{k, l\}$	1
3	$\{b, f\}$	1
4	$\{c, g\}$	2
5	$\{a, b\}$	2
6	$\{f, j\}$	2
7	$\{b, c\}$	3
8	$\{j, k\}$	3
9	$\{g, h\}$	3
10	$\{i, j\}$	3
11	$\{a, e\}$	3
	Total:	24

(a)   (b)

FIGURE 5   **A Minimum Spanning Tree Produced by Kruskal's Algorithm.**

thermore, since e_{k+1} was chosen by Prim's algorithm at the $(k+1)$st step, and e was also available at that step, the weight of e_{k+1} is less than or equal to the weight of e. From this observation it follows that T' is also a minimum spanning tree, since the sum of the weights of its edges does not exceed the sum of the weights of the edges of T. This contradicts the choice of k as the maximum integer so that a minimum spanning tree exists containing $e_1, \ldots, e_k$. Hence, $k = n - 1$, and $S = T$. It follows that Prim's algorithm produces a minimum spanning tree. ◁

It can be shown (see [CoLeRiSt01]) that to find a minimum spanning tree of a graph with e edges and v vertices, Kruskal's algorithm can be carried out using $O(e \log e)$ operations and Prim's algorithm can be carried out using $O(e \log v)$ operations. Consequently, it is preferable to use Kruskal's algorithm for graphs that are **sparse,** that is, where e is very small compared to $C(v, 2) = v(v - 1)/2$, the total number of possible edges in an undirected graph with v vertices. Otherwise, there is little difference in the complexity of these two algorithms.

Exercises

1. The roads represented by this graph are all unpaved. The lengths of the roads between pairs of towns are represented by edge weights. Which roads should be paved so that there is a path of paved roads between each pair of towns so that a minimum road length is paved? (*Note:* These towns are in Nevada.)

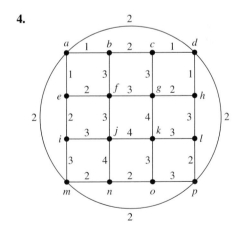

In Exercises 2–4 use Prim's algorithm to find a minimum spanning tree for the given weighted graph.

2.

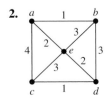

3.

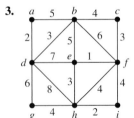

4.

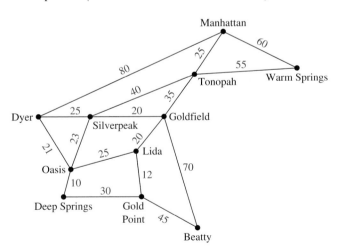

5. Use Kruskal's algorithm to design the communications network described at the beginning of the section.

6. Use Kruskal's algorithm to find a minimum spanning tree for the weighted graph in Exercise 2.

7. Use Kruskal's algorithm to find a minimum spanning tree for the weighted graph in Exercise 3.

8. Use Kruskal's algorithm to find a minimum spanning tree for the weighted graph in Exercise 4.

9. Find a connected weighted simple graph with the fewest edges possible that has more than one minimum spanning tree.

10. A **minimum spanning forest** in a weighted graph is a spanning forest with minimal weight. Explain how Prim's and Kruskal's algorithms can be adapted to construct minimum spanning forests.

A **maximum spanning tree** of a connected weighted undirected graph is a spanning tree with the largest possible weight.

11. Devise an algorithm similar to Prim's algorithm for constructing a maximum spanning tree of a connected weighted graph.

12. Devise an algorithm similar to Kruskal's algorithm for constructing a maximum spanning tree of a connected weighted graph.

13. Find a maximum spanning tree for the weighted graph in Exercise 2.

14. Find a maximum spanning tree for the weighted graph in Exercise 3.

15. Find a maximum spanning tree for the weighted graph in Exercise 4.

16. Find the second least expensive communications network connecting the five computer centers in the problem posed at the beginning of the section.

*17. Devise an algorithm for finding the second shortest spanning tree in a connected weighted graph.

*18. Show that an edge with smallest weight in a connected weighted graph must be part of any minimum spanning tree.

19. Show that there is a unique minimum spanning tree in a connected weighted graph if the weights of the edges are all different.

20. Suppose that the computer network connecting the cities in Figure 1 must contain a direct link between New York and Denver. What other links should be included so that there is a link between every two computer centers and the cost is minimized?

21. Find a spanning tree with minimal total weight containing the edges $\{e, i\}$ and $\{g, k\}$ in the weighted graph in Figure 3.

22. Describe an algorithm for finding a spanning tree with minimal weight containing a specified set of edges in a connected weighted undirected simple graph.

23. Express the algorithm devised in Exercise 22 in pseudocode.

Sollin's algorithm produces a minimum spanning tree from a connected weighted simple graph $G = (V, E)$ by successively adding groups of edges. Suppose that the vertices in V are ordered. This produces an ordering of the edges where $\{u_0, v_0\}$ precedes $\{u_1, v_1\}$ if u_0 precedes u_1 or if $u_0 = u_1$ and v_0 precedes v_1. The algorithm begins by simultaneously choosing the edge of least weight incident to each vertex. The first edge in the ordering is taken in the case of ties. This produces a graph with no simple circuits, that is, a forest of trees (Exercise 24 asks for a proof of this fact). Next, simultaneously choose for each tree in the forest the shortest edge between a vertex in this tree and a vertex in a different tree. Again the first edge in the ordering is chosen in the case of ties. (This produces a graph with no simple circuits containing fewer trees than were present before this step; see Exercise 24.) Continue the process of simultaneously adding edges connecting trees until $n - 1$ edges have been chosen. At this stage a minimum spanning tree has been constructed.

*24. Show that the addition of edges at each stage of Sollin's algorithm produces a forest.

25. Use Sollin's algorithm to produce a minimum spanning tree for the weighted graph shown in

   a) Figure 1.
   b) Figure 3.

*26. Express Sollin's algorithm in pseudocode.

**27. Prove that Sollin's algorithm produces a minimum spanning tree in a connected undirected weighted graph.

*28. Show that the first step of Sollin's algorithm produces a forest containing at least $\lceil n/2 \rceil$ edges.

*29. Show that if there are r trees in the forest at some intermediate step of Sollin's algorithm, then at least $\lceil r/2 \rceil$ edges are added by the next iteration of the algorithm.

*30. Show that no more than $\lfloor n/2^k \rfloor$ trees remain after the first step of Sollin's algorithm has been carried out and the second step of the algorithm has been carried out $k - 1$ times.

*31. Show that Sollin's algorithm requires at most $\log n$ iterations to produce a minimum spanning tree from a connected undirected weighted graph with n vertices.

32. Prove that Kruskal's algorithm produces minimum spanning trees.

Key Terms and Results

TERMS

tree: a connected undirected graph with no simple circuits

forest: an undirected graph with no simple circuits

rooted tree: a directed graph with a specified vertex, called the root, such that there is a unique path to any other vertex from this root

subtree: a subgraph of a tree that is also a tree

parent of v in a rooted tree: the vertex u such that (u, v) is an edge of the rooted tree

child of a vertex *v* in a rooted tree: any vertex with *v* as its parent

sibling of a vertex *v* in a rooted tree: a vertex with the same parent as *v*

ancestor of a vertex *v* in a rooted tree: any vertex on the path from the root to *v*

descendant of a vertex *v* in a rooted tree: any vertex that has *v* as an ancestor

internal vertex: a vertex that has children

leaf: a vertex with no children

level of a vertex: the length of the path from the root to this vertex

height of a tree: the largest level of the vertices of a tree

m-ary tree: a tree with the property that every internal vertex has no more than m children

full m-ary tree: a tree with the property that every internal vertex has exactly m children

binary tree: an m-ary tree with $m = 2$ (each child may be designated as a left or a right child of its parent)

ordered tree: a tree in which the children of each internal vertex are linearly ordered

balanced tree: a tree in which every vertex is at level h or $h - 1$, where h is the height of the tree

binary search tree: a binary tree in which the vertices are labeled with items so that a label of a vertex is greater than the labels of all vertices in the left subtree of this vertex and is less than the labels of all vertices in the right subtree of this vertex

decision tree: a rooted tree where each vertex represents a possible outcome of a decision and the leaves represent the possible solutions

prefix code: a code that has the property that the code of a character is never a prefix of the code of another character

minmax strategy: the strategy where the first player and second player move to positions represented by a child with maximum and minimum value, respectively

value of a vertex in a game tree: for a leaf, the payoff to the first player when the game terminates in the position represented by this leaf; for an internal vertex, the maximum or minimum of the values of its children, for an internal vertex at an even or odd level, respectively

tree traversal: a listing of the vertices of a tree

preorder traversal: a listing of the vertices of an ordered rooted tree defined recursively—the root is listed, followed by the first subtree, followed by the other subtrees in the order they occur from left to right

inorder traversal: a listing of the vertices of an ordered rooted tree defined recursively—the first subtree is listed, followed by the root, followed by the other subtrees in the order they occur from left to right

postorder traversal: a listing of the vertices of an ordered rooted tree defined recursively—the subtrees are listed in the order they occur from left to right, followed by the root

infix notation: the form of an expression (including a full set of parentheses) obtained from an inorder traversal of the binary tree representing this expression

prefix (or Polish) notation: the form of an expression obtained from a preorder traversal of the tree representing this expression

postfix (or reverse Polish) notation: the form of an expression obtained from a postorder traversal of the tree representing this expression

spanning tree: a tree containing all vertices of a graph

minimum spanning tree: a spanning tree with smallest possible sum of weights of its edges

greedy algorithm: an algorithm that optimizes by making the optimal choice at each step

RESULTS

A graph is a tree if and only if there is a unique simple path between any of its vertices.

A tree with n vertices has $n - 1$ edges.

A full m-ary tree with i internal vertices has $mi + 1$ vertices.

The relationships among the numbers of vertices, leaves, and internal vertices in a full m-ary tree (see Theorem 4 in Section 9.1).

There are at most m^h leaves in an m-ary tree of height h.

If an m-ary tree has l leaves, its height h is at least $\lceil \log_m l \rceil$. If the tree is also full and balanced, then its height is $\lceil \log_m l \rceil$.

Huffman coding: a procedure for constructing an optimal binary code for a set of symbols, given the frequencies of these symbols

depth-first search, or backtracking: a procedure for constructing a spanning tree by adding edges that form a path until this is not possible, and then moving back up the path until a vertex is found where a new path can be formed

breadth-first search: a procedure for constructing a spanning tree that successively adds all edges incident to the last set of edges added, unless a simple circuit is formed

Prim's algorithm: a procedure for producing a minimum spanning tree in a weighted graph that successively adds edges with minimal weight among all edges incident to a vertex already in the tree such that no edge produces a simple circuit when it is added

Kruskal's algorithm: a procedure for producing a minimum spanning tree in a weighted graph that successively adds edges of least weight that are not already in the tree such that no edge produces a simple circuit when it is added

Review Questions

1. a) Define a tree.   **b)** Define a forest.

2. Can there be two different simple paths between the vertices of a tree?

3. Give at least three examples of how trees are used in modeling.

4. a) Define a rooted tree and the root of such a tree.

b) Define the parent of a vertex and a child of a vertex in a rooted tree.

c) What are an internal vertex, a leaf, and a subtree in a rooted tree?

d) Draw a rooted tree with at least 10 vertices, where the degree of each vertex does not exceed 3. Identify the root, the parent of each vertex, the children of each vertex, the internal vertices, and the leaves.

5. a) How many edges does a tree with n vertices have?

b) What do you need to know to determine the number of edges in a forest with n vertices?

6. a) Define a full m-ary tree.

b) How many vertices does a full m-ary tree have if it has i internal vertices? How many leaves does the tree have?

7. a) What is the height of a rooted tree?

b) What is a balanced tree?

c) How many leaves can an m-ary tree of height h have?

8. a) What is a binary search tree?

b) Describe an algorithm for constructing a binary search tree.

c) Form a binary search tree for the words *vireo, warbler, egret, grosbeak, nuthatch,* and *kingfisher*.

9. a) What is a prefix code?

b) How can a prefix code be represented by a binary tree?

10. a) Define preorder, inorder, and postorder tree traversal.

b) Give an example of preorder, postorder, and inorder traversal of a binary tree of your choice with at least 12 vertices.

11. a) Explain how to use preorder, inorder, and postorder traversals to find the prefix, infix, and postfix forms of an arithmetic expression.

b) Draw the ordered rooted tree that represents $((x - 3) + ((x/4) + (x - y) \uparrow 3))$.

c) Find the prefix and postfix forms of the expression in part (b).

12. Show that the number of comparisons used by a sorting algorithm is at least $\lceil \log n! \rceil$.

13. a) Describe the Huffman coding algorithm for constructing an optimal code for a set of symbols, given the frequency of these symbols.

b) Use Huffman coding to find an optimal code for these symbols and frequencies: A: 0.2, B: 0.1, C: 0.3, D: 0.4.

14. Draw the game tree for nim if the starting position consists of two piles with one and four stones, respectively. Who wins the game if both players follow an optimal strategy?

15. a) What is a spanning tree of a simple graph?

b) Which simple graphs have spanning trees?

c) Describe at least two different applications that require that a spanning tree of a simple graph be found.

16. a) Describe two different algorithms for finding a spanning tree in a simple graph.

b) Illustrate how the two algorithms you described in (a) can be used to find the spanning tree of a simple graph, using a graph of your choice with at least eight vertices and 15 edges.

17. a) Explain how backtracking can be used to determine whether a simple graph can be colored using n colors.

b) Show, with an example, how backtracking can be used to show that a graph with a chromatic number equal to 4 cannot be colored with three colors, but can be colored with four colors.

18. a) What is a minimum spanning tree of a connected weighted graph?

b) Describe at least two different applications that require that a minimum spanning tree of a connected weighted graph be found.

19. a) Describe Kruskal's algorithm and Prim's algorithm for finding minimum spanning trees.

b) Illustrate how Kruskal's algorithm and Prim's algorithm are used to find a minimum spanning tree, using a weighted graph with at least eight vertices and 15 edges.

Supplementary Exercises

*1. Show that a simple graph is a tree if and only if it contains no simple circuits and the addition of an edge connecting two nonadjacent vertices produces a new graph that has exactly one simple circuit (where circuits that contain the same edges are not considered different).

*2. How many nonisomorphic rooted trees are there with six vertices?

3. Show that every tree with at least one edge must have at least two pendant vertices.

4. Show that a tree with n vertices that has $n - 1$ pendant vertices must be isomorphic to $K_{1,n-1}$.

5. What is the sum of the degrees of the vertices of a tree with n vertices?

*6. Suppose that $d_1, d_2, \ldots, d_n$ are n positive integers with sum $2n - 2$. Show that there is a tree that has n vertices so that the degrees of these vertices are $d_1, d_2, \ldots, d_n$.

7. Show that every tree is a planar graph.

8. Show that every tree is bipartite.

9. Show that every forest can be colored using two colors.

A **B-tree of degree** k is a rooted tree such that all its leaves are at the same level, its root has at least two and at most k children unless it is a leaf, and every internal vertex other than the root has at least $\lceil k/2 \rceil$, but no more than k, children. Computer files can be accessed efficiently when B-trees are used to represent them.

10. Draw three different B-trees of degree 3 with height 4.

*11. Give an upper bound and a lower bound for the number of leaves in a B-tree of degree k with height h.

*12. Give an upper bound and a lower bound for the height of a B-tree of degree k with n leaves.

The **binomial trees** $B_i, i = 0, 1, 2, \ldots$, are ordered rooted trees defined recursively:

Basis step: The binomial tree B_0 is the tree with a single vertex.

Recursive step: Let k be a nonnegative integer. To construct the binomial tree B_{k+1}, add a copy of B_k to a second copy of B_k by adding an edge that makes the root of the first copy of B_k the leftmost child of the second copy of B_k.

13. Draw B_k for $k = 0, 1, 2, 3, 4$.

14. How many vertices does B_k have? Prove that your answer is correct.

15. Find the height of B_k. Prove that your answer is correct.

16. How many vertices are there in B_k at depth j, where $0 \le j \le k$? Justify your answer.

17. What is the degree of the root of B_k? Prove that your answer is correct.

18. Show that the vertex of largest degree in B_k is the root.

A rooted tree T is called an S_k**-tree** if it satisfies this recursive definition. It is an S_0-tree if it has one vertex. For $k > 0$, T is an S_k-tree if it can be built from two S_{k-1}-trees by making the root of one the root of the S_k-tree and making the root of the other the child of the root of the first S_k-tree.

19. Draw an S_k-tree for $k = 0, 1, 2, 3, 4$.

20. Show that an S_k-tree has 2^k vertices and a unique vertex at level k. This vertex at level k is called the **handle.**

*21. Suppose that T is an S_k-tree with handle v. Show that T can be obtained from disjoint trees $T_0, T_1, \ldots, T_{k-1}$, where v is not in any of these trees, where T_i is an S_i-tree for $i = 0, 1, \ldots, k - 1$, by connecting v to r_0 and r_i to r_{i+1} for $i = 0, 1, \ldots, k - 2$.

The listing of the vertices of an ordered rooted tree in **level order** begins with the root, followed by the vertices at level 1 from left to right, followed by the vertices at level 2 from left to right, and so on.

22. List the vertices of the ordered rooted trees in Figures 3 and 9 of Section 9.3 in level order.

23. Devise an algorithm for listing the vertices of an ordered rooted tree in level order.

*24. Devise an algorithm for determining if a set of universal addresses can be the addresses of the leaves of a rooted tree.

25. Devise an algorithm for constructing a rooted tree from the universal addresses of its leaves.

A **cut set** of a graph is a set of edges such that the removal of these edges produces a subgraph with more connected components than in the original graph, but no proper subset of this set of edges has this property.

26. Show that a cut set of a graph must have at least one edge in common with any spanning tree of this graph.

A **cactus** is a connected graph in which no edge is in more than one simple circuit not passing through any vertex other than its initial vertex more than once or its initial vertex other than at its terminal vertex (where two circuits that contain the same edges are not considered different).

27. Which of these graphs are cacti?

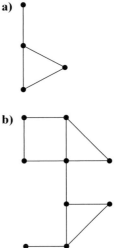

c)

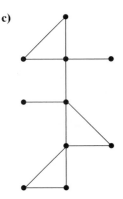

28. Is a tree necessarily a cactus?

29. Show that a cactus is formed if we add a circuit containing new edges beginning and ending at a vertex of a tree.

*30. Show that if every circuit not passing through any vertex other than its initial vertex more than once in a connected graph contains an odd number of edges, then this graph must be a cactus.

A **degree-constrained spanning tree** of a simple graph G is a spanning tree with the property that the degree of a vertex in this tree cannot exceed some specified bound. Degree-constrained spanning trees are useful in models of transportation systems where the number of roads at an intersection is limited, models of communications networks where the number of links entering a node is limited, and so on.

In Exercises 31–33 find a degree-constrained spanning tree of the given graph where each vertex has degree less than or equal to 3, or show that such a spanning tree does not exist.

31.

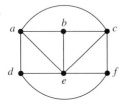

32.

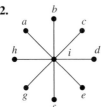

33.

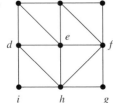

34. Show that a degree-constrained spanning tree of a simple graph in which each vertex has degree not exceeding 2 consists of a single Hamilton path in the graph.

35. A tree with n vertices is called **graceful** if its vertices can be labeled with the integers $1, 2, \ldots, n$ such that the absolute values of the difference of the labels of adjacent vertices are all different. Show that these trees are graceful.

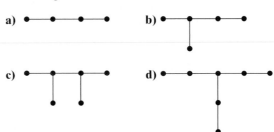

A **caterpillar** is a tree that contains a simple path such that every vertex not contained in this path is adjacent to a vertex in the path.

36. Which of the graphs in Exercise 35 are caterpillars?

37. How many nonisomorphic caterpillars are there with six vertices?

**38. a) Prove or disprove that all trees whose edges form a single path are graceful.

  b) Prove or disprove that all caterpillars are graceful.

39. Suppose that in a long bit string the frequency of occurrence of a 0 bit is 0.9 and the frequency of a 1 bit is 0.1 and bits occur independently.

  a) Construct a Huffman code for the four blocks of two bits, 00, 01, 10, and 11. What is the average number of bits required to encode a bit string using this code?

  b) Construct a Huffman code for the eight blocks of three bits. What is the average number of bits required to encode a bit string using this code?

40. Suppose that G is a directed graph with no circuits. Describe how depth-first search can be used to carry out a topological sort of the vertices of G.

*41. Suppose that e is an edge in a weighted graph that is incident to a vertex v so that the weight of e does not exceed the weight of any other edge incident to v. Show that there exists a minimum spanning tree containing this edge.

42. Three couples arrive at the bank of a river. Each of the wives is jealous and does not trust her husband when he is with one of the other wives (and perhaps with other people), but not with her. How can six people cross to the other side of the river using a boat that can hold no more than two people so that no husband is alone with a woman other than his wife? Use a graph theory model.

*43. Show that if no two edges in a weighted graph have the same weight, then the edge with least weight incident to a vertex v is included in every minimum spanning tree.

44. Find a minimum spanning tree of each of these graphs where the degree of each vertex in the spanning tree does not exceed 2.

a)

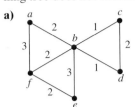

b)

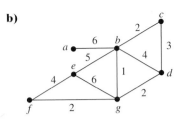

Computer Projects

WRITE PROGRAMS WITH THESE INPUT AND OUTPUT.

1. Given the adjacency matrix of an undirected simple graph, determine whether the graph is a tree.
2. Given the adjacency matrix of a rooted tree and a vertex in the tree, find the parent, children, ancestors, descendants, and level of this vertex.
3. Given the list of edges of a rooted tree and a vertex in the tree, find the parent, children, ancestors, descendants, and level of this vertex.
4. Given a list of items, construct a binary search tree containing these items.
5. Given a binary search tree and an item, locate or add this item to the binary search tree.
6. Given the ordered list of edges of an ordered rooted tree, find the universal addresses of its vertices.
7. Given the ordered list of edges of an ordered rooted tree, list its vertices in preorder, inorder, and postorder.
8. Given an arithmetic expression in prefix form, find its value.
9. Given an arithmetic expression in postfix form, find its value.
10. Given the frequency of symbols, use Huffman coding to find an optimal code for these symbols.

11. Given an initial position in the game of nim, determine an optimal strategy for the first player.
12. Given the adjacency matrix of a connected undirected simple graph, find a spanning tree for this graph using a depth-first search.
13. Given the adjacency matrix of a connected undirected simple graph, find a spanning tree for this graph using a breadth-first search.
14. Given a set of positive integers and a positive integer N, use backtracking to find a subset of these integers that have N as their sum.
*15. Given the adjacency matrix of an undirected simple graph, use backtracking to color the graph with three colors, if this is possible.
*16. Given a positive integer n, solve the n-queens problem using backtracking.
17. Given the list of edges and their weights of a weighted undirected connected graph, use Prim's algorithm to find a minimum spanning tree of this graph.
18. Given the list of edges and their weights of a weighted undirected connected graph, use Kruskal's algorithm to find a minimum spanning tree of this graph.

Computations and Explorations

USE A COMPUTATIONAL PROGRAM OR PROGRAMS YOU HAVE WRITTEN TO DO THESE EXERCISES.

1. Display all trees with six vertices.
2. Display a full set of nonisomorphic trees with seven vertices.
*3. Construct a Huffman code for the symbols with ASCII codes given the frequency of their occurrence in representative input.

4. Compute the number of different spanning trees of K_n for $n = 1, 2, 3, 4, 5, 6$. Conjecture a formula for the number of such spanning trees whenever n is a positive integer.
5. Compare the number of comparisons needed to sort lists of n elements for $n = 100, 1000,$ and $10,000$

where the elements are randomly selected positive integers, using the selection sort, the insertion sort, the merge sort, and the quick sort.

6. Compute the number of different ways n queens can be arranged on an $n \times n$ chessboard so that no two queens can attack each other for all positive integers n not exceeding 10.

*7. Find a minimum spanning tree of the graph that connects the capital cities of the 50 states in the United States to each other where the weight of each edge is the distance between the cities.

8. Draw the complete game tree for a game of checkers on a 4×4 board.

Writing Projects

RESPOND TO THESE WITH ESSAYS USING OUTSIDE SOURCES.

1. Explain how Cayley used trees to enumerate the number of certain types of hydrocarbons.

2. Define *AVL-trees* (sometimes also known as *height-balanced trees*). Describe how and why AVL-trees are used in a variety of different algorithms.

3. Define *quad trees* and explain how images can be represented using them. Describe how images can be rotated, scaled, and translated by manipulating the corresponding quad tree.

4. Define a *heap* and explain how trees can be turned into heaps. Why are heaps useful in sorting?

5. Describe dynamic algorithms for data compression based on letter frequencies as they change as characters are successively read, such as adaptive Huffman coding.

6. Explain how *alpha-beta pruning* can be used to simplify the computation of the value of a game tree.

7. Describe the techniques used by chess-playing programs such as Deep Blue.

8. Define the type of graph known as a *mesh of trees*.

Explain how this graph is used in applications to very large system integration and parallel computing.

9. Discuss the algorithms used in IP multicasting to avoid loops between routers.

10. Describe an algorithm based on depth-first search for finding the articulation points of a graph.

11. Describe an algorithm based on depth-first search to find the strongly connected components of a directed graph.

12. Describe the search techniques used by the crawlers and spiders in different search engines on the Web.

13. Describe an algorithm for finding the minimum spanning tree of a graph so that the maximum degree of any vertex in the spanning tree does not exceed a fixed constant k.

14. Compare and contrast some of the most important sorting algorithms in terms of their complexity and when they are used.

15. Discuss the history and origins of algorithms for constructing minimum spanning trees.

16. Describe algorithms for producing random trees.

Boolean Algebra

The circuits in computers and other electronic devices have inputs, each of which is either a 0 or a 1, and produce outputs that are also 0s and 1s. Circuits can be constructed using any basic element that has two different states. Such elements include switches that can be in either the on or the off position and optical devices that can either be lit or unlit. In 1938 Claude Shannon showed how the basic rules of logic, first given by George Boole in 1854 in his *The Laws of Thought,* could be used to design circuits. These rules form the basis for Boolean algebra. In this chapter we develop the basic properties of Boolean algebra. The operation of a circuit is defined by a Boolean function that specifies the value of an output for each set of inputs. The first step in constructing a circuit is to represent its Boolean function by an expression built up using the basic operations of Boolean algebra. We will provide an algorithm for producing such expressions. The expression that we obtain may contain many more operations than are necessary to represent the function. Later in the chapter we will describe methods for finding an expression with the minimum number of sums and products that represents a Boolean function. The procedures that we will develop, Karnaugh maps and the Quine–McCluskey method, are important in the design of efficient circuits.

10.1 Boolean Functions

INTRODUCTION

Boolean algebra provides the operations and the rules for working with the set $\{0, 1\}$. Electronic and optical switches can be studied using this set and the rules of Boolean algebra. The three operations in Boolean algebra that we will use most are complementation, the Boolean sum, and the Boolean product. The **complement** of an element, denoted with a bar, is defined by $\overline{0} = 1$ and $\overline{1} = 0$. The Boolean sum, denoted by $+$ or by *OR*, has the following values:

$$1 + 1 = 1, \qquad 1 + 0 = 1, \qquad 0 + 1 = 1, \qquad 0 + 0 = 0.$$

The Boolean product, denoted by $\cdot$ or by *AND,* has the following values:

$$1 \cdot 1 = 1, \qquad 1 \cdot 0 = 0, \qquad 0 \cdot 1 = 0, \qquad 0 \cdot 0 = 0.$$

When there is no danger of confusion, the symbol $\cdot$ can be deleted, just as in writing algebraic products. Unless parentheses are used, the rules of precedence for Boolean operators are: first, all complements are computed, followed by Boolean products, followed by all Boolean sums. This is illustrated in Example 1.

EXAMPLE 1   Find the value of $1 \cdot 0 + \overline{(0 + 1)}$.

Solution: Using the definitions of complementation, the Boolean sum, and the Boolean product, it follows that

$$(1 \cdot 0) + \overline{(0 + 1)} = 0 + \overline{1}$$
$$= 0 + 0$$
$$= 0. \qquad ◄$$

The complement, Boolean sum, and Boolean product correspond to the logical operators, ¬, ∨, and ∧, respectively, where 0 corresponds to F (false) and 1 corresponds to T (true). The results of Boolean algebra can be directly translated into results about propositions. Conversely, results about propositions can be translated into statements about Boolean algebra.

BOOLEAN EXPRESSIONS AND BOOLEAN FUNCTIONS

Let $B = \{0, 1\}$. Then $B^n = \{(x_1, x_2, \ldots, x_n) \mid x_i \in B \text{ for } 1 \leq i \leq n\}$ is the set of all possible n-tuples of 0s and 1s. The variable x is called a **Boolean variable** if it assumes values only from B, that is, if its only possible values are 0 and 1. A function from B^n to B is called a **Boolean function of degree n.**

EXAMPLE 2   The function $F(x, y) = x\overline{y}$ from the set of ordered pairs of Boolean variables to the set $\{0, 1\}$ is a Boolean function of degree 2 with $F(1, 1) = 0$, $F(1, 0) = 1$, $F(0, 1) = 0$, and $F(0, 0) = 0$. We display these values of F in Table 1.   ◄

Boolean functions can be represented using expressions made up from variables and Boolean operations. The **Boolean expressions** in the variables $x_1, x_2, \ldots, x_n$ are defined

TABLE 1

x	y	$F(x, y)$
1	1	0
1	0	1
0	1	0
0	0	0

Links

CLAUDE ELWOOD SHANNON (1916–2001)   Claude Shannon was born in Petoskey, Michigan, and grew up in Gaylord, Michigan. His father was a businessman and a probate judge, and his mother was a language teacher and a high school principal. Shannon attended the University of Michigan, graduating in 1936. He continued his studies at M.I.T., where he took the job of maintaining the differential analyzer, a mechanical computing device consisting of shafts and gears built by his professor, Vannevar Bush. Shannon's master's thesis, written in 1936, studied the logical aspects of the differential analyzer. This master's thesis presents the first application of Boolean algebra to the design of switching circuits; it is perhaps the most famous master's thesis of the twentieth century. He received his Ph.D. from M.I.T. in 1940. Shannon joined Bell Laboratories in 1940, where he worked on transmitting data efficiently. He was one of the first people to use bits to represent information. At Bell Laboratories he worked on determining the amount of traffic that telephone lines can carry. Shannon made many fundamental contributions to information theory. In the early 1950s he was one of the founders of the study of artificial intelligence. He joined the M.I.T. faculty in 1956, where he continued his study of information theory.

Shannon had an unconventional side. He is credited with inventing the rocket-powered Frisbee. He is also famous for riding a unicycle down the hallways of Bell Laboratories while juggling four balls. Shannon retired when he was 50 years old, publishing papers sporadically over the following 10 years. In his later years he concentrated on some pet projects, such as building a motorized pogo stick. One interesting quote from Shannon, published in *Omni Magazine* in 1987, is "I visualize a time when we will be to robots what dogs are to humans. And I am rooting for the machines."

TABLE 2					
x	y	z	xy	$\overline{z}$	$F(x, y, z) = xy + \overline{z}$
1	1	1	1	0	1
1	1	0	1	1	1
1	0	1	0	0	0
1	0	0	0	1	1
0	1	1	0	0	0
0	1	0	0	1	1
0	0	1	0	0	0
0	0	0	0	1	1

recursively as

$0, 1, x_1, x_2, \ldots, x_n$ are Boolean expressions;

if E_1 and E_2 are Boolean expressions, then $\overline{E}_1$, $(E_1 E_2)$, and $(E_1 + E_2)$ are Boolean expressions.

Each Boolean expression represents a Boolean function. The values of this function are obtained by substituting 0 and 1 for the variables in the expression. In Section 10.2 we will show that every Boolean function can be represented by a Boolean expression.

EXAMPLE 3   Find the values of the Boolean function represented by $F(x, y, z) = xy + \overline{z}$.

Solution: The values of this function are displayed in Table 2.   ◀

Note that we can represent a Boolean function graphically by distinguishing the vertices of the n-cube that correspond to the n-tuples of bits where the function has value 1.

EXAMPLE 4   The function $F(x, y, z) = xy + \overline{z}$ from B^3 to B from Example 3 can be represented by distinguishing the vertices that correspond to the five 3-tuples $(1, 1, 1)$, $(1, 1, 0)$, $(1, 0, 0)$, $(0, 1, 0)$, and $(0, 0, 0)$, where $F(x, y, z) = 1$, as shown in Figure 1. These vertices are displayed using solid black circles.   ◀

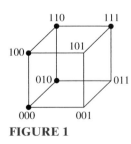
FIGURE 1

Boolean functions F and G of n variables are equal if and only if $F(b_1, b_2, \ldots, b_n) = G(b_1, b_2, \ldots, b_n)$ whenever $b_1, b_2, \ldots, b_n$ belong to B. Two different Boolean expressions that represent the same function are called **equivalent.** For instance, the Boolean expressions xy, $xy + 0$, and $xy \cdot 1$ are equivalent. The **complement** of the Boolean function F is the function $\overline{F}$, where $\overline{F}(x_1, \ldots, x_n) = \overline{F(x_1, \ldots, x_n)}$. Let F and G be Boolean functions of degree n. The **Boolean sum** $F + G$ and the **Boolean product** FG are defined by

$$(F + G)(x_1, \ldots, x_n) = F(x_1, \ldots, x_n) + G(x_1, \ldots, x_n),$$
$$(FG)(x_1, \ldots, x_n) = F(x_1, \ldots, x_n)G(x_1, \ldots, x_n).$$

A Boolean function of degree two is a function from a set with four elements, namely, pairs of elements from $B = \{0, 1\}$, to B, a set with two elements. Hence, there are 16

TABLE 3 The Boolean Functions of Degree Two.

x	y	F_1	F_2	F_3	F_4	F_5	F_6	F_7	F_8	F_9	F_{10}	F_{11}	F_{12}	F_{13}	F_{14}	F_{15}	F_{16}
1	1	1	1	1	1	1	1	1	1	0	0	0	0	0	0	0	0
1	0	1	1	1	1	0	0	0	0	1	1	1	1	0	0	0	0
0	1	1	1	0	0	1	1	0	0	1	1	0	0	1	1	0	0
0	0	1	0	1	0	1	0	1	0	1	0	1	0	1	0	1	0

different Boolean functions of degree two. In Table 3 we display the values of the 16 different Boolean functions of degree two, labeled $F_1, F_2, \ldots, F_{16}$.

EXAMPLE 5   How many different Boolean functions of degree n are there?

Solution: From the product rule for counting, it follows that there are 2^n different n-tuples of 0s and 1s. Since a Boolean function is an assignment of 0 or 1 to each of these 2^n different n-tuples, the product rule shows that there are 2^{2^n} different Boolean functions of degree n.   ◄

Table 4 displays the number of different Boolean functions of degrees one through six. The number of such functions grows extremely rapidly.

IDENTITIES OF BOOLEAN ALGEBRA

There are many identities in Boolean algebra. The most important of these are displayed in Table 5. (The reader should compare these identities to the logical equivalences in Table 5 of Section 1.2 and the set identities in Table 1 in Section 1.7. All are special cases of the same set of identities in a more abstract structure.) These identities are particularly useful in simplifying the design of circuits. Each of the identities in Table 5 can be proved using a table. We will prove one of the distributive laws in this way in the following example. The proofs of the remaining properties are left as exercises for the reader.

EXAMPLE 6   Show that the distributive law $x(y + z) = xy + xz$ is valid.

TABLE 4 The Number of Boolean Functions of Degree n.

Degree	*Number*
1	4
2	16
3	256
4	65,536
5	4,294,967,296
6	18,446,744,073,709,551,616

TABLE 5 **Boolean Identities.**	
Identity	**Name**
$\overline{\overline{x}} = x$	Law of the double complement
$x + x = x$ $x \cdot x = x$	Idempotent laws
$x + 0 = x$ $x \cdot 1 = x$	Identity laws
$x + 1 = 1$ $x \cdot 0 = 0$	Domination laws
$x + y = y + x$ $xy = yx$	Commutative laws
$x + (y + z) = (x + y) + z$ $x(yz) = (xy)x$	Associative laws
$x + yz = (x + y)(x + z)$ $x(y + z) = xy + xz$	Distributive laws
$\overline{(xy)} = \overline{x} + \overline{y}$ $\overline{(x + y)} = \overline{x}\,\overline{y}$	De Morgan's laws
$x + xy = x$ $x(x + y) = x$	Absorption laws
$x + \overline{x} = 1$	Unit property
$x\overline{x} = 0$	Zero property

Solution: The verification of this identity is shown in Table 6. The identity holds because the last two columns of the table agree. ◄

Identities in Boolean algebra can be used to prove further identities. We demonstrate this in Example 7.

TABLE 6							
x	y	z	$y + z$	xy	xz	$x(y + z)$	$xy + xz$
1	1	1	1	1	1	1	1
1	1	0	1	1	0	1	1
1	0	1	1	0	1	1	1
1	0	0	0	0	0	0	0
0	1	1	1	0	0	0	0
0	1	0	1	0	0	0	0
0	0	1	1	0	0	0	0
0	0	0	0	0	0	0	0

EXAMPLE 7

Prove the **absorption law** $x(x + y) = x$ using the other identities of Boolean algebra shown in Table 5. (This is called an absorption law since absorbing $x + y$ into x leaves x unchanged.)

Solution: The steps used to derive this identity and the law used in each step follow:

$$x(x + y) = (x + 0)(x + y) \quad \text{Identity law for the Boolean sum}$$
$$= x + 0 \cdot y \qquad\qquad \text{Distributive law of the Boolean sum over the Boolean product}$$
$$= x + y \cdot 0 \qquad\qquad \text{Commutative law for the Boolean product}$$
$$= x + 0 \qquad\qquad\quad \text{Domination law for the Boolean product}$$
$$= x. \qquad\qquad\qquad \text{Identity law for the Boolean sum} \qquad \blacktriangleleft$$

DUALITY

The identities in Table 5 come in pairs (except for the law of the double complement and the unit and zero properties). To explain the relationship between the two identities in each pair we use the concept of a dual. The **dual** of a Boolean expression is obtained by interchanging Boolean sums and Boolean products and interchanging 0s and 1s.

EXAMPLE 8
Find the duals of $x(y + 0)$ and $\overline{x} \cdot 1 + (\overline{y} + z)$.

Solution: Interchanging $\cdot$ signs and $+$ signs and interchanging 0s and 1s in these expressions produces their duals. The duals are $x + (y \cdot 1)$ and $(\overline{x} + 0)(\overline{y}z)$, respectively. $\qquad \blacktriangleleft$

The dual of a Boolean function F represented by a Boolean expression is the function represented by the dual of this expression. This dual function, denoted by F^d, does not depend on the particular Boolean expression used to represent F. An identity between functions represented by Boolean expressions remains valid when the duals of both sides of the identity are taken. (See Exercise 28 for the reason this is true.) This result, called the **duality principle,** is useful for obtaining new identities.

EXAMPLE 9
Construct an identity from the absorption law $x(x + y) = x$ by taking duals.

Solution: Taking the duals of both sides of this identity produces the identity $x + xy = x$, which is also called an absorption law and is shown in Table 5. $\qquad \blacktriangleleft$

THE ABSTRACT DEFINITION OF A BOOLEAN ALGEBRA

In this section we have focused on Boolean functions and expressions. However, the results we have established can be translated into results about propositions or results about sets. Because of this, it is useful to define Boolean algebras abstractly. Once it is shown that a particular structure is a Boolean algebra, then all results established about Boolean algebras in general apply to this particular structure.

Boolean algebras can be defined in several ways. The most common way is to specify the properties that operations must satisfy, as is done in Definition 1.

DEFINITION 1

A *Boolean algebra* is a set B with two binary operations $\vee$ and $\wedge$, elements 0 and 1, and a unary operation $^-$ such that these properties hold for all x, y, and z in B:

$$\left.\begin{array}{l} x \vee 0 = x \\ x \wedge 1 = x \end{array}\right\} \text{ Identity laws}$$

$$\left.\begin{array}{l} x \vee \overline{x} = 1 \\ x \wedge \overline{x} = 0 \end{array}\right\} \text{ Complement laws}$$

$$\left.\begin{array}{l} (x \vee y) \vee z = x \vee (y \vee z) \\ (x \wedge y) \wedge z = x \wedge (y \wedge z) \end{array}\right\} \text{ Associative laws}$$

$$\left.\begin{array}{l} x \vee y = y \vee x \\ x \wedge y = y \wedge x \end{array}\right\} \text{ Commutative laws}$$

$$\left.\begin{array}{l} x \vee (y \wedge z) = (x \vee y) \wedge (x \vee z) \\ x \wedge (y \vee z) = (x \wedge y) \vee (x \wedge z) \end{array}\right\} \text{ Distributive laws}$$

Using the laws given in Definition 1, it is possible to prove many other laws that hold for every Boolean algebra, such as idempotent and domination laws. (See Exercises 31–38.)

From our previous discussion, $B = \{0, 1\}$ with the *OR* and *AND* operations and the complement operator, satisfies all these properties. The set of propositions in n variables, with the $\vee$ and $\wedge$ operators, **F** and **T,** and the negation operator, also satisfies all the properties of a Boolean algebra, as can be seen from Table 5 in Section 1.2. Similarly, the set of subsets of a universal set U with the union and intersection operations, the empty set and the universal set, and the set complementation operator, is a Boolean algebra as can be seen by consulting Table 1 in Section 1.7. So, to establish results about each of Boolean expressions, propositions, and sets, we need only prove results about abstract Boolean algebras.

Boolean algebras may also be defined using the notion of a lattice, discussed in Chapter 7. Recall that a lattice L is a partially ordered set in which every pair of elements x, y has a least upper bound, denoted by $\text{lub}(x, y)$ and a greatest lower bound denoted by $\text{glb}(x, y)$. Given two elements x and y of L, we can define two operations $\vee$ and $\wedge$ on pairs of elements of L by $x \vee y = \text{lub}(x, y)$ and $x \wedge y = \text{glb}(x, y)$.

For a lattice L to be a Boolean algebra as specified in Definition 1, it must have two properties. First, it must be **complemented.** For a lattice to be complemented it must have a least element 0 and a greatest element 1 and for every element x of the lattice there must exist an element $\overline{x}$ such that $x \vee \overline{x} = 1$ and $x \wedge \overline{x} = 0$. Second, it must be **distributive.** This means that for every x, y, and z in $L, x \vee (y \wedge z) = (x \vee y) \wedge (x \vee z)$ and $x \wedge (y \vee z) = (x \wedge y) \vee (x \wedge z)$. Showing that a complemented, distributive lattice is a Boolean algebra is left as Exercise 39 at the end of this section.

Exercises

1. Find the values of these expressions.
  a) $1 \cdot \overline{0}$ **b)** $1 + \overline{1}$ **c)** $\overline{0} \cdot 0$ **d)** $\overline{(1 + 0)}$

2. Find the values, if any, of the Boolean variable x that satisfy these equations.

a) $x \cdot 1 = 0$
b) $x + x = 0$
c) $x \cdot 1 = x$
d) $x \cdot \overline{x} = 1$

3. Use a table to express the values of each of these Boolean functions.

a) $F(x, y, z) = \overline{x}y$
b) $F(x, y, z) = x + yz$
c) $F(x, y, z) = x\overline{y} + \overline{(xyz)}$
d) $F(x, y, z) = x(yz + \overline{y}\,\overline{z})$

4. Use a table to express the values of each of these Boolean functions.

a) $F(x, y, z) = \overline{z}$
b) $F(x, y, z) = \overline{x}y + \overline{y}z$
c) $F(x, y, z) = x\overline{y}z + \overline{(xyz)}$
d) $F(x, y, z) = \overline{y}(xz + \overline{x}\,\overline{z})$

5. Use a 3-cube Q_3 to represent each of the Boolean functions in Exercise 3 by displaying a black circle at each vertex that corresponds to a 3-tuple where this function has the value 1.

6. Use a 3-cube Q_3 to represent each of the Boolean functions in Exercise 4 by displaying a black circle at each vertex that corresponds to a 3-tuple where this function has the value 1.

7. What values of the Boolean variables x and y satisfy $xy = x + y$?

8. How many different Boolean functions are there of degree 7?

9. Prove the absorption law $x + xy = x$ using the other laws in Table 5.

☞ 10. Show that $F(x, y, z) = xy + xz + yz$ has the value 1 if and only if at least two of the variables x, y, and z have the value 1.

11. Show that $x\overline{y} + y\overline{z} + \overline{x}z = \overline{x}y + \overline{y}z + x\overline{z}$.

Exercises 12–21 deal with the Boolean algebra $\{0, 1\}$ with addition, multiplication, and complement defined at the beginning of this section. In each case, use a table as in Example 6.

12. Verify the law of the double complement.
13. Verify the idempotent laws.
14. Verify the identity laws.
15. Verify the domination laws.
16. Verify the commutative laws.
17. Verify the associative laws.
18. Verify the first distributive law in Table 5.
19. Verify De Morgan's laws.
20. Verify the unit property.
21. Verify the zero property.

The Boolean operator $\oplus$, called the *XOR* operator, is defined by $1 \oplus 1 = 0, 1 \oplus 0 = 1, 0 \oplus 1 = 1$, and $0 \oplus 0 = 0$.

22. Simplify these expressions.

a) $x \oplus 0$ b) $x \oplus 1$
c) $x \oplus x$ d) $x \oplus \overline{x}$

23. Show that these identities hold.

a) $x \oplus y = (x + y)\overline{(xy)}$
b) $x \oplus y = (x\overline{y}) + (\overline{x}y)$

24. Show that $x \oplus y = y \oplus x$.

25. Prove or disprove these equalities.

a) $x \oplus (y \oplus z) = (x \oplus y) \oplus z$
b) $x + (y \oplus z) = (x + y) \oplus (x + z)$
c) $x \oplus (y + z) = (x \oplus y) + (x \oplus z)$

26. Find the duals of these Boolean expressions.

a) $x + y$
b) $\overline{x}\,\overline{y}$
c) $xyz + \overline{x}\,\overline{y}\,\overline{z}$
d) $x\overline{z} + x \cdot 0 + \overline{x} \cdot 1$

*27. Suppose that F is a Boolean function represented by a Boolean expression in the variables $x_1, \ldots, x_n$. Show that $F^d(x_1, \ldots, x_n) = \overline{F(\overline{x}_1, \ldots, \overline{x}_n)}$.

*28. Show that if F and G are Boolean functions represented by Boolean expressions in n variables and $F = G$, then $F^d = G^d$, where F^d and G^d are the Boolean functions represented by the duals of the Boolean expressions representing F and G, respectively. (*Hint:* Use the result of Exercise 27.)

*29. How many different Boolean functions $F(x, y, z)$ are there so that $F(\overline{x}, \overline{y}, \overline{z}) = F(x, y, z)$ for all values of the Boolean variables x, y, and z?

*30. How many different Boolean functions $F(x, y, z)$ are there so that $F(\overline{x}, y, z) = F(x, \overline{y}, z) = F(x, y, \overline{z})$ for all values of the Boolean variables x, y, and z?

In Exercises 31–38, use the laws in Definition 1 to show that the stated properties hold in every Boolean algebra.

31. Show that in a Boolean algebra, the **idempotent laws** $x \vee x = x$ and $x \wedge x = x$ hold for every element x.

32. Show that in a Boolean algebra, every element x has a unique complement $\overline{x}$ such that $x \vee \overline{x} = 1$ and $x \wedge \overline{x} = 0$.

33. Show that in a Boolean algebra, the complement of the element 0 is the element 1 and vice versa.

34. Prove that in a Boolean algebra, the **law of the double complement** holds; that is, $\overline{\overline{x}} = x$ for every element x.

35. Show that **De Morgan's laws** hold in a Boolean algebra. That is, show that for all x and y, $\overline{(x \vee y)} = \overline{x} \wedge \overline{y}$ and $\overline{(x \wedge y)} = \overline{x} \vee \overline{y}$.

36. Show that in a Boolean algebra, the **modular properties** hold. That is, show that $x \wedge (y \vee (x \wedge z)) = (x \wedge y) \vee (x \wedge z)$ and $x \vee (y \wedge (x \vee z)) = (x \vee y) \wedge (x \vee z)$.

37. Show that in a Boolean algebra, if $x \vee y = 0$, then $x = 0$ and $y = 0$ and that if $x \wedge y = 1$, then $x = 1$ and $y = 1$.

38. Show that in a Boolean algebra, the **dual** of an identity, obtained by interchanging the $\vee$ and $\wedge$ operators and interchanging the elements 0 and 1, is also a valid identity.

39. Show that a complemented, distributive lattice is a Boolean algebra.

10.2 Representing Boolean Functions

Two important problems of Boolean algebra will be studied in this section. The first problem is: Given the values of a Boolean function, how can a Boolean expression that represents this function be found? This problem will be solved by showing that any Boolean function can be represented by a Boolean sum of Boolean products of the variables and their complements. The solution of this problem shows that every Boolean function can be represented using the three Boolean operators $\cdot$, $+$, and $^{-}$. The second problem is: Is there a smaller set of operators that can be used to represent all Boolean functions? We will answer this problem by showing that all Boolean functions can be represented using only one operator. Both of these problems have practical importance in circuit design.

SUM-OF-PRODUCTS EXPANSIONS

We will use examples to illustrate one important way to find a Boolean expression that represents a Boolean function.

EXAMPLE 1   Find Boolean expressions that represent the functions $F(x, y, z)$ and $G(x, y, z)$, which are given in Table 1.

TABLE 1

x	y	z	F	G
1	1	1	0	0
1	1	0	0	1
1	0	1	1	0
1	0	0	0	0
0	1	1	0	0
0	1	0	0	1
0	0	1	0	0
0	0	0	0	0

Solution: An expression that has the value 1 when $x = z = 1$ and $y = 0$, and the value 0 otherwise, is needed to represent F. Such an expression can be formed by taking the Boolean product of x, $\overline{y}$, and z. This product, $x\overline{y}z$, has the value 1 if and only if $x = \overline{y} = z = 1$, which holds if and only if $x = z = 1$ and $y = 0$.

To represent G, we need an expression that equals 1 when $x = y = 1$ and $z = 0$, or when $x = z = 0$ and $y = 1$. We can form an expression with these values by taking the Boolean sum of two different Boolean products. The Boolean product $xy\overline{z}$ has the value 1 if and only if $x = y = 1$ and $z = 0$. Similarly, the product $\overline{x}y\overline{z}$ has the value 1 if and only if $x = z = 0$ and $y = 1$. The Boolean sum of these two products, $xy\overline{z} + \overline{x}y\overline{z}$, represents G, since it has the value 1 if and only if $x = y = 1$ and $z = 0$ or $x = z = 0$ and $y = 1$.   ◀

Example 1 illustrates a procedure for constructing a Boolean expression representing a function with given values. Each combination of values of the variables for which the function has the value 1 leads to a Boolean product of the variables or their complements.

DEFINITION 1   A *literal* is a Boolean variable or its complement. A *minterm* of the Boolean variables $x_1, x_2, \ldots, x_n$ is a Boolean product $y_1 y_2 \cdots y_n$, where $y_i = x_i$ or $y_i = \overline{x}_i$. Hence, a minterm is a product of n literals, with one literal for each variable.

A minterm has the value 1 for one and only one combination of values of its variables. More precisely, the minterm $y_1 y_2 \cdots y_n$ is 1 if and only if each y_i is 1, and this occurs if and only if $x_i = 1$ when $y_i = x_i$ and $x_i = 0$ when $y_i = \overline{x}_i$.

EXAMPLE 2 Find a minterm that equals 1 if $x_1 = x_3 = 0$ and $x_2 = x_4 = x_5 = 1$, and equals 0 otherwise.

Solution: The minterm $\overline{x}_1 x_2 \overline{x}_3 x_4 x_5$ has the correct set of values. ◀

By taking Boolean sums of distinct minterms we can build up a Boolean expression with a specified set of values. In particular, a Boolean sum of minterms has the value 1 when exactly one of the minterms in the sum has the value 1. It has the value 0 for all other combinations of values of the variables. Consequently, given a Boolean function, a Boolean sum of minterms can be formed that has the value 1 when this Boolean function has the value 1, and has the value 0 when the function has the value 0. The minterms in this Boolean sum correspond to those combinations of values for which the function has the value 1. The sum of minterms that represents the function is called the **sum-of-products expansion** or the **disjunctive normal form** of the Boolean function.

EXAMPLE 3 Find the sum-of-products expansion for the function $F(x, y, z) = (x + y)\overline{z}$.

Solution: We will find the sum-of-products expansion of $F(x, y, z)$ in two ways. First, we will use Boolean identities to expand the product and simplify. We find that

$$
\begin{aligned}
F(x, y, z) &= (x + y)\overline{z} \\
&= x\overline{z} + y\overline{z} && \text{Distributive law} \\
&= x1\overline{z} + 1y\overline{z} && \text{Identity law} \\
&= x(y + \overline{y})\overline{z} + (x + \overline{x})y\overline{z} && \text{Unit property} \\
&= xy\overline{z} + x\overline{y}\,\overline{z} + xy\overline{z} + \overline{x}y\overline{z} && \text{Distributive law} \\
&= xy\overline{z} + x\overline{y}\,\overline{z} + \overline{x}y\,\overline{z}. && \text{Idempotent law}
\end{aligned}
$$

Second, we can construct the sum-of-products expansion by determining the values of F for all possible values of the variables x, y, and z. These values are found in Table 2. The sum-of-products expansion of F is the Boolean sum of three minterms corresponding to

TABLE 2					
x	y	z	$x + y$	$\overline{z}$	$(x + y)\overline{z}$
1	1	1	1	0	0
1	1	0	1	1	1
1	0	1	1	0	0
1	0	0	1	1	1
0	1	1	1	0	0
0	1	0	1	1	1
0	0	1	0	0	0
0	0	0	0	1	0

the three rows of this table that give the value 1 for the function. This gives

$$F(x, y, z) = xy\overline{z} + x\overline{y}\,\overline{z} + \overline{x}yz.$$ ◄

It is also possible to find a Boolean expression that represents a Boolean function by taking a Boolean product of Boolean sums. The resulting expansion is called the **conjunctive normal form** or **product-of-sums expansion** of the function. These expansions can be found from sum-of-products expansions by taking duals. How to find such expansions directly is described in Exercise 10 at the end of this section.

FUNCTIONAL COMPLETENESS

Every Boolean function can be expressed as a Boolean sum of minterms. Each minterm is the Boolean product of Boolean variables or their complements. This shows that every Boolean function can be represented using the Boolean operators $\cdot$, $+$, and $^-$. Since every Boolean function can be represented using these operators we say that the set $\{\cdot, +, ^-\}$ is **functionally complete.** Can we find a smaller set of functionally complete operators? We can do so if one of the three operators of this set can be expressed in terms of the other two. This can be done using one of De Morgan's laws. We can eliminate all Boolean sums using the identity

$$x + y = \overline{\overline{x}\,\overline{y}},$$

which is obtained by taking complements of both sides in the second De Morgan's law, given in Table 5 in Section 10.1, and then applying the double complementation law. This means that the set $\{\cdot, ^-\}$ is functionally complete. Similarly, we could eliminate all Boolean products using the identity

$$xy = \overline{\overline{x} + \overline{y}},$$

which is obtained by taking complements of both sides in the first De Morgan's law, given in Table 5 in Section 10.1, and then applying the double complementation law. Consequently $\{+, ^-\}$ is functionally complete. Note that the set $\{+, \cdot\}$ is not functionally complete, since it is impossible to express the Boolean function $F(x) = \overline{x}$ using these operators (see Exercise 19).

We have found sets containing two operators that are functionally complete. Can we find a smaller set of functionally complete operators, namely, a set containing just one operator? Such sets exist. Define two operators, the $|$ or **NAND** operator, defined by $1 \mid 1 = 0$ and $1 \mid 0 = 0 \mid 1 = 0 \mid 0 = 1$; and the $\downarrow$ or **NOR** operator, defined by $1 \downarrow 1 = 1 \downarrow 0 = 0 \downarrow 1 = 0$ and $0 \downarrow 0 = 1$. Both of the sets $\{\mid\}$ and $\{\downarrow\}$ are functionally complete. To see that $\{\mid\}$ is functionally complete, since $\{\cdot, ^-\}$ is functionally complete, all that we have to do is show that both of the operators $\cdot$ and $^-$ can be expressed using just the $|$ operator. This can be done as

$$\overline{x} = x \mid x,$$
$$xy = (x \mid y) \mid (x \mid y).$$

The reader should verify these identities (see Exercise 14). We leave the demonstration that $\{\downarrow\}$ is functionally complete for the reader (see Exercises 15 and 16).

Exercises

1. Find a Boolean product of the Boolean variables x, y, and z, or their complements, that has the value 1 if and only if

a) $x = y = 0, z = 1$.　　**b)** $x = 0, y = 1, z = 0$.
c) $x = 0, y = z = 1$.　　**d)** $x = y = z = 0$.

2. Find the sum-of-products expansions of these Boolean functions.

a) $F(x, y) = \bar{x} + y$　　**b)** $F(x, y) = x\,\bar{y}$
c) $F(x, y) = 1$　　　　**d)** $F(x, y) = \bar{y}$

3. Find the sum-of-products expansions of these Boolean functions.

a) $F(x, y, z) = x + y + z$
b) $F(x, y, z) = (x + z)y$
c) $F(x, y, z) = x$
d) $F(x, y, z) = x\,\bar{y}$

4. Find the sum-of-products expansions of the Boolean function $F(x, y, z)$ that equals 1 if and only if

a) $x = 0$.　　　　　**b)** $xy = 0$.
c) $x + y = 0$.　　　**d)** $xyz = 0$.

5. Find the sum-of-products expansion of the Boolean function $F(w, x, y, z)$ that has the value 1 if and only if an odd number of w, x, y, and z have the value 1.

6. Find the sum-of-products expansion of the Boolean function $F(x_1, x_2, x_3, x_4, x_5)$ that has the value 1 if and only if three or more of the variables x_1, x_2, x_3, x_4, and x_5 have the value 1.

Another way to find a Boolean expression that represents a Boolean function is to form a Boolean product of Boolean sums of literals. Exercises 7–11 are concerned with representations of this kind.

7. Find a Boolean sum containing either x or $\bar{x}$, either y or $\bar{y}$, and either z or $\bar{z}$ that has the value 0 if and only if

a) $x = y = 1, z = 0$.　　**b)** $x = y = z = 0$.
c) $x = z = 0, y = 1$.

8. Find a Boolean product of Boolean sums of literals that has the value 0 if and only if either $x = y = 1$

and $z = 0$, $x = z = 0$ and $y = 1$, or $x = y = z = 0$. [*Hint:* Take the Boolean product of the Boolean sums found in parts (a), (b), and (c) in Exercise 7.]

9. Show that the Boolean sum $y_1 + y_2 + \cdots + y_n$, where $y_i = x_i$ or $y_i = \bar{x}_i$, has the value 0 for exactly one combination of the values of the variables, namely, when $x_i = 0$ if $y_i = x_i$ and $x_i = 1$ if $y_i = \bar{x}_i$. This Boolean sum is called a **maxterm.**

10. Show that a Boolean function can be represented as a Boolean product of maxterms. This representation is called the **product-of-sums expansion or conjunctive normal form** of the function. (*Hint:* Include one maxterm in this product for each combination of the variables where the function has the value 0.)

11. Find the product-of-sums expansion of each of the Boolean functions in Exercise 3.

12. Express each of these Boolean functions using the operators $\cdot$ and $^-$.

a) $x + y + z$　　　　**b)** $x + \bar{y}(\bar{x} + z)$
c) $\overline{(x + \bar{y})}$　　　　**d)** $\bar{x}(x + \bar{y} + \bar{z})$

13. Express each of the Boolean functions in Exercise 12 using the operators $+$ and $^-$.

14. Show that

a) $\bar{x} = x \mid x$.　　　　**b)** $xy = (x \mid y) \mid (x \mid y)$.
c) $x + y = (x \mid x) \mid (y \mid y)$.

15. Show that

a) $\bar{x} = x \downarrow x$.
b) $xy = (x \downarrow x) \downarrow (y \downarrow y)$.
c) $x + y = (x \downarrow y) \downarrow (x \downarrow y)$.

16. Show that $\{ \downarrow \}$ is functionally complete using Exercise 15.

17. Express each of the Boolean functions in Exercise 3 using the operator $\mid$.

18. Express each of the Boolean functions in Exercise 3 using the operator $\downarrow$.

19. Show that the set of operators $\{+, \cdot\}$ is not functionally complete.

20. Are these sets of operators functionally complete?
a) $\{+, \oplus\}$　　**b)** $\{^-, \oplus\}$　　**c)** $\{\cdot, \oplus\}$

10.3 Logic Gates

INTRODUCTION

Links

Boolean algebra is used to model the circuitry of electronic devices. Each input and each output of such a device can be thought of as a member of the set $\{0, 1\}$. A computer, or other electronic device, is made up of a number of circuits. Each circuit can be designed

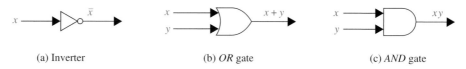

(a) Inverter   (b) *OR* gate   (c) *AND* gate

FIGURE 1   **Basic Types of Gates.**

FIGURE 2   **Gates with *n* Inputs.**

using the rules of Boolean algebra that were studied in Sections 10.1 and 10.2. The basic elements of circuits are called **gates.** Each type of gate implements a Boolean operation. In this section we define several types of gates. Using these gates, we will apply the rules of Boolean algebra to design circuits that perform a variety of tasks. The circuits that we will study in this chapter give output that depends only on the input, and not on the current state of the circuit. In other words, these circuits have no memory capabilities. Such circuits are called **combinational circuits** or **gating networks.**

We will construct combinational circuits using three types of elements. The first is an **inverter,** which accepts the value of one Boolean variable as input and produces the complement of this value as its output. The symbol used for an inverter is shown in Figure 1(a). The input to the inverter is shown on the left side entering the element, and the output is shown on the right side leaving the element.

The next type of element we will use is the ***OR* gate.** The inputs to this gate are the values of two or more Boolean variables. The output is the Boolean sum of their values. The symbol used for an *OR* gate is shown in Figure 1(b). The inputs to the *OR* gate are shown on the left side entering the element, and the output is shown on the right side leaving the element.

The third type of element we will use is the ***AND* gate.** The inputs to this gate are the values of two or more Boolean variables. The output is the Boolean product of their values. The symbol used for an *AND* gate is shown in Figure 1(c). The inputs to the *AND* gate are shown on the left side entering the element, and the output is shown on the right side leaving the element.

We will permit multiple inputs to *AND* and *OR* gates. The inputs to each of these gates are shown on the left side entering the element, and the output is shown on the right side. Examples of *AND* and *OR* gates with *n* inputs are shown in Figure 2.

COMBINATIONS OF GATES

Combinational circuits can be constructed using a combination of inverters, *OR* gates, and *AND* gates. When combinations of circuits are formed, some gates may share inputs. This is shown in one of two ways in depictions of circuits. One method is to use branchings that indicate all the gates that use a given input. The other method is to indicate this input separately for each gate. Figure 3 illustrates the two ways of showing gates with the same input values. Note also that output from a gate may be used as input by one or more other elements, as shown in Figure 3. Both drawings in Figure 3 depict the circuit that produces the output $xy + \overline{x}y$.

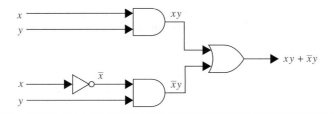

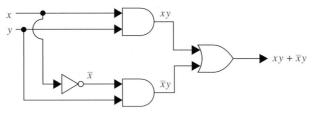

FIGURE 3   Two Ways to Draw the Same Circuit.

EXAMPLE 1   Construct circuits that produce the following outputs: (a) $(x + y)\bar{x}$, (b) $\bar{x}\,\overline{(y + \bar{z})}$, and (c) $(x + y + z)(\bar{x}\,\bar{y}\,\bar{z})$.

Solution: Circuits that produce these outputs are shown in Figure 4.   ◀

EXAMPLES OF CIRCUITS

We will give some examples of circuits that perform some useful functions.

EXAMPLE 2   A committee of three individuals decides issues for an organization. Each individual votes either yes or no for each proposal that arises. A proposal is passed if it receives at least two yes votes. Design a circuit that determines whether a proposal passes.

Solution: Let $x = 1$ if the first individual votes yes, and $x = 0$ if this individual votes no; let $y = 1$ if the second individual votes yes, and $y = 0$ if this individual votes no; let $z = 1$ if the third individual votes yes, and $z = 0$ if this individual votes no. Then a circuit must be designed that produces the output 1 from the inputs x, y, and z when two or more of x, y, and z are 1. One representation of the Boolean function that has these output values is $xy + xz + yz$ (see Exercise 10 in Section 10.1). The circuit that implements this function is shown in Figure 5.   ◀

EXAMPLE 3   Sometimes light fixtures are controlled by more than one switch. Circuits need to be designed so that flipping any one of the switches for the fixture turns the light on when it is off and turns the light off when it is on. Design circuits that accomplish this when there are two switches and when there are three switches.

Solution: We will begin by designing the circuit that controls the light fixture when two different switches are used. Let $x = 1$ when the first switch is closed and $x = 0$ when it is open, and let $y = 1$ when the second switch is closed and $y = 0$ when it is open. Let $F(x, y) = 1$ when the light is on and $F(x, y) = 0$ when it is off. We can arbitrarily decide that the light will be on when both switches are closed, so that $F(1, 1) = 1$.

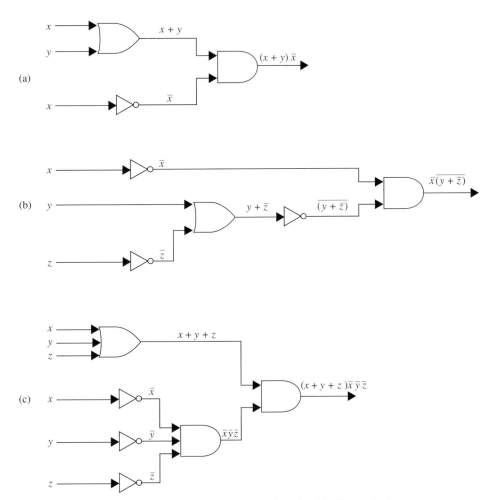

FIGURE 4 **Circuits that Produce the Outputs Specified in Example 1.**

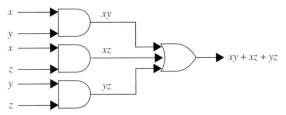

FIGURE 5 **A Circuit for Majority Voting.**

TABLE 1		
x	y	$F(x, y)$
1	1	1
1	0	0
0	1	0
0	0	1

This determines all the other values of F. When one of the two switches is opened, the light goes off, so $F(1, 0) = F(0, 1) = 0$. When the other switch is also opened, the light goes on, so that $F(0, 0) = 1$. Table 1 displays these values. We see that $F(x, y) = xy + \bar{x}\,\bar{y}$. This function is implemented by the circuit shown in Figure 6.

We will now design a circuit for three switches. Let x, y, and z be the Boolean variables that indicate whether each of the three switches is closed. We let $x = 1$ when the first

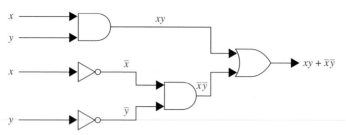

FIGURE 6   A Circuit for a Light Controlled by Two Switches.

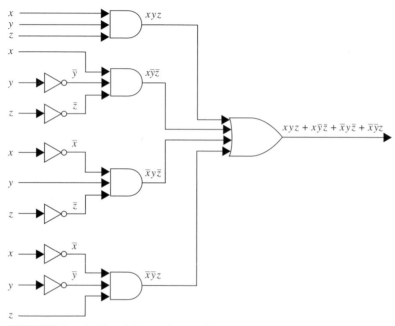

FIGURE 7   A Circuit for a Fixture Controlled by Three Switches.

TABLE 2			
x	y	z	$F(x, y, z)$
1	1	1	1
1	1	0	0
1	0	1	0
1	0	0	1
0	1	1	0
0	1	0	1
0	0	1	1
0	0	0	0

switch is closed, and $x = 0$ when it is open; $y = 1$ when the second switch is closed, and $y = 0$ when it is open; and $z = 1$ when the third switch is closed, and $z = 0$ when it is open. Let $F(x, y, z) = 1$ when the light is on and $F(x, y, z) = 0$ when the light is off. We can arbitrarily specify that the light be on when all three switches are closed so that $F(1, 1, 1) = 1$. This determines all other values of F. When one switch is opened, the light goes off so that $F(1, 1, 0) = F(1, 0, 1) = F(0, 1, 1) = 0$. When a second switch is opened, the light goes on so that $F(1, 0, 0) = F(0, 1, 0) = F(0, 0, 1) = 1$. Finally, when the third switch is opened, the light goes off again so that $F(0, 0, 0) = 0$. Table 2 shows the values of this function.

The function F can be represented by its sum-of-products expansion so that $F(x, y, z) = xyz + x\overline{y}\overline{z} + \overline{x}y\overline{z} + \overline{x}\,\overline{y}z$. The circuit shown in Figure 7 implements this function.   ◄

ADDERS

We will illustrate how logic circuits can be used to carry out addition of two positive integers from their binary expansions. We will build up the circuitry to do this addition

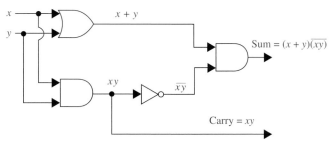

FIGURE 8   The Half Adder.

**TABLE 3
Input and
Output for
the Half
Adder.**

Input		Output	
x	y	s	c
1	1	0	1
1	0	1	0
0	1	1	0
0	0	0	0

**TABLE 4
Input and
Output for
the Full Adder.**

Input			Output	
x	y	c_i	s	c_{i+1}
1	1	1	1	1
1	1	0	0	1
1	0	1	0	1
1	0	0	1	0
0	1	1	0	1
0	1	0	1	0
0	0	1	1	0
0	0	0	0	0

from some component circuits. First, we will build a circuit that can be used to find $x + y$, where x and y are two bits. The input to our circuit will be x and y, since these each have the value 0 or the value 1. The output will consist of two bits, namely, s and c, where s is the sum bit and c is the carry bit. This circuit is called a **multiple output circuit** since it has more than one output. The circuit that we are designing is called the **half adder,** since it adds two bits, without considering a carry from a previous addition. We show the input and output for the half adder in Table 3. From Table 3 we see that $c = xy$ and that $s = x\overline{y} + \overline{x}y = (x + y)\overline{(xy)}$. Hence, the circuit shown in Figure 8 computes the sum bit s and the carry bit c from the bits x and y.

We use the **full adder** to compute the sum bit and the carry bit when two bits and a carry are added. The inputs to the full adder are the bits x and y and the carry c_i. The outputs are the sum bit s and the new carry c_{i+1}. The inputs and outputs for the full adder are shown in Table 4.

The two outputs of the full adder, the sum bit s and the carry c_{i+1}, are given by the sum-of-products expansions $xyc_i + x\overline{y}\,\overline{c}_i + \overline{x}y\overline{c}_i + \overline{x}\,\overline{y}c_i$ and $xyc_i + xy\overline{c}_i + x\overline{y}c_i + \overline{x}yc_i$, respectively. However, instead of designing the full adder from scratch, we will use half adders to produce the desired output. A full adder circuit using half adders is shown in Figure 9.

Finally, in Figure 10 we show how full and half adders can be used to add the two three-bit integers $(x_2x_1x_0)_2$ and $(y_2y_1y_0)_2$ to produce the sum $(s_3s_2s_1s_0)_2$. Note that s_3, the highest-order bit in the sum, is given by the carry c_2.

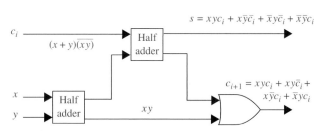

FIGURE 9   A Full Adder.

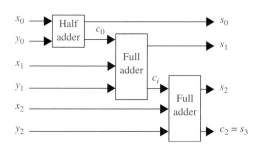

**FIGURE 10   Adding Two Three-Bit
Integers with Full and Half Adders.**

Exercises

In Exercises 1–5 find the output of the given circuit.

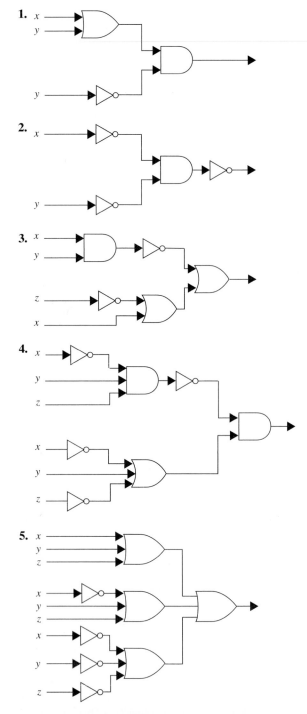

6. Construct circuits from inverters, *AND* gates, and *OR* gates to produce these outputs.

a) $\bar{x} + y$ **b)** $\overline{(x + y)x}$

c) $xyz + \bar{x}\,\bar{y}\,\bar{z}$ **d)** $\overline{(\bar{x} + z)(y + \bar{z})}$

7. Design a circuit that implements majority voting for five individuals.

8. Design a circuit for a light fixture controlled by four switches where flipping one of the switches turns the light on when it is off and turns it off when it is on.

9. Show how the sum of two five-bit integers can be found using full and half adders.

10. Construct a circuit for a half subtractor using *AND* gates, *OR* gates, and inverters. A **half subtractor** has two bits as input and produces as output a difference bit and a borrow.

11. Construct a circuit for a full subtractor using *AND* gates, *OR* gates, and inverters. A **full subtractor** has two bits and a borrow as input, and produces as output a difference bit and a borrow.

12. Use the circuits from Exercises 10 and 11 to find the difference of two four-bit integers, where the first integer is greater than the second integer.

*13. Construct a circuit that compares the two-bit integers $(x_1x_0)_2$ and $(y_1y_0)_2$, returning an output of 1 when the first of these numbers is larger and an output of 0 otherwise.

*14. Construct a circuit that computes the product of the two-bit integers $(x_1x_0)_2$ and $(y_1y_0)_2$. The circuit should have four output bits for the bits in the product.

Two gates that are often used in circuits are *NAND* and *NOR* gates. When *NAND* or *NOR* gates are used to represent circuits, no other types of gates are needed. The notation for these gates is as follows:

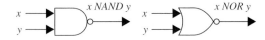

*15. Use *NAND* gates to construct circuits with these outputs.

a) $\bar{x}$ **b)** $x + y$ **c)** xy **d)** $x \oplus y$

*16. Use *NOR* gates to construct circuits for the outputs given in Exercise 15.

*17. Construct a half adder using *NAND* gates.

*18. Construct a half adder using *NOR* gates.

A **multiplexer** is a switching circuit that produces as output one of a set of input bits based on the value of control bits.

19. Construct a multiplexer using *AND* gates, *OR* gates, and inverters that has as input the four bits $x_0, x_1, x_2,$ and x_3 and the two control bits c_0 and c_1. Set up the circuit so that x_i is the output where i is the value of the two-bit integer $(c_1c_0)_2$.

The **depth** of a combinatorial circuit can be defined by specifying that the depth of the initial input is 0 and if a gate has n different inputs at depths $d_1, d_2, \ldots, d_n$, respectively, then its outputs have depth equal to $\max(d_1, d_2, \ldots, d_n) + 1$; this value is also defined to be the depth of the gate. The depth of a combinatorial circuit is the maximum depth of the gates in the circuit.

20. Find the depth of
  a) the circuit constructed in Example 2 for majority voting among three people.
  b) the circuit constructed in Example 3 for a light controlled by two switches.
  c) the half adder shown in Figure 8.
  d) the full adder shown in Figure 9.

10.4 Minimization of Circuits

INTRODUCTION

The efficiency of a combinational circuit depends on the number and arrangement of its gates. The process of designing a combinational circuit begins with the table specifying the output for each combination of input values. We can always use the sum-of-products expansion of a circuit to find a set of logic gates that will implement this circuit. However, the sum-of-products expansion may contain many more terms than are necessary. Terms in a sum-of-products expansion that differ in just one variable, so that in one term this variable occurs and in the other term the complement of this variable occurs, can be combined. For instance, consider the circuit that has output 1 if and only if $x = y = z = 1$ or $x = z = 1$ and $y = 0$. The sum-of-products expansion of this circuit is $xyz + x\overline{y}z$. The two products in this expansion differ in exactly one variable, namely, y. They can be combined as

$$xyz + x\overline{y}z = (y + \overline{y})(xz)$$
$$= 1 \cdot (xz)$$
$$= xz.$$

Hence, xz is a Boolean expression with fewer operators that represents the circuit. We show two different implementations of this circuit in Figure 1. The second circuit uses only one gate, whereas the first circuit uses three gates and an inverter.

This example shows that combining terms in the sum-of-products expansion of a circuit leads to a simpler expression for the circuit. We will describe two procedures that simplify sum-of-products expansions.

The goal of both procedures is to produce Boolean sums of Boolean products that represent a Boolean function with the fewest products of literals such that these products contain the fewest literals possible among all sums of products that represent a Boolean function. Finding such a sum of products is called **minimization of the Boolean function.** Minimizing a Boolean function makes it possible to construct a circuit for this function that uses the fewest gates and fewest inputs to the *AND* gates and *OR* gates in the circuit, among all circuits for the Boolean expression we are minimizing.

Until the early 1960s logic gates were individual components. To reduce costs it was important to use the fewest gates to produce a desired output. However, in the mid-1960s, integrated circuit technology was developed that made it possible to combine gates on a single chip. Even though it is now possible to build increasingly complex integrated circuits on chips at low cost, minimization of Boolean functions remains important. Reducing the number of gates on a chip can lead to a more reliable circuit and can reduce the cost to produce the chip. Also, minimization makes it possible to fit more circuits on the same chip. Furthermore, minimization reduces the number of inputs to gates in a circuit. This reduces the time used by a circuit to compute its output. Moreover, the

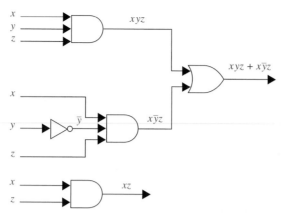

FIGURE 1   Two Circuits with the Same Output.

number of inputs to a gate may be limited because of the particular technology used to build logic gates.

The first procedure we will introduce, known as Karnaugh maps (or K-maps), was designed in the 1950s to help minimize circuits by hand. K-maps are useful in minimizing circuits with up to six variables, although they become rather complex even for five or six variables. The second procedure we will describe, the Quine–McCluskey method, was invented in the 1960s. It automates the process of minimizing combinatorial circuits and can be implemented as a computer program.

Unfortunately, minimizing Boolean functions with many variables is a computationally intensive problem. It has been shown that this problem is an NP-complete problem (see Section 2.3 and [Ka93]), so the existence of a polynomial-time algorithm for minimizing Boolean circuits is unlikely. The Quine–McCluskey method has exponential complexity. In practice, it can only be used when the number of literals does not exceed ten. Since the 1970s a number of newer algorithms have been developed for minimizing combinatorial circuits (see [Ha93] and [Ka93]). However, with the best algorithms yet devised, only circuits with no more than 25 variables can be minimized. Also, heuristic (or rule-of-thumb) methods can be used to substantially simplify, but not necessarily minimize, Boolean expressions with a larger number of literals.

KARNAUGH MAPS

To reduce the number of terms in a Boolean expression representing a circuit, it is necessary to find terms to combine. There is a graphical method, called a **Karnaugh map** or **K-map,** for finding terms to combine for Boolean functions involving a relatively small number of variables. The method we will describe was introduced by Maurice Karnaugh in 1953. His method is based on earlier work by E. W. Veitch. (This method is usually applied

MAURICE KARNAUGH (BORN 1924)   Maurice Karnaugh, born in New York City, received his B.S. from the City College of New York and his M.S. and Ph.D. from Yale University. He was a member of the technical staff at Bell Laboratories from 1952 until 1966 and Manager of Research and Development at the Federal Systems Division of AT&T from 1966 to 1970. In 1970 he joined IBM as a member of the research staff. Karnaugh has made fundamental contributions to the application of digital techniques in both computing and telecommunications. His current interests include knowledge-based systems in computers and heuristic search methods.

only when the function involves six or fewer variables.) K-maps give us a visual method for simplifying sum-of-products expansions; they are not suited for mechanizing this process. We will first illustrate how K-maps are used to simplify expansions of Boolean functions in two variables. We will continue by showing how K-maps can be used to minimize Boolean functions in three variables and then in four variables. Then we will describe the concepts that can be used to extend K-maps to minimize Boolean functions in more than four variables.

There are four possible minterms in the sum-of-products expansion of a Boolean function in the two variables x and y. A K-map for a Boolean function in these two variables consists of four cells, where a 1 is placed in the cell representing a minterm if this minterm is present in the expansion. Cells are said to be **adjacent** if the minterms that they represent differ in exactly one literal. For instance, the cell representing $\bar{x}y$ is adjacent to the cells representing xy and $\bar{x}\,\bar{y}$. The four cells and the terms that they represent are shown in Figure 2.

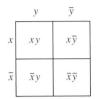

FIGURE 2
K-maps in Two Variables.

EXAMPLE 1   Find the K-maps for (a) $xy + \bar{x}y$, (b) $x\bar{y} + \bar{x}y$, and (c) $x\bar{y} + \bar{x}y + \bar{x}\,\bar{y}$.

Solution: We include a 1 in a cell when the minterm represented by this cell is present in the sum-of-products expansion. The three K-maps are shown in Figure 3.   ◀

We can identify minterms that can be combined from the K-map. Whenever there are 1s in two adjacent cells in the K-map, the minterms represented by these cells can be combined into a product involving just one of the variables. For instance, $x\bar{y}$ and $\bar{x}\,\bar{y}$ are represented by adjacent cells and can be combined into $\bar{y}$, since $x\bar{y} + \bar{x}\,\bar{y} = (x + \bar{x})\bar{y} = \bar{y}$. Moreover, if 1s are in all four cells, the four minterms can be combined into one term, namely, the Boolean expression 1 that involves none of the variables. We circle blocks of cells in the K-map that represent minterms that can be combined and then find the corresponding sum of products. The goal is to identify the largest possible blocks, and to cover all the 1s with the fewest blocks using the largest blocks first and always using the largest possible blocks.

EXAMPLE 2   Simplify the sum-of-products expansions given in Example 1.

Solution: The grouping of minterms is shown in Figure 4 using the K-maps for these expansions. Minimal expansions for these sums-of-products are (a) y, (b) $x\bar{y} + \bar{x}y$, and (c) $\bar{x} + \bar{y}$.   ◀

A K-map in three variables is a rectangle divided into eight cells. The cells represent the eight possible minterms in three variables. Two cells are said to be adjacent if the minterms that they represent differ in exactly one literal. One of the ways to form a K-map in three variables is shown in Figure 5(a). This K-map can be thought of as lying

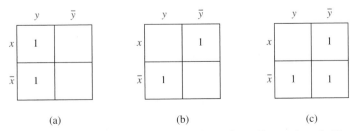

FIGURE 3   K-maps for the Sum-of-Products Expansions in Example 1.

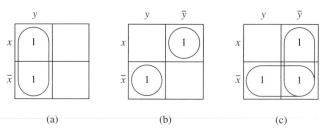

FIGURE 4   Simplifying the Sum-of-Products Expansion from Example 1.

on a cylinder, as shown in Figure 5(b). On the cylinder two cells have a common border if and only if they are adjacent.

To simplify a sum-of-products expansion in three variables, we use the K-map to identify blocks of minterms that can be combined. Blocks of two adjacent cells represent pairs of minterms that can be combined into a product of two literals; 2×2 and 4×1 blocks of cells represent minterms that can be combined into a single literal; and the block of all eight cells represents a product of no literals, namely, the function 1. In Figure 6, $1 \times 2, 2 \times 1, 2 \times 2, 4 \times 1$, and 4×2 blocks and the products they represent are shown.

The product of literals corresponding to a block of all 1s in the K-map is called an **implicant** of the function being minimized. It is called a **prime implicant** if this block of 1s is not contained in a larger block of 1s representing the product of fewer literals than in this product.

The goal is to identify the largest possible blocks in the map and cover all the 1s in the map with the least number of blocks, using the largest blocks first. The largest possible blocks are always chosen, but we must always choose a block if it is the only block of 1s covering a 1 in the K-map. Such a block represents an **essential prime implicant.** By covering all the 1s in the map with blocks corresponding to prime implicants we can express the sum of products as a sum of prime implicants. Note that there may be more than one way to cover all the 1s using the least number of blocks.

Example 3 illustrates how K-maps in three variables are used.

EXAMPLE 3   Use K-maps to minimize these sum-of-products expansions.

(a) $xy\overline{z} + x\overline{y}\,\overline{z} + \overline{x}yz + \overline{x}\,\overline{y}\,\overline{z}$
(b) $x\overline{y}z + x\overline{y}\,\overline{z} + \overline{x}yz + \overline{x}\,\overline{y}\,\overline{z} + \overline{x}\,\overline{y}z$
(c) $xyz + xy\overline{z} + x\overline{y}z + x\overline{y}\,\overline{z} + \overline{x}yz + \overline{x}\,\overline{y}z + \overline{x}\,\overline{y}\,\overline{z}$
(d) $xy\overline{z} + x\overline{y}\,\overline{z} + \overline{x}\,\overline{y}z + \overline{x}\,\overline{y}\,\overline{z}$

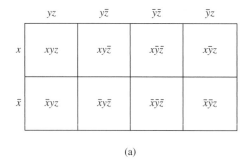

(a)

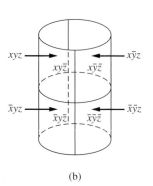

(b)

FIGURE 5   K-maps in Three Variables.

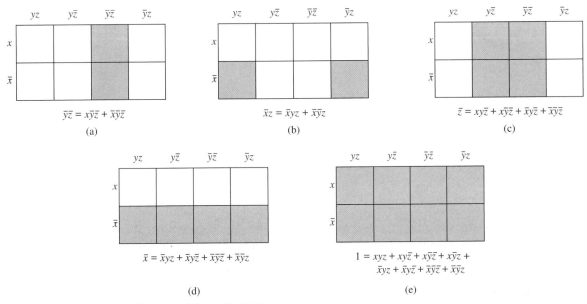

$$\overline{y}\,\overline{z} = x\overline{y}\,\overline{z} + \overline{x}\,\overline{y}\,\overline{z}$$

(a)

$$\overline{x}z = \overline{x}yz + \overline{x}\,\overline{y}z$$

(b)

$$\overline{z} = xy\overline{z} + x\overline{y}\,\overline{z} + \overline{x}y\overline{z} + \overline{x}\,\overline{y}\,\overline{z}$$

(c)

$$\overline{x} = \overline{x}yz + \overline{x}y\overline{z} + \overline{x}\,\overline{y}\,\overline{z} + \overline{x}\,\overline{y}z$$

(d)

$$1 = xyz + xy\overline{z} + x\overline{y}\,\overline{z} + x\overline{y}z +$$
$$\overline{x}yz + \overline{x}y\overline{z} + \overline{x}\,\overline{y}\,\overline{z} + \overline{x}\,\overline{y}z$$

(e)

FIGURE 6 **Blocks in K-maps in Three Variables.**

Solution: The K-maps for these sum-of-products expansions are shown in Figure 7. The grouping of blocks shows that minimal expansions into Boolean sums of Boolean products are (a) $x\overline{z} + \overline{y}\,\overline{z} + \overline{x}yz$, (b) $\overline{y} + \overline{x}z$, (c) $x + \overline{y} + z$, and (d) $x\overline{z} + \overline{x}\,\overline{y}$. In part (d) note that the prime implicants $x\overline{z}$ and $\overline{x}\,\overline{y}$ are essential prime implicants, but the prime implicant $\overline{y}\,\overline{z}$ is a prime implicant that is not essential since the cells it covers are covered by the other two prime implicants. ◀

A K-map in four variables is a square that is divided into 16 cells. The cells represent the 16 possible minterms in four variables. One of the ways to form a K-map in four variables is shown in Figure 8.

Two cells are adjacent if and only if the minterms they represent differ in one literal. Consequently, each cell is adjacent to four other cells. The K-map of a sum-of-products

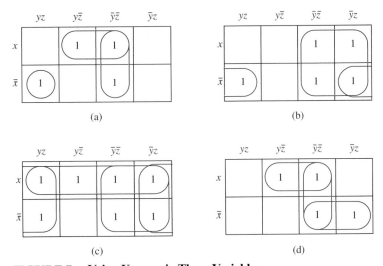

(a)

(b)

(c)

(d)

FIGURE 7 **Using K-maps in Three Variables.**

	yz	$y\bar{z}$	$\bar{y}\bar{z}$	$\bar{y}z$
wx	$wxyz$	$wxy\bar{z}$	$wx\bar{y}\bar{z}$	$wx\bar{y}z$
$w\bar{x}$	$w\bar{x}yz$	$w\bar{x}y\bar{z}$	$w\bar{x}\bar{y}\bar{z}$	$w\bar{x}\bar{y}z$
$\bar{w}\bar{x}$	$\bar{w}\bar{x}yz$	$\bar{w}\bar{x}y\bar{z}$	$\bar{w}\bar{x}\bar{y}\bar{z}$	$\bar{w}\bar{x}\bar{y}z$
$\bar{w}x$	$\bar{w}xyz$	$\bar{w}xy\bar{z}$	$\bar{w}x\bar{y}\bar{z}$	$\bar{w}x\bar{y}z$

FIGURE 8   K-maps in Four Variables.

expansion in four variables can be thought of as lying on a torus, so that adjacent cells have a common boundary (see Exercise 28). The simplification of a sum-of-products expansion in four variables is carried out by identifying those blocks of 2, 4, 8, or 16 cells that represent minterms that can be combined. Each cell representing a minterm must either be used to form a product using fewer literals, or be included in the expansion. In Figure 9 some examples of blocks that represent products of three literals, products of two literals, and a single literal are illustrated.

As is the case in K-maps in two and three variables, the goal is to identify the largest blocks of 1s in the map that correspond to the prime implicants and to cover all the 1s

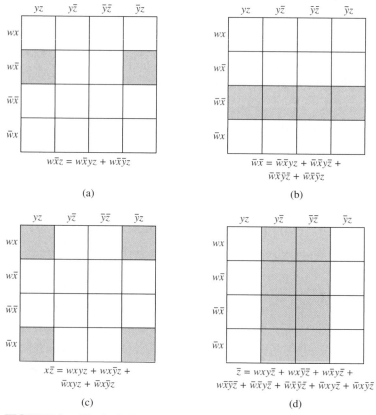

$w\bar{x}z = w\bar{x}yz + w\bar{x}\bar{y}z$

(a)

$\bar{w}\bar{x} = \bar{w}\bar{x}yz + \bar{w}\bar{x}y\bar{z} + \bar{w}\bar{x}\bar{y}\bar{z} + \bar{w}\bar{x}\bar{y}z$

(b)

$x\bar{z} = wxyz + wx\bar{y}z + \bar{w}xyz + \bar{w}x\bar{y}z$

(c)

$\bar{z} = wxy\bar{z} + wx\bar{y}\bar{z} + w\bar{x}y\bar{z} + w\bar{x}\bar{y}\bar{z} + \bar{w}\bar{x}y\bar{z} + \bar{w}\bar{x}\bar{y}\bar{z} + \bar{w}xy\bar{z} + \bar{w}x\bar{y}\bar{z}$

(d)

FIGURE 9   Blocks in K-maps in Four Variables.

using the fewest blocks needed, using the largest blocks first. The largest possible blocks are always used. Example 4 illustrates how K-maps in four variables are used.

EXAMPLE 4   Use K-maps to simplify these sum-of-products expansions.

    (a) $wxyz + wxy\overline{z} + wx\overline{y}\,\overline{z} + w\overline{x}yz + w\overline{x}\,\overline{y}z + w\overline{x}\,\overline{y}\,\overline{z} + \overline{w}x\overline{y}z +$
           $\overline{w}\,\overline{x}yz + \overline{w}\,\overline{x}y\overline{z}$

    (b) $wx\overline{y}\,\overline{z} + w\overline{x}yz + w\overline{x}y\overline{z} + w\overline{x}\,\overline{y}\,\overline{z} + \overline{w}xy\overline{z} + \overline{w}\,\overline{x}yz + \overline{w}\,\overline{x}\,\overline{y}\,\overline{z}$

    (c) $wxy\overline{z} + wx\overline{y}\,\overline{z} + w\overline{x}yz + w\overline{x}y\overline{z} + w\overline{x}\,\overline{y}\,\overline{z} + \overline{w}xyz + \overline{w}xy\overline{z} + \overline{w}x\overline{y}\,\overline{z} +$
           $\overline{w}x\,\overline{y}z + \overline{w}\,\overline{x}y\overline{z} + \overline{w}\,\overline{x}\,\overline{y}\,\overline{z}$

Solution: The K-maps for these expansions are shown in Figure 10. Using the blocks shown leads to the sum of products (a) $wyz + wx\overline{z} + w\overline{x}\,\overline{y} + \overline{w}\,\overline{x}y + \overline{w}x\overline{y}z$, (b) $\overline{y}\,\overline{z} + w\overline{x}y + \overline{x}\,\overline{z}$, and (c) $\overline{z} + \overline{w}x + w\overline{x}y$. The reader should determine whether there are other choices of blocks in each part that lead to different sums of products representing these Boolean functions.   ◀

K-maps can realistically be used to minimize Boolean functions with five or six variables, but beyond that, they are rarely used since they become extremely complicated. However, the concepts used in K-maps play an important role in newer algorithms. Furthermore, mastering these concepts helps you understand these newer algorithms and the computer-aided design (CAD) programs that implement them. As we develop these concepts, we will be able to illustrate them by referring back to our discussion of minimization of Boolean functions in three and in four variables.

The K-maps we used to minimize Boolean functions in two, three, and four variables are built using $2 \times 2, 2 \times 4$, and 4×4 rectangles, respectively. Furthermore, corresponding cells in the top row and bottom row and in the leftmost column and rightmost column in each of these cases are considered adjacent since they represent minterms differing in only one literal. We can build K-maps for minimizing Boolean functions in more than four variables in a similar way. We use a rectangle containing $2^{\lfloor n/2 \rfloor}$ rows and $2^{\lceil n/2 \rceil}$ columns. (These K-maps contain 2^n cells because $\lceil n/2 \rceil + \lfloor n/2 \rfloor = n$.) The rows and columns need to be positioned so that the cells representing minterms differing in just one literal are adjacent or are considered adjacent by specifying additional adjacencies of rows and columns. To help (but not entirely) achieve this, the rows and columns of a K-map are arranged using Gray codes (see Section 8.5), where we associate bit strings and products by specifying that a 1 corresponds to the appearance of a variable and

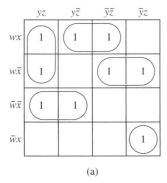

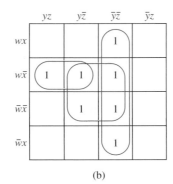

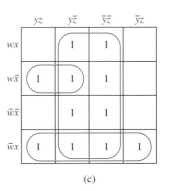

           (a)   (b)   (c)

FIGURE 10   **Using K-maps in Four Variables.**

a 0 with the appearance of its complement. For example, in a 10-dimensional K-map, the Gray code 01110 used to label a row corresponds to the product $\overline{x}_1 x_2 x_3 x_4 \overline{x}_5$.

EXAMPLE 5   The K-maps we used to minimize Boolean functions with four variables have two rows and two columns. Both the rows and the columns are arranged using the Gray code 11,10,00,01. The rows represent products wx, $w\overline{x}$, $\overline{w}\,\overline{x}$, and $\overline{w}x$, respectively, and the columns correspond to the products yz, $y\overline{z}$, $\overline{y}\,\overline{z}$, and $\overline{y}z$, respectively. Using Gray codes and considering cells adjacent in the first and last rows and in the first and last columns, we ensured that minterms that differ in only one variable are always adjacent.   ◄

EXAMPLE 6   To minimize Boolean functions in five variables we use K-maps with $2^3 = 8$ columns and $2^2 = 4$ rows. We label the four rows using the Gray code 11,10,00,01, corresponding to $x_1 x_2$, $x_1 \overline{x}_2$, $\overline{x}_1 \overline{x}_2$, and $\overline{x}_1 x_2$, respectively. We label the eight columns using the Gray code 111,110,100,101,001,000,010,011 corresponding to the terms $x_3 x_4 x_5$, $x_3 x_4 \overline{x}_5$, $x_3 \overline{x}_4 \overline{x}_5$, $x_3 \overline{x}_4 x_5$, $\overline{x}_3 \overline{x}_4 x_5$, $\overline{x}_3 \overline{x}_4 \overline{x}_5$, $\overline{x}_3 x_4 \overline{x}_5$, and $\overline{x}_3 x_4 x_5$, respectively. Using Gray codes to label columns and rows ensures that the minterms represented by adjacent cells differ in only one variable. However, to make sure all cells representing products that differ in only one variable are considered adjacent, we consider cells in the top and bottom rows to be adjacent, as well as cells in the first and eighth columns, the first and fourth columns, the second and seventh columns, the third and sixth columns, and the fifth and eighth columns (as the reader should verify).   ◄

To use a K-map to minimize a Boolean function in n variables, we first draw a K-map of the appropriate size. We place 1s in all cells corresponding to minterms in the sum-of-products expansion of this function. We then identify all prime implicants of the Boolean function. To do this we look for the blocks consisting of 2^k clustered cells all containing a 1, where $1 \le k \le n$. These blocks correspond to the product of $n - k$ literals. (Exercise 33 asks the reader to verify this.) Furthermore, a block of 2^k cells each containing a 1 not contained in a block of 2^{k+1} cells each containing a 1 represents a prime implicant. The reason that this implicant is a prime implicant is that no product obtained by deleting a literal is also represented by a block of cells all containing 1s.

EXAMPLE 7   A block of eight cells representing a product of two literals in a K-map for minimizing Boolean functions in five variables all containing 1s is a prime implicant if it is not contained in a larger block of 16 cells all containing 1s representing a single literal.   ◄

Once all prime implicants have been identified, the goal is to find the smallest possible subset of these prime implicants with the property that the cells representing these prime implicants cover all the cells containing a 1 in the K-map. We begin by selecting the essential prime implicants since each of these is represented by a block that covers a cell containing a 1 that is not covered by any other prime implicant. We add additional prime implicants to ensure that all 1s in the K-map are covered. When the number of variables is large, this last step can become exceedingly complicated.

DON'T CARE CONDITIONS

Links

In some circuits we care only about the output for some combinations of input values, because other combinations of input values are not possible or never occur. This gives us freedom in producing a simple circuit with the desired output since the output values for all those combinations that never occur can be arbitrarily chosen. The values of the function for these combinations are called ***don't care* conditions.** A *d* is used in a K-map to mark those combinations of values of the variables for which the function can be arbitrarily assigned. In the minimization process we can assign 1s as values to those combinations of the input values that lead to the largest blocks in the K-map. This is illustrated in Example 8.

EXAMPLE 8   One way to code decimal expansions using bits is to use the four bits of the binary expansion of each digit in the decimal expansion. For instance, 873 is encoded as 100001110011. This encoding of a decimal expansion is called a **binary coded decimal expansion.** Since there are 16 blocks of four bits and only 10 decimal digits, there are six combinations of four bits that are not used to encode digits. Suppose that a circuit is to be built that produces an output of 1 if the decimal digit is 5 or greater and an output of 0 if the decimal digit is less than 5. How can this circuit be simply built using *OR* gates, *AND* gates, and inverters?

Solution: Let $F(w, x, y, z)$ denote the output of the circuit, where $wxyz$ is a binary expansion of a decimal digit. The values of F are shown in Table 1. The K-map for F, with ds in the *don't care* positions, is shown in Figure 11(a). We can either include or exclude squares with ds from blocks. This gives us many possible choices for the blocks. For example, excluding all squares with ds and forming blocks as shown in Figure 11(b) produces the expression $w\overline{x}\,\overline{y} + \overline{w}xy + \overline{w}xz$. Including some of the ds and excluding others and forming blocks as shown in Figure 11(c) produces the expression $w\overline{x} + \overline{w}xy + x\overline{y}z$. Finally, including all the ds and using the blocks shown in Figure 11(d) produces the simplest sum-of-products expansion possible, namely, $F(x, y, z) = w + xy + xz$.   ◀

THE QUINE–McCLUSKEY METHOD

Links

We have seen that K-maps can be used to produce minimal expansions of Boolean functions as Boolean sums of Boolean products. However, K-maps are awkward to use when there are more than four variables. Furthermore, the use of K-maps relies on visual inspection to identify terms to group. For these reasons there is a need for a procedure for simplifying sum-of-products expansions that can be mechanized. The Quine–McCluskey

TABLE 1					
Digit	*w*	*x*	*y*	*z*	*F*
0	0	0	0	0	0
1	0	0	0	1	0
2	0	0	1	0	0
3	0	0	1	1	0
4	0	1	0	0	0
5	0	1	0	1	1
6	0	1	1	0	1
7	0	1	1	1	1
8	1	0	0	0	1
9	1	0	0	1	1

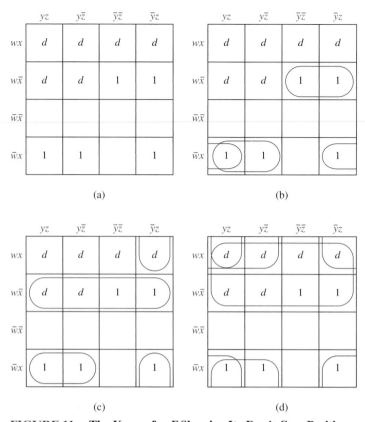

FIGURE 11   The K-map for _F_ Showing Its _Don't Care_ Positions.

method is such a procedure. It can be used for Boolean functions in any number of variables. It was developed in the 1950s by W. V. Quine and E. J. McCluskey, Jr. Basically, the Quine–McCluskey method consists of two parts. The first part finds those terms that are candidates for inclusion in a minimal expansion as a Boolean sum of Boolean products. The second part determines which of these terms to actually use. We will use Example 9 to illustrate how, by successively combining implicants into implicants with one fewer literal, this procedure works.

EXAMPLE 9   We will show how the Quine–McCluskey method can be used to find a minimal expansion equivalent to

$$xyz + x\overline{y}z + \overline{x}yz + \overline{x}\,\overline{y}z + \overline{x}\,\overline{y}\,\overline{z}.$$

EDWARD J. McCLUSKEY (BORN 1929)   Edward McCluskey attended Bowdoin College and M.I.T., where he received his doctorate in electrical engineering in 1956. He joined Bell Telephone Laboratories in 1955, remaining there until 1959. McCluskey was professor of electrical engineering at Princeton University from 1959 until 1966, also serving as director of the Computer Center at Princeton from 1961 to 1966. In 1967 he took a position as professor of computer science and electrical engineering at Stanford University, where he also served as director of the Digital Systems Laboratory from 1969 to 1978. McCluskey has worked in a variety of areas in computer science, including fault-tolerant computing, computer architecture, testing, and logic design. He is currently director of the Center for Reliable Computing at Stanford University. McCluskey is also an ACM Fellow.

TABLE 2		
Minterm	*Bit String*	*Number of 1s*
xyz	111	3
$x\overline{y}z$	101	2
$\overline{x}yz$	011	2
$\overline{x}\,\overline{y}z$	001	1
$\overline{x}\,\overline{y}\,\overline{z}$	000	0

We will represent the minterms in this expansion by bit strings. The first bit will be 1 if x occurs and 0 if $\overline{x}$ occurs. The second bit will be 1 if y occurs and 0 if $\overline{y}$ occurs. The third bit will be 1 if z occurs and 0 if $\overline{z}$ occurs. We then group these terms according to the number of 1s in the corresponding bit strings. This information is shown in Table 2.

Minterms that can be combined are those that differ in exactly one literal. Hence, two terms that can be combined differ by exactly one in the number of 1s in the bit strings that represent them. When two minterms are combined into a product, this product contains two literals. A product in two literals is represented using a dash to denote the variable that does not occur. For instance, the minterms $x\overline{y}z$ and $\overline{x}\,\overline{y}z$, represented by bit strings 101 and 001, can be combined into $\overline{y}z$, represented by the string –01. All pairs of minterms that can be combined and the product formed from these combinations are shown in Table 3.

Next, all pairs of products of two literals that can be combined are combined into one literal. Two such products can be combined if they contain literals for the same two variables, and literals for only one of the two variables differ. In terms of the strings representing the products, these strings must have a dash in the same position and must differ in exactly one of the other two slots. We can combine the products yz and $\overline{y}z$, represented by strings –11 and –01, into z, represented by the string – –1. We show all the combinations of terms that can be formed in this way in Table 3.

WILLARD VAN ORMAN QUINE (1908–2000) Willard Quine, born in Akron, Ohio, attended Oberlin College and later Harvard University, where he received his Ph.D. in philosophy in 1932. He became a Junior Fellow at Harvard in 1933 and was appointed to a position on the faculty there in 1936. He remained at Harvard his entire professional life, except for World War II, when he worked for the U.S. Navy decrypting messages from German submarines. Quine was always interested in algorithms, but not in hardware. He arrived at his discovery of what is now called the Quine–McCluskey method as a device for teaching mathematical logic, rather than as a method for simplifying switching circuits. Quine was one of the most famous philosophers of the twentieth century. He made fundamental contributions to the theory of knowledge, mathematical logic and set theory, and the philosophies of logic and language. His books, including *New Foundations of Mathematical Logic* published in 1937 and *Word and Object* published in 1960, have had profound impact. Quine retired from Harvard in 1978 but continued to commute from his home in Beacon Hill to his office there. He used his whole life the 1927 Remington typewriter on which he prepared his doctoral thesis. He even had an operation performed on this machine to add a few special symbols, removing the second period, the second comma, and the question mark. When asked whether he missed the question mark, he replied, "Well, you see, I deal in certainties." There is even a word *quine*, defined in the *New Hacker's Dictionary* as a program that generates a copy of its own source code as its complete output. Producing the shortest possible quine in a given programming language is a popular puzzle for hackers.

TABLE 3

			Step 1			Step 2	
	Term	*Bit String*		*Term*	*String*	*Term*	*String*
1	xyz	111	(1,2)	xz	1–1	(1,2,3,4)	z	– –1
2	$x\overline{y}z$	101	(1,3)	yz	–11		
3	$\overline{x}yz$	011	(2,4)	$\overline{y}z$	–01		
4	$\overline{x}\,\overline{y}z$	001	(3,4)	$\overline{x}z$	0–1		
5	$\overline{x}\,\overline{y}\,\overline{z}$	000	(4,5)	$\overline{x}\,\overline{y}$	00–		

TABLE 4

	xyz	$x\overline{y}z$	$\overline{x}yz$	$\overline{x}\,\overline{y}z$	$\overline{x}\,\overline{y}\,\overline{z}$
z	X	X	X	X	
$\overline{x}\,\overline{y}$				X	X

In Table 3 we also indicate which terms have been used to form products with fewer literals; these terms will not be needed in a minimal expansion. The next step is to identify a minimal set of products needed to represent the Boolean function. We begin with all those products that were not used to construct products with fewer literals. Next, we form Table 4, which has a row for each candidate product formed by combining original terms, and a column for each original term; and we put an X in a position if the original term in the sum-of-products expansion was used to form this candidate product. In this case, we say that the candidate product **covers** the original minterm. We need to include at least one product that covers each of the original minterms. Consequently, whenever there is only one X in a column in the table, the product corresponding to the row this X is in must be used. From Table 4 we see that both z and $\overline{x}\,\overline{y}$ are needed. Hence, the final answer is $z + \overline{x}\,\overline{y}$. ◀

As was illustrated in Example 9, the Quine–McCluskey method uses this sequence of steps to simplify a sum-of-products expression.

1. Express each minterm in n variables by a bit string of length n with a 1 in the ith position if x_i occurs and a 0 in this position if $\overline{x}_i$ occurs.

2. Group the bit strings according to the number of 1s in them.

3. Determine all products in $n - 1$ variables that can be formed by taking the Boolean sum of minterms in the expansion. Minterms that can be combined are represented by bit strings that differ in exactly one position. Represent these products in $n - 1$ variables with strings that have a 1 in the ith position if x_i occurs in the product, a 0 in this position if $\overline{x}_i$ occurs, and a dash in this position if there is no literal involving x_i in the product.

4. Determine all products in $n - 2$ variables that can be formed by taking the Boolean sum of the products in $n - 1$ variables found in the previous step. Products in $n - 1$ variables that can be combined are represented by bit strings that have a dash in the same position and differ in exactly one position.

5.   Continue combining Boolean products into products in fewer variables as long as possible.

6.   Find all the Boolean products that arose that were not used to form a Boolean product in one fewer literal.

7.   Find the smallest set of these Boolean products so that the sum of these products represents the Boolean function. This is done by forming a table showing which minterms are covered by which products. Every minterm must be covered by at least one product. The first step in using this table is to find all essential prime implicants. Each essential prime implicant must be included since it is the only prime implicant that covers one of the minterms. Once we have found essential prime implicants, we can simplify the table by eliminating the columns for minterms covered by this prime implicant. Furthermore, we can eliminate any prime implicants that cover a subset of minterms covered by another prime implicant (as the reader should verify). Moreover, we can eliminate from the table the column for a minterm if there is another minterm that is covered by a subset of the prime implicants that cover this minterm. This process of identifying essential prime implicants that must be included, followed by eliminating redundant prime implicants and identifying minterms that can be ignored, is iterated until the table does not change. At this point we use a backtracking procedure to find the optimal solution where we add prime implicants to the cover to find possible solutions, which we compare to the best solution found so far at each step.

A final example will illustrate how this procedure is used to simplify a sum-of-products expansion in four variables.

EXAMPLE 10   Use the Quine–McCluskey method to simplify the sum-of-products expansion $wxy\overline{z} + w\overline{x}yz + w\overline{x}y\overline{z} + \overline{w}xyz + \overline{w}x\overline{y}z + \overline{w}\,\overline{x}yz + \overline{w}\,\overline{x}\,\overline{y}z$.

Solution: We first represent the minterms by bit strings and then group these terms together according to the number of 1s in the bit strings. This is shown in Table 5. All the Boolean products that can be formed by taking Boolean sums of these products are shown in Table 6.

The only products that were not used to form products in fewer variables are $\overline{w}z$, $wy\overline{z}$, $w\overline{x}y$, and $\overline{x}yz$. In Table 7 we show the minterms covered by each of these products. To cover these minterms we must include $\overline{w}z$ and $wy\overline{z}$, since these products are the only

TABLE 5		
Term	*Bit String*	*Number of 1s*
$wxy\overline{z}$	1110	3
$w\overline{x}yz$	1011	3
$\overline{w}xyz$	0111	3
$w\overline{x}y\overline{z}$	1010	2
$\overline{w}x\overline{y}z$	0101	2
$\overline{w}\,\overline{x}yz$	0011	2
$\overline{w}\,\overline{x}\,\overline{y}z$	0001	1

TABLE 6			Step 1		Step 2	
	Term	Bit String	Term	String	Term	String
1	$wxy\overline{z}$	1110	(1,4)	$wy\overline{z}$	1–10	(3,5,6,7)	$\overline{w}z$	0 – –1
2	$w\overline{x}yz$	1011	(2,4)	$w\overline{x}y$	101–		
3	$\overline{w}xyz$	0111	(2,6)	$\overline{x}yz$	–011		
4	$w\overline{x}y\overline{z}$	1010	(3,5)	$\overline{w}xz$	01–1		
5	$\overline{w}x\overline{y}z$	0101	(3,6)	$\overline{w}yz$	0–11		
6	$\overline{w}\,\overline{x}yz$	0011	(5,7)	$\overline{w}\,\overline{y}z$	0–01		
7	$\overline{w}\,\overline{x}\,\overline{y}z$	0001	(6,7)	$\overline{w}\,\overline{x}z$	00–1		

TABLE 7	$wxy\overline{z}$	$w\overline{x}yz$	$\overline{w}xyz$	$w\overline{x}y\overline{z}$	$\overline{w}x\overline{y}z$	$\overline{w}\,\overline{x}yz$	$\overline{w}\,\overline{x}\,\overline{y}z$
$\overline{w}z$		X			X	X	X
$wy\overline{z}$	X			X			
$w\overline{x}y$		X		X			
$\overline{x}yz$		X				X	

products that cover $\overline{w}xyz$ and $wxy\overline{z}$, respectively. Once these two products are included, we see that only one of the two products left is needed. Consequently, we can take either $\overline{w}z + wy\overline{z} + w\overline{x}y$ or $\overline{w}z + wy\overline{z} + \overline{x}yz$ as the final answer. ◀

Exercises

1. a) Draw a K-map for a function in two variables and put a 1 in the cell representing $\overline{x}y$.

   b) What are the minterms represented by cells adjacent to this cell?

2. Find the sum-of-products expansions represented by each of these K-maps.

3. Draw the K-maps of these sum-of-products expansions in two variables.

   a) $x\overline{y}$ **b)** $xy + \overline{x}\,\overline{y}$

   c) $xy + x\overline{y} + \overline{x}y + \overline{x}\,\overline{y}$

4. Use a K-map to find a minimal expansion as a Boolean sum of Boolean products of each of these functions of the Boolean variables x and y.

   a) $\overline{x}y + \overline{x}\,\overline{y}$

   b) $xy + x\overline{y}$

   c) $xy + x\overline{y} + \overline{x}y + \overline{x}\,\overline{y}$

5. a) Draw a K-map for a function in three variables. Put a 1 in the cell that represents $\overline{x}y\overline{z}$.

   b) Which minterms are represented by cells adjacent to this cell?

6. Use K-maps to find simpler circuits with the same output as each of the circuits shown.

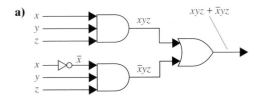

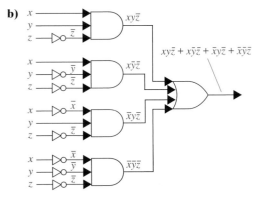

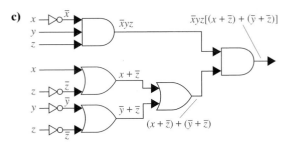

7. Draw the K-maps of these sum-of-products expansions in three variables.

a) $x\bar{y}\bar{z}$ **b)** $\bar{x}yz + \bar{x}\,\bar{y}\,\bar{z}$

c) $xyz + xy\bar{z} + \bar{x}y\bar{z} + \bar{x}\,\bar{y}z$

8. Construct a K-map for $F(x, y, z) = xz + yz + xy\bar{z}$. Use this K-map to find the implicants, prime implicants, and essential prime implicants of $F(x, y, z)$.

9. Construct a K-map for $F(x, y, z) = x\bar{z} + xyz + y\bar{z}$. Use this K-map to find the implicants, prime implicants, and essential prime implicants of $F(x, y, z)$.

10. Draw the 3-cube Q_3 and label each vertex with the minterm in the Boolean variables x, y, and z associated with the bit string represented by this vertex. For each literal in these variables indicate the 2-cube Q_2 that is a subgraph of Q_3 and represents this literal.

11. Draw the 4-cube Q_4 and label each vertex with the minterm in the Boolean variables $w, x, y,$ and z associated with the bit string represented by this vertex. For each literal in these variables, indicate which 3-cube Q_3 that is a subgraph of Q_4 represents this literal. Indicate which 2-cube Q_2 that is a subgraph of Q_4 represents the products $wz, \bar{x}y,$ and $\bar{y}\,\bar{z}$.

12. Use a K-map to find a minimal expansion as a Boolean sum of Boolean products of each of these functions in the variables $x, y,$ and z.

a) $\bar{x}yz + \bar{x}\,\bar{y}z$

b) $xyz + xy\bar{z} + \bar{x}yz + \bar{x}y\bar{z}$

c) $xy\bar{z} + x\bar{y}z + x\bar{y}\,\bar{z} + \bar{x}yz + \bar{x}\,\bar{y}z$

d) $xyz + x\bar{y}z + x\bar{y}\,\bar{z} + \bar{x}yz + \bar{x}y\bar{z} + \bar{x}\,\bar{y}\,\bar{z}$

13. a) Draw a K-map for a function in four variables. Put a 1 in the cell that represents $\bar{w}xy\bar{z}$.

b) Which minterms are represented by cells adjacent to this cell?

14. Use a K-map to find a minimal expansion as a Boolean sum of Boolean products of each of these functions in the variables $w, x, y,$ and z.

a) $wxyz + wx\bar{y}z + wx\bar{y}\,\bar{z} + w\bar{x}yz + w\bar{x}\,\bar{y}z$

b) $wxy\bar{z} + wx\bar{y}z + w\bar{x}yz + \bar{w}x\bar{y}z + \bar{w}\,\bar{x}yz + \bar{w}\,\bar{x}\,\bar{y}z$

c) $wxyz + wxy\bar{z} + wx\bar{y}z + w\bar{x}\,\bar{y}z + w\bar{x}\,\bar{y}\,\bar{z} + \bar{w}x\bar{y}z + \bar{w}\,\bar{x}y\bar{z} + \bar{w}\,\bar{x}\,\bar{y}z$

d) $wxyz + wxy\bar{z} + wx\bar{y}z + w\bar{x}yz + w\bar{x}y\bar{z} + \bar{w}xyz + \bar{w}\,\bar{x}yz + \bar{w}\,\bar{x}y\bar{z} + \bar{w}\,\bar{x}\,\bar{y}z$

15. Find the cells in a K-map for Boolean functions with five variables that correspond to each of these products.

a) $x_1x_2x_3x_4$ **b)** $\bar{x}_1x_3x_5$ **c)** x_2x_4

d) $\bar{x}_3\bar{x}_4$ **e)** x_3 **f)** $\bar{x}_5$

16. How many cells in a K-map for Boolean functions with six variables are needed to represent x_1, $\bar{x}_1x_6$, $\bar{x}_1x_2\bar{x}_6$, $x_2x_3x_4x_5$, and $x_1\bar{x}_2x_4\bar{x}_5$, respectively?

17. a) How many cells does a K-map in six variables have?

b) How many cells are adjacent to a given cell in a K-map in six variables?

18. Show that cells in a K-map for Boolean functions in five variables represent minterms that differ in exactly one literal if and only if they are adjacent or are in cells that become adjacent when the top and bottom rows and cells in the first and eighth columns, the first and fourth columns, the second and seventh columns, the third and sixth columns, and the fifth and eighth columns are considered adjacent.

19. Which rows and which columns of a 4×16 map for Boolean functions in six variables using the Gray codes 1111,1110,1010,1011,1001,1000,0000,0001,0011, 0010,0110,0111,0101,0100,1100,1101 to label the columns and 11,10,00,01 to label the rows need to be considered adjacent so that cells that represent minterms that differ in exactly one literal are considered adjacent?

*20. Use K-maps to find a minimal expansion as a Boolean sum of Boolean products of Boolean functions that have as input the binary code for each decimal digit and produce as output a 1 if and only if the digit corresponding to the input is

a) odd.   **b)** not divisible by 3.

c) not 4, 5, or 6.

*21. Suppose that there are five members on a committee, but that Smith and Jones always vote the opposite of Marcus. Design a circuit that implements majority voting of the committee using this relationship between votes.

22. Use the Quine–McCluskey method to simplify the sum-of-products expansions in Example 3.

23. Use the Quine–McCluskey method to simplify the sum-of-products expansions in Exercise 12.

24. Use the Quine–McCluskey method to simplify the sum-of-products expansions in Example 4.

25. Use the Quine–McCluskey method to simplify the sum-of-products expansions in Exercise 14.

*26. Explain how K-maps can be used to simplify product-of-sums expansions in three variables. (*Hint:* Mark with a 0 all the maxterms in an expansion and combine blocks of maxterms.)

27. Use the method from Exercise 26 to simplify the product-of-sums expansion $(x + y + z)(x + y + \bar{z})(x + \bar{y} + \bar{z})(x + \bar{y} + z)(\bar{x} + y + z)$.

*28. Draw a K-map for the 16 minterms in four Boolean variables on the surface of a torus.

29. Build a circuit using *OR* gates, *AND* gates, and inverters that produces an output of 1 if a decimal digit, encoded using a binary coded decimal expansion, is divisible by 3, and an output of 0 otherwise.

In Exercises 30–32 find a minimal sum-of-products expansion, given the K-map shown with *don't care* conditions indicated with *d*s.

30.

	yz	$y\bar{z}$	$\bar{y}\bar{z}$	$\bar{y}z$
wx	d	1	d	1
$w\bar{x}$		d	d	
$\bar{w}\bar{x}$		d	1	
$\bar{w}x$		1	d	

31.

	yz	$y\bar{z}$	$\bar{y}\bar{z}$	$\bar{y}z$
wx	1			1
$w\bar{x}$		d	1	
$\bar{w}\bar{x}$		1	d	
$\bar{w}x$	d			d

32.

	yz	$y\bar{z}$	$\bar{y}\bar{z}$	$\bar{y}z$
wx		d	d	1
$w\bar{x}$	d	d	1	d
$\bar{w}\bar{x}$				
$\bar{w}x$	1	1	1	d

33. Show that products of k literals correspond to 2^{n-k}-dimensional subcubes of the n-cube Q_n, where the vertices of the cube correspond to the minterms represented by the bit strings labeling the vertices, as described in Example 7 of Section 8.2.

Key Terms and Results

TERMS

Boolean variable: a variable that assumes only the values 0 and 1

$\bar{x}$ **(complement of x):** an expression with the value 1 when x has the value 0 and the value 0 when x has the value 1

$x \cdot y$ **(or xy) (Boolean product or conjunction of x and y):** an expression with the value 1 when both x and y have the value 1 and the value 0 otherwise

$x + y$ **(Boolean sum or disjunction of x and y):** an expression with the value 1 when either x or y, or both, has the value 1, and 0 otherwise

Boolean expressions: the expressions obtained recursively by specifying that $0, 1, x_1, \ldots, x_n$ are Boolean expressions and $\bar{E}_1$, $(E_1 + E_2)$, and $(E_1 E_2)$ are Boolean expressions if E_1 and E_2 are

dual of a Boolean expression: the expression obtained by interchanging + signs and · signs and interchanging 0s and 1s

Boolean function of degree n: a function from B^n to B where $B = \{0, 1\}$

Boolean algebra: a set B with two binary operations $\vee$ and $\wedge$, elements 0 and 1, and a complementation operator $^-$ that satisfies the identity, complement, associative, commutative, and distributive laws

literal of the Boolean variable x: either x or $\bar{x}$

minterm of $x_1, x_2, \ldots, x_n$: a Boolean product $y_1 y_2 \cdots y_n$ where each y_i is either x_i or $\bar{x}_i$

sum-of-products expansion (or disjunctive normal form): the representation of a Boolean function as a disjunction of minterms

functionally complete: a set of Boolean operators is called functionally complete if every Boolean function can be represented using these operators

$x \mid y$ **(or x NAND y):** the expression that has the value 0 when both x and y have the value 1 and the value 1 otherwise

$x \downarrow y$ **(or x NOR y):** the expression that has the value 0 when either x or y or both have the value 1 and the value 0 otherwise

inverter: a device that accepts the value of a Boolean variable as input and produces the complement of the input

OR **gate:** a device that accepts the values of two or more Boolean variables as input and produces their Boolean sum as output

AND **gate:** a device that accepts the values of two or more Boolean variables as input and produces their Boolean product as output

half adder: a circuit that adds two bits, producing a sum bit and a carry bit

full adder: a circuit that adds two bits and a carry, producing a sum bit and a carry bit

K-map for _n_ variables: a rectangle divided into 2^n cells where each cell represents a minterm in the variables

minimization of a Boolean function: representing a Boolean function as the sum of the fewest products of literals such that these products contain the fewest literals possible among all sums of products that represent this Boolean function

implicant of a Boolean function: a product of literals with the property that if this product has the value 1, then the value of this Boolean function is 1

prime implicant of a Boolean function: a product of literals that is an implicant of the Boolean function and no product obtained by deleting a literal is also an implicant of this function

essential prime implicant of a Boolean function: a prime implicant of the Boolean function that must be included in a minimization of this function

RESULTS

The identities for Boolean algebra (see Table 5 in Section 10.1).

An identity between Boolean functions represented by Boolean expressions remains valid when the duals of both sides of the identity are taken.

Every Boolean function can be represented by a sum-of-products expansion.

Each of the sets $\{+, {}^-\}$ and $\{\cdot, {}^-\}$ is functionally complete.

Each of the sets $\{\downarrow\}$ and $\{\mid\}$ is functionally complete.

The use of K-maps to minimize Boolean expressions.

The Quine–McCluskey method for minimizing Boolean expressions.

Review Questions

1. Define a Boolean function of degree n.
2. How many Boolean functions of degree two are there?
3. Give a recursive definition of the set of Boolean expressions.
4. **a)** What is the dual of a Boolean expression?
   b) What is the duality principle? How can it be used to find new identities involving Boolean expressions?
5. Explain how to construct the sum-of-products expansion of a Boolean function.
6. **a)** What does it mean for a set of operators to be functionally complete?
   b) Is the set $\{+, \cdot\}$ functionally complete?
   c) Are there sets of a single operator that are functionally complete?
7. Explain how to build a circuit for a light controlled by two switches using *OR* gates, *AND* gates, and inverters.
8. Construct a half adder using *OR* gates, *AND* gates, and inverters.
9. Is there a single type of logic gate that can be used to build all circuits that can be built using *OR* gates, *AND* gates, and inverters?
10. **a)** Explain how K-maps can be used to simplify sum-of-products expansions in three Boolean variables.
    b) Use a K-map to simplify the sum-of-products expansion $xyz + x\overline{y}z + x\overline{y}\,\overline{z} + \overline{x}yz + \overline{x}\,\overline{y}\,\overline{z}$.
11. **a)** Explain how K-maps can be used to simplify sum-of-products expansions in four Boolean variables.
    b) Use a K-map to simplify the sum-of-products expansion $wxyz + wxy\overline{z} + wx\overline{y}z + wx\overline{y}\,\overline{z} + w\overline{x}yz + w\overline{x}\,\overline{y}z + \overline{w}xyz + \overline{w}\,\overline{x}yz + \overline{w}\,\overline{x}y\overline{z}$.
12. **a)** What is a *don't care* condition?
    b) Explain how *don't care* conditions can be used to build a circuit using *OR* gates, *AND* gates, and inverters that produces an output of 1 if a decimal digit is 6 or greater, and an output of 0 if this digit is less than 6.
13. **a)** Explain how to use the Quine–McCluskey method to simplify sum-of-products expansions.
    b) Use this method to simplify $xy\overline{z} + x\overline{y}\,\overline{z} + \overline{x}yz + \overline{x}\,\overline{y}z + \overline{x}\,\overline{y}\,\overline{z}$.

Supplementary Exercises

1. For which values of the Boolean variables x, y, and z does
   a) $x + y + z = xyz$?
   b) $x(y + z) = x + yz$?
   c) $\overline{x}\,\overline{y}\,\overline{z} = x + y + z$?
2. Let x and y belong to $\{0, 1\}$. Does it necessarily follow that $x = y$ if there exists a value z in $\{0, 1\}$ such that
   a) $xz = yz$?   **b)** $x + z = y + z$?

c) $x \oplus z = y \oplus z$?   **d)** $x \downarrow z = y \downarrow z$?
e) $x \mid z = y \mid z$?

A Boolean function F is called **self-dual** if and only if $F(x_1, \ldots, x_n) = \overline{F(\overline{x}_1, \ldots, \overline{x}_n)}$.

3. Which of these functions are self-dual?

a) $F(x, y) = x$ **b)** $F(x, y) = xy + \overline{x}\,\overline{y}$
c) $F(x, y) = x + y$ **d)** $F(x, y) = xy + \overline{x}y$

4. Give an example of a self-dual Boolean function of three variables.

* 5. How many Boolean functions of degree n are self-dual?

We define the relation $\leq$ on the set of Boolean functions of degree n so that $F \leq G$ where F and G are Boolean functions if and only if $G(x_1, x_2, \ldots, x_n) = 1$ whenever $F(x_1, x_2, \ldots, x_n) = 1$.

6. Determine whether $F \leq G$ or $G \leq F$ for the following pairs of functions.

a) $F(x, y) = x, G(x, y) = x + y$
b) $F(x, y) = x + y, G(x, y) = xy$
c) $F(x, y) = \overline{x}, G(x, y) = x + y$

7. Show that if F and G are Boolean functions of degree n, then

a) $F \leq F + G$.   **b)** $FG \leq F$.

8. Show that if F, G, and H are Boolean functions of degree n, then $F + G \leq H$ if and only if $F \leq H$ and $G \leq H$.

* 9. Show that the relation $\leq$ is a partial ordering on the set of Boolean functions of degree n.

*10. Draw the Hasse diagram for the poset consisting of the set of the 16 Boolean functions of degree two (shown in Table 3 of Section 10.1) with the partial ordering $\leq$.

*11. For each of these equalities either prove it is an identity or find a set of values of the variables for which it does not hold.

a) $x \mid (y \mid z) = (x \mid y) \mid z$
b) $x \downarrow (y \downarrow z) = (x \downarrow y) \downarrow (x \downarrow z)$
c) $x \downarrow (y \mid z) = (x \downarrow y) \mid (x \downarrow z)$

Define the Boolean operator $\odot$ as follows: $1 \odot 1 = 1$, $1 \odot 0 = 0, 0 \odot 1 = 0$, and $0 \odot 0 = 1$.

12. Show that $x \odot y = xy + \overline{x}\,\overline{y}$.

13. Show that $x \odot y = \overline{(x \oplus y)}$.

14. Show that each of these identities holds.

a) $x \odot x = 1$ **b)** $x \odot \overline{x} = 0$
c) $x \odot y = y \odot x$

15. Is it always true that $(x \odot y) \odot z = x \odot (y \odot z)$?

*16. Determine whether the set $\{\odot\}$ is functionally complete.

*17. How many of the 16 Boolean functions in two variables x and y can be represented using only the given set of operators, variables x and y, and values 0 and 1?

a) $\{^-\}$ **b)** $\{\cdot\}$ **c)** $\{+\}$ **d)** $\{\cdot, +\}$

The notation for an **XOR gate,** which produces the output $x \oplus y$ from x and y, is as follows:

18. Determine the output of each of these circuits.

a)

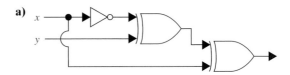

b)

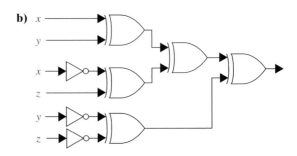

19. Show how a half adder can be constructed using fewer gates than are used in Figure 8 of Section 10.3 when *XOR* gates can be used in addition to *OR* gates, *AND* gates, and inverters.

20. Design a circuit that determines whether three or more of four individuals on a committee vote yes on an issue, where each individual uses a switch for the voting.

A **threshold gate** produces an output y that is either 0 or 1 given a set of input values for the Boolean variables $x_1, x_2, \ldots, x_n$. A threshold gate has a **threshold value** T, which is a real number, and **weights** $w_1, w_2, \ldots, w_n$, each of which is a real number. The output y of the threshold gate is 1 if and only if $w_1x_1 + w_2x_2 + \cdots + w_nx_n \geq T$. The threshold gate with threshold value T and weights $w_1, w_2, \ldots, w_n$ is represented by the following diagram. Threshold gates are useful in modeling in neurophysiology and in artificial intelligence.

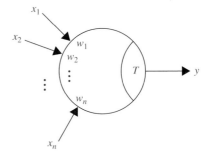

21. A threshold gate represents a Boolean function. Find a Boolean expression for the Boolean function represented by this threshold gate.

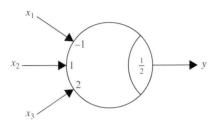

22. A Boolean function that can be represented by a threshold gate is called a **threshold function.** Show that each of these functions is a threshold function.

    a) $F(x) = \bar{x}$ **b)** $F(x, y) = x + y$
    c) $F(x, y) = xy$ **d)** $F(x, y) = x \mid y$
    e) $F(x, y) = x \downarrow y$
    f) $F(x, y, z) = x + yz$
    g) $F(w, x, y, z) = w + xy + z$
    h) $F(w, x, y, z) = wxz + x\bar{y}z$

*23. Show that $F(x, y) = x \oplus y$ is not a threshold function.
*24. Show that $F(w, x, y, z) = wx + yz$ is not a threshold function.

Computer Projects

WRITE PROGRAMS WITH THESE INPUT AND OUTPUT.

1. Given the values of two Boolean variables x and y, find the values of $x + y, xy, x \oplus y, x \mid y$, and $x \downarrow y$.

2. Construct a table listing the set of values of all 256 Boolean functions of degree three.

3. Given the values of a Boolean function in n variables, where n is a positive integer, construct the sum-of-products expansion of this function.

4. Given the table of values of a Boolean function, express this function using only the operators $\cdot$ and $\bar{}$.

5. Given the table of values of a Boolean function, express this function using only the operators $+$ and $\bar{}$.

*6. Given the table of values of a Boolean function, express this function using only the operator $\mid$.

*7. Given the table of values of a Boolean function, express this function using only the operator $\downarrow$.

8. Given the table of values of a Boolean function of degree three, construct its K-map.

9. Given the table of values of a Boolean function of degree four, construct its K-map.

**10. Given the table of values of a Boolean function, use the Quine–McCluskey method to find a minimal sum-of-products representation of this function.

11. Given a threshold value and a set of weights for a threshold gate and the values of the n Boolean variables in the input, determine the output of this gate.

12. Given a positive integer n, construct a random Boolean expression in n variables in disjunctive normal form.

Computations and Explorations

USE A COMPUTATIONAL PROGRAM OR PROGRAMS YOU HAVE WRITTEN TO DO THESE EXERCISES.

1. Compute the number of Boolean functions of degrees seven, eight, nine, and ten.

2. Construct a table of the Boolean functions of degree three.

3. Construct a table of the Boolean functions of degree four.

4. Express each of the different Boolean expressions in three variables in disjunctive normal form with just the *NAND* operator, using as few *NAND* operators as possible. What is the largest number of *NAND* operators required?

5. Express each of the different Boolean expressions in disjunctive normal form in four variables using just the *NOR* operator, with as few *NOR* operators as possible. What is the largest number of *NOR* operators required?

6. Randomly generate 10 different Boolean expressions in four variables and determine the average number of steps required to minimize them using the Quine–McCluskey method.

7. Randomly generate 10 different Boolean expressions in five variables and determine the average number of steps required to minimize them using the Quine–McCluskey method.

Writing Projects

RESPOND TO THESE QUESTIONS WITH ESSAYS USING OUTSIDE SOURCES.

1. Describe some of the early machines devised to solve problems in logic, such as the Stanhope Demonstrator, Jevons's Logic Machine, and the Marquand Machine.
2. Explain the difference between combinational circuits and sequential circuits. Then explain how *flip-flops* are used to build sequential circuits.
3. Define a *shift register* and discuss how shift registers are used. Show how to build shift registers using flip-flops and logic gates.
4. Show how *multipliers* can be built using logic gates.
5. Find out how logic gates are physically constructed. Discuss whether *NAND* and *NOR* gates are used in building circuits.
6. Explain how *dependency notation* can be used to describe complicated switching circuits.
7. Describe how multiplexers are used to build switching circuits.

8. Explain the advantages of using threshold gates to construct switching circuits. Illustrate this by using threshold gates to construct half and full adders.
9. Describe the concept of *hazard-free switching circuits* and give some of the principles used in designing such circuits.
10. Explain how to use K-maps to minimize functions of six variables.
11. Discuss the ideas used by newer methods for minimizing Boolean functions, such as Espresso. Explain how these methods can help solve minimization problems in as many as 25 variables.
12. Describe what is meant by the *functional decomposition* of a Boolean function of n variables and discuss procedures for decomposing Boolean functions into a composition of Boolean functions with fewer variables.

Modeling Computation

Computers can perform many tasks. Given a task, two questions arise. The first is: Can it be carried out using a computer? Once we know that this first question has an affirmative answer, we can ask the second question: How can the task be carried out? Models of computation are used to help answer these questions.

We will study three types of structures used in models of computation, namely, grammars, finite-state machines, and Turing machines. Grammars are used to generate the words of a language and to determine whether a word is in a language. Formal languages, which are generated by grammars, provide models for both natural languages, such as English, and for programming languages, such as Pascal, Fortran, Prolog, C, and Java. In particular, grammars are extremely important in the construction and theory of compilers. The grammars that we will discuss were first used by the American linguist Noam Chomsky in the 1950s.

Various types of finite-state machines are used in modeling. All finite-state machines have a set of states, including a starting state, an input alphabet, and a transition function that assigns a next state to every pair of a state and an input. The states of a finite-state machine give it limited memory capabilities. Some finite-state machines produce an output symbol for each transition; these machines can be used to model many kinds of machines, including vending machines, delay machines, binary adders, and language recognizers. We will also study finite-state machines that have no output but do have final states. Such machines are extensively used in language recognition. The strings that are recognized are those that take the starting state to a final state. The concepts of grammars and finite-state machines can be tied together. We will characterize those sets that are recognized by a finite-state machine and show that these are precisely the sets that are generated by a certain type of grammar.

Finally, we will introduce the concept of a Turing machine. We will show how Turing machines can be used to recognize sets. We will also show how Turing machines can be used to compute number-theoretic functions. We will discuss the Church–Turing thesis, which states that every effective computation can be carried out using a Turing machine.

11.1 Languages and Grammars

INTRODUCTION

Words in the English language can be combined in various ways. The grammar of English tells us whether a combination of words is a valid sentence. For instance, *the frog writes neatly* is a valid sentence, since it is formed from a noun phrase, *the frog,* made up of the article *the* and the noun *frog,* followed by a verb phrase, *writes neatly,* made up of the verb *writes* and the adverb *neatly*. We do not care that this is a nonsensical statement, because we are concerned only with the **syntax,** or form, of the sentence, and not its **semantics,** or meaning. We also note that the combination of words *swims quickly mathematics* is not a valid sentence because it does not follow the rules of English grammar.

The syntax of a **natural language,** that is, a spoken language, such as English, French, German, or Spanish, is extremely complicated. In fact, it does not seem possible to specify all the rules of syntax for a natural language. Research in the automatic translation of one language to another has led to the concept of a **formal language,** which, unlike a natural language, is specified by a well-defined set of rules of syntax. Rules of syntax are important not only in linguistics, the study of natural languages, but also in the study of programming languages.

We will describe the sentences of a formal language using a grammar. The use of grammars helps when we consider the two classes of problems that arise most frequently in applications to programming languages: (1) How can we determine whether a combination of words is a valid sentence in a formal language? (2) How can we generate the valid sentences of a formal language?

Before giving a technical definition of a grammar, we will describe an example of a grammar that generates a subset of English. This subset of English is defined using a list of rules that describe how a valid sentence can be produced. We specify that

1. a **sentence** is made up of a **noun phrase** followed by a **verb phrase;**
2. a **noun phrase** is made up of an **article** followed by an **adjective** followed by a **noun,** or
3. a **noun phrase** is made up of an **article** followed by a **noun;**
4. a **verb phrase** is made up of a **verb** followed by an **adverb,** or
5. a **verb phrase** is made up of a **verb;**
6. an article is *a*, or
7. an article is *the;*
8. an adjective is *large*, or
9. an adjective is *hungry;*
10. a noun is *rabbit*, or
11. a noun is *mathematician;*
12. a verb is *eats*, or
13. a verb is *hops;*
14. an adverb is *quickly*, or
15. an adverb is *wildly*.

From these rules we can form valid sentences using a series of replacements until no more rules can be used. For instance, we can follow the sequence of replacements:

sentence
noun phrase   verb phrase
article   adjective   noun   verb phrase
article   adjective   noun   verb   adverb
the **adjective   noun   verb   adverb**
the large **noun   verb   adverb**
the large rabbit **verb   adverb**
the large rabbit hops **adverb**
the large rabbit hops quickly

to obtain a valid sentence. It is also easy to see that some other valid sentences are: *a hungry mathematician eats wildly, a large mathematician hops, the rabbit eats quickly*, and so on. Also, we can see that *the quickly eats mathematician* is not a valid sentence.

PHRASE-STRUCTURE GRAMMARS

Before we give a formal definition of a grammar, we introduce a little terminology.

DEFINITION 1

> A *vocabulary* (or *alphabet*) V is a finite, nonempty set of elements called *symbols*. A *word* (or *sentence*) over V is a string of finite length of elements of V. The *empty string* or *null string,* denoted by λ, is the string containing no symbols. The set of all words over V is denoted by V^*. A *language over V* is a subset of V^*.

Note that λ, the empty string, is the string containing no symbols. It is different from $\emptyset$, the empty set. It follows that $\{\lambda\}$ is the set containing exactly one string, namely, the empty string.

Languages can be specified in various ways. One way is to list all the words in the language. Another is to give some criteria that a word must satisfy to be in the language. In this section we describe another important way to specify a language, namely, through the use of a grammar, such as the set of rules we gave in the introduction to this section. A grammar provides a set of symbols of various types and a set of rules for producing words. More precisely, a grammar has a **vocabulary** V, which is a set of symbols used to derive members of the language. Some of the elements of the vocabulary cannot be replaced by other symbols. These are called **terminals,** and the other members of the vocabulary, which can be replaced by other symbols, are called **nonterminals.** The sets of terminals and nonterminals are usually denoted by T and N, respectively. In the example given in the introduction of the section, the set of terminals is {*a, the, rabbit, mathematician, hops, eats, quickly, wildly*}, and the set of nonterminals is {**sentence, noun phrase, verb phrase, adjective, article, noun, verb, adverb**}. There is a special member of the vocabulary called the **start symbol,** denoted by S, which is the element of the vocabulary that we always begin with. In the example in the introduction, the start symbol is **sentence.** The rules that specify when we can replace a string from V^*, the set of all strings of elements in the vocabulary, with another string are called the **productions** of the grammar. We denote by $z_0 \to z_1$ the production that specifies that z_0 can be replaced by z_1 within a string. The productions in the grammar given in the introduction of this section were listed. The first production, written using this notation, is **sentence $\to$ noun phrase verb phrase.** We summarize with the following definition.

DEFINITION 2

> A *phrase-structure grammar* $G = (V, T, S, P)$ consists of a vocabulary V, a subset T of V consisting of terminal elements, a start symbol S from V, and a set of productions P. The set $V - T$ is denoted by N. Elements of N are called *nonterminal symbols*. Every production in P must contain at least one nonterminal on its left side.

EXAMPLE 1

Let $G = (V, T, S, P)$ where $V = \{a, b, A, B, S\}, T = \{a, b\}, S$ is the start symbol, and $P = \{S \to ABa, A \to BB, B \to ab, AB \to b\}$. G is an example of a phrase-structure grammar. ◀

We will be interested in the words that can be generated by the productions of a phrase-structure grammar.

DEFINITION 3

Let $G = (V, T, S, P)$ be a phrase-structure grammar. Let $w_0 = lz_0r$ (that is, the concatenation of l, z_0, and r) and $w_1 = lz_1r$ be strings over V. If $z_0 \rightarrow z_1$ is a production of G, we say that w_1 is *directly derivable* from w_0 and we write $w_0 \Rightarrow w_1$. If $w_0, w_1, \ldots, w_n$ are strings over V such that $w_0 \Rightarrow w_1, w_1 \Rightarrow w_2, \ldots, w_{n-1} \Rightarrow w_n$, then we say that w_n is *derivable from* w_0, and we write $w_0 \overset{*}{\Rightarrow} w_n$. The sequence of steps used to obtain w_n from w_0 is called a *derivation.*

EXAMPLE 2 The string $Aaba$ is directly derivable from ABa in the grammar in Example 1 since $B \rightarrow ab$ is a production in the grammar. The string $ababab$ is derivable from ABa since $ABa \Rightarrow Aaba \Rightarrow BBaba \Rightarrow Bababa \Rightarrow ababab$, using the productions $B \rightarrow ab$, $A \rightarrow BB, B \rightarrow ab$, and $B \rightarrow ab$ in succession. ◀

DEFINITION 4

Let $G = (V, T, S, P)$ be a phrase-structure grammar. The *language generated by G* (or the *language of G*), denoted by $L(G)$, is the set of all strings of terminals that are derivable from the starting state S. In other words,

$$L(G) = \{w \in T^* \mid S \overset{*}{\Rightarrow} w\}.$$

In Examples 3 and 4 we find the language generated by a phrase-structure grammar.

EXAMPLE 3 Let G be the grammar with vocabulary $V = \{S, A, a, b\}$, set of terminals $T = \{a, b\}$, starting symbol S, and productions $P = \{S \rightarrow aA, S \rightarrow b, A \rightarrow aa\}$. What is $L(G)$, the language of this grammar?

Solution: From the start state S we can derive aA using the production $S \rightarrow aA$. We can also use the production $S \rightarrow b$ to derive b. From aA the production $A \rightarrow aa$ can be used to derive aaa. No additional words can be derived. Hence $L(G) = \{b, aaa\}$. ◀

EXAMPLE 4 Let G be the grammar with vocabulary $V = \{S, 0, 1\}$, set of terminals $T = \{0, 1\}$, starting symbol S, and productions $P = \{S \rightarrow 11S, S \rightarrow 0\}$. What is $L(G)$, the language of this grammar?

Solution: From S we can derive 0 using $S \rightarrow 0$, or $11S$ using $S \rightarrow 11S$. From $11S$ we can derive either 110 or $1111S$. From $1111S$ we can derive 11110 and $111111S$. At any stage of a derivation we can either add two 1s at the end of the string or terminate the derivation by adding a 0 at the end of the string. We surmise that $L(G) = \{0, 110, 11110, 1111110, \ldots\}$, the set of all strings that begin with an even number of 1s and end with a 0. This can be proved using an inductive argument that shows that after n productions have been used, the only strings of terminals generated are those consisting of $n - 1$ or fewer concatenations of 11 followed by 0. (This is left as an exercise for the reader.) ◀

The problem of constructing a grammar that generates a given language often arises. Examples 5, 6, and 7 describe problems of this kind.

EXAMPLE 5   Give a phrase-structure grammar that generates the set $\{0^n 1^n \mid n = 0, 1, 2, \ldots\}$.

Solution: Two productions can be used to generate all strings consisting of a string of 0s followed by a string of the same number of 1s, including the null string. The first builds up successively longer strings in the language by adding a 0 at the start of the string and a 1 at the end. The second production replaces S with the empty string. The solution is the grammar $G = (V, T, S, P)$, where $V = \{0, 1, S\}, T = \{0, 1\}, S$ is the starting symbol, and the productions are

$$S \rightarrow 0S1$$
$$S \rightarrow \lambda.$$

The verification that this grammar generates the correct set is left as an exercise for the reader. ◄

Example 5 involved the set of strings made up of 0s followed by 1s, where the number of 0s and 1s are the same. Example 6 considers the set of strings consisting of 0s followed by 1s, where the number of 0s and 1s may differ.

EXAMPLE 6   Find a phrase-structure grammar to generate the set $\{0^m 1^n \mid m$ and n are nonnegative integers$\}$.

Solution: We will give two grammars G_1 and G_2 that generate this set. This will illustrate that two grammars can generate the same language.

The grammar G_1 has alphabet $V = \{S, 0, 1\}$, terminals $T = \{0, 1\}$, and productions $S \rightarrow 0S, S \rightarrow S1$, and $S \rightarrow \lambda$. G_1 generates the correct set, since using the first production m times puts m 0s at the beginning of the string, and using the second production n times puts n 1s at the end of the string. The details of this verification are left to the reader.

The grammar G_2 has alphabet $V = \{S, A, 0, 1\}$, terminals $T = \{0, 1\}$, and productions $S \rightarrow 0S, S \rightarrow 1A, S \rightarrow 1, A \rightarrow 1A, A \rightarrow 1, S \rightarrow \lambda$. The details that this grammar generates the correct set are left as an exercise for the reader. ◄

Sometimes a set that is easy to describe can be generated only by a complicated grammar. Example 7 illustrates this.

EXAMPLE 7   One grammar that generates the set $\{0^n 1^n 2^n \mid n = 0, 1, 2, 3, \ldots\}$ is $G = (V, T, S, P)$ with $V = \{0, 1, 2, S, A, B\}$, $T = \{0, 1, 2\}$, starting state S, and productions $S \rightarrow 0SAB, S \rightarrow \lambda, BA \rightarrow AB, 0A \rightarrow 01, 1A \rightarrow 11, 1B \rightarrow 12, 2B \rightarrow 22$. We leave it as an exercise for the reader to show that this statement is correct. The grammar given is the simplest type of grammar that generates this set, in a sense that will be made clear later in this section. The reader may wonder where this grammar came from, since it seems difficult to come up with this grammar from scratch. It may be comforting to know that this grammar can be systematically constructed using techniques from the theory of computation that are beyond the scope of this book. ◄

TYPES OF PHRASE-STRUCTURE GRAMMARS

Links

Phrase-structure grammars can be classified according to the types of productions that are allowed. We will describe the classification scheme introduced by Noam Chomsky. In Section 11.4 we will see that the different types of languages defined in this scheme correspond to the classes of languages that can be recognized using different models of computing machines.

A **type 0** grammar has no restrictions on its productions. A **type 1** grammar can have productions of the form $w_1 \rightarrow w_2$ where the length of w_2 is greater than the length of w_1 or of the form $w_1 \rightarrow \lambda$. A **type 2** grammar can have productions only of the form $w_1 \rightarrow w_2$, where w_1 is a single symbol that is not a terminal symbol. A **type 3** grammar can have productions only of the form $w_1 \rightarrow w_2$ with $w_1 = A$ and either $w_2 = aB$ or $w_2 = a$, where A and B are nonterminal symbols and a is a terminal symbol, or with $w_1 = S$ and $w_2 = \lambda$.

From these definitions we see that every type 3 grammar is a type 2 grammar, every type 2 grammar is a type 1 grammar, and every type 1 grammar is a type 0 grammar. Type 2 grammars are called **context-free grammars** since a nonterminal symbol that is the left side of a production can be replaced in a string whenever it occurs, no matter what else is in the string. A language generated by a type 2 grammar is called a **context-free language.** When there is a production of the form $lw_1r \rightarrow lw_2r$ (but not of the form $w_1 \rightarrow w_2$), the grammar is called type 1 or **context-sensitive** since w_1 can be replaced by w_2 only when it is surrounded by the strings l and r. Type 3 grammars are also called **regular grammars.** A language generated by a regular grammar is called **regular.** Section 11.4 deals with the relationship between regular languages and finite-state machines. The Venn diagram in Figure 1 shows the relationship among different types of grammars.

EXAMPLE 8   From Example 6 we know that $\{0^m1^n \mid m, n = 0, 1, 2, \ldots\}$ is a regular language, since it can be generated by a regular grammar, namely, the grammar G_2 in Example 6.   ◄

Context-free and regular grammars play an important role in programming languages. Context-free grammars are used to define the syntax of almost all programming languages. These grammars are strong enough to define a wide range of languages. Furthermore, efficient algorithms can be devised to determine whether and how a string can be generated. Regular grammars are used to search text for certain patterns and in lexical analysis, which is the process of transforming an input stream into a stream of tokens for use by a parser.

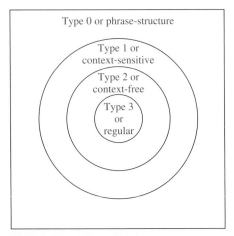

FIGURE 1   Types of Grammars.

EXAMPLE 9   It follows from Example 5 that $\{0^n 1^n \mid n = 0, 1, 2, \ldots\}$ is a context-free language, since the productions in this grammar are $S \rightarrow 0S1$ and $S \rightarrow \lambda$. However, it is not a regular language. This will be shown in Section 11.4. ◀

EXAMPLE 10   The set $\{0^n 1^n 2^n \mid n = 0, 1, 2, \ldots\}$ is a context-sensitive language, since it can be generated by a type 1 grammar, as Example 7 shows, but not by any type 2 language. (This is shown in Exercise 28 in the supplementary exercises at the end of the chapter.) ◀

Table 1 summarizes the terminology used to classify phrase-structure grammars.

DERIVATION TREES

A derivation in the language generated by a context-free grammar can be represented graphically using an ordered rooted tree, called a **derivation,** or **parse, tree.** The root of this tree represents the starting symbol. The internal vertices of the tree represent the nonterminal symbols that arise in the derivation. The leaves of the tree represent the terminal symbols that arise. If the production $A \rightarrow w$ arises in the derivation, where w is a word, the vertex that represents A has as children vertices that represent each symbol in w, in order from left to right.

EXAMPLE 11   Construct a derivation tree for the derivation of *the hungry rabbit eats quickly,* given in the introduction of this section.

Solution: The derivation tree is shown in Figure 2. ◀

TABLE 1 Types of Grammars.	
Type	*Restrictions on Productions $w_1 \rightarrow w_2$*
0	No restrictions
1	$l(w_1) < l(w_2)$, or $w_2 = \lambda$
2	$w_1 = A$ where A is a nonterminal symbol
3	$w_1 = A$ and $w_2 = aB$ or $w_2 = a$, where $A \in N, B \in N$, and $a \in T$, or $S \rightarrow \lambda$

AVRAM NOAM CHOMSKY (BORN 1928) Noam Chomsky, born in Philadelphia, is the son of a Hebrew scholar. He received his B.A., M.A., and Ph.D. in linguistics, all from the University of Pennsylvania. He was on the staff of the University of Pennsylvania from 1950 until 1951. In 1955 he joined the faculty at M.I.T., beginning his M.I.T. career teaching engineers French and German. Chomsky is currently the Ferrari P. Ward Professor of foreign languages and linguistics at M.I.T. He is known for his many fundamental contributions to linguistics, including the study of grammars. Chomsky is also widely known for his outspoken political activism.

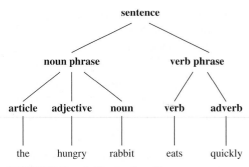

FIGURE 2   A Derivation Tree.

The problem of determining whether a string is in the language generated by a context-free grammar arises in many applications, such as in the construction of compilers. Two approaches to this problem are indicated in the following example.

EXAMPLE 12   Determine whether the word *cbab* belongs to the language generated by the grammar $G = (V, T, S, P)$ where $V = \{a, b, c, A, B, C, S\}$, $T = \{a, b, c\}$, S is the starting symbol, and the productions are

$$S \rightarrow AB$$
$$A \rightarrow Ca$$
$$B \rightarrow Ba$$
$$B \rightarrow Cb$$
$$B \rightarrow b$$
$$C \rightarrow cb$$
$$C \rightarrow b.$$

Solution: One way to approach this problem is to begin with S and attempt to derive *cbab* using a series of productions. Since there is only one production with S on its left-hand side, we must start with $S \Rightarrow AB$. Next we use the only production that has A on its left-hand side, namely $A \rightarrow Ca$, to obtain $S \Rightarrow AB \Rightarrow CaB$. Since *cbab* begins with the symbols *cb*, we use the production $C \rightarrow cb$. This gives us $S \Rightarrow Ab \Rightarrow CaB \Rightarrow cbaB$. We finish by using the production $B \rightarrow b$, to obtain $S \Rightarrow AB \Rightarrow CaB \Rightarrow cbaB \Rightarrow cbab$. The approach that we have used is called **top-down parsing,** since it begins with the starting symbol and proceeds by successively applying productions.

There is another approach to this problem, called **bottom-up parsing.** In this approach, we work backward. Since *cbab* is the string to be derived, we can use the production $C \rightarrow cb$, so that $Cab \Rightarrow cbab$. Then, we can use the production $A \rightarrow Ca$, so that $Ab \Rightarrow Cab \Rightarrow cbab$. Using the production $B \rightarrow b$ gives $AB \Rightarrow Ab \Rightarrow Cab \Rightarrow cbab$. Finally, using $S \rightarrow AB$ shows that a complete derivation for *cbab* is $S \Rightarrow AB \Rightarrow Ab \Rightarrow Cab \Rightarrow cbab$. ◀

BACKUS–NAUR FORM

There is another notation that is sometimes used to specify a type 2 grammar, called the **Backus–Naur form (BNF),** after John Backus, who invented it, and Peter Naur, who refined it for use in the specification of the programming language ALGOL. (Surprisingly, a notation quite similar to Backus–Naur form was used approximately 2500 years ago to describe the grammar of Sanskrit.) The Backus–Naur form is used to specify the syntactic rules of many computer languages, including Java. The productions in a type 2 grammar have a single nonterminal symbol as their left-hand side. Instead of listing all the productions separately, we can combine all those with the same nonterminal symbol on the left-hand side into one statement. Instead of using the symbol $\rightarrow$ in a production, we use the symbol $::=$. We enclose all nonterminal symbols in brackets, $\langle \rangle$, and we list all the right-hand sides of productions in the same statement, separating them by bars. For instance, the productions $A \rightarrow Aa$, $A \rightarrow a$, and $A \rightarrow AB$ can be combined into $\langle A \rangle ::= \langle A \rangle a \mid a \mid \langle A \rangle \langle B \rangle$.

Example 13 illustrates how the Backus–Naur form is used to describe the syntax of programming languages. Our example comes from the original use of Backus–Naur form in the description of ALGOL 60.

EXAMPLE 13

In ALGOL 60 an identifier (which is the name of an entity such as a variable) consists of a string of alphanumeric characters (that is, letters and digits) and must begin with a letter. We can use these rules in Backus–Naur to describe the set of allowable identifiers:

$\langle identifier \rangle ::= \langle letter \rangle \mid \langle identifier \rangle \langle letter \rangle \mid \langle identifier \rangle \langle digit \rangle$

$\langle letter \rangle ::= a \mid b \mid \cdots \mid y \mid z$ the ellipsis indicates that all 26 letters
                                                            are included

$\langle digit \rangle ::= 0 \mid 1 \mid 2 \mid 3 \mid 4 \mid 5 \mid 6 \mid 7 \mid 8 \mid 9$

JOHN BACKUS (BORN 1924) John Backus was born in Philadelphia and grew up in Wilmington, Delaware. He attended the Hill School in Pottstown, Pennsylvania. He needed to attend summer school every year since he disliked studying and was not a serious student. But he enjoyed spending his summers in New Hampshire where he attended summer school and amused himself with summer activities, including sailing. He obliged his father by enrolling at the University of Virginia to study chemistry. But he quickly decided chemistry was not for him, and in 1943 he entered the army, where he received medical training and worked in a neurosurgery ward in an army hospital. Ironically, Backus was soon diagnosed with a bone tumor in his skull and was fitted with a metal plate. His medical work in the army convinced him to try medical school, but he abandoned this after nine months because he disliked the rote memorization required. After dropping out of medical school, he entered a school for radio technicians because he wanted to build his own high fidelity set. A teacher in this school recognized his potential and asked him to help with some mathematical calculations needed for an article in a magazine. Finally, Backus found what he was interested in: mathematics and its applications. He enrolled at Columbia University, from which he received both a bachelor's and a master's degrees in mathematics. Backus joined IBM as a

programmer in 1950. He participated in the design and development of two of IBM's early computers. From 1954 to 1958 he led the IBM group that developed FORTRAN. Backus became a staff member at the IBM Watson Research Center in 1958. He was part of the committees that designed the programming language ALGOL, using what is now called the Backus–Naur form for the description of the syntax of this language. Later, Backus worked on the mathematics of families of sets and on a functional style of programming. Backus became an IBM Fellow in 1963, and he received the National Medal of Science in 1974 and the prestigious Turing Award from the Association of Computing Machinery in 1977.

PETER NAUR (BORN 1928) Peter Naur was born in Frederiksberg, near Copenhagen. As a boy he became interested in astronomy. Not only did he observe heavenly bodies, but he also computed the orbits of comets and asteroids. Naur attended Copenhagen University, receiving his degree in 1949. He spent 1950 and 1951 in Cambridge, where he used an early computer to calculate the motions of comets and planets. After returning to Denmark he continued working in astronomy but kept his ties to computing. In 1955 he served as a consultant to the building of the first Danish computer. In 1959 Naur made the switch from astronomy to computing as a full-time activity. His first job as a full-time computer scientist was participating in the development of the programming language ALGOL. From 1960 to 1967 he worked on the development of compilers for ALGOL and COBOL. In 1969 he became professor of computer science at Copenhagen University, where he has worked in the area of programming methodology.

For example, we can produce the valid identifier $x99a$ by using the first rule to replace ⟨*identifier*⟩ by ⟨*identifier*⟩⟨*letter*⟩, the second rule to obtain ⟨*identifier*⟩a, the first rule twice to obtain ⟨*identifier*⟩⟨*digit*⟩⟨*digit*⟩a, the third rule twice to obtain ⟨*identifier*⟩$99a$, the first rule to obtain ⟨*letter*⟩$99a$, and finally the second rule to obtain $x99a$. ◄

EXAMPLE 14   What is the Backus–Naur form of the grammar for the subset of English described in the introduction to this section?

Solution: The Backus–Naur form of this grammar is

⟨*sentence*⟩ ::= ⟨*noun phrase*⟩⟨*verb phrase*⟩
⟨*noun phrase*⟩ ::= ⟨*article*⟩⟨*adjective*⟩⟨*noun*⟩ | ⟨*article*⟩⟨*noun*⟩
⟨*verb phrase*⟩ ::= ⟨*verb*⟩⟨*adverb*⟩ | ⟨*verb*⟩
⟨*article*⟩ ::= *a* | *the*
⟨*adjective*⟩ ::= *large* | *hungry*
⟨*noun*⟩ ::= *rabbit* | *mathematician*
⟨*verb*⟩ ::= *eats* | *hops*
⟨*adverb*⟩ ::= *quickly* | *wildly* ◄

EXAMPLE 15   Give the Backus–Naur form for the production of signed integers in decimal notation. (A **signed integer** is a nonnegative integer preceded by a plus sign or a minus sign.)

Solution: The Backus–Naur form for a grammar that produces signed integers is

⟨*signed integer*⟩ ::= ⟨*sign*⟩⟨*integer*⟩
⟨*sign*⟩ ::= + | −
⟨*integer*⟩ ::= ⟨*digit*⟩ | ⟨*digit*⟩⟨*integer*⟩
⟨*digit*⟩ ::= 0 | 1 | 2 | 3 | 4 | 5 | 6 | 7 | 8 | 9 ◄

The Backus–Naur form, with a variety of extensions, is used extensively to specify the syntax of programming languages, such as Java and LISP; database languages, such as SQL; and markup languages, such as XML. Some extensions of the Backus–Naur form that are commonly used in the description of programming languages are introduced in the preamble to Exercise 28 at the end of this section.

Exercises

Exercises 1–3 refer to the grammar with start symbol **sentence,** set of terminals $T = \{the, sleepy, happy, tortoise, hare, passes, runs, quickly, slowly\}$, set of nonterminals $N = \{$**noun phrase, transitive verb phrase, intransitive verb phrase, article, adjective, noun, verb, adverb**$\}$, and productions:

sentence → **noun phrase   transitive verb phrase**
    noun phrase
sentence → **noun phrase   intransitive verb phrase**
noun phrase → **article   adjective   noun**
noun phrase → **article   noun**
transitive verb phrase → **transitive verb**
intransitive verb phrase → **intransitive verb   adverb**
intransitive verb phrase → **intransitive verb**
article → *the*

adjective → *sleepy*
adjective → *happy*
noun → *tortoise*
noun → *hare*
transitive verb → *passes*
intransitive verb → *runs*
adverb → *quickly*
adverb → *slowly*

1. Use the set of productions to show that each of these sentences is a valid sentence.
   a) *the happy hare runs*
   b) *the sleepy tortoise runs quickly*
   c) *the tortoise passes the hare*
   d) *the sleepy hare passes the happy tortoise*

2. Find five other valid sentences, besides those given in Exercise 1.

3. Show that *the hare runs the sleepy tortoise* is not a valid sentence.

*4. Let $V = \{S, A, B, a, b\}$ and $T = \{a, b\}$. Find the language generated by the grammar $\{V, T, S, P\}$ when the set P of productions consists of

   a) $S \to AB, A \to ab, B \to bb$.

   b) $S \to AB, S \to aA, A \to a, B \to ba$.

   c) $S \to AB, S \to AA, A \to aB, A \to ab, B \to b$.

   d) $S \to AA, S \to B, A \to aaA, A \to aa, B \to bB$, $B \to b$.

   e) $S \to AB, A \to aAb, B \to bBa, A \to \lambda, B \to \lambda$.

5. Construct a derivation of $0^3 1^3$ using the grammar given in Example 5.

6. Show that the grammar given in Example 5 generates the set $\{0^n 1^n \mid n = 0, 1, 2, \ldots\}$.

7. a) Construct a derivation of $0^2 1^4$ using the grammar G_1 in Example 6.

   b) Construct a derivation of $0^2 1^4$ using the grammar G_2 in Example 6.

8. a) Show that the grammar G_1 given in Example 6 generates the set $\{0^m 1^n \mid m, n = 0, 1, 2, \ldots\}$.

   b) Show that the grammar G_2 in Example 6 generates the same set.

9. Construct a derivation of $0^2 1^2 2^2$ in the grammar given in Example 7.

*10. Show that the grammar given in Example 7 generates the set $\{0^n 1^n 2^n \mid n = 0, 1, 2, \ldots\}$.

*11. Find a phrase-structure grammar for each of these languages.

   a) the set of all bit strings containing an even number of 0s and no 1s

   b) the set of all bit strings made up of a 1 followed by an odd number of 0s

   c) the set of all bit strings containing an even number of 0s and an even number of 1s

   d) the set of all strings containing 10 or more 0s and no 1s

   e) the set of all strings containing more 0s than 1s

   f) the set of all strings containing an equal number of 0s and 1s

   g) the set of all strings containing an unequal number of 0s and 1s

12. Construct phrase-structure grammars to generate each of these sets.

   a) $\{01^{2n} \mid n \geq 0\}$

   b) $\{0^n 1^{2n} \mid n \geq 0\}$

   c) $\{0^n 1^m 0^n \mid m \geq 0, n \geq 0\}$

13. Let $V = \{S, A, B, a, b\}$ and $T = \{a, b\}$. Determine whether $G = (V, T, S, P)$ is a type 0 grammar but not a type 1 grammar, a type 1 grammar but not a type 2 grammar, or a type 2 grammar but not a type 3 grammar if P, the set of productions, is

   a) $S \to aAB, A \to Bb, B \to \lambda$.

   b) $S \to aA, A \to a, A \to b$.

   c) $S \to ABa, AB \to a$.

   d) $S \to ABA, A \to aB, B \to ab$.

   e) $S \to bA, A \to B, B \to a$.

   f) $S \to aA, aA \to B, B \to aA, A \to b$.

   g) $S \to bA, A \to b, S \to \lambda$.

   h) $S \to AB, B \to aAb, aAb \to b$.

   i) $S \to aA, A \to bB, B \to b, B \to \lambda$.

   j) $S \to A, A \to B, B \to \lambda$.

14. A **palindrome** is a string that reads the same backward as it does forward, that is, a string w where $w = w^R$, where w^R is the reversal of the string w. Find a context-free grammar that generates the set of all palindromes over the alphabet $\{0, 1\}$.

*15. Let G_1 and G_2 be context-free grammars, generating the languages $L(G_1)$ and $L(G_2)$, respectively. Show that there is a context-free grammar generating each of these sets.

   a) $L(G_1) \cup L(G_2)$ **b)** $L(G_1)L(G_2)$

   c) $L(G_1)^*$

16. Find the strings constructed using the derivation trees shown here.

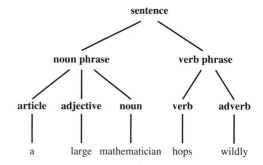

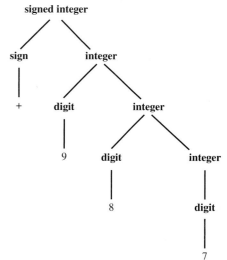

17. Construct derivation trees for the sentences in Exercise 1.

18. Let G be the grammar with $V = \{a, b, c, S\}$, $T = \{a, b, c\}$, starting symbol S, and productions $S \rightarrow abS$, $S \rightarrow bcS$, $S \rightarrow bbS$, $S \rightarrow a$, $S \rightarrow cb$. Construct derivation trees for
  a) *bcbba*.
  b) *bbbcbba*.
  c) *bcabbbbbcb*.

*19. Use top-down parsing to determine whether each of the following strings belongs to the language generated by the grammar in Example 12.
  a) *baba*   **b)** *abab*
  c) *cbaba*   **d)** *bbbcba*

*20. Use bottom-up parsing to determine whether the strings in Exercise 19 belong to the language generated by the grammar in Example 12.

21. Construct a derivation tree for -109 using the grammar given in Example 15.

22. a) Explain what the productions are in a grammar if the Backus–Naur form for productions is:

$$\langle expression \rangle ::= (\langle expression \rangle) \mid$$
$$\langle expression \rangle + \langle expression \rangle \mid$$
$$\langle expression \rangle * \langle expression \rangle \mid$$
$$\langle variable \rangle$$
$$\langle variable \rangle \quad ::= x \mid y$$

  b) Find a derivation tree for $(x * y) + x$ in this grammar.

23. a) Construct a phrase-structure grammar that generates all signed decimal numbers, consisting of a sign, either $+$ or $-$; a nonnegative integer; and a decimal fraction that is either the empty string or a decimal point followed by a positive integer, where initial zeros in an integer are allowed.
  b) Give the Backus–Naur form of this grammar.
  c) Construct a derivation tree for -31.4 in this grammar.

24. a) Construct a phrase-structure grammar for the set of all fractions of the form a/b, where a is a signed integer in decimal notation and b is a positive integer.
  b) What is the Backus–Naur form for this grammar?
  c) Construct a derivation tree for $+311/17$ in this grammar.

25. Give production rules in Backus–Naur form for an identifier if it can consist of
  a) one or more lowercase letters.
  b) at least three but no more than six lowercase letters.
  c) one to six uppercase or lowercase letters beginning with an uppercase letter.
  d) a lowercase letter, followed by a digit or an underscore, followed by three or four alphanumeric characters (lower or uppercase letters and digits).

26. Give production rules in Backus–Naur form for the name of a person if this name consists of a first name, which is a string of letters where only the first letter is uppercase; a middle initial; and a last name, which can be any string of letters.

27. Give production rules in Backus–Naur form that generate all identifiers in the C programming language. In C an identifier starts with a letter or an underscore (_) that is followed by one or more lowercase letters, uppercase letters, underscores, and digits.

Several extensions to Backus–Naur form are commonly used to define phrase-structure grammars. In one such extension, a question mark (?) indicates that the symbol, or group of symbols inside parentheses, to its left can appear zero or once (that is, it is optional), an asterisk (*) indicates that the symbol to its left can appear zero or more times, and a plus (+) indicates that the symbol to its left can appear one or more times. These extensions are part of **extended Backus–Naur form (EBNF),** and the symbols ?, *, and + are called **metacharacters.** In EBNF the brackets used to denote nonterminals are usually not shown.

28. Describe the set of strings defined by each of these sets of productions in EBNF.
  a) *string* $::= L+D?L+$
    $L ::= a \mid b \mid c$
    $D ::= 0 \mid 1$
  b) *string* $::= sign\ D+ \mid D+$
    $sign ::= + \mid -$
    $D ::= 0 \mid 1 \mid 2 \mid 3 \mid 4 \mid 5 \mid 6 \mid 7 \mid 8 \mid 9$
  c) *string* $::= L*(D+)?L*$
    $L ::= x \mid y$
    $D ::= 0 \mid 1$

29. Give production rules in extended Backus–Naur form that generate all decimal numerals consisting of an optional sign, a nonnegative integer, and a decimal fraction that is either the empty string or a decimal point followed by an optional positive integer optionally preceded by some number of zeros.

30. Give production rules in extended Backus–Naur form that generate a sandwich if a sandwich consists of a lower slice of bread; mustard or mayonnaise; optional lettuce; an optional slice of tomato; one or more slices of either turkey, chicken, or roast beef (in any combination); optionally some number of slices of cheese; and a top slice of bread.

31. Give production rules in extended Backus–Naur form for identifiers in the C programming language (see Exercise 27).

32. Describe how productions for a grammar in extended Backus–Naur form can be translated into a set of productions for the grammar in Backus–Naur form.

This is the Backus–Naur form that describes the syntax of expressions in postfix (or reverse Polish) notation.

⟨*expression*⟩ ::= ⟨*term*⟩ | ⟨*term*⟩⟨*term*⟩⟨*addOperator*⟩

⟨*addOperator*⟩ ::= + | −

⟨*term*⟩ ::= ⟨*factor*⟩ | ⟨*factor*⟩⟨*factor*⟩⟨*mulOperator*⟩

⟨*mulOperator*⟩ ::= ∗ | /

⟨*factor*⟩ ::= ⟨*identifier*⟩ | ⟨*expression*⟩

⟨*identifier*⟩ ::= a | b | $\cdots$ | z

33. For each of these strings, determine whether it is generated by the grammar given for postfix notation. If it is, find the steps used to generate the string

a) $abc*+$ **b)** $xy++$ **c)** $xy-z*$

d) $wxyz-*/$ **e)** $ade-*$

34. Use Backus–Naur form to describe the syntax of expressions in infix notation, where the set of operators and identifiers is the same as in the BNF for postfix expressions given in the preamble to Exercise 33, but parentheses must surround expressions being used as factors.

35. For each of these strings, determine whether it is generated by the grammar for infix expressions from Exercise 34. If it is, find the steps used to generate the string.

a) $x + y + z$ **b)** $a/b + c/d$

c) $m * (n + p)$ **d)** $+m - n + p - q$

e) $(m + n) * (p - q)$

36. Let G be a grammar and let R be the relation containing the ordered pair (w_0, w_1) if and only if w_1 is directly derivable from w_0 in G. What is the reflexive transitive closure of R?

11.2 Finite-State Machines with Output

INTRODUCTION

Many kinds of machines, including components in computers, can be modeled using a structure called a finite-state machine. Several types of finite-state machines are commonly used in models. All these versions of finite-state machines include a finite set of states, with a designated starting state, an input alphabet, and a transition function that assigns a next state to every state and input pair. Finite-state machines are used extensively in applications in computer science and data networking. For example, finite-state machines are the basis for programs for spell checking, grammar checking, indexing or searching large bodies of text, recognizing speech, transforming text using markup languages such as XML and HTML, and network protocols that specify how computers communicate.

In this section we will study those finite-state machines that produce output. We will show how finite-state machines can be used to model a vending machine, a machine that delays input, a machine that adds integers, and a machine that determines whether a bit string contains a specified pattern.

Before giving formal definitions, we will show how a vending machine can be modeled. A vending machine accepts nickels (5 cents), dimes (10 cents), and quarters (25 cents). When a total of 30 cents or more has been deposited, the machine immediately returns the amount in excess of 30 cents. When 30 cents has been deposited and any excess refunded, the customer can push an orange button and receive an orange juice or push a red button and receive an apple juice. We can describe how the machine works by specifying its states, how it changes states when input is received, and the output that is produced for every combination of input and current state.

The machine can be in any of seven different states s_i, $i = 0, 1, 2, \ldots, 6$, where s_i is the state where the machine has collected $5i$ cents. The machine starts in state s_0, with 0 cents received. The possible inputs are 5 cents, 10 cents, 25 cents, the orange button (O), and the red button (R). The possible outputs are nothing (n), 5 cents, 10 cents, 15 cents, 20 cents, 25 cents, an orange juice, and an apple juice.

TABLE 1 State Table for a Vending Machine.										
	Next State					**Output**				
	Input					*Input*				
State	**5**	**10**	**25**	**O**	**R**	**5**	**10**	**25**	**O**	**R**
s_0	s_1	s_2	s_5	s_0	s_0	n	n	n	n	n
s_1	s_2	s_3	s_6	s_1	s_1	n	n	n	n	n
s_2	s_3	s_4	s_6	s_2	s_2	n	n	5	n	n
s_3	s_4	s_5	s_6	s_3	s_3	n	n	10	n	n
s_4	s_5	s_6	s_6	s_4	s_4	n	n	15	n	n
s_5	s_6	s_6	s_6	s_5	s_5	n	5	20	n	n
s_6	s_6	s_6	s_6	s_0	s_0	5	10	25	OJ	AJ

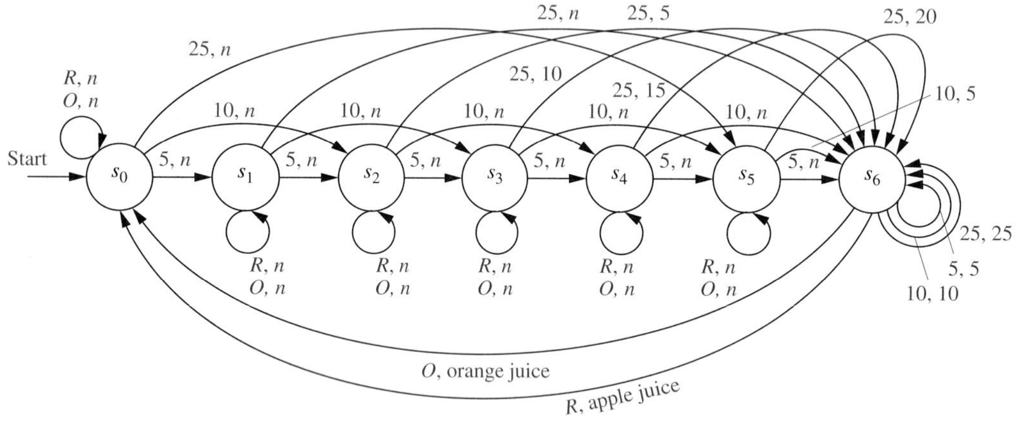

FIGURE 1   A Vending Machine.

We illustrate how this model of the machine works with this example. Suppose that a student puts in a dime followed by a quarter, receives 5 cents back, and then pushes the orange button for an orange juice. The machine starts in state s_0. The first input is 10 cents, which changes the state of the machine to s_2 and gives no output. The second input is 25 cents. This changes the state from s_2 to s_6, and gives 5 cents as output. The next input is the orange button, which changes the state from s_6 back to s_0 (since the machine returns to the start state) and gives an orange juice as its output.

We can display all the state changes and output of this machine in a table. To do this we need to specify for each combination of state and input the next state and the output obtained. Table 1 shows the transitions and outputs for each pair of a state and an input.

Another way to show the actions of a machine is to use a directed graph with labeled edges, where each state is represented by a circle, edges represent the transitions, and edges are labeled with the input and the output for that transition. Figure 1 shows such a directed graph for the vending machine.

FINITE-STATE MACHINES WITH OUTPUTS

We will now give the formal definition of a finite-state machine with output.

DEFINITION 1

A *finite-state machine* $M = (S, I, O, f, g, s_0)$ consists of a finite set S of *states,* a finite *input alphabet I*, a finite *output alphabet O*, a *transition function f* that assigns to each state and input pair a new state, an *output function g* that assigns to each state and input pair an output, and an initial state s_0.

Let $M = (S, I, O, f, g, s_0)$ be a finite-state machine. We can use a **state table** to represent the values of the transition function f and the output function g for all pairs of states and input. We previously constructed a state table for the vending machine discussed in the introduction to this section.

EXAMPLE 1   The state table shown in Table 2 describes a finite-state machine with $S = \{s_0, s_1, s_2, s_3\}$, $I = \{0, 1\}$, and $O = \{0, 1\}$. The values of the transition function f are displayed in the first two columns, and the values of the output function g are displayed in the last two columns.   ◀

Another way to represent a finite-state machine is to use a **state diagram,** which is a directed graph with labeled edges. In this diagram, each state is represented by a circle. Arrows labeled with the input and output pair are shown for each transition.

EXAMPLE 2   Construct the state diagram for the finite-state machine with the state table shown in Table 2.

Solution: The state diagram for this machine is shown in Figure 2.   ◀

EXAMPLE 3   Construct the state table for the finite-state machine with the state diagram shown in Figure 3.

Solution: The state table for this machine is shown in Table 3.   ◀

An input string takes the starting state through a sequence of states, as determined by the transition function. As we read the input string symbol by symbol (from left to right),

TABLE 2				
	f		g	
	Input		*Input*	
State	*0*	*1*	*0*	*1*
s_0	s_1	s_0	1	0
s_1	s_3	s_0	1	1
s_2	s_1	s_2	0	1
s_3	s_2	s_1	0	0

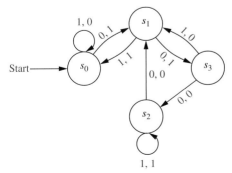

FIGURE 2   The State Diagram for the Finite-State Machine Shown in Table 2.

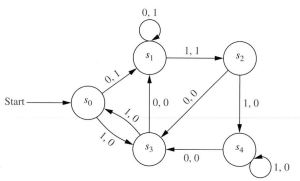

0, 1

s_1 1, 1 s_2

0, 1

Start —— s_0 0, 0   0, 0   1, 0

1, 0

1, 0 s_3 0, 0 s_4

1, 0

FIGURE 3   **A Finite-State Machine.**

	f		g	
	Input		**Input**	
State	**0**	**1**	**0**	**1**
s_0	s_1	s_3	1	0
s_1	s_1	s_2	1	1
s_2	s_3	s_4	0	0
s_3	s_1	s_0	0	0
s_4	s_3	s_4	0	0

TABLE 3

each input symbol takes the machine from one state to another. Because each transition produces an output, an input string also produces an output string.

Suppose that the input string is $x = x_1x_2\cdots x_k$. Then, reading this input takes the machine from state s_0 to state s_1, where $s_1 = f(s_0, x_1)$, then to state s_2, where $s_2 = f(s_1, x_2)$, and so on, ending at state $s_k = f(s_{k-1}, x_k)$. This sequence of transitions produces an output string $y_1y_2\cdots y_k$, where $y_1 = g(s_0, x_1)$ is the output corresponding to the transition from s_0 to s_1, $y_2 = g(s_1, x_2)$ is the output corresponding to the transition from s_1 to s_2, and so on. In general $y_j = g(s_{j-1}, x_j)$ for $j = 1, 2, \ldots, k$. Hence, we can extend the definition of the output function g to input strings so that $g(x) = y$, where y is the output corresponding to the input string x. This notation is useful in many applications.

EXAMPLE 4 Find the output string generated by the finite-state machine in Figure 3 if the input string is 101011.

Solution: The output obtained is 001000. The successive states and outputs are shown in Table 4. ◀

We can now look at some examples of useful finite-state machines. Examples 5, 6, and 7 illustrate that the states of a finite-state machine give it limited memory capabilities. The states can be used to remember the properties of the symbols that have been read by the machine. However, since there are only finitely many different states, finite-state machines cannot be used for some important purposes. This will be illustrated in Section 11.4.

EXAMPLE 5 An important element in many electronic devices is a *unit-delay machine,* which produces as output the input string delayed by a specified amount of time. How can a finite-state machine be constructed that delays an input string by one unit of time, that is, produces as output the bit string $0x_1x_2\cdots x_{k-1}$ given the input bit string $x_1x_2\cdots x_k$?

TABLE 4

Input	1	0	1	0	1	1	—
State	s_0	s_3	s_1	s_2	s_3	s_0	s_3
Output	0	0	1	0	0	0	—

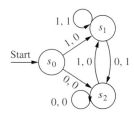

FIGURE 4   A Unit-Delay Machine.

Solution: A delay machine can be constructed that has two possible inputs, namely, 0 and 1. The machine must have a start state s_0. Since the machine has to remember whether the previous input was a 0 or a 1, two other states s_1 and s_2 are needed, where the machine is in state s_1 if the previous input was 1 and in state s_2 if the previous input was 0. An output of 0 is produced for the initial transition from s_0. Each transition from s_1 gives an output of 1, and each transition from s_2 gives an output of 0. The output corresponding to the input of a string $x_1 \cdots x_k$ is the string that begins with 0, followed by x_1, followed by $x_2, \ldots$, ending with x_{k-1}. The state diagram for this machine is shown in Figure 4. ◄

EXAMPLE 6   Produce a finite-state machine that adds two integers using their binary expansions.

Solution: When $(x_n \cdots x_1 x_0)_2$ and $(y_n \cdots y_1 y_0)_2$ are added, the following procedure (as described in Section 2.5) is followed. First, the bits x_0 and y_0 are added, producing a sum bit z_0 and a carry bit c_0. This carry bit is either 0 or 1. Then, the bits x_1 and y_1 are added, together with the carry c_0. This gives a sum bit z_1 and a carry bit c_1. This procedure is continued until the nth stage, where x_n, y_n, and the previous carry c_{n-1} are added to produce the sum bit z_n and the carry bit c_n, which is equal to the sum bit z_{n-1}.

A finite-state machine to carry out this addition can be constructed using just two states. For simplicity we assume that both the initial bits x_n and y_n are 0 (otherwise we have to make special arrangements concerning the sum bit z_{n+1}). The start state s_0 is used to remember that the previous carry is 0 (or for the addition of the rightmost bits). The other state, s_1, is used to remember that the previous carry is 1. Since the inputs to the machine are pairs of bits, there are four possible inputs. We represent these possibilities by 00 (when both bits are 0), 01 (when the first bit is 0 and the second is 1), 10 (when the first bit is 1 and the second is 0), and 11 (when both bits are 1). The transitions and the outputs are constructed from the sum of the two bits represented by the input and the carry represented by the state. For instance, when the machine is in state s_1 and receives 01 as input, the next state is s_1 and the output is 0, since the sum that arises is $0 + 1 + 1 = (10)_2$. The state diagram for this machine is shown in Figure 5. ◄

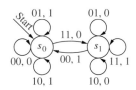

**FIGURE 5
A Finite-State
Machine for Addition.**

EXAMPLE 7   In a certain coding scheme, when three consecutive 1s appear in a message, the receiver of the message knows that there has been a transmission error. Construct a finite-state machine that gives a 1 as its output bit if and only if the last three bits received are all 1s.

Solution: Three states are needed in this machine. The start state s_0 remembers that the previous input value, if it exists, was not a 1. The state s_1 remembers that the previous input was a 1, but the input before the previous input, if it exists, was not a 1. The state s_2 remembers that the previous two inputs were 1s. An input of 1 takes s_0 to s_1, since now a 1, and not two consecutive 1s, has been read; it takes s_1 to s_2, since now two consecutive 1s have been read; and it takes s_2 to itself, since at least two consecutive 1s have been read. An input of 0 takes every state to s_0, since this breaks up any string of consecutive 1s. The output for the transition from s_2 to itself when a 1 is read is 1, since this combination of input and state shows that three consecutive 1s have been read. All other outputs are 0. The state diagram of this machine is shown in Figure 6. ◄

The machine in Figure 6 is an example of a **language recognizer,** because it produces an output of 1 if and only if the input string read so far has a specified property. Language recognition is an important application of finite-state machines.

TYPES OF FINITE-STATE MACHINES Many different kinds of finite-state machines have been developed to model computing machines. In this section we have given

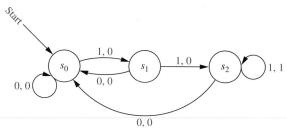

**FIGURE 6   A Finite-State Machine That Gives
an Output of 1 If and Only If the Input String Read So
Far Ends with 111.**

a definition of one type of finite-state machine. In the type of machine introduced in this section, outputs correspond to transitions between states. Machines of this type are known as **Mealy machines,** since they were first studied by G. H. Mealy in 1955. There is another important type of finite-state machine with output, where the output is determined only by the state. This type of finite-state machine is known as a **Moore machine,** since E. F. Moore introduced this type of machine in 1956. Moore machines are considered in a sequence of exercises at the end of this section.

In Example 7 we showed how a Mealy machine can be used for language recognition. However, another type of finite-state machine, giving no output, is usually used for this purpose. Finite-state machines with no output, also known as finite-state automata, have a set of final states and recognize a string if and only if it takes the start state to a final state. We will study this type of finite-state machine in Section 11.3.

Exercises

1. Draw the state diagrams for the finite-state machines with these state tables.

a)

	f		g	
	Input		**Input**	
State	**0**	**1**	**0**	**1**
s_0	s_1	s_0	0	1
s_1	s_0	s_2	0	1
s_2	s_1	s_1	0	0

b)

	f		g	
	Input		**Input**	
State	**0**	**1**	**0**	**1**
s_0	s_1	s_0	0	0
s_1	s_2	s_0	1	1
s_2	s_0	s_3	0	1
s_3	s_1	s_2	1	0

c)

	f		g	
	Input		**Input**	
State	**0**	**1**	**0**	**1**
s_0	s_0	s_4	1	1
s_1	s_0	s_3	0	1
s_2	s_0	s_2	0	0
s_3	s_1	s_1	1	1
s_4	s_1	s_0	1	0

2. Give the state tables for the finite-state machines with these state diagrams.

a)

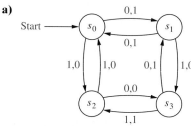

b)

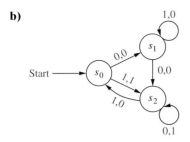

c)

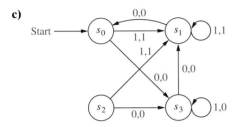

3. Given the finite-state machine shown in Example 2, determine the output for each of these input strings.

a) 0111   **b)** 11011011   **c)** 01010101010

4. Given the finite-state machine shown in Example 3, determine the output for each of these input strings.

a) 0000   **b)** 101010   **c)** 11011100010

5. Construct a finite-state machine that models a soda machine that accepts nickels, dimes, and quarters. The soda machine accepts change until 35 cents has been put in. It gives change back for any amount greater than 35 cents. Then the customer can push buttons to receive either a cola, a root beer, or a ginger ale.

6. Construct a finite-state machine that models a newspaper vending machine that has a door that can be opened only after either three dimes (and any number of other coins) or a quarter and a nickel (and any number of other coins) have been inserted. Once the door can be opened, the customer opens it and takes a paper, closing the door. No change is ever returned no matter how much extra money has been inserted. The next customer starts with no credit.

7. Construct a finite-state machine that delays an input string two bits, giving 00 as the first two bits of output.

8. Construct a finite-state machine that changes every other bit, starting with the second bit, of an input string, and leaves the other bits unchanged.

9. Construct a finite-state machine for the log on procedure for a computer, where the user logs in by entering a user identification number, which is considered to be a single input, and then a password, which is considered to be a single input. If the password is incorrect, the user is asked for the user identification number again.

10. Construct a finite-state machine for a combination lock that contains numbers 1 through 40 and that opens only when the correct combination, 10 right, 8 second left, 37 right, is entered. Each input is a triple consisting of a number, the direction of the turn, and the number of times the lock is turned in that direction.

11. Construct a finite-state machine for a toll machine that opens a gate after 25 cents, in nickels, dimes, or quarters, has been deposited. No change is given for overpayment, and no credit is given to the next driver when more than 25 cents has been deposited.

12. Construct a finite-state machine that gives an output of 1 if the number of input symbols read so far is divisible by 3 and an output of 0 otherwise.

13. Construct a finite-state machine that determines whether the input string has a 1 in the last position and a 0 in the third to the last position read so far.

14. Construct a finite-state machine that determines whether the input string read so far ends in at least five consecutive 1s.

15. Construct a finite-state machine that determines whether the word *computer* has been read as the last eight characters in the input read so far, where the input can be any string of English letters.

A **Moore machine** $M = (S, I, O, f, g, s_0)$ consists of a finite set of states, an input alphabet I, an output alphabet O, a transition function f that assigns a next state to every pair of a state and an input, an output function g that assigns an output to every state, and a starting state s_0. A Moore machine can be represented either by a table listing the transitions for each pair of state and input and the outputs for each state, or by a state diagram that displays the states, the transitions between states, and the output for each state. In the diagram, transitions are indicated with arrows labeled with the input, and the outputs are shown next to the states.

16. Construct the state diagram for the Moore machine with this state table.

	f		
	Input		
State	**0**	**1**	**g**
s_0	s_0	s_2	0
s_1	s_3	s_0	1
s_2	s_2	s_1	1
s_3	s_2	s_0	1

17. Construct the state table for the Moore machine with the state diagram shown here. Each input string to a Moore machine M produces an output string. In particular, the output corresponding to the input string $a_1 a_2 \cdots a_k$ is the string $g(s_0) g(s_1) \ldots g(s_k)$ where

$s_1 = f(s_{i-1}, a_i)$ for $i = 1, 2, \ldots, k$.

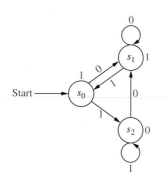

18. Find the output string generated by the Moore machine in Exercise 16 with each of these input strings.
  a) 0101 **b)** 111111 **c)** 11101110111
19. Find the output string generated by the Moore machine in Exercise 17 with each of the input strings in Exercise 18.
20. Construct a Moore machine that gives an output of 1 whenever the number of symbols in the input string read so far is divisible by 4.
21. Construct a Moore machine that determines whether an input string contains an even or odd number of 1s. The machine should give 1 as output if an even number of 1s are in the string and 0 as output if an odd number of 1s are in the string.

11.3 Finite-State Machines with No Output

INTRODUCTION

One of the most important applications of finite-state machines is in language recognition. This application plays a fundamental role in the design and construction of compilers for programming languages. In Section 11.2 we showed that a finite-state machine with output can be used to recognize a language, by giving an output of 1 when a string from the language has been read and a 0 otherwise. However, there are other types of finite-state machines that are specially designed for recognizing languages. Instead of producing output, these machines have final states. A string is recognized if and only if it takes the starting state to one of these final states.

SET OF STRINGS

Before discussing finite-state machines with no output, we will introduce some important background material on sets of strings. The operations that will be defined here will be used extensively in our discussion of language recognition by finite-state machines.

DEFINITION 1   Suppose that A and B are subsets of V^*, where V is a vocabulary. The *concatenation* of A and B, denoted by AB, is the set of all strings of the form xy where x is a string in A and y is a string in B.

EXAMPLE 1   Let $A = \{0, 11\}$ and $B = \{1, 10, 110\}$. Find AB and BA.

Solution: The set AB contains every concatenation of a string in A and a string in B. Hence, $AB = \{01, 010, 0110, 111, 1110, 11110\}$. The set BA contains every concatenation of a string in B and a string in A. Hence, $BA = \{10, 111, 100, 1011, 1100, 11011\}$. ◄

Note that it is not necessarily the case that $AB = BA$ when A and B are subsets of V^*, where V is an alphabet, as Example 1 illustrates.

From the definition of the concatenation of two sets of strings, we can define A^n, for $n = 0, 1, 2, \ldots$. This is done recursively by specifying that

$$A^0 = \{\lambda\},$$
$$A^{n+1} = A^n A \qquad \text{for } n = 0, 1, 2, \ldots.$$

EXAMPLE 2 Let $A = \{1, 00\}$. Find A^n for $n = 0, 1, 2,$ and 3.

Solution: We have $A^0 = \{\lambda\}$ and $A^1 = A^0 A = \{\lambda\} A = \{1, 00\}$. To find A^2 we take concatenations of pairs of elements of A. This gives $A^2 = \{11, 100, 001, 0000\}$. To find A^3 we take concatenations of elements in A^2 and A; this gives $A^3 = \{111, 1100, 1001, 10000, 0011, 00100, 00001, 000000\}$. ◀

DEFINITION 2

> Suppose that A is a subset of V^*. Then the *Kleene closure* of A, denoted by A^*, is the set consisting of concatenations of arbitrarily many strings from A. That is, $A^* = \bigcup_{k=0}^{\infty} A^k$.

EXAMPLE 3 What are the Kleene closures of the sets $A = \{0\}$, $B = \{0, 1\}$, and $C = \{11\}$?

Solution: The Kleene closure of A is the concatenation of the string 0 with itself an arbitrary finite number of times. Hence $A^* = \{0^n \mid n = 0, 1, 2, \ldots\}$. The Kleene closure of B is the concatenation of an arbitrary number of strings where each string is either 0 or 1. This is the set of all strings over the alphabet $V = \{0, 1\}$. That is, $B^* = V^*$. Finally, the Kleene closure of C is the concatenation of the string 11 with itself an arbitrary number of times. Hence, C^* is the set of strings consisting of an even number of 1s. That is, $C^* = \{1^{2n} \mid n = 0, 1, 2, \ldots\}$. ◀

FINITE-STATE AUTOMATA

We will now give a definition of a finite-state machine with no output. Such machines are also called **finite-state automata,** and that is the terminology we will use for them here. (*Note:* The singular of *automata* is *automaton*.) These machines differ from the finite-state machines studied in Section 11.2 in that they do not produce output, but they do have a set of final states. As we will see, they recognize strings that take the starting state to a final state.

STEPHEN COLE KLEENE (1909–1994) Stephen Kleene was born in Hartford, Connecticut. His mother, Alice Lena Cole, was a poet, and his father, Gustav Adolph Kleene, was an economics professor. Kleene attended Amherst College and received his Ph.D. from Princeton in 1934, where he studied under the famous logician Alonzo Church. Kleene joined the faculty of the University of Wisconsin in 1935, where he remained except for several leaves, including stays at the Institute for Advanced Study in Princeton. During World War II he was a navigation instructor at the Naval Reserve's Midshipmen's School and later served as the director of the Naval Research Laboratory. Kleene made significant contributions to the theory of recursive functions, investigating questions of computability and decidability, and proved one of the central results of automata theory. He served as the Acting Director of the Mathematics Research Center and as Dean of the College of Letters and Sciences at the University of Wisconsin. Kleene was a student of natural history. He discovered a previously undescribed variety of butterfly that is named after him. He was an avid hiker and climber. Kleene was also noted as a talented teller of anecdotes, using a powerful voice that could be heard several offices away.

DEFINITION 3

> A *finite-state automaton* $M = (S, I, f, s_0, F)$ consists of a finite set S of *states*, a finite *input alphabet I*, a transition function f that assigns a next state to every pair of state and input (so that $f : S \times I \to S$), an initial state s_0, and a subset F of S consisting of *final states*.

We can represent finite-state automata using either state tables or state diagrams. Final states are indicated in state diagrams by using double circles.

EXAMPLE 4   Construct the state diagram for the finite-state automaton $M = (S, I, f, s_0, F)$, where $S = \{s_0, s_1, s_2, s_3\}$, $I = \{0, 1\}$, $F = \{s_0, s_3\}$, and the transition function f is given in Table 1.

Solution: The state diagram is shown in Figure 1. Note that since both the inputs 0 and 1 take s_2 to s_0, we write 0,1 over the edge from s_2 to s_0.   ◄

The transition function f can be extended so that it is defined for all pairs of states and strings; that is, f can be extended to a function $f : S \times I^* \to S$. Let $x = x_1 x_2 \cdots x_k$ be a string in I^*. Then $f(s_1, x)$ is the state obtained by using each successive symbol of x, from left to right, as input, starting with state s_1. From s_1 we go on to state $s_2 = f(s_1, x_1)$, then to state $s_2 = f(s_2, x_2)$, and so on, with $f(s_1, x) = f(s_k, x_k)$.

A string x is said to be **recognized** or **accepted** by the machine $M = (S, I, f, s_0, F)$ if it takes the initial state s_0 to a final state, that is, $f(s_0, x)$ is a state in F. The **language recognized** or **accepted** by the machine M, denoted by $L(M)$, is the set of all strings that are recognized by M. Two finite-state automata are called **equivalent** if they recognize the same language.

EXAMPLE 5   Determine the language recognized by the finite-state automata M_1, M_2, and M_3 in Figure 2.

Solution: The only final state of M_1 is s_0. The strings that take s_0 to itself are those consisting of zero or more consecutive 1s. Hence, $L(M_1) = \{1^n \mid n = 0, 1, 2, \ldots\}$.

The only final state of M_2 is s_2. The only strings that take s_0 to s_2 are 1 and 01. Hence, $L(M_2) = \{1, 01\}$.

The final states of M_3 are s_0 and s_3. The only strings that take s_0 to itself are $\lambda, 0, 00, 000, \ldots$, that is, any string of zero or more consecutive 0s. The only strings

TABLE 1		
	f	
	Input	
State	**0**	**1**
s_0	s_0	s_1
s_1	s_0	s_2
s_2	s_0	s_0
s_3	s_2	s_1

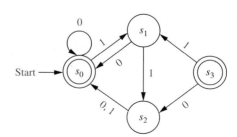

FIGURE 1   The State Diagram for a Finite-State Automaton.

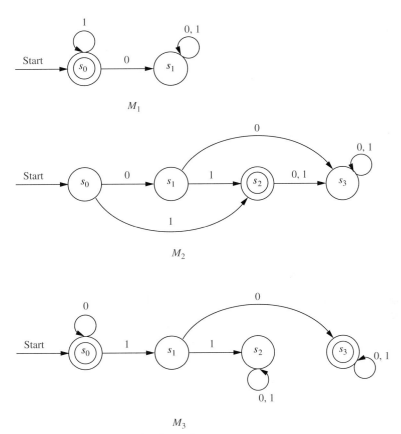

FIGURE 2   **Some Finite-State Automata.**

that take s_0 to s_3 are a string of zero or more consecutive 0s, followed by 10, followed by any string. Hence, $L(M_3) = \{0^n, 0^n 10x \mid n = 0, 1, 2, \ldots,$ and x is any string$\}$. ◀

The finite-state automata discussed so far are **deterministic,** since for each pair of state and input value there is a unique next state given by the transition function. There is another important type of finite-state automaton in which there may be several possible next states for each pair of input value and state. Such machines are called **nondeterministic.** Nondeterministic finite-state automata are important in determining which languages can be recognized by a finite-state automaton.

DEFINITION 4

Links

A *nondeterministic finite-state automaton* $M = (S, I, f, s_0, F)$ consists of a set S of states, an input alphabet I, a transition function f that assigns a set of states to each pair of state and input (so that $f : S \times I \rightarrow P(S)$), a starting state s_0, and a subset F of S consisting of the final states.

We can represent nondeterministic finite-state automata using state tables or state diagrams. When we use a state table, for each pair of state and input value we give a list of possible next states. In the state diagram we include an edge from each state to all possible next states, labeling edges with the input or inputs that lead to this transition.

EXAMPLE 6 Find the state diagram for the nondeterministic finite-state automaton with the state table shown in Table 2. The final states are s_2 and s_3.

Solution: The state diagram for this automaton is shown in Figure 3. ◀

EXAMPLE 7 Find the state table for the nondeterministic finite-state automaton with the state diagram shown in Figure 4.

Solution: The state table is given as Table 3. ◀

What does it mean for a nondeterministic finite-state automaton to recognize a string $x = x_1 x_2 \cdots x_k$? The first input symbol x_1 takes the starting state s_0 to a set S_1 of states. The next input symbol x_2 takes each of the states in S_1 to a set of states. Let S_2 be the union of these sets. We continue this process, including at a stage all states obtained using a state obtained at the previous stage and the current input symbol. We **recognize,** or **accept,** the string x if there is a final state in the set of all states that can be obtained from s_0 using x. The **language recognized** by a nondeterministic finite-state automaton is the set of all strings recognized by this automaton.

EXAMPLE 8 Find the language recognized by the nondeterministic finite-state automaton shown in Figure 4.

Solution: Since s_0 is a final state, and there is a transition from s_0 to itself when 0 is the input, the machine recognizes all strings consisting of zero or more consecutive 0s.

TABLE 2		
	f	
	Input	
State	**0**	**1**
s_0	s_0, s_1	s_3
s_1	s_0	s_1, s_3
s_2		s_0, s_2
s_3	s_0, s_1, s_2	s_1

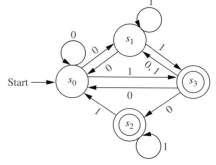

FIGURE 3   The Nondeterministic Finite-State Automaton with State Table Given in Table 2.

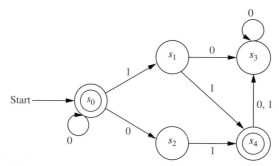

FIGURE 4   A Nondeterministic Finite-State Automaton.

TABLE 3		
	f	
	Input	
State	**0**	**1**
s_0	s_0, s_2	s_1
s_1	s_3	s_4
s_2		s_4
s_3	s_3	
s_4	s_3	s_3

Furthermore, since s_4 is a final state, any string that has s_4 in the set of states that can be reached from s_0 with this input string is recognized. The only such strings are strings consisting of zero or more consecutive 0s followed by 01 or 11. Since s_0 and s_4 are the only final states, the language recognized by the machine is $\{0^n, 0^n 01, 0^n 11 \mid n \geq 0\}$. ◀

One important fact is that a language recognized by a nondeterministic finite-state automaton is also recognized by a deterministic finite-state automaton. We will take advantage of this fact in the next section when we will determine which languages are recognized by finite-state automata.

THEOREM 1   If the language L is recognized by a nondeterministic finite-state automaton M_0, then L is also recognized by a deterministic finite-state automaton M_1.

Proof: We will describe how to construct the deterministic finite-state automaton M_1 that recognizes L from M_0, the nondeterministic finite-state automaton that recognizes this language. Each state in M_1 will be made up of a set of states in M_0. The start symbol of M_1 is $\{s_0\}$, which is the set containing the start state of M_0. The input set of M_1 is the same as the input set of M_0. Given a state $\{s_{i_1}, s_{i_2}, \ldots, s_{i_k}\}$ of M_1, the input symbol x takes this state to the union of the sets of next states for the elements of this set, that is, the union of the sets $f(s_{i_1}), f(s_{i_2}), \ldots, f(s_{i_k})$. The states of M_1 are all the subsets of S, the set of states of M_0, that are obtained in this way starting with s_0. (There are as many as 2^n states in the deterministic machine, where n is the number of states in the nondeterministic machine, since all subsets may occur as states, including the empty set, although usually far fewer states occur.) The final states of M_1 are those sets that contain a final state of M_0.

Suppose that an input string is recognized by M_0. Then one of the states that can be reached from s_0 using this input string is a final state (the reader should provide an inductive proof of this). This means that in M_1, this input string leads from $\{s_0\}$ to a set of states of M_0 that contains a final state. This subset is a final state of M_1, so this string is also recognized by M_1. Also, an input string not recognized by M_0 does not lead to any final states in M_0. (The reader should provide the details that prove this statement.) Consequently, this input string does not lead from $\{s_0\}$ to a final state in M_1.   ◁

EXAMPLE 9   Find a deterministic finite-state automaton that recognizes the same language as the nondeterministic finite-state automaton in Example 7.

Solution: The deterministic automaton shown in Figure 5 is constructed from the nondeterministic automaton in Example 7. The states of this deterministic automaton are subsets of the set of all states of the nondeterministic machine. The next state of a subset under an input symbol is the subset containing the next states in the nondeterministic machine of all elements in this subset. For instance, on input of 0, $\{s_0\}$ goes to $\{s_0, s_2\}$, since s_0 has transitions to itself and to s_2 in the nondeterministic machine; the set $\{s_0, s_2\}$ goes to $\{s_1, s_4\}$ on input of 1, since s_0 goes just to s_1 and s_2 goes just to s_4 on input of 1 in the nondeterministic machine; and the set $\{s_1, s_4\}$ goes to $\{s_3\}$ on input of 0, since s_1 and s_4 both go to just s_3 on input of 0 in the deterministic machine. All subsets that are obtained in this way are included in the deterministic finite-state machine. Note that the empty set is one of the states of this machine, since it is the subset containing all the next states of $\{s_3\}$ on input of 1. The start state is $\{s_0\}$, and the set of final states are all those that include s_0 or s_4.   ◀

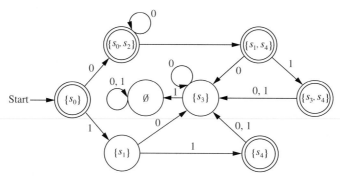

FIGURE 5 **A Deterministic Automaton Equivalent to the Nondeterministic Automaton in Example 7.**

Exercises

1. Let $A = \{0, 11\}$ and $B = \{00, 01\}$. Find each of these sets.

   a) AB **b)** BA **c)** A^2 **d)** B^3

2. Show that if A is a set of strings, then $A\emptyset = \emptyset A = \emptyset$.

3. Find all pairs of sets of strings A and B for which $AB = \{10, 111, 1010, 1000, 10111, 101000\}$.

4. Show that these equalities hold.

   a) $\{\lambda\}^* = \{\lambda\}$

   b) $(A^*)^* = A^*$ for every set of strings A

5. Describe the elements of the set A^* for these values of A.

   a) $\{10\}$ **b)** $\{111\}$ **c)** $\{0, 01\}$ **d)** $\{1, 101\}$

6. Let V be an alphabet, and let A and B be subsets of V^*. Show that $|AB| \le |A||B|$.

7. Let V be an alphabet, and let A and B be subsets of V^* with $A \subseteq B$. Show that $A^* \subseteq B^*$.

8. Suppose that A is a subset of V^* where V is an alphabet. Prove or disprove each of these statements.

   a) $A \subseteq A^2$ **b)** if $A = A^2$, then $\lambda \in A$

   c) $A\{\lambda\} = A$ **d)** $(A^*)^* = A^*$

   e) $A^*A = A^*$ **f)** $|A^n| = |A|^n$

9. Determine whether the string 11101 is in each of these sets.

   a) $\{0, 1\}^*$ **b)** $\{1\}^*\{0\}^*\{1\}^*$

   c) $\{11\}\{1\}^*\{01\}$ **d)** $\{11\}^*\{01\}^*$

   e) $\{111\}^*\{0\}^*\{1\}$ **f)** $\{111, 000\}\{00, 01\}$

10. Determine whether each of these strings is recognized by the deterministic finite-state automaton in Figure 1.

   a) 010   **b)** 1101   **c)** 1111110   **d)** 010101010

11. Determine whether all the strings in each of these sets are recognized by the deterministic finite-state automaton in Figure 1.

   a) $\{0\}^*$ **b)** $\{0\}\{0\}^*$ **c)** $\{1\}\{0\}^*$

   d) $\{01\}^*$ **e)** $\{0\}^*\{1\}^*$ **f)** $\{1\}\{0, 1\}^*$

In Exercises 12–16 find the language recognized by the given deterministic finite-state automaton.

12.

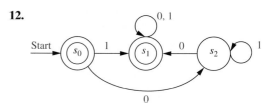

13.

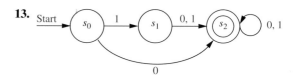

14.

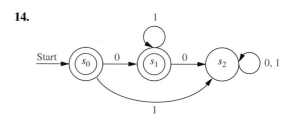

15.

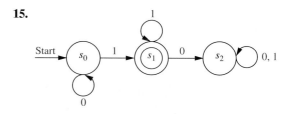

16.

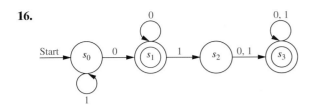

In Exercises 17–21 find the language recognized by the given nondeterministic finite-state automaton.

17.

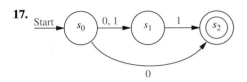

18.

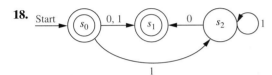

19.

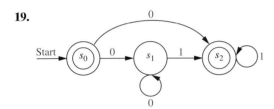

20.

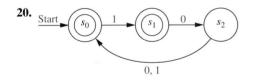

21.

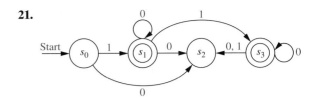

22. Find a deterministic finite-state automaton that recognizes the same language as the nondeterministic finite-state automaton in Exercise 17.

23. Find a deterministic finite-state automaton that recognizes the same language as the nondeterministic finite-state automaton in Exercise 18.

24. Find a deterministic finite-state automaton that recognizes the same language as the nondeterministic finite-state automaton in Exercise 19.

25. Find a deterministic finite-state automaton that recognizes the same language as the nondeterministic finite-state automaton in Exercise 20.

26. Find a deterministic finite-state automaton that recognizes the same language as the nondeterministic finite-state automaton in Exercise 21.

27. Find a deterministic finite-state automaton that recognizes each of these sets.

    a) $\{0\}$ **b)** $\{1, 00\}$ **c)** $\{1^n \mid n = 2, 3, 4, \ldots\}$

28. Find a nondeterministic finite-state automaton that recognizes each of the languages in Exercise 27, and has fewer states, if possible, than the deterministic automaton you found in that exercise.

*29. Show that there is no finite-state automaton that recognizes the set of bit strings containing an equal number of 0s and 1s.

11.4   Language Recognition

INTRODUCTION

We have seen that finite-state automata can be used as language recognizers. What sets can be recognized by these machines? Although this seems like an extremely difficult problem, there is a simple characterization of the sets that can be recognized by finite state automata. This problem was first solved in 1956 by the American mathematician Stephen Kleene. He showed that there is a finite-state automaton that recognizes a set if and only if this set can be built up from the null set, the empty string, and singleton strings by taking concatenations, unions, and Kleene closures, in arbitrary order. Sets that can be built up in this way are called **regular sets.**

Regular grammars were defined in Section 11.1. Because of the terminology used, it is not surprising that there is a connection between regular sets, which are the sets recognized by finite-state automata, and regular grammars. In particular, a set is regular if and only if it is generated by a regular grammar.

Finally, there are sets that cannot be recognized by any finite-state automata. We will give an example of such a set. We will briefly discuss more powerful models of computation, such as pushdown automata and Turing machines, at the end of this section.

REGULAR SETS

The regular sets are those that can be formed using the operations of concatenation, union, and Kleene closure in arbitrary order, starting with the empty set, the empty string, and singleton sets. We will see that the regular sets are those that can be recognized using a finite-state automaton. To define regular sets we first need to define regular expressions.

DEFINITION 1

The *regular expressions* over a set I are defined recursively by:

the symbol $\emptyset$ is a regular expression;
the symbol λ is a regular expression;
the symbol x is a regular expression whenever $x \in I$;
the symbols $(\mathbf{AB})$, $(\mathbf{A} \cup \mathbf{B})$, and $\mathbf{A}^*$ are regular expressions whenever $\mathbf{A}$
    and $\mathbf{B}$ are regular expressions.

Each regular expression represents a set specified by these rules:

$\emptyset$ represents the empty set, that is, the set with no strings;
λ represents the set $\{\lambda\}$, which is the set containing the empty string;
x represents the set $\{x\}$ containing the string with one symbol x;
$(\mathbf{AB})$ represents the concatenation of the sets represented by $\mathbf{A}$ and by $\mathbf{B}$;
$(\mathbf{A} \cup \mathbf{B})$ represents the union of the sets represented by $\mathbf{A}$ and by $\mathbf{B}$;
$\mathbf{A}^*$ represents the Kleene closure of the set represented by $\mathbf{A}$.

Sets represented by regular expressions are called **regular sets.** Henceforth regular expressions will be used to describe regular sets, so when we refer to the regular set $\mathbf{A}$, we will mean the regular set represented by the regular expression $\mathbf{A}$. The following example shows how regular expressions are used to specify regular sets.

EXAMPLE 1 What are the strings in the regular sets specified by the regular expressions $\mathbf{10}^*$, $(\mathbf{10})^*$, $\mathbf{0} \cup \mathbf{01}$, $\mathbf{0}(\mathbf{0} \cup \mathbf{1})^*$, and $(\mathbf{0}^*\mathbf{1})^*$?

Solution: The regular sets represented by these expressions are given in Table 1, as the reader should verify. ◀

KLEENE'S THEOREM

In 1956 Kleene proved that regular sets are the sets that are recognized by a finite-state automaton. Consequently, this important result is called Kleene's Theorem.

TABLE 1	
Expression	***Strings***
10*	a 1 followed by any number of 0s (including no zeros)
(10)*	any number of copies of 10 (including the null string)
0 ∪ 01	the string 0 or the string 01
0(0 ∪ 1)*	any string beginning with 0
(0*1)*	any string not ending with 0

THEOREM 1   **KLEENE'S THEOREM** A set is regular if and only if it is recognized by a finite-state automaton.

Kleene's Theorem is one of the central results in automata theory. We will prove the *only if* part of this theorem, namely, that every regular set is recognized by a finite-state automaton. The proof of the *if* part, that a set recognized by a finite-state automaton is regular, is left as an exercise for the reader.

Proof: Recall that a regular set is defined in terms of regular expressions, which are defined recursively. We can prove that every regular set is recognized by a finite-state automaton if we can do the following things.

1. Show that ∅ is recognized by a finite-state automaton.
2. Show that {λ} is recognized by a finite-state automaton.
3. Show that {*a*} is recognized by a finite-state automaton whenever *a* is a symbol in *I*.
4. Show that *AB* is recognized by a finite-state automaton whenever both *A* and *B* are.
5. Show that *A* ∪ *B* is recognized by a finite-state automaton whenever both *A* and *B* are.
6. Show that *A** is recognized by a finite-state automaton whenever *A* is.

We now consider each of these tasks. First, we show that ∅ is recognized by a nondeterministic finite-state automaton. To do this, all we need is an automaton with no final states. Such an automaton is shown in Figure 1(a).

Second, we show that {λ} is recognized by a finite-state automaton. To do this, all we need is an automaton that recognizes λ, the null string, but not any other string. This can be done by making the start state s_0 a final state and having no transitions so that

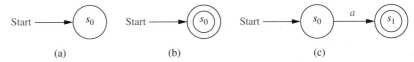

(a)   (b)   (c)

FIGURE 1   **Nondeterministic Finite-State Automata That Recognize Some Basic Sets.**

no other string takes s_0 to a final state. The nondeterministic automaton in Figure 1(b) shows such a machine.

Third, we show that $\{a\}$ is recognized by a nondeterministic finite-state automaton. To do this, we can use a machine with a starting state s_0 and a final state s_1. We have a transition from s_0 to s_1 when the input is a, and no other transitions. The only string recognized by this machine is a. This machine is shown in Figure 1(c).

Next, we show that AB and $A \cup B$ can be recognized by finite-state automata if A and B are languages recognized by finite-state automata. Suppose that A is recognized by $M_A = (S_A, I, f_A, s_A, F_A)$ and B is recognized by $M_B = (S_B, I, f_B, s_B, F_B)$.

We begin by constructing a finite-state machine $M_{AB} = (S_{AB}, I, f_{AB}, s_{AB}, F_{AB})$ that recognizes AB, the concatenation of A and B. We build such a machine by combining the machines for A and B in series, so a string in A takes the combined machine from s_A, the start state of M_A, to s_B, the start state of M_B. A string in B should take the combined machine from s_B to a final state of the combined machine. Consequently, we make the following construction. Let S_{AB} be $S_A \cup S_B$. The starting state s_{AB} is the same as s_A. The set of final states, F_{AB}, is the set of final states of M_B with s_{AB} included if and only if $\lambda \in A \cap B$. The transitions in M_{AB} include all transitions in M_A and in M_B, as well as some new transitions. For every transition in M_A that leads to a final state, we form a transition in M_{AB} from the same state to s_B, on the same input. In this way, a string in A takes M_{AB} from s_{AB} to s_B, and then a string in B takes s_B to a final state of M_{AB}. Moreover, for every transition from s_B we form a transition in M_{AB} from s_{AB} to the same state. Figure 2(a) contains an illustration of this construction.

We now construct a machine $M_{A \cup B} = \{S_{A \cup B}, I, f_{A \cup B}, s_{A \cup B}, F_{A \cup B}\}$ that recognizes $A \cup B$. This automaton can be constructed by combining M_A and M_B in parallel, using a new start state that has the transitions that both s_A and s_B have. Let $S_{A \cup B} = S_A \cup S_B \cup \{s_{A \cup B}\}$, where $s_{A \cup B}$ is a new state that is the start state of $M_{A \cup B}$. Let the set of final states $F_{A \cup B}$ be $F_A \cup F_B \cup \{s_{A \cup B}\}$ if $\lambda \in A \cup B$, and $F_A \cup F_B$ otherwise. The transitions in $M_{A \cup B}$ include all those in M_A and in M_B. Also, for each transition from s_A to a state s on input i we include a transition from $s_{A \cup B}$ to s on input i, and for each transition from s_B to a state s on input i we include a transition from $s_{A \cup B}$ to s on input i. In this way, a string in A leads from $s_{A \cup B}$ to a final state in the new machine, and a string in B leads from $s_{A \cup B}$ to a final state in the new machine. Figure 2(b) illustrates the construction of $M_{A \cup B}$.

Finally, we construct $M_{A^*} = (S_{A^*}, I, f_{A^*}, s_{A^*}, F_{A^*})$, a machine that recognizes A^*, the Kleene closure of A. Let S_{A^*} include all states in S_A and one additional state s_{A^*}, which is the starting state for the new machine. The set of final states F_{A^*} includes all states in F_A as well as the start state s_{A^*}, since λ must be recognized. To recognize concatenations of arbitrarily many strings from A, we include all the transitions in M_A, as well as transitions from s_{A^*} that match the transitions from s_A, and transitions from each final state that match the transitions from s_A. With this set of transitions, a string made up of concatenations of strings from A will take s_{A^*} to a final state when the first string in A has been read, returning to a final state when the second string in A has been read, and so on. Figure 2(c) illustrates the construction we used.   ◁

A nondeterministic finite-state automaton can be constructed for any regular set using the procedure described in this proof. We illustrate how this is done with Example 2.

EXAMPLE 2   Construct a nondeterministic finite-state automaton that recognizes the regular set **1*** $\cup$ **01**.

Solution: We begin by building a machine that recognizes **1***. This is done using the machine that recognizes **1** and then using the construction for M_{A^*} described in the

(a)

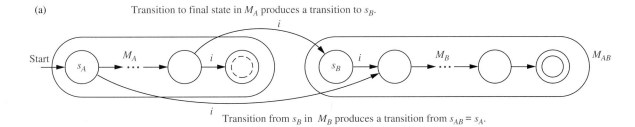

Start state is $s_{AB} = s_A$, which is final if s_A and s_B are final.   Final states include all finite states of M_B.

(b)

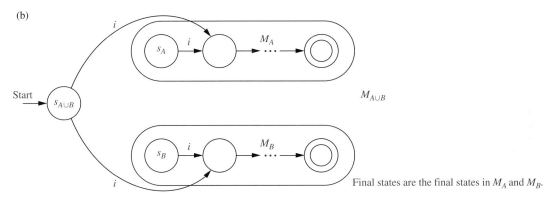

$s_{A \cup B}$ is the new start state, which is final if s_A or s_B is final.

(c)

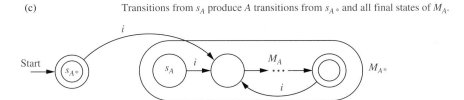

s_{A*} is the new start state, which is a final state.   Final states include all final states in M_A.

FIGURE 2   Building Automata to Recognize Concatenations, Unions, and Kleene Closures.

proof. Next, we build a machine that recognizes **01,** using machines that recognize **0** and **1** and the construction in the proof for M_{AB}. Finally, using the construction in the proof for $M_{A \cup B}$, we construct the machine for **1* ∪ 01.** The finite-state automata used in this construction are shown in Figure 3. The states in the successive machines have been labeled using different subscripts, even when a state is formed from one previously used in another machine. Note that the construction given here does not produce the simplest machine that recognizes **1* ∪ 01.** A much simpler machine that recognizes this set is shown in Figure 3(b). ◄

REGULAR SETS AND REGULAR GRAMMARS

In Section 11.1 we introduced phrase-structure grammars and defined different types of grammars. In particular we defined regular, or type 3, grammars, which are grammars of

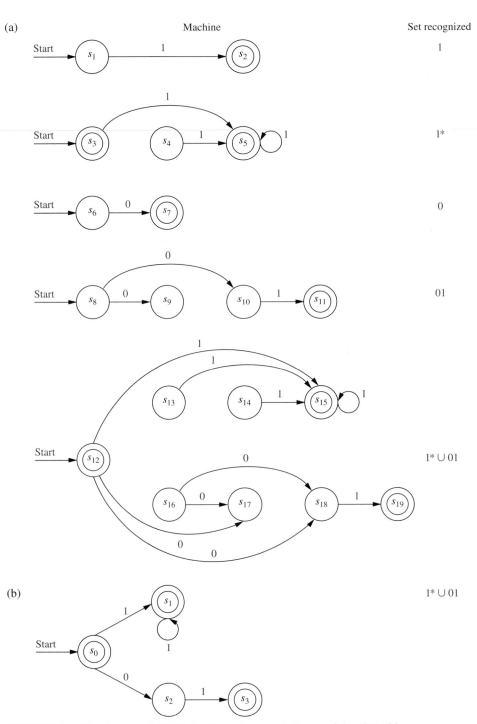

FIGURE 3 **Nondeterministic Finite-State Automata Recognizing $1^* \cup 01$.**

the form $G = (V, T, S, P)$ where each production is of the form $S \to \lambda$, $A \to a$, or $A \to aB$, where a is a terminal symbol, and A and B are nonterminal symbols. As the terminology suggests, there is a close connection between regular grammars and regular sets.

THEOREM 2   A set is generated by a regular grammar if and only if it is a regular set.

Proof: First we show that a set generated by a regular grammar is a regular set. Suppose that $G = (V, T, S, P)$ is a regular grammar generating the set $L(G)$. To show that $L(G)$ is regular we will build a nondeterministic finite-state machine $M = (S, I, f, s_0, F)$ that recognizes $L(G)$. Let S, the set of states, contain a state s_A for each nonterminal symbol A of G and an additional state s_F, which is a final state. The start state s_0 is the state formed from the start symbol S. The transitions of M are formed from the productions of G in the following way. A transition from s_A to s_F on input of a is included if $A \rightarrow a$ is a production, and a transition from s_A to s_B on input of a is included if $A \rightarrow aB$ is a production. The set of final states includes s_F and also includes s_0 if $S \rightarrow \lambda$ is a production in G. It is not hard to show that the language recognized by M equals the language generated by the grammar G, that is, the $L(M) = L(G)$. This can be done by determining the words that lead to a final state. The details are left as an exercise for the reader.   ◁

Before giving the proof of the converse, we illustrate how a nondeterministic machine is constructed that recognizes the same set as a regular grammar.

EXAMPLE 3   Construct a nondeterministic finite-state automaton that recognizes the language generated by the regular grammar $G = (V, T, S, P)$, where $V = \{0, 1, A, S\}$, $T = \{0, 1\}$, and the productions in P are $S \rightarrow 1A$, $S \rightarrow 0$, $S \rightarrow \lambda$, $A \rightarrow 0A$, $A \rightarrow 1A$, and $A \rightarrow 1$.

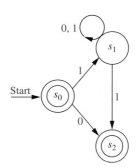

Solution: The state diagram for a nondeterministic finite-state automaton that recognizes $L(G)$ is shown in Figure 4. This automaton is constructed following the procedure described in the proof. In this automaton s_0 is the state corresponding to S, s_1 is the state corresponding to A, and s_2 is the final state.   ◀

FIGURE 4
A Nondeterministic
Finite-State
Automaton
Recognizing $L(G)$**.**

We now complete the proof of Theorem 2.

Proof: We now show that if a set is regular, then there is a regular grammar that generates it. Suppose that M is a finite-state machine that recognizes this set with the property that s_0, the starting state of M, is never the next state for a transition. (We can find such a machine by Exercise 14.) The grammar $G = (V, T, S, P)$ is defined as follows. The set V of symbols of G is formed by assigning a symbol to each state of S and each input symbol in I. The set T of terminal symbols of G is the symbols of G formed from the input symbols in I. The start symbol S is the symbol formed from the start state s_0. The set P of productions in G is formed from the transitions in M. In particular, if the state s goes to a final state under input a, then the production $A_s \rightarrow a$ is included in P, where A_s is the nonterminal symbol formed from the state s. If the state s goes to the state t on input a, then the production $A_s \rightarrow aA_t$ is included in P. The production $S \rightarrow \lambda$ is included in P if and only if $\lambda \in L(M)$. Since the productions of G correspond to the transitions of M and the productions leading to terminals correspond to transitions to final states, it is not hard to show that $L(G) = L(M)$. We leave the details as an exercise for the reader.   ◁

Example 4 illustrates the construction used to produce a grammar from an automaton that generates the language recognized by this automaton.

EXAMPLE 4

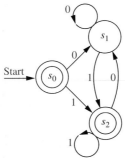

FIGURE 5   A Finite-State Automaton.

Find a regular grammar that generates the regular set recognized by the finite-state automaton shown in Figure 5.

Solution: The grammar $G = (V, T, S, P)$ generates the set recognized by this automaton where $V = \{S, A, B, 0, 1\}$; where the symbols S, A, and B correspond to the states s_0, s_1, and s_2, respectively; $T = \{0, 1\}$; S is the start symbol; and the productions are $S \rightarrow 0A$, $S \rightarrow 1B$, $S \rightarrow 1$, $S \rightarrow \lambda$, $A \rightarrow 0A$, $A \rightarrow 1B$, $A \rightarrow 1$, $B \rightarrow 0A$, $B \rightarrow 1B$, and $B \rightarrow 1$. ◄

A SET NOT RECOGNIZED BY A FINITE-STATE AUTOMATON

We have seen that a set is recognized by a finite-state automaton if and only if it is regular. We will now show that there are sets that are not regular by describing one such set. The technique used to show that this set is not regular illustrates an important method for showing that certain sets are not regular.

EXAMPLE 5

Show that the set $\{0^n 1^n \mid n = 0, 1, 2, \ldots\}$, made up of all strings consisting of a block of 0s followed by a block of an equal number of 1s, is not regular.

Solution: Suppose that this set were regular. Then there would be a deterministic finite-state automaton $M = (S, I, f, s_0, F)$ recognizing it. Let N be the number of states in this machine, that is, $N = |S|$. Since M recognizes all strings made up of a number of 0s followed by an equal number of 1s, M must recognize $0^N 1^N$. Let $s_0, s_1, s_2, \ldots, s_{2N}$ be the sequence of states that is obtained starting at s_0 and using the symbols of $0^N 1^N$ as input so that $s_1 = f(s_0, 0), s_2 = f(s_1, 0), \ldots, s_N = f(s_{N-1}, 0), s_{N+1} = f(s_N, 1), \ldots, s_{2N} = f(s_{2N-1}, 1)$. Note that s_{2N} is a final state.

Since there are only N states, the pigeonhole principle shows that at least two of the first $N + 1$ of the states, which are $s_0, \ldots, s_N$, must be the same. Say that s_i and s_j are two such identical states, with $0 \leq i < j \leq N$. This means that $f(s_i, 0^t) = s_j$ where $t = i - j$. It follows that there is a loop leading from s_i back to itself, obtained using the input 0 a total of t times, in the state diagram shown in Figure 6.

Now consider the input string $0^N 0^t 1^N = 0^{N+t} 1^N$. There are t more consecutive 0s at the start of this block than there are consecutive 1s that follow it. Since this string is not of the form $0^n 1^n$ (since it has more 0s that 1s), it is not recognized by M. Consequently, $f(s_0, 0^{N+t} 1^N)$ cannot be a final state. However, when we use the string $0^{N+t} 1^N$ as input, we end up in the same state as before, namely, s_{2N}. The reason for this is that the extra t 0s in this string take us around the loop from s_i back to itself an extra time, as shown in Figure 6. Then the rest of the string leads us to exactly the same state as before. This contradiction shows that $\{0^n 1^n \mid n = 1, 2, \ldots\}$ is not regular. ◄

MORE POWERFUL TYPES OF MACHINES

Finite-state automata are unable to carry out many computations. The main limitation of these machines is their finite amount of memory. This prevents them from recognizing languages that are not regular, such as $\{0^n 1^n \mid n = 0, 1, 2, \ldots\}$. Since a set is regular if and only if it is the language generated by a regular grammar, Example 5 shows that there

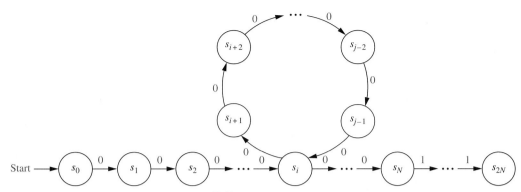

FIGURE 6   **The Path Produced by $0^N 1^N$.**

is no regular grammar that generates the set $\{0^n 1^n \mid n = 0, 1, 2, \ldots\}$. However, there is a context-free grammar that recognizes this set. Such a grammar was given in Example 5 in Section 11.1.

Because of the limitations of finite-state machines, it is necessary to use other, more powerful, models of computation. One such model is the **pushdown automaton.** A pushdown automaton includes everything in a finite-state automaton, as well as a stack, which provides unlimited memory. Symbols can be placed on the top or taken off the top of the stack. A set is recognized in one of two ways by a pushdown automaton. First, a set is recognized if the set consists of all the strings that produce an empty stack when they are used as input. Second, a set is recognized if it consists of all the strings that lead to a final state when used as input. It can be shown that a set is recognized by a pushdown automaton if and only if it is the language generated by a context-free grammar.

However, there are sets that cannot be expressed as the language generated by a context-free grammar. One such set is $\{0^n 1^n 2^n \mid n = 0, 1, 2, \ldots\}$. We will indicate why this set cannot be recognized by a pushdown automaton, but we will not give a proof, since we have not developed the machinery needed. (However, one method of proof is given in Exercise 28 of the supplementary exercises at the end of this chapter.) The stack can be used to show that a string begins with a sequence of 0s followed by an equal number of 1s by placing a symbol on the stack for each 0 (as long as only 0s are read), and removing one of these symbols for each 1 (as long as only 1s following the 0s are read). But once this is done, the stack is empty, and there is no way to determine that there are the same number of 2s in the string as 0s.

There are other machines called **linear bounded automata,** more powerful than pushdown automata, that can recognize sets such as $\{0^n 1^n 2^n \mid n = 0, 1, 2, \ldots\}$. In particular, linear bounded automata can recognize context-sensitive languages. However, these machines cannot recognize all the languages generated by phrase-structure grammars. To avoid the limitations of special types of machines, the model known as a **Turing machine,** named after the British mathematician Alan Turing, is used. A Turing machine is made up of everything included in a finite-state machine together with a tape, which is infinite in both directions. A Turing machine has read and write capabilities on the tape, and it can move back and forth along this tape. Turing machines can recognize all languages generated by phrase-structure grammars. In addition, Turing machines can model all the computations that can be performed on a computing machine. Because of their power, Turing machines are extensively studied in theoretical computer science. We will briefly study them in Section 11.5.

Exercises

1. Describe in words the strings in each of these regular sets.

a) 1*0
b) 1*00*
c) 111 ∪ 001
d) (1 ∪ 00)*
e) (00*1)*
f) (0 ∪ 1)(0 ∪ 1)*00

2. Determine whether 1011 belongs to each of these regular sets.

a) 10*1*
b) 0*(10 ∪ 11)*
c) 1(01)*1*
d) 1*01(0 ∪ 1)
e) (10)*(11)*
f) 1(00)*(11)*
g) (10)*1011
h) (1 ∪ 00)(01 ∪ 0)1*

3. Express each of these sets using a regular expression.

a) the set of strings of one or more 0s followed by a 1
b) the set of strings of two or more symbols followed by three or more 0s
c) the set of strings with either no 1 preceding a 0 or no 0 preceding a 1
d) the set of strings containing a string of 1s so that the number of 1s equals 2 modulo 3, followed by an even number of 0s

4. Construct deterministic finite-state automata that recognize these sets from I^*, where I is an alphabet.

a) ∅   **b)** {λ}   **c)** {a}, where $a \in I$

*5. Show that if A is a regular set, then A^R, the set of all reversals of strings in A, is also regular.

6. Find a finite-state automaton that recognizes
a) {λ, 0}.   **b)** {0, 11}.   **c)** {0, 11, 000}.

7. Using the constructions described in the proof of Kleene's Theorem, find nondeterministic finite-state automata that recognize each of these sets.
a) 0*1*   **b)** (0 ∪ 11)*   **c)** 01* ∪ 00*1

8. Construct a nondeterministic finite-state automaton that recognizes the language generated by the regular grammar $G = (V, T, S, P)$ where $V = \{0, 1, S, A, B\}$, $T = \{0, 1\}$, S is the start symbol, and the set of productions is

a) $S \to 0A, S \to 1B, A \to 0, B \to 0$.
b) $S \to 1A, S \to 0, S \to \lambda, A \to 0B, B \to 1B, B \to 1$.
c) $S \to 1B, S \to 0, A \to 1A, A \to 0B, A \to 1, A \to 0, B \to 1$.

In Exercises 9–11 construct a regular grammar $G = (V, T, S, P)$ that generates the language recognized by the given finite-state machine.

9.

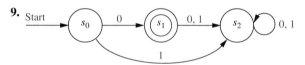

Links

ALAN MATHISON TURING (1912–1954) Alan Turing was born in London, although he was conceived in India, where his father was employed in the Indian Civil Service. As a boy, he was fascinated by chemistry, performing a wide variety of experiments, and by machinery. Turing attended Sherborne, an English boarding school. In 1931 he won a scholarship to King's College, Cambridge. After completing his dissertation, which included a rediscovery of the central limit theorem, a famous theorem in statistics, he was elected a fellow of his college. In 1935 Turing became fascinated with the decision problem, a problem posed by the great German mathematician Hilbert, which asked whether there is a general method that can be applied to any assertion to determine whether the assertion is true. Turing enjoyed running (later in life running as a serious amateur in competitions), and one day, while resting after a run, he discovered the key ideas needed to solve the decision problem. In his solution, he invented what is now called a **Turing machine** as the most general model of a computing machine. Using these machines, he found a problem, involving what he called computable numbers, that could not be decided using a general method.

From 1936 to 1938 Turing visited Princeton University to work with Alonzo Church, who had also solved Hilbert's decision problem. In 1939 Turing returned to King's College. However, at the outbreak of World War II, he joined the Foreign Office, performing cryptanalysis of German ciphers. His contribution to the breaking of the code of the Enigma, a mechanical German cipher machine, played an important role in winning the war.

After the war, Turing worked on the development of early computers. He was interested in the ability of machines to think, proposing that if a computer could not be distinguished from a person based on written replies to questions, it should be considered to be "thinking." He was also interested in biology, having written on morphogenesis, the development of form in organisms. In 1954 Turing committed suicide by taking cyanide, without leaving a clear explanation. Legal troubles related to a homosexual relationship and hormonal treatments mandated by the court to lessen his sex drive may have been factors in his decision to end his life.

10.

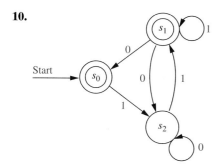

11.

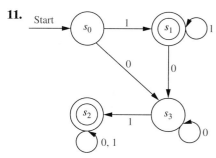

12. Show that the finite-state automaton constructed from a regular grammar in the proof of Theorem 2 recognizes the set generated by this grammar.

13. Show that the regular grammar constructed from a finite-state automaton in the proof of Theorem 2 generates the set recognized by this automaton.

14. Show that every nondeterministic finite-state automaton is equivalent to another such automaton that has the property that its starting state is never revisited.

*15. Let $M = (S, I, f, s_0, F)$ be a deterministic finite-state automaton. Show that the language recognized by M, $L(M)$, is infinite if and only if there is a word x recognized by M with $l(x) \geq |S|$.

*16. One important technique used to prove that certain sets are not regular is the **pumping lemma.** The pumping lemma states that if $M = (S, I, f, s_0, F)$ is a deterministic finite-state automaton and if x is a string in $L(M)$, the language recognized by M, with $l(x) \geq |S|$, then there are strings u, v, and w in $I*$ such that $x = uvw$, $l(uv) \leq |S|$ and $l(v) \geq 1$, and $uv^i w \in L(M)$ for $i = 0, 1, 2, \ldots$. Prove the pumping lemma. (*Hint:* Use the same idea as was used in Example 5.)

*17. Show that the set $\{0^{2n}1^n\}$ is not regular. You may use the pumping lemma given in Exercise 16.

*18. Show that the set $\{1^{n^2} \mid n = 0, 1, 2, \ldots\}$ is not regular. You may use the pumping lemma from Exercise 16.

*19. Show that the set of palindromes over $\{0, 1\}$ is not regular. You may use the pumping lemma given in Exercise 16. (*Hint:* Consider strings of the form $0^N 10^N$.)

**20. Show that a set recognized by a finite-state automaton is regular. (This is the *if* part of Kleene's Theorem.)

11.5 Turing Machines

INTRODUCTION

The finite-state automata studied earlier in this chapter cannot be used as general models of computation. They are limited in what they can do. For example, finite-state automata are able to recognize regular sets, but not able to recognize many easy-to-describe sets, including $\{0^n 1^n \mid n \geq 0\}$, which computers recognize using memory. We can use finite-state automata to compute relatively simple functions such as the sum of two numbers, but we cannot use them to compute functions that computers can, such as the product of two numbers. To overcome these deficiencies we can use a more powerful type of machine known as a Turing machine, after Alan Turing, the famous mathematician and computer scientist who invented them in the 1930s.

Basically, a Turing machine consists of a control unit, which at any step is in one of finitely many different states, together with a tape divided into cells, which is infinite in both directions. Turing machines have read and write capabilities on the tape as the control unit moves back and forth along this tape, changing states depending on the tape symbol read. Turing machines are more powerful than finite-state machines because they include memory capabilities that finite-state machines lack. We will show how to use Turing machines to recognize sets, including sets that cannot be recognized by finite-state machines. We will also show how to compute functions using Turing machines. Turing machines are the most general models of computation; essentially, they can do whatever a computer can do. Note that Turing machines are much more powerful than real computers, which have finite memory capabilities.

DEFINITION OF TURING MACHINES

We now give the formal definition of a Turing machine. Afterward we will explain how this formal definition can be interpreted in terms of a control head that can read and write symbols on a tape and move either right or left along the tape.

DEFINITION 1

A *Turing machine* $T = (S, I, f, s_0)$ consists of a finite set S of states, an alphabet I containing the blank symbol B, a partial function f from $S \times I$ to $S \times I \times \{R, L\}$, and a starting state s_0.

Recall from the preamble to Exercise 69 in Section 1.8 that a partial function is defined only for those elements in its domain of definition. This means that for some (state, symbol) pairs the partial function f may be undefined, but for a pair for which it is defined, there is a unique (state, symbol, direction) triple associated to this pair.

To interpret this definition in terms of a machine, consider a control unit and a tape divided into cells, infinite in both directions, having only a finite number of nonblank symbols on it at any given time, as pictured in Figure 1. The action of the Turing machine at each step of its operation depends on the value of the partial function f for the current state and tape symbol.

At each step, the control unit reads the current tape symbol x. If the control unit is in state s and if the partial function f is defined for the pair (s, x) with $f(s, x) = (s', x', d)$, the control unit:

1.   enters the state s',
2.   writes the symbol x' in the current cell, erasing x, and
3.   moves right one cell if $d = R$ or moves left one cell if $d = L$.

We write this step as the five-tuple (s, x, s', x', d). If the partial function f is undefined for the pair (s, x), then the Turing machine T will *halt*.

A common way to define a Turing machine is to specify a set of five-tuples of the form (s, x, s', x', d). The set of states and input alphabet is implicitly defined when such a definition is used.

At the beginning of its operation a Turing machine is assumed to be in the initial state s_0 and to be positioned over the leftmost nonblank symbol on the tape. If the tape is all

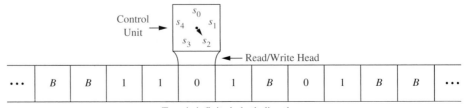

Tape is infinite in both directions.
Only finitely many nonblank cells at any time.

FIGURE 1   **A Representation of a Turing Machine.**

blank, the control head can be positioned over any cell. We will call the positioning of the control head over the leftmost nonblank tape symbol the *initial position* of the machine.

Example 1 illustrates how a Turing machine works.

EXAMPLE 1   What is the final tape when the Turing machine T defined by the seven five-tuples $(s_0, 0, s_0, 0, R)$, $(s_0, 1, s_1, 1, R)$, (s_0, B, s_3, B, R), $(s_1, 0, s_0, 0, R)$, $(s_1, 1, s_2, 0, L)$, (s_1, B, s_3, B, R), and $(s_2, 1, s_3, 0, R)$ is run on the tape shown in Figure 2(a)?

Solution: We start the operation with T in state s_0 and with T positioned over the leftmost nonblank symbol on the tape. The first step, using the five-tuple $(s_0, 0, s_0, 0, R)$, reads the 0 in the leftmost nonblank cell, stays in state s_0, writes a 0 in this cell, and moves one cell right. The second step, using the five-tuple $(s_0, 1, s_1, 1, R)$, reads the 1 in the current cell,

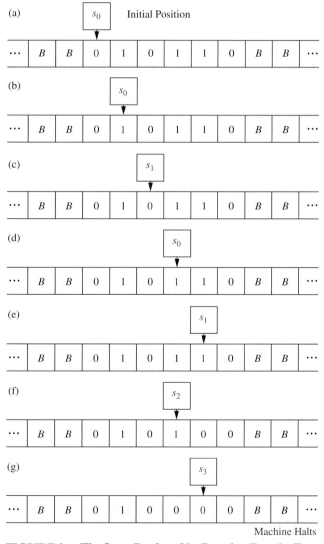

FIGURE 2   The Steps Produced by Running T on the Tape in Figure 1.

enters state s_1, writes a 1 in this cell, and moves one cell right. The third step, using the five-tuple $(s_1, 0, s_0, 0, R)$, reads the 0 in the current cell, enters state s_0, writes a 0 in this cell, and moves one cell right. The fourth step, using the five-tuple $(s_0, 1, s_1, 1, R)$, reads the 1 in the current cell, enters state s_1, writes a 1 in this cell, and moves right one cell. The fifth step, using the five-tuple $(s_1, 1, s_2, 0, L)$, reads the 1 in the current cell, enters state s_2, writes a 0 in this cell, and moves left one cell. The sixth step, using the five-tuple $(s_2, 1, s_3, 0, R)$, reads the 1 in the current cell, enters the state s_3, writes a 0 in this cell, and moves right one cell. Finally, in the seventh step, the machine halts because there is no five-tuple beginning with the pair $(s_3, 0)$ in the description of the machine. The steps are shown in Figure 2.

Note that T changes the first pair of consecutive 1s on the tape to 0s and then halts. ◄

USING TURING MACHINES TO RECOGNIZE SETS

Turing machines can be used to recognize sets. To do so requires that we define the concept of a *final state* as follows. A *final state* of a Turing machine T is a state that is not the first state in any five-tuple in the description of T using five-tuples (for example, state s_3 in Example 1).

We can now define what it means for a Turing machine to recognize a string. Given a string, we write consecutive symbols in this string in consecutive cells.

DEFINITION 2

Let V be a subset of an alphabet T. A Turing machine $T = (S, I, f, s_0)$ *recognizes* a string x in V^* if and only if T, starting in the initial position when x is written on the tape, halts in a final state. T is said to recognize a subset A of V^* if x is recognized by T if and only if x belongs to A.

Note that to recognize a subset A of V^* we can use symbols not in V. This means that the input alphabet I may include symbols not in V. These extra symbols are often used as markers (see Example 3).

When does a Turing machine T *not* recognize a string x in V^*? The answer is that x is not recognized if T does not halt or halts in a state that is not final when it operates on a tape containing the symbols of x in consecutive cells, starting in the initial position. (The reader should understand that this is one of many possible ways to define how to recognize sets using Turing machines.)

We illustrate this concept with Example 2.

EXAMPLE 2 Find a Turing machine that recognizes the set of bit strings that have a 1 as their second bit, that is, the regular set $(\mathbf{0} \cup \mathbf{1})\mathbf{1}(\mathbf{0} \cup \mathbf{1})^*$.

Solution: We want a Turing machine that, starting at the leftmost nonblank tape cell, moves right, and determines whether the second symbol is a 1. If the second symbol is 1, the machine should move into a final state. If the second symbol is not a 1, the machine should not halt or it should halt in a nonfinal state.

To construct such a machine, we include the five-tuples $(s_0, 0, s_1, 0, R)$ and $(s_0, 1, s_1, 1, R)$ to read in the first symbol and put the Turing machine in state s_1. Next, we include the five-tuples $(s_1, 0, s_2, 0, R)$ and $(s_1, 1, s_3, 1, R)$ to read in the second symbol and either move to state s_2 if this symbol is a 0, or to state s_3 if this symbol is a 1. We do not want to recognize strings that have a 0 as their second bit, so s_2 should not be a final state. We want s_3 to be a final state. So, we can include the five-tuple $(s_2, 0, s_2, 0, R)$. Since we do not want to recognize the empty string nor a string with one bit, we also include the five-tuples $(s_0, B, s_2, 0, R)$ and $(s_1, B, s_2, 0, R)$.

The Turing machine T consisting of the seven five-tuples listed here will terminate in the final state s_3 if and only if the bit string has at least two bits and the second bit of the input string is a 1. If the bit string contains fewer than two bits or if the second bit is not a 1, the machine will terminate in the nonfinal state s_2. ◀

Given a regular set, a Turing machine that always moves to the right can be built to recognize this set (as in Example 2). To build the Turing machine, first find a finite-state automaton that recognizes the set and then construct a Turing machine using the transition function of the finite-state machine, always moving to the right.

We will now show how to build a Turing machine that recognizes a nonregular set.

EXAMPLE 3 Find a Turing machine that recognizes the set $\{0^n 1^n \mid n \geq 1\}$.

Solution: To build such a machine, we will use an auxiliary tape symbol M as a marker. We have $V = \{0, 1\}$ and $I = \{0, 1, M\}$. We wish to recognize only a subset of strings in V^*. We will have one final state, s_6. The Turing machine successively replaces a 0 at the leftmost position of the string with an M and a 1 at the rightmost position of the string with an M, sweeping back and forth, terminating in a final state if and only if the string consists of a block of 0s followed by a block of the same number of 1s.

Although this is easy to describe and is easily carried out by a Turing machine, the machine we need to use is somewhat complicated. We use the marker M to keep track of the leftmost and rightmost symbols we have already examined. The five-tuples we use are: $(s_0, 0, s_1, M, R), (s_1, 0, s_1, 0, R), (s_1, 1, s_1, 1, R), (s_1, M, s_2, M, L),$ $(s_1, B, s_2, B, L), (s_2, 1, s_3, M, L), (s_3, 1, s_3, 1, L), (s_3, 0, s_4, 0, L), (s_3, M, s_5, M, R),$ $(s_4, 0, s_4, 0, L), (s_4, M, s_0, M, R),$ and (s_5, M, s_6, M, R). For example, the string 000111 would successively become $M00111, M0011M, MM011M, MM01MM, MMM1MM,$ $MMMMMM$ as the machine operates until it halts. Only the changes are shown, as most steps leave the string unaltered.

We leave it to the reader (Exercise 13 at the end of this section) to explain the actions of this Turing machine and to explain why it recognizes the set $\{0^n 1^n \mid n \geq 1\}$. ◀

It can be shown that a set can be recognized by a Turing machine if and only if it can be generated by a type 0 grammar, or in other words, if the set is generated by a phrase-structure grammar. The proof will not be presented here.

COMPUTING FUNCTIONS WITH TURING MACHINES

A Turing machine can be thought of as a computer that finds the values of a partial function. To see this, suppose that the Turing machine T, when given the string x as input, halts with the string y on its tape. We can then define $T(x) = y$. The domain of T is the set of strings for which T halts; $T(x)$ is undefined if T does not halt when given x as input. Thinking of a Turing machine as a machine that computes the values of a function on strings is useful, but how can we use Turing machines to compute functions defined on integers, on pairs of integers, on triples of integers, and so on?

To consider a Turing machine as a computer of functions from the set of k-tuples of nonnegative integers to the set of nonnegative integers (such functions are called **number-theoretic functions**), we need a way to represent k-tuples of integers on a tape. To do so, we use **unary representations** of integers. We represent the nonnegative integer n by a string of $n + 1$ 1s so that, for instance, 0 is represented by the string 1 and 5 is represented by the string 111111. To represent the k-tuple $(n_1, n_2, \ldots, n_k)$, we use a string of $n_1 + 1$ 1s, followed by an asterisk, followed by a string of $n_2 + 1$ 1s, followed by an asterisk, and so on, ending with a string of $n_k + 1$ 1s. For example, to represent the four-tuple $(2, 0, 1, 3)$ we use the string $111 * 1 * 11 * 1111$.

We can now consider a Turing machine T as computing a sequence of number-theoretic functions $T, T^2, \ldots, T^k, \ldots$. The function T^k is defined by the action of T on k-tuples of integers represented by unary representations of integers separated by asterisks.

EXAMPLE 4   Construct a Turing machine for adding two nonnegative integers.

Solution: We need to build a Turing machine T that computes the function $f(n_1, n_2) = n_1 + n_2$. The pair (n_1, n_2) is represented by a string of $n_1 + 1$ 1s followed by an asterisk followed by $n_2 + 1$ 1s. The machine T should take this as input and produce as output a tape with $n_1 + n_2 + 1$ 1s. One way to do this is as follows. The machine starts at the leftmost 1 of the input string, and carries out steps to erase this 1, halting if $n_1 = 0$ so that there are no more 1s before the asterisk, replaces the asterisk with the leftmost remaining 1, and then halts. We can use these five-tuples to do this: $(s_0, 1, s_1, B, R), (s_1, *, s_3, B, R),$ $(s_1, 1, s_2, B, R), (s_2, 1, s_2, 1, R),$ and $(s_2, *, s_3, 1, R)$. ◀

Unfortunately, constructing Turing machines to compute relatively simple functions can be extremely demanding. For example, one Turing machine for multiplying two non-negative integers found in many books has 31 five-tuples and 11 states. If it is challenging to construct Turing machines to compute even relatively simple functions, what hope do we have of building Turing machines for more complicated functions? One way to simplify this problem is to use a multitape Turing machine that uses more than one tape simultaneously and to build up multitape Turing machines for the composition of functions. It can be shown that for any multitape Turing machine there is a one-tape Turing machine that can do the same thing.

A function that can be computed by a Turing machine is called *computable*. It is fairly straightforward to show that there are number-theoretic functions that are not computable. However, it is not so easy to actually produce such a function. The *busy beaver function* defined in the preamble to Exercise 23 at the end of this section is an example of a noncomputable function. One way to show that the busy beaver function is not computable is to show that it grows faster than any computable function. (See Exercise 24.)

Links

DIFFERENT TYPES OF TURING MACHINES

There are many variations on the definition of a Turing machine. We can expand the capabilities of a Turing machine in a wide variety of ways. For example, we can allow a Turing machine to move right, left, or not at all at each step. We can allow a Turing machine to operate on multiple tapes, using $(2 + 3n)$-tuples to describe the Turing machine when n tapes are used. We can allow the tape to be two-dimensional, where at each step we move up, down, right, or left, not just right or left as we do on a one-dimensional tape. We can allow multiple tape heads that read different cells simultaneously. Furthermore, we can allow a Turing machine to be nondeterministic, by allowing a (state, tape symbol) pair to possibly appear as the first elements in more than one five-tuple of the Turing machine. We can also reduce the capabilities of a Turing machine in different ways. For example, we can restrict the tape to be infinite in only one dimension or we can restrict the tape alphabet to have only two symbols. All these variations of Turing machines have been studied in detail.

The crucial point is that no matter which of these variations we use, or even which combination of variations we use, we never increase or decrease the power of the machine. Anything that one of these variations can do can be done by the Turing machine defined in this section, and vice versa. The reason that these variations are useful is that sometimes they make doing some particular job much easier than if the Turing machine defined in Definition 1 were used. They never extend the capability of the machine.

THE CHURCH–TURING THESIS

Turing machines are relatively simple. They can have only finitely many states and they can read and write only one symbol at a time on a one-dimensional tape. But it turns out that Turing machines are extremely powerful. We have seen that Turing machines can be built to add numbers and to multiply numbers. Although it may be difficult to actually construct a Turing machine to compute a particular function that can be computed with an algorithm, such a Turing machine can always be found. This was the original goal of Turing when he invented his machines.

Furthermore, there is a tremendous amount of evidence for the **Church–Turing thesis,** which states that given any problem that can be solved with an effective algorithm, there is a Turing machine that can solve this problem. The reason this is called a *thesis* rather than a theorem is that the concept of solvability by an effective algorithm is informal and imprecise, as opposed to the notion of solvability by a Turing machine, which is formal and precise. Certainly, though, any problem that can be solved using a computer with a program written in any language, perhaps using an unlimited amount of memory, should be considered effectively solvable.

Many different formal theories have been developed to capture the notion of effective computability. These include Turing's theory and Church's lambda-calculus, as well as theories proposed by Kleene and by Post. These theories seem quite different on the surface. The surprising thing is that they can be shown to be equivalent by demonstrating that they define exactly the same class of functions. With this evidence, it seems that Turing's original ideas, formulated before the invention of modern computers, describe the ultimate capabilities of these machines.

ALONZO CHURCH (1903–1995) Alonzo Church was born in Washington, D.C. He studied at Göttingen under Hilbert and later in Amsterdam. He was a member of the faculty at Princeton University from 1927 until 1967 when he moved to UCLA. Church was one of the founding members of the Association for Symbolic Logic. He made many substantial contributions to the theory of computability, including his solution to the decision problem, his invention of the lambda-calculus, and, of course, his statement of what is now known as the Church–Turing thesis. Among Church's students were Stephen Kleene and Alan Turing. He published articles past his 90th birthday.

Exercises

1. Let T be the Turing machine defined by the five-tuples: $(s_0, 0, s_1, 1, R)$, $(s_0, 1, s_1, 0, R)$, $(s_0, B, s_1, 0, R)$, $(s_1, 0, s_2, 1, L)$, $(s_1, 1, s_1, 0, R)$, $(s_1, B, s_2, 0, L)$. For each of these initial tapes, determine the final tape when T halts, assuming that T begins in initial position.

a) ··· | B | B | 0 | 0 | 1 | 1 | B | B | ···

b) ··· | B | B | 1 | 0 | 1 | B | B | B | ···

c) ··· | B | B | 1 | 1 | B | 0 | 1 | B | ···

d) ··· | B | B | B | B | B | B | B | B | ···

2. Let T be the Turing machine defined by the five-tuples: $(s_0, 0, s_1, 0, R)$, $(s_0, 1, s_1, 0, L)$, $(s_0, B, s_1, 1, R)$, $(s_1, 0, s_2, 1, R)$, $(s_1, 1, s_1, 1, R)$, $(s_1, B, s_2, 0, R)$, $(s_2, B, s_3, 0, R)$. For each of these initial tapes, determine the final tape when T halts, assuming that T begins in initial position.

a) ··· | B | B | 0 | 1 | 0 | 1 | B | B | ···

b) ··· | B | B | 1 | 1 | 1 | B | B | B | ···

c) ··· | B | B | 0 | 0 | B | 0 | 0 | B | ···

d) ··· | B | B | B | B | B | B | B | B | ···

3. What does the Turing machine described by the five-tuples $(s_0, 0, s_0, 0, R)$, $(s_0, 1, s_1, 0, R)$, (s_0, B, s_2, B, R), $(s_1, 0, s_1, 0, R)$, $(s_1, 1, s_0, 1, R)$, and (s_1, B, s_2, B, R) do when given a bit string as input?

4. What does the Turing machine described by the five-tuples $(s_0, 0, s_1, B, R)$, $(s_0, 1, s_1, 1, R)$, $(s_1, 0, s_1, 0, R)$, $(s_1, 1, s_2, 1, R)$, $(s_2, 0, s_1, 0, R)$, $(s_2, 1, s_3, 0, L)$, $(s_3, 0, s_4, 0, R)$, and $(s_3, 1, s_4, 0, R)$ do when given a bit string as input?

5. Construct a Turing machine with tape symbols 0, 1, and B that replaces the first 0 with a 1 and does not change any of the other symbols on the tape.

6. Construct a Turing machine with tape symbols 0, 1, and B that, given a bit string as input, replaces all 0s on the tape with 1s and does not change any of the 1s on the tape.

7. Construct a Turing machine with tape symbols 0, 1, and B that, given a bit string as input, replaces all but the leftmost 1 on the tape with 0s and does not change any of the other symbols on the tape.

8. Construct a Turing machine with tape symbols 0, 1, and B that, given a bit string as input, replaces the first two consecutive 1s on the tape with 0s and does not change any of the other symbols on the tape.

9. Construct a Turing machine that recognizes the set of all bit strings that end with a 0.

10. Construct a Turing machine that recognizes the set of all bit strings that contain at least two 1s.

11. Construct a Turing machine that recognizes the set of all bit strings that contain an even number of 1s.

12. Show at each step the contents of the tape of the Turing machine in Example 3 starting with each of these strings.

    a) 0011   b) 00011   c) 101100   d) 000111

13. Explain why the Turing machine in Example 3 recognizes a bit string if and only if this string is of the form $0^n 1^n$ for some positive integer n.

*14. Construct a Turing machine that recognizes the set $\{0^{2n} 1^n \mid n \geq 0\}$.

*15. Construct a Turing machine that recognizes the set $\{0^n 1^n 2^n \mid n \geq 0\}$.

16. Construct a Turing machine that computes the function $f(n) = n + 2$ for all nonnegative integers n.

17. Construct a Turing machine that computes the function $f(n) = n - 3$ if $n \geq 3$ and $f(n) = 0$ for $n = 0, 1, 2$ for all nonnegative integers n.

18. Construct a Turing machine that computes the function $f(n) = n \bmod 3$.

19. Construct a Turing machine that computes the function $f(n) = 3$ if $n \geq 5$ and $f(n) = 0$ if $n = 0, 1, 2, 3$, or 4.

20. Construct a Turing machine that computes the function $f(n_1, n_2) = n_2 + 2$ for all pairs of nonnegative integers n_1 and n_2.

*21. Construct a Turing machine that computes the function $f(n_1, n_2) = \min(n_1, n_2)$ for all nonnegative integers n_1 and n_2.

22. Construct a Turing machine that computes the function $f(n_1, n_2) = n_1 + n_2 + 1$ for all nonnegative integers n_1 and n_2.

Let $B(n)$ be the maximum number of 1s that a Turing machine with n states with the alphabet $\{1, B\}$ may print on a tape that is initially blank. The problem of determining $B(n)$ for particular values of n is known as the **busy beaver problem.** This problem was first studied by Tibor Rado in 1962. Currently it is known that $B(2) = 4$, $B(3) = 6$, and $B(4) = 13$, but $B(n)$ is not known for $n \geq 5$.

*23. Show that $B(2)$ is at least 4 by finding a Turing machine with two states and alphabet $\{1, B\}$ that halts with four consecutive 1s on the tape.

**24. Show that the function $B(n)$ cannot be computed by any Turing machine. [*Hint:* Assume that there is a Turing machine that computes $B(n)$ in binary. Build a Turing machine T that, starting with a blank tape, writes n down in binary, computes $B(n)$ in binary, and converts $B(n)$ from binary to unary. Show that for sufficiently large n, the number of states of T is less than $B(n)$, leading to a contradiction.]

Key Terms and Results

TERMS

alphabet (or vocabulary): a set that contains elements used to form strings

language: a subset of the set of all strings over an alphabet

phrase-structure grammar (V, T, S, P): a description of a language containing an alphabet V, a set of terminal symbols T, a start symbol S, and a set of productions P

the production $w \to w_1$: w can be replaced by w_1 whenever it occurs in a string in the language

$w_1 \Rightarrow w_2$ (w_2 is directly derivable from w_1): w_2 can be obtained from w_1 using a production to replace a string in w_1 with another string

$w_1 \overset{*}{\Rightarrow} w_2$ (w_2 is derivable from w_1): w_2 can be obtained from w_1 using a sequence of productions to replace strings by other strings

type 0 grammar: any phrase-structure grammar

type 1 grammar: a phrase-structure grammar in which every production is of the form $w_1 \to w_2$, where $l(w_1) \le l(w_2)$ or $w_2 = \lambda$

type 2, or context-free, grammar: a phrase-structure grammar in which every production is of the form $A \to w_1$, where A is a nonterminal symbol

type 3, or regular, grammar: a phrase-structure grammar where every production is of the form $A \to aB$, $A \to a$, or $S \to \lambda$, where A and B are nonterminal symbols, S is the start symbol, and a is a terminal symbol

derivation (or parse) tree: an ordered rooted tree where the root represents the starting symbol of a type 2 grammar, internal vertices represent nonterminals, leaves represent terminals, and the children of a vertex are the symbols on a right side of a production, in order from left to right, where the symbol represented by the parent is on the left-hand side

Backus–Naur form: a description of a context-free grammar in which all productions having the same nonterminal as their left-hand side are combined with the different right-hand sides of these productions, each separated by a bar, with nonterminal symbols enclosed in angular brackets and the symbol $\to$ replaced by ::=

finite-state machine (S, I, O, f, g, s_0) (or a Mealy machine): a six-tuple containing a set S of states, an input alphabet I, an output alphabet O, a transition function f that assigns a next state to every pair of a state and an input, an output function g that assigns an output to every pair of a state and an input, and a starting state s_0

AB (concatenation of A and B): the set of all strings formed by concatenating a string in A and a string in B in that order

A^* (Kleene closure of A): the set of all strings made up by concatenating arbitrarily many strings from A

deterministic finite-state automaton (S, I, f, s_0, F): a five-tuple containing a set S of states, an input alphabet I, a transition function f that assigns a next state to every pair of a state and an input, a starting state s_0, and a set of final states F

nondeterministic finite-state automaton (S, I, f, s_0, F): a five-tuple containing a set S of states, an input alphabet I, a transition function f that assigns a set of possible next states to every pair of a state and an input, a starting state s_0, and a set of final states F

language recognized by an automaton: the set of input strings that take the start state to a final state of the automaton

regular expression: an expression defined recursively by specifying that $\emptyset, \lambda$, and x, for all x in the input alphabet, are regular expressions, and that (AB), $(A \cup B)$, and $(A)^*$ are regular expressions when A and B are regular expressions

regular set: a set defined by a regular expression (see page 766)

Turing machine $T = (S, I, f, s_0)$: a four-tuple consisting of a finite set S of states, an alphabet I containing the blank symbol B, a partial function f from $S \times I$ to $S \times I \times \{R, L\}$, and a starting state s_0

RESULTS

For any nondeterministic finite-state automaton there is a deterministic finite-state automaton that recognizes the same set.

Kleene's Theorem: A set is regular if and only if there is a finite-state automaton that recognizes it.

A set is regular if and only if it is generated by a regular grammar.

Review Questions

1. a) Define a phrase-structure grammar.
   b) What does it mean for a string to be derivable from a string w by a phrase-structure grammar G?

2. a) What is the language generated by a phrase-structure grammar G?
   b) What is the language generated by the grammar G with vocabulary $\{S, 0, 1\}$, set of terminals $T = \{0, 1\}$, starting symbol S, and productions $S \rightarrow 000S, S \rightarrow 1$?
   c) Give a phrase-structure grammar that generates the set $\{01^n \mid n = 0, 1, 2, \ldots\}$.

3. a) Define a type 1 grammar.
   b) Give an example of a grammar that is not a type 1 grammar.
   c) Define a type 2 grammar.
   d) Give an example of a grammar that is not a type 2 grammar but is a type 1 grammar.
   e) Define a type 3 grammar.
   f) Give an example of a grammar that is not a type 3 grammar but is a type 2 grammar.

4. a) Define a regular grammar.
   b) Define a regular language.
   c) Show that the set $\{0^m 1^n \mid m, n = 0, 1, 2, \ldots\}$ is a regular language.

5. a) What is Backus–Naur form?
   b) Give an example of the Backus–Naur form of the grammar for a subset of English of your choice.

6. a) What is a finite-state machine?
   b) Show how a vending machine that accepts only quarters and dispenses a soft drink after 75 cents has been deposited can be modeled using a finite-state machine.

7. a) What is the Kleene closure of a set of strings?
   b) Find the Kleene closure of the set $\{11, 0\}$.

8. a) Define a finite-state automaton.
   b) What does it mean for a string to be recognized by a finite-state automaton?

9. a) Define a nondeterministic finite-state automaton.
   b) Show that given a nondeterministic finite-state automaton, there is a deterministic finite-state automaton that recognizes the same language.

10. a) Define the set of regular expressions over a set I.
    b) Explain how regular expressions are used to represent regular sets.

11. State Kleene's Theorem.

12. Show that a set is generated by a regular grammar if and only if it is a regular set.

13. Give an example of a set not recognized by a finite-state automaton. Show that no finite-state automaton recognizes it.

14. Define a Turing machine.

15. Describe how Turing machines are used to recognize sets.

16. Describe how Turing machines are used to compute number-theoretic functions.

Supplementary Exercises

*1. Find a phrase-structure grammar that generates each of these languages.
   a) the set of bit strings of the form $0^{2n} 1^{3n}$, where n is a nonnegative integer
   b) the set of bit strings with twice as many 0s as 1s
   c) the set of bit strings of the form w^2, where w is a bit string

*2. Find a phrase-structure grammar that generates the set $\{0^{2^n} \mid n \geq 0\}$.

For Exercises 3 and 4, let $G = (V, T, S, P)$ be the context-free grammar with $V = \{(,), S, A, B\}, T = \{(,)\}$, starting symbol S, and productions $S \rightarrow A, A \rightarrow AB, A \rightarrow B$, $B \rightarrow (A)$, and $B \rightarrow (), S \rightarrow \lambda$.

3. Construct the derivation trees of these strings.
   a) (())   b) ()(())   c) ((()()))

*4. Show that $L(G)$ is the set of all balanced strings of parentheses, defined in the preamble to Supplementary Exercise 57 in Chapter 3.

A context-free grammar is **ambiguous** if there is a word in $L(G)$ with two derivations that produce different derivation trees, considered as ordered, rooted trees.

5. Show that the grammar $G = (V, T, S, P)$ with $V = \{0, S\}, T = \{0\}$, starting state S, and productions $S \rightarrow 0S, S \rightarrow S0$, and $S \rightarrow 0$ is ambiguous by constructing two different derivation trees for 0^3.

6. Show that the grammar $G = (V, T, S, P)$ with $V = \{0, S\}, T = \{0\}$, starting state S, and productions $S \rightarrow 0S$ and $S \rightarrow 0$ is unambiguous.

7. Suppose that A and B are finite subsets of V^*, where V is an alphabet. Is it necessarily true that $|AB| = |BA|$?

8. Prove or disprove each of these statements for subsets A, B, and C of V^*, where V is an alphabet.

   a) $A(B \cup C) = AB \cup AC$
   b) $A(B \cap C) = AB \cap AC$
   c) $(AB)C = A(BC)$
   d) $(A \cup B)^* = A^* \cup B^*$

9. Suppose that A and B are subsets of V^*, where V is an alphabet. Does it follow that $A \subseteq B$ if $A^* \subseteq B^*$?

10. What set of strings with symbols in the set $\{0, 1, 2\}$ is represented by the regular expression $(\mathbf{2}^*)(\mathbf{0} \cup (\mathbf{12}^*))^*$?

The **star height** $h(\mathbf{E})$ of a regular expression over the set I is defined recursively by

$$h(\emptyset) = 0;$$
$$h(\mathbf{x}) = 0 \text{ if } \mathbf{x} \in I;$$
$$h((\mathbf{E_1} \cup \mathbf{E_2})) = h((\mathbf{E_1}\mathbf{E_2})) = \max(h(\mathbf{E_1}), h(\mathbf{E_2}))$$
$$\text{if } \mathbf{E_1} \text{ and } \mathbf{E_2} \text{ are regular expressions;}$$
$$h(\mathbf{E}^*) = h(\mathbf{E}) + 1 \text{ if } \mathbf{E} \text{ is a regular expression.}$$

11. Find the star height of each of these regular expressions.

   a) $\mathbf{0}^*\mathbf{1}$
   b) $\mathbf{0}^*\mathbf{1}^*$
   c) $(\mathbf{0}^*\mathbf{01})^*$
   d) $((\mathbf{0}^*\mathbf{1})^*)^*$
   e) $(\mathbf{010}^*)(\mathbf{1}^*\mathbf{01}^*)^*((\mathbf{01})^*(\mathbf{10})^*)^*$
   f) $(((((\mathbf{0}^*)\mathbf{1})^*\mathbf{0})^*)\mathbf{1})^*$

*12. For each of these regular expressions find a regular expression that represents the same language with minimum star height.

   a) $(\mathbf{0}^*\mathbf{1}^*)^*$
   b) $(\mathbf{0}(\mathbf{01}^*\mathbf{0})^*)^*$
   c) $(\mathbf{0}^* \cup (\mathbf{01})^* \cup \mathbf{1}^*)^*$

13. Construct a finite-state machine with output that produces an output of 1 if the bit string read so far as input contains four or more 1s. Then construct a deterministic finite-state automaton that recognizes this set.

14. Construct a finite-state machine with output that produces an output of 1 if the bit string read so far as input contains four or more consecutive 1s. Then construct a deterministic finite-state automaton that recognizes this set.

15. Construct a finite-state machine with output that produces an output of 1 if the bit string read so far as input ends with four or more consecutive 1s. Then construct a deterministic finite-state automaton that recognizes this set.

16. A state s' in a finite-state machine is said to be **reachable** from state s if there is an input string x such that $f(s, x) = s'$. A state s is called **transient** if there is no nonempty input string x with $f(s, x) = s$. A state s is called a **sink** if $f(s, x) = s$ for all input strings x. Answer these questions about the finite-state machine with the state diagram illustrated here.

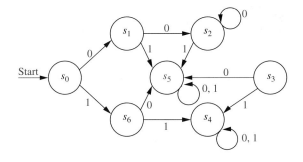

   a) Which states are reachable from s_0?
   b) Which states are reachable from s_2?
   c) Which states are transient?
   d) Which states are sinks?

*17. Suppose that S, I, and O are finite sets such that $|S| = n$, $|I| = k$, and $|O| = m$.

   a) How many different finite-state machines (Mealy machines) $M = (S, I, O, f, g, s_0)$ can be constructed, where the starting state s_0 can be arbitrarily chosen?
   b) How many different Moore machines $M = (S, I, O, f, g, s_0)$ can be constructed, where the starting state s_0 can be arbitrarily chosen?

*18. Suppose that S and I are finite sets such that $|S| = n$ and $|I| = k$. How many different finite-state automata $M = (S, I, f, s_0, F)$ are there where the starting state s_0 and the subset F of S consisting of final states can be chosen arbitrarily

   a) if the automata are deterministic?
   b) if the automata may be nondeterministic? (*Note:* This includes deterministic automata.)

19. Construct a deterministic finite-state automaton that is equivalent to the nondeterministic automaton with the state diagram shown here.

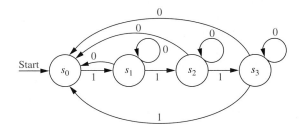

20. What is the language recognized by the automaton in Exercise 19?

21. Construct finite-state automata that recognize these sets.

   a) $\mathbf{0}^*(\mathbf{10})^*$
   b) $(\mathbf{01} \cup \mathbf{111})^*\mathbf{10}^*(\mathbf{0} \cup \mathbf{1})$
   c) $(\mathbf{001} \cup (\mathbf{11})^*)^*$

*22. Find regular expressions that represent the set of all strings of 0s and 1s

  a) made up of blocks of even numbers of 1s interspersed with odd numbers of 0s.

  b) with at least two consecutive 0s or three consecutive 1s.

  c) with no three consecutive 0s or two consecutive 1s.

*23. Show that if A is a regular set, then so is $\overline{A}$.

*24. Show that if A and B are regular sets, then so is $A \cap B$.

*25. Find finite-state automata that recognize these sets of strings of 0s and 1s.

  a) the set of all strings that start with no more than three consecutive 0s and contain at least two consecutive 1s.

  b) the set of all strings with an even number of symbols that do not contain the pattern 101.

  c) the set of all strings with at least three blocks of two or more 1s and at least two 0s.

*26. Show that $\{0^{2^n} \mid n \in \mathbf{N}\}$ is not regular. You may use the pumping lemma given in Exercise 16 of Section 11.4.

*27. Show that $\{1^p \mid p \text{ is prime}\}$ is not regular. You may use the pumping lemma given in Exercise 16 of Section 11.4.

*28. There is a result for context-free languages analogous to the pumping lemma for regular sets. Suppose that $L(G)$ is the language recognized by a context-free language G. This result states that there is a constant N such that if z is a word in $L(G)$ with $l(w) \geq N$, then z can be written as $uvwxy$ where $l(vwx) \leq N$, $l(vx) \geq 1$, and $uv^i wx^i y$ belongs to $L(G)$ for $i = 0, 1, 2, 3, \ldots$. Use this result to show that there is no context-free grammar G with $L(G) = \{0^n 1^n 2^n \mid n = 0, 1, 2, \ldots\}$.

Computer Projects

WRITE PROGRAMS WITH THESE INPUT AND OUTPUT.

1. Given the productions in a phrase-structure grammar, determine which type of grammar this is in the Chomsky classification scheme.

*2. Given the productions of a context-free grammar and a string, produce a derivation tree for this string if it is in the language generated by this grammar.

3. Given the state table of a Moore machine and an input string, produce the output string generated by the machine.

4. Given the state table of a Mealy machine and an input string, produce the output string generated by the machine.

5. Given the state table of a deterministic finite-state automaton and a string, decide whether this string is recognized by the automaton.

6. Given the state table of a nondeterministic finite-state automaton and a string, decide whether this string is recognized by the automaton.

*7. Given the state table of a nondeterministic finite-state automaton, construct the state table of a deterministic finite-state automaton that recognizes the same language.

**8. Given a regular expression, construct a nondeterministic finite-state automaton that recognizes the set that this expression represents.

9. Given a regular grammar, construct a finite-state automaton that recognizes the language generated by this grammar.

10. Given a finite-state automaton, construct a regular grammar that generates the language recognized by this automaton.

*11. Given a Turing machine, find the output string produced by a given input string.

Computations and Explorations

USE A COMPUTATIONAL PROGRAM OR PROGRAMS YOU HAVE WRITTEN TO DO THESE EXERCISES.

1. Solve the busy beaver problem for two states by testing all possible Turing machines with two states and alphabet $\{1, B\}$.

*2. Solve the busy beaver problem for three states by testing all possible Turing machines with three states and alphabet $\{1, B\}$.

**3. Find a busy beaver machine with four states by testing all possible Turing machines with four states and alphabet {1, *B*}.

**4. Make as much progress as you can toward finding a busy beaver machine with five states.

**5. Make as much progress as you can toward finding a busy beaver machine with six states.

Writing Projects

RESPOND TO THESE QUESTIONS WITH ESSAYS USING OUTSIDE SOURCES.

1. Describe how the growth of certain types of plants can be modeled using a Lidenmeyer system. Such a system uses a grammar with productions modeling the different ways plants can grow.

2. Describe the Backus–Naur form (and extended Backus–Naur form) rules used to specify the syntax of a programming language, such as Java, LISP, or ADA, or the database language SQL.

3. Explain how finite-state machines are used by spell-checkers.

4. Explain how finite-state machines are used in the study of network protocols.

5. Explain how finite-state machines are used in speech recognition programs.

6. Explain the concept of minimizing finite-state automata. Give an algorithm that carries out this minimization.

7. Give the definition of cellular automata. Explain their applications. Use the Game of Life as an example.

8. Define a pushdown automaton. Explain how pushdown automata are used to recognize sets. Which sets are recognized by pushdown automata? Provide an outline of a proof justifying your answer.

9. Define a linear-bounded automaton. Explain how linear-bounded automata are used to recognize sets. Which sets are recognized by linear-bounded automata? Provide an outline of a proof justifying your answer.

10. Look up Turing's original definition of what we now call a Turing machine. What was his motivation for defining these machines?

11. Describe the concept of the universal Turing machine. Explain how such a machine can be built.

12. Explain the kinds of applications in which nondeterministic Turing machines are used instead of deterministic Turing machines.

13. Show that a Turing machine can simulate any action of a nondeterministic Turing machine.

14. Show that a set is recognized by a Turing machine if and only if it is generated by a phrase-structure grammar.

15. Describe the basic concepts of the lambda-calculus and explain how it is used to study computability of functions.

16. Show that a Turing machine as defined in this chapter can do anything a Turing machine with *n* tapes can do.

17. Show that a Turing machine with a tape infinite in one direction can do anything a Turing machine with a tape infinite in both directions can do.

A-1 Exponential and Logarithmic Functions

I n this appendix we review some of the basic properties of exponential functions and logarithms. These properties are used throughout the text. Students requiring further review of this material should consult precalculus or calculus books, such as those mentioned in the Suggested Readings.

EXPONENTIAL FUNCTIONS

Let n be a positive integer, and let b be a fixed positive real number. The function $f_b(n) = b^n$ is defined by

$$f_b(n) = b^n = b \cdot b \cdot b \cdot \cdots \cdot b,$$

where there are n factors of b multiplied together on the right-hand side of the equation.

We can define the function $f_b(x) = b^x$ for all real numbers x using techniques from calculus. The function $f_b(x) = b^x$ is called the **exponential function to the base b.** We will not discuss how to find the values of exponential functions to the base b when x is not an integer.

Two of the important properties satisfied by exponential functions are given in Theorem 1. Proofs of these and other related properties can be found in calculus texts.

THEOREM 1   Let b be a real number. Then
1. $b^{x+y} = b^x b^y$, and
2. $(b^x)^y = b^{xy}$.

We display the graphs of some exponential functions in Figure 1.

LOGARITHMIC FUNCTIONS

Suppose that b is a real number with $b > 1$. Then the exponential function b^x is strictly increasing (a fact shown in calculus). It is a one-to-one correspondence from the set of real numbers to the set of nonnegative real numbers. Hence, this function has an inverse, called the **logarithmic function to the base b.** In other words, if b is a real number greater than 1 and x is a positive real number, then

$$b^{\log_b x} = x.$$

The value of this function at x is called the **logarithm of x to the base b.**
From the definition, it follows that

$$\log_b b^x = x.$$

We give several important properties of logarithms in Theorem 2.

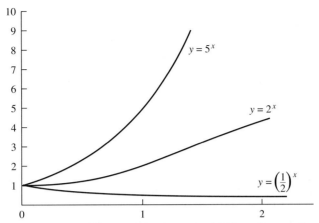

FIGURE 1 **Graphs of the Exponential Functions to the Bases $\frac{1}{2}$, 2, and 5.**

THEOREM 2   Let b be a real number greater than 1. Then

1. $\log_b(xy) = \log_b x + \log_b y$ whenever x and y are positive real numbers, and

2. $\log_b(x^y) = y \log_b x$ whenever x is a positive real number.

Proof: Since $\log_b(xy)$ is the unique real number with $b^{\log_b(xy)} = xy$, to prove part 1 it suffices to show that $b^{\log_b x + \log_b y} = xy$. By part 1 of Theorem 1, we have

$$b^{\log_b x + \log_b y} = b^{\log_b x} b^{\log_b y}$$
$$= xy.$$

To prove part 2, it suffices to show that $b^{y \log_b x} = x^y$. By part 2 of Theorem 1, we have

$$b^{y \log_b x} = (b^{\log_b x})^y$$
$$= x^y. \qquad \qquad \triangleleft$$

The following theorem relates logarithms to two different bases.

THEOREM 3   Let a and b be real numbers greater than 1, and let x be a positive real number. Then

$$\log_a x = \log_b x / \log_b a.$$

Proof: To prove this result, it suffices to show that

$$b^{\log_a x \cdot \log_b a} = x.$$

By part 2 of Theorem 1, we have

$$b^{\log_a x \cdot \log_b a} = (b^{\log_b a})^{\log_a x}$$
$$= a^{\log_a x}$$
$$= x.$$

This completes the proof. $\triangleleft$

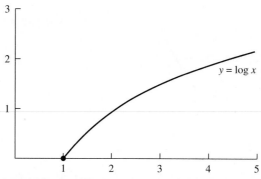

FIGURE 2   The Graph of $f(x) = \log x$.

Since the base used most often for logarithms in this text is $b = 2$, the notation $\log x$ is used throughout the test to denote $\log_2 x$.

The graph of the function $f(x) = \log x$ is displayed in Figure 2. From Theorem 3, when a base b other than 2 is used, a function that is a constant multiple of the function $\log x$, namely, $(1/\log b) \log x$, is obtained.

Exercises

1. Express each of the following quantities as powers of 2.

  a) $2 \cdot 2^2$
  b) $(2^2)^3$
  c) $2^{(2^2)}$

2. Find each of the following quantities.

  a) $\log_2 1024$
  b) $\log_2 1/4$
  c) $\log_4 8$

3. Suppose that $\log_4 x = y$. Find each of the following quantities.

  a) $\log_2 x$

  b) $\log_8 x$
  c) $\log_{16} x$

☞ **4.** Let a, b, and c be positive real numbers. Show that $a^{\log_b c} = c^{\log_b a}$.

5. Draw the graph of $f(x) = b^x$ if b is

  a) 3.
  b) 1/3.
  c) 1.

6. Draw the graph of $f(x) = \log_b x$ if b is

  a) 4.
  b) 100.
  c) 1000.

A-2 Pseudocode

Links

The algorithms in this text are described both in English and in **pseudocode.** Pseudocode is an intermediate step between an English language description of the steps of a procedure and a specification of this procedure using an actual programming language. The advantages of using pseudocode include the simplicity with which it can be written and understood and the ease of producing actual computer code (in a variety of programming languages) from the pseudocode.

This appendix describes the format and syntax of the pseudocode used in the text. This pseudocode is designed so that its basic structure resembles that of Pascal, which is currently the most commonly taught programming language. However, the pseudocode we use will be a lot looser than a formal programming language since a lot of English language descriptions of steps will be allowed.

This appendix is not meant for formal study. Rather, it should serve as a reference guide for students when they study the descriptions of algorithms given in the text and when they write pseudocode solutions to exercises.

PROCEDURE STATEMENTS

The pseudocode for an algorithm begins with a **procedure** statement that gives the name of an algorithm, lists the input variables, and describes what kind of variable each input is. For instance, the statement

procedure *maximum*(L: list of integers)

is the first statement in the pseudocode description of the algorithm, which we have named *maximum,* that finds the maximum of a list L of integers.

ASSIGNMENTS AND OTHER TYPES OF STATEMENTS

An assignment statement is used to assign values to variables. In an assignment statement the left-hand side is the name of the value and the right-hand side is an expression that involves constants, variables that have been assigned values, or functions defined by procedures. The right-hand side may contain any of the usual arithmetic operations. However, in the pseudocode in this book it may include any well-defined operation, even if this operation can be carried out only by using a large number of statements in an actual programming language.

The symbol := is used for assignments. Thus, an assignment statement has the form

---

variable := *expression*

For example, the statement

---

max := *a*

---

assigns to the variable *max* the value of *a*. A statement such as

---

$x :=$ largest integer in the list L

---

can also be used. This sets x equal to the largest integer in the list L. To translate this statement into an actual programming language would require more than one statement. Also, the instruction

---

interchange *a* and *b*

---

can be used to interchange *a* and *b*. We could also express this one statement with several assignment statements (see Exercise 2 at the end of this appendix), but for simplicity, we will often prefer this abbreviated form of pseudocode.

BLOCKS OF STATEMENTS

Statements can be grouped into blocks to carry out complicated procedures. These blocks are set off using **begin** and **end** statements, and the statements in the block are all indented the same amount.

---

begin
    statement 1
    statement 2
    statement 3
    .
    .
    .
    statement n
end

The statements in the block are executed sequentially.

COMMENTS

In the pseudocode in this book, statements enclosed in curly braces are not executed. Such statements serve as comments or reminders that help explain how the procedure works. For instance, the statement

$\{x$ is the largest element in $L\}$

can be used to remind the reader that at that point in the procedure the variable x equals the largest element in the list L.

CONDITIONAL CONSTRUCTIONS

The simplest form of the conditional construction that we will use is

if condition **then** statement

or

if condition **then**
begin
     block of statements
end

Here, the condition is checked, and if it is true, then the statement given is carried out. For example, in Algorithm 1 in Section 2.1, which finds the maximum of a set of integers, we use a conditional statement to check whether $max < a_i$ for each variable; if it is, we assign the value of a_i to max.

Often, we require the use of a more general type of construction. This is used when we wish to do one thing when the indicated condition is true, but another when it is false. We use the construction

if condition **then** statement 1
else statement 2

Note that either one or both of statement 1 and statement 2 can be replaced with a block of statements. Sometimes, we require the use of an even more general form of a conditional. The general form of the conditional construction that we will use is

> **if** condition 1 **then** statement 1
> **else if** condition 2 **then** statement 2
> **else if** condition 3 **then** statement 3
>
>   .
>   .
>   .
>
> **else if** condition n **then** statement n
> **else** statement $n + 1$

When this construction is used, if condition 1 is true, then statement 1 is carried out, and the program exits this construction. In addition, if condition 1 is false, the program checks whether condition 2 is true; if it is, statement 2 is carried out, and so on. Thus, if none of the first $n - 1$ conditions hold, but condition n does, statement n is carried out. Finally, if none of condition 1, condition 2, condition 3, $\ldots$, condition n is true, then statement $n + 1$ is executed. Note that any of the $n + 1$ statements can be replaced by a block of statements.

LOOP CONSTRUCTIONS

There are two types of loop construction in the pseudocode in this book. The first is the "for" construction, which has the form

> **for** *variable* $:=$ *initial value* **to** *final value*
>   statement

or

> **for** *variable* $:=$ *initial value* **to** *final value*
> **begin**
>   block of statements
> **end**

Here, at the start of the loop, *variable* is assigned *initial value* if *initial value* is less than or equal to *final value,* and the statements at the end of this construction are carried out with this value of *variable*. Then *variable* is increased by one, and the statement, or the statements in the block, are carried out with this new value of *variable*. This is repeated until *variable* reaches *final value*. After the instructions are carried out with *variable* equal to *final value,* the algorithm proceeds to the next statement. When *initial value* exceeds *final value,* none of the statements in the loop is executed.

We can use the "for" loop construction to find the sum of the positive integers from 1 to n with the following pseudocode.

$sum := 0$
for $i := 1$ **to** n
    $sum := sum + i$

Also, the more general "for" statement, of the form

for all elements with a certain property

is used in this text. This means that the statement or block of statements that follow are carried out successively for the elements with the given property.

The second type of loop construction that we will use is the "while" construction. This has the form

while condition
        statement

or

while condition
begin
        block of statements
end

When this construction is used, the condition given is checked, and if it is true, the statements that follow are carried out, which may change the values of the variables that are part of the condition. If the condition is still true after these instructions have been carried out, the instructions are carried out again. This is repeated until the condition becomes false. As an example, we can find the sum of the integers from 1 to n using the following block of pseudocode including a "while" construction.

$sum := 0$
while $n > 0$
begin
        $sum := sum + n$
        $n := n - 1$
end

Note that any "for" construction can be turned into a "while" construction (see Exercise 3 at the end of this appendix). However, it is often easier to understand the "for" construction. So, when it makes sense, we will use the "for" construction in preference to the corresponding "while" construction.

USING PROCEDURES IN OTHER PROCEDURES

We can use a procedure from within another procedure (or within itself in a recursive program) simply by writing the name of this procedure followed by the inputs to this procedure. For instance,

$max(L)$

will carry out the procedure *max* with the input list L. After all the steps of this procedure have been carried out, execution carries on with the next statement in the procedure.

Exercises

1. What is the difference between the following blocks of two assignment statements?

   $a := b$

   $b := c$

and

   $b := c$

   $a := b$

2. Give a procedure using assignment statements to interchange the values of the variables x and y. What is the minimum number of assignment statements needed to do this?

3. Show how a loop of the form

   for i := *initial value* **to** *final value*

      statement

can be written using the "while" construction.

Suggested Readings

Among the resources available for learning more about the topics covered in this book are printed materials and relevant websites. Printed resources are described in this section of suggested readings. Readings are listed by chapter and keyed to particular topics of interest. Some general references also deserve special mention. A book you may find particularly useful is the *Handbook of Discrete and Combinatorial Mathematics* by Rosen [Ro00a], a comprehensive reference book. Additional applications of discrete mathematics can be found in Michaels and Rosen [MiRo91]. Deeper coverage of many topics in computer science, including those discussed in this book, can be found in Gruska [Gr97]. Biographical information about many of the mathematicians and computer scientists mentioned in this book can be found in Gillispie [Gi70].

To find pertinent websites, consult the links found on the website for this book. Its address is **www.mhhe.com/rosen**.

CHAPTER 1

An entertaining way to study logic is to read Lewis Carroll's book [Ca78]. General references for logic include Mendelson [Me64], Stoll [St74], and Suppes [Su87]. A comprehensive treatment of logic in discrete mathematics can be found in Gries and Schneider [GrSc93]. System specifications are discussed in Ince [In93]. Smullyan's knights and knaves puzzles were introduced in [Sm78]. He has written many fascinating books on logic puzzles including [Sm92] and [Sm98]. Prolog is discussed in depth in Nilsson and Maluszynski [NiMa95] and in Clocksin and Mellish [ClMe94]. The basics of proofs are covered in Cupillari [Cu01], Solow [So01], Velleman [Ve94], and Wolf [Wo98]. Lin and Lin [LiLi81] is an easily read text on sets and their applications. Axiomatic developments of set theory can be found in Halmos [Ha60], Monk [Mo69], and Stoll [St74]. Brualdi [Br77] and Reingold, Nievergelt, and Deo [ReNiDe77] contain introductions to multisets. Fuzzy sets and their application to expert systems and artificial intelligence are treated in Negoita [Ne85] and Zimmerman [Zi91]. Calculus books, such as Apostol [Ap67], Spivak [Sp94], and Thomas and Finney [ThFi96], contain discussions of functions. Discussions of the mathematical foundations needed for computer science can be found in Arbib, Kfoury, and Moll [ArKfMo80], Bobrow and Arbib [BoAr74], Beckman [Be80], and Tremblay and Manohar [TrMa75]. Biographical information on many of the math-

ematicians and computer scientists mentioned in this book can be found in Gillispie [Gi70].

CHAPTER 2

The articles by Knuth [Kn77] and Wirth [Wi84] are accessible introductions to the subject of algorithms. Extensive material on big-O estimates of functions can be found in Knuth [Kn97a]. General references for algorithms and their complexity include Aho, Hopcroft, and Ullman [AhHoUl74], Baase and Van Gelder [BaGe99], Cormen, Leierson, Rivest, and Stein [CoLeRiSt01], Gonnet [Go84], Goodman and Hedetniemi [GoHe77], Harel [Ha87], Horowitz and Sahni [HoSa82], Kreher and Stinson [KrSt98], the famous series of books by Knuth on the art of computer programming [Kn97a,b,98], Kronsjö [Kr87], Pohl and Shaw [PoSh81], Purdom and Brown [PuBr85], Rawlins [Ra92], Sedgewick [Se88], Wilf [Wi86], and Wirth [Wi76]. Sorting and searching algorithms and their complexity are studied in detail in Knuth [98]. References for number theory include Hardy and Wright [HaWr79], LeVeque [Le77], Rosen [Ro00], and Stark [St78]. Applications of number theory to cryptography are covered in Denning [De82], Menezes, van Oorschot, and Vanstone [MeOoVa97], Rosen [Ro00], Seberry and Pieprzyk [SePi89], Sinkov [Si66], and Stinson [St02]. The RSA public-key system was introduced in Rivest, Shamir, and Adleman [RiShAd78]. Algorithms for computer arithmetic are discussed in Knuth [Kn97b] and Pohl and Shaw [PoSh81]. Matrices and their operations are covered in all linear algebra books, such as Curtis [Cu84] and Strang [St88].

CHAPTER 3

The science and art of constructing proofs is discussed in a delightful way in three books by Pólya: [Po62, 71, 90]. The best printed source of information about integer sequences is Sloan and Plouffe [SlPl95]. Stanat and McAllister [StMc77] has a thorough section on countability. An accessible introduction to mathematical induction can be found in the English translation (from the Russian original) of Sominskii [So61]. Problems involving the covering of some or all of the squares of a chessboard using L-shaped pieces or other types of pieces are treated by Golomb [Go94]. Books that contain thorough treatments of mathematical induction and recursive definitions include Liu [Li85], Sahni [Sa85], Stanat and McAllister [StMc77], and Tremblay and Manohar [TrMa75]. The Ackermann function, introduced in 1928 by W. Ackermann, arises in the theory of recursive function (see Beckman [Be80] and McNaughton [Mc82], for instance) and in the analysis of the complexity of certain set theoretic algorithms (see Tarjan [Ta83]). Recursion is studied in Roberts [Ro86], Rohl [Ro84a], and Wand [Wa80]. Discussions of program correctness and the logical machinery used to prove that programs are correct can be found in Alagic and Arbib [AlAr78], Anderson [An79], Backhouse [Ba86], Sahni [Sa85], and Stanat and McAllister [StMc77].

CHAPTER 4

General references for counting techniques and their applications include Anderson [An74], Berman and Fryer [BeFr72], Bogart [Bo86], Bose and Manvel [BoMa86], Brualdi [Br97], Cohen [Co78], Grimaldi [Gr94], Liu [Li68], Pólya, Tarjan, and Woods [PoTaWo83], Riordan [Ri58], Roberts [Ro84], Tucker [Tu02], and Williamson [Wi85]. Vilenkin [Vi71] contains a selection of combinatorial problems and their solutions. A selection of more difficult combinatorial problems can be found in Lovász [Lo79]. Information about Internet protocol addresses and datagrams can be found in Comer [Co00]. Applications of the pigeonhole principle can be found in Brualdi [Br77], Liu [Li85], and Roberts [Ro84]. A wide selection of combinatorial identities can be found in Riordan [Ri68]. Combinatorial algorithms, including algorithms for generating permutations and combinations, are described by Even [Ev73], Lehmer [Le64], and Reingold, Nievergelt, and Deo [ReNiDe77].

CHAPTER 5

Useful references for discrete probability theory include Feller [Fe68], Nabin [Na00], and Ross [Ro01]. Ross [Ro02], which focuses on the application of probability theory to computer science, provides examples of average case complexity analysis and covers the probabilistic method. Aho and Ullman [AhUl95] includes a discussion of various aspects of probability theory important in computer science, including programming applications of probability. The probabilistic method is discussed in a chapter in Aigner and Ziegler [AiZi01], a monograph devoted to clever, insightful, and brilliant proofs, that is, proofs that Paul Erdős described as coming from "The Book." Extensive coverage of the probabilistic method can be found in Alon and Spencer [AlSp00].

CHAPTER 6

Many different models using recurrence relations can be found in Roberts [Ro84] and Tucker [Tu02]. Exhaustive treatments of linear homogeneous recurrence relations with constant coefficients, and related inhomogeneous recurrence relations, can be found in Brualdi [Br77], Liu [Li68], and Mattson [Ma93]. Divide-and-conquer algorithms and their complexity are covered in Roberts [Ro84] and Stanat and McAllister [StMc77]. Descriptions of fast multiplication of integers and matrices can be found in Aho, Hopcroft, and Ullman [AhHoUl74] and Knuth [Kn81]. An excellent introduction to generating functions can be found in Pólya, Tarjan, and Woods [PoTaWo83]. Generating functions are studied in detail in Brualdi [Br77], Cohen [Co78], Graham, Knuth, and Patashnik [GrKnPa94], Grimaldi [Gr94], and Roberts [Ro84]. Additional applications of the principle of inclusion–exclusion can be found in Liu [Li85, Li68], Roberts [Ro84], and Ryser [Ry63].

CHAPTER 7

General references for relations, including treatments of equivalence relations and partial orders, include Bobrow and Arbib [BoAr74], Grimaldi [Gr94], Sanhi [Sa85], and Tremblay and Manohar [TrMa75]. Discussions of relational models for databases are given in Date [Da82] and Aho and Ullman [AhUl95]. The original papers by Roy and Warshall for finding transitive closures can be found in [Ro59] and [Wa62], respectively. Directed graphs are studied in Chartrand and Lesniak [ChLe86], Gross and Yellen [GrYe99], Robinson and Foulds [RoFo80], Roberts [Ro84], and Tucker [Tu85]. The application of lattices to information flow is treated in Denning [De82].

CHAPTER 8

General references for graph theory include Behzad and Chartrand [BeCh71], Chartrand and Lesniak [ChLe96], Bondy and Murty [BoMu76], Chartrand and Oellermann [ChOe93], Graver and Watkins [GrWa77], Gross and Yellen [GrYe99], Harary [Ha69], Ore [Or63], Roberts [Ro84], Tucker [Tu85], Wilson [Wi85], and Wilson and Watkins [WiWa90]. A wide variety of applications of graph theory can be found in Chartrand [Ch77], Deo [De74], Foulds [Fo92], Roberts [Ro84, Ro76], Wilson and Beineke [WiBe79], and McHugh [Mc90]. Applications involving large graphs, including the Web graph, are discussed in Hayes [Ha00a, 00b]. A comprehensive description of algorithms in graph theory can be found in Gibbons [Gi85]. Other references for algorithms in graph theory include Buckley and Harary [BuHa90], Chartrand and Oellermann [ChOe93], Chachra, Ghare, and Moore [ChGhMo79], Even [Ev73, Ev79], Hu [Hu82], and Reingold, Nievergelt, and Deo [ReNiDe77]. A translation of Euler's original paper on the Königsberg bridge problem can be found in [Eu53]. Dijkstra's algorithm is studied in Gibbons [Gi85], Liu [Li85], and Reingold, Nievergelt, and Deo [ReNiDe77]. Dijkstra's original paper can be found in [Di59]. A proof of Kuratowski's theorem can be found in Harary [Ha69] and Liu [Li68]. Crossing numbers and thicknesses of graphs are studied in Chartrand and Lesniak [ChLe96]. References for graph coloring and the four-color theorem are included in Barnette [Ba83] and Saaty and Kainen [SaKa86]. The original conquest of the four-color theorem is reported in Appel and Haken [ApHa76]. Applications of graph coloring are described by Roberts [Ro84]. The history of graph theory is covered in Biggs, Lloyd, and Wilson [BiLlWi86]. Interconnection networks for parallel processing are discussed in Akl [Ak89] and Siegel and Hsu [SiHs88].

CHAPTER 9

Trees are studied in Deo [De74], Grimaldi [Gr94], Knuth [Kn97a], Roberts [Ro84], and Tucker [Tu85]. The use of trees in computer science is described by Gotlieb and Gotlieb [GoGo78], Horowitz and Sahni [HoSa82], and Knuth [Kn97a, 98]. Roberts [Ro84] covers applications of trees to many different areas. Prefix codes and Huffman coding are covered in Hamming [Ha80]. Backtracking is an old technique; its use to solve maze puzzles can be found in the 1891 book by Lucas [Lu91]. An extensive discussion of how to solve problems using backtracking can be found in Reingold, Nievergelt, and Deo [ReNiDe77]. Gibbons [Gi85] and Reingold, Nievergelt, and Deo [ReNiDe77] contain discussions of algorithms for constructing spanning trees and minimal spanning trees. The background and history of algorithms for finding minimal spanning trees is covered in Graham and Hell [GrHe85]. Prim and Kruskal described their algorithms for finding minimal spanning trees in [Pr57] and [Kr56], respectively. Sollin's algorithm is an example of an algorithm well suited for parallel processing; although Sollin never published a description of it, his algorithm has been described by Even [Ev73] and Goodman and Hedetniemi [GoHe77].

CHAPTER 10

Boolean algebra is studied in Hohn [Ho66], Kohavi [Ko86], and Tremblay and Manohar [TrMa75]. Applications of Boolean algebra to logic circuits and switching circuits are described by Hayes [Ha93], Hohn [Ho66], Katz [Ka93], and Kohavi [Ko86]. The original papers dealing with the minimization of sum-of-products expansions using maps are Karnaugh [Ka53] and Veitch [Ve52]. The Quine-McCluskey method was introduced in McCluskey [Mc56] and Quine [Qu52, Qu55]. Threshold functions are covered in Kohavi [Ko78].

CHAPTER 11

General references for languages and automata theory include Denning, Dennis, and Qualitz [DeDeQu81], Hopcroft and Ullman [HoUl79], Hopkin and Moss [HoMo76], Lewis and Papadimitriou [LePa84], and McNaughton [Mc82]. Mealy machines and Moore machines were originally introduced in Mealy [Me55] and Moore [Mo56]. The original proof of Kleene's theorem can be found in [Kl56]. Powerful models of computation, including pushdown automata and Turing machines, are discussed in Brookshear [Br89], Hennie [He77], Hopcroft and Ullman [HoUl79], Hopkin and Moss [HoMo76], Martin [Ma91], and Wood [Wo87]. Barwise and Etchemendy [BaEt93] is an excellent introduction to Turing machines. Interesting articles about the history and application of Turing machines and related machines can be found in Herken [He88]. Busy beaver machines were first introduced by Rado in [Ra62], and information about them can be found in Dewdney [De84,93], the article by Brady in Herken [He88], and in Wood [Wo87].

APPENDIXES

Detailed treatments of exponential and logarithmic functions can be found in calculus books such as Apostol [Ap67], Spivak [Sp80], and Thomas and Finney [ThFi92]. The pseudocode described in Appendix 2 resembles Pascal. Pohl and Shaw [PoSh81] uses a similar type of language to describe algorithms. Wirth [Wi76] describes how to use algorithms and data structures to build programs using Pascal. Also, Rohl [Ro84] shows how to implement recursive programs using Pascal.

REFERENCES

[AhHoUl74] A. V. Aho, J. E. Hopcroft, and J. D. Ullman, *The Design and Analysis of Computer Algorithms,* Addison-Wesley, Reading, MA, 1974.

[AhUl95] Alfred V. Aho and Jeffrey D. Ullman, *Foundations of Computer Science,* C ed., Computer Science Press, New York, 1995.

[AiZi01] Martin Aigner and Günter M. Ziegler, *Proofs from THE BOOK,* 2d ed., Springer, Berlin, 2001.

[Ak89] S. G. Akl, *The Design and Analysis of Parallel Algorithms,* Prentice Hall, Englewood Cliffs, NJ, 1989.

[AlAr78] S. Alagic and M. A. Arbib, *The Design of Well-Structured and Correct Programs,* Springer-Verlag, New York, 1978.

[AlSp00] Noga Alon and Joel H. Spencer, *The Probabilistic Method,* 2d ed., Wiley, New York, 2000.

[An74] I. Anderson, *A First Course in Combinatorial Mathematics,* Clarendon, Oxford, England, 1974.

[An79] R. B. Anderson, *Proving Programs Correct,* Wiley, New York, 1979.

[Ap67] T. M. Apostol, *Calculus,* vol. I, 2d ed., Wiley, New York, 1967.

[ApHa76] K. Appel and W. Haken, "Every Planar Map Is 4-colorable," *Bulletin of the AMS,* 82 (1976), 711–712.

[ArKfMo80] M. A. Arbib, A. J. Kfoury, and R. N. Moll, *A Basis for Theoretical Computer Science,* Springer-Verlag, New York, 1980.

[BaGe99] S. Baase and A. Van Gelder, *Computer Algorithms: Introduction to Design and Analysis,* 3d ed., Addison-Wesley, Reading, MA, 1999.

[Ba86] R. C. Backhouse, *Program Construction and Verification,* Prentice-Hall, Englewood Cliffs, NJ, 1986.

[Ba83] D. Barnette, *Map Coloring, Polyhedra, and the Four-Color Problem,* Mathematical Association of America, Washington, D.C., 1983.

[BaEt93] Jon Barwise and John Etchemendy, *Turing's World 3.0 for the Macintosh,* CSLI Publications, Stanford, CA, 1993.

[Be80] F. S. Beckman, *Mathematical Foundations of Programming,* Addison-Wesley, Reading, MA, 1980.

[BeCh71] M. Behzad and G. Chartrand, *Introduction to the Theory of Graphs,* Allyn & Bacon, Boston, 1971.

[Be86] J. Bentley, *Programming Pearls,* Addison-Wesley, Reading, MA, 1986.

[BeFr72] G. Berman and K. D. Fryer, *Introduction to Combinatorics,* Academic Press, New York, 1972.

[BiLlWi86] N. L. Biggs, E. K. Lloyd, and R. J. Wilson, *Graph Theory 1736–1936,* Clarendon, Oxford, England, 1986.

[BoAr74] L. S. Bobrow and M. A. Arbib, *Discrete Mathematics,* Saunders, Philadelphia, 1974.

[Bo86] K. P. Bogart, *Introductory Combinatorics,* Wiley, New York, 1986.

[BoMu76] J. A. Bondy and U. S. R. Murty, *Graph Theory with Applications,* North-Holland, New York, 1976.

[BoMa86] R. C. Bose and B. Manvel, *Introduction to Combinatorial Theory,* Wiley, New York, 1986.

[Br89] J. G. Brookshear, *Theory of Computation,* Benjamin Cummings, Redwood City, CA, 1989.

[Br97] R. A. Brualdi, *Introductory Combinatorics,* 3d ed., Prentice-Hall, Englewood Cliffs, NJ, 1997.

[BuHa90] F. Buckley and F. Harary, *Distance in Graphs,* Addison-Wesley, Redwood City, CA, 1990.

[Ca78] L. Carroll, *Symbolic Logic,* Crown, New York, 1978.

[ChGhMo79] V. Chachra, P. M. Ghare, and J. M. Moore, *Applications of Graph Theory Algorithms,* North-Holland, New York, 1979.

[Ch77] G. Chartrand, *Graphs as Mathematical Models,* Prindle, Weber & Schmidt, Boston, 1977.

[ChLe96] G. Chartrand and L. Lesniak, *Graphs and Digraphs,* 3d ed., Wadsworth, Belmont, CA, 1996.

[ChOe93] G. Chartrand and O. R. Oellermann, *Applied Algorithmic Graph Theory,* McGraw-Hill, New York, 1993.

[ClMe94] W. F. Clocksin and C. S. Mellish, *Programming in Prolog,* 4th ed., Springer-Verlag, New York, 1994.

[Co78] D. I. A. Cohen, *Basic Techniques of Combinatorial Theory,* Wiley, New York, 1978.

[Co00] D. Comer, Internetworking with TCP/IP, *Principles, Protocols, and Architecture,* vol. 1, 4th ed., Prentice-Hall, Englewood Cliffs, NJ, 2000.

[CoLeRiSt01] T. H. Cormen, C. E. Leierson, R. L. Rivest, and Stein, *Introduction to Algorithms,* 2d ed., MIT Press, Cambridge, MA, 2001.

[CrPo01] Richard Crandall and Carl Pomerance, *Prime Numbers: a Computational Perspective,* Springer-Verlag, New York, 2001.

[Cu01] Antonella Cupillari, *The Nuts and Bolts of Proofs,* 2d ed., Academic Press, San Diego, 2001.

[Cu84] C. W. Curtis, *Linear Algebra,* Springer-Verlag, New York, 1984.

[Da82] C. J. Date, *An Introduction to Database Systems,* 3d ed., Addison-Wesley, Reading, MA, 1982.

[De82] D. E. R. Denning, *Cryptography and Data Security,* Addison-Wesley, Reading, MA, 1982.

[DeDeQu81] P. J. Denning, J. B. Dennis, and J. E. Qualitz, *Machines, Languages, and Computation,* Prentice-Hall, Englewood Cliffs, NJ, 1981.

[De74] N. Deo, *Graph Theory with Applications to Engineering and Computer Science,* Prentice-Hall, Englewood Cliffs, NJ, 1974.

[De84] A. K. Dewdney, "Computer Recreations," *Scientific American,* 251, no. 2 (August 1984), 19–23; 252, no. 3 (March 1985), 14–23; 251, no. 4 (April 1985), 20–30.

[De93] A. K. Dewdney, *The New Turing Omnibus: Sixty-Six Excursions in Computer Science,* W. H. Freeman, New York, 1993.

[Di59] E. Dijkstra, "Two Problems in Connexion with Graphs," *Numerische Mathematik,* 1 (1959), 269–271.

[Eu53] L. Euler, "The Koenigsberg Bridges," *Scientific American,* 189, no. 1 (July 1953), 66–70.

[Ev73] S. Even, *Algorithmic Combinatorics,* Macmillan, New York, 1973.

[Ev79] S. Even, *Graph Algorithms,* Computer Science Press, Rockville, MD, 1979.

[Fe68] W. Feller, *An Introduction to Probability Theory and Its Applications,* vol. 1, 3d ed., Wiley, New York, 1968.

[Fo92] L. R. Foulds, *Graph Theory Applications,* Springer-Verlag, New York, 1992.

[GaJo79] Michael R. Garey and David S. Johnson, *Computers and Intractability: A Guide to NP-Completeness,* Freeman, New York, 1979.

[Gi85] A. Gibbons, *Algorithmic Graph Theory,* Cambridge University Press, Cambridge, England, 1985.

[Gi70] C. C. Gillispie, ed., *Dictionary of Scientific Biography,* Scribner's, New York, 1970.

[Go94] S. W. Golomb, *Polyominoes,* Princeton University Press, Princeton, NJ, 1994.

[Go84] G. H. Gonnet, *Handbook of Algorithms and Data Structures,* Addison-Wesley, London, 1984.

[GoHe77] S. E. Goodman and S. T. Hedetniemi, *Introduction to the Design and Analysis of Algorithms,* McGraw-Hill, New York, 1977.

[GoGo78] C. C. Gotlieb and L. R. Gotlieb, *Data Types and Structures,* Prentice-Hall, Englewood Cliffs, NJ, 1978.

[GrHe85] R. L. Graham and P. Hell, "On the History of the Minimum Spanning Tree Problem," *Annals of the History of Computing,* 7 (1985), 43–57.

[GrKnPa94] R. L. Graham, D. E. Knuth, and O. Patashnik, *Concrete Mathematics,* 2d ed., Addison-Wesley, Reading, MA, 1994.

[GrRoSp90] Ronald L. Graham, Bruce L. Rothschild, and Joel H. Spencer, *Ramsey Theory,* 2d ed., Wiley, New York, 1990.

[GrWa77] J. E. Graver and M. E. Watkins, *Combinatorics with Emphasis on the Theory of Graphs,* Springer-Verlag, New York, 1977.

[GrSc93] D. Gries and F. B. Schneider, *A Logical Approach to Discrete Math,* Springer-Verlag, New York, 1993.

[Gr94] R. P. Grimaldi, *Discrete and Combinatorial Mathematics,* 3d ed., Addison-Wesley, Reading, MA, 1994.

[GrYe99] J. L. Gross and J. Yellen, *Graph Theory and Its Applications,* CRC Press, Boca Raton, FL, 1999.

[Gr90] Jerrold W. Grossman, *Discrete Mathematics: An Introduction to Concepts, Methods, and Applications,* Macmillan, New York, 1990.

[Gr97] J. Gruska, *Foundations of Computing,* International Thomsen Computer Press, London, 1997.

[Ha60] P. R. Halmos, *Naive Set Theory,* D. Van Nostrand, New York, 1960.

[Ha80] R. W. Hamming, *Coding and Information Theory,* Prentice-Hall, Englewood Cliffs, NJ, 1980.

[Ha69] F. Harary, *Graph Theory,* Addison-Wesley, Reading, MA, 1969.

[HaWr79] G. H. Hardy and E. M. Wright, *An Introduction to the Theory of Numbers,* 5th ed., Oxford University Press, Oxford, England, 1979.

[Ha87] D. Harel, *Algorithmics, The Spirit of Computing,* Addison-Wesley, Reading, MA, 1987.

[Ha00a] Brian Hayes, "Graph-Theory in Practice, Part I," *American Scientist,* 88, no. 1 (2000), 9–13.

[Ha00b] Brian Hayes, "Graph-Theory in Practice, Part II," *American Scientist,* 88, no. 2 (2000), 104–109.

[Ha93] John P. Hayes, *Introduction to Digital Logic Design,* Addison-Wesley, Reading, MA, 1993.

[He77] F. Hennie, *Introduction to Computability,* Addison-Wesley, Reading, MA, 1977.

[He88] R. Herken, *The Universal Turing Machine, A Half-Century Survey,* Oxford University Press, New York, 1988.

[Ho99] D. Hofstadter, *Gödel, Escher, Bach: An Internal Golden Braid,* Basic Books, New York, 1999.

[Ho66] F. E. Hohn, *Applied Boolean Algebra,* 2d ed., Macmillan, New York, 1966.

[HoUl79] J. E. Hopcroft and J. D. Ullman, *Introduction to Automata Theory, Languages, and Computation,* Addison-Wesley, Reading, MA, 1979.

[HoMo76] D. Hopkin and B. Moss, *Automata,* Elsevier, North-Holland, New York, 1976.

[HoSa82] E. Horowitz and S. Sahni, *Fundamentals of Computer Algorithms,* Computer Science Press, Rockville, MD, 1982.

[Hu82] T. C. Hu, *Combinatorial Algorithms,* Addison-Wesley, Reading, MA, 1982.

[In93] D. C. Ince, *An Introduction to Discrete Mathematics, Formal System Specification, and Z,* 2d ed., Oxford, New York, 1993.

[Ka53] M. Karnaugh, "The Map Method for Synthesis of Combinatorial Logic Circuits," *Transactions of the AIEE,* part I, 72 (1953), 593–599.

[Ka93] Randy H. Katz, *Contemporary Logic Design,* Addison-Wesley, Reading, MA, 1993.

[Kl56] S. C. Kleene, "Representation of Events by Nerve Nets," in *Automata Studies,* 3–42, Princeton University Press, Princeton, NJ, 1956.

[Kn77] D. E. Knuth, "Algorithms," *Scientific American,* 236, no. 4 (April 1977), 63–80.

[Kn97a] D. E. Knuth, *The Art of Computer Programming, Vol. I: Fundamental Algorithms,* 3d ed., Addison-Wesley, Reading, MA, 1997a.

[Kn97b] D. E. Knuth, *The Art of Computer Programming, Vol. II: Seminumerical Algorithms,* 3d ed., Addison-Wesley, Reading, MA, 1997b.

[Kn98] D. E. Knuth, *The Art of Computer Programming, Vol. III: Sorting and Searching,* 2d ed., Addison-Wesley, Reading, MA, 1998.

[Ko86] Z. Kohavi, *Switching and Finite Automata Theory,* 2d ed., McGraw-Hill, New York, 1986.

[KrSt98] Donald H. Kreher and Douglas R. Stinson, *Combinatorial Algorithms: Generation, Enumeration, and Search,* CRC Press, Boca Raton, FL 1998.

[Kr87] L. Kronsjö, *Algorithms: Their Complexity and Efficiency,* 2d ed., Wiley, New York, 1987.

[Kr56] J. B. Kruskal, "On the Shortest Spanning Subtree of a Graph and the Traveling Salesman Problem," *Proceedings of the AMS,* 1 (1956), 48–50.

[Le64] D. H. Lehmer, "The Machine Tools of Combinatorics," in E. F. Beckenbach (ed.), *Applied Combinatorial Mathematics,* Wiley, New York, 1964.

[Le77] W. J. LeVeque, *Fundamentals of Number Theory,* Addison-Wesley, Reading, MA, 1977.

[LePa81] H. R. Lewis and C. H. Papadimitriou, *Elements of the Theory of Computation,* Prentice-Hall, Englewood Cliffs, NJ, 1981.

[LiLi81] Y. Lin and S. Y. T. Lin, *Set Theory with Applications,* 2d ed., Mariner, Tampa, FL, 1981.

[Li68] C. L. Liu, *Introduction to Combinatorial Mathematics,* McGraw-Hill, New York, 1968.

[Li85] C. L. Liu, *Elements of Discrete Mathematics,* 2d ed., McGraw-Hill, New York, 1985.

[Lo79] L. Lovász, *Combinatorial Problems and Exercises,* North-Holland, Amsterdam, 1979.

[Lu91] E. Lucas, *Récreations Mathématiques,* Gauthier-Villars, Paris, 1891.

[Ma91] J. C. Martin, *Introduction to Languages and the Theory of Computation,* McGraw-Hill, New York, 1991.

[Ma93] H. F. Mattson, Jr., *Discrete Mathematics with Applications,* Wiley, New York, 1993.

[Mc56] E. J. McCluskey, Jr., "Minimization of Boolean Functions," *Bell System Technical Journal,* 35 (1956), 1417–1444.

[Mc90] J. A. McHugh, *Algorithmic Graph Theory,* Prentice-Hall, Englewood Cliffs, NJ, 1990.

[Mc82] R. McNaughton, *Elementary Computability, Formal Languages, and Automata,* Prentice-Hall, Englewood Cliffs, NJ, 1982.

[Me55] G. H. Mealy, "A Method for Synthesizing Sequential Circuits," *Bell System Technical Journal,* 34 (1955), 1045–1079.

[Me64] E. Mendelson, *Introduction to Mathematical Logic,* Van Nostrand Reinhold, New York, 1964.

[MeOoVa97] A. J. Menezes, P. C. van Oorschot, S. A. Vanstone, *Handbook of Applied Cryptography,* CRC Press, Boca Raton, FL, 1997.

[MiRo91] J. G. Michaels and K. H. Rosen, *Applications of Discrete Mathematics,* McGraw-Hill, New York, 1991.

[Mo69] J. R. Monk, *Introduction to Set Theory,* McGraw-Hill, New York, 1969.

[Mo56] E. F. Moore, "Gedanken-Experiments on Sequential Machines," in *Automata Studies,* 129–153, Princeton University Press, Princeton, NJ, 1956.

[Na00] Paul J. Nabin, *Duelling Idiots and Other Probability Puzzlers,* Princeton University Press, Princeton, NJ, 2000.

[Ne85] C. V. Negoita, *Expert Systems and Fuzzy Systems,* Benjamin Cummings, Menlo Park, CA, 1985.

[NiMa95] Ulf Nilsson and Jan Maluszynski, *Logic, Programming, and Prolog,* 2d ed., Wiley, Chichester, England, 1995.

[Or63] O. Ore, *Graphs and Their Uses,* Mathematical Association of America, Washington, D.C., 1963.

[Pe87] A. Pelc, "Solution of Ulam's Problem on Searching with a Lie," *Journal of Combinatorial Theory,* Series A, 44 (1987), 129–140.

[PoSh81] I. Pohl and A. Shaw, *The Nature of Computation: An Introduction to Computer Science,* Computer Science Press, Rockville, MD, 1981.

[Po62] George Pólya, *Mathematical Discovery,* Vols. 1 and 2, Wiley, New York, 1962.

[Po71] George Pólya, *How to Solve It,* Princeton University Press, Princeton, NJ, 1971.

[Po90] George Pólya, *Mathematics and Plausible Reasoning,* Princeton University Press, Princeton, NJ, 1990.

[PoTaWo83] G. Pólya, R. E. Tarjan, and D. R. Woods, *Notes on Introductory Combinatorics,* Birkhäuser, Boston, 1983.

[Pr57] R. C. Prim, "Shortest Connection Networks and Some Generalizations," *Bell System Technical Journal,* 36 (1957), 1389–1401.

[PuBr85] P. W. Purdom, Jr. and C. A. Brown, *The Analysis of Algorithms,* Holt, Rinehart & Winston, New York, 1985.

[Qu52] W. V. Quine, "The Problem of Simplifying Truth Functions," *American Mathematical Monthly,* 59 (1952), 521–531.

[Qu55] W. V. Quine, "A Way to Simplify Truth Functions," *American Mathematical Monthly,* 62 (1955), 627–631.

[Ra62] T. Rado, "On Non-Computable Functions," *Bell System Technical Journal,* (May 1962), 877–884.

[Ra92] Gregory J. E. Rawlins, *Compared to What? An Introduction to the Analysis of Algorithms,* Computer Science Press, New York, 1992.

[ReNiDe77] E. M. Reingold, J. Nievergelt, and N. Deo, *Combinatorial Algorithms: Theory and Practice,* Prentice-Hall, Englewood Cliffs, NJ, 1977.

[Ri58] J. Riordan, *An Introduction to Combinatorial Analysis,* Wiley, New York, 1958.

[Ri68] J. Riordan, *Combinatorial Identities,* Wiley, New York, 1968.

[RiShAd78] R. Rivest, A. Shamir, and L. Adleman, "A Method for Obtaining Digital Signatures and Public-Key Cryptosystems," *Communications of the Association for Computing Machinery,* 31, no. 2 (1978), 120–128.

[Ro86] E. S. Roberts, *Thinking Recursively*, Wiley, New York, 1986.

[Ro76] F. S. Roberts, *Discrete Mathematics Models*, Prentice-Hall, Englewood Cliffs, NJ, 1976.

[Ro84] F. S. Roberts, *Applied Combinatorics*, Prentice-Hall, Englewood Cliffs, NJ, 1984.

[RoFo80] D. F. Robinson and L. R. Foulds, *Digraphs: Theory and Techniques*, Gordon and Breach, New York, 1980.

[Ro84a] J. S. Rohl, *Recursion via Pascal*, Cambridge University Press, Cambridge, England, 1984.

[Ro00] K. H. Rosen, *Elementary Number Theory and Its Applications*, 4th ed., Addison-Wesley, Reading, MA, 2000.

[Ro00a] K. H. Rosen, *Handbook of Discrete and Combinatorial Mathematics*, CRC Press, Boca Raton, FL, 2000.

[Ro01] Sheldon M. Ross, *A First Course in Probability Theory*, 6th ed., Prentice-Hall, Englewood Cliffs, NJ, 2001.

[Ro02] Sheldon M. Ross, *Probability Models for Computer Science*, Harcourt/Academic Press, 2002.

[Ro59] B. Roy, "Transitivité et Connexité," *C.R. Acad. Sci. Paris*, 249 (1959), 216.

[Ry63] H. Ryser, *Combinatorial Mathematics*, Mathematical Association of America, Washington, D.C., 1963.

[SaKa86] T. L. Saaty and P. C. Kainen, *The Four-Color Problem: Assaults and Conquest*, Dover, New York, 1986.

[Sa85] S. Sahni, *Concepts in Discrete Mathematics*, Camelot, Minneapolis, 1985.

[SePi89] J. Seberry and J. Pieprzyk, *Cryptography: An Introduction to Computer Security*, Prentice-Hall, Englewood Cliffs, NJ, 1989.

[Se88] R. Sedgewick, *Algorithms*, 2d ed., Addison-Wesley, Reading, MA, 1988.

[SiHs88] H. J. Siegel and W. T. Hsu, "Interconnection Networks," in *Computer Architectures*, V. M. Milutinovic (ed.), North-Holland, New York, 1988, pp. 225–264.

[Si66] A. Sinkov, *Elementary Cryptanalysis*, Mathematical Association of America, Washington, DC, 1966.

[SlPl95] N. J. A. Sloane and S. Plouffe, *The Encyclopedia of Integer Sequences*, Academic Press, New York, 1995.

[Sm78] Raymond Smullyan, *What Is the Name of This Book?: The Riddle of Dracula and Other Logical Puzzles*, Prentice-Hall, Englewood Cliffs, NJ, 1978.

[Sm92] Raymond Smullyan, *Lady or the Tiger? And Other Logic Puzzles Including a Mathematical Novel That Features Godel's Great Discovery*, Times Book, New York, 1992.

[Sm98] Raymond Smullyan, *The Riddle of Scheherazade: And Other Amazing Puzzles, Ancient & Modern*, Harvest Book, Fort Washington, PA, 1998.

[So01] Daniel Solow, *How to Read and Do Proofs: An Introduction to Mathematical Thought Processes*, Wiley, New York, 2001.

[So61] I. S. Sominskii, *Method of Mathematical Induction*, Blaisdell, New York, 1961.

[Sp94] M. Spivak, *Calculus*, 3d ed., Publish or Perish, Wilmington, DE, 1994.

[StMc77] D. Stanat and D. F. McAllister, *Discrete Mathematics in Computer Science*, Prentice-Hall, Englewood Cliffs, NJ, 1977.

[St78] H. M. Stark, *An Introduction to Number Theory*, MIT Press, Cambridge, MA, 1978.

[St02] Douglas R. Stinson, *Cryptography, Theory and Practice*, 2d ed., Chapman and Hall/CRC, Boca Raton, FL, 2002.

[St94] P. K. Stockmeyer, "Variations on the Four-Post Tower of Hanoi Puzzle," *Congressus Numerantrum* 102 (1994), 3–12.

[St74] R. R. Stoll, *Sets, Logic, and Axiomatic Theories*, 2d ed., W. H. Freeman, San Francisco, 1974.

[St88] G. W. Strang, *Linear Algebra and Its Applications*, 3d ed., Harcourt Brace Jovanovich, San Diego, 1988.

[Su87] P. Suppes, *Introduction to Logic*, D. Van Nostrand, Princeton, NJ, 1987.

[Ta83] R. E. Tarjan, *Data Structures and Network Algorithms*, Society for Industrial and Applied Mathematics, Philadelphia, 1983.

[ThFi96] G. B. Thomas and R. L. Finney, *Calculus and Analytic Geometry*, 9th ed., Addison-Wesley, Reading, MA, 1996.

[TrMa75] J. P. Tremblay and R. P. Manohar, *Discrete Mathematical Structures with Applications to Computer Science*, McGraw-Hill, New York, 1975.

[Tu02] Alan Tucker, *Applied Combinatorics*, 4th ed., Wiley, New York, 2002.

[Ve52] E. W. Veitch, "A Chart Method for Simplifying Truth Functions," *Proceedings of the ACM* (1952), 127–133.

[Ve94] David Velleman, *How to Prove It: A Structured Approach*, Cambridge University Press, New York, 1994.

[Vi71] N. Y. Vilenkin, *Combinatorics*, Academic Press, New York, 1971.

[Wa80] M. Wand, *Induction, Recursion, and Programming*, North-Holland, New York, 1980.

[Wa62] S. Warshall, "A Theorem on Boolean Matrices," *Journal of the ACM*, 9 (1962), 11–12.

[Wi86] Herbert S. Wilf, *Algorithms and Complexity*, Prentice-Hall, Englewood Cliffs, NJ, 1986.

[Wi85] S. G. Williamson, *Combinatorics for Computer Science*, Computer Science Press, Rockville, MD, 1985.

[Wi85] R. J. Wilson, *Introduction to Graph Theory*, 3d ed., Longman, Essex, England, 1985.

[WiBe79] R. J. Wilson and L. W. Beineke, *Applications of Graph Theory*, Academic Press, London, 1979.

[WiWa90] R. J. Wilson and J. J. Watkins, *Graphs, An Introductory Approach*, Wiley, New York, 1990.

[Wi76] N. Wirth, *Algorithms + Data Structures = Programs*, Prentice-Hall, Englewood Cliffs, NJ, 1976.

[Wi84] N. Wirth, "Data Structures and Algorithms," *Scientific American,* 251 (September 1984), 60–69.

[Wo98] Robert S. Wolf, *Proof, Logic, and Conjecture: The Mathematician's Toolbox,* New York, 1998.

[Wo87] D. Wood, *Theory of Computation,* Harper & Row, New York, 1987.

[Zi91] H. J. Zimmermann, *Fuzzy Set Theory and Its Applications,* 2d ed., Kluwer, Boston, 1991.

Answers to Odd-Numbered Exercises

CHAPTER 1

SECTION 1.1

1. a) Yes, T **b)** Yes, F **c)** Yes, T **d)** Yes, F **e)** No **f)** No **g)** Yes, T **3. a)** Today is not Thursday. **b)** There is pollution in New Jersey. **c)** $2 + 1 \neq 3$. **d)** The summer in Maine is not hot or it is not sunny. **5. a)** Sharks have not been spotted near the shore. **b)** Swimming at the New Jersey shore is allowed, and sharks have been spotted near the shore. **c)** Swimming at the New Jersey shore is not allowed, or sharks have been spotted near the shore. **d)** If swimming at the New Jersey shore is allowed, then sharks have not been spotted near the shore. **e)** If sharks have not been spotted near the shore, then swimming at the New Jersey shore is allowed. **f)** If swimming at the New Jersey shore is not allowed, then sharks have not been spotted near the shore. **g)** Swimming at the New Jersey shore is allowed if and only if sharks have not been spotted near the shore. **h)** Swimming at the New Jersey shore is not allowed, and either swimming at the New Jersey shore is allowed or sharks have not been spotted near the shore. (Note that we were able to incorporate the parentheses by using the word "either" in the second half of the sentence.) **7. a)** $p \land q$ **b)** $p \land \neg q$ **c)** $\neg p \land \neg q$ **d)** $p \lor q$ **e)** $p \to q$ **f)** $(p \lor q) \land (p \to \neg q)$ **g)** $q \leftrightarrow p$ **9. a)** $\neg p$ **b)** $p \land \neg q$ **c)** $p \to q$ **d)** $\neg p \to \neg q$ **e)** $p \to q$ **f)** $q \land \neg p$ **g)** $q \to p$ **11. a)** $r \land \neg p$ **b)** $\neg p \land q \land r$ **c)** $r \to (q \leftrightarrow \neg p)$ **d)** $\neg q \land \neg p \land r$ **e)** $(q \to (\neg r \land \neg p)) \land ((\neg r \land \neg p) \to q)$ **f)** $(p \land r) \to \neg q$ **13. a)** False **b)** True **c)** True **d)** True **e)** True **f)** True **g)** False **h)** True **15. a)** Inclusive *or*: It is allowable to take discrete mathematics if you have had calculus or computer science, or both. Exclusive *or*: It is allowable to take discrete mathematics if you have had calculus or computer science, but not if you have had both. Most likely the inclusive *or* is intended. **b)** Inclusive *or*: You can take the rebate, or you can get a low-interest loan, or you can get both the rebate and a low-interest loan. Exclusive *or*: You can take the rebate, or you can get a low-interest loan, but you cannot get both the rebate and a low-interest loan. Most likely the exclusive *or* is intended. **c)** Inclusive *or*: You can order two items from column A and none from column B, or three items from column B and none from column A, or five items including two from column A and three from column B. Exclusive *or*: You can

order two items from column A or three items from column B, but not both. Almost certainly the exclusive *or* is intended. **d)** Inclusive *or*: More than 2 feet of snow or windchill below -100, or both, will close school. Exclusive *or*: More than 2 feet of snow or windchill below -100, but not both, will close school. Certainly the inclusive *or* is intended. **17. a)** If the wind blows from the northeast, then it snows. **b)** If it stays warm for a week, then the apple trees will bloom. **c)** If the Pistons win the championship, then they beat the Lakers. **d)** If you get to the top of Long's Peak, then you must have walked 8 miles. **e)** If you are world-famous, then you will get tenure as a professor. **f)** If you drive more than 400 miles, then you will need to buy gasoline. **g)** If your guarantee is good, then you must have bought your CD player less than 90 days ago. **19. a)** You buy an ice cream cone if and only if it is hot outside. **b)** You win the contest if and only if you hold the only winning ticket. **c)** You get promoted if and only if you have connections. **d)** Your mind will decay if and only if you watch television. **e)** The train runs late if and only if it is a day I take the train. **21. a)** Converse: "I will ski tomorrow only if it snows today." Contrapositive: "If I do not ski tomorrow, then it will not have snowed today." Inverse: "If it does not snow today, then I will not ski tomorrow." **b)** Converse: "If I come to class, then there will be a quiz." Contrapositive: "If I do not come to class, then there will not be a quiz." Inverse: "If there is not going to be a quiz, then I don't come to class." **c)** Converse: "A positive integer is a prime if it has no divisors other than 1 and itself." Contrapositive: "If a positive integer has a divisor other than 1 and itself, then it is not prime." Inverse: "If a positive integer is not prime, then it has a divisor other than 1 and itself."

23. a)

p	$\neg p$	$p \land \neg p$
T	F	F
F	T	F

b)

p	$\neg p$	$p \lor \neg p$
T	F	T
F	T	T

c)

p	q	$\neg q$	$p \lor \neg q$	$(p \lor \neg q) \to q$
T	T	F	T	T
T	F	T	T	F
F	T	F	F	T
F	F	T	T	F

d)

p	q	p ∨ q	p ∧ q	(p ∨ q) → (p ∧ q)
T	T	T	T	T
T	F	T	F	F
F	T	T	F	F
F	F	F	F	T

e)

p	q	p → q	¬q	¬p	¬q → ¬p	(p → q) ↔ (¬q → ¬p)
T	T	T	F	F	T	T
T	F	F	T	F	F	T
F	T	T	F	T	T	T
F	F	T	T	T	T	T

f)

p	q	p → q	q → p	(p → q) → (q → p)
T	T	T	T	T
T	F	F	T	T
F	T	T	F	F
F	F	T	T	T

25. For parts **(a)**, **(b)**, **(c)**, **(d)**, and **(f)** we have this table.

p	q	(p ∨ q) → (p ⊕ q)	(p ⊕ q) → (p ∧ q)	(p ∨ q) ⊕ (p ∧ q)	(p ↔ q) ⊕ (¬p ↔ q)	(p ⊕ q) → (p ⊕ ¬q)
T	T	F	T	F	T	T
T	F	T	F	T	T	F
F	T	T	F	T	T	F
F	F	T	T	F	T	T

For part **(e)** we have this table.

p	q	r	¬p	¬r	p ↔ q	¬p ↔ ¬r	(p ↔ q) ⊕ (¬p ↔ ¬r)
T	T	T	F	F	T	T	F
T	T	F	F	T	T	F	T
T	F	T	F	F	F	T	T
T	F	F	F	T	F	F	F
F	T	T	T	F	F	F	F
F	T	F	T	T	F	T	T
F	F	T	T	F	T	F	T
F	F	F	T	T	T	T	F

27.

p	q	p → ¬q	¬p ↔ q	(p → q)∨ (¬p → q)	(p → q)∧ (¬p → q)	(p ↔ q)∨ (¬p ↔ q)	(¬p ↔ ¬q) ↔ (p ↔ q)
T	T	F	F	T	T	T	T
T	F	T	T	T	F	T	T
F	T	T	T	T	T	T	T
F	F	T	F	T	F	T	T

29.

p	q	r	p → (¬q ∨ r)	¬p → (q → r)	(p → q)∨ (¬p → r)	(p → q)∧ (¬p → r)	(p ↔ q)∨ (¬q ↔ r)	(¬p ↔ ¬q) ↔ (q ↔ r)
T	T	T	T	T	T	T	T	T
T	T	F	F	T	T	T	T	F
T	F	T	T	T	T	T	F	T
T	F	F	T	T	T	T	F	F
F	T	T	T	T	T	T	F	F
F	T	F	T	F	T	F	T	T
F	F	T	T	T	T	T	T	F
F	F	F	T	T	T	T	T	T

31.

p	q	r	s	p ↔ q	r ↔ s	(p ↔ q) ↔ (r ↔ s)
T	T	T	T	T	T	T
T	T	T	F	T	F	F
T	T	F	T	T	F	F
T	T	F	F	T	T	T
T	F	T	T	F	T	F
T	F	T	F	F	F	T
T	F	F	T	F	F	T
T	F	F	F	F	T	F
F	T	T	T	F	T	F
F	T	T	F	F	F	T
F	T	F	T	F	F	T
F	T	F	F	F	T	F
F	F	T	T	T	T	T
F	F	T	F	T	F	F
F	F	F	T	T	F	F
F	F	F	F	T	T	T

33. a) Bitwise *OR* is 111 1111; bitwise *AND* is 000 0000; bitwise *XOR* is 111 1111. **b)** Bitwise *OR* is 1111 1010; bitwise *AND* is 1010 0000; bitwise *XOR* is 0101 1010. **c)** Bitwise *OR* is 10 0111 1001; bitwise *AND* is 00 0100 0000; bitwise *XOR* is 10 0011 1001. **d)** Bitwise *OR* is 11 1111 1111; bitwise *AND* is 00 0000 0000; bitwise *XOR* is 11 1111 1111. **35.** 0.2, 0.6 **37.** 0.8, 0.6 **39. a)** The 99th statement is true and the rest are false. **b)** Statements 1 through 50 are all true and statements 51 through 99 are all false. **c)** This cannot happen; it is a paradox, showing that these cannot be statements. **41.** "If I were to ask you whether the right branch leads to the ruins, would you answer yes?" **43. a)** $q \rightarrow p$ **b)** $q \wedge \neg p$ **c)** $q \rightarrow p$ **d)** $\neg q \rightarrow \neg p$ **45.** Not consistent **47.** Consistent **49.** NEW **AND** JERSEY **AND** BEACHES, (JERSEY **AND** BEACHES) **NOT** NEW **51.** *A* is a knight and *B* is a knave. **53.** *A* is a knight and *B* is a knight. **55.** *A* is a knave and *B* is a knight. **57.** In order of decreasing salary: Fred, Maggie, Janice **59.** The detective can determine that the butler and cook are lying but cannot determine whether the gardener is telling the truth or whether the handyman is telling the truth. **61.** The Japanese man owns the zebra, and the Norwegian drinks water.

SECTION 1.2

1. The equivalences follow by showing that the appropriate pairs of columns of this table agree.

p	p ∧ T	p ∨ F	p ∧ F	p ∨ T	p ∨ p	p ∧ p
T	T	T	F	T	T	T
F	F	F	F	T	F	F

3. a)

p	q	p ∨ q	q ∨ p
T	T	T	T
T	F	T	T
F	T	T	T
F	F	F	F

b)

p	q	p ∧ q	q ∧ p
T	T	T	T
T	F	F	F
F	T	F	F
F	F	F	F

5.

p	q	r	q ∨ r	p ∧ (q ∨ r)	p ∧ q	p ∧ r	(p ∧ q) ∨ (p ∧ r)
T	T	T	T	T	T	T	T
T	T	F	T	T	T	F	T
T	F	T	T	T	F	T	T
T	F	F	F	F	F	F	F
F	T	T	T	F	F	F	F
F	T	F	T	F	F	F	F
F	F	T	T	F	F	F	F
F	F	F	F	F	F	F	F

7. a)

p	q	p ∧ q	(p ∧ q) → p
T	T	T	T
T	F	F	T
F	T	F	T
F	F	F	T

b)

p	q	p ∨ q	p → (p ∨ q)
T	T	T	T
T	F	T	T
F	T	T	T
F	F	F	T

c)

p	q	¬p	p → q	¬p → (p → q)
T	T	F	T	T
T	F	F	F	T
F	T	T	T	T
F	F	T	T	T

d)

p	q	p ∧ q	p → q	(p ∧ q) → (p → q)
T	T	T	T	T
T	F	F	F	T
F	T	F	T	T
F	F	F	T	T

e)

p	q	p → q	¬(p → q)	¬(p → q) → p
T	T	T	F	T
T	F	F	T	T
F	T	T	F	T
F	F	T	F	T

f)

p	q	$p \rightarrow q$	$\neg(p \rightarrow q)$	$\neg q$	$\neg(p \rightarrow q) \rightarrow \neg q$
T	T	T	F	F	T
T	F	F	T	T	T
F	T	T	F	F	T
F	F	T	F	T	T

9. In each case we will show that if the hypothesis is true, then the conclusion is also. **a)** If the hypothesis $p \wedge q$ is true, then by the definition of conjunction, the conclusion p must also be true. **b)** If the hypothesis p is true, by the definition of disjunction, the conclusion $p \vee q$ is also true. **c)** If the hypothesis $\neg p$ is true, that is, if p is false, then the conclusion $p \rightarrow q$ is true. **d)** If the hypothesis $p \wedge q$ is true, then both p and q are true so that the conclusion $p \rightarrow q$ is also true. **e)** If the hypothesis $\neg(p \rightarrow q)$ is true, then $p \rightarrow q$ is false, so that the conclusion p is true (and q is false). **f)** If the hypothesis $\neg(p \rightarrow q)$ is true, then $p \rightarrow q$ is false, so that p is true and q is false. Hence, the conclusion $\neg q$ is true. **11.** That the fourth column of the truth table shown is identical to the first column proves part **(a)**, and that the sixth column is identical to the first column proves part **(b)**.

p	q	$p \wedge q$	$p \vee (p \wedge q)$	$p \vee q$	$p \wedge (p \vee q)$
T	T	T	T	T	T
T	F	F	T	T	T
F	T	F	F	T	F
F	F	F	F	F	F

13. The only way this implication can be false is when $\neg q \wedge (p \rightarrow q)$ is true and $\neg p$ is false. For $\neg p$ to be false, p must be true. For $\neg q \wedge (p \rightarrow q)$ to be true, $\neg q$ must be true, so q is false. Since p is true, this makes $p \rightarrow q$ false, which is impossible. **15.** These are not logically equivalent since when p, q, and r are all false, $(p \rightarrow q) \rightarrow r$ is false, but $p \rightarrow (q \rightarrow r)$ is true. **17.** The proposition $\neg p \leftrightarrow q$ is true when $\neg p$ and q have the same truth values, which means that p and q have different truth values. Similarly, $p \leftrightarrow \neg q$ is true in exactly the same cases. Therefore, these two expressions are logically equivalent. **19.** The proposition $\neg(p \leftrightarrow q)$ is true when $p \leftrightarrow q$ is false, which means that p and q have different truth values. Since this is precisely when $\neg p \leftrightarrow q$ is true, the two expressions are logically equivalent. **21.** For $(p \rightarrow r) \wedge (q \rightarrow r)$ to be false, one of the two implications must be false, which happens exactly when r is false and at least one of p and q is true. But these are precisely the cases in which $p \vee q$ is true and r is false, which is precisely when $(p \vee q) \rightarrow r$ is false. Since the two propositions are false in exactly the same situations, they are logically equivalent. **23.** For $(p \rightarrow r) \vee (q \rightarrow r)$ to be false, both of the two implications must be false, which happens exactly when r is false and both p and q are true. But this is precisely the case in which $p \wedge q$ is true and r is

false, which is precisely when $(p \wedge q) \rightarrow r$ is false. Since the two propositions are false in exactly the same situations, they are logically equivalent. **25.** This fact was observed in Section 1 when the biconditional was first defined. Each of these is true precisely when p and q have the same truth values. **27.** Each of these is true precisely when p and q have opposite truth values. **29.** The last column is all Ts.

p	q	r	$p \rightarrow q$	$q \rightarrow r$	$(p \rightarrow q) \wedge (q \rightarrow r)$	$p \rightarrow r$	$(p \rightarrow q) \wedge (q \rightarrow r) \rightarrow (p \rightarrow r)$
T	T	T	T	T	T	T	T
T	T	F	T	F	F	F	T
T	F	T	F	T	F	T	T
T	F	F	F	T	F	F	T
F	T	T	T	T	T	T	T
F	T	F	T	F	F	T	T
F	F	T	T	T	T	T	T
F	F	F	T	T	T	T	T

31. If we take duals twice, every $\vee$ changes to an $\wedge$ and then back to an $\vee$, every $\wedge$ changes to an $\vee$ and then back to an $\wedge$, every **T** changes to an **F** and then back to a **T**, every **F** changes to a **T** and then back to an **F**. Hence, $(s^*)^* = s$. **33.** Let p and q be equivalent compound propositions involving only the operators $\wedge, \vee$, and $\neg$, and **T** and **F**. Note that $\neg p$ and $\neg q$ are also equivalent. Use De Morgan's laws as many times as necessary to push negations in as far as possible within these compound propositions, changing $\vee$s to $\wedge$s, and vice versa, and changing **T**s to **F**s, and vice versa. This shows that $\neg p$ and $\neg q$ are the same as p^* and q^* except that each atomic proposition p_i within them is replaced by its negation. From this we can conclude that p^* and q^* are equivalent since $\neg p$ and $\neg q$ are. **35.** $(p \wedge q \wedge \neg r) \vee (p \wedge \neg q \wedge r) \vee (\neg p \wedge q \wedge r)$ **37.** Given a compound proposition p, form its truth table and then write down a proposition q in disjunctive normal form that is logically equivalent to p. Since q involves only $\neg$, $\wedge$, and $\vee$, this shows that these three operators form a functionally complete set. **39.** By Exercise 37, given a compound proposition p, we can write down a proposition q that is logically equivalent to p and involves only $\neg$, $\wedge$, and $\vee$. By De Morgan's law we can eliminate all the $\wedge$s by replacing each occurrence of $p_1 \wedge p_2 \wedge \cdots \wedge p_n$ with $\neg(\neg p_1 \vee \neg p_2 \vee \cdots \vee \neg p_n)$. **41.** $\neg(p \wedge q)$ is true when either p or q, or both, are false, and is false when both p and q are true. Since this was the definition of $p \mid q$, the two compound propositions are logically equivalent. **43.** $\neg(p \vee q)$ is true when both p and q are false, and is false otherwise. Since this was the definition of $p \downarrow q$, the two are logically equivalent. **45.** $((p \downarrow p) \downarrow q) \downarrow ((p \downarrow p) \downarrow q)$ **47.** This follows immediately from the truth table or definition of $p \mid q$. **49.** 16 **51.** If the database is open, then either the system is in its initial state or the monitor is put in a closed state. **53.** All nine **55.** To determine whether c is a tautology apply an algorithm for satisfiability to $\neg c$. If the algorithm says that $\neg c$ is satisfiable, then we report that c is not a tautology, and if the algorithm says that $\neg c$ is not satisfiable, then we report that c is a tautology.

SECTION 1.3

1. a) T **b)** T **c)** F **3. a)** T **b)** F **c)** F **d)** F **5. a)** There is a student who spends more than 5 hours every weekday in class. **b)** Every student spends more than 5 hours every weekday in class. **c)** There is a student who does not spend more than 5 hours every weekday in class. **d)** No student spends more than 5 hours every weekday in class. **7. a)** Every comedian is funny. **b)** Every person is a funny comedian. **c)** There exists a person such that if she or he is a comedian, then she or he is funny. **d)** Some comedians are funny. **9. a)** $\exists x(P(x) \wedge Q(x))$ **b)** $\exists x(P(x) \wedge \neg Q(x))$ **c)** $\forall x(P(x) \vee Q(x))$ **d)** $\forall x \neg(P(x) \vee Q(x))$ **11. a)** T **b)** T **c)** F **d)** F **e)** T **f)** F **13. a)** True **b)** True **c)** True **d)** True **15. a)** True **b)** False **c)** True **d)** False **17. a)** $P(0) \vee P(1) \vee P(2) \vee P(3) \vee P(4)$ **b)** $P(0) \wedge P(1) \wedge P(2) \wedge P(3) \wedge P(4)$ **c)** $\neg P(0) \vee \neg P(1) \vee \neg P(2) \vee \neg P(3) \vee \neg P(4)$ **d)** $\neg P(0) \wedge \neg P(1) \wedge \neg P(2) \wedge \neg P(3) \wedge \neg P(4)$ **e)** $\neg(P(0) \vee P(1) \vee P(2) \vee P(3) \vee P(4))$ **f)** $\neg(P(0) \wedge P(1) \wedge P(2) \wedge P(3) \wedge P(4))$ **19. a)** $P(1) \vee P(2) \vee P(3) \vee P(4) \vee P(5)$ **b)** $P(1) \wedge P(2) \wedge P(3) \wedge P(4) \wedge P(5)$ **c)** $\neg(P(1) \vee P(2) \vee P(3) \vee P(4) \vee P(5))$ **d)** $\neg(P(1) \wedge P(2) \wedge P(3) \wedge P(4) \wedge P(5))$ **e)** $(P(1) \wedge P(2) \wedge P(4) \wedge P(5)) \vee (\neg P(1) \vee \neg P(2) \vee \neg P(3) \vee \neg P(4) \vee \neg P(5))$ **21.** Let $C(x)$ be the propositional function "x is in your class." **a)** $\exists x H(x)$ and $\exists x(C(x) \wedge H(x))$, where $H(x)$ is "x can speak Hindi" **b)** $\forall x F(x)$ and $\forall x(C(x) \rightarrow F(x))$, where $F(x)$ is "x is friendly" **c)** $\exists x \neg B(x)$ and $\exists x(C(x) \wedge \neg B(x))$, where $B(x)$ is "x was born in California" **d)** $\exists x M(x)$ and $\exists x(C(x) \wedge M(x))$, where $M(x)$ is "x has been in a movie" **e)** $\forall x \neg L(x)$ and $\forall x(C(x) \rightarrow \neg L(x))$, where $L(x)$ is "x has taken a course in logic programming" **23.** Let $P(x)$ be "x is perfect"; let $F(x)$ be "x is your friend"; and let the universe of discourse be all people. **a)** $\forall x \neg P(x)$ **b)** $\neg \forall x P(x)$ **c)** $\forall x(F(x) \rightarrow P(x))$ **d)** $\exists x(F(x) \wedge P(x))$, assuming that this means "at least one" of your friends is perfect **e)** $\forall x(F(x) \wedge P(x))$ or $(\forall x F(x)) \wedge (\forall x P(x))$ **f)** $(\forall x \neg F(x)) \vee (\exists x \neg P(x))$ **25.** Let $Y(x)$ be the propositional function that x is in your school or class, as appropriate. **a)** If we let $V(x)$ be "x has lived in Vietnam," then we have $\exists x V(x)$ if the universe is just your schoolmates, or $\exists x(Y(x) \wedge V(x))$ if the universe is all people. If we let $D(x, y)$ mean that person x has lived in country y, then we can rewrite this last one as $\exists x(Y(x) \wedge D(x, \text{Vietnam}))$. **b)** If we let $H(x)$ be "x can speak Hindi," then we have $\exists x \neg H(x)$ if the universe is just your schoolmates, or $\exists x(Y(x) \wedge \neg H(x))$ if the universe is all people. If we let $S(x, y)$ mean that person x

can speak language y, then we can rewrite this last one as $\exists x(Y(x) \wedge \neg S(x, \text{Hindi}))$. **c)** If we let $J(x)$, $P(x)$, and $C(x)$ be the propositional functions asserting x's knowledge of Java, Prolog, and C++, respectively, then we have $\exists x(J(x) \wedge P(x) \wedge C(x))$ if the universe is just your schoolmates, or $\exists x(Y(x) \wedge J(x) \wedge P(x) \wedge C(x))$ if the universe is all people. If we let $K(x, y)$ mean that person x knows programming language y, then we can rewrite this last one as $\exists x(Y(x) \wedge K(x, \text{Java}) \wedge K(x, \text{Prolog}) \wedge K(x, \text{C++}))$. **d)** If we let $T(x)$ be "x enjoys Thai food," then we have $\forall x T(x)$ if the universe is just your classmates, or $\forall x(Y(x) \rightarrow T(x))$ if the universe is all people. If we let $E(x, y)$ mean that person x enjoys food of type y, then we can rewrite this last one as $\forall x(Y(x) \rightarrow E(x, \text{Thai}))$. **e)** If we let $H(x)$ be "x plays hockey," then we have $\exists x \neg H(x)$ if the universe is just your classmates, or $\exists x(Y(x) \wedge \neg H(x))$ if the universe is all people. If we let $P(x, y)$ mean that person x plays game y, then we can rewrite this last one as $\exists x(Y(x) \wedge \neg P(x, \text{hockey}))$. **27.** Let $T(x)$ mean that x is a tautology and $C(x)$ mean that x is a contradiction. **a)** $\exists x T(x)$ **b)** $\forall x(C(x) \rightarrow T(\neg x))$ **c)** $\exists x \exists y(\neg T(x) \wedge \neg C(x) \wedge \neg T(y) \wedge \neg C(y) \wedge T(x \vee y))$ **d)** $\forall x \forall y((T(x) \wedge T(y)) \rightarrow T(x \wedge y))$ **29. a)** $Q(0, 0, 0) \wedge Q(0, 1, 0)$ **b)** $Q(0, 1, 1) \vee Q(1, 1, 1) \vee Q(2, 1, 1)$ **c)** $\neg Q(0, 0, 0) \vee \neg Q(0, 0, 1)$ **d)** $\neg Q(0, 0, 1) \vee \neg Q(1, 0, 1) \vee \neg Q(2, 0, 1)$ **31. a)** Let $T(x)$ be the predicate that x can learn new tricks, and let the universe of discourse be old dogs. Original is $\exists x T(x)$. Negation is $\forall x \neg T(x)$: "No old dogs can learn new tricks." **b)** Let $C(x)$ be the predicate that x knows calculus, and let the universe of discourse be rabbits. Original is $\neg \exists x C(x)$. Negation is $\exists x C(x)$: "There is a rabbit that knows calculus." **c)** Let $F(x)$ be the predicate that x can fly, and let the universe of discourse be birds. Original is $\forall x F(x)$. Negation is $\exists x \neg F(x)$: "There is a bird who cannot fly." **d)** Let $T(x)$ be the predicate that x can talk, and let the universe of discourse be dogs. Original is $\neg \exists x T(x)$. Negation is $\exists x T(x)$: "There is a dog that talks." **e)** Let $F(x)$ and $R(x)$ be the predicates that x knows French and knows Russian, respectively, and let the universe of discourse be people in this class. Original is $\neg \exists x(F(x) \wedge R(x))$. Negation is $\exists x(F(x) \wedge R(x))$: "There is someone in this class who knows French and Russian." **33. a)** There is no counterexample. **b)** $x = 0$ **c)** $x = 2$ **35. a)** $\forall x((F(x, 25{,}000) \vee S(x, 25)) \rightarrow E(x))$, where $E(x)$ is "Person x qualifies as an elite flyer in a given year," $F(x, y)$ is "Person x flies more than y miles in a given year," and $S(x, y)$ is "Person x takes more than y flights in a given year" **b)** $\forall x(((M(x) \wedge T(x, 3)) \vee (\neg M(x) \wedge T(x, 3.5))) \rightarrow Q(x))$, where $Q(x)$ is "Person x qualifies for the marathon," $M(x)$ is "Person x is a man," and $T(x, y)$ is "Person x has run the marathon in less than y hours" **c)** $M \rightarrow ((H(60) \vee (H(45) \wedge T)) \wedge \forall y G(\text{B}, y))$, where M is the proposition "The student received a masters degree," $H(x)$ is "The student took at least x course hours," T is the proposition "The student wrote a thesis," and $G(x, y)$ is "The person got grade x or higher in course y"

d) $\exists x\,((T(x, 21) \land G(x, 4.0))$, where $T(x, y)$ is "Person x took more than y credit hours" and $G(x, p)$ is "Person x earned grade point average p" (we assume that we are talking about one given semester)   **37. a)** If there is a printer that is both out of service and busy, then some job has been lost.   **b)** If every printer is busy, then there is a job in the queue.   **c)** If there is a job that is both queued and lost, then some printer is out of service.   **d)** If every printer is busy and every job is queued, then some job is lost.   **39. a)** $(\exists x\, F(x, 10)) \to \exists x\, S(x)$, where $F(x, y)$ is "Disk x has more than y kilobytes of free space," and $S(x)$ is "Mail message x can be saved"   **b)** $(\exists x\, A(x)) \to \forall x(Q(x) \to T(x))$, where $A(x)$ is "Alert x is active," $Q(x)$ is "Message x is queued," and $T(x)$ is "Message x is transmitted"   **c)** $\forall x((x \neq \text{main console}) \to T(x))$, where $T(x)$ is "The diagnostic monitor tracks the status of system x"   **d)** $\forall x(\neg L(x) \to B(x))$, where $L(x)$ is "The host of the conference call put participant x on a special list" and $B(x)$ is "Participant x was billed"   **41.** Not always   **43.** Both statements are true precisely when at least one of $P(x)$ and $Q(x)$ is true for at least one value of x in the universe of discourse.   **45. a)** If A is true, then both sides are logically equivalent to $\forall x\, P(x)$. If A is false, the left-hand side is clearly false. Furthermore, for every x, $P(x) \land A$ is false, so the right-hand side is false. Hence, the two sides are logically equivalent.   **b)** If A is true, then both sides are logically equivalent to $\exists x\, P(x)$. If A is false, the left-hand side is clearly false. Furthermore, for every x, $P(x) \land A$ is false, so $\exists x(P(x) \land A)$ is false. Hence, the two sides are logically equivalent.   **47.** To show these are not logically equivalent, let $P(x)$ be the statement "x is positive," and let $Q(x)$ be the statement "x is negative" with universe of discourse the set of integers. Then $\exists x\, P(x) \land \exists x\, Q(x)$ is true, but $\exists x(P(x) \land Q(x))$ is false.   **49. a)** True   **b)** False, unless the universe of discourse consists of just one element   **c)** True   **51. a)** Yes   **b)** No   **c)** juana, kiko   **d)** math273, cs301   **e)** juana, kiko
53. `sibling(X,Y) :- mother(M,X),`
`mother(M,Y), father(F,X), father(F,Y)`
55. a) $\forall x(P(x) \to \neg Q(x))$ **b)** $\forall x(Q(x) \to R(x))$
c) $\forall x(P(x) \to \neg R(x))$ **d)** The conclusion does not follow. There may be vain professors, since the premises do not rule out the possibility that there are other vain people besides ignorant ones.   **57. a)** $\forall x(P(x) \to \neg Q(x))$ **b)** $\forall x(R(x) \to \neg S(x))$ **c)** $\forall x(\neg Q(x) \to S(x))$
d) $\forall x(P(x) \to \neg R(x))$ **e)** The conclusion follows. Suppose x is a baby. Then by the first premise, x is illogical, so by the third premise, x is despised. The second premise says that if x could manage a crocodile, then x would not be despised. Therefore, x cannot manage a crocodile.

SECTION 1.4

1. a) For every real number x there exists a real number y such that x is less than y.   **b)** For every real number x and real number y, if x and y are both nonnegative, then their product is nonnegative.   **c)** For every real number x and real number y, there exists a real number z such that $xy = z$.   **3. a)** There is some student in your class who has sent a message to some student in your class.   **b)** There is some student in your class who has sent a message to every student in your class.   **c)** Every student in your class has sent a message to at least one student in your class.   **d)** There is a student in your class who has been sent a message by every student in your class.   **e)** Every student in your class has been sent a message from at least one student in your class.   **f)** Every student in the class has sent a message to every student in the class.   **5. a)** Sarah Smith has visited www.att.com.   **b)** At least one person has visited www.imdb.org.   **c)** Jose Orez has visited at least one website.   **d)** There is a website that both Ashok Puri and Cindy Yoon have visited.   **e)** There is a person besides David Belcher who has visited all the websites that David Belcher has visited.   **f)** There are two different people who have visited exactly the same websites.   **7. a)** Abdallah Hussein does not like Japanese cuisine.   **b)** Some student at your school likes Korean cuisine, and everyone at your school likes Mexican cuisine.   **c)** There is some cuisine that either Monique Arsenault or Jay Johnson likes.   **d)** For every pair of distinct students at your school, there is some cuisine that at least one them does not like.   **e)** There are two students at your school who like exactly the same set of cuisines.   **f)** For every pair of students at your school, there is some cuisine about which they have the same opinion (either they both like it or they both do not like it).   **9. a)** $\forall x L(x, \text{Jerry})$ **b)** $\forall x \exists y L(x, y)$
c) $\exists y \forall x L(x, y)$ **d)** $\forall x \exists y \neg L(x, y)$ **e)** $\exists x \neg L(\text{Lydia}, x)$
f) $\exists x \forall y \neg L(y, x)$ **g)** $\exists x (\forall y L(y, x) \land \forall z((\forall w L(w, z)) \to z = x))$ **h)** $\exists x \exists y (x \neq y \land L(\text{Lynn}, x) \land L(\text{Lynn}, y) \land \forall z(L(\text{Lynn}, z) \to (z = x \lor z = y)))$ **i)** $\forall x L(x, x)$
j) $\exists x \forall y (L(x, y) \leftrightarrow x = y)$ **11. a)** $A(\text{Lois}, \text{Professor Michaels})$ **b)** $\forall x(S(x) \to A(x, \text{Professor Gross}))$
c) $\forall x(F(x) \to (A(x, \text{Professor Miller}) \lor A(\text{Professor Miller}, x)))$ **d)** $\exists x(S(x) \land \forall y(F(y) \to \neg A(x, y)))$ **e)** $\exists x(F(x) \land \forall y(S(y) \to \neg A(y, x)))$ **f)** $\forall y(F(y) \to \exists x(S(x) \lor A(x, y)))$
g) $\exists x(F(x) \land \forall y((F(y) \land (y \neq x)) \to A(x, y)))$ **h)** $\exists x(S(x) \land \forall y(F(y) \to \neg A(y, x)))$ **13. a)** $\neg M(\text{Chou}, \text{Koko})$
b) $\neg M(\text{Arlene}, \text{Sarah}) \land \neg T(\text{Arlene}, \text{Sarah})$ **c)** $\neg M(\text{Deborah}, \text{Jose})$ **d)** $\forall x M(x, \text{Ken})$ **e)** $\forall x \neg T(x, \text{Nina})$
f) $\forall x((T, x, \text{Avi}) \lor M(x, \text{Avi}))$ **g)** $\exists x \forall y(y \neq x \to M(x, y))$ **h)** $\exists x \forall y(y \neq x \to (M(x, y) \lor T(x, y)))$
i) $\exists x \exists y(x \neq y \land M(x, y) \land M(y, x))$ **j)** $\exists x M(x, x)$
k) $\exists x \forall y(x \neq y \to (\neg M(x, y) \land \neg T(y, x)))$ **l)** $\forall x(\exists y(x \neq y \land (M(y, x) \lor T(x, y))))$ **m)** $\exists x \exists y(x \neq y \land M(x, y) \land T(y, x))$ **n)** $\exists x \exists y(x \neq y \land \forall z((z \neq x \land z \neq y) \to (M(x, z) \lor M(y, z) \lor T(x, z) \lor T(y, z))))$ **15. a)** $\forall x P(x)$, where $P(x)$ is "x needs a course in discrete mathematics" and the universe of discourse consists of all computer science students   **b)** $\exists x P(x)$, where $P(x)$ is "x owns a personal computer" and the universe consists of all students in this class   **c)** $\forall x \exists y P(x, y)$, where $P(x, y)$ is "x has taken

y," the universe of discourse for x consists of all students in this class, and the universe of discourse for y consists of all computer science classes   **d)** $\exists x \exists y P(x, y)$, where $P(x, y)$ and universes of discourse are the same as in part (c)   **e)** $\forall x \forall y P(x, y)$, where $P(x, y)$ is "x has been in y," the universe of discourse for x consists of all students in this class, and the universe of discourse for y consists of all buildings on campus   **f)** $\exists x \exists y \forall z (P(z, y) \rightarrow Q(x, z))$, where $P(z, y)$ is "z is in y" and $Q(x, z)$ is "x has been in z"; the universe of discourse for x consists of all students in the class, the universe of discourse for y consists of all buildings on campus, and the universe of discourse of z consists of all rooms.   **g)** $\forall x \forall y \exists z (P(z, y) \wedge Q(x, z))$, with same environment as in part (f)   **17. a)** $\forall u \exists m (A(u, m) \wedge \forall n(n \neq m \rightarrow \neg A(u, n)))$, where $A(u, m)$ means that user u has access to mailbox m **b)** $\exists p \forall e (H(e) \wedge S(p, \text{ running})) \rightarrow S(\text{kernel, working correctly})$, where $H(e)$ means that error condition e is in effect and $S(x, y)$ means that the status of x is y **c)** $\forall u \forall s (E(s, .edu) \rightarrow A(u, s))$, where $E(s, x)$ means that website s has extension x, and $A(u, s)$ means that user u can access website s **d)** $\exists x \exists y (x \neq y \wedge \forall z ((\forall s \, M(z, s)) \leftrightarrow (z = x \vee z = y)))$, where $M(a, b)$ means that system a monitors remote server b **19. a)** $\forall x \forall y ((x < 0) \wedge (y < 0) \rightarrow (x + y < 0))$ **b)** $\neg \forall x \forall y ((x > 0) \wedge (y > 0) \rightarrow (x - y > 0))$ **c)** $\forall x \forall y \, (x^2 + y^2 \geq (x + y)^2)$ **d)** $\forall x \forall y \, (|xy| = |x||y|)$ **21.** $\forall x \exists a \exists b \exists c \exists d \, ((x > 0) \rightarrow x = a^2 + b^2 + c^2 + d^2)$, where the universe of discourse consists of all integers   **23. a)** $\forall x \forall y ((x < 0) \wedge (y < 0) \rightarrow (xy > 0))$ **b)** $\forall x (x - x = 0)$ **c)** $\forall x \exists a \exists b (a \neq b \wedge \forall c (c^2 = x \leftrightarrow (c = a \vee c = b)))$ **d)** $\forall x ((x < 0) \rightarrow \neg \exists y(x = y^2))$ **25. a)** There is a multiplicative identity for the real numbers.   **b)** The product of two negative real numbers is always a positive real number.   **c)** There exist real numbers x and y such that x^2 exceeds y but x is less than y.   **d)** The real numbers are closed under the operation of addition.   **27. a)** True   **b)** True   **c)** True   **d)** True   **e)** True   **f)** False   **g)** False   **h)** True   **i)** False   **29. a)** $P(1, 1) \wedge P(1, 2) \wedge P(1, 3) \wedge P(2, 1) \wedge P(2, 2) \wedge P(2, 3) \wedge P(3, 1) \wedge P(3, 2) \wedge P(3, 3)$ **b)** $P(1, 1) \vee P(1, 2) \vee P(1, 3) \vee P(2, 1) \vee P(2, 2) \vee P(2, 3) \vee P(3, 1) \vee P(3, 2) \vee P(3, 3)$ **c)** $(P(1, 1) \wedge P(1, 2) \wedge P(1, 3)) \vee (P(2, 1) \wedge P(2, 2) \wedge P(2, 3)) \vee (P(3, 1) \wedge P(3, 2) \wedge P(3, 3))$ **d)** $(P(1, 1) \vee P(2, 1) \vee P(3, 1)) \wedge (P(1, 2) \vee P(2, 2) \vee P(3, 2)) \wedge (P(1, 3) \vee P(2, 3) \vee P(3, 3))$ **31. a)** $\exists x \forall y \exists z \neg T(x, y, z)$ **b)** $\exists x \forall y \neg P(x, y) \wedge \exists x \forall y \, \neg Q(x, y)$ **c)** $\exists x \forall y (\neg P(x, y) \vee \forall z \, \neg R(x, y, z))$ **d)** $\exists x \forall y (P(x, y) \wedge \neg Q(x, y))$ **33. a)** $\exists x \exists y \neg P(x, y)$ **b)** $\exists y \forall x \neg P(x, y)$ **c)** $\exists y \exists x (\neg P(x, y) \wedge \neg Q(x, y))$ **d)** $(\forall x \forall y P(x, y)) \vee (\exists x \exists y \neg Q(x, y))$ **e)** $\exists x (\forall y \exists z \neg P(x, y, z) \vee \forall z \exists y \neg P(x, y, z))$ **35. a)** There is someone in this class such that for every two different math courses, these are not the two and only two math courses this person has taken.   **b)** Every person has either visited Libya or has not visited a country other than Libya.   **c)** Someone has climbed every mountain in the Himalayas.   **d)** There is someone who has neither been in a movie with Kevin Bacon nor has been in a movie with someone who has been in a movie with Kevin Bacon.   **37. a)** $x = 2, y = -2$ **b)** $x = -4$ **c)** $x = 17, y = -1$ **39.** $\forall x \forall y \forall z ((x \cdot y) \cdot z = x \cdot (y \cdot z))$ **41. a)** True   **b)** False   **c)** True   **43.** $\neg(\exists x \forall y P(x, y)) \leftrightarrow \forall x(\neg \forall y P(x, y)) \leftrightarrow \forall x \exists y \neg P(x, y)$ **45. a)** Suppose that $\forall x P(x) \wedge \exists x Q(x)$ is true. Then $P(x)$ is true for all x and there is an element y for which $Q(y)$ is true. Since $P(x) \wedge Q(y)$ is true for all x and there is a y for which $Q(y)$ is true, $\forall x \exists y (P(x) \wedge Q(y))$ is true. Conversely, suppose that the second proposition is true. Let x be an element in the universe of discourse. There is a y such that $Q(y)$ is true, so $\exists x Q(x)$ is true. Since $\forall x P(x)$ is also true, it follows that the first proposition is true.   **b)** Suppose that $\forall x P(x) \vee \exists x Q(x)$ is true. Then either $P(x)$ is true for all x, or there exists a y for which $Q(y)$ is true. In the former case, $P(x) \vee Q(y)$ is true for all x, so $\forall x \exists y (P(x) \vee Q(y))$ is true. In the latter case, $Q(y)$ is true for a particular y, so $P(x) \vee Q(y)$ is true for all x and consequently $\forall x \exists y (P(x) \vee Q(y))$ is true. Conversely, suppose that the second proposition is true. If $P(x)$ is true for all x, then the first proposition is true. If not, $P(x)$ is false for some x, and for this x there must be a y such that $P(x) \vee Q(y)$ is true. Hence, $Q(y)$ must be true, so $\exists y Q(y)$ is true. It follows that the first proposition must hold.   **47.** We will show how an expression can be put into prenex normal form (PNF) if subexpressions in it can be put into PNF. Then, working from the inside out, any expression can be put in PNF. (To formalize the argument, it is necessary to use the method of structural induction that will be discussed in Section 3.4.) By Exercise 39 of Section 1.2, we can assume that the proposition uses only $\vee$ and $\neg$ as logical connectives. Now note that any proposition with no quantifiers is already in PNF. (This is the basis case of the argument.) Now suppose that the proposition is of the form $Qx P(x)$, where Q is a quantifier. Since $P(x)$ is a shorter expression than the original proposition, we can put it into PNF. Then Qx followed by this PNF is again in PNF and is equivalent to the original proposition. Next, suppose that the proposition is of the form $\neg P$. If P is already in PNF, we slide the negation sign past all the quantifiers using the equivalences in Table 2 in Section 1.3. Finally, assume that proposition is of the form $P \vee Q$, where each of P and Q is in PNF. If only one of P and Q has quantifiers, then we can use Exercise 44 in Section 1.3 to bring the quantifier in front of both. If both P and Q have quantifiers, we can use Exercise 43 in Section 1.3, Exercise 44, or part (b) of Exercise 45 to rewrite $P \vee Q$ with two quantifiers preceding the disjunction of a proposition of the form $R \vee S$, and then put $R \vee S$ into PNF.   **49.** $\exists x P(x) \wedge \forall x \forall y ((P(x) \wedge P(y)) \rightarrow x = y)$ **51.** $\forall L \exists \epsilon \forall N \exists n (n > N \wedge |a_n - L| \geq \epsilon)$ **53.** $\forall \epsilon (\forall N \exists n (n > N \wedge a_n > L - \epsilon) \wedge \exists N \forall n (n > N \rightarrow a_n \leq L + \epsilon))$

SECTION 1.5

1. a) Addition   **b)** Simplification   **c)** Modus ponens   **d)** Modus tollens   **e)** Hypothetical syllogism   **3.** Let w

be "Randy works hard," let d be "Randy is a dull boy," and let j be "Randy will get the job." The hypotheses are w, $w \rightarrow d$, and $d \rightarrow \neg j$. Using modus ponens and the first two hypotheses, d follows. Using modus ponens and the last hypothesis, $\neg j$, which is the desired conclusion, "Randy will not get the job," follows. **5.** Universal instantiation is used to conclude that "If Socrates is a man, then Socrates is mortal." Modus ponens is then used to conclude that Socrates is mortal. **7. a)** Valid conclusions are "I did not take Tuesday off," "I took Thursday off," "It rained on Thursday." **b)** "I did not eat spicy foods and it did not thunder" is a valid conclusion. **c)** "I am clever" is a valid conclusion. **d)** "Ralph is not a CS major" is a valid conclusion. **e)** "That you buy lots of stuff is good for the U.S. and is good for you" is a valid conclusion. **f)** "Mice gnaw their food" and "Rabbits are not rodents" are valid conclusions. **9. a)** Let $c(x)$ be "x is in this class," $j(x)$ be "x knows how to write programs in JAVA," and $h(x)$ be "x can get a high-paying job." The premises are $c(\text{Doug})$, $j(\text{Doug})$, $\forall x (j(x) \rightarrow h(x))$. Using universal instantiation and the last premise, $j(\text{Doug}) \rightarrow h(\text{Doug})$ follows. Applying modus ponens to this conclusion and the second premise, $h(\text{Doug})$ follows. Using conjunction and the first premise, $c(\text{Doug}) \wedge h(\text{Doug})$ follows. Finally, using existential generalization, the desired conclusion, $\exists x (c(x) \wedge h(x))$ follows. **b)** Let $c(x)$ be "x is in this class," $w(x)$ be "x enjoys whale watching," and $p(x)$ be "x cares about ocean pollution." The premises are $\exists x (c(x) \wedge w(x))$ and $\forall x (w(x) \rightarrow p(x))$. From the first premise, $c(y) \wedge w(y)$ for a particular person y. Using simplification, $w(y)$ follows. Using the second premise and universal instantiation, $w(y) \rightarrow p(y)$ follows. Using modus ponens, $p(y)$ follows, and by conjunction, $c(y) \wedge p(y)$ follows. Finally, by existential generalization, the desired conclusion, $\exists x (c(x) \wedge p(x))$, follows. **c)** Let $c(x)$ be "x is in this class," $p(x)$ be "x owns a PC," and $w(x)$ be "x can use a word-processing program." The premises are $c(\text{Zeke})$, $\forall x (c(x) \rightarrow p(x))$, and $\forall x (p(x) \rightarrow w(x))$. Using the second premise and universal instantiation, $c(\text{Zeke}) \rightarrow p(\text{Zeke})$ follows. Using the first premise and modus ponens, $p(\text{Zeke})$ follows. Using the third premise and universal instantiation, $p(\text{Zeke}) \rightarrow w(\text{Zeke})$ follows. Finally, using modus ponens, $w(\text{Zeke})$, the desired conclusion, follows. **d)** Let $j(x)$ be "x is in New Jersey," $f(x)$ be "x lives within 50 miles of the ocean," and $s(x)$ be "x has seen the ocean." The premises are $\forall x (j(x) \rightarrow f(x))$ and $\exists x (j(x) \wedge \neg s(x))$. The second hypothesis and existential instantiation imply that $j(y) \wedge \neg s(y)$ for a particular person y. By simplification, $j(y)$ for this person y. Using universal instantiation and the first premise, $j(y) \rightarrow f(y)$, and by modus ponens, $f(y)$ follows. By simplification, $\neg s(y)$ follows from $j(y) \wedge \neg s(y)$. So $f(y) \wedge \neg s(y)$ follows by conjunction. Finally, the desired conclusion, $\exists x (f(x) \wedge \neg s(x))$, follows by existential generalization. **11. a)** Correct, using universal instantiation and modus ponens **b)** Invalid; fallacy

of affirming the conclusion **c)** Invalid; fallacy of denying the hypothesis **d)** Correct, using universal instantiation and modus tollens **13. a)** Fallacy of affirming the conclusion **b)** Fallacy of begging the question **c)** Valid argument using modus tollens **d)** Fallacy of denying the hypothesis **15.** We know that *some x* exists that makes $H(x)$ true, but we cannot conclude that Lola is one such x. **17.** The proposition is vacuously true since 0 is not a positive integer. Vacuous proof. **19.** $P(1)$ is true since $(a + b)^1 = a + b \geq a^1 + b^1 = a + b$. Direct proof. **21. a)** Assume that n is odd, so $n = 2k + 1$ for some integer k. Then $n^3 + 5 = 2(4k^3 + 6k^2 + 3k + 3)$. Since $n^3 + 5$ is two times some integer, it is even. **b)** Suppose that $n^3 + 5$ is odd and n is odd. Since n is odd and the product of two odd numbers is odd, it follows that n^2 is odd and then that n^3 is odd. But then $5 = (n^3 + 5) - n^3$ would have to be even since it is the difference of two odd numbers. Therefore, the supposition that $n^3 + 5$ and n were both odd is wrong. **23.** Let $n = 2k + 1$ and $m = 2l + 1$ be odd integers. Then $n + m = 2(k + l + 1)$ is even. **25.** Suppose that r is rational and i is irrational and $s = r + i$ is rational. Then by Example 18, $s + (-r) = i$ is rational, which is a contradiction. **27.** Since $\sqrt{2} \cdot \sqrt{2} = 2$ is rational and $\sqrt{2}$ is irrational, the product of two irrational numbers is not necessarily irrational. **29.** Indirect proof: If $1/x$ were rational, then by definition $1/x = p/q$ for some integers p and q with $q \neq 0$. Since $1/x$ cannot be 0 (if it were, then we'd have the contradiction $1 = x \cdot 0$ by multiplying both sides by x), we know that $p \neq 0$. Now $x = 1/(1/x) = 1/(p/q) = q/p$ by the usual rules of algebra and arithmetic. Hence x can be written as the quotient of two integers with the denominator nonzero. Thus by definition, x is rational. **31.** If there were 9 or fewer days on each day of the week, this would account for at most $9 \cdot 7 = 63$ days. But we chose 64 days. This contradiction shows that at least 10 of the days must be on the same day of the week. **33.** If $x \leq y$, then $\max(x, y) + \min(x, y) = y + x = x + y$. If $x \geq y$, then $\max(x, y) + \min(x, y) = x + y$. Since these are the only two cases, the equality always holds. **35.** There are four cases. *Case 1:* $x \geq 0$ and $y \geq 0$. Then $|x| + |y| = x + y = |x + y|$. *Case 2:* $x < 0$ and $y < 0$. Then $|x| + |y| = -x + (-y) = -(x + y) = |x + y|$ since $x + y < 0$. *Case 3:* $x \geq 0$ and $y < 0$. Then $|x| + |y| = x + (-y)$. If $x \geq -y$, then $|x + y| = x + y$. But since $y < 0$, $-y > y$, so that $|x| + |y| = x + (-y) > x + y = |x + y|$. If $x < -y$, then $|x + y| = -(x + y) = -x + (-y)$. But since $x \geq 0$, $x \geq -x$, so that $|x| + |y| = x + (-y) \geq -x + (-y) = |x + y|$. *Case 4:* $x < 0$ and $y \geq 0$. Identical to Case 3 with the roles of x and y reversed. **37.** Without loss of generality we can assume that n is nonnegative, since the fourth power of an integer and the fourth power of its negative are the same. Following the hint given in Exercise 36, we divide an arbitrary positive integer n by 10, obtaining a quotient k and remainder l, whence $n = 10k + l$, and l is an integer between 0 and 9, inclusive. Then we compute n^4 in each of these ten cases. We get the following values, where X is some integer that is a multiple of 10, whose exact value we do not care about. $(10k + 0)^4 = 10,000k^4 = 10,000k^4 + 0$, $(10k + 1)^4 = 10,000k^4 + X \cdot k^3 + X \cdot k^2 + X \cdot k + 1$, $(10k + 2)^4 = 10,000k^4 + X \cdot k^3 + X \cdot k^2 + X \cdot k + 16$, $(10k + 3)^4 = 10,000k^4 + X \cdot k^3 + X \cdot k^2 + X \cdot k + 81$, $(10k + 4)^4 = 10,000k^4 + X \cdot k^3 + X \cdot k^2 + X \cdot k + 256$, $(10k + 5)^4 = 10,000k^4 + X \cdot k^3 + X \cdot k^2 + X \cdot k + 625$, $(10k + 6)^4 = 10,000k^4 + X \cdot k^3 + X \cdot k^2 + X \cdot k + 1296$, $(10k + 7)^4 = 10,000k^4 + X \cdot k^3 + X \cdot k^2 + X \cdot k + 2401$, $(10k + 8)^4 = 10,000k^4 + X \cdot k^3 + X \cdot k^2 + X \cdot k + 4096$, $(10k + 9)^4 = 10,000k^4 + X \cdot k^3 + X \cdot k^2 + X \cdot k + 6561$. Since each coefficient indicated by X is a multiple of 10, the corresponding term has no effect on the ones digit of the answer. Therefore the ones digits are 0, 1, 6, 1, 6, 5, 6, 1, 6, 1, respectively, so it is always a 0, 1, 5, or 6. **39.** First, assume that n is odd, so that $n = 2k + 1$ for some integer k. Then $5n + 6 = 5(2k + 1) + 6 = 10k + 11 = 2(5k + 5) + 1$. Hence $5n + 6$ is odd. To prove the converse, suppose that n is even, so that $n = 2k$ for some integer k. Then $5n + 6 = 10k + 6 = 2(5k + 3)$, so that $5n + 6$ is even. Hence n is odd if and only if $5n + 6$ is odd. **41.** This proposition is true. Suppose that m is neither 1 nor -1. Then mn has a factor m larger than 1. On the other hand, $mn = 1$, and 1 has no such factor. Hence $m = 1$ or $m = -1$. In the first case $n = 1$, and in the second case $n = -1$, since $n = 1/m$. **43.** We prove that all these are equivalent to x being even. If x is even, then $x = 2k$ for some integer k. Therefore $3x + 2 = 3 \cdot 2k + 2 = 6k + 2 = 2(3k + 1)$, which is even, since it has been written in the form $2t$, where $t = 3k + 1$. Similarly, $x + 5 = 2k + 5 = 2k + 4 + 1 = 2(k + 2) + 1$, so $x + 5$ is odd; and $x^2 = (2k)^2 = 2(2k^2)$, so x^2 is even. For the converses, we will use an indirect proof. So assume that x is not even; thus x is odd and we can write $x = 2k + 1$ for some integer k. Then $3x + 2 = 3(2k + 1) + 2 = 6k + 5 = 2(3k + 2) + 1$, which is odd (i.e., not even), since it has been written in the form $2t + 1$, where $t = 3k + 2$. Similarly, $x + 5 = 2k + 1 + 5 = 2(k + 3)$, so $x + 5$ is even (i.e., not odd). That x^2 is odd was already proved in Example 14. **45.** We give indirect proofs of (i) $\to$ (ii), (ii) $\to$ (i), (i) $\to$ (iii), and (iii) $\to$ (i). For the first of these, suppose that $3x + 2$ is rational, namely, equal to p/q for some integers p and q with $q \neq 0$. Then we can write $x = ((p/q) - 2)/3 = (p - 2q)/(3q)$, where $3q \neq 0$. This shows that x is rational. For the second implication, suppose that x is rational, namely, equal to p/q for some integers p and q with $q \neq 0$. Then we can write $3x + 2 = (3p + 2q)/q$, where $q \neq 0$. This shows that $3x + 2$ is rational. For the third implication, suppose that $x/2$ is rational, namely, equal to p/q for some integers p and q with $q \neq 0$. Then we can write $x = 2p/q$, where $q \neq 0$. This shows that x is rational. And for the fourth implication, suppose that x is rational, namely, equal to p/q for some integers p and q with $q \neq 0$. Then we can write $x/2 = p/(2q)$, where $2q \neq 0$. This shows that $x/2$ is ra-

tional. **47.** No **49.** $10,001, 10,002, \ldots, 10,100$ are all nonsquares, since $100^2 = 10,000$ and $101^2 = 10,201$; constructive **51.** $8 = 2^3$ and $9 = 3^2$ **53. a)** This statement asserts the existence of x with a certain property. If we let $y = x$, then we see that $P(x)$ is true. If y is anything other than x, then $P(x)$ is not true. Thus x is the unique element that makes P true. **b)** The first clause here says that there is an element that makes P true. The second clause says that whenever two elements both make P true, they are in fact the same element. Together this says that P is satisfied by exactly one element. **c)** This statement asserts the existence of an x that makes P true and has the further property that whenever we find an element that makes P true, that element is x. In other words, x is the unique element that makes P true. **55.** The equation $|a - c| = |b - c|$ is equivalent to the disjunction of two equations: $a - c = b - c$ or $a - c = -b + c$. The first of these is equivalent to $a = b$, which contradicts the assumptions made in this problem, so the original equation is equivalent to $a - c = -b + c$. By adding $b + c$ to both sides and dividing by 2, we see that this equation is equivalent to $c = (a + b)/2$. Thus there is a unique solution. Furthermore, this c is an integer, because the sum of the odd integers a and b is even. **57.** We are being asked to solve $n = (k - 2) + (k + 3)$ for k. Using the usual, reversible, rules of algebra, we see that this equation is equivalent to $k = (n - 1)/2$. In other words, this is the one and only value of k that makes our equation true. Since n is odd, $n - 1$ is even, so k is an integer. **59.** If x is itself an integer, then we can take $n = x$ and $\epsilon = 0$. No other solution is possible in this case, since if the integer n is greater than x, then n is at least $x + 1$, which would make $\epsilon \geq 1$. If x is not an integer, then round it up to the next integer, and call that integer n. We let $\epsilon = n - x$. Clearly $0 \leq \epsilon < 1$, this is the only ϵ that will work with this n, and n cannot be any larger, since ϵ is constrained to be less than 1. **61.** Let p be "It is raining"; let q be "Yvette has her umbrella"; let r be "Yvette gets wet." Assumptions are $\neg p \vee q$, $\neg q \vee \neg r$, and $p \vee \neg r$. Resolution on the first two gives $\neg p \vee \neg r$. Resolution on this and the third assumption gives $\neg r$, as desired. **63.** Assume that this proposition is satisfiable. Using resolution on the first two clauses enables us to conclude $q \vee q$; in other words, we know that q has to be true. Using resolution on the last two clauses enables us to conclude $\neg q \vee \neg q$; in other words, we know that $\neg q$ has to be true. This is a contradiction. So this proposition is not satisfiable. **65.** Let $x = 2$ and $y = \sqrt{2}$. If $x^y = 2^{\sqrt{2}}$ is irrational, we are done. If not, then let $x = 2^{\sqrt{2}}$ and $y = \sqrt{2}/4$. Then $x^y = (2^{\sqrt{2}})^{\sqrt{2}/4} = 2^{\sqrt{2} \cdot (\sqrt{2})/4} = 2^{1/2} = \sqrt{2}$. **67.** Suppose that $p_1 \to p_4 \to p_2 \to p_5 \to p_3 \to p_1$. To prove that one of these propositions implies any of the others, just use hypothetical syllogism repeatedly. **69.** Every domino placed on a chessboard covers exactly one white and one black square. Hence a set of dominos covers exactly the same number of white squares and black squares.

Since removing opposite corners leaves a board with two more black squares than white squares or two more white squares than black squares, no set of dominos can cover the board with opposite corners removed. **71.** We will give a proof by contradiction. Suppose that $a_1, a_2, \ldots, a_n$ are all less than A, where A is the average of these numbers. Then $a_1 + a_2 + \cdots + a_n < nA$. Dividing both sides by n shows that $A = (a_1 + a_2 + \cdots + a_n)/n < A$, which is a contradiction. **73.** We will show that the four statements are equivalent by showing that (i) implies (ii), (ii) implies (iii), (iii) implies (iv), and (iv) implies (i). First, assume that n is even. Then $n = 2k$ for some integer k. Then $n + 1 = 2k + 1$, so that $n + 1$ is odd. This shows that (i) implies (ii). Next, suppose that $n + 1$ is odd, so that $n + 1 = 2k + 1$ for some integer k. Then $3n + 1 = 2n + (n + 1) = 2(n + k) + 1$, which shows that $3n + 1$ is odd, showing that (ii) implies (iii). Next, suppose that $3n + 1$ is odd, so that $3n + 1 = 2k + 1$ for some integer k. Then $3n = (2k + 1) - 1 = 2k$, so that $3n$ is even. This shows that (iii) implies (iv). Finally, suppose that n is not even. Then n is odd, so $n = 2k + 1$ for some integer k. Then $3n = 3(2k + 1) = 6k + 3 = 2(3k + 1) + 1$, so that $3n$ is odd. This completes an indirect proof that (iv) implies (i). **75.** By the second premise, there is some lion that does not drink coffee. Let Leo be such a creature. By simplification we know that Leo is a lion. By modus ponens we know from the first premise that Leo is fierce. Hence Leo is fierce and does not drink coffee. By the definition of the existential quantifier, there exist fierce creatures that do not drink coffee, that is, some fierce creatures do not drink coffee. **77.** Valid

SECTION 1.6

1. a) $\{-1, 1\}$ **b)** $\{1, 2, 3, 4, 5, 6, 7, 8, 9, 10, 11\}$
c) $\{0, 1, 4, 9, 16, 25, 36, 49, 64, 81\}$ **d)** $\emptyset$ **3. a)** Yes **b)** No
c) No **5. a)** Yes **b)** No **c)** Yes **d)** No **e)** No **f)** No
7. a) False **b)** False **c)** False **d)** True **e)** False **f)** False
g) True **9. a)** True **b)** True **c)** False **d)** True **e)** True
f) False **11.** Suppose that $x \in A$. Since $A \subseteq B$, this implies that $x \in B$. Since $B \subseteq C$, we see that $x \in C$. Since $x \in A$ implies that $x \in C$, it follows that $A \subseteq C$.
13. a) 1 **b)** 1 **c)** 2 **d)** 3 **15. a)** $\{\emptyset, \{a\}\}$ **b)** $\{\emptyset, \{a\}, \{b\},$ $\{a, b\}\}$ **c)** $\{\emptyset, \{\emptyset\}, \{\{\emptyset\}\}, \{\emptyset, \{\emptyset\}\}\}$ **17. a)** 8 **b)** 16 **c)** 2
19. a) $\{(a, y), (b, y), (c, y), (d, y), (a, z), (b, z), (c, z), (d, z)\}$
b) $\{(y, a), (y, b), (y, c), (y, d), (z, a), (z, b), (z, c), (z, d)\}$
21. The set of triples (a, b, c), where a is an airline and b and c are cities. **23.** $\emptyset \times A = \{(x, y) \mid x \in \emptyset$ and $y \in A\} = \emptyset = \{(x, y) \mid x \in A$ and $y \in \emptyset\} = A \times \emptyset$ **25.** mn
27. a) The square of a real number is never -1. True
b) There exists an integer whose square is 2. False **c)** The square of every integer is positive. False **d)** There is a real number equal to its own square. True **29.** We must show that $\{\{a\}, \{a, b\}\} = \{\{c\}, \{c, d\}\}$ if and only if $a = c$ and $b = d$. The "if" part is immediate. So assume these two sets are equal. First, consider the case when $a \neq b$. Then $\{\{a\}, \{a, b\}\}$ contains exactly two elements, one of which contains one element. Thus $\{\{c\}, \{c, d\}\}$ must have

the same property, so $c \neq d$ and $\{c\}$ is the element containing exactly one element. Hence, $\{a\} = \{c\}$, which implies that $a = c$. Also, the two-element sets $\{a, b\}$ and $\{c, d\}$ must be equal. Since $a = c$ and $a \neq b$, it follows that $b = d$. Second, suppose that $a = b$. Then $\{\{a\}, \{a, b\}\} = \{\{a\}\}$, a set with one element. Hence $\{\{c\}, \{c, d\}\}$ has only one element, which can happen only when $c = d$, and the set is $\{\{c\}\}$. It then follows that $a = c$ and $b = d$. **31.** Let $S = \{a_1, a_2, \ldots, a_n\}$. Represent each subset of S with a bit string of length n, where the ith bit is 1 if and only if $a_i \in S$. To generate all subsets of S, list all 2^n bit strings of length n (for instance, in increasing order), and write down the corresponding subsets.

SECTION 1.7

1. a) The set of students who live within 1 mile of school and who walk to classes. **b)** The set of students who live within 1 mile of school or who walk to classes (or who do both). **c)** The set of students who live within 1 mile of school but do not walk to classes. **d)** The set of students who walk to classes but live more than 1 mile away from school. **3. a)** $\{0, 1, 2, 3, 4, 5, 6\}$ **b)** $\{3\}$
c) $\{1, 2, 4, 5\}$ **d)** $\{0, 6\}$ **5.** $\overline{\overline{A}} = \{x \mid \neg(x \in \overline{A})\} = \{x \mid \neg(\neg x \in A)\} = \{x \mid x \in A\} = A$. **7. a)** $A \cup B = \{x \mid x \in A \vee x \in B\} = \{x \mid x \in B \vee x \in A\} = B \cup A$; **b)** $A \cap B = \{x \mid x \in A \wedge x \in B\} = \{x \mid x \in B \wedge x \in A\} = B \cap A$.
9. Suppose $x \in A \cap (A \cup B)$. Then $x \in A$ and $x \in A \cup B$ by the definition of intersection. Since $x \in A$, we have proved that the left-hand side is a subset of the right-hand side. Conversely, let $x \in A$. Then by the definition of union, $x \in A \cup B$ as well. Therefore $x \in A \cap (A \cup B)$ by the definition of intersection, so the right-hand side is a subset of the left-hand side. **11. a)** $x \in \overline{(A \cup B)} \equiv x \notin (A \cup B) \equiv \neg(x \in A \vee x \in B) \equiv \neg(x \in A) \wedge \neg(x \in B) \equiv x \notin A \wedge x \notin B \equiv x \in \overline{A} \wedge x \in \overline{B} \equiv x \in \overline{A} \cap \overline{B}$.

b)

A	B	$A \cup B$	$\overline{(A \cup B)}$	$\overline{A}$	$\overline{B}$	$\overline{A} \cap \overline{B}$
1	1	1	0	0	0	0
1	0	1	0	0	1	0
0	1	1	0	1	0	0
0	0	0	1	1	1	1

13. a) $x \in \overline{A \cap B \cap C} \equiv x \notin A \cap B \cap C \equiv x \notin A \vee x \notin B \vee x \notin C \equiv x \in \overline{A} \vee x \in \overline{B} \vee x \in \overline{C} \equiv x \in \overline{A} \cup \overline{B} \cup \overline{C}$.
b)

A	B	C	$A \cap B \cap C$	$\overline{(A \cap B \cap C)}$	$\overline{A}$	$\overline{B}$	$\overline{C}$	$\overline{A} \cup \overline{B} \cup \overline{C}$
1	1	1	1	0	0	0	0	0
1	1	0	0	1	0	0	1	1
1	0	1	0	1	0	1	0	1
1	0	0	0	1	0	1	1	1
0	1	1	0	1	1	0	0	1
0	1	0	0	1	1	0	1	1
0	0	1	0	1	1	1	0	1
0	0	0	0	1	1	1	1	1

15. Both sides equal $\{x \mid x \in A \wedge x \notin B\}$. **17. a)** $x \in A \cup (B \cup C) \equiv (x \in A) \vee (x \in (B \cup C)) \equiv (x \in A) \vee (x \in B \vee x \in C) \equiv (x \in A \vee x \in B) \vee (x \in C) \equiv x \in (A \cup B) \cup C$ **b)** same as part (a) with $\cup$ replaced by $\cap$ and $\vee$ replaced by $\wedge$ **c)** $x \in A \cup (B \cap C) \equiv (x \in A) \vee (x \in (B \cap C)) \equiv (x \in A) \vee (x \in B \wedge x \in C) \equiv (x \in A \vee x \in B) \wedge (x \in A \vee x \in C) \equiv x \in (A \cup B) \cap (A \cup C)$ **19. a)** $\{4, 6\}$ **b)** $\{0, 1, 2, 3, 4, 5, 6, 7, 8, 9, 10\}$ **c)** $\{4, 5, 6, 8, 10\}$ **d)** $\{0, 2, 4, 5, 6, 7, 8, 9, 10\}$ **21. a)** $B \subseteq A$ **b)** $A \subseteq B$ **c)** $A \cap B = \emptyset$ **d)** nothing, since this is always true **e)** $A = B$ **23.** $A \subseteq B \equiv \forall x (x \in A \rightarrow x \in B) \equiv \forall x (x \notin B \rightarrow x \notin A) \equiv \forall x (x \in \overline{B} \rightarrow x \in \overline{A}) \equiv \overline{B} \subseteq \overline{A}$ **25.** The set of students who are computer science majors but not mathematics majors or who are mathematics majors but not computer science majors. **27.** An element is in $(A \cup B) - (A \cap B)$ if it is in the union of A and B but not in the intersection of A and B, which means that it is in either A or B but not in both A and B. This is exactly what it means for an element to belong to $A \oplus B$. **29. a)** $A \oplus A = (A - A) \cup (A - A) = \emptyset \cup \emptyset = \emptyset$ **b)** $A \oplus \emptyset = (A - \emptyset) \cup (\emptyset - A) = A \cup \emptyset = A$ **c)** $A \oplus U = (A - U) \cup (U - A) = \emptyset \cup \overline{A} = \overline{A}$ **d)** $A \oplus \overline{A} = (A - \overline{A}) \cup (\overline{A} - A) = A \cup \overline{A} = U$ **31.** $B = \emptyset$ **33.** Yes. Suppose that $x \in A$ but $x \notin B$. If $x \in C$, then $x \notin A \oplus C$ but $x \in B \oplus C$, a contradiction. If $x \notin C$ then $x \in A \oplus C$ but $x \notin B \oplus C$, a contradiction. Hence, $A \subseteq B$. Similarly, $B \subseteq A$, so that $A = B$. **35.** Yes. **37. a)** $\{1, 2, 3, \ldots, n\}$ **b)** $\{1\}$ **39. a)** A_n **b)** $\{0, 1\}$ **41. a)** $\{1, 2, 3, 4, 7, 8, 9, 10\}$ **b)** $\{2, 4, 5, 6, 7\}$ **c)** $\{1, 10\}$ **43.** The bit in the ith position of the bit string of the difference of two sets is 1 if the ith bit of the first string is 1 and the ith bit of the second string is 0, and is 0 otherwise. **45. a)** 11 1110 0000 0000 0000 0000 $\vee$ 01 1100 1000 0000 0100 0101 0000 = 11 1110 1000 0000 0100 0101 0000, representing $\{a, b, c, d, e, g, p, t, v\}$ **b)** 11 1110 0000 0000 0000 0000 0000 $\wedge$ 01 1100 1000 0000 0100 0101 0000 = 01 1100 0000 0000 0000 0000, representing $\{b, c, d\}$ **c)** (11 1110 0000 0000 0000 0000 0000 $\vee$ 00 0110 0110 0001 1000 0110 0110) $\wedge$ (01 1100 1000 0000 0100 0101 0000 $\vee$ 00 1010 0010 0000 1000 0010 0111) = 11 1110 0110 0001 1000 0110 0110 $\wedge$ 01 1110 1010 0000 1100 0111 0111 = 01 1110 0010 0000 1000 0110 0110, representing $\{b, c, d, e, i, o, t, u, x, y\}$ **d)** 11 1110 0000 0000 0000 0000 0000 $\vee$ 01 1100 1000 0000 0100 0101 0000 $\vee$ 00 1010 0010 0000 1000 0010 0111 $\vee$ 00 0110 0110 0001 1000 0110 0110 = 11 1110 1110 0001 1100 0111 0111, representing $\{a, b, c, d, e, g, h, i, n, o, p, t, u, v, x, y, z\}$ **47. a)** $\{1, 2, 3, \{1, 2, 3\}\}$ **b)** $\{\emptyset\}$ **c)** $\{\emptyset, \{\emptyset\}\}$ **d)** $\{\emptyset, \{\emptyset\}, \{\emptyset, \{\emptyset\}\}\}$ **49. a)** $\{3 \cdot a, 3 \cdot b, 1 \cdot c, 4 \cdot d\}$ **b)** $\{2 \cdot a, 2 \cdot b\}$ **c)** $\{1 \cdot a, 1 \cdot c\}$ **d)** $\{1 \cdot b, 4 \cdot d\}$ **e)** $\{5 \cdot a, 5 \cdot b, 1 \cdot c, 4 \cdot d\}$ **51.** $\overline{F} = \{0.4\,\text{Alice}, 0.1\,\text{Brian}, 0.6\,\text{Fred}, 0.9\,\text{Oscar}, 0.5\,\text{Rita}\}$, $\overline{R} = \{0.6\,\text{Alice}, 0.2\,\text{Brian}, 0.8\,\text{Fred}, 0.1\,\text{Oscar}, 0.3\,\text{Rita}\}$ **53.** $F \cap R = \{0.4\,\text{Alice}, 0.8\,\text{Brian}, 0.2\,\text{Fred}, 0.1\,\text{Oscar}, 0.5\,\text{Rita}\}$

SECTION 1.8

1. a) $f(0)$ is not defined. **b)** $f(x)$ is not defined for $x < 0$. **c)** $f(x)$ is not well-defined since there are two distinct values assigned to each x. **3. a)** Not a function **b)** A function **c)** Not a function **5. a)** The set of integers **b)** The set of even nonnegative integers **c)** The set of nonnegative integers not exceeding 7 **d)** The set of squares of integers $= \{0, 1, 4, 9, 16, \ldots\}$ **7. a)** Domain $\mathbf{Z}^+ \times \mathbf{Z}^+$; range $\mathbf{Z}^+$ **b)** Domain $\mathbf{Z}^+$; range $\{0, 1, 2, 3, 4, 5, 6, 7, 8, 9\}$ **c)** Domain the set of bit strings; range $\mathbf{N}$ **d)** Domain the set of bit strings; range $\mathbf{N}$ **9. a)** 1 **b)** 0 **c)** 0 **d)** -1 **e)** 3 **f)** -1 **g)** 2 **h)** 1 **11.** Only the function in part (a) **13.** Only the functions in parts (a) and (d) **15. a)** Onto **b)** Not onto **c)** Onto **d)** Not onto **e)** Onto **17. a)** The function $f(x)$ with $f(x) = 3x + 1$ when $x \geq 0$ and $f(x) = -3x + 2$ when $x < 0$ **b)** $f(x) = |x| + 1$ **c)** The function $f(x)$ with $f(x) = 2x + 1$ when $x \geq 0$ and $f(x) = -2x$ when $x < 0$ **d)** $f(x) = x^2 + 1$ **19. a)** Yes **b)** No **c)** Yes **d)** No **21.** Suppose that f is strictly decreasing. This means that $f(x) > f(y)$ whenever $x < y$. To show that g is strictly increasing, suppose that $x < y$. Then $g(x) = 1/f(x) < 1/f(y) = g(y)$. Conversely, suppose that g is strictly increasing. This means that $g(x) < g(y)$ whenever $x < y$. To show that f is strictly decreasing, suppose that $x < y$. Then $f(x) = 1/g(x) > 1/g(y) = f(y)$. **23. a)** $f(S) = \{0, 1, 3\}$ **b)** $f(S) = \{0, 1, 3, 5, 8\}$ **c)** $f(S) = \{0, 8, 16, 40\}$ **d)** $f(S) = \{1, 12, 33, 65\}$ **25. a)** Let x and y be distinct elements of A. Since g is one-to-one, $g(x)$ and $g(y)$ are distinct elements of B. Since f is one-to-one, $f(g(x)) = (f \circ g)(x)$ and $f(g(y)) = (f \circ g)(y)$ are distinct elements of C. Hence, $f \circ g$ is one-to-one. **b)** Let $y \in C$. Since f is onto, $y = f(b)$ for some $b \in B$. Now since g is onto, $b = g(x)$ for some $x \in A$. Hence, $y = f(b) = f(g(x)) = (f \circ g)(x)$. It follows that $f \circ g$ is onto. **27.** No. For example, suppose that $A = \{a\}$, $B = \{b, c\}$, and $C = \{d\}$. Let $g(a) = b$, $f(b) = d$, and $f(c) = d$. f and $f \circ g$ are onto, but g is not. **29.** $(f + g)(x) = x^2 + x + 3$, $(fg)(x) = x^3 + 2x^2 + x + 2$. **31.** f is one-to-one since $f(x_1) = f(x_2) \equiv ax_1 + b = ax_2 + b \equiv ax_1 = ax_2 \equiv x_1 = x_2$. f is onto since $f((y - b)/a) = y$. $f^{-1}(y) = (y - b)/a$. **33.** Let $f(1) = a$, $f(2) = a$. Let $S = \{1\}$ and $T = \{2\}$. Then $f(S \cap T) = f(\emptyset) = \emptyset$, but $f(S) \cap f(T) = \{a\} \cap \{a\} = \{a\}$. **35. a)** $\{x \mid 0 \leq x < 1\}$ **b)** $\{x \mid -1 \leq x < 2\}$ **c)** $\emptyset$ **37.** $f^{-1}(\overline{S}) = \{x \in A \mid f(x) \notin S\} = \overline{\{x \in A \mid f(x) \in S\}} = \overline{f^{-1}(S)}$ **39.** Let $x = \lfloor x \rfloor + \epsilon$, where ϵ is a real number with $0 \leq \epsilon < 1$. If $\epsilon < \frac{1}{2}$, then $\lfloor x \rfloor - 1 < x - \frac{1}{2} < \lfloor x \rfloor$, so $\lceil x - \frac{1}{2} \rceil = \lfloor x \rfloor$ and this is the integer closest to x. If $\epsilon > \frac{1}{2}$, then $\lfloor x \rfloor < x - \frac{1}{2} < \lfloor x \rfloor + 1$, so $\lceil x - \frac{1}{2} \rceil = \lfloor x \rfloor + 1$ and this is the integer closest to x. If $\epsilon = \frac{1}{2}$, then $\lceil x - \frac{1}{2} \rceil = \lfloor x \rfloor$, which is the smaller of the two integers that surround x and are the same distance from x. **41.** Write the real number

x as $\lfloor x \rfloor + \epsilon$, where ϵ is a real number with $0 \le \epsilon < 1$. Since $\epsilon = x - \lfloor x \rfloor$, it follows that $0 \le -\lfloor x \rfloor < 1$. The first two inequalities, $x - 1 < \lfloor x \rfloor$ and $\lfloor x \rfloor \le x$, follow directly. For the other two inequalities, write $x = \lceil x \rceil - \epsilon'$, where $0 \le \epsilon' < 1$. Then $0 \le \lceil x \rceil - x < 1$, and the desired inequality follows.   **43. a)** If $x < n$, since $\lfloor x \rfloor \le x$, it follows that $\lfloor x \rfloor < n$. Suppose that $x \ge n$. By the definition of the floor function, it follows that $\lfloor x \rfloor \ge n$. This means that if $\lfloor x \rfloor < n$, then $x < n$.   **b)** If $n < x$, then since $x \le \lceil x \rceil$, it follows that $n \le \lceil x \rceil$. Suppose that $n \ge x$. By the definition of the ceiling function, it follows that $\lceil x \rceil \le n$. This means that if $n < \lceil x \rceil$, then $n < x$.   **45.** If n is even, then $n = 2k$ for some integer k. Thus $\lfloor n/2 \rfloor = \lfloor k \rfloor = k = n/2$. If n is odd, then $n = 2k + 1$ for some integer k. Thus $\lfloor n/2 \rfloor = \lfloor k + \frac{1}{2} \rfloor = k = (n - 1)/2$.   **47.** Assume that $x \ge 0$. The left-hand side is $\lceil -x \rceil$ and the right-hand side is $-\lfloor x \rfloor$. If x is an integer, then both sides equal $-x$. Otherwise, let $x = n + \epsilon$, where n is a natural number and ϵ is a real number with $0 \le \epsilon < 1$. Then $\lceil -x \rceil = \lceil -n - \epsilon \rceil = -n$ and $-\lfloor x \rfloor = -\lfloor n + \epsilon \rfloor = -n$ also. When $x < 0$, the equation also holds since it can be obtained by substituting $-x$ for x.   **49.** $\lceil b \rceil - \lfloor a \rfloor - 1$ **51. a)** 1   **b)** 3   **c)** 126   **d)** 3600   **53. a)** 100   **b)** 256   **c)** 1030   **d)** 30,200

55.

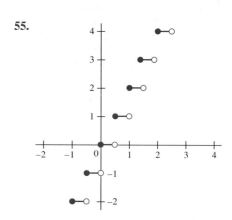

57.

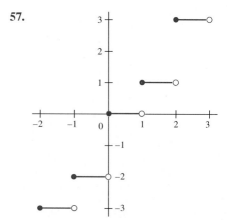

59. a)

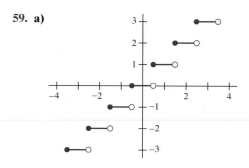

b)

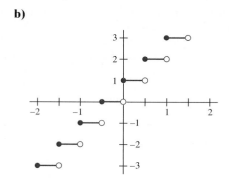

c)

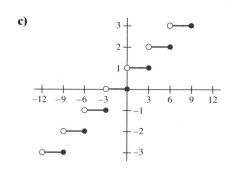

d)

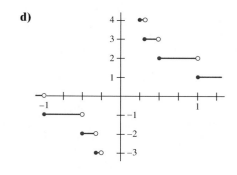

e)

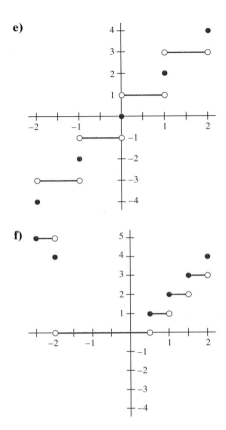

f)

g) See part (a).

61. $f^{-1}(y) = (y-1)^{1/3}$

63. a) $f_{A \cap B}(x) = 1 \equiv x \in A \cap B \equiv x \in A$ and $x \in B \equiv f_A(x) = 1$ and $f_B(x) = 1 \equiv f_A(x) f_B(x) = 1$ **b)** $f_{A \cup B}(x) = 1 \equiv x \in A \cup B \equiv x \in A$ or $x \in B \equiv f_A(x) = 1$ or $f_B(x) = 1 \equiv f_A(x) + f_B(x) - f_A(x) f_B(x) = 1$ **c)** $f_{\overline{A}}(x) = 1 \equiv x \in \overline{A} \equiv x \notin A \equiv f_A(x) = 0 \equiv 1 - f_A(x) = 1$ **d)** $f_{A \oplus B}(x) = 1 \equiv x \in A \oplus B \equiv (x \in A$ and $x \notin B)$ or $(x \notin A$ and $x \in B) \equiv f_A(x) + f_B(x) - 2 f_A(x) f_B(x) = 1$ **65. a)** True; since $\lfloor x \rfloor$ is already an integer, $\lceil \lfloor x \rfloor \rceil = \lfloor x \rfloor$. **b)** False; $x = \frac{1}{2}$ is a counterexample. **c)** True; if x or y is an integer, then using property 4b in Table 1 of Section 1.8, the difference is 0. If neither x nor y is an integer, then $x = n + \epsilon$ and $y = m + \delta$, where n and m are integers and ϵ and δ are positive real numbers less than 1. Then $m + n < x + y < m + n + 2$, so $\lceil x + y \rceil$ is either $m + n + 1$ or $m + n + 2$. Therefore, the given expression is either $(n+1) + (m+1) - (m+n+1) = 1$ or $(n+1) + (m+1) - (m+n+2) = 0$, as desired. **d)** False; $x = \frac{1}{4}$ and $y = 3$ is a counterexample. **e)** False; $x = \frac{1}{2}$ is a counterexample. **67. a)** If x is a positive integer, then the two sides are equal. So suppose that $x = n^2 + m + \epsilon$, where n^2 is the largest perfect square less than x, m is a nonnegative integer, and $0 < \epsilon \le 1$. Then both $\sqrt{x}$ and $\sqrt{\lfloor x \rfloor} = \sqrt{n^2 + m}$ are between n and $n + 1$, so both sides equal n. **b)** If x is a positive integer, then the two sides are

equal. So suppose that $x = n^2 - m - \epsilon$, where n^2 is the smallest perfect square greater than x, m is a nonnegative integer, and ϵ is a real number with $0 < \epsilon \le 1$. Then both $\sqrt{x}$ and $\sqrt{\lceil x \rceil} = \sqrt{n^2 - m}$ are between $n - 1$ and n. Therefore, both sides of the equation equal n. **69. a)** Domain is $\mathbf{Z}$; codomain is $\mathbf{R}$; domain of definition is the set of nonzero integers; the set of values for which f is undefined is $\{0\}$; not a total function. **b)** Domain is $\mathbf{Z}$; codomain is $\mathbf{Z}$; domain of definition is $\mathbf{Z}$; set of values for which f is undefined is $\emptyset$; total function. **c)** Domain is $\mathbf{Z} \times \mathbf{Z}$; codomain is $\mathbf{Q}$; domain of definition is $\mathbf{Z} \times (\mathbf{Z} - \{0\})$; set of values for which f is undefined is $\mathbf{Z} \times \{0\}$; not a total function. **d)** Domain is $\mathbf{Z} \times \mathbf{Z}$; codomain is $\mathbf{Z}$; domain of definition is $\mathbf{Z} \times \mathbf{Z}$; set of values for which f is undefined is $\emptyset$; total function. **e)** Domain is $\mathbf{Z} \times \mathbf{Z}$; codomain is $\mathbf{Z}$; domain of definitions is $\{(m, n) \mid m > n\}$; set of values for which f is undefined is $\{(m, n) \mid m \le n\}$; not a total function. **71.** It is clear from the formula that the range of values the function takes on for a fixed value of $m + n$, say $m + n = x$, is $(x-2)(x-1)/2 + 1$ through $(x-2)(x-1)/2 + (x-1)$, since m can assume the values $1, 2, 3, \ldots, (x-1)$ under these conditions, and the first term in the formula is a fixed positive integer when $m + n$ is fixed. To show that this function is one-to-one and onto, we merely need to show that the range of values for $x + 1$ picks up precisely where the range of values for x left off, i.e., that $f(x-1, 1) + 1 = f(1, x)$. We have $f(x-1, 1) + 1 = \frac{(x-2)(x-1)}{2} + (x-1) + 1 = \frac{x^2 - x + 2}{2} = \frac{(x-1)x}{2} + 1 = f(1, x)$.

SUPPLEMENTARY EXERCISES

1. a) $q \to p$ **b)** $q \wedge p$ **c)** $\neg q \vee \neg p$ **d)** $q \leftrightarrow p$ **3. a)** The proposition cannot be false unless $\neg p$ is false, so that p is true. If p is true and q is true then $\neg q \wedge (p \to q)$ is false and the implication is true. If p is true and q is false then $p \to q$ is false so that $\neg q \wedge (p \to q)$ is false and the implication is true. **b)** The proposition cannot be false unless q is false. If q is false and p is true then $(p \vee q) \wedge \neg p$ is false, and the implication is true. If q is false and p is false then $(p \vee q) \wedge \neg p$ is false and the implication is true. **5.** $(p \wedge q \wedge r \wedge \neg s) \vee (p \wedge q \wedge \neg r \wedge s) \vee (p \wedge \neg q \wedge r \wedge s) \vee (\neg p \wedge q \wedge r \wedge s)$ **7.** Translating these statements into symbols, using the obvious letters, we have $\neg t \to \neg g$, $\neg g \to \neg q$, $r \to q$, and $\neg t \wedge r$. Assume the statements are consistent. The fourth statement tells us that $\neg t$ must be true. Therefore by modus ponens with the first statement, we know that $\neg g$ is true, hence (from the second statement) that $\neg q$ is true. Also, the fourth statement tells us that r must be true, and so again modus ponens (third statement) makes q true. This is a contradiction: $q \wedge \neg q$. Thus the statements are inconsistent. **9.** Brenda **11. a)** F **b)** T **c)** F **d)** T **e)** F **f)** T **13.** $\forall x \exists y \exists z (y \ne z \wedge \forall w (P(w, x) \leftrightarrow (w =$

$y \lor w = z)))$ **15.** Suppose that $\exists x(P(x) \to Q(x))$ is true. Then either $Q(x_0)$ is true for some x_0, in which case $\forall x\, P(x) \to \exists x\, Q(x)$ is true; or $P(x_0)$ is false for some x_0, in which case $\forall x\, P(x) \to \exists x\, Q(x)$ is true. Conversely, suppose that $\exists x(P(x) \to Q(x))$ is false. That means that $\forall x(P(x) \land \neg Q(x))$ is true, which implies $\forall x\, P(x)$ and $\forall x(\neg Q(x))$. This latter proposition is equivalent to $\neg \exists x\, Q(x)$. Thus, $\forall x\, P(x) \to \exists x\, Q(x)$ is false. **17.** No **19.** $\forall x \forall z \exists y\, T(x, y, z)$, where $T(x, y, z)$ is the statement that student x has taken class y in department z, where the universes of discourse are the set of students in the class, the set of courses at this university, and the set of departments in the school of mathematical sciences. **21. a)** $\overline{A}$ **b)** $A \cap B$ **c)** $A - B$ **d)** $\overline{A} \cap \overline{B}$ **e)** $A \oplus B$ **23.** The second use of existential instantiation cannot use the letter c, since that letter has already been used for something else. There is no guarantee that the x that makes P true is the same as the x that makes Q true, so we can't call them both c. **25.** Let x be given. We must show that $\neg R(x) \to P(x)$, so assume that $\neg R(x)$ is true. We apply universal instantiation to the second hypothesis and then take the contrapositive and apply De Morgan's law to obtain $\neg R(x) \to (P(x) \lor \neg Q(x))$. By modus ponens, this gives us $P(x) \lor \neg Q(x)$. The first hypothesis gives $P(x) \lor Q(x)$ by universal instantiation. Combining these by resolution, we have $P(x)$, as desired. **27.** We give an indirect proof that if $\sqrt{x}$ is rational, then x is rational, assuming throughout that $x \geq 0$. Suppose that $\sqrt{x} = p/q$ is rational, $q \neq 0$. Then $x = (\sqrt{x})^2 = p^2/q^2$ is also rational (q^2 is again nonzero). **29.** We can give a constructive proof by letting $m = 10^{500} + 1$. Then $m^2 = (10^{500} + 1)^2 > (10^{500})^2 = 10^{1000}$. **31.** 23 cannot be written as the sum of eight cubes. **33.** 223 cannot be written as the sum of 36 fifth powers. **35.** Yes **37. a)** $A \cap \overline{A} = \{x \mid x \in A \land x \notin A\} = \emptyset$ **b)** $A \cup \overline{A} = \{x \mid x \in A \lor x \notin A\} = U$ **39.** $A - (A - B) = A - (A \cap \overline{B}) = A \cap \overline{(A \cap \overline{B})} = A \cap (\overline{A} \cup B) = (A \cap \overline{A}) \cup (A \cap B) = \emptyset \cup (A \cap B) = A \cap B$ **41.** Let $A = \{1\}, B = \emptyset, C = \{1\}$. Then $(A - B) - C = \emptyset$ but $A - (B - C) = \{1\}$. **43.** No. For example, let $A = B = \{a, b\}, C = \emptyset$, and $D = \{a\}$. Then $(A - B) - (C - D) = \emptyset - \emptyset = \emptyset$, but $(A - C) - (B - D) = \{a, b\} - \{b\} = \{a\}$. **45. a)** $|\emptyset| \leq |A \cap B| \leq |A| \leq |A \cup B| \leq |U|$ **b)** $|\emptyset| \leq |A - B| \leq |A \oplus B| \leq |A \cup B| \leq |A| + |B|$ **47. a)** yes, no **b)** yes, no **c)** f has inverse with $f^{-1}(a) = 3, f^{-1}(b) = 4, f^{-1}(c) = 2, f^{-1}(d) = 1$; g has no inverse **49.** Let $f(a) = f(b) = 1, f(c) = f(d) = 2, S = \{a, c\}, T = \{b, d\}$. Then $f(S \cap T) = f(\emptyset) = \emptyset$, but $f(S) \cap f(T) = \{1, 2\} \cap \{1, 2\} = \{1, 2\}$. **51.** The equation is true if and only if the sum of the fractional parts of x and y is less than 1.

CHAPTER 2

SECTION 2.1
1. $max := 1, i := 2, max := 8, i := 3, max := 12, i := 4, i := 5, i := 6, i := 7, max := 14, i := 8, i := 9, i := 10, i := 11$

3. procedure $sum(a_1, \ldots, a_n$: integers)
  $sum := a_1$
  for $i := 2$ **to** n
    $sum := sum + a_i$
  $\{sum$ has desired value$\}$

5. procedure $duplicates(a_1, a_2, \ldots, a_n$: integers in nondecreasing order)
  $k := 0$ $\{$this counts the duplicates$\}$
  $j := 2$
  while $j \leq n$
  begin
    if $a_j = a_{j-1}$ **then**
    begin
      $k := k + 1$
      $c_k := a_j$
      while $(j \leq n$ and $a_j = c_k)$
        $j := j + 1$
    end
    $j := j + 1$
  end $\{c_1, c_2, \ldots, c_k$ is the desired list$\}$

7. procedure *last even location*$(a_1, a_2, \ldots, a_n$: integers)
  $k := 0$
  for $i := 1$ **to** n
    if a_i is even **then** $k := i$
  end $\{k$ is the desired location (or 0 if there are no evens)$\}$

9. procedure *palindrome check*$(a_1 a_2 \ldots a_n$: string)
  $answer := $ **true**
  for $i := 1$ **to** $\lfloor n/2 \rfloor$
    if $a_i \neq a_{n+1-i}$ **then** $answer := $ **false**
  end $\{answer$ is true iff string is a palindrome$\}$

11. procedure *interchange*$(x, y$: real numbers)
  $z := x$
  $x := y$
  $y := z$
The minimum number of assignments needed is three.
13. Linear search: $i := 1, i := 2, i := 3, i := 4, i := 5, i := 6, i := 7, location := 7$; binary search: $i := 1, j := 8, m := 4, i := 5, m := 6, i := 7, m := 7, j := 7, location := 7$
15. procedure *insert*$(x, a_1, a_2, \ldots, a_n$: integers)
  $\{$the list is in order: $a_1 \leq a_2 \leq \cdots \leq a_n\}$
  $a_{n+1} := x + 1$
  $i := 1$
  while $x > a_i$
    $i := i + 1$
  for $j := 0$ **to** $n - i$
    $a_{n-j+1} := a_{n-j}$
  $a_i := x$
  $\{x$ has been inserted into correct position$\}$
17. procedure *first largest*$(a_1, \ldots, a_n$: integers)
  $max := a_1$
  $location := 1$
  for $i := 2$ **to** n
  begin
    if $max < a_i$ **then**
    begin
      $max := a_i$

$location := i$
   end
end

19. procedure *mean-median-max-min*(a, b, c: integers)
$mean := (a + b + c) / 3$
{the six different orderings of a, b, c with respect
   to $\geq$ will be handled separately}
if $a > b$ **then**
begin
   if $b > c$ **then**
   $median := b$; $max := a$; $min := c$
end
   ...
(The rest of the algorithm is similar.)

21. procedure *first-three* ($a_1, a_2, \ldots, a_n$: integers)
if $a_1 > a_2$ **then** interchange a_1 and a_2
   if $a_2 > a_3$ **then** interchange a_2 and a_3
   if $a_1 > a_2$ **then** interchange a_1 and a_2

23. procedure *onto*(f: function from A to B where
   $A = \{a_1, \ldots, a_n\}$, $B = \{b_1, \ldots, b_m\}$, $a_1, \ldots, a_n$,
   $b_1, \ldots, b_m$ are integers)
for $i := 1$ **to** m
   $hit(b_i) := 0$
$count := 0$
for $j := 1$ **to** n
   if $hit(f(a_j)) = 0$ **then**
   begin
     $hit(f(a_j)) := 1$
     $count := count + 1$
   end
if $count = m$ **then** $onto :=$ **true**
else $onto :=$ **false**

25. procedure *ones*(a: bit string, $a = a_1 a_2 \cdots a_n$)
$ones := 0$
for $i := 1$ **to** n
begin
   if $a_i := 1$ **then**
     $ones := ones + 1$
   end {*ones* is the number of ones in the bit
strings a}

27. procedure *ternary search*(s: integer, $a_1, a_2, \ldots, a_n$:
   increasing integers)
$i := 1$
$j := n$
while $i < j - 1$
begin
   $l = \lfloor (i + j)/3 \rfloor$
   $u = \lfloor 2(i + j)/3 \rfloor$
   if $x > a_u$ **then** $i := u + 1$
   else if $x > a_l$ **then**
   begin
     $i := l + 1$
     $j := u$
   end
   else $j := l$
end

if $x = a_i$ **then** $location := i$
else if $x = a_j$ **then** $location := j$
else $location := 0$
{$location$ is the subscript of the term equal to x
   (0 if not found)}

29. procedure *find a mode*($a_1, a_2, \ldots, a_n$: nondecreasing
   integers)
$modecount := 0$
$i := 1$ **while** $i \leq n$
begin
   $value := a_i$
   $count := 1$
   while $i \leq n$ and $a_i = value$
   begin
     $count := count + 1$
     $i := i + 1$
   end
   if $count > modecount$ **then**
   begin
     $modecount := count$
     $mode := value$
   end
end
{*mode* is the first value occurring most often}

31. procedure *find duplicate*($a_1, a_2, \ldots, a_n$: integers)
$location := 0$
$i := 2$
while $i \leq n$ and $location = 0$
begin
   $j := 1$
   while $j < i$ and $location = 0$
   if $a_i = a_j$ **then** $location := i$
   else $j := j + 1$
   $i := i + 1$
end
{*location* is the subscript of the first value that repeats
a previous value in the sequence}

33. procedure *find decrease*($a_1, a_2, \ldots, a_n$: positive
   integers)
$location := 0$
$i := 2$
while $i \leq n$ and $location = 0$
   if $a_i < a_{i-1}$ **then** $location := i$
   else $i := i + 1$
{*location* is the subscript of the first value less than the
immediately preceding one}

35. At the end of the first pass: 1, 3, 5, 4, 7; at the end of
the second pass: 1, 3, 4, 5, 7; at the end of the third pass: 1,
3, 4, 5, 7; at the end of the fourth pass: 1, 3, 4, 5, 7

37. procedure *better bubblesort* ($a_1, \ldots, a_n$: integers)
$i := 1$; $done :=$ **false**
while ($i < n$ and $done =$ **false**)
begin
   $done :=$ **true**
   for $j := 1$ **to** $n - i$

if $a_j > a_{j+1}$ **then**
   begin
      interchange a_j and a_{j+1}
      done := **false**
   end
   $i := i + 1$
end $\{a_1, \ldots, a_n$ is in increasing order$\}$

39. At the end of the first, second, and third passes: $1, 3, 5, 7, 4$; at the end of the fourth pass: $1, 3, 4, 5, 7$ **41. a)** $1, 5, 4, 3, 2$; $1, 2, 4, 3, 5$; $1, 2, 3, 4, 5$; $1, 2, 3, 4, 5$ **b)** $1, 4, 3, 2, 5$; $1, 2, 3, 4, 5$; $1, 2, 3, 4, 5$; $1, 2, 3, 4, 5$ **c)** $1, 2, 3, 4, 5$; $1, 2, 3, 4, 5$; $1, 2, 3, 4, 5$; $1, 2, 3, 4, 5$ **43.** We carry out the linear search algorithm given as Algorithm 2 in this section, except that we replace $x \neq a_i$ by $x < a_i$, and we replace the **else** clause with **else** *location* := $n + 1$. **45.** $2 + 3 + 4 + \cdots + n = (n^2 + n - 2)/2$ **47.** Find the location for the 2 in the list 3 (one comparison), and insert it in front of the 3, so the list now reads $2, 3, 4, 5, 1, 6$. Find the location for the 4 (compare it to the 2 and then the 3), and insert it, leaving $2, 3, 4, 5, 1, 6$. Find the location for the 5 (compare it to the 3 and then the 4), and insert it, leaving $2, 3, 4, 5, 1, 6$. Find the location for the 1 (compare it to the 3 and then the 2 and then the 2 again), and insert it, leaving $1, 2, 3, 4, 5, 6$. Find the location for the 6 (compare it to the 3 and then the 4 and then the 5), and insert it, giving the final answer $1, 2, 3, 4, 5, 6$.

49. procedure *binary insertion sort*$(a_1, a_2, \ldots, a_n$: real
        numbers with $n \geq 2$)
 for $j := 2$ **to** n
 begin
   $\{$binary search for insertion location $i\}$
   left := 1
   right := $j - 1$
   while *left* < *right*
   begin
      middle := $\lfloor (left + right)/2 \rfloor$
      if $a_j > a_{middle}$ **then** *left* := *middle* + 1
      else *right* := *middle*
   end
   if $a_j < a_{left}$ **then** $i := left$ **else** $i := left + 1$
   $\{$insert a_j in location i by moving a_i through a_{j-1}
      toward back of list$\}$
   $m := a_j$
   for $k := 0$ **to** $j - i - 1$
      $a_{j-k} := a_{j-k-1}$
   $a_i := m$
 end $\{a_1, a_2, \ldots, a_n$ are sorted$\}$

51. The variation from Exercise 50 **53. a)** Two quarters, one penny **b)** Two quarters, one dime, one nickel, four pennies **c)** A three quarters, one penny **d)** Two quarters, one dime **55.** Greedy algorithm uses fewest coins in parts **(a)**, **(c)**, and **(d)**. **a)** Two quarters, one penny **b)** Two quarters, one dime, nine pennies **c)** Three quarters, one penny **d)** Two quarters, one dime **57. a)** The

variable f will give the finishing time of the talk last selected, starting out with f equal to the time the hall becomes available. Order the talks in increasing order of the ending times, and start at the top of the list. At each stage of the algorithm, go down the list of talks from where it left off, and find the first one whose starting time is not less than f. Schedule that talk and update f to record its finishing time. **b)** the 9:00–9:45 talk, the 9:50–10:15 talk, the 10:15–10:45 talk, the 11:00–11:15 talk

SECTION 2.2

1. The choices of C and k are not unique. **a)** $C = 1, k = 10$ **b)** $C = 4, k = 7$ **c)** No **d)** $C = 5, k = 1$ **e)** $C = 1, k = 0$ **f)** $C = 1, k = 2$ **3.** $x^4 + 9x^3 + 4x + 7 \leq 4x^4$ for all $x > 9$; witnesses $C = 4, k = 9$. **5.** $(x^2 + 1)/(x + 1) = x - 1 + 2/(x + 1) < x$ for all $x > 1$; witnesses $C = 1, k = 1$. **7.** The choices of C and k are not unique. **a)** $n = 3$, $C = 3, k = 1$ **b)** $n = 3, C = 4, k = 1$ **c)** $n = 1, C = 2$, $k = 1$ **d)** $n = 0, C = 2, k = 1$ **9.** $x^2 + 4x + 17 \leq 3x^3$ for all $x > 17$, so $x^2 + 4x + 17$ is $O(x^3)$, with witnesses $C = 3, k = 17$. However, if x^3 were $O(x^2 + 4x + 17)$, then $x^3 \leq C(x^2 + 4x + 17) \leq 3Cx^2$ for some C, for all sufficiently large x, which implies that $x \leq 3C$ for all sufficiently large x, which is impossible. Hence x^3 is not $O(x^2 + 4x + 17)$. **11.** $3x^4 + 1 \leq 4x^4 = 8(x^4/2)$ for all $x > 1$, so $3x^4 + 1$ is $O(x^4/2)$, with witnesses $C = 8$, $k = 1$. Also $x^4/2 \leq 3x^4 + 1$ for all $x > 0$, so $x^4/2$ is $O(3x^4 + 1)$, with witnesses $C = 1, k = 0$. **13.** Since $2^n \leq 3^n$ for all $n > 0$, it follows that 2^n is $O(3^n)$, with witnesses $C = 1, k = 0$. However, if 3^n were $O(2^n)$, then for some C, $3^n \leq C \cdot 2^n$ for all sufficiently large n. This says that $C \geq (3/2)^n$ for all sufficiently large n, which is impossible. Hence 3^n is not $O(2^n)$. **15.** All functions for which there exist real numbers k and C with $|f(x)| \leq C$ for $x > k$. These are the functions $f(x)$ that are bounded for all sufficiently large x. **17.** There are constants C_1, C_2, k_1, and k_2 such that $|f(x)| \leq C_1|g(x)|$ for all $x > k_1$ and $|g(x)| \leq C_2|h(x)|$ for all $x > k_2$. Hence, for $x > \max(k_1, k_2)$ it follows that $|f(x)| \leq C_1|g(x)| \leq C_1C_2|h(x)|$. This shows that $f(x)$ is $O(h(x))$. **19. a)** $O(n^3)$ **b)** $O(n^5)$ **c)** $O(n^3 \cdot n!)$ **21. a)** $O(n^2 \log n)$ **b)** $O(n^2(\log n)^2)$ **c)** $O(n^{2^n})$ **23. a)** neither $\Theta(x^2)$ nor $\Omega(x^2)$ **b)** $\Theta(x^2)$ and $\Omega(x^2)$ **c)** neither $\Theta(x^2)$ nor $\Omega(x^2)$ **d)** $\Omega(x^2)$, but not $\Theta(x^2)$ **e)** $\Omega(x^2)$, but not $\Theta(x^2)$ **f)** $\Omega(x^2)$ and $\Theta(x^2)$ **25.** If $f(x)$ is $\Theta(g(x))$, then there exist constants C_1 and C_2 with $C_1|g(x)| \leq |f(x)| \leq C_2|g(x)|$. It follows that $|f(x)| \leq C_2|g(x)|$ and $|g(x)| \leq (1/C_1)|f(x)|$ for $x > k$. Thus, $f(x)$ is $O(g(x))$ and $g(x)$ is $O(f(x))$. Conversely, suppose that $f(x)$ is $O(g(x))$ and $g(x)$ is $O(f(x))$. Then there are constants C_1, C_2, k_1, and k_2 such that $|f(x)| \leq C_1|g(x)|$ for $x > k_1$ and $|g(x)| \leq C_2|f(x)|$ for $x > k_2$. We can assume that $C_2 > 0$ (we can always make C_2 larger). Then we have $(1/C_2)|g(x)| \leq |f(x)| \leq C_1|g(x)|$ for $x > \max(k_1, k_2)$. Hence, $f(x)$ is $\Theta(g(x))$. **27.** If $f(x)$ is $\Theta(g(x))$, then $f(x)$ is both $O(g(x))$ and

$\Omega(g(x))$. Hence there are positive constants C_1, k_1, C_2, and k_2 such that $|f(x)| \leq C_2|g(x)|$ for all $x > k_2$ and $|f(x)| \geq C_1|g(x)|$ for all $x > k_1$. It follows that $C_1|g(x)| \leq |f(x)| \leq C_2|g(x)|$ whenever $x > k$, where $k = \max(k_1, k_2)$. Conversely, if there are positive constants C_1, C_2, and k such that $C_1|g(x)| \leq |f(x)| \leq C_2|g(x)|$ for $x > k$, then taking $k_1 = k_2 = k$ shows that $f(x)$ is both $O(g(x))$ and $\Theta(g(x))$.

29.

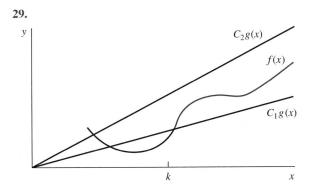

31. If $f(x)$ is $\Theta(1)$, then $|f(x)|$ is bounded between positive constants C_1 and C_2. In other words, $f(x)$ cannot grow larger than a fixed bound or smaller than the negative of this bound and must not get closer to 0 than some fixed bound. **33.** Since $f(x)$ is $O(g(x))$ there are constants C and l such that $|f(x)| \leq C|g(x)|$ for $x > l$. Hence $|f^k(x)| \leq C^k|g^k(x)|$ for $x > l$, so $f^k(x)$ is $O(g^k(x))$ by taking the constant to be C^k. **35.** Since $f(x)$ and $g(x)$ are increasing and unbounded, we can assume $f(x) \geq 1$ and $g(x) \geq 1$ for sufficiently large x. There are constants C and k with $f(x) \leq Cg(x)$ for $x > k$. This implies that $\log f(x) \leq \log C + \log g(x) < 2\log g(x)$ for sufficiently large x. Hence $\log f(x)$ is $O(\log g(x))$. **37.** By definition there are positive constraints $C_1, C_1', C_2, C_2', k_1, k_1', k_2$, and k_2' such that $f_1(x) \geq C_1|g(x)|$ for all $x > k_1$, $f_1(x) \leq C_1'|g(x)|$ for all $x > k_1'$, $f_2(x) \geq C_2|g(x)|$ for all $x > k_2$, and $f_2(x) \leq C_2'|g(x)|$ for all $x > k_2'$. Adding the first and third inequalities shows that $f_1(x) + f_2(x) \geq (C_1 + C_2)|g(x)|$ for all $x > k$ where $k = \max(k_1, k_2)$. Adding the second and fourth inequalities shows that $f_1(x) + f_2(x) \leq (C_1' + C_2')|g(x)|$ for all $x > k'$ where $k' = \max(k_1', k_2')$. Hence $f_1(x) + f_2(x)$ is $\Theta(g(x))$. This is no longer true if f_1 and f_2 can assume negative values. **39.** This is false. Let $f_1 = x^2 + 2x$, $f_2(x) = x^2 + x$, and $g(x) = x^2$. Then $f_1(x)$ and $f_2(x)$ are both $O(g(x))$, but $(f_1 - f_2)(x)$ is not. **41.** Take $f(n)$ to be the function with $f(n) = n$ if n is an odd positive integer and $f(n) = 1$ if n is an even positive integer and $g(n)$ to be the function with $g(n) = 1$ if n is an odd positive integer and $g(n) = n$ if n is an even positive integer. **43.** There are positive constants $C_1, C_2, C_1', C_2', k_1, k_1', k_2$, and k_2' such that $|f_1(x)| \geq C_1|g_1(x)|$ for all $x > k_1$, $|f_1(x)| \leq C_1'|g_1(x)|$ for all $x \geq k_1'$, $|f_2(x)| > C_2|g_2(x)|$ for all $x > k_2$, and

$|f_2(x)| \leq C_2'|g_2(x)|$ for all $x > k_2'$. Since f_2 and g_2 are never zero, the last two inequalities can be rewritten as $|1/f_2(x)| \leq (1/C_2)|1/g_2(x)|$ for all $x > k_2$ and $|1/f_2(x)| \geq (1/C_2')|1/g_2(x)|$ for all $x > k_2'$. Multiplying the first and rewritten fourth inequalities shows that $|f_1(x)/f_2(x)| \geq (C_1/C_2')|g_1(x)/g_2(x)|$ for all $x > \max(k_1, k_2')$, and multiplying the second and rewritten third inequalities gives $|f_1(x)/f_2(x)| \leq (C_1'/C_2)|g_1(x)/g_2(x)|$ for all $x > \max(k_1', k_2)$. It follows that f_1/f_2 is big-Theta of g_1/g_2. **45.** There exist positive constants $C_1, C_2, k_1, k_2, k_1', k_2'$ such that $|f(x,y)| \leq C_1|g(x,y)|$ for all $x > k_1$ and $y > k_2$ and $|f(x,y)| \geq C_2|g(x,y)|$ for all $x > k_1'$ and $y > k_2'$. **47.** $(x^2 + xy + x\log y)^3 < (3x^2y^3) = 27x^6y^3$ for $x > 1$ and $y > 1$, since $x^2 < x^2y$, $xy < x^2y$, and $x\log y < x^2y$. Hence $(x^2 + xy + x\log y)^3$ is $O(x^6y^3)$. **49.** For all positive real numbers x and y, $\lfloor xy \rfloor \leq xy$. Hence, $\lfloor xy \rfloor$ is $O(xy)$ from the definition, taking $C = 1$ and $k_1 = k_2 = 0$. **51. a)** $\lim_{x\to\infty} x^2/x^3 = \lim_{x\to\infty} 1/x = 0$ **b)** $\lim_{x\to\infty} \frac{x\log x}{x^2} = \lim_{x\to\infty} \frac{\log x}{x} = \lim_{x\to\infty} \frac{1}{x\ln 2} = 0$ (using L'Hôpital's rule) **c)** $\lim_{x\to\infty} \frac{x^2}{2^x} = \lim_{x\to\infty} \frac{2x}{2^x \cdot \ln 2} = \lim_{x\to\infty} \frac{2}{2^x \cdot (\ln 2)^2} = 0$ (using L'Hôpital's rule) **d)** $\lim_{x\to\infty} \frac{x^2+x+1}{x^2} = \lim_{x\to\infty} \left(1 + \frac{1}{x} + \frac{1}{x^2}\right) = 1 \neq 0$

53.

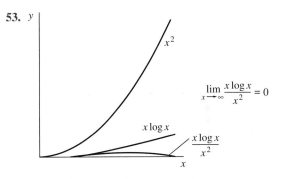

55. No. Take $f(x) = 1/x^2$ and $g(x) = 1/x$. **57. a)** Since $\lim_{x\to\infty} f(x)/g(x) = 0$, $|f(x)|/|g(x)| < 1$ for sufficiently large x. Hence, $|f(x)| < |g(x)|$ for $x > k$ for some constant k. Therefore, $f(x)$ is $O(g(x))$. **b)** Let $f(x) = g(x) = x$. Then $f(x)$ is $O(g(x))$, but $f(x)$ is not $o(g(x))$ since $f(x)/g(x) = 1$. **59.** Since $f_2(x)$ is $o(g(x))$, from Exercise 57(a) it follows that $f_2(x)$ is $O(g(x))$. By Corollary 1, we have $f_1(x) + f_2(x)$ is $O(g(x))$. **61.** We can easily show that $(n-i)(i+1) \geq n$ for $i = 0, 1, \ldots, n-1$. Hence, $(n!)^2 = (n \cdot 1)((n-1) \cdot 2) \cdot ((n-2) \cdot 3) \cdots (2 \cdot (n-1)) \cdot (1 \cdot n) \geq n^n$. Therefore, $2\log n! \geq n\log n$. **63. a)** Asymptotic **b)** Not asymptotic **c)** Asymptotic **d)** Asymptotic **e)** Not asymptotic **f)** Not asymptotic **g)** Not asymptotic

SECTION 2.3

1. $2n - 1$ **3.** Linear **5.** $O(n)$ **7. a)** *power* := 1, $y := 1$; $i := 1$, *power* := 2, $y := 3$; $i := 2$, *power*

$:= 4$, $y := 15$ **b)** $2n$ multiplications and n additions
9. a) $2^{10^9} \approx 10^{3 \times 10^8}$ **b)** 10^9 **c)** 3.96×10^7 **d)** 3.16×10^4
e) 29 **f)** 12 **11. a)** 36 years **b)** 13 days **c)** 19 minutes **13.** The average number of comparisons is
$(3n + 4)/2$. **15.** $O(\log n)$ **17.** $O(n)$ **19.** $O(n^2)$
21. $O(n)$ **23.** $O(n)$ **25.** $O(\log n)$ comparisons; $O(n^2)$
swaps **27. a)** doubles **b)** increases by 1

SECTION 2.4

1. a) Yes **b)** No **c)** Yes **d)** No **3.** Suppose that $a \mid b$.
Then there exists an integer k such that $ka = b$. Since
$a(ck) = bc$ it follows that $a \mid bc$. **5.** If $a \mid b$ and
$b \mid a$, there are integers c and d such that $b = ac$ and
$a = bd$. Hence $a = acd$. Since $a \neq 0$ it follows that
$cd = 1$. Thus either $c = d = 1$ or $c = d = -1$.
Hence either $a = b$ or $a = -b$. **7.** Since $ac \mid bc$
there is an integer k such that $ack = bc$. Hence $ak = b$, so $a \mid b$. **9. a)** $2, 5$ **b)** $-11, 10$ **c)** $34, 7$ **d)** $77, 0$
e) $0, 0$ **f)** $0, 3$ **g)** $-1, 2$ **h)** $4, 0$ **11. a)** $2^3 \cdot 11$ **b)** $2 \cdot
3^2 \cdot 7$ **c)** 3^6 **d)** $7 \cdot 11 \cdot 13$ **e)** $11 \cdot 101$ **f)** $2 \cdot 3^3 \cdot 5 \cdot 7 \cdot 13 \cdot 37$
13. $2^8 \cdot 3^4 \cdot 5^2 \cdot 7$ **15.** Suppose that $\log_2 3 = a/b$ where
$a, b \in \mathbf{Z}^+$ and $b \neq 0$. Then $2^{a/b} = 3$, so that $2^a = 3^b$.
This violates the Fundamental Theorem of Arithmetic.
Hence $\log_2 3$ is irrational. **17.** $1, 7, 11, 13, 17, 19, 23, 29$
19. a) Yes **b)** No **c)** Yes **d)** Yes **21.** If $a \bmod m =
b \bmod m$, then a and b have the same remainder when divided by m. Hence $a = q_1 m + r$ and $b = q_2 m + r$, where
$0 \leq r < m$. It follows that $a - b = (q_1 - q_2)m$ so that
$m \mid (a - b)$. It follows that $a \equiv b \pmod{m}$. **23.** Suppose that n is not prime so that $n = ab$, where a and
b are integers greater than 1. Since $a > 1$, by the identity in the hint, $2^a - 1$ is a factor of $2^n - 1$ that is greater
than 1, and the second factor in this identity is also greater
than 1. Hence $2^n - 1$ is not prime. **25. a)** 2 **b)** 4 **c)** 12
27. $\phi(p^k) = p^k - p^{k-1}$ **29. a)** $3^5 \cdot 5^3$ **b)** 1 **c)** 23^{17}
d) $41 \cdot 43 \cdot 53$ **e)** 1 **f)** 1111 **31. a)** $2^{11} \cdot 3^7 \cdot 5^9 \cdot 7^3$
b) $2^9 \cdot 3^7 \cdot 5^5 \cdot 7^3 \cdot 11 \cdot 13 \cdot 17$ **c)** 23^{31} **d)** $41 \cdot 43 \cdot 53$
e) $2^{12} 3^{13} 5^{17} 7^{21}$ **f)** undefined **33.** There is some b with
$(b - 1)k < n \leq bk$. Hence $(b - 1)k \leq n - 1 < bk$. Divide
by k to obtain $b - 1 < n/k \leq b$ and $b - 1 \leq (n - 1)/k < b$.
Hence $\lceil n/k \rceil = b$ and $\lfloor (n-1)/k \rfloor = b - 1$. **35.** $x \bmod m$ if
$x \bmod m \leq \lceil m/2 \rceil$ and $(x \bmod m) - m$ if $x \bmod m > \lceil m/2 \rceil$
37. a) 1 **b)** 2 **c)** 3 **d)** 9 **39. a)** No **b)** No **c)** Yes
d) No **41.** Since $\min(x, y) + \max(x, y) = x + y$, the exponent of p_i in the prime factorization of $\gcd(a, b) \cdot
\text{lcm}(a, b)$ is the sum of the exponents of p_i in the prime
factorizations of a and b. **43.** Let $m = tn$. Since $a \equiv
b \pmod{m}$ there exists an integer s such that $a = b + sm$.
Hence $a = b + (st)n$, so that $a \equiv b \pmod{n}$. **45.** Let
$m = c = 2, a = 0$, and $b = 1$. Then $0 = ac \equiv bc = 2
\pmod 2$, but $0 = a \not\equiv b = 1 \pmod 2$. **47.** Since $a \equiv
b \pmod{m}$, there exists an integer s such that $a = b + sm$,
so that $a - b = sm$. Then $a^k - b^k = (a - b)(a^{k-1} +
a^{k-2}b + \cdots + ab^{k-2} + b^{k-1})$, $k \geq 2$, is also a multiple of

m. It follows that $a^k \equiv b^k \pmod{m}$. **49. a)** 7, 19, 7, 7,
18, 0 **b)** Take the next available space **mod** 31. **51.** 2,
6, 7, 10, 8, 2, 6, 7, 10, 8, ... **53. a)** GR QRW SDVV JR
b) QB ABG CNFF TB **c)** QX UXM AHJJ ZX **55.** 0
57. The check digit of the ISBN for this book is valid since
$1 \cdot 0 + 2 \cdot 0 + 3 \cdot 7 + 4 \cdot 2 + 5 \cdot 4 + 6 \cdot 2 + 7 \cdot 4 + 8 \cdot 3 + 9 \cdot 4 + 10 \cdot 6 \equiv
0 \pmod{11}$. **59. a)** $a_n = 1$ if n is prime and $a_n = 0$ otherwise. **b)** a_n is the smallest prime factor of n with $a_1 = 1$.
c) a_n is the number of positive divisors of n. **d)** $a_n = 1$
if n has no divisors that are perfect squares greater than 1
and $a_n = 0$ otherwise. **e)** a_n is the largest prime less than
or equal to n. **f)** a_n is the product of the first $n - 1$ primes.

SECTION 2.5

1. a) 1110 0111 **b)** 1 0001 1011 0100 **c)** 1 0111 1101 0110
1100 **3. a)** 31 **b)** 513 **c)** 341 **d)** 26,896 **5. a)** 1000
0000 1110 **b)** 1 0011 0101 1010 1011 **c)** 1010 1011 1011
1010 **d)** 1101 1110 1111 1010 1100 1110 1101 **7.** 1010
1011 1100 1101 1110 1111 **9.** $(B7B)_{16}$ **11.** Adding up
to three leading 0s if necessary, write the binary expansion
as $(\dots b_{23}b_{22}b_{21}b_{20}b_{13}b_{12}b_{11}b_{10}b_{03}b_{02}b_{01}b_{00})_2$. The value of
this numeral is $b_{00} + 2b_{01} + 4b_{02} + 8b_{03} + 2^4 b_{10} + 2^5 b_{11} +
2^6 b_{12} + 2^7 b_{13} + 2^8 b_{20} + 2^9 b_{21} + 2^{10} b_{22} + 2^{11} b_{23} + \cdots$, which
we can rewrite as $b_{00} + 2b_{01} + 4b_{02} + 8b_{03} + (b_{10} + 2b_{11} +
4b_{12} + 8b_{13}) \cdot 2^4 + (b_{20} + 2b_{21} + 4b_{22} + 8b_{23}) \cdot 2^8 + \cdots$.
Now $(b_{i3}b_{i2}b_{i1}b_{i0})_2$ translates into the hexadecimal digit
h_i. So our number is $h_0 + h_1 \cdot 2^4 + h_2 \cdot 2^8 + \cdots =
h_0 + h_1 \cdot 16 + h_2 \cdot 16^2 + \cdots$, which is the hexadecimal
expansion $(\dots h_1 h_1 h_0)_{16}$. **13.** Group together blocks of
three binary digits, adding up to two initial 0s if necessary,
and translate each block of three binary digits into a single octal digit. **15.** $(111011100101011010001)_2$, $(1273)_8$
17. Convert the given octal numeral to binary using Exercise 14, then convert from binary to hexadecimal using Example 6. **19.** 27 **21. a)** 6 **b)** 3 **c)** 11 **d)** 3 **e)** 40
f) 12 **23.** 8 **25.** The binary expansion of the integer is
the unique such sum. **27.** Let $a = (a_{n-1}a_{n-2} \cdots a_1 a_0)_{10}$.
Then $a = 10^{n-1}a_{n-1} + 10^{n-2}a_{n-2} + \cdots + 10a_1 + a_0 \equiv
a_{n-1} + a_{n-2} + \cdots + a_1 + a_0 \pmod 3$, since $10^j \equiv 1 \pmod 3$ for
all nonnegative integers j. It follows that $3 \mid a$ if and only
if 3 divides the sum of the decimal digits of a. **29.** Let
$a = (a_{n-1}a_{n-2} \cdots a_1 a_0)_2$. Then $a = a_0 + 2a_1 + 2^2 a_2 +
\cdots + 2^{n-1}a_{n-1} \equiv a_0 - a_1 + a_2 - a_3 + \cdots \pm a_{n-1} \pmod 3$.
It follows that a is divisible by 3 if and only if the sum
of the binary digits in the even-numbered positions minus the sum of the binary digits in the odd-numbered
positions is divisible by 3. **31. a)** -6 **b)** 13 **c)** -14
d) 0 **33.** The one's complement of the sum is found by
adding the one's complements of the two integers except
that a carry in the leading bit is used as a carry to the
last bit of the sum. **35.** If $m \geq 0$, then the leading bit
a_{n-1} of the one's complement expansion of m is 0 and
the formula reads $m = \sum_{i=0}^{n-2} a_i 2^i$. This is correct since
the right-hand side is the binary expansion of m. When
m is negative, the leading bit a_{n-1} of the one's comple-

ment expansion of m is 1. The remaining $n - 1$ bits can be obtained by subtracting $-m$ from $111 \ldots 1$ (where there are $n - 1$ 1s), since subtracting a bit from 1 is the same as complementing it. Hence, the bit string $a_{n-2} \ldots a_0$ is the binary expansion of $(2^{n-1} - 1) - (-m)$. Solving the equation $(2^{n-1}-1)-(-m) = \sum_{i=0}^{n-2} a_i 2^i$ for m gives the desired equation, since $a_{n-1} = 1$.   **37. a)** -7 **b)** 13   **c)** -15 **d)** -1
39. To obtain the two's complement representation of the sum of two integers, add their two's complement representations (as binary integers are added) and ignore any carry out of the leftmost column. However, the answer is invalid if an overflow has occurred. This happens when the leftmost digits in the two's complement representation of the two terms agree and the leftmost digit of the answer differs.   **41.** If $m \geq 0$, then the leading bit a_{n-1} is 0 and the formula reads $m = \sum_{i=0}^{n-2} a_i 2^i$. This is correct since the right-hand side is the binary expansion of m. If $m < 0$, its two's complement expansion has 1 as its leading bit and the remaining $n - 1$ bits are the binary expansion of $2^{n-1} - (-m)$. This means that $(2^{n-1}) - (-m) = \sum_{i=0}^{n-2} a_i 2^i$. Solving for m gives the desired equation, since $a_{n-1} = 1$.
43. $4n$
45. procedure *Cantor*$(x$: positive integer$)$
  $n := 1; f := 1$
  while $(n + 1) \cdot f \leq x$
  begin
    $n := n + 1$
    $f := f \cdot n$
  end
  $y := x$
  while $n > 0$
  begin
    $a_n := \lfloor y/f \rfloor$
    $y := y - a_n \cdot f$
    $f := f/n$
    $n := n - 1$
  end $\{x = a_n n! + a_{n-1}(n-1)! + \cdots + a_1 1!\}$
47. First step: $c = 0, d = 0, s_0 = 1$; second step: $c = 0$, $d = 1, s_1 = 0$; third step: $c = 1, d = 1, s_2 = 0$; fourth step: $c = 1, d = 1, s_3 = 0$; fifth step: $c = 1, d = 1, s_4 = 1$; sixth step: $c = 1, s_5 = 1$
49. procedure *subtract*$(a, b$: positive integers, $a > b$,
    $a = (a_{n-1}a_{n-2}\cdots a_1 a_0)_2$,
    $b = (b_{n-1}b_{n-2}\cdots b_1 b_0)_2$
    $B := 0$ $\{B$ is the borrow$\}$
    for $j := 0$ **to** $n - 1$
    begin
      if $a_j \geq b_j + B$ **then**
      begin
        $s_j := a_j - b_j - B$
        $B := 0$
      end
      else
      begin
        $s_j := a_j + 2 - b_j - B$

    $B := 1$
    end
  end $\{(s_{n-1}s_{n-2}\cdots s_1 s_0)_2$ is the difference$\}$
51. procedure *compare*$(a, b$: positive integers and
    $a = (a_n a_{n-1}\cdots a_1 a_0)_2$,
    $b = (b_n b_{n-1}\cdots b_1 b_0)_2$
    $k := n$
    while $a_k = b_k$ and $k > 0$
    $k := k - 1$
    if $a_k = b_k$ **then** print "a equals b"
    if $a_k > b_k$ **print** "a is greater than b"
    if $a_k < b_k$ **then** **print** "a is less than b"

53. $O(\log n)$ **55.** The only time-consuming part of the algorithm is the **while** loop, which is iterated q times. The work done inside is a subtraction of integers no bigger than $|a|$, which has $\log |a|$ bits. The result now follows from Example 8.

SECTION 2.6

1. a) $1 = (-1) \cdot 10 + 1 \cdot 11$ **b)** $1 = 21 \cdot 21 + (-10) \cdot 44$
c) $12 = (-1) \cdot 36 + 48$ **d)** $1 = 13 \cdot 55 + (-21) \cdot 34$
e) $3 = 11 \cdot 213 + (-20) \cdot 117$ **f)** $223 = 1 \cdot 0 + 1 \cdot 223$ **g)** $1 = 37 \cdot 2347 + (-706) \cdot 123$ **h)** $2 = 1128 \cdot 3454 + (-835) \cdot 4666$
i) $1 = 2468 \cdot 9999 + (-2221) \cdot 11111$ **3.** $15 \cdot 7 = 105 \equiv 1 \pmod{26}$ **5.** 7   **7.** 52   **9.** Suppose that b and c are both inverses of a modulo m. Then $ba \equiv 1 \pmod{m}$ and $ca \equiv 1 \pmod{m}$. Hence $ba \equiv ca \pmod{m}$. Since $\gcd(a, m) = 1$ it follows by Theorem 2 that $b \equiv c \pmod{m}$.
11. $x \equiv 8 \pmod 9$ **13.** Let $m' = m/\gcd(c, m)$. Since all the common factors of m and c are divided out of m to obtain m', it follows that m' and c are relatively prime. Since m divides $(ac - bc) = (a - b)c$, it follows that m' divides $(a - b)c$. By Lemma 1, we see that m' divides $a - b$, so that $a \equiv b \pmod{m'}$.   **15.** Suppose that $x^2 \equiv 1 \pmod p$. Then p divides $x^2 - 1 = (x + 1)(x - 1)$. By Lemma 2 it follows that $p \mid (x + 1)$ or $p \mid (x - 1)$, so that $x \equiv -1 \pmod p$ or $x \equiv 1 \pmod p$.   **17. a)** Suppose that $ia \equiv ja \pmod p$, where $1 \leq i < j < p$. Then p divides $ja - ia = a(j - i)$. By Theorem 1, since a is not divisible by p, p divides $j - i$, which is impossible since $j - i$ is a positive integer less than p.   **b)** By (a), since no two of $a, 2a, \ldots, (p - 1)a$ are congruent modulo p, each must be congruent to a different number from 1 to $p - 1$. It follows that $a \cdot 2a \cdot 3a \cdot \cdots \cdot (p-1) \cdot a \equiv 1 \cdot 2 \cdot 3 \cdot \cdots \cdot (p-1)$ $\pmod p$. It follows that $(p - 1)! \cdot a^{p-1} \equiv (p - 1)! \pmod p$.
c) By Wilson's theorem and (b), if p does not divide a, it follows that $(-1) \cdot a^{p-1} \equiv -1 \pmod p$. Hence $a^{p-1} \equiv 1 \pmod p$.   **d)** If $p \mid a$, then $p \mid a^p$. Hence $a^p \equiv a \equiv 0 \pmod p$. If p does not divide a, then $a^{p-1} \equiv a \pmod p$, by (c). Multiplying both sides of this congruence by a gives $a^p \equiv a \pmod p$.   **19.** all integers of the form $323 + 330k$, where k is an integer   **21.** all integers of the form $16 + 252k$, where k is an integer   **23.** Suppose that p is a prime appearing in the prime factoriza-

tion of $m_1 m_2 \cdots m_n$. Since the m_is are relatively prime, p is a factor of exactly one of the m_is, say m_j. Since m_j divides $a - b$, it follows that $a - b$ has the factor p in its prime factorization to a power at least as large as the power to which it appears in the prime factorization of m_j. It follows that $m_1 m_2 \cdots m_n$ divides $a - b$, so $a \equiv b$ (mod $m_1 m_2 \cdots m_n$). **25.** $x \equiv 1 \pmod 6$ **27. a)** By Fermat's Little Theorem, we have $2^{10} \equiv 1 \pmod{11}$. Hence $2^{340} = (2^{10})^{34} \equiv 1^{34} \equiv 1 \pmod{11}$. **b)** Since $32 \equiv 1 \pmod{31}$, it follows that $2^{340} = (2^5)^{68} \equiv 32^{68} \equiv 1^{68} \equiv 1 \pmod{31}$. **c)** Since 11 and 31 are relatively prime, and $11 \cdot 31 = 341$, it follows by (a) and (b) and Exercise 23 that $2^{340} \equiv 1 \pmod{341}$. **29. a)** 3, 4, 8 **b)** 983 **31.** First, $2047 = 23 \cdot 89$ is composite. Write $2047 - 1 = 2046 = 2 \cdot 1023$, so $s = 1$ and $t = 1023$ in the definition. Then $2^{1023} = (2^{11})^{93} \equiv 2048^{93} \equiv 1^{93} \equiv 1 \pmod{2047}$, as desired. **33.** We must show that $b^{2820} \equiv 1 \pmod{2821}$ for all b relatively prime to 2821. Note that $2821 = 7 \cdot 13 \cdot 31$, and if $\gcd(b, 2821) = 1$, then $\gcd(b, 7) = \gcd(b, 13) = \gcd(b, 31) = 1$. Using Fermat's Little Theorem we find that $b^6 \equiv 1 \pmod 7$, $b^{12} \equiv 1 \pmod{13}$, and $b^{30} \equiv 1 \pmod{31}$. It follows that $b^{2820} = (b^6)^{470} \equiv 1 \pmod 7$, $b^{2820} = (b^{12})^{235} \equiv 1 \pmod{13}$, and $b^{2820} = (b^{30})^{94} \equiv 1 \pmod{31}$. By Exercise 23 (or the Chinese Remainder Theorem) it follows that $b^{2820} \equiv 1 \pmod{2821}$, as desired. **35. a)** If we multiply out this expression, we get $n = 1296m^3 + 396m^2 + 36m + 1$. Clearly $6m \mid n - 1$, $12m \mid n - 1$, and $18m \mid n - 1$. Therefore, the conditions of Exercise 34 are met, and we conclude that n is a Carmichael number. **b)** Letting $m = 51$ gives $n = 172{,}947{,}529$. **37.** $0 = (0, 0)$, $1 = (1, 1)$, $2 = (2, 2)$, $3 = (0, 3)$, $4 = (1, 4)$, $5 = (2, 0)$, $6 = (0, 1)$, $7 = (1, 2)$, $8 = (2, 3)$, $9 = (0, 4)$, $10 = (1, 0)$, $11 = (2, 1)$, $12 = (0, 2)$, $13 = (1, 3)$, $14 = (2, 4)$ **39.** We have $m_1 = 99$, $m_2 = 98$, $m_3 = 97$, and $m_4 = 95$, so $m = 99 \cdot 98 \cdot 97 \cdot 95 = 89{,}403{,}930$. We find that $M_1 = m/m_1 = 903{,}070$, $M_2 = m/m_2 = 912{,}285$, $M_3 = m/m_3 = 921{,}690$, and $M_4 = m/m_4 = 941{,}094$. Using the Euclidean algorithm, we compute that $y_1 = 37$, $y_2 = 33$, $y_3 = 24$, and $y_4 = 4$ are inverses of M_k modulo m_k for $k = 1, 2, 3, 4$, respectively. It follows that the solution is $65 \cdot 903{,}070 \cdot 37 + 2 \cdot 912{,}285 \cdot 33 + 51 \cdot 921{,}690 \cdot 24 + 10 \cdot 941{,}094 \cdot 4 = 3{,}397{,}886{,}480 \equiv 537{,}140 \pmod{89{,}403{,}930}$. **41.** By Exercise 40 it follows that $\gcd(2^b - 1, (2^a - 1) \bmod (2^b - 1)) = \gcd(2^b - 1, 2^{a \bmod b} - 1)$. Since the exponents involved in the calculation are b and $a \bmod b$, the same as the quantities involved in computing $\gcd(a, b)$, the steps used by the Euclidean algorithm to compute $\gcd(2^a - 1, 2^b - 1)$ run in parallel to those used to compute $\gcd(a, b)$ and show that $\gcd(2^a - 1, 2^b - 1) = 2^{\gcd(a,b)} - 1$. **43.** Suppose that q is an odd prime with $q \mid 2^p - 1$. By Fermat's Little Theorem, $q \mid 2^{q-1} - 1$. From Exercise 41, $\gcd(2^p - 1, 2^{Q-1} - 1) = 2^{\gcd(p,q-1)} - 1$. Since q is a common divisor of $2^p - 1$ and $2^{q-1} - 1$, $\gcd(2^p - 1, 2^{q-1} - 1) > 1$. Hence, $\gcd(p, q - 1) = p$, since the only other possibility, namely, $\gcd(p, q - 1) = 1$, gives us $\gcd(2^p - 1, 2^{q-1} - 1) = 1$. Hence $p \mid q - 1$, and therefore

there is a positive integer m such that $q - 1 = mp$. Since q is odd, m must be even, say, $m = 2k$, and so every prime divisor of $2^p - 1$ is of the form $2kp + 1$. Furthermore, the product of numbers of this form is also of this form. Therefore, all divisors of $2^p - 1$ are of this form. **45.** Suppose we know both $n = pq$ and $(p - 1)(q - 1)$. To find p and q, first note that $(p-1)(q-1) = pq - p - q + 1 = n - (p+q) + 1$. From this we can find $s = (p + q)$. Since $q = s - p$, we have $n = p(s - p)$. Hence $p^2 - ps + n = 0$. We now can use the quadratic formula to find p. Once we have found p, we can find q since $q = n/p$. **47.** SILVER **49.** $34 \cdot 144 + (-55) \cdot 89 = 1$ **51. procedure** *extended Euclidean*(a, b: positive integers)

```
x := a
y := b
oldolds := 1
olds := 0
oldoldt := 0
oldt := 1
while y ≠ 0
begin
    q := x div y
    r := x mod y
    x := y
    y := r
    s := oldolds − q · olds
    t := oldoldt − q · oldt
    oldolds := olds
    oldoldt := oldt
    olds := s
    oldt := t
end {gcd(a, b) is x, and (oldolds)a + (oldoldt)b = x}
```

53. Assume that s is a solution of $x^2 \equiv a \pmod p$. Then since $(-s)^2 = s^2$, $-s$ is also a solution. Furthermore, $s \not\equiv -s \pmod p$. Otherwise, $p \mid 2s$, which implies that $p \mid s$, and this implies, using the original assumption, that $p \mid a$, which is a contradiction. Furthermore, if s and t are incongruent solutions modulo p, then since $s^2 \equiv t^2 \pmod p$, $p \mid (s^2 - t^2)$. This implies that $p \mid (s + t)(s - t)$, and by Lemma 2, $p \mid (s - t)$ or $p \mid (s + t)$, so $s \equiv t \pmod p$ or $s \equiv -t \pmod p$. Hence, there are at most two solutions. **55.** The value of $\left(\frac{a}{p}\right)$ depends only on whether a is a quadratic residue modulo p, that is, whether $x^2 \equiv a \pmod p$ has a solution. Since this only depends on the equivalence class of a modulo p, it follows that $\left(\frac{a}{p}\right) = \left(\frac{b}{p}\right)$ if $a \equiv b \pmod p$. **57.** By Exercise 56, $\left(\frac{a}{p}\right)\left(\frac{b}{p}\right) = a^{(p-1)/2} b^{(p-1)/2} = (ab)^{(p-1)/2} \equiv \left(\frac{ab}{p}\right) \pmod p$. **59.** $x \equiv 8, 13, 22$ or $27 \pmod{35}$

SECTION 2.7

1. a) 3×4 **b)** $\begin{bmatrix} 1 \\ 4 \\ 3 \end{bmatrix}$ **c)** $[2 \quad 0 \quad 4 \quad 6]$ **d)** 1

e) $\begin{bmatrix} 1 & 2 & 1 \\ 1 & 0 & 1 \\ 1 & 4 & 3 \\ 3 & 6 & 7 \end{bmatrix}$ **3. a)** $\begin{bmatrix} 1 & 11 \\ 2 & 18 \end{bmatrix}$ **b)** $\begin{bmatrix} 2 & -2 & -3 \\ 1 & 0 & 2 \\ 9 & -4 & 4 \end{bmatrix}$

c) $\begin{bmatrix} -4 & 15 & -4 & 1 \\ -3 & 10 & 2 & -3 \\ 0 & 2 & -8 & 6 \\ 1 & -8 & 18 & -13 \end{bmatrix}$ **5.** $\begin{bmatrix} 9/5 & -6/5 \\ -1/5 & 4/5 \end{bmatrix}$

7. $\mathbf{0} + \mathbf{A} = [0 + a_{ij}] = [a_{ij} + 0] = \mathbf{0} + \mathbf{A}$ **9.** $\mathbf{A} + (\mathbf{B} + \mathbf{C}) = [a_{ij} + (b_{ij} + c_{ij})] = [(a_{ij} + b_{ij}) + c_{ij}] = (\mathbf{A} + \mathbf{B}) + \mathbf{C}$ **11.** The number of rows of $\mathbf{A}$ equals the number of columns of $\mathbf{B}$, and the number of columns of $\mathbf{A}$ equals the number of rows of $\mathbf{B}$. **13.** $\mathbf{A}(\mathbf{BC}) = \left[\sum_q a_{iq} \left(\sum_r b_{qr} c_{rl} \right) \right] = \left[\sum_q \sum_r a_{iq} b_{qr} c_{rl} \right] = \left[\sum_r \sum_q a_{iq} b_{qr} c_{rl} \right] = \left[\sum_r \left(\sum_q a_{iq} b_{qr} \right) c_{rl} \right] = (\mathbf{AB})\mathbf{C}$ **15.** $\mathbf{A}^n = \begin{bmatrix} 1 & n \\ 0 & 1 \end{bmatrix}$ **17. a)** Let $\mathbf{A} = [a_{ij}]$ and $\mathbf{B} = [b_{ij}]$. Then $(\mathbf{A} + \mathbf{B}) = [a_{ij} + b_{ij}]$. We have $(\mathbf{A} + \mathbf{B})^t = [a_{ji} + b_{ji}] = [a_{ji}] + [b_{ji}] = \mathbf{A}^t + \mathbf{B}^t$. **b)** Using the same notation as in (a), we have $[\mathbf{B}^t\mathbf{A}^t = \left[\sum_q b_{qi} a_{jq} \right] = \left[\sum_q a_{jq} b_{qi} \right] = (\mathbf{AB})^t$, since the (i, j)th entry is the (j, i)th entry of $\mathbf{AB}$. **19.** The result follows since $\begin{bmatrix} a & b \\ c & d \end{bmatrix} \begin{bmatrix} d & -b \\ -c & a \end{bmatrix} = \begin{bmatrix} ad - bc & 0 \\ 0 & ad - bc \end{bmatrix} = (ad - bc)\mathbf{I}_2 = \begin{bmatrix} d & -b \\ -c & a \end{bmatrix} \begin{bmatrix} a & b \\ c & d \end{bmatrix}$. **21.** $\mathbf{A}^n(\mathbf{A}^{-1})^n = \mathbf{A}(\mathbf{A} \cdots (\mathbf{A}(\mathbf{A}\mathbf{A}^{-1})\mathbf{A}^{-1}) \cdots \mathbf{A}^{-1})\mathbf{A}^{-1}$ by the associative law. Since $\mathbf{A}\mathbf{A}^{-1} = \mathbf{I}$, working from the inside shows that $\mathbf{A}^n(\mathbf{A}^{-1})^n = \mathbf{I}$. Similarly $(\mathbf{A}^{-1})^n\mathbf{A}^n = \mathbf{I}$. Therefore $(\mathbf{A}^n)^{-1} = (\mathbf{A}^{-1})^n$. **23.** There are m_2 multiplications used to find each of the $m_1 m_3$ entries of the product. Hence $m_1 m_2 m_3$ multiplications are used. **25.** $\mathbf{A}_1((\mathbf{A}_2\mathbf{A}_3)\mathbf{A}_4)$ **27.** $x_1 = 1, x_2 = -1, x_3 = -2$

29. a) $\begin{bmatrix} 1 & 1 & 1 \\ 1 & 1 & 1 \\ 1 & 0 & 1 \end{bmatrix}$ **b)** $\begin{bmatrix} 0 & 0 & 1 \\ 1 & 0 & 0 \\ 0 & 0 & 1 \end{bmatrix}$ **c)** $\begin{bmatrix} 1 & 1 & 1 \\ 1 & 1 & 1 \\ 1 & 0 & 1 \end{bmatrix}$

31. a) $\begin{bmatrix} 1 & 0 & 0 \\ 1 & 1 & 0 \\ 1 & 0 & 1 \end{bmatrix}$ **b)** $\begin{bmatrix} 1 & 0 & 0 \\ 1 & 0 & 1 \\ 1 & 1 & 0 \end{bmatrix}$ **c)** $\begin{bmatrix} 1 & 0 & 0 \\ 1 & 1 & 1 \\ 1 & 1 & 1 \end{bmatrix}$

33. a) $\mathbf{A} \vee \mathbf{B} = [a_{ij} \vee b_{ij}] = [b_{ij} \vee a_{ij}] = \mathbf{B} \vee \mathbf{A}$ **b)** $\mathbf{A} \wedge \mathbf{B} = [a_{ij} \wedge b_{ij}] = [b_{ij} \wedge a_{ij}] = \mathbf{B} \wedge \mathbf{A}$ **35. a)** $\mathbf{A} \vee (\mathbf{B} \wedge \mathbf{C}) = [a_{ij}] \vee [b_{ij} \wedge c_{ij}] = [a_{ij} \vee (b_{ij} \wedge c_{ij})] = [(a_{ij} \vee b_{ij}) \wedge (a_{ij} \vee c_{ij})] = [a_{ij} \vee b_{ij}] \wedge [a_{ij} \vee c_{ij}] = (\mathbf{A} \vee \mathbf{B}) \wedge (\mathbf{A} \vee \mathbf{C})$ **b)** $\mathbf{A} \wedge (\mathbf{B} \vee \mathbf{C}) = [a_{ij}] \wedge [b_{ij} \vee c_{ij}] = [a_{ij} \wedge (b_{ij} \vee c_{ij})] = [(a_{ij} \wedge b_{ij}) \vee (a_{ij} \wedge c_{ij})] = [a_{ij} \wedge b_{ij}] \vee [a_{ij} \wedge c_{ij}] = (\mathbf{A} \wedge \mathbf{B}) \vee (\mathbf{A} \wedge \mathbf{C})$ **37.** $\mathbf{A} \odot (\mathbf{B} \odot \mathbf{C}) = \left[\bigvee_q a_{iq} \wedge \left(\bigvee_r (b_{qr} \wedge c_{rl}) \right) \right] = \left[\bigvee_q \bigvee_r (a_{iq} \wedge b_{qr} \wedge c_{rl}) \right] = \left[\bigvee_r \bigvee_q (a_{iq} \wedge b_{qr} \wedge c_{rl}) \right] = \left[\bigvee_r \left(\bigvee_q (a_{iq} \wedge b_{qr}) \right) \wedge c_{rl} \right] = (\mathbf{A} \odot \mathbf{B}) \odot \mathbf{C}$

SUPPLEMENTARY EXERCISES

1. a) procedure *last max*($a_1, \ldots, a_n$: integers)

```
max := a_1
last := 1
i := 2
while i ≤ n
begin
    if a_i ≥ max then
    begin
        max := a_i
        last := i
    end
    i := i + 1
end {last is the location of final occurrence of
    largest integer in list}
```

b) $2n - 1 = O(n)$ comparisons

3. a) procedure *pair zeros*($b_1 b_2 \cdots b_n$: bit string,

```
    n ≥ 2)
x := b_1
y := b_2
k := 2
while (k < n and (x ≠ 0 or y ≠ 0))
begin
    k := k + 1
    x := y
    y := b_k
end
if (x = 0 and y = 0) then print "YES"
else print "NO"
```

b) $O(n)$ comparisons

5. a) and **b)**

procedure *smallest and largest*($a_1, a_2, \ldots, a_n$: integers)

```
min := a_1
max := a_1
for i := 2 to n
begin
    if a_i < min then min := a_i
    if a_i > max then max := a_i
end {min is the smallest integer among the input, and max
    is the largest}
```

c) $2n - 2$

7. Before any comparisons are done, there is a possibility that each element could be the maximum and a possibility that it could be the minimum. This means that there are $2n$ different possibilities, and $2n - 2$ of them have to be eliminated through comparisons of elements, since we need to find the unique maximum and the unique minimum. We classify comparisons of two elements as "virgin" or "nonvirgin," depending on whether or not both elements being compared have been in any previous comparison. A virgin comparison eliminates the possibility that the larger one is the minimum and that the smaller one is the maximum; thus each virgin comparison eliminates two possibilities, but it clearly cannot do more. A nonvirgin comparison must be between two elements that are still in the running to be the maximum or two elements that are still in the run-

ning to be the minimum, and at least one of these elements must *not* be in the running for the other category. For example, we might be comparing x and y, where all we know is that x has been eliminated as the minimum. If we find that $x > y$ in this case, then only one possibility has been ruled out—we now know that y is not the maximum. Thus in the worst case, a nonvirgin comparison eliminates only one possibility. (The cases of other nonvirgin comparisons are similar.) Now there are at most $\lfloor n/2 \rfloor$ comparisons of elements that have not been compared before, each removing two possibilities; they remove $2\lfloor n/2 \rfloor$ possibilities altogether. Therefore we need $2n - 2 - 2\lfloor n/2 \rfloor$ more comparisons that, as we have argued, can remove only one possibility each, in order to find the answers in the worst case, since $2n - 2$ possibilities have to be eliminated. This gives us a total of $2n - 2 - 2\lfloor n/2 \rfloor + \lfloor n/2 \rfloor$ comparisons in all. But $2n - 2 - 2\lfloor n/2 \rfloor + \lfloor n/2 \rfloor = 2n - 2 - \lfloor n/2 \rfloor = 2n - 2 + \lceil -n/2 \rceil = \lceil 2n - n/2 \rceil - 2 = \lceil 3n/2 \rceil - 2$, as desired. **9.** At end of first pass: 3, 1, 4, 5, 2, 6; at end of second pass: 1, 3, 2, 4, 5, 6; at end of third pass: 1, 2, 3, 4, 5, 6; fourth pass finds nothing to exchange and algorithm terminates **11.** There are possibly as many as n passes through the list, and each pass uses $O(n)$ comparisons. Thus there are $O(n^2)$ comparisons in all. **13.** Since $\log n < n$, we have $(n \log n + n^2)^3 \le (n^2 + n^2)^3 \le (2n^2)^3 = 8n^6$ for all $n > 0$. This proves that $(n \log n + n^2)^3$ is $O(n^6)$, with witnesses $C = 8$ and $k = 0$. **15.** $O(x^2 2^x)$ **17.** Note that $\frac{n!}{2^n} = \frac{n}{2} \cdot \frac{n-1}{2} \cdots \frac{3}{2} \cdot \frac{2}{2} \cdot \frac{1}{2} > \frac{n}{2} \cdot 1 \cdot 1 \cdots 1 \cdot \frac{1}{2} = \frac{n}{4}$. **19.** $5, 22, -12, -29$ **21.** Since $ac \equiv bc \pmod{m}$ there is an integer k such that $ac = bc + km$. Hence $a - b = km/c$. Since $a - b$ is an integer, $c \mid km$. Letting $d = \gcd(m, c)$, write $c = de$. Since no factor of e divides m/d, it follows that $d \mid m$ and $e \mid k$. Thus $a - b = (k/e)(m/d)$, where $k/e \in \mathbf{Z}$ and $m/d \in \mathbf{Z}$. Therefore $a \equiv b \pmod{m/d}$. **23.** 1 **25.** 1 **27.** $(a_n a_{n-1} \cdots a_1 a_0)_{10} = \sum_{k=0}^{n} 10^k a_k \equiv \sum_{k=0}^{n} a_k \pmod 9$ since $10^k \equiv 1 \pmod 9$ for every nonnegative integer k. **29. a)** Not mutually relatively prime **b)** Mutually relatively prime **c)** Mutually relatively prime **d)** Mutually relatively prime **31. a)** The decryption function is $g(q) = \bar{a}(q - b) \bmod 26$, where $\bar{a}$ is an inverse of a modulo 26. **b)** PLEASE SEND MONEY **33.** $x \equiv 28$ (mod 30) **35.** $\mathbf{A}^{4n} = \begin{bmatrix} 1 & 0 \\ 0 & 1 \end{bmatrix}$, $\mathbf{A}^{4n+1} = \begin{bmatrix} 0 & 1 \\ -1 & 0 \end{bmatrix}$, $\mathbf{A}^{4n+2} = \begin{bmatrix} -1 & 0 \\ 0 & -1 \end{bmatrix}$, $\mathbf{A}^{4n+3} = \begin{bmatrix} 0 & -1 \\ 1 & 0 \end{bmatrix}$, for $n \ge 0$.
37. Suppose that $\mathbf{A} = \begin{bmatrix} a & b \\ c & d \end{bmatrix}$. Let $\mathbf{B} = \begin{bmatrix} 0 & 1 \\ 0 & 0 \end{bmatrix}$. Since $\mathbf{AB} = \mathbf{BA}$, it follows that $c = 0$ and $a = d$. Let $\mathbf{B} = \begin{bmatrix} 0 & 0 \\ 1 & 0 \end{bmatrix}$. Since $\mathbf{AB} = \mathbf{BA}$, it follows that $b = 0$. Hence $\mathbf{A} = \begin{bmatrix} a & 0 \\ 0 & a \end{bmatrix} = a\mathbf{I}$.
39. procedure *triangular matrix multiplication*$(\mathbf{A}, \mathbf{B}$: upper triangular $n \times n$ matrices, $\mathbf{A} = [a_{ij}]$, $\mathbf{B} = [b_{ij}])$

```
for i := 1 to n
begin
  for j := i to n
  begin
    c_ij := 0
    for k := i to j
      c_ij := c_ij + a_ik b_kj
  end
end
```

41. $(\mathbf{AB})(\mathbf{B}^{-1}\mathbf{A}^{-1}) = \mathbf{A}(\mathbf{BB}^{-1})\mathbf{A}^{-1} = \mathbf{AIA}^{-1} = \mathbf{AA}^{-1} = \mathbf{I}$. Similarly, $(\mathbf{B}^{-1}\mathbf{A}^{-1})(\mathbf{AB}) = \mathbf{I}$. Hence $(\mathbf{AB})^{-1} = \mathbf{B}^{-1}\mathbf{A}^{-1}$. **43. a)** Let $\mathbf{A} \odot \mathbf{0} = [b_{ij}]$. Then $b_{ij} = (a_{i1} \wedge 0) \vee \cdots \vee (a_{ip} \wedge 0) = 0$. Hence $\mathbf{A} \odot \mathbf{0} = \mathbf{0}$. Similarly $\mathbf{0} \odot \mathbf{A} = \mathbf{0}$. **b)** $\mathbf{A} \vee \mathbf{0} = [a_{ij} \vee 0] = [a_{ij}] = \mathbf{A}$. Hence $\mathbf{A} \vee \mathbf{0} = \mathbf{A}$. Similarly $\mathbf{0} \vee \mathbf{A} = \mathbf{A}$. **c)** $\mathbf{A} \wedge \mathbf{0} = [a_{ij} \wedge 0] = [0] = \mathbf{0}$. Hence $\mathbf{A} \wedge \mathbf{0} = \mathbf{0}$. Similarly $\mathbf{0} \wedge \mathbf{A} = \mathbf{0}$. **45.** We assume that someone has chosen a positive integer less than 2^n, which we are to guess. We ask the person to write the number in binary, using leading 0s if necessary to make it n bits long. We then ask "Is the first bit a 1?" "Is the second bit a 1?" "Is the third bit a 1?" and so on. After we know the answers to these n questions, we will know the number, because we will know its binary expansion.

CHAPTER 3

SECTION 3.1

1. Since every second integer is divisible by 2, the product is divisible by 2. Since every third integer is divisible by 3, the product is divisible by 3. Therefore the product has both 2 and 3 in its prime factorization and is therefore divisible by $3 \cdot 2 = 6$. **3.** $n^4 - 1 = (2k + 1)^4 - 1 = 16k^4 + 32k^3 + 24k^2 + 8k = 8k(k + 1)(2k^2 + 2k + 1)$. One of k or $k + 1$ is even, so 16 divides $n^4 - 1$. **5.** Since $5^4 = 625$, both x and y must be less than 5. Then $x^4 + y^4 \le 4^4 + 4^4 = 512 < 625$. **7.** If $3x^2 - 1 = 8y$, then $3x^2 \equiv 1 \pmod 8$. If x is an odd integer, then by Exercise 2, $x^2 \equiv 1 \pmod 8$, so $3x^2 \equiv 3 \pmod 8$. If x is even, then $x^2 \equiv 0$ or 4 $\pmod 8$, so $3x^2 \equiv 0$ or 4 $\pmod 8$. This contradiction implies that there are no integer solutions to the given equation. **9.** If $x^4 - 3 = 16y^4$, then $x^4 \equiv 3 \pmod{16}$. By Exercise 3, the fourth power of an odd integer is congruent to 1 modulo 16. The fourth power of an even integer is congruent to 0 modulo 16. Thus x^4 cannot be congruent to 3 modulo 16, and the equation has no solutions. **11.** The harmonic mean of distinct positive real numbers a and b is always less than their geometric mean. To prove $2ab/(a + b) < \sqrt{ab}$, multiply both sides by $(a + b)/(2\sqrt{ab})$ to obtain the equivalent inequality $\sqrt{ab} < (a + b)/2$, which is proved in Example 1. **13.** The parity (oddness or evenness) of the sum of the numbers written on the board never changes, since $j + k$ and $|j - k|$ have the same parity (and at each step we reduce the sum by $j + k$ but increase it by $|j - k|$). Therefore the integer at the end of the process must have the same parity as $1 + 2 + \cdots + (2n) = n(2n + 1)$, which is odd since n is odd. **15.** $n = 1601$ is a counterexample. **17.** Suppose that $3^{1/3} = a/b$ where $a, b \in \mathbf{Z}, b \neq 0$, and $\gcd(a, b) = 1$. Then $3 = a^3/b^3$, so that $3b^3 = a^3$. Hence $3 \mid a^3$, which can happen only if $3 \mid a$. Let $a = 3m$. Then $3b^3 = 27m^3$, or $b^3 = 9m^3$. Thus $3 \mid b^3$, which shows that $3 \mid b$. This is a contradiction of the assumption that $\gcd(a, b) = 1$. **19.** By the rational root test from high school algebra, any rational number that satisfies this polynomial with integer coefficients is of the form p/q, where p is a factor of the constant term (1), and q is a factor of the leading coefficient (1). So the only possible rational roots are ± 1; since neither of them is a root, there are no rational roots. **21.** $a^2 \equiv b^2 \pmod{p}$ if and only if $p \mid (a^2 - b^2) = (a + b)(a - b)$. By the uniqueness of prime factorization, this is equivalent to $p \mid (a - b)$ or $p \mid (a + b)$, which is the same as $a \equiv b \pmod{p}$ or $a \equiv -b \pmod{p}$. **23.** Both sides are nonnegative integers less than m, congruent to ab modulo m, so they must be equal. **25.** 3, 5, and 7 are primes of the desired form. **27.** The integer $p_1 p_2 \cdots p_n p_{n+1}$, where p_j is the jth prime, is divisible by $n + 1$ distinct primes. **29.** If not, then suppose that $q_1, q_2, \ldots, q_n$ are all the primes of the form $6k + 5$. Let $Q = 6q_1 q_2 \cdots q_n - 1$. Note that Q is of the form $6k + 5$, where $k = q_1 q_2 \cdots q_n - 1$. Let $Q = p_1 p_2 \cdots p_t$ be the prime factorization of Q. No p_i is 2, 3, or any q_j, because the remainder when Q is divided by 2 is 1, by 3 is 2, and by q_j is $q_j - 1$. All odd primes other than 3 are of the form $6k + 1$ or $6k + 5$, and the product of primes of the form $6k + 1$ is also of this form. Therefore at least one of the p_i's must be of the form $6k + 5$, a contradiction. **31.** We prove this by considering the three cases of $n = 3k, n = 3k + 1$, and $n = 3k + 2$ for some integer k. If n is a multiple of 3, then the left-hand side is $\lfloor \frac{x}{3} \rfloor + \lfloor \frac{x+n}{3} \rfloor + \lfloor \frac{x+2n}{3} \rfloor = \lfloor \frac{x}{3} \rfloor + \lfloor \frac{x}{3} \rfloor + \frac{n}{3} + \lfloor \frac{x}{3} \rfloor + \frac{2n}{3} = 3\lfloor \frac{x}{3} \rfloor + n$, and the right-hand side of the desired equality is also $3\lfloor \frac{x}{3} \rfloor + n$. If $n = 3k + 1$, then the left-hand side is $\lfloor \frac{x}{3} \rfloor + \lfloor \frac{x+3k+1}{3} \rfloor + \lfloor \frac{x+6k+2}{3} \rfloor = \lfloor \frac{x}{3} \rfloor + \lfloor \frac{x+1}{3} \rfloor + k + \lfloor \frac{x+2}{3} \rfloor + 2k = \lfloor \frac{x}{3} \rfloor + \lfloor \frac{x+1}{3} \rfloor + \lfloor \frac{x+2}{3} \rfloor + n - 1$, and the right-hand side is $\lfloor x \rfloor + n - 1$. By Exercise 68 in Section 1.8, these are equal. The third case is similar. **33. a)** Without loss of generality, assume that the x sequence is already sorted into nondecreasing order, since we can relabel the indices. There are only a finite number of possible orderings for the y sequence, so if we can show that we can increase the sum (or at least keep it the same) whenever we find y_i and y_j that are out of order (i.e., $i < j$ but $y_i > y_j$) by switching them, then we will have shown that the sum is largest when the y sequence is in nondecreasing order. Indeed, if we perform the swap, then we have added $x_i y_j + x_j y_i$ to the sum and subtracted $x_i y_i + x_j y_j$. The net effect is to have added $x_i y_j + x_j y_i - x_i y_i - x_j y_j = (x_j - x_i)(y_i - y_j)$, which is nonnegative by our ordering assumptions. **b)** similar to part (a) **35.** Let $x = n + (r/m) + \epsilon$, where n is an integer, r is a nonnegative integer less than m, and ϵ is

a real number with $0 \leq \epsilon < 1/m$. The left-hand side is $\lfloor nm + r + m\epsilon \rfloor = nm + r$. On the right-hand side, the terms $\lfloor x \rfloor$ through $\lfloor x + (m + r - 1)/m \rfloor$ are all just n and the terms from $\lfloor x + (m - r)/m \rfloor$ on are all $n + 1$. Therefore, the right-hand side is $(m - r)n + r(n + 1) = nm + r$, as well. **37.** Recall that a nonconstant polynomial can take on the same value only a finite number of times. Thus f can take on the values 0 and ± 1 only finitely many times, so if there is not some y such that $f(y)$ is composite, then there must be some x_0 such that $\pm f(x_0)$ is prime, say p. Look at $f(x_0 + kp)$. When we plug $x_0 + kp$ in for x in the polynomial and multiply it out, every term will contain a factor of p except for the terms that form $f(x_0)$. Therefore $f(x_0 + kp) = f(x_0) + mp = (m + 1)p$ for some integer m. As k varies, this value can be 0, p, or $-p$ only finitely many times; therefore it must be a composite number for some values of k. **39.** Assume that every even integer greater than 2 is the sum of two primes, and let n be an integer greater than 5. If n is odd, write $n = 3 + (n - 3)$ and decompose $n - 3 = p + q$ into the sum of two primes; if n is even, then write $n = 2 + (n - 2)$ and decompose $n - 2 = p + q$ into the sum of two primes. For the converse, assume that every integer greater than 5 is the sum of three primes, and let n be an even integer greater than 2. Write $n + 2$ as the sum of three primes, one of which is necessarily 2, so $n + 2 = 2 + p + q$, whence $n = p + q$. **41. a)** $6 \to 3 \to 10 \to 5 \to 16 \to 8 \to 4 \to 2 \to 1$ **b)** $7 \to 22 \to 11 \to 34 \to 17 \to 52 \to 26 \to 13 \to 40 \to 20 \to 10 \to 5 \to 16 \to 8 \to 4 \to 2 \to 1$ **c)** $17 \to 52 \to 26 \to 13 \to 40 \to 20 \to 10 \to 5 \to 16 \to 8 \to 4 \to 2 \to 1$ **d)** $21 \to 64 \to 32 \to 16 \to 8 \to 4 \to 2 \to 1$ **43.** $1 + 2 + 3 = 6$; $1 + 2 + 4 + 7 + 14 = 28$; $1 + 2 + 4 + 8 + 16 + 31 + 62 + 124 + 248 = 496$ **45.** The sum of the proper divisors is $(1 + 2 + 4 + 8 + \cdots + 2^{p-1}) + (2^p - 1)(1 + 2 + 4 + 8 + \cdots + 2^{p-2}) = 2^p - 1 + (2^p - 1)(2^{p-1} - 1) = (2^p - 1)(1 + 2^{p-1} - 1) = (2^p - 1)2^{p-1}$. **47.** $(3, 5, 0) \to (3, 2, 3) \to (6, 2, 0) \to (6, 0, 2) \to (1, 5, 2) \to (1, 4, 3)$ **49.** Suppose we had a program S that could tell whether a program with its given input ever prints the digit 1. Here is an algorithm for solving the halting problem: Given a program P and its input I, construct a program P', which is just like P but never prints anything (even if P did print something) except that if and when it is about to halt, it prints a 1 and halts. Then P halts on an input if and only if P' ever prints a 1 on that same input. Feed P' and I to S, and that will tell us whether or not P halts on input I. Since we know that the halting problem is in fact not solvable, we have a contradiction. Therefore no such program S exists. **51.** Run the two programs concurrently and wait for one to halt.

SECTION 3.2

1. a) 3 **b)** -1 **c)** 787 **d)** 2639 **3. a)** $a_0 = 2, a_1 = 3$, $a_2 = 5, a_3 = 9$ **b)** $a_0 = 1, a_1 = 4, a_2 = 27, a_3 = 256$

c) $a_0 = 0, a_1 = 0, a_2 = 1, a_3 = 1$ **d)** $a_0 = 0, a_1 = 1$, $a_2 = 2, a_3 = 3$ **5. a)** 2, 5, 8, 11, 14, 17, 20, 23, 26, 29 **b)** 1, 1, 1, 2, 2, 2, 3, 3, 3, 4 **c)** 1, 1, 3, 3, 5, 5, 7, 7, 9, 9 **d)** $-1, -2, -2, 8, 88, 656, 4912, 40064, 362368$, 3627776 **e)** 3, 6, 12, 24, 48, 96, 192, 384, 768, 1536 **f)** 1, 1, 2, 3, 5, 8, 13, 21, 34, 55 **g)** 1, 2, 2, 3, 3, 3, 3, 4, 4, 4 **h)** 3, 3, 5, 4, 4, 3, 5, 5, 4, 3 **7.** Each term could be twice the previous term; the nth term could be obtained from the previous term by adding $n - 1$; the terms could be the positive integers that are not multiples of 3; there are infinitely many other possibilities. **9. a)** One 1 and one 0, followed by two 1s and two 0s, followed by three 1s and three 0s, and so on **b)** The positive integers are listed in increasing order with each even positive integer listed twice. **c)** The terms in odd-numbered locations are the successive powers of 2; the terms in even-numbered locations are all 0. **d)** $a_n = 3 \cdot 2^{n-1}$ **e)** $a_n = 15 - 7(n - 1) = 22 - 7n$ **f)** $a_n = (n^2 + n + 4)/2$ **g)** $a_n = 2n^3$ **h)** $a_n = n! + 1$ **11.** Among the integers $1, 2, \ldots, a_n$, where a_n is the nth positive integer not a perfect square, the nonsquares are $a_1, a_2, \ldots, a_n$ and the squares are $1^2, 2^2, \ldots, k^2$, where k is the integer with $k^2 < n+k < (k+1)^2$. Consequently, $a_n = n+k$, where $k^2 < a_n < (k+1)^2$. To find k, first note that $k^2 < n+k < (k+1)^2$, so $k^2 + 1 \le n + k \le (k + 1)^2 - 1$. Hence $(k - \frac{1}{2})^2 + \frac{3}{4} = k^2 - k + 1 \le n \le k^2 + k = (k + \frac{1}{2})^2 - \frac{1}{4}$. It follows that $k - \frac{1}{2} < \sqrt{n} < k + \frac{1}{2}$, so $k = \{\sqrt{n}\}$ and $a_n = n+k = n+\{\sqrt{n}\}$. **13. a)** 20 **b)** 11 **c)** 30 **d)** 511 **15. a)** 1533 **b)** 510 **c)** 4923 **d)** 9842 **17. a)** 21 **b)** 78 **c)** 18 **d)** 18 **19.** $\sum_{j=1}^{n}(a_j - a_{j-1}) = a_n - a_0$ **21. a)** n^2 **b)** $n(n + 1)/2$ **23.** 15150 **25.** $\frac{n(n+1)(2n+1)}{3} + \frac{n(n+1)}{2} + (n+1)(m - (n+1)^2 + 1)$, where $n = \lfloor \sqrt{m} \rfloor - 1$ **27. a)** 0 **b)** 1680 **c)** 1 **d)** 1024 **29.** 34 **31. a)** Countable, $-1, -2, -3, -4, \ldots$ **b)** Countable, $0, 2, -2, 4, -4, \ldots$ **c)** Uncountable **d)** Countable, $0, 7, -7, 14, -14, \ldots$ **33.** Assume that $A - B$ is countable. Then, since $A = (A - B) \cup B$, the elements of A can be listed in a sequence by alternating elements of $A - B$ and elements of B. This contradicts the uncountability of A. **35.** Assume that B is countable. Then the elements of B can be listed as $b_1, b_2, b_3, \ldots$. Since A is a subset of B, taking the subsequence of $\{b_n\}$ that contains the terms that are in A gives a listing of the elements of A. Since A is uncountable, this is impossible. **37.** Suppose that $A_1, A_2, A_3, \ldots$ are countable sets. Since A_i is countable, we can list its elements in a sequence as $a_{i1}, a_{i2}, a_{i3}, \ldots$. The elements of the set $\bigcup_{i=1}^{n} A_i$ can be listed by listing all terms a_{ij} with $i + j = 2$, then all terms a_{ij} with $i + j = 3$, then all terms a_{ij} with $i + j = 4$, and so on. **39.** There are a finite number, namely, 2^m, bit strings of length m. The set of all bit strings is the union of the bit strings of lengths m over $m = 0, 1, 2, \ldots$. Since the union of a countable number of countable sets is countable, there are a countable number of bit strings. **41.** For any finite alphabet there are a finite number of strings of length n, whenever n is a positive integer. It follows by the result of

Exercise 37 that there are only a finite number of strings from any given finite alphabet. Since the set of all computer programs in a particular language is a subset of the set of all strings of a finite alphabet, which is a countable set, by the result of Exercise 34, it is itself a countable set. **43.** Exercise 41 shows that there are only a countable number of computer programs. Consequently, there are only a countable number of computable functions. Since, as Exercise 42 shows, there are an uncountable number of functions, not all functions are computable. **45.** Given a positive integer x, we show that there is exactly one positive rational number m/n (in lowest terms) such that $K(m/n) = x$. From the prime factorization of x, read off the m and n such that $K(m/n) = x$. The primes that occur to even powers are the primes that occur in the prime factorization of m, with the exponents being half the corresponding exponents in x; and the primes that occur to odd powers are the primes that occur in the prime factorization of n, with the exponents being half of one more than the exponents in x.

SECTION 3.3

1. $n(n + 1)$ **3.** Let $P(n)$ be "$\sum_{j=0}^{n} 3 \cdot 5^j = 3(5^{n+1} - 1)/4$." *Basis step:* $P(0)$ is true since $\sum_{j=0}^{0} 3 \cdot 5^j = 3 = 3(5^1 - 1)/4$. *Inductive step:* Assume that $\sum_{j=0}^{k} 3 \cdot 5^j = 3(5^{k+1} - 1)/4$. Then $\sum_{j=0}^{k+1} 3 \cdot 5^j = (\sum_{j=0}^{k} 3 \cdot 5^j) + 3 \cdot 5^{k+1} = 3(5^{k+1} - 1)/4 + 3 \cdot 5^{k+1} = 3(5^{k+1} + 4 \cdot 5^{k+1} - 1)/4 = 3(5^{k+2} - 1)/4$. **5.** By examining small values of n we make the conjecture that $P(n)$ is true where $P(n)$ is the statement "$\sum_{j=1}^{n} 1/2^j = (2^n - 1)/2^n$." *Basis step:* $P(1)$ is true since $\frac{1}{2} = (2^1 - 1)/2^1$. *Inductive step:* Assume that $\sum_{j=1}^{k} 1/2^j = (2^k - 1)/2^k$. Then $\sum_{j=1}^{k+1} \frac{1}{2^j} = (\sum_{j=1}^{k} \frac{1}{2^j}) + \frac{1}{2^{k+1}} = \frac{2^k-1}{2^k} + \frac{1}{2^{k+1}} = \frac{2^{k+1}-2+1}{2^{k+1}} = \frac{2^{k+1}-1}{2^{k+1}}$. **7.** Let $P(n)$ be "$\sum_{j=1}^{n} j^2 = n(n + 1)(2n + 1)/6$." *Basis step:* $P(1)$ is true since $\sum_{1}^{1} j^2 = 1 = 1(1 + 1)(2 \cdot 1 + 1)/6$. *Inductive step:* Assume that $\sum_{j=1}^{k} j^2 = k(k+1)(2k+1)/6$. Then $\sum_{j=1}^{k+1} j^2 = (\sum_{j=1}^{k} j^2) + (k + 1)^2 = k(k+1)(2k+1)/6 + (k+1)^2 = (k+1)[2k^2 + k + 6k + 6]/6 = (k+1)(k+2) \cdot (2k+3)/6 = (k+1)((k+1)+1)(2(k+1)+1)/6$. **9.** Let $P(n)$ be "$1^2 + 3^2 + \cdots + (2n + 1)^2 = (n + 1)(2n + 1)(2n + 3)/3$." *Basis step:* $P(0)$ is true since $1^2 = 1 = (0 + 1)(2 \cdot 0 + 1)(2 \cdot 0 + 3)/3$. *Inductive step:* Assume that $P(k)$ is true. Then $1^2 + 3^2 + \cdots + (2k+1)^2 + (2(k+1)+1)^2 = (k+1)(2k+1)(2k+3)/3 + (2k+3)^2 = (2k+3)[(k+1)(2k+1)/3 + (2k+3)] = (2k+3)(2k^2 + 9k + 10)/3 = (2k+3)(2k+5)(k+2)/3 = ((k+1)+1)(2(k+1)+1) \cdot (2(k+1)+3)/3$. **11.** Let $P(n)$ be "$1 + nh \le (1 + h)^n, h > -1$." *Basis step:* $P(0)$ is true since $1 + 0 \cdot h = 1 \le 1 = (1 + h)^0$. *Inductive step:* Assume $1 + kh \le (1 + h)^k$. Then since $(1+h) > 0$, $(1+h)^{k+1} = (1+h)(1+h)^k \ge (1+h)(1+kh) = 1 + (k + 1)h + kh^2 \ge 1 + (k + 1)h$. **13.** Let $P(n)$ be "$2^n > n^2$." *Basis step:* $P(5)$ is true since $2^5 = 32 > 25 = 5^2$. *Inductive step:* Assume that $P(k)$ is true, that is, $2^k > k^2$.

Then $2^{k+1} = 2 \cdot 2^k > k^2 + k^2 > k^2 + 4k \geq k^2 + 2k + 1 = (k+1)^2$ since $k > 4$.   **15.** Let $P(n)$ be "$1 \cdot 2 + 2 \cdot 3 + \cdots + n(n+1) = n(n+1)(n+2)/3$." *Basis step:* $P(1)$ is true since $1 \cdot 2 = 2 = 1(1+1)(1+2)/3$. *Inductive step:* Assume that $P(k)$ is true. Then $1 \cdot 2 + 2 \cdot 3 + \cdots + k(k+1) + (k+1)(k+2) = [k(k+1)(k+2)/3] + (k+1)(k+2) = (k+1)(k+2)[(k/3)+1] = (k+1)(k+2)(k+3)/3$.   **17.** Let $P(n)$ be "$1^2 - 2^2 + 3^2 - \cdots + (-1)^{n-1}n^2 = (-1)^{n-1}n(n+1)/2$." *Basis step:* $P(1)$ is true since $1^2 = 1 = (-1)^0 1^2$. *Inductive step:* Assume that $P(k)$ is true. Then $1^2 - 2^2 + 3^2 - \cdots + (-1)^{k-1}k^2 + (-1)^k(k+1)^2 = (-1)^{k-1}k(k+1)/2 + (-1)^k(k+1)^2 = (-1)^k(k+1)[-k/2 + (k+1)] = (-1)^k(k+1)[(k/2)+1] = (-1)^k(k+1)(k+2)/2$. **19.** Let $P(n)$ be "a postage of n cents can be formed using 3-cent and 5-cent stamps." *Basis step:* $P(8)$ is true since 8 cents postage can be formed with one 3-cent and one 5-cent stamp. *Inductive step:* Assume that $P(k)$ is true, that is, postage of k cents can be formed. We will show how to form postage of $k+1$ cents. By the inductive hypothesis postage of k cents can be formed. If this included a 5-cent stamp, replace this with two 3-cent stamps to obtain $k+1$ cents postage. Otherwise, only 3-cent stamps were used and $k \geq 9$. Remove three of these 3-cent stamps and replace them with two 5-cent stamps to obtain $k+1$ cents postage. **21.** Let $P(n)$ be "$n^5 - n$ is divisible by 5." *Basis step:* $P(0)$ is true since $0^5 - 0 = 0$ is divisible by 5. *Inductive step:* Assume that $P(k)$ is true, that is, $k^5 - 5$ is divisible by 5. Then $(k+1)^5 - (k+1) = (k^5 + 5k^4 + 10k^3 + 10k^2 + 5k + 1) - (k+1) = (k^5 - k) + 5(k^4 + 2k^3 + 2k^2 + k)$ is also divisible by 5, since both terms in this sum are divisible by 5.   **23.** Let $P(n)$ be the proposition that $(2n-1)^2 - 1$ is divisible by 8. The basis case $P(1)$ is true since $8 \mid 0$. Now assume that $P(k)$ is true. Since $((2(k+1)-1)^2 - 1) = ((2k-1)^2 - 1) + 8k$, $P(k+1)$ is true since both terms on the right-hand side are divisible by 8. This shows that $P(n)$ is true for all positive integers so that $m^2 - 1$ is divisible by 8 whenever m is an odd positive integer.   **25.** Let $P(n)$ be the statement that a set with n elements has $n(n-1)/2$ two-element subsets. $P(2)$, the basis case, is true, since a set with two elements has one subset with two elements—namely, itself—and $2(2-1)/2 = 1$. Now assume that $P(k)$ is true. Let S be a set with $k+1$ elements. Choose an element a in S and let $T = S - \{a\}$. A two-element subset of S either contains a or it does not. Those subsets not containing a are the subsets of two elements of T; by the inductive hypothesis there are $k(k-1)/2$ of these. There are k subsets of two elements of S that contain a, since such a subset contains a and one of the k elements in T. Hence there are $k(k-1)/2 + k = (k+1)k/2$ two-element subsets of S. This completes the inductive proof.   **27.** Let $P(n)$ be the statement that $1^4 + 2^4 + 3^4 + \cdots + n^4 = n(n+1)(2n+1)(3n^2 + 3n - 1)/30$. $P(1)$ is true since $1 \cdot 2 \cdot 3 \cdot 5/30 = 1$. Assume that $P(k)$ is true. Then $(1^4 + 2^4 + 3^4 + \cdots + k^4) + (k+1)^4 = k(k+1)(2k+1)(3k^2 + 3k - 1)/30 + (k+1)^4 = ((k+1)/30)(k(2k+1)(3k^2 + 3k - 1) + 30(k+1)^3) = ((k+1)/30)(6k^4 + 39k^3 + 91k^2 + 89k + 30) = ((k+1)/30)(k+2)(2k+3)(3(k+1)^2 + 3(k+1) - 1)$. This

demonstrates that $P(k+1)$ is true.   **29.** By inspection we find that the inequality $2n + 3 \leq 2^n$ does not hold for $n = 0, 1, 2, 3$. Let $P(n)$ be the proposition that this inequality holds for the positive integer n. $P(4)$, the basis case, is true since $2 \cdot 4 + 3 = 11 \leq 16 = 2^4$. For the inductive step assume that $P(k)$ is true. Then, by the inductive hypothesis, $2(k+1) + 3 = (2k+3) + 2 < 2^k + 2$. But since $k \geq 1$, $2^k + 2 \leq 2^k + 2^k = 2^{k+1}$. This shows that $P(k+1)$ is true.   **31. a)** The postages that can be formed using 5-cent and 6-cent stamps are 5 cents, 6 cents, 10 cents, 11 cents, 12 cents, 15 cents, 16 cents, 17 cents, 18 cents, and all postages of 20 cents or more.   **b)** We will prove that all postages of 20 cents or more can be formed using 5-cent and 6-cent stamps. Let $P(n)$ be the statement that postage of n cents can be formed. $P(20)$ is true since postage of 20 cents can be formed using four 5-cent stamps. Now assume that $P(k)$ is true. If a 5-cent stamp was used to form postage of n cents, then replace it by a 6-cent stamp to form postage of $k+1$ cents. Otherwise, if only 6-cent stamps were used, since $k \geq 20$ at least four 6-cent stamps were used. Replace four 6-cent stamps by five 5-cent stamps to get postage of $k+1$ cents. Hence $P(k+1)$ is true. This completes the proof by mathematical induction.   **c)** Let $P(n)$ be as in (b). The basis cases are $P(20)$, $P(21)$, $P(22)$, $P(23)$, and $P(24)$. These are true since postage of 20 cents, 21 cents, 22 cents, 23 cents, and 24 cents can be formed using four 5-cent stamps, three 5-cent stamps and one 6-cent stamp, two 5-cent stamps and two 6-cent stamps, one 5-cent and three 6-cent stamps, and four 6-cent stamps, respectively. Now assume that $P(j)$ is true for $20 \leq j \leq k$, where $k \geq 24$. Since $k + 1 \geq 25$, it follows that $k - 4 \geq 20$, so that by the inductive hypothesis postage of $k - 4$ can be formed. Add a 5-cent stamp to obtain postage of $k+1$ cents, showing that $P(k+1)$ is true. This completes the proof by strong induction.   **33.** All multiples of \$10 greater than or equal to \$40 can be formed as well as \$20. Let $P(n)$ be the statement that $10n$ dollars can be formed. $P(4)$ is true since \$40 can be formed using two \$20s. Now assume that $P(k)$ is true with $k \geq 4$. If a \$50 bill is used to form $10k$ dollars, replace it by three \$20 bills to obtain $10(k+1)$ dollars. Otherwise, at least two \$20 bills were used since $10k$ is at least \$40. Replace two \$20 bills with a \$50 bill to obtain \$10$(k+1)$. This shows that $P(k+1)$ is true.   **35.** *Basis step:* There are four base cases. If $n = 1 = 4 \cdot 0 + 1$, then clearly the second player wins. If there are two, three, or four matches ($n = 4 \cdot 0 + 2$, $n = 4 \cdot 0 + 3$, or $n = 4 \cdot 1$), then the first player can win by removing all but one match. *Inductive step:* Assume the strong inductive hypothesis, that in games with k or fewer matches, the first player can win if $k \equiv 0$, 2 or 3 (mod 4) and the second player can win if $k \equiv 1$ (mod 4). Suppose we have a game with $k+1$ matches, with $k \geq 4$. If $k+1 \equiv 0$ (mod 4), then the first player can remove three matches, leaving $k - 2$ matches for the other player. Since $k - 2 \equiv 1$ (mod 4), by the inductive hypothesis, this is a game that the second player at that point (who is the first player in our game) can win. Similarly, if $k + 1 \equiv 2$ (mod 4), then the first player can remove one match; and if $k + 1 \equiv 3$ (mod 4), then the first player can remove two matches. Finally, if $k + 1 \equiv 1$ (mod 4), then the first player must leave $k, k - 1$, or $k - 2$ matches for the other player. Since $k \equiv 0$ (mod 4), $k - 1 \equiv 3$ (mod 4), and $k - 2 \equiv 2$ (mod 4), by the inductive hypothesis, this is a game that the first player at that point (who is the second player in our game) can win. **37.** *Basis step:* For $n = 1$, the left-hand side is just $\frac{1}{1}$ which is 1. For $n = 2$, there are three nonempty subsets $\{1\}$, $\{2\}$, and $\{1, 2\}$, so the left-hand side is $\frac{1}{1} + \frac{1}{2} + \frac{1}{1\cdot 2} = 2$. *Inductive step:* Assume that the statement is true for k. The set of the first $k + 1$ positive integers has many nonempty subsets, but they fall into three categories: a nonempty subset of the first k positive integers together with $k + 1$, a nonempty subset of the first k positive integers, or just $\{k + 1\}$. By the inductive hypothesis, the sum of the first category is k. For the second category, we can factor out $1/(k + 1)$ from each term of the sum and what remains is just k by the inductive hypothesis, so this part of the sum is $k/(k + 1)$. Finally, the third category simply yields $1/(k + 1)$. Hence, the entire summation is $k + k/(k + 1) + 1/(k + 1) = k + 1$. **39.** We use the notation (i, j) to mean the square in row i and column j and use induction on $i + j$ to show that every square can be reached by the knight. *Basis step:* There are six base cases, for the cases when $i + j \le 2$. The knight is already at $(0, 0)$ to start, so the empty sequence of moves reaches that square. To reach $(1, 0)$, the knight moves $(0, 0) \to (2, 1) \to (0, 2) \to (1, 0)$. Similarly, to reach $(0, 1)$, the knight moves $(0, 0) \to (1, 2) \to (2, 0) \to (0, 1)$. Note that the knight has reached $(2, 0)$ and $(0, 2)$ in the process. For the last basis step there is $(0, 0) \to (1, 2) \to (2, 0) \to (0, 1) \to (2, 2) \to (0, 3) \to (1, 1)$. *Inductive step:* Assume the inductive hypothesis, that the knight can reach any square (i, j) for which $i + j = k$, where k is an integer greater than 1. We must show how the knight can reach each square (i, j) when $i + j = k + 1$. Since $k + 1 \ge 3$, at least one of i and j is at least 2. If $i \ge 2$, then by the inductive hypothesis, there is a sequence of moves ending at $(i - 2, j + 1)$, since $i - 2 + j + 1 = i + j - 1 = k$; from there it is just one step to (i, j); similarly, if $j \ge 2$. **41.** *Basis step:* The base cases $n = 0$ and $n = 1$ are true since the derivative of x^0 is 0 and the derivative $x^1 = x$ is 1. *Inductive step:* Using the product rule, the inductive hypothesis, and the base case shows that $\frac{d}{dx}x^{k+1} = \frac{d}{dx}(x \cdot x^k) = x \cdot \frac{d}{dx}x^k + x^k \frac{d}{dx}x = x \cdot kx^{k-1} + x^k \cdot 1 = kx^k + x^k = (k+1)x^k$. **43.** Let $P(n)$ be the statement that $\mathbf{AB}^n = \mathbf{B}^n\mathbf{A}$. $P(1)$ is true since $\mathbf{AB} = \mathbf{BA}$. Now assume that $P(n)$ is true. Then $\mathbf{AB}^{n+1} = \mathbf{AB}^n\mathbf{B} = \mathbf{B}^n\mathbf{AB} = \mathbf{B}^n\mathbf{BA} = \mathbf{B}^{n+1}\mathbf{A}$. It follows that $P(n + 1)$ is true. **45.** Let $P(n)$ be "$(A_1 \cup A_2 \cup \cdots \cup A_n) \cap B = (A_1 \cap B) \cup (A_2 \cap B) \cup \cdots \cup (A_n \cap B)$." *Basis step:* $P(1)$ is trivially true. *Inductive step:* Assume that $P(k)$ is true. Then $(A_1 \cup A_2 \cup \cdots \cup A_k \cup A_{k+1}) \cap B = [(A_1 \cup A_2 \cup \cdots \cup$

$A_k) \cup A_{k+1}] \cap B = [(A_1 \cup A_2 \cup \cdots \cup A_k) \cap B] \cup (A_{k+1} \cap B) = [(A_1 \cap B) \cup (A_2 \cap B) \cup \cdots \cup (A_k \cap B)] \cup (A_{k+1} \cap B) = (A_1 \cap B) \cup (A_2 \cap B) \cup \cdots \cup (A_k \cap B) \cup (A_{k+1} \cap B)$. **47.** Let $P(n)$ be "$\overline{\bigcup_{k=1}^{n} A_k} = \bigcap_{k=1}^{n} \overline{A_k}$." *Basis step:* $P(1)$ is trivially true. *Inductive step:* Assume that $P(k)$ is true. Then $\overline{\bigcup_{j=1}^{k+1} A_j} = \overline{\left(\bigcup_{j=1}^{k} A_j\right) \cup A_{k+1}} = \overline{\left(\bigcup_{j=1}^{k} A_j\right)} \cap \overline{A_{k+1}} = \left(\bigcap_{j=1}^{k} \overline{A_j}\right) \cap \overline{A_{k+1}} = \bigcap_{j=1}^{k+1} \overline{A_j}$. **49.** Let $P(n)$ be "$[(p_1 \to p_2) \wedge (p_2 \to p_3) \wedge \cdots \wedge (p_{n-1} \to p_n)] \to [(p_1 \wedge \cdots \wedge p_{n-1}) \to p_n]$." *Basis step:* $P(2)$ is true since $(p_1 \to p_2) \to (p_1 \to p_2)$ is a tautology. *Inductive step:* Assume $P(k)$ is true. To show $[(p_1 \to p_2) \wedge \cdots \wedge (p_{k-1} \to p_k) \wedge (p_k \to p_{k+1})] \to [(p_1 \wedge \cdots \wedge p_{k-1} \wedge p_k) \to p_{k+1}]$ is a tautology, assume the hypothesis of this implication is true. Since both the hypothesis and $P(k)$ are true, it follows that $(p_1 \wedge \cdots \wedge p_{k-1}) \to p_k$ is true. Since this is true, and since $p_k \to p_{k+1}$ is true (it is part of the assumption) it follows by hypothetical syllogism that $(p_1 \wedge \cdots \wedge p_{k-1}) \to p_{k+1}$ is true. The weaker statement $(p_1 \wedge \cdots \wedge p_{k-1} \wedge p_k) \to p_{k+1}$ follows from this. **51.** The two sets do not overlap if $n + 1 = 2$. In fact, the implication $P(1) \to P(2)$ is false. **53.** The error is in going from the base case $n = 0$ to the next case, $n = 1$; we cannot write 1 as the sum of two smaller natural numbers. **55.** Assume that the well-ordering property holds. Suppose that $P(1)$ is true and that the implication $(P(1) \wedge P(2) \wedge \cdots \wedge P(n)) \to P(n + 1)$ is true for every positive integer $n \ge 1$. Let S be the set of integers n for which $P(n)$ is false. We will show $S = \emptyset$. Assume that $S \ne \emptyset$. Then by the well-ordering property there is a least integer m in S. We know that m cannot be 1 because $P(1)$ is true. Since $n = m$ is the least integer such that $P(n)$ is false, $P(1), P(2), \ldots, P(m - 1)$ are true, and $m - 1 \ge 1$. Since $(P(1) \wedge P(2) \wedge \cdots \wedge P(m-1)) \to P(m)$ is true, it follows that $P(m)$ must also be true, which is a contradiction. Hence $S = \emptyset$. **57.** Let $P(n)$ be "$H_{2^n} \le 1 + n$." *Basis step:* $P(0)$ is true since $H_{2^0} = H_1 = 1 \le 1 + 0$. *Inductive step:* Assume that $H_{2^k} \le 1 + k$. Then $H_{2^{k+1}} = H_{2^k} + \sum_{j=2^k+1}^{2^{k+1}} \frac{1}{j} \le 1 + k + 2^k \left(\frac{1}{2^{k+1}}\right) < 1 + k + 1 = 1 + (k + 1)$. **59.** Let $P(n)$ be "$1/\sqrt{1} + 1/\sqrt{2} + 1/\sqrt{3} + \cdots + 1/\sqrt{n} > 2\left(\sqrt{n+1} - 1\right)$." *Basis step:* $P(1)$ is true since $1 > 2\left(\sqrt{2} - 1\right)$. *Inductive step:* Assume that $P(k)$ is true. Then $1 + 1/\sqrt{2} + \cdots + 1/\sqrt{k} + 1/\sqrt{k+1} > 2\left(\sqrt{k+1} - 1\right) + 1/\sqrt{k+1}$. If we show that $2\left(\sqrt{k+1} - 1\right) + 1/\sqrt{k+1} > 2\left(\sqrt{k+2} - 1\right)$, it follows that $P(k + 1)$ is true. This inequality is equivalent to $2\left(\sqrt{k+2} - \sqrt{k+1}\right) < 1/\sqrt{k+1}$, which is equivalent to $2\left(\sqrt{k+2} - \sqrt{k+1}\right)\left(\sqrt{k+2} + \sqrt{k+1}\right) < \sqrt{k+1}/\sqrt{k+1} + \sqrt{k+2}/\sqrt{k+1}$. This is equivalent to $2 < 1 + \sqrt{k+2}/\sqrt{k+1}$, which is clearly true. **61.** We will first prove the result when n is a power of 2, that is, if $n = 2^k$, $k = 1, 2, \ldots$. Let $P(k)$ be the statement $A \ge G$, where A and G are the arithmetic and geometric means of a set of $n = 2^k$ positive real numbers. *Basis* *step:* $k = 1$ and $n = 2^1 = 2$. Note that $(\sqrt{a_1} - \sqrt{a_2})^2 \geq 0$. Expanding this shows that $a_1 - 2\sqrt{a_1 a_2} + a_2 \geq 0$, that is, $(a_1 + a_2)/2 \geq (a_1 a_2)^{1/2}$. *Inductive step:* Assume that $P(k)$ is true, with $n = 2^k$. We will show that $P(k + 1)$ is true. We have $2^{k+1} = 2n$. Now $(a_1 + a_2 + \cdots + a_{2n})/(2n) = ((a_1 + a_2 + \cdots + a_n)/n + (a_{n+1} + a_{n+2} + \cdots + a_{2n})/n)/2$ and similarly $(a_1 a_2 \cdots a_{2n})^{1/(2n)} = [(a_1 \cdots a_n)^{1/n} (a_{n+1} \cdots a_{2n})^{1/n}]^{1/2}$. To simplify the notation, let $A(x, y, \ldots)$ and $G(x, y, \ldots)$ denote the arithmetic mean and geometric mean of $x, y, \ldots$, respectively. Also, if $x \leq x'$, $y \leq y'$, and so on, then $A(x, y, \ldots) \leq A(x', y', \ldots)$ and $G(x, y, \ldots) \leq G(x', y', \ldots)$. Hence $A(a_1, \ldots, a_{2n}) = A(A(a_1, \ldots, a_n), A(a_{n+1}, \ldots, a_{2n})) \geq A(G(a_1, \ldots, a_n), G(a_{n+1}, \ldots, a_{2n})) \geq G(G(a_1, \ldots, a_n), G(a_{n+1}, \ldots, a_{2n})) = G(a_1, \ldots, a_{2n})$. This finishes the proof for powers of 2. Now if n is not a power of 2, let m be the next higher power of 2, and let $a_{n+1}, \ldots, a_m$ all equal $A(a_1, \ldots, a_n) = \bar{a}$. Then we have $((a_1 a_2 \cdots a_n)\bar{a}^{m-n})^{1/m} \leq A(a_1, \ldots, a_m)$, since m is a power of 2. Since $A(a_1, \ldots, a_m) = \bar{a}$, it follows that $(a_1 \cdots a_n)^{1/m}\bar{a}^{1-n/m} \leq \bar{a}^{n/m}$. Raising both sides to the (m/n)th power gives $G(a_1, \ldots, a_n) \leq A(a_1, \ldots, a_n)$. **63.** There is nothing to prove for the basis case when $n = 1$. Now assume the inductive hypothesis. Suppose that $p \mid a_1 a_2 \cdots a_k a_{k+1}$. Note that $\gcd(p, a_1, a_2, \ldots, a_k) = 1$ or p. If it is 1, by Lemma 1 in Section 2.6, $p \mid a_{k+1}$. If it is p, then $p \mid a_1 a_2 \cdots a_k$, so that by the inductive hypothesis, $p \mid a_i$ for some $i \leq k$. This completes the proof. **65.** Following the hint, we first want to show that if $w^2 + x^2 + y^2 + z^2 = 2wxyz$, then all the variables must be even. Clearly either none, two, or all must be even, since the right-hand side is even. Suppose all the variables were odd. Then the left-hand side is $1 + 1 + 1 + 1 \equiv 0 \pmod 4$ (see Exercise 2 in Section 3.1). But the right-hand side is congruent to 2 modulo 4 in this case, a contradiction. Next suppose that two of the variables were odd and two were even. Then the left-hand side is 2 modulo 4 while the right-hand side is 0 modulo 4, another contradiction. So we conclude that all the variables are even, and we make the substitution $w = 2a, x = 2b, y = 2c, z = 2d$ and simplify to obtain $a^2 + b^2 + c^2 + d^2 = 8abcd$. We repeat the argument modulo 8 to conclude $a = 2e, b = 2f, c = 2g, d = 2h$, whence $e^2 + f^2 + g^2 + h^2 = 32efgh$, and so on forever. Thus each of our original variables must have arbitrarily large powers of 2 as factors, an impossibility. **67.** Let $P(n)$ be the statement that if $x_1, x_2, \ldots, x_n$ are n distinct real numbers, then $n - 1$ multiplications are used to find the product of these numbers no matter how parentheses are inserted in the product. We will prove that $P(n)$ is true using the second principle of mathematical induction. The basis case $P(1)$ is true since $1 - 1 = 0$ multiplications are required to find the product of x_1, a product with only one factor. Suppose that $P(k)$ is true for $1 \leq k \leq n$. The last multiplication used to find the product of the $n + 1$ distinct real numbers $x_1, x_2, \ldots, x_n, x_{n+1}$ is a multiplication of

the product of the first k of these numbers for some k and the product of the last $n + 1 - k$ of them. By the inductive hypothesis, $k - 1$ multiplications are used to find the product of k of the numbers, no matter how parentheses were inserted in the product of these numbers, and $n - k$ multiplications are used to find the product of the other $n + 1 - k$ of them, no matter how parentheses were inserted in the product of these numbers. Since one more multiplication is required to find the product of all $n + 1$ numbers, the total number of multiplications used equals $(k - 1) + (n - k) + 1 = n$. Hence $P(n + 1)$ is true. The proof is complete.

69.

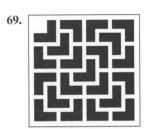

71. Let $P(n)$ be the statement that every $2^n \times 2^n \times 2^n$ chessboard with a $1 \times 1 \times 1$ cube removed can be covered by tiles that are $2 \times 2 \times 2$ cubes each with a $1 \times 1 \times 1$ cube removed. The basis step, $P(1)$, holds since one tile coincides with the solid to be tiled. Now assume that $P(k)$ holds. Now consider a $2^{k+1} \times 2^{k+1} \times 2^{k+1}$ cube with a $1 \times 1 \times 1$ cube removed. Split this object into eight pieces using planes parallel to its faces and running through its center. The missing $1 \times 1 \times 1$ piece occurs in one of these eight pieces. Now position one tile with its center at the center of the large object so that the missing $1 \times 1 \times 1$ cube lies in the octant in which the large object is missing a $1 \times 1 \times 1$ cube. This creates eight $2^k \times 2^k \times 2^k$ cubes, each missing a $1 \times 1 \times 1$ cube. By the inductive hypothesis we can fill each of these eight objects with tiles. Putting these tilings together produces the desired tiling.

73.

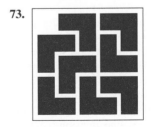

75. Assume that $a = dq + r = dq' + r'$ such that $0 \leq r < d$ and $0 \leq r' < d$. Then $d(q - q') = r' - r$. It follows that d divides $r' - r$. Since $-d < r' - r < d$, we have $r' - r = 0$. Hence $r' = r$. It follows that $q = q'$. **77.** This is a paradox caused by self-reference. The answer is clearly "no." There are a finite number of English words, so only a finite number of strings of 15 words or less; therefore only a finite number of positive integers can be so described, not all of them.

SECTION 3.4

1. a) $f(1) = 3$, $f(2) = 5$, $f(3) = 7$, $f(4) = 9$ **b)** $f(1) = 3, f(2) = 9, f(3) = 27, f(4) = 81$ **c)** $f(1) = 2$, $f(2) = 4$, $f(3) = 16$, $f(4) = 65{,}536$ **d)** $f(1) = 3$, $f(2) = 13, f(3) = 183, f(4) = 33{,}673$ **3. a)** $f(2) = -1$, $f(3) = 5, f(4) = 2, f(5) = 17$ **b)** $f(2) = -4, f(3) = 32$, $f(4) = -4096$, $f(5) = 536{,}870{,}912$ **c)** $f(2) = 8$, $f(3) = 176$, $f(4) = 92{,}672$, $f(5) = 25{,}764{,}174{,}848$ **d)** $f(2) = -\frac{1}{2}$, $f(3) = -4$, $f(4) = \frac{1}{8}$, $f(5) = -32$ **5. a)** Not valid **b)** $f(n) = 1 - n$. *Basis step:* $f(0) = 1 = 1 - 0$. *Inductive step:* if $f(k) = 1 - k$, then $f(k + 1) = f(k) - 1 = 1 - k - 1 = 1 - (k + 1)$. **c)** $f(n) = 4 - n$ if $n > 0$, and $f(0) = 2$. *Basis step:* $f(0) = 2$ and $f(1) = 3 = 4 - 1$. *Inductive step* (with $k \geq 1$): $f(k + 1) = f(k) - 1 = (4 - k) - 1 = 4 - (k + 1)$. **d)** $f(n) = 2^{\lfloor (n+1)/2 \rfloor}$. *Basis step:* $f(0) = 1 = 2^{\lfloor (0+1)/2 \rfloor}$ and $f(1) = 2 = 2^{\lfloor (1+1)/2 \rfloor}$. *Inductive step* (with $k \geq 1$): $f(k + 1) = 2f(k - 1) = 2 \cdot 2^{\lfloor k/2 \rfloor} = 2^{\lfloor k/2 \rfloor + 1} = 2^{\lfloor ((k+1)+1)/2 \rfloor}$. **e)** $f(n) = 3^n$. *Basis step:* Trivial. *Inductive step:* For odd n, $f(n) = 3f(n - 1) = 3 \cdot 3^{n-1} = 3^n$; and for even $n > 1$, $f(n) = 9f(n - 2) = 9 \cdot 3^{n-2} = 3^n$. **7.** There are many possible correct answers. We will supply relatively simple ones. **a)** $a_{n+1} = a_n + 6$ for $n \geq 1$ and $a_1 = 6$ **b)** $a_{n+1} = a_n + 2$ for $n \geq 1$ and $a_1 = 3$ **c)** $a_{n+1} = 10a_n$ for $n \geq 1$ and $a_1 = 10$ **d)** $a_{n+1} = a_n$ for $n \geq 1$ and $a_1 = 5$ **9.** $F(0) = 0, F(n) = F(n - 1) + n$ for $n \geq 1$ **11.** $P_m(0) = 0, P_m(n + 1) = P_m(n) + m$ **13.** Let $P(n)$ be "$f_1 + f_3 + \cdots + f_{2n-1} = f_{2n}$." *Basis step:* $P(1)$ is true since $f_1 = 1 = f_2$. *Inductive step:* Assume that $P(k)$ is true. Then $f_1 + f_3 + \cdots + f_{2k-1} + f_{2k+1} = f_{2k} + f_{2k+1} = f_{2k+2} = f_{2(k+1)}$. **15.** *Basis step:* $f_0 f_1 + f_1 f_2 = 0 \cdot 1 + 1 \cdot 1 = 1^2 = f_2^2$. *Inductive step:* Assume that $f_0 f_1 + f_1 f_2 + \cdots + f_{2k-1} f_{2k} = f_{2k}^2$. Then $f_0 f_1 + f_1 f_2 + \cdots + f_{2k-1} f_{2k} + f_{2k} f_{2k+1} + f_{2k+1} f_{2k+2} = f_{2k}^2 + f_{2k} f_{2k+1} + f_{2k+1} f_{2k+2} = f_{2k}(f_{2k} + f_{2k+1}) + f_{2k+1} f_{2k+2} = f_{2k} f_{2k+2} + f_{2k+1} f_{2k+2} = (f_{2k} + f_{2k+1}) f_{2k+2} = f_{2k+2}^2$. **17.** The number of divisions used by the Euclidean algorithm to find $\gcd(f_{n+1}, f_n)$ is 0 for $n = 0$, 1 for $n = 1$, and $n - 1$ for $n \geq 2$. To prove this result for $n \geq 2$ we use mathematical induction. For $n = 2$, one division shows that $\gcd(f_3, f_2) = \gcd(2, 1) = \gcd(1, 0) = 1$. Now assume that $k - 1$ divisions are used to find $\gcd(f_{k+1}, f_k)$. To find $\gcd(f_{k+2}, f_{k+1})$, first divide f_{k+2} by f_{k+1} to obtain $f_{k+2} = 1 \cdot f_{k+1} + f_k$. After one division we have $\gcd(f_{k+2}, f_{k+1}) = \gcd(f_{k+1}, f_k)$. By the inductive hypothesis it follows that exactly $k - 1$ more divisions are required. This shows that k divisions are required to find $\gcd(f_{k+2}, f_{k+1})$, finishing the inductive proof. **19.** $|A| = -1$. Hence $|A^n| = (-1)^n$. It follows that $f_{n+1} f_{n-1} - f_n^2 = (-1)^n$. **21. a)** Proof by induction. *Basis step:* For $n = 1$, $\max(-a_1) = -a_1 = -\min(a_1)$. For $n = 2$, there are two

cases. If $a_2 \geq a_1$ then $-a_1 \geq -a_2$, so $\max(-a_1, -a_2) = -a_1 = -\min(a_1, a_2)$. If $a_2 < a_1$, then $-a_1 < -a_2$, so $\max(-a_1, -a_2) = -a_2 = -\min(a_1, a_2)$. *Inductive step:* Assume true for k with $k \geq 2$. Then $\max(-a_1, -a_2, \ldots, -a_k, -a_{k+1}) = \max(\max(-a_1, \ldots, -a_k), -a_{k+1}) = \max(-\min(a_1, \ldots, a_k), -a_{k+1}) = -\min(\min(a_1, \ldots, a_k), a_{k+1}) = -\min(a_1, \ldots, a_{k+1})$. **b)** Proof by mathematical induction. *Basis step:* For $n = 1$, the result is the identity $a_1 + b_1 = a_1 + b_1$. For $n = 2$, first consider the case in which $a_1 + b_1 \geq a_2 + b_2$. Then $\max(a_1 + b_1, a_2 + b_2) = a_1 + b_1$. Also note that $a_1 \leq \max(a_1, a_2)$ and $b_1 \leq \max(b_1, b_2)$, so that $a_1 + b_1 \leq \max(a_1, a_2) + \max(b_1, b_2)$. Therefore $\max(a_1 + b_1, a_2 + b_2) = a_1 + b_1 \leq \max(a_1, a_2) + \max(b_1, b_2)$. The case with $a_1 + b_1 < a_2 + b_2$ is similar. *Inductive step:* Assume that the result is true for k. Then $\max(a_1 + b_1, a_2 + b_2, \ldots, a_k + b_k, a_{k+1} + b_{k+1}) = \max(\max(a_1 + b_1, a_2 + b_2, \ldots, a_k + b_k), a_{k+1} + b_{k+1}) \leq \max(\max(a_1, a_2, \ldots, a_k) + \max(b_1, b_2, \ldots, b_k), a_{k+1} + b_{k+1}) \leq \max(\max(a_1, a_2, \ldots, a_k), a_{k+1}) + \max(\max(b_1, b_2, \ldots, b_k), b_{k+1}) = \max(a_1, a_2, \ldots, a_k, a_{k+1}) + \max(b_1, b_2, \ldots, b_k, b_{k+1})$. **c)** Same as (b), but replace every occurrence of "max" by "min" and invert each inequality. **23.** $5 \in S$, and $x + y \in S$ if $x, y \in S$. **25. a)** $0 \in S$, and if $x \in S$, then $x + 2 \in S$ and $x - 2 \in S$. **b)** $2 \in S$, and if $x \in S$, then $x + 3 \in S$. **c)** $1 \in S, 2 \in S, 3 \in S, 4 \in S$, and if $x \in S$, then $x + 5 \in S$. **27. a)** $(0, 1)$, $(1, 1)$, $(2, 1)$; $(0, 2)$, $(1, 2)$, $(2, 2)$, $(3, 2)$, $(4, 2)$; $(0, 3)$, $(1, 3)$, $(2, 3)$, $(3, 3)$, $(4, 3)$, $(5, 3)$, $(6, 3)$; $(0, 4)$, $(1, 4)$, $(2, 4)$, $(3, 4)$, $(4, 4)$, $(5, 4)$, $(6, 4)$, $(7, 4)$, $(8, 4)$ **b)** Let $P(n)$ be the statement that $a \leq 2b$ whenever $(a, b) \in S$ is obtained by n applications of the recursive step. *Basis step:* $P(0)$ is true, since the only element of S obtained with no applications of the recursive step is $(0, 0)$, and indeed $0 \leq 2 \cdot 0$. *Inductive step:* Assume that $a \leq 2b$ whenever $(a, b) \in S$ is obtained by k or fewer applications of the recursive step, and consider an element obtained with $k + 1$ applications of the recursive step. Since the final application of the recursive step to an element (a, b) must be applied to an element obtained with fewer applications of the recursive step, we know that $a \leq 2b$. Add $0 \leq 2$, $1 \leq 2$, and $2 \leq 2$, respectively, to obtain $a \leq 2(b + 1)$, $a + 1 \leq 2(b + 1)$, and $a + 2 \leq 2(b + 1)$, as desired. **c)** This holds for the basis step, since $0 \leq 0$. If this holds for (a, b), then it also holds for the elements obtained from (a, b) in the recursive step, since adding $0 \leq 2$, $1 \leq 2$, and $2 \leq 2$, respectively, to $a \leq 2b$ yields $a \leq 2(b + 1)$, $a + 1 \leq 2(b + 1)$, and $a + 2 \leq 2(b + 1)$. **29. a)** Define S by $(1, 1) \in S$, and if $(a, b) \in S$, then $(a + 2, b) \in S$, $(a, b + 2) \in S$, and $(a + 1, b + 1) \in S$. All elements put in S satisfy the condition, because $(1, 1)$ has an even sum of coordinates, and if (a, b) has an even sum of coordinates, then so do $(a + 2, b)$, $(a, b + 2)$, and $(a + 1, b + 1)$. Conversely, we show by induction on the sum of the coordinates that if $a + b$ is even, then $(a, b) \in S$. If the sum is 2, then $(a, b) = (1, 1)$, and the basis step put (a, b) into S. Otherwise the sum is at least 4, and at least one of $(a-2, b)$, $(a, b-2)$, and $(a-1, b-1)$ must have positive integer coordinates whose sum is an even number smaller than $a + b$, and therefore must be in S. Then one application of the recursive step shows that $(a, b) \in S$. **b)** Define S by $(1, 1)$, $(1, 2)$, and $(2, 1)$ are in S, and if $(a, b) \in S$, then $(a+2, b)$ and $(a, b+2)$ are in S. To prove that our definition works, we note first that $(1, 1), (1, 2)$, and $(2, 1)$ all have an odd coordinate, and if (a, b) has an odd coordinate, then so do $(a + 2, b)$ and $(a, b + 2)$. Conversely, we show by induction on the sum of the coordinates that if (a, b) has at least one odd coordinate, then $(a, b) \in S$. If $(a, b) = (1, 1)$ or $(a, b) = (1, 2)$ or $(a, b) = (2, 1)$, then the basis step put (a, b) into S. Otherwise either a or b is at least 3, so at least one of $(a - 2, b)$ and $(a, b - 2)$ must have positive integer coordinates whose sum is smaller than $a + b$, and therefore must be in S. Then one application of the recursive step shows that $(a, b) \in S$. **c)** $(1, 6) \in S$ and $(2, 3) \in S$, and if $(a, b) \in S$, then $(a + 2, b) \in S$ and $(a, b + 6) \in S$. To prove that our definition works, we note first that $(1, 6)$ and $(2, 3)$ satisfy the condition, and if (a, b) satisfies the condition, then so do $(a + 2, b)$ and $(a, b + 6)$. Conversely we show by induction on the sum of the coordinates that if (a, b) satisfies the condition, then $(a, b) \in S$. For sums 5 and 7, the only points are $(1, 6)$, which the basis step put into S, $(2, 3)$, which the basis step put into S, and $(4, 3) = (2+2, 3)$, which is in S by one application of the recursive definition. For a sum greater than 7, either $a \geq 3$, or $a \leq 2$ and $b \geq 9$, in which case either $(a - 2, b)$ or $(a, b - 6)$ must have positive integer coordinates whose sum is smaller than $a + b$ and satisfy the condition for being in S. Then one application of the recursive step shows that $(a, b) \in S$. **31.** If x is a set or a variable representing a set, then x is a well-formed formula. If x and y are well-formed formulae, then so are $\overline{x}, (x \cup y), (x \cap y)$, and $(x - y)$. **33. a)** If $x \in D = \{0, 1, 2, 3, 4, 5, 6, 7, 8, 9\}$, then $m(x) = x$; if $s = tx$, where $t \in D^*$ and $x \in D$, then $m(s) = \min(m(s), x)$. **b)** Let $t = wx$, where $w \in D^*$ and $x \in D$. If $w = \lambda$, then $m(st) = m(sx) = \min(m(s), x) = \min(m(s), m(x))$ by the recursive step and the basis step of the definition of m. Otherwise, $m(st) = m((sw)x) = \min(m(sw), x)$ by the definition of m. Now $m(sw) = \min(m(s), m(w))$ by the inductive hypothesis of the structural induction, so $m(st) = \min(\min(m(s), m(w)), x) = \min(m(s), \min(m(w), x))$ by the meaning of min. But $\min(m(w), x) = m(wx) = m(t)$ by the recursive step of the definition of m. Thus $m(st) = \min(m(s), m(t))$. **35.** $\lambda^R = \lambda$ and $(ux)^R = xu^R$ for $x \in \Sigma, u \in \Sigma^*$. **37.** $w^0 = \lambda$ and $w^{n+1} = ww^n$. **39.** When the string consists of n 0s followed by n 1s for some nonnegative integer n **41.** Let $P(i)$ be "$l(w^i) = i \cdot l(w)$." $P(0)$ is true since $l(w^0) = 0 = 0 \cdot l(w)$. Assume $P(i)$ is true. Then $l(w^{i+1}) = l(ww^i) = l(w) + l(w^i) = l(w) + i \cdot l(w) = (i + 1) \cdot l(w)$. **43.** *Basis step:* For

the full binary tree consisting of just a root the result is true since $n(T) = 1$ and $h(T) = 0$, and $1 \geq 2 \cdot 0 + 1$. *Inductive step:* Assume that $n(T_1) \geq 2h(T_1) + 1$ and $n(T_2) \geq 2h(T_2) + 1$. By the recursive definitions of $n(T)$ and $h(T)$, we have $n(T) = 1 + n(T_1) + n(T_2)$ and $h(T) = 1 + \max(h(T_1), h(T_2))$. Therefore $n(T) = 1 + n(T_1) + n(T_2) \geq 1 + 2h(T_1) + 1 + 2h(T_2) + 1 \geq 1 + 2 \cdot \max(h(T_1), h(T_2)) + 2 = 1 + 2(\max(h(T_1), h(T_2)) + 1) = 1 + 2h(T)$. **45.** *Basis step:* $a_{0,0} = 0 = 0 + 0$. *Inductive step:* Assume that $a_{m',n'} = m' + n'$ whenever (m', n') is less than (m, n) in the lexicographic ordering of $\mathbf{N} \times \mathbf{N}$. If $n = 0$ then $a_{m,n} = a_{m-1,n} + 1 = m - 1 + n + 1 = m + n$. If $n > 0$, then $a_{m,n} = a_{m,n-1} + 1 = m + n - 1 + 1 = m + n$. **47. a)** $P_{m,m} = P_m$ since a number exceeding m cannot be used in a partition of m. **b)** Since there is only one way to partition 1, namely, $1 = 1$, it follows that $P_{1,n} = 1$. Since there is only one way to partition m into 1s, $P_{m,1} = 1$. When $n > m$ it follows that $P_{m,n} = P_{m,m}$ since a number exceeding m cannot be used. $P_{m,m} = 1 + P_{m,m-1}$ since one extra partition, namely, $m = m$, arises when m is allowed in the partition. $P_{m,n} = P_{m,n-1} + P_{m-n,n}$ if $m > n$ since a partition of m into integers not exceeding n either does not use any ns and hence is counted in $P_{m,n-1}$ or else uses an n and a partition of $m - n$, and hence is counted in $P_{m-n,n}$. **c)** $P_5 = 7, P_6 = 11$ **49.** Let $P(n)$ be "$A(n, 2) = 4$." *Basis step:* $P(1)$ is true since $A(1, 2) = A(0, A(1, 1)) = A(0, 2) = 2 \cdot 2 = 4$. *Inductive step:* Assume that $P(n)$ is true, that is, $A(n, 2) = 4$. Then $A(n + 1, 2) = A(n, A(n + 1, 1)) = A(n, 2) = 4$. **51. a)** 16 **b)** 65,536 **53.** Use a double induction argument to prove the stronger statement: $A(m, k) > A(m, l)$ when $k > l$. *Basis step:* When $m = 0$ the statement is true since $k > l$ implies that $A(0, k) = 2k > 2l = A(0, l)$. *Inductive step:* Assume that $A(m, x) > A(m, y)$ for all nonnegative integers x and y with $x > y$. We will show that this implies that $A(m + 1, k) > A(m + 1, l)$ if $k > l$. *Basis steps:* When $l = 0$ and $k > 0$, $A(m + 1, l) = 0$ and either $A(m + 1, k) = 2$ or $A(m + 1, k) = A(m, A(m + 1, k - 1))$. If $m = 0$, this is $2A(1, k - 1) = 2^k$. If $m > 0$ this is greater than 0 by the inductive hypothesis. In all cases $A(m + 1, k) > 0$, and in fact, $A(m + 1, k) \geq 2$. If $l = 1$ and $k > 1$ then $A(m + 1, l) = 2$ and $A(m + 1, k) = A(m, A(m + 1, k - 1))$, with $A(m + 1, k - 1) \geq 2$. Hence by the inductive hypothesis, $A(m, A(m + 1, k - 1)) \geq A(m, 2) > A(m, 1) = 2$. *Inductive step:* Assume that $A(m + 1, r) > A(m + 1, s)$ for all $r > s, s = 0, 1, \ldots, l$. Then if $k + 1 > l + 1$ it follows that $A(m+1, k+1) = A(m, A(m+1, k)) > A(m, A(m+1, k)) = A(m + 1, l + 1)$. **55.** From Exercise 54 it follows that $A(i, j) \geq A(i - 1, j) \geq \cdots \geq A(0, j) = 2j \geq j$. **57.** Let $P(n)$ be "$F(n)$ is well-defined." Then $P(0)$ is true since $F(0)$ is specified. Assume that $P(k)$ is true for all $k < n$. Then $F(n)$ is well-defined at n since $F(n)$ is given in terms of $F(0), F(1), \ldots, F(n - 1)$. So $P(n)$ is true for all integers n. **59. a)** The value of $F(1)$ is ambigu-

ous.   **b)** $F(2)$ is not defined, since $F(0)$ is not defined.
c) $F(3)$ is ambiguous and $F(4)$ is not defined since $F(\frac{4}{3})$
makes no sense.   **d)** The definition of $F(1)$ is ambiguous
since both the second and third clause seem to apply.
e) $F(2)$ cannot be computed since trying to compute
$F(2)$ gives $F(2) = 1 + F(F(1)) = 1 + F(2)$.   **61. a)** 1
b) 2   **c)** 3   **d)** 3   **e)** 4   **f)** 4   **g)** 5   **63.** $f_0^*(n) = \lceil n/a \rceil$
65. $f_2^*(n) = \lceil \log \log n \rceil$ for $n \geq 2$, $f_2^*(1) = 0$

SECTION 3.5

1. procedure *mult*(*n*: positive integer, *x*: integer)
  if $n = 1$ **then** $mult(n, x) := x$
  else $mult(n, x) := x + mult(n - 1, x)$
3. procedure *sum of odds* (*n*: positive integer)
  if $n = 1$ **then** *sum of odds* $(n) := 1$
  else *sum of odds*$(n) := $*sum of odds* $(n - 1) + 2n - 1$
5. procedure *smallest*($a_1, \ldots, a_n$: integers)
  if $n = 1$ **then** $smallest(a_1, \ldots, a_n) = a_1$
  else $smallest(a_1, \ldots, a_n) :=$
    $\min(smallest(a_1, \ldots, a_{n-1}), a_n)$
7. procedure *modfactorial*(*n, m*: positive integers)
  if $n = 1$ **then** $modfactorial(n, m) := 1$
  else $modfactorial(n, m) :=$
    $(n \cdot modfactorial(n - 1, m))$ **mod** m
9. procedure gcd(*a, b*: nonnegative integers)
  {$a < b$ assumed to hold}
  if $a = 0$ **then** $gcd(a, b) := b$
  else if $a = b - a$ **then** $gcd(a, b) := a$
  else if $a < b - a$ **then** $gcd(a, b) := gcd(a, b - a)$
  else $gcd(a, b) := gcd(b - a, a)$
11. procedure *multiply*(*x, y* : nonnegative integers)
  if $y = 0$ **then** $multiply(x, y) := 0$
  else if y is even **then**
      $multiply(x, y) := 2x \cdot multiply(x, y/2)$
  else $multiply(x, y) := 2x \cdot multiply(x, (y - 1)/2) + x$
13. If $n = 1$, then $nx = x$, and the algorithm correctly returns x. Assume that the algorithm correctly computes kx.
To compute $(k + 1)x$ it recursively computes the product
of $k + 1 - 1 = k$ and x, and then adds x. By the inductive hypothesis, it computes that product correctly, so the
answer returned is $kx + x = (k + 1)x$, which is correct.
15. If $n = 0$, then the first line of the program correctly
returns the function value 1. Assume that the program
works correctly for $n = k$. If $n = k + 1$, then the **else**
clause is executed, and the value a times *power*(a, k) is
returned. Since the latter is a^k by the inductive hypothesis, this equals $a \cdot a^k = a^{k+1}$, which is correct.   **17.** n
multiplications versus 2^n **19.** $O(\log n)$ versus n
21. procedure *a*(*n*: nonnegative integer)
  if $n = 0$ **then** $a(n) := 1$
  else if $n = 1$ **then** $a(n) := 2$
  else $a(n) := a(n - 1) * a(n - 2)$

23. Iterative
25. procedure *iterative*(*n*: nonnegative integer)
  if $n = 0$ **then** $z := 1$
  else if $n = 1$ **then** $z := 2$
  else
  begin
   $x := 1$
   $y := 2$
   $z := 3$
   for $i := 1$ **to** $n - 2$
   begin
    $w := x + y + z$
    $x := y$
    $y := z$
    $z := w$
   end
  end
  {z is the nth term of the sequence}
27. We first give a recursive procedure and then an iterative procedure.
procedure *r*(*n*: nonnegative integer)
if $n < 3$ **then** $r(n) := 2n + 1$
else $r(n) = r(n - 1) \cdot (r(n - 2))^2 \cdot (r(n - 3))^3$

procedure *i*(*n* : nonnegative integer)
if $n = 0$ **then** $z := 1$
else if $n = 1$ **then** $z := 3$
else
begin
 $x := 1$
 $y := 3$
 $z := 5$
 for $i := 1$ **to** $n - 2$
 begin
  $w := z * y^2 * x^3$
  $x := y$
  $y := z$
  $z := w$
 end
end
{z is the nth term of the sequence}
The iterative version is more efficient.
29. procedure *reverse* (*w*: bit string)
  $n := \text{length}(w)$
  if $n \leq 1$ **then** $reverse(w) := w$
  else $reverse(w) :=$
   $substr(w, n, n)reverse(substr(w, 1, n - 1))$
  {$substr(w, a, b)$ is the substring of w consisting of
   the symbols in the ath through bth positions}
31. procedure A(*m, n*: nonnegative integers)
  if $m = 0$ **then** $A(m, n) := 2n$
  else if $n = 0$ **then** $A(m, n) := 0$
  else if $n = 1$ **then** $A(m, n) := 2$
  else $A(m, n) := A(m - 1, A(m, n - 1))$

33.

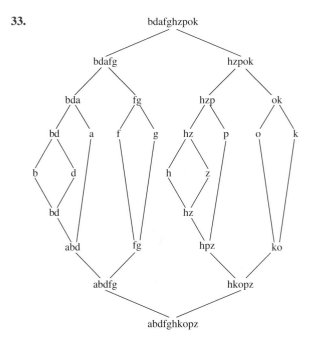

35. Let the two lists be $1, 2, \ldots, m - 1, m + n - 1$ and m, $m + 1, \ldots, m + n - 2, m + n$, respectively.   **37.** If $n = 1$, then the algorithm does nothing, which is correct, since a list with one element is already sorted. Assume that the algorithm works correctly for $n = 1$ through $n = k$. If $n = k + 1$, then the list is split into two lists, L_1 and L_2. By the inductive hypothesis, *mergesort* correctly sorts each of these sublists; furthermore *merge* correctly merges two sorted lists into one, since with each comparison, the smallest element in $L_1 \cup L_2$ not yet put into L is put there. **39.** $O(n)$ **41.** 6   **43.** $O(n^2)$

SECTION 3.6

1. Suppose that $x = 0$. The program segment first assigns the value 1 to y and then assigns the value $x + y = 0 + 1 = 1$ to z.   **3.** Suppose that $y = 3$. The program segment assigns the value 2 to x and then assigns the value $x + y = 2 + 3 = 5$ to z. Since $y = 3 > 0$ it then assigns the value $z + 1 = 5 + 1 = 6$ to z.

5.
$$(p \wedge condition1)\{S_1\}q$$
$$(p \wedge \neg condition1 \wedge condition2)\{S_2\}q$$

.

.

.

$$(p \wedge \neg condition1 \wedge \neg condition2$$
$$\cdots \wedge \neg condition(n - 1)\{S_n\}q$$

$$\therefore p\{\textbf{if } condition1 \textbf{ then } S_1;$$
$$\textbf{else if } condition2 \textbf{ then } S_2; \ldots; \textbf{ else } S_n\}q$$

7. We will show that p : "*power* $= x^{i-1}$ and $i \leq n + 1$" is a loop invariant. Note that p is true initially, since before the loop starts, $i = 1$ and *power* $= 1 = x^0 = x^{1-1}$. Next, we must show that if p is true and $i \leq n$ after an execution of the loop, then p remains true after one more execution. The loop increments i by 1. Hence, since $i \leq n$ before this pass, $i \leq n + 1$ after this pass. Also the loop assigns *power*$\cdot x$ to *power*. By the inductive hypothesis we see that *power* is assigned the value $x^{i-1} \cdot x = x^i$. Hence p remains true. Furthermore, the loop terminates after n traversals of the loop with $i = n + 1$ since i is assigned the value 1 prior to entering the loop, is incremented by 1 on each pass, and the loop terminates when $i > n$. Consequently, at termination *power* $= x^n$, as desired.   **9.** Suppose that p is "m and n are integers." Then if the condition $n < 0$ is true, $a = -n = |n|$ after S_1 is executed. If the condition $n < 0$ is false, then $a = n = |n|$ after S_1 is executed. Hence $p\{S_1\}q$ is true where q is $p \wedge (a = |n|)$. Since S_2 assigns the value 0 to both k and x, it is clear that $q\{S_2\}r$ is true where r is $q \wedge (k = 0) \wedge (x = 0)$. Suppose that r is true. Let $P(k)$ be "$x = mk$ and $k \leq a$." We can show that $P(k)$ is a loop invariant for the loop in S_3. $P(0)$ is true, since before the loop is entered $x = 0 = m \cdot 0$ and $0 \leq a$. Now assume $P(k)$ is true and $k < a$. Then $P(k + 1)$ is true since x is assigned the value $x + m = mk + m = m(k + 1)$. The loop terminates when $k = a$, and at that point $x = ma$. Hence $r\{S_3\}s$ is true where s is "$a = |n|$ and $x = ma$." Now assume that s is true. Then if $n < 0$ it follows that $a = -n$, so $x = -mn$. In this case S_4 assigns $-x = mn$ to *product*. If $n > 0$ then $x = ma = mn$, so S_4 assigns mn to *product*. Hence $s\{S_4\}t$ is true.   **11.** Suppose that the initial assertion p is true. Then since $p\{S\}q_0$ is true, q_0 is true after the segment S is executed. Since $q_0 \to q_1$ is true, it also follows that q_1 is true after S is executed. Hence $p\{S\}q_1$ is true.   **13.** We will use the proposition p, "$\gcd(a, b) = \gcd(x, y)$ and $y \geq 0$," as the loop invariant. Note that p is true before the loop is entered, since at that point $x = a, y = b$, and y is a positive integer, using the initial assertion. Now assume that p is true and $y > 0$; then the loop will be executed again. Inside the loop, x and y are replaced by y and x **mod** y, respectively. By Lemma 1 of Section 2.5, $\gcd(x, y) = \gcd(y, x \textbf{ mod } y)$. Therefore, after execution of the loop, the value of $\gcd(x, y)$ is the same as it was before. Moreover, since y is the remainder, it is at least 0. Hence p remains true, so that it is a loop invariant. Furthermore, if the loop terminates, then $y = 0$. In this case, we have $\gcd(x, y) = x$, the final assertion. Therefore the program, which gives x as its output, has correctly computed $\gcd(a, b)$. Finally, we can prove the loop must terminate, since each iteration causes the value of y to decrease by at least 1. Therefore, the loop can be iterated at most b times.

SUPPLEMENTARY EXERCISES

1. $2^{2^5} + 1 = 4294967297 = 641 \cdot 6700417$ **3.** Proof by contradiction: Suppose that $\sqrt[3]{7} = p/q$, where p and q are positive integers with no common factor. Then $p^3 = 7q^3$. Therefore 7 is a factor of p^3, and hence a factor of p (if it were not, then the prime factorization of p^3 could not contain any 7s). Write $p = 7k$ for some positive integer k. This gives $49k^3 = q^3$. Hence 7 is a factor of q (by the same reasoning as before). But this contradicts the fact that p and q had no nontrivial common factors. Therefore $\sqrt[3]{7}$ is not a rational number. **5.** The only possible summands are $1^4 = 1, 2^4 = 16, 3^4 = 81, 4^4 = 256$, and $5^4 = 625$. Clearly no two of these sum to 1000. **7.** Let $z = x - y$, so that $x = y + z$. Then the inequality we are trying to prove becomes $|z| \geq |y + z| - |y|$, which is equivalent to $|y| + |z| \geq |y + z|$. But this is the triangle inequality proved as Exercise 35 of Section 1.5. **9.** Let $P(x)$ be as in the hint. Then $P(x)$ is a polynomial (of degree $n - 1$, in fact); and if $x = x_m$, then $\prod_{i \neq j} (x - x_j)/(x_i - x_j) = 0$ unless $i = m$. Thus $P(x_m) = \prod_{j \neq m} y_m (x_m - x_j)/(x_m - x_j) = 1 \cdot y_m = y_m$. **11.** Let $P(n)$ be "$1 \cdot 1 + 2 \cdot 2 + \cdots + n \cdot 2^{n-1} = (n-1)2^n + 1$." *Basis step:* $P(1)$ is true since $1 \cdot 1 = 1 = (1-1)2^1 + 1$. *Inductive step:* Assume that $P(k)$ is true. Then $1 \cdot 1 + 2 \cdot 2 + \cdots + k \cdot 2^{k-1} + (k+1) \cdot 2^k = (k-1)2^k + 1 + (k+1)2^k = 2k \cdot 2^k + 1 = ((k+1) - 1)2^{k+1} + 1$. **13.** Let $P(n)$ be "$1/1 \cdot 4 + \cdots + 1/[(3n-2)(3n+1)] = n/(3n+1)$." *Basis step:* $P(1)$ is true since $1/1 \cdot 4 = 1/4$. *Inductive step:* Assume $P(k)$ is true. Then $1/1 \cdot 4 + \cdots + 1/[(3k-2)(3k+1)] + 1/[(3k+1)(3k+4)] = k/(3k+1) + 1/[(3k+1)(3k+4)] = [k(3k+4)+1]/[(3k+1)(3k+4)] = [(3k+1)(k+1)]/[(3k+1)(3k+4)] = (k+1)/(3k+4)$. **15.** Let $P(n)$ be "$2^n > n^3$." *Basis step:* $P(10)$ is true since $1024 > 1000$. *Inductive step:* Assume $P(k)$ is true. Then $(k+1)^3 = k^3 + 3k^2 + 3k + 1 \leq k^3 + 9k^2 \leq k^3 + k^3 = 2k^3 < 2 \cdot 2^k = 2^{k+1}$. **17.** Let $P(n)$ be "$a - b$ is a factor of $a^n - b^n$." *Basis step:* $P(1)$ is trivially true. Assume $P(k)$ is true. Then $a^{k+1} - b^{k+1} = a^{k+1} - ab^k + ab^k - b^{k+1} = a(a^k - b^k) + b^k(a - b)$. Then since $a - b$ is a factor of $a^k - b^k$ and $a - b$ is a factor of $a - b$, it follows that $a - b$ is a factor of $a^{k+1} - b^{k+1}$. **19.** Let $P(n)$ be "$a + (a+d) + \cdots + (a+nd) = (n+1)(2a+nd)/2$." *Basis step:* $P(1)$ is true since $a + (a + d) = 2a + d = 2(2a+d)/2$. *Inductive step:* Assume that $P(k)$ is true. Then $a + (a+d) + \cdots + (a+kd) + (a+(k+1)d) = (k+1)(2a+kd)/2 + a + (k+1)d = \frac{1}{2}[2ak + 2a + k^2d + kd + 2a + 2kd + 2d] = \frac{1}{2}[2ak + 4a + k^2d + 3kd + 2d] = \frac{1}{2}(k+2)(2a+(k+1)d)$. **21.** *Basis step:* This is true for $n = 1$, since $5/6 = 10/12$. *Inductive step:* Assume that the equation holds for $n = k$, and consider $n = k + 1$. Then $\sum_{i=1}^{k+1} \frac{i+4}{i(i+1)(i+2)} = \sum_{i=1}^{k} \frac{i+4}{i(i+1)(i+2)} + \frac{k+5}{(k+1)(k+2)(k+3)} = \frac{k(3k+7)}{2(k+1)(k+2)} + \frac{k+5}{(k+1)(k+2)(k+3)}$ (by the inductive hypothesis) $= \frac{1}{(k+1)(k+2)} \cdot (\frac{k(3k+7)}{2} + \frac{k+5}{k+3}) = \frac{1}{2(k+1)(k+2)(k+3)} \cdot (k(3k+7)(k+3) + 2(k+5)) = \frac{1}{2(k+1)(k+2)(k+3)} \cdot (3k^3 + 16k^2 + 23k + 10) = \frac{1}{2(k+1)(k+2)(k+3)} \cdot (3k+10)(k+1)^2 = \frac{1}{2(k+2)(k+3)} \cdot (3k+10)(k+1) = \frac{(k+1)(3(k+1)+7)}{2((k+1)+1)((k+1)+2)}$, as desired. **23.** *Basis step:* The statement is true for $n = 1$, since the derivative of $g(x) = xe^x$ is $x \cdot e^x + e^x = (x+1)e^x$ by the product rule. *Inductive step:* Assume that the statement is true for $n = k$, i.e., the kth derivative is given by $g^{(k)} = (x+k)e^x$. Differen-

tiating by the product rule gives the $(k+1)$st derivative: $g^{(k+1)} = (x+k)e^x + e^x = (x + (k+1))e^x$, as desired. **25.** We will use strong induction to show that f_n is even if $n \equiv 0 \pmod 3$ and is odd otherwise. *Basis step:* This follows since $f_0 = 0$ is even and $f_1 = 1$ is odd. *Inductive step:* Assume that if $j \leq k$, then f_j is even if $j \equiv 0 \pmod 3$ and is odd otherwise. Now suppose $k + 1 \equiv 0 \pmod 3$. Then $f_{k+1} = f_k + f_{k-1}$ is even since f_k and f_{k-1} are both odd. If $k + 1 \equiv 1 \pmod 3$, then $f_{k+1} = f_k + f_{k-1}$ is odd since f_k is even and f_{k-1} is odd. Finally, if $k + 1 \equiv 2 \pmod 3$, then $f_{k+1} = f_k + f_{k-1}$ is odd, since f_k is odd and f_{k-1} is even. This completes the inductive proof. **27.** Let $P(n)$ be the statement that $f_k f_n + f_{k+1} f_{n+1} = f_{n+k+1}$ for every nonnegative integer k. *Basis step:* This consists of showing that $P(0)$ and $P(1)$ both hold. $P(0)$ is true since $f_k f_0 + f_{k+1} f_1 = f_{k+1} \cdot 0 + f_{k+1} \cdot 1 = f_{k+1}$. Since $f_k f_1 + f_{k+1} f_2 = f_k + f_{k+1} = f_{k+2}$, it follows that $P(1)$ is true. *Inductive step:* Now assume that $P(j)$ holds. Then, by the inductive hypothesis and the recursive definition of the Fibonacci numbers, it follows that $f_{k+1} f_{j+1} + f_{k+2} f_{j+2} = f_k(f_{j-1} + f_j) + f_{k+1}(f_j + f_{j+1}) = (f_k f_{j-1} + f_{k+1} f_j) + (f_k f_j + f_{k+1} f_{j+1}) = f_{j-1+k+1} + f_{j+k+1} = f_{j+k+2}$. This shows that $P(j+1)$ is true and completes the proof. **29.** Let $P(n)$ be the statement $l_0^2 + l_1^2 + \cdots + l_n^2 = l_n l_{n+1} + 2$. *Basis step:* $P(0)$ and $P(1)$ both hold since $l_0^2 = 2^2 = 2 \cdot 1 + 2 = l_0 l_1 + 2$ and $l_0^2 + l_1^2 = 2^2 + 1^2 = 1 \cdot 3 + 2 = l_1 l_3 + 2$. *Inductive step:* Assume that $P(k)$ holds. Then by the inductive hypothesis $l_0^2 + l_1^2 + \cdots + l_k^2 + l_{k+1}^2 = l_k l_{k+1} + 2 + l_{k+1}^2 = l_{k+1}(l_k + l_{k+1}) + 2 = l_{k+1} l_{k+2} + 2$. This shows that $P(k+1)$ holds and completes the proof. **31.** Let $P(n)$ be the statement that the identity holds for the integer n. *Basis step:* $P(1)$ is obviously true. *Inductive step:* Assume that $P(k)$ is true. Then $\cos(k+1)x + i\sin(k+1)x = \cos(kx+x) + i\sin(kx + x) = \cos kx \cos x - \sin kx \sin x + i(\sin kx \cos x + \cos kx \sin x) = \cos x(\cos kx + i\sin kx)(\cos x + i\sin x) = (\cos x + i\sin x)^k(\cos x + i\sin x) = (\cos x + i\sin x)^{k+1}$. It follows that $P(k+1)$ is true, completing the proof. **33.** Rewrite the right-hand side as $2^{n+1}(n^2 - 2n + 3) - 6$. For $n = 1$ we have $2 = 4 \cdot 2 - 6$. Assume that the equation holds for $n = k$, and consider $n = k + 1$. Then $\sum_{j=1}^{k+1} j^2 2^j = \sum_{j=1}^{k} j^2 2^j + (k+1)^2 2^{k+1} = 2^{k+1}(k^2 - 2k + 3) - 6 + (k^2 + 2k + 1)2^{k+1}$ (by the inductive hypothesis) $= 2^{k+1}(2k^2 + 4) - 6 = 2^{k+2}(k^2 + 2) - 6 = 2^{k+2}((k+1)^2 - 2(k+1) + 3) - 6$. **35.** When $n = 1$ the statement is vacuously true. Assume that the statement is true for $n = k$, and consider $k + 1$ people standing in a line, with a woman first and a man last. If the kth person is a woman, then we have that woman standing in front of the man at the end. If the kth person is a man, then the first k people in line satisfy the conditions of the inductive hypothesis for the first k people in line, so again we can conclude that there is a woman directly in front of a man somewhere in the line. **37.** *Basis step:* When $n = 1$ there is one circle, and we can color the inside blue and the outside red to satisfy the conditions. *Inductive step:* Assume the inductive hypothesis that if there are k circles, then the regions can be 2-colored such that no regions with a common boundary have the same color, and consider a situation with $k + 1$ circles. Remove one of the circles, producing a picture with k circles, and invoke the inductive hypothesis to color it in the prescribed manner. Then replace the removed circle and change the color of every region inside this circle. The resulting figure satisfies the condition, since if two regions have a common boundary, then either that boundary involved the new circle, in which case the regions on either side used to be the same region and now the inside portion is different from the outside, or else the boundary did not involve the new circle, in which case the regions are colored differently because they were colored differently before the new circle was restored. **39.** If $n = 1$ then the equation reads $1 \cdot 1 = 1 \cdot 2/2$, which is true. Assume that the equation is true for n and consider it for $n + 1$. Then $\sum_{j=1}^{n+1}(2j-1)\left(\sum_{k=j}^{n+1}\frac{1}{k}\right) = \sum_{j=1}^{n}(2j-1)\left(\sum_{k=j}^{n+1}\frac{1}{k}\right) + (2(n+1)-1)\cdot\frac{1}{n+1} = \sum_{j=1}^{n}(2j-1)\left(\frac{1}{n+1}+\sum_{k=j}^{n}\frac{1}{k}\right) + \frac{2n+1}{n+1} = \left(\frac{1}{n+1}\sum_{j=1}^{n}(2j-1)\right) + \left(\sum_{j=1}^{n}(2j-1)\sum_{k=j}^{n}\frac{1}{k}\right) + \frac{2n+1}{n+1} = \left(\frac{1}{n+1}\cdot n^2\right) + \frac{n(n+1)}{2} + \frac{2n+1}{n+1}$ (by the inductive hypothesis) $= \frac{2n^2+n(n+1)^2+(4n+2)}{2(n+1)} = \frac{2(n+1)^2+n(n+1)^2}{2(n+1)} = \frac{(n+1)(n+2)}{2}$. **41. a)** 92 **b)** 91 **c)** 91 **d)** 91 **e)** 91 **f)** 91 **43.** The basis step is incorrect since $n \neq 1$ for the sum shown. **45.** Let $P(n)$ be "the plane is divided into $n^2 - n + 2$ regions by n circles if every two of these circles have two common points but no three have a common point." *Basis step:* $P(1)$ is true since a circle divides the plane into $2 = 1^2 - 1 + 2$ regions. *Inductive step:* Assume that $P(k)$ is true, that is, k circles with the specified properties divide the plane into k^2-k+2 regions. Suppose that a $(k + 1)$st circle is added. This circle intersects each of the other k circles in two points so that these points of intersection form $2k$ new arcs, each of which splits an old region. Hence there are $2k$ regions split, which shows that there are $2k$ more regions than there were previously. Hence $k+1$ circles satisfying the specified properties divide the plane into $k^2 - k + 2 + 2k = (k^2 + 2k + 1) - (k + 1) + 2 = (k + 1)^2 - (k + 1) + 2$ regions. **47.** Suppose $\sqrt{2}$ were rational. Then $\sqrt{2} = a/b$, where a and b are positive integers. It follows that the set $S = \{n\sqrt{2} \mid n \in \mathbf{N}\} \cap \mathbf{N}$ is a nonempty set of positive integers, since $b\sqrt{2} = a$ belongs to S. Let t be the least element of S, which exists by the well-ordering property. Then $t = s\sqrt{2}$ for some integer s. We have $t - s = s\sqrt{2}-s = s(\sqrt{2}-1)$, so that $t - s$ is a positive integer since $\sqrt{2} > 1$. Hence $t - s$ belongs to S. This is a contradiction since $t - s = s\sqrt{2}-s < s$. Hence $\sqrt{2}$ is irrational. **49.** Suppose that the well-ordering property were false. Let S be a nonempty set of nonnegative integers that has no least element. Let $P(n)$ be the statement "$i \notin S, i = 0, 1, \ldots, n$." $P(0)$ is true because if $0 \in S$ then S has a least element, namely, 0. Now suppose that $P(n)$ is true. Thus $0 \notin S, 1 \notin S, \ldots, n \notin S$. Clearly $n + 1$ cannot be in S, for if it were, it would be

its least element. Thus $P(n + 1)$ is true. So by the principle of mathematical induction, $n \notin S$ for all nonnegative integers n. Thus $S = \emptyset$, a contradiction. **51. a)** Let $d = \gcd(a_1, a_2, \ldots, a_n)$. Then d is a divisor of each a_i and so must be a divisor of $\gcd(a_{n-1}, a_n)$. Hence d is a common divisor of $a_1, a_2, \ldots, a_{n-2}$, and $\gcd(a_{n-1}, a_n)$. To show it is the greatest common divisor of these numbers, suppose that c is a common divisor of them. Then c is a divisor of a_i for $i = 1, 2, \ldots, n - 2$ and a divisor of $\gcd(a_{n-1}, a_n)$, so that it is a divisor of a_{n-1} and a_n. Hence c is a common divisor of $a_1, a_2, \ldots, a_{n-1}, a_n$. Hence it is a divisor of d, the greatest common divisor of $a_1, a_2, \ldots, a_n$. It follows that d is the greatest common divisor, as claimed. **b)** If $n = 2$, apply the Euclidean algorithm. Otherwise, apply the Euclidean algorithm to a_{n-1} and a_n, obtaining $d = \gcd(a_{n-1}, a_n)$, and then apply the algorithm recursively to $a_1, a_2, \ldots, a_{n-2}, d$. **53.** $f(n) = n^2$. Let $P(n)$ be "$f(n) = n^2$." *Basis step:* $P(1)$ is true since $f(1) = 1 = 1^2$, which follows from the definition of f. *Inductive step:* Assume $f(n) = n^2$. Then $f(n+1) = f((n+1)-1)+2(n+1)-1 = f(n)+2n+1 = n^2+2n+1 = (n+1)^2$. **55. a)** $\lambda, 0, 1, 00, 01, 11, 000, 001, 011, 111, 0000, 0001, 0011, 0111, 1111, 00000, 00001, 00011, 00111, 01111, 11111$ **b)** $S = \{\alpha\beta \mid \alpha$ is a string of m 0s and β is a string of n 1s, $m \geq 0, n \geq 0\}$ **57.** Apply the first recursive step to λ to get $() \in B$. Apply the second recursive step to this string to get $()() \in B$. Apply the first recursive step to this string to get $(()()) \in B$. By Exercise 60, $(())$ is not in B because the number of left parentheses does not equal the number of right parentheses. **59.** $\lambda, (), (()), ()()$ **61. a)** 0 **b)** -2 **c)** 2 **d)** 0 **63. procedure** *generate*(n: nonnegative integer)
  if n is odd **then**
  begin
    $S := S(n - 1); T := T(n - 1)$
  end
  else if $n = 0$ **then**
  begin
    $S := \emptyset; T := \{\lambda\}$
  end
  else
  begin
    $T_1 := T(n - 2); S_1 := S(n - 2)$
    $T := T_1 \cup \{(x) \mid x \in T_1 \cup S_1$ and $l(x) = n - 2\}$
    $S := S_1 \cup \{xy \mid x \in T_1$ and $y \in T_1 \cup S_1$
    and $l(xy) = n\}$
  end $\{T \cup S$ is the set of balanced strings of length at most $n\}$

65. If $x \leq y$ initially, $x := y$ is not executed, so $x \leq y$ is a true final assertion. If $x > y$ initially, then $x := y$ is executed, so $x \leq y$ is again a true final assertion.

67. procedure *zerocount*($a_1, a_2, \ldots, a_n$: list of integers)
  if $n = 1$ **then**
    if $a_1 = 0$ **then** *zerocount*($a_1, a_2, \ldots, a_n$) := 1
    else *zerocount*($a_1, a_2, \ldots, a_n$) := 0
  else

if $a_n = 0$ **then** $zerocount(a_1, a_2, \ldots, a_n) :=$
 $\quad zerocount(a_1, a_2, \ldots, a_{n-1}) + 1$
else $zerocount(a_1, a_2, \ldots, a_n) :=$
 $\quad zerocount(a_1, a_2, \ldots, a_{n-1})$

69. We will prove that $a(n)$ is a natural number and $a(n) \leq n$. This is true for the base case $n = 0$, since $a(0) = 0$. Now assume that $a(n-1)$ is a natural number and $a(n-1) \leq n-1$. Then $a(a(n-1))$ is a applied to a natural number less than or equal to $n-1$. Hence $a(a(n-1))$ is also a natural number less than or equal to $n-1$. Therefore $n - a(a(n-1))$ is n less some natural number less than or equal to $n-1$, which is a natural number less than or equal to n.   **71.** From Exercise 70, $a(n) = \lfloor (n+1)\mu \rfloor$ and $a(n-1) = \lfloor n\mu \rfloor$. Since $\mu < 1$, these two values are equal or they differ by 1. First suppose that $\mu n - \lfloor \mu n \rfloor < 1 - \mu$. This is equivalent to $\mu(n+1) < 1 + \lfloor \mu n \rfloor$. If this is true, then $\lfloor \mu(n+1) \rfloor = \lfloor \mu n \rfloor$. On the other hand, if $\mu n - \lfloor \mu n \rfloor \geq 1 - \mu$, then $\mu(n+1) \geq 1 + \lfloor \mu n \rfloor$, so $\lfloor \mu(n+1) \rfloor = \lfloor \mu n \rfloor + 1$, as desired.   **73.** $f(0) = 1, m(0) = 0; f(1) = 1, m(1) = 0; f(2) = 2, m(2) = 1; f(3) = 2, m(3) = 2; f(4) = 3, m(4) = 2; f(5) = 3, m(5) = 3; f(6) = 4, m(6) = 4; f(7) = 5, m(7) = 4; f(8) = 5, m(8) = 5; f(9) = 6, m(9) = 6$ **75.** The last occurrence of n is in the position for which the total number of 1s, 2s, $\ldots$, n's all together is that position number. But since a_k is the number of occurrences of k, this is just $\sum_{k=1}^{n} a_k$, as desired. Since $f(n)$ is the sum of the first n terms of the sequence, $f(f(n))$ is the sum of the first $f(n)$ terms of the sequence. But since $f(n)$ is the last term whose value is n, this means that the sum is the sum of all terms of the sequence whose value is at most n. Since there are a_k terms of the sequence whose value is k, this sum is $\sum_{k=1}^{n} k \cdot a_k$, as desired   **77.** Suppose that $f(n)$ and $g(n)$ are both in $\mathcal{L}$ and eventually positive. Then $\ln f(n)$ and $\ln g(n)$ are in $\mathcal{L}$, so by Exercise 76, $\ln f(n) + \ln g(n) = \ln(f(n)g(n)) \in \mathcal{L}$. Therefore, $e^{\ln(f(n)g(n))} = f(n)g(n) \in \mathcal{L}$. Similarly, $\ln f(n) - \ln g(n) = \ln(f(n)/g(n)) \in \mathcal{L}$, so $e^{\ln(f(n)/g(n))} = f(n)/g(n) \in \mathcal{L}$.

Now suppose that $f(n)$ and $g(n)$ are both in $\mathcal{L}$ and neither is identically zero. If $f(n)$ and $g(n)$ are both eventually negative, then $-f(n)$ and $-g(n)$ are both eventually positive and also in $\mathcal{L}$, so $f(n)g(n) = (-f(n))(-g(n))$ and $f(n)/g(n) = (-f(n))/(-g(n))$ are in $\mathcal{L}$. If one of the functions, say, $f(n)$, is eventually positive and the other eventually negative, then $f(n)g(n) = -(f(n))(-g(n)) \in \mathcal{L}$ and similarly for the quotient.   **79.** If $f(n) \in \mathcal{L}$ and is eventually positive, then $\ln f(n) \in \mathcal{L}$, so by the result of Exercise 77, $\frac{1}{2} \ln f(n) = \ln(\sqrt{f(n)}) \in \mathcal{L}$. Hence $\sqrt{f(n)} = e^{\ln \sqrt{f(n)}} \in \mathcal{L}$.

CHAPTER 4

SECTION 4.1

1. a) 5850 **b)** 343   **3. a)** 4^{10} **b)** 5^{10} **5.** 42   **7.** 26^3 **9.** 676   **11.** 2^8 **13.** $n + 1$ (counting the empty string)   **15.** 475,255 (counting the empty string)   **17.** 1,321,368,961   **19. a)** 128 **b)** 450 **c)** 9 **d)** 675 **e)** 450 **f)** 450 **g)** 225 **h)** 75   **21. a)** 990 **b)** 500 **c)** 27   **23.** 3^{50} **25.** 52,457,600   **27.** 20,077,200   **29. a)** 37,822,859,361 **b)** 8,204,716,800 **c)** 40,159,050,880 **d)** 12,113,640,000 **e)** 171,004,205,215 **f)** 72,043,541,640 **g)** 6,230,721,635 **h)** 223,149,655   **31. a)** 0 **b)** 120 **c)** 720 **d)** 2520   **33. a)** 2 if $n = 1$, 2 if $n = 2$, 0 if $n \geq 3$ **b)** 2^{n-2} for $n > 1$; 1 if $n = 1$ **c)** $2(n-1)$ **35.** $(n+1)^m$ **37.** If n is even $2^{n/2}$; if n is odd $2^{(n+1)/2}$ **39. a)** 240 **b)** 480 **c)** 360   **41.** 352   **43.** 147   **45.** 33   **47.** 7,104,000,000,000   **49.** 18   **51.** 17   **53.** 22   **55.** Let $P(m)$ be the sum rule for m tasks. For the basis case take $m = 2$. This is just the sum rule for two tasks. Now assume that $P(m)$ is true. Consider $m + 1$ tasks, $T_1, T_2, \ldots, T_m, T_{m+1}$, which can be done in $n_1, n_2, \ldots, n_m, n_{m+1}$ ways, respectively, such that no two of these tasks can be done at the same time. To do one of these tasks, we can either do one of the first m of these, or do task T_{m+1}. By the sum rule for two tasks, the number of ways to do this is the sum of the number of ways to do one of the first m tasks, plus n_{m+1}. By the inductive hypothesis this is $n_1 + n_2 + \cdots + n_m + n_{m+1}$, as desired.   **57.** $n(n-3)/2$

SECTION 4.2

1. Since there are six classes, but only five weekdays, the pigeonhole principle shows that at least two classes must be held on the same day.   **3. a)** 3 **b)** 14   **5.** Since there are four possible remainders when an integer is divided by 4, the pigeonhole principle implies that given five integers, at least two have the same remainder.   **7.** Let $a, a+1, \ldots, a+n-1$ be the integers in the sequence. The integers $(a + i) \bmod n, i = 0, 1, 2, \ldots, n - 1$, are distinct, since $0 < (a+j) - (a+k) < n$ whenever $0 \leq k < j \leq n-1$. Since there are n possible values for $(a + i) \bmod n$, and there are n different integers in the set, each of these values is taken on exactly once. It follows that there is exactly one integer in the sequence that is divisible by n.   **9.** 4951   **11.** The midpoint of the segment joining the points (a, b, c) and (d, e, f) is $((a+d)/2, (b+e)/2, (c+f)/2)$. It has integer coefficients if and only if a and d have the same parity, b and e have the same parity, and c and f have the same parity. Since there are eight possible triples of parity [such as (*even, odd, even*)], by the pigeonhole principle at least two of the nine points have the same triple of parities. The midpoint of the segment joining two such points has integer coefficients.   **13. a)** Group the first eight positive integers into four subsets of two integers each so that the integers of each subset add up to 9: $\{1, 8\}, \{2, 7\}, \{3, 6\}$, and $\{4, 5\}$. If five integers are selected from the first eight positive integers, by the pigeonhole principle at least two of them come from the same subset. Two such integers have a sum of 9, as desired. **b)** No. Take $\{1, 2, 3, 4\}$, for example.   **15.** 4   **17.** 21,251   **19. a)** If there were fewer than 9 freshmen, fewer than 9 sophomores, and fewer than

9 juniors in the class, there would be no more than 8 of each of these three class standings, for a total of at most 24 students, contradicting the fact that there are 25 students in the class. **b)** If there were fewer than 3 freshmen, fewer than 19 sophomores, and fewer than 5 juniors, then there would be at most 2 freshmen, at most 18 sophomores, and at most 4 juniors, for a total of at most 24 students. This contradicts the fact that there are 25 students in the class.
21. $4, 3, 2, 1, 8, 7, 6, 5, 12, 11, 10, 9, 16, 15, 14, 13$
23. procedure $long(a_1, \ldots, a_n$: positive integers)
$\quad$ {first find longest increasing subsequence}
$\quad max := 0; set := 00 \ldots 00\{n$ bits}
$\quad$ **for** $i := 1$ **to** 2^n
$\quad$ **begin**
$\quad\quad last := 0; count := 0, OK := $ **true**
$\quad\quad$ **for** $j := 1$ **to** n
$\quad\quad$ **begin**
$\quad\quad\quad$ **if** $set(j) = 1$ **then**
$\quad\quad\quad$ **begin**
$\quad\quad\quad\quad$ **if** $a_j > last$ **then** $last := a_j$
$\quad\quad\quad\quad count := count + 1$
$\quad\quad\quad$ **end**
$\quad\quad\quad$ **else** $OK := false$
$\quad\quad$ **end**
$\quad\quad$ **if** $count > max$ **then**
$\quad\quad$ **begin**
$\quad\quad\quad max := count$
$\quad\quad\quad best := set$
$\quad\quad$ **end**
$\quad\quad set := set + 1$ (binary addition)
$\quad$ **end** {max is length and $best$ indicates the sequence}
$\quad$ {repeat for decreasing subsequence with only
$\quad\quad$ changes being $a_j < last$ instead of $a_j > last$
$\quad\quad$ and $last := \infty$ instead of $last := 0$}
25. By symmetry we need prove only the first statement. Let A be one of the people. Either A has at least four friends, or A has at least six enemies among the other nine people (since $3 + 5 < 9$). Suppose, in the first case, that $B, C, D,$ and E are all A's friends. If any two of these are friends with each other, then we have found three mutual friends. Otherwise $\{B, C, D, E\}$ is a set of four mutual enemies. In the second case, let $\{B, C, D, E, F, G\}$ be a set of enemies of A. By Example 11, among $B, C, D, E, F,$ and G there are either three mutual friends or three mutual enemies, who form, with A, a set of four mutual enemies.
27. We need to show two things: that if we have a group of n people, then among them we must find either a pair of friends or a subset of n of them all of whom are mutual enemies; and that there exists a group of $n - 1$ people for which this is not possible. For the first statement, if there is any pair of friends, then the condition is satisfied, and if not, then every pair of people are enemies, so the second condition is satisfied. For the second statement, if we have a group of $n - 1$ people all of whom are enemies of each other, then there is neither a pair of friends nor a subset of n of them all of whom are mutual enemies.

29. There are 6,432,816 possibilities for the three initials and a birthday. So, by the generalized pigeonhole principle, there are at least $\lceil 34,000,000/6,432,816 \rceil = 6$ people who share the same initials and birthday. **31.** 18 **33.** Since there are six computers, the number of other computers a computer is connected to is an integer between 0 and 5, inclusive. However, 0 and 5 cannot both occur. To see this, note that if some computer is connected to no others, then no computer is connected to all five others, and if some computer is connected to all five others, then no computer is connected to no others. Hence, by the pigeonhole principle, since there are at most five possibilities for the number of computers a computer is connected to, there are at least two computers in the set of six connected to the same number of others. **35.** Label the computers C_1 through C_{100}, and label the printers P_1 through P_{20}. If we connect C_k to P_k for $k = 1, 2, \ldots, 20$ and connect each of the computers C_{21} through C_{100} to *all* the printers, then we have used a total of $20 + 80 \cdot 20 = 1620$ cables. Clearly this is sufficient, because if computers C_1 through C_{20} need printers, then they can use the printers with the same subscripts, and if any computers with higher subscripts need a printer instead of one or more of these, then they can use the printers that are not being used, since they are connected to all the printers. Now we must show that 1619 cables is not enough. Since there are 1619 cables and 20 printers, the average number of computers per printer is $1619/20$, which is less than 81. Therefore some printer must be connected to fewer than 81 computers. That means it is connected to 80 or fewer computers, so there are 20 computers that are not connected to it. If those 20 computers all needed a printer simultaneously, then they would be out of luck, since they are connected to at most the 19 other printers. **37.** Let a_i be the number of matches completed by hour i. Then $1 \leq a_1 < a_2 < \cdots < a_{75} \leq 125$. Also $25 \leq a_1 + 24 < a_2 + 24 < \cdots < a_{75} + 24 \leq 149$. There are 150 numbers $a_1, \ldots, a_{75}, a_1 + 24, \ldots, a_{75} + 24$. By the pigeonhole principle, at least two are equal. Since all the a_is are distinct and all the $(a_i + 24)$s are distinct, it follows that $a_i = a_j + 24$ for some $i > j$. Thus, in the period from the $(j + 1)$st to the ith hour, there are exactly 24 matches. **39.** Use the generalized pigeonhole principle, placing the $|S|$ objects $f(s)$ for $s \in S$ in $|T|$ boxes, one for each element of T. **41.** Let d_j be $jx - N(jx)$, where $N(jx)$ is the integer closest to jx for $1 \leq j \leq n$. Each d_j is an irrational number between $-1/2$ and $1/2$. We will assume that n is even; the case where n is odd is messier. Consider the n intervals $\{x \mid j/n < x < (j+1)/n\}$, $\{x \mid -(j+1)/n < x < -j/n\}$ for $j = 0, 1, \ldots, (n/2) - 1$. If d_j belongs to the interval $\{x \mid 0 < x < 1/n\}$ or to the interval $\{x \mid -1/n < x < 0\}$ for some j, we are done. If not, since there are $n - 2$ intervals and n numbers d_j, the pigeonhole principle tells us that there is an interval $\{x \mid (k-1)/n < x < k/n\}$ containing d_r and d_s with $r < s$. The proof can be finished by showing that $(s - r)x$ is within $1/n$ of its nearest integer. **43. a)** Assume that $i_k \leq n$

for all k. Then by the generalized pigeonhole principle, at least $\lceil (n^2+1)/n \rceil = n+1$ of the numbers $i_1, i_2, \ldots, i_{n^2+1}$ are equal.   **b)** If $a_{k_j} < a_{k_{j+1}}$, then the subsequence consisting of a_{k_j}, followed by the increasing subsequence of length $i_{k_{j+1}}$ starting at $a_{k_{j+1}}$ contradicts the fact that $i_{k_j} = i_{k_{j+1}}$. Hence $a_{k_j} > a_{k_{j+1}}$.   **c)** If there is no increasing subsequence of length greater than n, then parts (a) and (b) apply. Therefore we have $a_{k_{n+1}} > a_{k_n} > \cdots > a_{k_2} > a_{k_1}$, a decreasing sequence of length $n+1$.

SECTION 4.3

1. $abc, acb, bac, bca, cab, cba$ **3.** 720   **5. a)** 120   **b)** 720
c) 8   **d)** 6720   **e)** 40,320   **f)** 3,628,800   **7.** 15,120
9. 1320   **11. a)** 210   **b)** 386   **c)** 848   **d)** 252   **13.** $2(n!)^2$
15. 65,780   **17.** $2^{100} - 5051$ **19. a)** 1024   **b)** 45   **c)** 176
d) 252   **21. a)** 120   **b)** 24   **c)** 120   **d)** 24   **e)** 6
f) 0   **23.** 609,638,400   **25. a)** 94,109,400   **b)** 941,094
c) 3,764,376   **d)** 90,345,024   **e)** 114,072   **f)** 2328   **g)** 24
h) 79,727,040   **i)** 3,764,376   **j)** 109,440   **27. a)** 12,650
b) 303,600   **29. a)** 37,927   **b)** 18,915   **31. a)** 122,523,030
b) 72,930,375   **c)** 223,149,655   **d)** 100,626,625   **33.** 54,600
35. 45   **37.** 912   **39.** 11,232,000   **41.** 13   **43.** 873

SECTION 4.4

1. $x^4 + 4x^3y + 6x^2y^2 + 4xy^3 + y^4$ **3.** $x^6 + 6x^5y + 15x^4y^2 + 20x^3y^3 + 15x^2y^4 + 6xy^5 + y^6$ **5.** 101   **7.** $-2^{10}\binom{19}{9} = -94,595,072$ **9.** $-2^{101}3^{99}\binom{200}{99}$ **11.** $(-1)^{(200-k)/3}\binom{100}{(200-k)/3}$ if $k \equiv 2 \pmod 3$ and $-100 \le k \le 200$; 0 otherwise
13. 1 9 36 84 126 126 84 36 9 1   **15.** The sum of *all* the positive numbers $\binom{n}{k}$, as k runs from 0 to n, is 2^n, so each one of them is no bigger than this sum.   **17.** $\binom{n}{k} = \frac{n(n-1)(n-2)\cdots(n-k+1)}{k(k-1)(k-2)\cdots 2} \le \frac{n \cdot n \cdots n}{2 \cdot 2 \cdots 2} = n^k/2^{k-1}$
19. $\binom{n}{k-1} + \binom{n}{k} = \frac{n!}{(k-1)!(n-k+1)!} + \frac{n!}{k!(n-k)!} = \frac{n!}{k!(n-k+1)!} \cdot [k + (n - k + 1)] = \frac{(n+1)!}{k!(n+1-k)!} = \binom{n+1}{k}$ **21. a)** We show that each side counts the number of ways to choose from a set with n elements a subset with k elements and a distinguished element of that set. For the left-hand side, first choose the k-set [this can be done in $\binom{n}{k}$ ways] and then choose one of the k elements in this subset to be the distinguished element (this can be done in k ways). For the right-hand side, first choose the distinguished element out of the entire n-set [this can be done in n ways], and then choose the remaining $k-1$ elements of the subset from the remaining $n-1$ elements of the set (this can be done in $\binom{n-1}{k-1}$ ways).
b) $k\binom{n}{k} = k \cdot \frac{n!}{k!(n-k)!} = \frac{n \cdot (n-1)!}{(k-1)!(n-k)!} = n\binom{n-1}{k-1}$ **23.** $\binom{n+1}{k} = \frac{(n+1)!}{k!(n+1-k)!} = \frac{(n+1)}{k} \cdot \frac{n!}{(k-1)!(n-(k-1))!} = (n+1)\binom{n}{k-1}/k$. This identity together with $\binom{n}{0} = 1$ gives a recursive definition.
25. $\binom{2n}{n+1} + \binom{2n}{n} = \binom{2n+1}{n+1} = \frac{1}{2}\left(\binom{2n+1}{n+1} + \binom{2n+1}{n+1}\right) = \frac{1}{2}\left(\binom{2n+1}{n} + \binom{2n+1}{n}\right) = \frac{1}{2}\left(\binom{2n+2}{n+1}\right)$ **27. a)** $\binom{n+r+1}{r}$ counts the number of ways to choose a sequence of r 0s and $n+1$ 1s by choosing the positions of the 0s. Alternately, suppose that the $(j+1)$st term is the last term equal to 1, so $n \le j \le n+r$. Once we have determined where the last 1 is, we decide

where the 0s are to be placed in the j spaces before the last 1. There are n 1s and $j - n$ 0s in this range. By the sum rule it follows that there are $\sum_{j=n}^{n+r}\binom{j}{j-n} = \sum_{k=0}^{r}\binom{n+k}{0}$ ways to do this.   **b)** Let $P(r)$ be the statement to be proved. The basis step is the equation $\binom{n}{0} = \binom{n+1}{0}$, which is just $1 = 1$. Assume that $P(r)$ is true. Then $\sum_{k=0}^{r+1}\binom{n+k}{k} = \sum_{k=0}^{r}\binom{n+k}{k} + \binom{n+r+1}{r+1} = \binom{n+r+1}{r} + \binom{n+r+1}{r+1} = \binom{n+r+2}{r+1}$, using the inductive hypothesis and Pascal's identity.   **29.** We can choose the leader first in n different ways. We can then choose the rest of the committee in 2^{n-1} ways. Hence there are $n2^{n-1}$ ways to choose the committee and its leader. Meanwhile, the number of ways to select a committee with k people is $\binom{n}{k}$. Once we have chosen a committee with k people, there are k ways to choose its leader. Hence there are $\sum_{k=1}^{n} k\binom{n}{k}$ ways to choose the committee and its leader. Hence $\sum_{k=1}^{n} k\binom{n}{k} = n2^{n-1}$.
31. Let the set have n elements. From Corollary 2 we have $\binom{n}{0} - \binom{n}{1} + \binom{n}{2} - \cdots + (-1)^n\binom{n}{n} = 0$. It follows that $\binom{n}{0} + \binom{n}{2} + \binom{n}{4} + \cdots = \binom{n}{1} + \binom{n}{3} + \binom{n}{5} + \cdots$. The left-hand side gives the number of subsets with an even number of elements, and the right-hand side gives the number of subsets with an odd number of elements.   **33. a)** A path of the desired type consists of m moves to the right and n moves up. Each such path can be represented by a bit string of length $m + n$ with m 0s and n 1s, where a 0 represents a move to the right and a 1 a move up.   **b)** The number of bit strings of length $m + n$ containing exactly n 1s equals $\binom{m+n}{n} = \binom{m+n}{m}$ since such a string is determined by specifying the positions of the n 1s or by specifying the positions of the m 0s.   **35.** By Exercise 33 the number of paths of length n of the type described in that exercise equals 2^n, the number of bit strings of length n. On the other hand, a path of length n of the type described in Exercise 33 must end at a point that has n as the sum of its coordinates, say $(n - k, k)$ for some k between 0 and n, inclusive. By Exercise 33, the number of such paths ending at $(n - k, k)$ equals $\binom{n-k+k}{k} = \binom{n}{k}$. Hence $\sum_{k=0}^{n}\binom{n}{k} = 2^n$.
37. By Exercise 33 the number of paths from $(0, 0)$ to $(n + 1, r)$ of the type described in that exercise equals $\binom{n+r+1}{r}$. But such a path starts by going j steps vertically for some j with $0 \le j \le r$. The number of these paths beginning with j vertical steps equals the number of paths of the type described in Exercise 33 that go from $(1, j)$ to $(n+1, r)$. This is the same as the number of such paths that go from $(0, 0)$ to $(n, r - j)$, which by Exercise 33 equals $\binom{n+r-j}{r-j}$. Since $\sum_{j=0}^{r}\binom{n+r-j}{r-j} = \sum_{k=0}^{r}\binom{n+k}{k}$, it follows that $\sum_{k=1}^{r}\binom{n+k}{k} = \binom{n+r-1}{r}$.   **39. a)** $\binom{n+1}{2}$ **b)** $\binom{n+2}{3}$ **c)** $\binom{2n-2}{n-1}$ **d)** $\binom{n-1}{\lfloor (n-1)/2 \rfloor}$ **e)** Largest odd entry in nth row of Pascal's triangle   **f)** $\binom{3n-3}{n-1}$

SECTION 4.5

1. 243   **3.** 26^6 **5.** 125   **7.** 35   **9. a)** 1716   **b)** 50,388
c) 2,629,575   **d)** 330   **e)** 9,724   **11.** 9   **13.** 4,504,501
15. a) 10,626   **b)** 1,365   **c)** 11,649   **d)** 106   **17.** 2,520

19. 302,702,400   **21.** 3003   **23.** 7,484,400   **25.** 30,492
27. $C(59, 50)$ **29.** 35   **31.** 83,160   **33.** 63
35. 19,635   **37.** 210   **39.** 27,720   **41.** $52!/(7!^5 17!)$
43. $24 \cdot 13^4/(52 \cdot 51 \cdot 50 \cdot 49)$ **45. a)** $C(k + n - 1, n)$
b) $(k + n - 1)!/(k - 1)!$ **47.** There are $C(n, n_1)$ ways
to choose n_1 objects for the first box. Once these objects
are chosen, there are $C(n - n_1, n_2)$ ways to choose objects
for the second box. Similarly, there are $C(n - n_1 - n_2, n_3)$
ways to choose objects for the third box. Continue in
this way until there is $C(n - n_1 - n_2 - \cdots - n_{k-1}, n_k) =$
$C(n_k, \ln_k) = 1$ way to choose the objects for the last box
(since $n_1 + n_2 + \cdots + n_k = n$). By the product rule, the number of ways to make the entire assignment is $C(n, n_1)C(n -$
$n_1, n_2)C(n - n_1 - n_2, n_3) \cdots C(n - n_1 - n_2 - \cdots - n_{k-1}, n_k)$,
which equals $n!/(n_1!n_2! \cdots n_k!)$, as straightforward simplification shows.   **49. a)** Since $x_1 \le x_2 \le \cdots \le x_r$, it follows
that $x_1 + 0 < x_2 + 1 < \cdots < x_r + r - 1$. The inequalities are
strict since $x_j + j - 1 < x_{j+1} + j$ as long as $x_j \le x_{j+1}$. Since
$1 \le x_j \le n + r - 1$, this sequence is made up of r distinct
elements from T.   **b)** Suppose that $1 \le x_1 < x_2 < \cdots <$
$x_r \le n+r-1$. Let $y_k = x_k -(k-1)$. Then it is not hard to see
that $y_k \le y_{k+1}$ for $k = 1, 2, \ldots, r - 1$ and that $1 \le y_k \le n$
for $k = 1, 2, \ldots r$. It follows that $\{y_1, y_2, \ldots, y_r\}$ is an r-
combination with repetition allowed of S.   **c)** From (a)
and (b) it follows that there is a one-to-one correspondence of r-combinations with repetition allowed of S and
r-combinations of T, a set with $n + r - 1$ elements. We
conclude that there are $C(n + r - 1, r)$ r-combinations
with repetitions allowed of S.   **51.** 5   **53.** The terms
in the expansion are of the form $x_1^{n_1} x_2^{n_2} \cdots x_m^{n_m}$, where
$n_1 + n_2 + \cdots + n_m = n$. Such a term arises from choosing
the x_1 in n_1 factors, the x_2 in n_2 factors,..., and the x_m in
n_m factors. This can be done in $C(n; n_1, n_2, \ldots, n_m)$ ways,
since a choice is a permutation of n_1 labels "1," n_2 labels
"2," ...and n_m labels "m."   **55.** 2520

SECTION 4.6

1. a) 2134   **b)** 54132   **c)** 12534   **d)** 45312   **e)** 6714253
f) 31542678   **3.** 14532, 15432, 21345, 23451, 23514, 31452,
31542, 43521, 45213, 45321   **5.** 1234, 1243, 1324, 1342,
1423, 1432, 2134, 2143, 2314, 2341, 2413, 2431, 3124, 3142,
3214, 3241, 3412, 3421, 4123, 4132, 4213, 4231, 4312, 4321
7. $\{1, 2, 3\}, \{1, 2, 4\}, \{1, 2, 5\}, \{1, 3, 4\}, \{1, 3, 5\}, \{1, 4, 5\},$
$\{2, 3, 4\}, \{2, 3, 5\}, \{2, 4, 5\}, \{3, 4, 5\}$ **9.** The bit string representing the next largest r-combination must differ from
the bit string representing the original one in position i
since positions $i + 1, \ldots, r$ are occupied by the largest possible numbers. Also $a_i + 1$ is the smallest possible number
we can put in position i if we want a combination greater
than the original one. Then $a_i + 2, \ldots, a_i + r - i + 1$ are the
smallest allowable numbers for positions $i + 1$ to r. Thus
we have produced the next r-combination.   **11.** 123, 132,
213, 231, 312, 321, 124, 142, 214, 241, 412, 421, 125, 152, 215,
251, 512, 521, 134, 143, 314, 341, 413, 431, 135, 153, 315,
351, 513, 531, 145, 154, 415, 451, 514, 541, 234, 243, 324, 342,

423, 432, 235, 253, 325, 352, 523, 532, 245, 254, 425, 452, 524,
542, 345, 354, 435, 453, 534, 543   **13.** We will show it is
a bijection by showing it has an inverse. Given a positive
integer less than $n!$, let $a_1, a_2, \ldots, a_{n-1}$ be its Cantor digits.
Put n in position $n - a_{n-1}$, so clearly a_{n-1} is the number of
integers less than n that follow n in the permutation. Then
put $n - 1$ in free position $(n - 1) - a_{n-2}$, where we have
numbered the free positions $1, 2, \ldots, n - 1$ (excluding the
position that n is already in). Continue until 1 is placed in
the only free position left. Since we have constructed an
inverse, the correspondence is a bijection.
15. procedure *Cantor permutation*$(n, i$: integers with
  $n \ge 1$ and $0 \le i < n!)$
  $x := n$
  for $j := 1$ **to** n
   $p_j := 0$
  for $k := 1$ **to** $n - 1$
  begin
   $c := \lfloor x/(n - k)! \rfloor; x := x - c(n - k)!; h := n$
   while $p_h \ne 0$
    $h := h - 1$
   for $j := 1$ **to** c
   begin
    $h := h - 1$
    while $p_h \ne 0$
     $h := h - 1$
   end
   $p_h := n - k + 1$
  end
  $h := 1$
  while $p_h \ne 0$
   $h := h + 1$
  $p_h := 1$
  $\{p_1 p_2 \cdots p_n$ is the permutation corresponding to $i\}$

SUPPLEMENTARY EXERCISES

1. a) 151,200   **b)** 1,000,000   **c)** 210   **d)** 5005   **3.** 3^{100}
5. 24,600   **7. a)** 4,060   **b)** 2688   **c)** 25,009,600   **9. a)** 192
b) 301   **c)** 300   **d)** 300   **11.** 639   **13.** The maximum possible sum is 240, and the minimum possible
sum is 15. So the number of possible sums is 226. Since
there are 252 subsets with five elements of a set with
10 elements, by the pigeonhole principle it follows that
at least two have the same sum.   **15. a)** 50   **b)** 50
c) 14   **d)** 5   **17.** Let $a_1, a_2, \ldots, a_m$ be the integers, and
let $d_i = \sum_{j=1}^{i} a_j$. If $d_i \equiv 0 \pmod{m}$ for some i, we are
done. Otherwise $d_1 \bmod m, d_2 \bmod m, \ldots, d_m \bmod m$ are
m integers with values in $\{1, 2, \ldots, m - 1\}$. By the pigeonhole principle $d_k = d_l$ for some $1 \le k < l \le m$. Then
$\sum_{j=k+1}^{l} a_j = d_l - d_k \equiv 0 \pmod{m}$.   **19.** The decimal expansion of the rational number a/b can be obtained by
division of b into a, where a is written with a decimal point
and an arbitrarily long string of 0s following it. The basic
step is finding the next digit of the quotient, namely, $\lfloor r/b \rfloor$,
where r is the remainder with the next digit of the div-

idend brought down. The current remainder is obtained from the previous remainder by subtracting b times the previous digit of the quotient. Eventually the dividend has nothing but 0s to bring down. Furthermore, there are only b possible remainders. Thus, at some point, by the pigeonhole principle, we will have the same situation as had previously arisen. From that point onward, the calculation must follow the same pattern. In particular, the quotient will repeat. **21. a)** 125,970 **b)** 20 **c)** 141,120,525 **d)** 141,120,505 **e)** 177,100 **f)** 141,078,021 **23. a)** 10 **b)** 8 **c)** 7 **25.** 3^n **27.** $C(n+2, r+1) = C(n+1, r+1) + C(n+1, r) = 2C(n+1, r+1) - C(n+1, r+1) + C(n+1, r) = 2C(n+1, r+1) - (C(n, r+1) + C(n, r)) + (C(n, r) + C(n, r-1)) = 2C(n+1, r+1) - C(n+1, r+1) + C(n, r-1)$ **29.** Substitute $x = 1$ and $y = 3$ into the binomial theorem. **31.** $C(n+1, 5)$ **33.** 3,491,888,400 **35.** 5^{24} **37. a)** 45 **b)** 57 **c)** 12 **39. a)** 386 **b)** 56 **c)** 512 **41.** 0 if $n < m$; $C(n-1, n-m)$ if $n \geq m$ **43. procedure** *next permutation* (n: positive integer,
$\quad a_1, a_2, \ldots, a_r$: positive integers not exceeding
$\quad n$ with $a_1 a_2 \cdots a_r \neq nn \cdots n$)
$\quad i := r$
$\quad$**while** $a_i = n$
$\quad$**begin**
$\quad\quad a_i := 1$
$\quad\quad i := i - 1$
$\quad$**end**
$\quad a_i := a_i + 1$
$\quad \{a_1 a_2 \cdots a_r$ is the next permutation in lexicographic order$\}$
45. We must show that if there are $R(m, n-1) + R(m-1, n)$ people at a party, then there must be at least m mutual friends or n mutual enemies. Consider one person; let's call him Jerry. Then there are $R(m-1, n) + R(m, n-1) - 1$ other people at the party, and by the pigeonhole principle there must be at least $R(m-1, n)$ friends of Jerry or $R(m, n-1)$ enemies of Jerry among these people. First let's suppose there are $R(m-1, n)$ friends of Jerry. By the definition of R, among these people we are guaranteed to find either $m-1$ mutual friends or n mutual enemies. In the former case these $m-1$ mutual friends together with Jerry are a set of m mutual friends; and in the latter case we have the desired set of n mutual enemies. The other situation is similar: Suppose there are $R(m, n-1)$ enemies of Jerry; we are guaranteed to find among them either m mutual friends or $n-1$ mutual enemies. In the former case we have the desired set of m mutual friends, and in the latter case these $n-1$ mutual enemies together with Jerry are a set of n mutual enemies.

CHAPTER 5

SECTION 5.1

1. 1/13 **3.** 1/2 **5.** 1/2 **7.** 1/64 **9.** 47/52 **11.** $1/C(52, 5)$ **13.** $1 - (C(48, 5)/C(52, 5))$

15. $C(13, 2)C(4, 2)C(4, 2)C(44, 1)/C(52, 5)$ **17.** $10,240/C(52, 5)$ **19.** $1,302,540/C(52, 5)$ **21.** 1/64 **23.** 8/25 **25. a)** $1/C(50, 6) = 1/15,890,700$ **b)** $1/C(52, 6) = 1/20,358,520$ **c)** $1/C(56, 6) = 1/32,468,436$ **d)** $1/C(60, 6) = 1/50,063,860$ **27. a)** 139,128/319,865 **b)** 212,667/511,313 **c)** 151,340/386,529 **d)** 163,647/446,276 **29.** $1/C(100, 8)$ **31.** 3/100 **33. a)** 1/7,880,400 **b)** 1/8,000,000 **35. a)** 9/19 **b)** 81/361 **c)** 1/19 **d)** 1,889,568/2,476,099 **e)** 48/361 **37.** Three dice **39.** The door the contestant chooses is chosen at random without knowing where the prize is, but the door chosen by the host is not chosen at random, since he always avoids opening the door with the prize. This makes any argument based on symmetry invalid. **41. a)** 671/1296 **b)** $1 - 35^{24}/36^{24}$; no **c)** Yes

SECTION 5.2

1. $p(T) = 1/4$, $p(H) = 3/4$ **3.** $p(1) = p(3) = p(5) = p(6) = 1/16$; $p(2) = p(4) = 3/8$ **5.** 9/49 **7. a)** 1/2 **b)** 1/2 **c)** 1/3 **d)** 1/4 **e)** 1/4 **9. a)** 1/26! **b)** 1/26 **c)** 1/2 **d)** 1/26 **e)** 1/650 **f)** 1/15,600 **11.** Clearly $p(E \cup F) \geq p(E) = 0.7$. Also, $p(E \cup F) \leq 1$. If we apply Theorem 2 from Section 5.1, we can rewrite this as $p(E) + p(F) - p(E \cap F) \leq 1$, or $0.7 + 0.5 - p(E \cap F) \leq 1$. Solving for $p(E \cap F)$ gives $p(E \cap F) \geq 0.2$. **13.** Since $p(E \cup F) = p(E) + p(F) - p(E \cap F)$ and $p(E \cup F) \leq 1$, it follows that $1 \geq p(E) + p(F) - p(E \cap F)$. From this inequality we conclude that $p(E) + p(F) \leq 1 + p(E \cap F)$. **15.** We will use mathematical induction to prove that the inequality holds for $n \geq 2$. Let $P(n)$ be the statement that $p(\bigcup_{j=1}^{n} E_j) \leq \sum_{j=1}^{n} p(E_j)$. *Basis step:* $P(2)$ is true since $p(E_1 \cup E_2) = p(E_1) + p(E_2) - p(E_1 \cap E_2) \leq p(E_1) + p(E_2)$. *Inductive step:* Assume that $P(k)$ is true. Using the basis case and the inductive hypothesis, it follows that $p(\bigcup_{j=1}^{k+1} E_j) \leq p(\bigcup_{j=1}^{k} E_j) + p(E_{k+1}) \leq \sum_{j=1}^{k+1} p(E_j)$. This shows that $P(k+1)$ is true, completing the proof by mathematical induction. **17.** Since $E \cup \overline{E}$ is the entire sample space S, the event F can be split into two disjoint events: $F = S \cap F = (E \cup \overline{E}) \cap F = (E \cap F) \cup (\overline{E} \cap F)$, using the distributive law. Therefore $p(F) = p((E \cap F) \cup (\overline{E} \cap F)) = p(E \cap F) + p(\overline{E} \cap F)$, since these two events are disjoint. Subtracting $p(E \cap F)$ from both sides, using the fact that $p(E \cap F) = p(E) \cdot p(F)$ (the hypothesis that E and F are independent), and factoring, we have $p(F)(1 - p(E)) = p(\overline{E} \cap F)$. Since $1 - p(E) = p(\overline{E})$, this says that $p(\overline{E} \cap F) = p(\overline{E}) \cdot p(F)$, as desired. **19.** (assuming that births are independent and the probability of a birth in each month is 1/12) **a)** 1/12 **b)** $1 - \frac{11}{12} \cdot \frac{10}{12} \cdots \cdots \frac{13-n}{12}$ **c)** 5 **21.** 614 **23.** 1/4 **25.** 3/8 **27. a)** Not independent **b)** Not independent **c)** Not independent **29.** 3/16 **31. a)** $1/32 = 0.03125$ **b)** $0.49^5 \approx 0.02825$ **c)** 0.03795012 **33. a)** 5/8 **b)** 0.627649 **c)** 0.6431 **35. a)** p^n **b)** $1 - p^n$ **c)** $p^n + n \cdot p^{n-1} \cdot (1 - p)$ **d)** $1 - [p^n + n \cdot p^{n-1} \cdot (1 - p)]$ **37.** $p(\bigcup_{i=1}^{\infty} E_i)$ is the sum of $p(s)$ for each outcome s in $\bigcup_{i=1}^{\infty} E_i$. Since the E_i's are pairwise disjoint, this is the sum of the probabilities of all the outcomes in any of the E_i's, which is what $\sum_{i=1}^{\infty} p(E_i)$ is. (We can rearrange the summands and still get the same answer because this series converges absolutely.) **39. a)** $\overline{E} = \bigcup_{j=1}^{\binom{m}{k}} F_j$, so the given inequality now follows from Boole's inequality (Exercise 15). **b)** The probability that a particular player not in the jth set beats all k of the players in the jth set is $(1/2)^k = 2^{-k}$. Therefore the probability that this player does not do so is $1 - 2^{-k}$, so the probability that all $m - k$ of the players not in the jth set are unable to boast of a perfect record against everyone in the jth set is $(1 - 2^{-k})^{m-k}$. That is precisely $p(F_j)$. **c)** The first inequality follows immediately, since all the summands are the same and there are $\binom{m}{k}$ of them. If this probability is less than 1, then it must be possible that $\overline{E}$ fails, i.e., that E happens. So there is a tournament that meets the conditions of the problem as long as the second inequality holds. **d)** $m \geq 21$ for $k = 2$, and $m \geq 91$ for $k = 3$

41. procedure *probabilistic prime*(n, k)
   composite := **false**
   $i := 0$
   while *composite* = **false** and $i < k$
   begin
     $i := i + 1$
     choose b uniformly at random with $1 < b < n$
     apply Miller's test to base b
     if n fails the test **then** *composite* = **true**
   end
   if *composite* = **true then print** ("composite")
   else print ("probably prime")

SECTION 5.3

1. By Theorem 2, with $p = 1/2$ and $n = 5$, we see that the expected number of heads is 2.5. **3.** 5/3 **5.** 336/49 **7.** 170 **9.** $(4n + 6)/3$ **11.** $50{,}700{,}551/10{,}077{,}696 \approx$ 5.03 **13.** 6 **15.** $p(X \geq j) = \sum_{k=j}^{\infty} p(X = k) = \sum_{k=j}^{\infty} (1 - p)^{k-1} p = p(1 - p)^{j-1} \sum_{k=0}^{\infty} (1 - p)^k = p(1 - p)^{j-1}/(1 - (1-p)) = (1-p)^{j-1}$ **17.** 2302 **19.** $(7/2) \cdot 7 \neq$ 329/12 **21.** $p + (n - 1)p(1 - p)$ **23.** 5/2 **25. a)** 0 **b)** 1 **27.** 1/100 **29.** $E(X)/a = \sum_r (r/a) \cdot p(X = r) \geq \sum_{r \geq a} 1 \cdot p(X = r) = p(X \geq a)$ **31. a)** 10/11 **b)** 0.9984 **33. a)** Each of the $n!$ permutations occurs with probability $1/n!$, so $E(X)$ is the number of comparisons, averaged over all these permutations. **b)** Even if the algorithm continues $n - 1$ rounds, X will be at most $n(n-1)/2$. It follows from the formula for expectation that $E(X) \leq n(n-1)/2$. **c)** The algorithm proceeds by comparing adjacent elements and then swapping them if necessary. Thus the only way that inverted elements can become uninverted is for them to be compared and swapped. **d)** Since $X(P) \geq I(P)$ for all P, it follows from the definition of expectation that $E(X) \geq E(I)$. **e)** This summation counts

1 for every instance of an inversion. **f)** This follows from Theorem 3. **g)** By Theorem 2 with $n = 1$, the expectation of $I_{j,k}$ is the probability that a_k precedes a_j in the permutation. This is clearly 1/2 by symmetry. **h)** The summation in part (f) consists of $C(n, 2) = n(n-1)/2$ terms, each equal to 1/2, so the sum is $n(n - 1)/4$. **i)** From part (a) and part (b) we know that $E(X)$, the object of interest, is at most $n(n - 1)/2$, and from part (d) and part (h) we know that $E(X)$ is at least $n(n-1)/4$, both of which are $\Theta(n^2)$. **35.** 1 **37.** $V(X+Y) = E((X+Y)^2) - E(X+Y)^2 = E(X^2 + 2XY + Y^2) - (E(X) + E(Y))^2 = E(X^2) + 2E(XY) + E(Y^2) - E(X)^2 - 2E(X)E(Y) - E(Y)^2 = E(X^2) - E(X)^2 + 2(E(XY) - E(X)E(Y)) + E(Y^2) - E(Y)^2 = V(X) + 2\operatorname{Cov}(X, Y) + V(Y)$ **39.** $((n - 1)/n)^m$ **41.** $(n - 1)^m/n^{m-1}$

SUPPLEMENTARY EXERCISES

1. $1/109{,}668$ **3. a)** $1/C(52, 13)$ **b)** $4/C(52, 13)$ **c)** $2{,}944{,}656/C(52, 13)$ **d)** $35{,}335{,}872/C(52, 13)$ **5. a)** 9/2 **b)** 21/4 **7. a)** 9 **b)** 21/2 **9. a)** 8 **b)** 49/6 **11. a)** $n/2^{n-1}$ **b)** $p(1 - p)^{k-1}$, where $p = n/2^{n-1}$ **c)** $2^{n-1}/n$ **13.** $\frac{(m-1)(n-1)+\gcd(m, n)-1}{mn-1}$ **15 a)** 2/3 **b)** 2/3 **17.** 1/32 **19. a)** The probability that one wins 2^n dollars is $1/2^n$, since that happens precisely when the player gets $n - 1$ tails followed by a head. The expected value of the winnings is therefore the sum of 2^n times $1/2^n$ as n goes from 1 to infinity. Since each of these terms is 1, the sum is infinite. In other words, one should be willing to wager any amount of money and expect to come out ahead in the long run. **b)** \$9, \$9 **21. a)** 1/3 when $S = \{1, 2, 3, 4, 5, 6, 7, 8, 9, 10, 11, 12\}$, $A = \{1, 2, 3, 4, 5, 6, 7, 8, 9\}$, and $B = \{1, 2, 3, 4\}$; 1/12 when $S = \{1, 2, 3, 4, 5, 6, 7, 8, 9, 10, 11, 12\}$, $A = \{4, 5, 6, 7, 8, 9, 10, 11, 12\}$, and $B = \{1, 2, 3, 4\}$ **b)** 1 when $S = \{1, 2, 3, 4, 5, 6, 7, 8, 9, 10, 11, 12\}$, $A = \{4, 5, 6, 7, 8, 9, 10, 11, 12\}$, and $B = \{1, 2, 3, 4\}$; 3/4 when $S = \{1, 2, 3, 4, 5, 6, 7, 8, 9, 10, 11, 12\}$, $A = \{1, 2, 3, 4, 5, 6, 7, 8, 9\}$, and $B = \{1, 2, 3, 4\}$ **23. a)** $p(E_1 \cap E_2) = p(E_1)p(E_2)$, $p(E_1 \cap E_3) = p(E_1)p(E_3)$, $p(E_2 \cap E_3) = p(E_2)p(E_3)$, $p(E_1 \cap E_2 \cap E_3) = p(E_1)p(E_2)p(E_3)$ **b)** Yes **c)** Yes **d)** $2^n - n - 1$ **25.** The events $E \cap F_i, i = 1, 2, \ldots, n$, are mutually exclusive and cover all the conditions under which E occurs. Therefore, $p(E) = \sum_{i=1}^{n} p(E \cap F_i)$. By the definition of conditional probability, we know that $p(E \cap F_i) = p(E \mid F_i)p(F_i)$ and $p(F_i \mid E) = p(E \cap F_i)/p(E)$. Hence $p(F_j \mid E) = \frac{p(E \cap F_j)}{p(E)} = \frac{p(E \mid F_j)p(F_j)}{\sum_{i=1}^{n} p(E \cap F_i)} = \frac{p(E \mid F_j)p(F_j)}{\sum_{i=1}^{n} p(E \mid F_i)p(F_i)}$. **27.** $V(aX + b) = E((aX + b)^2) - E(aX + b)^2 = E(a^2X^2 + 2abX + b^2) - (aE(X) + b)^2 = E(a^2X^2) + E(2abX) + E(b^2) - (a^2E(X)^2 + 2abE(X) + b^2) = a^2E(X^2) + 2abE(X) + b^2 - a^2E(X)^2 - 2abE(X) - b^2 = a^2(E(X^2) - E(X)^2) = a^2V(X)$ **29.** To count every element in the sample space exactly once, we must include every element in each of the sets and then take away the double counting of the elements in the intersections. Thus $p(E_1 \cup E_2 \cup \cdots \cup E_m) = p(E_1) + p(E_2) + \cdots + p(E_m) - p(E_1 \cap E_2) - p(E_1 \cap E_3) - \cdots - p(E_1 \cap E_m) - p(E_2 \cap E_3) - p(E_2 \cap E_4) - \cdots - p(E_2 \cap E_m) - \cdots -$

$p(E_{m-1} \cap E_m) = qm - (m(m-1)/2)r$, since $C(m, 2)$ terms are being subtracted. But $p(E_1 \cup E_2 \cup \cdots \cup E_m) = 1$, so we have $qm - (m(m-1)/2)r = 1$. Since $r \geq 0$, this equation tells us that $qm \geq 1$, so $q \geq 1/m$. Since $q \leq 1$, this equation also implies that $(m(m-1)/2)r = qm - 1 \leq m - 1$, from which it follows that $r \leq 2/m$. **31. a)** We purchase the cards until we have gotten one of each type. That means we have purchased X cards in all. On the other hand, that also means that we purchased X_0 cards until we got the first type we got, and then purchased X_1 more cards until we got the second type we got, and so on. Thus X is the sum of the X_j's. **b)** Once j distinct types have been obtained, there are $n - j$ new types available out of a total of n types available. Since it is equally likely that we get each type, the probability of success on the next purchase (getting a new type) is $(n - j)/n$. **c)** This follows immediately from the definition of geometric distribution, the definition of X_j, and part (b). **d)** From part (c) it follows that $E(X_j) = n/(n-j)$. Thus by the linearity of expectation and part (a), we have $E(X) = E(X_0) + E(X_1) + \cdots + E(X_{n-1})$ $= \frac{n}{n} + \frac{n}{n-1} + \cdots + \frac{n}{1} = n\left(\frac{1}{n} + \frac{1}{n-1} + \cdots + \frac{1}{1}\right)$. **e)** About 224.46

CHAPTER 6

SECTION 6.1

1. a) 2, 12, 72, 432, 2592 **b)** 2, 4, 16, 256, 65,536 **c)** 1, 2, 5, 11, 26 **d)** 1, 1, 6, 27, 204 **e)** 1, 2, 0, 1, 3 **3. a)** 6, 17, 49, 143, 421 **b)** $49 = 5 \cdot 17 - 6 \cdot 6$, $143 = 5 \cdot 49 - 6 \cdot 17$, $421 = 5 \cdot 143 - 6 \cdot 49$ **c)** $5a_{n-1} - 6a_{n-2} = 5(2^{n-1} + 5 \cdot 3^{n-1}) - 6(2^{n-2} + 5 \cdot 3^{n-2}) = 2^{n-2}(10 - 6) + 3^{n-2}(75 - 30) = 2^{n-2} \cdot 4 + 3^{n-2} \cdot 9 \cdot 5 = 2^n + 3^n \cdot 5 = a_n$ **5. a)** Yes **b)** No **c)** No **d)** Yes **e)** Yes **f)** Yes **g)** No **h)** No **7. a)** $a_{n-1} + 2a_{n-2} + 2n - 9 = -(n-1) + 2 + 2(-(n-2) + 2) + 2n - 9 = -n + 2 = a_n$ **b)** $a_{n-1} + 2a_{n-2} + 2n - 9 = 5(-1)^{n-1} - (n-1) + 2 + 2(5(-1)^{n-2} - (n-2) + 2) + 2n - 9 = 5(-1)^{n-2}(-1 + 2) - n + 2 = a_n$ **c)** $a_{n-1} + 2a_{n-2} + 2n - 9 = 3(-1)^{n-1} + 2^{n-1} - (n-1) + 2 + 2(3(-1)^{n-2} + 2^{n-2} - (n-2) + 2) + 2n - 9 = 3(-1)^{n-2}(-1 + 2) + 2^{n-2}(2 + 2) - n + 2 = a_n$ **d)** $a_{n-1} + 2a_{n-2} + 2n - 9 = 7 \cdot 2^{n-1} - (n-1) + 2 + 2(7 \cdot 2^{n-2} - (n-2) + 2) + 2n - 9 = 2^{n-2}(7 \cdot 2 + 2 \cdot 7) - n + 2 = a_n$ **9. a)** $a_n = 2 \cdot 3^n$ **b)** $a_n = 2n + 3$ **c)** $a_n = 1 + n(n+1)/2$ **d)** $a_n = n^2 + 4n + 4$ **e)** $a_n = 1$ **f)** $a_n = (3^{n+1} - 1)/2$ **g)** $a_n = 5n!$ **h)** $a_n = 2^n n!$ **11. a)** $a_n = 3a_{n-1}$ **b)** 5,904,900 **13. a)** $a_n = n + a_{n-1}$, $a_0 = 0$ **b)** $a_{12} = 78$ **c)** $a_n = n(n+1)/2$ **15.** $B(k) = (1 + (0.07/12))B(k-1) - 100$, with $B(0) = 5000$ **17.** Let $P(n)$ be "$H_n = 2^n - 1$." *Basis step:* $P(1)$ is true since $H_1 = 1$. *Inductive step:* Assume that $H_n = 2^n - 1$. Then since $H_{n+1} = 2H_n + 1$, it follows that $H_{n+1} = 2(2^n - 1) + 1 = 2^{n+1} - 1$. **19. a)** $a_n = 2a_{n-1} + a_{n-5}$ for $n \geq 5$ **b)** $a_0 = 1, a_1 = 2, a_3 = 8, a_4 = 16$ **c)** 1217 **21.** 9494 **23. a)** $a_n = a_{n-1} + a_{n-2} + 2^{n-2}$ for $n \geq 2$

b) $a_0 = 0, a_1 = 0$ **c)** 94 **25. a)** $a_n = a_{n-1} + a_{n-2} + a_{n-3}$ for $n \geq 3$ **b)** $a_0 = 1, a_1 = 2, a_2 = 4$ **c)** 81 **27. a)** $a_n = a_{n-1} + a_{n-2}$ for $n \geq 2$ **b)** $a_0 = 1, a_1 = 1$ **c)** 34 **29. a)** $a_n = 2a_{n-1} + 2a_{n-2}$ for $n \geq 2$ **b)** $a_0 = 1, a_1 = 3$ **c)** 448 **31. a)** $a_n = 2a_{n-1} + a_{n-2}$ for $n \geq 2$ **b)** $a_0 = 1, a_1 = 3$ **c)** 239 **33. a)** $a_n = 2a_{n-1}$ for $n \geq 2$ **b)** $a_1 = 3$ **c)** 96 **35. a)** $a_n = a_{n-1} + a_{n-2}$ for $n \geq 2$ **b)** $a_0 = 1, a_1 = 1$ **c)** 89 **37. a)** $R_n = n + R_{n-1}, R_0 = 1$ **b)** $R_n = n(n+1)/2 + 1$ **39. a)** $S_n = S_{n-1} + (n^2 - n + 2)/2$, $S_0 = 1$ **b)** $S_n = (n^3 + 5n + 6)/6$ **41.** 64 **43. a)** $a_n = 2a_{n-1} + 2a_{n-2}$ **b)** $a_0 = 1, a_1 = 3$ **c)** 1224 **45.** Clearly $S(m, 1) = 1$ for $m \geq 1$. If $m \geq n$, then a function that is not onto from the set with m elements to the set with n elements can be specified by picking the size of the range, which is an integer between 1 and $n - 1$ inclusive, picking the elements of the range, which can be done in $C(n, k)$ ways, and picking an onto function onto the range, which can be done in $S(m, k)$ ways. Hence there are $\sum_{k=1}^{n-1} C(n, k)S(m, k)$ functions that are not onto. But there are n^m functions altogether, so $S(m, n) = n^m - \sum_{k=1}^{n-1} C(n, k)S(m, k)$. **47. a)** $C_5 = C_0C_4 + C_1C_3 + C_2C_2 + C_3C_1 + C_4C_0 = 1 \cdot 14 + 1 \cdot 5 + 2 \cdot 2 + 5 \cdot 1 + 14 \cdot 1 = 42$ **b)** $C(10, 5)/6 = 42$ **49.** $J(1) = 1, J(2) = 1, J(3) = 3, J(4) = 1, J(5) = 3, J(6) = 5, J(7) = 7, J(8) = 1, J(9) = 3, J(10) = 5, J(11) = 7, J(12) = 9, J(13) = 11, J(14) = 13, J(15) = 15, J(16) = 1$ **51.** First, suppose that the number of people is even, say, $2n$. After going around the circle once and returning to the first person, since the people at locations with even numbers have been eliminated, there are exactly n people left and the person currently at location i is the person who was originally at location $2i - 1$. Therefore, the survivor [originally in location $J(2n)$] is now in location $J(n)$; this was the person who was at location $2J(n) - 1$. Hence, $J(2n) = 2J(n) - 1$. Similarly, when there are an odd number of people, say $2n + 1$, then after going around the circle once and then eliminating person 1, there are n people left and the person currently at location i is the person who was at location $2i + 1$. Therefore, the survivor will be the player currently occupying location $J(n)$, namely, the person who was originally at location $2J(n) + 1$. Hence, $J(2n + 1) = 2J(n) + 1$. The base case is $J(1) = 1$. **53.** 73, 977, 3617 **55.** These nine moves solve the puzzle: Move disk 1 from peg 1 to peg 2; move disk 2 from peg 1 to peg 3; move disk 1 from peg 2 to peg 3; move disk 3 from peg 1 to peg 2; move disk 4 from peg 1 to peg 4; move disk 3 from peg 2 to peg 4; move disk 1 from peg 3 to peg 2; move disk 2 from peg 3 to peg 4; move disk 1 from peg 2 to peg 4. To see that at least nine moves are required, first note that at least seven moves are required no matter how many pegs are present: three to unstack the disks, one to move the largest disk 4, and three more moves to restack them. At least two other moves are needed, since to move disk 4 from peg 1 to peg 4 the other three disks must be on pegs 2 and 3, so at least one move is needed to restack them and one move to unstack them.

57. The base cases are obvious. If $n > 1$, the algorithm consists of three stages. In the first stage, by the inductive hypothesis, $R(n-k)$ moves are used to transfer the smallest $n-k$ disks to peg 2. Then using the usual three-peg Tower of Hanoi algorithm, it takes $2^k - 1$ moves to transfer the rest of the disks (the largest k disks) to peg 4, avoiding peg 2. Then again by induction, it takes $R(n-k)$ moves to transfer the smallest $n-k$ disks to peg 4; all the pegs are available for this, since the largest disks, now on peg 4, do not interfere. This establishes the recurrence relation. **59.** First note that $R(n) = \sum_{j=1}^{n} (R(j) - R(j-1))$ (which follows since the sum is telescoping and $R(0) = 0$). By Exercise 58, this is the sum of $2^{k'-1}$ for this range of values of j. Therefore, the sum is $\sum_{i=1}^{k} i 2^{i-1}$, except that if n is not a triangular number, then the last few values when $i = k$ are missing, and that is what the final term in the given expression accounts for. **61.** By Exercise 59, $R(n)$ is no larger than $\sum_{i=1}^{k} i 2^{i-1}$. It can be shown that this sum equals $(k + 1)2^k - 2^{k+1} + 1$, so it is no greater than $(k + 1)2^k$. Since $n > k(k - 1)/2$, the quadratic formula can be used to show that $k < 1 + \sqrt{2n}$ for all $n > 1$. Therefore $R(n)$ is bounded above by $(1 + \sqrt{2n} + 1)2^{1+\sqrt{2n}} < 8\sqrt{n}2^{\sqrt{2n}}$ for all $n > 2$. Hence $R(n)$ is $O(\sqrt{n}2^{\sqrt{2n}})$. **63. a)** 0 **b)** 0 **c)** 2 **d)** $2^{n-1} - 2^{n-2}$ **65.** $a_n - 2\nabla a_n + \nabla^2 a_n = a_n - 2(a_n - a_{n-1}) + (\nabla a_n - \nabla a_{n-1}) = -a_n + 2a_{n-1} + ((a_n - a_{n-1}) - (a_{n-1} - a_{n-2})) = -a_n + 2a_{n-1} + (a_n - 2a_{n-1} + a_{n-2}) = a_{n-2}$ **67.** $a_n = a_{n-1} + a_{n-2} = (a_n - \nabla a_n) + (a_n - 2\nabla a_n + \nabla^2 a_n) = 2a_n - 3\nabla a_n + \nabla^2 a_n$, or $a_n = 3\nabla a_n - \nabla^2 a_n$

SECTION 6.2

1. a) Degree 3 **b)** No **c)** Degree 4 **d)** No **e)** No **f)** Degree 2 **g)** No **3. a)** $a_n = 3 \cdot 2^n$ **b)** $a_n = 2$ **c)** $a_n = 3 \cdot 2^n - 2 \cdot 3^n$ **d)** $a_n = 6 \cdot 2^n - 2 \cdot n2^n$ **e)** $a_n = n(-2)^{n-1}$ **f)** $a_n = 2^n - (-2)^n$ **g)** $a_n = (1/2)^{n+1} - (-1/2)^{n+1}$ **5.** $a_n = \frac{1}{\sqrt{5}} \left(\frac{1+\sqrt{5}}{2} \right)^{n+1} - \frac{1}{\sqrt{5}} \left(\frac{1-\sqrt{5}}{2} \right)^{n+1}$ **7.** $(2^{n+1} + (-1)^n)/3$ **9. a)** $P_n = 1.2P_{n-1} + 0.45P_{n-2}, P_0 = 100,000, P_1 = 120,000$ **b)** $P_n = (250,000/3)(3/2)^n + (50,000/3)(-3/10)^n$ **11. a)** *Basis step:* For $n = 1$ we have $1 = 0 + 1$, and for $n = 2$ we have $3 = 1 + 2$. *Inductive step:* Assume true for $k \leq n$. Then $L_{n+1} = L_n + L_{n-1} = f_{n-1} + f_{n+1} + f_{n-2} + f_n = (f_{n-1} + f_{n-2}) + (f_{n+1} + f_n) = f_n + f_{n+2}$. **b)** $L_n = \left(\frac{1+\sqrt{5}}{2} \right)^n + \left(\frac{1-\sqrt{5}}{2} \right)^n$ **13.** $a_n = 8(-1)^n - 3(-2)^n + 4 \cdot 3^n$ **15.** $a_n = 5 + 3(-2)^n - 3^n$ **17.** Let $a_n = C(n, 0) + C(n-1, 1) + \cdots + C(n-k, k)$ where $k = \lfloor n/2 \rfloor$. First, assume that n is even, so that $k = n/2$, and the last term is $C(k, k)$. By Pascal's identity we have $a_n = 1 + C(n-2, 0) + C(n-2, 1) + C(n-3, 1) + C(n-3, 2) + \cdots + C(n-k, k-2) + C(n-k, k-1) + 1 = 1 + C(n-2, 1) + C(n-3, 2) + \cdots + C(n-k, k-1) + C(n-2, 0) + C(n-3, 1) + \cdots + C(n-k, k-2) + 1 = a_{n-1} + a_{n-2}$ since $\lfloor (n-1)/2 \rfloor = k - 1 = \lfloor (n-2)/2 \rfloor$. A similar calculation works when n is odd. Hence $\{a_n\}$ satisfies the recurrence relation $a_n = a_{n-1} + a_{n-2}$ for all

positive integers $n, n \geq 2$. Also, $a_1 = C(1, 0) = 1$ and $a_2 = C(2, 0) + C(1, 1) = 2$, which are f_2 and f_3. It follows that $a_n = f_{n+1}$ for all positive integers n. **19.** $a_n = (n^2 + 3n + 5)(-1)^n$ **21.** $(a_{1,0} + a_{1,1}n + a_{1,2}n^2 + a_{1,3}n^3) + (a_{2,0} + a_{2,1}n + a_{2,2}n^2)(-2)^n + (a_{3,0} + a_{3,1}n)3^n + a_{4,0}(-4)^n$ **23. a)** $3a_{n-1} + 2^n = 3(-2)^n + 2^n = 2^n(-3+1) = -2^{n+1} = a_n$ **b)** $a_n = \alpha 3^n - 2^{n+1}$ **c)** $a_n = 3^{n+1} - 2^{n+1}$ **25. a)** $A = -1$, $B = -7$ **b)** $a_n = \alpha 2^n - n - 7$ **c)** $a_n = 11 \cdot 2^n - n - 7$ **27. a)** $p_3 n^3 + p_2 n^2 + p_1 n + p_0$ **b)** $n^2 p_0 (-2)^n$ **c)** $n^2(p_1 n + p_0)2^n$ **d)** $(p_2 n^2 + p_1 n + p_0)4^n$ **e)** $n^2(p_2 n^2 + p_1 n + p_0)(-2)^n$ **f)** $n^2(p_4 n^4 + p_3 n^3 + p_2 n^2 + p_1 n + p_0)2^n$ **g)** p_0 **29. a)** $a_n = \alpha 2^n + 3^{n+1}$ **b)** $a_n = -2 \cdot 2^n + 3^{n+1}$ **31.** $a_n = \alpha 2^n + \beta 3^n - n \cdot 2^{n+1} + 3n/2 + 21/4$ **33.** $a_n = (\alpha + \beta n + n^2 + n^3/6)2^n$ **35.** $a_n = -4 \cdot 2^n - n^2/4 - 5n/2 + 1/8 + (39/8)3^n$ **37.** $a_n = n(n+1)(n+2)/6$ **39. a)** $1, -1, i, -i$ **b)** $a_n = \frac{1}{4} - \frac{1}{4}(-1)^n + \frac{2+i}{4}i^n + \frac{2-i}{4}(-i)^n$ **41. a)** Using the formula for f_n, we see that $\left| f_n - \frac{1}{\sqrt{5}} \left(\frac{1+\sqrt{5}}{2} \right)^n \right| = \left| \frac{1}{\sqrt{5}} \left(\frac{1-\sqrt{5}}{2} \right)^n \right| < 1/\sqrt{5} < 1/2$ This means that f_n is the integer closest to $\frac{1}{\sqrt{5}} \left(\frac{1+\sqrt{5}}{2} \right)^n$. **b)** Less when n is even; greater when n is odd **43.** $a_n = f_{n-1} + 2f_n - 1$ **45. a)** $a_n = 3a_{n-1} + 4a_{n-2}, a_0 = 2, a_1 = 6$ **b)** $a_n = (4^{n+1} + (-1)^n)/5$ **47. a)** $a_n = 2a_{n+1} + (n-1)10,000$ **b)** $a_n = 70,000 \cdot 2^{n-1} - 10,000n - 10,000$ **49.** $a_n = 5n^2/12 + 13n/12 + 1$ **51.** See Chapter 11, Section 5 in [Ma93]. **53.** $6^n \cdot 4^{n-1}/n$

SECTION 6.3

1. 14 **3.** The first step is $(1110)_2 (1010)_2 = (2^4 + 2^2)(11)_2 (10)_2 + 2^2((11)_2 - (10)_2)((10)_2 - (10)_2) + (2^2 + 1)(10)_2 \cdot (10)_2$. The product is $(10001100)_2$. **5.** $C = 50,665C + 729 = 33,979$ **7. a)** 2 **b)** 4 **c)** 7 **9. a)** 79 **b)** 48,829 **c)** 30,517,579 **11.** $O(\log n)$ **13.** $O(n^{\log_3 2})$ **15.** 5 **17. a)** *Basis step:* If the sequence has just one element, then the one person on the list is the winner. *Recursive step:* Divide the list into two parts—the first half and the second half—as equally as possible. Apply the algorithm recursively to each half to come up with at most two names. Then run through the entire list to count the number of occurrences of each of those names to decide which, if either, is the winner. **b)** $O(n \log n)$ **19. a)** $f(n) = f(n/2) + 2$ **b)** $O(\log n)$ **21. a)** 7 **b)** $O(\log n)$ **23. a)** **procedure** *largest sum*$(a_1, \ldots, a_n)$
    best := 0 {empty subsequence has sum 0}
    for $i := 1$ **to** n
    begin
      sum := 0
      for $j := i + 1$ **to** n
      begin
        sum := *sum* + a_j
        if *sum* > *best* **then** *best* := *sum*
      end
    end
    {*best* is the maximum possible sum of numbers in
      the list}

b) $O(n^2)$ **c)** We divide the list into a first half and a second half and apply the algorithm recursively to find the largest sum of consecutive terms for each half. The largest sum of consecutive terms in the entire sequence is either one of these two numbers or the sum of a sequence of consecutive terms that crosses the middle of the list. To find the largest possible sum of a sequence of consecutive terms that crosses the middle of the list, we start at the middle and move forward to find the largest possible sum in the second half of the list, and move backward to find the largest possible sum in the first half of the list; the desired sum is the sum of these two quantities. The final answer is then the largest of this sum and the two answers obtained recursively. The base case is that the largest sum of a sequence of one term is the larger of that number and 0. **d)** $11, 9, 14$ **e)** $S(n) = 2S(n/2) + n$, $C(n) = 2C(n/2) + n + 2, S(1) = 0, C(1) = 1$ **f)** $O(n \log n)$, better than $O(n^2)$ **25.** $(1, 6)$ and $(3, 6)$ at distance 2 **27.** The algorithm is essentially the same as the algorithm given in Example 12. The central strip still has width $2d$ but we need to consider just two boxes of size $d \times d$ rather than eight boxes of size $(d/2) \times (d/2)$. The recurrence relation is the same as the recurrence relation in Example 12, except that the coefficient 7 is replaced by 1. **29.** With $k = \log_b n$, it follows that $f(n) = a^k f(1) + \sum_{j=0}^{k-1} a^j c(n/b^j)^d = a^k f(1) + \sum_{j=0}^{k-1} cn^d = a^k f(1) + kcn^d = a^{\log_b n} f(1) + c(\log_b n)n^d = n^{\log_b a} f(1) + cn^d \log_b n = n^d f(1) + cn^d \log_b n$. **31.** Let $k = \log_b n$ where n is a power of b. *Basis step:* If $n = 1$ and $k = 0$, then $c_1 n^d + c_2 n^{\log_b a} = c_1 + c_2 = b^d c/(b^d - a) + f(1) + b^d c/(a - b^d) = f(1)$. *Inductive step:* Assume true for k, where $n = b^k$. Then for $n = b^{k+1}, f(n) = af(n/b) + cn^d = a\{[b^d c/(b^d - a)](n/b)^d + [f(1) + b^d c/(a - b^d)] \cdot (n/b)^{\log_b a}\} + cn^d = b^d c/(b^d - a)n^d a/b^d + [f(1) + (b^d c/(a - b^d))]n^{\log_b a} + cn^d = n^d[ac/(b^d - a) + c(b^d - a)/(b^d - a)] + [f(1) + b^d c/(a - b^d c)]n^{\log_b a} = [(b^d c)/(b^d - a)]n^d + [f(1) + b^d c/(a - b^d)]n^{\log_b a}$. **33.** If $a > b^d$, then $\log_b a > d$, so the second term dominates, giving $O(n^{\log_b a})$. **35.** $O(n^{\log_4 5})$ **37.** $O(n^3)$

SECTION 6.4

1. $f(x) = 2(x^6 - 1)/(x - 1)$ **3. a)** $f(x) = 2x(1 - x^6)/(1 - x)$ **b)** $x^3/(1 - x)$ **c)** $x/(1 - x^3)$ **d)** $2/(1 - 2x)$ **e)** $(1 + x)^7$ **f)** $2/(1 + x)$ **g)** $(1/(1 - x)) - x^2$ **h)** $x^3/(1 - x)^2$ **5. a)** $5/(1 - x)$ **b)** $1/(1 - 3x)$ **c)** $2x^3/(1 - x)$ **d)** $(3 - x)/(1 - x)^2$ **e)** $(1 + x)^8$ **f)** $1/(1 - x)^5$ **7. a)** $a_0 = -64, a_1 = 144, a_2 = -108, a_3 = 27$, and $a_n = 0$ for all $n \geq 4$ **b)** The only nonzero coefficients are $a_0 = 1, a_3 = 3, a_6 = 3, a_9 = 1$. **c)** $a_n = 5^n$ **d)** $a_n = (-3)^{n-3}$ for $n \geq 3$, and $a_0 = a_1 = a_2 = 0$ **e)** $a_0 = 8, a_1 = 3, a_2 = 2, a_n = 0$ for odd n greater than 2 and $a_n = 1$ for even n greater than 2 **f)** $a_n = 1$ if n is a positive multiple 4, $a_n = -1$ if $n < 4$, and $a_n = 0$ otherwise **g)** $a_n = n - 1$ for $n \geq 2$ and $a_0 = a_1 = 0$ **h)** $a_n = 2^{n+1}/n!$ **9. a)** 6 **b)** 3 **c)** 9 **d)** 0 **e)** 5 **11. a)** 1024 **b)** 11

c) 66 **d)** 292,864 **e)** 20,412 **13.** 10 **15.** 50 **17.** 20 **19.** $f(x) = 1/((1 - x)(1 - x^2)(1 - x^5)(1 - x^{10}))$ **21.** 15 **23. a)** $x^4(1 + x + x^2 + x^3)^2/(1 - x)$ **b)** 6 **25. a)** The coefficient of x^r in the power series expansion of $1/((1 - x^3)(1 - x^4)(1 - x^{20}))$ **b)** $1/(1 - x^3 - x^4 - x^{20})$ **c)** 7 **d)** 3224 **27. a)** 3 **b)** 29 **c)** 29 **d)** 242 **29. a)** 10 **b)** 49 **c)** 2 **d)** 4 **31. a)** $G(x) - a_0 - a_1 x - a_2 x^2$ **b)** $G(x^2)$ **c)** $x^4 G(x)$ **d)** $G(2x)$ **e)** $\int_0^x G(t)dt$ **f)** $G(x)/(1 - x)$ **33.** $a_k = 2 \cdot 3^k - 1$ **35.** $a_k = 18 \cdot 3^k - 12 \cdot 2^k$ **37.** $a_k = k^2 + 8k + 20 + (6k - 18)2^k$ **39.** Let $G(x) = \sum_{k=0}^{\infty} f_k x^k$. After shifting indices of summation and adding series, we see that $G(x) - xG(x) - x^2 G(x) = f_0 + (f_1 - f_0)x + \sum_{k=2}^{\infty}(f_k - f_{k-1} - f_{k-2})x^k = 0 + x + \sum_{k=2}^{\infty} 0x^k$. Hence $G(x) - xG(x) - x^2 G(x) = x$. Solving for $G(x)$ gives $G(x) = x/(1 - x - x^2)$. By the method of partial fractions, it can be shown that $x/(1 - x - x^2) = (1/\sqrt{5})[1/(1 - \alpha x) - 1/(1 - \beta x)]$ where $\alpha = (1 + \sqrt{5})/2$ and $\beta = (1 - \sqrt{5})/2$. Using the fact that $1/(1 - \alpha x) = \sum_{k=0}^{\infty} \alpha^k x^k$, it follows that $G(x) = (1/\sqrt{5}) \cdot \sum_{k=0}^{\infty}(\alpha^k - \beta^k)x^k$. Hence $f_k = (1/\sqrt{5}) \cdot (\alpha^k - \beta^k)$. **41. a)** Let $G(x) = \sum_{n=0}^{\infty} C_n x^n$ be the generating function for $\{C_n\}$. Then $G(x)^2 = \sum_{n=0}^{\infty}(\sum_{k=0}^{n} C_k C_{n-k})x^n = \sum_{n=1}^{\infty}(\sum_{k=0}^{n-1} C_k C_{n-1-k})x^{n-1} = \sum_{n=1}^{\infty} C_n x^{n-1}$. Hence $xG(x)^2 = \sum_{n=1}^{\infty} C_n x^n$, which implies that $xG(x)^2 - G(x) + 1 = 0$. Applying the quadratic formula shows that $G(x) = \frac{1 \pm \sqrt{1 - 4x}}{2x}$. We choose the minus sign in this formula since the choice of the plus sign leads to a division by zero. **b)** By Exercise 40, $(1 - 4x)^{-1/2} = \sum_{n=0}^{\infty} \binom{2n}{n} x^n$. Integrating term by term (which is valid by a theorem from calculus) shows that $\int_0^x (1 - 4t)^{-1/2}dt = \sum_{n=0}^{\infty} \frac{1}{n+1}\binom{2n}{n}x^{n+1} = x\sum_{n=0}^{\infty} \frac{1}{n+1}\binom{2n}{n}x^n$. Since $\int_0^x (1 - 4t)^{-1/2}dt = \frac{1 - \sqrt{1 - 4x}}{2} = xG(x)$, equating coefficients shows that $C_n = \frac{1}{n+1}\binom{2n}{n}$. **43.** Applying the Binomial Theorem to the equality $(1 + x)^{m+n} = (1 + x)^m(1 + x)^n$, shows that $\sum_{r=0}^{m+n} C(m + n, r)x^r = \sum_{r=0}^{m} C(m, r)x^r \cdot \sum_{r=0}^{n} C(n, r)x^r = \sum_{r=0}^{m+n}(\sum_{k=0}^{r} C(m, r - k)C(n, k))x^r$. Comparing coefficients gives the desired identity. **45. a)** $2e^x$ **b)** e^{-x} **c)** e^{3x} **d)** $xe^x + e^x$ **e)** $(e^x - 1)/x$ **47. a)** $a_n = (-1)^n$ **b)** $a_n = 3 \cdot 2^n$ **c)** $a_n = 3^n - 3 \cdot 2^n$ **d)** $a_n = (-2)^n$ for $n \geq 2, a_1 = -3$, $a_0 = 2$ **e)** $a_n = (-2)^n + n!$ **f)** $a_n = (-3)^n + n! \cdot 2^n$ for $n \geq 2, a_0 = 1, a_1 = -2$ **g)** $a_n = 0$ if n is odd and $a_n = n!/(n/2)!$ if n is even **49. a)** $a_n = 6a_{n-1} + 8^{n-1}$ for $n \geq 1, a_0 = 1$ **b)** The general solution of the associated linear homogeneous recurrence relation is $a_n^{(h)} = \alpha 6^n$. A particular solution is $a_n^{(p)} = \frac{1}{2} \cdot 8^n$. Hence the general solution is $a_n = \alpha 6^n + \frac{1}{2} \cdot 8^n$. Using the initial condition, it follows that $\alpha = \frac{1}{2}$. Hence $a_n = (6^n + 8^n)/2$. **c)** Let $G(x) = \sum_{k=0}^{\infty} a_k x^k$. Using the recurrence relation for $\{a_k\}$, it can be shown that $G(x) - 6xG(x) = (1 - 7x)/(1 - 8x)$. Hence $G(x) = (1 - 7x)/((1 - 6x)(1 - 8x))$. Using partial fractions, it follows that $G(x) = (1/2)/(1 - 6x) + (1/2)/(1 - 8x)$. With the help of Table 1, it follows that $a_n = (6^n + 8^n)/2$. **51.** $\frac{1}{1-x} \cdot \frac{1}{1-x^2} \cdot \frac{1}{1-x^3} \cdots$ **53.** $(1 + x)(1 + x)^2(1 + x)^3 \cdots$ **55.** The generating functions obtained in Exercises 52 and 53 are equal since $(1 + x)(1 + x^2)(1 + x^3) \cdots =$

$\frac{1-x^2}{1-x} \cdot \frac{1-x^4}{1-x^2} \cdot \frac{1-x^6}{1-x^3} \cdots = \frac{1}{1-x} \cdot \frac{1}{1-x^3} \cdot \frac{1}{1-x^5} \cdots$. **57. a)** $G_X(1) = \sum_{k=0}^{\infty} p(X = k) \cdot 1^k = \sum_{k=0}^{\infty} P(X = k) = 1$ **b)** $G'_X(1) = \frac{d}{dx} \sum_{k=0}^{\infty} p(X = k) \cdot x^k |_{x=1} = \sum_{k=0}^{\infty} p(X = k) \cdot k \cdot x^{k-1} |_{x=1} = \sum_{k=0}^{\infty} p(X = k) \cdot k = E(X)$ **c)** $G''_X(1) = \frac{d^2}{dx^2} \sum_{k=0}^{\infty} p(X = k) \cdot x^k |_{x=1} = \sum_{k=0}^{\infty} p(X = k) \cdot k(k - 1) \cdot x^{k-2} |_{x=1} = \sum_{k=0}^{\infty} p(X = k) \cdot (k^2 - k) = V(X) + E(X)^2 - E(X)$. Combining this with part (b) gives the desired results. **59. a)** $G(x) = p^m/(1 - qx)^m$ **b)** $V(x) = mq/p^2$

SECTION 6.5

1. a) 30 **b)** 29 **c)** 24 **d)** 18 **3.** 1% **5. a)** 300 **b)** 150 **c)** 175 **d)** 100 **7.** 492 **9.** 974 **11.** 55 **13.** 248 **15.** 50,138 **17.** 234 **19.** $|A_1 \cup A_2 \cup A_3 \cup A_4 \cup A_5| = |A_1| + |A_2| + |A_3| + |A_4| + |A_5| - |A_1 \cap A_2| - |A_1 \cap A_3| - |A_1 \cap A_4| - |A_1 \cap A_5| - |A_2 \cap A_3| - |A_2 \cap A_4| - |A_2 \cap A_5| - |A_3 \cap A_4| - |A_3 \cap A_5| - |A_4 \cap A_5| + |A_1 \cap A_2 \cap A_3| + |A_1 \cap A_2 \cap A_4| + |A_1 \cap A_2 \cap A_5| + |A_1 \cap A_3 \cap A_4| + |A_1 \cap A_3 \cap A_5| + |A_1 \cap A_4 \cap A_5| + |A_2 \cap A_3 \cap A_4| + |A_2 \cap A_3 \cap A_5| + |A_2 \cap A_4 \cap A_5| + |A_3 \cap A_4 \cap A_5| - |A_1 \cap A_2 \cap A_3 \cap A_4| - |A_1 \cap A_2 \cap A_3 \cap A_5| - |A_1 \cap A_2 \cap A_4 \cap A_5| - |A_1 \cap A_3 \cap A_4 \cap A_5| - |A_2 \cap A_3 \cap A_4 \cap A_5| + |A_1 \cap A_2 \cap A_3 \cap A_4 \cap A_5|$ **21.** $|A_1 \cup A_2 \cup A_3 \cup A_4 \cup A_5 \cup A_6| = |A_1| + |A_2| + |A_3| + |A_4| + |A_5| + |A_6| - |A_1 \cap A_2| - |A_1 \cap A_3| - |A_1 \cap A_4| - |A_1 \cap A_5| - |A_1 \cap A_6| - |A_2 \cap A_3| - |A_2 \cap A_4| - |A_2 \cap A_5| - |A_2 \cap A_6| - |A_3 \cap A_4| - |A_3 \cap A_5| - |A_3 \cap A_6| - |A_4 \cap A_5| - |A_4 \cap A_6| - |A_5 \cap A_6|$ ⋯ **23.** $p(E_1 \cup E_2 \cup E_3) = p(E_1) + p(E_2) + p(E_3) - p(E_1 \cap E_2) - p(E_1 \cap E_3) - p(E_2 \cap E_3) + p(E_1 \cap E_2 \cap E_3)$ **25.** 4972/71,295 **27.** $p(E_1 \cup E_2 \cup E_3 \cup E_4 \cup E_5) = p(E_1) + p(E_2) + p(E_3) + p(E_4) + p(E_5) - p(E_1 \cap E_2) - p(E_1 \cap E_3) - p(E_1 \cap E_4) - p(E_1 \cap E_5) - p(E_2 \cap E_3) - p(E_2 \cap E_4) - p(E_2 \cap E_5) - p(E_3 \cap E_4) - p(E_3 \cap E_5) - p(E_4 \cap E_5) + p(E_1 \cap E_2 \cap E_3) + p(E_1 \cap E_2 \cap E_4) + p(E_1 \cap E_2 \cap E_5) + p(E_1 \cap E_3 \cap E_4) + p(E_1 \cap E_3 \cap E_5) + p(E_1 \cap E_4 \cap E_5) + p(E_2 \cap E_3 \cap E_4) + P(E_2 \cap E_3 \cap E_5) + p(E_2 \cap E_4 \cap E_5) + p(E_3 \cap E_4 \cap E_5)$ **29.** $p\left(\bigcup_{i=1}^{n} E_i\right) = \sum_{1 \le i \le n} p(E_i) - \sum_{1 \le i < j \le n} p(E_i \cap E_j) + \sum_{1 \le i < j < k \le n} p(E_i \cap E_j \cap E_k) - \cdots + (-1)^{n+1} p\left(\bigcap_{i=1}^{n} E_i\right)$

SECTION 6.6

1. 75 **3.** 6 **5.** 46 **7.** 9875 **9.** 540 **11.** 2100 **13.** 1854 **15. a)** $D_{100}/100!$ **b)** $100D_{99}/100!$ **c)** $C(100,2)/100!$ **d)** 0 **e)** 1/100! **17.** 2,170,680 **19.** By Exercise 18 we have $D_n - nD_{n-1} = -(D_{n-1} - (n - 1)D_{n-2})$. Iterating, we have $D_n - nD_{n-1} = -(D_{n-1} - (n - 1)D_{n-2}) = -(-(D_{n-2} - (n - 2)D_{n-3})) = D_{n-2} - (n - 2)D_{n-3} = \cdots = (-1)^n(D_2 - 2D_1) = (-1)^n$ since $D_2 = 1$ and $D_1 = 0$. **21.** When n is odd **23.** $\phi(n) = n - \sum_{i=1}^{m} \frac{n}{p_i} + \sum_{1 \le i < j \le m} \frac{n}{p_i p_j} - \cdots \pm \frac{n}{p_1 p_2 \cdots p_m} = n \prod_{i=1}^{m} \left(a - \frac{1}{p_i}\right)$ **25.** 4 **27.** There are n^m functions from a set with m elements to a set with n elements, $C(n, 1)(n-1)^m$ functions from a set with m elements to a set with n elements that miss exactly one element, $C(n, 2)(n - 2)^m$ functions from a set with m elements to a set with n elements that miss exactly two elements, and

so on, with $C(n, n - 1)1^m$ functions from a set with m elements to a set with n elements that miss exactly $n - 1$ elements. Hence by the principle of inclusion–exclusion there are $n^m - C(n, 1)(n - 1)^m + C(n, 2)(n - 2)^m - \cdots + (-1)^{n-1}C(n, n - 1)1^m$ onto functions.

SUPPLEMENTARY EXERCISES

1. a) $A_n = 4A_{n-1}$ **b)** $A_1 = 40$ **c)** $A_n = 10 \cdot 4^n$ **3. a)** $M_n = M_{n-1} + 160,000$ **b)** $M_1 = 186,000$ **c)** $M_n = 160,000n + 26,000$ **d)** $T_n = T_{n-1} + 160,000n + 26,000$ **e)** $T_n = 80,000n^2 + 106,000n$ **5. a)** $a_n = a_{n-2} + a_{n-3}$ **b)** $a_1 = 0, a_2 = 1, a_3 = 1$ **c)** $a_{12} = 12$ **7. a)** 2 **b)** 5 **c)** 8 **d)** 16 **9.** $a_n = 2^n$ **11.** $a_n = 2 + 4n/3 + n^2/2 + n^3/6$ **13.** $a_n = a_{n-2} + a_{n-3}$ **15.** $O(n^4)$ **17.** $O(n)$ **19. a)** $18n + 18$ **b)** 18 **c)** 0 **21.** $\Delta(a_n b_n) = a_{n+1}b_{n+1} - a_n b_n = a_{n+1}(b_{n+1} - b_n) + b_n(a_{n+1} - a_n) = a_{n+1}\Delta b_n + b_n \Delta a_n$ **23. a)** Let $G(x) = \sum_{n=0}^{\infty} a_n x^n$. Then $G'(x) = \sum_{n=1}^{\infty} n a_n x^{n-1} = \sum_{n=0}^{\infty} (n + 1)a_{n+1} x^n$. Therefore, $G'(x) - G'(x) = \sum_{n=0}^{\infty} ((n + 1)a_{n+1} - a_n)x^n = \sum_{n=0}^{\infty} x^n/n! = e^x$, as desired. That $G(0) = a_0 = 1$ is given. **b)** We have $(e^{-x}G(x))' = e^{-x}G'(x) - e^{-x}G(x) = e^{-x}(G'(x) - G(x)) = e^{-x} \cdot e^x = 1$. Hence $e^{-x}G(x) = x + c$ where c is a constant. Consequently, $G(x) = xe^x + ce^x$. Since $G(0) = 1$, it follows that $c = 1$. **c)** We have $G(x) = \sum_{n=0}^{\infty} (x^{n+1}/n!) + \sum_{n=0}^{\infty} (x^n/n!) = \sum_{n=1}^{\infty} (x^n/(n - 1)!) + \sum_{n=0}^{\infty} (x^n/n!)$. Therefore, $a_n = 1/(n-1)! + 1/n!$ for all $n \ge 1$, and $a_0 = 1$. **25.** 7 **27.** 110 **29.** 0 **31. a)** 19 **b)** 65 **c)** 122 **d)** 167 **e)** 168 **33.** $D_{n-1}/(n - 1)!$ **35.** 11/32

CHAPTER 7

SECTION 7.1

1. a) $\{(0, 0), (1, 1), (2, 2), (3, 3)\}$ **b)** $\{(1, 3), (2, 2), (3, 1), (4, 0)\}$ **c)** $\{(1, 0), (2, 0), (2, 1), (3, 0), (3, 1), (3, 2), (4, 0), (4, 1), (4, 2), (4, 3)\}$ **d)** $\{(1, 0), (1, 1), (1, 2), (1, 3), (2, 0), (2, 2), (3, 0), (3, 3), (4, 0)\}$ (assuming that 0 does not divide 0) **e)** $\{(0, 1), (1, 0), (1, 1), (1, 2), (1, 3), (2, 1), (2, 3), (3, 1), (3, 2), (4, 1), (4, 3)\}$ **f)** $\{(1, 2), (2, 1), (2, 2)\}$ **3. a)** Transitive **b)** Reflexive, symmetric, transitive **c)** Symmetric **d)** Antisymmetric **e)** Reflexive, symmetric, antisymmetric, transitive **f)** None of these properties **5. a)** Reflexive, transitive **b)** Symmetric **c)** Symmetric **d)** Symmetric **7. a)** Symmetric **b)** Symmetric, transitive **c)** Symmetric **d)** Reflexive, symmetric, transitive **e)** Reflexive, transitive **f)** Reflexive, symmetric, transitive **g)** Antisymmetric **h)** Antisymmetric, transitive **9.** (c), (d), (f) **11. a)** Not irreflexive **b)** Not irreflexive **c)** Not irreflexive **d)** Not irreflexive **13.** Yes, for instance $\{(1, 1)\}$ on $\{1, 2\}$ **15.** $(a, b) \in R$ if and only if a is taller than b **17.** (a) **19.** None **21.** $\forall a \forall b ((a, b) \in R \rightarrow (b, a) \notin R)$ **23.** 2^{mn} **25. a)** $\{(a, b) \mid b$ divides

a} **b)** {(a, b) | a does not divide b} **27.** The graph of f^{-1} **29. a)** {(a, b) | a is required to read or has read b} **b)** {(a, b) | a is required to read and has read b} **c)** {(a, b) | either a is required to read b but has not read it or a has read b but is not required to} **d)** {(a, b) | a is required to read b but has not read it} **e)** {(a, b) | a has read b but is not required to} **31.** $S \circ R = $ {(a, b) | a is a parent of b and b has a sibling}, $R \circ S = $ {(a, b) | a is an aunt or uncle of b} **33. a)** $\mathbf{R}^2$ **b)** R_6 **c)** R_3 **d)** R_3 **e)** $\emptyset$ **f)** R_1 **g)** R_4 **h)** R_4 **35. a)** R_1 **b)** R_2 **c)** R_3 **d)** $\mathbf{R}^2$ **e)** R_3 **f)** $\mathbf{R}^2$ **g)** $\mathbf{R}^2$ **h)** $\mathbf{R}^2$ **37.** b got his or her doctorate under someone who got his or her doctorate under a; there is a sequence of $n + 1$ people, starting with a and ending with b, such that each is the advisor of the next person in the sequence **39. a)** {(a, b) | $a - b \equiv 0, 3, 4, 6, 8,$ or $9 \pmod{12}$} **b)** {(a, b) | $a \equiv b \pmod{12}$} **c)** {(a, b) | $a - b \equiv 3, 6,$ or $9 \pmod{12}$} **d)** {(a, b) | $a - b \equiv 4$ or $8 \pmod{12}$} **e)** {(a, b) | $a - b \equiv 3, 4, 6, 8,$ or $9 \pmod{12}$} **41.** 8 **43. a)** 65,536 **b)** 32,768 **45. a)** $2^{n(n+1)/2}$ **b)** $2^n 3^{n(n-1)/2}$ **c)** $3^{n(n-1)/2}$ **d)** $2^{n(n-1)}$ **e)** $2^{n(n-1)/2}$ **f)** $2^{n^2} - 2 \cdot 2^{n(n-1)}$ **47.** There may be no such b. **49.** If R is symmetric and $(a, b) \in R$, then $(b, a) \in R$, so that $(a, b) \in R^{-1}$. Hence $R \subseteq R^{-1}$. Similarly $R^{-1} \subseteq R$. So $R = R^{-1}$. Conversely, if $R = R^{-1}$ and $(a, b) \in R$, then $(a, b) \in R^{-1}$, so that $(b, a) \in R$. Thus R is symmetric. **51.** R is reflexive if and only if $(a, a) \in R$ for all $a \in A$ if and only if $(a, a) \in R^{-1}$ [since $(a, a) \in R$ if and only if $(a, a) \in R^{-1}$] if and only if R^{-1} is reflexive. **53.** Use mathematical induction. The result is trivial for $n = 1$. Assume R^n is reflexive and transitive. By Theorem 1, $R^{n+1} \subseteq R$. To see that $R \subseteq R^{n+1} = R^n \circ R$, let $(a, b) \in R$. By the inductive hypothesis, $R^n = R$ and hence is reflexive. Thus $(b, b) \in R^n$. Therefore $(a, b) \in R^{n+1}$. **55.** Use mathematical induction. The result is trivial for $n = 1$. Assume R^n is reflexive. Then $(a, a) \in R^n$ for all $a \in A$ and $(a, a) \in R$. Thus $(a, a) \in R^n \circ R = R^{n+1}$ for all $a \in A$. **57.** No, for instance, take $R = \{(1, 2), (2, 1)\}$.

SECTION 7.2

1. {$(1, 2, 3), (1, 2, 4), (1, 3, 4), (2, 3, 4)$} **3.** (Nadir, 122, 34, Detroit, 08:10), (Acme, 221, 22, Denver, 08:17), (Acme, 122, 33, Anchorage, 08:22), (Acme, 323, 34, Honolulu 08:30), (Nadir, 199, 13, Detroit, 08:47), (Acme, 222, 22, Denver, 09:10), (Nadir, 322, 34, Detroit, 09:44) **5.** Airline and flight number, airline and departure time **7. a)** Yes **b)** No **c)** No **9. a)** Social Security number **b)** There are no two people with the same name who happen to have the same street address. **c)** There are no two people with the same name living together. **11.** (Nadir, 122, 34, Detroit, 08: 10), (Nadir, 199, 13, Detroit, 08: 47), (Nadir, 322, 34, Detroit, 09: 44) **13.** (Nadir, 122, 34, Detroit, 08: 10), (Nadir, 199, 13, Detroit, 08: 47), (Nadir, 322, 34, Detroit, 09: 44), (Acme, 221, 22, Denver, 08: 17), (Acme, 222, 22, Denver, 09: 10) **15.** $P_{3,5,6}$

17.

Airline	Destination
Nadir	Detroit
Acme	Denver
Acme	Anchorage
Acme	Honolulu

19.

Supplier	Part_number	Project	Quantity	Color_code
23	1092	1	2	2
23	1101	3	1	1
23	9048	4	12	2
31	4975	3	6	2
31	3477	2	25	2
32	6984	4	10	1
32	9191	2	80	4
33	1001	1	14	8

21. Both sides of this equation pick out the subset of R consisting of those n-tuples satisfying both conditions C_1 and C_2. **23.** Both sides of this equation pick out the set of n-tuples that are in R, are in S, and satisfy condition C. **25.** Both sides of this equation pick out the m-tuples consisting of i_1th, i_2th, $\ldots, i_m$th components of n-tuples in either R or S. **27.** Let $R = \{(a, b)\}$ and $S = \{(a, c)\}, n = 2, m = 1,$ and $i_1 = 1; P_1(R - S) = \{(a)\},$ but $P_1(R) - P_1(S) = \emptyset$. **29. a)** J_2 followed by $P_{1,3}$ **b)** (23, 1), (23, 3), (31, 3), (32, 4)

SECTION 7.3

1. a) $\begin{bmatrix} 1 & 1 & 1 \\ 0 & 0 & 0 \\ 0 & 0 & 0 \end{bmatrix}$ **b)** $\begin{bmatrix} 0 & 1 & 0 \\ 1 & 1 & 0 \\ 0 & 0 & 1 \end{bmatrix}$

c) $\begin{bmatrix} 1 & 1 & 1 \\ 0 & 1 & 1 \\ 0 & 0 & 1 \end{bmatrix}$ **d)** $\begin{bmatrix} 0 & 0 & 1 \\ 0 & 0 & 0 \\ 1 & 0 & 0 \end{bmatrix}$

3. a) (1, 1), (1, 3), (2, 2), (3, 1), (3, 3) **b)** (1, 2), (2, 2), (3, 2) **c)** (1, 1), (1, 2), (1, 3), (2, 1), (2, 3), (3, 1), (3, 2), (3, 3) **5.** The relation is not irreflexive if and only if there is at least one loop in the digraph. **7. a)** Reflexive, symmetric, transitive **b)** Antisymmetric, transitive **c)** Symmetric **9. a)** 4950 **b)** 9900 **c)** 99 **d)** 100 **e)** 1 **11.** Change each 0 to a 1 and each 1 to a 0.

13. a) $\begin{bmatrix} 0 & 1 & 1 \\ 1 & 1 & 0 \\ 1 & 0 & 1 \end{bmatrix}$ **b)** $\begin{bmatrix} 1 & 0 & 0 \\ 0 & 0 & 1 \\ 0 & 1 & 0 \end{bmatrix}$ **c)** $\begin{bmatrix} 1 & 1 & 1 \\ 1 & 1 & 1 \\ 1 & 1 & 1 \end{bmatrix}$

15. a) $\begin{bmatrix} 0 & 0 & 1 \\ 1 & 1 & 0 \\ 0 & 1 & 1 \end{bmatrix}$ **b)** $\begin{bmatrix} 1 & 1 & 0 \\ 0 & 1 & 1 \\ 1 & 1 & 1 \end{bmatrix}$ **c)** $\begin{bmatrix} 0 & 1 & 1 \\ 1 & 1 & 1 \\ 1 & 1 & 1 \end{bmatrix}$

17. $n^2 - k$

19. a)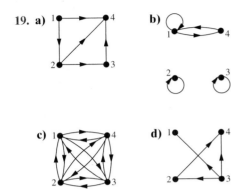
b)

c) **d)**

9. a) 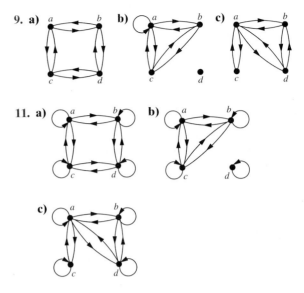 **b)** **c)**

11. a) **b)**

c)

21. For simplicity we have indicated pairs of edges between the same two vertices in opposite directions by using a double arrowhead, rather than drawing two separate lines.

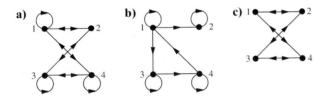

a) **b)** **c)**

23. $\{(a, b), (a, c), (b, c), (c, b)\}$ **25.** $(a, c), (b, a), (c, d),$ (d, b) **27.** $\{(a, a), (a, b), (a, c), (b, a), (b, b), (b, c), (c, a),$ $(c, b), (d, d)\}$ **29.** The relation is asymmetric if and only if the directed graph has no loops and no closed paths of length 2. **31.** Exercise 23: irreflexive. Exercise 24: reflexive, antisymmetric, transitive. Exercise 25: irreflexive, antisymmetric. **33.** Reverse the direction on every edge in the digraph for R. **35.** Proof by mathematical induction. *Basis step:* Trivial for $n = 1$. *Inductive step:* Assume true for k. Since $R^{k+1} = R^k \circ R$, its matrix is $\mathbf{M}_R \odot \mathbf{M}_{R^k}$. By the inductive hypothesis this is $\mathbf{M}_R \odot \mathbf{M}_R^{[k]} = \mathbf{M}_R^{[k+1]}$.

13. The symmetric closure of R is $R \cup R^{-1}$. $\mathbf{M}_{R \cup R^{-1}} = \mathbf{M}_R \vee \mathbf{M}_{R^{-1}} = \mathbf{M}_R \vee \mathbf{M}_R^t$. **15.** Only when R is irreflexive, in which case it is its own closure. **17.** $a, a, a, a; b,$ $c, c, b; c, b, c, c; c, c, c, b, c; c, c, c, c; d, d, e, e, d; e, d, e, e; e,$ $e, d, e; e, e, e, e$ **19. a)** $\{(1, 1), (1, 5), (2, 3), (3, 1), (3, 2),$ $(3, 3), (3, 4), (4, 1), (4, 5), (5, 3), (5, 4)\}$ **b)** $\{(1, 1), (1, 2),$ $(1, 3), (1, 4), (2, 1), (2, 5), (3, 1), (3, 3), (3, 4), (3, 5), (4, 1),$ $(4, 2), (4, 3), (4, 4), (5, 1), (5, 3), (5, 5)\}$ **c)** $\{(1, 1), (1, 3),$ $(1, 4), (1, 5), (2, 1), (2, 2), (2, 3), (2, 4), (3, 1), (3, 2), (3, 3),$ $(3, 4), (3, 5), (4, 1), (4, 3), (4, 4), (4, 5), (5, 1), (5, 2), (5, 3),$ $(5, 4), (5, 5)\}$ **d)** $\{(1, 1), (1, 2), (1, 3), (1, 4), (1, 5), (2, 1),$ $(2, 3), (2, 4), (2, 5), (3, 1), (3, 2), (3, 3), (3, 4), (3, 5), (4, 1),$ $(4, 2), (4, 3), (4, 4), (4, 5), (5, 1), (5, 2), (5, 3), (5, 4), (5, 5)\}$ **e)** $\{(1, 1), (1, 2), (1, 3), (1, 4), (1, 5), (2, 1), (2, 2), (2, 3), (2, 4),$ $(2, 5), (3, 1), (3, 2), (3, 3), (3, 4), (3, 5), (4, 1), (4, 2), (4, 3),$ $(4, 4), (4, 5), (5, 1), (5, 2), (5, 3), (5, 4), (5, 5)\}$ **f)** $\{(1, 1),$ $(1, 2), (1, 3), (1, 4), (1, 5), (2, 1), (2, 2), (2, 3), (2, 4), (2, 5),$ $(3, 1), (3, 2), (3, 3), (3, 4), (3, 5), (4, 1), (4, 2), (4, 3), (4, 4),$ $(4, 5), (5, 1), (5, 2), (5, 3), (5, 4), (5, 5)\}$ **21. a)** If there is a student c who shares a class with a and a class with b **b)** If there are two students c and d such that a and c share a class, c and d share a class, and d and b share a class **c)** If there is a sequence $s_0, \dots, s_n$ of students with $n \geq 1$ such that $s_0 = a, s_n = b,$ and for each $i = 1, 2, \dots, n,$ s_i and s_{i-1} share a class **23.** The result follows from $(R^*)^{-1} = (\bigcup_{n=1}^{\infty} R^n)^{-1} = \bigcup_{n=1}^{\infty} (R^n)^{-1} = \bigcup_{n=1}^{\infty} R^n = R^*$.

25. a) $\begin{bmatrix} 1 & 1 & 1 & 1 \\ 1 & 1 & 1 & 1 \\ 1 & 1 & 1 & 1 \\ 1 & 1 & 1 & 1 \end{bmatrix}$ **b)** $\begin{bmatrix} 0 & 0 & 0 & 0 \\ 1 & 0 & 1 & 1 \\ 1 & 0 & 1 & 1 \\ 1 & 0 & 1 & 1 \end{bmatrix}$

c) $\begin{bmatrix} 0 & 1 & 1 & 1 \\ 0 & 0 & 1 & 1 \\ 0 & 0 & 0 & 1 \\ 0 & 0 & 0 & 0 \end{bmatrix}$ **d)** $\begin{bmatrix} 1 & 1 & 1 & 1 \\ 1 & 1 & 1 & 1 \\ 1 & 1 & 1 & 1 \\ 1 & 1 & 1 & 1 \end{bmatrix}$

SECTION 7.4

1. a) $\{(0, 0), (0, 1), (1, 1), (1, 2), (2, 0), (2, 2), (3, 0), (3, 3)\}$ **b)** $\{(0, 1), (0, 2), (0, 3), (1, 0), (1, 1), (1, 2), (2, 0), (2, 1),$ $(2, 2), (3, 0)\}$ **3.** $\{(a, b) \mid a$ divides b or b divides $a\}$

5. 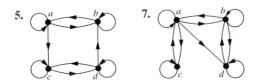 **7.**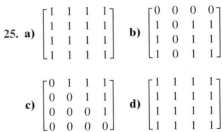

27. Answers same as for Exercise 25. **29. a)** $\{(1, 1),$ $(1, 2), (1, 4), (2, 2), (3, 3), (4, 1), (4, 2), (4, 4)\}$ **b)** $\{(1, 1),$ $(1, 2), (1, 4), (2, 1), (2, 2), (2, 4), (3, 3), (4, 1), (4, 2), (4, 4)\}$ **c)** $\{(1, 1),(1, 2),(1, 4),(2, 1),(2, 2),(2, 4),(3, 3),(4, 1),(4, 2),$ $(4, 4)\}$ **31.** Algorithm 1: $O(n^{3.8})$; Algorithm 2: $O(n^3)$ **33.** Initialize with $\mathbf{A} := \mathbf{M}_R \vee \mathbf{I}_n$ and loop only for $i := 2$ to $n - 1$. **35. a)** Since R is reflexive, every relation containing it must also be reflexive. **b)** Both $\{(0, 0), (0, 1), (0, 2),$ $(1, 1), (2, 2)\}$ and $\{(0, 0), (0, 1), (1, 0), (1, 1), (2, 2)\}$ contain R and have an odd number of elements, but neither is a subset of the other.

SECTION 7.5

1. a) Equivalence relation **b)** Not reflexive, not transitive **c)** Equivalence relation **d)** Not transitive **e)** Not symmetric, not transitive **3. a)** Equivalence relation **b)** Not transitive **c)** Not reflexive, not symmetric, not transitive **d)** Equivalence relation **e)** Not reflexive, not transitive **5. a)** $(x, x) \in R$ since $f(x) = f(x)$. Hence R is reflexive. $(x, y) \in R$ if and only if $f(x) = f(y)$, which holds if and only if $f(y) = f(x)$ if and only if $(y, x) \in R$. Hence R is symmetric. If $(x, y) \in R$ and $(y, z) \in R$, then $f(x) = f(y)$ and $f(y) = f(z)$. Hence $f(x) = f(z)$. Thus $(x, z) \in R$. It follows that R is transitive. **b)** The sets $f^{-1}(b)$ for b in the range of f. **7.** Let x be a string of length 3 or more. Because x agrees with itself in the first three bits, $(x, x) \in R$. Hence R is reflexive. Suppose that $(x, y) \in R$. Then x and y agree in the first three bits. Hence y and x agree in the first three bits. Thus $(y, x) \in R$. If (x, y) and (y, z) are in R then x and y agree in the first three bits, as do y and z. Hence x and z agree in the first three bits. Hence $(x, z) \in R$. It follows that R is transitive. **9.** The statement p is equivalent to q means that p and q have the same entries in their truth tables. R is reflexive, since p has the same truth table as p. R is symmetric, for if p and q have the same truth table, then q and p have the same truth table. If p and q have the same entries in their truth tables and q and r have the same entries in their truth tables, then p and r also do, so R is transitive. **11. a)** This follows from Exercise 5, where the function f from the set of differentiable functions (from $\mathbf{R}$ to $\mathbf{R}$) to the set of functions (from $\mathbf{R}$ to $\mathbf{R}$) is the differentiation operator. **b)** The set of all functions of the form $g(x) = x^2 + C$ for some constant C **13.** This follows from Exercise 5, where the function f from the set of all URLs to the set of all Web pages is the function that assigns to each URL the Web page for that URL. **15.** No **17.** No **19.** R is reflexive since a bit string s has the same number of 1s as itself. R is symmetric since s and t having the same number of 1s implies that t and s do. R is transitive since s and t having the same number of 1s, and t and u having the same number of 1s implies that s and u have the same number of 1s. **21. a)** The sets of people of the same age **b)** The sets of people with the same two parents **23.** The set of all bit strings with exactly two

1s. **25. a)** $\{s \mid s$ is a bit string of length 3$\}$ **b)** $\{s1 \mid s$ is a bit string of length 3$\}$ **c)** $\{s11 \mid s$ is a bit string of length 3$\}$ **d)** $\{s10101 \mid s$ is a bit string of length 3$\}$ **27.** $\{6n + k \mid n \in \mathbf{Z}\}$ for $k \in \{0, 1, 2, 3, 4, 5\}$ **29. a)** No **b)** Yes **c)** Yes **d)** No **31.** (a), (c), (e) **33.** (b), (d), (e) **35. a)** $\{(0, 0), (1, 1), (1, 2), (2, 1), (2, 2), (3, 3), (3, 4),$ $(3, 5), (4, 3), (4, 4), (4, 5), (5, 3), (5, 4), (5, 5)\}$ **b)** $\{(0, 0),$ $(0, 1), (1, 0), (1, 1), (2, 2), (2, 3), (3, 2), (3, 3), (4, 4), (4, 5),$ $(5, 4), (5, 5)\}$ **c)** $\{(0, 0), (0, 1), (0, 2), (1, 0), (1, 1), (1, 2),$ $(2, 0), (2, 1), (2, 2), (3, 3), (3, 4), (3, 5), (4, 3), (4, 4), (4, 5),$ $(5, 3), (5, 4), (5, 5)\}$ **d)** $\{(0, 0), (1, 1), (2, 2), (3, 3), (4, 4),$ $(5, 5)\}$ **37.** $[0]_6 \subseteq [0]_3, [1]_6 \subseteq [1]_3, [2]_6 \subseteq [2]_3, [3]_6 \subseteq [0]_3,$ $[4]_6 \subseteq [1]_3, [5]_6 \subseteq [2]_3$ **39.** Let A be a set in the first partition. Pick a particular element x of A. The set of all bit strings of length 16 that agree with x on the last eight bits is one of the sets in the second partition, and clearly every string in A is in that set. **41.** $\{(a, a), (a, b),$ $(a, c), (b, a), (b, b), (b, c), (c, a), (c, b), (c, c), (d, d), (d, e),$ $(e, d), (e, e)\}$ **43. a)** $\mathbf{Z}$ **b)** $\{n + \frac{1}{2} \mid n \in \mathbf{Z}\}$ **45. a)** R is reflexive since any coloring can be obtained from itself via a 360-degree rotation. To see that R is symmetric and transitive, use the fact that each rotation is the composition of two reflections and conversely the composition of two reflections is a rotation. Hence, (C_1, C_2) belongs to R if and only if C_2 can be obtained from C_1 by a composition of reflections. So if (C_1, C_2) belongs to R, so does (C_2, C_1) since the inverse of the composition of reflections is also a composition of reflections (in the opposite order). Hence, R is symmetric. To see that R is transitive, suppose (C_1, C_2) and (C_2, C_3) belong to R. Taking the composition of the reflections in each case yields a composition of reflections, showing that (C_1, C_3) belongs to R. **b)** We express colorings with sequences of length four, with r and b denoting red and blue, respectively. We list letters denoting the colors of the upper left square, upper right square, lower left square, and lower right square, in that order. The equivalence classes are: $\{rrrr\}$, $\{bbbb\}$, $\{rrrb, rrbr, rbrr, brrr\}$, $\{bbbr, bbrb, brbb, rbbb\}$, $\{rbbr, brrb\}$, $\{rrbb, brbr, bbrr, rbrb\}$. **47.** 5 **49.** Yes **51.** R **53.** First form the reflexive closure of R, then form the symmetric closure of the reflexive closure, and finally form the transitive closure of the symmetric closure of the reflexive closure. **55.** $p(0) = 1, p(1) = 1,$ $p(2) = 2, p(3) = 5, p(4) = 15, p(5) = 52, p(6) = 203,$ $p(7) = 877, p(8) = 4140, p(9) = 21147, p(10) = 115975$

SECTION 7.6

1. a) Yes **b)** No **c)** Yes **d)** No **3.** No **5.** Yes **7. a)** $\{(0, 0), (1, 0), (1, 1), (2, 0), (2, 1), (2, 2)\}$ **b)** $(\mathbf{Z}, \leq)$ **c)** $(P(\mathbf{Z}), \subseteq)$ **d)** $(\mathbf{Z}^+,$ "is a multiple of") **9. a)** $\{0\}$ and $\{1\}$, for instance **b)** 4 and 6, for instance **11. a)** $(1, 1, 2) < (1, 2, 1)$ **b)** $(0, 1, 2, 3) < (0, 1, 3, 2)$ **c)** $(0, 1, 1, 1, 0) < (1, 0, 1, 0, 1)$ **13.** $0 < 0001 < 001 <$ $01 < 010 < 0101 < 011 < 11$

15.

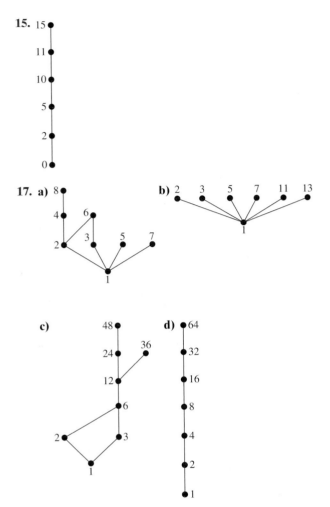

17. a)

b)

c)

d)

19. $(a, b), (a, c), (a, d), (b, c), (b, d), (a, a), (b, b), (c, c),$ (d, d) **21.** $(a, a), (a, g), (a, d), (a, e), (a, f), (b, b), (b, g),$ $(b, d), (b, e), (b, f), (c, c), (c, g), (c, d), (c, e), (c, f), (g, d),$ $(g, e), (g, f), (g, g), (d, d), (e, e), (f, f)$ **23.** $(\emptyset, \{a\}),$ $(\emptyset, \{b\}), (\emptyset, \{c\}), (\{a\}, \{a, b\}), (\{a\}, \{a, c\}), (\{b\}, \{a, b\}),$ $(\{b\}, \{b, c\}), (\{c\}, \{a, c\}), (\{c\}, \{b, c\}), (\{a, b\}, \{a, b, c\}),$ $(\{a, c\}, \{a, b, c\})$ $(\{b, c\}, \{a, b, c\})$ **25.** Let $(S, \preccurlyeq)$ be a finite poset. We will show that this poset is the reflexive transitive closure of its covering relation. Suppose that (a, b) is in the reflexive transitive closure of the covering relation. Then $a = b$ or $a \prec b$, so $a \preccurlyeq b$, or else there is a sequence $a_1, a_2, \ldots, a_n$ such that $a \prec a_1 \prec a_2 \prec \cdots \prec a_n \prec b$, in which case again $a \preccurlyeq b$ by the transitivity of $\preccurlyeq$. Conversely, suppose that $a \prec b$. If $a = b$ then (a, b) is in the reflexive transitive closure of the covering relation. If $a \prec b$ and there is no z such that $a \prec z \prec b$, then (a, b) is in the covering relation and therefore in its reflexive transitive closure. Otherwise, let $a \prec a_1 \prec a_2 \prec \cdots \prec a_n \prec b$ be a longest possible sequence of this form (which exists since the poset is finite). Then no intermediate elements can be inserted, so each pair $(a, a_1), (a_1, a_2), \ldots, (a_n, b)$

is in the covering relation, so again (a, b) is in its reflexive transitive closure.   **27. a)** 24, 45 **b)** 3, 5 **c)** No **d)** No **e)** 15, 45 **f)** 15 **g)** 15, 5, 3 **h)** 15 **29. a)** $\{1, 2\}, \{1, 3, 4\}, \{2, 3, 4\}$ **b)** $\{1\}, \{2\}, \{4\}$ **c)** No **d)** No **e)** $\{2, 4\}, \{2, 3, 4\}$ **f)** $\{2, 4\}$ **g)** $\{3, 4\}, \{4\}$ **h)** $\{3, 4\}$ **31.** Since $(a, b) \preccurlyeq (a, b)$, $\preccurlyeq$ is reflexive. If $(a_1, a_2) \preccurlyeq (b_1, b_2)$ and $(a_1, a_2) \neq (b_1, b_2)$, either $a_1 \prec b_1$, or $a_1 = b_1$ and $a_2 \prec b_2$. In either case, (b_1, b_2) is not less than or equal to (a_1, a_2). Hence $\preccurlyeq$ is antisymmetric. Suppose that $(a_1, a_2) \prec (b_1, b_2) \prec (c_1, c_2)$. Then if $a_1 \prec b_1$ or $b_1 \prec c_1$, we have $a_1 \prec c_1$, so $(a_1, a_2) \prec (c_1, c_2)$, but if $a_1 = b_1 = c_1$, then $a_2 \prec b_2 \prec c_2$, which implies that $(a_1, a_2) \prec (c_1, c_2)$. Hence $\preccurlyeq$ is transitive.   **33.** Since $(s, t) \leq (s, t)$, $\preccurlyeq$ is reflexive. If $(s, t) \preccurlyeq (u, v)$ and $(u, v) \preccurlyeq (s, t)$, then $s \preccurlyeq u \preccurlyeq s$ and $t \preccurlyeq v \preccurlyeq t$; hence $s = u$ and $t = v$. Hence $\preccurlyeq$ is antisymmetric. Suppose that $(s, t) \preccurlyeq (u, v) \preccurlyeq (w, x)$. Then $s \preccurlyeq u, t \preccurlyeq v, u \preccurlyeq w$, and $v \preccurlyeq x$. It follows that $s \preccurlyeq w$ and $t \preccurlyeq x$. Hence $(s, t) \preccurlyeq (w, x)$. Hence $\preccurlyeq$ is transitive. **35. a)** Suppose that x is maximal and that y is the largest element. Then $x \preccurlyeq y$. Since x is not less than y, it follows that $x = y$. By Exercise 34(a) y is unique. Hence x is unique.   **b)** Suppose that x is minimal and that y is the smallest element. Then $x \succcurlyeq y$. Since x is not greater than y, it follows that $x = y$. By Exercise 34(b) y is unique. Hence x is unique.   **37. a)** Yes **b)** No **c)** Yes **39.** Use mathematical induction. Let $P(n)$ be "Every subset with n elements from a lattice has a least upper bound and a greatest lower bound." *Basis step:* $P(1)$ is true since the least upper bound and greatest lower bound of $\{x\}$ is x. *Inductive step:* Assume that $P(k)$ is true. Let S be a set with $k + 1$ elements. Let $x \in S$ and $S' = S - \{x\}$. Since S' has k elements, by the inductive hypothesis, it has a least upper bound y and a greatest lower bound a. Now since we are in a lattice, there are elements $z = \text{lub}(x, y)$ and $b = \text{glb}(x, a)$. We are done if we can show that z is the least upper bound of S and b is the greatest lower bound of S. To show that z is the least upper bound of S, first note that if $w \in S$, then $w = x$ or $w \in S'$. If $w = x$ then $w \preccurlyeq z$ since z is the least upper bound of x and y. If $w \in S'$, then $w \preccurlyeq z$ because $w \preccurlyeq y$ since y is the least upper bound of S' and $y \preccurlyeq z$ since $z = \text{lub}(x, y)$. To see that z is the least upper bound of S suppose that u is an upper bound of S. Note that such an element u must be an upper bound of x and y, but since $z = \text{lub}(x, y)$, it follows that $z \preccurlyeq u$. We omit the similar argument that b is the greatest lower bound of S. **41. a)** No. **b)** Yes. **c)** (*Proprietary, {Cheetah,Puma}*), (*Restricted, {Cheetah,Puma}*), (*Registered, {Cheetah, Puma}*), (*Proprietary, {Cheetah,Puma,Impala}*), (*Restricted, {Cheetah,Puma,Impala}*), (*Registered, {Cheetah,Puma,Impala}*)   **d)** (*Nonproprietary, {Impala, Puma}*), (*Proprietary, {Impala, Puma}*), (*Restricted, {Impala, Puma}*), (*Nonproprietary, {Impala}*), (*Proprietary, {Impala}*), (*Restricted, {Impala}*), (*Nonproprietary, {Puma}*), (*Proprietary, {Puma}*),

(*Restricted*, {*Puma*}), (*Nonproprietary*, ∅), (*Proprietary*, ∅), (*Restricted*, ∅) **43.** Let Π be the set of all partitions of a set S with $P_1 \preccurlyeq P_2$ if P_1 is a refinement of P_2, that is, if every set in P_1 is a subset of a set in P_2. First, we show that $(\Pi, \preccurlyeq)$ is a poset. Since $P \preccurlyeq P$ for every partition P, $\preccurlyeq$ is reflexive. Now suppose that $P_1 \preccurlyeq P_2$ and $P_2 \preccurlyeq P_1$. Let $T \in P_1$. Since $P_1 \preccurlyeq P_2$, there is a set $T' \in P_2$ such that $T \subseteq T'$. Since $P_2 \preccurlyeq P_1$ there is a set $T'' \in P_1$ such that $T' \subseteq T''$. It follows that $T \subseteq T''$. But since P_1 is a partition, $T = T''$, which implies that $T = T'$ since $T \subseteq T' \subseteq T''$. Thus $T \in P_2$. By reversing the roles of P_1 and P_2 it follows that every set in P_2 is also in P_1. Hence $P_1 = P_2$ and $\preccurlyeq$ is antisymmetric. Next, suppose that $P_1 \preccurlyeq P_2$ and $P_2 \preccurlyeq P_3$. Let $T \in P_1$. Then there is a set $T' \in P_2$ such that $T \subseteq T'$. Since $P_2 \preccurlyeq P_3$ there is a set $T'' \in P_3$ such that $T' \subseteq T''$. This means that $T \subseteq T''$. Hence $P_1 \preccurlyeq P_3$. It follows that $\preccurlyeq$ is transitive. The greatest lower bound of the partitions P_1 and P_2 is the partition P whose subsets are the nonempty sets of the form $T_1 \cap T_2$ where $T_1 \in P_1$ and $T_2 \in P_2$. We omit the justification of this statement here. The least upper bound of the partitions P_1 and P_2 is the partition that corresponds to the equivalence relation in which $x \in S$ is related to $y \in S$ if there is a sequence $x = x_0, x_1, x_2, \ldots, x_n = y$ for some nonnegative integer n such that for each i from 1 to n, x_{i-1} and x_i are in the same element of P_1 or of P_2. We omit the details that this is an equivalence relation and the details of the proof that this is the least upper bound of the two partitions. **45.** By Exercise 39 there is a least upper bound and a greatest lower bound for the entire finite lattice. By definition these elements are the greatest and least elements, respectively. **47.** The least element of a subset of $\mathbf{Z}^+ \times \mathbf{Z}^+$ is that pair that has the smallest possible first coordinate, and, if there is more than one such pair, that pair among those that has the smallest second coordinate. **49.** If x is an integer in a decreasing sequence of elements of this poset, then at most $|x|$ elements can follow x in the sequence, namely, integers whose absolute values are $|x|-1, |x|-2, \ldots, 1, 0$. Therefore there can be no infinite decreasing sequence. This is not a totally ordered set, since 5 and -5, for example, are incomparable. **51.** To find which of two rational numbers is larger, write them with a positive common denominator and compare numerators. To show that this set is dense, suppose $x < y$ are two rational numbers. Then their average, i.e., $(x + y)/2$ is a rational number between them. **53.** Let $(S, \preccurlyeq)$ be a partially ordered set. It is enough to show that every nonempty subset of S contains a least element if and only if there is no infinite decreasing sequence of elements $a_1, a_2, a_3, \ldots$ in S (i.e., where $a_{i+1} \prec a_i$ for all i). An infinite decreasing sequence of elements clearly has no least element. Conversely, let A be any nonempty subset of S that has no least element. Since A is nonempty, choose $a_1 \in A$. Since a_1 is not the least element of A, choose $a_2 \in A$ with $a_2 \prec a_1$. Since a_2 is not the least element of A, choose $a_3 \in A$ with $a_3 \prec a_2$. Continue in this manner, producing an infinite decreasing sequence in S.

55. $a \prec b \prec c \prec d \prec e \prec f \prec g \prec h \prec i \prec j \prec k \prec l \prec m$ **57.** $C \prec A \prec B \prec D \prec E \prec F \prec G$ **59.** Determine user needs $\prec$ Write functional requirements $\prec$ Set up test sites $\prec$ Develop system requirements $\prec$ Write documentation $\prec$ Develop module $A \prec$ Develop module $B \prec$ Develop module $C \prec$ Integrate modules $\prec \alpha$ test $\prec \beta$ test $\prec$ Completion

SUPPLEMENTARY EXERCISES

1. a) Irreflexive (we do not include the empty string), symmetric **b)** Irreflexive, symmetric **c)** Irreflexive, antisymmetric, transitive **3.** $((a, b), (a, b)) \in R$ since $a + b = a + b$. Hence R is reflexive. If $((a, b), (c, d)) \in R$ then $a + d = b + c$, so that $c + b = d + a$. It follows that $((c, d), (a, b)) \in R$. Hence R is symmetric. Suppose that $((a, b), (c, d))$ and $((c, d), (e, f))$ belong to R. Then $a + d = b + c$ and $c + f = d + e$. Adding these two equations and subtracting $c + d$ from both sides gives $a + f = b + e$. Hence $((a, b), (e, f))$ belongs to R. Hence R is transitive. **5.** Suppose that $(a, b) \in R$. Since $(b, b) \in R$ it follows that $(a, b) \in R^2$. **7.** Yes, yes **9.** Yes, yes **11.** Two records with identical keys in the projection would have identical keys in the original. **13.** $(\Delta \cup R)^{-1} = \Delta^{-1} \cup R^{-1} = \Delta \cup R^{-1}$ **15. a)** $R = \{(a, b), (a, c)\}$. The transitive closure of the symmetric closure of R is $\{(a, a), (a, b), (a, c), (b, a), (b, b), (b, c), (c, a), (c, b), (c, c)\}$ and is different from the symmetric closure of the transitive closure of R, which is $\{(a, b), (a, c), (b, a), (c, a)\}$. **b)** Suppose (a, b) is in the symmetric closure of the transitive closure of R. We must show that (a, b) is in the transitive closure of the symmetric closure of R. We know that at least one of (a, b) and (b, a) is in the transitive closure of R. Hence, there is either a path from a to b in R or a path from b to a in R (or both). In the former case, there is a path from a to b in the symmetric closure of R. In the latter case, we can form a path from a to b in the symmetric closure of R by reversing the directions of all the edges in a path from b to a, going backward. Hence (a, b) is in the transitive closure of the symmetric closure of R. **17.** The closure of S with respect to property **P** is a relation with property **P** that contains R since $R \subseteq S$. Hence the closure of S with respect to property **P** contains the closure of R with respect to property **P**. **19.** Use the basic idea of Warshall's algorithm, except let $w_{ij}^{[k]}$ equal the length of the longest path from v_i to v_j using interior vertices with subscripts not exceeding k, and equal to -1 if there is no such path. To find $w_{ij}^{[k]}$ from the entries of $\mathbf{W}_{k-1}$, determine for each pair (i, j) whether there are paths from v_i to v_k and from v_k to v_j using no vertices labeled greater than k. If either $w_{ik}^{[k-1]}$ or $w_{kj}^{[k-1]}$ is -1, then such a pair of paths does not exist, so set $w_{ij}^{[k]} = w_{ij}^{[k-1]}$. If such a pair of paths exists, then there are two possibilities. If $w_{kk}^{[k-1]} > 0$, there are paths of arbitrary long length from v_i to v_j, so set $w_{ij}^{[k]} = \infty$. If $w_{kk}^{[k-1]} = 0$, set $w_{ij}^{[k-1]} = \max(w_{ij}^{[k-1]}, w_{ik}^{[k-1]} + w_{kj}^{[k-1]})$. (Initially take $\mathbf{W}_0 = \mathbf{M}_R$.) **21.** 25 **23.** Since $A_i \cap B_j$ is a subset of A_i and of B_j, the collection of subsets is a refinement of each of the given partitions. We must show that it is a partition. By construction, each of these sets is nonempty. To see that their union is S, suppose that $s \in S$. Since P_1 and P_2 are partitions of S, there are sets A_i and B_j such that $s \in A_i$ and $s \in B_j$. Therefore $s \in A_i \cap B_j$. Hence the union of these sets is S. To see that they are pairwise disjoint, note that unless $i = i'$ and $j = j'$, $(A_i \cap B_j) \cap (A_{i'} \cap B_{j'}) = (A_i \cap A_{i'}) \cap (B_j \cap B_{j'}) = \emptyset$. **25.** The subset relation is a partial ordering on any collection of sets, since it is reflexive, antisymmetric, and transitive. Here the collection of sets is $\mathbf{R}(S)$. **27.** Find recipe $\prec$ Buy seafood $\prec$ Buy groceries $\prec$ Wash shellfish $\prec$ Cut ginger and garlic $\prec$ Clean fish $\prec$ Steam rice $\prec$ Cut fish $\prec$ Wash vegetables $\prec$ Chop water chestnuts $\prec$ Make garnishes $\prec$ Cook in wok $\prec$ Arrange on platter $\prec$ Serve **29. a)** The only antichain with more than one element is $\{c, d\}$. **b)** The only antichains with more than one element are $\{b, c\}, \{c, e\}$, and $\{d, e\}$. **c)** The only antichains with more than one element are $\{a, b\}, \{a, c\}, \{b, c\}$, $\{a, b, c\}, \{d, e\}, \{d, f\}, \{e, f\}, \{d, e, f\}$. **31.** Let $(S, \preccurlyeq)$ be a finite poset, and let A be a maximal chain. Since $(A, \preccurlyeq)$ is also a poset it must have a minimal element m. Suppose that m is not minimal in S. Then there would be an element a of S with $a \prec m$. However, this would make the set $A \cup \{a\}$ a larger chain than A. To show this, we must show that a is comparable with every element of A. Since m is comparable with every element of A and m is minimal, it follows that $m \prec x$ when x is in A and $x \neq m$. Since $a \prec m$ and $m \prec x$, the transitive law shows that $a \prec x$ for every element of A. **33.** Let aRb denote that a is a descendant of b. By Exercise 32, if no set of $n + 1$ people none of whom is a descendant of any other (an antichain) exists, then $k \le n$, so the set can be partitioned into $k \le n$ chains. By the pigeonhole principle, at least one of these chains contains at least $m + 1$ people. **35.** We prove by contradiction that if S has no infinite decreasing sequence and $\forall x((\forall y(y \prec x \to P(y))) \to P(x))$, then $P(x)$ is true for all $x \in S$. If it does not hold that $P(x)$ is true for all $x \in S$, let x_1 be an element of S such that $P(x_1)$ is not true. Then by the implication already given, it must be the case that $\forall y(y \prec x_1 \to P(y))$ is not true. This means that there is some x_2 with $x_2 \prec x_1$ such that $P(x_2)$ is not true. Again invoking the implication, we get an $x_3 \prec x_2$ such that $P(x_3)$ is not true, and so on forever. This contradicts the well-foundedness of our poset. Therefore, $P(x)$ is true for all $x \in S$. **37.** Suppose that R is a quasi-ordering. Since R is reflexive, if $a \in A$, then $(a, a) \in R$. This implies that $(a, a) \in R^{-1}$. Hence $a \in R \cap R^{-1}$. It follows that $R \cap R^{-1}$ is reflexive. $R \cap R^{-1}$ is symmetric for any relation R since, for any relation R, if $(a, b) \in R$ then $(b, a) \in R^{-1}$ and vice versa. To show that $R \cap R^{-1}$ is transitive, suppose that $(a, b) \in R \cap R^{-1}$ and $(b, c) \in R \cap R^{-1}$. Since $(a, b) \in R$ and $(b, c) \in R$, $(a, c) \in R$, since R is transitive. Similarly, since $(a, b) \in R^{-1}$ and $(b, c) \in R^{-1}$, $(b, a) \in R$ and $(c, b) \in R$, so that $(c, a) \in R$ and $(a, c) \in R^{-1}$. Hence

$(a, c) \in R \cap R^{-1}$. It follows that $R \cap R^{-1}$ is an equivalence relation. **39. a)** Since $\mathrm{glb}(x, y) = \mathrm{glb}(y, x)$ and $\mathrm{lub}(x, y) = \mathrm{lub}(y, x)$, it follows that $x \wedge y = y \wedge x$ and $x \vee y = y \vee x$. **b)** Using the definition, $(x \wedge y) \wedge z$ is a lower bound of x, y, and z that is greater than every other lower bound. Since x, y, and z play interchangeable roles, $x \wedge (y \wedge z)$ is the same element. Similarly, $(x \vee y) \vee z$ is an upper bound of x, y, and z that is less than every other upper bound. Since x, y, and z play interchangeable roles, $x \vee (y \vee z)$ is the same element. **c)** To show that $x \wedge (x \vee y) = x$ it is sufficient to show that x is the greatest lower bound of x and $x \vee y$. Note that x is a lower bound of x and since $x \vee y$ is by definition greater than x, x is a lower bound for it as well. Therefore, x is a lower bound. But any lower bound of x has to be less than x, so x is the greatest lower bound. The second statement is the dual of the first; we omit its proof. **d)** x is a lower, and an upper, bound for itself and itself, and the greatest, and least, such bound. **41. a)** Since 1 is the only element greater than or equal to 1, it is the only upper bound for 1 and therefore the only possible value of the least upper bound of x and 1. **b)** Since $x \preccurlyeq 1$, x is a lower bound for both x and 1 and no other lower bound can be greater than x, so $x \wedge 1 = x$. **c)** Since $0 \preccurlyeq x$, x is an upper bound for both x and 0 and no other bound can be less than x, so $x \vee 0 = x$. **d)** Since 0 is the only element less than or equal to 0, it is the only lower bound for 0 and therefore the only possible value of the greatest lower bound of x and 0. **43.** $L = (S, \subseteq)$ where $S = \{\emptyset, \{1\}, \{2\}, \{3\}, \{1, 2\}, \{2, 3\}, \{1, 2, 3\}\}$ **45.** Yes. **47.** The complement of a subset $X \subseteq S$ is its complement $S - X$. To prove this, note that $X \vee (S - X) = 1$ and $X \wedge (S - X) = 0$ since $X \cup (S - X) = S$ and $X \cap (S - X) = \emptyset$.

CHAPTER 8

SECTION 8.1

1. a)

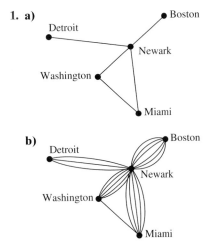

c)

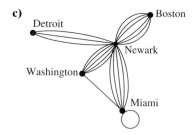

d)

e)

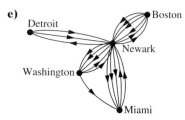

3. Simple graph   **5.** Pseudograph   **7.** Directed graph
9. Directed multigraph

11. a)

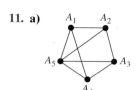

b)

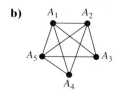

c)

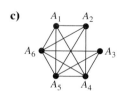

13.

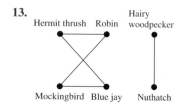

15.

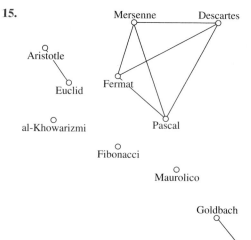

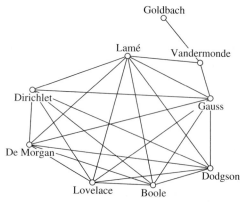

17.

19.

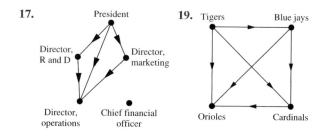

21. We find the telephone numbers in the call graph for February that are not present in the call graph for January and vice versa. For each number we find, we make a list of the numbers they called or were called by using the edges in the call graph. We examine these lists to find new telephone numbers in February that had similar calling patterns to defunct telephone numbers in January.   **23.** We use the graph model that has e-mail addresses as vertices and for each message sent, an edge from the e-mail address of the sender to the e-mail address of the recipient. For each e-mail address, we can make a list of other addresses they sent messages to and a list of other address from which they received messages. If two e-mail addresses had almost the same pattern, we conclude that these addresses might have belonged to the same person who had recently changed his or her e-mail address.   **25.** Let the set of vertices be a set of people, and two vertices are joined by an edge if the two people were ever married. Ignoring complications, this graph has the property that there are two types of vertices (men and women), and every edge joins vertices of opposite types.

27.

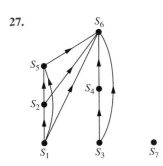

29. Represent people in the group by vertices. Put a directed edge into the graph for every pair of vertices. Label the edge from the vertex representing A to the vertex representing B with a + (plus) if A likes B, a − (minus) if A dislikes B, and a 0 if A is neutral about B.

SECTION 8.2

1. $v = 6$; $e = 6$; deg $(a) = 2$, deg$(b) = 4$, deg$(c) = 1$, deg$(d) = 0$, deg$(e) = 2$, deg$(f) = 3$; c is pendant; d is isolated.   **3.** $v = 9$; $e = 12$; deg$(a) = 3$, deg$(b) = 2$, deg$(c) = 4$, deg$(d) = 0$, deg$(e) = 6$, deg$(f) = 0$; deg$(g) = 4$; deg$(h) = 2$; deg$(i) = 3$; d and f are isolated. **5.** No, since the sum of the degrees of the vertices cannot be odd.   **7.** $v = 4$; $e = 7$; deg$^-(a) = 3$, deg$^-(b) = 1$, deg$^-(c) = 2$, deg$^-(d) = 1$, deg$^+(a) = 1$, deg$^+(b) = 2$, deg$^+(c) = 1$, deg$^+(d) = 3$.   **9.** 5 vertices, 13 edges; deg$^-(a) = 6$, deg$^+(a) = 1$, deg$^-(b) = 1$, deg$^+(b) = 5$, deg$^-(c) = 2$, deg$^+(c) = 5$, deg$^-(d) = 4$, deg$^+(d) = 2$, deg$^-(e) = 0$, deg$^+(e) = 0$.

11.

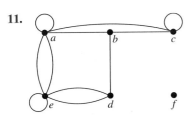

13. The number of collaborators v has; someone who has never collaborated; someone who has just one collaborator   **15.** In the directed graph deg$^-(v) =$ number of calls v received, deg$^+(v) =$ number of calls v made; in the undirected graph, deg(v) is the number of calls either made or received by v.   **17.** (deg$^+(v)$, deg$^-(v)$) is the win–loss record of v.   **19.** Bipartite   **21.** Not bipartite   **23.** Not bipartite   **25. a)** n vertices, $n(n-1)/2$ edges   **b)** n vertices, n edges   **c)** $n + 1$ vertices, $2n$ edges   **d)** $m + n$

vertices, mn edges   **e)** 2^n vertices, $n2^{n-1}$ edges   **27.** 7

29. a) Yes

b) No, sum of degrees is odd.   **c)** No   **d)** No, sum of degrees is odd.

e) Yes

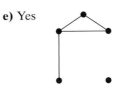

f) No, sum of degrees is odd.   **31.** 17

33.

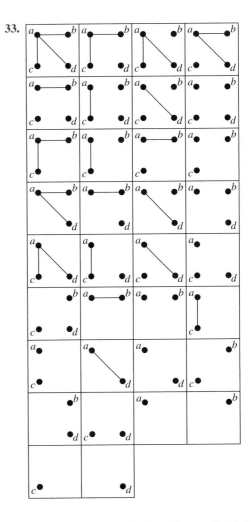

35. a) For all $n \geq 1$ **b)** For all $n \geq 3$ **c)** For $n = 3$ **d)** For all $n \geq 0$ **37.** 5

39.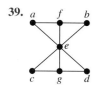

41. a) The graph with n vertices and no edges **b)** The disjoint union of K_m and K_n **c)** The graph with vertices $\{v_1, \ldots, v_n\}$ with an edge between v_i and v_j unless $i \equiv j \pm 1 \pmod{n}$ **d)** The graph whose vertices are represented by bit strings of length n with an edge between two vertices if the associated bit strings differ in more than one bit **43.** $v(v-1)/2 - e$ **45.** The union of G and $\overline{G}$ contains an edge between each pair of the n vertices. Hence this union is K_n.

47. Exercise 7:

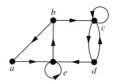

Exercise 8:   Exercise 9:

49. A directed graph $G = (V, E)$ is its own converse if and only if it satisfies the condition $(u, v) \in E$ if and only if $(v, u) \in E$. But this is precisely the condition that the associated relation must satisfy to be symmetric.

51. $P(0, 0) \quad P(0, 1) \quad P(0, 2)$

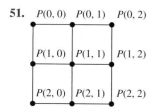

53. We can connect $P(i, j)$ and $P(k, l)$ by using $|i - k|$ hops to connect $P(i, j)$ and $P(k, j)$ and $|j - l|$ hops to connect $P(k, j)$ and $P(k, l)$. Hence the total number of hops required to connect $P(i, j)$ and $P(k, l)$ does not exceed $|i - k| + |j - l|$. This is less than or equal to $m + m = 2m$, which is $O(m)$.

SECTION 8.3

1.

Vertex	Adjacent Vertices
a	b, c, d
b	a, d
c	a, d
d	a, b, c

3.

Vertex	Terminal Vertices
a	a, b, c, d
b	d
c	a, b
d	b, c, d

5. $\begin{bmatrix} 0 & 1 & 1 & 1 \\ 1 & 0 & 0 & 1 \\ 1 & 0 & 0 & 1 \\ 1 & 1 & 1 & 0 \end{bmatrix}$

where vertices are listed in alphabetical order

7. $\begin{bmatrix} 1 & 1 & 1 & 1 \\ 0 & 0 & 0 & 1 \\ 1 & 1 & 0 & 0 \\ 0 & 1 & 1 & 1 \end{bmatrix}$ **9. a)** $\begin{bmatrix} 0 & 1 & 1 & 1 \\ 1 & 0 & 1 & 1 \\ 1 & 1 & 0 & 1 \\ 1 & 1 & 1 & 0 \end{bmatrix}$

b) $\begin{bmatrix} 0 & 1 & 1 & 1 & 1 \\ 1 & 0 & 0 & 0 & 0 \\ 1 & 0 & 0 & 0 & 0 \\ 1 & 0 & 0 & 0 & 0 \\ 1 & 0 & 0 & 0 & 0 \end{bmatrix}$ **c)** $\begin{bmatrix} 0 & 0 & 1 & 1 & 1 \\ 0 & 0 & 1 & 1 & 1 \\ 1 & 1 & 0 & 0 & 0 \\ 1 & 1 & 0 & 0 & 0 \\ 1 & 1 & 0 & 0 & 0 \end{bmatrix}$

d) $\begin{bmatrix} 0 & 1 & 0 & 1 \\ 1 & 0 & 1 & 0 \\ 0 & 1 & 0 & 1 \\ 1 & 0 & 1 & 0 \end{bmatrix}$ **e)** $\begin{bmatrix} 0 & 1 & 0 & 1 & 1 \\ 1 & 0 & 1 & 0 & 1 \\ 0 & 1 & 0 & 1 & 1 \\ 1 & 0 & 1 & 0 & 1 \\ 1 & 1 & 1 & 1 & 0 \end{bmatrix}$

f) $\begin{bmatrix} 0 & 1 & 1 & 0 & 1 & 0 & 0 & 0 \\ 1 & 0 & 0 & 1 & 0 & 1 & 0 & 0 \\ 1 & 0 & 0 & 1 & 0 & 0 & 1 & 0 \\ 0 & 1 & 1 & 0 & 0 & 0 & 0 & 1 \\ 1 & 0 & 0 & 0 & 0 & 1 & 1 & 0 \\ 0 & 1 & 0 & 0 & 1 & 0 & 0 & 1 \\ 0 & 0 & 1 & 0 & 1 & 0 & 0 & 1 \\ 0 & 0 & 0 & 1 & 0 & 1 & 1 & 0 \end{bmatrix}$

11. 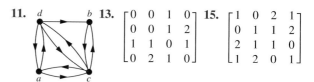 **13.** $\begin{bmatrix} 0 & 0 & 1 & 0 \\ 0 & 0 & 1 & 2 \\ 1 & 1 & 0 & 1 \\ 0 & 2 & 1 & 0 \end{bmatrix}$ **15.** $\begin{bmatrix} 1 & 0 & 2 & 1 \\ 0 & 1 & 1 & 2 \\ 2 & 1 & 1 & 0 \\ 1 & 2 & 0 & 1 \end{bmatrix}$

17. 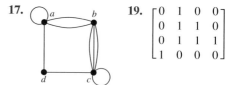 **19.** $\begin{bmatrix} 0 & 1 & 0 & 0 \\ 0 & 1 & 1 & 0 \\ 0 & 1 & 1 & 1 \\ 1 & 0 & 0 & 0 \end{bmatrix}$

21. $\begin{bmatrix} 1 & 1 & 2 & 1 \\ 1 & 0 & 0 & 2 \\ 1 & 0 & 1 & 1 \\ 0 & 2 & 1 & 0 \end{bmatrix}$ **23.** **25.** Yes

27. Exercise 13: $\begin{bmatrix} 1 & 0 & 0 & 0 & 0 \\ 0 & 1 & 1 & 1 & 0 \\ 1 & 1 & 0 & 0 & 1 \\ 0 & 0 & 1 & 1 & 1 \end{bmatrix}$

Exercise 14: $\begin{bmatrix} 1 & 1 & 1 & 1 & 0 & 0 & 0 & 0 \\ 1 & 1 & 1 & 0 & 1 & 0 & 0 & 0 \\ 0 & 0 & 0 & 0 & 1 & 1 & 1 & 1 \\ 0 & 0 & 0 & 1 & 0 & 1 & 1 & 1 \end{bmatrix}$

Exercise 15: $\begin{bmatrix} 1 & 1 & 1 & 1 & 0 & 0 & 0 & 0 & 0 & 0 \\ 0 & 0 & 0 & 0 & 1 & 1 & 1 & 1 & 0 & 0 \\ 0 & 1 & 1 & 0 & 0 & 1 & 0 & 0 & 1 & 0 \\ 0 & 0 & 0 & 1 & 0 & 0 & 1 & 1 & 0 & 1 \end{bmatrix}$

29. $\deg(v)$ – number of loops at v; $\deg^-(v)$ **31.** 2 if e is not a loop, 1 if e is a loop

33. a) $\begin{bmatrix} 1 & 1 & \cdots & 1 & 0 & \cdots & 0 \\ 1 & 0 & \cdots & 0 & 1 & \cdots & 0 \\ 0 & 1 & \cdots & 0 & 1 & \cdots & 0 \\ \vdots & \vdots & & \vdots & \vdots & & \vdots \\ 0 & 0 & \cdots & 0 & 0 & \cdots & 1 \\ 0 & 0 & \cdots & 1 & 0 & \cdots & 1 \end{bmatrix}$

b) $\begin{bmatrix} 1 & 0 & \cdots & 0 & 1 \\ 1 & 1 & \cdots & 0 & 0 \\ 0 & 1 & \cdots & 0 & 0 \\ \vdots & \vdots & & \vdots & \vdots \\ 0 & 0 & \cdots & 1 & 0 \\ 0 & 0 & \cdots & 1 & 1 \end{bmatrix}$

c) $\begin{bmatrix} 0 & 0 & \cdots & 0 & 1 & 1 & \cdots & 1 \\ & & & & 1 & 0 & \cdots & 0 \\ & & \mathbf{B} & & 0 & 1 & \cdots & 0 \\ & & & & \vdots & \vdots & & \vdots \\ & & & & 0 & 0 & \cdots & 1 \end{bmatrix}$

where **B** is the answer to (b)

d) $\begin{bmatrix} 1 & 1 & \cdots & 1 & 0 & \cdots & 0 \\ 0 & 0 & \cdots & 0 & 1 & \cdots & 0 \\ \vdots & \vdots & & \vdots & \vdots & & \vdots \\ 0 & 0 & \cdots & 0 & 0 & \cdots & 1 \\ 1 & 0 & & 0 & 1 & \cdots & 0 \\ 0 & 1 & & 0 & 0 & & \\ \vdots & \vdots & & \vdots & \vdots & & \\ 0 & 0 & \cdots & 1 & 0 & \cdots & 1 \end{bmatrix}$

35. Isomorphic   **37.** Isomorphic   **39.** Isomorphic   **41.** Not isomorphic   **43.** Isomorphic   **45.** G is isomorphic to itself by the identity function, so isomorphism is reflexive. Suppose that G is isomorphic to H. Then there exists a one-to-one correspondence f from G to H that preserves adjacency and nonadjacency. It follows that f^{-1} is a one-to-one correspondence from H to G that preserves adjacency and nonadjacency. Hence isomorphism is symmetric. If G is isomorphic to H and H is isomorphic to K, then there are one-to-one correspondences f and g from G to H and from H to K that preserve adjacency and nonadjacency. It follows that $g \circ f$ is a one-to-one correspondence from G to K that preserves adjacency and nonadjacency. Hence isomorphism is transitive.   **47.** All zeros   **49.** Label the vertices in order so that all of the vertices in the first set of the partition of the vertex set come first. Since no edges join vertices in the same set of the partition, the matrix has the desired form.   **51.** C_5 **53.** $n = 5$ only

55. 4   **57. a)** Yes   **b)** No   **c)** No   **59.** $G = (V_1, E_1)$ is isomorphic to $H = (V_2, E_2)$ if and only if there exist functions f from V_1 to V_2 and g from E_1 to E_2 such that each is a one-to-one correspondence and for every edge e in E_1 the endpoints of $g(e)$ are $f(v)$ and $f(w)$ where v and w are the endpoints of e.   **61.** Yes   **63.** Yes   **65.** If f is an isomorphism from a directed graph G to a directed graph H, then f is also an isomorphism from G^c to H^c. To see this note that (u, v) is an edge of G^c if and only if (v, u) is an edge of G if and only if $(f(v), f(u))$ is an edge of H if and only if $(f(u), f(v))$ is an edge of H^c.   **67.** The product is $[a_{ij}]$ where a_{ij} is the number of edges from v_i to v_j when $i \neq j$ and a_{ii} is the number of edges incident to v_i.   **69.** The graphs in Exercise 41 provide a devil's pair.

SECTION 8.4

1. a) Path of length 4; not a circuit; not simple   **b)** Not a path   **c)** Not a path   **d)** Simple circuit of length 5   **3.** No   **5.** No   **7.** Maximal sets of people with the property that for any two of them, we can find a string of acquaintances that takes us from one to the other   **9.** If a person has Erdős number n, then there is a path of length n from that person to Erdős in the collaboration graph, so by definition, that means that that person is in the same component as Erdős. If a person is in the same component as Erdős, then there is a path from that person to Erdős, and the length of the shortest such path is that person's Erdős number.   **11.** The maximal sets of phone numbers for which it is possible to find directed paths between every two different numbers in the set   **13 a)** $\{a, b, f\}, \{c, d, e\}$ **b)** $\{a, b, c, d, e, h\}, \{f\}, \{g\}$ **c)** $\{a, b, d, e, f, g, h, i\}, \{c\}$ **15. a)** 2   **b)** 7   **c)** 20   **d)** 61   **17. a)** 3   **b)** 0   **c)** 27   **d)** 0   **19. a)** 1   **b)** 0   **c)** 2   **d)** 1   **e)** 5   **f)** 3   **21.** R is reflexive by definition. Assume that $(u, v) \in R$; then there is a path from u to v. Then $(v, u) \in R$ since there is a path from v to u, namely, the path from u to v traversed backward. Assume that $(u, v) \in R$ and $(v, w) \in R$; then there are paths from u to v and from v to w. Putting these two paths together gives a path from u to w. Hence $(u, w) \in R$. It follows that R is transitive.   **23.** c **25.** b, c, e, i **27.** If a vertex is pendant it is clearly not a cut vertex. So an endpoint of a cut edge that is a cut vertex is not pendant. Removal of a cut edge produces a graph with more connected components than in the original graph. If an endpoint of a cut edge is not pendant, the connected component it is in after the removal of the cut edge contains more than just this vertex. Consequently, removal of that vertex, and all edges incident to it, including the original cut edge, produces a graph with more connected components than were in the original graph. Hence an endpoint of a cut edge that is not pendant is a cut vertex.   **29.** Assume there exists a connected graph G with at most one vertex that is not a cut vertex. Define the distance between the vertices u and v, denoted by $d(u, v)$, to be the length of the shortest path between u and v in G. Let s and t be vertices in G such that $d(s, t)$ is a maximum. Either s or t (or both) is a cut vertex, so without loss of generality suppose that s is a cut vertex. Let w belong to the connected component that does not contain t of the graph obtained by deleting s and all edges incident to s from G. Since every path from w to t contains s, $d(w, t) > d(s, t)$, which is a contradiction. **31. a)** Denver–Chicago, Boston–New York **b)** Seattle–Portland, Portland–San Francisco, Salt Lake City–Denver, New York–Boston, Boston–Burlington, Boston–Bangor. **33.** A set of people who collectively influence everyone (directly or indirectly); {Deborah} **35.** An edge cannot connect two vertices in different connected components. Since there are at most $C(n_i, 2)$ edges in the connected component with n_i vertices, it follows that there are at most $\sum_{i=1}^{k} C(n_i, 2)$ edges in the graph. **37.** Suppose that G is not connected. Then it has a component of k vertices for some k, $1 \leq k \leq n - 1$. The most edges G could have is $C(k, 2) + C(n - k, 2) = (k(k - 1) + (n - k)(n - k - 1))/2 = k^2 - nk + (n^2 - n)/2$. This quadratic function of f is minimized at $k = n/2$ and maximized at $k = 1$ or $k = n - 1$. Hence if G is not connected, the number of edges does not exceed the value of this function at 1 and at $n - 1$, namely, $(n - 1)(n - 2)/2$. **39. a)** 1 **b)** 2 **c)** 6 **d)** 21 **41.** 2 **43.** Let the paths P_1 and P_2 be $u = x_0, x_1, \ldots, x_n = v$, respectively. Since P_1 and P_2 do not contain the same set of edges, they must eventually diverge. If this happens only after one of them has ended, the rest of the other path is a simple circuit from v to v. Otherwise, we can suppose that $x_0 = y_0, x_1 = y_1, \ldots, x_t = y_t$, but $x_{t+1} \neq y_{t+1}$. Follow the path y_1, y_{t+1}, y_{t+2}, and so on, until it once again encounters a vertex on P_1. Once we are back on P_1, follow it, forward or backward as necessary, back to x_t. Since $x_i = y_i$, this forms a circuit that must be simple since no edge among the x_ks can be repeated and no edge among the x_ks can equal one of the Y_is that we used. **45.** The graph G is connected if and only if every off-diagonal entry of $\mathbf{A} + \mathbf{A}^2 + \mathbf{A}^3 + \cdots + \mathbf{A}^{n-1}$ is positive where $\mathbf{A}$ is the adjacency matrix of G. **47.** If the graph is bipartite, say with parts A and B, then the vertices in every path must alternately lie in A and B. Therefore a path that starts in A, say, will end in B after an odd number of steps and in A after an even number of steps. Since a circuit ends at the same vertex where it starts, the length must be even. Conversely, suppose that all circuits have even length; we must show that the graph is bipartite. We can assume that the graph is connected, because if it is not, then we can just work on one component at a time. Let v be a vertex of the graph, and let A be the set of all vertices to which there is a path of odd length starting at v, and let B be the set of all vertices to which there is a path of even length starting at v. Since the component is connected, every vertex lies in A or B. No vertex can lie in both A and B, since then following the odd-length path from v to that vertex and then back along the even-length path from that vertex to v would produce

an odd circuit, contrary to the hypothesis. Thus the set of vertices has been partitioned into two sets. To show that every edge has endpoints in different parts, suppose that xy is an edge where $x \in A$. Then the odd-length path from v to x followed by xy produces an even-length path from v to y, so $y \in B$. (Similarly if $x \in B$.)

SECTION 8.5

1. Neither **3.** No Euler circuit; $a, e, c, e, b, e, d, b, a, c, d$ **5.** $a, b, c, d, c, e, d, b, e, a, e, a$ **7.** $a, i, h, g, d, e, f, g, c, e, h, d, c, a, b, i, c, b, h, a$ **9.** No, A still has odd degree. **11.** When the graph in which vertices represent intersections and edges streets has an Euler path **13.** Yes **15.** No **17.** If there is an Euler path, as we follow it each vertex except the starting and ending vertices must have equal in-degree and out-degree, since whenever we come to a vertex along an edge, we leave it along another edge. The starting vertex must have out-degree 1 larger than its in-degree, since we use one edge leading out of this vertex and whenever we visit it again we use one edge leading into it and one leaving it. Similarly, the ending vertex must have in-degree 1 greater than its out-degree. Since the Euler path with directions erased produces a path between any two vertices, in the underlying undirected graph, the graph is weakly connected. Conversely, suppose the graph meets the degree conditions stated. If we add one more edge from the vertex of deficient out-degree to the vertex of deficient in-degree, then the graph has every vertex with equal in-degree and out-degree. Since the graph is still weakly connected, by Exercise 16 this new graph has an Euler circuit. Now delete the added edge to obtain the Euler path. **19.** Neither **21.** No Euler circuit; $a, d, e, d, b, a, e, c, e, b, c, b, e$ **23.** Neither **25.** Follow the same procedure as Algorithm 1, taking care to follow the directions of edges. **27. a)** $n = 2$ **b)** None **c)** None **d)** $n = 1$ **29.** Exercise 1:1 time; Exercises 2–7: 0 times **31.** a, b, c, d, e, a is a Hamilton circuit. **33.** No Hamilton circuit exists, because once a purported circuit has reached e it would have nowhere to go. **35.** No Hamilton circuit exists, because every edge in the graph is incident to a vertex of degree 2 and therefore must be in the circuit. **37.** a, b, c, f, d, e is a Hamilton path. **39.** f, e, d, a, b, c is a Hamilton path. **41.** No Hamilton path exists. There are eight vertices of degree 2, and only two of them can be end vertices of a path. For each of the other six, their two incident edges must be in the path. It is not hard to see that if there is to be a Hamilton path, exactly one of the inside corner vertices must be an end, and that this is impossible. **43.** $a, b, c, f, i, h, g, d, e$ is a Hamilton path. **45.** $m = n \geq 2$ **47. a)** (i) No, (ii) No, (iii) Yes **b)** (i) No, (ii) No, (iii) Yes **c)** (i) Yes, (ii) Yes, (iii) Yes **d)** (i) Yes, (ii) Yes, (iii) Yes **49.** The result is trivial for $n = 1$: code is 0, 1. Assume we have a Gray code of order n. Let $c_1, \ldots, c_k$, $k = 2^n$ be such a code. Then $0c_1, \ldots, 0c_k, 1c_k, \ldots, 1c_l$ is a Gray code of order $n + 1$.

51. **procedure** *Fleury* ($G = (V, E)$: connected multigraph
with the degree of all vertices even, $V = \{v_l, \ldots, v_n\}$)

$v := v_1$

circuit $:= v$

$H := G$

while H has edges

begin

$e :=$ first edge with endpoint V in H (with respect to
listing of V) such that e is not a cut edge of H, if one
exists, and simply the first edge in H with endpoint
v otherwise

$w :=$ other endpoint of e

circuit $:=$ *circuit* with edge e, w added

$v := w$

$H := H - e$

end {*circuit* is an Euler circuit}

53. If G has an Euler circuit, it is also an Euler path. If
not, add an edge between the two vertices of odd degree and apply the algorithm to get an Euler circuit. Then
delete the new edge. **55.** Suppose $G = (V, E)$ is a bipartite graph with $V = V_1 \cup V_2$, where no edge connects a vertex in V_1 and a vertex in V_2. Suppose that V
has a Hamilton circuit. Such a circuit must be of the form
$a_1, b_1, a_2, b_2, \ldots, a_k, b_k, a_1$, where $a_i \in V_1$ and $b_i \in V_2$ for
$i = 1, 2, \ldots, k$. Since the Hamilton circuit visits each ver-
tex exactly once, except for v_1, where it begins and ends,
the number of vertices in the graph equals $2k$, an even
number. Hence, a bipartite graph with an odd number of
vertices cannot have a Hamilton circuit.

57.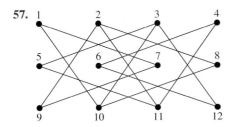

59. We represent the squares of a 3×4 chessboard as follows:

1	2	3	4
5	6	7	8
9	10	11	12

A knight's tour can be made by following the moves 8, 10,
1, 7, 9, 2, 11, 5, 3, 12, 6, 4 **61.** We represent the squares
of a 4×4 chessboard as follows:

1	2	3	4
5	6	7	8
9	10	11	12
13	14	15	16

There are only two moves from each of the four corner
squares. If we include all the edges 1–10, 1–7, 16–10, and
16–7, a circuit is completed too soon, so at least one of these
edges must be missing. Without loss of generality, assume
the path starts 1–10, 10–16, and 16–7. Now the only moves
from square 3 are to squares 5, 10, and 12 and square 10 already has two incident edges. Therefore, 3–5 and 3–12 must
be in the Hamilton circuit. Similarly, edges 8–2 and 8–15
must be in the circuit. Now the only moves from square 9
are to squares 2, 7, and 15. If there were edges from square
9 to both squares 2 and 15, a circuit would be completed
too soon. Therefore the edge 9–7 must be in the circuit giving square 7 its full complement of edges. But now square
14 is forced to be joined to squares 5 and 12, completing a circuit too soon (5–14–12–3–5). This contradiction
shows that there is no knight's tour on the 4×4 board.
63. Since there are mn squares on an $m \times n$ board, if both
m and n are odd, there are an odd number of squares.
Since by Exercise 62 the corresponding graph is bipartite,
by Exercise 55 it has no Hamilton circuit. Hence there is
no reentrant knight's tour. **65. a)** If G does not have
a Hamilton circuit, continue as long as possible adding
missing edges one at a time in such a way that we do not
obtain a graph with a Hamilton circuit. This cannot go on
forever, because once we've formed the complete graph
by adding all missing edges, there is a Hamilton circuit.
Whenever the process stops, we have obtained a (necessarily noncomplete) graph H with the desired property.
b) Add one more edge to H. This produces a Hamilton
circuit, which uses the added edge. The path consisting of
this circuit with the added edge omitted is a Hamilton path
in H. **c)** Clearly v_1 and v_n are not adjacent in H, since H
has no Hamilton circuit. Therefore they are not adjacent
in G. But the hypothesis was that the sum of the degrees of
vertices not adjacent in G was at least n. This inequality can
be rewritten as $n - \deg(v_n) \leq \deg(v_1)$. But $n - \deg(v_n)$ is just
the number of vertices not adjacent to v_n. **d)** Since there
is no vertex following v_n in the Hamilton path, v_n is not in S.
Each one of the $\deg(v_1)$ vertices adjacent to v_1 gives rise to
an element of S, so S contains $\deg(v_1)$ vertices. **e)** By part
(c) there are at most $\deg(v_1) - 1$ vertices other than v_n not
adjacent to v_n, and by part (d) there are $\deg(v_1)$ vertices
in S, none of which is v_n. Therefore at least one vertex of S is
adjacent to v_n. By definition, if v_k is this vertex, then H contains edges $v_k v_n$ and $v_1 v_{k+1}$, where $1 < k < n - 1$. **f)** Now
$v_1, v_2, \ldots, v_{k-1}, v_k, v_n, v_{n-1}, \ldots, v_{k+1}, v_1$ is a Hamilton circuit in H, contradicting the construction of H. Therefore

our assumption that G did not originally have a Hamilton circuit is wrong, and our proof by contradiction is complete.

SECTION 8.6

1. a) Vertices are the stops, edges join adjacent stops, weights are the times required to travel between adjacent stops. **b)** Same as (a), except weights are distances between adjacent stops. **c)** Same as (a), except weights are fares between stops. **3.** 16 **5.** Exercise 2: *a, b, e, d, z* Exercise 3: *a, c, d, e, g, z* Exercise 4: *a, b, e, h, l, m, p, s, z* **7. a)** *a, c, d* **b)** *a, c, d, f* **c)** *c, d, f* **e)** *b, d, e, g, z* **9. a)** Direct **b)** Via New York **c)** Via Atlanta and Chicago **d)** Via New York **11. a)** Via Chicago **b)** Via Chicago **c)** Via Los Angeles **d)** Via Chicago **13. a)** Via Chicago **b)** Via Chicago **c)** Via Los Angeles **d)** Via Chicago **15.** Do not stop the algorithm when z is added to the set S. **17. a)** Via Woodbridge, via Woodbridge and Camden **b)** Via Woodbridge, via Woodbridge and Camden **19.** For instance, sightseeing tours, street cleaning

21.

	a	*b*	*c*	*d*	*e*	*z*
a	4	3	2	8	10	13
b	3	2	1	5	7	10
c	2	1	2	6	8	11
d	8	5	6	4	2	5
e	10	7	8	2	4	3
z	13	10	11	5	3	6

23. $O(n^3)$ **25.** *a-c-b-d-a* (or the same circuit starting at some point but traversing the vertices in the same or reverse order) **27.** San Francisco–Denver–Detroit–New York–Los Angeles–San Francisco (or the same circuit starting at some other point but traversing the vertices in the same or reverse order)
29. Consider this graph:

The circuit *a-b-a-c-a* visits each vertex at least once (and the vertex *a* twice) and has total weight 6. Every Hamilton circuit has total weight 103.

SECTION 8.7

1. Yes **3.**

5. No

7. Yes

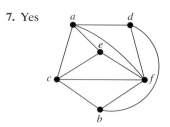

9. No **11.** A triangle is formed by the planar representation of the subgraph of K_5 consisting of the edges connecting v_1, v_2, and v_3. The vertex v_4 must be placed either within the triangle or outside of it. We will consider only the case when v_4 is inside the triangle; the other case is similar. Drawing the three edges from v_1, v_2, and v_3 to v_4 forms four regions. No matter which of these four regions v_5 is in, it is possible to join it to only three, and not all four, of the other vertices. **13.** 8 **15.** Since there are no loops or multiple edges and no simple circuits of length 3, and the degree of the unbounded region is at least 4, each region has degree at least 4. Thus $2e \geq 4r$, or $r \leq e/2$. But $r = e - v + 2$, so we have $e - v + 2 \leq e/2$, which implies that $e \leq 2v - 4$. **17.** As in Corollary 2, we have $2e \geq 5r$ and $r = e - v + 2$. Thus $e - v + 2 \leq 2e/5$, which implies that $e \leq (5/3)v - (10/3)$. **19.** Only (a) and (c) **21.** Not homeomorphic to $K_{3,3}$ **23.** Planar **25.** Nonplanar **27. a)** 1 **b)** 3 **c)** 9 **d)** 2 **e)** 4 **f)** 16 **29.** Draw $K_{m,n}$ as described in the hint. The number of crossings is four times the number in the first quadrant. The vertices on the x-axis to the right of the origin are $(1, 0), (2, 0), \ldots, (m/2, 0)$ and the vertices on the y-axis above the origin are $(0, 1), (0, 2), \ldots, (0, n/2)$. We obtain all crossings by choosing any two distinct numbers a and b with $1 \leq a < b \leq m/2$ and two distinct numbers r and s with $1 \leq r < s \leq n/2$; we get exactly one crossing in the graph between the edge connecting $(a, 0)$ and $(0, s)$ and the edge connecting $(b, 0)$ and $(0, r)$. Hence the number of crossings in the first quadrant is $C\left(\frac{m}{2}, 2\right) \cdot C\left(\frac{n}{2}, 2\right) = \frac{(m/2)(m/2-1)}{2} \cdot \frac{(n/2)(n/2-1)}{2}$. Hence the total number of crossings is $4 \cdot mn(m-2)(n-2)/64 = mn(m-2)(n-2)/16$. **31. a)** 2 **b)** 2 **c)** 2 **d)** 2 **e)** 2 **f)** 2 **33.** The formula is valid for $n \leq 4$. If $n > 4$, by Exercise 32 the thickness of K_n is at least $C(n, 2)/(3n - 6) = (n + 1 + \frac{2}{n-2})/6$ rounded up. Since this quantity is never an integer, it equals the next largest integer rounded down, which equals $\lfloor (n + 7)/6 \rfloor$. **35.** This follows from Exercise 34 since $K_{m,n}$ has mn edges and $m + n$ vertices and has no triangles since it is bipartite.

37.

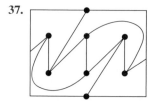

SECTION 8.8

1. Four colors   **3.** Three colors

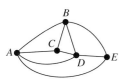

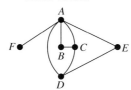

5. 3 **7.** 3 **9.** 2 **11.** 3 **13.** Graphs with no edges **15.** 3 if n is even, 4 if n is odd **17.** Period 1: Math 115, Math 185; period 2: Math 116, CS 473; period 3: Math 195, CS 101; period 4: CS 102; period 5: CS 273 **19.** 5 **21.** Exercise 5: 3 Exercise 6: 6 Exercise 7: 3 Exercise 8: 4 Exercise 9: 3 Exercise 10: 6 Exercise 11: 4 **23.** 5 **25.** The set of vertices with one of the colors is one of the parts, and the set of vertices with the other color is the other part. Since no edge can join vertices of the same color, there are no edges between vertices in the same part. **27.** Color 1: e, f, d; color 2: c, a, i, g; color 3: h, b, j **29.** Color C_6 **31. a)** 6 **b)** 7 **c)** 9 **d)** 11 **33.** Represent frequencies by colors and zones by vertices. Join two vertices with an edge if the zones these vertices represent interfere with one another. Then a k-tuple coloring is precisely an assignment of frequencies that avoids interference. **35.** We use induction on the number of vertices of the graph. Every graph with five or fewer vertices can be colored with five or fewer colors, since each vertex can get a different color. That takes care of the basis case(s). So we assume that all graphs with k vertices can be 5-colored and consider a graph G with $k + 1$ vertices. By Corollary 2 in Section 8.7, G has a vertex v with degree at most 5. Remove v to form the graph G'. Since G' has only k vertices, we 5-color it by the inductive hypothesis. If the neighbors of v do not use all five colors, then we can 5-color G by assigning to v a color not used by any of its neighbors. The difficulty arises if v has five neighbors, and each has a different color in the 5-coloring of G'. Suppose that the neighbors of v, when considered in clockwise order around v, are $a, b, c, m,$ and p. (This order is determined by the clockwise order of the curves representing the edges incident to v.) Suppose that the colors of the neighbors are azure, blue, chartreuse, magenta, and purple, respectively. Consider the azure-chartreuse subgraph (i.e., the vertices in G colored azure or chartreuse and all the edges between them). If a and c are not in the same component of this graph, then in the component containing a we can interchange these two colors (make the azure vertices chartreuse and vice versa), and G' will still be properly colored. That makes a chartreuse, so we can now color v azure, and G has been properly colored. If a and c are in the same component, then there is a path of vertices alternately colored azure and chartreuse joining a and c. This path together with edges av and vc divides

the plane into two regions, with b in one of them and m in the other. If we now interchange blue and magenta on all the vertices in the same region as b, we will still have a proper coloring of G', but now blue is available for v. In this case, too, we have found a proper coloring of G. This completes the inductive step, and the theorem is proved.

SUPPLEMENTARY EXERCISES

1. 2500 **3.** Yes **5.** Yes **7.** $\sum_{i=1}^{m} n_i$ vertices, $\sum_{t<j} n_i n_j$ edges

9. a)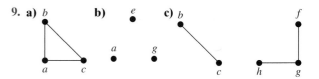

11. Complete subgraphs containing the following sets of vertices: $\{b, c, e, f\}$, $\{a, b, g\}$, $\{a, d, g\}$, $\{d, e, g\}$, $\{b, e, g\}$ **13.** Complete subgraphs containing the following sets of vertices: $\{b, c, d, j, k\}$, $\{a, b, j, k\}$, $\{e, f, g, i\}$, $\{a, b, i\}$, $\{a, i, j\}$, $\{b, d, e\}$, $\{b, e, i\}$, $\{b, i, j\}$, $\{g, h, i\}$, $\{h, i, j\}$ **15.** $\{c, d\}$ is a minimum dominating set.

17. a)

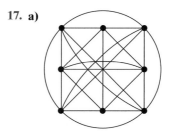

b)

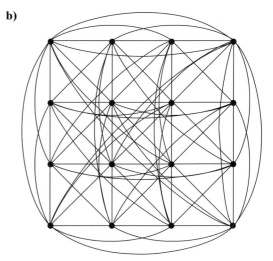

19. a) 1 **b)** 2 **c)** 3 **21. a)** A path from u to v in a graph G induces a path from $f(u)$ to $f(v)$ in an isomorphic graph H. **b)** Suppose f is an isomorphism from

G to H. If $v_0, v_1, \ldots, v_n, v_0$ is a Hamilton circuit in G, then $f(v_0), f(v_1), \ldots, f(v_n), f(v_0)$ must be a Hamilton circuit in H since it is still a circuit and $f(v_i) \neq f(v_j)$ for $0 \leq i < j \leq n$. **c)** Suppose f is an isomorphism from G to H. Then if $v_0, v_1, \ldots, v_n, v_0$ is an Euler circuit in G, then $f(v_0), f(v_1), \ldots, f(v_n), f(v_0)$ must be an Euler circuit in H since it is a circuit that contains each edge exactly once. **d)** Two isomorphic graphs must have the same crossing number since they can be drawn exactly the same way in the plane. **e)** Suppose f is an isomorphism from G to H. Then v is isolated in G if and only if $f(v)$ is isolated in H. Hence the graphs must have the same number of isolated vertices. **f)** Suppose f is an isomorphism from G to H. If G is bipartite, then the vertex set of G can be partitioned into V_1 and V_2 with no edge connecting a vertex of V_1 with one in V_2. Then the vertex set of H can be partitioned into $f(V_1)$ and $f(V_2)$ with no edge connecting a vertex in $f(V_1)$ and one in $f(V_2)$. **23.** 3 **25. a)** Yes **b)** No **27.** No **29.** Yes **31.** If e is a cut edge with endpoints u and v, then if we direct e from u to v, there will be no path in the directed graph from v to u, or else e would not have been a cut edge. Similar reasoning works if we direct e from v to u. **33.** $n-1$ **35.** Let the vertices represent the chickens. We include the edge (u, v) in the graph if and only if chicken u dominates chicken v. **37. a)** 4 **b)** 2 **c)** 3 **d)** 4 **e)** 4 **f)** 2 **39. a)** Suppose that $G = (V, E)$. Let $a, b \in V$. We must show that the distance between a and b in $\overline{G}$ is at most 2. If $\{a, b\} \notin E$ this distance is 1, so assume $\{a, b\} \in E$. Since the diameter of G is greater than 3, there are vertices u and v such that the distance in G between u and v is greater than 3. Either u or v, or both, is not in the set $\{a, b\}$. Assume that u is different from both a and b. Either $\{a, u\}$ or $\{b, u\}$ belongs to E; otherwise a, u, b would be a path in $\overline{G}$ of length 2. So, without loss of generality, assume $\{a, u\} \in E$. Thus v cannot be a or b, and by the same reasoning either $\{a, v\} \in E$ or $\{b, v\} \in E$. In either case, this gives a path of length less than or equal to 3 from u to v in G, a contradiction. **b)** Suppose $G = (V, E)$. Let $a, b \in V$. We must show that the distance between a and b in $\overline{G}$ does not exceed 3. If $\{a, b\} \notin E$, the result follows, so assume that $\{a, b\} \in E$. Since the diameter of G is greater than or equal to 3, there exist vertices u and v such that the distance in G between u and v is greater than or equal to 3. Either u or v, or both, is not in the set $\{a, b\}$. Assume u is different from both a and b. Either $\{a, u\} \in E$ or $\{b, u\} \in E$; otherwise a, u, b is a path of length 2 in $\overline{G}$. So, without loss of generality, assume $\{a, u\} \in E$. Thus v is different from a and from b. If $\{a, v\} \in E$, then u, a, v is a path of length 2 in G, so $\{a, v\} \notin E$ and thus $\{b, v\} \in E$ (or else there would be a path a, v, b of length 2 in $\overline{G}$). Hence $\{u, b\} \notin E$; otherwise u, b, v is a path of length 2 in G. Thus, a, v, u, b is a path of length 3 in $\overline{G}$, as desired.

41. a, b, e, z **43.** a, c, b, d, e, z **45.** If G is planar, then since $e \leq 3v - 6$, G has at most 27 edges. (If G is not connected it has even fewer edges.) Similarly, $\overline{G}$ has at most 27 edges. But the union of G and $\overline{G}$ is K_{11}, which has 55 edges, and $55 > 27 + 27$. **47.** Suppose that G is colored with k colors and has independence number i. Since each color class must be an independent set, each color class has no more than i elements. Thus there are at most ki vertices. **49. a)** By Theorem 2 of Section 5.2, the probability of selecting exactly m edges is $C(n, m)p^m(1 - p)^{n-m}$. **b)** By Theorem 2 in Section 5.3, the expected value is np. **c)** To generate a labeled graph G, as we apply the process to pairs of vertices, the random number x chosen must be less than or equal to $1/2$ when G has an edge between that pair of vertices and greater than $1/2$ when G has no edge there. Hence the probability of making the correct choice is $1/2$ for each edge and $1/2^{C(n,2)}$ overall. Hence all labeled graphs are equally likely. **51.** Suppose P is monotone increasing. If the property of not having P were not retained whenever edges are removed from a simple graph, there would be a simple graph G not having P and another simple graph G' with the same vertices but with some of the edges of G missing which has P. But P is monotone increasing, so since G' has P, so does G obtained by adding edges to G'. This is a contradiction. The proof of the converse is similar.

CHAPTER 9

SECTION 9.1

1. (a),(c),(e) **3. a)** a **b)** a,b,c,d,f,h,j,q,t **c)** $e,g,i,$ k,l,m,n,o,p,r,s,u **d)** q,r **e)** c **f)** p **g)** f,b,a **h)** $e,$ f, l, m, n **5.** No **7.** level 0: a; level 1: b, c, d; level 2: e through k (in alphabetical order); level 3: l through r; level 4: s, t; level 5: u **9. a)** The entire tree **b)** $c, g, h, o,$ p and the four edges cg, ch, ho, hp **c)** e alone **11. a)** 1 **b)** 2 **13. a)** 3 **b)** 9 **15.** The "only if" part is Theorem 2 and the definition of a tree. Suppose G is a connected simple graph with n vertices and $n - 1$ edges. If G is not a tree, it contains, by Exercise 14, an edge whose removal produces a graph G', which is still connected. If G' is not a tree, remove an edge to produce a connected graph G''. Repeat this procedure until the result is a tree. This requires at most $n - 1$ steps since there are only $n - 1$ edges. By Theorem 2, the resulting graph has $n - 1$ edges since it has n vertices. It follows that no edges were deleted, so that G was already a tree. **17.** 9999 **19.** 2000 **21.** 999 **23.** 1,000,000 dollars **25.** No such tree exists by Theorem 4 since it is impossible for $m = 2$ or $m = 84$.

27. Complete binary tree of height 4:

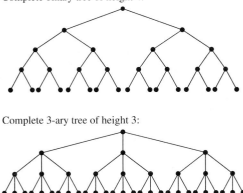

Complete 3-ary tree of height 3:

path cannot contain both u and v or else there would be a simple circuit. In fact, this path from c to w leaves P and does not return to P once it, possibly, follows part of P toward either u or v. Without loss of generality, assume this path does not follow P toward u. Then the path from u to c to w is simple, and of length more than e, a contradiction. Hence u and v are adjacent. Now since any two centers are adjacent, if there were more than two centers, T would contain K_3, a simple circuit, as a subgraph, which is a contradiction.

45.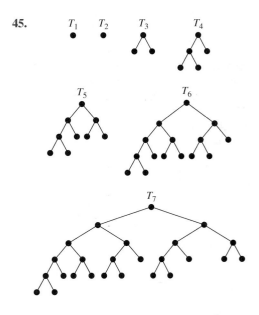

29. a) By Theorem 3 it follows that $n = mi + 1$. Since $i + l = n$, we have $l = n - i$, so that $l = (mi + 1) - i = (m - 1)i + 1$. **b)** We have $n = mi + 1$ and $i + l = n$. Hence $i = n - l$. It follows that $n = m(n - l) + 1$. Solving for n gives $n = (ml - 1)/(m - 1)$. From $i = n - l$ we obtain $i = [(ml - 1)/(m - 1)] - l = (l - 1)/(m - 1)$. **31.** $n - t$ **33. a)** 1 **b)** 3 **c)** 5 **35. a)** The parent directory **b)** A subdirectory or contained file **c)** A subdirectory or contained file in the same parent directory **d)** All directories in the path name **e)** All subdirectories and files continued in the directory or a subdirectory of this directory, and so on **f)** The length of the path to this directory or file **g)** The depth of the system, i.e., the length of the longest path **37.** Let $n = 2^k$, where k is a positive integer. If $k = 1$, there is nothing to prove since we can add two numbers with $n - 1 = 1$ processor in $\log 2 = 1$ step. Assume we can add $n = 2^k$ numbers in $\log n$ steps using a tree-connected network of $n - 1$ processors. Let $x_1, x_2, \ldots, x_{2n}$ be $2n = 2^{k+1}$ numbers that we wish to add. The tree-connected network of $2n - 1$ processors consists of the tree-connected network of $n - 1$ processors together with two new processors as children of each leaf. In one step we can use the leaves of the larger network to find $x_1 + x_2, x_3 + x_4, \ldots, x_{2n-1} + x_{2n}$, giving us n numbers, which, by the inductive hypothesis, we can add in $\log n$ steps using the rest of the network. Since we have used $\log n + 1$ steps and $\log(2n) = \log 2 + \log n = 1 + \log n$, this completes the proof. **39.** c only **41.** c and h **43.** Suppose a tree T has at least two centers. Let u and v be distinct centers, both with eccentricity e, with u and v not adjacent. Since T is connected, there is a simple path P from u to v. Let c be any other vertex on this path. Since the eccentricity of c is at least e, there is a vertex w such that the unique simple path from c to w has length at least e. Clearly this

47. The statement is that *every* tree with n vertices has a path of length $n - 1$, and it was shown only that there exists a tree with n vertices having a path of length $n - 1$.

SECTION 9.2

1.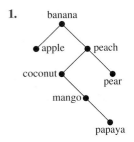

3. a) 3 **b)** 1 **c)** 4 **d)** 5

5.

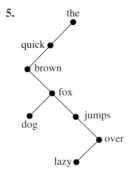

7. At least $\lceil \log_3 4 \rceil = 2$ weighings are needed, since there are only four outcomes (since it is not required to determine whether the coin is lighter or heavier). In fact, two weighings suffice. Begin by weighing coin 1 against coin 2. If they balance, weigh coin 1 against coin 3. If coin 1 and coin 3 are the same weight, coin 4 is the counterfeit coin, and if they are not the same weight, then coin 3 is the counterfeit coin. If coin 1 and coin 2 are not the same weight, again weigh coin 1 against coin 3. If they balance, coin 2 is the counterfeit coin; if they do not balance, coin 1 is the counterfeit coin. **9.** At least $\lceil \log_3 13 \rceil = 3$ weighings are needed. In fact, three weighings suffice. Start by putting coins 1, 2, and 3 on the left-hand side of the balance and coins 4, 5, and 6 on the right-hand side. If equal, apply Example 2 to coins 1, 2, 7, 8, 9, 10, 11, and 12. If unequal, apply Example 2 to 1, 2, 3, 4, 5, 6, 7, and 8. **11.** The least number is five. Call the elements $a, b, c,$ and d. First compare a and b; then compare c and d. Without loss of generality, assume that $a < b$ and $c < d$. Next compare a and c. Whichever is smaller is the smallest element of the set. Again without loss of generality, suppose $a < c$. Finally, compare b with both c and d to completely determine the ordering. **13.** The first two steps are shown in the text. After 22 has been identified as the second largest element, we replace the leaf 22 by $-\infty$ in the tree and recalculate the winner in the path from the leaf where 22 used to be up to the root. Next, we see that 17 is the third largest element, so we repeat the process: replace the leaf 17 by $-\infty$ and recalculate. Next, we see that 14 is the fourth largest element, so we repeat the process: replace the leaf 14 by $-\infty$ and recalculate. Next, we see that 11 is the fifth largest element, so we repeat the process: replace the leaf 11 by $-\infty$ and recalculate. The process continues in this manner. We determine that 9 is the sixth largest element, 8 is the seventh largest element, and 3 is the eighth largest element. The trees produced in all steps, except the second to last, are shown here.

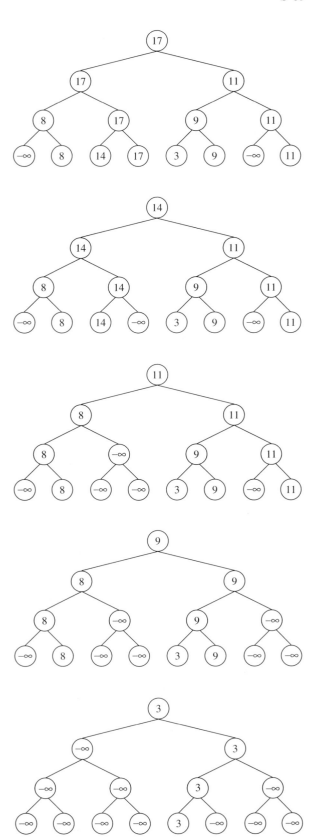

15. The value of a vertex is the list element currently there, and the label is the name (i.e., location) of the leaf responsible for that value.

procedure *tournament sort*$(a_1, \ldots, a_n)$
$k := \lceil \log n \rceil$
build a binary tree of height k
for $i := 1$ **to** n
  set the value of the ith leaf to be a_i and its label to be itself
for $i := n + 1$ **to** 2^k
  set the value of the ith leaf to be $-\infty$ and its label to be itself
for $i := k - 1$ **downto** 0
  for each vertex v at level i
    set the value of v to the larger of the values of its children and its label to be the label of the child with the larger value
for $i := 1$ **to** n
begin
  $c_i :=$ value at the root
  let v be the label of the root
  set the value of v to be $-\infty$
  while the label at the root is still v
  begin
    $v := parent(v)$
    set the value of v to the larger of the values of its children and its label to be the label of the child with the larger value
  end
end $\{c_1, \ldots, c_n$ is the list in nonincreasing order$\}$

17. $k - 1$, where $n = 2^k$ **19. a)** Yes **b)** No **c)** Yes **d)** Yes   **21.** $a: 000, e: 001, i: 01, k: 1100, o: 1101, p: 11110, u:$ 11111   **23.** a: 11; b: 101; c: 100; d: 01; e: 00; 2.25 bits (Note: This coding depends on how ties are broken, but the average number of bits is always the same.)   **25.** There are four possible answers in all, the one shown here and three more obtained from this one by swapping t and v and/or swapping u and w.

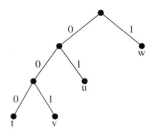

27. A:0001;   B:101001;   C:11001;   D:00000;   E:100; F:001100;   G:001101;   H:0101;   I:0100;   J:110100101; K:1101000; L:00001; M:10101; N:0110; O:0010; P:101000; Q:1101001000; R:1011; S:0111; T:111; U:00111; V:110101;

W:11000; X:11010011; Y:11011; Z:1101001001   **29.** A:2; E:1; N:010; R:011; T:02; Z:00   **31.** n **33.** Since the tree is rather large, we have indicated in some places to "see text." Refer to Figure 9; the subtree rooted at these square or circle vertices is exactly the same as the corresponding subtree in Figure 9. First player wins.

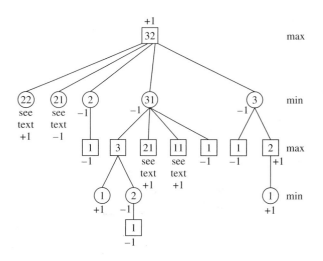

35. a) \$1 **b)** \$3 **c)** −\$3   **37.** See the figures shown next. **a)** 0 **b)** 0 **c)** 1 **d)** This position cannot have occurred in a game; this picture is impossible.

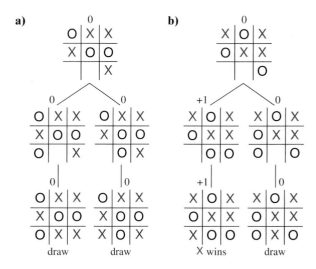

c)

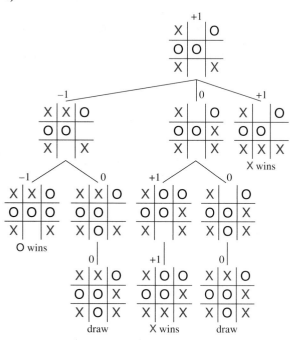

of the root in the first tree whose subtrees are not shown because of space limitations. They should be thought of as spliced into the first picture.

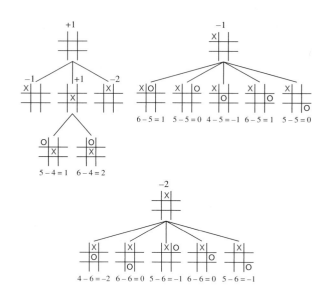

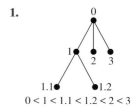

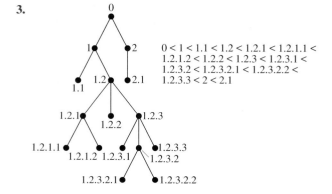

39. Proof by strong induction: *Basis step:* When there are $n = 2$ stones in each pile, if first player takes two stones from a pile, then second player takes one stone from the remaining pile and wins. If first player takes one stone from a pile, then second player takes two stones from the other pile and wins. *Inductive step:* Assume inductive hypothesis that second player can always win if the game starts with two piles of j stones for all $2 \leq j \leq k$, where $k \geq 2$, and consider a game with two piles containing $k + 1$ stones each. If first player takes all the stones from one of the piles, then second player takes all but one stone from the remaining pile and wins. If first player takes all but one stone from one of the piles, then second player takes all the stones from the other pile and wins. Otherwise first player leaves j stones in one pile, where $2 \leq j \leq k$, and $k + 1$ stones in the other pile. Second player takes the same number of stones from the larger pile, also leaving j stones there. At this point the game consists of two piles of j stones each. By the inductive hypothesis, the second player in that game, who is also the second player in our actual game, can win, and the proof by strong induction is complete. **41.** 7; 49 **43.** Value of tree is 1. Note: The second and third trees are the subtrees of the two children

SECTION 9.3

1.

$$0 < 1 < 1.1 < 1.2 < 2 < 3$$

3.

$0 < 1 < 1.1 < 1.2 < 1.2.1 < 1.2.1.1 <$
$1.2.1.2 < 1.2.2 < 1.2.3 < 1.2.3.1 <$
$1.2.3.2 < 1.2.3.2.1 < 1.2.3.2.2 <$
$1.2.3.3 < 2 < 2.1$

5. No   **7.** a, b, d, e, f, g, c

9. $a, b, e, k, l, m, f, g, n, r, s, c, d, h, o, i, j, p, q$

11. $d, b, i, e, m, j, n, o, a, f, c, g, k, h, p, l$

13. d, f, g, e, b, c, a

15. $k, l, m, e, f, r, s, n, g, b, c, o, h, i, p, q, j, d, a$

17. a)

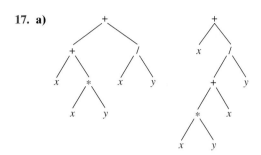

b) $+ + x * x y / x y, + x / + * x y x y$ **c)** $x x y * + x y / +, x x y * x + y / +$ **d)** $((x + (x * y)) + (x / y)), (x + (((x * y) + x) / y))$

19. a)

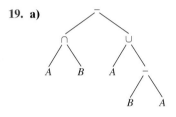

b) $- \cap A B \cup A - B A$ **c)** $A B \cap A B A - \cup -$

d) $((A \cap B) - (A \cup (B - A)))$ **21.** 14   **23. a)** 1   **b)** 1

c) 4   **d)** 2205

25.

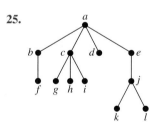

27. Use mathematical induction. The result is trivial for a list with one element. Assume the result is true for a list with n elements. For the inductive step, start at the end. Find the sequence of vertices at the end of the list starting with the last leaf, ending with the root, each vertex being

the last child of the one following it. Remove this leaf and apply the inductive hypothesis.   **29.** c, d, b, f, g, h, e, a in each case   **31.** Proof by mathematical induction. Let $S(X)$ and $O(X)$ represent the number of symbols and number of operators in the well-formed formula X, respectively. The statement is true for well-formed formulae of length 1, since they have 1 symbol and 0 operators. Assume the statement is true for all well-formed formulae of length less than n. A well-formed formula of length n must be of the form $* X Y$ where $*$ is an operator and X and Y are well-formed formulae of length less than n. Then by the inductive hypothesis $S(* X Y) = S(X) + S(Y) = (O(X) + 1) + (O(Y) + 1) = O(X) + O(Y) + 2$. Since $O(* X Y) = 1 + O(X) + O(Y)$, it follows that $S(* X Y) = O(* X Y) + 1$.   **33.** $x y + z x \circ + x \circ, x y z + + y x + +,$ $x y x y \circ \circ x y \circ \circ z \circ +, x z \times z z + \circ, y y y y \circ \circ \circ, z x + y z + \circ,$ for instance

SECTION 9.4

1. $m - n + 1$

3.

 **5.**

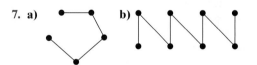

7. a)

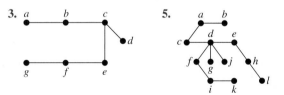

b)

c)

d)   **e)**   **f)**

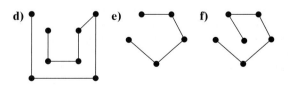

9.

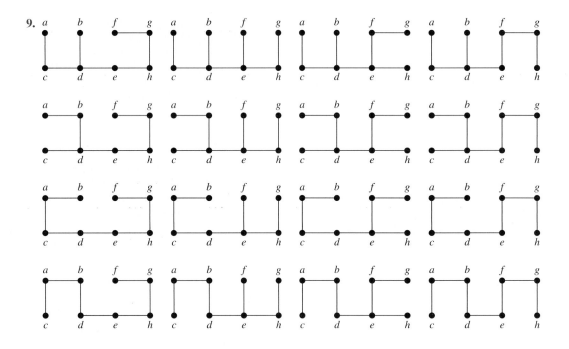

11. a) 3 **b)** 16 **c)** 4 **d)** 5

13.

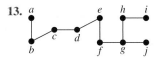

15.

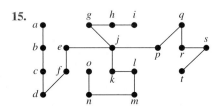

17. a) A path of length 6 **b)** A path of length 5 **c)** A path of length 6 **d)** Depends on order chosen to visit the vertices; may be a path of length 7 **19.** With breadth-first search, the initial vertex is the middle vertex, and the n spokes are added to the tree as this vertex is processed. Thus the resulting tree is $K_{1,n}$. With depth-first search, we start at the vertex in the middle of the wheel and visit a neighbor—one of the vertices on the rim. From there we move to an adjacent vertex on the rim, and so on all the way around until we have reached every vertex. Thus the resulting spanning tree is a path of length n. **21.** With breadth-first search, we fan out from a vertex of degree m to all the vertices of degree n as the first step. Next a vertex of degree n is

processed, and the edges from it to all the remaining vertices of degree m are added. The result is a $K_{1,n-1}$ and a $K_{1,m-1}$ with their centers joined by an edge. With depth-first search, we travel back and forth from one partite set to the other until we can go no further. If $m = n$ or $m = n - 1$, then we get a path of length $m + n - 1$. Otherwise, the path ends while some vertices in the larger partite set have not been visited, so we back up one link in the path to a vertex v and then successively visit the remaining vertices in that set from v. The result is a path with extra pendant edges coming out of one end of the path. **23.** A possible set of flights to discontinue are: Boston–New York, Detroit–Boston, Boston–Washington, New York–Washington, New York–Chicago, Atlanta–Washington, Atlanta–Dallas, Atlanta–Los Angeles, Atlanta–St. Louis, St. Louis–Dallas, St. Louis–Detroit, St. Louis–Denver, Dallas–San Diego, Dallas–Los Angeles, Dallas–San Francisco, San Diego–Los Angeles, Los Angeles–San Francisco, San Francisco–Seattle. **25.** Trees **27.** Proof by induction on the length of the path: If the path has length 0, then the result is trivial. If the length is 1, then u is adjacent to v, so u is at level 1 in the breadth-first spanning tree. Assume that the result is true for paths of length l. If the length of a path is $l + 1$, let u' be the next-to-last vertex in a shortest path from v to u. By the inductive hypothesis, u' is at level l in the breadth-first spanning tree. If u were at a level not exceeding l, then clearly the length of the shortest path from v to u would also not exceed l. So u has not been added to the breadth-first spanning tree yet after the vertices of level l have been added. Since u is adjacent to u', it will be added at level $l + 1$ (although the edge connecting u' and u is not necessarily added).   **29. a)** No solution

b)   **c)**

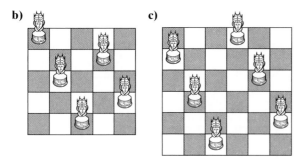

31. Start at a vertex and proceed along a path without repeating vertices as long as possible, allowing the return to the start after all vertices have been visited. When it is impossible to continue along a path, backtrack and try another extension of the current path.   **33.** Take the union of the spanning trees of the connected components of G. They are disjoint so the result is a forest.   **35.** $m - n + c$ **37.** Use depth-first search on each component.   **39.** If an edge uv is not followed while we are processing vertex u during the depth-first search process, then it can only be the case that the vertex v had already been visited. There are two cases. If vertex v was visited after we started processing u, then, since we are not finished processing u yet, v must appear in the subtree rooted at u (and hence must be a descendant of u). On the other hand, if the processing of v had already begun before we started processing u, then why wasn't this edge followed at that time? It must be that we had not finished processing v, in other words, that we are still forming the subtree rooted at v, so u is a descendant of v, and hence v is an ancestor of u.   **41.** Certainly these two procedures produce the identical spanning trees if the graph we are working with is a tree itself, since in this case there is only one spanning tree (the whole graph). This is the only case in which that happens, however. If the original graph has any other edges, then by Exercise 39 they must be back edges and hence join a vertex to an ancestor or descendant, whereas by Exercise 40, they must connect vertices at the same level or at levels that differ by 1. Clearly these two possibilities are mutually exclusive. Therefore there can be no edges other than tree edges if the two spanning trees are to be the same.   **43.** Since the edges not in the spanning tree are not followed in the process, we can ignore them. Thus we can assume that the graph was a rooted tree to begin with. The base case is trivial (there is only one vertex), so we assume the inductive hypothesis that breadth-first search applied to trees with n vertices have their vertices visited in order of their level in the tree and consider a tree T with $n + 1$ vertices. The last vertex to be visited during breadth-first search of this tree, say v, is the one that was added last to the list

of vertices waiting to be processed. It was added when its parent, say u, was being processed. We must show that v is at the lowest (bottom-most, i.e., numerically greatest) level of the tree. Suppose not; say vertex x, whose parent is vertex w, is at a lower level. Then w is at a lower level than u. Clearly v must be a leaf, since any child of v could not have been seen before v is seen. Consider the tree T' obtained from T by deleting v. By the inductive hypothesis, the vertices in T' must be processed in order of their level in T' (which is the same as their level in T, and the absence of v in T' has no effect on the rest of the algorithm). Therefore u must have been processed before w, and therefore v would have joined the waiting list before x did, a contradiction. Therefore v is at the bottom-most level of the tree, and the proof is complete.   **45.** We modify the pseudocode given in Algorithm 2 by initializing m to be 0 at the beginning of the algorithm, and adding the statements "$m := m + 1$" and "assign m to vertex v" after the statement that removes vertex v from L.   **47.** If a directed edge uv is not followed while we are processing its tail u during the depth-first search process, then it can only be the case that its head v had already been visited. There are three cases. If vertex v was visited after we started processing u, then, since we are not finished processing u yet, v must appear in the subtree rooted at u (and hence must be a descendant of u), so we have a forward edge. Otherwise, the processing of v must have already begun before we started processing u. If it had not yet finished (i.e., we are still forming the subtree rooted at v), then u is a descendant of v, and hence v is an ancestor of u (we have a back edge). Finally, if the processing of v had already finished, then by definition we have a cross edge.   **49.** Let T be the spanning tree constructed in Figure 3 and T_1, T_2, T_3, and T_4 the spanning trees in Figure 4. Denote by $d(T', T'')$ the distance between T' and T''. Then $d(T, T_1) = 6$, $d(T, T_2) = 4$, $d(T, T_3) = 4$, $d(T, T_4) = 2$, $d(T_1, T_2) = 4$, $d(T_1, T_3) = 4$, $d(T_1, T_4) = 6$, $d(T_2, T_3) = 4$, $d(T_2, T_4) = 2$, and $d(T_3, T_4) = 4$.   **51.** Suppose $e_1 = \{u, v\}$ is as specified. Then $T_2 \cup \{e_1\}$ contains a simple circuit C containing e_1. The graph $T_1 - \{e_1\}$ has two connected components; the endpoints of e_1 are in different components. Travel C from u in the direction opposite to e_1 until you come to the first vertex in the same component as v. The edge just crossed is e_2. Clearly $T_2 \cup \{e_1\} - \{e_2\}$ is a tree, since e_2 was on C. Also $T_1 - \{e_1\} \cup \{e_2\}$ is a tree, since e_2 reunited the two components.

53. Exercise 18:   Exercise 19:   Exercise 20:

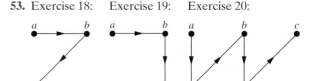

Exercise 21:   **Exercise 22:**   **Exercise 23:**

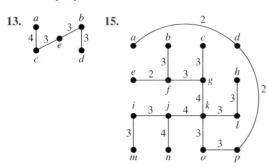

55. First construct an Euler circuit in the directed graph. Then delete from this circuit every edge that goes to a vertex previously visited.   **57.** According to Exercise 56, a directed graph contains a circuit if and only if there are any back edges. We can detect back edges as follows. Add a marker on each vertex v to indicate what its status is: not yet seen (the initial situation), seen (i.e., put into T) but not yet finished (i.e., *visit(v)* has not yet terminated), or finished (i.e., *visit(v)* has terminated). A few extra lines in Algorithm 1 will accomplish this bookkeeping. Then to determine whether a directed graph has a circuit, we just have to check when looking at edge uv whether the status of v is "seen." If that ever happens, then we know there is a circuit; if not, then there is no circuit.

SECTION 9.5

1. Deep Springs–Oasis, Oasis–Dyer, Oasis–Silverspeak, Silverspeak–Goldfield, Lida–Gold Point, Gold Point–Beatty, Lida–Goldfield, Goldfield–Tonopah, Tonopah–Manhattan, Tonopah–Warm Springs   **3.** $\{e, f\}$, $\{c, f\}$, $\{e, h\}$, $\{h, i\}$, $\{b, c\}$, $\{b, d\}$, $\{a, d\}$, $\{g, h\}$

5.

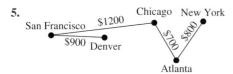

7. $\{e, f\}, \{a, d\}, \{h, i\}, \{b, d\}, \{c, f\}, \{e, h\}, \{b, c\}, \{g, h\}$

9.

11. Instead of choosing minimum-weight edges at each stage, choose maximum-weight edges at each stage with the same properties.

13.   **15.**

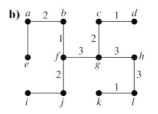

17. First find a minimum spanning tree T of the graph G with n edges. Then for $i = 1$ to $n - 1$, only delete the ith edge of T from G and find a minimum spanning tree of the remaining graph. Pick the one of these $n - 1$ trees with the shortest length.   **19.** If all edges have different weights a contradiction is obtained in the proof that Prim's algorithm works when an edge e_{k+1} is added to T and an edge e is deleted, instead of possibly producing another spanning tree.

21.

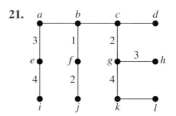

23. Same as Kruskal's algorithm, except start with $T :=$ this set of edges and iterate from $i = 1$ to $n - 1 - s$ where s is the number of edges you start with.

25. a)

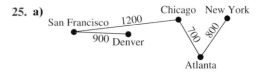

b)

27. By Exercise 24, at each stage of Sollin's algorithm a forest results. Hence after $n - 1$ edges are chosen, a tree results. It remains to show that this tree is a minimum spanning tree. Let T be a minimum spanning tree with as many edges in common with Sollin's tree S as possible. If $T \neq S$, then there is an edge $e \in S - T$ added at some stage in the algorithm, where prior to that stage all edges in S are also in T. $T \cup \{e\}$ contains a unique simple circuit. Find an edge $e' \in S - T$ and an edge $e'' \in T - S$ on this circuit and "adjacent" when viewing the trees of this stage as "supervertices." Then by the algorithm, $w(e') \leq w(e'')$. So replace T by $T - \{e''\} \cup \{e'\}$ to produce a minimum spanning tree closer to S than T was.   **29.** Each of the r trees is joined to at least one other tree by a new edge. Hence there are at most $r/2$ trees in the result (each new tree contains two or more old trees). To accomplish this, we need to add $r - (r/2) = r/2$ edges. Since the number of edges added is integral, it is at least $\lceil r/2 \rceil$.   **31.** If $k \geq \log n$, then $n/2^k \leq 1$, so $\lceil n/2^k \rceil = 1$, so by Exercise 30 the algorithm is finished after at most $\log n$ iterations.

SUPPLEMENTARY EXERCISES

1. Suppose T is a tree. Then clearly T has no simple circuits. If we add an edge e connecting two nonadjacent vertices u and v, then obviously a simple circuit is formed, since when e is added to T the resulting graph has too many edges to be a tree. The only simple circuit formed is made up of the edge e together with the unique path P in T from v to u. Suppose T satisfies the given conditions. All that is needed is to show that T is connected, since there are no simple circuits in the graph. Assume that T is not connected. Then let u and v be in separate connected components. Adding $e = \{u, v\}$ does not satisfy the conditions. **3.** Suppose that a tree T has n vertices of degrees $d_1, d_2, \ldots, d_n$, respectively. Since $2e = \sum_{i=1}^{n} d_i$ and $e = n - 1$, we have $2(n - 1) = \sum_{i=1}^{n} d_i$. Since each $d_i \geq 1$, it follows that $2(n - 1) = n + \sum_{i=1}^{n}(d_i - 1)$, or that $n - 2 = \sum_{i=1}^{n}(d_i - 1)$. Hence at most $n - 2$ of the terms of this sum can be 1 or more. Hence, at least two of them are 0. It follows that $d_i = 1$ for at least two values of i. **5.** $2n - 2$ **7.** T has no circuits, so it cannot have a subgraph homeomorphic to $K_{3,3}$ or K_5. **9.** Color each connected component separately. For each of these connected components, first root the tree, then color all vertices at even levels red and all vertices at odd levels blue. **11.** Upper bound: k^h; lower bound: $2\lceil k/2 \rceil^{h-1}$

13.

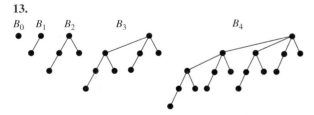

B_0 B_1 B_2 B_3 B_4

15. Since B_{k+1} is formed from two copies of B_k, one shifted down one level, the height increases by 1 as k increases by 1. Since B_0 had height 0, it follows by induction that B_k has height k. **17.** Since the root of B_{k+1} is the root of B_k with one additional child (namely the root of the other B_k), the degree of the root increases by 1 as k increases by 1. Since B_0 had a root with degree 0, it follows by induction that B_k has a root with degree k.

19.

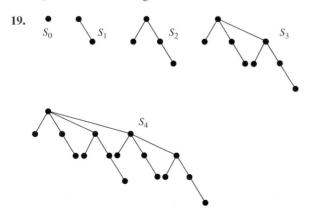

S_0 S_1 S_2 S_3

S_4

21. Use mathematical induction. The result is trivial for $k = 0$. Suppose it is true for $k - 1$. T_{k-1} is the parent tree for T. By induction, the child tree for T can be obtained from $T_0, \ldots, T_{k-2}$ in the manner stated. The final connection of r_{k-2} to r_{k-1} is as stated in the definition of S_k-tree.

23. procedure *level*(T: ordered rooted tree with root r)
　queue := sequence consisting of just the root r
　while *queue* contains at least one term
　begin
　　v := first vertex in queue
　　list v
　　remove v from queue and put children of v onto
　　　the end of queue
　end

25. Build the tree by inserting a root for the address 0, and then inserting a subtree for each vertex labeled i, for i a positive integer, built up from subtrees for each vertex labeled $i.j$ for j a positive integer, and so on. **27. a)** Yes **b)** No **c)** Yes **29.** The resulting graph has no edge that is in more than one simple circuit of the type described. Hence it is a cactus.

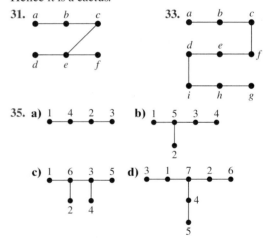

31. a　b　c　　d　e　f

33. a　b　c　d　e　f　i　h　g

35. a) 1　4　2　3　**b)** 1　5　3　4　2

c) 1　6　3　5　2　4　**d)** 3　1　7　2　6　4　5

37. 6 **39. a)** 0 for 00, 11 for 01, 100 for 10, 101 for 11 (exact coding depends on how ties were broken, but all versions are equivalent); $0.645n$ for string of length n **b)** 0 for 000, 100 for 001, 101 for 010, 110 for 100, 11100 for 011, 11101 for 101, 11110 for 110, 11111 for 111 (exact coding depends on how ties were broken, but all versions are equivalent); $0.53\overline{26}n$ for string of length n **41.** Let G' be the graph obtained by deleting from G the vertex v and all edges incident to v. A minimum spanning tree of G can be obtained by taking an edge of minimal weight incident to v together with a minimum spanning tree of G'. **43.** Suppose that edge e is the edge of least weight incident to vertex v, and suppose that T is a spanning tree that does not include e. Add e to T, and delete from the simple circuit formed thereby the other edge of the circuit that contains v. The result will be a spanning tree of strictly smaller weight (since the deleted edge has weight greater than the weight of e). This is a contradiction, so T must include e.

CHAPTER 10

SECTION 10.1

1. a) 1 **b)** 1 **c)** 0 **d)** 0

3. a)

x	y	z	$\overline{x}y$
1	1	1	0
1	1	0	0
1	0	1	0
1	0	0	0
0	1	1	1
0	1	0	1
0	0	1	0
0	0	0	0

b)

x	y	z	$x + yz$
1	1	1	1
1	1	0	1
1	0	1	1
1	0	0	1
0	1	1	1
0	1	0	0
0	0	1	0
0	0	0	0

c)

x	y	z	$x\overline{y} + \overline{x\,yz}$
1	1	1	0
1	1	0	1
1	0	1	1
1	0	0	1
0	1	1	1
0	1	0	1
0	0	1	1
0	0	0	1

d)

x	y	z	$x(yz + \overline{y}\,\overline{z})$
1	1	1	1
1	1	0	0
1	0	1	0
1	0	0	1
0	1	1	0
0	1	0	0
0	0	1	0
0	0	0	0

5. a) **b)** **c)** **d)**

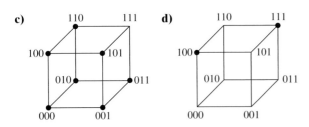

7. (0,0) and (1,1) **9.** $x + xy = x \cdot 1 + xy = x(1 + y) = x(y + 1) = x \cdot 1 = x$

11.

x	y	z	$x\overline{y}$	$y\overline{z}$	$\overline{x}z$	$x\overline{y} + y\overline{z} + \overline{x}z$	$\overline{x}y$	$\overline{y}z$	$x\overline{z}$	$\overline{x}y + \overline{y}z + x\overline{z}$
1	1	1	0	0	0	0	0	0	0	0
1	1	0	0	1	0	1	0	0	1	1
1	0	1	1	0	0	1	0	1	0	1
1	0	0	1	0	0	1	0	0	1	1
0	1	1	0	0	1	1	1	0	0	1
0	1	0	0	1	0	1	1	0	0	1
0	0	1	0	0	1	1	0	1	0	1
0	0	0	0	0	0	0	0	0	0	0

13.

x	$x + x$	$x \cdot x$
0	0	0
1	1	1

15.

x	$x + 1$	$x \cdot 0$
0	1	0
1	1	0

17.

x	y	z	$y + z$	$x + (y + z)$	$x + y$	$(x + y) + z$	yz	$x(yz)$	xy	$(xy)z$
1	1	1	1	1	1	1	1	1	1	1
1	1	0	1	1	1	1	0	0	1	0
1	0	1	1	1	1	1	0	0	0	0
1	0	0	0	1	1	1	0	0	0	0
0	1	1	1	1	1	1	1	0	0	0
0	1	0	1	1	1	1	0	0	0	0
0	0	1	1	1	0	1	0	0	0	0
0	0	0	0	0	0	0	0	0	0	0

19.

x	y	xy	$\overline{(xy)}$	$\overline{x}$	$\overline{y}$	$\overline{x} + \overline{y}$	$x + y$	$\overline{(x + y)}$	$\overline{x}\,\overline{y}$
1	1	1	0	0	0	0	1	0	0
1	0	0	1	0	1	1	1	0	0
0	1	0	1	1	0	1	1	0	0
0	0	0	1	1	1	1	0	1	1

21. $0 \cdot \overline{0} = 0 \cdot 1 = 0; 1 \cdot \overline{1} = 1 \cdot 0 = 0$

23.

x	y	$x \oplus y$	$x + y$	xy	$\overline{(xy)}$	$(x + y)\overline{(xy)}$	$x\overline{y}$	$\overline{x}y$	$x\overline{y} + \overline{x}y$
1	1	0	1	1	0	0	0	0	0
1	0	1	1	0	1	1	1	0	1
0	1	1	1	0	1	1	0	1	1
0	0	0	0	0	1	0	0	0	0

25. a) True, as a table of values can show **b)** False; take $x = 1, y = 1, z = 1$, for instance **c)** False; take $x = 1, y = 1, z = 0$, for instance **27.** By De Morgan's laws, the complement of an expression is like the dual except that the complement of each variable has been taken. **29.** 16 **31.** By the domination, distributive, and identity laws, $x \vee x = (x \vee x) \wedge 1 = (x \vee x) \wedge (x \vee \overline{x}) = x \vee (x \wedge \overline{x}) = x \vee 0 = x$. Similarly, $x \wedge x = (x \wedge x) \vee 0 = (x \wedge x) \vee (x \wedge \overline{x}) = x \wedge (x \vee \overline{x}) = x \wedge 1 = x$. **33.** Since $0 \vee 1 = 1$ and $0 \wedge 1 = 0$ by the identity and commutative laws, it follows that $\overline{0} = 1$. Similarly, since $1 \vee 0 = 1$ and $1 \wedge 0 = 1$, it follows that $\overline{1} = 0$. **35.** First, note that $x \wedge 0 = 0$ and $x \vee 1 = 1$ for all x, as can easily be proved. To prove the first identity, it is sufficient to show that $(x \vee y) \vee (\overline{x} \wedge \overline{y}) = 1$ and $(x \vee y) \wedge (\overline{x} \wedge \overline{y}) = 0$. By the associative, commutative, distributive, domination, and identity laws, $(x \vee y) \vee (\overline{x} \wedge \overline{y}) = y \vee (x \vee (\overline{x} \wedge \overline{y})) = y \vee ((x \vee \overline{x}) \wedge (x \vee \overline{y})) = y \vee (1 \wedge (x \vee \overline{y})) = y \vee (x \vee \overline{y}) = (y \vee \overline{y}) \vee x = 1 \vee x = 1$ and $(x \vee y) \wedge (\overline{x} \wedge \overline{y}) = \overline{y} \wedge (\overline{x} \wedge (x \vee y)) = \overline{y} \wedge ((\overline{x} \wedge x) \vee (\overline{x} \wedge y)) = \overline{y} \wedge (0 \vee (\overline{x} \wedge y)) = \overline{y} \wedge (\overline{x} \wedge y) = \overline{x} \wedge (y \wedge \overline{y}) = \overline{x} \wedge 0 = 0$. The second identity is proved in a similar way. **37.** Using the hypotheses, Exercise 31, and the distributive law it follows

that $x = x \vee 0 = x \vee (x \vee y) = (x \vee x) \vee y = x \vee y = 0$.
Similarly, $y = 0$. To prove the second statement, note that
$x = x \wedge 1 = x \wedge (x \wedge y) = (x \wedge x) \wedge y = x \wedge y = 1$.
Similarly, $y = 1$.   **39.** Use Exercises 39 and 41 in the
Supplementary Exercises in Chapter 7 and the definition
of a complemented, distributed lattice to establish the five
pairs of laws in the definition.

SECTION 10.2

1. a) $\overline{x}\,\overline{y}z$ **b)** $\overline{x}y\overline{z}$ **c)** $\overline{x}yz$ **d)** $\overline{x}\,\overline{y}\,\overline{z}$ **3. a)** $xyz + xy\overline{z} +$
$x\overline{y}z + x\overline{y}\,\overline{z} + \overline{x}yz + \overline{x}y\overline{z} + \overline{x}\,\overline{y}z$ **b)** $xyz + xy\overline{z} + \overline{x}yz$ **c)** $xyz +$
$xy\overline{z} + x\overline{y}z + x\overline{y}\,\overline{z}$ **d)** $x\overline{y}z + x\overline{y}\,\overline{z}$ **5.** $wxy\overline{z} + wx\overline{y}z + w\overline{x}yz +$
$\overline{w}xyz + \overline{w}x\overline{y}\,\overline{z} + \overline{w}\,\overline{x}\,\overline{y}z + \overline{w}\,\overline{x}y\overline{z} + w\overline{x}\,\overline{y}\,\overline{z}$ **7. a)** $\overline{x} + \overline{y} + z$
b) $x + y + z$ **c)** $x + \overline{y} + z$ **9.** $y_1 + y_2 + \cdots + y_n = 0$
if and only if $y_i = 0$ for $i = 1, 2, \ldots, n$. This holds if and
only if $x_i = 0$ when $y_i = x_i$ and $x_i = 1$ when $y_i = \overline{x_i}$.
11. a) $x + y + z$ **b)** $(x + y + z)(x + y + \overline{z})(x + \overline{y} + z)(\overline{x} +$
$y + z)(\overline{x} + y + \overline{z})$ **c)** $(x + y + z)(x + y + \overline{z})(x + \overline{y} + z)(x + \overline{y} + \overline{z})$
d) $(x + y + z)(x + y + \overline{z})(x + \overline{y} + z)(x + \overline{y} + \overline{z})(\overline{x} + \overline{y} + z)$
$(\overline{x} + \overline{y} + \overline{z})$ **13. a)** $\overline{x + y + z}$ **b)** $\overline{x + (y + \overline{(\overline{x} + z)})}$
c) $\overline{(x + \overline{y})}$ **d)** $\overline{(x + \overline{(x + \overline{y} + \overline{z})})}$

15. a)

x	$\overline{x}$	$x \downarrow x$
1	0	0
0	1	1

b)

x	y	xy	$x \downarrow x$	$y \downarrow y$	$(x \downarrow x) \downarrow (y \downarrow y)$
1	1	1	0	0	1
1	0	0	0	1	0
0	1	0	1	0	0
0	0	0	1	1	0

c)

x	y	$x + y$	$(x \downarrow y)$	$(x \downarrow y) \downarrow (x \downarrow y)$
1	1	1	0	1
1	0	1	0	1
0	1	1	0	1
0	0	0	1	0

17. a) $(((x \mid x) \mid (y \mid y)) \mid ((x \mid x) \mid (y \mid y))) \mid (z \mid z)$
b) $(((x \mid x) \mid (z \mid z)) \mid y) \mid (((x \mid x) \mid (z \mid z)) \mid y)$ **c)** x
d) $(x \mid (y \mid y)) \mid (x \mid (y \mid y))$ **19.** It is impossible to
represent $\overline{x}$ using $+$ and $\cdot$ since there is no way to get the
value 0 if the input is 1.

SECTION 10.3

1. $(x + y)\overline{y}$ **3.** $\overline{(xy)} + (\overline{z} + x)$ **5.** $(x + y + z) + (\overline{x} + y +$
$z) + (\overline{x} + \overline{y} + \overline{z})$

7.

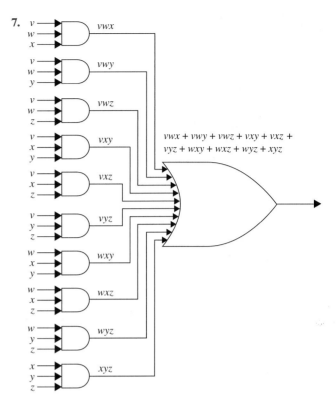

$vwx + vwy + vwz + vxy + vxz +$
$vyz + wxy + wxz + wyz + xyz$

9.

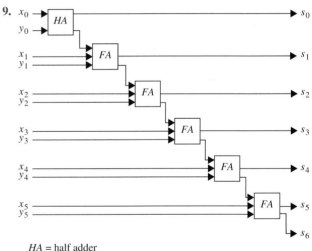

HA = half adder
FA = full adder

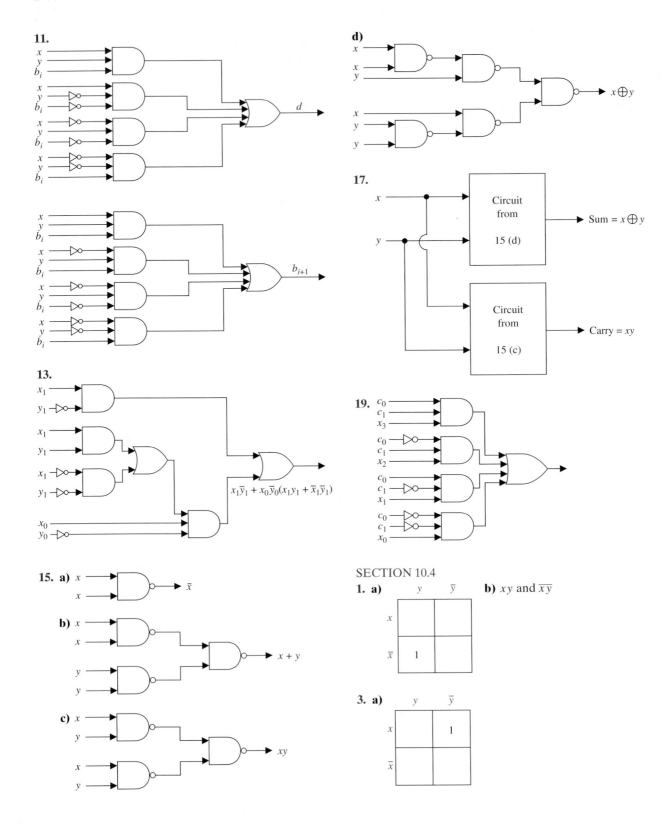

11.

d)

17.

Circuit from 15 (d) → Sum = $x \oplus y$

Circuit from 15 (c) → Carry = xy

13.

$x_1\bar{y}_1 + x_0\bar{y}_0(x_1y_1 + \bar{x}_1\bar{y}_1)$

19.

15. a) $x \rightarrow \bar{x}$

b) $x + y$

c) xy

SECTION 10.4

1. a)

	y	$\bar{y}$
x		
$\bar{x}$	1	

b) xy and $\overline{x}\,\overline{y}$

3. a)

	y	$\bar{y}$
x		1
$\bar{x}$		

b)

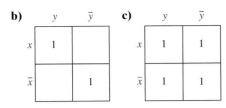

c)

5. a)

b) $\overline{x}yz, \overline{x}\,\overline{y}\,\overline{z}, xy\overline{z}$

7. a)

b)

c)

9. Implicants: $xyz, xy\overline{z}, x\,\overline{y}\,\overline{z}, \overline{x}y\overline{z}, xy, x\overline{z}, y\overline{z}$; prime implicants: $xy, x\overline{z}, y\overline{z}$; essential prime implicants: $xy, x\overline{z}, y\overline{z}$

11. The 3-cube on the right corresponds to w; the 3-cube given by the top surface of the whole figure represents x; the 3-cube given by the back surface of the whole figure represents y; the 3-cube given by the right surfaces of both the left and the right 3-cube represents z. In each case, the opposite 3-face represents the complemented literal. The 2-cube that represents wz is the right face of the 3-cube on the right; the 2-cube that represents $\overline{x}y$ is bottom rear; the 2-cube that represents $\overline{y}\,\overline{z}$ is front left.

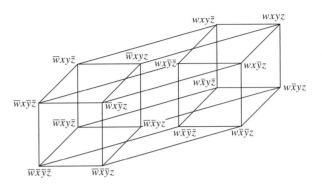

13. a)

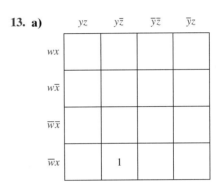

b) $\overline{w}xyz, \overline{w}\,\overline{x}\,y\overline{z}, \overline{w}x\overline{y}\,\overline{z}, \overline{w}x\overline{y}\,\overline{z}$

15. a)

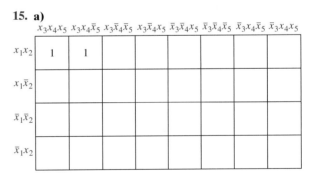

b)

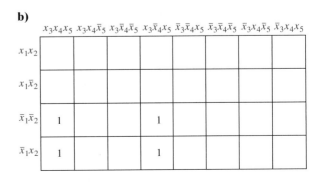

c)

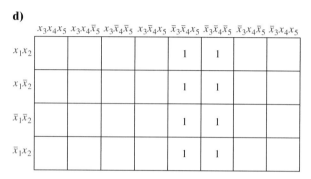

	$x_3x_4x_5$	$x_3x_4\bar{x}_5$	$x_3\bar{x}_4\bar{x}_5$	$x_3\bar{x}_4x_5$	$\bar{x}_3\bar{x}_4x_5$	$\bar{x}_3\bar{x}_4\bar{x}_5$	$\bar{x}_3x_4\bar{x}_5$	$\bar{x}_3x_4x_5$
x_1x_2	1	1					1	1
$x_1\bar{x}_2$								
$\bar{x}_1\bar{x}_2$								
$\bar{x}_1x_2$	1	1					1	1

d)

	$x_3x_4x_5$	$x_3x_4\bar{x}_5$	$x_3\bar{x}_4\bar{x}_5$	$x_3\bar{x}_4x_5$	$\bar{x}_3\bar{x}_4x_5$	$\bar{x}_3\bar{x}_4\bar{x}_5$	$\bar{x}_3x_4\bar{x}_5$	$\bar{x}_3x_4x_5$
x_1x_2					1	1		
$x_1\bar{x}_2$					1	1		
$\bar{x}_1\bar{x}_2$					1	1		
$\bar{x}_1x_2$					1	1		

e)

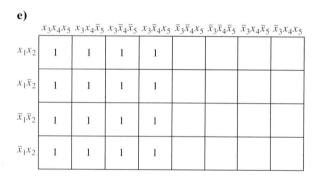

	$x_3x_4x_5$	$x_3x_4\bar{x}_5$	$x_3\bar{x}_4\bar{x}_5$	$x_3\bar{x}_4x_5$	$\bar{x}_3\bar{x}_4x_5$	$\bar{x}_3\bar{x}_4\bar{x}_5$	$\bar{x}_3x_4\bar{x}_5$	$\bar{x}_3x_4x_5$
x_1x_2	1	1	1	1				
$x_1\bar{x}_2$	1	1	1	1				
$\bar{x}_1\bar{x}_2$	1	1	1	1				
$\bar{x}_1x_2$	1	1	1	1				

f)

	$x_3x_4x_5$	$x_3x_4\bar{x}_5$	$x_3\bar{x}_4\bar{x}_5$	$x_3\bar{x}_4x_5$	$\bar{x}_3\bar{x}_4x_5$	$\bar{x}_3\bar{x}_4\bar{x}_5$	$\bar{x}_3x_4\bar{x}_5$	$\bar{x}_3x_4x_5$
x_1x_2		1	1			1	1	
$x_1\bar{x}_2$		1	1			1	1	
$\bar{x}_1\bar{x}_2$		1	1			1	1	
$\bar{x}_1x_2$		1	1			1	1	

17. a) 64 **b)** 6 **19.** Rows 1 and 4 are considered adjacent. The pairs of columns considered adjacent are: columns 1 and 4, 1 and 12, 1 and 16, 2 and 11, 2 and 15, 3 and 6, 3 and 10, 4 and 9, 5 and 8, 5 and 16, 6 and 15, 7 and 10, 7 and 14, 8 and 13, 9 and 12, 11 and 14, 13 and 16.

21.

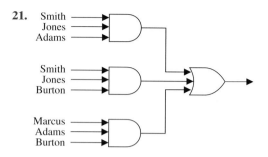

Smith
Jones
Adams

Smith
Jones
Burton

Marcus
Adams
Burton

23. a) $\bar{x}z$ **b)** y **c)** $x\bar{z} + \bar{x}z + \bar{y}z$ **d)** $xz + \bar{x}y + \bar{y}\bar{z}$
25. a) $wxz + wx\bar{y} + w\bar{y}z + w\bar{x}yz$ **b)** $x\bar{y}z + \bar{w}\,\bar{y}z + wxy\bar{z} + w\bar{x}yz + \bar{w}\,\bar{x}yz$ **c)** $\bar{y}z + wxy + w\bar{x}\,\bar{y} + \bar{w}\,\bar{x}yz$ **d)** $wy + yz + \bar{x}y + wxz + \bar{w}\,\bar{x}z$ **27.** $x(y + z)$

29.

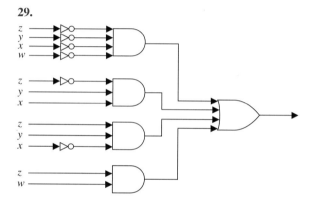

z
y
x
w

z
y
x

z
y
x

z
w

31. $\bar{x}\,\bar{z} + xz$ **33.** We use induction on n. If $n = 1$, then we are looking at a line segment, labeled 0 at one end and 1 at the other end. The only possible value of k is also 1, and if the literal is x_1, then the subcube we have is the 0-dimensional subcube consisting of the endpoint labeled 1, and if the literal is $\bar{x}_1$, then the subcube we have is the 0-dimensional subcube consisting of the endpoint labeled 0. Now assume that the statement is true for n; we must show that it is true for $n + 1$. If the literal x_{n+1} (or its complement) is not part of the product, then by the inductive hypothesis, the product when viewed in the setting of n variables corresponds to an $(n - k)$-dimensional subcube of the n-dimensional cube, and the Cartesian product of that subcube with the line segment $[0, 1]$ gives us a subcube one dimension higher in our given $(n + 1)$-dimensional cube, namely having dimension $(n + 1) - k$, as desired. On the other hand, if the literal x_{n+1} (or its complement) is part of the product, then the product of the remaining $k - 1$ literals corresponds to a subcube of dimension $n - (k - 1) = (n + 1) - k$ in the n-dimensional cube, and that slice, at either the 1-end or the 0-end in the last variable, is the desired subcube.

SUPPLEMENTARY EXERCISES

1. a) $x = 0, y = 0, z = 0; x = 1, y = 1, z = 1$ **b)** $x = 0,$
$y = 0, z = 0; x = 0, y = 0, z = 1; x = 0, y = 1,$
$z = 0; x = 1, y = 0, z = 1; x = 1, y = 1, z = 0;$
$x = 1, y = 1, z = 1$ **c)** No values **3. a)** Yes **b)** No
c) No **d)** Yes **5.** $2^{2^{n-1}}$ **7. a)** If $F(x_1, \ldots, x_n) = 1$,
then $(F+G)(x_1, \ldots, x_n) = F(x_1, \ldots, x_n)+G(x_1, \ldots, x_n) =$
1 by the dominance law. Hence $F \leq F + G$. **b)** If
$(FG)(x_1, \ldots, x_n) = 1$, then $F(x_1, \ldots, x_n) \cdot G(x_1, \ldots, x_n) =$
1. Hence $F(x_1, \ldots, x_n) = 1$. It follows that $FG \leq F$.
9. Since $F(x_1, \ldots, x_n) = 1$ implies that $F(x_1, \ldots, x_n) = 1$,
$\leq$ is reflexive. Suppose that $F \leq G$ and $G \leq F$. Then
$F(x_1, \ldots, x_n) = 1$ if and only if $G(x_1, \ldots, x_n) = 1$. This
implies that $F = G$. Hence $\leq$ is antisymmetric. Suppose
that $F \leq G \leq H$. Then if $F(x_1, \ldots, x_n) = 1$, it follows that
$G(x_1, \ldots, x_n) = 1$, which implies that $H(x_1, \ldots, x_n) = 1$.
Hence $F \leq H$, so that $\leq$ is transitive. **11. a)** $x = 1, y =$
$0, z = 0$ **b)** $x = 1, y = 0, z = 0$ **c)** $x = 1, y = 0, z = 0$

13.

x	y	$x \odot y$	$x \oplus y$	$\overline{(x \oplus y)}$
1	1	1	0	1
1	0	0	1	0
0	1	0	1	0
0	0	1	0	1

15. Yes, as a truth table shows. **17. a)** 6 **b)** 5 **c)** 5 **d)** 6

19.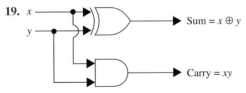

21. $x_3 + x_2\overline{x_1}$ **23.** Suppose it were with weights a and b.
Then there would be a real number T such that $xa + yb \geq T$
for $(1,0)$ and $(0,1)$, but with $xa + yb < T$ for $(0,0)$ and $(1,1)$.
Hence $a \geq T, b \geq T, 0 < T$, and $a + b < T$. Thus a and b
are positive, which implies that $a + b > a \geq T$, a contradiction.

CHAPTER 11

SECTION 11.1

1. a) sentence ⇨ **noun phrase   intransitive verb phrase**
⇨ **article   adjective noun   intransitive verb phrase** ⇨
article   adjective   noun   intransitive verb ⇨ . . . (after
3 steps). . . ⇨ *the happy hare runs*.
b) sentence ⇨ **noun phrase   intransitive verb phrase** ⇨
article   adjective   noun   intransitive verb phrase ⇨
article   adjective   noun   intransitive verb
adverb . . . (after 4 steps). . . ⇨ *the sleepy tortoise runs*
quickly

c) sentence ⇨ **noun phrase   transitive verb phrase**
noun phrase ⇨ **article   noun   transitive verb phrase**
noun phrase ⇨ **article   noun   transitive verb   noun**
phrase ⇨ **article   noun   transitive verb   article   noun**
⇨ . . . (after 4 steps). . . ⇨ *the tortoise passes the hare*

d) sentence ⇨ **noun phrase   transitive verb phrase**
noun phrase ⇨ **article   adjective   noun   transitive verb**
phrase   noun phrase ⇨ **article   adjective   noun**
transitive verb   noun phrase ⇨ **article   adjective   noun**
transitive verb   article   adjective   noun ⇨ . . . (after 6
steps). . . ⇨ *the sleepy hare passes the happy tortoise*

3. The only way to get a noun, such as tortoise, at the end
is to have a noun phrase at the end, which can be achieved
only via the production **sentence → noun phrase   transi-**
tive verb phrase   noun phrase. However, **transitive verb**
phrase → transitive verb → *passes*, and this sentence does
not contain *passes*.

5. $S \Rightarrow 0S1 \Rightarrow 00S11 \Rightarrow 000S111 \Rightarrow 000111$

7. a) $S \Rightarrow 0S \Rightarrow 00S \Rightarrow 00S1 \Rightarrow 00S11 \Rightarrow 00S111 \Rightarrow 00S1111 \Rightarrow$
001111 **b)** $S \Rightarrow 0S \Rightarrow 00S \Rightarrow 001A \Rightarrow 0011A \Rightarrow 00111A \Rightarrow$
001111

9. $S \Rightarrow 0SAB \Rightarrow 00SABAB \Rightarrow 00ABAB \Rightarrow 00AABB \Rightarrow$
$001ABB \Rightarrow 0011BB \Rightarrow 00112B \Rightarrow 001122$

11. a) $S \rightarrow 00S, S \rightarrow \lambda$ **b)** $S \rightarrow 10A, A \rightarrow 00A, A \rightarrow \lambda$
c) $S \rightarrow AAS, S \rightarrow BBS, AB \rightarrow BA, BA \rightarrow AB, S \rightarrow$
$\lambda, A \rightarrow 0, B \rightarrow 1$ **d)** $S \rightarrow 0000000000A, A \rightarrow 0A, A \rightarrow$
λ **e)** $S \rightarrow AS, S \rightarrow ABS, S \rightarrow A, AB \rightarrow BA, BA \rightarrow$
$AB, A \rightarrow 0, B \rightarrow 1$ **f)** $S \rightarrow ABS, S \rightarrow \lambda, AB \rightarrow$
$BA, BA \rightarrow AB, A \rightarrow 0, B \rightarrow 1$ **g)** $S \rightarrow ABS, S \rightarrow$
$T, S \rightarrow U, T \rightarrow AT, T \rightarrow A, U \rightarrow BU, U \rightarrow B, AB \rightarrow$
$BA, BA \rightarrow AB, A \rightarrow 0, B \rightarrow 1$ **13. a)** Type 2, not type
3 **b)** Type 3, not type 2 **c)** Type 0, not type 1 **d)** Type
2, not type 3 **e)** Type 2 **f)** Type 0, not type 1 **g)** Type
3 **h)** Type 0, not type 1 **i)** Type 2, not type 3 **j)** Type 2,
not type 3 **15.** Let S_1 and S_2 be the start symbols of G_1
and G_2, respectively. Let S be a new start symbol. **a)** Add
S and productions $S \rightarrow S_1$ and $S \rightarrow S_2$. **b)** Add S and
production $S \rightarrow S_1S_2$. **c)** Add S and production $S \rightarrow \lambda$
and $S \rightarrow S_1S$.

17. a)

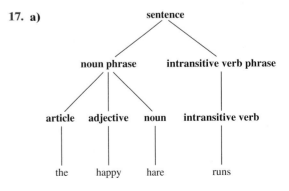

b)

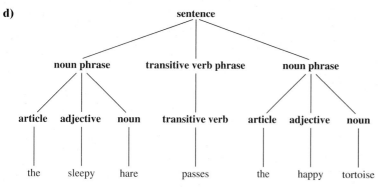

c)

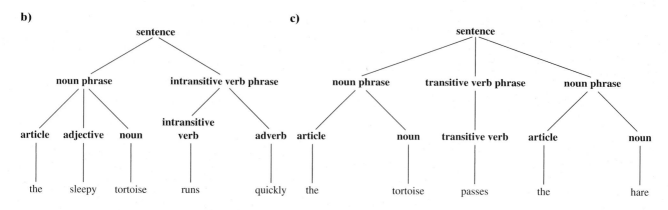

d)

19. a) Yes **b)** No **c)** Yes **d)** No

21.

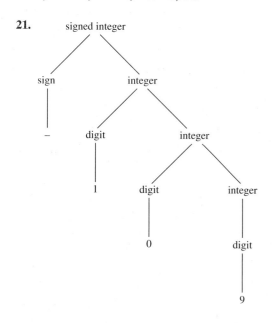

23. a) $S \rightarrow \langle sign \rangle \langle integer \rangle$
$S \rightarrow \langle sign \rangle \langle integer \rangle \,.\, \langle positive\ integer \rangle$
$\langle sign \rangle \rightarrow +$
$\langle sign \rangle \rightarrow -$
$\langle integer \rangle \rightarrow \langle digit \rangle$
$\langle integer \rangle \rightarrow \langle integer \rangle \langle digit \rangle$
$\langle digit \rangle \rightarrow i, i = 1, 2, 3, 4, 5, 6, 7, 8, 9, 0$
$\langle positive\ integer \rangle \rightarrow \langle integer \rangle \langle nonzero\ digit \rangle$
$\langle positive\ integer \rangle \rightarrow \langle nonzero\ digit \rangle \langle integer \rangle$
$\langle positive\ integer \rangle \rightarrow \langle integer \rangle \langle nonzero\ digit \rangle \langle integer \rangle$
$\langle positive\ integer \rangle \rightarrow \langle nonzero\ digit \rangle$
$\langle nonzero\ digit \rangle \rightarrow i, i = 1, 2, 3, 4, 5, 6, 7, 8, 9$

b) $\langle signed\ decimal\ number \rangle ::= \langle sign \rangle \langle integer \rangle \mid$
$\quad \langle sign \rangle \langle integer \rangle \,.\, \langle positive\ integer \rangle$
$\langle sign \rangle ::= + \mid -$
$\langle integer \rangle ::= \langle digit \rangle \mid \langle integer \rangle \langle digit \rangle$
$\langle digit \rangle ::= 0 \mid 1 \mid 2 \mid 3 \mid 4 \mid 5 \mid 6 \mid 7 \mid 8 \mid 9$
$\langle nonzero\ digit \rangle ::= 1 \mid 2 \mid 3 \mid 4 \mid 5 \mid 6 \mid 7 \mid 8 \mid 9$
$\langle positive\ integer \rangle ::= \langle integer \rangle \langle nonzero\ digit \rangle \mid$
$\quad \langle nonzero\ digit \rangle \langle integer \rangle \mid$
$\quad \langle integer \rangle \langle nonzero\ integer \rangle \langle integer \rangle \mid \langle nonzero\ digit \rangle$

c)

25. a) $\langle identifier \rangle ::= \langle lcletter \rangle \mid \langle identifier \rangle \langle lcletter \rangle$
$\langle lcletter \rangle ::= a \mid b \mid c \mid \ldots \mid z$

b) $\langle identifier \rangle ::= \langle lcletter \rangle \langle lcletter \rangle \langle lcletter \rangle \mid \langle lcletter \rangle \langle lcletter \rangle \langle lcletter \rangle \langle lcletter \rangle \mid$
$\qquad \langle lcletter \rangle \langle lcletter \rangle \langle lcletter \rangle \langle lcletter \rangle \langle lcletter \rangle \mid$
$\qquad \langle lcletter \rangle \langle lcletter \rangle \langle lcletter \rangle \langle lcletter \rangle \langle lcletter \rangle \langle lcletter \rangle$
$\langle lcletter \rangle ::= a \mid b \mid c \mid \ldots \mid z$

c) $\langle identifier \rangle ::= \langle ucletter \rangle \mid \langle ucletter \rangle \langle letter \rangle \mid \langle ucletter \rangle \langle letter \rangle \langle letter \rangle \mid$
$\qquad \langle ucletter \rangle \langle letter \rangle \langle letter \rangle \langle letter \rangle \mid \langle ucletter \rangle \langle letter \rangle \langle letter \rangle \langle letter \rangle \langle letter \rangle \mid$
$\qquad \langle ucletter \rangle \langle letter \rangle \langle letter \rangle \langle letter \rangle \langle letter \rangle \langle letter \rangle$
$\langle letter \rangle ::= \langle lcletter \rangle \mid \langle ucletter \rangle$
$\langle lcletter \rangle ::= a \mid b \mid c \mid \ldots \mid z$
$\langle ucletter \rangle ::= A \mid B \mid C \mid \ldots \mid Z$

d) $\langle identifier \rangle ::= \langle lcletter \rangle \langle digitorus \rangle \langle alphanumeric \rangle \langle alphanumeric \rangle \langle alphanumeric \rangle \mid$
$\qquad \langle lcletter \rangle \langle digitorus \rangle \langle alphanumeric \rangle \langle alphanumeric \rangle \langle alphanumeric \rangle \langle alphanumeric \rangle$
$\langle digitorus \rangle ::= \langle digit \rangle \mid \_$
$\langle alphanumeric \rangle ::= \langle letter \rangle \mid \langle digit \rangle$
$\langle letter \rangle ::= \langle lcletter \rangle \mid \langle ucletter \rangle$
$\langle lcletter \rangle ::= a \mid b \mid c \mid \ldots \mid z$
$\langle ucletter \rangle ::= A \mid B \mid C \mid \ldots \mid Z$
$\langle digit \rangle ::= 0 \mid 1 \mid 2 \mid \ldots \mid 9$

27. $\langle identifier \rangle ::= \langle letterorus \rangle \mid \langle identifier \rangle \langle symbol \rangle$
$\langle letterorus \rangle ::= \langle letter \rangle \mid \_$
$\langle symbol \rangle ::= \langle letterorus \rangle \mid \langle digit \rangle$
$\langle letter \rangle ::= \langle lcletter \rangle \mid \langle ucletter \rangle$
$\langle lcletter \rangle ::= a \mid b \mid c \mid \ldots \mid z$
$\langle ucletter \rangle ::= A \mid B \mid C \mid \ldots \mid Z$
$\langle digit \rangle ::= 0 \mid 1 \mid 2 \mid \ldots \mid 9$

29. $numeral ::= sign?\ nonzerodigit\ digit* \ decimal? \mid sign?\ 0\ decimal?$
$sign ::= + \mid -$
$nonzerodigit ::= 1 \mid 2 \mid \ldots \mid 9$
$digit ::= 0 \mid nonzerodigit$
$decimal ::= .digit*$

31. $identifier ::= letterorus\ symbol*$
$letterorus ::= letter \mid \_$
$symbol ::= letterorus \mid digit$
$letter ::= lcletter \mid ucletter$
$lcletter ::= a \mid b \mid c \mid \ldots \mid z$
$ucletter ::= A \mid B \mid C \mid \ldots \mid Z$
$digit ::= 0 \mid 1 \mid 2 \mid \ldots \mid 9$

33. a) ⟨*expression*⟩

⟨*term*⟩⟨*term*⟩⟨*addOperator*⟩

⟨*factor*⟩⟨*factor*⟩⟨*factor*⟩⟨*mulOperator*⟩⟨*addOperator*⟩

⟨*identifier*⟩⟨*identifier*⟩⟨*identifier*⟩⟨*mulOperator*⟩⟨*addOperator*⟩

a b c ∗ +

b) Not generated

c) ⟨*expression*⟩

⟨*term*⟩

⟨*factor*⟩⟨*factor*⟩⟨*mulOperator*⟩

⟨*expression*⟩⟨*factor*⟩⟨*mulOperator*⟩

⟨*term*⟩⟨*term*⟩⟨*addOperator*⟩⟨*factor*⟩⟨*mulOperator*⟩

⟨*factor*⟩⟨*factor*⟩⟨*addOperator*⟩⟨*factor*⟩⟨*mulOperator*⟩

⟨*identifier*⟩⟨*identifier*⟩⟨*addOperator*⟩⟨*identifier*⟩⟨*mulOperator*⟩

x y − *z* ∗

d) ⟨*expression*⟩

⟨*term*⟩

⟨*factor*⟩⟨*factor*⟩⟨*mulOperator*⟩

⟨*factor*⟩⟨*expression*⟩⟨*mulOperator*⟩

⟨*factor*⟩⟨*term*⟩⟨*mulOperator*⟩

⟨*factor*⟩⟨*factor*⟩⟨*factor*⟩⟨*mulOperator*⟩⟨*mulOperator*⟩

⟨*factor*⟩⟨*factor*⟩⟨*expression*⟩⟨*mulOperator*⟩⟨*mulOperator*⟩

⟨*factor*⟩⟨*factor*⟩⟨*term*⟩⟨*term*⟩⟨*addOperator*⟩⟨*mulOperator*⟩⟨*mulOperator*⟩

⟨*factor*⟩⟨*factor*⟩⟨*factor*⟩⟨*factor*⟩⟨*addOperator*⟩⟨*mulOperator*⟩⟨*mulOperator*⟩

⟨*identifier*⟩⟨*identifier*⟩⟨*identifier*⟩⟨*identifier*⟩⟨*addOperator*⟩⟨*mulOperator*⟩⟨*mulOperator*⟩

w x y z − ∗ /

e) ⟨*expression*⟩

⟨*term*⟩

⟨*factor*⟩⟨*factor*⟩⟨*mulOperator*⟩

⟨*factor*⟩⟨*expression*⟩⟨*mulOperator*⟩

⟨*factor*⟩⟨*term*⟩⟨*term*⟩⟨*addOperator*⟩⟨*mulOperator*⟩

⟨*factor*⟩⟨*factor*⟩⟨*factor*⟩⟨*addOperator*⟩⟨*mulOperator*⟩

⟨*identifier*⟩⟨*identifier*⟩⟨*identifier*⟩⟨*addOperator*⟩⟨*mulOperator*⟩

a d e − ∗

35. a) Not generated

b) ⟨*expression*⟩

⟨*term*⟩⟨*addOperator*⟩⟨*term*⟩

⟨*factor*⟩⟨*mulOperator*⟩⟨*factor*⟩⟨*addOperator*⟩⟨*factor*⟩⟨*mulOperator*⟩⟨*factor*⟩

⟨*identifier*⟩⟨*mulOperator*⟩⟨*identifier*⟩⟨*addOperator*⟩⟨*identifier*⟩⟨*mulOperator*⟩⟨*identifier*⟩

a/*b* + *c*/*d*

c) ⟨*expression*⟩

⟨*term*⟩

⟨*factor*⟩⟨*mulOperator*⟩⟨*factor*⟩

⟨*factor*⟩⟨*mulOperator*⟩(⟨*expression*⟩)

⟨*factor*⟩⟨*mulOperator*⟩(⟨*term*⟩⟨*addOperator*⟩⟨*term*⟩)

⟨*factor*⟩⟨*mulOperator*⟩(⟨*factor*⟩⟨*addOperator*⟩⟨*factor*⟩)

⟨*identifier*⟩⟨*mulOperator*⟩(⟨*identifier*⟩⟨*addOperator*⟩⟨*identifier*⟩)

m ∗ (*n* + *p*)

d) Not generated

e) ⟨*expression*⟩

⟨*term*⟩

⟨*factor*⟩⟨*mulOperator*⟩⟨*factor*⟩

(⟨*expression*⟩)⟨*mulOperator*⟩(⟨*expression*⟩)

(⟨*term*⟩⟨*addOperator*⟩⟨*term*⟩)⟨*mulOperator*⟩(⟨*term*⟩⟨*addOperator*⟩⟨*term*⟩)

(⟨*factor*⟩⟨*addOperator*⟩⟨*factor*⟩)⟨*mulOperator*⟩(⟨*factor*⟩⟨*addOperator*⟩⟨*factor*⟩)

(⟨*identifier*⟩⟨*addOperator*⟩⟨*identifier*⟩)⟨*mulOperator*⟩(⟨*identifier*⟩⟨*addOperator*⟩⟨*identifier*⟩)

(*m* + *n*) ∗ (*p* − *q*)

SECTION 11.2

1. a)

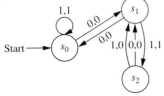

b)

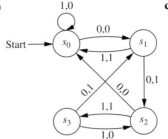

c)

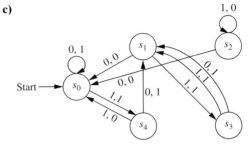

3. a) 1100   **b)** 00110110   **c)** 11111111111

5.

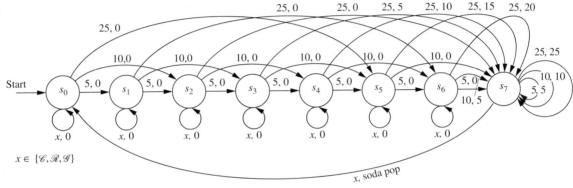

$\mathscr{C}$ = cola
$\mathscr{R}$ = root beer
$\mathscr{G}$ = ginger ale

7.

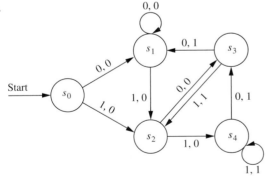

9.

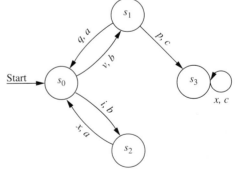

v = Valid ID a = "Enter user ID"
i = Invalid ID b = "Enter password"
p = Valid password c = Prompt
q = Invalid password x = Any input

11.

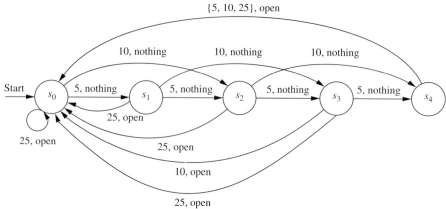

13.

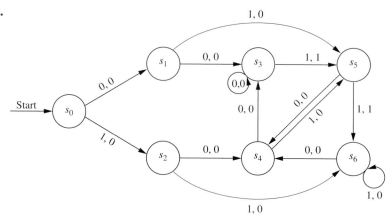

15.

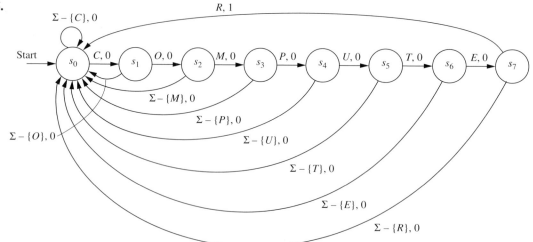

17.

	f		
	Input		
State	**0**	**1**	***g***
s_0	s_1	s_2	1
s_1	s_1	s_0	1
s_2	s_1	s_2	0

19. a) 11111
b) 1000000
c) 100011001100

21.

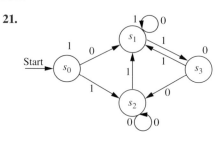

SECTION 11.3

1. a) $\{000, 001, 1100, 1101\}$ **b)** $\{000, 0011, 010, 0111\}$
c) $\{00, 011, 110, 1111\}$ **d)** $\{000000, 000001, 000100, 000101,$
$010000, 010001, 010100, 010101\}$ **3.** $A = \{1, 101\}$,
$B = \{0, 11, 000\}; A = \{10, 111, 1010, 1000, 10111, 101000\}$,
$B = \{\lambda\}; A = \{\lambda, 10\}$, $B = \{10, 111, 1000\}$ or $A = \{\lambda\}$,
$B = \{10, 111, 1010, 1000, 10111, 101000\}$ **5. a)** The set
of all strings consisting of zero or more consecutive bit
pairs 10   **b)** The set of all strings consisting of all 1s such
that the number of 1s is divisible by 3, including the null
string   **c)** The set of all strings in which every 1 is immediately preceded by a 0   **d)** The set of all strings that
begin and end with a 1 and have at least two 1s between
every pair of 0s   **7.** A string is in A^* if and only if it
is a concatenation of an arbitrary number of strings in
A. Since each string in A is also in B, it follows that a
string in A^* is also a concatenation of strings in B. Hence
$A^* \subseteq B^*$.   **9. a)** Yes **b)** Yes **c)** Yes **d)** No **e)** Yes
f) Yes   **11.**   **a)** Yes **b)** Yes **c)** No **d)** No **e)** No
f) No   **13.** $\{0, 10, 11\}\{0, 1\}^*$ **15.** $\{0^m 1^n \mid m \geq 0 \text{ and } n \geq 1\}$ **17.** $\{0, 01, 11\}$ **19.** $\{\lambda, 0\} \cup \{0^m 1^n \mid m \geq 1, n \geq 1\}$
21. $\{10^n \mid n \geq 0\} \cup \{10^n 10^m \mid n, m \geq 0\}$

23.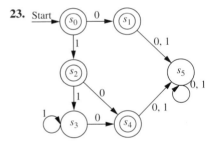

25. Add a non-final state s_3 with transitions to s_3 from s_0
on input 0, from s_1 on input 1, and from s_3 on input 0 or 1.

27. a)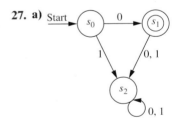

b)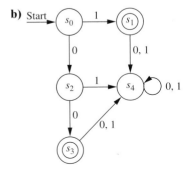

c)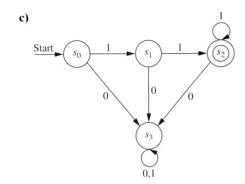

29. Suppose that M is a finite-state automaton that accepts
the set of bit strings containing an equal number of 0s and
1s. Suppose M has n states. Consider the string $0^{n+1}1^{n+1}$.
By the pigeonhole principle, as M processes this string, it
must encounter the same state more than once as it reads
the first $n + 1$ 0s; so let s be a state it hits at least twice.
Then k 0s in the input takes M from state s back to itself
for some positive integer k. But then M ends up exactly
at the same place after reading $0^{n+1+k}1^{n+1}$ as it will after
reading $0^{n+1}1^{n+1}$. Therefore, since M accepts $0^{n+1}1^{n+1}$ it
also accepts $0^{n+k+1}1^{n+1}$, which is a contradiction.

SECTION 11.4

1. a) Any number of 1s followed by a 0   **b)** Any number
of 1s followed by one or more 0s   **c)** 111 or 001   **d)** A
string of any number of 1s or 00s or some of each in a
row   **e)** λ or a string that ends with a 1 and has one or
more 0s before each 1   **f)** A string of length at least 3 that
ends with 00   **3. a)** 00^*1 **b)** $(0 \cup 1)(0 \cup 1)(0 \cup 1)^*0000^*$
c) $0^*1^* \cup 1^*0^*$ **d)** $11(111)^*(00)^*$ **5.** Use an inductive
proof. If the regular expression for A is $\emptyset$, λ, or x, the result
is trivial. Otherwise, suppose the regular expression for A
is $\mathbf{BC}$. Then $A = BC$ where B is the set generated by $\mathbf{B}$ and
C is the set generated by $\mathbf{C}$. By the inductive hypothesis
there are regular expressions $\mathbf{B}'$ and $\mathbf{C}'$ that generate B^R
and C^R, respectively. Since $A^R = (BC)^R = C^R B^R$, $\mathbf{C}'\mathbf{B}'$ is
a regular expression for A^R. If the regular expression for
A is $\mathbf{B} \cup \mathbf{C}$, then the regular expression for A is $\mathbf{B}' \cup \mathbf{C}'$ since
$(B \cup C)^R = (B^R) \cup (C^R)$. Finally, if the regular expression
for A is $\mathbf{B}^*$, then it is easy to see that $(\mathbf{B}')^*$ is a regular
expression for A^R.

7. a)

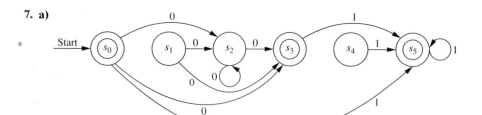

b)

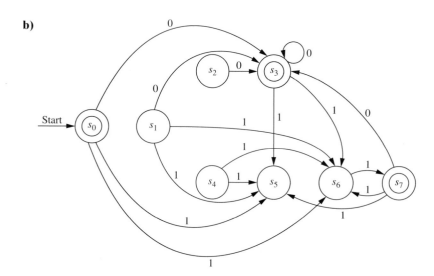

c)

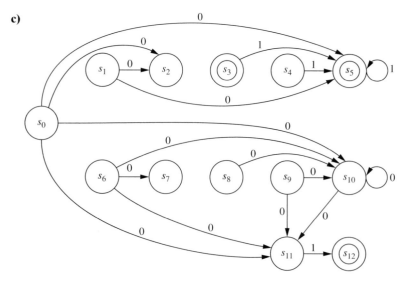

9. $S \to 0A, S \to 1B, S \to 0, A \to 0B, A \to 1B, B \to 0B, B \to 1B$ **11.** $S \to 0C, S \to 1A, S \to 1, A \to 1A, A \to 0C, A \to 1, B \to 0B, B \to 1B, B \to 0, B \to 1, C \to 0C, C \to 1B, C \to 1.$ **13.** This follows since input that leads to a final state in the automaton corresponds uniquely to a derivation in the grammar. **15.** The "only if" part is clear since I is finite. For the "if" part let the states be $s_{i_0}, s_{i_1}, s_{i_2}, \ldots, s_{i_n}$, where $n = l(x)$. Since $n \geq |S|$, some state is repeated by the pigeonhole principle. Let y be the part of x that causes the loop, so that $x = uyv$ and y sends s_j to s_j, for some j. Then $uy^k v \in L(M)$ for all k. Hence $L(M)$ is infinite. **17.** Suppose that $L = \{0^{2n}1^n\}$ were regular. Let S be the set of states of a finite-state machine recognizing this set. Let $z = 0^{2n}1^n$ where $3n \geq |S|$. Then by the pumping lemma, $z = 0^{2n}1^n = uvw, l(v) \geq 1$, and $uv^i w \in \{0^{2n}1^n \mid n \geq 0\}$. Obviously v cannot contain both 0 and 1, since v^2 would then contain 10. So v is all 0s or all 1s, and hence $uv^2 w$ contains too many 0s or too many 1s, so it is not in L. This contradiction shows that L is not regular. **19.** Suppose that the set of palindromes over $\{0, 1\}$ were regular. Let S be the set of states of a finite-state machine recognizing this set. Let $z = 0^n 10^n$, where $n > |S|$. Apply the pumping lemma to get $uv^i w \in L$ for all nonnegative integers i where $l(v) \geq 1$, and $l(uv) \leq |S|$, and $z = 0^n 10^n = uvw$. Then v must be a string of 0s (since $|n| > |S|$), so $uv^2 w$ is not a palindrome. Hence the set of palindromes is not regular.

SECTION 11.5

1. a) The nonblank portion of the tape contains the string 1111 when the machine halts. **b)** The nonblank portion of the tape contains the string 011 when the machine halts. **c)** The nonblank portion of the tape contains the string 00001 when the machine halts. **d)** The nonblank portion of the tape contains the string 00 when the machine halts. **3.** If the tape contains at least one 1, the machine changes every other 1 to a 0 starting at the first 1, and halts when it reaches the first blank symbol. If the tape is blank initially the machine halts without changing the tape. If the nonblank portion of the tape contains all 0s, the machine moves successively through these 0s and halts.
5. $(s_0, 0, s_1, 1, R), (s_0, 1, s_0, 1, R)$
7. $(s_0, 0, s_0, 0, R), (s_0, 1, s_1, 1, R), (s_1, 0, s_1, 0, R), (s_1, 1, s_1, 0, R)$
9. $(s_0, 0, s_1, 0, R), (s_0, 1, s_0, 0, R), (s_1, 0, s_1, 0, R), (s_1, 1, s_0, 0, R), (s_1, B, s_2, B, R)$
11. $(s_0, 0, s_0, 0, R), (s_0, 1, s_1, 1, R), (s_1, 0, s_1, 0, R), (s_1, 1, s_0, 1, R), (s_0, B, s_2, B, R)$
13. If the input string is blank or starts with a 1 the machine halts in nonfinal state s_0. Otherwise, the initial 0 is changed to an M and the machine skips past all the intervening 0s and 1s until it either comes to the end of the input string or else comes to an M. At this point, it backs up

one square and enters state s_2. Since the acceptable strings must have a 1 at the right for each 0 at the left, there must be a 1 here if the string is acceptable. Therefore, the only transition out of s_2 occurs when this square contains a 1. If it does, the machine replaces it with an M and makes its way back to the left; if it does not the machine halts in nonfinal state s_2. On its way back, it stays in s_3 as long as it sees 1s, then stays in s_4 as long as it sees 0s. Eventually either it encouters a 1 while in s_4 at which point it halts without accepting or else it reaches the rightmost M that had been written over a 0 at the start of the string. If it is in s_3 when this happens, then there are no more 0s in the string, so it had better be the case that there are no more 1s either; this is accomplished by the transitions (s_3, M, s_5, M, R) and (s_5, M, s_6, M, R), and s_6 is a final state. Otherwise the machine halts in nonfinal state s_5. If it is in s_4 when this M is encountered, things start all over again, except now the string will have had its leftmost remaining 0 and its rightmost remaining 1 replaced by Ms. So the machine moves, staying in state s_4, to the leftmost remaining 0 and goes back into state s_0 to repeat the process.
15. $(s_0, B, s_9, B, L), (s_0, 0, s_1, 0, L), (s_1, B, s_2, E, R), (s_2, M, s_2, M, R), (s_2, 0, s_3, M, R), (s_3, 0, s_3, 0, R), (s_3, M, s_3, M, R), (s_3, 1, s_4, M, R), (s_4, 1, s_4, 1, R), (s_4, M, s_4, M, R), (s_4, 2, s_5, M, R), (s_5, 2, s_5, 2, R), (s_5, B, s_6, B, L), (s_6, M, s_8, M, L), (s_6, 2, s_7, 2, L), (s_7, 0, s_7, 0, L), (s_7, 1, s_7, 1, L), (s_7, 2, s_7, 2, L), (s_7, M, s_7, M, L), (s_7, E, s_2, E, R), (s_8, M, s_8, M, L), (s_8, E, s_9, E, L)$ where M and E are markers, with E marking the left end of the input
17. $(s_0, 1, s_1, B, R), (s_1, 1, s_2, B, R), (s_2, 1, s_3, B, R), (s_3, 1, s_4, 1, R), (s_1, B, s_4, 1, R), (s_2, B, s_4, 1, R), (s_3, B, s_4, 1, R)$
19. $(s_0, 1, s_1, B, R), (s_1, 1, s_2, B, R), (s_1, B, s_6, B, R), (s_2, 1, s_3, B, R), (s_2, B, s_6, B, R), (s_3, 1, s_4, B, R), (s_3, B, s_6, B, R), (s_4, 1, s_5, B, R), (s_4, B, s_6, B, R), (s_6, B, s_{10}, 1, R), (s_5, 1, s_5, B, R), (s_5, B, s_7, 1, R), (s_7, B, s_8, 1, R), (s_8, B, s_9, 1, R), (s_9, B, s_{10}, 1, R)$
21. $(s_0, 0, s_0, 0, R), (s_0, *, s_5, B, R), (s_3, *, s_3, *, L), (s_3, 0, s_3, 0, L), (s_3, 1, s_3, 1, L), (s_3, B, s_0, B, R), (s_5, 1, s_5, B, R), (s_5, 0, s_5, B, R), (s_5, B, s_6, B, L), (s_6, B, s_6, B, L), (s_6, 0, s_7, 1, L), (s_7, 0, s_7, 1, L), (s_0, 1, s_1, 0, R), (s_1, 1, s_1, 1, R), (s_1, *, s_2, *, R), (s_2, 0, s_2, 0, R), (s_2, 1, s_3, 0, L), (s_2, B, s_4, B, L), (s_4, 0, s_4, 1, L), (s_4, *, s_8, B, L), (s_8, 0, s_8, B, L), (s_8, 1, s_8, B, L)$
23. $(s_0, B, s_1, 1, L), (s_0, 1, s_1, 1, R), (s_1, B, s_0, 1, R)$

SUPPLEMENTARY EXERCISES

1. a) $S \to 00S111, S \to \lambda$ **b)** $S \to AABS, AB \to BA, BA \to AB, A \to 0, B \to 1, S \to \lambda$ **c)** $S \to ET, T \to 0TA, T \to 1TB, T \to \lambda, 0A \to A0, 1A \to A1, 0B \to B0, 1B \to B1, EA \to E0, EB \to E1, E \to \lambda$

3.

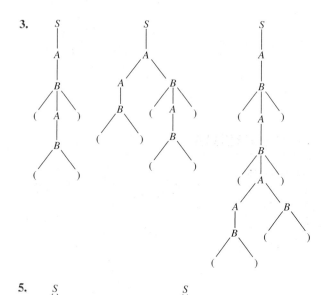

5.

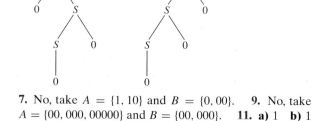

7. No, take $A = \{1, 10\}$ and $B = \{0, 00\}$. **9.** No, take $A = \{00, 000, 00000\}$ and $B = \{00, 000\}$. **11. a)** 1 **b)** 1 **c)** 2 **d)** 3 **e)** 2 **f)** 4

13.

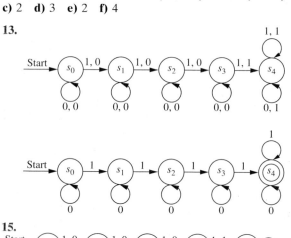

15.

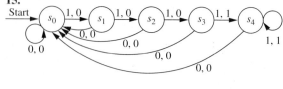

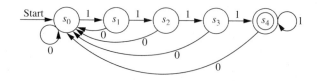

17. a) $n^{nk+1}m^{nk}$ **b)** $n^{nk+1}m^n$

19.

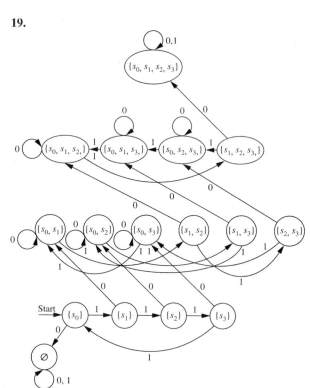

21. a)

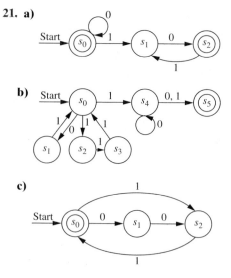

b)

c)

23. Construct the deterministic finite automaton for A with states S and final states F. For $\overline{A}$ use the same automaton but with final states $S - F$.

25. a)

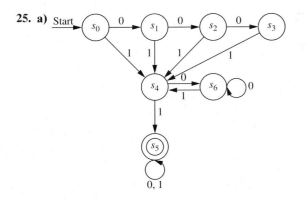

b)

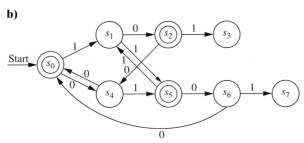

c)

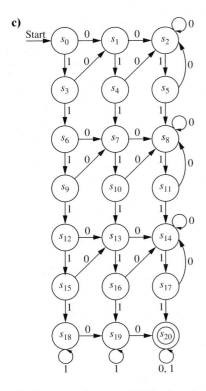

27. Suppose that $L = \{1^p \mid p \text{ is prime}\}$ is regular, and let S be the set of states in a finite-state machine recognizing L. Let $z = 1^p$ where p is a prime with $p > |S|$ (such a prime exists since there are infinitely many primes). By the pumping lemma it must be possible to write $z = uvw$ with $l(uv) \le |S|, l(v) \ge 1$ and for all nonnegative integers

i, $uv^i w \in L$. Since z is a string of all 1s, $u = 1^a$, $v = 1^b$, and $w = 1^c$, where $a + b + c = p, a + b \le n$, and $b \ge 1$. This means that $uv^i w = 1^a 1^{bi} 1^c = 1^{(a+b+c)+b(i-1)} = 1^{p+b(i-1)}$. Now take $i = p + 1$. Then $uv^i w = 1^{p(1+b)}$. Since $p(1 + b)$ is not prime, $uv^i w \notin L$, which is a contradiction.

APPENDIXES

APPENDIX 1
1. a) 2^3 **b)** 2^6 **c)** 2^4 **3. a)** $2y$ **b)** $2y/3$ **c)** $y/2$

5.

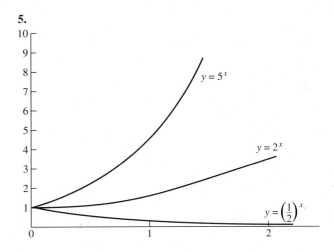

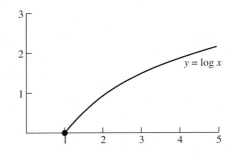

APPENDIX 2
1. After the first block is executed, a has been assigned the original value of b and b has been assigned the original value of c, whereas after the second block is executed, b is assigned the original value of c and a the original value of c as well.   **3.** The following **while** construction does the same thing.

```
i := initial value
while i ≤ final value
begin
   statement
   i := i + 1
end
```

Photo Credits

Chapter 1

Page 2: ©National Library of Medicine; **p. 4:** Library of Congress; **p. 13:** Courtesy Indiana University Archives; **p. 15:** University Archives. Department of Rare Books and Special Collections. Princeton University Library; **p. 23:** By permission of the London Mathematical Society; **p. 26 top:** ©Mary Evans Picture Library; **p. 26 bottom:** Courtesy, Harvard University Art Museums, Cambridge, MA; **p. 29:** ©Bettmann/Corbis Images; **p. 37:** ©The Granger Collection; **p. 78, p. 79:** Library of Congress; **p. 80:** By permission of the President and Council of the Royal Society; **p. 83:** Scientists and Inventors Portrait File Collection, Archives Center, National Museum of American History.

Chapter 2

Page 120: From *The Scientists Of The Ancient World,* by Anderson/Stephenson; **p. 90:** Illustration by Marie le Glatin Keis; **p. 133:** Bachmann, Paul. Die Arithmetik Der Quadratischen Formen. Verlag und Druck von BG Teubner. Leipzig. Berlin. 1923; **p. 134:** Smith Collection, Rare Book and Manuscript Library, Columbia University; **p. 135:** Stanford University News Service; **p. 157:** ©Mary Evans Picture Library; **p. 162:** Smithsonian Institution. Photo No. 46, 834-M; **p. 177:** Scientists and Inventors Portrait File Collection, Archives Center, National Museum of American History; **p. 189:** From A History of Science, Technology and Philosophy in the 16th and 17th Centuries, by Abraham Wolf. Reprinted by permission of HarperCollins Publishers Ltd. ©Abraham Wolf; **p. 192:** Courtesy Ronald L. Rivest; **p. 193:** ©Kristen Tsolis; **p. 194:** ©Courtesy Leonard Adleman.

Chapter 3

Page 229: Courtesy, Neil Sloane and AT&T Shannon Lab; **p. 259:** ©The Granger Collection; **p. 260:** ©Mary Evans Picture Library; **p. 286:** Courtesy Tony Hoare; **p. 293:** ©Matthew Naythons/TimePix; **p. 296, p. 297:** Master and Fellows of Trinity College Cambridge.

Chapter 4

Page 313: ©The Granger Collection; **p. 318:** Courtesy, Mrs. Jane Burch; **p. 330:** National Library of Medicine.

Chapter 5

Page 356: From A Concise History of Mathematics by Dirk Struik, 1967, Dover Publications; **p. 369:** ©The Granger Collection; **p. 391:** Smith Collection, Rare Book and Manuscript Library, Columbia University.

Chapter 7

Page 503: Courtesy Stephen Warshall; **p. 533:** By permission of the President and Council of the Royal Society; **p. 520:** ©Deutsche Museum; **p. 579:** Scientists and Inventors Portrait File Collection, Archives Center, National Museum of American History; **p. 584:** Master and Fellows of Trinity College Cambridge; **p. 592:** Julius Petersen: Metoder og Teorier, Edition no. 6, Det Schoenbergske Forlag, Copenhagen 1912. Photo courtesy of Bjarne Toft; **p. 595:** Courtesy, Edsger W. Dijkstra; **p. 609:** Courtesy, the Archive of Polish Mathematicians, at the Institute of Mathematics of the Polish Academy of Sciences. Reproduced with permission of the Institute; **p. 614:** By permission of the London Mathematical Society.

Chapter 9

Page 636: Master and Fellows of Trinity College Cambridge; **p. 650:** Courtesy, California State University, Santa Cruz; **p. 670:** Courtesy, the Archive of Polish Mathematicians, at the Institute of Mathematics of the Polish Academy of Sciences. Reproduced with permission of the Institute; **p. 689:** Courtesy, Robert Clay Prim; **p. 690:** Courtesy, Joseph B. Kruskal. Photo by Rachel Kruskal.

Chapter 10

Page 702: ©MIT Museum; **p. 720:** Courtesy, Maurice Karnaugh; **p. 727:** Courtesy, Edward J. McCluskey; **p. 728:** Courtesy, The Quine Family.

Chapter 11

Page 745: ©Donna Coveney/MIT; **p. 747 top:** Courtesy IBM, Almaden Research Center; **p. 747 bottom:** Courtesy Peter Naur; **p. 759:** From: *I Have A Photographic Memory* (1987), Plate 506, published by The American Mathematical Society. Photo Courtesy, Paul Halmos; **p. 774:** ©Science Photo Library/Photo Researchers; **p. 781:** University Archives. Department of Rare Books and Special Collections. Princeton University Library.

Index of Biographies

Adleman, Leonard, 192
Aristotle, 2
Augusta, Ada (Countess of Lovelace), 25
Bachmann, Paul Gustav Heinrich, 133
Backus, John, 747
Bernoulli, James, 369
Boole, George, 4
Cantor, Georg, 78
Carmichael, Robert Daniel, 190
Carroll, Lewis (Charles Dodgson), 37
Cayley, Arthur, 636
Chebyshev, Pafnuty Lvovich, 391
Chomsky, Avram Noam, 745
Church, Alonzo, 781
De Morgan, Augustus, 23
Descartes, René, 83
Dijkstra, Edsger Wybe, 595
Dirichlet, G. Lejeune, 313
Dodgson, Charles (Lewis Carroll), 37
Eratosthenes, 459
Erdős, Paul, 533
Euclid, 177
Euler, Leonhard, 579
Fermat, Pierre de, 189
Fibonacci, 259
Gauss, Karl Friedrich, 162
Goldbach, Christian, 221
Hamilton, William Rowan, 584
Hardy, Godfrey Harold, 296
Hasse, Helmut, 520
Hoare, C. Anthony R., 286
Huffman, David A., 650
Karnaugh, Maurice, 720

Kempe, Alfred Bray, 614
al-Khowarizmi, Abu Ja'far Mohammed ibn Musa, 120
Kleene, Stephen Cole, 759
Knuth, Donald E., 135
Kruskal, Joseph Bernard, 690
Kuratowski, Kazimierz, 609
Lamé, Gabriel, 260
Landau, Edmund, 134
Laplace, Pierre Simon, 356
Łukasiewicz, Jan, 670
McCarthy, John, 294
McCluskey, Edward J., 728
Mersenne, Marin, 157
Naur, Peter, 747
Pascal, Blaise, 330
Peirce, Charles Sanders, 29
Petersen, Julius Peter Christian, 592
Prim, Robert Clay, 689
Quine, Willard van Orman, 729
Ramanujan, Srinivasa, 297
Ramsey, Frank Plumpton, 318
Rivest, Ronald, 192
Russell, Bertrand, 79
Shamir, Adi, 192
Shannon, Claude Elwood, 702
Sheffer, Henry Maurice, 26
Sloane, Neil, 229
Smullyan, Raymond, 13
Tukey, John Wilder, 15
Turing, Alan Mathison, 774
Vandermonde, Alexandre-Théophile, 332
Venn, John, 80
Warshall, Stephen, 503

Index

Absorption laws
  for Boolean algebra, 705, 706
  for lattices, 535
  for propositions, 24
  for sets, 89
Abstract definition of Boolean
      algebra, 706–707
Academic papers, collaboration in, 542
Academy, Plato's, 2
Accepted language, 760–763
Accepted strings, 760
Ackermann, Wilhelm, 273, B-2
Ackermann's function, 273, B-2
Acquaintanceship graph, 541
  paths in, 568
Actors, 541
Adapting existing proofs, 216–217
Adders, 716–717
  full, 717, 735
  half, 717, 734
Addition
  of functions, 98–99
  of geometric progressions, 243
  of integers, 172–174
  of matrices, 197–198
  of multisets, 96
  of subsets, 684
Addition rule of inference, 58
Address system, universal, 660
Adjacency list, 557
Adjacency matrices, 557–559, 622
  counting paths between vertices
      by, 574–575
Adjacent cells, 721
Adjacent vertices, 546, 547, 622
Adjective, 740
Adleman, Leonard, 191, 192
Adverb, 740
Affine transformation, 166
Airline system, weighted graphs
      modeling, 593–594
Alexander the Great, 2
Algebra, Boolean, 4, 26, 701–738
  abstract definition of, 706–707
  definition of, 734
  identities of, 704–706, 707
ALGOL, 746
ALGOL 60, 747
Algorithm(s), 119, 120–131, B-1, B-2
  approximation, 600–601
  for base b expansion, 169, 171
  for binary search tree, 645–646
  for Boolean product of zero-one
      matrices, 202–203
  bubble sort, 126
  change-making, 128
  for coloring graphs, 618, 621
  complexity of, 144–152
    average-case, 146–147, 149, 206, 384–385
    of binary search algorithms, 146
    of breadth-first search algorithms, 681

of bubble sort, 147
computational, 144
  of Dijkstra's algorithm, 599
constant, 148
of depth-first search algorithms, 679
exponential, 148
factorial, 148
of finding maximum element of set, 145
of insertion sort, 147–148
linear, 148
of linear search algorithms, 145–147
logarithmic, 148
of merge sort, 430
polynomial, 148
of searching algorithms, 145–147, 148
of sorting algorithms, 147–148, 647–649
space, 144, 206
time, 144, 145–148, 149, 206
worst-case, 146, 147–148, 149, 206
computer time used by, 150–151
for computing **div** and **mod**, 175–176
for computing x^n (mod m), 176
definition of, 120, 206
Dijkstra's, 595–599, 623, B-3
divide-and-conquer, 426, 465
division, 158–159, 207
estimating number of operations of, 132–140
Euclidean, 177–179, 207, 260–261
  extended, 182, 196
for Euler circuits, 581–582
for Euler paths, 582–583
for evaluating polynomials, 151–152
for fast integer multiplication, 427–428, B-2
for fast matrix multiplication, 428
for Fibonacci numbers, 278–279
for finding maximum element in
      sequence, 121–122, 209
for finding minimum element in sequence, 209
Fleury's, 582, 591
Floyd's, 602–603
for generating combinations, 346–348, B-2
for generating permutations, 344–346, B-2
in graph theory, B-3
greedy, 127–129, 172, 248–249, 651, 689, 695
history of word, 120
Huffman coding, 650–651
inorder traversal, 666
insertion sort, 127
for integer addition, 172–174
for integer multiplication,
      174–175, 427–428
for integer operations, 172–176
integers and, 169–181
iterative, for Fibonacci numbers, 279
Kruskal's, 690–691, 692, 695, B-3
for matrix multiplication, 199–200
merge sort, 280
for minimum spanning trees, 689–693
for modular exponentiation, 176
Monte Carlo, 373–375
NAUTY, 563

optimal, 152
parallel, 552
postorder traversal, 668
preorder traversal, 666
Prim's, 689, 690, 695, B-3
probabilistic, 355, 373–375, 394
properties of, 122
recursive, 274–284, 291
  binary search, 276
  for computing a^n, 275
  for computing greatest common
      divisor, 276
  correctness of, 275
  for factorials, 277
  for Fibonacci numbers, 278
  linear search, 276
  modular exponentiation, 275–276
searching, 123–125, 206, 207
  binary, 123–125, 206, 207, 426
  Boolean, 12
  breadth-first, 680–681, 695
  complexity of, 145–147, 148
  depth-first, 677–680, 695
    applications with, 681–684
    in directed graphs, 684–685
  linear, 123, 206, 207
    average-case complexity of, 384
    complexity of, 145–147, 148
    recursive, 276
  recursive sequential, 276–277
  ternary, 130
serial, 552
shortest-path, 595–599
Sollin's, 694, B-3
sorting, 125–127, 206
  complexity of, 147–148, 647–649
  topological, 525–527, 531
  for spanning trees, 689–693
  topological sorting, 526
  for transitive closure, 502
  traversal, 660–668
  Warshall's, 503–506, 531, B-2
Alice in Wonderland (Carroll), 37
Allen, Layman, 117
Alphabet, 741, 783
Alpha-beta pruning, 655, 700
Alphanumeric character, 306
Ambiguous grammar, 784
Analytic Engine, 26
Ancestors of vertex, 634, 695
AND, 12, 14–15
AND gate, 713, 734
Antecedent, 6
Antichain, 534
Antisymmetric relation, 475–476, 531
  representing
    using digraphs, 494
    using matrices, 491
Appel, Kenneth, 615

Application(s)
  with backtracking, 681–684
  with congruences, 163–166
  with graphs colorings, 618–619
  with inclusion-exclusion, 457–465
  with number theory, 181–196
  with pigeonhole principle, 316–318
  with trees, 644–659
Approximation algorithm, 600–601
Argument(s), 38
  Cantor diagonalization, 235
  valid, 58–59
Aristotle, 2, 3
Arithmetic
  computer, with large integers, 187–189
  Fundamental Theorem of, 155, 207, 250
  matrix, 197–199
  modular, 161–163
Arithmetic mean, 215, 256
Arithmeticorum Libri Duo (Maurolico), 240
Arithmetic progression, 226, 290
Array
  linear, 552, 553
  two-dimensional, 552
Arrow, Peirce, 27
Ars Conjectandi (Bernoulli), 369
Article, 740
Articulation points, 571
Artificial intelligence, 293, B-1
  fuzzy logic in, 18
  fuzzy sets in, 96–97
Art of Computer Programming, The
    (Knuth), 125, 135
Assertion
  final, 285, 291
  initial, 285, 291
Assignment of jobs, 464
Assignments, frequency, 618
Assignment statements, A-4–A-5
Associated homogenous recurrence relations,
  419
Associative laws
  for Boolean algebra, 705, 707
  for lattices, 535
  for propositions, 23, 24
  for sets, 89
Asymmetric relation, 480
Asymptotic functions, 144
Asynchronous transfer mode (ATM), 106–107
ATM cells, 106–107
AT&T network, 571
Augusta, Ada (Countess of Lovelace), 23, 25
Automated Theorem Proving, 117
Automaton
  finite-state, 759–764
    deterministic, 761, 763, 783
    nondeterministic, 761–763, 767–769,
      771, 783
    regular sets recognized by, 766–769
    set not recognized by, 772
  linear bounded, 773
  pushdown, 773
  theory, 751–773
Average-case complexity
  of algorithms, 146–147, 149, 206, 384–385
  of bubble sort variant, 393
  of insertion sort, 385
  of linear search, 384
  of quick sort, 393–394
AVL-trees, 700
Axioms, 56, 77, 112

Babbage, Charles, 26
Bachmann, Paul Gustav Heinrich, 133, 134
Back edges, 678
Backtracking, 677–680, 695, B-3
  applications with, 681–684
  in directed graphs, 684–685
  solving the *n*-queens problem, 683
Backus, John, 746, 747
Backus–Naur form (BNF), 746–748, 783
Backward differences, 412–413
Backward reasoning, 214
Bacon, Kevin, 569
Bacon number, 569
Bacteria, 409, 467
Balanced rooted trees, 640, 695
Balanced strings of parentheses, 295
Balanced ternary expansion, 180
Bandwidth of graph, 627
Base *b*
  Miller's test for, 195
  strong pseudoprime to, 195
Base *b* expansion, 169, 171
Baseball, 309
Base conversion, 170
BASIC, counting variable names, 306
Basis, vertex, 577
Basis step, 240, 261, 267, 291, 639, 655
  of recursive definition of functions, 257–258
  of recursive definition of sets, 261–262
Bayes' formula, 397
Beaver, busy, 780, 782
Begging the question, 72
Begin statements, A-6
Bell, E. T., 516
Bell numbers, 516
Bernoulli, James, 368, 369
Bernoulli family tree, 632
Bernoulli's inequality, 253
Bernoulli trial, 368–370, 381, 394
Biconditional, 8–9, 112
  logical equivalences for, 24
Bidirectional bubble sort, 125, 209
Big-Omega notation, 134, 140–142
Big-*O* notation, 132–138
Big-Theta notation, 134, 140–142
Bijective function, 101, 113
Binary coded decimal, 181
Binary coded decimal expansion, 727
Binary digit, 112
  origin of term, 14, 15
Binary expansion, 169, 207
Binary insertion sort, 125, 131
Binary relation, 471, 530
Binary search algorithm, 123–125, 206, 207, 426
  complexity of, 146, 148
  recursive, 276–277
Binary search trees, 644–646, 695
Binary trees, 264–265, 634, 635
  definition of, 695
  extended, 265
  full, 265–266
    height of, 269
    number of vertices of, 269
  with prefix code, 649
Binding variables, 33–34
Binomial coefficients, 322, 327–334, 349
  extended, 438
Binomial distribution, 368–370
Binomial expression, 327
Binomial Theorem, 327–330, 349
  Extended, 439

Binomial trees, 697
Bipartite graphs, 549–551, 622
Birthday problem, 371–372, 377
Bit, 112
  origin of term, 14, 15
Bit operations, 14, 112
Bit strings, 14, 93–94, 112
  counting, 302
    without consecutive 0s, 406–407
  decoding, 649
  generating next largest, 347
  length of, 14, 406–407
Bitwise operators, 14–15, 112
Blocks of statements, A-5
BNF (Backus–Naur form), 746–748, 783
Bonferroni's inequality, 377
Book number of graph, 629
Boole, George, 3, 4, 701
Boolean algebra, 4, 26, 701–738
  abstract definition of, 706–707, B-3
  definition of, 734
  identities of, 704–706, 707
Boolean expressions, 702–704, 734
  dual of, 706, 734
Boolean functions, 701–708, 734
  dual of, 706
  functionally complete set of operators for, 711
  implicant of, 722, 735
  minimization of, 719–734, 735
  representing, 709–712
  self-dual, 736
  threshold, 737
Boolean power (of zero-one matrices), 203–204
Boolean product, 701–702, 703, 709–711, 734
  of zero-one matrices, 202–203, 207
Boolean searches, 12
Boolean sum, 701–702, 703, 709–711, 734
Boolean variable, 14, 112, 702, 709, 734
Boole's inequality, 377
Boruvka, Otakar, 691
Bots, 685
Bottom-up parsing, 746
Bound
  greatest lower, 531
    of poset, 523
  least upper, 56, 531
    of poset, 523
  lower
    of lattice, 535
    of poset, 523, 531
  upper, 56
    of lattice, 535
    of poset, 523, 531
Bounded lattices, 535
Bound variable, 33
Boxes, distributing objects into, 341
Breadth-first search, 680–681, 695
Bridge, in graphs, 571
Bridge problem, Königsberg, 577–578, 581, 583
B-tree of degree *k*, 697
Bubble sort, 125–126, 207, 393
  bidirectional, 125, 209
  varient of, 393
  worst-case complexity of, 147
Busy beaver function, 780, 782
Butane, 637
Bytes, 170

Cactus, 697
CAD programs, 725
Caesar, Julius, 165

Caesar cipher, 165–166
Calculus
  predicate, 30
  propositional, 3
Call graphs, 542
  connected components of, 571
Canterbury Puzzles, The (Dudeney), 406
Cannibals, 19
Cantor, Georg, 77, 78, 235
Cantor diagonalization argument, 235
Cantor digits, 348
Cantor expansion, 181, 348
Cardinality, of set, 81, 113, 233–236
Cards, deck of, 357–358
Carmichael, Robert Daniel, 190
Carmichael numbers, 190–191, 207
Carroll, Lewis (Charles Dodgson), 37–38, 43–44,
  B-1
Carry, 173
Cartesian products, 82–84
Cases, proofs by, 67, 112, 215–216
Catalan numbers, 408
Caterpillar, 698
Cayley, Arthur, 631, 636
Ceiling function, 105–107, 113
Cells, adjacent, 721
Center, 643
Chain, 534
Chain letter, 640
Change-making algorithm, 128–129
Characteristic equation, 414
Characteristic function, 110–111
Characteristic roots, 414
Chebyshev, Pafnuty Lvovich, 391
Chebyshev's inequality, 390–392, 394
Cheese, in Reve's puzzle, 406
Chen, J. R., 221
Chess, game tree for, 655
Chessboard
  knight's tour on, 591–592
  n-queens problem on, 683
  squares controlled by queen on, 626
  tiling, mathematical induction for, 247–248
Chickens, 627
Child of vertex, 634, 635, 695
Chinese meal, preparing, 534
Chinese postman problem, 583, 629
Chinese Remainder Theorem, 185–187, 207
Chocolate bar, 254
Chomsky, Avram Noam, 739, 743, 745
Chomsky classification of grammars, 743–744
Chromatic number, 614, 623
  edge, 620
Church, Alonzo, 781
Church–Turing thesis, 781
Cipher
  affine, 166
  Caesar, 165–166
  RSA, 191–194
  shift, 166
Circuits
  combinational, 713–716
    depth of, 719
    examples of, 714–716
    minimization of, 719–734
  in directed graph, 498, 531
  Euler, 577–583, 622, 623
  in graph, 567
  Hamilton, 583–588, 622
  multiple output, 717
  simple, 567

Circular reasoning, 72, 112
Circular relation, 533
Civil War, U.S., 29
Class
  congruence, 162
  equivalence, 509–510, 531
    definition of, 509
    and partitions, 510–512
    representative of, 509
  NP, 149
  P, 149
Class A addresses, 307
Class B addresses, 307
Class C addresses, 307
Class D addresses, 307
Class E addresses, 307
Clauses, 60
Clique, 625
Closest-pair problem, 431–433
Closure, Kleene, 759, 783
Closures of relations, 496–507, 531
  reflexive, 497, 531
  symmetric, 497, 531
  transitive, 496, 499–502, 531
    computing, 503–506
Coast Survey, U.S., 29
COBOL, 747
Codes, Gray, 587–588, 725
Codes, prefix, 649–651, 695
Codeword enumeration, 407–408
Coding, Huffman, 650–651, 695, B-3
Codomain of function, 98, 111, 113
Coefficient(s)
  binomial, 322, 327–334, 349
    extended, 438
  constant
    linear homogenous recurrent relations
      with, 413, 414–419, 465
    linear nonhomogenous recurrent relations
      with, 419–425, 465
  multinomial, 344
Coins, change using fewest, 128–129
Collaboration graphs, 542
  paths in, 568–569
Collatz problem, 221–222
Collision, 164
  in hashing functions, probability of, 372–373
Coloring
  of graphs, 613–621, 623, 682
  of maps, 613–614
Column of matrix, 197
Combinational circuits, 713–716
  depth of, 719
  examples of, 714–716
  minimization of, 719–734
Combinations of events, 358–360, 365–366
Combinatorial identity, 323–326
Combinatorial proof, 323, 349
Combinatorics, 301, 349
Comments in pseudocode, A-6
Common difference, 226
Common divisor, greatest, 159–160
Common multiple, least, 160–161
Common ratio, 226
Commutative laws
  for Boolean algebra, 705, 707
  for lattices, 535
  for propositions, 24
  for sets, 89
Comparable elements in a poset, 517, 531
Compatible total ordering, 525, 531

Complement, 701, 734
  of Boolean function, 703
  double, law of, in Boolean algebra, 705, 708
  of fuzzy set, 96–97
  one's, 180
  of set, 88, 113
  two's, 180–181
Complementary event, 358
Complementary graph, 556
Complementary relation, 480
Complementation law, for sets, 89
Complemented lattice, 535, 707
Complement laws, for sets, 89
Complete bipartite graph, 550–551, 622
Complete graphs, 548, 622
Complete *m*-ary tree, 643
Complete *m*-partite graph, 625
Complexity of algorithms, 144–152
  average-case, 146–147, 149, 206, 384–385
  of binary search algorithms, 146
  of Boolean product of zero-one matrices, 204
  of breadth-first search algorithms, 681
  of bubble sort, 147
  computational, 144
  of computing **div** and **mod,** 175
  constant, 148
  of depth-first search algorithms, 679
  of Dijkstra's algorithm, 599
  of Euclidean algorithm, 179, 260–261
  exponential, 148
  factorial, 148
  of fast integer multiplication, 427, 430
  of fast matrix multiplication, 428, 431
  of finding maximum element of set, 145
  of insertion sort, 147–148
  of integer addition, 174
  of integer multiplication, 175, 427
  linear, 148
  of linear search algorithms, 145–147
  logarithmic, 148
  of matrix multiplication, 154–155, 337
  of merge sort, 430
  of modular exponentiation, 177
  polynomial, 148
  of searching algorithms, 145–147, 148
  of sorting algorithms, 147–148, 647–649
  space, 144, 206
  time, 144, 145–148, 149, 206
  worst-case, 146, 147–148, 149, 206
Complexity of merge sort, 430
Components of graphs, connected, 570–571
  strongly, 572, 622
Composite integers, 155, 189–190, 207, 218–219
  consecutive, 218–219
Composite key, 484, 531
Composite of relations, 478–479, 530
Composition rule, 286
Compositions of functions, 103–104, 113
Compound interest, 403
Compound propositions, 3, 9–10, 21, 22, 111
  consistent, 112
  disjunctive normal form of, 27
  dual of, 27
  satisfiable, 28
  well-formed formula for, 263
Computable function, 238, 780
Computation, models of, 739–787
Computational complexity of algorithms, 144
  of Dijkstra's algorithm, 599
Computer-aided design (CAD) programs, 725

Computer arithmetic, with large
    integers, 187–189
Computer file systems, 637–638
Computer network, 538
  with diagnostic lines, 539
  interconnection networks, 552–553
  local area networks, 551
  multicasting over, 676–677
  with multiple lines, 538
  with multiple one-way lines, 540
  with one-way lines, 539
Computer programming languages, 60, 98
Computer representation, of sets, 93–94
Computer time used by algorithms, 150–151
Concatenation, 263, 758–759, 783
Conclusion, 6, 38, 58–59
  fallacy of affirming, 60
Concurrent processing, 543
Conditional constructions, A-7–A-8
Conditional probability, 363, 366–367, 394
Conditional statement, 6
  for program correctness, 286–288
Conditions
  don't care, 726–727
  initial, 402, 465
Congruence class, 162
Congruence modulo m, 161–162, 207, 509,
    510, 531
  inverse of, 184, 207
Congruences, 161–166
  linear, 184–185, 207
Conjecture, 56, 112
  $3x + 1$, 221–222
  and counterexamples, 219–222
  de Polignac's, 224
  Frame's, 412
  Goldbach's, 220–221
  and proof, 217–219
  twin prime, 221
Conjunction, 4–5, 111, 734
  distributive law of disjunction over, 22
Conjunction rule of inference, 58
Conjunctive normal form, 711, 712
Connected components of graphs, 570–571
  strongly, 572, 622
Connected graphs, 567–577
  directed, 571–572
  planar simple, 606–609
  strongly, 571
  undirected, 569–571
  weakly, 571
Connecting vertices, 546
Connectives, 4
Connectivity relation, 499–501, 531
Consecutive composite integers, 218–219
Consecutive terms, finding largest sum of, 434
Consequence, 6
Consistency, of system specifications, 11–12
Consistent compound propositions, 112
Constant coefficients
  linear homogenous recurrent relations with,
    413, 414–419, 465
  linear nonhomogenous recurrent relations
    with, 419–425, 465
Constant complexity, 148
Constructions
  conditional, A-6–A-7
  loop, A-7–A-8
Constructive existence proof, 69, 112
Contain, 78
Context-free grammars, 744, 783

Context-free language, 744
Context-sensitive grammars, 744
Contingency, 21, 112
Contradiction, 21, 112
  proofs by, 66–67, 112
Contrapositive, of implication, 8, 111
Converse
  of directed graph, 556
  of implication, 8, 111
Cookie, 338
Corollary, 56, 112
Correctness
  of programs, 284–290, 291
    conditional statements for, 286–288
    loop invariants for, 288–289
    partial, 285
    program verification for, 285
    rules of inference for, 285–286
  of recursive algorithms, 275
Correspondences, one-to-one, 101, 113, 234, 235
Countability, of rational numbers, 234, 238
Countable set, 234, 291
Counterexamples, 31, 70–71, 112
  conjecture and, 219–222
Counting, 301–354, 401–470
  basic principles of, 302–306
  bit strings, 302
    without consecutive 0s, 406–407
  Boolean functions, 704
  combinations, 322–324
  derangements, 462, 465
  by distributing objects into boxes, 341
  functions, 303
  generating functions for, 439–450
  Internet addresses, 307–308
  one-to-one functions, 303
  onto functions, 460–461, 465
  passwords, 306–307
  paths between vertices, 574–575
  permutations, 321–322
  pigeonhole principle and, 313–320
  reflexive relations, 477
  relations, 474
  with repetition, 335
  subsets of finite set, 304
  telephone numbers, 303–304
  tree diagrams for, 309–310
  variable names, 306
Course in Pure Mathematics, A (Hardy), 296
Covariance, 394
Covering relation, 528
Covers, 528
CPM (Critical Path Method), 536
C programming language, 312
Crawlers, 685
Cricket, 296
Critical Path Method (CPM), 536
Crossing number, 613
Cryptography, 165–166
  private key, 191, 207
  public key, 181, 191–194, 207
Cunningham numbers, 158
Cut edge, 571
Cut set of graph, 697
Cut vertices, 571
Cycle
  in directed graph, 498, 531
  with n vertices, 548, 622
Cylinder, 722
Czekanowski, Jan, 691

Database
  composite key of, 484
  intension of, 484
  primary key of, 483
  records in, 483
  relational model of, 483–484, 531
Database query language, 487–488
Data compression, 650
Datagrams, 312
de Bruijn sequences, 629
Decimal, binary coded, 181
Decision trees, 646–649, 695
Decryption, 166, 181, 207
  RSA, 193
Deep Blue, 656, 700
Definition
  domain of, 111
  recursive, 213, 256–274
Degree
  of linear homogenous recurrence
    relations, 413
  of membership, 96
  of n-ary relations, 482
  of region, 608
  of vertex in undirected graph, 546, 622
Degree-constrained spanning tree, 698
Delay machine, 754–755
de Méré, Chevalier, 362
De Morgan, Augustus, 21, 23, 26, 615
De Morgan's laws
  for Boolean algebra, 705, 708, 711
  for propositions, 21, 24
  proving by mathematical induction, 246–247
  for sets, 89
Dense poset, 530
Dependency notation, 738
de Polignac's conjecture, 224
Depth-first search, 677–680, 695
  applications with, 681–684
  in directed graphs, 684–685
Depth of combinatorial circuit, 719
Derangements, 461–464, 465
  number of, 462, 465
Derivable form, 742, 783
Derivation, 742, 783
Derivation trees, 745–746, 783
Descartes, René, 83
Descendants of vertex, 634, 695
Descent, infinite, 252
Detachment, law of, 57
Deterministic finite-state automata, 761, 763, 783
Deviation, standard, 388
Devil's pair, 567
Diagonalization argument, 235
Diagonal matrix, 205
Diagonal relation, 497
Diagrams
  Hasse, 521–522, 523, 524, 531
  state, for finite-state machine, 753
  tree, 309–310, 349
  Venn, 79, 80, 81, 87, 88, 113
Diameter of graph, 627
Dice, 356–357
Die, 379, 389
  dodecahedral, 396
  octahedral, 396
Difference
  backward, 412–413
  common, 226
  forward, 467

Difference—*Cont.*
  of multisets, 96
  of sets, 88, 113
    symmetric, 95, 113
Difference equation, 413
Digits
  binary, 112
    origin of term, 14, 15
  Cantor, 348
Digraphs, 531, 539, 622
  circuit (cycle) in, 498, 531
  connectedness in, 571–572
  converse of, 556
  depth-first search in, 684–685
  edges of, 539, 547–548, 558
  Euler circuit of, 579
  loops in, 493, 531, 539
  paths in, 498–499, 531, 622
  representing relations using, 493–494
  self-converse, 626
  vertex of, 538, 547, 558
Dijkstra, Edsger Wybe, 595
Dijkstra's algorithm, 595–599, 623
Dimes, 128–129
Diophantus, 189, 220
Dirac, G. A., 586
Dirac's Theorem, 586
Directed edges, 540, 547–548, 622
Directed graphs, 531, 539, 622
  circuit (cycle) in, 498, 531
  connectedness in, 571–572
  converse of, 556
  depth-first search in, 684–685
  edges of, 539, 547–548, 558
  Euler circuit of, 579
  loops in, 493, 531, 539
  paths in, 498–499, 531, 622
  representing relations using, 493–494
  self-converse, 626
  vertex of, 538, 547, 558
Directed multigraph, 540, 622
  paths in, 568
Directly derivable form, 742, 783
Direct proof, 63–64, 112
Dirichlet, G. Lejeune, 313
Dirichlet drawer principle, 313
Discours (Descartes), 83
Discourse, universe of, 30, 112
Discrete mathematics, definition of, xix
Discrete probability, 355–399, 394
  assigning, 363–365
  of collision in hashing functions, 372–373
  of combinations of events, 358–360, 365–366
  conditional, 363, 366–367, 394
  finite, 356–358
  Laplace's definition of, 356
Disjoint set, 87
Disjunction, 4–5, 111, 734
  associative law of, 23
  distributive law of, over conjunction, 22
Disjunctive normal form
  for Boolean variables, 710
  for propositions, 27
Disjunctive syllogism, 58
Distance
  between distinct vertices, 627
  between spanning trees, 688
Distributing objects into boxes, 341
Distribution
  binomial, 368–370
  probability, 363

of random variable, 370, 394
  geometric, 386, 394
  uniform, 364, 394
Distributive lattice, 535, 707
Distributive laws
  for Boolean algebra, 704, 705, 707
  for propositions, 22, 24
  for sets, 89, 91
div function, 158–159, 175–176
Divide-and-conquer algorithms, 426, 465
Divide-and-conquer recurrence relations,
    426–433
Dividend, 158
Divides, 153
Divisibility relation, 516
Division
  of integers, 153–168
  trial, 158
Division algorithm, 158–159, 207
Divisor, 158
  greatest common, 159–160, 182–185, 207
DNA sequencing, 583
Dodecahedral die, 396
Dodgson, Charles (Lewis Carroll), 37–38, 43–44
Domain
  of definition, 111
  of function, 98, 111, 113
  of *n*-ary relation, 483
  of relation, 482
  universe of, 30, 112
Dominating set, 625
Domination laws
  for Boolean algebra, 705
  for propositions, 24
  for sets, 89
Dominos, 239
Don't care conditions, 726–727
Double complement, law of, in Boolean algebra,
    705, 708
Double negation law, for propositions, 24
Double summations, 231
Dual
  of Boolean expression, 706, 734
  of Boolean function, 706
  of compound proposition, 27
  of poset, 528
Dual graph, 613
Duality in lattice, 536
Duality principle, for Boolean identities, 706, 708
Dudeney, Henry, 406

EBNF (Extended Backus–Naur form), 750
Eccentricity of vertex, 643
Ecology, niche overlap graph in, 541
Edge chromatic number, 620
Edge coloring, 620
Edges
  back, 678
  cut, 571
  directed, 540, 547–548, 622
  of directed graph, 539, 547–548, 558
  of directed multigraph, 540
  endpoints of, 546
  incident, 546, 622
  of multigraph, 538
  multiple, 538, 540, 622
  of pseudograph, 539
  of simple graph, 538, 557
  tree, 678
  undirected, 540, 622
  of undirected graph, 540, 546
Egyptian (unit) fraction, 293–294

Einstein, Albert, 20
Elementary subdivision, 609, 623
Elements, 78, 112
  comparable, in partially ordered set, 517, 531
  equivalent, 508
  fixed, 394
  greatest, of partially ordered set, 522, 531
  incomparable, in partially ordered
    set, 517, 531
  least, of partially ordered set, 522, 531
  of matrix, 197
  maximal, of partially ordered set, 522, 531
  minimal, of partially ordered set, 522,
    525–526, 531
Elements (Euclid), 156, 177
Elements of Mathematical Logic (Łukasiewicz),
    670
Elkies, Noam, 220
Ellipses (...), 78
Empty set, 80, 112
Empty string, 226, 291, 741
Encryption, 165–166, 181, 207
  RSA, 191, 192–193
Encyclopedia of Integers Sequences, The (Sloane
    and Plouffe), 228
Endpoints of edge, 546
End statements, A-5–A-6
End vertex, 493, 547
English sentences
  translating logical expressions to, 42, 44–45
  translating to logical expressions, 10–11,
    35–37, 45–47
Entry of matrix, 197
Enumeration, 301, 349
  codeword, 407–408
Equality, of sets, 79, 113
Equation
  characteristic, 414
  difference, 413
Equivalence, proofs of, 67–68
Equivalence classes, 509–510, 531
  definition of, 509
  and partitions, 510–512
  representative of, 509
Equivalence relations, 507–516, 531
  definition of, 508
Equivalent Boolean expressions, 703
Equivalent elements, 508
Equivalent finite-state automata, 760
Equivalent propositions, 8, 20–28, 112
Eratosthenes, 459
  sieve of, 155, 458–460, 465
Erdős, Paul, 533, 569
Erdős number, 533, 568, 569
Erdős Number Project, 569
Errors, in proofs, 71–73
Essential prime implicant, 722, 735
Euclid, 156, 177
Euclidean algorithm, 177–179, 207, 260–261
  extended, 182, 196
Euler, Leonhard, 219, 220, 537, 578, 579
Euler circuits, 577–583, 622, 623
Euler *φ*-function, 167
Euler paths, 577–583, 622, 623
Euler's formula, 606–609, 623
Evaluation functions, 656
Even integers, 63
Event(s), 356
  combinations of, 358–360, 365–366
  independent, 362, 363, 367–368, 394
  mutually independent, 397

Eventually negative function, 297
Eventually positive function, 297
Exams, scheduling, 618, 619
Exclusion rule in recursive definition, 262
Exclusive or, 5, 111
Exercises in this book, key to, xx
Existence proofs, 69, 112
    constructive, 69, 112
    nonconstructive, 69, 112
Existential generalization, 62
Existential instantiation, 61, 62
Existential quantification, 32, 112
    negation of, 34
Existential quantifier, 32–33
Expansion
    balanced ternary, 180
    base b, 169, 171
    binary, 169, 207
    binary coded decimal, 727
    Cantor, 181, 348
    hexadecimal, 170, 207
    octal, 170, 207
    one's complement, 180
    two's complement, 180–181
Expectations, linearity of, 382–384
Expected number of inversions, 383
Expected values, 363, 379–381, 394
    in hatcheck problem, 383
    of inversions in permutation, 383–384
    linearity of, 382–384, 394
Experiment, 356
Exponential complexity, 148
Exponential functions, A-1, A-2
Exponential generating function, 449
Exponentiation, modular, 175–177
    recursive, 275
Expression(s)
    binomial, 327
    Boolean, 702–704, 734
    infix form of, 669
    logical
        translating English sentences into, 10–11, 35–37, 45–47
        translating to English sentences, 42, 44–45
    postfix form of, 670
    prefix form of, 669
    regular, 766, 783
Extended Backus–Naur form (EBNF), 750
Extended binary trees, 265
    recursive definition of, 265
Extended binomial coefficients, 438
Extended Binomial Theorem, 439
Extended Euclidean algorithm, 182, 196

Factor, of integers, 153
Factorial complexity, 148
Factorial function, 108
Factorials, recursive procedure for, 277
Factorization into primes, 155–156
    for finding greatest common divisor, 160
    for finding least common multiple, 160–161
    uniqueness of, 183–184
Facts, 56
Failure, 368
Fallacy, 56, 60–61, 112
    of affirming conclusion, 60
    of begging the question, 72, 112
    of circular reasoning, 72, 112
    of denying hypothesis, 61
Family trees, 631, 632

Fast multiplication
    of integers, 427–428
    of matrices, 428
Fermat, Pierre de, 189, 220, 252
Fermat's Last Theorem, 220, 260
Fermat's Little Theorem, 189, 207
Fibonacci, 259, 403
Fibonacci numbers, 259, 658
    formula for, 415
    and Huffman coding, 658
    iterative algorithm for, 279
    rabbits and, 403–404
    recursive algorithms for, 278
Fibonacci trees, rooted, 644
Fields, 483
Final assertion, 285, 291
Final exams, scheduling, 618, 619
Final value, A-7
Finite probability, 356–358
Finite sets, 81, 113, 234, 451, 465
    subsets of
        counting, 304
        number of, 244–245
    union of three, number of elements in, 453–454, 465
    union of two, number of elements in, 451–453, 465
Finite-state automata, 759–764
    deterministic, 761, 763, 783
    nondeterministic, 761–763, 767–769, 771, 783
    regular sets recognized by, 766–769
    set not recognized by, 772
Finite-state machines, 739, 751–765, 783
    with no output, 758–765
    with outputs, 753–756
First difference, 412
First forward difference, 467
Fixed elements, 394
Fixture controlled by three switches, circuit for, 714–716
Fleury's algorithm, 582, 591
Flip-flops, 738
Floor function, 105–107, 113
Floyd's algorithm, 602–603
Forests, 632, 633
    definition of, 694
    minimum spanning, 694
    spanning, 687
Form(s)
    Backus–Naur, 746–748, 783
    conjunctive normal, 711, 712
    disjunctive normal
        for Boolean variables, 710
        for propositions, 27
    infix, 669
    postfix, 670
    prefix, 669
    prenex normal, 56
Formal language, 740
Formal power series, 437
Formula(s)
    Bayes', 397
    Euler's, 606–609, 623
    for Fibonacci numbers, 415
    summation, 233
    well-formed, 263
        for compound propositions, 263
        of operators and operands, 263–264
FORTRAN, 747
Forward differences, 467
Forward reasoning, 214

Four Color Theorem, 614–615, 623
Fraction, unit (Egyptian), 293–294
Frame's conjecture, 412
Free variable, 33, 112
Frend, Sophia, 23
Frend, William, 26
Frequency assignments, 618
Frisbee, rocket-powered, 702
Full adder, 717, 735
Full binary trees, 265–266
    height of, 269
    number of vertices of, 269
    recursive definition of, 266
Full m-ary tree, 634, 639, 695
    complete, 643
Full subtractor, 718
Function(s), 97–111, 113
    Ackermann's, 273
    addition of, 98–99
    asymptotic, 144
    bijective, 101, 113
    Boolean, 701–708, 734
        dual of, 706
        functionally complete set of operators for, 711
        implicant of, 722, 735
        minimization of, 719–734, 735
        representing, 709–712
        self-dual, 736
        threshold, 737
    busy beaver, 780, 782
    ceiling, 105–107, 113
    characteristic, 110–111
    codomain of, 98, 111, 113
    compositions of, 103–104, 113
    computable, 238, 780
    counting, 303
    definition of, 97, 113
    domain of, 98, 111, 113
    Euler ϕ, 167
    evaluation, 656
    eventually negative, 297
    eventually positive, 297
    exponential, A-1, A-2
    factorial, 108
    floor, 105–107, 113
    generating, 435–450, 465
        for counting, 439–450
        exponential, 449
        probability, 450
        for proving identities, 446
        for recurrence relations, 444–450
    graphs of, 104–105
    greatest integer, 105–107
    growth of, 131–144
    growth of combinations of, 138–140
    hashing, 163–164
        collision in, probability of, 372–373
    identity, 102
    INT, 110
    inverse, 102–103, 113
    invertible, 103
    iterated, 274
    iterated logarithm, 273
    logarithmic, A-1–A-3
    logarithmic-exponential, 296–297
    McCarthy 91, 294
    multiplication of, 98–99
    number-theoretic, 780
    one-to-one (injective), 99–100, 101, 113
        counting, 303

Function(s)—*Cont.*
  onto (surjective), 100–101, 113
    number of, 460–461, 465
  partial, 111
  product of, 98–99
  propositional, 28, 29, 112
    existential quantification of, 32, 112
    universal quantification of, 30, 112
  range of, 98, 113
  recursive, 97, 257–261, 291
  as relations, 472–473
  strictly decreasing, 100
  strictly increasing, 100
  sum of, 98–99
  threshold, 737
  total, 111
  transition, 753
  Turing machines computing, 779–780
Functional completeness, 711
Functional decomposition, 738
Functionally complete set of operators
    for Boolean functions, 711, 734
    for propositions, 27
Fundamental Theorem of Arithmetic, 155,
    207, 250
Fuzzy logic, 18, B-1
Fuzzy sets, 96–97, B-1
  complement of, 96–97
  intersection of, 97
  union of, 97

Gambling, 355
Game trees, 651–656, 695
Gates, logic, 712–719
  AND, 713, 734
  combination of, 713–716
  NAND, 718
  NOR, 718
  OR, 713, 734
  threshold, 736
  XOR, 736
Gating networks, 713–716
  depth of, 719
  examples of, 714–716
  minimization of, 719–734
Gauss, Karl Friedrich, 157, 162
GCD (greatest common divisor), 159–160,
    182–185, 207
Generalization
  existential, 62
  universal, 61, 62
Generalized induction, 269–270
Generalized pigeonhole principle, 314–316, 349
Generating combinations, 336–339
Generating functions, 435–450, 465
  for counting, 439–450
  exponential, 449
  probability, 450
  for proving identities, 446
  for recurrence relations, 444–450
Generating permutations, 335
Geometric distribution, 386, 394
Geometric mean, 215, 256
Geometric progression(s), 226, 230, 290
  sums of, 243
Geometric series, 230
Giant strongly connected components
    (GSCC), 572
GIMPS (Great Internet Mersenne Prime
    Search), 157
Gödel, Escher, Bach (Hofstader), 295
Goldbach, Christian, 220, 221

Goldbach's conjecture, 220–221
Golomb's self-generating sequence, 296
Google, 685
Graceful trees, 698
Grammars, 739–751
  ambiguous, 784
  Backus–Naur form of, 746–748, 783
  context-free (type 2), 744, 783
  context-sensitive, 744
  phrase-structure, 741–745, 783
  productions of, 741, 783
  regular (type 3), 744, 766, 769–772, 783
  type 0, 744, 783
  type 1, 744, 783
Graphs, 537–630
  acquaintanceship, 541
  bandwidth of, 627
  bipartite, 549–551, 622
  book number of, 629
  call, 542
    connected components of, 571
  collaboration, 542
  coloring, 613–621, 623, 682
  complementary, 556
  complete, 548, 622
  complete bipartite, 550–551, 622
  complete *m*-partite, 625
  connected components of, 570–571, 622
  connectedness in, 567–577
  converse of, 556
  crossing number, 613
  cut set of, 697
  diameter of, 627
  directed, 531, 539, 622
    circuit (cycle) in, 498, 531
    connectedness in, 571–572
    converse of, 556
    depth-first search in, 684–685
    edges of, 539, 547–548, 558
    Euler circuit of, 579
    loops in, 493, 531, 539
    paths in, 498–499, 531, 622
    representing relations using, 493–494
    self-converse, 626
    vertex of, 538, 547, 558
  directed multigraph, 540, 622
  dual, 613
  of functions, 104–105
  Hollywood, 541
  homeomorphic, 609–611, 623
  independence number of, 627
  influence, 541, 542
  intersection, 544
  isomorphic, 557, 560–563, 622
    paths and, 572–573
  monotone decreasing property of, 628
  monotone increasing property of, 628
  multigraphs, 538, 622
  niche overlap, 541
  nonplanar, 609–611
  n-regular, 556
  orientable, 626
  paths in, 567–569
  Petersen, 591
  planar, 603–613, 623
  planar representation of, 604
  precedence, 543
  pseudograph, 539, 540, 622
  radius of, 627
  random, 627 (exercise 49)
  regular, 556, 622

representing, 557–567
  self-complementary, 566
  simple, 538, 548–551, 622
    coloring of, 614
    connected planar, 606–609
    crossing number of, 613
    edges of, 538, 557
    isomorphic, 560–563, 622
    orientation of, 626
    paths in, 567
    random, 627–628
    self-complementary, 566
    with spanning trees, 675–676
    thickness of, 613
    vertices of, 538, 557
  sparse, 693
  strongly connected directed, 571
  subgraph of, 553, 622
  terminology of, 545–556
  thickness of, 613
  undirected, 540, 546–547, 622
    connectedness in, 569–571
    Euler circuit of, 578
    Euler path of, 582
    orientation of, 626
    paths in, 567, 622
    underlying, 548, 622
  union of, 554, 622
  very large scale integration (VLSI), 629
  weakly connected, 571
  Web, 543
    strongly connected components of, 572
  weighted, 623
    shortest path between, 593–599
    traveling salesman problem with, 599–601
  wheel, 548, 622
Gray, Frank, 588
Gray codes, 587–588, 725
"Greater than or equal" relation, 516
Greatest common divisor (GCD), 159–160,
    182–185, 207
Greatest element of poset, 522, 531
Greatest integer function, 105–107, 113
Greatest lower bound, 531
  of poset, 523
Great Internet Mersenne Prime Search
    (GIMPS), 157
Greedy algorithm for scheduling talks
  optimality of, 248
Greedy algorithms, 127–129, 172, 248–249, 651,
    689, 695
Greedy change-making algorithm, 128
Growth of functions, 131–144
GSCC (giant strongly connected components),
    572
Guare, John, 568
Guthrie, Francis, 615

Hadamard, Jacques, 158
Haken, Wolfgang, 615
Half adder, 717, 734
Half subtractor, 718
Halting problem, 149, 222–223, 291
Hamilton, William Rowan, 583, 584, 615
Hamilton circuits, 583–588, 622
Hamilton paths, 583–588, 622
Hamilton's "Voyage Round the World" Puzzle,
    583–584
Handle, 697
Handshaking Theorem, 546

Hanoi, tower of, 404–406
Hardy, Godfrey Harold, 69, 296, 297
Harmonic mean, 223
Harmonic number(s), 144
  inequality of, 243–244
Harmonic series, 244
Hashing functions, 163–164
  collision in, probability of, 372–373
Hasse, Helmut, 520
Hasse diagrams, 521–522, 523, 524, 531
Hasse's algorithm, 221–222
Hatcheck problem, 383, 461–462
Hazard-free switching circuits, 738
Heaps, 700
Heawood, Percy, 615
Height, star, 785
Height-balanced trees, 700
Height of full binary tree, 269
Height of rooted tree, 640–641, 695
Hexadecimal expansion, 170, 207
Hexagon identity, 333
Hoare, C. Anthony R., 285, 286
Hoare triple, 285
Hofstader, Douglas, 295
Hollywood graph, 541
  paths in, 569
Homeomorphic graphs, 609–611, 623
Hops, 552
Horner's method, 151
Horses, same color, 255
Host number (hostid), 307
Hotbot, 685
House, scheduling building tasks, 530
HTML, 751
Huffman, David A., 650
Huffman coding, 650–651, 695
  variations of, 651
Hummingbirds, 38
Hybrid topology for local area network, 551
Hydrocarbons, trees modeling, 636–637
Hypercube, 553
Hypothesis, 6
  fallacy of denying, 61
  inductive, 240
Hypothetical syllogism, 58

"Icosian Game," 583, 584
Idempotent laws
  for Boolean algebra, 705, 708
  for lattices, 535
  for propositions, 24
  for sets, 89
Identifier, 747
Identities
  combinatorial, 328–333
Identity
  hexagon, 333
  Pascal's, 330–331, 349
  proving, generating functions for, 446
  Vandermonde's, 331–332
Identity function, 102
Identity laws
  for Boolean algebra, 704–706, 707
  for propositions, 24
  for sets, 89
Identity matrix, 201, 207
"If and only if" statement, 8, 9
If then statement, 7, 9, 30, A-6–A-7
Image, 98, 113
  inverse, of set, 110
  of set, 99

Implicant, 722
  prime, 722, 735
    essential, 722, 735
Implications, 6–9, 112
  contrapositive of, 8
  converse of, 8
  inverse of, 8
  logical equivalences for, 24
Incidence matrices, 559–560, 622
Incident edge, 546, 622
Inclusion-exclusion principle, 87, 308–309,
    451–457, 465
  alternative form of, 457–458
  applications with, 457–465
Inclusion relation, 516
Inclusive or, 5
Incomparable elements in poset, 517, 531
Increasing function, 100
Increment, 164
In-degree of vertex, 547, 622
Independence number, 627
Independent events, 362, 363, 367–368, 394
Independent random variables, 387–388, 394
Independent set of vertices, 627
Index, of summation, 229
Index registers, 619
Indicator random variable, 392
Indirect proof, 64, 112
Indistinguishable objects, permutations
    with, 339–340
Induced subgraph, 625
Induction, mathematical, 23, 238–256
  generalized, 269–270
  principle of, 291
  proofs by, 240–249
  second principle of, 249–251
  strong, 249–251, 275, 291
  structural, 257, 266–269, 291
  validity of, 251–253
  well-ordered, 517–518, 531
Inductive definitions, 213, 256–274
  of factorials, 277
  of functions, 97, 257–261, 291
  of sets, 261–266, 291
  of strings, 262
  of structures, 261–266
Inductive hypothesis, 240
Inductive step, 240, 291, 639, 655
Inequality
  Bernoulli's, 253
  Bonferroni's, 377
  Boole's, 377
  Chebyshev's, 390–392, 394
  of harmonic numbers, 243–244
  Markov's, 393
  proving by mathematical induction, 241, 244
  triangle, 75
Inference, rules of, 56, 57–58, 60, 112
  for program correctness, 285–286
  for statements, 61–63
Infinite descent, 252
Infinite series, 233
Infinite set, 82, 113, 234
Infinitude of primes, 156
Infix form, 669
Infix notation, 668–672, 695
Influence graphs, 541, 542
Information flow, lattice model of, 525
Initial assertion, 285, 291
Initial conditions, 402, 465
Initial position of a Turing machine, 777

Initial state of a finite-state machine, 753
Initial term, 226
Initial value, A-7
Initial vertex, 493, 547
Injection, 99–100
Injective (one-to-one) function, 99–100, 101, 113
  counting, 303
Inorder traversal, 662–664, 665, 668, 695
Input alphabet, 753, 760
Insertion sort, 125, 126–127, 207
  average-case complexity of, 385
  worst-case complexity of, 147–148
Instantiation
  existential, 61, 62
  universal, 61, 62
Integers, 78, 119
  addition of, algorithms for, 172–174
  applications with, 181–196
  composite, 155, 189–190, 207, 218–219
  division of, 153–168
  even, 63
  finding maximum element in finite sequence
    of, 121–122
  greatest common divisor of, 159–160
  large, computer arithmetic with, 187–189
  least common multiple of, 160–161
  multiplication of
    algorithms for, 174–175
    fast, 427–428
  odd, 63
  partition of, 273
  perfect, 167
  pseudoprimes, 189–191, 195, 207
  representations of, 169–172, 180–181
  set of, 78
  squarefree, 464
  sum of the first n, 237, 245
Integer sequences, 227–228
Intension, of database, 484
Interconnection networks for parallel
    computation, 552–553
Interest, compound, 403
Interior vertices, 503
Internal vertices, 634, 695
International Standard Book Number (ISBN),
    168
Internet
  search engines on, 685
  searching, 12
Internet addresses, counting, 307–308
Internet datagram, 312
Internet Movie Database, 541
Internet Protocol (IP) multicasting, 676–677
Intersection
  of fuzzy sets, 97
  graph, 544
  of multisets, 96
  of sets, 86–87, 91–93, 113
Intersection graph, 544
INT function, 110
Intractable problem, 149
Invariant for graph isomorphism, 561, 622
Invariants, loop, 288–289, 291
Inverse, 530
  of functions, 102–103, 113
  of implication, 8, 112
  of matrix, 205
  modulo m, 184, 207
Inverse image of set, 110
Inverse relation, 480

Inversions, in permutation, expected number of, 383–384
Inverter, 713, 734
Invertible function, 103
Invertible matrix, 205
IP multicasting, 676–677
Irrationality of $\sqrt{2}$, 252
Irrational numbers, 167
Irreflexive relation, 480
ISBN (International Standard Book Number), 168
Isobutane, 637
Isolated vertex, 546, 622
Isomorphic graphs, 557, 560–563
    paths and, 572–573
Isomorphism of graphs, 557, 560–563
Iterated function, 274
Iterated logarithm, 273
Iteration, 273, 274, 291
Iterative algorithm, for Fibonacci numbers, 279
Iterative procedure, 277
    for factorials, 278
Iwaniec, Henryk, 221

Java, 98
Jigsaw puzzle, 294
Jobs, assignment of, 464
Join, in lattice, 535
Join of n-ary relations, 485–486, 531
Join of zero-one matrices, 202, 207
Joint authorship, 542
Josephus, Flavius, 412
Josephus problem, 412
Jug, 225

Kakutani's problem, 222
Kaliningrad, Russia, 577
Karnaugh, Maurice, 720, 721
Karnaugh maps, 720–726, 735
Kempe, Alfred Bray, 614, 615
Key
    composite, 484, 531
    for hashing, 163
    primary, 483–484, 531
al-Khowarizmi, Abu Ja'far Mohammed ibn Musa, 120
Kissing problem, 229
Kleene, Stephen Cole, 759, 765
Kleene closure, 759, 783
Kleene's Theorem, 766–769, 783
K-maps, 720–726, 735
Kneiphof Island, 578
Knights and knaves, 12–13
Knight's tour, 591–592
    reentrant, 591–592
Knuth, Donald E., 125, 134, 135, 140, B-1, B-2, B-3
Königsberg bridge problem, 577–578, 581, 583
Kruskal, Joseph Bernard, 690, 691
Kruskal's algorithm, 690–691, 692, 695
k-tuple graph coloring, 621
Kuratowski, Kazimierz, 609
Kuratowski's Theorem, 609–611, 623

Labeled tree, 643
Lagarias, Jeffrey, 222
Lamé, Gabriel, 260
Lamé's Theorem, 260
Landau, Edmund, 134
Landau symbol, 134
Lander, L. J., 219

Language, 739–751, 783
    context-free, 744
    formal, 740
    generated by grammar, 742
    natural, 740
    recognized by finite-state automata, 760–763, 765–775, 783
    regular, 744
Language recognizer, 755
Laplace, Pierre Simon, 355, 356
Laplace's definition of probability, 356
Large numbers, law of, 369
Lattice model of information flow, 525
Lattices, 524–525, 531
    absorption laws for, 535
    associative laws for, 535
    bounded, 535
    commutative laws for, 535
    complemented, 535, 707
    distributive, 535, 707
    duality in, 536
    idempotent laws for, 535
    join in, 535
    meet in, 535
    modular, 536
Law(s)
    absorption
        for Boolean algebra, 705, 706
        for lattices, 535
        for propositions, 24
        for sets, 89
    associative
        for Boolean algebra, 705, 707
        for lattices, 535
        for propositions, 23, 24
        for sets, 89
    commutative
        for Boolean algebra, 705, 707
        for lattices, 535
        for propositions, 24
        for sets, 89
    complement, for sets, 89
    complementation, for sets, 89
    De Morgan's
        for Boolean algebra, 705, 708, 711
        for propositions, 21, 24
        proving by mathematical induction, 246–247
        for sets, 89
    of detachment, 57
    distributive
        for Boolean algebra, 704, 705, 707
        for propositions, 22, 24
        for sets, 89, 91
    domination
        for Boolean algebra, 705
        for propositions, 24
        for sets, 89
    of double complement, in Boolean algebra, 705, 708
    idempotent
        for Boolean algebra, 705, 708
        for lattices, 535
        for propositions, 24
        for sets, 89
    identity
        for Boolean algebra, 704–706, 707
        for propositions, 24
        for sets, 89
    of large numbers, 369
    negation, for propositions, 24

Laws of Thought, The (Boole), 3, 4, 701
Leaf, 634, 695
Least common multiple (LCM), 160–161, 207
Least element of poset, 522, 531
Least upper bound, 56, 531
    of poset, 523
Left child of vertex, 635
Left subtree, 635
Lemma, 56, 112
    pumping, 775
Length of bit string, 14, 406–407
Length of path
    in directed graph, 498
    in weighted graph, 594
Length of string, 226
    recursive definition of, 263
"Less than or equals" relation, 517, 519
Level of vertex, 640, 695
Level order of vertex, 697
Leveraging proof by cases, 215
Lexicographic ordering, 270, 344–348, 518–520
Liber Abaci (Fibonacci), 259, 403–404
Light fixture controlled by three switches, circuit for, 714–716
Limit, definition of, 47
Limit superior, 56
Linear array, 552, 553
Linear bounded automata, 773
Linear combination, gcd as, 182
Linear complexity, 148
Linear congruences, 184–185, 207
Linear congruential method, 164
Linear equations, simultaneous, 205–206
Linear homogenous recurrence relations, 413, 414–419, 465
Linearity of expectations, 382–384, 394
Linearly ordered set, 517, 531
Linear nonhomogeneous recurrence relations, 419–425, 465
Linear ordering, 517, 531
Linear search algorithm, 123, 206, 207
    average-case complexity of, 384
    complexity of, 145–147, 148
    recursive, 276
Lists, merging two, 282
Literal, 709, 734
Little-o notation, 143
Littlewood, John E., 296
Loan, 410
Lobsters, 423
Local area networks, 551
Logarithm, iterated, 273
Logarithmic complexity, 148
Logarithmic-exponential functions, 296–297
Logarithmic functions, A-1–A-3
Logic, 1–56, 111–112
    fuzzy, 18
    predicate, 30
        rules of inference, 61–62
    propositional, 3
        rules of inference for, 57–58
Logical connectives, 4
Logical equivalences, 21–26, 112
Logical expressions
    translating English sentences into, 10–11, 35–37, 45–47
    translating to English sentences, 42, 44–45
Logical operators, 111
    functionally complete, 27
    precedence of, 9–10

Logic gates, 712–719
    AND, 713, 734
    combination of, 713–716
    NAND, 718
    NOR, 718
    OR, 713, 734
    threshold, 736
Logic Problem, 76
Logic programming, 39–40
Logic puzzles, 12–14
Long-distance telephone network, 542
Longest monotone subsequence, 317–318
Loop
    in directed graphs, 493, 531, 539, 622
    nested, 50
    in pseudocode, A-4–A-9
    in undirected graphs, 539
Loop constructions, A-7–A-8
Loop invariants, 288–289, 291
Lottery, 357
Lovelace, Countess of (Ada Augusta), 23, 25
Lower bound
    of lattice, 535
    of poset, 523, 531
Lower limit of summation, 229
Lucas, Edouard, 404
Lucas–Lehmer test, 157
Lucas numbers, 293, 423
Lucky numbers, 470
Łukasiewicz, Jan, 669, 670
Lyceum, 2
Lycos, 685

McCarthy, John, 294
McCarthy 91 function, 294
McCluskey, Edward J., 728
Machines
    delay, 754–755
    finite-state, 739, 751–765, 783
        with no output, 758–765
        with outputs, 753–756
    Mealy, 756
    Moore, 756, 757
    Turing, 739, 773, 774, 775–783
        computing functions with, 779–780
        definition of, 776–778
        sets recognized by, 778–779
        types of, 780–781
    vending, 751–752
Macro, 46
MAD Magazine, 135
Magic tricks, 13
Majority voting, circuit for, 714, 715
Maps
    K-, 720–726
    coloring of, 613–614
    Karnaugh, 720–726
Markov's inequality, 393
Markup languages, 751
m-ary tree, 634, 695
    complete, 643
    full, 634, 639
    height of, 640–641
Master Theorem, 430
Mathematical Analysis of Logic, The (Boole), 4
Mathematical induction, 23, 238–256
    generalized, 269–270
    principle of, 291
    proofs by, 240–249
    strong, 249–251, 291
    structural, 257, 266–269, 291
    validity of, 251–253

Mathematician's Apology, A (Hardy), 296
Mathematics, devoting one's life to, 296
Matrix-chain multiplication, 200
Matrix (matrices), 119, 196–206
    addition of, 197–198
    adjacency, 557–559, 622
        counting paths between vertices
            by, 574–575
    Boolean product of, 202–203, 207
    column of, 197
    definition of, 207
    diagonal, 205
    identity, 201, 207
    incidence, 559–560, 622
    inverse of, 205
    invertible, 205
    multiplication of, 198–199
        algorithms for, 199–200
        fast, 428
    powers of, 201
    product of, 198
    representing relations using, 490–492
    row of, 197
    sparse, 558
    square, 201
    symmetric, 201, 202, 207
    transposes of, 201, 207
    upper triangular, 210
    zero-one, 202–204, 207
        Boolean product of, 202–203
        join of, 202, 207
        meet of, 202, 207
        representing relations using, 490–492
        of transitive closure, 501–502
        Warshall's algorithm and, 503–506
Maurolico, Francesco, 240
Maximal element of poset, 522, 531
Maximum, of sequence, 426–427
Maximum element
    in finite sequence, 121–122
    time complexity of finding, 145
Maximum satisfiability problems, 398
Maximum spanning tree, 694
Maxterm, 712
Mealy, G. H., 756
Mealy machines, 756, 783
Mean, 130
    arithmetic, 215, 256
    deviation from, 391
    geometric, 215, 256
    harmonic, 223
    quadratic, 223
Median, 130
Meet, in lattice, 535
Meet of zero-one matrices, 202, 207
Meigu, Guan, 583
Members, 78, 112
Membership, degree of, 96
Membership tables, 91, 113
Ménages, problème des, 470
Merge sort, 125, 279–283, 291, 427
    complexity of, 430
    recursive, 280
Merging two lists, 282
Mersenne, Marin, 157
Mersenne primes, 157, 207, 218
Mesh network, 552, 553
Mesh of trees, 700
Metacharacters, 750
Metafont, 134
Method(s)
    Critical Path, 536

Horner's, 151
    of infinite descent, 252
    linear congruential, 164
    probabilistic, 375–376, 394
    of proof, 56–77, 112
    Quine–McCluskey, 720, 727–732
Methods of Proving Theorem, 63–69
Miller's test for base *b*, 195, 374
Minimal element of poset, 522, 525–526, 531
Minimization
    of Boolean functions, 719–734, 735
    of combinational circuits, 719–734
Minimum, of sequence, 426–427
Minimum dominating set, 625
Minimum spanning forest, 694
Minimum spanning trees, 688–694, 695
Minmax strategy, 654, 695
Minterm, 709, 734
Mistakes, in proofs, 71–73
mod function, 158–159, 161–163, 175–176
Mode, 130
Modeling
    computation, 739–787
    with recurrence relations, 403–409
    with trees, 636–638
Modular arithmetic, 161–163
Modular exponentiation, 175–177
    recursive, 275
Modular lattice, 536
Modular properties, in Boolean algebra, 708
Modulus, 164
Modus ponens, 57, 58
Modus tollens, 58
Mohammed's scimitars, 581
Molecules, trees modeling, 636–637
Monks, 405–406
Monotone decreasing property of graph, 628
Monotone increasing property of graph, 628
Monte Carlo algorithms, 373–375
Montmort, Pierre Raymond de, 462, 470
Monty Hall Three Door Puzzle, 360
Moore, E. F., 756, B-3
Moore machine, 756, 757, B-3
Motorized pogo stick, 702
m-tuple, 485
Muddy children puzzle, 13–14
Multicasting, 676–677
Multigraphs, 538, 622
    directed, 540
    Euler circuit of, 581
    Euler path of, 582
Multinomial coefficient, 344
Multinomial Theorem, 344
Multiple
    of integers, 153
    least common, 160–161, 207
Multiple edges, 538, 540, 622
Multiple output circuit, 717
Multiplexer, 718–719
Multiplication
    of functions, 98–99
    of integers
        algorithms for, 174–175
        fast, 427–428
    of matrices, 198–199
        algorithms for, 199–200
        fast, 428
    matrix-chain, 200
Multiplicity of membership in multisets, 96
Multiplier, 164, 738
Multisets, 96

Mutually independent events, 397
Mutually relative prime, 210

Naive set theory, 77
Namagiri, 297
NAND, 27, 711, 734
NAND gate, 718
n-ary relations, 482–489, 531
  domain of, 483
  operations on, 484–487
Natural language, 740
Natural numbers, 78
Naur, Peter, 746, 747
NAUTY, 563
n-cubes, 549, 622
Necessary, 6
  and sufficient, 9
Necessary and sufficient conditions for
  Euler circuits, 579–582
  Euler paths, 582–583
Negating nested quantifiers, 47–48
Negating quantifiers, 35
Negation
  of existential quantification, 34
  of propositions, 3–4, 111
  of universal quantification, 34
Negation laws, for propositions, 24
Negation operator, 4
Neighbors in graphs, 546
Nested loops, 50
Nested quantifiers, 44–56
  negating, 47–48
Network
  computer, 538
    with diagnostic lines, 539
    interconnection networks, 552–553
    local area networks, 551
    multicasting over, 676–677
    with multiple lines, 538
    with multiple one-way lines, 540
    with one-way lines, 539
  gating, 713–716
    depth of, 719
    examples of, 714–716
    minimization of, 719–734
  tree-connected, 638
Network number (netid), 307
New Jersey, crags, 229
Newton–Pepys problem, 399
Niche overlap graph, 541
Nickels, 128–129
Nim, 653, 655, 656
$n \log n$ complexity, 148
Nodes, 493
Nonconstructive existence proof, 69, 112
Nondeterministic finite-state automata,
  761–763, 767–769, 771, 783
Nonplanar graphs, 609–611
Nonterminals, 741
NOR, 27, 711, 734
NOR gate, 718
NOT, 12
Notation
  big-O, 132–138
  big-Omega, 134, 140–142
  big-Theta, 134, 140–142
  dependency, 738
  infix, 668–672, 695
  little-o notation, 143
  one's complement, 180
  Polish, 669, 695

postfix, 668–672, 695
prefix, 668–672, 695
  well-formed formula in, 674
  for products, 237
  reverse Polish, 670, 695
  set builder, 79
  summation, 229
  two's complement, 180–181
Noun, 740
Noun phrase, 740
Nova, 220
NP-complete problems, 149, 586, 720
n-queens problem, 683
n-regular graph, 556
n-tuples, 483–485
  ordered, 82–83
Null set, 80, 112
Null string, 226, 291, 741
Number(s)
  Bacon, 569
  Bell, 516
  Carmichael, 190–191, 207
  Catalan, 408
  chromatic, 614, 623
    edge, 620
  crossing, 613
  Cunningham, 158
  Erdős, 533, 568, 569
  Fibonacci, 259, 658
    formula for, 415
    iterative algorithm for, 279
    rabbits and, 403–404
    recursive algorithms for, 278
  harmonic, 144
    inequality of, 243–244
  independence, 627
  irrational, 167
  large, law of, 369
  Lucas, 293, 423
  lucky, 470
  natural, 78
  perfect, 224
  pseudorandom, 164–165, 212
  Ramsey, 318
  rational, 78, 112
    countability of, 234, 238
  real, 78
  Ten Most Wanted, 158
  Ulam, 292
Numbering plan, 303–304
Number-theoretic functions, 780
Number theory, 153–196

Object, 77–78
Octahedral die, 396
Octal expansion, 170, 207
Odd integer, 63
Odlyzko, Andrew, 229
One's complement expansion, 180
One-to-one correspondence, 101, 113, 234, 235
One-to-one (injective) function, 99–100, 101, 113
  counting, 303
"Only if" statement, 9
Onto (surjective) function, 100–101, 113
  number of, 460–461, 465
Open problems, 220
Operands, well-formed formula of, 263–264
Operations
  bit, 14, 112
  estimating number of, of algorithms, 132–140
  integer, algorithms for, 172–176

on n-ary relations, 484–487
on set, 86–97
Operators
  bitwise, 14–15, 112
  functionally complete set of, for
    propositions, 27
  logical, 111
    functionally complete, 27
    precedence of, 9–10
  negation, 4
  selection, 484, 531
  well-formed formula of, 263–264
Optimal algorithm, 152, 248
Optimization problems, 127–129
OR, 12, 14–15
Oracle of Bacon, 569
Order, 141
  of quantifiers, 49–51
  same, 133
Ordered n-tuples, 82–83
Ordered pairs, 83
Ordered rooted tree, 635, 695, 697
Ordering
  lexicographic, 270, 344–348, 518–520
  linear, 517
  partial, 516–530, 531
  quasi-ordering, 535
  total, 517, 531
Ordinary generating function, 436
Ore, O., 586
Ore's Theorem, 586, 592
Organizational tree, 637
OR gate, 713, 734
Orientable graphs, 626
Orientation of undirected graph, 626
Out-degree of vertex, 547, 622
Output alphabet, 753
Outputs
  finite-state machines with, 753–756
  finite-state machines without, 758–765

Pairs
  devil's, 567
  ordered, 83
Pairwise relatively prime integers, 160, 207
Palindrome, 130, 311, 749
Paradox, 77, 112
  barber, 19
  Russell's, 86
  St. Petersburg, 396
Parallel algorithms, 552
Parallel edges, 538
Parallel processing, 151, 552
  tree-connected, 638
Parentheses, balanced strings of, 295
Parent of vertex, 634, 694
Parent relation, 478
Parkin, T. R., 219
Parse tree, 745–746, 783
Parsing
  bottom-up, 746
  top-down, 746
Partial correctness, 285
Partial function, 111
Partially ordered set, 516, 531
  antichain in, 534
  chain in, 534
  comparable elements in, 517, 531
  dense, 530
  dual of, 528
  greatest element of, 522, 531
  Hasse diagram of, 520–522, 523, 524, 531

incomparable elements in, 517, 531
  least element of, 522, 531
  lower bound of, 523, 531
  maximal element of, 522, 531
  minimal element of, 522, 525–526, 531
  upper bound of, 523, 531
  well-founded, 530
Partial orderings, 516–530, 531
  compatible total ordering from, 531
Partition, 450
  of positive integer, 273
  refinement of, 515
  of set, 511–512, 531
Pascal, 98, 121, 189
Pascal, Blaise, 330, 355, 362
Pascal's identity, 330–331, 349
Pascal's triangle, 331, 349
Passwords, 301, 306–307
Paths, 567–569
  in acquantanceship graphs, 568
  in collaboration graphs, 568–569
  counting between vertices, 574–575
  in directed graphs, 498–499, 531, 622
  in directed multigraphs, 568
  Euler, 577–583, 622, 623
  and graph isomorphism, 572–573
  Hamilton, 583–588, 622
  in Hollywood graph, 569
  of length n, 567
  length of, in weighted graph, 594
  shortest, 593–603, 623
  in simple graphs, 567
  in undirected graphs, 567, 622
Payoff, 652
Pecking order, 627
Peirce, Charles Sanders, 27, 29
Peirce arrow, 27
Pelc, A., 435
Pendant vertex, 546, 622
Pennies, 128–129
Perfect integers, 167
Perfect number, 224
Peripatetics, 2
Permutations, 321–322, 349
  generating, 344–346
  with indistinguishable objects, 339–340
  inversions in, expected number of, 383–384
  with repetition, 335
PERT (Program Evaluation and Review
    Technique), 536
Petersen, Julius Peter Christian, 592
Petersen graph, 591
Phrase-structure grammars, 741–745, 783
Pigeonhole principle, 313–320, 349
  applications of, 316–318
  generalized, 314–316, 349
Planar graphs, 603–613, 623
Plato, 2
Plato's Academy, 2
Plouffe, Simon, 228
PNF (prenex normal form), 55
Pogo stick, motorized, 702
Pointer, position of, digital representation
    of, 587–588
Poisonous snakes, 650
Polish notation, 669, 695
Pólya, George, 223
Polynomial complexity, 148
Polynomials, rook, 470
Polyominoes, 299
Poset, 516, 531
  antichain in, 534
  chain in, 534

comparable elements in, 517, 531
  dense, 530
  dual of, 528
  greatest element of, 522, 531
  Hasse diagram of, 520–522, 523, 524, 531
  incomparable elements in, 517, 531
  least element of, 522, 531
  lower bound of, 523, 531
  maximal element of, 522, 531
  minimal element of, 522, 525–526, 531
  upper bound of, 523, 531
  well-founded, 530
Positive integers, 78, 273
Positive rational numbers, countability of,
    234, 238
Postfix form, 670
Postfix notation, 668–672, 695
Postorder traversal, 663–664, 666, 667, 668, 695
Postulates, 56
Power, Boolean (of zero-one matrices), 203–204
Powers
  of matrices, 201
  of relation, 479, 530
Power series, 437–439
  formal, 437
Power set, 82, 113
Precedence graphs, 543
Precedence of logical operators, 10
Predicate, 28, 29
Predicate calculus, 30
Prefix codes, 649–651, 695
Prefix form, 669
Prefix notation, 668–672, 695
  well-formed formula in, 674
Pregel River, 577–578
Pre-image, 98, 113
Premise, 6, 38
Prenex normal form (PNF), 55
Preorder traversal, 661–662, 666, 695
Prim, Robert Clay, 689, 691
Primality testing, probabilistic, 374–375
Primary key, 483–484, 531
Prime factorization, 155–156
  for finding greatest common divisor, 160
  for finding least common multiple, 160–161
  uniqueness of, 183–184
Prime implicant, 722, 735
  essential, 722, 735
Prime Number Theorem, 157
Primes, 154–158
  definition of, 207
  distribution of, 157
  infinitude of, 156
  Mersenne, 157, 207, 218
  mutually relative, 210
  pseudoprimes, 189–191, 195, 207
  relatively, 160, 207
  sieve of Eratosthenes and, 155, 458–460
  twin, 221
Prim's algorithm, 689, 690, 695
Principia Mathematica (Whitehead and
    Russell), 26, 79
Principle(s)
  of counting, 302–306
  duality, for Boolean identities, 706, 708
  of inclusion-exclusion, 87, 308–309,
      451–457, 465
    alternative form of, 457–458
    applications of, 457–465
  of mathematical induction, 291

pigeonhole, 313–320, 349
    applications of, 316–318
    generalized, 314–316, 349
  of well-founded induction, 534
  of well-ordered induction, 517–518
Private key cryptography, 191, 207
Probabilistic algorithms, 355, 373–375, 394
Probabilistic method, 375–376, 394
Probabilistic primality testing, 374–375
Probabilistic reasoning, 360
Probability, discrete, 355–399
  assigning, 363–365
  of collision in hashing functions, 372–373
  of combinations of events, 358–360, 365–366
  conditional, 363, 366–367, 394
  finite, 356–358
  Laplace's definition of, 356
Probability distribution, 363
Probability generating function, 450
Probability theory, 355, 362–378
Problem(s)
  birthday, 371–372, 377
  busy beaver, 780, 782
  Chinese postman, 583, 629
  closest-pair, 431–433
  Collatz, 221–222
  halting, 149, 222–223, 291
  hatcheck, 383, 461–462
  intractable, 149
  Josephus, 412
  Kakutani's, 222
  kissing, 229
  Königsberg bridge problem, 577–578, 581, 583
  maximum satisfiability, 398
  Newton-Pepys, 399
  NP-complete, 149, 586, 720
  n-queens, 683
  open, 220
  optimization, 127–129
  P, 149
  scheduling, 131
    final exams, 618, 619
    greedy algorithm for, 248–249
  searching, 123–125
  shortest-path, 593–603, 623
  solvable, 149
  Syracuse, 222
  tractable, 149
  traveling salesman, 586, 595, 599–601, 623
  Ulam's, 221–222, 435
  unsolvable, 149
  utilities-and-houses, 603–605
Problème de rencontres, 462, 470
Problème des ménages, 470
procedure statements, A-4
Processing
  concurrent, 543
  parallel, 151, 552
    tree-connected, 638
Product
  Boolean, 701–702, 703, 709–711, 734
    of zero-one matrices, 202–203, 207
  Cartesian, 82–84
  notation for, 237
Productions of grammar, 741, 783
Product-of-sums expansion, 711, 712
Product rule, 302–304, 349
Program correctness, 284–290, 291
  conditional statements for, 286–288
  loop invariants for, 288–289
  partial, 285

Program correctness—*Cont.*
  program verification for, 285
  rules of inference for, 285–286
Program Evaluation and Review Technique
  (PERT), 536
Programming, logic, 39–40
Programming languages, 60, 98, 312
Program verification, 285
Progression(s)
  arithmetic, 226, 290
  geometric, 226, 230, 290
    sums of, 243
Projection of n-ary relations, 485, 531
Prolog, 39–40, 60
Prolog facts, 39
Prolog rules, 39
Proof(s)
  adapting existing, 216–217
  by Cantor diagonalization, 235
  by cases, 67, 112, 215–216
  combinatorial, 323, 349
  conjecture and, 217–219
  constructive existence, 69, 122
  by contradiction, 66–67, 112
  definition of, 56, 112
  direct, 63–64, 112
  of equivalence, 67–68
  existence, 69, 112
  indirect, 64, 112
  by infinite descent, 252
  by mathematical induction, 240–249
  methods of, 56–77, 112
  mistakes in, 71–73
  nonconstructive existence, 69, 112
  by strong induction, 249
  by structural induction, 267
  trivial, 64, 112
  uniqueness, 69–70, 112
  vacuous, 64, 112
Proof strategies, 65, 214–225
Proper subset, 81, 113
Property, well-ordering, 251–253, 291
Proposition(s), 2–6, 56
  compound, 3, 9–10, 21, 22, 111
    consistent, 112
    disjunctive normal form of, 27
    dual of, 27
    satisfiable, 28
    well-formed formulae, 263
  conjunction of, 4–5, 111
  definition of, 2, 111
  disjunction of, 4–5, 111
  equivalent, 8, 20–28, 112
  negation of, 3–4, 111
  rules of inference for, 57–58
  truth value of, 3
Propositional equivalences, 20–28
Propositional function, 28, 29, 112
  existential quantification of, 32, 112
  universal quantification of, 30, 112
Propositional logic, 3
Pseudocode, 121, A-4–A-9
Pseudograph, 539, 540, 622
Pseudoprimes, 189–191, 195, 207
Pseudorandom numbers, 164–165, 212
Public key cryptography, 181, 191–194, 207
Pumping lemma, 775
  for context free languages, 786
Pure multiplicative generator, 165
Pushdown automata, 773, B-3

Puzzle(s)
  Birthday Problem, 371–372
  Icosian, 583, 584
  jigsaw, 294
  logic, 12–14
  Monty Hall Three Door, 360
  muddy children, 13–14
  Reve's, 406
  Tower of Hanoi, 404–406
  "Voyage Round the World," 583–584
  zebra, 20

Quadratic mean, 223
Quadratic residue, 196
Quad trees, 700
Quality control, 373–374
Quantification
  existential, 32, 112
    negation of, 34
  as loops, 50
  universal, 30, 112
    negation of, 34
Quantified statements
  rules of inference for, 61
Quantifiers, 30–33
  existential, 32–33
  negating, 35
  nested, 44–56
    negating, 47–48
  order of, 49–51
  scope of, 33
  Theorems involving, 68–69
  "uniqueness," 46
  universal, 30–32
  using set notation with, 84–85
Quarters, 128–129
Quasi-ordering, 535
Queens on chessboard, 626, 683
Question, begging, 72, 112
Quick sort, 125, 284
  average case complexity, 393–394
Quine, Willard van Orman, 728, 729
Quine-McCluskey method, 720, 727–732
Quotient, 158

Rabbits, 403–404
Radius of graph, 627
Rado, Tibor, 782
Ramanujan, Srinivasa, 69, 297
Ramaré, O., 221
Ramsey, Frank Plumpton, 318
Ramsey number, 318, 375
Ramsey theory, 318
Random, at, 364
Random simple graph, 627–628
Random variables, 363, 370–371
  covariance of, 394
  definition of, 394
  distribution of, 370, 394
    geometric, 386
  expected values of, 363, 379–381, 394
  independent, 387–388, 394
  indicator, 392
  standard deviation of, 388
  variance of, 379, 388–390, 394
Range of function, 98, 113
Ratio, common, 226
Rational numbers, 78, 112
  countability of, 234, 238
r-combination, 322, 349
Reachable states, 785

Real numbers, 78
  uncountability of, 235–236
Reasoning
  backward, 214
  circular, 72, 112
  forward, 214
  probabilistic, 360
Recognized language, 760–763, 783
Recognized strings, 760
Recognizer, language, 755
Record in a database, 483
Recurrence relations, 401–413
  associated homogenous, 419
  definition of, 402, 465
  divide-and-conquer, 426–433
  initial condition for, 402, 465
  linear homogenous, 413, 414–419, 465
  linear nonhomogenous, 419–425, 465
  modeling with, 403–409
  simultaneous, 424
  solution of, 402
  solving, 413–425
    using generating functions for, 444–450
Recursion, 256
Recursive algorithms, 274–284, 291
  for binary search, 276
  for computing exponentials, 275
  for computing factorials, 277
  for computing Fibonacci numbers, 278
  for computing greatest common divisor, 276
  correctness of, 275
  for linear search, 276
  for merge search, 276
  for modular exponentiation, 275–276
Recursive definitions, 213, 256–274
  of extended binary trees, 265
  of factorials, 277
  of full binary trees, 266
  of functions, 97, 257–261, 291
  of rooted trees, 264
  of sets, 261–266, 291
  of strings, 262
  of structures, 261–266
Recursive merge sort, 280
Recursive modular exponentiation, 275
Recursive sequential search algorithm, 276–277
Recursive step, 261, 267
Reentrant knight's tour, 591–592
Refinement, of partition, 515
Reflexive closure of relation, 497, 531
Reflexive relation, 474–475, 530
  representing
    using digraphs, 493
    using matrices, 490
Regions of planar representation
  graphs, 606, 623
Regular expressions, 766, 783
Regular grammars, 744, 766, 769–772, 783
Regular graph, 556, 622
Regular language, 744
Regular sets, 765, 766, 767, 769–772, 783
Relation(s), 471–536
  antisymmetric, 475–476, 531
  asymmetric, 480
  binary, 471, 530
  circular, 533
  closures of, 496–507, 531
    reflexive, 497, 531
    symmetric, 497, 531
    transitive, 496, 499–502, 531
      computing, 503–506

combining, 477–479
complementary, 480
composite of, 478–479, 530
connectivity, 499–501, 531
counting, 474
covering, 528
definition of, 471, 473
diagonal, 497
difference of, 477
divide-and-conquer recurrence, 426
divisibility, 516
domains of, 482
equivalence, 507–516, 531
functions as, 472–473
"greater than or equal," 516
inclusion, 516
intersection of, 477
inverse, 480, 530
irreflexive, 480
"less than or equals," 517, 519
n-ary, 482–489, 531
    domain of, 483
    operations on, 484–487
parent, 478
paths in, 498–499
powers of, 479, 530
properties of, 474–477
recurrence, 401
reflexive, 474–475, 530
representing
    using digraphs, 493–494
    using matrices, 490–492
on set, 84, 473–474, 530
symmetric, 475–476, 531
transitive, 476–477, 479, 531
union of, 477
Relational database model, 483–484, 531, B-2
Relatively prime integers, 160, 207
    mutually, 210
    pairwise, 160
Remainder, 158, 161, 207
Rencontres, 462, 470
Repetition
    combinations with, 336–339
    permutations with, 335
    sampling with, 358
Replacement
    sampling with, 358
    sampling without, 358
Representation
    binary, 169
    hexadecimal, 170
    octal, 170
    unary, 780
Representative, of equivalence class, 509
Representing Boolean functions, 709–712
Representing relations
    using directed graphs, 493–494
    using directed matrices, 490–492
Residue, quadratic, 197
Resolution, 58, 60
Results, 56
Reversal of string, 272
Reverse Polish notation, 670, 695
Reve's puzzle, 406
Right child of vertex, 635
Right subtree, 635
Ring topology for local area network, 551
Rivest, Ronald, 191, 192
Rocket-powered Frisbee, 702
Rook polynomials, 470

Rooted Fibonacci trees, 644
Rooted spanning tree, 688
Rooted trees, 264, 633–635
    balanced, 640, 695
    binomial, 697
    B-tree of degree k, 697
    decision trees, 646–649
    definition of, 694
    height of, 640, 695
    level order of vertices of, 697
    ordered, 635, 695, 697
    recursive definition of, 264
    S_k-tree, 697
Root of a tree, 264
Roots, characteristic, 414
Round-robin tournaments, 252, 378, 542
Row of matrix, 197
Roy, Bernard, 503
Roy-Warshall algorithm, 503–506
r-permutation, 321, 349
RSA cryptosystem, 191–194
RSA decryption, 193
RSA encryption, 191, 192–193
Rule
    product, 302
    sum, 305
Rules of inference, 56, 57–58, 60, 112
    for program correctness, 285–286
    for propositions, 57–58
    for quantified statements, 61–63
Run, 392
Russell, Bertrand, 77, 79
Russell's paradox, 86

St. Petersburg Paradox, 396
Same order, 133
Samos, Michael, 431
Sample space, 356
Sampling
    with replacement, 358
    without replacement, 358
Sanskrit, 746
Satisfiability problem, maximum, 398
Satisfiable compound propositions, 28
Saturated hydrocarbons, trees modeling, 636–637
Scheduling problems, 131
    final exams, 618, 619
    greedy algorithm for, 248–249
    software projects, 530
    tasks, 526–527
Scimitars, Mohammed's, 581
Scope of quantifier, 33
Search engines, 685
Searching, 123–125, 206, 207
    binary, 123–125, 206, 207, 426
    Boolean, 12
    breadth-first, 680–681, 695
    complexity of, 145–147, 148
    depth-first, 677–680, 695
        applications with, 681–684
        in directed graphs, 684–685
    linear, 123, 206, 207
        average-case complexity of, 384
        complexity of, 145–147, 148
        recursive, 276
    recursive sequential, 276–277
    ternary, 130
    Web, 12
Search trees, binary, 644–646, 695

Second principle of mathematical induction, 249–251, 275, 291
Seed, 164
Selection operator, 484, 531
Selection sort, 125, 130–131
Self-complementary graph, 566
Self-converse directed graph, 626
Self-dual, 736
Self-generating sequences, 295, 296
Semantics, 740
Sentence, 740, 741
Sequence(s), 225–228, 290
    de Bruijn, 629
    definition of, 225
    finding maximum and minimum of, 426–427
    generating functions for, 436
    Golomb's self-generating, 296
    integer, 227–228
    strictly decreasing, 317
    strictly increasing, 317
Sequential search, 123
Serial algorithms, 552
Series
    geometric, 230
    harmonic, 244
    infinite, 233
    power, 437–439
        formal, 437
Set(s), 77–97, 112–113, B-1
    cardinality of, 81, 113, 233–236
    complement of, 88, 113
    computer representation of, 93–94
    countable, 234, 291
    cut, 697
    definition of, 77, 112
    description of, 78–79
    difference of, 88, 113
        symmetric, 95, 113
    disjoint, 87
    dominating, 625
    elements of, 78, 112
    empty (null), 80, 112
    equal, 79, 113
    finite, 81, 113, 234, 451, 465
        subsets of
            counting, 304
            number of, 244–245
        union of three, number of elements in, 453–454, 465
        union of two, number of elements in, 451–453, 465
    fuzzy, 96–97, B-1
        complement of, 96–97
        intersection of, 97
        union of, 97
    identities of, 89–91
    image of, 99
    infinite, 82, 113, 234
    intersection of, 86–87, 91–93, 113
    inverse image of, 110
    linearly ordered, 517, 531
    members of, 78, 112
    not recognized by finite-set automata, 772
    operations on, 86–97
    partially ordered, 516, 531
        antichain in, 534
        chain in, 534
        comparable elements in, 517, 531
        dense, 530
        dual of, 528
        greatest element of, 522, 531

Set(s)—*Cont.*
  Hasse diagram of, 520–522, 523, 524, 531
  incomparable elements in, 517, 531
  least element of, 522, 531
  lower bound of, 523, 531
  maximal element of, 522, 531
  minimal element of, 522, 525–526, 531
  upper bound of, 523, 531
  well-founded, 530
  partition of, 511–512, 531
  power, 82, 113
  recognized by finite-set automata, 766–769
  recognized by Turing machines, 778–779
  recursively defined, 261–266, 291
  regular, 765, 766, 767, 769–772, 783
  relation on, 84, 473–474, 530
  singleton, 80
  successor of, 96
  symmetric difference of, 95, 113
  totally ordered, 517, 531
  uncountable, 235–236, 291
  union of, 86, 91–93, 113
  universal, 79, 113
  well-ordered, 295, 517, 531
Set builder notation, 79
Set notation, with quantifiers, 84–85
Sex, 367
Shaker sort, 125, 209
Shamir, Adi, 191, 192
Shannon, Claude Elwood, 701, 702
Sheffer, Henry Maurice, 26, 27
Sheffer stroke, 26, 27
Shift cipher, 166
Shifting, 174
Shift register, 738
Shortest-path algorithm, 595–599
Shortest-path problems, 593–603, 623
Sibling of vertex, 634, 695
Sieve of Eratosthenes, 155, 458–460, 465
Simple circuit, 567
Simple graphs, 538, 548–551, 622
  coloring of, 614
  connected planar, 606–609
  crossing number of, 613
  edges of, 538, 557
  isomorphic, 560–563, 622
  orientation of, 626
  paths in, 567
  random, 627–628
  self-complementary, 566
  with spanning trees, 675–676
  thickness of, 613
  vertices of, 538, 557
Simplification rule of inference, 58
Simultaneous linear equations, 205–206
Singleton set, 80
Sink, 785
Six Degrees of Separation (Guare), 568
S_k-tree, 697
Sloane, Neil, 229
Smullyan, Raymond, 12, 13
Snakes, poisonous, 650
Sneakers (movie), 192
Socks, 319
Software project, ordering tasks of, 530
Software systems, 11
Sollin's algorithm, 694
Solution of a recurrence relation, 402

Solving
  counting problems using generating
    functions, 439–444
  recurrence relations using generating
    functions, 444–445
Sort
  binary insertion, 125, 131
  bubble, 125–126, 207
    bidirectional, 125, 209
    worst-case complexity of, 147
  insertion, 125, 126–127, 207
    average-case complexity of, 385
    worst-case complexity of, 147–148
  merge, 125, 279–283, 291, 427
    complexity of, 430
    recursive, 280
  quick, 125, 284
  selection, 125, 130–131
  shaker, 125, 209
  tournament, 125, 657
Sorting algorithms, 125–127, 206
  complexity of, 147–148, 647–649
  topological, 525–527, 531
Space, sample, 356
Space complexity of algorithms, 144, 206
Spanning forest, 687
  minimum, 694
Spanning trees, 674–688, 695
  building
    by breadth-first search, 680–681
    by depth-first search, 677–680
  definition of, 675
  degree-constrained, 698
  distance between, 688
  in IP multicasting, 676–677
  maximum, 694
  minimum, 688–694, 695
  rooted, 688
Sparse graphs, 693
Sparse matrix, 558
Specifications, system, 11
Spiders, web, 685
SQL, 487–488
Squarefree integer, 464
Square matrix, 201
Squirrel, 541
Stairs, climbing, 410
Standard deviation, 388
Star height, 785
Star topology for local area network, 551
Starting state, 753, 760
Start symbol, 741
State
  final, 760
  initial, 753, 760
  start, 753, 760
State diagram for finite-state machine, 753
Statement(s)
  assignment, A-4–A-5
  begin, A-5
  blocks of, A-5
  conditional, 6
    for program correctness, 286–288
  end, A-5
  "if and only if," 8, 9
  if then, 7, 9, 30, A-6–A-7
  procedure, A-4
  rules of inference for, 61–63
States, 753
  reachable, 785
  transient, 785

State table for finite-state machine, 753
Strategies
  minmax, 654, 695
  proof, 214–225
S_k-tree, 697
Strictly decreasing function, 100
Strictly decreasing sequence, 317
Strictly increasing function, 100
Strictly increasing sequence, 317
Strings, 226, 291
  concatenation of, 263, 758–759, 783
  empty, 226, 291, 741
  length of, 226
    recursive definition of, 263
  lexicographic ordering of, 518–519
  of parentheses, balanced, 295
  recognized (accepted), 760
  recursively defined, 262
  reversal, 272
  structural induction and, 268
  ternary, 410
Strings, bit, 14, 93–94, 112
  counting, 302
    without consecutive 0s, 406–407
  decoding, 649
  generating next largest, 347
  length of, 14, 406–407
Stroke, Sheffer, 27
Strong components of a directed graph, 572
Strong induction, 249–251, 275, 291
Strongly connected components of
    graphs, 572, 622
Strongly connected graphs, 571
Structural induction, 257, 266–269, 291
Structured query language (SQL), 487–488
Structures, recursively defined, 261–266
Subgraph, 553, 622
  induced, 625
Subsequence, 317
Subsets, 80, 113
  of finite set
    counting, 304
    number of, 244–245
  proper, 81, 113
  sums of, 684
Subtractors
  full, 718
  half, 718
Subtree, 634, 635, 694
Success, 368
Successor, of set, 96
Sufficient, 6
  necessary and, 9
Sum(s)
  Boolean, 701–702, 703, 709–711, 734
  of first *n* positive integers, 75, 78, 193, 328
  of functions, 98–99
  of geometric progressions, 243
  of matrices, 197–198
  of multisets, 96
  of subsets, 684
Summation notation, 229
Summations, 229–233
  double, 231
  formulae, 231–233
  index of, 229
  lower limit of, 229
  upper limit of, 229
Sum-of-products expansions, 709–711, 734
  simplifying, 719–734
Sum rule for counting, 305, 349

Sun-Tsu, 186
Surjective (onto) function, 100–101, 113
    number of, 460–461, 465
Switching circuits, 738
Syllogism
    disjunctive, 58
    hypothetical, 58
Symbol(s), 741
    Landau, 134
    nonterminal, 741
    start, 741
    terminal, 741
Symbolic Logic (Carroll), 38, 43–44
Symbolic Logic (Venn), 80
Symmetric closure of relation, 497, 531
Symmetric difference, of sets, 95, 113
Symmetric matrix, 201, 202, 207
Symmetric relation, 475–476, 531
    representing
        using digraphs, 494
        using matrices, 491
Syntax, 740
Syracuse problem, 222
System specifications, 11–12
    consistent, 11

Tables, 483
    membership, 91
    state, 753
    truth, 3
Tautology, 21, 25, 57–58, 60, 112
Tee-shirts, 310
Telephone call graph, 571
Telephone calls, 542
Telephone network, 542, 571
Telephone number, 542
Telephone numbering plan, 303–304
Telescoping sum, 237
Ten Most Wanted Numbers, 158
Term, initial, 226
Terminals, 741
Terminal vertex, 493, 547
Ternary expansion, balanced, 180
Ternary search algorithm, 130
Ternary string, 410
Test
    Lucas–Lehmar, 157
    Miller's, 374
    probabilistic primality, 374
TeX, 134
Theorem(s), 56, 112
    Binomial, 327–330, 349
    Chinese Remainder, 185–187, 207
    Dirac's, 586
    Extended Binomial, 439
    Fermat's Last, 220, 260
    Fermat's Little, 189, 207
    Four Color, 614–615, 623
    Fundamental Theorem of Arithmetic, 155,
        207, 250
    Handshaking, 546
    Kleene's, 766–769, 783
    Kuratowski's, 609–611, 623
    Lamé's, 260
    Master, 430
    Methods of Proving, 63–69
    Multinomial Theorem, 344
    Ore's, 586, 592
    Prime Number, 157
    Proving, Automated, 117
    Wilson's, 194

Theory Ramsay, 318
Thesis, Church–Turing, 781
Thickness of a graph, 613
$3x + 1$ conjecture, 221–222
Threshold function, 737
Threshold gate, 736
Threshold value, 736
Tic-tac-toe, 653, 654
Time complexity of algorithms, 144, 145–148,
    149, 206
Top-down parsing, 746
Topological sorting, 525–527, 531
Topology for local area network
    hybrid, 551
    ring, 551
    star, 551
Torus, 613
Total function, 111
Totally ordered set, 517, 531
Total ordering, 517, 531
    compatible, 525, 531
Tournament, 627
    round-robin, 378, 542
Tournament sort, 125, 657
Tower of Hanoi, 404–406
Tractable problem, 149
Transfer mode, asynchronous, 106–107
Transient state, 785
Transition function, 753
Transitive closure of relation, 496, 499–502, 531
    computing, 503–506
Transitive relation, 476–477, 479, 531
    representing, using digraphs, 494
Translating
    English sentences to logical expressions,
        10–11, 35–37, 45–47
    logical expressions to English sentences, 42,
        44–45
    system specifications into logical
        expressions, 11
Transposes of matrices, 201, 207
"Traveler's Dodecahedron," 583–584
Traveling salesman problem, 586, 595,
    599–601, 623
Traversal of tree, 660–674, 695
    inorder, 662–664, 665, 668, 695
    level order, 697
    postorder, 663–664, 666, 667, 668, 695
    preorder, 661–662, 666, 695
Tree(s), 631–700
    applications of, 644–659
    AVL, 700
    binary, 264–265, 634, 635
        extended, 265
        full, 265–266
    binary search, 644–646, 695
    binomial, 697
    caterpillar, 698
    decision, 646–649, 695
    definition of, 631, 694
    derivation, 745–746, 783
    family, 631, 632
    Fibonacci, rooted, 644
    full m-ary, 634, 639, 695
    game, 651–656
    graceful, 698
    height-balanced, 700
    labeled, 643
    m-ary, 634, 695
        complete, 643
        height of, 640–641
    mesh of, 700

as models, 636–638
    properties of, 638–641
    quad, 700
    rooted, 264, 633–635
        balanced, 640, 695
        binomial, 697
        B-tree of degree k, 697
        decision trees, 646–649
        definition of, 694
        Fibonacci, 644
        height of, 640, 695
        level order of vertices of, 697
        ordered, 635, 695, 697
        S_k-tree, 697
    rooted Fibonacci, 644
    spanning, 674–688, 695
        building
            by breadth-first search, 680–681
            by depth-first search, 677–680
        definition of, 675
        degree-constrained, 698
        distance between, 688
        in IP multicasting, 676–677
        maximum, 694
        minimum, 688–694, 695
        rooted, 688
Tree-connected network, 638
Tree-connected parallel processors, 638
Tree diagrams, 309–310, 349
Tree edges, 678
Tree traversal, 660–674, 695
    inorder, 662–664, 665, 668, 695
    postorder, 663–664, 666, 667, 668, 695
    preorder, 661–662, 666, 695
Trial division, 158
Triangle, Pascal's, 331, 349
Triangle inequality, 75
Trivial proof, 64, 112
Truth table, 3–6, 111
    for biconditional statement, 9
    for conjunction of two propositions, 5
    for disjunction of two propositions, 5
    for exclusive or of two propositions, 6
    for implication, 6
    for logical equivalences, 22
    for negation of proposition, 3
Truth value, 111
    of implication, 6
    of proposition, 3
Tukey, John Wilder, 14, 15
Turing, Alan, 149, 222, 773, 775
Turing, Alan Mathison, 774
Turing machines, 739, 773, 774, 775–783
    computing functions with, 779–780
    definition of, 776–778
    sets recognized by, 778–779
    types of, 780–781
Twin prime conjecture, 221
Twin primes, 221
Two-dimensional array, 552
Two's complement expansion, 180–181
Type 0 grammar, 744, 783
Type 1 grammar, 744, 783
Type 2 grammar, 744, 783
Type 3 grammar, 744, 766, 769–772, 783

Ulam, Stanislaw, 435
Ulam numbers, 292
Ulam's problem, 221–222, 435
Unary representations, 780
Uncountable set, 235–236, 291
Undefined function, 111
Underlying undirected graph, 548, 622

Undirected edges, 540, 622
  of simple graph, 538
Undirected graphs, 540, 546–547, 622
  connectedness in, 569–571
  Euler circuit of, 578
  Euler path of, 582
  orientation of, 626
  paths in, 567, 622
  underlying, 548, 622
Unicasting, 676–677
Unicycle, 702
Uniform distribution, 364, 394
Union
  of fuzzy sets, 97
  of graphs, 554, 622
  of multisets, 96
  of sets, 86, 91–93, 113
  of three finite sets, number of elements in,
      453–454, 465
  of two finite sets, number of elements in,
      451–453, 465
Uniqueness proofs, 69–70, 112
"Uniqueness quantifier," 46
Unit-delay machine, 754–755
United States Coast Survey, 29
Unit (Egyptian) fraction, 293–294
Unit property, in Boolean algebra, 705
Universal address system, 660
Universal generalization, 61, 62
Universal instantiation, 61, 62
Universal quantification, 30, 112
  negation of, 34
Universal quantifier, 30–32
Universal set, 79, 113
Universe of discourse (domain), 30, 112
Unsolvable problem, 149
Upper bound, 56
  of lattice, 535
  of poset, 523, 531
Upper limit of summation, 229
Upper triangular matrix, 210
U.S. Coast Survey, 29
Utilities-and-houses problem, 603–605

Vacuous proof, 64, 112
Valeé-Poussin, Charles-Jean-
      Gustave-Nicholas de la, 158
Valid arguments, 58–59
Value(s)
  expected, 363, 379–381, 394
    in hatcheck problem, 383
    of inversions in permutation, 383–384
    linearity of, 382–384, 394
  final, A-7
  initial, A-8
  threshold, 736
  of tree, 654
  of vertex in game tree, 654–655, 695
Vandermonde, Alexandre-Théophile, 332

Vandermonde's identity, 331–332
Variable(s)
  binding, 33–34
  Boolean, 14, 112, 702, 709, 734
  bound, 33
  free, 33, 112
  random, 363, 370–371
    covariance of, 394
    definition of, 394
    distribution of, 370, 394
      geometric, 386
    expected values of, 363, 379–381, 394
    independent, 387–388, 394
    indicator, 392
    standard deviation of, 388
    variance of, 379, 388–390, 394
Variance, 379, 388–390, 394
Veitch, E. W., 721
Vending machine, 751–752
Venn, John, 79, 80
Venn diagrams, 79, 80, 81, 87, 88, 113
Verb, 740
Verb phrase, 740
Vertex basis, 577
Vertex (vertices), 493
  adjacent, 546, 547, 622
  ancestor of, 634, 695
  child of, 634, 635, 695
  connecting, 546
  counting paths between, 574–575
  cut, 571
  degree of, in undirected graph, 546, 622
  descendant of, 634, 695
  of directed graph, 539, 547, 558
  of directed multigraph, 540
  distance between, 627
  eccentricity of, 643
  end, 493, 547
  in-degree of, 547, 622
  independent set of, 627
  initial, 493, 547
  interior, 503
  internal, 634, 695
  isolated, 546, 622
  level of, 640, 695
  level order of, 697
  of multigraph, 538
  number of, of full binary tree, 269
  out-degree of, 547, 622
  parent of, 634, 694
  pendant, 546, 622
  of pseudograph, 539
  sibling of, 634, 695
  of simple graph, 538, 557
  terminal, 493, 547
  of undirected graph, 540, 546, 547
    degree of, 546, 622
  value of, in game tree, 654–655, 695

Very large scale integration (VLSI), 629
Vocabulary, 741, 783
"Voyage Round the World" Puzzle, 583–584

Warshall, Stephen, 503
Warshall's algorithm, 503–506, 531
Weakly connected digraphs, 571
Web crawlers, 543
Web graph, 543, 685
  strongly connected components of, 572
Web pages, 12, 543, 685
Web page searching, 12
Website for this book, xvii–xviii
Web spiders, 685
Weighted graphs, 623
  minimum spanning tree for, 689, 690, 691
  shortest path between, 593–599
  traveling salesman problem with, 599–601
Well-formed expressions, 267
Well-formed formula, 263
  for compound propositions, 263
  of operators and operands, 263–264
  in prefix notation, 674
  structural induction and, 267–268
Well-founded induction, principle of, 534
Well-founded poset, 530
Well-ordered induction, 531
  principle of, 517–518
Well-ordered set, 295, 517, 531
Well-ordering property, 251–253, 291
WFF'N PROOF, The Game of Modern Logic
      (Allen), 76, 117
Wheels, 548, 622
Whitehead, Alfred North, 79
Wiles, Andrew, 189, 220
Wilson's Theorem, 194
Witnesses, 132, 206
Word, 741
World Cup soccer tournament, 326
World's record, for twin primes, 221
World Wide Web graph, 543
Worst-case complexity of algorithms, 146,
      147–148, 149, 206

XML, 751
XOR, 14–15
XOR gate, 736

Zebra puzzle, 20
Zero-one matrices, 202–204, 207
  Boolean product of, 202–203
  join of, 202, 207
  meet of, 202, 207
  representing relations using, 490–492
  of transitive closure, 501–502
  Warshall's algorithm and, 503–506
Zero property, in Boolean algebra, 705
Ziegler's Giant Bar, 135
Zodiac, signs of, 351

LIST OF SYMBOLS

TOPIC	SYMBOL	MEANING	PAGE
COUNTING	$P(n, r)$	number of r-permutations of a set with n elements	321
	$C(n, r)$	number of r-combinations of a set with n elements	322
	$\binom{n}{r}$	binomial coefficient n choose r	322
	$C(n; n_1, n_2, \ldots, n_m)$	multinomial coefficient	344
	$p(E)$	probability of E	356
	$p(E \mid F)$	conditional probability of E given F	366
	$E(X)$	expected value of the random variable X	379
	$V(X)$	variance of the random variable X	388
	C_n	Catalan number	408
	$N(P_{i_1} \cdots P_{i_n})$	number of elements having properties $P_{i_j}, j = 1, \ldots, n$	457
	$N(P'_{i_1} \cdots P'_{i_n})$	number of elements not having properties $P_{i_j}, j = 1, \ldots, n$	457
	D_n	number of derangements of n objects	461
RELATIONS	$S \circ R$	composite of the relations R and S	478
	R^n	nth power of the relation R	479
	R^{-1}	inverse relation	480
	s_C	select operator for condition C	484
	$P_{i_1, i_2, \ldots, i_m}$	projection	485
	$J_p(R, S)$	join	486
	Δ	diagonal relation	497
	R^*	connectivity relation of R	499
	$[a]_R$	equivalence class of a with respect to R	509
	$[a]_m$	congruence class modulo m	510
	(S, R)	poset consisting of the set S and partial ordering R	516
	$a \prec b$	a is less than b	516
	$a \succ b$	a is greater than b	516
	$a \preccurlyeq b$	a is less than or equal to b	517
	$a \succcurlyeq b$	a is greater than or equal to b	517
GRAPHS AND TREES	(u, v)	directed edge	493
	$\{u, v\}$	undirected edge	538
	$G = (V, E)$	graph with vertex set V and edge set E	538
	$\deg(v)$	degree of the vertex v	546
	$\deg^-(v)$	in-degree of the vertex v	547
	$\deg^+(v)$	out-degree of the vertex v	547
	K_n	complete graph on n vertices	548
	C_n	cycle of size n	548
	W_n	wheel of size n	548
	Q_n	n-cube	549
	$K_{m,n}$	complete bipartite graph of size m, n	550